Antennas

Editors
Naser Ojaroudi Parchin
Edinburgh Napier University
Edinburgh
UK

Chan Hwang See
Edinburgh Napier University
Edinburgh
UK

Raed A. Abd-Alhameed
University of Bradford
Bradford
UK

Editorial Office
MDPI
St. Alban-Anlage 66
4052 Basel, Switzerland

This is a reprint of articles from the Topic published online in the open access journals *Electronics* (ISSN 2079-9292), *Future Internet* (ISSN 1999-5903), and *Sensors* (ISSN 1424-8220) (available at: https://www.mdpi.com/topics/Antennas).

For citation purposes, cite each article independently as indicated on the article page online and as indicated below:

Lastname, A.A.; Lastname, B.B. Article Title. *Journal Name* **Year**, *Volume Number*, Page Range.

ISBN 978-3-7258-0671-3 (Hbk)
ISBN 978-3-7258-0672-0 (PDF)
doi.org/10.3390/books978-3-7258-0672-0

Contents

About the Editors

Naser Ojaroudi Parchin

Naser Ojaroudi Parchin (SMIEEE) is currently an Assistant Professor (lecturer) at Edinburgh Napier University, Edinburgh, UK. He obtained his Ph.D. in Electrical Engineering from the University of Bradford, UK, where he was a Research Assistant at the Faculty of Engineering and Informatics and worked as a Research Fellow in the SATNEX V project, funded by the European Space Agency. From 2018 to 2020, Naser was a Marie Curie Research Fellow in the H2020-ITN-SECRET project funded by the EU Commission. His research interests include phased arrays, MIMO systems, smartphone antennas, SAR/user impact, 5G antennas, implementable and biomedical sensors, millimeter-wave/terahertz components, metamaterials and metasurfaces, microwave filters, reconfigurable structures, and wireless propagation. Dr. Naser is the author and co-author of several books/book chapters and more than 350 technical journal and conference papers. He is a member of the Institute of IEEE, the MCAA, the EurAAP, and the EPC. He is also an active reviewer for various high-ranking journals and publishers such as *IEEE Transactions*, *IEEE Access/Letters*, IET, Wiley, Springer, Elsevier, MDPI, etc. He has been appointed as a Guest Editor and Topic Board Member for several Special Issues. He is also the recipient and co-recipient of various awards and grants for research publications, such as the Research Development Fund, a Research & Innovation Grant, MDPI travel awards, selected journal papers, and best paper awards at the URSI Symposium 2019, UK URSI Festival 2020, and IMDC 2021. Naser has been endorsed as a UK Talent by the Royal Academy of Engineering. He was included in the World's Top Scientists list in 2016, 2020, 2021, 2022, and 2023. His papers have more than 7400 citations with a 49 h-index, reported by Google Scholar.

Chan Hwang See

Chan Hwang See received a first-class B.Eng. Honours degree in Electronic, Telecommunication and Computer Engineering and a Ph.D. degree from the University of Bradford, UK. Currently, he is an Associate Professor in the School of Computing, Engineering & the Built Environment, Edinburgh Napier University, UK. Previously, he was Head of Electrical Engineering & Mathematics from 2019 to 2022. His research interests cover the Internet of Things (IoTs), wireless sensor network system design, acoustic/microwave sensors, antennas, microwave wireless power transfer, computational electromagnetics, and microwave circuits. He has published over 150 peer-reviewed journal articles and over 150 conference papers. He is a co-author of one book and five book chapters. He was named in the recent 2022 and 2023 edition of the Stanford University List of World Top 2% Scientists. Dr. See was awarded an IEEE Malaysia AP/MTT/EMC Joint Chapter—Best Paper Award in 2020. He was also a recipient of two Young Scientist Awards from the International Union of Radio Science (URSI, US) and the Asia-Pacific Radio Science Conference (AP-RASC, Japan) in 2008 and 2010, respectively. He was awarded a certificate of excellence for his successful knowledge transfer partnership (KTP) with Yorkshire Water on the design and implementation of a wireless sensor system for sewerage infrastructure monitoring in 2009. Dr. See is a Chartered Engineer (CEng), Senior Member of the Institute of Electrical and Electronics Engineers (SMIEEE), and Fellow of the Institution of Engineering and Technology (FIET). He is also a Fellow of The Higher Education Academy (FHEA), a full member of the EPSRC Review College, an Associate Editor for *IEEE Access*, and an Editor for the *Journal of Electronics and Electrical Engineering, Scientific Reports, Peerj Computer Science*, and *PLoS ONE*. According to the Web of Science, he has completed over 500 verified reviews and over 250 verified editor records.

Raed A. Abd-Alhameed

Raed A. Abd-Alhameed (Senior Member, IEEE) has been a Professor Research Visitor with Wrexham University, Wales, since 2009, covering the wireless and communications research areas. He is currently a Professor of electromagnetic and radiofrequency engineering with the University of Bradford, U.K. He is also a Principal Investigator for several funded applications to EPSRCs, Innovate U.K., and the British Council, and the Leader of several successful knowledge transfer programs, such as with Arris (previously known as Pace plc), Yorkshire Water plc, Harvard Engineering plc, IETG Ltd., Seven Technologies Group, Emkay Ltd., and Two World Ltd. He has also been a co-investigator in several funded research projects. He has published over 800 academic journal articles and conference papers; in addition, he has coauthored eight books and several book chapters, including seven patents. His research interests include computational methods and optimizations, wireless and mobile communications, sensor design, EMC, beam steering antennas, beam forming, energy-efficient PAs, and RF predistorter design applications. He was a recipient of the Business Innovation Award for his successful knowledge transfer program with the companies Pace and Datong on the design and implementation of MIMO sensor systems and an antenna array design for service localizations. He is the chair of several successful workshops, including the workshop "energy-efficient and reconfigurable transceivers: an approach toward energy conservation and CO_2 reduction", which addresses the biggest challenges for future wireless systems. He has been the General Chair of the IMDC-IST International Conference since 2020. He has been a Co-Editor of *Electronics* (MDPI) since June 2019. Since 2009, he has been a Guest Editor of *IET Science, Measurement and Technology*.

sensors

Editorial

Editorial on Antennas

Naser Ojaroudi Parchin [1,*], Chan Hwang See [1] and Raed A. Abd-Alhameed [2,3]

1 School of Computing, Engineering and the Built Environment, Edinburgh Napier University, Edinburgh EH10 5DT, UK; c.see@napier.ac.uk
2 Faculty of Engineering and Informatics, University of Bradford, Bradford BD7 1DP, UK; r.a.a.abd@bradford.ac.uk
3 Department of Information and Communication Engineering, Basrah University College of Science and Technology, Basrah 61004, Iraq
* Correspondence: n.ojaroudiparchin@napier.ac.uk

1. Introduction

In the ever-evolving landscape of modern wireless communication systems, the escalating demand for seamless connectivity has propelled the imperative for avant garde, versatile, and high-performance antennas to unprecedented heights. These antennas, standing at the forefront of future communication systems, serve as linchpins that can potentially revolutionise the overall performance of communication networks [1]. The exponential surge in connectivity needs has instigated a notable surge in dedicated research endeavours, underscoring the paramount importance of advancing antenna technologies. The adaptability of these antennas, which is essential to meet the diverse and evolving requirements of wireless services, necessitates not only careful consideration but also a continuous stream of innovative approaches [2]. Navigating the dynamic terrain of wireless communication, antennas are required to embody a plethora of features to ensure efficacy in both current and future systems. These encompass expansive and multi-frequency coverage, compact form factors, meticulously defined radiation patterns, multi-mode operational capabilities, cost-effectiveness in fabrication, energy efficiency, streamlined integration and assembly processes, and conformity to ever-evolving standards [3]. The symbiotic integration of Multiple Input Multiple Output (MIMO) configurations and phased array arrangements, fortified by adaptive and smart antennas, stands as the linchpin for significantly augmenting system capacity, thereby adeptly meeting the burgeoning demands of the unfolding epoch of wireless networks [4].

As we delve into the vast expanse of antennas and their applications, we recognise their pivotal role in shaping the contemporary technological landscape. The advent of 5G and the looming prospect of 6G technologies underscore the escalating role of antennas in shaping the future of communication [5]. From applications in wearable devices and biomedical wireless communication to their indispensable role in the Internet of Things (IoT) and smart cities, antennas are catalysts propelling the world towards unprecedented connectivity and technological advancement. They are the silent architects of our interconnected future, laying the foundation for the seamless exchange of information in the realms of autonomous vehicles, intelligent infrastructure, and beyond [6]. In this continuous journey of antenna evolution, we are witnessing not just technological progress but a transformative force shaping how we connect, communicate, and innovate. Antennas, in their myriad forms, have become integral components in diverse applications. Their role extends beyond mere conduits of signals; they are the lifelines of global connectivity, enabling progress and ushering in an era of unparalleled interconnectedness [7]. The relentless pursuit of excellence in antenna design and technology is not merely a shared ethos but a driving force propelling the evolution of wireless technologies into uncharted territories. The future promises antennas that seamlessly operate across a broader range of frequencies, with increased energy efficiency and reduced interference. These antennas are not just

Citation: Parchin, N.O.; See, C.H.; Abd-Alhameed, R.A. Editorial on Antennas. *Sensors* **2023**, *23*, 9643. https://doi.org/10.3390/s23249643

Received: 16 November 2023
Accepted: 24 November 2023
Published: 6 December 2023

The proposed design offers a comprehensive solution for spectrum utilisation efficiency in cognitive radio applications.

The work by Ayaz et al. (contribution 15) introduces a conformal cylindrical phased array antenna excited with composite right-/left-handed (CRLH) phase shifters. The novel aspect involves embedding magneto-static field-responsive micron-sized particles into the CRLH phase shifter structure, enabling variable phase shifts without increasing the insertion loss or phase error. The proposed antenna, operating in the C-band (5–6 GHz), exhibits low insertion loss and phase error, making it suitable for a printed and flexible electronics design. The prototype of the cylindrical phased array, with the particle-embedded CRLH phase shifters, demonstrates a close agreement between simulated and measured results, presenting a promising solution for conformal array applications.

The research by Nurhayati et al. (contribution 16) delves into the design of a 1×2 MIMO Palm Tree Coplanar Vivaldi Antenna in the E-Plane, aiming to overcome challenges related to mutual coupling and grating lobes. The authors employ diverse patch structures, such as Back Cut Palm Tree (BCPT) and Horizontal Wave Structure Palm Tree (HWSPT), demonstrating superior return loss and mutual scattering. Additionally, their incorporation of Metamaterial Lens Palm Tree (MLPT) into radar applications further extends the antenna's utility.

Bohao Tang et al. (Contribution 17) present an Evolutionary Computation approach for the Sparse Synthesis Optimization of Concentric Circular Antenna Arrays (CCAAs). By introducing hybrid solution initialisation and crossover methods, the proposed algorithm optimises CCAAs to reduce sidelobes while turning off specific antennas, addressing challenges related to overhead and excessive sidelobes in these arrays.

In contribution 18, by Ning Zhang et al., a dual-polarisation dipole antenna for a cylindrical phased array in Ku-Band is introduced. The dual-layer structured antenna, composed of butterfly shaped dipoles, demonstrates improved isolation between ports and effective scanning capabilities. The proposed design, applied in a 32-element cylinder array, indicates its potential for conformal devices in Ku-band frequencies.

Marcellin Atemkeng et al. (contribution 19) focus on the expansion of the African Very Long Baseline Interferometry (AVN) Network, particularly in the central African region. Analysing the scientific impact of additional antennas in countries like Cameroon and Chad, the paper emphasises the economic and human capital impacts of radio interferometers, contributing to the broader success of the AVN project.

In contribution 20, by Madiha Farasat et al., a simple yet effective approach for scattering suppression in multiband base station antennas is presented. By introducing a novel horizontal and vertical radiating element, the authors successfully mitigate high band pattern distortions, providing improved return loss and comparable pattern performance over the entire frequency band.

Irfan Ullah et al. (contribution 21) explore adaptive beamforming patterns of microstrip patch antenna arrays on flexible surfaces, emphasising the importance of conformal and self-adapting beamforming in the era of wireless spectrum growth. Their work offers insights into efficient and robust conformal phased-array antennas with multiple beamforming capabilities.

Chunli Wang et al. (contribution 22) propose novel coplanar meta-surface-based substrate-integrated waveguide antennas for K-Band beam scanning. Their innovative designs leverage coplanar rhombus- and hexagon-shaped meta-surfaces, providing low reflection and wide bandwidth, thus paving the way for highly directive scanning radiation in mm-Wave applications.

In contribution 23, by Aiting Wu et al., a compact four-port MIMO antenna for UWB applications is introduced. Utilising a polarisation diversity approach, the authors achieve a compact design with small dimensions and demonstrate promising performance in terms of impedance bandwidth, isolation, and gain, making it suitable for UWB applications.

Eunice Oluwabunmi Owoola et al. (contribution 24) propose an advanced marine predator algorithm (AMPA) to optimise non-uniform CAA beam patterns. The algorithm effectively optimises amplitude current and inter-element spacing for CAAs with varying element counts, achieving superior peak sidelobe level (SLL) suppression and convergence rates compared to other algorithms.

In contribution 25, by Bancha Luadang et al., a portable Yagi–Uda-based directional antenna for digital terrestrial television (DTT) is proposed. Simulations and tests show an impedance bandwidth of 75.93%, gain from 2.69 to 4.84 dBi, and a unidirectional radiation pattern. The radome has minimal impact, and outdoor/indoor tests yield power measurements at 514 MHz of 38.4 dBμV (−70.4 dBm) and 26.6 dBμV (−82.2 dBm), with carrier-to-noise ratios (C/N) of 11.6 dB and 10.9 dB.

Boasting a compact size of 30 × 18 × 1.6 mm^3, the ultra-wideband (UWB) multiple-input, multiple-output (MIMO) antenna by Weidong Mu et al. (contribution 26) exhibits flower-shaped radiating components, providing high isolation and excellent performance across the entire operation band of 4.3–15.63 GHz. The proposed design holds promise for diverse UWB applications.

In contribution 27, by Abdul Wajid et al., a dual-band wearable patch antenna with split-ring resonator (SRR)- and electromagnetic bandgap (EBG)-based designs is presented. The SRR-based antenna demonstrates improved gain, surface wave suppression, and compactness, showcasing its potential for wearable sensor networks and IoT applications.

Contribution 28 by Xianjin Yi et al. introduces a dual-band high-gain shared-aperture antenna that integrates Fabry–Perot and reflect-array mechanisms. Operating in both the S-band and X-band, the antenna achieves impressive gains with good isolation between the two frequency bands, making it an attractive candidate for high-performance communication systems.

The paper by Rozenn Allanic et al. (contribution 29) introduces polarisation-reconfigurable patch antennas using semiconductor-distributed doped areas (ScDDAs). The co-design method presented enables the optimisation of both the antenna and the ScDDAs, offering a practical and efficient solution for polarisation reconfiguration.

Contribution 30 by Yuefei Yan et al. addresses the growing importance of electromechanical coupling in high-frequency communication base station antennas. The authors establish a comprehensive channel capacity model, considering factors like positional shift, attitude deflection, and temperature change, providing insights crucial for the design and manufacture of advanced communication systems.

The research by Fengan Li et al. (contribution 31) introduces a novel metasurface comprising complementary units, enabling multi-band dual polarisation conversion. The design not only achieves remarkable frequency bands for linear and linear-to-circular polarisation conversion, but also shows radar cross-section (RCS) reduction capabilities, extending its application to multiple microwave frequency bands.

In the realm of 5G communication systems, Syed Aftab Naqvi and team (contribution 32) present a dual-band metamaterial-based absorber operating at 24 GHz and 28 GHz. Addressing the challenges of massive MIMO techniques, their absorber design enhances gain and spatial multiplexing while effectively isolating adjacent antennas. This innovative absorber holds promise for compact 5G devices by preventing unwanted interactions between antennas.

Kerlos Atia Abdalmalak and colleagues (contribution 33) contribute a unique feeding method for linear dielectric resonator antenna (DRA) arrays. By utilising standing waves and discrete metallic patches, the authors achieve a high-gain DRA array with low losses, presenting a cost-effective and compact design. The proposed 3D-printed structure demonstrates high efficiency, making it a noteworthy advancement in feeding techniques.

In contribution 34, Zhiyi Li et al. present a low-profile wideband magnetoelectric (ME) dipole antenna. The design involves the intricate bending of structures to achieve a reduced antenna height, while maintaining wideband properties. The relative bandwidth for VSWR < 2.0 and low boresight gain drop make this antenna suitable for applications requiring limited height and wideband characteristics.

Chao Ni et al. (contribution 35) propose a filtering slot antenna using characteristic mode analysis (CMA). Their approach involves analysing and designing characteristic magnetic currents to achieve a wide filtering bandwidth with stable gain. The fabricated prototype validates the design process, showing promising results for applications where stable gain and wide filtering are crucial.

Niamat Hussain et al. (contribution 36) contribute a conformal frequency-reconfigurable antenna catering to smart portable devices. With a focus on flexibility and multi-frequency operation, the antenna employs a coplanar waveguide-fed slotted circular patch. The integration of a frequency-reconfigurable element enhances its adaptability to various operating bands, making it suitable for modern wireless devices.

Addressing the demands of 5G millimetre-wave communications, Hussain Askari and colleagues (Contribution 37) present a wideband, circularly polarised magnetoelectric (CP ME) dipole antenna operating at 28 GHz. The unique geometry, including metallic plates and hook-shaped strips, enables stable gain and wideband characteristics. The antenna's compact footprint makes it well-suited for 5G smart devices and sensors.

Shenko Chura Aredo et al. (contribution 38) tackle the challenges associated with massive MIMO systems, proposing hardware-efficient solutions with optimal antenna selection. By evaluating low-resolution digital-to-analogue conversion and antenna selection techniques, the authors aim to reduce power consumption and enhance energy efficiency in massive MIMO systems.

The study by Ruisi Ge et al. (contribution 39) explores the influence of conformal metasurfaces on passive beam steering. The research introduces a passive approach to beam steering, utilising conformal metasurfaces on conventional patch antennas. The simplicity of the proposed system, combined with its passive nature, holds promise for low-power consumption beam steering systems.

The research from Faxiao Sun et al. (contribution 40) introduces a novel rotating shutter antenna designed for ultra-low-frequency (ULF) communication, with a focus on its potential application as a transmitter for magnetic induction (MI) underground communication systems. The authors employ advanced simulations and experiments to validate the antenna's performance, showcasing its ability to generate 2FSK signals in the ULF band. The proposed rotating shutter antenna holds promise for enhancing communication in challenging environments, such as underground spaces.

The article by Kazuhiro Honda (contribution 41) delves into the over-the-air testing of a massive multiple-input multiple-output (MIMO) antenna. The paper presents an innovative testing method involving a full-rank channel matrix created through a fading emulator with a minimal number of scatterers. By virtually positioning scatterers through antenna rotation, the study demonstrates a practical approach to assessing the performance of massive MIMO systems. This methodology offers insights into the real-world functionality of large-scale antenna arrays.

Marek Garbaruk et al. (contribution 42) propose a planar four-element ultrawideband (UWB) antenna array designed for the 6–8.5 GHz UWB frequency band. The symmetrical structure and elliptical-shaped radiators, fed by a stripline excitation network, contribute to uniform power distribution. With measured gains ranging from 6.4 to 10.8 dBi, the UWB antenna array exhibits favourable impedance matching. This research addresses the demand for high-performance antennas in the European Commission's designated UWB frequency band.

Shimaa A. M. Soliman et al. (contribution 43) offer insights into the analysis and design of an X-band reflect-array antenna tailored for a medium Earth orbit (MEO) remote sensing satellite system. The study explores various reflect-array configurations, including broadside and tilted pencil beam options, optimising the antenna for a nearly constant response across the coverage area. The use of a Yagi–Uda array and a genetic algorithm (GA) optimisation method demonstrates an efficient design process for achieving a flat-top radiation pattern.

In contribution 44, Raffaele Moretta et al. address the challenge of efficiently sampling the field radiated by a circumference arc source. The paper introduces a methodology to determine the minimum number of basis functions required for accurate representation and proposes an interpolation formula that optimally exploits non-redundant field samples. This work contributes to advancing the efficiency of field measurement techniques, crucial for applications where acquisition time is a limiting factor.

Mingcong Xie et al. (contribution 45) tackle the complex problem of aperture-level simultaneous transmit and receive (ALSTAR) with a digital phased array. The authors propose an adaptive random group quantum brainstorming optimisation (ARGQBSO) algorithm to simplify the array design, addressing the isolation between transmit and receive apertures. This innovative algorithm demonstrates robust performance, reducing complexity and enhancing overall efficiency in ALSTAR systems.

Hamdi Bilel and Aguili Taoufik (contribution 46) present a novel formulation based on Floquet spectral analysis for the almost-periodic modulation of massive finite and infinite strongly coupled arrays. The study has significant implications for applications such as dense-massive-MIMO, intelligent-surfaces, and future wireless technologies (5G and 6G). The numerical methods adopted, including the method of moments, pave the way for the advanced modelling of antenna structures in small areas with a large number of elements.

Collectively, these contributions demonstrate the breadth and depth of contemporary antenna research, addressing challenges and pushing the boundaries of what is possible in diverse applications. The innovative methodologies and findings presented in this Topic contribute to the ongoing evolution of antenna technologies, playing a vital role in shaping the future of wireless communication and sensing systems.

3. Conclusions

The odyssey through contributions 1 to 46 in this Topic has not only illuminated the cutting edge of antenna research but has also unfurled a vibrant tapestry of groundbreaking discoveries that significantly redefine the boundaries of what was once deemed possible in the realms of wireless communication, sensing systems, and beyond. Each contribution stands as a testament to the collective ingenuity and tireless efforts of researchers dedicated to pushing the frontiers of knowledge. The diverse array of antennas, from the non-linear chaotic optimisation algorithm to the innovative Floquet spectral analysis, shows the versatility and adaptability of antenna technologies. This collection encapsulates the dynamic and ever-evolving nature of the field, demonstrating how research endeavours continue to reshape our understanding and utilisation of antennas in various applications. As we traverse this intellectual landscape, it becomes evident that the relentless pursuit of excellence is not merely a shared ethos but a driving force propelling the evolution of wireless technologies into uncharted territories. In essence, the odyssey through these contributions serves as a compass, guiding us toward a future where antennas will play an even more pivotal role in shaping the landscape of global connectivity, sensing, and communication.

As we navigate the frontiers of antenna advancements, the prospect of future research beckons with enticing challenges and unexplored possibilities. One promising avenue lies in further enhancing the adaptability and robustness of antennas for emerging applications such as the Internet of Things (IoT) and smart cities. Exploring sustainable materials and fabrication techniques to reduce environmental impact is another crucial direction. Additionally, the integration of artificial intelligence and machine learning algorithms for the autonomous optimisation of antenna parameters holds immense potential [9]. The advent of 6G technologies and beyond will likely demand antennas that can seamlessly operate across a broader range of frequencies, with increased energy efficiency and reduced interference. The quest for antennas capable of supporting massive MIMO systems and intelligent surfaces represents a frontier in which interdisciplinary collaboration and innovation will play a pivotal role. In this ever-evolving field, embracing the challenges of real-world deployment scenarios, including the influence of environmental factors and dynamic interference, will be essential for pushing the boundaries of what can be achieved with antenna technologies [10]. As we stand at the precipice of a new era in wireless communication, the contributions within this collection not only mark significant milestones but also lay down the gauntlet for the exciting journey that lies ahead.

We wish to express our sincere gratitude to the esteemed authors whose remarkable contributions have profoundly enriched this MDPI Topic. Their invaluable research has not only propelled the field of antennas forward significantly, but has also contributed to the broader landscape of knowledge in impactful ways. Our deepest thanks extend to the diligent reviewers whose insightful comments and constructive feedback have played a pivotal role in elevating the quality of the articles presented here. Their expertise and unwavering dedication have been instrumental in ensuring the robustness and precision of the published works.

Acknowledgment goes to the dedicated editorial board and the supportive editorial offices of MDPI's *Electronics*, *Future Internet*, *Sensors*, and *Telecom* journals. Their consistent guidance and tireless assistance throughout the entire publication process have been indispensable, contributing significantly to the realisation of this comprehensive collection. We are confident that readers will find these papers not only informative but also enlightening, providing valuable insights into the dynamic and evolving realm of antennas. Looking ahead, we eagerly anticipate continued collaboration and future contributions from researchers in the field. Together, we aim to further advance and explore new frontiers in antenna technology. To all involved, we extend our heartfelt appreciation for your remarkable efforts and unwavering dedication to the advancement of knowledge in this field.

Conflicts of Interest: The authors declare no conflict of interest.

List of Contributions

1. Zhao, K.; Liu, Y.; Hu, K. Optimal Pattern Synthesis of Linear Array Antennas Using the Nonlinear Chaotic Grey Wolf Algorithm. *Electronics* **2023**, *12*, 4087; https://doi.org/10.3390/electronics12194087.
2. Wang, Z.; Zhang, R.; Song, W.; Xie, B.; Lin, X.; Li, H.; Tian, L. W-Band Broadband Circularly Polarized Reflectarray Antenna. *Electronics* **2023**, *12*, 3849; https://doi.org/10.3390/electronics12183849.
3. Wang, D.; Liu, Y.; Liang, J. Design of a Crossed Dielectric Resonator-Loaded, Dual-Band Dual-Polarized Differential Patch Antenna with Improved Port Isolation and Gain. *Electronics* **2023**, *12*, 3570; https://doi.org/10.3390/electronics12173570.
4. Abd Elrahman, S.; Elkhawaga, A.; Hussein, A.; Shaalan, A. Linear Antenna Array Sectorized Beam Scanning Approaches Using Element Position Perturbation in the Azimuth Plane. *Sensors* **2023**, *23*, 6557; https://doi.org/10.3390/s23146557.
5. Yassin, M.; Hussein, K.; Abbasi, Q.; Imran, M.; Mohassieb, S. Flexible Antenna with Circular/Linear Polarization for Wideband Biomedical Wireless Communication. *Sensors* **2023**, *23*, 5608; https://doi.org/10.3390/s23125608.
6. Noor, S.; Jusoh, M.; Sabapathy, T.; Rambe, A.; Vettikalladi, H.; M. Albishi, A.; Himdi, M. A Patch Antenna with Enhanced Gain and Bandwidth for Sub-6 GHz and Sub-7 GHz 5G Wireless Applications. *Electronics* **2023**, *12*, 2555; https://doi.org/10.3390/electronics12122555.
7. López-Álvarez, C.; López-Martín, M.; Rodríguez-González, J.; Ares-Pena, F. Maximizing Antenna Array Aperture Efficiency for Footprint Patterns. *Sensors* **2023**, *23*, 4982; https://doi.org/10.3390/s23104982.
8. Mingle, S.; Kampouridou, D.; Feresidis, A. Multi-Layer Beam Scanning Leaky Wave Antenna for Remote Vital Signs Detection at 60 GHz. *Sensors* **2023**, *23*, 4059; https://doi.org/10.3390/s23084059.
9. Teodorani, L.; Vernì, F.; Giordanengo, G.; Gaffoglio, R.; Vecchi, G. Experimental Demonstration of Beam Scanning of Dual-Metasurface Antenna. *Electronics* **2023**, *12*, 1833; https://doi.org/10.3390/electronics12081833.

10. Han, J.; Jia, D.; Du, B.; Han, G.; Jia, Y.; Zhao, Z. Design of Broadband Low-RCS Array Antennas Based on Characteristic Mode Cancellation. *Electronics* **2023**, *12*, 1536; https://doi.org/10.3390/electronics12071536.
11. Abbas, M.; Allam, A.; Gaafar, A.; Elhennawy, H.; Sree, M. Compact UWB MIMO Antenna for 5G Millimeter-Wave Applications. *Sensors* **2023**, *23*, 2702; https://doi.org/10.3390/s23052702.
12. Wang, R.; Liao, D.; Yang, F. Antipodal Linearly Tapered Slot Antenna with Quasi-Hemispherical Pattern Using Subwavelength Elements. *Electronics* **2023**, *12*, 628; https://doi.org/10.3390/electronics12030628.
13. Cai, L.; Tong, K. A Single-Fed Wideband Circularly Polarized Cross-Fed Cavity-Less Magneto-Electric Dipole Antenna. *Sensors* **2023**, *23*, 1067; https://doi.org/10.3390/s23031067.
14. Srikar, D.; Nella, A.; Mamidi, R.; Babu, A.; Das, S.; Lavadiya, S.; Algarni, A.; El-Shafai, W. A Novel Integrated UWB Sensing and 8-Element MIMO Communication Cognitive Radio Antenna System. *Electronics* **2023**, *12*, 330; https://doi.org/10.3390/electronics12020330.
15. Ayaz, M.; Iftikhar, A.; Braaten, B.; Khalil, W.; Ullah, I. A Composite Right/Left-Handed Phase Shifter-Based Cylindrical Phased Array with Reinforced Particles Responsive to Magneto-Static Fields. *Electronics* **2023**, *12*, 306; https://doi.org/10.3390/electronics12020306.
16. Nurhayati, N.; Setijadi, E.; de Oliveira, A.; Kurniawan, D.; Yasin, M. Design of 1 × 2 MIMO Palm Tree Coplanar Vivaldi Antenna in the E-Plane with Different Patch Structure. *Electronics* **2023**, *12*, 177; https://doi.org/10.3390/electronics12010177.
17. Tang, B.; Cai, L.; Yang, S.; Xu, J.; Yu, Y. Evolutionary Computation for Sparse Synthesis Optimization of CCAAs: An Enhanced Whale Optimization Algorithm Method. *Future Internet* **2022**, *14*, 347; https://doi.org/10.3390/fi14120347.
18. Zhang, N.; Xue, Z.; Zheng, P.; Gao, L.; Liu, J. Design of a Dual-Polarization Dipole Antenna for a Cylindrical Phased Array in Ku-Band. *Electronics* **2022**, *11*, 3796; https://doi.org/10.3390/electronics11223796.
19. Atemkeng, M.; Okouma, P.; Maina, E.; Ianjamasimanana, R.; Zambou, S. Radio Astronomical Antennas in the Central African Region to Improve the Sampling Function of the VLBI Network in the SKA Era?. *Sensors* **2022**, *22*, 8466; https://doi.org/10.3390/s22218466.
20. Farasat, M.; Thalakotuna, D.; Hu, Z.; Yang, Y. A Simple and Effective Approach for Scattering Suppression in Multiband Base Station Antennas. *Electronics* **2022**, *11*, 3423; https://doi.org/10.3390/electronics11213423.
21. Ullah, I.; Braaten, B.; Iftikhar, A.; Nikolaou, S.; Anagnostou, D. Beamforming with 1 × N Conformal Arrays. *Sensors* **2022**, *22*, 6616; https://doi.org/10.3390/s22176616.
22. Wang, C.; Gao, D.; Liang, L.; Wang, Y. Coplanar Meta-Surface-Based Substrate-Integrated Waveguide Antennas with Broadband and Low Reflections for K-Band Beam Scanning. *Sensors* **2022**, *22*, 6353; https://doi.org/10.3390/s22176353.
23. Wu, A.; Zhao, M.; Zhang, P.; Zhang, Z. A Compact Four-Port MIMO Antenna for UWB Applications. *Sensors* **2022**, *22*, 5788; https://doi.org/10.3390/s22155788.
24. Owoola, E.; Xia, K.; Ogunjo, S.; Mukase, S.; Mohamed, A. Advanced Marine Predator Algorithm for Circular Antenna Array Pattern Synthesis. *Sensors* **2022**, *22*, 5779; https://doi.org/10.3390/s22155779.
25. Luadang, B.; Pukraksa, R.; Janpangngern, P.; Pookkapund, K.; Dentri, S.; Kosulvit, S.; Phongcharoenpanich, C. Portable Wideband Directional Antenna Scheme with Semicircular Corrugated Reflector for Digital Television Reception. *Sensors* **2022**, *22*, 5338; https://doi.org/10.3390/s22145338.
26. Mu, W.; Lin, H.; Wang, Z.; Li, C.; Yang, M.; Nie, W.; Wu, J. A Flower-Shaped Miniaturized UWB-MIMO Antenna with High Isolation. *Electronics* **2022**, *11*, 2190; https://doi.org/10.3390/electronics11142190.

electronics

MDPI

Article

Floquet Spectral Almost-Periodic Modulation of Massive Finite and Infinite Strongly Coupled Arrays: Dense-Massive-MIMO, Intelligent-Surfaces, 5G, and 6G Applications

Hamdi Bilel * and Aguili Taoufik

Communication System Laboratory Sys'Com, National Engineering School of Tunis (ENIT), University of Tunis El Manar, BP 37, Le Belvédère, Tunis 1002, Tunisia; taoufik.aguili@gmail.com
* Correspondence: hbilel.enit@gmail.com; Tel.: +32-466422407

Abstract: In this study, we introduce a new formulation based on Floquet (Fourier) spectral analysis combined with a spectral modulation technique (and its spatial form) to study strongly coupled sublattices predefined in the infinite and large finite extent of almost-periodic antenna arrays (e.g., metasurfaces). This analysis is very relevant for dense-massive-MIMO, intelligent-surfaces, 5G, and 6G applications (used for very small areas with a large number of elements such as millimeter and terahertz waves applications). The numerical method that is adopted to model the structure is the method of moments simplified by equivalent circuits MoM GEC. Other numerical methods (such as the ASM-array scanning method and the windowing Fourier method) used this analysis in their kernel to treat periodic and pseudo-periodic (or quasi-periodic) arrays.

Keywords: Floquet analysis; MoM method; almost-periodic antenna arrays; Fourier analysis; strong mutual coupling; dense massive MIMO; mm and THz waves; 5G and 6G applications

Citation: Bilel, H.; Taoufik, A. Floquet Spectral Almost-Periodic Modulation of Massive Finite and Infinite Strongly Coupled Arrays: Dense-Massive-MIMO, Intelligent-Surfaces, 5G, and 6G Applications. *Electronics* **2022**, *11*, 36. https://doi.org/10.3390/electronics11010036

Academic Editors: Raed A. Abd-Alhameed, Naser Ojaroudi Parchin and Chan Hwang See

Received: 20 November 2021
Accepted: 20 December 2021
Published: 23 December 2021

1. Introduction

Antenna arrays, and in particular dense (or massive) coupled almost-periodic antenna arrays, have been of great interest in telecommunications and RF electronic applications (such as dense-massive-MIMO, smart-surfaces, 5G, and 6G applications) [1–4], including those used for very small surfaces with large numbers of elements such as millimeter and terahertz array applications. Therefore, the spectrum analysis based on a Fourier transformation (in the Floquet domain) is proposed to simplify the EM calculation on an elementary cell surrounded by periodic walls, as explained in [1–7] (in other research, they use periodic Green's functions) [8–19]. In the bibliography and recent studies, only spatial modulation techniques have been proposed to study periodic systems with large sizes [20–23]. Except in our case, a Fourier spectral analysis is presented to introduce a spectral modulation technique and its spatial equivalent (Fourier and Fourier inverse) to study strongly coupled sub arrays in an infinite and large finite almost-periodic support [24–27]. In this context, several numerical methods such as FDTD and FEM and other integral methods like the method of moments and full-wave methods [28] are proposed to resolve the given problem. In our work, we are interested only in the method of moments combined with equivalent circuits and Floquet analysis to study the suggested structure with the principle of modulation. This work is divided into four parts: we start with an explication of the almost-periodic modal (or spectral) modulation and its spatial equivalent to examine strongly coupled cells [29–33]. Then, we applied MoM-GEC as a numerical method to solve the proposed problem [1–7,34–36]. Next, several numerical results are presented to confirm the validity of the approach. Finally, some conclusions are established.

2. Almost-Periodic Modulation to Study Strongly Coupled Arrays

The concept is a signal-processing concept for a filter with a periodic spectral response, as shown in [30–32]. Its response is described as an impulse response function that is given by:

$$\int_{-\infty}^{\infty} K(\Omega - \alpha)u(\alpha)\, d\alpha = V(\Omega) \tag{1}$$

Its Fourier representation of Equation (1) (in the spatial domain) yields to

$$H(x)U(x) = V(x) \tag{2}$$

From the transformations by way of analogy, which we took into account, we note that x is a spatial coordinate and $\alpha \in [-\frac{\pi}{d_x}, \frac{\pi}{d_x}]$ is a spectral coordinate in the Brillouin domain and d_x is a spatial period. Note that Ω is a spectral coordinate, as α, where $H(x)$ is the Fourier transform of $K(\Omega)$ and is defined as the optical (or optoelectronic [30,31]) transfer function; $U(x)$ and $V(x)$ are the Fourier transforms of $u(\alpha)$ and $V(\Omega)$, respectively. More details are provided in [30].

2.1. Modulation in Infinite Almost-Periodic Arrays

Let us consider $f_\alpha(x)$ as a spectral periodic response for an infinite array that is written [4,33]:

$$f_\alpha(x) = \sum_{n=-N}^{N=+\infty} f_n(x)e^{+jn\alpha d_x} \tag{3}$$

α is a continuous Floquet mode $\alpha \in [-\frac{\pi}{d_x}, \frac{\pi}{d_x}]$; x (or $x_n = x(n) = nd_x$) represents the position in the spatial domain (a spatial distribution) ; and n is the position index in the periodic lattice.

with

$$f_n(x) = \frac{d}{2\pi}\int_{-\frac{\pi}{d}}^{\frac{\pi}{d}} f_\alpha(x)e^{-j\alpha n d_x}\, d\alpha \tag{4}$$

Considering f_0 is built from x_0, f_1 is identically constructed from $x_0 \pm d_x$. f_0 is a weight for x_0. In the same way, f_1 is a weight for $x_1 = x_0 \pm d_x$. Now, we can generalize the construction towards n elements. Then, $f_n(x) = f_0(x - nd_x)$, $n \in \mathbb{Z}$. $f_n(x)$ is a periodic function [34].

Now, let us put the given spectral modulation [30]:

$$U^{\infty}_{mod}(\alpha) = u(\alpha)f_\alpha(x) = u(\alpha)\sum_{n=-N}^{N=+\infty} f_n(x)e^{+j\alpha n d_x} \tag{5}$$

Next, we are considering: $TF(U^{\infty}_{mod}(\alpha)) = U^{\infty}_{mod}(x)$

It is possible to take $TF^{-1}(U^{\infty}_{mod}(\alpha)) = U^{\infty}_{mod}(x)$ (it depends the Fourier notation); As a result,

$$U^{\infty}_{mod}(x) = \sum_{n=-N}^{N=+\infty} f_n U(x - nd_x) \tag{6}$$

A simple demonstration from (5) to (6) is provided:

$$\begin{aligned} U^{\infty}_{mod}(x) &= TF^{-1}(U^{\infty}_{mod}(\alpha)) \\ &= TF^{-1}(u(\alpha)\textstyle\sum_{n=-N}^{N=+\infty} f_n e^{+j\alpha x}) \\ &= \textstyle\sum_{n=-N}^{N=+\infty} f_n TF^{-1}(u(\alpha)e^{+j\alpha x}) \\ &= \textstyle\sum_{n=-N}^{N=+\infty} f_n(\frac{d_x}{2\pi}\int_{-\frac{\pi}{d_x}}^{\frac{\pi}{d_x}} u(\alpha)e^{+j\alpha x}e^{-j\alpha n d_x}\, d\alpha) \end{aligned} \tag{7}$$

$$= \sum_{n=-N}^{N=+\infty} f_n\left(\frac{d_x}{2\pi}\int_{-\frac{\pi}{d_x}}^{\frac{\pi}{d_x}} u(\alpha)e^{+j\alpha(x-nd_x)}\,d\alpha\right)$$
$$= \sum_{n=-N}^{N=+\infty} f_n U(x-nd_x)$$

Equation (7) is established referring to the theorem of aliasing given in [37] and Equation (6.107) of [38].

Finally, a spatial modulation is derived from Equation (2) [24,30]

$$V_{mod}^{\infty}(x) = H(x)U_{mod}^{\infty}(x) = H(x)\sum_{n=-N}^{N=+\infty} f_n U(x-nd_x) \quad (8)$$

Notice that our old published work [34] on rectangular pulse functions can be a special case of this process of analysis and development, where $f_{n,\alpha} = \frac{W}{2d}\,\mathrm{sinc}((\frac{2n\pi}{d}+\alpha)\frac{w}{2})$ are the Fourier series coefficients of the periodic pulse train (in the presence of Floquet modes), as described in [39].

For more details, following the pulse (or impulse) trains, we can introduce the Floquet phases as follows [4,33]:

$$f_\alpha(x) = \sum_{n=-N}^{N=+\infty} rect(x-nd_x)e^{+j\alpha x} \quad (9)$$
$$= \sum_{n=-N}^{N=+\infty} rect_n(x)e^{+j\alpha x}$$

Note that: $\sum_{n=-N}^{N=+\infty} rect_n(x) = \sum_{n=-N}^{N=+\infty} f_n e^{+j\frac{2n\pi}{d_x}x}$ (for a standard rectangular pulse train), which explains how to recover (reconstruct) a pulse train by means the Fourier series, as shown in Figure 1 and explained in [39,40] (see the subsection on periodic pulse and impulse trains in [39]). By adding the Floquet contribution $e^{+j\alpha x}$, we obtain the definition $f_\alpha(x) = \sum_{n=-N}^{N=+\infty} rect(x-nd_x)e^{+j\alpha x}$ (see Figure 1).

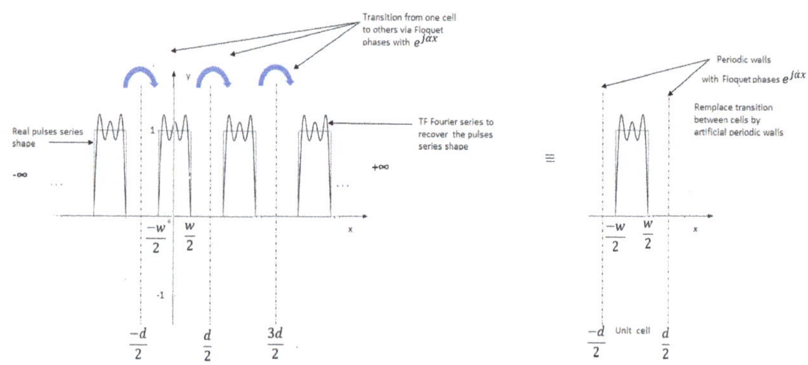

Figure 1. Construction of $f_\alpha(x)$ as given in Equations (9)–(11).

Rect is a rectangular function with a width W.

Then, we can develop a series of rectangle functions into a series of Fourier functions, which allows us to write: (see Figure 1 and [34])

$$f_\alpha(x) = \sum_{n=-N}^{N=+\infty} f_{n,\alpha}e^{+j\frac{2n\pi}{d_x}x}e^{+j\alpha x} = \sum_{n=-N}^{N=+\infty} f_{n,\alpha}e^{+j(\frac{2n\pi}{d_x}+\alpha)x} \quad (10)$$

we can note: $K_{x,n} = \frac{2n\pi}{d_x} + \alpha$ as a wavenumber that leads to decompose

$$f_\alpha(x) = \sum_{n=-N}^{N=+\infty} f_{n,\alpha} e^{+jK_{x,n}x} \tag{11}$$

With

$$\begin{aligned} f_{n,\alpha} &= \frac{1}{d_x} \int_{-\frac{d}{2}}^{\frac{d}{2}} f_\alpha(x) e^{-jK_{x,n}x}\, dx \\ &= \frac{1}{d_x} \int_{-\frac{w}{2}}^{\frac{w}{2}} e^{-jK_{x,n}x}\, dx \\ &= \frac{W}{2d_x} \operatorname{sinc}(K_{x,n} \tfrac{w}{2}) = \frac{W}{2d_x} \operatorname{sinc}((\tfrac{2n\pi}{d_x} + \alpha)\tfrac{w}{2}) \end{aligned} \tag{12}$$

Equation (12) is proven based on the example (Example 3.17) of [39].

Now let us apply the superposition theorem of Equation (10) (based on Floquet states) to generate the spatial solution [9,19]: it is also called modulation (see Equation (7))

$$\begin{aligned} &\frac{d_x}{2\pi} \int_{-\frac{\pi}{d_x}}^{\frac{\pi}{d_x}} f_\alpha(x) e^{-j\alpha x}\, d\alpha = TF^{-1}(f_\alpha(x)) \\ &= \frac{d_x}{2\pi} \int_{-\frac{\pi}{d_x}}^{\frac{\pi}{d_x}} (\textstyle\sum_{n=-N}^{N=+\infty} f_{n,\alpha} e^{+j(\frac{2n\pi}{d_x}+\alpha)x}) e^{-j\alpha x}\, d\alpha \\ &= \frac{d_x}{2\pi} \int_{-\frac{\pi}{d_x}}^{\frac{\pi}{d_x}} (\textstyle\sum_{n=-N}^{N=+\infty} f_{n,\alpha} e^{+j(\frac{2n\pi}{d_x})x})\, d\alpha \\ &= \frac{d_x}{2\pi} \int_{-\frac{\pi}{d_x}}^{\frac{\pi}{d_x}} \textstyle\sum_{n=-N}^{N=+\infty} rect_n(x)\, d\alpha \\ &= \frac{d_x}{2\pi} \int_{-\frac{\pi}{d_x}}^{\frac{\pi}{d_x}} \textstyle\sum_{n=-N}^{N=+\infty} rect(x - nd_x)\, d\alpha \\ &= \frac{d_x}{2\pi} \textstyle\sum_{n=-N}^{N=+\infty} rect(x - nd_x) \int_{-\frac{\pi}{d_x}}^{\frac{\pi}{d_x}} d\alpha \\ &= \textstyle\sum_{n=-N}^{N=+\infty} rect(x - nd) \\ &= \textstyle\sum_{n=-N}^{N=+\infty} rect(x - nd_x) \\ &\text{The spatial solution is a periodic pulses series} \\ &\text{and is similar to} = U_{mod}^{\infty}(x) \text{ when}(u(\alpha) = 1) \end{aligned} \tag{13}$$

What we get in (13) is similar to Equation (14) of our published work [34] (and Equation (4) of the WATANABE reference [21]). Additionally, it is of the same type as Equation (6) and the expansion that follows in Equation (7). Then, a spatial modulation that was performed in Equation (8) follows.

2.2. Modulation in Finite Almost-Periodic Arrays

As previously explained in [4,6], the interactions between cells in the spectral domain for periodic finite arrays are governed by a discrete phase law such that $\alpha_p = \frac{2\pi p d_x}{D} = \frac{2\pi p}{N_x}$ (with $-\frac{N_x}{2} \leq p \leq -\frac{N_x}{2} - 1$), which comes from a rule-of-three math reasoning.

For a large period of finite arrays, $D \longrightarrow 2\pi$ (2π is the hole interval of phases).

For a local period of finite arrays, $pd_x \longrightarrow \alpha_p =?$ (is the spectral contribution for one cell in position pd_x) (p is the index position in a finite array, and d is the local period).

So, $\Longrightarrow \alpha_p = \frac{2\pi p}{N_x}$, where N_x is the total number of elements in finite array, and p is the index position [13].

Figure 2 explains how to discretize the phases from the infinite case to the finite case. According to the same Figure 2, each cell interacts spectrally with its neighbors through the continuous Floquet modes α (or the phase shift $e^{+j\alpha x}$) in the infinite case and α_p (or the phase shift $e^{+j\alpha_p x}$) in the finite case.

This allows writing the spectral solution as the sum of discrete Floquet states [4,33],

$$f_{\alpha_p}(x) = \frac{1}{\sqrt{N_x}} \sum_{n=-\frac{N_x}{2}}^{\frac{N_x}{2}-1} f_n(x) e^{+j\alpha_p n d_x} \tag{14}$$

and

$$f_n(x) = \frac{1}{\sqrt{N_x}} \sum_{p=-\frac{N_x}{2}}^{\frac{N_x}{2}-1} f_{\alpha_p}(x) e^{-j\alpha_p n d_x} \tag{15}$$

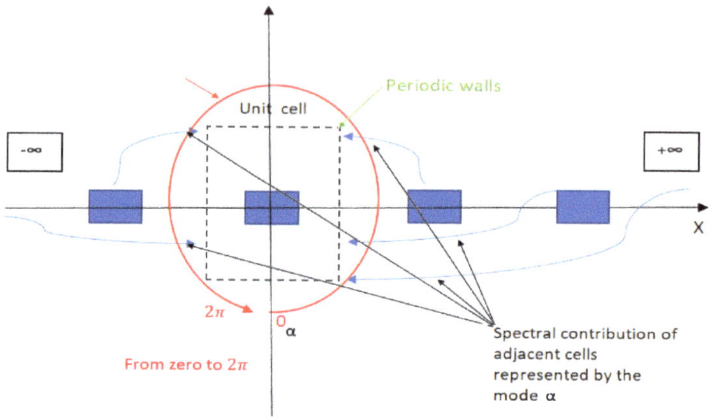

Contribution of continuous Floquet modes α in an infinite array

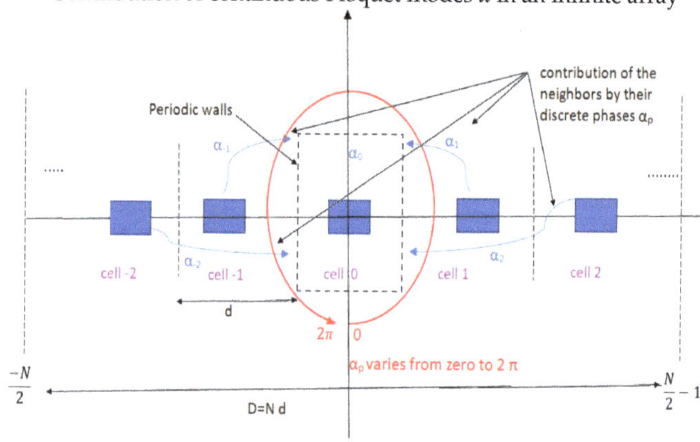

Contribution of discrete Floquet modes α_p in a finite array.

Figure 2. Spectral representation of the interactions of a unit cell with its neighbors (infinite and finite cases) (valid for strong coupling interaction by using Floquet phases).

In the same way, $f_n(x) = f_0(x - nd_x)$ with $-\frac{N_x}{2} \leq n \leq -\frac{N_x}{2} - 1$, and we can rewrite the spectral modulation law for a discrete Floquet mode [30],

$$\begin{aligned} U_{mod}^{Finite}(\alpha_p) &= u(\alpha_p) f_{\alpha_p}(x) \\ &= u(\alpha_p) \frac{1}{\sqrt{N_x}} \sum_{n=-\frac{N_x}{2}}^{\frac{N_x}{2}-1} f_n(x) e^{+j\alpha_p n d_x} \end{aligned} \tag{16}$$

Thus, the DFT is written as $DFT(U_{mod}^{Finite}(\alpha_p)) = U_{mod}^{Finite}(x_i)$ with $x_i = id_x$ and $-\frac{N_x}{2} \leq i \leq -\frac{N_x}{2} - 1$

from which

$$U_{mod}^{Finite}(x) = \frac{1}{\sqrt{N_x}} \sum_{n=-\frac{N_x}{2}}^{\frac{N_x}{2}-1} f_n U(x - nd_x) \tag{17}$$

Eventually

$$V_{mod}^{Finite}(x) = H(x)U_{mod}^{Finite}(x) \quad (18)$$
$$= H(x)\frac{1}{\sqrt{N_x}}\sum_{n=-\frac{N_x}{2}}^{\frac{N_x}{2}-1} f_n U(x - nd_x)$$

3. MoM-GeC Modelization Based on Floquet Analysis

The problem formulation is already explained in [1–7]. It explains the development of the method of moments combined with Floquet spectral analysis to determine the electromagnetic performance in the presence of strong mutual couplings such as input impedance, surface current, surface electric field, radiated field, and directivity... etc.

4. Numerical Results

A part of our results was presented in [4,6,34,35]. Let us now display the other obtained results.

This approach can be applied to frequency-modulated continuous-wave (FMCW) radar antennas (to scan the radiation beam produced by very small areas of an antenna system) as well as to antennas that are massively placed in a coupled almost-periodic antenna array. The provided antenna example can be used to show how to model a 77 GHz (2 × 4) antenna array for frequency-modulated continuous-wave (FMCW) radar applications. The availability of antennas and antenna arrays in and on vehicles has become ordinary with the inclusion of remote crash-recognition-and-aversion systems, as well as lane-departure alerting systems. The two frequency bands appropriate for these systems are approximately 24 GHz and 77 GHz, respectively. In this model, we consider the microstrip patch antenna as a phased-array radiator. The dielectric substrate is air. According to Figures 3 and 4, the patch antenna has its first resonance (parallel resonance) at 24.52 and 77.9 GHz (after adjusting the length of the used antenna) . It is a common practice to shift this resonance to 24 and 77 GHz by scaling the length of the planar dipole antenna, as described in [35].

The next stage is to reconfirm the reflection coefficient of the planar antenna dipole, as depicted in Figure 5 and Figure 1 of [34]. The purpose of this check is to consider a good impedance match. It is very common to highlight the value as a limit value for the calculation of the bandwidth of the antenna. The deepest minima at 24 GHz and 77 GHz indicated a good fit with 120π. The bandwidths of the antennas are roughly 1 GHz and 2 GHz, respectively. Thus, the spectrum bands are 23.5 GHz to 25.5 GHz and 76.5 GHz to 77.5 GHz. Finally, in terms of input impedances and S parameters, a good comparison between the adopted MoM GeC method and the references [41,42] is obtained.

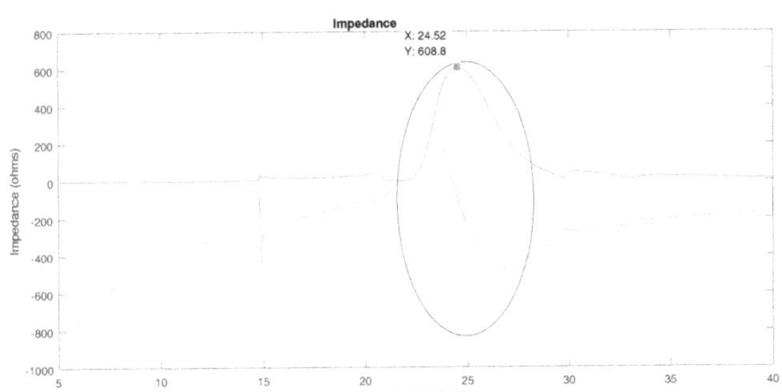

Figure 3. Impedance variation against frequency band around 24 GHz: obtained by the MoM-GEC method.

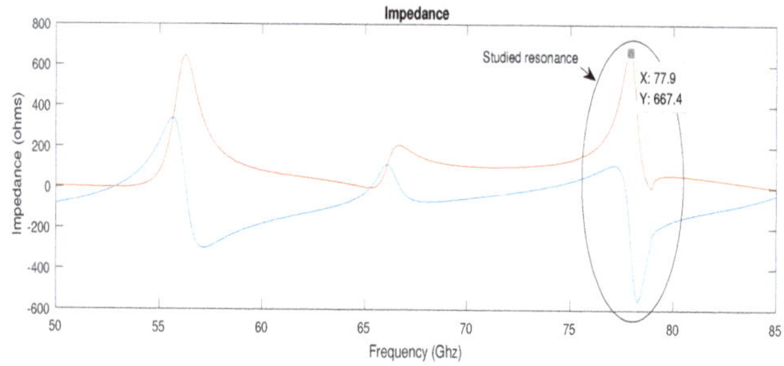

Figure 4. Impedance variation against frequency band around 77 GHz: obtained by the MoM GEC method.

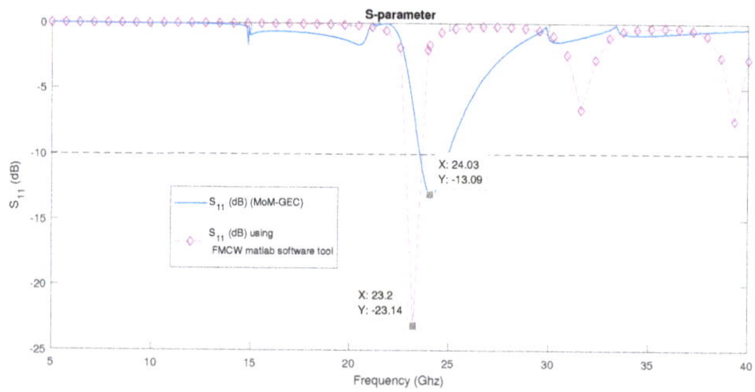

Figure 5. S_{11}-parameter (dB) variation against frequency band around 24 GHz: a comparison between the MoM-GEC and a MATLAB tool (see references [41,42]).

Therefore, the radiation pattern response for a small 2 × 4 antenna array is proposed based on the Floquet analysis (via the superposition theorem). Let us assume that the radar antenna system operates at 77 GHz with a bandwidth of 700 MHz. The following Figure 1 of [35] shows the spatial-radiation pattern of the resulting planar antenna using the superposition theorem of 2 × 4 Floquet radiation examples. Then, the sum of the discrete Floquet radiation patterns assigned to the FMCW radar permits the prediction of the global spatial-radiation pattern (what is called spatial modulation). A good comparison of the given numerical radiation pattern and the radiation obtained by patch array and cosine array is presented in Figure 3 of [35]. After validating the FMCW radar, we propose to evaluate the approach for a very large number of elements that uses the same frequency band at 24 and 77 GHz (for example, a lattice of 100 elements). Figure 6 gives an example of the superposition theorem (or a spatial modulation) for a large array to generate a spatial radiation pattern through the addition of the radiation patterns of Floquet states. After that, Figure 4 of [35] presents the variation of the spatial-radiation pattern (obtained using Floquet analysis) in the function of steering angles for 100 antenna elements that are distributed in a uni-dimensional configuration (for 5G application). In the same way, Figure 7 and Figure 5 of [35] show the variation of the 3D radiation pattern against different steering angles that are described with ($\theta_0 = 45°$, $\phi_0 = 0°$), and ($\theta_0 = 90°$, $\phi_0 = 0°$), in Cartesian coordinates and (u,v) space, respectively. Following the same study, Tables 1 and 2 show the directivity values for each Floquet state and the superposition theorem (in the two cases of the FMCW radar and the 5G application), with two different steering angles and at the frequencies of 24 GHz and 77 GHz. From Figure 1 of [35] and Figure 6, we can see that the radiation patterns are nearly identical and verify the condition where the directivities are similar for both the Floquet states and the superposition theorem. This is why, in both Tables 1 and 2, we find that all directivity values are identical (for the Floquet states as well as for the superposition), even when we change the steering angle. Knowing that generally, the directivity value close to 20 dB satisfies a narrow beam angle of about 20 degrees, as shown graphically in Figure 7a.

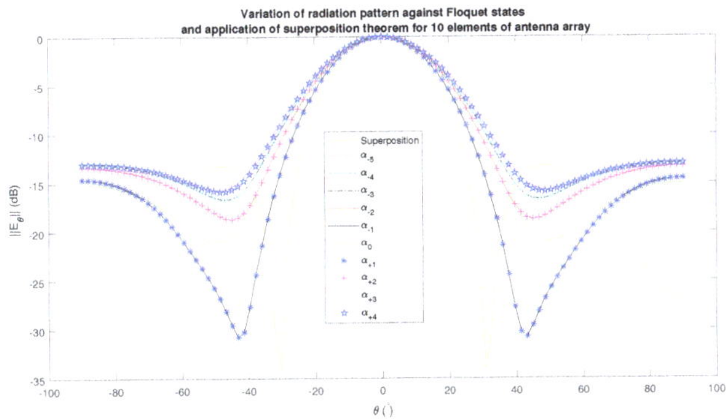

Figure 6. Variation of radiation pattern against Floquet states and application of superposition theorem for 10 elements of antenna array (uni-dimentionnal configuration) at 77 GHz: obtained by the MoM-GEC method.

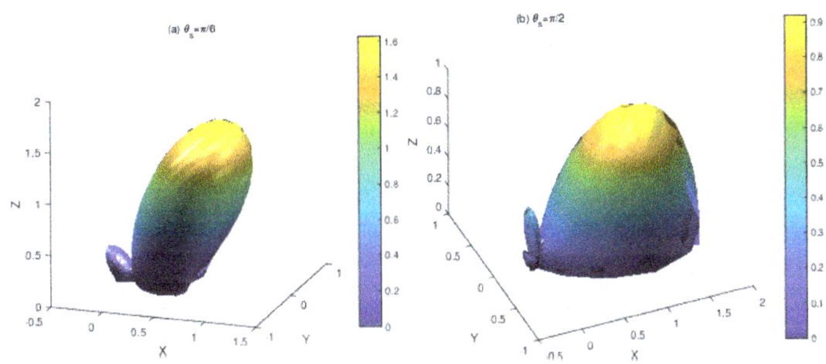

Figure 7. Variation of radiation pattern against Floquet states and application of superposition theorem for 100 elements of antenna array (uni-dimentionnal configuration) at 77 GHz: obtained by the MoM-GEC method.

Table 1. Directivity versus some Floquet states (considering 100 antenna arrays) and the superposition theorem (or the modulation as explained in Formula (13), which transformed to study a finite array) for $\phi_s = 0$, $\theta_s = 30°$ steering angles (used for 5G application).

Floquet States	Directivity Values (dB) at 24 GHz
$(\alpha_{-49}, \beta = 0)$	23.0200
$(\alpha_{-10}, \beta = 0)$	22.5953
$(\alpha_{+20}, \beta = 0)$	22.8874
$(\alpha_{+30}, \beta = 0)$	23.2055
Superposition	23.3009
Floquet States	**Directivity Values (dB) at 77 GHz**
$(\alpha_{-49}, \beta = 0)$	22.8771
$(\alpha_{-10}, \beta = 0)$	24.1083
$(\alpha_{+20}, \beta = 0)$	23.4323
$(\alpha_{+30}, \beta = 0)$	23.1353
Superposition	23.5783

Table 2. Directivity versus Floquet states and the superposition theorem (or the modulation as explained in Formula (13), which transformed to study a finite array) for $\phi_s = 0,\ \theta_s = 45°$ steering angles (FMCW radar application).

Floquet States	Directivity Values (dB) at 24 GHz
$(\alpha_{-2}, \beta_{-1})$	21.7249
$(\alpha_{-1}, \beta_{-1})$	26.7351
(α_{0}, β_{-1})	21.1290
$(\alpha_{+1}\beta_{-1})$	20.8615
(α_{-2}, β_{0})	20.6344
(α_{-1}, β_{0})	21.0667
(α_{0}, β_{0})	14.8990
(α_{+1}, β_{0})	21.0737
Superposition	21.0455
Floquet States	**Directivity Values (dB) at 77 GHz**
$(\alpha_{-2}, \beta_{-1})$	22.3762
$(\alpha_{-1}, \beta_{-1})$	18.0573
(α_{0}, β_{-1})	21.9229
$(\alpha_{+1}\beta_{-1})$	22.1662
(α_{-2}, β_{0})	20.9345
(α_{-1}, β_{0})	21.3088
(α_{0}, β_{0})	13.7990
(α_{+1}, β_{0})	21.3088
Superposition	21.7022

5. Conclusions

In this article, we illustrated the principle of Floquet spectral modulation based on the Fourier analysis (and its spatial form) to study almost-periodic sub-arrays (with finite size) in the presence of strong mutual coupling interaction, defined on infinite support (or a really large finite size). This study is very useful for the new generation of technologies based on millimeter and terahertz waves in phased arrays, for example, in dense-massive-MIMO, smart-surfaces, 5G, and 6G applications. In future work, we are interested to investigate the randomly modulated almost-periodic arrays (also in the presence of strong coupling).

Author Contributions: Conceptualization, H.B.; software, H.B. and A.T.; formal analysis, H.B. and A.T.; investigation, H.B. and A.T.; writing—original draft preparation, H.B. and A.T.; writing—review and editing, H.B. and A.T. All authors have read and agreed to the published version of the manuscript.

Funding: This project received funding from the Laboratory Sys'Com-ENIT (LR-99- ES21)-National Engineering School of Tunis ENIT, Tunis, Tunisia, 1002.

Data Availability Statement: Not applicable.

Acknowledgments: I would like to thank each of the Ammar Bouallegue (ENIT Tunis-El Manar University), Christophe Craeye (UCLouvain University), Junwu TAO (N7-University of Toulouse) and especiallyHenri Baudrand (N7-University of Toulouse), who died recently, for their help in the achievement of this work.

Conflicts of Interest: The authors declare no conflict of interest.

References

1. Mekkioui, Z.; Baudrand, H. *Bi-Periodic Centered-Fed Microstrip Leaky-Wave Antenna (LWA) Analysis by a Source Modal Decomposition in Spectral Domain*; IET Microwaves Antennas and Propagation: Sarrebruck, Germany, 2009; pp. 1141–1149.
2. Baudrand, H.; Titaouine, M.; Raveu, N.; Fontgland, G. Electromagnetic modeling of planar almost periodic structures. In Proceedings of the SBMOI/IEEE MTT-S International Microwave and Optoelectronics Conference, Belem, Brazil, 3–6 November 2009; pp. 427–431.
3. Azizi, M.K.; Latrach, L.; Raveu, N.; Gharsallah, A.; Baudrand, H. A new approach of almost periodic lumped elements circuits by an iterative method using auxiliary sources. *Am. Appl. Sci.* **2013**, *10*, 1457–1472. [CrossRef]
4. Hamdi, B.; Aguili, T.; Raveu, N.; Baudrand, H. Calculation of the mutual coupling parameters and their effects in 1-D planar almost periodic structures. *Prog. Electromagn. Res.* **2014**, *59*, 269–289. [CrossRef]
5. Hamdi, B.; Aguili, T.; Tao, J. *Modélisation des Circuits Presque-Périodiques*; Éditions Universitaires Européennes: Sarrebruck, Germany, 2019; 156p, ISBN 6138476387, ISBN 978-6138476382.
6. Hamdi, B.; Aguili, T.; Baudrand, H. Floquet Modal Analysis To Modelize and Study 2-D Planar Almost Periodic Structures In Finite And Infinite Extent With Coupled Motifs. *Prog. Electromagn. Res. B* **2015**, *62*, 63–86. [CrossRef]
7. Latifa, N.B.; Aguili, T. Synthesis and Optimization of Almost Periodic Antennas Using Floquet Modal Analysis and MoM-GEC Method. *J. Electromagn. Anal. Appl.* **2019**, *11*, 1–16. [CrossRef]
8. Ishimaru, A.; Coe, R.J.; Miller, G.E.; Green, W.P. Finite periodic structure approach to large scanning array problems. *IEEE Trans. Antennas Propagat.* **1985**, *33*, 1213–1220. [CrossRef]
9. Valerio, G.; Baccarelli, P.; Burghignoli, P.; Galli, A.; Rodrguez-Berral, R.; Mesa, F. Analysis of periodic shielded microstrip lines excited by nonperiodic sources through the array scanning method. *Radio Sci.* **2008**, *43*, 1–15. [CrossRef]
10. Rodrguez-Berral, R.; Mesa, F.; Baccarelli, P.; Burghignoli, P. Excitation of a periodic microstrip line by an aperiodic delta-gap source. *IEEE Trans. Antennas Propagat. Lett.* **2009**, *8*, 641–644. [CrossRef]
11. Dardenne, X.; Craeye, C. Application of the Array Scanning Method with Windowing to the Analysis of Finite Rectangular Periodic Structures. In Proceedings of the 2005 18th International Conference on Applied Electromagnetics and Communications, Dubrovnik, Croatia, 12–14 October 2005; pp. 1–4. [CrossRef]
12. Craeye, C.; González-Ovejero, D. A review on array mutual coupling analysis. *Radio Sci.* **2012**, *46*, 1–25. [CrossRef]
13. Craeye, C. Expoitation of Infinte-Array Results for Accurate Solution of Finite Widebands Arrays. In Proceedings of the URSI Electromagnetic Theory Symposium (EMTS'07), San Diego, CA, USA, 27–31 May 2019.
14. Mesa, F.; di Nallo, C.; Jackson, D.R. The theory of surface-wave and space-wave leaky-mode excitation on microstrip lines. *IEEE Trans. Microw. Theory Tech.* **1999**, *47*, 207–215. [CrossRef]
15. Capolino, F. *Theory and Phenomena of Metamaterials*; CRC Press: Boca Raton, FL, USA, 2017.
16. Skrivervik, K.; Mosig, L.R. Finite phased array of microstrip patch antennas: The infinite array approach. *IEEE Trans. Antennas Propagat.* **1992**, *40*, 579–582. [CrossRef]
17. Skrivervik, K.; Mosig, J.R. *Réseaux Périodiques D'antennes Microruban*; Thèse n° 1032; EPFL: Lausanne, Switzerland, 1992.
18. Sze, K.Y.; Shafai, L. Reflection properties of infinite periodic arrays of rectangular conducting patches. *Can. J. Electr. Comput.-Eng.-Rev. Can. Genie Electr. Inform.* **1999**, *24*, 27–33.
19. Vardaxoglou, J.C. *Frequency Selective Surfaces, Analysis and Design*; John Wiley and Sons: Hoboken, NJ, USA, 1997.
20. De Sabata, A.; Matekovits, L.; Lipan, O. Band pattern of commensurate modulated periodic structures. *IET Microw. Antennas Propag.* **2017**, *11*, 1303–1307. [CrossRef]
21. Salazar-Arrieta, J.d.J.; Halevi, P. Wave Propagation in Electric Periodic Structure in Space with Modulation in Time (2D+ 1). I. Theory. *arXiv* **2021**, arXiv:2105.07105.
22. Patel, A.M.; Grbic, A. A Printed Leaky-Wave Antenna Based on a Sinusoidally-Modulated Reactance Surface. *IEEE Trans. Antennas Propag.* **2011**, *59*, 2087–2096. [CrossRef]
23. Oliner, A.; Hessel, A. Guided waves on sinusoidally-modulated reactance surfaces. *IRE Trans. Antennas Propag.* **1959**, *7*, 201–208. [CrossRef]
24. Watanabe, K. Study on Spectral-Domain Formulation of Electromagnetic Scattering by Periodic Strip Array with Period Modulation. In Proceedings of the 2019 IEEE Conference on Antenna Measurements & Applications (CAMA), Kuta, Indonesia, 1 October 2019; pp. 165–166. [CrossRef]
25. Watanabe, K. Spectral-domain approach to electromagnetic scattering from imperfectly periodic structures. In Proceedings of the 2010 International Conference on Mathematical Methods in Electromagnetic Theory, Kyiv, Ukraine, 6–8 September 2010; pp. 1–6.
26. Watanabe, K.; Yasumoto, K. Two-dimensional electromagnetic scattering of non-plane incident waves by periodic structures. *Prog. Electromagn. Res.* **2007**, *74*, 241–271. [CrossRef]
27. Bhattacharyya, A.K.*Phased Array Antennas: Floquet Analysis, Synthesis, BFNs and Active Array Systems*; John Wiley & Sons: Hoboken, NJ, USA, 2006.
28. Abdallah, Y.; Menudier, C.; Thevenot, M.; Monediere, T. Investigations of the effects of mutual coupling in reflectarray antennas. *IEEE Antennas Propag. Mag.* **2013**, *55*, 49–61. [CrossRef]
29. Appendix B: Almost-Periodic Functions and Their Spectra. 2009. Available online: https://rd.springer.com/content/pdf/bbm:978-1-84882-181-1%2F1.pdf (accessed on 20 November 2021).

domain, analog domain, and digital domain [4]. Yet the methods in the propagation domain normally achieved only 30~40 dB of isolation in a narrow bandwidth [5] by antenna separation, antenna orientation, and placing absorptive shielding. Due to the use of active attenuators, phase shifters, and delayers, the analog-domain cancellation methods introduced inestimable non-linear distortion and they are often power-intensive, expensive, and physically large [6]. Digital-domain methods include digital self-interference cancellation, transmitting and receiving beamforming, channel estimation, and fingerprint signals. These approaches will be flexible, efficient, and reliable if the signal of interest is effectively received within the dynamic range of ADCs and DACs. Ahmed et al. proposed an all-digital nonlinear estimation and self-interference cancellation technology to improve the isolation above 20 dB in a single-receiving and single-transmitting full-duplex system [7]. Qiu et al. verified that adaptive beamforming provides a high isolation between the transmitter and the receiver [8]. They optimized the beamformers using the linear constrained minimum variance algorithm to provide at least 110 dB of isolation, without affecting target detection. Liang et al. proposed a method to realize adaptive transmit beamforming together with digital SIC [9], which provides positive inspiration for the advancement of digital cancellation.

Aperture-level simultaneous transmit and receive (ALSTAR) based on a digital phased array was proposed by the MIT Lincoln Laboratory in 2016 [10]. ALSTAR integrates digital cancellation and transmit and receive beamforming technologies to achieve extremely high isolation, termed effective isotropic isolation (EII). Since the theory is immature, a lot of work is still needed to extend the method to practice. For instance, effective isotropic radiated power (EIRP) and effective isotropic sensitivity (EIS) of the system should be taken into account to ensure a high emission efficiency and receiver performance. The increase of target parameters complicates the design of the STAR array. In this case, the designers wish to make a trade between EII, EIRP, and EIS to meet the requirements of different scenarios. The authors in [11] used an alternate optimization (AO) to design the EII of the ALSTAR array and then loaded a weight diagonally in the noise covariance matrix to reveal the relationship between EII and the gain. Yet this approach only analyzes the connection between them, and does not study how to trade EII, EIRP, and EIS. Furthermore, the AO algorithm needs to construct identities of two objective functions for alternate iteration updates in two directions, and perform a beamformer optimization at every elevation angle, which increases the computational complexity and hardware cost. This may limit the further promotion and application of ALSTAR.

This paper is dedicated to the improvement of the overall performance of the ALSTAR array and the reduction of complexity. We trade EII, EIRP, and EIS by proposing the weight w_f to enhance the overall performance and optimize a set of transmitting and receiving beamformers independent of the elevation angle in order to reduce complexity. We found that the swarm intelligence optimization methods have the potential to achieve these goals. They have been successful in the fields such as beamforming-based pattern synthesis [12], array optimization [13,14], DC brushless motor efficiency problems [15], Loney's solenoid problem [16], and stock index forecasting [17]. Extensive literature reveals that compared to traditional particle swarm optimization (PSO), genetic algorithm (GA), and differential evolution (DE), the brainstorm optimization (BSO) algorithm [18] has the characteristics of fast convergence, excellent robustness, and a strong global optimization ability in solving non-convex, multi-objective, and multi-modal optimization problems. This algorithm is quite suitable for the design of the ALSTAR array, but there is still space for improvement in the BSO algorithm [19].

In the classic BSO algorithm, each individual in the population may become a potential solution, which corresponds to a new idea in the process of human brainstorming. The process can be summarized in three steps. First of all, a blocking method similar to k-means is used to group individuals in the population, and the individual with the best fitness value will be the center of each block. Next, a new individual will be obtained by interacting information between individuals in one or more blocks. At last, the logarithmic sigmoid function with Gaussian random is used as the step size to update the individual. Correspondingly, the drawbacks of the classic BSO algorithm include k-means grouping with complex calculations, poor individual creation methods, and suboptimal update mechanisms.

In view of these shortcomings of BSO, we propose the adaptive random grouping quantum BSO (ARGQBSO) algorithm and apply it to design the ALSTAR array. The improvements of ARGQBSO are made by presetting initial value, random grouping, dynamic probability function, and quantum computing. Experimental results show that the preset initial value shortens the search range and speeds up the convergence of the algorithm. Random grouping reduces the complexity of the algorithm. The dynamic probability function and quantum update improve the accuracy of the algorithm. In terms of algorithm architecture, our innovations are mainly manifested in two aspects.

(1) According to the demands of the ALSTAR array, the weight w_f is put forward to trade EII, EIRP, and EIS. Its significance is to enable the performance of the ALSTAR array to meet the needs of EII, EIRP, and EIS in various scenarios.
(2) The proposed ARGQBSO algorithm aims to achieve digital self-interference cancellation and adaptive beamforming. By proposing preset initial values and improving random grouping, dynamic probability functions, and quantum updates, the algorithm is a better balance in solving accuracy, solution time, and robustness.
(3) The beamformer optimized by ARGQBSO is independent of an angle and can be applied to any scanning angle. Its advantage is that the resources of the digital chip are greatly saved.

The remainder of this paper is organized as follows. In Section 2, the signal model and optimization model of the ALSTAR array are given. Section 3 introduces the origin of the ARGQBSO algorithm and detailed improvement measures. Section 4 discusses and analyzes the results of the six algorithms in the ALSTAR application, and verifies the competitiveness of the proposed algorithm. Finally, the conclusion and future work are provided in Section 5.

2. System Model

2.1. Signal Model

The ALSTAR architecture is shown in Figure 1, the symbol $t \in \mathbb{Q}^{J \times 1}$ means the transmit signal vector and $r \in \mathbb{Q}^{K \times 1}$ denotes the received signal vector. The parameters $w_t \in \mathbb{Q}^{J \times 1}$, $w_r \in \mathbb{Q}^{K \times 1}$, and $w_c \in \mathbb{Q}^{J \times 1}$ are the weights of transmit beamforming, receive beamforming, and adaptive cancellation filters, respectively. $H_m \in \mathbb{Q}^{K \times J}$ and $H_o \in \mathbb{Q}^{J \times J}$ are the characteristic matrix of the coupled channel and the observation channel, respectively. J and K are the number of transmitting channels and receiving channels, respectively. The signal x at time index n can be written as follows:

$$x(n) = w_t \cdot t(n) + n_t(n) \tag{1}$$

where $t(n)$ is the expected signal to be transmitted and $E\left[|t(n)|^2\right] = 1$. $n_t \in \mathbb{Q}^{J \times 1}$ is the complex additive white gaussian noise with zero-mean. The signal y is mainly composed of two parts, one is the self-interference signal coupled by the transmit signal through the coupling channel, and the other is the signal of interest s.

$$y(n) = H_m \cdot x(n) + s(n) \tag{2}$$

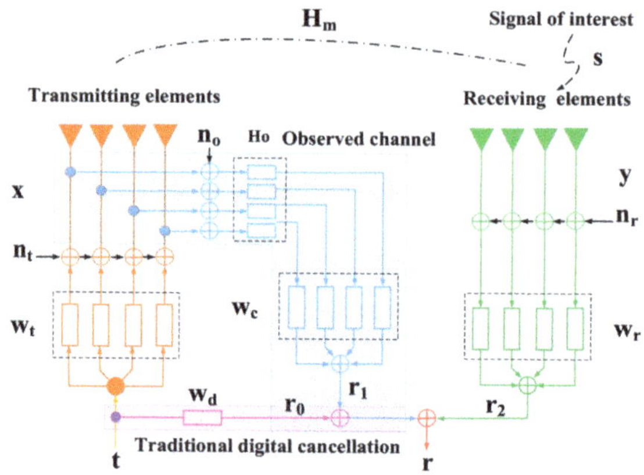

Figure 1. Block diagram of the ALSTAR array cancellation architecture.

The receiving noise $n_r \in \mathbb{Q}^{K \times 1}$ is mixed with the received signal y. After being processed by the receiving beamformer, they are accepted by the receiver. Therefore, the received signal can be shown as follows:

$$r_2(n) = w_r^H \cdot (n_r(n) + y(n)) \tag{3}$$

In the case of the ALSTAR architecture, the final received signal $r(n)$ after the cancellation can be expressed as follows:

$$r(n) = r_2(n) - r_1(n) \tag{4}$$

where $r_1(n)$ is the reference signal after passing through the observation channel, and its expression can be noted as follows:

$$r_1(n) = w_c^H[H_o(x(n) + n_o(n))] \tag{5}$$

Assuming ${w_c}^H = w_r^H \cdot H_m \cdot H_o^{-1}$ and combining the Equations (4) and (5), we can get

$$r(n) = w_r^H[n_r(n) + s(n) - H_m \cdot n_o(n)] \tag{6}$$

where $r(n)$ is composed of three parts: receiving noise $n_r(n)$, the signal of interest $s(n)$, and observation noise $n_o(n)$. The signal $n_o \in \mathbb{Q}^{J \times 1}$ is additive white gaussian noise, which obeys the normal distribution, i.e.,

$$n_o(n) \sim N(0, N_o) \tag{7}$$

where N_o is equal to $diag(w_t \cdot w_t^H)/\rho_r$ and ρ_r denotes the receive dynamic range. Apparently, mitigation of the observation noise and emission noise can be achieved by optimizing the transmitting beamforming weight w_t and receiving the beamforming weight w_r.

2.2. Metrics and Optimization Problems

This article optimizes the ALSTAR array design from the perspective of selecting the three best parameters, w_t, w_r, and w_f, to improve the overall performance and simplify the design. For the directional system, the isolation between the transmitter and the receiver changes with each position in the three-dimensional space, so EII is used to accurately measure the isolation between the directional systems, which is defined as follows:

$$\text{EII} = \text{EIRP}/\text{EIS} \tag{8}$$

where EIRP and EIS describe the performance of transmitters and receivers, respectively, and their expressions are shown:

$$\text{EIRP} = P_t \cdot G_{t\_total} = P_t \cdot \sum_{i=1}^{J} w_{t_i} g_{t_i}(\theta,\varphi) u_{t_i}(\theta,\varphi) \tag{9}$$

$$\text{EIS} = P_r / G_{r\_total} = (P_{n_r} + P_{n_o}) / \sum_{i=1}^{K} w_{r_i} g_{r_i}(\theta,\varphi) u_{r_i}(\theta,\varphi) \tag{10}$$

where u_t and u_r are the manifold vectors of the transmitting and receiving array, respectively, and g_{t_i} and g_{r_i} are gain of the transmitting and receiving elements, respectively. θ and φ represent the elevation angle and azimuth angle, respectively. The scalars of P_{n_r} and P_{n_o} are the receiver and observation noise, respectively.

$$P_{n_r} = (H_m diag(w_t w_t^H) H_m^H)/\rho_r \tag{11}$$

$$P_{n_o} = w_r^H \cdot \left\{ [diag(H_m w_t w_t^H H_m^H)]/\rho_r + [diag(H_m diag(w_t w_t^H) H_m^H)]/\rho_r \rho_t + \vartheta^2 \cdot eye(J) \right\} \cdot w_r \tag{12}$$

where the symbol ϑ^2 is the noise floor. From Equations (8)–(10), it can be seen that the EII, EIRP, and EIS of the array are closely related to the transmit beamforming weight w_t and receive beamforming weight w_r. Thus, we optimize the transmit and receive beamformers to obtain the desired EII, EIRP, and EIS. In frequency-modulated continuous-wave radars, it is possible to lower EII to achieve higher EIRP and gain. To achieve this, the weight $w_f = \left[w_{f\ EII}, w_{f\ EIRP}, w_{f\ EIS} \right]$ is proposed to trade between the EII, EIRP, and EIS in this paper. Therefore, the ALSTAR array can be designed via optimizing three parameters (i.e., w_t, w_r, and w_f) to achieve a higher EII with the premise of relatively high radiation efficiency and better receiving gain. The value ranges of them are as follows:

$$\begin{cases} 0 \le |w_t| \le 1;\ 0 \le \angle w_t \le 2\pi \\ 0 \le |w_r| \le 1;\ 0 \le \angle w_r \le 2\pi \\ \left\| w_f \right\|_1 = 1 \end{cases} \tag{13}$$

as the EII, EIRP, and EIS of the ALSTAR array have a theoretical upper bound, their optimization results must be smaller than it. The scope of EII, EIRP, and EIS is given as follows:

$$\begin{cases} \text{EII} \le \frac{P_t \cdot G_{t\_total} \cdot G_{r\_total}}{\vartheta^2} \\ \text{EIRP} \le P_t \cdot G_{t\_total} \\ \text{EIS} \le \frac{\vartheta^2}{G_{r\_total}} \end{cases} \tag{14}$$

thus, the objective function or fitness function complying with the above conditions can be provided as follows:

$$Fitness = \min(w_f \times fitness^T) \pm \zeta \tag{15}$$

where $fitness$ signifies the total objective function value, which consists of three parts, as shown in Equation (16). The coefficient w_f indicates the weight of each part. The interaction of the three goals can be achieved by modifying their weights. As the fault tolerance value, ζ is used to promote the iterative process of the algorithm.

$$\begin{aligned} fitness &= abs(\text{EII} - \tfrac{P_t \cdot G_{t\_total} \cdot G_{r\_total}}{\vartheta^2}) \\ &+ abs(\text{EIRP} - P_t \cdot G_{t\_total}) \\ &+ abs(\text{EIS} - \tfrac{\vartheta^2}{G_{r\_total}}) \end{aligned} \tag{16}$$

3. Our Proposed Algorithm

In response to the problems of traditional BSO algorithms, we consider amending the BSO from the presetting initial value, grouping strategy, individual creation, and individual update to enhance the convergence speed, global search capability, and reduce the operating overhead. We call the improved algorithm the adaptive random grouping quantum brainstorming optimization (ARGQBSO), which is a hybrid algorithm that combines quantum computing and classical BSO. Its performance-improvement is mainly reflected in four aspects.

3.1. Preset Initial Value

The initial value of the transmitting and receiving beamformer is directly related to the convergence speed of the algorithm. We reduce the search space of the algorithm without reducing the performance by setting an appropriate initial value. In the original BSO algorithm, the initial range of each idea (individual) is a random number of [0, 1]. According to the principle of maximum entropy, the maximum EIRP value can be obtained when the w_t is uniformly distributed. When $w_t = 0$, the vintage EIS can be obtained. In order to find a result that meets the actual needs among EII, EIRP, and EII, the optimal solution range of w_t must appear in $(0, \frac{1}{\hbar})$. $\hbar$ is the number of transmitting elements. Obviously, by setting the initial value in this way, the search space has been reduced $1 - \frac{1}{\hbar}$. In addition, the initial value of the transmit and receive beamforming weights should be related to the number of array elements and the array manifold vector. Its expression is as follows:

$$Population = [\frac{1}{\hbar}u_t; \frac{1}{\lambda}u_r] * X_{\min} + (X_{\max} - X_{\min}).* rand(D, N) \tag{17}$$

where the symbol $\hbar$ and λ represent the number of transmitting and receiving elements. $X_{\max}$ and $X_{\min}$ imply the upper boundary and lower boundary of the population, respectively, D implies the dimension of the population, and N denotes the size of the population.

3.2. Random Grouping

Random grouping is used to replace the original k-means, which avoids calculating the distance between different individuals and reduces the calculation. The population random grouping process is shown in Figure 2.

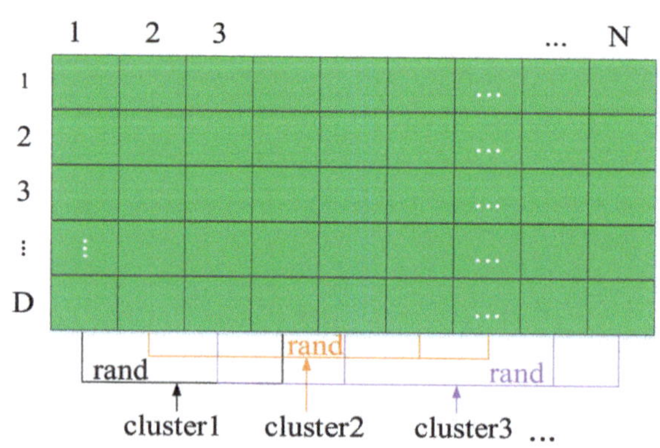

Figure 2. Schematic diagram of random grouping.

D depends on the number of elements. The cluster represents the same kind of individual population. The specific process is as follows: ℓ_1 individuals are randomly selected from populations N, and recorded as cluster1. Similarly, randomly select ℓ_2 individuals from the remaining $(N - \ell_1)$ individual species and record them as cluster2, and so on. It should be noted that there ℓ is a $1 \times n_c$ vector, and $\|\ell\|_1 = n_c, n_c$ represents the number of clusters.

3.3. Dynamic Probability Function

In the original BSO, there are four ways of individual creation [20], two of which create individuals based on a single cluster center, and the others create individuals based on two cluster centers. The individuals created by the former surround their clusters and have strong local search characteristics, while the individuals created by the latter may appear in the entire space and means the global search. The dynamic probability function p_0 improves the local and global search performance of ARGQBSO, making the algorithm focus on the global search in the early stage and the local search in the later stage to enhance the convergence speed. The dynamic probability function p_3 is used in the later stage of the algorithm to avoid missing the global best solution. Their expressions are as follows:

$$\begin{cases} p_{0,t+1} = \exp(-\gamma(t/t_{\max})^k) \times p_{0,t} \\ p_{3,t+1} = \exp(\gamma(t/t_{\max})^k) \times p_{3,t} \end{cases} \tag{18}$$

The meaning of t is the current number of iterations; $t_{\max}$ is the maximum number of iterations; $p_{0,t}$, $p_{3,t}$, and $p_{0,t+1}$, $p_{3,t+1}$ are the probabilities of the present and next generations, respectively. γ and κ are positive integers.

3.4. Quantum Update

The quantum behavior mechanism is introduced into all individuals so that every individual is transformed from the original classical state to the quantum state. At this time, individuals in the quantum state pass through the quantum revolving gate to update iteratively.

Different from [16], in this paper, the individual's quantum behavior only takes effect in the later stage of the algorithm, and the quantum state is transformed through a dynamic quantum spin gate. As the value of the dynamic probability function is already quite small in the later stage of the algorithm, the individuals created at this time are distributed almost in the center of a cluster, and diversity is completely lost. This is highly unfavorable for solving infinite domain and multi-extreme problems like ALSTAR. In particular, the noise, channel, and transmit power may change with the environment, if the ALSTAR system is running. To automatically adapt to these accidents, a dynamic quantum revolving gate is introduced in the later stage of the algorithm to enhance the global convergence ability in the later stage to avoid missing the best solution. The individuals in the quantum state follow the Equation (19) to update.

$$\begin{array}{l} R\_n_{t+1,D} = R_{t,D} + sign(R_{t,D} - globe_{t,D}) \times \Delta\Theta \\ \Delta\Theta = \exp(-\varepsilon \times \frac{I_t}{I_{\max}}) \times \Theta_0 \end{array} \tag{19}$$

where $R_{t,D}$ and $R_{t+1,D}$ represent contemporary and next-generation individuals, respectively; $globe_{t,D}$ is the current global optimal individual; $\Delta\Theta$ is the dynamic revolving gate; and Θ_0 represents the initial rotation angle. The flowchart of the ARGQBSO algorithm is shown in Figure 3. The scalars p_1 and p_2 are random numbers of [0, 1]. The variables p_0 and p_3 are dynamic probability functions, respectively.

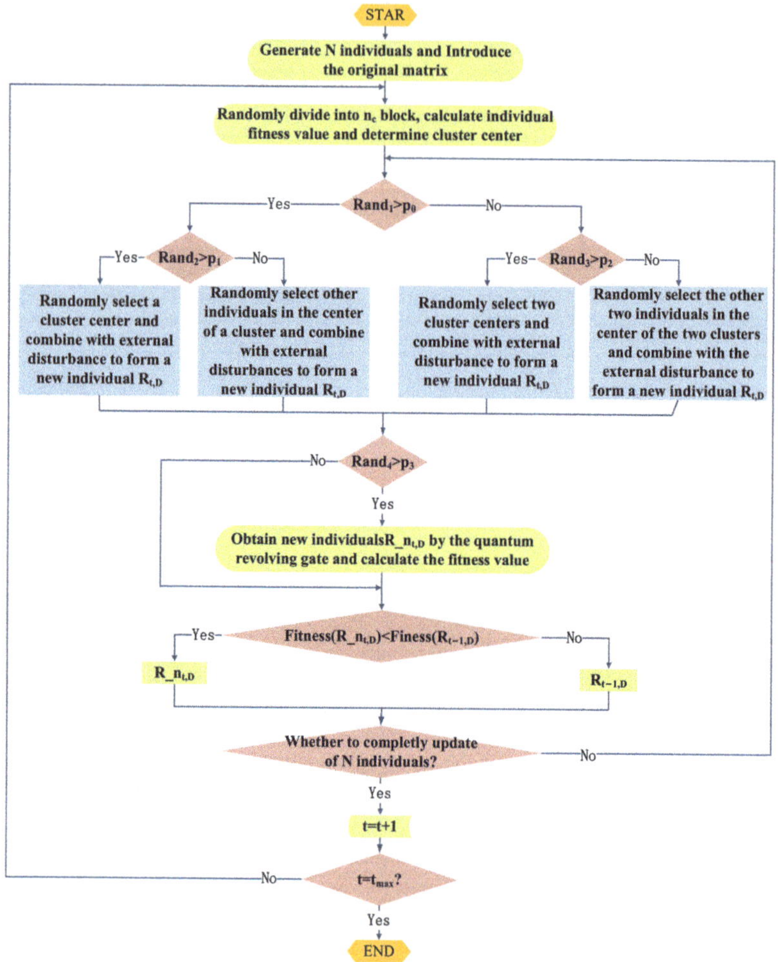

Figure 3. The flowchart of the ARGQBSO algorithm.

4. Simulation Results

The performance of the ALSTAR array is measured by the three objectives of EII, EIRP, and EIS. We minimize the fitness function to obtain the best transmit beamforming weight w_t and receive beamforming weigh w_r. The optimization model is given in the second part. As the characteristics of the coupling matrix and antenna gain are directly linked to EII, EIRP, and EIS of the ALSTAR from Equations (8)–(12), it is necessary to design a phased array with high isolation and high gain. Subsequently, based on the coupling matrix of the designed phased array and the pattern data of each element, the effects of the preset initial value, random grouping, dynamic probability function, and quantum update on the proposed algorithm are analyzed in detail. In addition, by comparing the performance of six commonly used optimization algorithms on the ALSTAR array design, the competitiveness of the ARGQBSO algorithm is verified. Finally, we explore the EII, EIRP, and EIS under extreme differences w_f using ARGQBSO.

4.1. Phased Array with High Isolation

The array element adopts the microstrip antenna fed by slot coupling, and its structures are shown in Figure 4. The microstrip patch is divided into several small pieces in the horizontal and vertical directions to form the series capacitor periodic array loaded with metamaterial patch elements [21]. In addition, loading a reflector on the bottom of the antenna enhances the directivity and front-to-back ratio, without deteriorating the matching performance. The proposed antenna is simulated in ANSYS HFSS, and its optimized parameters are $L = 34.2$ mm; $L_p = 22.5$ mm; $L_{p0} = 5.8$ mm; $L_{p1} = 4.6$ mm; $L_f = 16.8$ mm $L_{s0} = 8.8$ mm; $L_{s1} = 14$ mm; $W = 34.2$ mm; $W_{p0} = 3.2$ mm; $W_{p1} = 23.2$ mm; $W_f = 2.1$ mm; $W_s = 1.6$ mm; $H = 5.235$ mm; $g = 0.83$ mm.

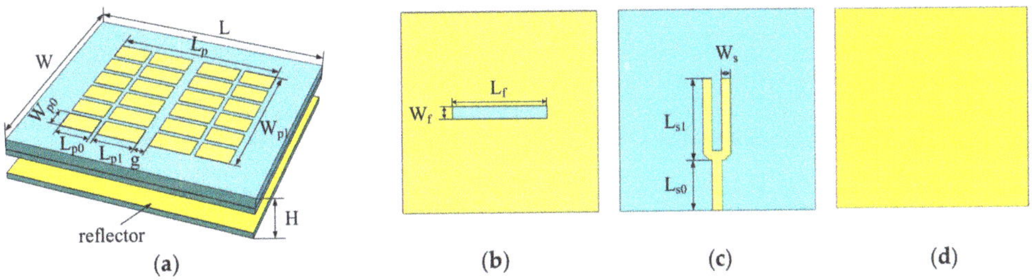

Figure 4. Broadband antenna model: (**a**) 3D view; (**b**) gap structure; (**c**) feed-network; (**d**) metal reflector.

The STAR array is shown in Figure 5. It is composed of eight designed broadband antenna elements evenly arranged at a pitch of 0.55 λ_0. λ_0 is the wavelength in free space. The simulation results of the phased array are shown in Figure 6. Part of the simulation data is shown in Appendix A. The impedance bandwidth of this antenna is 6.92~13.45 GHz under the condition that the port reflection coefficient is less than −10 dB, and the isolation between the adjacent antennas is about 25 dB in the whole X-band.

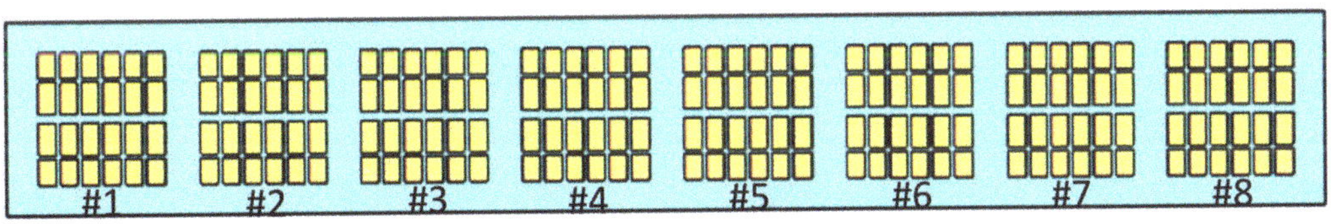

Figure 5. Schematic diagram of the broadband digital phased array structure.

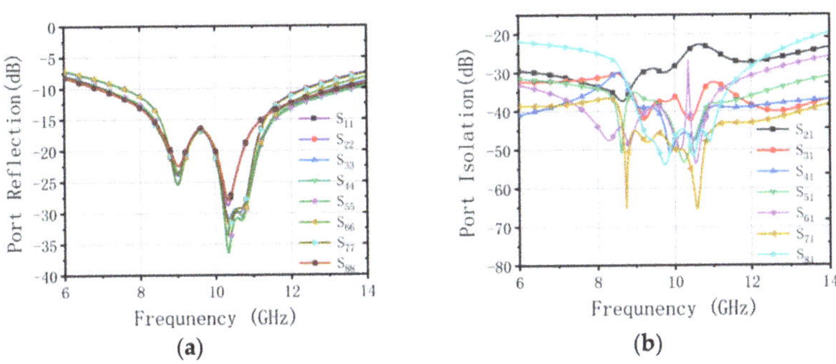

Figure 6. *Cont.*

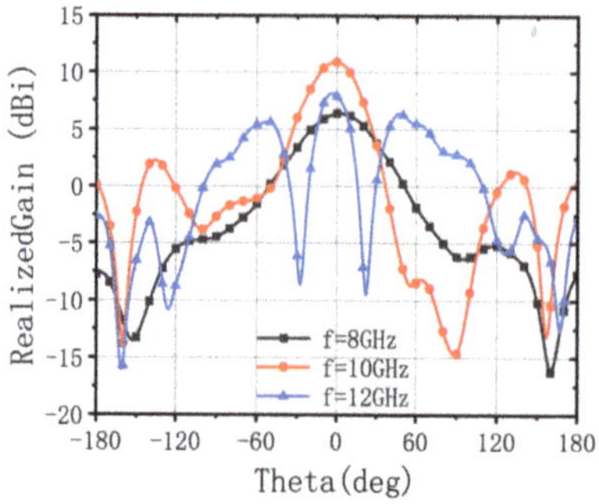

(c)

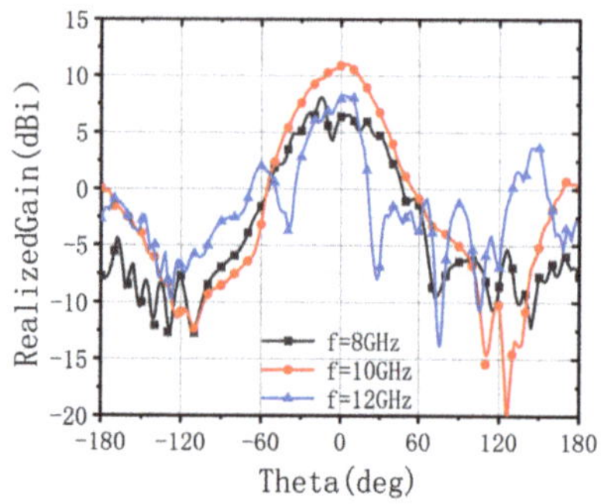

(d)

Figure 6. (**a**) The port reflection parameters of the array. (**b**) Port isolation parameters of the array. (**c**) E-plane pattern of the first element. (**d**) H-plane pattern of the first element.

In the ALSTAR array, we assume that the 1~4 elements on the left of the phased array are transmitting antennas and the 5~8 are receiving antennas. The coupling matrix H_m is used to describe the state of electromagnetic waves from the transmitter to the receiver and it is a $K \times J$ matrix. K and J represent the number of receiving and transmitting antennas, respectively. H_m is written as follows:

$$H_m = \begin{bmatrix} S_{51} & S_{52} & S_{53} & S_{54} \\ S_{61} & S_{62} & S_{63} & S_{64} \\ S_{71} & S_{72} & S_{73} & S_{74} \\ S_{81} & S_{82} & S_{83} & S_{84} \end{bmatrix} \tag{20}$$

In practice, adaptive beamforming and digital cancellation requires a great estimate of the mutual coupling channel in the ALSTAR array in order to consider the coupling caused by the time-varying environment or external interference. Yet it can be considered that the coupling matrix varies slowly in most scenarios. For example, there are not many fast-varying scenes in communications. In the radar, we mainly focus on specific targets, and may not be scene varies. In this case, the slow vary of the coupling matrix can be ignored for the time being.

4.2. Algorithm Performance Analysis

4.2.1. Analysis of the Role of Improved Operations

In order to expose the effects of preset initial values, random grouping, dynamic probability functions, and quantum updates on the AGRQBSO algorithm, we separately analyzed the benefits of each operation. Before that, we incorporated the algorithm into the metrics and optimization model of the ALSTAR array established in Section 2.2. The ALSTAR antenna adopts the phased array designed above, and the system's dynamic range of transmitting and receiving channels is $\rho_t = 40$ dB and $\rho_r = 80$ dB, respectively. The noise floor of the receiving channel is −81 dB, which is obtained by the 1000 MHz bandwidth channel with a 3 dB noise figure. The objective function and the boundary range of the parameters are given in Section 2. The experiments were executed in MATLAB software (Version: R2021a) and all comparative experiments were executed on a desktop PC with an Intel Core i7-8700 CPU processor @ 3.20 GHz, 16GB RAM, under the Windows10 64-bit OS.

We first analyzed the contribution of the preset initial value operation to the ARGQBSO algorithm. As mentioned earlier, the preset initial value reduces the search space of the algorithm and can speed up the convergence of the algorithm. Figure 7a shows the average result of the convergence curve with or without preset initial value operation by repeating the experiment 200 times. It can be seen that since the search space is compressed $1 - \frac{1}{h}$, the algorithm using preset initial values only takes 132 iterations to the state of convergence. Compared with the original scheme, the number of iterations is reduced by 29. In addition, the initial fitness value of the algorithm using the preset initial value is 4.8 dB smaller than the original algorithm. This hints that the preset the initial value places the algorithm in a better position in the actual stage and speeds up the convergence.

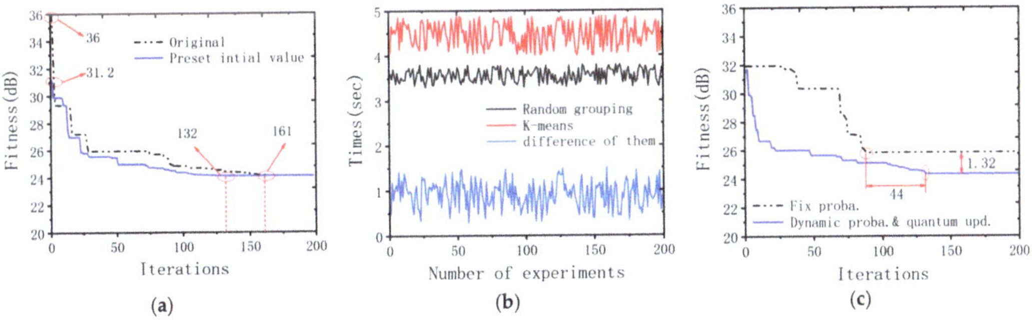

Figure 7. (**a**) Comparison of the fitness value with/without preset initial value. (**b**) The influence curve of grouping method on algorithm running time. (**c**) Comparison of the fitness value of fixed probability and dynamic probability density function and quantum update.

The contribution of the random grouping to the ARGQBSO algorithm is mainly to reduce the complexity of the clustering. In order to intuitively feel the reduction of the algorithm complexity by the random grouping, Figure 7b shows the running time of the algorithm using random grouping and K-means grouping in 200 experiments. The result of the experiment implies that the time of the random grouping algorithm is reduced in the range of 0.3145 to 1.4852 s.

The usefulness of the dynamic probability function to the algorithm is mainly reflected in the adjustment of the individual generation mode. By controlling the proportion of individuals participating in the global and local search, the global search is strengthened in the early stage of the algorithm to find a better position, and the local search is strengthened in the later stage of the algorithm to speed up the convergence speed.

The essence of the quantum update mechanism is to alter the evolution step length of the newly generated individual, but it is different from the individual update step length of the original BSO. It is a dynamic quantum rotation update, and the generated individuals surround the current global optimum. The contribution of dynamic probability density function and quantum update to the algorithm is reflected in the accuracy of the solution. Figure 7c shows the average results of 200 experiments. The algorithm using dynamic probability density function and quantum update is 1.32 dB smaller than the fitness value of the fixed probability, indicating that the accuracy of the former is better than the latter.

On the other hand, the number of iterations that the fixed probability algorithm uses to reach the convergence state is 44 less than that of the dynamic probability function and quantum update algorithm. This shows that the fixed probability algorithm has an insufficient global convergence ability and is easy to fall into the local optimal region.

4.2.2. Comparison of ARGQBSO with Other Algorithms

We explore the comprehensive performance of ARGQBSO, BSO, PSO, AO, GA, and DE in terms of the objective function value, convergence speed, and running time. What is different from [8] is that the w_t and w_r obtained by the ARGQBSO can be applied to all elevation angles, while there is no need for a set of w_t and w_r at every elevation angle. The advantage of this scheme is to reduce the processing time of beamforming and the calculation cost of the DSP chip.

considerable adjustment range makes it possible for ALSTAR to provide multi-scenario applications. On this basis, we can design an ALSTAR array to realize the trade of EII, EIRP, and EIS by changing the parameter w_f for different scenarios. This is to prevent engineers from considering only the isolation, while ignoring the performance of the transmitter and receiver itself when they are designing the STAR array.

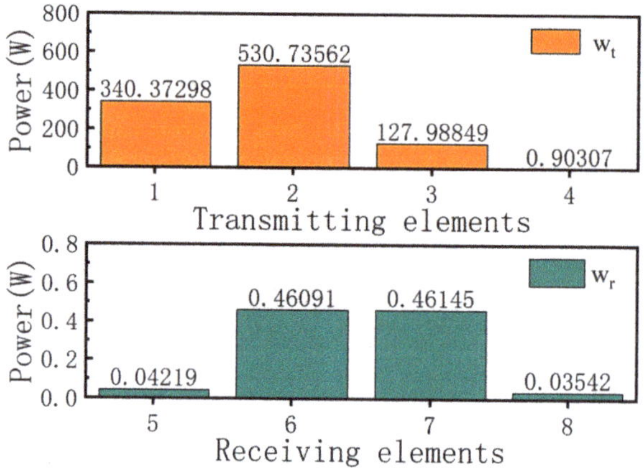

Figure 9. Transmit and receive beamforming vector.

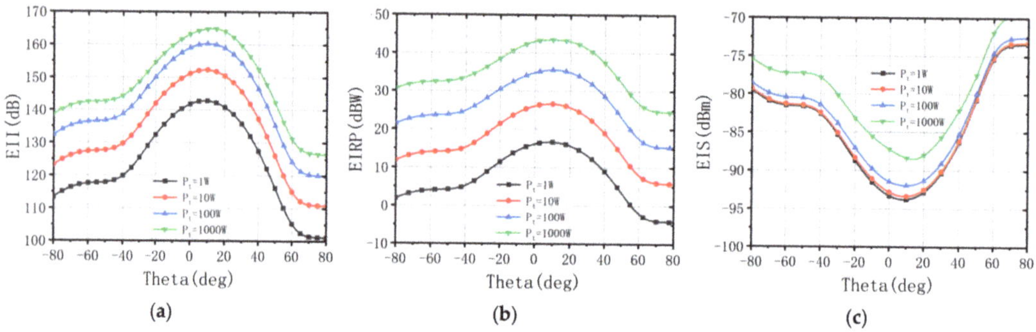

Figure 10. (**a**) EII with different transmit power. (**b**) EIRP with different transmit power. (**c**) EIS with different transmit power.

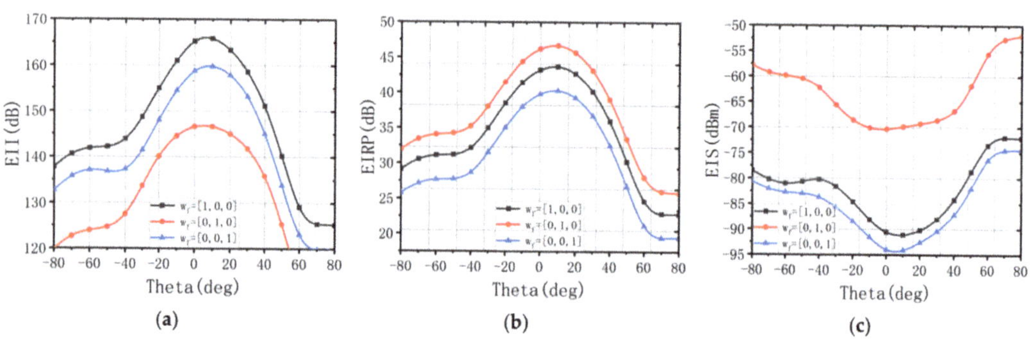

Figure 11. (**a**) EII with different w_f. (**b**) EIRP with different w_f. (**c**) EIS with different w_f.

5. Conclusions and Future Work

A robust design for the ALSTAR array was proposed in this paper. The transmit and receive beamforming weights obtained by the ARGQBSO algorithm are independent of the scanning angles, which reduces the computational complexity and maintains the excellent overall performance of the ALSTAR array. It expands the scope of application of transmitting and receiving beamforming weights and has a high robustness. ARGQBSO is an improved version of the original BSO algorithm. Its improvements include four aspects: initial value, random grouping mechanism, dynamic probability function, and quantum computing. Experimental results show that the preset initial value shortens the search range and speeds up the convergence Random grouping reduces the complexity of the algorithm. Dynamic probability function and quantum update improve the accuracy of the algorithm. In addition, the solution time, accuracy, and robustness of the proposed algorithm are claimed as a better trade-off, compared to other synthesis approaches. The simulated results based on an eight-element phased array indicate that the array achieves 166.8 dB of peak EII, 47.1 dBW of peak EIRP, and −94.6 dBm of peak EIS at $P_t = 1000$ W with $w_f = [1,0,0], w_f = [0,1,0], w_f = [0,0,1]$, respectively. In addition, EII fluctuates between 146.7~166.8 dB, EIRP fluctuates between 40.2~47.1 dBW, and EIS fluctuates between −70.3~−94.6 dBm. The results verify that the ARGQBSO is competitive in the ALSTAR application, and it trades among EII, EIRP, and EIS to satisfy the needs of different scenarios. The ARGQBSO algorithm provides engineers with a concise way to design massive transmit and receive arrays, and to achieve a superior overall performance of the ALSTAR array.

In the future, ARGQBSO will be used to solve more optimization problems in engineering design and other fields. Furthermore, some other strategies will be developed to further optimize the performance of the proposed algorithm.

Author Contributions: Conceptualization, M.X., Y.T. and X.W.; methodology, M.X.; software, D.H.; validation, M.X., Y.T. and D.H.; formal analysis, M.X.; investigation, M.X.; resources, X.W.; data curation, M.X.; writing—original draft preparation, M.X.; writing—review and editing, Y.T. and D.H.; visualization, M.X.; supervision, X.W.; project administration, X.W.; funding acquisition, Y.T. All authors have read and agreed to the published version of the manuscript.

Funding: This research was funded in part by the Guangdong Natural Science Foundation under Grant 2019A1515011622 and the Foundation under Grant 2019-JCJQ-JJ-411.

Institutional Review Board Statement: Not applicable.

Informed Consent Statement: Not applicable.

Data Availability Statement: Not applicable.

Acknowledgments: The authors are sincerely grateful to the editor and anonymous reviewers for their insightful and constructive comments that helped us improve this work greatly.

Conflicts of Interest: The authors declare that they have no known competing financial interests or personal relationships that could have appeared to influence the work reported in this paper.

Appendix A

The pattern of array elements 2–4 is shown in Additional Figure A1. Due to the high isolation performance of the designed phased array, the pattern characteristics of each array element are almost similar, which is conducive to beam forming. In addition, as the array elements numbered 5–8 are symmetrical to the center of array elements 1–4, their pattern data are almost the same, and they are not listed here.

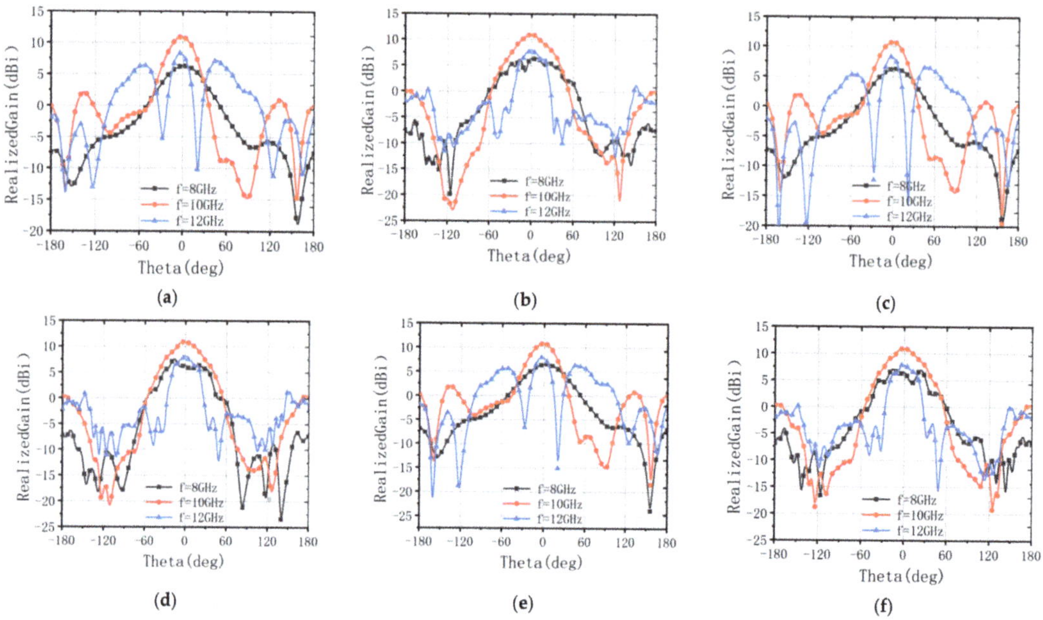

Figure A1. (**a**) E-Pattern of the second element. (**b**) H-Pattern of the second element. (**c**) E-Pattern of the third element. (**d**) H-Pattern of the third element. (**e**) E-Pattern of the fourth element. (**f**) H-Pattern of the fourth element.

References

1. Nwankwo, C.D.; Zhang, L.; Quddus, A.; Imran, M.A.; Tafazolli, R. A survey of self-interference management techniques for single frequency full duplex systems. *IEEE Access* **2017**, *6*, 30242–30268. [CrossRef]
2. Riihonen, T.; Korpi, D.; Rantula, O.; Rantanen, H.; Saarelainen, T.; Valkama, M. Inband full-duplex radio transceivers: A paradigm shifts in tactical communications and electronic warfare? *IEEE Commun. Mag.* **2017**, *55*, 30–36. [CrossRef]
3. Kolodziej, K.E.; Doane, J.P.; Perry, B.T.; Herd, J.S. Adaptive Beamforming for Multi-Function In-Band Full-Duplex Applications. *IEEE Wirel. Commun.* **2021**, *28*, 28–35. [CrossRef]
4. Alves, H.; Riihonen, T.; Suraweera, H.A. *Full-Duplex Communications for Future Wireless Networks*, 1st ed.; Springer: Singapore, 2020; pp. 61–72.
5. Makar, G.; Tran, N.; Karacolak, T. A high-isolation monopole array with ring hybrid feeding structure for in-band full-duplex systems. *IEEE Antennas Wirel. Propag. Lett.* **2016**, *16*, 356–359. [CrossRef]
6. Li, R.; Masmoudi, A.; Le-Ngoc, T. Self-interference cancellation with nonlinearity and phase-noise suppression in full-duplex systems. *IEEE Trans. Veh. Technol.* **2017**, *67*, 2118–2129. [CrossRef]
7. Ahmed, E.; Eltawil, A.M. All-digital self-interference cancellation technique for full-duplex systems. *IEEE Trans. Wirel. Commun.* **2015**, *14*, 3519–3532. [CrossRef]
8. Qiu, J.; Yao, Y.; Wu, G. Research on aperture-level simultaneous transmit and receive. *J. Eng.* **2019**, *2019*, 8006–8008. [CrossRef]
9. Liang, D.; Xiao, P.; Chen, G.; Ghoraishi, M.; Tafazolli, R. Digital self-interference cancellation for Full-Duplex MIMO systems. In *Proceedings of the International Wireless Communications and Mobile Computing Conference (IWCMC), Dubrovnik, Croatia, 24–28 August 2015*; IEEE: New York, NY, USA, 2015.
10. Doane, J.P.; Kolodziej, K.E.; Perry, B.T. Simultaneous transmit and receive with digital phased arrays. In *Proceedings of the IEEE International Symposium on Phased Array Systems and Technology (PAST), Waltham, MA, USA, 18–21 October 2016*; IEEE: New York, NY, USA, 2017.
11. Cummings, I.T.; Doane, J.P.; Schulz, T.J.; Havens, T.C. Aperture-level simultaneous transmit and receive with digital phased arrays. *IEEE Trans. Signal Processing* **2020**, *68*, 1243–1258. [CrossRef]
12. Jiang, X.; Xie, M.C.; Wei, J.; Wang, K.P.; Peng, L. 3D-MIMO beamforming realized by AQPSO algorithm for cylindrical conformal phased array. *IET Microw. Antennas Propag.* **2019**, *13*, 2701–2705. [CrossRef]
13. Sharp, C.; DuPont, B. Wave energy converter array optimization: A genetic algorithm approach and minimum separation distance study. *Ocean. Eng.* **2018**, *163*, 148–156. [CrossRef]

14. Subhashini, K.R.; Satapathy, J.K. Development of an enhanced ant lion optimization algorithm and its application in antenna array synthesis. *Appl. Soft Comput.* **2017**, *59*, 153–173. [CrossRef]
15. Duan, H.; Li, S.; Shi, Y. Predator–prey brain storm optimization for DC brushless motor. *IEEE Trans. Magn.* **2013**, *49*, 5336–5340. [CrossRef]
16. Duan, H.; Li, C. Quantum-behaved brain storm optimization approach to solving Loney's solenoid problem. *IEEE Trans. Magn.* **2014**, *51*, 1–7. [CrossRef]
17. Sun, Y. A hybrid approach by integrating brain storm optimization algorithm with grey neural network for stock index forecasting. *Abstr. Appl. Anal.* **2014**, *2014*, 759862. [CrossRef]
18. Shi, Y. Brain storm optimization algorithm. In *Proceedings of the International Conference in Swarm Intelligence, Chongqing, China, 12–15 June 2011*; Springer: Berlin/Heidelberg, Germany, 2011.
19. Cheng, S.; Qin, Q.; Chen, J.; Shi, Y. Brain storm optimization algorithm: A review. *Artif. Intell. Rev.* **2016**, *46*, 445–458. [CrossRef]
20. Li, C.; Hu, D.; Song, Z.; Yang, F.; Luo, Z.; Fan, J.; Liu, P.X. A vector grouping learning brain storm optimization algorithm for global optimization problems. *IEEE Access* **2018**, *6*, 78193–78213. [CrossRef]
21. Liu, W.; Chen, Z.N.; Qing, X. Metamaterial-based low-profile broadband aperture-coupled grid-slotted patch antenna. *IEEE Trans. Antennas Propag.* **2015**, *63*, 3325–3329. [CrossRef]
22. Poli, R.; Kennedy, J.; Blackwell, T. Particle swarm optimization. *Swarm Intell.* **2007**, *1*, 33–57. [CrossRef]
23. Maruyama, T.; Igarashi, H. An effective robust optimization based on genetic algorithm. *IEEE Trans. Magn.* **2008**, *44*, 990–993. [CrossRef]
24. Hui, S.; Suganthan, P.N. Ensemble and arithmetic recombination-based speciation differential evolution for multimodal optimization. *IEEE Trans. Cybern.* **2015**, *46*, 64–74. [CrossRef] [PubMed]

electronics

MDPI

Article

Optimal Field Sampling of Arc Sources via Asymptotic Study of the Radiation Operator

Raffaele Moretta *, Giovanni Leone, Fortuna Munno and Rocco Pierri

Dipartimento di Ingegneria, Università della Campania "Lugi Vanvitelli", Via Roma 29, 81031 Aversa, Italy; giovanni.leone@unicampania.it (G.L.); fortuna.munno@unicampania.it (F.M.); rocco.pierri@unicampania.it (R.P.)
* Correspondence: raffaele.moretta@unicampania.it; Tel.: +39-334-2470823

Abstract: In this paper, the question of how to efficiently sample the field radiated by a circumference arc source is addressed. Classical sampling strategies require the acquisition of a redundant number of field measurements that can make the acquisition time prohibitive. For such reason, the paper aims at finding the minimum number of basis functions representing the radiated field with good accuracy and at providing an interpolation formula of the radiated field that exploits a non-redundant number of field samples. To achieve the first task, the number of relevant singular values of the radiation operator is computed by exploiting a weighted adjoint operator. In particular, the kernel of the related eigenvalue problem is first evaluated asymptotically; then, a warping transformation and a proper choice of the weight function are employed to recast such a kernel as a convolution and bandlimited function of sinc type. Finally, the number of significant singular values of the radiation operator is found by invoking the Slepian–Pollak results. The second task is achieved by exploiting a Shannon sampling expansion of the reduced field. The analysis is developed for both the far and the near fields radiated by a 2D scalar arc source observed on a circumference arc.

Keywords: field sampling; number of degrees of freedom (NDF); singular values decomposition (SVD); conformal source

Citation: Moretta, R.; Leone, G.; Munno, F.; Pierri, R. Optimal Field Sampling of Arc Sources via Asymptotic Study of the Radiation Operator. *Electronics* **2022**, *11*, 270. https://doi.org/10.3390/electronics11020270

Academic Editors: Naser Ojaroudi Parchin, Chan Hwang See and Raed A. Abd-Alhameed

Received: 13 December 2021
Accepted: 11 January 2022
Published: 14 January 2022

1. Introduction

The question of sampling the field radiated by a source or the one scattered by an object is a classical research topic of the electromagnetics literature [1–9].

On one hand, a proper sampling of the radiated/scattered field allows representing the field from the knowledge of its samples in a discrete and finite number of points. On the other, it allows acquiring independent information to address the correspondent inverse source/inverse scattering problem [10,11].

In this paper, the attention is limited to the sampling of the radiated field E which is linked to the source current J by a linear operator T called *radiation operator*.

1.1. Literature Review

Classical sampling schemes of the radiated field for the case of planar [12], cylindrical [13] and spherical scanning [14] were proposed. Despite this, such schemes do not take into account explicitly the source shape; for such reason, they require collecting a number of field measurements that may be significantly higher than the number of degrees of freedom (NDF) of the radiated field [15,16]. The latter represents the number of independent parameters required to represent the radiated field with good accuracy and, at the same time, the minimum number of field measurements required to reconstruct the source current stably.

The acquisition of a number of field samples larger than the number of degrees of freedom affects badly the acquisition time and also the processing time to interpolate the field samples or to retrieve the source current. For such a reason, it is of great interest to devise a sampling scheme that exploits a non-redundant number of field measurements.

To reduce the number of measurements, over the years different sampling schemes have been proposed. A first strategy is based on an adaptive procedure that increases the sampling rate only when the measured field oscillates faster [17]. In [18], an efficient sampling scheme suitable for any source enclosed in an ellipsoid is devised by exploiting reasoning on the local bandwidth of the reduced field. Such a sampling scheme can be used also for the planar disk and spherical sources that can be seen as particular cases of the ellipsoidal source.

Recently, a new method that exploits the point spread function (PSF) in the observation domain has been proposed in [19]. Such a method relies on the idea that two adjacent points are independent if the main lobes of the PSFs centered in such points do not overlap and, hence, are distinguishable from each other.

Other methods recast the question of efficiently sampling the radiated field as a sensor selection problem and choose the optimal sampling points in such a way that the radiation operator and its discrete counterpart exhibit the same relevant singular values. Such a goal can be achieved by exploiting a numerical procedure that optimizes a metric related to the singular values [20–22] or, alternatively, by an analytical study and a proper discretization of the radiation operator [23–25].

1.2. Goal of the Paper

Here, with reference to a 2D geometry consisting of a circumference arc source whose radiated field is collected on a circumference observation arc, the minimum number of measurements required to discretize the radiated field without loss of information is first determined. From the mathematical point of view, this implies an evaluation of the NDF of the source over the assigned observation domain. This task is performed by computing analytically the number of relevant singular values of the radiation operator with the asymptotic approach proposed in [25]. Next, an efficient interpolation formula of the radiated field is found by exploiting a sampling representation of the left singular functions of the radiation operator.

Let us remark that the optimal locations of the sampling points and an efficient interpolation formula of the field radiated by a circumference arc source are provided also in [19] by a numerical procedure. Here, instead, the optimal sampling points and the basis functions used in the interpolation stage of the radiated field are analytically found. This allows highlighting the key role played by the geometric parameters of the problem.

The paper is organized as follows. In Section 2, the geometry of the problem and the explicit expression of the radiation operator in the case of an observation domain in far-field and near-field is provided. In Section 3, an outline of our sampling strategy is shown. In Section 4, the NDF and an efficient interpolation formulation of the far field are derived. In Section 5, all the results of Section 4 are extended to the case of an observation domain in near field. In Section 6, for sake of comparison, a sampling scheme based on a uniform sampling step is considered and the number of measurements points saved by our non-uniform sampling scheme with respect to the uniform case is estimated. In Section 7, a numerical validation of our analytical results on the NDF and the field sampling is shown. Conclusions follow in Section 8.

2. Geometry of the Problem

Consider the 2D scalar geometry depicted in Figure 1 where the y-axis represents the direction of invariance. An electric current $\boldsymbol{J}(\phi) = J(\phi)\, \boldsymbol{i}_y$ is supported over an arc of circumference of radius a spanning the interval $SD = [-\phi_{max}, \phi_{max}]$.

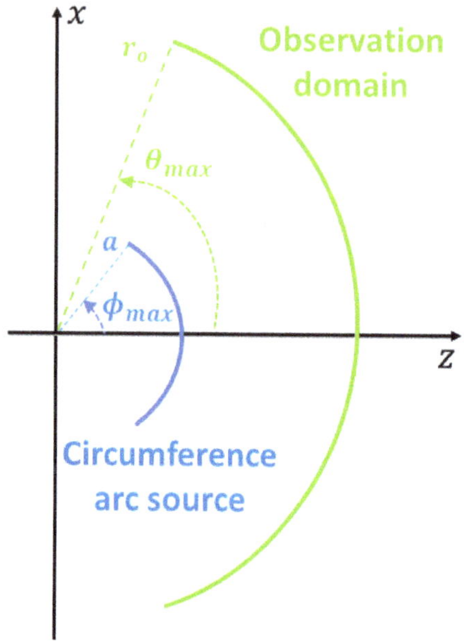

Figure 1. Geometry of the problem.

The electric field $\boldsymbol{E}(r,\theta) = E(r,\theta)\, \boldsymbol{i}_y$ radiated by such source is observed on an arc of circumference of radius $r_o > a$ that subtends an angular sector $OD = [-\theta_{max}, \theta_{max}]$.

For the considered geometry, the radiation operator T is defined as

$$J \in L_2(SD) \rightarrow E \in L_2(OD) \tag{1}$$

where $L_2(SD)$ and $L_2(OD)$ denotes the set of square integrable functions on SD and OD, respectively. Apart some unessential factors, such operator can be explicitly written as

$$TJ = a \int_{-\phi_{max}}^{\phi_{max}} g(\phi,\theta)\; J(\phi)\; d\phi \tag{2}$$

where the $2D$ Green *function* $g(\phi,\theta)$ is given by

$$g(\phi,\theta) = \begin{cases} \frac{e^{-j\beta R(\phi,\theta)}}{\sqrt{\beta R(\phi,\theta)}} & if\;\; r_o > a + \lambda \\ e^{j\beta a \cos(\theta-\phi)} & if\;\; r_o > \frac{4a^2}{\lambda} \end{cases} \tag{3}$$

with

$$R(\phi,\theta) = \sqrt{r_o^2 + a^2 - 2a\, r_o \cos(\phi-\theta)} \tag{4}$$

Since the radiation operator T is linear and compact, its singular values decomposition can be introduced. The latter is provided by the triple $\{u_n, v_n, \sigma_n\}$ where the right singular functions $\{u_n\}$ and the left singular functions $\{v_n\}$ represent a set of basis functions for the density current J and the radiated field E, instead, $\{\sigma_n\}$ stand for the singular values. As well known, the right and the left singular functions are related by the following equations

$$Tu_n = \sigma_n v_n \qquad T^{+} v_n = \sigma_n u_n \tag{5}$$

where $T^{\dagger}$ stands for the adjoint operator. Hereafter, the adjoint operator is not defined as usual since a weight function $p(\theta,\phi)$ is introduced in its definition. In other words, $T^{\dagger}: v_n \in L_2(OD) \rightarrow u_n \in L_2(SD)$ is defined as

$$T^{\dagger}v_n(\theta) = a \int_{-\theta_{max}}^{\theta_{max}} p(\theta,\phi)\; g^*(\phi,\theta)\; v_n(\theta)d\theta \tag{6}$$

with g^* denoting the conjugate of the Green function.

It is worth noting that the use of a weighted adjoint affects only the shape of the singular values behavior of T but not the critical index after which they abrupt decay. For such reason, it can be used to estimate the most significant singular values of the radiation operator.

3. Outline of the Sampling Strategy

In this section, the methodology followed in the paper is described and an outline of such methodology is sketched in the block diagram of Figure 2.

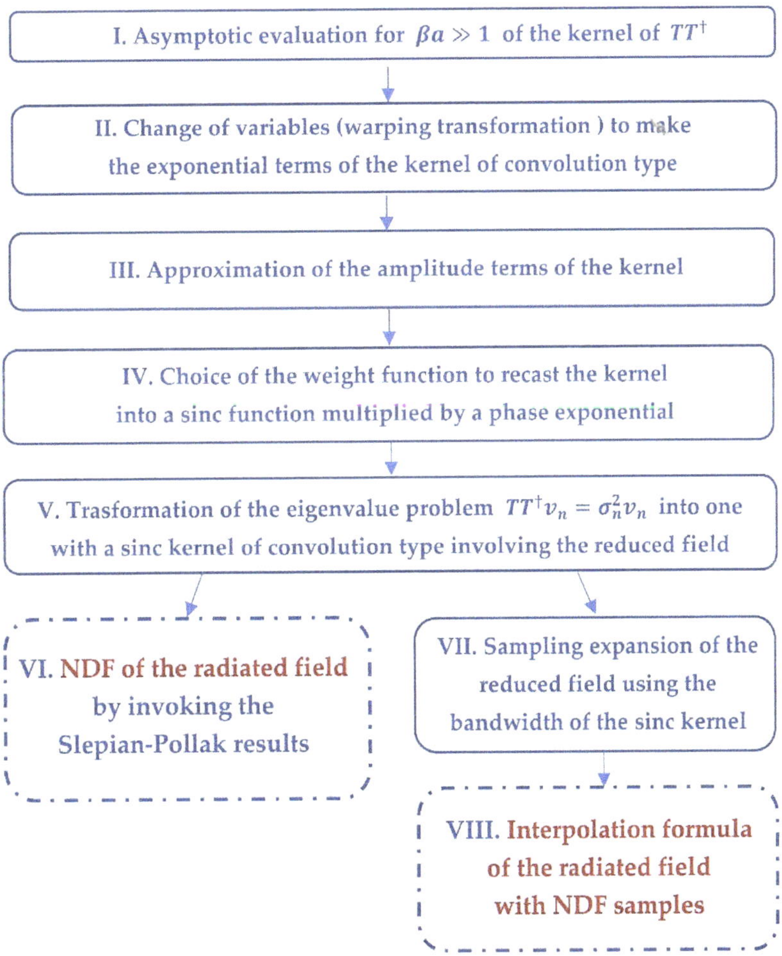

Figure 2. Block diagram of the study.

The first aim of the paper is to provide a closed form expression of the NDF of the radiated field. The NDF is estimated by evaluating the number of relevant singular values of the radiation operator. In particular, since the eigenvalues of the auxiliary operator $TT^{\dagger}$ are the square of the singular values of T, the number of relevant singular values of the radiation operator will be evaluated by studying the eigenvalue problem $TT^{\dagger}v_n = \sigma_n^2 v_n$. The latter can be explicitly written as

$$a^2 \int_{-\theta_{max}}^{\theta_{max}} K(\theta_o, \theta)\, v_n(\theta)\, d\theta = \sigma_n^2\, v_n(\theta_o) \tag{7}$$

where the kernel $K(\theta_o, \theta)$ is given by

$$K(\theta_o, \theta) = \int_{-\phi_{max}}^{\phi_{max}} p(\phi, \theta)\, g(\phi, \theta_o)\, g^*(\phi, \theta) d\phi \tag{8}$$

Here, such a kernel is estimated by exploiting an asymptotic approach. After, by exploiting a change of variables and proper choice of the weight function, such a kernel is recast like a sinc kernel of convolution type multiplied by a phase exponential. Finally, by redefining the eigenfunctions, the eigenvalues problem $TT^{\dagger}v_n = \sigma_n^2 v_n$ is rewritten in a new form with a purely sinc kernel. This allows exploiting the Slepian–Pollak theory to estimate the number of relevant eigenvalues of $TT^{\dagger}$ which, as said before, provides an estimation of NDF of the radiated field.

The second aim of the paper is to provide the optimal sampling where the field must be collected and an interpolation formula of the radiated field that exploits a non-redundant number of field samples. To achieve this goal, the Shannon sampling theorem is adopted to derive a sampling representation of the reduced field in the warped variable. Then, starting from it, an efficient interpolation formula of the radiated field is easily obtained.

The study is developed not only for an observation arc in far-field but also for a near-field configuration.

4. Optimal Sampling of the Far-Field

In this section, all the steps illustrated in Figure 2 are detailed to compute the NDF of the far-field and to provide an interpolation formula that exploits a non-redundant number of field samples.

4.1. Asymptotic Study of the Operator $TT^{\dagger}$ in Far Zone

I. The kernel of the auxiliary operator $TT^{\dagger}$ is particularized to the far-zone by substituting in (8) the far zone Green function. It follows that

$$K(\theta_o, \theta) = \int_{-\phi_{max}}^{\phi_{max}} p(\phi, \theta)\, e^{j\,\beta\, a\, \Psi(\phi, \theta, \theta_o)}\, d\phi \tag{9}$$

with

$$\Psi(\phi, \theta_o, \theta) = \cos(\theta_o - \phi) - \cos(\theta - \phi) \tag{10}$$

For $\theta_o = \theta$, the value of the kernel can be easily obtained by evaluating the integral $\int_{-\phi_{max}}^{\phi_{max}} p(\phi, \theta)\, d\phi$.

For $\theta_o \neq \theta$, the integral in (9) can be asymptotically evaluated if the condition $\beta a \gg 1$ is satisfied. The choice of the asymptotic technique is related to the presence/absence of stationary points in the phase function $\Psi(\phi, \theta, \theta_o)$. In Appendix A, it is shown that if the following condition holds

$$\theta_{max} + \phi_{max} \leq \frac{\pi}{2} \tag{11}$$

then no stationary points fall into the set $[-\phi_{max}, \phi_{max}]$. In such a case, the kernel in (9) can be asymptotically evaluated by considering only the contribution by the endpoints $\phi = \phi_{max}$ and $\phi = -\phi_{max}$ [26]. Accordingly, for each $\theta_o \neq \theta$ it results that

$$K(\theta_o, \theta) \approx \frac{p(\phi_{max},\theta)}{j\beta a\, \Psi'(\phi_{max},\theta_o,\theta)} e^{j\beta a\, \Psi(\phi_{max},\theta_o,\theta)} - \frac{p(-\phi_{max},\theta)}{j\beta a\, \Psi'(-\phi_{max},\theta_o,\theta)} e^{j\beta a\, \Psi(-\phi_{max},\theta_o,\theta)} \tag{12}$$

where Ψ' denotes the partial derivative of Ψ with respect to ϕ, hence, $\Psi'(\phi,\theta_o,\theta) = \sin(\theta_o - \phi) - \sin(\theta - \phi)$.

II. The kernel of $TT^{\dagger}$ in the variables (θ_o, θ) is not convolution and this does not allow to find easily its eigenvalues. To make $TT^{\dagger}$ more similar to a convolution operator, it is first recast as

$$\begin{aligned} K(\theta_o, \theta) &\approx \frac{1}{j\beta a} e^{j\frac{\beta a}{2}[\Psi(\phi_{max},\theta_0,\theta)+\Psi(-\phi_{max},\theta_o,\theta)]} \left(\frac{p(\phi_{max},\theta)}{\Psi'(\phi_{max},\theta_o,\theta)} e^{j\frac{\beta a}{2}[\Psi(\phi_{max},\theta_o,\theta)-\Psi(-\phi_{max},\theta_o,\theta)]} \right. \\ &\left. - \frac{p(-\phi_{max},\theta)}{\Psi'(-\phi_{max},\theta_o,\theta)} e^{-j\frac{\beta a}{2}[\Psi(\phi_{max},\theta_o,\theta)-\Psi(-\phi_{max},\theta_o,\theta)]} \right) \end{aligned} \tag{13}$$

Then, the following variables

$$\overline{u}(\theta) = \frac{\cos(\theta - \phi_{max}) - \cos(\theta + \phi_{max})}{2} \tag{14}$$

$$\overline{w}(\theta) = \frac{\cos(\theta - \phi_{max}) + \cos(\theta + \phi_{max})}{2} \tag{15}$$

are introduced. Equations (14) and (15) allow rewriting the kernel of $TT^{\dagger}$ as

$$K(\overline{u}_o, \overline{u}) \approx \frac{1}{j\beta a} e^{j\beta a(\overline{w}(\overline{u}_o) - \overline{w}(\overline{u}))} \left(\frac{p(\phi_{max},\overline{u})}{\Psi'(\phi_{max},\overline{u}_o,\overline{u})} e^{j\beta a(\overline{u}_o - \overline{u})} - \frac{p(-\phi_{max},\overline{u})}{\Psi'(-\phi_{max},\overline{u}_o,\overline{u})} e^{-j\beta a(\overline{u}_o - \overline{u})} \right) \tag{16}$$

III. At this juncture, the kernel function has still an intricate structure. However, by expanding $\Psi'(\phi, \overline{u}_o, \overline{u})$ with respect to the variable $\overline{u}_o$ in a Taylor series stopped at the first order, one obtains

$$\Psi'(\phi_{max}, \overline{u}_o, \overline{u}) \approx \cos(\theta(\overline{u}) - \phi_{max})(\overline{u}_o - \overline{u}) \left.\frac{d\theta}{d\overline{u}_o}\right|_{\overline{u}_o = \overline{u}} \tag{17}$$

$$\Psi'(-\phi_{max}, \overline{u}_o, \overline{u}) \approx \cos(\theta(\overline{u}) + \phi_{max})(\overline{u}_o - \overline{u}) \left.\frac{d\theta}{d\eta_o}\right|_{\overline{u}_o = \overline{u}} \tag{18}$$

Hence, taking into account of Equations (17) and (18), the kernel can be approximated as

$$K(\overline{u}_o, \overline{u}) \approx \frac{1}{j\beta a} \frac{1}{\frac{d\theta}{d\overline{u}}} \frac{e^{j\beta a(\overline{w}(\overline{u}_o) - \overline{w}(\overline{u}))}}{\overline{u}_o - \overline{u}} \left(\frac{p(\phi_{max},\overline{u})}{\cos(\theta(\overline{u}) - \phi_{max})} e^{j\beta a(\overline{u}_o - \overline{u})} - \frac{p(-\phi_{max},\overline{u})}{\cos(\theta(\overline{u}) + \phi_{max})} e^{-j\beta a(\overline{u}_o - \overline{u})} \right) \tag{19}$$

IV. Now, the weight function $p(\phi, \overline{u})$ must be chosen. The best choice for $p(\phi, \overline{u})$ is to fix it in such a way that the eigenvalues of $TT^{\dagger}$ are known in closed form. Such goal can be reached by choosing

$$p(\phi, \overline{u}) = \cos(\theta(\overline{u}) - \phi) \tag{20}$$

Then, the kernel can be recast as

$$K(\overline{u}_o, \overline{u}) \approx 2 \frac{1}{\frac{d\theta}{d\overline{u}}} e^{j\beta a(\overline{w}(\overline{u}_o) - \overline{w}(\overline{u}))}\, sinc\,(\beta a\,(\overline{u}_o - \overline{u})) \tag{21}$$

Then, (36) can be rewritten as

$$K(\theta_o,\theta) = \frac{1}{\beta}\int_{-\phi_{max}}^{\phi_{max}} p(\phi,\theta)\ A(\phi,\theta_o,\theta)\ e^{-j\beta a\ \Phi(\phi,\theta_o,\theta)}\ d\phi \tag{37}$$

At $\theta_o = \theta$, the kernel of $TT^\dagger$ can be evaluated by computing the integral $\frac{1}{\beta}\int_{-\phi_{max}}^{\phi_{max}} p(\phi,\theta)\ A(\phi,\theta,\theta)\ d\phi$.

For $\theta_o \neq \theta$, if the hypothesis $\beta a \gg 1$ is fulfilled, the integral (37) can be asymptotically evaluated. To establish if stationary phase points appear in the phase function, the equation $\Phi'(\phi,\theta_o,\theta) = 0$ must be solved for ϕ. The latter can be explicitly written as

$$\Phi'(\phi,\theta,\theta_o) = \frac{r_o \sin(\theta-\phi)}{R(\phi,\theta)} - \frac{r_o \sin(\theta_o-\phi)}{R(\phi,\theta_o)} = 0 \tag{38}$$

Unfortunately, the previous equation cannot be analytically solved. For such reason, here, the attention is limited to all those cases where the geometrical parameters are such that no stationary points appear in the set $[-\phi_{max},\phi_{max}]$. In particular, through a numerical analysis, it has been shown that

- fixing the source angle ϕ_{max}, the θ interval for which no stationary points appear on the source increases with the ratio r_o/a.
- fixing the ratio r_o/a, the θ interval for which no stationary points appear on the source decreases with the source angle ϕ_{max}.

This behavior can be observed in the tables of the Appendix B. From such tables, it is evident that in near zone the condition for the lack of stationary points in $\Phi\ (\phi,\theta_o,\theta)$ is given by

$$\theta_{max} + \phi_{max} \leq C\left(\frac{r_o}{a}\right) \tag{39}$$

where C is a function depending on the ratio $\frac{r_o}{a}$ whose values are reported in Table 1.

Table 1. Values of C in terms of $\frac{r_o}{a}$.

$\frac{r_o}{a}$	$C(\frac{r_o}{a})$
1.4	0.70 (40°)
1.6	0.87 (50°)
2	1.05 (60°)
4	1.22 (70°)
8	1.40 (80°)
15	1.48 (85°)

Accordingly, $C(\frac{r_o}{a})$ is a monotonic function and its diagram is shown in Figure 3.

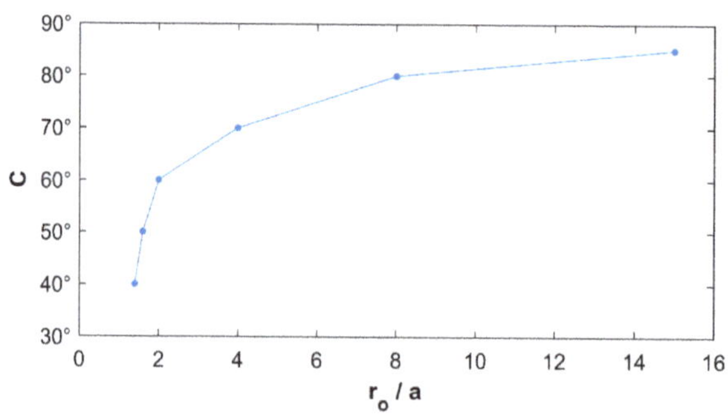

Figure 3. Diagram in degrees of the function C in terms of $\frac{r_o}{a}$.

For all the configurations in which no stationary points appear in the phase function Φ, the integral in (37) can be evaluated by considering only the contributions by the endpoints. Then, it results that for each $\theta_o \neq \theta$

$$K(\theta_o,\theta) \approx -\frac{1}{j\beta^2 a}\left[p(\phi_{max},\theta)\frac{A(\phi_{max},\theta_o,\theta)}{\Phi'(\phi_{max},\theta_o,\theta)}e^{-j\beta a\ \Phi(\phi_{max},\theta_o,\theta)} -p(-\phi_{max},\theta)\frac{A(-\phi_{max},\theta_o,\theta)}{\Phi'(-\phi_{max},\theta_o,\theta)}e^{-j\beta a\ \Phi(-\phi_{max},\theta_o,\theta)}\right] \tag{40}$$

II. In order to recast the kernel in a form more similar to a known convolution kernel, let us rewrite it as

$$K(\theta_o,\theta) \approx -\frac{1}{j\beta^2 a}\, e^{-j\,\frac{\beta a}{2}\,(\Phi(\phi_{max},\theta,\theta_0)+\,\Phi(-\phi_{max},\theta,\theta_0))} \left(\frac{p(\phi_{max},\theta)\ A(\phi_{max},\theta,\theta_o)}{\Phi'(\phi_{max},\theta,\theta_o)}e^{j\,\frac{\beta a}{2}\,(\Phi(-\phi_{max},\theta,\theta_0)-\Phi(+\phi_{max},\theta,\theta_0))} -\frac{p(-\phi_{max},\theta)A(-\phi_{max},\theta,\theta_o)}{\Phi'(-\phi_{max},\theta,\theta_o)}e^{-j\,\frac{\beta a}{2}\,(\Phi(-\phi_{max},\theta,\theta_0)-\Phi(\phi_{max},\theta,\theta_0))}\right) \tag{41}$$

Then, let us introduce the following functions

$$\eta(\theta) = \frac{R(-\phi_{max},\theta) - R(\phi_{max},\theta)}{2a} \tag{42}$$

$$\gamma(\theta) = \frac{R(-\phi_{max},\theta) + R(\phi_{max},\theta)}{2a} \tag{43}$$

which allow recasting the kernel as

$$K(\eta_o,\eta) \approx -\frac{1}{j\beta^2 a}\, e^{-j\ \beta a\ (\gamma(\eta_o)-\gamma(\eta))}\left(\frac{p(\phi_{max},\eta)\ A(\phi_{max},\eta_o,\eta)}{\Phi'(\phi_{max},\eta_o,\eta)}e^{j\ \beta a\ (\eta_o-\eta)} -\frac{p(-\phi_{max},\eta)A(-\phi_{max},\eta_o,\eta)}{\Phi'(-\phi_{max},\eta_o,\eta)}e^{-j\ \beta a\ (\eta_o-\eta)}\right) \tag{44}$$

III. The kernel function (44) has still an intricate structure. However, if the numerator and the denominator of amplitude term $\frac{A}{\Phi'}$ are expanded with respect to the variable η_o in a Taylor series truncated to the first order, the amplitude term can be simplified as below

$$\frac{A(\phi,\eta_o,\eta)}{\Phi'(\phi,\eta_o,\eta)} \approx \frac{R^2(\eta,\phi)}{r_o\left[\ a\ r_o\sin^2(\theta(\eta)-\phi)-R^2(\eta,\phi)\cos(\theta(\eta)-\phi\ )\right]}\,\frac{1}{(\eta_o-\eta)\ \frac{d\theta}{d\eta_o}\Big|_{\eta_o=\eta}} \tag{45}$$

IV. At this juncture, if the weight function $p(\eta,\phi)$ is chosen as

$$p(\eta,\phi) = \frac{r_o\ \left[\ a\ r_o\sin^2(\theta(\eta)-\phi) - R^2(\eta,\phi)\cos(\theta(\eta)-\phi\ )\right]}{R^2(\eta,\phi)} \tag{46}$$

the kernel can be recast as

$$K(\eta_o,\eta) \approx -\frac{2}{\beta}\frac{1}{\frac{d\theta}{d\eta}}\, e^{-j\ \beta a\ (\gamma(\eta_o)-\gamma(\eta))} sinc(\beta a(\eta-\eta_o) \tag{47}$$

Accordingly, in the variable (η,η_o) the eigenvalues problem in (8) can be expressed as

$$-\frac{2a^2}{\beta}\int_{\eta(-\theta_{max})}^{\eta(\theta_{max})}\frac{1}{\frac{d\theta}{d\eta}}\, e^{-j\ \beta a\ (\gamma(\eta_o)-\gamma(\eta))}\ sinc(\beta a(\eta-\eta_o)\ v_n(\eta)\ \frac{d\theta}{d\eta}\ d\eta = \sigma_n^2\ v_n(\eta_o) \tag{48}$$

from which follows that

$$-\frac{2a^2}{\beta}\int_{\eta(-\theta_{max})}^{\eta(\theta_{max})} e^{-j\ \beta a\ (\gamma(\eta_o)-\gamma(\eta))}\ sinc(\beta a(\eta-\eta_o)\ v_n(\eta)\ d\eta = \sigma_n^2\ v_n(\eta_o) \tag{49}$$

V. Now, by fixing

$$\widetilde{v}_n(\eta_o) = e^{\,j\,\beta a\,\gamma(\eta_o)}\,v_n(\eta_o), \tag{50}$$

the eigenvalues problem (49) can be rewritten in the simple and nice form

$$-\frac{2a^2}{\beta}\int_{\eta(-\theta_{max})}^{\eta(\theta_{max})} sinc\,(\beta a\,(\eta_o-\eta))\,\widetilde{v}_n(\eta)\,d\eta = \sigma_n^2\,\widetilde{v}_n(\eta_o) \tag{51}$$

5.2. NDF Estimation and Interpolation of the near Field

VI. The eigenvalues of (51) can be found by resorting again to [27]. Accordingly, it is possible to state that the eigenvalues of (51) exhibit a step-like behavior with the knee occurring at the index

$$N = \left[\,2\frac{\beta a}{\pi}\,\eta(\theta_{max})\,\right] \tag{52}$$

This number provides an estimation of the NDF of the near-field. Let us highlight that, when no stationary points appear in the integral (38), the NDF of the far-field radiated by a circumference arc source has the same mathematical expression of that of the near-field radiated by a strip source (see Equation (25) in [21]).

VII. At this juncture an interpolation formula of the near-field is provided. Since the set of basis functions $\{\widetilde{v}_n(\eta_o)\}$ are bandlimited functions with a bandwidth βa, for each $n \in N$ $\widetilde{v}_n(\eta_o)$ can be expressed through the following truncated sampling series

$$\widetilde{v}_n(\eta_o) \approx \sum_{m\,\in\,I} \widetilde{v}_n(m\,\Delta\eta)\,sinc(\beta a\,\eta_o - m\pi) \tag{53}$$

where

- $\Delta\eta = \frac{\pi}{\beta a}$;
- I is the set containing all those index m such that $m\,\Delta\eta \in [\eta(-\theta_{max}), \eta(\theta_{max})]$.

The set of functions $\{\widetilde{v}_n(\eta_o)\}$ represent a basis for the reduced field

$$\widetilde{E}(\eta_o) = e^{\,j\,\beta a\,\gamma(\eta_o)}\,E(\eta_o) \tag{54}$$

Accordingly, it can be expressed by the following truncated sampling expansion

$$\widetilde{E}(\eta_o) \approx \sum_{m\,\in\,I} \widetilde{E}(m\,\Delta\eta)\,sinc(\beta a\,\eta_o - m\pi) \tag{55}$$

VIII. At this juncture, taking into account of (54) and (55), the near field can be approximated as below

$$E(\eta_o) \approx e^{-j\,\beta a\,\gamma(\eta_o)} \sum_{m\,\in\,I} \widetilde{E}(m\,\Delta\eta)\,sinc(\,\beta a\,\eta_o - m\pi\,) \tag{56}$$

Since $\widetilde{E}(m\,\Delta\eta) = e^{\,j\,\beta a\,\gamma(m\,\Delta\eta)}\,E(m\,\Delta\eta)$, Equation (56) can be rewritten as

$$E(\eta_o) \approx e^{-j\,\beta a\,\gamma(\eta_o)} \sum_{m\,\in\,I} E(m\,\Delta\eta)\,e^{\,j\,\beta a\,\gamma(m\Delta\eta)}\,sinc(\beta a\,\eta_o - m\pi) \tag{57}$$

the previous equation provides an interpolation formula of the near field based on a Shannon sampling series of the reduced field. The number of field samples used by (57) is equal to the Shannon number which is given by

$$N_{SH} = 2\left[\frac{\eta(\theta_{max})}{\Delta\eta}\right] + 1 = 2\left[\frac{\beta a}{\pi}\,\eta(\theta_{max})\right] + 1 \tag{58}$$

accordingly, also in the case of observation domain in near field, the Shannon number is exactly equal to the NDF. From Equation (57), it can be noted that the optimal locations of the sampling points in the variable η_o are given by

$$\eta_{om} = m\frac{\pi}{\beta a} \tag{59}$$

hence, in the variable θ_o the optimal sampling points of the near-field can be found by solving numerically the equation

$$\sqrt{r_o^2 + a^2 - 2a\, r_o \cos(-\phi_{max} - \theta_o)} - \sqrt{r_o^2 + a^2 - 2a\, r_o \cos(\phi_{max} - \theta_o)} = m\frac{2\pi}{\beta} \tag{60}$$

naturally, since the transformation $\eta_o = \eta(\theta_o)$ is nonlinear, the uniform sampling in the variable η_o is mapped into a non-uniform sampling in the variable θ_o. In particular, the optimal field samples are denser for small values of $|\theta_o|$ whereas their step is larger when $|\theta_o|$ approaches to θ_{max}.

6. Comparison between Non-Uniform and Uniform Sampling

In the previous section, a non-uniform sampling scheme for the far field and the near field has been shown. Here, for sake of comparison, a more standard sampling scheme based on a uniform sampling is recalled from the literature. It is well known that the field radiated by a source enclosed in a circle of radius a can be expressed in a series of Fourier harmonics or periodic Dirichlet functions. In particular, if the observation domain is a full circumference ($\theta \in [-2\pi, 2\pi]$), the number of terms of such series can be truncated to $N_{upper} = 2[\beta a] + 1$ where $[\beta a]$ stands for the integer part of βa [29]. Instead, if the radiated field is observed on a limited angular interval $[-\theta_{max}, \theta_{max}]$, a number of terms

$$N_{upper} = 2\left[\frac{\beta a}{\pi}\theta_{max}\right] + 1 \tag{61}$$

is sufficient to represent the radiated field with good accuracy [10].

Accordingly, if the observation domain is a limited arc extending on the angular sector $[-\theta_{max}, \theta_{max}]$, the field radiated by a source enclosed in a circle of radius a can be expressed as below

$$E(\theta_o) \approx \sum_{m=N_o}^{N_o} E(\theta_m)\, D_{M_o}(\theta_o - \theta_{om}) \tag{62}$$

where

- $N_o = \left[\frac{\beta a}{\pi}\theta_{max}\right]$
- $D_{N_o}(\theta_o) = \frac{1}{2N_o+1}\frac{\sin\left(\frac{\pi}{\theta_{max}}\left(N_o+\frac{1}{2}\right)\theta_o\right)}{\sin\left(\frac{\pi}{\theta_{max}}\frac{1}{2}\theta_o\right)}$ is the Dirichlet function
- $\{\theta_{om}\}$ are the sampling points uniformly spaced over the observation arc, hence, $\theta_{om} = -\theta_{max} + m\Delta\theta$ with $\Delta\theta = \frac{2\theta_{max}}{N_{upper}}$ and $m \in \{1, 2, \ldots, N_{upper}\}$.

Now, it is possible to quantify the percentual reduction of field samples P of the present non-uniform sampling scheme when it is compared with the uniform sampling strategies. The latter is given by

$$P = \left(1 - \frac{N_{SH}}{N_{upper}}\right) \cdot 100\% \tag{63}$$

Accordingly, it results that the percentual reduction of field samples for the far-field and the near field can be approximated as below

$$P \approx \begin{cases} [1 - sin\phi_{max}\, sinc(\theta_{max})] \cdot 100\% & \text{for far} - \text{field} \\ \left(1 - \frac{\eta(\theta_{\max})}{\theta_{max}}\right) \cdot 100\% & \text{for near} - \text{field} \end{cases}. \tag{64}$$

7. Numerical Validation

In this section, the NDF evaluation and the interpolation formula of the radiated field provided in Sections 4 and 5 are validated by a numerical analysis. Moreover, the non-uniform sampling strategies developed for the far field and the near field are compared with the uniform sampling strategies described in Section 6. In such comparison, the misfit between the exact field and its approximation provided by the interpolation is measured by the relative error

$$e = \frac{||E - E_{int}||}{||E||} \tag{65}$$

with $||\cdot||$ denoting the Euclidean norm. In all the cases, the exact field E will be that provided by Equation (2) while the interpolated field E_{int} will be provided by (32), (57) or (62), according to the type of considered sampling (either non-uniform or uniform).

The numerical validation of the analytical results is provided in two subsections: the first concerning the far field sampling, the second one regarding the near-field sampling.

7.1. Far-Field Sampling Validation

In this section, a numerical check of the analytical results for the far field is provided. A circumference arc source of radius $a = 20\lambda$ spanning the interval $[-\phi_{max}, \phi_{max}] = [-35°, 35°]$ is considered. The density current of the source is chosen as

$$J(\phi) = e^{-j\beta a \cos(\theta^* - \phi)} \tag{66}$$

where $\theta^* = 15°$. As is well known, such current radiates an electric field focusing at $\theta = \theta^*$. The far field is observed on a circumference arc spanning the interval $[-\theta_{max}, \theta_{max}] = [-50°, 50°]$.

In Figure 4, the actual singular values of the radiation operator T are compared with the ones obtained by considering the weighted adjoint.

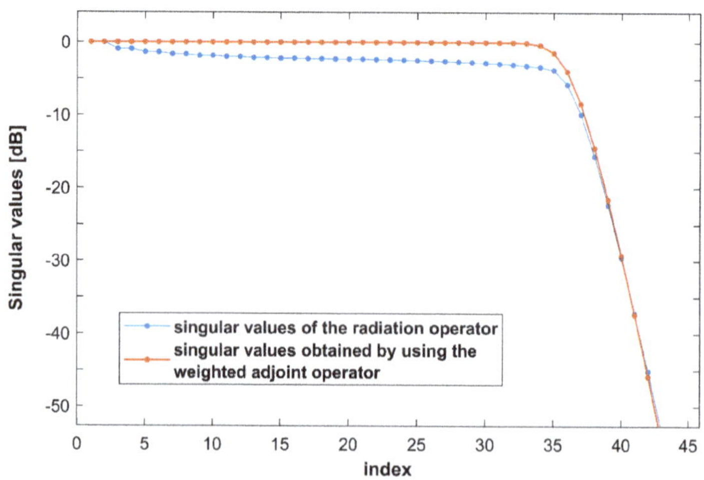

Figure 4. Comparison between the singular values of the radiation operator and those obtained by introducing the weighted adjoint. The diagram refers to the configuration $a = 20\lambda$, $\phi_{max} = 35°$, $\theta_{max} = 50°$.

As it is clear from Figure 4, the use of a weighted adjoint changes the behavior of the singular values but not the index at which they decay abruptly. The latter is predicted by (27) which, in the considered test case, returns $N = 35$ in perfect agreement with the diagrams in Figure 4. In Figure 5, the optimal sampling points of the far-field in the variable θ are shown. They are non-uniformly arranged along the observation domain. In particular, the sampling step is minimum around the direction $\theta = 0°$, whereas it increases by moving towards the directions $\theta = \pm\theta_{max}$.

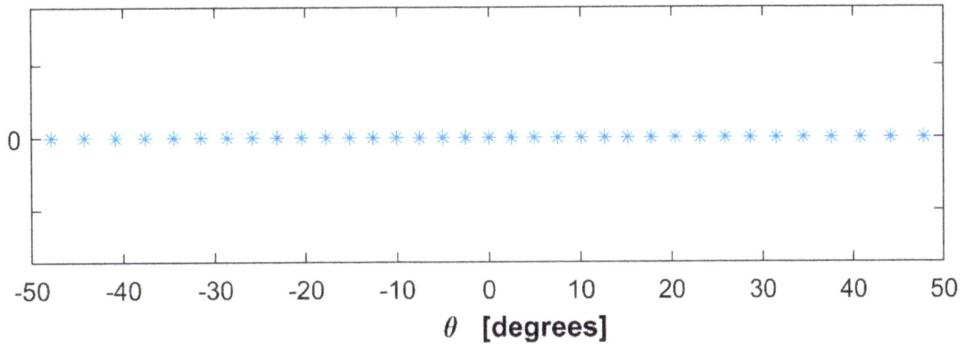

Figure 5. Optimal position of the far-field samples in the variable θ. The diagram refers to the configuration $a = 20\lambda$, $\phi_{max} = 35°$, $\theta_{max} = 50°$.

In Figure 5, the exact far-field computed by means of the equation $E(\theta) = TJ(\phi)$ is compared with the field returned by the interpolation Formula (32).

As can be seen from Figure 6, despite the interpolation Formula (32) exploits a number of samples that is as low as possible (only $N_{SH} = 35$ non-uniform field samples are used for the interpolation), the interpolated field agrees very well with the exact field and $e = 0.028$.

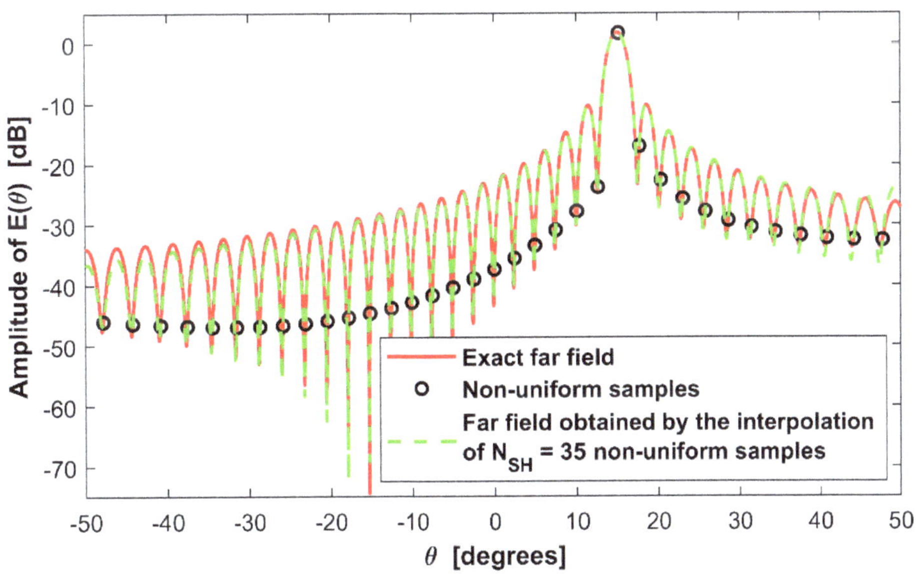

Figure 6. Comparison between the far field computed by the radiation model in (2) and the far field returned by the interpolation Formula (32). The diagram refers to the configuration $a = 20\lambda$, $\phi_{max} = 35°$, $\theta_{max} = 50°$.

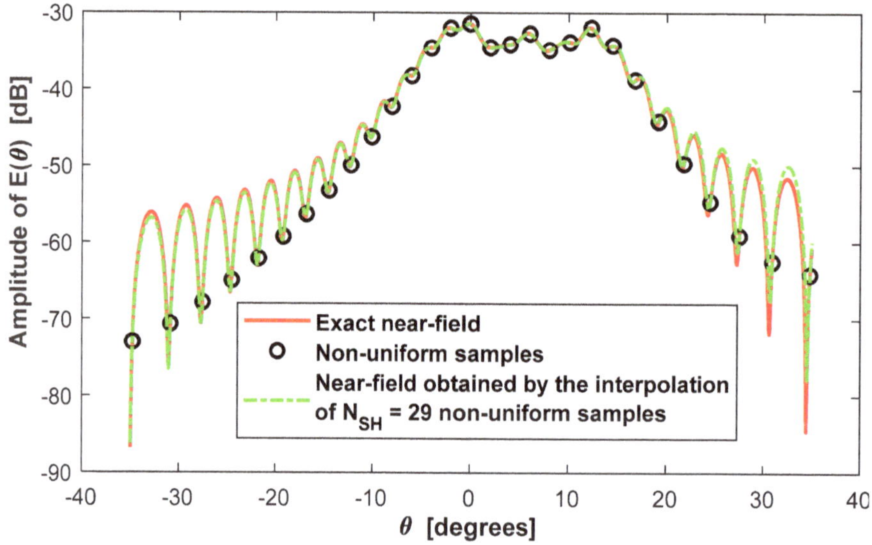

Figure 10. Comparison between the far-field computed by the radiation model in (3) and the far-field returned by the interpolation Formula (57). The diagram refers to the configuration $a = 20\lambda$, $\phi_{max} = 25°$, $r_o = 40\lambda$, $\theta_{max} = 35°$.

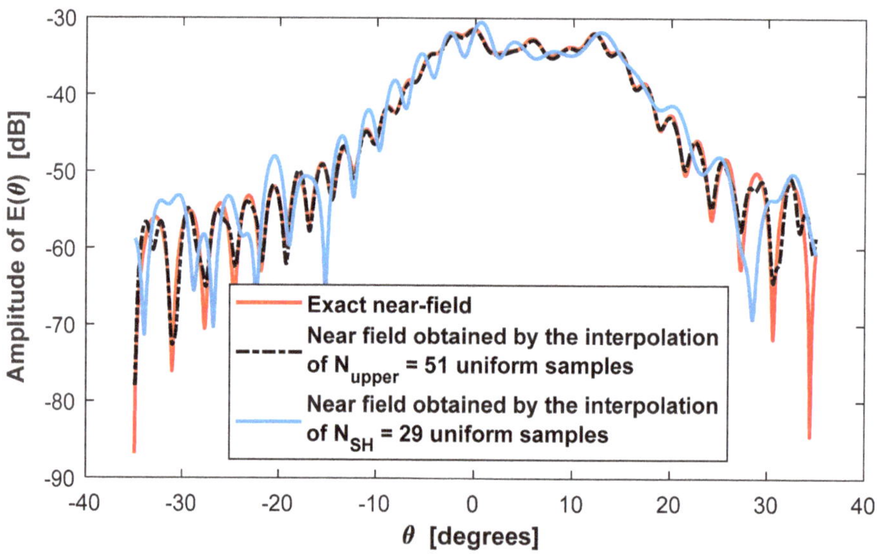

Figure 11. Comparison between the exact near field, the near field obtained by the interpolation of $N_{SH} = 29$ uniform samples and the near field obtained by the interpolation of $N_{upper} = 51$ uniform samples. The diagram refers to the configuration $a = 20\lambda$, $\phi_{max} = 25°$, $r_o = 40\lambda$, $\theta_{max} = 35°$.

As can be seen from Figure 11, the interpolation of the near field obtained with only $N_{SH} = 29$ uniform field samples is not very accurate and the relative error is equal to 0.294. On the contrary, the interpolation obtained with $N_{upper} = 71$ approximates well the near field with a relative error $e = 0.034$. Accordingly, also for the near field the non-uniform sampling scheme allows to achieve the same accuracy with a lower number of field samples.

In the considered example, the use of the non-uniform sampling strategy allows to reach $P = 43\%$.

8. Conclusions

In this paper, an optimal sampling strategy of the field radiated by a 2D current supported over a circumference arc source has been developed. In particular, through an analytical study of the relevant singular values of the radiation operator, the minimum number of sampling points required to sample the radiated field without loss of information has been first found. Then, starting from a sampling representation of the reduced field, an interpolation formula of the radiated field that exploits a non-redundant number of field samples has been provided. The developed sampling strategy allows us to reach the same accuracy in the field interpolation of the uniform sampling scheme with a lower number of field measurements. This is very important in practical cases since a reduction of the number of field measurements allows reducing the acquisition time in near-field testing techniques which is dominated by the mechanical positioning of the field probe. The only limitation of the developed sampling method is the fulfillment of condition (11) for the far-field and condition (39) for the near-field.

Future developments concern the extension of the proposed sampling strategy to the cases of other conformal sources with a different shape and more realistic scenarios involving 3D geometries. Moreover, another possible extension regards the case of phaseless measurements [30–32].

Author Contributions: Conceptualization, R.P., G.L. and R.M.; methodology, R.M. and R.P.; software, R.M. and F.M.; validation, R.M. and F.M.; formal analysis, R.M., G.L. and F.M.; investigation, R.M. and G.L.; resources, G.L. and R.P.; data curation, R.M. and F.M.; writing—original draft preparation, R.M. and G.L.; writing—review and editing, R.M., G.L. and F.M.; visualization, R.M.; supervision, R.P. and G.L.; project administration, G.L.; funding acquisition, G.L. and R.P. All authors have read and agreed to the published version of the manuscript.

Funding: This work was funded by the Italian Ministry of University and Research through PRIN 2017 Program.

Data Availability Statement: Data supporting reported results are generated during the study.

Conflicts of Interest: The authors declare no conflict of interest.

Appendix A

In this appendix the mathematical condition (11) ensuring the absence of stationary points in the phase function $\Psi(\phi, \theta_o, \theta) = \cos(\theta_o - \phi) - \cos(\theta - \phi)$ is derived. A possible stationary point ϕ_s is solution of the equation $\Psi(\phi_s, \theta_o, \theta) = 0$, that is

$$\sin(\theta_o - \phi_s) - \sin(\theta - \phi_s) = 0 \tag{A1}$$

By resorting to sum-to-product identity for trigonometric functions (A1) recasts as

$$2\sin\left(\frac{\theta_o - \theta}{2}\right)\cos\left(\phi_s - \frac{\theta_o + \theta}{2}\right) = 0 \tag{A2}$$

which, excluding the case $\theta_o = \theta$, is generally verified when

$$\phi_s = \frac{\theta_o + \theta + \pi}{2} + m\pi, \quad m = 0 \pm 1, \pm 2, \dots \tag{A3}$$

Thus, the interval $[-\phi_{max}, \phi_{max}]$ is devoid of all the stationary points when they fall outside this interval, namely when

$$|\phi_s| > \phi_{max} \tag{A4}$$

From (A3) one can note that once m is fixed, since θ and θ_0 are at most equal to $\pi/2$, the values of ϕ_s fall into the interval $[m\pi, (m+1)\pi]$. Limiting the analysis to the interval $[-\pi, \pi]$, condition (A4) translates into the condition

$$\frac{\theta + \theta_0 + \pi}{2} > \phi_{max} \cup \frac{\theta + \theta_0 - \pi}{2} < -\phi_{max} \,, \tag{A5}$$

where $m = 0$ in the left condition obtained for positive values of ϕ_s and $m = -1$ in the right condition for negative values of ϕ_s. The left condition rewrites as

$$\theta + \theta_0 > 2\phi_{max} - \pi \,, \tag{A6}$$

and, to be always verified, the smallest possible value of $\theta + \theta_0$ should be larger than $2\phi_{max} - \pi$, that is

$$-2\theta_{max} > 2\phi_{max} - \pi \tag{A7}$$

Analogously, the right condition is always satisfied when the larger possible value of $\theta + \theta_0$ (that is $2\theta_{max}$) is less than $\pi - 2\phi_{max}$. Hence, both the conditions lead to the final condition.

$$\theta_{max} + \phi_{max} < \frac{\pi}{2} \tag{A8}$$

Appendix B

In this appendix, with reference to three different values of ϕ_{max}, the limit angle θ_{max} under which no stationary points appear in the phase function $\Phi\ (\phi, \theta_o, \theta)$ is found by a numerical analysis. In particular, the cases $\phi_{max} = \{0.35\ (20^\circ),\ 0.52\ (30^\circ),\ 0.7\ (40^\circ)\}$ are respectively considered in Tables A1–A3.

Table A1. Maximum value of θ_{max} such that no stationary points appear in $\Phi\ (\phi, \theta_o, \theta)$ when $\phi_{max} = 0.35\ (20^\circ)$.

r_o/a	**Maximum Value of θ_{max} Such That no Stationary Points Appear in $\Phi\ (\phi, \theta, \theta_o)$**
1.4	0.35 (20°)
1.6	0.52 (30°)
2	0.70 (40°)
4	0.87 (50°)
8	1.05 (60°)
15	1.13 (65°)

Table A2. Maximum value of θ_{max} such that no stationary points appear in $\Phi\ (\phi, \theta_o, \theta)$ when $\phi_{max} = 0.52\ (30^\circ)$.

r_o/a	**Maximum Value of θ_{max} Such That no Stationary Points Appear in $\Phi\ (\phi, \theta, \theta_o)$**
1.4	0.17 (10°)
1.6	0.35 (20°)
2	0.52 (30°)
4	0.70 (40°)
8	0.87 (50°)
15	0.96 (55°)

Table A3. Maximum value of θ_{max} such that no stationary points appear in Φ (ϕ, θ_o, θ) when $\phi_{max} = 0.7$ (40°).

r_o/a	Maximum Value of θ_{max} Such That no Stationary Points Appear in Φ (ϕ, θ, θ_o)
1.6	0.17 (10°)
2	0.35 (20°)
4	0.52 (30°)
8	0.70 (40°)
15	$\pi/4$ (45°)

The results in the Tables A1–A3 have been obtained by numerically solving the stationary condition (39) with respect to ϕ for each value of r_o/a and for each couple (θ_o, θ).

References

1. Rahmat-Samii, Y.; Cheung, R. Non-uniform sampling techniques for antenna applications. *IEEE Trans. Antennas Propag.* **1987**, *35*, 268–279. [CrossRef]
2. Wang, J.; Yarovoy, A. Sampling design of synthetic volume arrays for three-dimensional microwave imaging. *IEEE Trans. Comp. Imag.* **2018**, *4*, 648–660. [CrossRef]
3. Mézières, N.; Fuchs, B.; Le Coq, L.; Lerat, J.M.; Contreres, R.; Le Fur, G. On the Antenna Position to Improve the Radiation Pattern Characterization. *IEEE Trans. Antennas Propag.* **2021**, *69*, 5335–5344. [CrossRef]
4. Migliore, M.D. Near field antenna measurement sampling strategies: From linear to nonlinear interpolation. *Electronics* **2018**, *7*, 257. [CrossRef]
5. Foged, L.J.; Saccardi, F.; Mioc, F.; Iversen, P.O. Spherical Near Field Offset Measurements using Downsampled Acquisition and Advanced NF/FF Transformation Algorithm. In Proceedings of the 10th European Conference on Antennas and Propagation (EuCAP), Davos, Switzerland, 10–15 April 2016.
6. D'Agostino, F.; Ferrara, F.; Gennarelli, C.; Guerriero, R.; Migliozzi, M. Fast and Accurate Far-Field Prediction by Using a Reduced Number of Bipolar Measurements. *IEEE Antennas Wirel. Propag. Lett.* **2017**, *16*, 2939–2942.
7. Hofmann, B.; Neitz, O.; Eibert, T. On the minimum number of samples for sparse recovery in spherical antenna near-field measurements. *IEEE Trans. Antennas Propag.* **2019**, *67*, 7597–7610. [CrossRef]
8. Kim, Y.H. Greedy sensor selection based on QR factorization. *EURASIP J. Adv. Signal Process.* **2021**, *1*, 1–13. [CrossRef]
9. Behjoo, H.R.; Pirhadi, A.; Asvadi, R. Optimal Sampling in Spherical Near-Field Antenna Measurements by Utilizing the Information Content of Spherical Wave Harmonics. *IEEE Trans. Antennas Propag.* **2021**, 1. [CrossRef]
10. Leone, G.; Maisto, M.A.; Pierri, R. Inverse Source of Circumference Geometries: SVD Investigation Based on Fourier Analysis. *Progr. Electromagn. Res. M* **2018**, *76*, 217–230. [CrossRef]
11. Solimene, R.; Brancaccio, A.; Romano, J.; Pierri, R. Localizing Thin Metallic Cylinders by a 2.5-D Linear Distributional Approach: Experimental Results. *IEEE Trans. Antennas Propag.* **2008**, *56*, 2630–2637. [CrossRef]
12. Joy, E.; Paris, D. Spatial sampling and filtering in near-field measurements. *IEEE Trans. Antennas Propag.* **1972**, *20*, 253–261. [CrossRef]
13. Leach, W.; Paris, D. Probe compensated near-field measurements on a cylinder. *IEEE Trans. Antennas Propag.* **1973**, *21*, 435–445. [CrossRef]
14. Bucci, O.M.; Gennarelli, C.; Saverese, C. Optimal Interpolation of radiated fields over a sphere. *IEEE Trans. Antennas Propag.* **1991**, *39*, 1633–1643. [CrossRef]
15. Piestun, R.; Miller, D.A.B. Electromagnetic degrees of freedom of an optical system. *J. Opt. Soc. Amer. A* **2000**, *17*, 892–902. [CrossRef] [PubMed]
16. Pierri, R.; Moretta, R. NDF of the near-zone field on a line perpendicular to the source. *IEEE Access* **2021**, *9*, 91649–91660. [CrossRef]
17. Qureshi, M.A.; Schmidt, C.H.; Eibert, T.F. Adaptive Sampling in Spherical and Cylindrical Near-Field Antenna Measurements. *IEEE Antennas Propag. Mag.* **2013**, *55*, 243–249. [CrossRef]
18. Bucci, O.M.; Gennarelli, C.; Savarese, C. Representation of Electromagnetic Fields over Arbitrary Surfaces by a Finite and Nonredundant Number of Samples. *IEEE Trans. Antennas Propag.* **1998**, *46*, 351–359. [CrossRef]
19. Leone, G.; Munno, F.; Solimene, R.; Pierri, R. A PSF Approach to Far Field Discretization for Conformal Sources. *IEEE Trans. Comp. Imag.* **2022**. Available online: https://www.techrxiv.org/articles/preprint/A_PSF_Approach_to_Far_Field_Discretization_for_Conformal_Sources/17708579 (accessed on 12 December 2021).
20. Capozzoli, A.; Curcio, C.; Liseno, A. Different Metrics for Singular Value Optimization in Near-Field Antenna Characterization. *Sensors* **2021**, *21*, 2122. [CrossRef]
21. Joshi, S.; Boyd, S. Sensor selection via convex optimization. *IEEE Trans. Signal Process.* **2008**, *57*, 451–462. [CrossRef]

22. Jiang, C.; Soh, Y.; Li, H. Sensor placement by maximal projection on minimum eigenspace for linear inverse problems. *IEEE Trans. Signal Process.* **2016**, *64*, 5595–5610. [CrossRef]
23. Solimene, R.; Maisto, M.A.; Pierri, R. Sampling approach for singular system computation of a radiation operator. *JOSA A* **2019**, *36*, 353–361. [CrossRef] [PubMed]
24. Maisto, M.A.; Pierri, R.; Solimene, R. Near-Field Warping Sampling Scheme for Broad-Side Antenna Characterization. *Electronics* **2020**, *9*, 1047. [CrossRef]
25. Pierri, R.; Moretta, R. Asymptotic Study of the Radiation Operator for the Strip Current in Near Zone. *Electronics* **2020**, *9*, 911. [CrossRef]
26. Bleistein, N.; Handelsman, R.A. *Asymptotic Expansions of Integrals*; Dover: New York, NY, USA, 1986.
27. Slepian, D.; Pollack, H.O. Prolate spheroidal wave functions, Fourier analysis, and uncertainty—I. *Bell Syst. Tech. J.* **1961**, *40*, 43. [CrossRef]
28. Khare, K.; George, N. Sampling theory approach to prolate spheroidal wavefunctions. *J. Phys. A Math. Gen.* **2003**, *36*, 10011. [CrossRef]
29. Devaney, A. *Mathematical Foundations of Imaging, Tomography and Wavefield Inversion*; Cambridge University Press: Cambridge, UK, 2012.
30. Pierri, R.; Leone, G.; Moretta, R. The Dimension of Phaseless Near-Field Data by Asymptotic Investigation of the Lifting Operator. *Electronics* **2021**, *10*, 1658. [CrossRef]
31. Pierri, R.; Moretta, R. On the Sampling of the Fresnel Field Intensity over a Full Angular Sector. *Electronics* **2021**, *10*, 832. [CrossRef]
32. Rodríguez Varela, F.; Fernandez Álvarez, J.; Galocha Iragüen, B.; Sierra Castañer, M.; Breinbjerg, O. Numerical and Experimental Investigation of Phaseless Spherical Near-Field Antenna Measurements. *IEEE Trans. Antennas Propag.* **2021**, *69*, 8830–8841. [CrossRef]

sensors

MDPI

Article

Analysis and Design of an X-Band Reflectarray Antenna for Remote Sensing Satellite System

Shimaa A. M. Soliman *, Eman M. Eldesouki and Ahmed M. Attiya †

Microwave Engineering Department, Electronics Research Institute, Joseph Tito St., Huckstep, El Nozha, Cairo 11843, Egypt; eman@eri.sci.eg (E.M.E.); attiya@eri.sci.eg (A.M.A.)
* Correspondence: shimaa_megahed@eri.sci.eg
† Ahmed M. Attiya, Member, IEEE.

Abstract: This paper presents the analysis and design of an X-band reflectarray. The proposed antenna can be used for a medium Earth orbit (MEO) remote sensing satellite system in the 8.5 GHz band. To obtain a nearly constant response along the coverage area of this satellite system, the proposed antenna was designed with a flat-top radiation pattern with a beam width of around 29° for the required MEO system. In addition, broadside pencil beam and tilted pencil beam reflectarrays were also investigated. The feeding element of the proposed reflectarray antennas is a Yagi–Uda array. The amplitude and phase distribution of the fields due to the feeding element on the aperture of the reflectarray antenna are obtained directly by numerical simulation without introducing any approximation. The required phase distribution along the aperture of the reflectarray to obtain the required flat-top radiation pattern is obtained using the genetic algorithm (GA) optimization method. The reflecting elements of the reflectarray are composed of stacked circular patches. This stacked configuration was found to be appropriate for obtaining a wide range of reflection phase shift, which is required to implement the required phase distribution on the reflectarray aperture. The antenna was fabricated and measured for verification.

Keywords: reflectarray antenna; flat-top radiation pattern; remote sensing satellite system; genetic algorithm

Citation: Soliman, S.A.M.; Eldesouki, E.M.; Attiya, A.M. Analysis and Design of an X-Band Reflectarray Antenna for Remote Sensing Satellite System. *Sensors* **2022**, *22*, 1166. https://doi.org/10.3390/s22031166

Academic Editors: Raed A. Abd-Alhameed, Naser Ojaroudi Parchin and Chan Hwang See

Received: 30 December 2021
Accepted: 26 January 2022
Published: 3 February 2022

1. Introduction

Recently, satellite communication systems in low and medium Earth orbit (LEO/MEO) have experienced rapid development. Satellites are characterized by their design-and-deployment cost, power consumption, and down-link bandwidth [1]. In order to increase the down-link data rate of the satellite, a high gain antenna with a low profile, light weight, and small volume, in addition to a cheap assembly process, is required. These prerequisites can be obtained by utilizing a reflectarray antenna, which comprises a spatial feed and a planar structure. Reflectarray antennas are based on focusing the incident fields from an antenna feeding element to obtain the required radiation pattern by compensating for the phase differences between the reflectarray elements. There are different types of reflectarrays, such as planar microstrip reflectarrays [2–6] and dielectric resonator reflectarrays [7–9]. The main feature of the reflectarray is that its radiation characteristics can be manipulated by tuning the geometrical dimensions of its unit cells [10–12].

Although reflectarrays for high-gain pencil beam patterns in a certain direction can be easily designed using analytical equations [13], the synthesis of shaped or contoured beams is a challenging problem. These shaped or contoured beams are required in many satellite communication systems for better power management [14]. In order to generate a specific radiation pattern, different algorithms have been used to synthesize and optimize the phase distribution on the reflectarray elements. The optimization methods used to synthesis the reflectarray pattern are classified into two main categories: local search algorithms,

$\Delta\varphi_{m,n} = \beta_x + \beta_y$. For the case of a reflectarray, the magnitude of the excitation at the elements $|A_{m,n}|$ is determined by the amplitude distribution of the fields due to the feeding element on the aperture of the reflectarray. On the other hand, the total phase at the *mn*th element is the summation of the phase of the field distribution of the feeding element plus the phase distribution of the reflection coefficient on the aperture of the reflectarray. The key point in the design of the reflectarray antenna is to determine the required total phase distribution for the obtained amplitude distribution by the feeding element to obtain the required radiation pattern. The next step is the implementation of the reflecting elements of the reflectarray antenna to verify this total phase distribution. For simple radiation patterns such as a broadside pencil beam or a tilted pencil beam, this phase distribution can be obtained analytically in a closed form as follows:

$$\Delta\varphi_{m,n} = 2\pi N' - k\left(r_{m,n} - \vec{R}\cdot\hat{r_o}\right) \quad (4)$$

where $N' = 1, 2, 3, \ldots$, $r_{m,n}$ is the distance from the feed to each array element, $\hat{R}$ is the position vector from each element to the array center $(0,0,0)$ and $\hat{r_o}$ is the position vector in the direction of the main beam of the reflectarray. However, for reflectarray antennas having beams with more complicated shapes, this phase distribution is obtained using optimization techniques.

3. Feeding Antenna

In this section, the analysis and design of the feeding antenna for the proposed reflectarray are discussed. The proposed feeding antenna is a Yagi–Uda antenna, as shown in Figure 2a. It consists of a fed dipole antenna inserted between two parasitic elements: a director and a reflector element. The reflector and the director elements enhance the radiation in the direction of the aperture of the reflectarray. Typically, the feed element length L_2 is usually around 0.45–0.49λ, while the director length L_3 is approximately 0.4λ to 0.45λ. In addition, the reflector length L_1 is slightly greater than the fed element. The separation between the elements d_s is found to be around 0.1 λ. The radius of these wire elements (a) is set to around approximately 0.025 λ. The proposed Yagi–Uda antenna is simulated using HFSS. Parametric studies are performed through EM simulation for setting the optimum values of the dimensional parameters of an *X*-band Yagi–Uda antenna at the operating frequency of 8.5 GHz. The required performance of the feeding antenna of the reflectarray in the present case comprises input matching below -10 dB and forward to backward radiation of more than 10 dB. The optimum values of the proposed Yagi–Uda antenna are $L_1 = 20.9$ mm, $L_2 = 16.5$ mm, $L_3 = 10.8$ mm, $d_s = 3.9$ mm, and $a = 0.5$ mm.

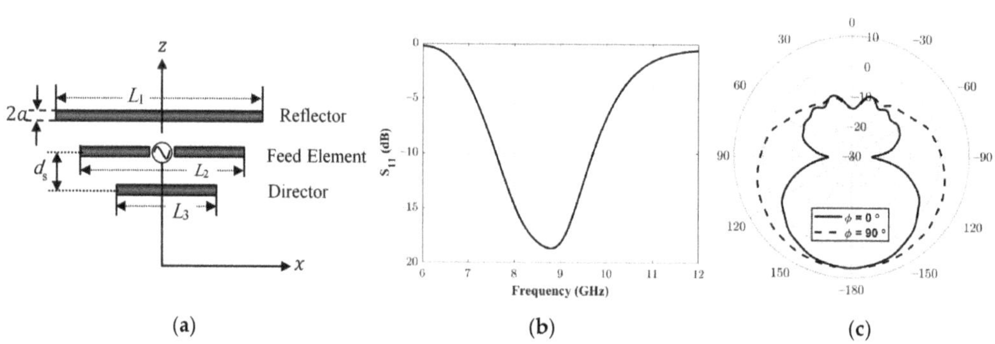

Figure 2. Proposed Yagi–Uda feeding antenna. (**a**) Geometry, (**b**) Simulated Reflection coefficient, (**c**) Simulated total gain pattern.

Figure 2b shows the simulated $|S_{11}|$ for the feed antenna. It can be noted that $|S_{11}|$ at the required operating frequency 8.5 GHz is less than -15 dB, which represents a good matching. The simulated realized gain pattern is shown in Figure 2c. The peak gain is obtained in the –ve z direction towards the aperture of the reflectarray. The peak gain is greater than 6 dBi and the backward radiation is less than −12 dBi; thus, the front to back ratio is around 18 dB. Thus, the proposed Yagi–Uda antenna is suitable for the proposed requirements for the feeding antenna of the reflectarray.

4. Field Distribution of the Feeding Antenna on the Aperture of the Reflectarray

In order to obtain the required beam pattern of the reflectarray, the amplitudes and phase distribution on the plane of the reflectarray should be determined. In previous studies of reflectarray antennas, the field distribution is presented as a simple analytical approximation based on $(\cos\theta)^q$, where the value of q is chosen to obtain the corresponding approximate radiation pattern of the feeding antenna. In addition, the phase distribution of the field on the aperture of the reflectarray is calculated in terms of the distance from the center of the feeding point to each point on the aperture of the reflectarray. In this paper, this approximation is replaced by directly calculating this field distribution numerically using the commercially EM simulation software HFSS. The advantage of this method is that it does not require any assumptions. Figure 3a shows the 2D distribution of the magnitude of the complex total field on an aperture located at a distance $F = 352.9$ mm from the center of the fed element of the Yagi–Uda antenna. It should be noted that this is only a calculation plane and it does not represent any additional boundary to the simulation problem. This plane and the feeding antenna are included inside a common radiation boundary in the simulation process. This distribution can be presented as a radial function of the magnitude around the z-axis. Figure 3b shows the 1D distribution of the normalized amplitudes along the x-axis at the plane of the reflectarray. In addition, one can also obtain the corresponding phase distribution using the argument for this complex field, as shown in Figure 3b. These amplitude and phase distributions are discretized along the proposed grid of the reflectarray to obtain the amplitude and phase of the incident field on each element of the reflectarray.

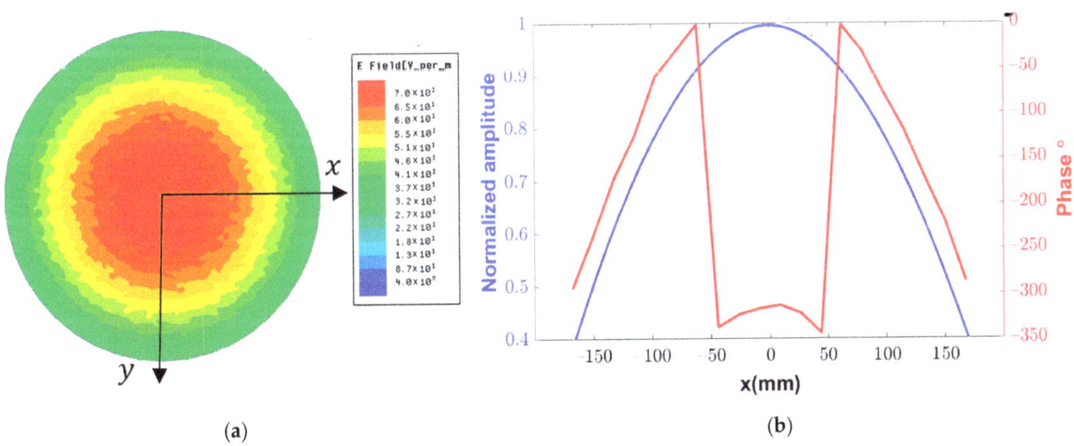

Figure 3. Complex field distribution of the feeding element at the plane of the reflectarray: (**a**) 2D representation, (**b**) normalized amplitude and phase along the x-axis.

The next step is to use this amplitude distribution to find the required phase on each element to obtain the required radiation pattern. Then, it is required to design each element to introduce a phase reflection added to the phase distribution of the feeding element, such that the total phase on this element equals the required phase, to obtain the required radiation pattern.

For the case of a pencil beam radiation pattern, the required total phase distribution on the aperture of the reflectarray antenna can be obtained analytically, as shown in Equation (4). Thus, the reflection phase of each reflecting element can be obtained by subtracting the phase of the feeding element from the required phase distribution of the aperture of the reflectarray. For the present reflectarray structure mentioned in Section 2 and the proposed Yagi–Uda feeding element, the required reflection phases on the aperture of the reflectarray for both broadside and tilted pencil beams with a tilting angle 15° are shown in Figure 4a,b, respectively. It should be noted that the total number of reflecting elements is equal to 316 unit cells, arranged uniformly in a planar grid with $M \times N = 20 \times 20$ elements.

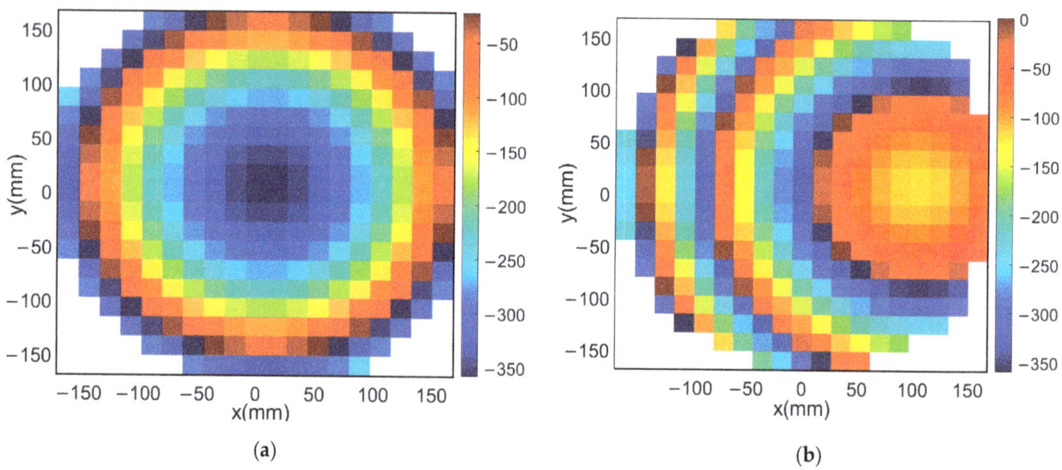

Figure 4. Phase distribution of the pencil beam reflectarray: (**a**) broadside beam, (**b**) tilted beam with a tilting angle of 15°.

By comparison, for the case of a shaped beam, such as a flat-top beam, it is required to determine the required phase distribution using an optimization algorithm because it cannot be determined directly using Equations (1)–(4), as in the case of broadside or tilted pencil beams. However, the phase distribution of the broadside pencil beam can be considered as a good starting point for the proposed optimization process to obtain the corresponding phase distribution for the flat-top beam.

5. Flat-Top Pattern Synthesized Using a Genetic Algorithm

In order to start the optimization process, the requirements of the flat-top pattern that can be applied in the optimization procedure should be first introduced. To obtain a nearly constant communication link along the coverage angle, an antenna with a flat-top pattern is required. The maximum coverage angle θ_{max} is defined from the Earth-satellite geometry shown in Figure 5, as:

$$\theta_{max} = \cos^{-1}\left(\frac{d_{max}}{h_o + R_e}\right) \tag{5}$$

where R_e is the radius of the Earth and h_o is the vertical distance from the satellite to the Earth's surface. d_{max} is maximum distance from the satellite to the Earth given by the trigonometric equation as:

$$d_{max} = \sqrt{(R_e + h_o)^2 - {R_e}^2} \tag{6}$$

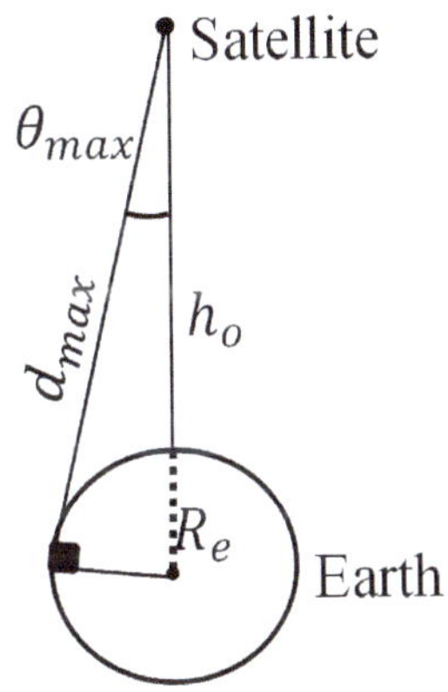

Figure 5. Geometry of the satellite coverage of the MEO satellite system.

The constraints on the required radiation patterns are considered by using appropriate masks. The requirement of the flat-top normalized pattern is given by means of two mask templates, as shown in Figure 6, which impose the minimum and maximum values that the far field must achieve. Thus, if $AF(\theta_i)_{\text{dB}}$ is the normalized array factor of the flat-top beam in dB, it should fulfil the following:

$$Mask_l(\theta_i)_{\text{dB}} \leq AF(\theta_i)_{\text{dB}} \leq Mask_u(\theta_i)_{\text{dB}} \tag{7}$$

where the upper mask shapes the normalized radiation pattern in the angular span $-\theta_{max} \leq \theta_i \leq \theta_{max}$ with an amplitude of 0 dB. This span is assumed as the transition region of the satellite. For the other θ directions, outside of the main beam the SLL limit is assumed to be below −20 dB. The definition of the upper mask of the flat-top beam pattern, as illustrated in Figure 6, is given in dB as follows:

$$Mask_u(\theta_i)_{\text{dB}} = \begin{cases} SLL & -90^o \leq \theta_i \leq -\theta_{max} \\ 0 & -\theta_{max} \leq \theta_i \leq \theta_{max} \\ SLL & \theta_{max} \leq \theta_i \leq 90^o \end{cases} \tag{8}$$

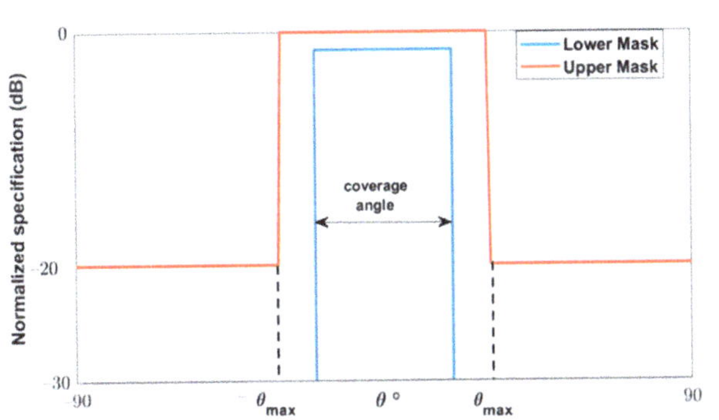

Figure 6. Required mask for the normalized radiation pattern.

The lower mask is mostly used to control the allowable ripple level (ARL) of the shaped beam, which is −3 dB for the flat-top beam. The lower mask in dB is given as:

$$Mask_l(\theta_i)_{\text{dB}} = \begin{cases} < -30 & \theta_i \geq \theta_{max} - \Delta\theta \\ ARL & -\theta_{max} + \Delta\theta \leq \theta_i \leq \theta_{max} - \Delta\theta \\ < -30 & \theta_i \leq -\theta_{max} + \Delta\theta \end{cases} \tag{9}$$

where $\Delta\theta$ is the allowable angle between the upper and the lower mask.

To achieve the flat-top radiation pattern, the phase distribution on the reflectarray elements should be determined through an optimization of a properly defined cost function. The cost function is defined as the error between the obtained normalized array factor AF and the required upper and lower masks. The cost function is normalized for N_{total} angles for both the upper and lower limits of the mask as follows:

$$cost = \frac{\varepsilon_u(\theta_i) + \varepsilon_l(\theta_i)}{N_{total}} \tag{10}$$

where $\varepsilon_u(\theta_i)$ and $\varepsilon_l(\theta_i)$ are given by Equations (11) and (12), respectively:

$$\varepsilon_u(\theta_i) = \sum_{i}^{N_{total}} [AF(\theta_i)_{\mathrm{dB}} - Mask_u(\theta_i)_{\mathrm{dB}}] \frac{[1 + \mathrm{sgn}(AF(\theta_i)_{\mathrm{dB}} - Mask_u(\theta_i)_{\mathrm{dB}})]}{2} \tag{11}$$

$$\varepsilon_l(\theta_i) = \sum_{i}^{N_{total}} [Mask_l(\theta_i)_{\mathrm{dB}} - AF(\theta_i)_{\mathrm{dB}}] \frac{[1 + \mathrm{sgn}(Mask_l(\theta_i)_{\mathrm{dB}} - AF(\theta_i)_{\mathrm{dB}})]}{2} \tag{12}$$

where $AF(\theta_i)_{\mathrm{dB}} = 20\log(|AF(\theta_i)|)$ and $\mathrm{sgn}(x) = 1$ for $x > 0$ and $\mathrm{sgn}(x) = -1$ for $x < 0$. Thus, a better match between the obtained pattern and the required pattern is obtained for the minimum value of this cost function.

After determining the required radiation pattern for the proposed MEO satellite communication system and the amplitude distribution of the fields due to the feeding element, the corresponding phase distribution on the aperture of the reflectarray must be determined. This phase distribution is obtained using GA. In GA, the optimization starts with an initial population comprising a number of candidate solutions (designated as chromosomes). These parents are controlled using different factors (combination, crossover, or mutation) to make a new set of chromosomes for the next generation. During the advancement of the arrangement, chromosomes are reviewed with respect to the enhancement of the fitness between the obtained radiation pattern and the required mask. The higher-positioned chromosomes are chosen to proceed to the next generation. Once the new generation is formed, the fitness of its chromosomes is estimated and the process continues until the convergence condition is satisfied. The algorithm stops when the value of the fitness function for the best point in the current population is less than or equal to the fitness limit. The important basic genetic algorithm steps are presented in Figure 7.

The main problem when applying GA is the large number of optimization variables, which correspond to all reflecting elements on the reflectarray. This large number of variables requires a large computational time, which affects the overall convergence of the optimization process. However, because the proposed flat-top radiation pattern and amplitude distribution of the fields due to the feeding element are radially symmetric around the z-axis, as shown in Figure 3, the required phase distribution should be also be radially symmetric around the z-axis. Thus, the number of the unknown variables can be reduced by taking into consideration this symmetry. For the case of a circular reflectarray as shown in Figure 1, the reflecting elements can be arranged into four image-symmetric quarters. Each quarter can also be divided into two symmetric halves, such that each column in the quarter would be the same as the corresponding perpendicular raw in the same quarter as shown in Figure 8.

Using this approach, it is possible to reduce the number of unknowns in the optimization process to one-eighth of the number of reflectarray elements. This significant reduction reduces the convergence computational time and also improves the resulting convergence. Figure 9 shows the phase distribution obtained by using GA to obtain the required flat-top radiation pattern. Moreover, the obtained flat-top radiation pattern is shown in Figure 10. It can be noted that the obtained flat-top radiation pattern almost satisfies the required conditions of the maximum coverage angle of 29°, $\Delta\theta$ of 5°, and ARL of −3 dB.

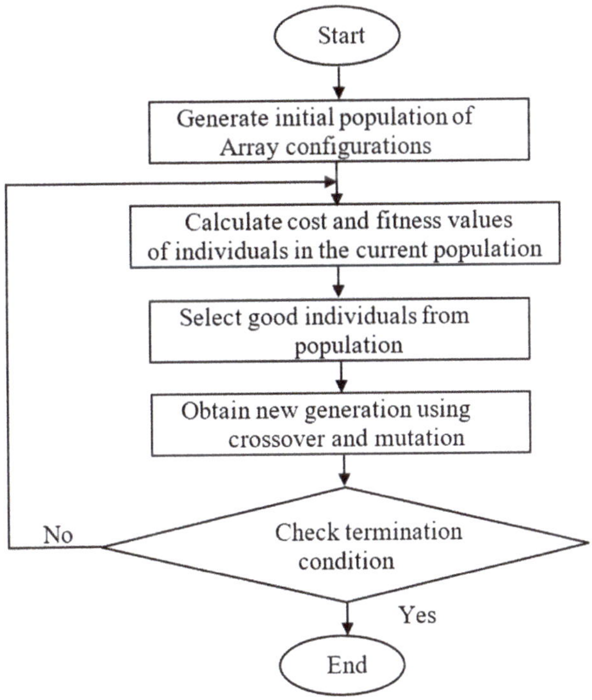

Figure 7. Flowchart of a general GA approach.

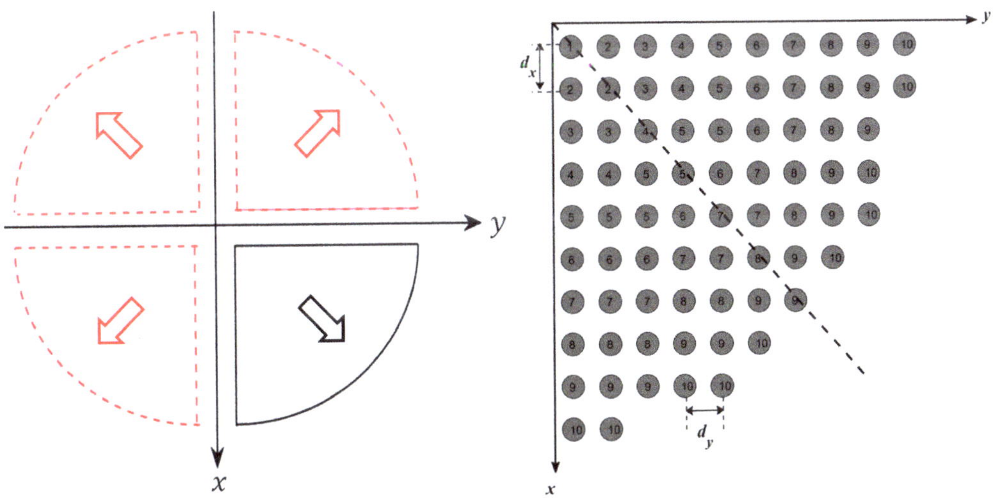

Figure 8. Phase symmetry on half of the elements of the proposed reflectarray antenna.

was fabricated and measured to validate the flat-top pattern obtained by GA and that obtained by numerical calculation. Figure 17b–d shows the fabricated layers of the flat-top reflectarray antenna.

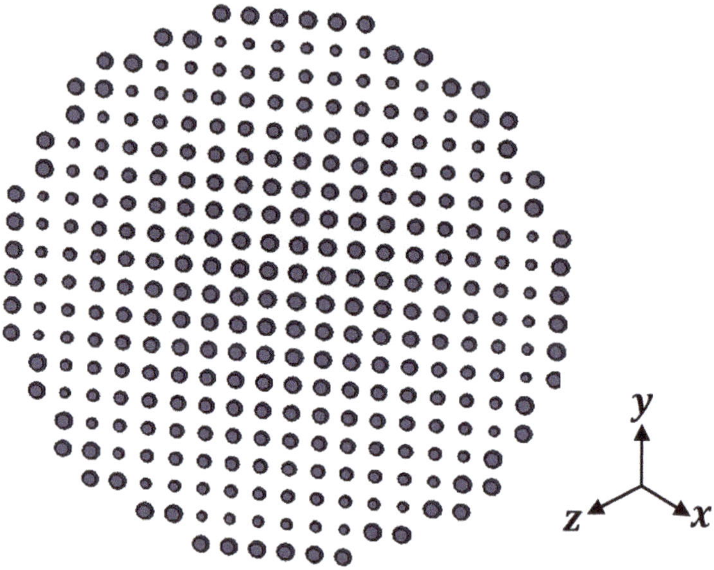

Figure 13. Simulation layout of the reflecting elements for the broadside pencil beam pattern.

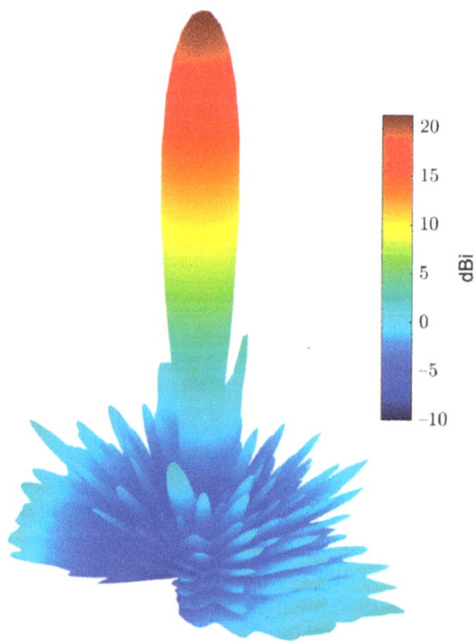

Figure 14. 3D radiation pattern of the simulated broadside pencil beam reflectarray antenna.

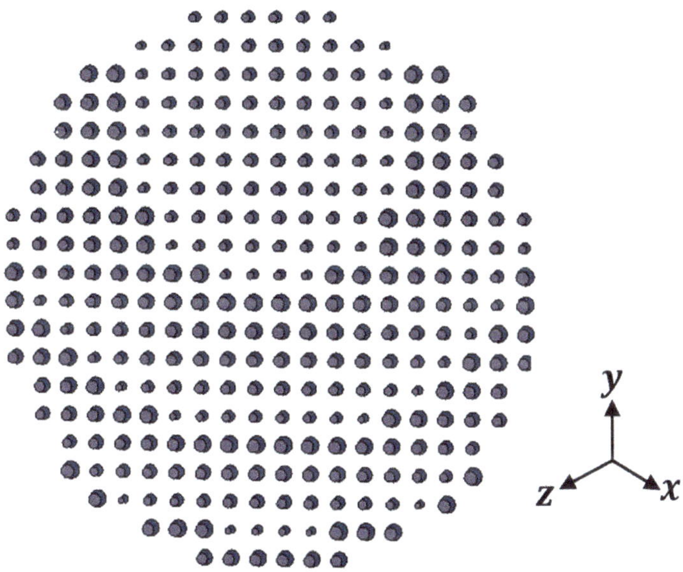

Figure 15. Simulation layout of the reflecting elements for the tilted beam pattern.

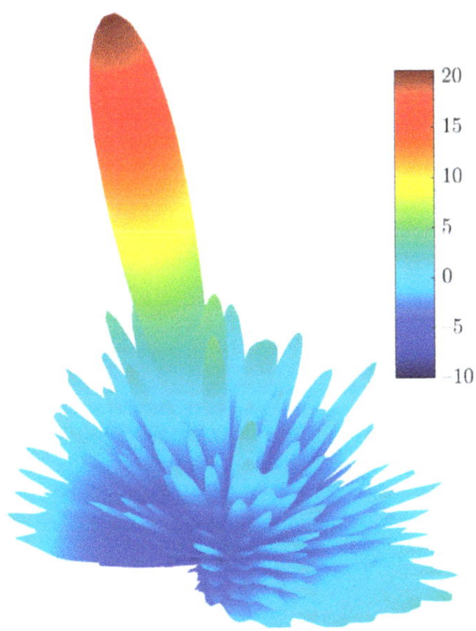

Figure 16. 3D radiation pattern of the simulated tilted beam reflectarray antenna.

Moreover, a Yagi–Uda antenna was fabricated, as shown in Figure 18a, to complete the structure of the reflectarray. The frequency response of the reflection coefficient magnitude of the fabricated Yagi–Uda antenna was measured using a vector network analyzer (VNA; Rhode and Schwartz model ZVA67), as shown in Figure 18b. The excellent matching of the fabricated antenna at the operating frequency of 8.5 GHz can be noted in Figure 18c.

(a) (b) (c) (d)

Figure 17. Reflecting elements for the flat-top pattern: (**a**) simulated layout, (**b**) fabricated upper layer, (**c**) fabricated bottom layer, and (**d**) fabricated ground plane.

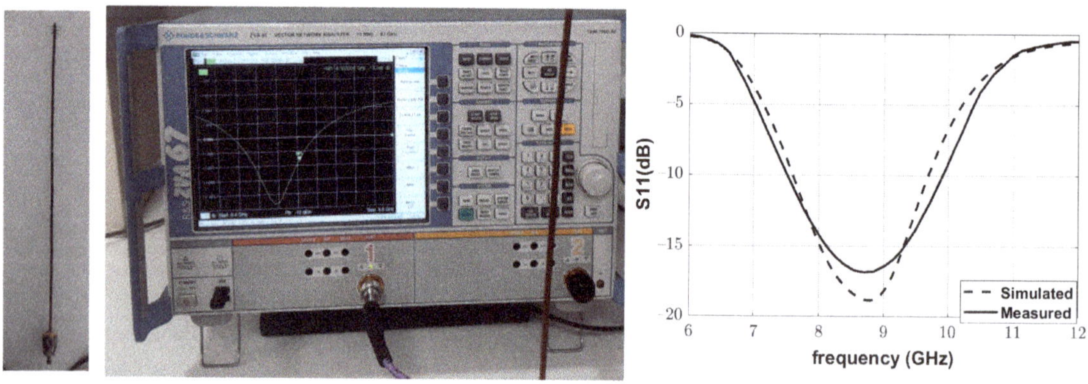

(a) (b) (c)

Figure 18. Measurement of the reflection coefficient of the fabricated Yagi–Uda antenna: (**a**) the fabricated Yagi–Uda antenna, (**b**) Yagi–Uda antenna connected to the Rhode and Schwartz model ZVA67 VNA, and (**c**) simulated and measured magnitude of the reflection coefficient.

Figure 19 shows the complete reflectarray antenna with its feeding antenna. The radiation patterns of the fabricated antenna were measured inside as anechoic chamber, as shown in Figure 20. Figure 21 shows the measured normalized radiation pattern of this reflectarray antenna compared to the radiation pattern obtained by GA for the required phase distribution. It can be noted that the obtained radiation pattern satisfies the required mask to a good extent. The slight differences in the obtained radiation pattern can be explained due to the alignment and the fabrication accuracy. In addition, Figure 22 shows the 3D radiation pattern of the flat-top beam with peak gain of around 8 dBi.

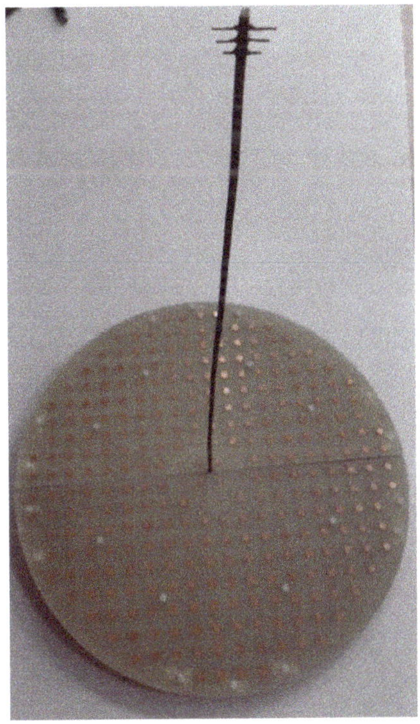

Figure 19. Fabricated prototype of the proposed reflectarray antenna.

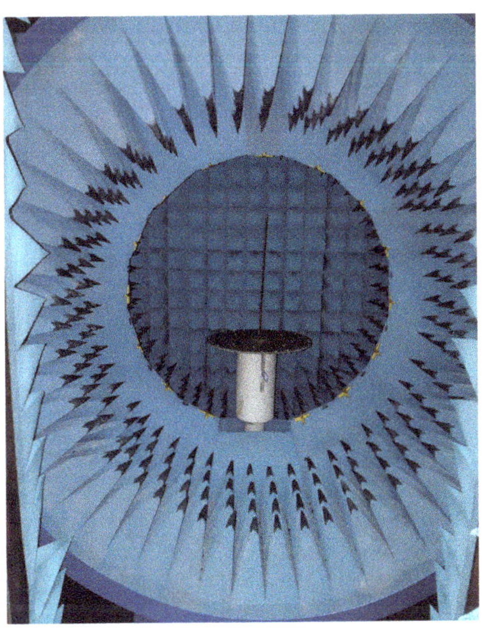

Figure 20. Fabricated antenna inside the anechoic chamber for the radiation pattern measurement.

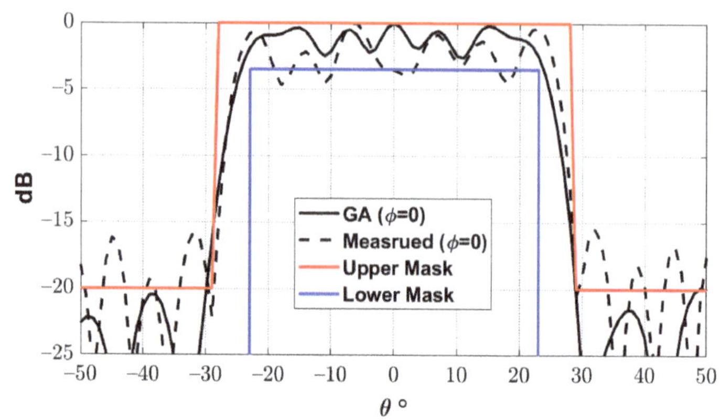

Figure 21. Measured and optimized radiation pattern of the proposed reflectarray antenna.

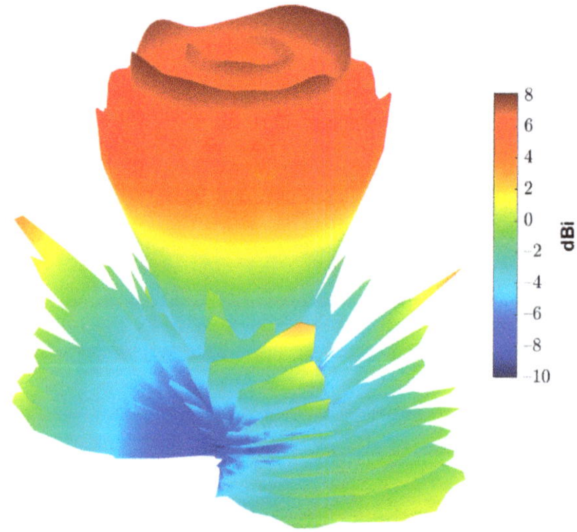

Figure 22. Three-dimensional radiation pattern of the proposed reflectarray antenna.

8. Conclusions

This paper presents the analysis and design of a flat-top reflectarray antenna for an MEO satellite system for remote sensing at an X-band frequency of 8.5 GHz. The feeding antenna is a Yagi–Uda antenna. The amplitude and phase of the field distribution due to the feeding antenna at the aperture of the reflectarray antenna are obtained numerically without introducing any approximation. This field distribution is used to obtain the required reflection phase distribution to obtain the required flat-top radiation pattern. This phase distribution is obtained using genetic algorithm optimization. The initial phase distribution of the optimization process is taken to be the phase distribution of a broadside pencil beam, which is obtained analytically. The problem of the tilted pencil beam is also investigated using analytical calculations. These reflection phase distributions are implemented using periodic reflecting elements. The reflecting element is composed of stacked circular patches on a grounded double-layered dielectric substrate. The reflection phases of these reflecting elements are adjusted by controlling the diameter of the stacked circular patches. These reflecting elements are arranged according to the required phase distribution for the cases of broadside pencil beam, tilted pencil beam, and flat-top beam. The complete reflectarray systems for the three cases are investigated. Details of simulation steps are discussed. The flat-top beam reflectarray was fabricated and measured for verification. Good agreements between the obtained results and the simulated results were obtained. The results of the reflection coefficient of the designed and fabricated feeding antenna showed a good agreement. Moreover, the obtained radiation pattern of the complete reflectarray antenna was found to satisfy the required radiation mask to a good extent.

Author Contributions: S.A.M.S., E.M.E. and A.M.A. have contributed to the conceptualization, methodology, software, validation, formal analysis, investigation, resources, data curation, writing, review, editing and visualizations. All authors have read and agreed to the published version of the manuscript.

Funding: This research received no external funding.

Institutional Review Board Statement: Not applicable.

Informed Consent Statement: Not applicable.

Data Availability Statement: Data sharing not applicable.

Conflicts of Interest: The authors declare no conflict of interest.

References

1. Rahmat-Samii, Y.; Densmore, A.C. Technology trends and challenges of antennas for satellite communication systems. *IEEE Trans. Antennas Propag.* **2014**, *63*, 1191–1204. [CrossRef]
2. Imaz-Lueje, B.; Prado, D.R.; Arrebola, M.; Pino, M.R. Reflectarray antennas: A smart solution for new generation satellite mega-constellations in space communications. *Sci. Rep.* **2020**, *10*, 21554. [CrossRef] [PubMed]
3. Prado, D.R.; Arrebola, M.; Pino, M.R. Reflectarray pattern optimization for advanced wireless communications. In *Advances in Array Optimization*; IntechOpen: London, UK, 2019.
4. Martinez-De-Rioja, D.; Martinez-De-Rioja, E.; Rodriguez-Vaqueiro, Y.; Encinar, J.A.; Pino, A. Multibeam reflectarrays in Ka-band for efficient antenna farms onboard broadband communication satellites. *Sensors* **2020**, *21*, 207. [CrossRef] [PubMed]
5. Dahri, M.H.; Jamaluddin, M.H.; Seman, F.C.; Abbasi, M.I.; Ashyap, A.Y.I.; Kamarudin, M.R.; Hayat, O. A Novel Asymmetric Patch Reflectarray Antenna with Ground Ring Slots for 5G Communication Systems. *Electronics* **2020**, *9*, 1450. [CrossRef]
6. Su, W.; Luo, W.; Nie, Z.; Liu, W.-W.; Cao, Z.-H.; Wang, Z. A Wideband Folded Reflectarray Antenna Based on Single-Layered Polarization Rotating Metasurface. *IEEE Access* **2020**, *8*, 158579–158584. [CrossRef]
7. Zhao, X.; Wei, F.; Li, B.; Shi, X. Design of circularly polarized dielectric resonator reflectarray antenna. In Proceedings of the 2018 Asia-Pacific Microwave Conference (APMC), Kyoto, Japan, 6–9 November 2018.
8. Sun, Y.-X.; Leung, K.W. Millimeter-wave substrate-based dielectric reflectarray. *IEEE Antennas Wirel. Propag. Lett.* **2018**, *17*, 2329–2333. [CrossRef]
9. Zhang, S. Three-dimensional printed millimetre wave dielectric resonator reflectarray. *IET Microw. Antennas Propag.* **2017**, *11*, 2005–2009. [CrossRef]
10. Han, C.; Zhang, Y.; Yang, Q. A novel single-layer unit structure for broadband reflectarray antenna. *IEEE Antennas Wirel. Propag. Lett.* **2016**, *16*, 681–684. [CrossRef]
11. Li, W.; Gao, S.; Zhang, L.; Luo, Q.; Cai, Y. An ultra-wide-band tightly coupled dipole reflectarray antenna. *IEEE Trans. Antennas Propag.* **2017**, *66*, 533–540. [CrossRef]
12. Bodur, H.; Çimen, S. Reflectarray antenna design with double cutted ring element for X-band applications. *Microw. Opt. Technol. Lett.* **2020**, *62*, 3248–3254. [CrossRef]
13. Huang, J.; Encinar, J.A. *Reflectarray Antennas*; John Wiley & Sons: Hoboken, NJ, USA, 2008.
14. Henderson, K.Q.; Ghalichechian, N. Triangular and Rectangular Lattices for Cosecant-squared Shaped Beam Reflectarrays. *IEEE Antennas Wirel. Propag. Lett.* **2021**, *20*, 2058–2062. [CrossRef]
15. Liu, C.; Li, C.; Yang, B. Improved alternating projection algorithm for pattern synthesis with dual polarised conformal arrays. *IET Microw. Antennas Propag.* **2020**, *14*, 891–896. [CrossRef]
16. Quijano, J.L.A.; Vecchi, G. Alternating adaptive projections in antenna synthesis. *IEEE Trans. Antennas Propag.* **2009**, *58*, 727–737. [CrossRef]
17. Rahmat-Samii, Y.; Michielssen, E. Electromagnetic optimization by genetic algorithms. *Microw. J.* **1999**, *42*, 232.
18. Ares-Pena, F.J.; González, J.A.R.; Villanueva-Lopez, E.; Rengarajan, S.R. Genetic algorithms in the design and optimization of antenna array patterns. *IEEE Trans. Antennas Propag.* **1999**, *47*, 506–510. [CrossRef]
19. Haupt, R.L. Genetic algorithm applications for phased arrays. *Appl. Comput. Electromagn. Soc. J.* **2006**, *21*, 325.
20. Sallam, T.; Attiya, A.M. Low sidelobe cosecant-squared pattern synthesis for large planar array using genetic algorithm. *Prog. Electromagn. Res.* **2020**, *93*, 23–34. [CrossRef]

of breast tumors consisting of a moveable array of improved negative-index ultrawideband antenna sensor is presented in [9]. An active slot antenna integrating a low-noise amplifier for tissue sensing arrays was proposed in [10]. A beam scanning technique was developed with a time delay for UWB arrays in [11]. The design of a grid array antenna for automotive radar sensors constructed using astroid unit cells, characterized by high gain and bandwidth, was developed in [12]. In [13], the bandwidth enhancement and frequency scanning for a UWB array antenna utilizing the novel technique of band-pass filter integration for wireless vital signs monitoring and vehicle navigation sensors was presented. In [14] a compact eight-element antenna array was presented for UWB pulsed radar in a highly reflective metallic environment. This Vivaldi antenna and array were characterized in terms of the transient energy patterns and the signal fidelity given in terms of the off-angle signal correlation.

Ultrawideband antenna arrays for MIMO UWB systems are also very popular in the literature. Their concept often differs from the concept of classic antenna arrays, but they are in fact systems containing many radiators and can be considered as antenna arrays. A compact printed UWB slot antenna consisting of two modified coplanar waveguides for MIMO diversity applications was described in [15]. Another example of antenna with two coplanar stripline-fed staircase-shaped radiating elements, with high isolation, is proposed for portable UWB MIMO systems in [16]. In [17] an ultra-compact frequency reconfigurable UWB MIMO antenna with four radiators that are capable of rejecting WLAN signals on demand by activating the PIN diodes is presented. A generic design method of spiral MIMO antenna arrays for short-range ultrawideband imaging application and its focusing property were discussed in [18]. In [19], the authors developed a compact uniplanar 4-port MIMO antenna array with rejecting band and polarization diversity. Several fabricated prototypes of 8 × 4 tightly coupled dipoles in linearly polarized phased arrays were presented in [20]. Tightly coupled dipoles in 11 × 11 UWB arrays with integrated baluns were also discussed in [21]. In [22] UWB MIMO array installed around a polystyrene block in the 3D-octagonal arrangement system was proposed for 3D non-planar applications. In [23], a UWB metasurface-based beam-switching antenna system was proposed. A four-element planar UWB antenna requiring no decoupling circuit for MIMO system was developed in [24]. In [25] an ultrawideband MIMO antenna system composed of two radiating elements with an improved isolation by using slotted stubs is presented. An eight-element UWB MIMO antenna with a deployed inductor capacitor stub on the ground plane for 3G/4G/5G networks was proposed in [26]. Another design of MIMO UWB antenna composed of two offset microstrip-fed elements with a band-notched function was analyzed in [27]. Finally, a wideband neutralization line was proposed in [28] to reduce the mutual coupling of a compact MIMO UWB antenna. Moreover, in the case of antenna array designing, an additional problem of optimizing the ratio between the power deposited over a given area and the whole transmitted power may arise. The synthesis of fields able to maximize the power radiated in an arbitrary portion of the visible space was discussed in [29]. Effective antenna arrays, including UWB arrays, can therefore be utilized for high beam efficiency transmissions. This overview presents that there is a great interest in the design of UWB antenna arrays and many design approaches can be found.

In this paper there are presented constructions of two variants of the four-element UWB planar antenna array designed for modern UWB radio systems. The antenna design and simulated and measured results are shown. Each array variant contains a stripline power divider, which is an excitation network providing a uniform distribution of excitations of all radiators. These feeding networks contain in their structure a few staircase-shaped sections. A new concept in this design is placing a metallic layer with the radiators between two layers of dielectric in the stripline structure. That results in getting a fully two-layer symmetrical structure. At the same time, in the part of the array containing the excitation network, its shielding is obtained. The antenna arrays have been designed to operate in the main band (6–8.5 GHz) allowed by the European Commission for use by UWB systems in Europe [2]. The obtained results were compared with the results achieved for a single prototype UWB antenna, with the same dimensions of the radiator and the length of excitation section, which was a reference antenna in the design of four-element UWB arrays.

The paper is organized as follows: the design of the proposed UWB prototype antenna and two variants of the four-element antenna array are presented in Section 2. The results of simulation and measurements are discussed in Section 3. The conclusion of the work is presented in Section 4.

2. Antenna Design

In order to design the 4-element antenna array considered in this work, a single UWB prototype antenna was firstly designed. Its design specifies the dimensions of the radiator and the length of the excitation line connecting the antenna input with the radiator. It was decided to choose an elliptical-shaped radiator due to the possibility of obtaining a good impedance matching in a wide frequency operation range.

Figure 1 shows the structure of a single UWB prototype antenna with the orientation of the adopted coordinate system. This single antenna contains an elliptical-shaped radiator with axis lengths of 14 mm for a shorter (horizontal) axis and 16 mm for a longer (vertical) one, fed by the stripline.

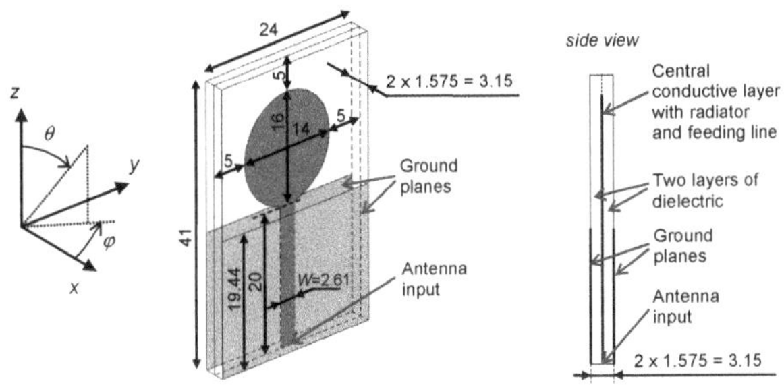

Figure 1. Geometry of proposed single prototype UWB antenna (dimensions in mm).

A two-layer, symmetrical structure of the antenna was obtained. Between the two layers of the dielectric there is a central conductive middle layer, while on the outer dielectric sides there are two identical areas of ground planes placed exactly one above the other in the plane of the antenna surface (*yz*-plane). The elliptical radiator is connected to the central conductor of the stripline structure and is located between two dielectric layers each 1.575 mm thick, $\varepsilon_r = 2.2$ (for fabrication of the considered antennas Rogers Duroid 5880 substrate with the same parameters was used). The total thickness of this two-layer structure is 3.15 mm. A 50 Ω feeding line is employed to feed the radiator. The calculated width of the excitation line is $W = 2.61$ mm, and its length equals 20 mm. The width of this excitation line was calculated in the stripline impedance calculator integrated with software used for a computer simulation. This line is shielded on both sides by two external ground planes with a slightly shorter length, equal to 19.44 mm. This shift of the upper border of the reference ground plane with respect to the lower end of the radiator has a positive effect on the impedance matching of the prototype antenna.

The boundaries of the whole single UWB antenna in its upper part (containing the radiator) are moved by 5 mm from the boundaries of the radiator. The overall dimensions of the single antenna are 24 mm × 41 mm. The antenna input is at its bottom edge. In the computer simulation a port dedicated for stripline structures was employed as the antenna input. The SMA connector mounted to fabricated antennas was not included in computer models. The presented dimensions of the radiator and the whole structure of the single antenna provide a nearly omnidirectional shaped radiation pattern and a good impedance matching.

Based on the structure of the single UWB prototype antenna, two variants of four-element UWB planar antenna array have been developed (Figure 2). The four-element arrays presented in this paper contain four radiators of the same elliptical shape and the same orientation as the radiator of the single prototype antenna. The radiators are positioned in a straight line along the y-axis at equal distances (the surfaces of all antennas are oriented in the yz-plane, as the single prototype antenna shown in Figure 1). Their positioning and orientation in relation to the top and bottom border of the entire array structure (in relation to the substrate boundaries) is analogous to the prototype UWB antenna. The boundaries of the entire antenna array structure are moved by 5 mm upwards and in the left and right directions from the first and the last radiator.

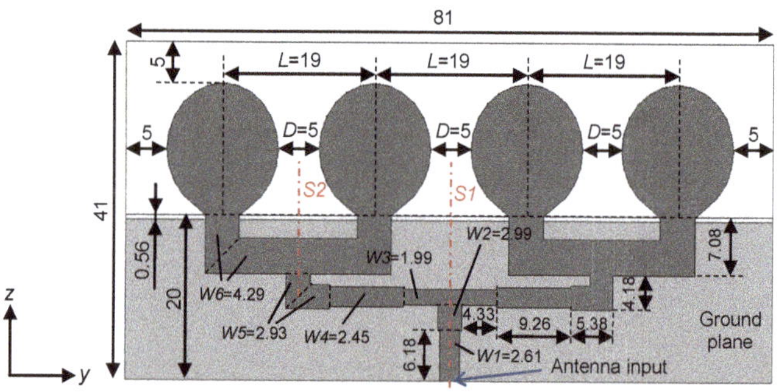

(a)

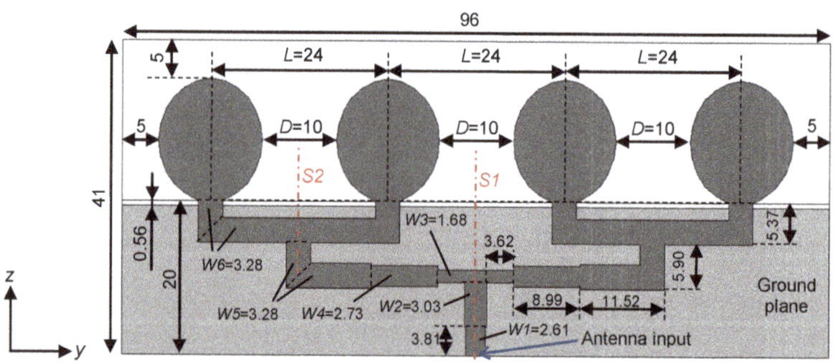

(b)

Figure 2. Geometry of proposed four-element antenna arrays in yz-plane (dimensions in mm): (**a**) Shorter variant of UWB antenna array; (**b**) Longer variant of UWB antenna array.

For two considered antenna array variants the centers of the radiators are positioned in the distance: L = 19 mm (shorter variant) and L = 24 mm (longer variant). The distances between the radiators' edges on the y-axis are thus D = 5 mm and D = 10 mm, respectively. The electrical length between the centers of neighboring radiating elements varies throughout the operation frequency range from 0.38λ at 6 GHz to 0.54λ at 8.5 GHz for the shorter variant and from 0.48λ at 6 GHz to 0.68λ at 8.5 GHz for the longer variant (λ is the free-space wavelength). The overall dimensions of both variants of the four-element UWB antenna arrays are 81 mm × 41 mm for the shorter variant and 96 mm × 41 mm for the longer variant. The single input of each array variant is placed in the middle of the bottom edge of each array variant structure. Ports dedicated to stripline structures were used again as the arrays' inputs in the computer analysis without modeling SMA connectors. The length of the excitation section measured along the z-axis, from the bottom of the ground plane to the bottom edges of the radiators, is 20 mm. The length of the ground planes in this direction is shortened by 0.54 mm to 19.44 mm as for the single prototype antenna.

The radiating elements are fed by symmetrical multi-section broadband power dividers which were developed as the stripline structures, as for the single antenna. Figure 2 shows the structures of the central conducting layer of both variants of the four-element arrays, including the radiators with the feeding network. In the beginning part of the feeding network structure, staircase-shaped sections can be distinguished. In the design of the entire array structure, this approach resulted in a better impedance matching of the array in the operating frequency band. The widths of all subsequent sections of feeding network are marked as *W1*–*W6*. All remaining dimensions fully describing the geometry of the designed networks have also been denoted (those dimensions that result from the symmetry of the array structure have not been marked).

The uniform power distribution of the radiators is adopted in the development of both variants of excitation networks. This power distribution allows the direction of the maximum radiation to be perpendicular to the surface of the antenna array, both in the positive and negative direction of *x*-axis. Two feeding structures for both array variants include dividing parts that provide a uniform power division between all corresponding sections and radiators. They also provide equal electrical lengths from antenna arrays inputs to all radiators.

In the area where the section of width *W2* passes into two lines of width *W3*, the input power is divided equally into two parts. It is similar in places where sections with *W5* widths change into sections with *W6* widths. The uniform power distribution of the radiators was obtained under the condition of full symmetry of the designed arrays structures. In the design of both variants of antenna arrays there is the symmetry of the feeding network and also the symmetry of the entire structure of the antenna array with respect to the *z*-axis passing through the center of the first section of width *W1* to which the input port is connected (a symmetry line *S1* in Figure 2). At the same time, the extension of the axis of symmetry of the section *W5* in the direction of the *z*-axis (a symmetry line *S2* in Figure 2) is located exactly halfway between the first and the second radiator of the array on both sides of the overall structure, respectively.

All dimensions of both variants of feeding networks were determined in the optimization process of their whole geometry. The impedance of the first sections connected to the array input sockets was set as 50 Ω (*W1* = 2.61 mm for the selected substrate as for the single prototype antenna), the other sections with widths from *W2* to *W6* were optimized. The adopted optimization criterion was to obtain $|S_{11}| < -10$ dB in the 6–8.5 GHz operation band. Computer simulations of the analyzed structures were carried out in the IE3D/Hyper Lynx 3D EM program, a 3D electromagnetic simulator using method of moments (MoM).

It can be noticed that the determined widths of the subsequent sections of the feeding network of both array variants are relatively wide. This is due to the values of parameters of the substrate that was selected for the antennas' design and their fabrication. The use of a thinner dielectric or a dielectric with a higher value of ε_r would result in narrower widths of the successive sections. In that case the entire structure of the feeding networks should be optimized to achieve the desired level of impedance matching.

The single antenna and two four-element antenna arrays were fabricated by the author. SMA-connectors were used as inputs. The center conductive layer with a mosaic of radiators and feeding excitation network was etched on one side of the first dielectric layer. Its other external side, considering the structure of the entire antenna array, contains the first ground reference plane. The second layer of dielectric contains only the second ground plane on its external side. After soldering the SMA-connector, both dielectric layers were tightly stuck together. The metallization thickness of all conducting layers is 35 μm, however in the computer project and simulation the thickness was assumed as 0 mm.

Figure 3a,b presents the central metallic layers of both array variants and photos of the ready-made antennas are shown in Figure 3c.

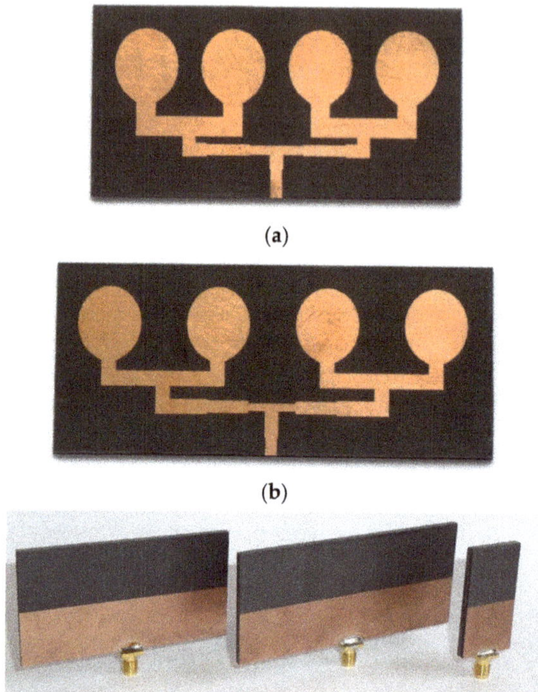

(**a**)

(**b**)

(**c**)

Figure 3. Photographs of proposed antennas: (**a**) Central metallic layer of shorter variant of four-element antenna array; (**b**) Central metallic layer of longer variant of four-element antenna array; (**c**) Fabricated antennas.

3. Simulated and Measured Results

The experimental verification of S-parameters and radiation patterns of the proposed UWB prototype antenna and both variants of the four-element antenna arrays were conducted. The simulated and measured results were compared.

The measurements of $|S_{11}|$ were made with the Agilent N5230A vector network analyzer. The results of the simulations and measurements of $|S_{11}|$ of the proposed antennas are shown in Figure 4. In the case of the single prototype UWB antenna a good impedance matching was achieved. The simulated $|S_{11}|$ is less than -21.5 dB and measured less than -16.6 dB. Considering the four-element antenna arrays, the impedance matching is slightly worse. The shorter variant of the antenna array shows the simulated $|S_{11}|$ less than -10.5 dB and measured less than -8.7 dB, although the simulated $|S_{11}|$ is larger than -10 dB only in part of the frequency band above 8.1 GHz. For the longer variant of the four-element antenna array the simulated $|S_{11}|$ is less than -9.7 dB and measured is also less than -9.7 dB. There is a good agreement between the simulated and measured results.

The radiation characteristics were measured in an anechoic chamber in the 6–8.5 GHz frequency band with 0.2 GHz frequency steps. The photography of the wider variant of the four-element UWB antenna array during the measurements in the anechoic chamber is presented in Figure 5. The radiation patterns were measured in the horizontal xy-plane (the measured antenna was mounted vertically in yz-plane, as a computer model presented in Figure 1). During each measurement procedure two orthogonal components of the gain were measured in E-plane and H-plane and then total gain was calculated.

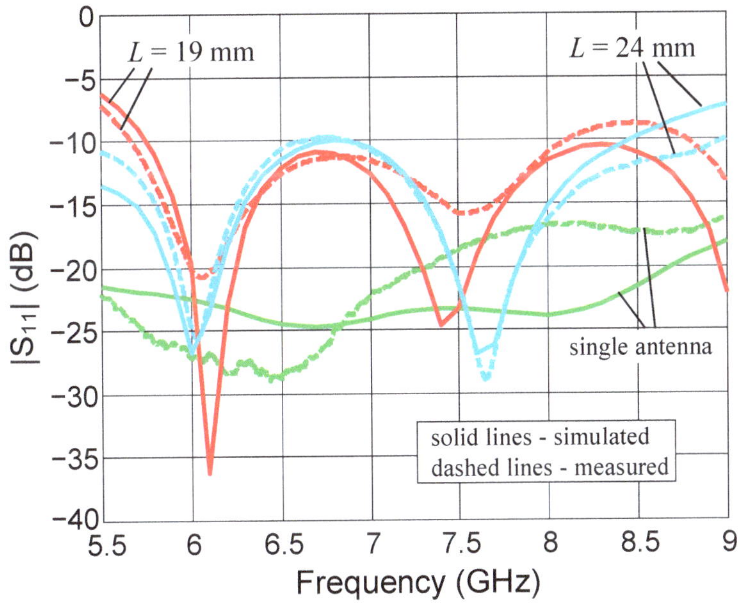

Figure 4. Simulated and measured $|S_{11}|$ of proposed antennas.

Figure 5. Longer variant of four-element antenna array in anechoic chamber during measurements.

Figure 6 shows the simulated and measured frequency characteristics of the gain determined in the direction perpendicular to the antenna surface (both in the positive and negative direction of the x-axis due to the array symmetry). The simulated gain (solid line in Figure 6) of the single UWB prototype antenna has a value in the range 1–2.1 dBi in the frequency operating band 6–8.5 GHz while the measured gain (dashed line in Figure 6) differs from the simulated one by a maximum of 1.3 dB. In the case of the four-element antenna arrays, the simulated gain for the shorter variant changes in the operating band in the range 6.7–8.9 dBi and for the wider variant in the range 8.5–9.3 dBi. The measured gain has slightly higher values. For corresponding UWB arrays the measured ranges of gain are: 6.4–10.6 dBi and 8.5–10.8 dBi. The differences between the simulated and measured results are caused mostly by errors during measurements of the radiation patterns. The difference in the simulated gain of proposed arrays in Figure 6 decreases from 1.8 dB at 6 GHz to 0.3 dB at 8.5 GHz and the measured from 2.1 dB to 0.4 dB, respectively.

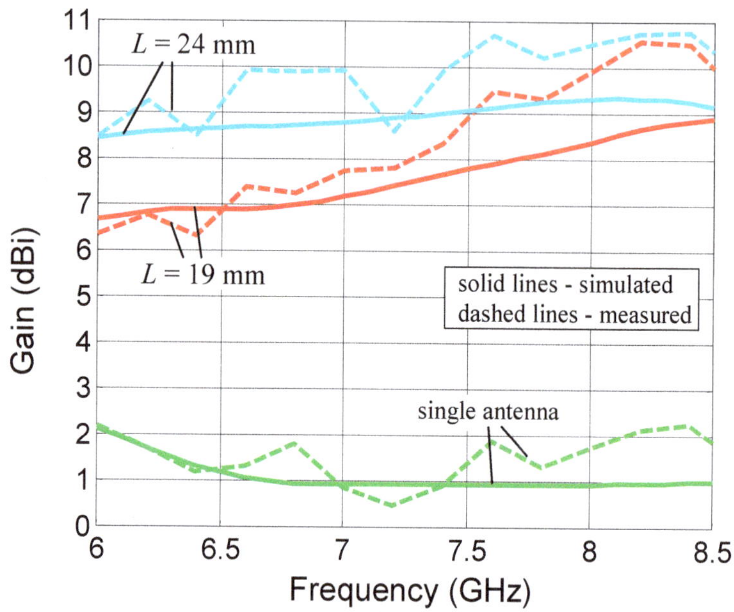

Figure 6. Simulated and measured gain of proposed antennas in direction perpendicular to antenna surface (along +/− *x*-axis).

Figure 7 shows the simulated and measured radiation patterns in the *xy*-plane at three frequencies from the operating band of the antenna arrays: 6, 7.2 and 8.4 GHz. A high agreement of the simulated and measured characteristics is observed. Due to the double symmetry of the designed antenna array structures, the gain characteristics are symmetrical in the antenna alignment *yz*-plane and also in the *xz*-plane. There is a noticeable rise in the value of each antenna array gain with increasing the frequency and with increasing the spacing *L* between radiators. It is also observable that the growth of the spacing between the radiators does not increase the gain very much. The increase of the gain is inhibited by the growth of the sidelobes, especially visible in the upper part of the frequency operation band. The same relationship is observed in the results presented in Figure 6. Considering the maximum gain for both variants of the array in Figure 7 it can be observed that the largest difference between the simulated and measured gain is 0.3 dB at 6 GHz, 0.4 dB at 7.2 GHz and 1.7 dB at 8.4 GHz. This difference rises with an increase in frequency. It may occur mostly from two factors. Firstly, in the computer simulation (MoM), the meshing was performed automatically and is the same for all analyzed frequencies. Thus, in cases of increases in frequency, the size of the cells in relation to the wavelength decreases, which may turn into a slight decrease in the precision of the calculation. Secondly, the measurement uncertainty is difficult to estimate precisely. It may be influenced, among others, by errors in the calibration of the measurement system or by the influence of a rigid coaxial cable used for mounting antennas during measurements (visible in Figure 5).

The simulated and measured radiation patterns of the designed UWB antenna arrays are characterized by a fairly high level of sidelobes, especially noticeable in the upper part of the frequency operation band for the wider array variant. The level of sidelobes can be minimized by the design of the feeding network that provides a heterogeneous distribution of excitations at the inputs of the radiators, which will be the subject of the future research.

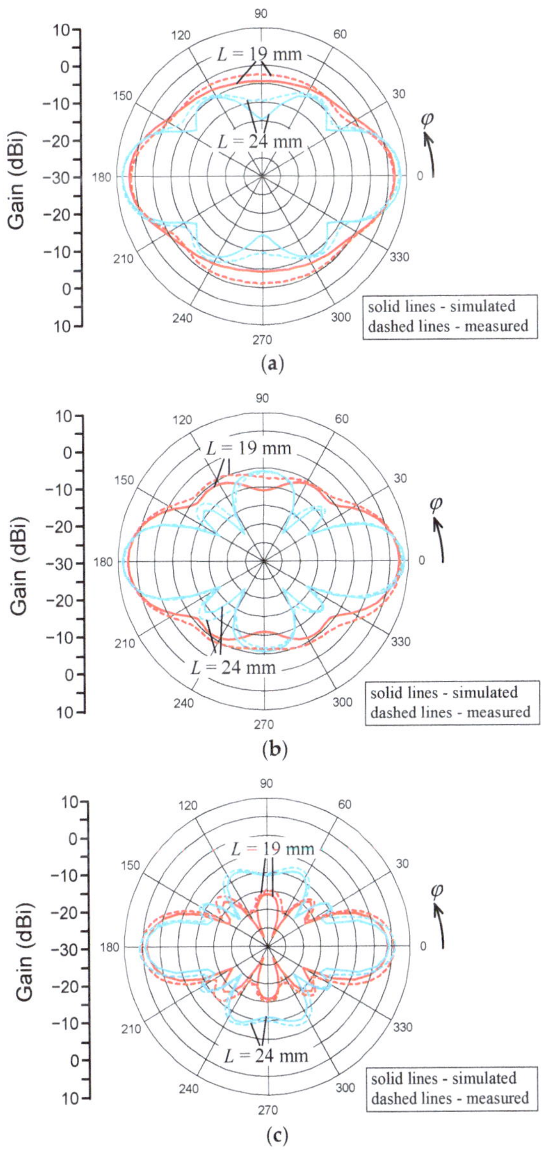

Figure 7. Simulated and measured radiation pattern of the proposed arrays in *xy*-plane at different frequencies: (**a**) 6 GHz; (**b**) 7.2 GHz; (**c**) 8.4 GHz.

4. Conclusions

In this paper the four-element UWB antenna array in two variants of the spacing between the radiating elements is proposed. Both designed variants are excited by multi-section stripline feeding network with the uniform power distribution. The structure of the feeding network contains staircase-shaped sections that improve impedance matching. Two external ground planes fully shield the structure of the excitation network. The characteristics of $|S_{11}|$ and radiation patterns are presented, both simulated and measured. For comparison, the $|S_{11}|$ characteristics are also shown for the UWB single prototype antenna, which has the same sized radiator as the antenna arrays.

The simulated and measured results for both variants of the UWB antenna arrays present a similar value of $|S_{11}|$; less than -10 dB in almost the whole operation band. The single UWB antenna shows better impedance matching than the proposed antenna arrays, but the measured gain is less by 4.2–9.1 dB depending on the frequency and the array spacing variant. The wider variant of the antenna array shows a slightly higher gain than the shorter one. The simulation and measurement results for all considered structures show a good agreement.

Funding: This research was realized in Bialystok University of Technology, Poland, and supported by the Polish Ministry of Education and Science under Rector's Project WZ/WE-IA/1/2020.

Conflicts of Interest: The author declares no conflict of interest.

References

1. Federal Communications Commission. *The Federal Communications Commission Revision of Part 15 of the Commission's Rules Regarding Ultrawideband Transmission Systems from 3.1 to 10.6 GHz*; Federal Communications Commission: Washington, DC, USA, 2002.
2. The European Commission. *Commission Implementing Decision (EU) 2019/785 on the Harmonisation of Radio Spectrum for Equipment Using Ultra-Wideband Technology in the Union and Repealing Decision 2007/131/EC (Notified under Document C(2019) 3461)*; European Union: Brussels, Belgique, 2019.
3. Agrawall, N.P.; Kumar, G.; Ray, K.P. Wide-Band Planar Monopole Antennas. *IEEE Trans. Antennas Propag.* **1998**, *46*, 294–295. [CrossRef]
4. Liang, J.; Chiau, C.C.; Chen, X.; Parini, C.G. Study of a Printed Circular Disc Monopole Antenna for UWB Systems. *IEEE Trans. Antennas Propag.* **2005**, *53*, 3500–3504. [CrossRef]
5. Huang, C.-Y.; Hsia, W.-C. Planar elliptical antenna for ultra-wideband communications. *Electron. Lett.* **2005**, *41*, 296–297. [CrossRef]
6. Klemm, M.; Troester, G. Textile UWB Antennas for Wireless Body Area Networks. *IEEE Trans. Antennas Propag.* **2006**, *54*, 3192–3197. [CrossRef]
7. Sugitani, T.; Kubota, S.; Toya, A.; Xiao, X.; Kikkawa, T. A Compact 4 × 4 Planar UWB Antenna Array for 3-D Breast Cancer Detection. *IEEE Antennas Wireless Propag. Lett.* **2013**, *12*, 733–736. [CrossRef]
8. Hossain, A.; Islam, M.T.; Islam, M.T.; Chowdhury, M.E.H.; Rmili, H.; Samsuzzaman, M. A Planar Ultrawideband Patch Antenna Array for Microwave Breast Tumor Detection. *Materials* **2020**, *13*, 4918. [CrossRef]
9. Islam, M.T.; Samsuzzaman, M.; Islam, M.T.; Kibria, S. Experimental Breast Phantom Imaging with Metamaterial-Inspired Nine-Antenna Sensor Array. *Sensors* **2018**, *18*, 4427. [CrossRef]
10. Foroutan, F.; Nikolova, N.K. UWB Active Antenna for Microwave Breast Imaging Sensing Arrays. *IEEE Antennas Wirel. Propag. Lett.* **2019**, *18*, 1951–1955. [CrossRef]
11. Liao, C.H.; Hsu, P.; Chang, D.C. Energy Patterns of UWB Antenna Arrays with Scan Capability. *IEEE Trans. Antennas Propag.* **2011**, *59*, 1140–1147. [CrossRef]
12. Gulam Nabi Alsath, M.; Lawrance, L.; Kanagasabai, M. Bandwidth-Enhanced Grid Array Antenna for UWB Automotive Radar Sensors. *IEEE Trans. Antennas Propag.* **2015**, *63*, 5215–5219. [CrossRef]
13. Rahman, M.; NaghshvarianJahromi, M.; Mirjavadi, S.S.; Hamouda, A.M. Bandwidth Enhancement and Frequency Scanning Array Antenna Using Novel UWB Filter Integration Technique for OFDM UWB Radar Applications in Wireless Vital Signs Monitoring. *Sensors* **2018**, *18*, 3155. [CrossRef]
14. Maunder, A.; Mousavi, P. Application of UWB Arrays for Material Identification of Multilayer Media in Metallic Tanks. *IEEE Trans. Antennas Propag.* **2015**, *63*, 4901–4909. [CrossRef]
15. Gao, P.; He, S.; Wei, X.; Xu, Z.; Wang, N.; Zheng, Y. Compact Printed UWB Diversity Slot Antenna with 5.5-GHz Band-Notched Characteristics. *IEEE Antennas Wireless Propag. Lett.* **2014**, *13*, 376–379. [CrossRef]
16. Roshna, T.K.; Deepak, U.; Sajitha, V.R.; Vasudevan, K.; Mohanan, P. A Compact UWB MIMO Antenna with Reflector to Enhance Isolation. *IEEE Trans. Antennas Propag.* **2015**, *63*, 1873–1877. [CrossRef]
17. Khan, M.S.; Iftikhar, A.; Shubair, R.M.; Capobianco, A.-D.; Asif, S.M.; Braaten, B.D.; Anagnostou, D.E. Ultra-Compact Reconfigurable Band Reject UWB MIMO Antenna with Four Radiators. *Electronics* **2020**, *9*, 584. [CrossRef]
18. Cheng, Q.; Liu, Y.; Zhang, H.; Hao, Y. A Generic Spiral MIMO Array Design Method for Short-Range UWB Imaging. *IEEE Antennas Wireless Propag. Lett.* **2020**, *19*, 851–855. [CrossRef]
19. Gómez-Villanueva, R.; Jardón-Aguilar, H. Compact UWB Uniplanar Four-Port MIMO Antenna Array with Rejecting Band. *IEEE Antennas Wireless Propag. Lett.* **2019**, *18*, 2543–2547. [CrossRef]
20. Mansouri Moghaddam, S.; Yang, J.; Uz Zaman, A. Fully-Planar Ultrawideband Tightly-Coupled Array (FPU-TCA) With Integrated Feed for Wide-Scanning Millimeter-Wave Applications. *IEEE Trans. Antennas Propag.* **2020**, *68*, 6591–6601. [CrossRef]
21. Zhong, J.; Johnson, A.; Alwan, E.A.; Volakis, J.L. Dual-Linear Polarized Phased Array With 9:1 Bandwidth and 60° Scanning Off Broadside. *IEEE Trans. Antennas Propag.* **2019**, *67*, 1996–2001. [CrossRef]

22. Shabbir, T.; Saleem, R.; Al-Bawri, S.S.; Shafique, M.F.; Islam, M.T. Eight-Port Metamaterial Loaded UWB-MIMO Antenna System for 3D System-in-Package Applications. *IEEE Access* **2020**, *8*, 106982–106992. [CrossRef]
23. Swain, R.; Naik, D.K.; Panda, A.K. Low-loss ultra-wideband beam switching metasurface antenna in X-band. *IET Microw. Antennas Propag.* **2020**, *14*, 1216–1221. [CrossRef]
24. Sipal, D.; Abegaonkar, M.P.; Koul, S.K. Easily Extendable Compact Planar UWB MIMO Antenna Array. *IEEE Antennas Wireless Propag. Lett.* **2017**, *16*, 2328–2331. [CrossRef]
25. Altaf, A.; Iqbal, A.; Smida, A.; Smida, J.; Althuwayb, A.A.; Kiani, S.H.; Alibakhshikenari, M.; Falcone, F.; Limiti, E. Isolation Improvement in UWB-MIMO Antenna System Using Slotted Stub. *Electronics* **2020**, *9*, 1582. [CrossRef]
26. Khan, M.S.; Iftikhar, A.; Shubair, R.M.; Capobianco, A.-D.; Braaten, B.D.; Anagnostou, D.E. Eight-Element Compact UWB-MIMO/Diversity Antenna with WLAN Band Rejection for 3G/4G/5G Communications. *IEEE Open J. Antennas Propag.* **2020**, *1*, 196–206. [CrossRef]
27. Kang, L.; Li, H.; Wang, X.; Shi, X. Compact Offset Microstrip-Fed MIMO Antenna for Band-Notched UWB Applications. *IEEE Antennas Wireless Propag. Lett.* **2015**, *14*, 1754–1757. [CrossRef]
28. Zhang, S.; Pedersen, G.F. Mutual Coupling Reduction for UWB MIMO Antennas with a Wideband Neutralization Line. *IEEE Antennas Wireless Propag. Lett.* **2015**, *15*, 166–169. [CrossRef]
29. Morabito, A.F.; Lagana, A.R.; Isernia, T. Optimizing Power Transmission in Given Target Areas in the Presence of Protection Requirements. *IEEE Antennas Wirel. Propag. Lett.* **2014**, *14*, 44–47. [CrossRef]

sensors

Technical Note

Over-the-Air Testing of a Massive MIMO Antenna with a Full-Rank Channel Matrix

Kazuhiro Honda

Graduate School of Engineering, Toyama University, 3190 Gofuku, Toyama 930-8555, Japan; hondak@eng.u-toyama.ac.jp; Tel.: +81-76-445-6759

Abstract: This paper presents an over-the-air testing method in which a full-rank channel matrix is created for a massive multiple-input multiple-output (MIMO) antenna system utilizing a fading emulator with a small number of scatterers. In the proposed method, in order to mimic a fading emulator with a large number of scatterers, the scatterers are virtually positioned by rotating the massive MIMO antenna. The performance of a 64-element quasi-half-wavelength dipole circular array antenna was evaluated using a two-dimensional fading emulator. The experimental results reveal that a large number of available eigenvalues are obtained from the channel response matrix, confirming that the proposed method, which utilizes a full-rank channel matrix, can be used to assess a massive MIMO antenna system.

Keywords: massive multiple-input multiple-output (MIMO) antenna; over-the-air (OTA) testing; channel matrix; full-rank; fading emulator

Citation: Honda, K. Over-the-Air Testing of a Massive MIMO Antenna with a Full-Rank Channel Matrix. *Sensors* **2022**, *22*, 1240. https://doi.org/10.3390/s22031240

Academic Editors: Naser Ojaroudi Parchin, Chan Hwang See and Raed A. Abd-Alhameed

Received: 23 December 2021
Accepted: 5 February 2022
Published: 6 February 2022

1. Introduction

Global commercial services for ultra-high speed fifth-generation (5G) mobile communication using multiple-input multiple-output (MIMO) systems are currently available [1,2]. One of the possible solutions to significantly enhance the channel capacity of MIMO systems is to utilize a large number of antenna elements for both the base station (BS) and the mobile station (MS). Such a system is called a massive MIMO system [3].

Most of the activity so far undertaken in developing massive MIMO systems has been directed toward providing large-scale MIMO antennas at the BS, with antennas comprising more than 100 antenna elements [4,5], and there are few reports of doing something similar at the MS [6]. The author is currently developing a method to achieve a large-scale MIMO antenna system that maintains an invariable channel capacity over the full-azimuth at the MS, such as, for example, a connected ground-based or flying car [7].

The usual technique for evaluating the performance of MIMO antennas with multipath fading channels is to do Monte Carlo simulation where several scatterers are placed on a circle [8,9]. This is known as the Clarke model or ring model. Using this model, the number of scatterers required to simulate the full-rank property of the channel matrix for a massive MIMO system is greater than the number of subchannels.

To analyze the capability of the developed antenna [7], the author proposed a Monte Carlo simulation with randomly arranged scatterers [10]. A small number of differently positioned scatterers were set for each BS antenna using random numbers, confirming that the channel matrix created can achieve full-rank status similar to conventional Monte Carlo simulation. However, a large number of scatterers are necessary to emulate a lot of channels.

A legitimate manner of assessing the performance of a fabricated massive MIMO antenna is to test it in the field [11]. However, with field testing, the measurements are neither repeatable nor controllable, and, moreover, the measurement process is considerably time-consuming and labor-intensive. Hence, over-the-air (OTA) testing, which evaluates the ability of a MIMO antenna by reproducing a realistic multipath radio propagation environment in the laboratory, is essential.

A proper OTA testing method for massive MIMO antennas is required to accelerate the development and optimization of the antenna [12]. As many massive MIMO BS antennas have been developed, radiating evaluation methods for massive MIMO BS antennas using fading emulators have also been investigated [13,14]. Many BS antennas comprise a two-dimensional planar array antenna with a number of patch antennas. Consequently, a large number of scatterers are placed in a limited direction with respect to the OTA apparatus and scatterers are selected from among them for the assessment.

On the other hand, an OTA testing method for massive MIMO MS antennas is not currently available. A fading emulator with a small number of scatterers placed on a circle has been adopted for proper OTA testing of a handset MS comprising a few MIMO antenna elements [15,16]. In the standard method utilizing a fading emulator, the arrangement of the scatterers depends on the number of subchannels, indicating that a large number of scatterers are necessary to assess the performance of a massive MIMO system with a full-rank channel matrix. Therefore, because of the size and cost of the equipment, the standard method is not effective for massive MIMO MS antennas.

The author proposed an OTA evaluation method for a massive MIMO antenna that creates a full-rank channel matrix [17]. The results of the Monte Carlo simulation, which included simulation of the proposed OTA testing method, revealed that even though the channel model comprised a limited number of scatterers, a full-rank channel matrix can be created. However, an experimental verification of this has not been done.

This paper presents an experimental verification of the proposed method utilizing a two-dimensional fading emulator with a small number of scatterers. In the OTA measurements, the massive MIMO MS antenna is located at the center of the fading emulator and is rotated depending on the measured channel response at the BS antenna, in such a way that a small number of scatterers are equivalent to a much larger number of scatterers. The experimental results showed that the number of available eigenvalues is greater than that obtained with the previous method.

2. Measurement Method of the Full-Rank Channel

In a previous Monte Carlo simulation with a uniform arrangement of scatterers, which represented the secondary wave source, $N \times M$ channel responses were calculated to form Equation (1).

$$\mathbf{H}_S = \begin{bmatrix} h_{11} & h_{12} & \cdots & h_{1K_m} & \cdots & h_{1M} \\ h_{21} & h_{22} & \cdots & h_{2K_m} & \cdots & h_{2M} \\ \vdots & \vdots & \ddots & \vdots & \ddots & \vdots \\ h_{N1} & h_{N2} & \cdots & h_{NK_m} & \cdots & h_{NM} \end{bmatrix} \tag{1}$$

where K_m indicates the number of actual scatterers. M and N denote the number of elements in the BS and MS, respectively. Furthermore, the author assumed that the number of elements at the BS is equal to that at the MS, that is, $M = N$.

In Equation (1), because all the signals from the M elements at BS overlap with each other at the same location, the number of columns that satisfy linear independence is equal to K_m in the channel model. Consequently, assuming that K_m is less than M, Equation (1) is transformed into Equation (2) using diagonalization.

$$\mathbf{H}_S = \begin{bmatrix} h_{11} & h_{12} & \cdots & h_{1K_m} & 0 & \cdots & 0 \\ h_{21} & h_{22} & \cdots & h_{2K_m} & 0 & \cdots & 0 \\ \vdots & \vdots & \ddots & \vdots & \vdots & \ddots & \vdots \\ h_{N1} & h_{N2} & \cdots & h_{NK_m} & 0 & \cdots & 0 \end{bmatrix} \tag{2}$$

Therefore, the eigenvalue vector obtained using singular value decomposition (SVD) is denoted by Equation (3).

$$\mathbf{\Lambda} = \begin{bmatrix} \lambda_1 & \lambda_2 & \cdots & \lambda_{K_m} & 0 & \cdots & 0 \end{bmatrix} \tag{3}$$

The rank of Equation (2) equals K_m which is less than M, indicating a rank-deficient status. Consequently, the number of eigenvalues, that is, the number of channels, observed is only K_m. Hence, a large number of scatterers, greater than M, are necessary to obtain full-rank status with rank ($\mathbf{H}_S$) = M.

The method of randomly arranged scatterers, in which a limited number of scatterers are arranged to simulate each BS element, was proposed [10]. The required number of scatterers to generate the Rayleigh fading environment for one BS element is small. However, the total number of scatterers is the product of M and K_m. Consequently, for OTA testing of a massive MIMO system incorporating the method of randomly arranged scatterers, a small number of scatterers must be selected from among the large number of scatterers on the circle of the fading emulator. Otherwise, the actual scatterers need to be relocated for each BS. Hence, the OTA testing implemented using the method of randomly arranged scatterers is extremely labor-intensive process compared with the standard OTA testing method.

The author proposed an OTA testing method in which the scatterers are virtually formed emulating a large number of scatterers [17]. Figure 1 shows the configuration of the proposed fading emulator to enable a full-rank channel matrix for a massive MIMO antenna. In Figure 1, K_m, that is, the number of scatterers in the 1st set, is 14 which is sufficient to produce a Rayleigh fading environment. The 2nd to S-th sets of scatterers are virtually placed, where S indicates the number of scatterer sets.

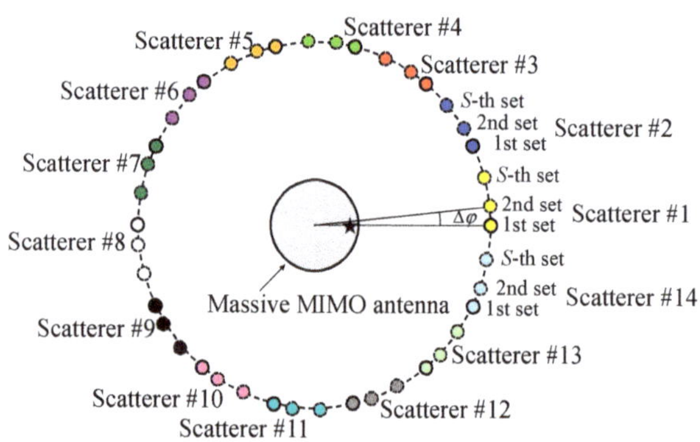

Figure 1. Configuration of the scatterers for evaluating the massive MIMO system.

The angular intervals between the i-th and $(i + 1)$-th sets of scatterers, $\Delta\varphi$, as shown in Figure 1, are the same. Therefore, each set of scatterers is formed by rotating the 1st set of scatterers. Consequently, the placement of each of scatterer differs, and a large number of scatterers can be emulated, with the expectation that measurement with a full-rank property of the channel matrix can be achieved.

The most important parameter in the measurement is the number of scatterer sets which depends on the number of actual scatterers. Independent paths via each BS antenna are generated by orthogonal initial phase sets at the actual scatterers. Hence, the maximum number of BS elements emulated in accordance with each set of scatterers is the same as the number of actual scatterers. In order to achieve a full-rank matrix, S must be adjusted to be greater than M divided by K_m. Another important parameter is $\Delta\varphi$. If $\Delta\varphi$ is small, the possibility of generating different paths from adjacent sets of scatterers is small. In this paper, $\Delta\varphi$ is set to equal angular intervals, and it is calculated as follows:

$$\Delta\varphi = \frac{360}{K_m}\frac{1}{S} \quad (4)$$

There are two ways to construct the fading emulator embodied using the proposed method. One of the possible ways is to rotate the turn rail on which the actual scatterers are located, as shown in the insert in Figure 2a, which is the same structure shown in Figure 1. Another is that the massive MIMO antenna, which is placed at the center of the turntable, is rotated, as illustrated on the left hand side of the inset in Figure 2b. In this case, the rotation target, that is, the massive MIMO antenna, is different to that of Figure 1, but Figures 1 and 2b have the same benefit of achieving a full-rank channel matrix measurement, as explained in below.

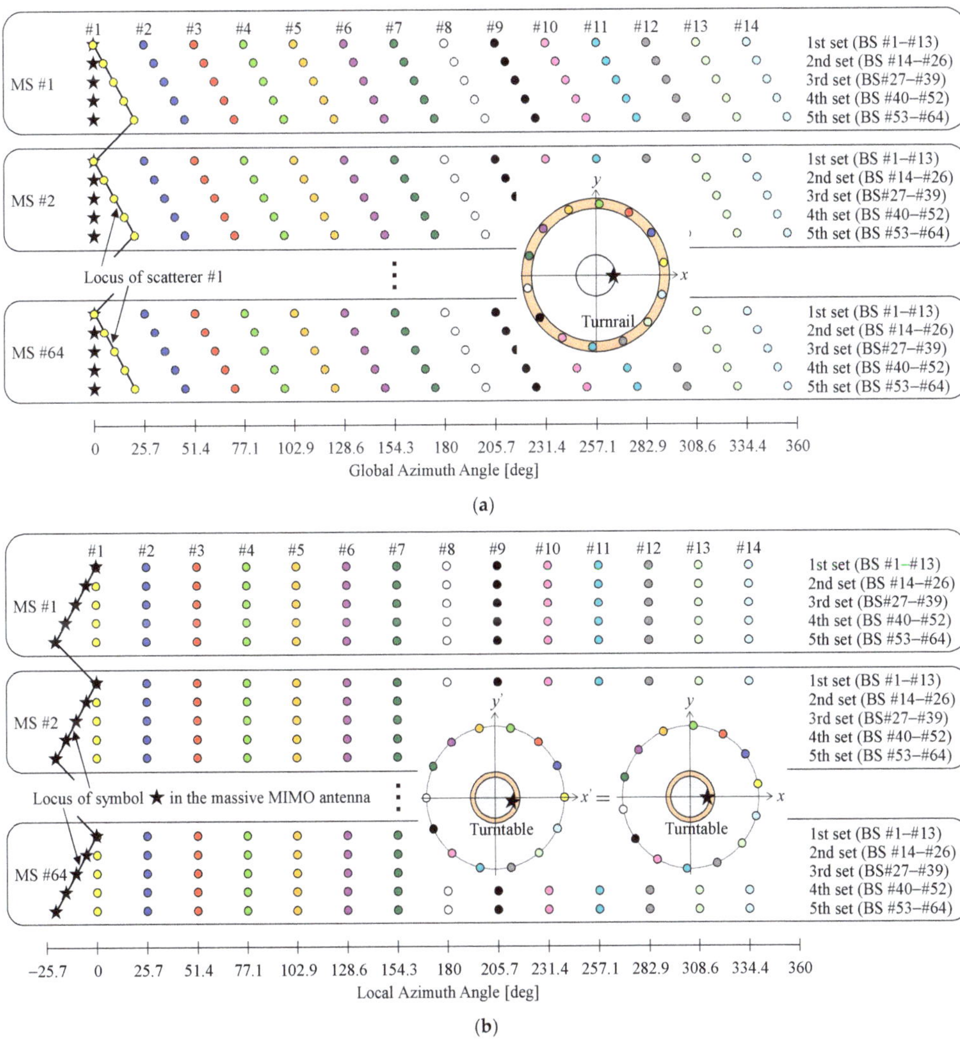

Figure 2. Massive MIMO OTA apparatus: (**a**) with the turn rail; (**b**) with the turntable.

It is known that with multiple probe antenna based methods, such as those with fading emulators, the repeatability and controllability of the radio propagation environment are superior to those obtained with other OTA testing methods, such as reverberation chamber based methods or two-stage methods [18]. In OTA assessment using a fading emulator, the MIMO channel response between the m-th BS antenna element and n-th MS antenna element is measured individually, taking advantage of the high time correlation characteristics of the apparatus.

Figure 2 shows the relationship between the channel response measured, the azimuth angle of the actual scatterers, and the azimuth angle of the massive MIMO antenna in the case of a 64 × 64 MIMO system, with K_m = 14, and S = 5. The symbols in Figure 2 are associated with Figure 1. The circles indicate the positions of the scatterers, whereas the star symbol denotes the azimuth angle of the massive MIMO antenna, which starts from 0° in the measurements.

In Figure 2a, the massive MIMO MS antenna is fixed at the center of the fading emulator, and the actual scatterers are moved by rotating the turn rail. Accordingly, the global azimuth angle of the scatterers is varied depending on the measured channel response for the m-th BS antenna. The black line indicates the locus of scatterer #1.

The channel responses from BS #1 to BS #13 are measured with the 1st set of scatterers in place. Then, the turn rail is rotated by $\Delta\varphi$, and the channel responses from BS #14 to BS #26 are measured with the 2nd set of scatterers in place. By repeating this procedure, the channel responses of the n-th MS antenna are fulfilled. Moreover, this method applies to all MS antennas, resulting in a full-rank channel response matrix.

In contrast, in Figure 2b, the actual scatterers remain stationary, and the turntable with the massive MIMO MS antenna is rotated. Consequently, the local azimuth angle of the massive MIMO MS antenna is changed corresponding to the measured channel response for the m-th BS antenna. The black line shows the locus of the star symbol expressing the angle of the massive MIMO antenna. However, the global azimuth angle of the MIMO antenna is fixed during OTA testing. When the local azimuth angle is transformed so that the global azimuth angle is 0°, as shown on the right hand side of the inset in Figure 2b, it becomes the same as in Figure 2a. Eventually, the actual scatterers are virtually positioned in different locations.

The channel responses from BS #1 to BS #13 are measured. Then, the turntable is rotated by $-\Delta\varphi$, and the channel responses from BS #14 to BS #26 are obtained. By repeating this operation, the channel responses of the n-th MS antenna are satisfied. Furthermore, this is done for all MS antennas, demonstrating that the channel matrix is full-rank status.

3. Results and Discussion

3.1. Analytical Results

To verify the proposed OTA method, Monte Carlo simulation of a massive MIMO antenna was conducted. The massive MIMO antenna comprises a 64-element quasi-half-wavelength dipole MIMO circular array antenna at 5 GHz to exclude the effect caused by electromagnetic mutual coupling. The array antenna was arranged with equal angular intervals. The radiation pattern of the massive MIMO antenna was calculated by the method of moments.

Figure 3 shows the number of channels as a function of the total number of scatterers K_n, that is the number of scatterers located to perform all measurements. In Figure 3, the black circles represent the analytical outcome with the proposed method as a function of S, whereas the blue rhombuses are those of the randomly arranged scatterers method, in which the scatterers on each BS were randomly selected from among all the scatterers. The red line represents the theoretical value, as expressed in Equation (3). K_m is set to 14.

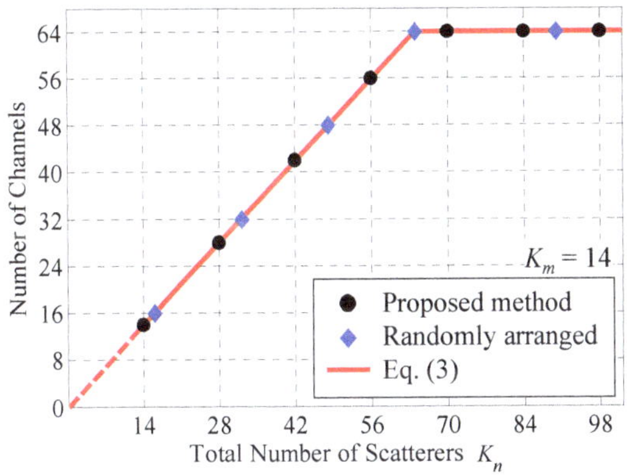

Figure 3. Number of channels vs. number of total scatterers.

It can be seen from Figure 3 that the proposed method has the same effect as the method of randomly arranged scatterers. When K_n is greater than 64, the number of channels equals 64. In contrast, when K_n is less than 64, the number of channels is understood to be just K_n. Hence, full-rank status can be achieved by rotating the actual scatterers multiple times, even with a small number of scatterers on the OTA apparatus.

Figure 4 shows the average of the 64th eigenvalues through all snapshots as a function of the angular interval between the i-th and $(i + 1)$-th sets of scatterers with S as a parameter [17]. The green, red, and blue curves indicate the cases where S is 3, 5, and 7, respectively. K_m is set to 14.

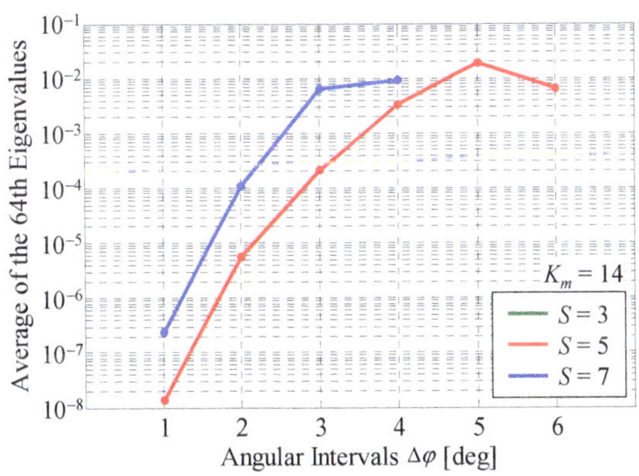

Figure 4. Average of the 64th eigenvalues.

As can be seen in Figure 4, there is no green curve for $S = 3$ because the total number of scatterers is only 42 (=14 × 3), indicating a rank-deficient status. In contrast, when S is greater than 5, the average of the 64th eigenvalues is confirmed, indicating full-rank status.

The average of the 64th eigenvalues is increased with increasing the angular interval. This is because, when $\Delta\varphi$ is small, there is high correlation coefficient between the incoming waves due to the closely spaced arrangement of the scatterers. On the other hand, the independence of the incoming wave is greater as $\Delta\varphi$ is increased i.e., the eigenvalues are larger. But the average of the 64th eigenvalues is reduced for more large angular intervals such as $\Delta\varphi = 6°$ with $S = 5$. In this case, the azimuth angle of scatterer #1 in the 5th set is 24°, which is close to that of scatterer #2 in the 1st set, 25.7°. Thus, all the scatterers including the virtual scatterers should be set to equal angular intervals.

3.2. Experimental Results

This subsection is devoted to verification of the proposed method utilizing the fading emulator with a small number of scatterers. The channel response matrix of the 64 × 64 MIMO system was measured using a two-dimensional fading emulator with a uniform incident wave distribution in azimuth. In millimeter wave 5G communications, the distance between the BS and MS is smaller than previous communication systems because the path loss between the BS and MS is large, resulting in an environment in which line-of-sight propagation or a cluster power distribution is assumed. In the sub-6 GHz frequency bands, the propagation environment is like the cluster or uniform power distributions found in previous communication systems. In this paper, a uniform power distribution, with which a sufficient number of paths in a propagation environment are expected, is considered.

Figure 5 shows the configuration of the massive MIMO-OTA apparatus that embodies the proposed method using the turntable, as shown in Figure 2b. The inserted lower right photo is a bird's eye view of the fading emulator. A 64-element quasi half-wavelength dipole MIMO circular array antenna was placed on a turntable located at the center of the fading emulator. The radius of the massive MIMO antenna was 20 cm. Fourteen scatterers, comprising vertically polarized half-wavelength sleeve dipole antennas, were set at equal angular intervals on a circle of radius 120 cm. The frequency was set to 5 GHz.

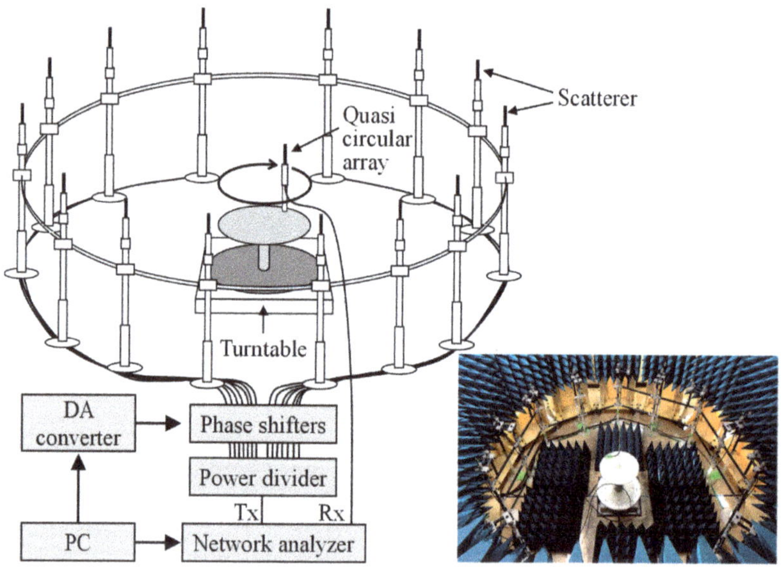

Figure 5. Configuration of the massive MIMO-OTA apparatus.

Figure 6 shows the cumulative distribution function (CDF) of the instantaneous eigenvalues obtained through SVD for a measured channel response matrix utilizing a fading emulator with a small number of scatterers. Figure 6a,b show the results for the previous method without rotation of the massive MIMO antenna and for the proposed method with rotation of the massive MIMO antenna, respectively. In the proposed method, S is set to 5 and the incremental rotation angle, $\Delta\varphi$, as illustrated in Figure 1, is 5.1°, that is, the angular intervals between the scatterers are equal.

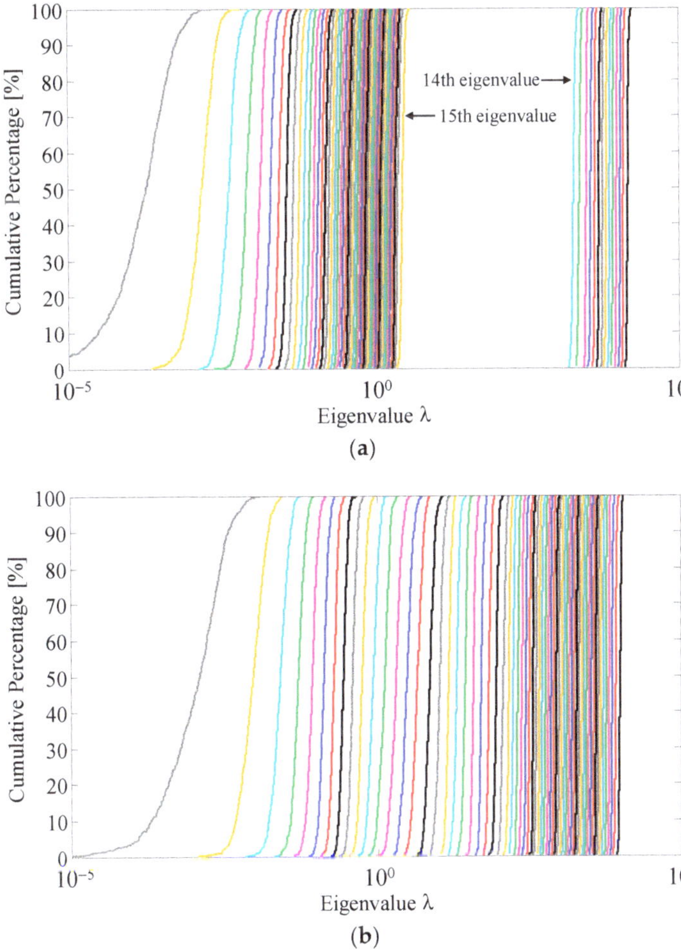

Figure 6. CDF characteristics of the eigenvalues: (**a**) previous method; (**b**) proposed method.

As shown in Figure 6a, there is a large interval between the 14th and 15th eigenvalues. Therefore, the previous OTA testing method can emulate only 14 channels with the same K_m. However, the 15th and subsequent eigenvalues are observed. This is due to the fact that the measured channel matrix includes noise and that the time correlation is 0.999, not absolutely 1. In contrast, in Figure 6b, an extremely dense eigenvalue distribution is observed because five scatterer sets have the same effect as 70 scatterers. Consequently, a full-rank channel matrix for a massive MIMO system can be implemented using the proposed method.

Figure 7a shows the average eigenvalue distribution with S as a parameter. The green, red, and blue curves denote the cases where S is 3, 5, and 7, respectively. For comparison, the previous method is depicted by the black curve in the graph. When the actual scatterers were rotated, $\Delta\varphi$ was set to equal angular intervals between the i-th and $(i + 1)$-th sets. As the angular interval between the scatterers is 25.7° with K_m set to 14, $\Delta\varphi$ is 5.1° with S equal to 5.

is concluded from these results that OTA testing incorporating the proposed method is a valid approach for obtaining a full-rank channel matrix for a massive MIMO system.

4. Conclusions

This paper presents an OTA evaluation method in which a full-rank channel matrix is created for a massive MIMO MS antenna utilizing a fading emulator with a small number of scatterers. The massive MIMO MS antenna is placed at the center of a turntable which is rotated; the total number of scatterers can be determined by controlling the rotation of the massive MIMO antenna. The experimental results reveal that a full-rank channel matrix for a massive MIMO antenna system can be obtained embodying the proposed method. This is a valuable tool for assessing the high MIMO channel capacity of a massive MIMO antenna.

Future work may include verification of the proposed method for the cluster propagation environment assumed in 5G mobile communication systems.

Funding: This research received no external funding.

Institutional Review Board Statement: Not applicable.

Informed Consent Statement: Not applicable.

Data Availability Statement: Not applicable.

Conflicts of Interest: The author declares no conflict of interest.

References

1. Dahlman, E.; Mildh, G.; Parkvall, S.; Peisa, J.; Sachs, J.; Selén, Y.; Sköld, J. 5G Wireless Access: Requirements and Realization. *IEEE Commun. Mag.* **2014**, *52*, 42–47. [CrossRef]
2. Andrews, J.G.; Buzzi, S.; Choi, W.; Hanly, S.V.; Lozano, A.; Soong, A.C.; Zhang, J.C. What Will 5G Be? *IEEE J. Sel. Areas Commun.* **2014**, *32*, 1065–1082. [CrossRef]
3. Larsson, E.G.; Edfors, O.; Tufvesson, F.; Marzetta, T.L. Massive MIMO for Next Generation Wireless Systems. *IEEE Commun. Mag.* **2014**, *52*, 186–195. [CrossRef]
4. Mao, C.X.; Gao, S.; Wang, Y. Broadband High-Gain Beam-Scanning Antenna Array for Millimeter-Wave Applications. *IEEE Trans. Antennas Propag.* **2017**, *65*, 4864–4868. [CrossRef]
5. Kibaroglu, K.; Sayginer, M.; Phelps, T.; Rebeiz, G.M. A 64-Element 28-GHz Phased-Array Transceiver With 52-dBm EIRP and 8–12-Gb/s 5G Link at 300 Meters Without Any Calibration. *IEEE Trans. Microw. Theory Tech.* **2018**, *66*, 5796–5811. [CrossRef]
6. Li, Y.; Sim, C.-Y.-D.; Luo, Y.; Yang, G. 12-port 5G massive MIMO antenna array in sub-6GHz mobile handset for LTE bands 42/43/46 applications. *IEEE Access* **2017**, *6*, 344–354. [CrossRef]
7. Honda, K.; Fukushima, T.; Ogawa, K. Full-Azimuth Beam Steering MIMO Antenna Arranged in a Daisy Chain Array Structure. *Micromachines* **2020**, *11*, 871. [CrossRef] [PubMed]
8. Shiu, D.S.; Foschini, G.J.; Gans, M.J.; Kahn, J.M. Fading correlation and its effect on the capacity of multielement antenna systems. *IEEE Trans. Commun.* **2020**, *48*, 502–513. [CrossRef]
9. Kyösti, P.; Jämsä, T.; Nuutinen, J.P. Channel modelling for multiprobe over-the-air MIMO testing. *Int. J. Antennas Propag.* **2012**, *2012*, 615954. [CrossRef]
10. Fukushima, T.; Narukawa, N.; Ogawa, K.; Honda, K. Randomly Arranged Scatterers Monte-Carlo Simulation for a Large-Scale MIMO Antenna. In Proceedings of the 2019 IEEE Asia-Pacific Microwave Conference (APMC 2019), Singapore, 10–13 December 2019; pp. 278–280.
11. Taniguchi, R.; Nishimori, K.; Kataoka, R.; Kameyama, K.; Kitao, K.; Tran, N.; Imai, T. Evaluation of Massive MIMO Considering Real Propagation Characteristics in the 20-GHz Band. *IEEE Trans. Antennas Propag.* **2017**, *65*, 6703–6711. [CrossRef]
12. Fan, W.; Carton, I.; Kyosti, P.; Karstensen, A.; Jamsa, T.; Gustafsson, M.; Pedersen, G.F. A Step Toward 5G in 2020: Low-cost OTA performance evaluation of massive MIMO base stations. *IEEE Antennas Propag. Mag.* **2017**, *59*, 38–47. [CrossRef]
13. Kyösti, P.; Hentilä, L.; Fan, W.; Lehtomäki, J.; Latva-Aho, M. On Radiated Performance Evaluation of Massive MIMO Devices in Multiprobe Anechoic Chamber OTA Setups. *IEEE Trans. Antennas Propag.* **2018**, *66*, 5485–5497. [CrossRef]
14. Gao, H.; Wang, W.; Wu, Y.; Liu, Y.; Pedersen, G.F. A Virtual Over-the-Air Method for 5G Massive MIMO Base Station Testing With Flexible Virtual Probes. *IEEE Access* **2019**, *7*, 108474–108485. [CrossRef]
15. Sakata, T.; Yamamoto, A.; Ogawa, K.; Iwai, H.; Takada, J.; Sakaguchi, K. A Spatial Fading Emulator for Evaluation of MIMO Antennas in a Cluster Environment. *IEICE Trans. Commun.* **2014**, *E97-B*, 2127–2135. [CrossRef]
16. Honda, K. A Method of Implementing a 4 $\times$ 4 Correlation Matrix for Evaluating the Uplink Channel Properties of MIMO Over-The-Air Apparatus. *Sensors* **2021**, *21*, 6184. [CrossRef] [PubMed]

17. Honda, K.; Ogawa, K. Over-The-Air Apparatus for Large-Scale MIMO Antennas to Create the Full-Rank Channel Matrix. In Proceedings of the 2020 International Symposium on Antennas and Propagation (ISAP 2020), Osaka, Japan, 25–28 January 2021; pp. 547–548.
18. 3G Partnership Project. *Measurement of Radiated Performance for Multiple Input Multiple Output (MIMO) and Multi-Antenna Reception for High Speed Packet Access (HSPA) and LTE Terminals*; TR 37.976 v16.0.0; 3GPP: Sophia Antipolis, France, 2020. Available online: https://portal.3gpp.org/desktopmodules/Specifications/SpecificationDetails.aspx?specificationId=2636 (accessed on 12 December 2021).

electronics

MDPI

Article

Research on Ultra-Low-Frequency Communication Based on the Rotating Shutter Antenna

Faxiao Sun [1,2,3], Feng Zhang [1,2,*], Xiaoya Ma [1,2,3], Zhaoqian Gong [1,2], Yicai Ji [1,2,3] and Guangyou Fang [1,2,3]

[1] Aerospace Information Research Institute, Chinese Academy of Sciences, Beijing 100190, China; sunfaxiao19@mails.ucas.ac.cn (F.S.); maxiaoya20@mails.ucas.ac.cn (X.M.); zqgong@mail.ie.ac.cn (Z.G.); ycji@mail.ie.ac.cn (Y.J.); gyfang@mail.ie.ac.cn (G.F.)

[2] Key Laboratory of Electromagnetic Radiation and Sensing Technology, Chinese Academy of Sciences, Beijing 100190, China

[3] School of Electronic, Electrical and Communication Engineering, University of Chinese Academy of Sciences, Beijing 100049, China

* Correspondence: zhangfeng002723@aircas.ac.cn

Abstract: This paper proposes a rotating shutter antenna that can directly generate 2FSK signals in the ULF band and it is expected to be used as the transmitter for magnetic induction (MI) underground communication systems. The antenna was modeled using ANSYS Maxwell and the magnetic field distribution was simulated. The results show that the interaction between the high-permeability shutter and the mutual cancellation of magnets decreased the transmitting magnetic moment of the antenna. A prototype antenna was manufactured and the time and frequency properties of the measured B_z field were the same as the simulated results, while the magnitude of the measured signal was larger. The propagation characteristics of the antenna in air–soil–rock were simulated using FEKO and the results show that the signal strength was greater than 1 fT at a depth of 450 m from the antenna whose magnetic moment as 1 Am^2. The relationship between different magnetic components and azimuth could be used to enhance the signal strength. The formula of the B_z field was derived using the measured magnitude versus distance and the path loss was also analyzed. Finally, the 2FSK modulation property of the antenna was verified by indoor communication experiments with a code rate of 12.5 bps in the ULF band.

Keywords: ultra-low-frequency communication; mechanical antenna; layered media; rotating shutter antenna

Citation: Sun, F.; Zhang, F.; Ma, X.; Gong, Z.; Ji, Y.; Fang, G. Research on Ultra-Low-Frequency Communication Based on the Rotating Shutter Antenna. *Electronics* **2022**, *11*, 596. https://doi.org/10.3390/electronics11040596

Academic Editor: Naser Ojaroudi Parchin

Received: 6 January 2022
Accepted: 14 February 2022
Published: 15 February 2022

1. Introduction

The electromagnetic waves at radio frequencies have short wavelengths and poor penetrating capabilities, which make them lose part of their energy in a complex electromagnetic environment [1]. For example, the permittivity and conductivity of soil have a great relationship with the water content, which causes significant attenuation of high-frequency electromagnetic waves. This leads to the fact that radio-frequency communication cannot be widely used in underground communication. To overcome this problem, magnetic induction communication technology with ultra-low-frequency magnetic fields is widely used in underground communication, such as Through-The-Earth communication [2]. The permeability of non-magnetic materials is almost the same, which provides a relatively stable channel for the propagation of magnetic field signals and the stable channel enables the magnetic induction communication to avoid the shortcomings of multipath propagation, large propagation delay and high bit error rate of acoustic transmission through the ground [3].

Normally, magnetic inductive communication systems use coils to generate and receive magnetic field signals in near-field zone [4]. However, in the ultra-low-frequency band, the size of the coil is also very large, which brings great inconvenience to the placement and layout of the system [5,6].

To overcome the shortcomings of traditional low-frequency antennas, such as large size, low efficiency and high-power consumption, in 2017, the US Department of Defense Advanced Research Project Agency (DARPA) proposed the AMEBA plan [7]. Different from traditional antennas, AMEBA mainly generates alternating electromagnetic waves directly through electrical charges or magnetic moments in mechanical motion, which is a process of converting mechanical energy into electromagnetic energy [8,9]. As a kind of mechanical antenna, the rotating permanent magnet antenna (RPMA) drives the permanent magnet to rotate so as to obtain an alternating magnetic field. It is likely to be used as the transmitting antenna in the magnetic induction communication system.

After AMEBA was proposed, many researchers have become interested in RPMA and began to study it, including field generation and measurement [10,11] and radiation power analysis [12]. Skyler Selvin et al. analyzed the efficiency of RPMA and proposed a method to build an RPMA array to improve antenna efficiency [13]. Srinivas Prasad M.N. et al. proposed a prototype consisting of an array of magnetic pendulums in oscillatory motion at ULF. The transmission efficiency of the magnetic pendulums array is higher than a bare coil through the measurement in the near field [14,15].

The communication system which uses RPMA as the transmitter mostly adopts direct antenna modulation (DAM). For example, we can obtain an OOK signal by controlling the driver switch and an FSK signal by controlling the speed of motion [16–20]. Besides, Refs. [21–24] proposed that amplitude and phase modulation can be achieved by using an external modulator. However, these methods require additional energy to control the modulator in real time, which is difficult to implement.

To effectively avoid the limitation of the motor to the antenna's operating frequency range, we used a rotating shutter antenna as our transmitting antenna, which could obtain a magnetic field signal that is four times the rotating frequency. ANSYS Maxwell was used to analyze the magnetic field distribution of the rotating shutter antenna and the results show that the interaction between the high-permeability shutter and the mutual cancellation of magnets decreased the transmitting magnetic moment. A prototype antenna was manufactured and the experimental results are in great agreement with the simulated results, while the magnitude of the measured signal was larger. FEKO was used to analyze the propagation characteristics of the RPMA in air–soil–rock media and the simulated results show that the B_z-field at a depth of 450 m away from the antenna on ground with a magnetic moment of 1 Am^2 was 1 fT. The formula of the B_z-field was derived using measured magnitude versus distance and the path loss was also analyzed. The formula shows that the path loss of the rotating shutter antenna in free space was 303 dB at a distance of 570 m, while the signal strength was 1 fT. The prototype antenna was also used to carry out indoor communication experiments with a code rate of 12.5 bps in the ULF. This paper is organized as follows: In Section 2, the principle and simulation of the rotating shutter antenna are presented. The simulation of RPMA in layered media is shown in Section 3. In Section 4, the prototype and the experimental results are presented. Finally, Section 5 draws the conclusion.

2. Principle and Simulation of the Rotating Shutter Antenna

The radiation principle of the rotating permanent magnet antenna can be summarized as follows: the motor drives the magnet to rotate so that its static magnetic field is converted to an alternating magnetic field. The corresponding relationship between the signal frequency and the motor speed n can be expressed as

$$f = \frac{n}{60} \tag{1}$$

When the speed n of the drive motor is greater than 18,000 rpm, the frequency of the signal obtained is greater than 300 Hz, which is difficult for common servo motors on the market.

The rotating shutter antenna was proposed by M. Golkowski in 2018 [25] and the prototype mainly consists of magnets and a shutter made of soft magnetic materials, such as permalloy, whose relative permeability is significant. Assuming that the number of openings is N, the antenna needs two N magnets of different polarities which are alternately placed to form a circle. During the working process, the shutter rotates to block the permanent magnets of different polarities to generate an altering magnetic field, whose frequency is N times the rotating fundamental frequency. The relationship between frequency and speed can be expressed as

$$f = \frac{n}{60} \cdot N \tag{2}$$

Compared with RPMA, the rotating shutter antenna can reduce the speed to n/N. The block diagram of the antenna principle when $N = 4$ is shown in Figure 1.

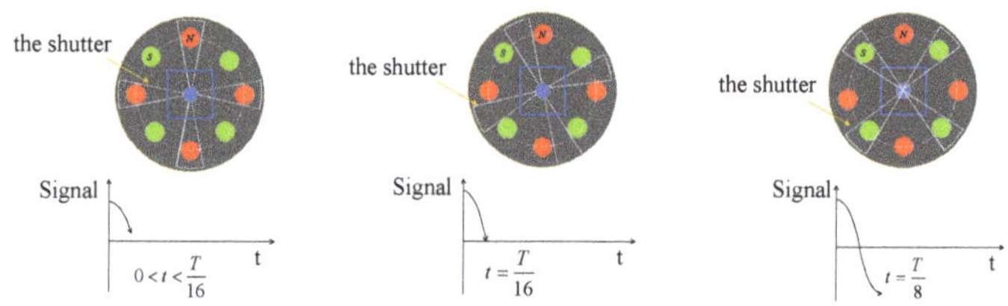

Figure 1. Block diagram of the rotating shutter antenna when $N = 4$.

ANSYS Maxwell was used to analyse a rotating shutter antenna with $N = 4$; the parameters of the simulation are shown in Table 1. A free-space sphere with a radius of 4 m was established and the boundary adopted natural boundary conditions, which made the magnetic field continuous inside and outside the simulation area. The magnetic field produced by the permanent magnets was used as excitation and the effects of eddy current induced on the shutter were considered. Regardless of mechanical damping, a band area with a set rotation speed of 4800 rpm was used to simulate the rotation of the shutter, which corresponded to a fundamental rotation frequency of 80 Hz, so the frequency of the magnetic field signal was, theoretically, 320 Hz.

Table 1. The parameters of the simulation.

Parameters	Values	Parameters	Values
radius of the magnet	10 mm	magnetic remanence	1.47 T
height of the magnet	100 mm	thickness of the shutter	1.5 mm
openings of the shutter	4	rotating speed	4800 rpm

We placed a point at the position of 3 m in the axial direction of the rotating shutter antenna and obtained the B_z-field. The simulated results are shown in Figure 2. Figure 2a shows the signal with an amplitude of 3.5 nT at the point 3 m away from the transmitter, which is much smaller than the theoretical value. The corresponding spectrogram is shown in Figure 2b, which represents B_z-field strength as a function of time and frequency. It can be clearly seen the signal whose frequency is 320 Hz in the spectrogram and multiple harmonics are visible.

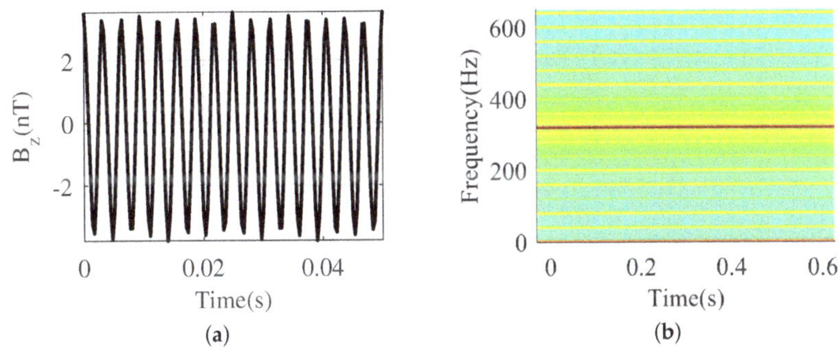

Figure 2. (**a**) Field at the point 3 m away from the shutter antenna. (**b**) Spectrogram of field.

The magnetic field intensity distribution on the rotating shutter at different times is shown in Figure 3. It is clear that a large part of the energy was concentrated on the shutter, which may be one of the reasons for the field strength being smaller than the theoretical calculation.

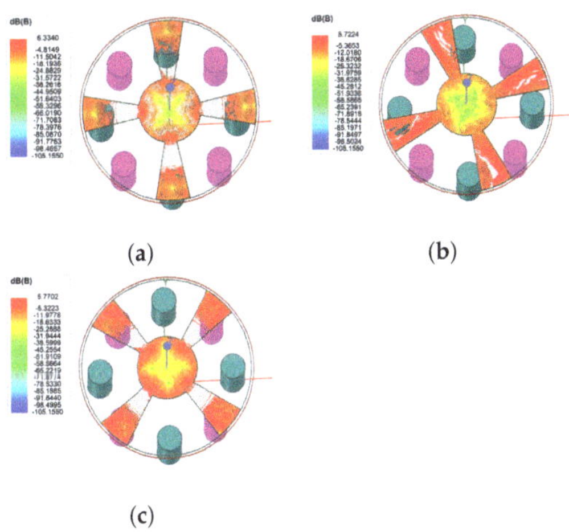

Figure 3. B-field distribution on the rotating shutter at different times: (**a**) t = 0 s; (**b**) t = 0.667 ms; (**c**) t = 1.67 ms.

The side view and top view of the distribution of the B-field by intercepting different planes in the simulation are shown in Figure 4. The fast attenuation of the signal can be ascribed to both the interaction of the high-permeability shutter and the mutual cancellation of magnets.

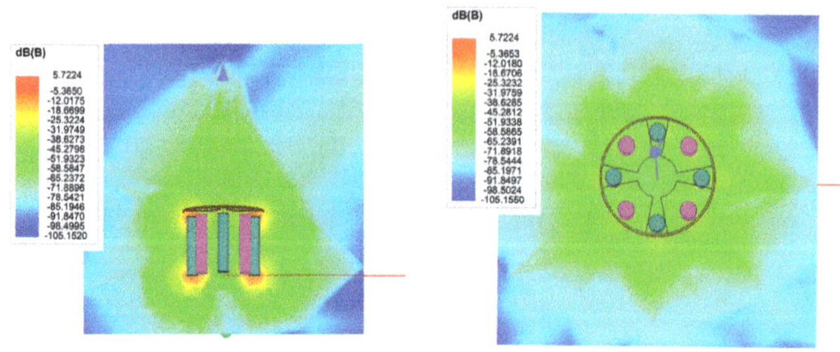

(**a**) (**b**)

Figure 4. B-field distribution from different viewing angles: (**a**) side view; (**b**) top view.

3. Propagation Characteristics in Layered Media

It is promising that RPMA can be used as the transmitting antenna in magnetic inductive underground communication and it is particularly important to study the radiation characteristics of RPMA in layered media. Layered media cause interface emission loss and transmission medium loss because of different permittivity and conductivity.

This paper used FEKO to construct an equivalent model of a rotating permanent magnet antenna that rotated vertically on the ground and analyzed its magnetic field characteristics in three-layer media, such as air–soil–rock. The simulation parameters are shown in Table 2.

Table 2. The parameters of the simulation.

	Parameters	Values
antenna	frequency	320 Hz
	moment	1 Am^2
soil	depth	1500 m
	permittivity	10
	conductivity	0.01 S/m
rock	permittivity	10
	conductivity	0.01 S/m

The curves of magnetic flux intensity versus distance are shown in Figure 5 and it is clear that the maximum penetrating depth could reach 450 m for magnetic induction communication.

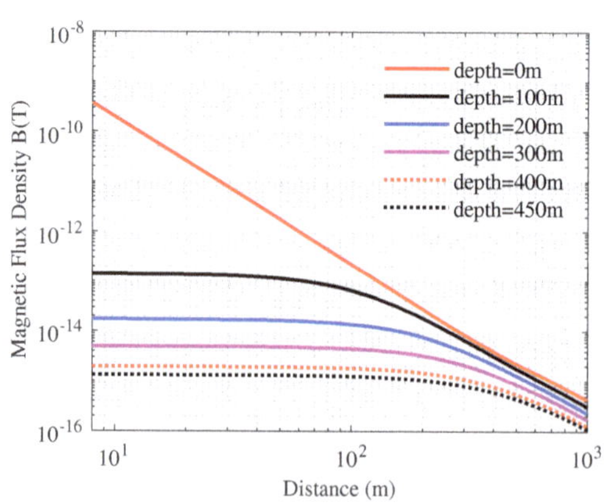

Figure 5. Change in magnetic flux density with the distance at different depths.

The curves of B_r and B_φ with azimuth at different depths shown in Figure 6 show that, at the same depth, the phase between the two magnetic field components was constant. If we used a two-axis magnetic field sensor to receive, we could obtain an enhanced signal by shifting the φ-direction component with a constant phase and then adding it to the r-direction component.

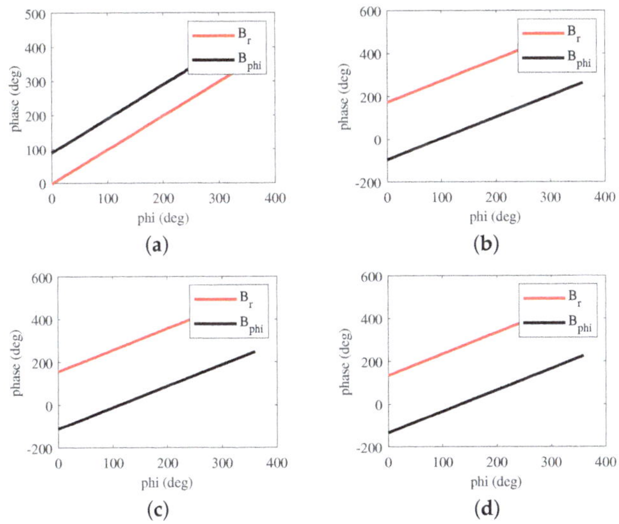

Figure 6. Change in B_r and B_φ with azimuth at different depths: (**a**) depth = 0 m; (**b**) depth = 150 m; (**c**) depth = 300 m; (**d**) depth = 450 m.

4. Experimental Results

The experimental results are presented in two sub-sections, i.e., (a) measurement of the rotating shutter antenna and (b) communication based on the rotating shutter antenna.

4.1. Measurement of the Rotating Shutter Antenna

A prototype, as shown in Figure 7, with the same size as shown in Table 2 was manufactured and magnets were fixed on an aluminum plate through an aluminum alloy cover. The YASKAWA servo AC motors with model SGM7J-01AFC6S were used as the driver, which could provide a rated torque of 637 mN·m and a rated power of 200 W. The rotation axis of the shutter antenna and the ULF receiver shown in Figure 8 were placed in line to measure the B_z-field. The distance between the antenna and the receiver was 3 m, as that in simulation. In the communication experiment, the information symbols were sent to the servo through software named MPE720 Ver. 7 to control the rotating shutter antenna in real time.

The portable ULF receiver, shown in Figure 8, consisted of a magnetic sensor and an NI collector. The conversion coefficient of the magnetic sensor was 8 mV/nT, so we could directly convert the received voltage signal into field strength. The NI collector was utilized to digitize and store the received signal with a sampling rate of 3 Ksps.

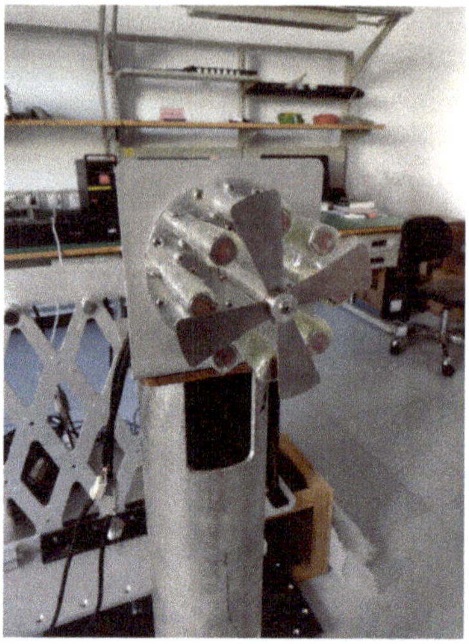

Figure 7. The rotating shutter antenna.

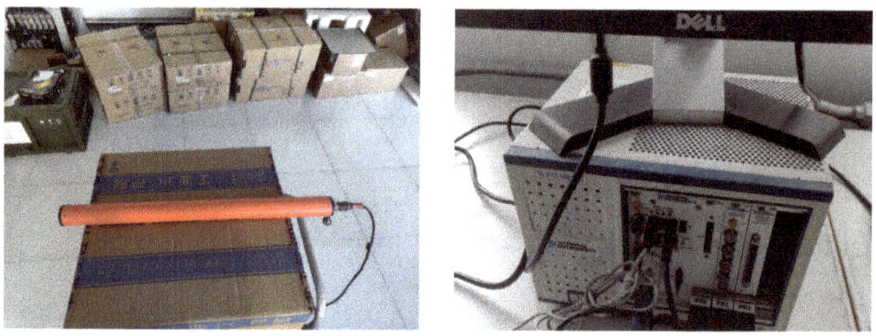

(a) (b)

Figure 8. ULF receiver: (**a**) magnetic sensor; (**b**) NI sampler.

The measured result shown in Figure 9 shows that, when the motor speed was set to 4800 rpm, the frequency of the signal obtained was in good agreement with the simulated result, but the amplitude was larger than that obtained in the simulation, which may be ascribed to the fact that the material in the simulation was not consistent with our prototype and the metal substances in the experimental environment may also have had a certain influence on the signal strength. From the spectrogram of the field shown in Figure 9b, we can clearly see the 320 Hz signal.

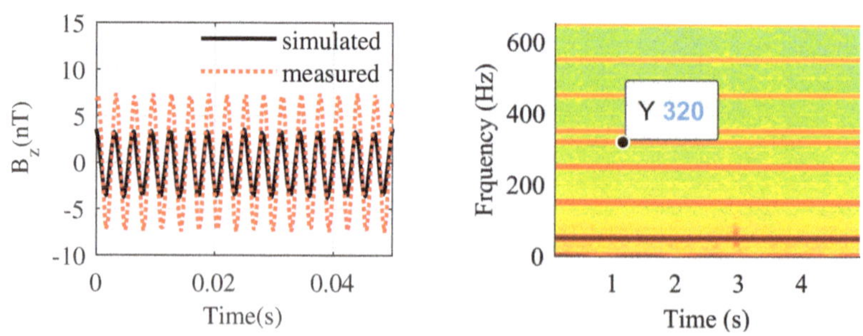

(a) (b)

Figure 9. (**a**) Field at the point 3 m away from the shutter antenna. (**b**) Spectrogram of field.

We measured the B_z-field by placing the receiving antenna along the axis direction of the rotating shutter antenna and changing the distance between them (such as 0.6 m, 1.2 m, ..., 4.8 m and 5.4 m); the variation in the field strength with distance is shown in Figure 10. The propagation characteristic curve was fitted using MATLAB and the result is also shown in Figure 10. The fitting curve can be expressed as

$$B_z(\mathrm{nT}) = \frac{188}{r^3} \tag{3}$$

The intensity of the B_z-field is inversely proportional to the third power of the distance, which is in line with the attenuation law of the near-field signal. However, different from the rotating permanent magnet antennas ($B_z = \frac{\mu m}{2\pi r^3}$), the expression of the magnetic flux density of the rotating shutter antenna shown in Equation (3) cannot find the similar law related to the magnetic moment.

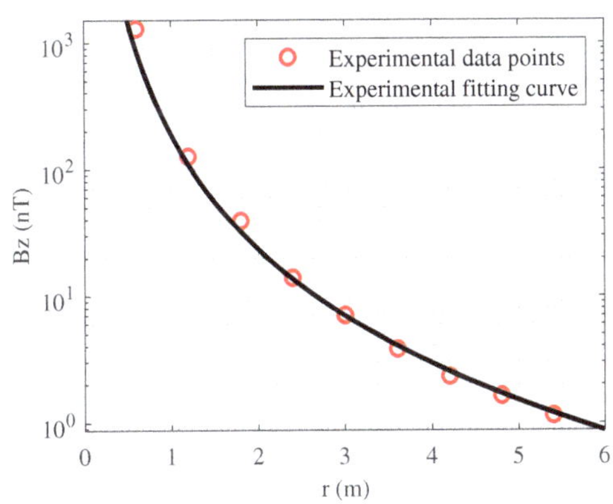

Figure 10. The change in B_z-field with distance.

In MI communication, the rotating shutter antenna is used as the transmitter. The path loss can be defined as

$$PL = 20\lg\frac{B_r}{B_z} \approx 137.4 + 60\lg(r) \tag{4}$$

where B_r is the remanence of the permanent magnet and B_z is given in Equation (3).

The path loss determined by the propagation range was simulated, as shown in Figure 11. The point with path loss of 303 dB (the received magnetic field was about 1 fT for $B_r = 1.4\mathrm{T}$) is marked with red lines in Figure 11. If the receiver can handle magnetic field signals of 1 fT, the MI communication distance of up to 570 m could be achieved. The antenna prototype shown in Figure 7 was equivalent to an RPMA with a transmitting magnetic moment of 0.94 Am^2. If the antenna was used as the transmitting antenna in the magnetic induction communication, a two-unit antenna array would be required to make the maximum detection depth reach 450 m. The synchronization between the elements would also be difficult. Searching for a new method to enhance magnetic moments will be the focus of our research work.

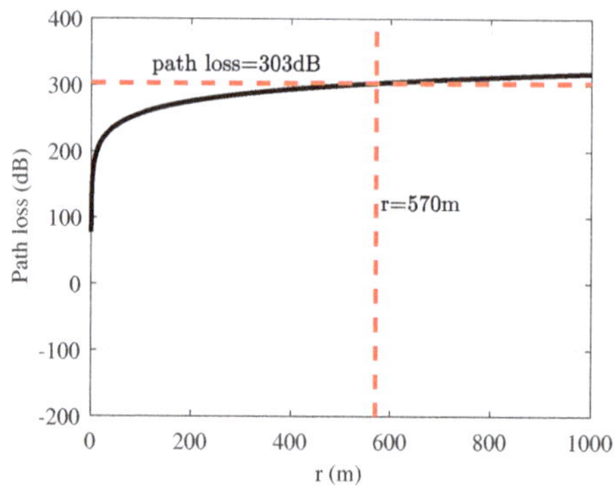

Figure 11. Path loss versus propagation range.

4.2. Communication Based on the Rotating Shutter Antenna

We changed the distance between the transmitter and the receiver antenna to 4.5 m, while the other experimental parameters were the same as those mentioned above. Using control software to rotate the motor between 4800 rpm and 4950 rpm, we theoretically obtained 320 Hz and 330 Hz signals, which can represent different binary information symbols such as '1' and '0'. The time required for the motor to switch between different speeds was set to 100 ms and the time for a constant rotation was 80 ms, which determined the code rate of communication.

The modulated signal after bandpass filtering is shown in Figure 12a. Its corresponding spectrum, shown in Figure 12b, demonstrates that the frequency of the 2FSK signal did not reach the theoretical value, which may have been due to insufficient control accuracy. In addition, there was a very obvious frequency component between the frequency spectra of the 303.7 Hz and 316.2 Hz signals, which was introduced during the motor speed switching process. This problem could be solved by increasing the constant speed time of the motor, but this would reduce the communication code rate.

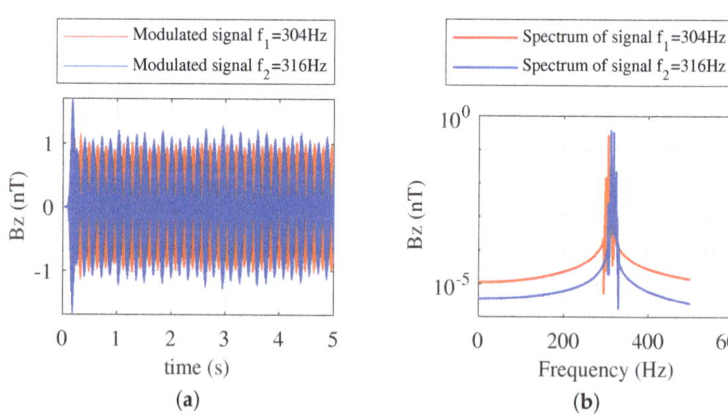

Figure 12. (**a**) Modulated signals at the point 5.4 m away from the rotating shutter antenna; (**b**) spectrum of signals.

It can be clearly observed from the spectrogram shown in Figure 13a that the signals with frequencies of 303.7 Hz and 316.2 Hz appeared alternately in sequence, which means that the information symbol '1010101...' was received. The received signal is processed by filtering out the power–frequency signal, passing through bandpass filtering and extracting the envelope of the filtered signal; then, the information symbol shown in Figure 13b can be obtained through amplitude judgment. It was found, from the demodulated information, that there were 12.5 symbols in one second, which means that the code rate of communication was 12.5 bps.

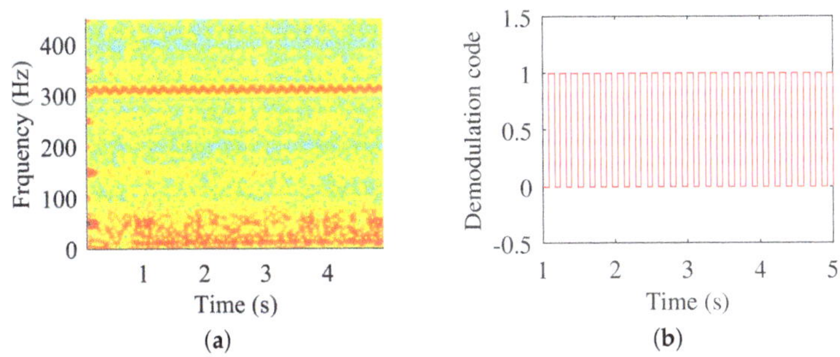

Figure 13. (**a**) Spectrogram of 2FSK signal; (**b**) demodulated code.

Table 3 summarizes the performance comparison between the previously proposed systems and the system adopted in this paper. It can be found that the frequency of the signal, the communication rate and the communication distance could not be taken into account simultaneously in the existing research study. Our system mainly considers a higher communication code rate, but the communication range was not the longest, which is also a focus for our future research work.

Table 3. Performance comparison of rotating permanent magnet antennas.

Reference	Frequency	Data Rate	Detection Distance
Ref. [10]	500 Hz	/	<130 m
Ref. [11]	22 Hz	8 bps	0.7 m
Ref. [14]	1.03 kHz	2 bps	30 m
Ref. [17]	100 Hz	0.25 bps	5 m
Ref. [20]	30 Hz	5 bps	100 m
Ref. [22]	40 Hz	<2 bps	5 m
Ref. [23]	50 Hz	40 bps	0.25 m
Our work	320 Hz	12.5 bps	5 m

Since the antenna operates in its near-field region, the signal strength decays with the third power of the distance, which limits the maximum communication distance of the communication system. Similarly, due to the low carrier frequency of the system, the data rate cannot be too high, which limits the diversity of communication. For example, the communication rate is only suitable for message transmission. In addition, an array with more shutter antennas would be a good choice to increase the magnetic moment, but assuring the control accuracy of the motor would be necessary and difficult to achieve.

5. Conclusions

This paper used ANSYS Maxwell to analyze the magnetic field distribution of the rotating shutter antenna and the results show that the interaction between the high-permeability shutter and the mutual cancellation of magnets decreased the transmitting magnetic moment. A prototype was manufactured; the measured results are in great agreement with the simulated results, while the magnitude of the measured signal was larger. We used FEKO to analyze the propagation characteristics of RPMA in air–soil–rock and the simulated results show that the signal strength was greater than 1 fT at a depth of 450 m from the antenna, whose moment was 1 Am^2. Since the phase difference between B_r and B_φ remained constant at $\pm 90^\circ$, we could use a multi-directional receiver to receive the multi-directional field components; then, the signal strength could be enhanced by shifting the phases of the B_r-field component and adding it to another component, such as a B_φ-field component. The formula shows that the path loss of the rotating shutter antenna at 570 m was 303 dB, which means that the signal strength was 1 fT. The rotating shutter antenna was equivalent to an RPMA with a transmitting magnetic moment of 0.94 Am^2. If the antenna was used as the transmitting antenna in the magnetic induction communication, a two-element antenna array would be required to make the maximum detection depth reach 450 m. The synchronization between the elements would make the control of the system very difficult. Finding a new method to increase the transmitting magnetic moment and the synchronization of the elements will be the focus of our next research studies. FSK modulation could be realized by controlling the motion state of the motor and the constant time of the motor determined the symbol rate. The prototype was used to achieve indoor communication with a code rate of 12.5 bps in the ULF.

Author Contributions: Conceptualization, F.S., X.M. and F.Z.; methodology, F.S. and X.M.; software, F.S. and Z.G.; validation, F.S. and F.Z.; formal analysis, F.S.; investigation, F.S. and X.M.; resources, Y.J.; data curation, F.Z. and Z.G.; writing—original draft preparation, F.S.; writing—review and editing, F.Z. and Y.J.; visualization, F.S.; supervision, G.F.; project administration, Y.J.; funding acquisition, G.F. All authors have read and agreed to the published version of the manuscript.

Funding: This research study was funded by the National Natural Science Foundation of China, grant number 61827803.

Institutional Review Board Statement: Not applicable.

Informed Consent Statement: Not applicable.

Data Availability Statement: Not applicable.

Conflicts of Interest: The authors declare no conflict of interest.

References

1. Kilinc, D.; Akan, O.B. Nanoscale magneto-inductive communication. In Proceedings of the 2013 Asilomar Conference on Signals, Systems and Computers, Pacific Grove, CA, USA, 3–6 November 2013; pp. 1061–1065.
2. Yan, L.; Waynert, J.; Sunderman, C.; Damiano, N. Statistical analysis and modeling of VLF/ELF noise in coal mines for through-the-earth wireless communications. In Proceedings of the 2014 IEEE Industry Application Society Annual Meeting, Vancouver, BC, Canada, 5–9 October 2014; pp. 1–5.
3. Domingo, M.C. Magnetic Induction for Underwater Wireless Communication Netwvorks. *IEEE Trans. Antennas Propag.* **2012**, *60*, 2929–2939. [CrossRef]
4. Sun, Z.; Akyildiz, I.F. Magnetic Induction Communications for Wireless Underground Sensor Networks. *IEEE Trans. Antennas Propag.* **2010**, *58*, 2426–2435. [CrossRef]
5. Gulbahar, B.; Akan, O.B. A communication theoretical modeling and analysis of underwater magneto-inductive wireless channels. *IEEE Trans. Wireless Commun.* **2010**, *11*, 3326–3334. [CrossRef]
6. Akyildiz, I.F.; Wang, P.; Sun, Z. Realizing underwater communication through magnetic induction. *IEEE Commun. Mag.* **2015**, *53*, 42–48. [CrossRef]
7. Ding, H. DARPA mechanical antenna project may set off a military communications revolution. *Conmilit* **2017**, *26*, 71–73.
8. Bickford, J.A.; McNabb, R.S.; Ward, P.A.; Freeman, D.K.; Weinberg, M.S. Low frequency mechanical antennas: Electrically short transmitters from mechanically-actuated dielectrics. In Proceedings of the 2017 IEEE International Symposium on Antennas and Propagation & USNC/URSI National Radio Science Meeting, San Diego, CA, USA, 9–14 July 2017; pp. 1475–1476.
9. Manteghi, M. A navigation and positining system for unmanned underwater vehicles based on a mechanical antenna. In Proceedings of the 2017 IEEE International Symposium on Antennas and Propagation & USNC/URSI National Radio Science Meeting, San Diego, CA, USA, 9–14 July 2017; pp. 1997–1998.
10. Burch, H.C.; Garraud, A.; Mitchell, M.F.; Moore , R.C.; Arnold, D.P. Experimental Generation of ELF Radio Signals Using a Rotating Magnet. *IEEE Trans. Antennas Propag.* **2018**, *66*, 6265–6272. [CrossRef]
11. Fawole, O.C.; Tabib-Azar, M. An Electromechanically Modulated Permanent Magnet Antenna for Wireless Communication in Harsh Electromagnetic Environments. *IEEE Trans. Antennas Propag.* **2017**, *65*, 6927–6936. [CrossRef]
12. Bickford, J.A.; Duwel, A.E.; Weinberg, M.S.; McNabb, R.S.; Freeman, D.K.; Ward, P.A. Performance of Electrically Small Conventional and Mechanical Antennas. *IEEE Trans. Antennas Propag.* **2019**, *67*, 2209–2223. [CrossRef]
13. Selvin, S.; Srinivas, P.M.N.; Huang, Y.; Wang, E. Spinning magnet antenna for VLF transmitting. In Proceedings of the 2017 IEEE International Symposium on Antennas and Propagation & USNC/URSI National Radio Science Meeting, San Diego, CA, USA, 9–14 July 2017; pp. 1477–1478.

14. Srinivas, P.M.N.; Tok, R.U.; Wang, Y.E. Magnetic pendulum arrays for ULF transmission. In Proceedings of the 2018 IEEE International Symposium on Antennas and Propagation & USNC/URSI National Radio Science Meeting, Boston, MA, USA, 8–13 July 2018; pp. 71–72.
15. Srinivas, P.M.N.; Fereidoony, F.; Wang, Y.E. 2D Stacked magnetic pendulum arrays for efficient ULF transmission. In Proceedings of the 2020 IEEE International Symposium on Antennas and Propagation and North American Radio Science Meeting, Montreal, QC, Canada, 5–10 July 2020; pp.1305–1306.
16. Liu, Y.; Hou, M.; Gong, S. A Rotating Permanent magnet transmitter for magnetic induction communication in RF-impenetrable environment. In Proceedings of the 2020 IEEE MTT-S International Conference on Numerical Electromagnetic and Multiphysics Modeling and Optimization (NEMO), Hangzhou, China, 7–9 December 2020; pp.1–3.
17. Glickstein, J.S.; Liang, J.; Choi, S.; Madanayake, A.; Mandal, S. Power-efficient ELF wireless communications using electro-mechanical transmitters. *IEEE Access* **2020**, *8*, 2455–2471. [CrossRef]
18. Tarek, M.T.B.; Dharmasena, S.; Madanayake, A.; Choi, S.; Glickstein, J.; Liang, J.; Mandal, S. Power-efficient data modulation for all-mechanical ULF/VLF transmitters. In Proceedings of the 2018 IEEE 61st International Midwest Symposium on Circuits and Systems (MWSCAS), Windsor, ON, Canada, 5–8 August 2018; pp.759–762.
19. Cao, J.; Liu, Y.; Gong, S. Low frequency mechanical antenna for underwater communication. In Proceedings of the 2019 International Conference on Electronic Engineering and Informatics (EEI), Nanjing, China, 8–10 November 2019; pp.140–142.
20. Liu, Y.; Gong, S.; Liu, Q.; Hou, M. A Mechanical Transmitter for Undersea Magnetic Induction Communication. *IEEE Trans. Antennas Propag.* **2021**, *69*, 6391–6400. [CrossRef]
21. Strachen, N.D.; Booske, J.H.; Behdad, N. Mechanical Super-Low Frequency (SLF) transmitter using electrically-modulated reluctance. In Proceedings of the 2018 IEEE International Symposium on Antennas and Propagation & USNC/URSI National Radio Science Meeting, Boston, MA, USA, 8–13 July 2018; pp.67–68.
22. Zhang, J.; Song, Z.; Zhang, D.; Xi, X. Amplitude modulation method of the mechanically rotating-based antenna. *Electron. Lett.* **2020**, *56*, 321–323 [CrossRef]
23. Barani, N.; Sarabandi, K. A phase modulation scheme for super-low frequency handheld mechanical antennas. In Proceedings of the 2020 IEEE International Symposium on Antennas and Propagation and North American Radio Science Meeting, Montreal, QC, Canada, 5–10 July 2020; pp.467–468.
24. Barani, N.; Sarabandi, K. A frequency multiplier and phase modulation approach for mechanical antennas operating at Super Low Frequency (SLF) Band. In Proceedings of the 2019 IEEE International Symposium on Antennas and Propagation and USNC-URSI Radio Science Meeting, Atlanta, GA, USA, 7–12 July 2019; pp. 2169–2170.
25. Gołkowski, M.; Park, J.; Bittle, J.; Babaiahgari, B.; Rorrer, R.A.L.; Celinski, Z. Novel mechanical magnetic shutter antenna for ELF /VLF radiation. In Proceedings of the 2018 IEEE International Symposium on Antennas and Propagation & USNC/URSI National Radio Science Meeting, Boston, MA, USA, 8–13 July 2018; pp. 65–66.

Figure 1. Overview of the conformal metasurface.

2. Materials and Methods

2.1. Proposed Conformal Metasurface Antenna System

The proposed conformal metasurface antenna system consists of two parts: 1. the conformal metasurface, and 2. the conventional patch antenna. As shown in Figure 2, a conformal metasurface was placed near a conventional patch antenna. The conformal metasurface acted as a reflector of a patch antenna. With radio waves traveling with different distances and angles, the conformal metasurface was divided into multiple subwavelength segments. The angle transformation from conformal into planar waves can be based on the geometric method. The shortest distance in a desired direction can be used as a reference, i.e., $R_0 + S_0$. The phase difference between rays S_1 and S_2 can be regarded as [34]:

$$\Delta\varphi = k_0[(|R_1|+|S_1|) - (|R_0|+|S_0|)] \quad (1)$$

where $\Delta\varphi$ is the phase difference and k_0 is the free-space wave vector. From (1) the phase difference can be controlled by manipulating the distance of different waves. One way to achieve that is by changing the curvature of the metasurface. As Figure 2 shows, when changing the curvature of the conformal surface radius from 40 mm to 88 mm, the direction of the beam is expected to change.

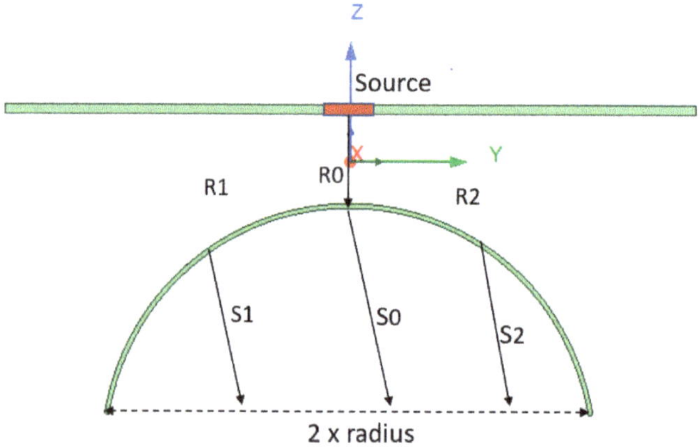

(**a**)

Figure 2. *Cont.*

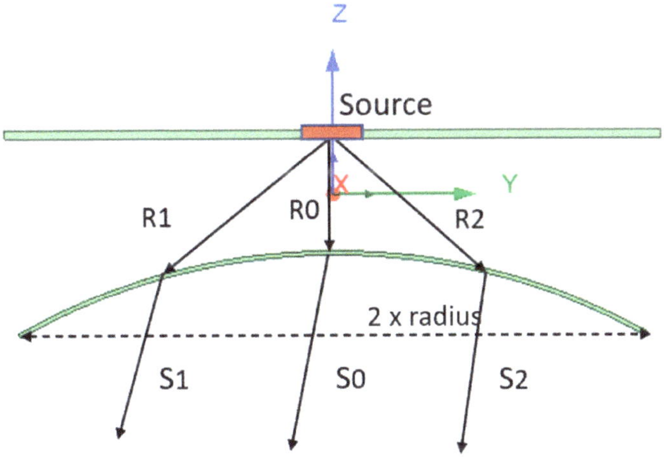

(**b**)

Figure 2. Conformal metasurface antenna proposal: (**a**) proposed conformal metasurface with radius of 40 mm. (**b**) proposed conformal metasurface with radius of 88 mm.

To further verify this design idea, simulation was performed using the finite element solver HFSS. The layout of the patch antenna and metasurface is shown in Figures 3 and 4. Figure 3a shows the layout of the conventional patch antenna. This patch antenna was designed with a center frequency of 2.4 GHz and was printed on a 1.6 mm thick FR4 substrate with dielectric constant of 4.04 [37]. The goal here is to have the most conventional and cost-effective patch antenna to verify the effectiveness of the conformal metasurface. The substrate size was designed as the same size as the conformal metasurface path.

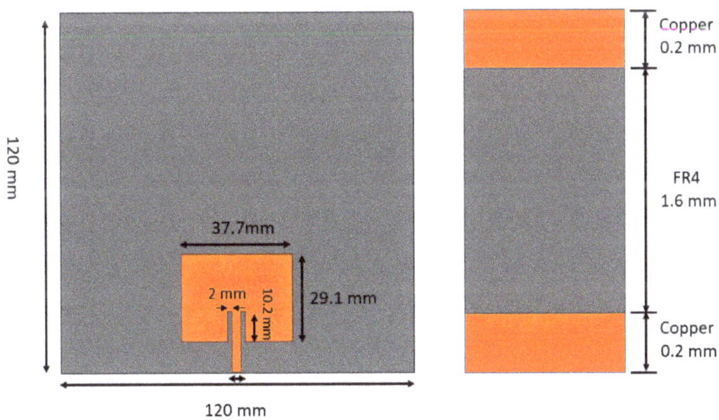

(**a**) (**b**)

Figure 3. Patch Antenna layout: (**a**) Top view; (**b**) Side View.

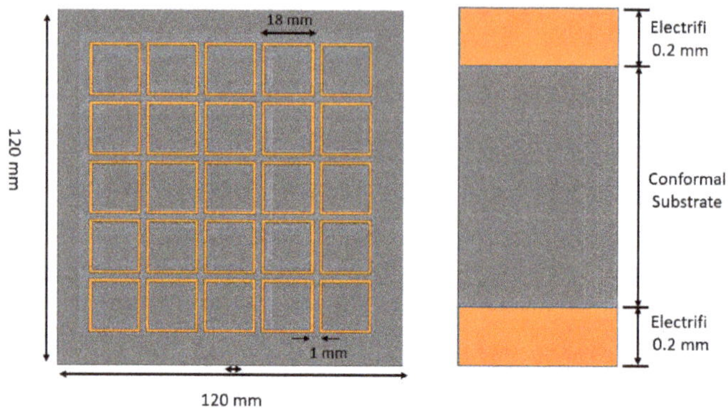

(a) (b)

Figure 4. Conformal metasurface layout: (**a**) Top view; (**b**) Side View.

The layout of the conformal metasurface is presented in Figure 4a. The layout consisted of a 5 × 5 unit cell square with a size of 18 mm. The overall length of the substrate was 120 mm. The gap between each unit square was 1 mm. Three different conformal materials were simulated as substrate: 1. Rogers 5870 conformal metasurface, 2. 3D Printing Ninja Flex conformal metasurface, and 3. Panasonic Felios F775 flexible PCB conformal metasurface. With each material came a different permittivity value and firmness of the substrate. The specific property comparison of the conformal materials is shown in Table 1:

Table 1. Comparison of the conformal material's properties.

Material	Permitivity
Rogers 5870	2.33 [38]
3D Printing Ninja Flex	3.7 [39]
Panasonic Felios F775	3.2 [40]

2.2. HFSS Simulation Result

The proposed design was simulated using the finite element solver HFSS. The conformal metasurface was set to wrap around a cylindrical non model object in HFSS. The curvature of this metasurface was changed by changing the cylindrical object's radius from 43 mm to 88 mm. All other physical design parameters remained the same. A 3D polar plot, a 2D gain polar plot, and a 2D far field gain rectangular plot were generated by HFSS at 2.4 GHz. Three targeted materials: Rogers 5870, 3D Printing Ninja Flex, and Panasonic Felios F775 were used as conformal substrate material, the permittivity of those materials was set to 2.33, 3.7, and 3.2 based on Table 1.

As shown in the Figure 5, for the Rogers 5870 conformal metasurface, the overall gain increased from 1.12 dB to 3.718 dB. The phase difference between the 43 mm radius and 88 mm radius was 25 degrees. Figure 6 shows the simulation results of the Ninja Flex metasurface influence. Ninja Flex is a 3D printing material that can be used as a substrate for an antenna. Previous research [39] demonstrates that this can be used as a substrate for a patch antenna. The overall gain was increased from 1.31 dB to 3.12 dB when the radius changed from 43 mm to 88 mm; and the phase difference between the largest and smallest radius was 19 degrees. Figure 7 shows the simulation results of the Panasonic Felios F775 conformal metasurface influence. The overall gain was increased from 1.26 dB to 2.51 dB when the radius changed from 43 mm to 88 mm.

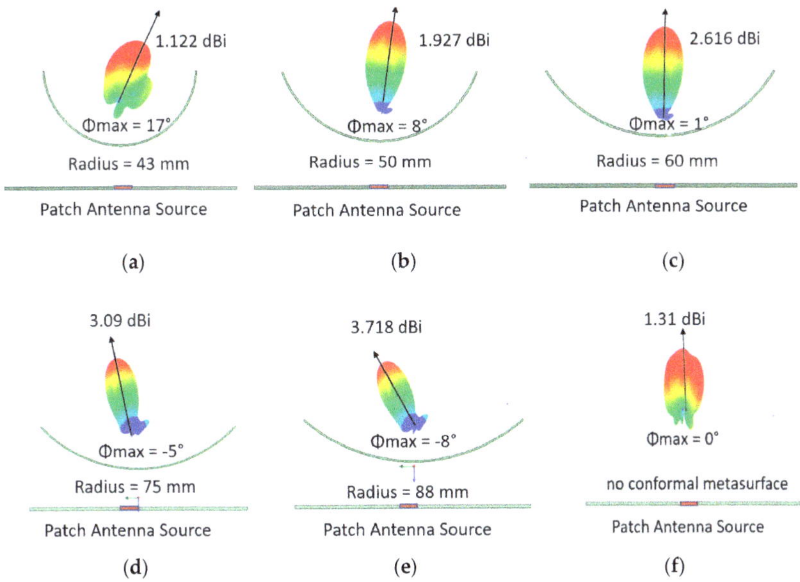

Figure 5. Simulation result: comparison of different radii for Rogers 5870 conformal metasurface on patch antenna at 2.45 GHz. (**a**) r = 43 mm; (**b**) r = 50 mm; (**c**) r = 60 mm; (**d**) r = 75 mm; (**e**) r = 88 mm. (**f**) no conformal metasurface.

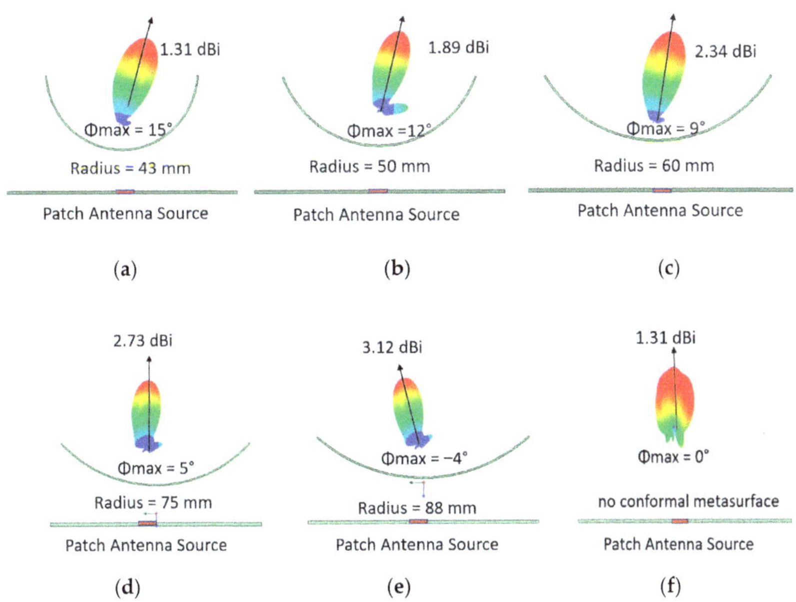

Figure 6. Effect of different radii for Ninja Flex conformal metasurface on patch antenna at 2.45 GHz. (**a**) r = 43 mm; (**b**) r = 50 mm; (**c**) r = 60 mm; (**d**) r = 75 mm; (**e**) r = 88 mm. (**f**) no conformal metasurface.

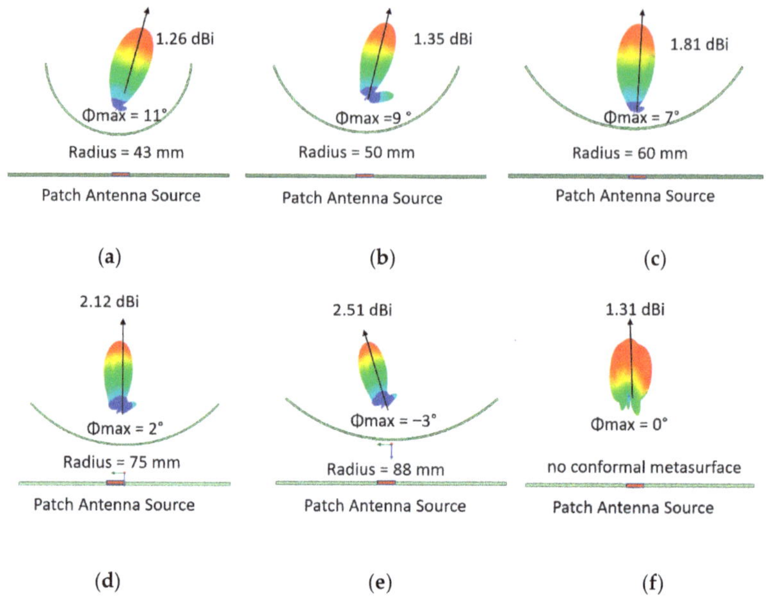

Figure 7. Effect of different radii for Panasonic Felios F775 conformal metasurface on patch antenna at 2.45 GHz. (**a**) r = 43 mm; (**b**) r = 50 mm; (**c**) r = 60 mm; (**d**) r = 75 mm; (**e**) r = 88 mm. (**f**) no conformal metasurface.

Comparing three different materials, it is clear the overall gain increased when the radius of the conformal surface increased. When the conformal metasurface was not placed near the patch antenna, the overall gain was shown to be 1.31 dB, when adding the conformal metasurface, the overall gain increased and phase shifted.

Overall, the Rogers 5870 conformal metasurface was shown to have a better ability to shift the phase while the Panasoic Felios F775 had the lowest in simulation. On the other hand, the Panasoic Felios F775 had the best ability to bend, while Rogers was more difficult to bend than the other two.

3. Fabrication and Measurement

3.1. Fabrication

There are many materials and techniques can be utilized to fabricate metasurfaces. Most of those materials and techniques can be divided into three categories: all-dielectric, printed layers, and all-metal. All-dielectric metasurfaces were fabricated by low loss and high permittivity dielectric structures [41]. It was utilized for many applications such as broadband optical beam splitters [25], electromagnetic band gap (EBG) resonator antennas [42], and ultralow profile lens antenna [43].

A metasurface can also be made of printed materials as explained in [44]. Printed layer metasurface are known for their low cost and fast prototyping time for complex structure. Studies such as [32] demonstrated the wide adaptability for low-cost project. There are also some studies utilizing all-metal structure such as [45]. In this paper, inexpensive thin layers of metal sheet were used to fabricate frequency-selective surface (FSS), and those surfaces could be manufactured in large quantity with low cost.

In this work, the goal is to explore different conformal metasurfaces for passive beam steering. Most applications for passive beam steering require fast prototyping time and relatively low cost. While conformal antennas are easy to manipulate gain by changing their shape, they are harder to manufacture than traditional patch antennas.

In this work, three manufacture methods were utilized to manufacture the conformal metasurface. The Rogers 5870 conformal metasurface was fabricated by laser direct method. The Ninja Flex conformal metasurface was fabricated by 3D printing. The Panasoic Felios F775 is a flex pcb material. For the Rogers 5870 conformal metasurface shown in Figure 8a, a laser-direct engraving method was used to fabricate it. The conductive layer was etched away, leaving only designed square rings on the surface. A LPFK laser engraving machine was used for this printing. The thickness of this antenna was 0.7874 mm (0.031 inch) which made the metasurface conformal. There are some challenges that come with this method. Mainly, the surface area is large, some of the surface area is not totally smooth, but it should not influence the performance of the conformal metasurface.

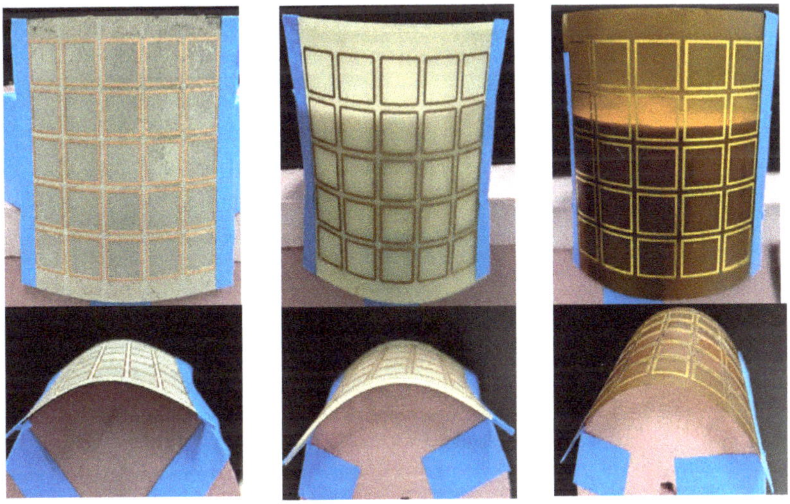

(**a**) (**b**) (**c**)

Figure 8. Fabricated conformal metasurface: (**a**) Laser-direct engraving Rogers 5870; (**b**) 3D printed Ninja Flex; (**c**) The Panasoic Felios F775 flexible PCB.

The Ninja Flex 3D printed conformal metasurface is shown in Figure 8b. Additive manufactured by 3D printing has made a lot of progress recently [46–48]. The growing demands for low-cost and complex 3D structures make 3D printing a great option for many researchers. Ninja Flex is a flexible 3D printing filament which allows for conformal surfaces to be fabricated by a 3D printer. Previous research [39] indicates it can be a good substrate material for conformal surfaces. The permittivity of the material was reported as 3.7 [39]. For the 3D printing process, the infill was set as 100% to get a better result. The squareconductors for designed ring were Electrifi. As discussed in [49], utilizing Ninja Flex and Electrifi to print this design requires great care due to the difference in filament extrusion temperatures. Furthermore, Electrifi conducting filament loses some of its conductivity when dispensed at higher temperatures. However, when a lower print temperature is used, extrusion is inconsistent and may produce voids (open circuits).

To achieve high-quality extrusion at low temperatures, the factory-default print nozzle (0.4 mm) was replaced with a larger 1.0 mm diameter nozzle, and the extrusion temperature was reduced to 160 °C. To print at this lower temperature, g-code command M302 P1 was used to disable the printer's cold extrusion checking.

The Panasoic Felios F775 conformal metasurface is shown in Figure 9c. Compared with the other two materials, this conformal metasurface has the best ability to bend. The prototype was outsourced to Oshpark which is very cost effective. This printed design sample comes with great uniformity for the substrate and shows a great potential.

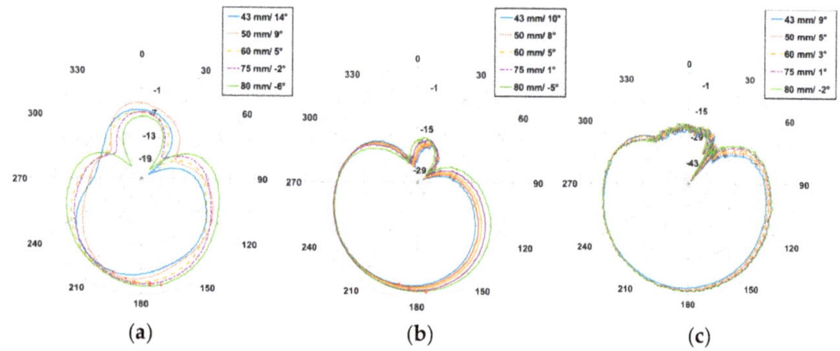

Figure 12. Measured radiation pattern polar plot in: (**a**) Rogers 5870. (**b**) Ninja Flex 3D printing. (**c**) Panasoic Felios F775 flexible PCB.

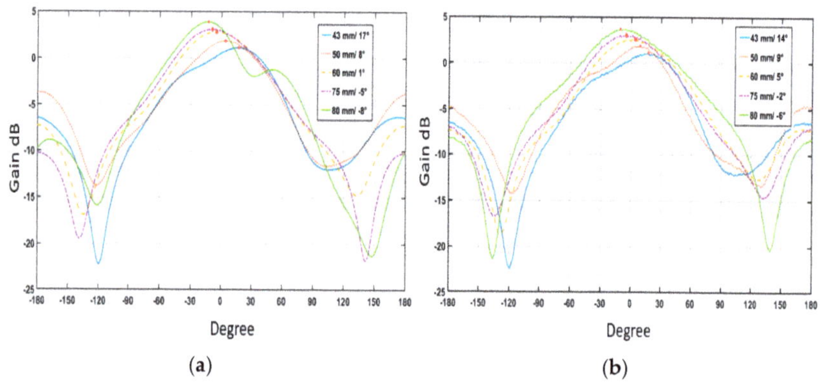

Figure 13. Simulation and measurement results for Rogers 5870 conformal metasurface. (**a**) Simulation. (**b**) Measurement.

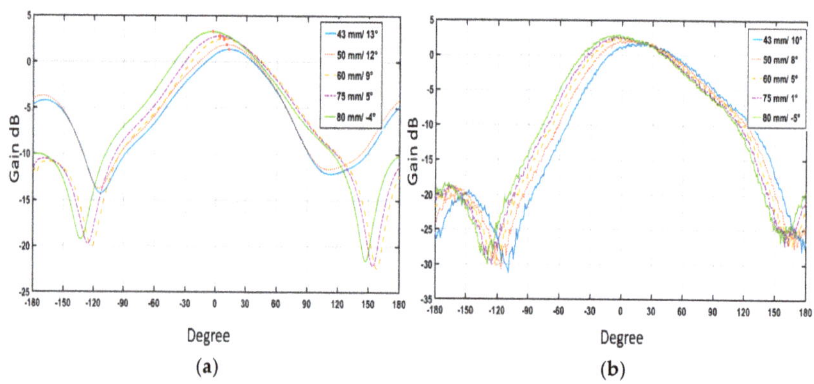

Figure 14. Simulation and measurement results for Ninja Flex 3D printing conformal metasurface. (**a**) Simulation. (**b**) Measurement.

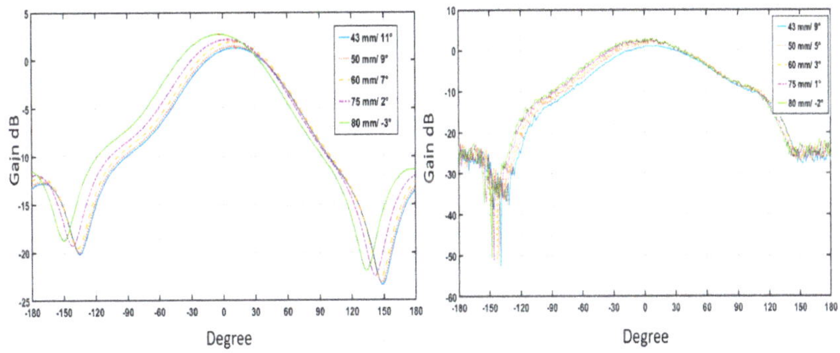

(a) (b)

Figure 15. Simulation and measurement results for Panasonic Felios F775 conformal metasurface: (**a**) Simulation. (**b**) Measurement.

Table 2 shows the gain and phase comparison between each material. The above simulation and measurement results verified the effectiveness of the passive conformal metasurface.

Table 2. Simulation and measurement results for gain and phase.

Material	Radius	Simulated G_{max}	Measured G_{max}	Simulated P_{max}	Measured P_{max}
Rogers 5870	43 mm	1.122 dB	1.21 dB	17°	14°
	50 mm	1.927 dB	1.86 dB	8°	9°
	60 mm	2.616 dB	2.53 dB	1°	5°
	75 mm	3.09 dB	2.98 dB	−5°	−2°
	88 mm	3.718 dB	3.47 dB	−8°	−6°
Ninja Flex	43 mm	1.31 dB	1.38 dB	13°	10°
	50 mm	1.89 dB	1.74 dB	12°	8°
	60 mm	2.34 dB	2.14 dB	9°	5°
	75 mm	2.73 dB	2.76 dB	5°	1°
	88 mm	3.12 dB	2.89 dB	−4°	−5°
Panasoic Felios	43 mm	1.26 dB	1.13 dB	11°	9°
	50 mm	1.35 dB	1.31 dB	9°	5°
	60 mm	1.81 dB	1.74 dB	7°	3°
	75 mm	2.12 dB	2.07 dB	2°	1°
	88 mm	2.51 dB	2.31 dB	−3°	−2°

4. Discussions

The purpose of this work is to experiment different conformal material for beam shifting and gain improvement. The conventional patch antenna used here was a very common 2.4 GHz FR4 patch antenna with low cost. The overall gain without the conformal metasurface was about 1.31 dB. By simply placing a passive conformal metasurface at a fixed distance, the main lobe of the patch antenna was shifted. Furthermore, beam shifting could be changed by increasing the radius of the conformal metasurface. Comparing the measurement results of the three prototypes, the Rogers 5870 substrate had the highest gain improvement of 2.26 dB. The max phase angle also shifted 20° when changing the radius from 43 mm to 88 mm. Ninja Flex had a 1.41 dB gain improvement and 15° beam shifting. Panasoic Felios had a 1.18 dB gain improvement and the max phase angle shifted 11°. On the other hand, Panasoic Felios is the most flexible one among the three fabricated conformal metasurfaces. It is a good option for gain improvement of compact antenna designs. Three different manufacturing methods were explored for each conformal metasurface material. Ninja Flex 3D printing material had a lower cost compared to the

Rogers 5870 substrate making it a good option for fast prototyping. It also came with challenges to adjust temperatures in case losing the conductivity of Electrifi.

Compared with active conformal metasurface, which the beam can shift up to 60 degrees [34], this passive conformal metasurface has narrower beam shifting ability up to 20 degrees. However, active conformal metasurfaces rely on phase control circuitry and beam-forming networks interrelated with active phase shifters. Passive conformal metasurface is more cost effective, simpler to manufacture and easier to integrate.

To improve related research in the future, more complex structure such as antenna array can be added as radiating elements to improve the overall gain. The conformal metasurface can extract as a model for future research. Furthermore, the relationship between the side lobe and radius when bending can be further studied.

5. Conclusions

In conclusion, this work demonstrates that the conformal metasurface can be used to steer the beam from a conventional patch antenna without using active phase shifters, a beam-forming network, or complex structures. The proposed passive conformal metasurface was placed near a conventional patch antenna as a parasitic surface. Three conformal materials were experimented and evaluated to demonstrate the beam steering ability and efficiency. The proposed designs were simulated, fabricated, and measured. The radiation pattern shows up to a 20° phase shift. Furthermore, different fabrication methods were explored. The proposed approach can be further extended to higher frequencies, enabling future work such as low power consumption millimeter-wave beam steering systems.

Author Contributions: Conceptualization, R.G.; Methodology, R.G.; software, R.G.; validation, R.G., R.S.; fabrication, R.S.; measurement, R.G., R.S.; formal analysis, R.G.; investigation, R.G., R.S., B.B.; resources, B.B.; writing—original draft preparation, R.G.; writing—review and editing, R.G., B.B., R.S., supervision, B.B.; project administration, B.B.; funding acquisition, B.B. All authors have read and agreed to the published version of the manuscript.

Funding: This research received no external funding.

Data Availability Statement: Data are available upon reasonable request from the corresponding authors.

Conflicts of Interest: The authors declare no conflict of interest.

References

1. Andrews, J.G.; Buzzi, S.; Choi, W.; Hanly, S.V.; Lozano, A.; Soong, A.C.; Zhang, J.C. What will 5G be? *IEEE J. Sel. Areas Commun.* **2014**, *32*, 1065–1082. [CrossRef]
2. Navarro, A.; Chang, K. *Integrated Active Antennas and Spatial Power Combining*; Wiley: New York, NY, USA, 1996.
3. Yan, W.; Yuan, X.; He, Z.-Q.; Kuai, X. Passive Beamforming and Information Transfer Design for Reconfigurable Intelligent Surfaces Aided Multiuser MIMO Systems. *IEEE J. Sel. Areas Commun.* **2020**, *38*, 1793–1808. [CrossRef]
4. Hu, S.; Rusek, F.; Edfors, O. Beyond Massive MIMO: The Potential of Data Transmission with Large Intelligent Surfaces. *IEEE Trans. Signal Process.* **2018**, *66*, 2746–2758. [CrossRef]
5. Zhang, Q.; Guo, H.; Liang, Y.-C.; Yuan, X. Constellation learningbased signal detection for ambient backscatter communication systems. *IEEE J. Sel. Areas Commun.* **2019**, *37*, 452–463. [CrossRef]
6. Wan, X.; Qi, M.Q.; Chen, T.Y.; Cui, T.J. Field-programmable beam reconfiguring based on digitally-controlled coding metasurface. *Sci. Rep.* **2016**, *6*, 20663. [CrossRef]
7. Zhang, L.; Chen, X.Q.; Liu, S.; Zhang, Q.; Zhao, J.; Dai, J.Y.; Bai, G.D.; Wan, X.; Cheng, Q.; Castaldi, G.; et al. Space-time-coding digital metasurfaces. *Nat. Commun.* **2018**, *9*, 4334. [CrossRef]
8. Pal, A.; Mehta, A.; Goonesinghe, H.; Mirshekar-Syahkal, D.; Nakano, H. Conformal Beam-Steering Antenna Controlled by a Raspberry Pi for Sustained High-Throughput Applications. *IEEE Trans. Antennas Propag.* **2017**, *66*, 918–926. [CrossRef]
9. Hashemi, M.R.M.; Yang, S.-H.; Wang, T.; Sepúlveda, N.; Jarrahi, M. Electronically-Controlled Beam-Steering through Vanadium Dioxide Metasurfaces. *Sci. Rep.* **2016**, *6*, 35439. [CrossRef]
10. Orazbayev, B.; Beruete, M.; Khromova, I. Tunable beam steering enabled by graphene metamaterials. *Opt. Express* **2016**, *24*, 8848–8861. [CrossRef]
11. Cui, T.J.; Qi, M.Q.; Wan, X.; Zhao, J.; Cheng, Q. Coding metamaterials, digital metamaterials and programmable metamaterials. *Light Sci. Appl.* **2014**, *3*, e218. [CrossRef]

12. Liu, S.; Cui, T.J. Concepts, working principles, and applications of coding and programmable metamaterials. *Adv. Opt. Mater.* **2017**, *5*, 1700624. [CrossRef]
13. Zhu, W.M.; Liu, A.Q.; Zhang, W.; Tao, J.F.; Bourouina, T.; Teng, J.H.; Zhang, X.H.; Wu, Q.Y.; Tanoto, H.; Guo, H.C.; et al. Po-larization dependent state to polarization independent state change in THz metamaterials. *Appl. Phys. Lett.* **2011**, *99*, 221102. [CrossRef]
14. Pitchappa, P.; Ho, C.P.; Dhakar, L.; Lee, C. Microelectromechanically reconfigurable interpixelated metamaterial for inde-pendent tuning of multiple resonances at terahertz spectral region. *Optica* **2015**, *2*, 571–578. [CrossRef]
15. Qin, P.-Y.; Song, L.-z.; Guo, Y.J. Beam Steering Conformal Transmitarray Employing Ultra-Thin Triple-Layer Slot Ele-ments. *IEEE Trans. Antennas Propag.* **2019**, *67*, 5390–5398. [CrossRef]
16. Wu, Y.F.; Cheng, Y.J. Proactive Conformal Antenna Array for Near-Field Beam Focusing and Steering Based on Curved Substrate Integrated Waveguide. *IEEE Trans. Antennas Propag.* **2019**, *67*, 2354–2363. [CrossRef]
17. Shi, S.; Bai, J.; Schneider, G.J.; Zhang, Y.; Nelson, R.; Wilson, J.; Schuetz, C.; Grund, D.W.; Prather, D.W. Conformal Wideband Optically Addressed Transmitting Phased Array with Photonic Receiver. *J. Light. Technol.* **2014**, *32*, 3468–3477. [CrossRef]
18. Huang, G.-L.; Li, J.; Yuan, T.; Sim, C.-Y.-D. Recent Progress in Practical Waveguide-Based Antennas and Passive Com-ponents With Additive Manufacturing Technology. In Proceedings of the2018 IEEE Asia-Pacific Conference on Antennas and Propagation (APCAP), Auckland, New Zealand, 5–8 August 2018; pp. 70–71.
19. Zaghloul, A.I.; Kilic, O. Hybrid Beam Former for Distributed-Aperture Electronically Steered Arrays. *IEEE Antennas Wirel. Propag. Lett.* **2010**, *9*, 678–681. [CrossRef]
20. Huang, C.; Pan, W.; Ma, X.; Zhao, B.; Cui, J.; Luo, X. Using Reconfigurable Transmitarray to Achieve Beam-Steering and Polarization Manipulation Applications. *IEEE Trans. Antennas Propag.* **2015**, *63*, 4801–4810. [CrossRef]
21. Singh, K.; Afzal, M.U.; Kovaleva, M.; Esselle, K.P. Controlling the Most Significant Grating Lobes in Two-Dimensional Beam-Steering Systems with Phase-Gradient Metasurfaces. *IEEE Trans. Antennas Propag.* **2019**, *68*, 1389–1401. [CrossRef]
22. Lin, F.H.; Chen, Z.N. Low-Profile Wideband Metasurface Antennas Using Characteristic Mode Analysis. *IEEE Trans. Antennas Propag.* **2017**, *65*, 1706–1713. [CrossRef]
23. Nguyen, Q.; Zaghloul, A.I. Design of Beam Steering Patch Arrays Using Self-Phased Metasurface Pixels. In Proceedings of the 2020 IEEE International Symposium on Antennas and Propagation and North American Radio Science Meeting, Montreal, QC, Canada, 5–10 July 2020; pp. 909–910.
24. Turpin, J.P.; Bossard, J.A.; Morgan, K.L.; Werner, D.H.; Werner, P.L. Reconfigurable and Tunable Metamaterials: A Review of the Theory and Applications. *Int. J. Antennas Propag.* **2014**, *2014*, 429837. [CrossRef]
25. Lalbakhsh, A.; Esselle, K.P. Directivity Improvement of a Fabry-Perot Cavity Antenna by Enhancing Near Field Charac-Teristic. In Proceedings of the 2016 17th International Symposium on Antenna Technology and Applied Electromagnetics (ANTEM), Montreal, QC, Canada, 10–13 July 2016; pp. 1–2.
26. Lalbakhsh, A.; Afzal, M.U.; Esselle, K.P.; Smith, S.; Zeb, B.A. Single-Dielectric Wideband Partially Reflecting Surface with Variable Reflection Components for Realization of a Compact High-Gain Resonant Cavity Antenna. *IEEE Trans. Antennas Propag.* **2019**, *67*, 1916–1921. [CrossRef]
27. Lalbakhsh, A.; Afzal, M.U.; Esselle, K.P. Multiobjective Particle Swarm Optimization to Design a Time-Delay Equalizer Metasur-face for an Electromagnetic Band-Gap Resonator Antenna. *IEEE Antennas Wirel. Propag. Lett.* **2017**, *16*, 912–915. [CrossRef]
28. Das, P.; Mandal, K.; Lalbakhsh, A. Single-layer polarization-insensitive frequency selective surface for beam reconfigurability of monopole antennas. *J. Electromagn. Waves Appl.* **2020**, *34*, 86–102. [CrossRef]
29. Lalbakhsh, A.; Afzal, M.U.; Esselle, K.P.; Smith, S. Design of an Artificial Magnetic Conductor Surface Using an Evolutionary Algorithm. In Proceedings of the 2017 International Conference on Electromagnetics in Advanced Applications (ICEAA), Verona, Italy, 11–15 September 2017; pp. 885–887.
30. Paul, G.S.; Mandal, K.; Lalbakhsh, A. Single-layer ultra-wide stop-band frequency selective surface using inter-connected square rings. *Aeu-Int. J. Electron. Commun.* **2021**, *132*, 153630. [CrossRef]
31. Lalbakhsh, A.; Afzal, M.U.; Hayat, T.; Esselle, K.P.; Mandal, K. All-metal wideband metasurface for near-field transformation of medium-to-high gain electromagnetic sources. *Sci. Rep.* **2021**, *11*, 9421. [CrossRef]
32. Afzal, M.U.; Matekovits, L.; Esselle, K.P.; Lalbakhsh, A. Beam-Scanning Antenna Based on Near-Electric Field Phase Transforma-tion and Refraction of Electromagnetic Wave Through Dielectric Structures. *IEEE Access* **2020**, *8*, 199242–199253. [CrossRef]
33. Kim, Y.-B.; Lim, S.; Lee, H.L. Electrically Conformal Antenna Array with Planar Multipole Structure for 2-D Wide Angle Beam Steering. *IEEE Access* **2020**, *8*, 157261–157269. [CrossRef]
34. Li, H.; Ma, C.; Shen, F.; Xu, K.; Ye, D.; Huangfu, J.; Li, C.; Ran, L.; Denidni, T.A. Wide-Angle Beam Steering Based on an Active Conformal Metasurface Lens. *IEEE Access* **2019**, *7*, 185264–185272. [CrossRef]
35. Psychoudakis, D.; Volakis, J.L. Conformal Asymmetric Meandered Flare (AMF) Antenna for Body-Worn Applications. *IEEE Antennas Wirel. Propag. Lett.* **2009**, *8*, 931–934. [CrossRef]
36. Karthikeya, G.S.; Agnihotri, N.; Siddiq, S.S.; Mehul, K.S.; Thyagaraj, T. A Conformal UHF Antenna for Cargo Helicopter Belly. In Proceedings of the 2016 IEEE 5th Asia-Pacific Conference on Antennas and Propagation (APCAP), Kaohsiung, Taiwan, 26–29 July 2016; pp. 285–286.
37. Isola Group. 370HR. Available online: https://www.isola-group.com/pcb-laminates-prepreg/370hr-laminate-prepreg/ (ac-cessed on 2 December 2021).

38. RT/duroid®5870/5880 High Frequency Laminates. Available online: https://rogerscorp.com/-/media/project/rogerscorp/documents/advanced-electronics-solutions/english/data-sheets/rt-duroid-5870---5880-data-sheet.pdf/ (accessed on 2 December 2021).
39. Striker, R.; Mitra, D.; Braaten, B.D. Permittivity of 3D Printed NinjaFlex Filament for use in Conformal Antenna Designs up to 20 GHz. In Proceedings of the 2020 IEEE International Conference on Electro Information Technology (EIT), Chicago, IL, USA, 31 July–1 August 2020; pp. 224–227.
40. Panasonic Felios R-F775. Available online: https://docs.oshpark.com/resources/flex-substrate-Panasonic-Felios-F775.pdf/ (accessed on 2 December 2021).
41. Monti, A.; Alu, A.; Toscano, A.; Bilotti, F. Surface Impedance Modeling of All-Dielectric Metasurfaces. *IEEE Trans. Antennas Propag.* **2019**, *68*, 1799–1811. [CrossRef]
42. Lalbakhsh, A.; Afzal, M.; Esselle, K.; Smith, S. A High-Gain Wideband EBG Resonator Antenna for 60 GHz Unlicenced Frequency Band. In Proceedings of the 12th European Conference on Antennas and Propagation (EuCAP 2018), London, UK, 9–13 April 2018. [CrossRef]
43. Wang, Z.; Wu, H.; Chen, J.; Wu, Z.; Feng, Y. An ultralow-profile lens antenna based on all-dielectric metasur-faces. In Proceedings of the 2016 IEEE 5th Asia-Pacific Conference on Antennas and Propagation (APCAP), Kaohsiung, Taiwan, 26–29 July 2016; pp. 367–368.
44. Adibi, S.; Honarvar, M.A.; Lalbakhsh, A. Gain Enhancement of Wideband Circularly Polarized UWB Antenna Using FSS. *Radio Sci.* **2020**, *56*, 1–8. [CrossRef]
45. Lalbakhsh, A.; Afzal, M.U.; Esselle, K.P.; Smith, S.L. All-Metal Wideband Frequency-Selective Surface Bandpass Filter for TE and TM polarizations. *IEEE Trans. Antennas Propag.* **2022**, 1. [CrossRef]
46. Zhang, S.; Cadman, D.; Vardaxoglou, J.Y.C. Additively Manufactured Profiled Conical Horn Antenna with Dielectric Loading. *IEEE Antennas Wirel. Propag. Lett.* **2018**, *17*, 2128–2132. [CrossRef]
47. Dorle, A.; Gillard, R.; Menargues, E.; Van Der Vorst, M.; De Rijk, E.; Martin-Iglesias, P.; Garcia-Vigueras, M. Additive Manufacturing of Modulated Triple-Ridge Leaky-Wave Antenna. *IEEE Antennas Wirel. Propag. Lett.* **2018**, *17*, 2123–2127. [CrossRef]
48. Su, Z.; Klionovski, K.; Bilal, R.M.; Shamim, A. A Dual Band Additively Manufactured 3-D Antenna on Package with Near-Isotropic Radiation Pattern. *IEEE Trans. Antennas Propag.* **2018**, *66*, 3295–3305. [CrossRef]
49. Roy, S.; Qureshi, M.B.; Asif, S.; Sajal, S.; Braaten, B.D. A study of microstrip transmission lines on substrates created using additive manufacturing and flexible or semi-rigid filaments. In Proceedings of the 2017 IEEE International Conference on Electro Information Technology (EIT), Lincoln, NE, USA, 14–17 May 2017; pp. 200–205. [CrossRef]

sensors

MDPI

Article

Hardware Efficient Massive MIMO Systems with Optimal Antenna Selection

Shenko Chura Aredo [1,*], Yalemzewd Negash [2], Yihenew Wondie Marye [2], Hailu Belay Kassa [3], Kevin T. Kornegay [3,*] and Feyisa Debo Diba [4]

1 School of Electrical and Computer Engineering, Hawassa University, Hawaasa 05, Ethiopia
2 School of Electrical and Computer Engineering, Addis Ababa University, Addis Ababa 1176, Ethiopia; yalemzewdn@yahoo.com (Y.N.); yihenew.wondie@aait.edu.et (Y.W.M.)
3 Department of Electrical & Computer Engineering, Morgan State University, Baltimore, MD 21251, USA; hailu.kassa@morgan.edu
4 Scool of Electrical Engineering and Computing, Adama Science and Technology University, Adama 1024, Ethiopia; feyisa2006@yahoo.com
* Correspondence: duskaanoo@gmail.com (S.C.A.); kevin.kornegay@morgan.edu (K.T.K.)

Abstract: An increase in the number of transmit antennas (M) poses an equivalent rise in the number of Radio Frequency (RF) chains associated with each antenna element, particularly in digital beamforming. The chain exhibits a substantial amount of power consumption accordingly. Hence, to alleviate such problems, one of the potential solutions is to reduce the number of RFs or to minimize their power consumption. In this paper, low-resolution Digital to Analogue Conversion (DAC) and transmit antenna selection at the downlink are evaluated to favour reducing the total power consumption and achieving energy efficiency in mMIMO with reasonable complexity. Antenna selection and low-resolution DAC techniques are proposed to leverage massive MIMO systems in free space and Close In (CI) path-loss models. The simulation results show that the power consumption decreases with antenna selection and low-resolution DAC. Then, the system achieves more energy efficiency than without low-resolution of DAC and full array utilization.

Keywords: antenna selection; beamforming; Digital to Analogue Conversion; energy efficiency; massive MIMO; mmWave

Citation: Aredo, S.C.; Negash, Y.; Marye, Y.W.; Kassa, H.B.; Kornegay, K.T.; Diba, F.D. Hardware Efficient Massive MIMO Systems with Optimal Antenna Selection. *Sensors* **2022**, *22*, 1743. https://doi.org/10.3390/s22051743

Academic Editors: Naser Ojaroudi Parchin, Chan Hwang See and Raed A. Abd-Alhameed

Received: 5 January 2022
Accepted: 16 February 2022
Published: 23 February 2022

1. Introduction

Massive MIMO (mMIMO) is a large-scale MIMO device that is becoming more common in wireless communications and which scales up traditional MIMO by orders of magnitude [1]. It considers multi-user MIMO in which a base station has hundreds and thousands of antennas supporting multiple single-antenna terminals at the same time and frequency resources.

A device with a large number of antenna elements increases the connection reliability, spectral quality, and radiated energy efficiency. Each antenna element is linked to a single RF chain at the base station, which comprises mixers, analogue-to-digital converters (ADC), and amplifiers [2]. Furthermore, the increase in the number of antennas and associated RF chains at the base station will result in physical restrictions, complexity, and expense [3]. According to [4], RF chains are responsible for approximately 50–80% of a base station's total transceiver power consumption.

The hardware complexity and power consumption of DACs increase exponentially with the number of quantization bits as the Base Station (BS) antenna elements increase. Thus, using a low-resolution DAC is a promising option [5]. The energy consumption of the power amplifier is also influenced by conversion from analogue to digital and digital to analogue (ADC/DAC), phase shifters, and power amplifiers. Though the digital beamforming system provides a high data rate, the energy consumption becomes excessive since the transceiver system uses the same number of antennas as the chains.

In contrast, a hybrid beamforming system uses fewer RF components, which can be used to offer comparable spectral efficiency to a digital beamforming system while being more energy-efficient [6]. Even though hybrid beamforming is the solution as a technique employing a small number of RF chains, cutting down some numbers among the entire array is one of the questions left as an open issue. Due to this, working on low-resolution DAC and antenna selection has been used as one of the power reduction techniques for a system with an extensive array of RF components.

Most recent literature studies have concentrated on performance analysis for large MIMO uplinks using analogue-to-digital converters with limited resolution. In [7], the effect of signal detection schemes on uplink MIMO systems' energy efficiency with low-resolution analogue-to-digital converters were evaluated. There have been an increasing number of studies for the case of downlink transmission with low-resolution DACs. The Energy Efficiency (EE) of hybrid transmitters with DACs quantized based on additive quantization noise was explored in [8–11].

Consequently, a sub-optimal method was utilized to build an optimum hybrid precoder based on the Additive Quantization Noise Model (AQNM). It also compares quantized digital precoders to hybrid ones with a wholly or partially linked phase-shifting network of active/passive phase-shifters. The challenges of downlink precoding for multi-user MIMO on a narrow-band system with low-resolution DACs at a BS are investigated in [9].

Nonetheless, most researchers have proposed low-resolution DAC for hybrid beamforming, where limited baseband units are used and with low power demand; there should be an equivalent solution for digital beamforming. Since digital beamforming is known for its high capacity at the expense of increased power consumption, we propose antenna selection with low-resolution DAC as a viable option for addressing the inherent hardware complexities and power consumption.

Over the last few decades, various antenna selection techniques and algorithms have been investigated for the classic MIMO. In [12], basic selection algorithms for realistic detectors were used to examine error rate-based performance evaluations. The studies in [13–15] promoted capacity-oriented selection criteria like the greedy algorithm and convex optimization. The authors in [16] presented an antenna selection technique (AS) with a minimal level of complexity that picks antennas that minimize constructive user interference. When the transmitter uses precoders in conjunction with a matched filter, the suggested AS algorithm outperforms systems that use a more complicated channel inversion method. The work in [17,18] aimed to remove the destructive portion of the interference, which was established by the connection between the substreams of a modulated Phase Shift Keying (PSK) scheme.

According to the authors in [19], singular value decomposition was utilized to offer a new Euclidean Distance based Antenna Selection technique (EDAS) for antenna selection in spatial modulation systems that has lower computational complexity than exhaustive search. Furthermore, the Symbol Error Rate (SER) approaches a full search when the number of received antennas grows. Therefore, in comparison with the past and current research trends, the authors of [20] stated that there is still considerable interest in mmWave-based massive MIMO antenna selection with manageable complexity, more energy efficiency, and optimal data rates in recent years.

In this paper, a system with transmit antenna selection for massive MIMO-enabled BS is considered after low-resolution DAC is applied. The procedure is divided into three parts: First, the EE of an entire array device is evaluated at the cell edge using a fixed power allocation technique. In this case, the optimal number of BS antennas ($M^{\star}$) at which the EE reaches its maximum is determined among the total number of BS antennas (M) using the initial access condition. Second, the minimum Signal-to-Noise Ratio (SNR) to be found at the cell edge is used as a threshold value to search the optimal number further when users move from the cell edge to outskirts or centre positions. In this scenario, $M^{\star}$ is considered to be a maximum number of elements.

Due to the position changes, $M^{\star}$ is transformed to (M^{o}), representing the number of selected antennas while the transmit power adaptively changes due to distance changes in mobility. Following the determination of M^{o}, the subset of antennas with the best channel conditions are chosen, and EE is assessed using spatial selectivity at sub-6 GHz and mmWave frequency ranges. Finally, EE is evaluated by integrating a selection algorithm with a low-resolution DAC.

The main contributions of this paper are stated as follows:

- We introduce an energy-efficient downlink antenna selection technique for mobile and static users. The proposed technique considers two-phase selection:
  1. Optimal number of BS antennas ($M^{\star}$) at which the energy efficiency graph becomes maximum and starts declining, is determined as, $M^{\star} = M(EE_i = EE_{max})$. For this, the following assumptions are used:
     - A maximum number users a BS can support is assumed, and all users are to be at the cell edge distances.
     - All BS antennas and RF components are employed to determine total downlink power consumption according to (34).
     - The channel is assumed to be random, and we consider fixed SNR ($\bar{\gamma}_k$), which is the average of least SNR values from several random channel generations for cell edge as in (18). A minimum SNR value is considered to accommodate the worst-case in which the channel is in deep fading.
  2. Next, user mobility-based selection is made. In this case, our selection algorithm incorporates the exhaustive searching method to select a group of elements with the best channel gain as in (21) and (23). The double section also reduces the number of search combinations and computational complexity. Again, since double selection using algorithms one and two minimizes the number of RF components directly associated with the antenna elements in the case of digital beamforming, the power consumption is substantially reduced and makes the system energy efficient.

     In comparison to prior methods, our proposed algorithm lowers the computational complexity of the transceiver system.
- We design a heuristic and simple formulation of antenna selection to evaluate the performance for mMIMO at sub-6 GHz and mmWave bands with CI and FS path-loss models.
- We introduce an energy-efficient and optimal DAC resolution algorithm for massive MIMO systems.
- Finally, by integrating our novel algorithms, the effect of selection on the EE was evaluated with low resolution and typical DAC.

The rest of the work is structured as follows: A system model for mMIMO beamforming and array geometry is defined in Section 2. After the propagation model is explained in Section 3, antenna selection and power consumption models are followed in Sections 4 and 5, where results and analysis are presented. Finally, our conclusions are drawn in Section 6.

2. System Model and Description

A downlink massive MIMO system with digital beamforming is considered. As shown in Figure 1, the system consists of digital switches that are associated with each BS antenna serving simultaneous users with multiple data streams. Both DAC resolution and antenna selection are performed on the basis of digital beamforming. As shown in Figure 1, the BS has a Uniform Rectangular Array (URA) geometry with $\lambda/2$ spaced M antennas, where λ is the signal wavelength, and mutual coupling between antenna elements is ignored. Inside the cell, k single-antenna devices transmit data to the BS simultaneously using the same time-frequency resources.

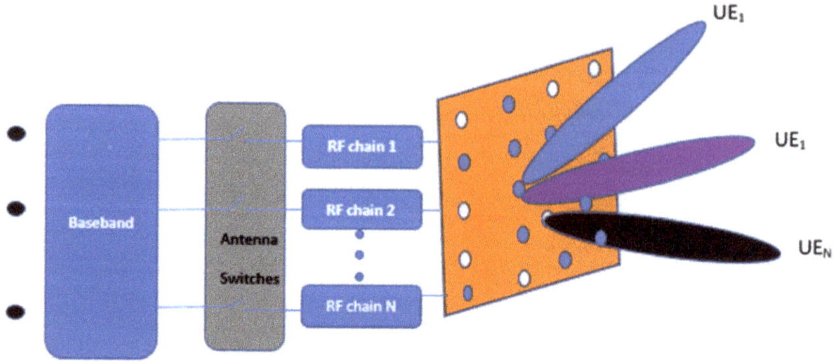

Figure 1. Digital beamforming URA configuration.

3. Propagation Model and Analysis

Unlike typical low-frequency channels, mmWave channel propagation characteristics are no longer affected by Rayleigh fading, and its spatial selectivity is restricted due to significant path loss in free space [20]. As a result, the channel vector for $\hat{w}$th user and $\hat{m}$th antenna element is given by the product of Equations (2)–(5) as follows:

$$h_{\hat{w},\hat{m}} = \vec{G}\Omega\widetilde{A}\varpi, \tag{1}$$

where Equations (2)–(5) are as stated below.

$$\vec{G} = \sqrt{p_k}\begin{bmatrix} \hat{\mathcal{F}}_{rx,\hat{w},\bar{\beta}}(\bar{\beta}_k, ZoA, \delta_k, AoA) \\ \hat{\mathcal{F}}_{rx,\hat{w},\bar{\beta}}(\bar{\beta}_k, ZoA, \delta_k, AoA) \end{bmatrix}^T \tag{2}$$

$$\Omega = \begin{bmatrix} e^{j\Phi_k^{\bar{\beta}\bar{\beta}}} & \sqrt{\Theta_k^{-1}}e^{j\Phi_k^{\bar{\beta}\delta}} \\ \sqrt{\Theta_k^{-1}}e^{j\Phi_k^{\bar{\beta}\delta}} & e^{j\Phi_k^{\phi\delta}} \end{bmatrix} \tag{3}$$

$$\widetilde{A} = \begin{bmatrix} \hat{\mathcal{F}}_{tx,\hat{m},\bar{\beta}}(\bar{\beta}_k, ZoD, \delta_k, AoD) \\ \hat{\mathcal{F}}_{tx,\hat{m},\phi}(\bar{\beta}_k, ZoD, \delta_k, AoD) \end{bmatrix} \tag{4}$$

$$\varpi = e^{(2\pi\lambda^{-1}(\hat{r}^T_{rx,k}\cdot\hat{d}_{rx,\hat{w}}))}e^{(2\pi\lambda^{-1}(\hat{r}^T_{rx,k}\cdot\hat{d}_{tx,\hat{m}}))} \tag{5}$$

3.1. Channel Model

The power associated with the kth user terminal is denoted by p_k, and the field patterns for $\hat{m}$th BS antenna and $\hat{w}$th user terminal in the direction of elevation angle, δ and azimuth angle, $\bar{\beta}$ are given by $\hat{\mathcal{F}}_{tx,\hat{m},\delta}$, $\hat{\mathcal{F}}_{tx,\hat{m},\bar{\beta}}$, $\hat{\mathcal{F}}_{rx,\hat{w},\delta}$, and $\hat{\mathcal{F}}_{rx,\hat{w},\bar{\beta}}$ for both elevation and azimuth directions, respectively. The arrival–departure of elevation and azimuth angles are represented as $\delta_k, AoA/\delta_k, AoD$, and $\bar{\beta}_k, ZoA/\bar{\beta}_k, ZoD$ for elevation, and $\bar{\beta}_k, ZoA$, $\bar{\beta}_k, ZoD$, δ_k, AoA, and δ_k, AoD for azimuth locations and Θ_k in (3) is the cross-polarization energy ratio.

For four distinct polarization combinations, the random starting phases are $\hat{\Phi}^{\bar{\beta}\bar{\beta}}$, $\hat{\Phi}^{\bar{\beta}\delta}$, $\hat{\Phi}^{\delta\bar{\beta}}$ and $\hat{\Phi}^{\delta\delta}$. The spherical unit vectors of the transmitter and receiver, given in Cartesian coordinates, are denoted by the symbols $\hat{r}_{tx,k}$ and $\hat{r}_{rx,k}$ where $\hat{r}_{i,k,\ell} = [sin\bar{\beta}cos\delta \ sin\bar{\beta}sin\delta \ cos\bar{\beta}]^T$ for $i \in [tx, rx]$. The position vectors of the transmit and receive antenna components are $\hat{d}_{tx,\hat{w}}$ and $\hat{d}_{rx,\hat{w}}$, respectively, and therefore the uplink channel vector for a single terminal is formulated as

$$h_{\hat{w}} = [h_{\hat{w}1} \ h_{\hat{w}2} \ldots h_{\hat{w}\hat{m}}]. \tag{6}$$

The index k represents a user, and w represents the index of the antenna element of kth user. Since a user has only a single antenna in multi-user MIMO or massive MIMO systems, wth user antenna element can be used for the kth user. However, a user terminal can have more than one antenna in LTE or classical MIMO systems. Hence, the wth antenna of a user is the same to kth user.

3.2. Array Steering Vector

The array steering vectors for uplink and downlink are provided by $\nu_r(\bar{\beta}^r_k, \delta^r_k)$ and $\nu_t(\bar{\beta}^t_k, \delta^t_k)$, respectively. According to the URA antenna geometry of Figure 1, the azimuth-elevation plane is also given by

$$\hat{A}_{URA}(\bar{\beta}, \delta) = \hat{\upsilon}_{az}(\delta, \bar{\beta})\hat{\upsilon}_{el}(\delta, \bar{\beta})^T, \tag{7}$$

where $\nu_{az}(\delta, \bar{\beta})$ and $\nu_{el}(\delta, \bar{\beta})$ are steering vectors. The maximized uplink received signal due to beamforming weight and the general forms of both planes are stated in (8) and (9).

$$\varphi_k = \nu_k(\bar{\beta}, \delta)x = \sum_{k=1}^{K} \varphi_k \hat{s}_k \tag{8}$$

The general form of both azimuth and elevation planes can further be expressed as [21]

$$\hat{\upsilon}_i(\delta, \bar{\beta}) = \frac{1}{\sqrt{M_i}}[1, e^{j\varrho_i}, \ldots, e^{j(M_i-1)\varrho_i}]^T, i \in [az, el]. \tag{9}$$

where M_i represents the number of BS antennas in the azimuth and elevation planes, $\varrho_{az} = \xi d sin(\bar{\beta}) cos(\delta)$, $\varrho_{el} = \xi d sin(\bar{\beta}) sin(\delta)$, $\xi = 2\pi/\lambda$, d is the distance between the elements and λ is wave length of the signal for M number of antennas. The array form of URA steering vector's component is $\frac{1}{M_{az}M_{el}}$, where M_{az} and M_{el} are the azimuth and elevation directions of the superimposed signal components of the antenna array, respectively.

$$\nu(\delta, \bar{\beta}) = f_1[1, \ldots, e^{2j\varrho_{az}}, \ldots, e^{j(M_{az}-1)\varrho_{az}+(M_{el}-1)\varrho_{el}}]^T, \tag{10}$$

where $f_1 = \frac{1}{M_{az}M_{el}}$. In digital beamforming, a single radio frequency chain is required for each antenna element, resulting in high power consumption and complex architecture, where an RF chain includes a down-converter, low-noise amplifier, digital to analogue converter, analogue to digital converter, and others. Therefore, for many BS antennas, the power consumption of mixed-signal components, such as high-resolution DACs and ADCs is higher, which makes the assignment of a separate RF chain for each antenna highly inefficient.

Though hybrid beamforming is one way to attain this goal, by deploying beamforming in both the digital and analogue schemes, its capacity is less than digital. Since beamforming is accomplished at the baseband frequency in digital [22], this paper aims to reduce power overheads in digital beamforming due to several RF chain components, which account for a large amount of power consumption in cellular communication.

Figure 1 presents a digital beamforming transmitter with RF URA configurations. As shown in the figure, equal number of RF chains and number of antennas are present in digital beamforming. In every symbol duration, the vector form of the data symbol can be stated as $\ddot{d} = [\ddot{d}_1, \ddot{d}_2, \ldots, \ddot{d}_k]$. Then, multiplying kth user symbol by its beamforming vector $b_k \in \mathbb{C}^{N\times 1}$ to form a transmit vector that corresponds to the kth user as $x_k = b_k\ddot{d}_k$, where N is the number of RF chain components. Finally, the added transmitted matrix, $\bar{\ddot{d}} = \sum_{i=1}^{k} x_i$, pass through the RF chain components consisting of a DAC for converting the digital samples to analogue signals and transmitting them to all antennas via power amplifiers.

At the transmitter, several DACs are utilized to convert the transmission to the analogue domain. For the quantization noise of the DACs, a linear model approximation is adopted as [6]

$$Q(u_i) = \sqrt{1 - \rho b_i} + \epsilon_i, \quad (11)$$

where $Q(.)$ is a uniform scalar quantization for the ith RF chain and $\rho b_i = \frac{\pi\sqrt{3}}{2}2^{-2b_i}$ is the quantization distortion parameter for bit resolution equal to b_i. The input u is assumed to be Gaussian distributed, ϵ_i is the quantization noise independent of u, and $\epsilon \sim CN(0, \sigma^2)$, where σ^2 is the noise variance. Extending the argument to the massive MIMO scenario, this is approximated as

$$Q(u) \approx \; cu + \epsilon, \quad (12)$$

where c is a diagonal matrix with values based on each RF chain's DAC resolution, and $\epsilon \in CN(0, \sigma^2 I)$, where I is $N \times N$ identity matrix. Approximating the linear quantizer, we may find $x_k \in C^{MN \times 1}$, and, as a result, the broadcast signal can be stated as follows:

$$x = H_N Q(u) = H_N(cu \; + \; \epsilon) = H_N \; cu + H_N \; \epsilon, \quad (13)$$

where M and N are the total numbers of antennas and RF chain components, respectively, and H_N is the channel matrix due to the number of antennas equal to the number of RF chains, N. The total power consumption of DAC as a function of bits is expressed as

$$P_{dac}(b) = c_1 f b + c_2(2^b), \quad (14)$$

where c_1 and c_2 are static and dynamic power consumption of DAC, and f and b are operating frequency and random number of bits, respectively.

3.3. Signal Model

Before user equipment establishes a physical connection with a BS, an initial access procedure uses reference signals. These reference signals are used for channel estimation and equalization. For our antenna selection process, we use Sounding Reference Signal (SRS) to be transmitted from a mobile terminal(uplink). SRS is used to provide the channel characteristics of a user to a BS so that the BS uses this information to allocate resources for user terminals.

In (15), the uplink reference signal that is received by a BS is represented as

$$Y_j = \sum_{i=1}^{k_s} h_{ij} x_i w_i + z_j, \quad (15)$$

where j represents any antenna element on the BS, k_s is the number of signal sources, h_{ij} is the channel between k and j, element z_j is additive noise at the BS, and w_i is beamforming weight. After channel estimation is done using SRS of the uplink, a BS starts sending data to user equipments through physical downlink shared channel. Therefore, the downlink signal to be received at the uplink is given by

$$y_k = \sqrt{\frac{p_t M}{K}} H_k W_k x_k + \sqrt{\frac{p_t M}{K}} \sum_{i=1, j \neq k}^{K} H_k W_i x_i + n_k, \quad (16)$$

where the first term in the right side is the desired signal, the second term is multi user interferences, and the third is user's additive noise. H_k, W_k and x_k are the channel matrix, beamformer weight and downlink signal corresponding to k user, respectively.

Therefore, disregarding the interference part in Equation (16) due to assumption of a single BS designed with precoding or digital beamforming, the capacity at the cell edge that is to be maintained as a threshold is given by

$$\varsigma_{th} = B\sum_{k=1}^{K} log_2(1 + \frac{p_r|H_k W_k|^2}{N_o B}). \quad (17)$$

where p_r is the received signal power, and, considering large and small scale fading, it can be evaluated as

$$\varsigma_{th} = B\sum_{k=1}^{K} log_2(1 + \frac{p_t|H_k W_k|^2}{\Gamma(D_{max})N_o B}) = B\sum_{k=1}^{K} log_2(1 + \bar{\gamma}_k) \quad (18)$$

where $\bar{\gamma}_k$ is the minimum SNR as a function of the total transmit power of a BS and path loss at the cell edge ($\Gamma(D_{max})$ (which is calculated as in (19)), N_o is the noise spectral density, and B is the channel band width. This threshold capacity also depends on $|H_k W_k|^2$, which changes with the number of users and BS antennas.

3.4. Mobile Location and Positioning

Identifying a mobile position in today's cellular networks is a critical problem. The angle of Arrival (AoA), Time of Arrival (ToA), and Global Positioning System (GPS) are among the techniques used. In general, there are three methods for determining the location of a mobile terminal: satellite positioning, cellular network-based positioning, and indoor positioning. The trilateration method is used to calculate a mobile's location using a relative position of a BS. Unlike the triangulation process, which requires the angle of each user for position tracking, only the distance between the BS and each user is required in this case [23].

3.5. Close-In (CI) Path Loss Model

The CI model is based on Friis and Bullington's fundamental radio propagation concepts, wherein the values are 2 for free space and 4 for the asymptotic two-ray ground bouncing model. This provides insight into path loss as a function of the environment since base station towers are tall and inter-site distances for specific frequency bands are several kilometres. Previous Ultra High Frequency (UHF)/microwave models employed a close-in standard distance of 1 km or 100 m [24].

The CI 1 m reference distance, as proposed in [25], is a suitable recommended norm that relates the real transmit power or PL to a usable distance of 1 m. Standardization to a 1 m reference distance simplifies dimension and model comparisons, provides a consistent description of the Path Loss Exponent (PLE), and allows for quick and straightforward route loss estimates without the need for a calculator [26].

Using power control mechanisms, user terminals nearer to the BS are allocated lower power than those on the outskirts to control interference and fairness. The CI path-loss model is a generic frequency model that explains large-scale path loss at all applicable frequencies in a specific context. The dynamic range of signals perceived by users in a commercial system becomes significantly lower than the equation for the CI model, which is formulated as [24]

$$PL_{CI}(.)_{dB} = PL_{FS}(f, 1m) + 10nlog_{10}(d) + \chi_{\sigma}^{CI}, \quad (19)$$

where $n = \frac{\sum(\bar{D}A)}{\sum(\bar{D}^2)}$, $A = PL_{CI}(.)_{dB} - PL_{FS}(f, 1m)$, $\bar{D} = 10log_{10}d$ denotes a single model criterion, the PLE, d is the transmitter-receiver separation distance d starting from 1 m, and

(.) represents frequency and distance parameters. The free space path loss, $PL_{FS}(f,1)_{dB}$ at a 1 m distance from a station and carrier frequency f is given as

$$PL_{FS}(f,1m)[dB] = 20log10(4\pi/\lambda), \tag{20}$$

where λ is the wavelength of the signal. The CI model includes an intrinsic frequency interdependence of path loss in the 1 m PL_{FS} value, and it only has one parameter compared to the Alpha-Beta-Gamma (ABG) or (α, β, and γ) model, where α and γ are coefficients showing the dependence of path loss on distance and frequency, respectively, and β is an optimized offset value for path loss in dB. $\hat{\sigma}^{CI} = \sqrt{\Sigma \chi_{\hat{\sigma}}^{CI^2}/T}$, where T is the number of data points and $\chi_{\hat{\sigma}}^{CI}$ is large-scale signal fluctuations due to the CI pathloss model.

4. Antenna Selection and Power Model

4.1. Antenna Selection

The number of antennas to be chosen is determined by adjusting an appropriate amount of transmit power to be radiated during the selection process. The number of antenna elements to be selected accounts for the system complexity [27–29].

Trilateration is used to pinpoint a user's position so that the main beam can focus only on the target region, thus, reducing leakage. The transmit power adapts the user's location changes as a function of distance due to user mobility. In this case, instead of using all the arrays and wasting resources, a certain number of transmit antennas can be decreased adaptively as user nodes become closer to the centre of the BS. For this, we consider the minimum SNR at the cell edge as a threshold value. In contrast to when allocating maximum transmit power based on edge distance, the BS only allocates power proportional to the reduced distance, resulting in only a few antennas being activated. Therefore, the first optimal number of antennas for cell-edge users $(M^{\star})$ is selected as follows

$$M^{\star} = \max_{\iota \in M,\ \varsigma \in [\varsigma_{th}, \varsigma_{max}]} (EE(M(\iota))) \tag{21}$$

where

$$EE(M(\iota)) = \frac{log_2\Big(real(det(I + (\frac{\bar{\gamma}_k}{M(\iota)})))\Big)}{\wp_{tot}(\iota)}, \tag{22}$$

and $\wp_{tot}(\iota)$ is found using (34) and $\iota \in [1, M]$, where $\wp_{tot}(\iota)$ and $M(\iota)$ are the total power and total number of antennas as a function of each iteration. Next, adaptive selection follows as the users move from cell edge to cell centre or the outskirts and given by

$$M^{o} = \frac{\Big(\Sigma(\Gamma(r))/k\Big)M^{\star}}{\Gamma(D_{max})}. \tag{23}$$

Substituting for $\Gamma(D_{max})$ from (18),

$$M^{o} = \frac{\Big(\Sigma(\Gamma(r))/k\Big)M^{\star}\bar{\gamma}_k N_o B}{p_t|H_k W_k|^2}. \tag{24}$$

Finally, using factorial permutations as $\binom{M^{\star}}{M^{o}} = \frac{M^{\star}!}{M^{o}!(M^{\star} - M^{o})!}$, the antennas with the best channel gains are chosen from the list, and this minimizes complexity as compared to $\frac{M!}{M^{o}!(M - M^{o})!}$.

The terms $\Gamma(r)$ and $\Gamma(D_{max})$ in (23) refer to the CI path loss of users at any arbitrary distance and maximum distance, respectively. The number of RF chains N becomes the same as the number of antennas to be selected M^o in the case of changes in channel condition or user position, in this case, in particular for digital beamforming. The selection process for the whole system is stated in algorithms one and two below.

In addition to selecting branches with the best channel gain and pilot test at the initial access, this section also aims to demonstrate the computational complexity at the receiving end in floating operation points (flops). The complexity order $\mathcal{O}$, which is the number of times the capacity evaluation with the best channel gains, is formulated as $\mathcal{O}\left(\binom{M}{M_{selected}}\right)$ and its complexity level is shown in Table 1.

Table 1. Complexity level measurement of the algorithms.

Algorithm 2	Algorithm 1 + 2	Algorithm 1 + 2 + 3
$\left(\hat{n}\binom{M}{M^o}\right)$	$\left(\hat{n}\binom{M}{M^o-\ell^*}\right)$	$\left(\hat{n}\binom{M}{M^o-\ell^*}\right)+\sqcup$

The table states the combinational permutation of the algorithms, which we compare with that of [21], which accounts for $\hat{n}\binom{M}{M^o}$, where $\hat{n}=M^2+2M^{o2}+M^o$ and ℓ^* and $\sqcup$ are the deducted elements due to selection and additional iterations due to incorporation of low resolution DAC, respectively. According to [28,29], the computational complexity due to the selection process is shown in (27) and (28), respectively.

$$\mathcal{O}_1(.)=16n^3+\hat{n}^2(24M^2+40M+24-24M^{o2}-24M^o), \tag{25}$$

$$\mathcal{O}_2(.)=\hat{e}+20(M^2+M-M^{o2}-M^o), \tag{26}$$

$$\mathcal{O}_3(.)=\mathcal{O}_1(.)+\mathcal{O}_2(.), \tag{27}$$

$$\mathcal{O}_4(.)=\hat{n}(M^{o2}(M^3(M+1)), \tag{28}$$

where $\hat{e}=\hat{n}(34M^2+44M-36M^{o2}-34M^o)$ and (.) denotes (M,M^o).

The above algorithm selects the optimal number of antennas considering cell edge users at initial access utilizing the reference signal. The reference signal is only used for connection setup and determining the optimal number of antennas for static users at the cell edge. Therefore, the optimal number of antennas M^* to be used is found using at a point in which the EE graph starts declining. All the components use only a limited amount of power, which is assumed to be 15% of its connection power.

4.2. Capacity and Power Consumption Model

According to [19], the down-link sum capacity is given by

$$\varsigma_{full}=log_2\left[det\left(I+\frac{\bar{\gamma}}{M}HH^{\ddagger}\right)\right], \tag{29}$$

where $\bar{\gamma}$ is the average SNR per receiver, I is the $k\times k$ identity matrix, superscript $\ddagger$ denotes the Hermitian transpose, and H is the channel matrix of entire antenna array elements. The antennas that maximize the capacity are now selected so that

$$\varsigma_{sel}=\max_{s(\bar{H})} log_2\left[det\left(I_k+\frac{\bar{\gamma}}{M^o}\bar{H}\bar{H}^{\ddagger}\right)\right], \tag{30}$$

where $\bar{H}$ is created by deleting $M-M^o$ rows from H, and $s(\bar{H})$ denotes the set of all possible $\bar{H}$ (whose cardinality is $\binom{M}{M^o}$), and K is the total number of user terminals.

$$\varsigma_{DPC_\iota} = \max_{P_\iota} \log_2 det\left(I + \bar{\gamma}k(\bar{H}\ddagger^{(M^o)})P_\iota \bar{H}^{(M^o)}\right), \quad (31)$$

where $P\iota$ denotes a fraction of the transmit power for a single antenna element, and K is the total number of user terminals. Throughput will be further enhanced with Dirty Paper Coding (DPC) by selecting antennas with better gains after iteration over all possible combinations. In the process of selecting those columns from a full array of H_ι, an $M \times M$ matrix $\bigtriangledown$ with a binary diagonal element is introduced.

$$\bigtriangledown_i = \begin{cases} 1, & \text{Select} \\ 0, & \text{otherwise,} \end{cases} \quad (32)$$

showing whether the ith element is chosen and achieving $\sum_{i=1}^{L} = M^o$. According to Sylvester's identity, det $(I + UJ) = det(I + JU)$, and the DPC total rate can be rewritten as in (31) as

$$\varsigma_{DPC_\iota} = \max_{P_\iota} \log_2 det(1 + \bar{\gamma}kP_\iota \bar{H}_\iota \bigtriangledown (\bar{H}_\iota)^\ddagger), \quad (33)$$

where $\sum_{i=1}^{k} P_{\iota,i} = 1$. The desired $\bigtriangledown$ is discovered by increasing the average DPC sum rate.

We can categorize the total power before and after low resolution DAC and selection as

$$\wp_{tot} = P_{LO} + P_{PA} + M(2P_{DAC} + 2P_M + 2P_{LF} + P_{HB}) \quad (34)$$

$$\vec{\wp}_{tot} = P_{LO} + P_{PA} + M(2\vec{P}_{DAC} + 2P_M + 2P_{LF} + P_{HB}) \quad (35)$$

$$\wp^o_{tot} = P_{LO} + P_{PA} + M^o(2P_{DAC} + 2P_M + 2P_{LF} + P_{HB}) \quad (36)$$

$$\vec{\wp^o}_{tot} = P_{LO} + P_{PA} + M^o(2\vec{P}_{DAC} + 2P_M + 2P_{LF} + P_{HB}) \quad (37)$$

where $\wp_{tot}$, $\vec{\wp}_{tot}$, $\wp^o_{tot}$, and $\vec{\wp^o}_{tot}$ are the total power before both low resolution and selection, after low resolution and no selection, with selection and no resolution, and with both selection and low resolution, respectively. Again, P_{LO} denotes the power consumption of the local oscillator, and P_{PA} is the power consumption of the power amplifier. P_M, P_{LF}, and P_{HB} are the mixer power, low pass filter power, and hybrid with buffer power, respectively.

$$EE_{FA} = \left(\frac{\varsigma}{\wp_{tot}}\right)_M \approx \frac{log_2\left[det\left(I + \frac{\bar{\gamma}}{M}HH^\ddagger\right)\right]}{\wp_{tot}}. \quad (38)$$

$$EE_{Se} = \left(\frac{\varsigma}{\wp_{tot}}\right)_{M_{selected}} \approx \frac{log_2\left[det\left(I + \frac{\bar{\gamma}}{M^{o/*}}HH^\ddagger\right)\right]}{\wp^{selected}_{tot}}. \quad (39)$$

$$EE_{alg_{1+2}} \approx \frac{log_2\left[det\left(I + \frac{\bar{\gamma}}{\left(\frac{M\sum_{i=1}^{k}\alpha_t/k}{p_t}\right)}HH^\ddagger\right)\right]}{\wp^o_{tot}}. \quad (40)$$

$$EE_{alg_{1+2+3}} \approx \frac{log_2\left[det\left(I + \frac{\bar{\gamma}}{\left(\frac{M\sum_{i=1}^{k}\alpha_t/k}{p_t}\right)}HH^\ddagger\right)\right]}{\vec{\wp^o}_{tot}}. \quad (41)$$

$$log_2\left[det\left(I + \frac{\bar{\gamma}}{(\frac{M\sum_{i=1}^{K}\alpha_t/k}{p_t})}HH^{\ddagger}\right)\right]$$
$$= \frac{}{P_{LO} + P_{PA} + M^o(2\vec{P}_{DAC} + 2P_M + 2P_{LF} + P_{HB})}. \quad (42)$$

$$EE_{DPC} = \frac{\varsigma_{DPC_l}}{\vec{\wp}^{o}{}_{tot}} \quad (43)$$

where $\varsigma_{DPC_l} = \max_{P_l} \log_2 det\left(I + \bar{\gamma}k(\bar{H}^{\ddagger(M^o)})P_l\bar{H}^{(M^o)}\right)$. In the *EE* equations listed above, (38) and (39) belong to energy evaluations for full array antennas and partially selected elements based on the respective criteria. In contrast, (40)–(42) incorporate Algorithms 1–3 with low resolution DAC and DPC-enabled transmitters.

Algorithm 1: Initial access-based optimal number selection algorithm.

Input: $D_{max}, M, f, \bar{\gamma}, p_t, B$
Output: $EE, M^{\star}$, Maximum EE
begin
  $\wp_{tot} = 0, \varsigma = 0$ `/* Initializing total power and capacity for service signal not to affect the first addition in the iteration. This is the same to when a BS is at down or off state */`
  $\bar{\gamma}_k = \frac{p_t|H_kW_k|^2}{\Gamma(D_{max})N_oB}$
  `/* SNR equals mimimum received power over noise at the cell edge user */`
  for $\iota = 1 : length\ (M)$ **do**
    $\wp_{tot(\iota)} \leftarrow P_{LO} + P_{PA} + \iota(2P_{DAC} + 2P_M + 2P_{LF} + P_{HB})$ `/* Computing total power consumtion as the number of antenna elements increases iteratively */`
    $\varsigma(\iota) = log_2\left(real(det(I + (\frac{\bar{\gamma}_k}{\iota})))\right)$ `/* Computing capacity as the number of BS antenna elements increases iteratively */`
    $EE(\iota) \leftarrow \frac{\varsigma(\iota)}{\wp_{tot}(\iota)}$ `/* Computing energy efficiency as the number of antenna elements increases iteratively */`
  if $EE(\iota) \geq EE(M)$ **then**
    $M^{\star} \leftarrow \iota(find(EE = max(EE)))$ `/* Selecting the number of antennas at wich EE reaches its maximum (first optimal M) */`
  $M^{*} \leftarrow M$ `/* Full array utilization (no selection) */`

Algorithm 2: Number and element selection after reduced distance.

Input: $d_{min}, M^{\star}, \Gamma(D_{max})$, p_t /* Minimum distance, first selected number of antennas for cell edge users, cell edge pathloss (from algorithm one), and total transmit power */

Output: *EE*, M^o

begin

  $\varsigma_{max} = 0$ /* Setting the capacity value to zero; assuming the BS is at off state */

  re-trilateration: **for** $i \in k$ **do**

    $r_k \leftarrow D_{max}$ /* Distance update of a user from cell edge to outskirts or centre of the cell due to mobility */

    if $r_k \neq d_{min}$ **then**

      $P_{rmin} \leftarrow p_t/\Gamma(D_{max})$ /* Minimum received power at the cell edge */

      $P_r(k) \leftarrow \Gamma(r)Pr_{min}$ /* Received power at k user and $\Gamma(r)$ is the pathloss at a reduced distance */

      $M_1^o = \dfrac{M^{\star}\sum_{i=1}^{k} p_t/k}{\wp_{tot}}$ /* finding an optimal number of antennas transmit as a function of the maximum transmit power, total number of users, and total power. */

      $M_2^o \leftarrow \dfrac{round\left(\sum\Gamma(r)/k\right)M^{\star}}{\Gamma(D_{max})}$ /* where Γ is path loss at respective distance */

    if $M_1^o \neq M_2^o$ **then**

      goto re-trilateration

      $M_1^o \leftarrow M_2^o$

      for $\bar{\gamma}_o = 1 : length(M^o)$ **do**

        /* Assuming the range of SNR values from 1 to length of adaptively selected number of antennas, where $\bar{\gamma}_o$ represents average SNR as a function of selected number of antennas, M^o. */

        $H = \dfrac{\Psi}{\sqrt{M^o}}$; /* $\Psi = rand(k; M^{\star}) + jrand(k, M^{\star})$; this is random channel matrix for M antennas and k users and H is the nurmallized channel with the selected branches */

        $i = 1$;

        for $\iota = 1 : M^o$ **do**

          $H_c = [H :; [M_\iota^o \; M^{\star} - \iota]]$ /* Selecting a column with maximum channel gain */

          $\varsigma_{DPC}(M^o(\iota) = log_2(real(det(I + \bar{\gamma}_o * H_c * H_c')))$ /* Computing DPC capacity with selected columns with maximum gain. */

        $\varsigma_{max} = \varsigma_{max} + max(\varsigma_{DPC}(M^o(\iota))$ /* Selecting maximum Capacity among the capacity array elements foung in the iteration where $M^o(\iota)$, where $\iota \in [1, M^o]$ */

      $\varsigma(\bar{\gamma}_o) \leftarrow \varsigma_{max}$

    $EE = \dfrac{\varsigma(\bar{\gamma}_o)}{\wp_{tot}^o}$ /* Calculating energy efficiency with total capacity and tital power as a function of selected branches. $\wp_{tot}^o$ is computed according to (36) */

Algorithm 3: Proposed algorithm for low resolution DAC.

Input: p_t, σ, D_{max}, f
  /* Total transmit power, Bandwidth, maximum radius and operating frequency */
Output: ς, EE
$\hat{n} = randi([0,1],[1,n])$ /* Generating random number of bits */
$\varsigma_{DAC} = \hat{n}/\tau$ /* Calculating capacity of DAC with n random bits to be generated at time τ, */
if $\varsigma_{DAC} < \varsigma_{th}$ **then**
  $b = 1$ /* Initialize bit counter */
  $M^s \leftarrow 2^b, \varsigma_M^s = \varsigma_{DAC} * log_2 M^s$ /* Capacity after modulation with symbol M^s */
  $P_{DAC} \leftarrow c_1 f_t + c_2 2^b$ /* P_{DAC} is DAC power before lowering the resolution */
  while $\varsigma_{DAC} log_2(2^b) < \varsigma_{th}$ **do**
    $b \leftarrow b + 1$ /* increment the bits by one */
    $\varsigma(b) = \varsigma_{DAC} log_2(2^b)$
    if $\varsigma(b) > \varsigma_{th}$ **then**
      continue   /* skip the iteration */
  $EE = \frac{\varsigma}{P_{DAC}}$
while $\frac{b}{\tau} > \varsigma_{th}$ **do**
  $b = b - 1$ $\vec{P}_{DAC} \leftarrow c_1 f_t + c_2 2^b$ /* $\vec{P}_{DAC}$ is DAC power after lowering the resolution */
  if b==0 **then**
    break
  $\varsigma_{dac}(b) = \frac{b}{\tau}$,
  $EE(b) = \varsigma_{dac}/\vec{P}_{DAC}$,
  if $\frac{b}{\tau} < \varsigma_{th}$ **then**
    Continue   /* Ending iteraton */

The procedures of the above two algorithms are summarized as follows:

- In algorithm one, the selection is made with dynamic or nondeterministic channel conditions. The input for selection is a pilot signal, and the number is identified based on the optimal value $M^\star$ of the EE curve.
- In algorithm two, the maximum number of antennas M^* is transformed to M^o, which was obtained as a new number in algorithm one.
- After identifying the minimum received signal at the cell edge and an optimal number of antennas ($M^\star$) using the initial access condition in algorithm one, the number of antennas is adaptively reduced. Instead of reducing the transmit power when users move from the cell edge area to the cell centre positions, in this case, the number of antennas is reduced, maintaining the minimum required SNR for connection quality. Then, after finding the average SNR, the number is changed from M^* to M^o.
- After identifying M^o, the branches with best channel gains are selected from $M^\star$ iteratively, and the DPC capacity is computed.

The general idea behind algorithm three is stated as follows:

- Assuming fixed SNR at the cell edge, the minimum capacity is obtained and considered as a threshold capacity. Here, the free space propagation model is considered, and small-scale fading is ignored.

- Through randomly generated bits, the capacity of DAC is calculated and compared with the threshold.
- If the capacity of DAC, which was obtained with random bits, is greater than the threshold capacity, then DAC down resolution is done by iteratively decreasing the number of bits before analogue conversion is done. This reduction in bits ultimately reduces the operating power consumption of DAC according to (14). This DAC power directly affects the total system power consumption according to (37).
- Finally, the EE is computed as a function of the capacity and total power with low resolution. Throughout the evaluation process in this paper, the following parameters in Table 2 are used.

Table 2. Simulation parameters.

Parameter	Description	Value
c_1	Static power of DAC	9×10^{-12}
c_2	Dynamic power of DAC	1.5×10^{-5}
B	Channel bandwidth	40 MHz
D_{max}	Cell edge distance	200 m
d_{min}	Minimum distance	3 m
f	Operating frequency	38 GHz
$\bar{\gamma}$	Total average SNR at the cell edge	$\frac{Pr_{min}}{N_o B}$
c	Speed of light	3×10^8 m/s
$\Gamma(D_{max})$	Path loss at cell edge distance	$(\frac{4\pi D_{max} f}{c})^2$
τ	Random bit generation time interval	5 ms
p_A	Amplifier power	0.05 mW
p_M	Mixer power	0.04 mW
P_{LO}	Local oscillator power	0.01 mW
$\chi_{\bar{\sigma}}^{CI}$	Large-scale signal fluctuations due to the CI pathloss model	4.4 dB
P_{LP}	Low pass filter power	0.012 mW
P_{HB}	Hybrid with buffer power	0.033 mW

5. Results and Analysis

In this section, the simulation results with different scenarios are discussed. Figure 2 illustrates the relationship between distortion and the number of bits or symbols. As the number of bits or symbols increases, distortion decreases logarithmically, and this accounts for the increase in bit resolution. As shown from the figure, the distortion reaches its minimum point when the number of symbols is 7 when the evaluation is at the bit level, and it lasts until around 65 at the symbol level. Figure 3 shows the effect of the number of symbols on quantization. Quantization is the inverse of distortion, which increases logarithmically with the number of symbols. As the number of bits is mapped to each constellation point in any digital modulation scheme, the number of symbols increases exponentially and directly affects the quantization level.

The relation between DAC power consumption and energy efficiency is illustrated in Figure 4. The EE is evaluated through a range of power values from 0.3 to 3.48 mW. As the power consumption of the DAC increases due to an increase in the number of bits entering to DAC as in (14); then, the energy efficiency decreases exponentially according to the EE equation in algorithm two. This happens if the increase in bits affects the total power consumption more than the capacity. Moreover, an increase in the total power consumption with maintained minimum capacity leads to a decrease in the EE. Again, the graph shows that applying low resolution in DAC lowers power consumption and so that the EE increases.

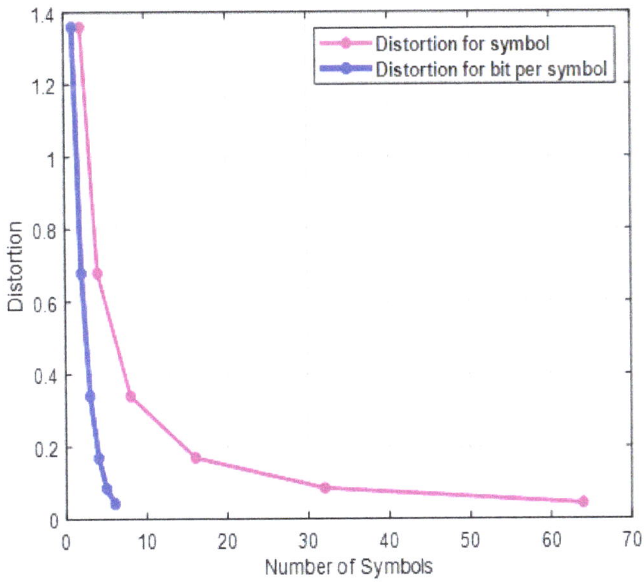

Figure 2. Distortion versus number of symbols with $k = 5$ or 10 and $M = 64$.

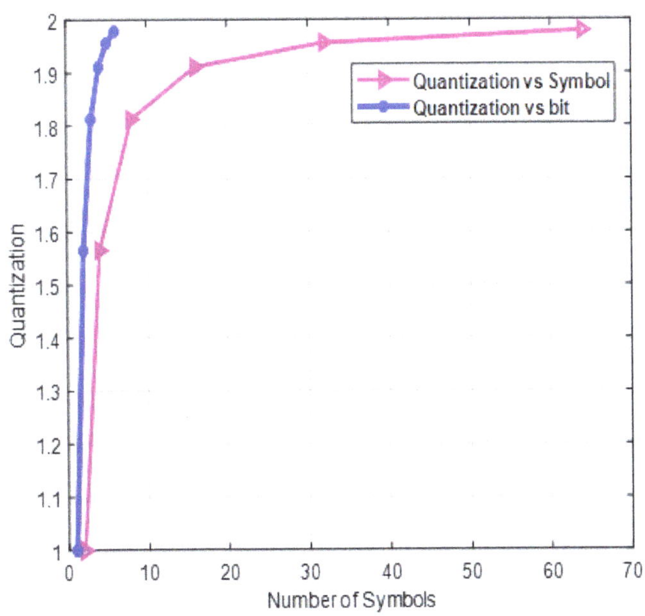

Figure 3. Quantization versus number of symbols with BPSK modulation.

The relationship between energy efficiency and the number of antennas considering the low-resolution DAC and without resolution case is shown in Figure 5. As the number of antennas increases, the energy efficiency also increases, and when low-resolution DAC is applied, the energy efficiency becomes higher than without resolution. The peak EE with low-resolution is 14.8 and 10.2 Mbits/J with and without low resolution, respectively. The EE begins to decline after the peak point because the increasing number of antennas on total power consumption exceeds the increase in capacity, which is achieved due to the increasing resolution.

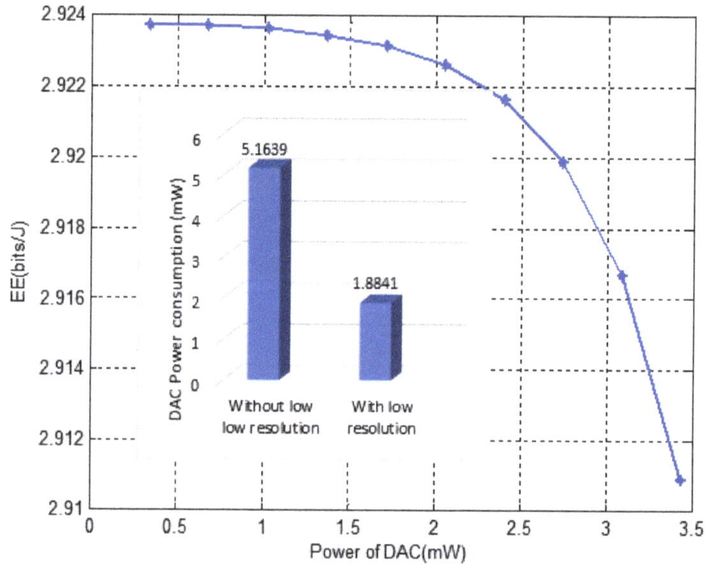

Figure 4. EE evaluation with different DAC power values, and evaluation of the total power consumption with and without low DAC resolution at randomly generated bits.

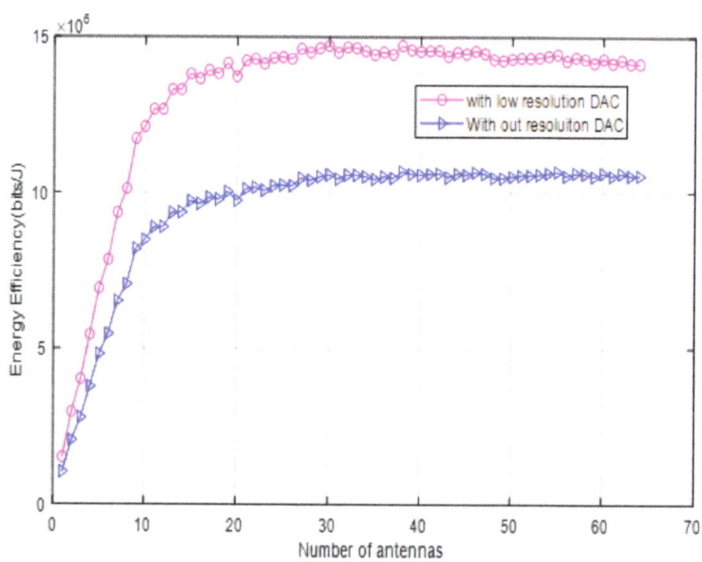

Figure 5. Energy efficiency as a function of the number of BS antennas when the number of terminals $k = 30$.

The trade-off between spectral and energy efficiency for the given number of users with and without low resolution is shown in Figure 6. Both spectral and energy efficiency increase together for a fixed bandwidth up to the optimal energy efficiency point. For this, we evaluated the energy efficiency for a different number of user scenarios; for example, $k = 5$ or 10. For a smaller number of users, the diminishing rate is faster than that of a larger number. Hence, the EE graph starts declining at 55 and 59 bps/Hz of SE for five users with low-resolution DAC and without, respectively.

High energy efficiency is achieved with low-resolution DAC for the given spectral efficiency. The maximum energy efficiency point is 14 Mbits/J, which is when the spectral efficiency reaches 115 b/s/Hz with ten users and without resolution. It becomes 10.5 Mbits/J for the same number of users and spectral efficiency. Figure 7 depicts the relationship between energy efficiency, k, and M in a massive MIMO system with statistical and instantaneous SNR values. For cell edge users in LoS conditions, the outcome is evaluated using procedures of algorithm one. While the energy efficiency increases with the increase of M at first, it begins to decline at some point as M continues to grow, according to the simulation.

In this figure, statistical and instantaneous or fixed SNR are also compared for the same k. It has been demonstrated that fixed SNR outperforms for small M and under-performs for large M. EE has also been shown as the number of user terminals increases. However, due to random channel conditions, EE exhibits different optimal points. Moreover, the optimal threshold of EE for each configuration varies according to the number of users and SNR modalities.

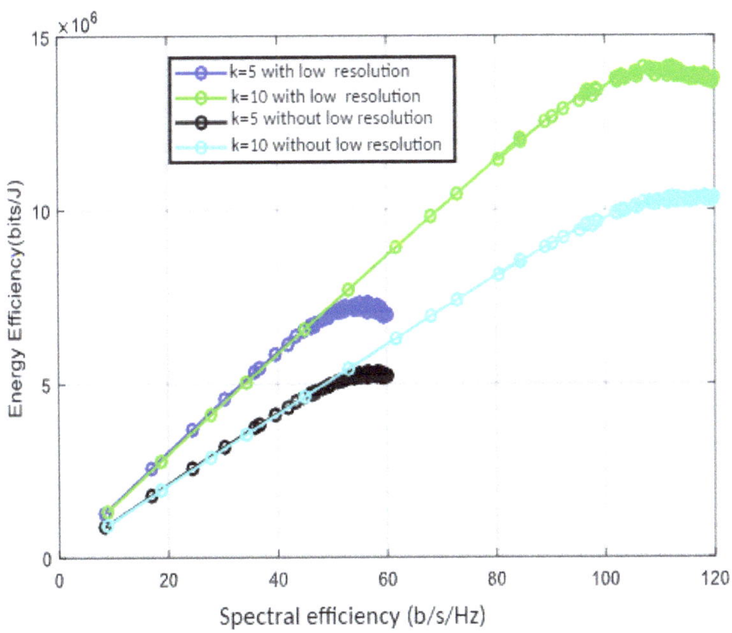

Figure 6. Energy efficiency versus spectral efficiency with without low resolution DAC.

Figures 8 and 9 present the results according to the proposed algorithm by combining the three scenarios, and we compared the performance of each at CI and FSPL using mmWave and sub-6 GHz frequency ranges. The first scenario entails locating $M^{\star}$ from the entire array at the indoor cell edge, locating M^{o} according to (23), and finally evaluating capacity values as $\binom{M}{M^{o}}$ using combinational permutation. At initial access, equal power allocation among all BS antenna elements and the point at which the EE graph starts diminishing is evaluated using a reference signal. Then, the number of antennas is used as a baseline for our further considerations.

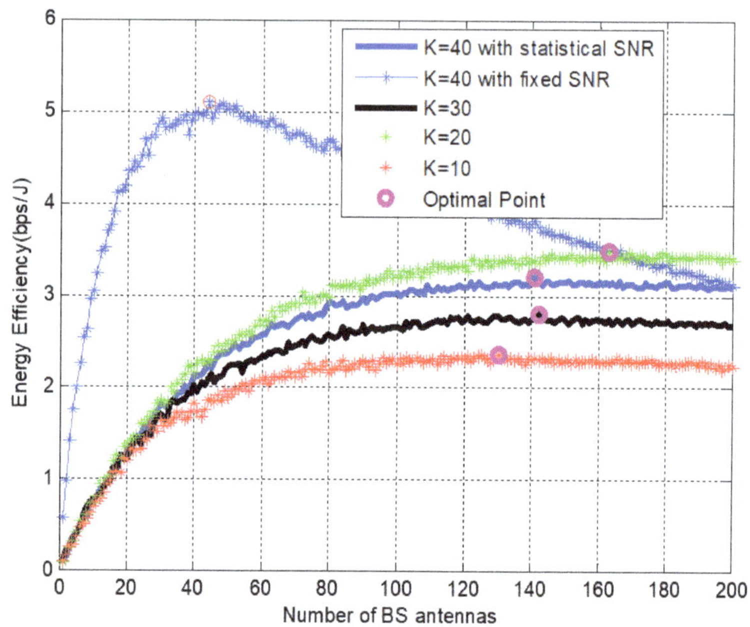

Figure 7. Energy efficiency as a function of M base station antennas with $P_t = 20$ mW.

Before the energy efficiency evaluation process, we performed an analysis of teh free space and CI path-loss models according to their formulations stated in (19) and (20). Accordingly, the FS model provides higher data rates due to obstruction freedom; however, CI is more realistic than FS in practical scenarios. Based on this intuition, we applied an antenna selection algorithm to both, and the results show that a minimal number of antennas are selected in free space compared to CI.

When CI path loss is applied to mmWave and sub-6 GHz frequency ranges and compared for fixed total system power, CI with sub-6 GHz is more energy-efficient than mmWave. Although the high-frequency signal carries larger data than the low-frequency signal, as frequency increases, the blockage due to different impairments also exhibits low wavelength, which negatively affects the received signal. Low received signal accounts for a low data rate at the receiver, so EE is degraded compared to CI. Finally, we found that the FS path loss with the DPC precoder changes the graph from a logarithmic to almost linear because only a few antenna elements were selected compared to CI.

The effect of the transmit power on the EE is depicted in Figure 9. We evaluated EE as a function of BS antennas at different power levels for full array and selection implementations. The system's performance was also evaluated with and without the non-linear preceding, which showed that antenna selection with the minimum SNR significantly improved the energy efficiency with less transmit power and a DPC precoder.

Figure 10 illustrates EE with a number, complex, and random element selection and finally compares it with full array utilization or no selection. It can be observed that random number selection followed by complex selection shows better performance than no selection or full array. However, when it is compared with the number of selections made with our proposed algorithm and complex selection, random selection still outperforms for a smaller number of antennas employed and under-performs for large numbers of antennas.

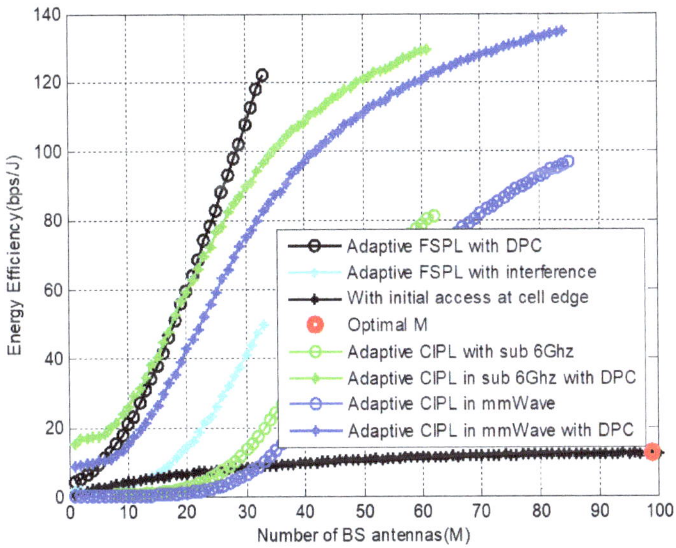

Figure 8. Energy efficiency evaluation as a function of the number of BS antennas with at mmWave frequency, f = 38 GHz and sub-6 GHz, and f = 2.5 GHz.

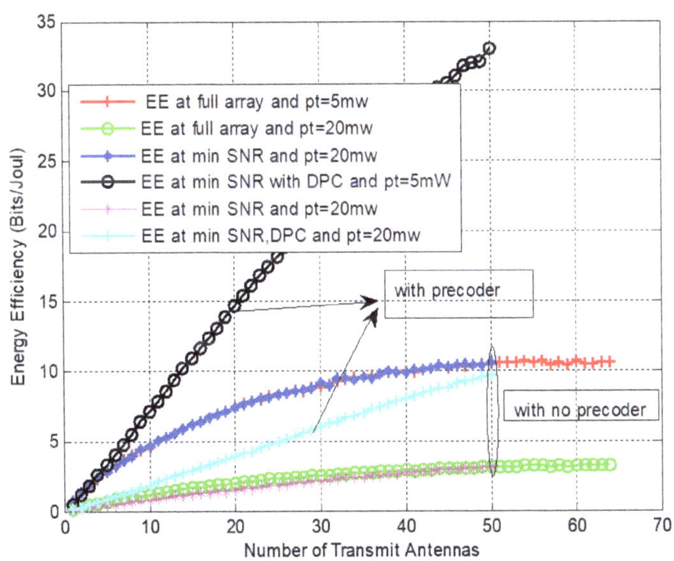

Figure 9. Energy efficiency evaluation as a function of number of BS antennas with at mmWave frequency, f = 38 GHz, and M = 64.

The relationship of energy efficiency with low-resolution DAC and selection are shown in Figure 11. The EE can be enhanced by applying a low-resolution DAC algorithm even if the full array is utilized. We also evaluated the random selection after finding the optimal antennas and applying low DAC resolution. The results show that applying low DAC resolution still enhances EE as it has a significant role in minimizing the total power consumption.

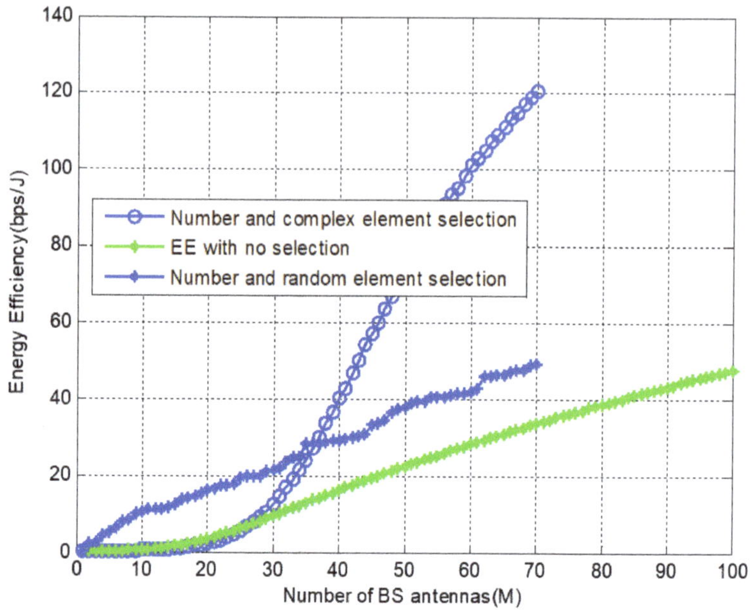

Figure 10. Energy efficiency evaluation with random and complex selection at f = 38 GHz and M = 100.

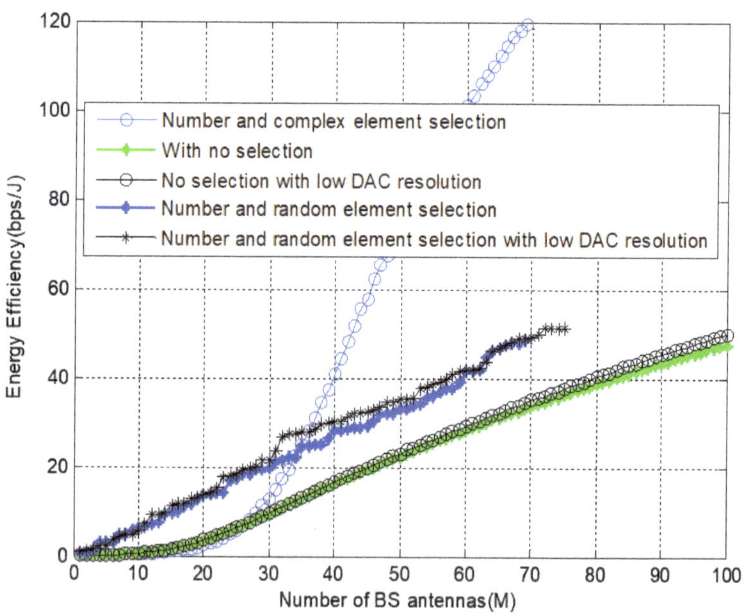

Figure 11. Energy efficiency with and without low resolution DAC at f = 38 GHz and M = 100.

Figure 12 presents the complex nature of the proposed selection algorithm and compares it with selected literature that used similar techniques. Complexity, in this case, is the number of iterations primary and nested loops to happen while selection is made to identify the branch with better channel gain among the entire array. Hence, the result also illustrates the system's complexity when selection incorporates low-resolution DAC according to table one. From the graph, we can observe that random selection is the least complex even though it has a lower capacity than complex selection. This is because the selection is made irrespective of the channel gain, which plays a crucial role in enhancing the capacity and complexity. For random selection, the number of iterations to select M antennas is only one as it has no combination with the channel branches.

Our proposed algorithm is also compared with [21,28,29], which are among the simplest and following similar approaches to the best of our knowledge. The complexity order of each is [21] our proposed technique [28,29] and random selection according to (27) and (28). We also found that the proposed algorithm is more energy-efficient than random at the cost of some complexity, which is less than that of [21].

Moreover, since the energy efficiency of the proposed technique has been shown to surpass random selection and full array utilization or no selection in Figure 11. Figure 12 is to show only how complexity costs while working for energy efficiency; however, the rate of the effect and trade-off, including the EE of the aforementioned literature, is left as future work. Therefore, the selection technique meets our main goal of proposing an energy-efficient system at the cost of some complexity.

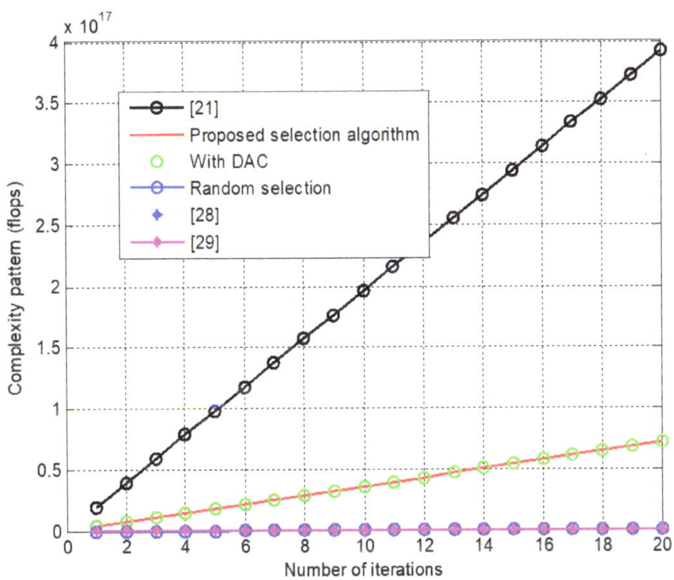

Figure 12. Computationalx complexity of selection algorithms with adaptive ℓ^* and $M = 64$.

6. Conclusions

In this paper, energy-efficient antenna selection techniques were presented. In particular, we proposed adaptive transmit antenna selection strategies for downlink systems to minimize the power consumption of RF chain components associated with each antenna element. The selection process was categorized into two parts to reduce the complexity arising from several iterations. In the first part, we considered only cell edge users and found the average minimum SNR value from multiple generations of random channels to find the number of antennas at which the EE curve reached its maximum and began declining.

The optimal number of antennas obtained through this process was used as a baseline for further selection while users moved to the centre of the cell. The selection depended on the distance and channel condition between the users and a BS in the second case. The number of antennas to be selected adaptively changed with channel and user distance variations. We also proposed low-resolution DAC to further minimize the system's total power consumption to enhance EE.

We evaluated the system's performance at sub-6 GHz and mmWave frequencies with CI and free space propagation models. Furthermore, we compared the proposed antenna selection with and without low-resolution DAC. The results show that selecting only a few antennas instead of employing all the arrays improved the EE by reducing the total power consumption. Furthermore, we demonstrated that applying non-linear precoders, such as DPC, further improved the EE by enhancing the system's capacity. However, the combined average EE was found to surpass selection without low resolution at the cost of some complexity.

Author Contributions: Formal analysis, Y.W.M.; Funding acquisition, K.T.K.; Investigation, H.B.K.; Methodology, Y.N.; Resources, F.D.D.; Writing—original draft, S.C.A. The authors equally contributed to the reported research and writing of the paper. All authors have read and agreed to the published version of the manuscript.

Funding: This fund was granted by Morgan State University.

Institutional Review Board Statement: Not applicable.

Informed Consent Statement: Not applicable.

Data Availability Statement: Not applicable.

Conflicts of Interest: In the collection, analyses, or interpretation of data; in the writing of the manuscript, or in the decision to publish the results. The authors declare no conflict of interest.

Abbreviations

The following abbreviations are used in this manuscript:

AoA	Angle of Arrival
AoD	Angle of Departure
BS	Base Station
CI	Close In
DAC	Digital to Analogue Conversion
EE	Energy Efficiency
mMIMO	massive Multiple Input Multiple Output
RF	Radio Frequency
SE	Spectral Efficiency
SNR	Signal-to-Noise Ratio
ZoD	Azimuth Angle of Departure

References

1. Larsson, E.G.; Tufvesson, F.; Edfors, O.; Marzetta, T.L. Massive MIMO for next-generation wireless systems. *IEEE Commun. Mag.* **2014**, *52*, 186–195. [CrossRef]
2. Heath, R.W.; Sandhu, S.; Paulraj, A. Antenna selection for spatial multiplexing systems with linear receivers. *IEEE Commun. Lett.* **2001**, *5*, 142–144. [CrossRef]
3. Rusek, F.; Persson, D.; Lau, B.K.; Larsson, E.G.; Marzetta, T.L.; Edfors, O.; Tufvesson, F. Scaling up MIMO: Opportunities and Challenges with Very Large Arrays. *IEEE Signal Process. Mag.* **2013**, *30*, 40–60. [CrossRef]
4. Correia, L.; Zeller, D.; Blume, O.; Ferling, D.; Jading, Y.; Godor, I.; Auer, G.; van der Perre, L. Challenges and enabling technologies for energy-aware mobile radio Networks. *IEEE Commun. Mag.* **2010**, *48*, 6672. [CrossRef]
5. Ribeiro, L.; Schwarz, S.; Rupp, M.; de Almeida, A.L.F. Energy efficiency of mmWave massive MIMO precoding with low-resolution DACs. *IEEE J. Sel. Top. Signal Process.* **2018**, *12*, 298–312. [CrossRef]
6. Vlachos, E.; Kaushik, A.; Thompson, J. Energy-efficient transmitter with low resolution DACs for massive MIMO with partially connected hybrid architecture. In Proceedings of the IEEE 2018 87th Vehicular Technology Conference (VTC Spring), Porto, Portugal, 3–6 June 2018; pp. 1–5.
7. Ding, Q.; Deng, Y.; Gao, X. Spectral and energy efficiency of hybrid precoding for mmWave massive MIMO with low-resolution ADCs/DACs. *IEEE Access* **2019**, *7*, 186529–186537. [CrossRef]
8. Moghadam, N.N.; Fodor, G.; Bengtsson, M.; Love, D.J. On the energy efficiency of MIMO hybrid beamforming for millimeter-wave systems with nonlinear power amplifiers. *IEEE Trans. Wirel. Commun.* **2018**, *17*, 7208–7221. [CrossRef]

9. Jacobsson, S.; Duris, G.; Coldrey, M.; Goldstein, T.; Studer, C. Quantized precoding for massive MU-MIMO. *IEEE Trans. Commun.* **2017**, *65*, 4670–4684. [CrossRef]
10. Rabie, E.; Shokair, S. A Proposed Efficient Hybrid Precoding Algorithm for millimetre-wave Massive MIMO 5G Networks. *Wirel. Pers. Commun.* **2020**, *112*, 1–16.
11. Albreem, M.A.; Al Habbash, A.H.; Abu-Hudrouss, A.M.; Ikki, S.S. Overview of precoding techniques for massive MIMO. *IEEE Access* **2021**, *9*, 60764–60801. [CrossRef]
12. Yang, P.; Zhu, J.; Chen, Z. Antenna selection for MIMO system based on pattern recognition. *Digit. Commun. Netw.* **2019**, *5*, 34–39. [CrossRef]
13. Gharavi-Alkhansari, M.; Gershman, A. Fast Antenna Subset Selection in MIMO Systems. *IEEE Trans. Signal Process.* **2004**, *52*, 339–347. [CrossRef]
14. Dua, A.; Medepalli, K.; Paulraj, A.J. Receive Antenna Selection in MIMO Systems Using Convex Optimization. *IEEE Trans. Wirel. Commun.* **2006**, *5*, 2353–2357. [CrossRef]
15. Wang, B.; Hui, T.; Leong, M.S. Global and Fast Receiver Antenna Selection for IMO Systems. *IEEE Trans. Common.* **2006**, *58*, 2505–2510. [CrossRef]
16. Xu, Z.; Sfar, S.; Blum, R.S. Analysis of MIMO systems with receive antenna selection in spatially correlated Rayleigh fading channels. *IEEE Trans. Veh. Technol.* **2009**, *58*, 251262.
17. Masouros, C.; Alsusa, E. Dynamic linear precoding for the exploitation of known interference in MIMO broadcast systems. *IEEE Trans. Wirel. Commun.* **2009**, *8*, 1396–1404. [CrossRef]
18. Gesbert, M. Soft linear precoding for the downlink of DS/CDMA communication systems. *IEEE Trans. Veh. Technol.* **2010**, *59*, 203215.
19. Xiang, E.; Liu, J.; Tufvesson, F. Antenna selection in measured massive MIMO channels using convex optimization. In Proceedings of the 2013 IEEE Globecom Workshops (GC Wkshps), Atlanta, GA, USA, 9–13 December 2013.
20. Zhang, X.; Zhao, F. Hybrid Precoding Algorithm for Millimeter-Wave Massive MIMO Systems with Sub connection Structures. *Wirel. Commun. Mob. Comput.* **2021**, *2021*, 5532939. [CrossRef]
21. Husbands, R.; Ahmed, Q.; Wang, J. Transmit Antenna Selection for Massive MIMO: A Knapsack Problem Formulation. In Proceedings of the 2017 IEEE International Conference on Communications (ICC), Paris, France, 21–25 May 2017.
22. Bogale, T.E.; Le, L.B.; Haghighat, A.; Vandendorpe, L. On the Number of RF Chains and Phase Shifters, and Scheduling Design with Hybrid Analog-Digital Beamforming. *IEEE Trans. Wirel. Commun.* **2016**, *15*, 3311–3326. [CrossRef]
23. Kanhere, O.; Rappaport, T.S. Outdoor sub-THz Position Location and Tracking using Field Measurements at 142 GHz. In Proceedings of the 2021 IEEE International Conference on Communications (ICC), Montreal, QC, Canada, 14–23 June 2021; pp. 1–6.
24. Rappaport, T. *Wireless Communications: Principles and Practice*, 2nd ed.; Prentice-Hall: Upper Saddle River, NJ, USA, 2002.
25. Molisch, A.F.; Win, M.Z.; Winters, J.H. Capacity of MIMO Systems with Antenna Selection. *IEEE Trans. Wirel. Commun.* **2005**, *4*, 1759–1772. [CrossRef]
26. Rappaport, T. Wideband millimeter-wave propagation measurements and chan-new models for future wireless communication system design (Invited Paper). *IEEE Trans. Commun.* **2015**, *63*, 30293056. [CrossRef]
27. Challita, F.; Lienard, M.; Gaillot, P.; Pardo, M. Evaluation of an Antenna Selection Strategy for Reduced Massive MIMO Complexity. *Radio Sci.* **2021**, *56*, e2020RS007242. [CrossRef]
28. Yu, W.; Wang, T.; Wang, S. Multi-Label Learning-Based Antenna Selection in Massive MIMO Systems. *IEEE Trans. Veh. Technol.* **2021**, *70*, 7255–7260. [CrossRef]
29. Kim, S. Efficient Transmit Antenna Subset Selection for Multiuser Space–Time Line Code Systems. *Sensors* **2021**, *21*, 2690. [CrossRef]

sensors

MDPI

Communication

A Wideband Circularly Polarized Magnetoelectric Dipole Antenna for 5G Millimeter-Wave Communications

Hussain Askari [1,†], Niamat Hussain [2,†], Md. Abu Sufian [1], Sang Min Lee [3] and Nam Kim [1,*]

[1] Department of Information and Communication, Chungbuk National University, Cheongju 28644, Korea; askarihussain01@gmail.com (H.A.); abu.sufian@chungbuk.ac.kr (M.A.S.)
[2] Department of Smart Device Engineering, Sejong University, Seoul 05006, Korea; niamathussain@sejong.ac.kr
[3] Department of Corporate Support Centre, Korea National University of Transportation, Chungju-si 27469, Korea; leesm@ut.ac.kr
* Correspondence: namkim@chungbuk.ac.kr
† These authors contributed equally to this work.

Abstract: In this paper, a wideband circularly polarized (CP) magnetoelectric (ME) dipole antenna operating at 28 GHz band was proposed for 5G millimeter-wave (mm-wave) communications. The antenna geometry included two metallic plates with extended hook-shaped strips at its principal diagonal position, and two corners of truncated metallic plates at the secondary diagonal position. The pair of metallic vias connected the modified strips to the ground plane to create the magnetic dipole. The L-shaped probe feed between the strips was used to excite the antenna. The antenna showed stable gain and wideband characteristics. The simulated and measured results showed that the proposed CP ME dipole antenna had an overlapping ($|S_{11}| < -10$ dB impedance and 3 dB axial ratio) bandwidth of 18.1% (25–30 GHz), covering the frequency bands dedicated for 5G new radio communications. Moreover, an average gain of 8 dBic was achieved by the antenna throughout the operating bandwidth. The measured data verified the design concept, and the proposed antenna had a small footprint of $0.83\ \lambda_o \times 0.83\ \lambda_o \times 0.125\ \lambda_o$ (λ_o is free space wavelength at the lowest operating frequency), suitable for its application in 5G smart devices and sensors.

Keywords: magnetoelectric; dipole; circularly polarized; 5G antenna; 28 GHz

Citation: Askari, H.; Hussain, N.; Sufian, M.A.; Lee, S.M.; Kim, N. A Wideband Circularly Polarized Magnetoelectric Dipole Antenna for 5G Millimeter-Wave Communications. *Sensors* **2022**, *22*, 2338. https://doi.org/10.3390/s22062338

Academic Editors: Raed A. Abd-Alhameed, Chan Hwang See and Naser Ojaroudi Parchin

Received: 16 February 2022
Accepted: 16 March 2022
Published: 17 March 2022

1. Introduction

The twenty-first century has experienced tremendous development in the field of wireless communication. The fast evolving fifth-generation (5G) technology has brought new advancements which have posed a great challenge for antenna researchers and engineers. Large channel capacity and high data rates should be achieved to meet the demands of the 5G communication link. In this scenario, the millimeter-wave (mm-wave) band is used to fulfill the gigabit high-data transmission, and addresses the lack of wider spectral resource in the currently allocated band below 6 GHz [1]. Recently, the third-generation partnership project (3GPP) has allocated new radio (NR) frequency bands in the mm-wave range from 24.25 to 29.5 GHz, which are known as n257 and n258. Figure 1 shows the mm-wave frequency band distribution adopted by the leading 5G countries in the world [2].

In addition, circularly polarized (CP) antennas have been a unique choice for stable communication due to their resilience to multipath interference and polarization mismatching between receiving and transmitting antennas. These features make CP antennas the desirable candidate for numerous applications including satellite communication, radar communication, randomly oriented RFID (radio frequency identification) tags, GPS (global positioning system), and sensors [3–9]. Therefore, the design of CP antennas has always been a hot topic for antenna designers, compared with linear polarized (LP) antennas. In the literature, different kinds of CP antennas have been reported for mm-wave applications, including patch antenna, aperture antenna, slot antenna, cavity antenna, and magneto-electric (ME) dipole antenna. The ME dipole antenna is the most popular type of antenna due to its complementary performance.

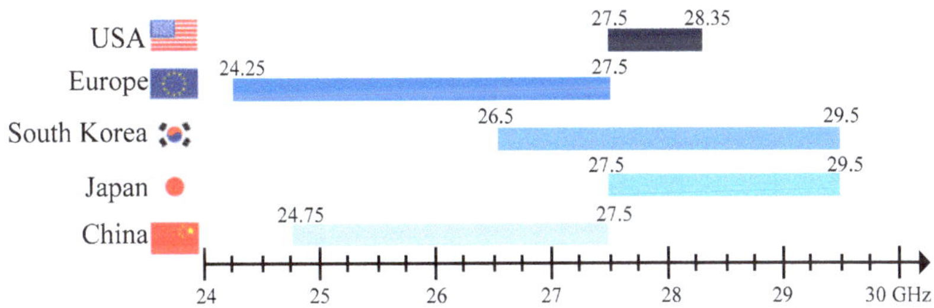

Figure 1. Global snapshot of the 5G millimeter-wave spectrum.

The concept of the first ever complimentary antenna was proposed by Chlavin in 1954 [10]. Since then, many structures were proposed based on a slot/dipole or a slot/monopole combination. In 2006, Luk and Wong introduced a complementary antenna based on patch and dipole antenna combinations for the first time, achieving wideband performance [11]. The idea used a pair of copper patches parallel to the ground as an electric dipole and a pair of copper shorted patches perpendicular to the ground as a magnetic dipole. The antenna showed linear polarization with wide bandwidth, stable gain, and low levels of cross-polarization. In the course of designing ME dipole antennas [12,13], less attention has been paid to CP ME dipole based antennas. In the past few years, some CP antennas based on ME dipole structures have been reported [8,14–21]. In [14], an ME dipole structure was reported to use a pair of bowtie patch antennas and a pair of trapezoidal-shaped dipoles fed by a Wilkinson power divider to achieve a 3 dB AR bandwidth of 33%. Moreover, to ensure the CP radiation was presented, an ME dipole antenna was integrated with a crossed dipole fed by double phased delay rings in [15]. The proposed antenna achieved 27.67% of 3 dB AR bandwidth. In [16], substrate integrated waveguide (SIW) and aperture coupled feeding was used to excite an ME dipole antenna for mm-wave applications. This antenna achieved a 3 dB AR bandwidth of 12.8%. However, the above-mentioned reported ME dipole antennas showed limited AR bandwidth and had complex feeding structures. The wideband ME dipole antennas investigated in [17,18] had the advantage of a wider 3 dB AR bandwidth of 71.5% and 47.7%, respectively. However, the drawback of these antennas was their huge volume which limits their usage in modern electronic devices and sensors. On the other hand, the CP ME antennas designed at 60 GHz band offer wide AR bandwidth/high gain at the expense of large antenna size [19–21]. In [22], the AR bandwidth of the ME dipole antenna was significantly improved (53%) by employing a novel crossed feeding structure. This design has a high antenna profile (33.3 mm antenna height) and limited gain (6.6 dBic). A wideband circularly polarized ME dipole array fed by a complementary SIW power distribution phase shifter was presented [23]. The single-element antenna gain was 7dBic, while the AR bandwidth was restricted to 9.7%. To mitigate increased atmospheric losses in mm-wave bands, the gain of the wideband CP ME dipole antennas was increased in [24,25]. Moreover, the ME dipole antenna offered the advantages of low-profile and wide bandwidth with LP radiation [26]. In summary, the existing CP ME dipole antennas are suffering from either narrowband operation or high antenna profiles.

In this paper, a compact, wideband CP ME antenna operating at 28 GHz band is presented for 5G communications. The antenna's CP bandwidth covers 25–30 GHz band (18.1%), covering the frequency bands dedicated for 5G new radio communications. Moreover, the antenna offers stable gain (average 8 dBic) and radiation patterns, with the advantages of a small footprint ($0.9\ \lambda_o \times 0.9\ \lambda_o \times 0.14\ \lambda_o$) for its applications in 5G smart devices and sensors.

The rest of the manuscript has been structured as follows: Section 2 explains the detailed analysis and design of the proposed circularly polarized ME dipole antenna. The simulated and measured results have been presented in Section 3, while the paper has been concluded in Section 4.

2. CP ME Antenna Design and Analysis

2.1. Antenna Geometry

The schematics of the proposed CP ME dipole antenna are shown in Figure 2. The antenna is composed of two metallic strips with an extended hook shape on its principal diagonal position, two L-shaped metallic strips on secondary diagonal position, and four sets of via holes. The modified metallic strips above the ground plane act as two planar electric dipoles. The four sets of via holes, each containing three metallic plated vias, are shorted between the modified patches and the ground plane. The metallic vias and the ground plane between them act as a magnetic dipole. The antenna is fed by an L-shaped probe to achieve low cross-polarization and back radiation levels [27,28]. The metallic patch of L-shaped probe acts as a coplanar waveguide (CPW) feed between the planar dipoles, which work as coplanar grounds. A square size (10 mm × 10 mm) Rogers 5880 with a thickness of 1.5 mm was used as the substrate of the antenna. The proposed ME dipole antenna's optimized dimensions are summarized in Table 1.

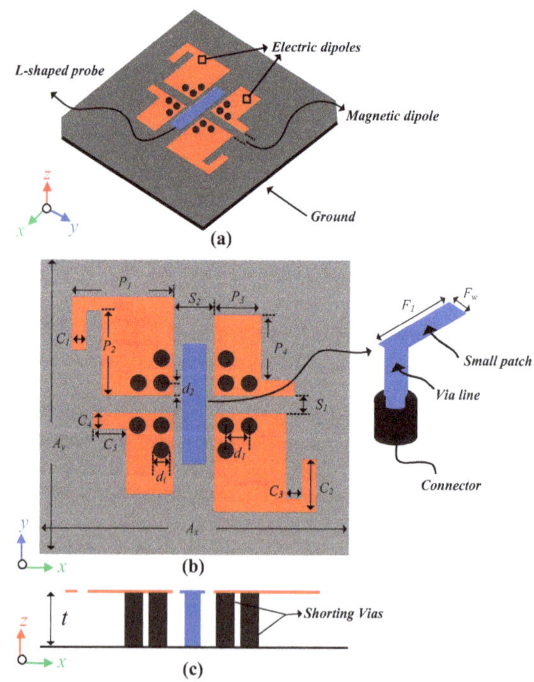

Figure 2. The geometry of the proposed CP ME antenna: (a) 3D view, (b) top view, and (c) side view.

Table 1. Optimized dimensions of the proposed CP ME dipole antenna.

Parameter	Value (mm)	Parameter	Value (mm)	Parameter	Value (mm)
t	1.5	P_4	1.8	C_4	0.45
F_1	2.5	S_1	0.5	C_5	1
F_w	0.7	S_2	1	d_i	0.5
P_1	3	C_1	0.4	d_1	0.7
P_2	2.4	C_2	1.5	d_2	0.35
P_3	1.4	C_3	0.5	A_x, A_y	10

During the antenna design, we found that the antenna offered optimum performance under the given constraints of the optimized parameters. Changing any parameter may disturb the AR or impedance bandwidth, or even both. In particular, varying the length of the patch (F_1) changed the $|S_{11}|$ resonance of the antenna, however, the AR resonance did not change significantly. The distance between the monopole and metallic strips (S_2) and S_1 was sensitive to both $|S_{11}|$ and AR bandwidth. A small change may deteriorate the overlapping bandwidth. Similarly, all other parameters, especially S_1, C_1, C_2, C_3, C_4, and C_5 should be carefully tuned to achieve the best possible usable bandwidth.

2.2. CP Mechanism

The CP mechanism of the proposed ME dipole antenna can easily be understood by visualizing the surface current distributions on coplanar electric dipoles, as shown in Figure 3. The surface current distributions were examined and recorded at 28 GHz for different time phases of 0°, 90°, 180°, and 270°. It can be noticed that the vector representation of surface current rotates in an anti-clockwise direction as the phase shifts from 0° to 90°, 180°, and 270°. The resultant vector of these surface current distributions makes a quasi-loop due to the rotational symmetry of this antenna, a necessary condition for CP radiation [29–31]. Thus, the proposed ME dipole antenna generates right-hand circular polarization (RHCP) in the positive z-axis direction.

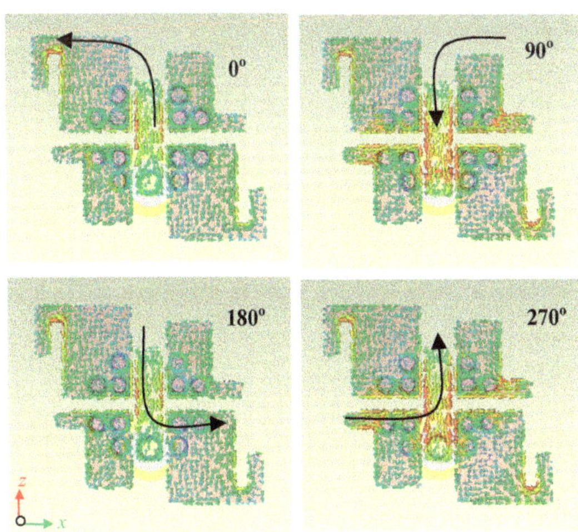

Figure 3. Surface current distributions of the proposed CP ME dipole antenna for different time phases at 28 GHz.

The origin of CP radiations in the antenna is further explained by considering the degenerated modes which are excited in the planar electric dipoles and L-shaped microstrip patch antenna, separately. At time $t = 0$, the aperture of the shorted L-shaped patch acquires maximum surface current, and similarly for the time $t = T/2$, but at this latter time, the direction of surface current is opposite. Therefore, this causes induction of equivalent and orthogonal magnetic current in the aperture of the same L-shaped patch along the x-axis, which causes a 90 degree phase difference with respect to electric surface currents. In time t = T/4 and 3T/4, the planar electric dipoles attain maximum and opposite directed electric surface currents at their edges. It can be noted that the direction of electric surface currents on planar dipoles is also along the x-axis. Hence, the equivalent electric and magnetic currents are directed in the same direction and are in phase. This is how the proposed antenna can generate the CP radiation.

3. Results and Discussion

The proposed CP ME dipole antenna's prototype was fabricated and measured to verify its performance, as shown in Figure 4. Figure 4a shows the fabricated prototype of the proposed antenna. The setup for the antenna far field measurements is illustrated in Figure 4b. The antenna was measured in a multi-probe 360 degree scanning anechoic chamber. Overall, the simulated results showed good agreement with the measured results.

Figure 4. Photographs of the proposed CP ME dipole antenna: (**a**) fabricated prototype and (**b**) far field measurement setup.

3.1. S-Parameters and Axial Ratio

The measurement result for the S-parameter $|S_{11}|$ was obtained using a network analyzer (Rohde and Schwarz ZVA 40) in open air condition. The little difference in simulated and measured results, as shown in Figure 5a, was due to connector/cable losses. The proposed antenna has an impedance bandwidth for $|S_{11}| < -10$ dB was 24.6%, that is, from 24.1 to 31.0 GHz. The simulated and measured 3 dB AR bandwidth of the proposed antenna was 18.1%, ranging from 25 to 30 GHz, as shown in Figure 5b. Thus, the overlapping operating bandwidth of the antenna covered the important band spectrum proposed for 5G mm-wave applications.

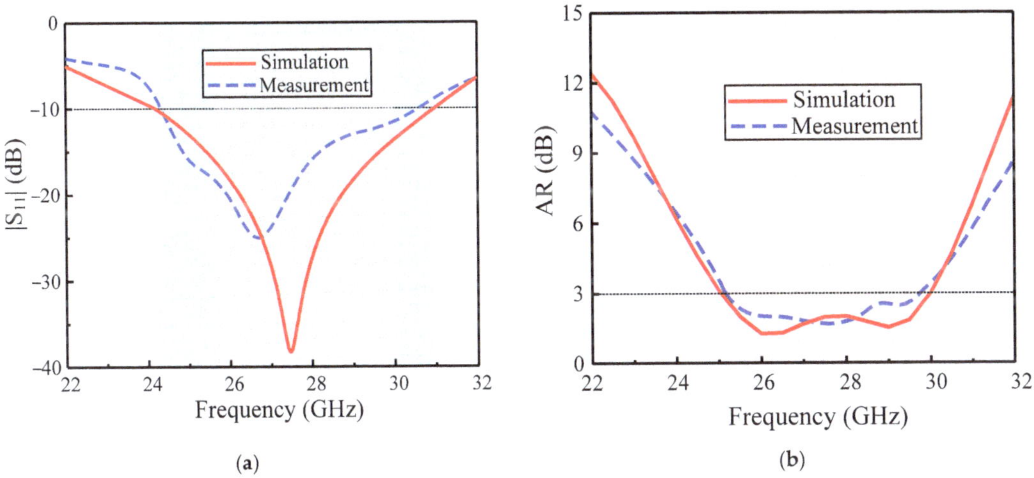

Figure 5. Proposed CP ME dipole antenna: (**a**) S-parameter $|S_{11}|$ and (**b**) axial ratio.

3.2. Broadside Gain and Efficiency

The simulated and measured broadside gain of the proposed antenna is shown in Figure 6a. The antenna showed a stable broadside gain of 8 dBic with a deviation of only ± 0.5 dBic within the frequency range of interest. Moreover, the antenna also offered high efficiency, both total and radiation efficiency, due to good antenna impedance matching (Figure 6b). The simulated radiation efficiency of the antenna was more than 96%, which was in close agreement with its measured value. The measured total efficiency was observed to be a little lower than its simulated values (88%) due to possible cable/connector and substrate losses.

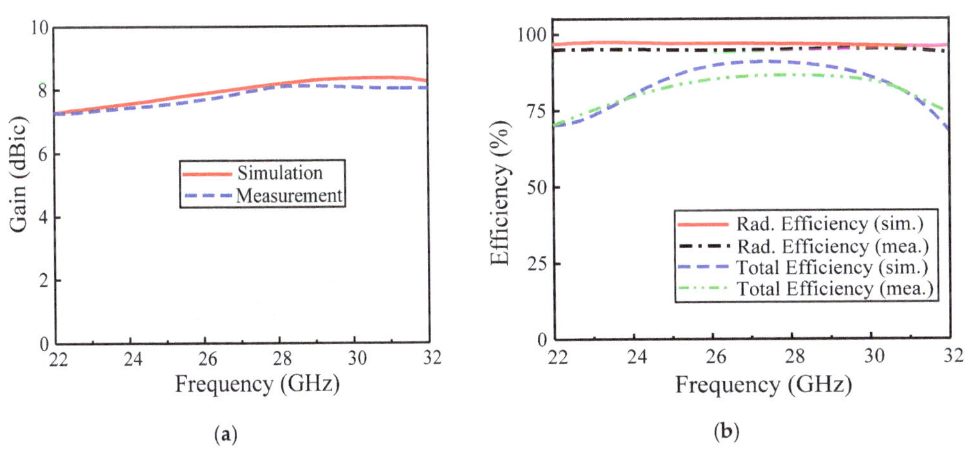

Figure 6. Proposed CP ME dipole antennas: (**a**) broadside gain, (**b**) radiation efficiency and total efficiency.

3.3. Radiation Patterns

The measured and simulated radiation patterns of the proposed antenna are shown in Figure 7. The radiations patterns were analyzed in both the *xz*- and *yz*-principal planes for 28.5 GHz and 29.5 GHz frequencies. Since the LHCP radiations were minor, the antenna had stable RHCP radiation patterns. At 28.5 GHz, the antenna showed a gain of 8.2 dBic, a front to back ratio of 26.9 dB, and an angular beamwidth (3 dB) of 61.1 and 74.3 in *xz*- and *yz*-planes, respectively. At 29.5 GHz, the antenna showed a gain of 8.3 dBic, a front to back ratio of 26.7 dB, and angular beamwidth (3 dB) of 58 and 76.5 in *xz*- and *yz*- planes, respectively.

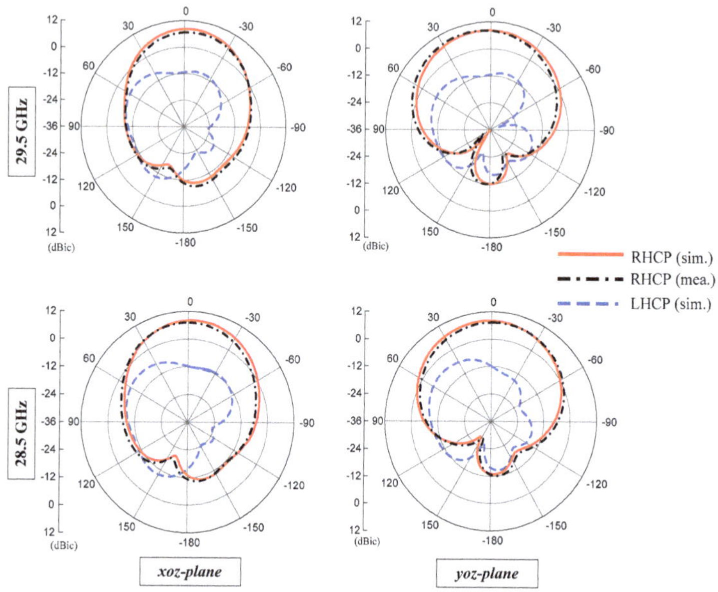

Figure 7. Radiation patterns of the proposed CP ME dipole antenna for different frequencies.

3.4. Performance Comparison

Table 2 shows the comparison of the proposed antenna with the similar state-of-the-art CP ME dipole antennas. To ensure a fair comparison, different performance metrics including antenna design geometry, antenna volume, frequency of operation, polarization, AR bandwidths, and the peak gain value are considered. It can be seen from the table that the proposed antenna demonstrated outstanding performance in terms of overall size, AR bandwidth, and peak gain. Many interesting designs of CP antennas based on ME dipole structure have been reported [8,16–25]. The proposed antenna is the smallest among its competitors with comparable gain and CP bandwidth. The antennas presented in [8,16] have smaller gain narrow bandwidth. The antennas presented in [17–19] have superiority in terms of bandwidth at the expense of larger antenna size, although, their gain performance is comparable with our proposed antenna. The antenna design in [20] has the advantages of higher gain (10.4 dBi), however, it has the limitations of narrower bandwidth and larger antenna profile. Similarly, the mm-wave ME CP antenna in [21] has the merits of wide bandwidth but has lower gain and a larger size. Although the design developed in [22] offers a very wide AR bandwidth due to its novel feeding mechanism, it has limited gain and 3D geometry with a high antenna profile. The CP ME antenna arrays offer wideband and high gain due to their increased number of radiating elements, which, of course, increases the antenna size and complex feeding networks [23–25]. It is noted that our antenna consists of a single element and has a stable gain (average 8 dBic) with a deviation of only ±0.5 dBic. In conclusion, the proposed antenna outperforms the existing CP ME dipole antennas with its highest gain of 8.5 dBic, wide 3 dB AR bandwidth (18.1%), and a small volume of only 0.83 λ_o × 0.83 λ_o × 0.125 $\lambda_{o.}$ Since MIMO antennas with increased gain are the key requirements for the 5G systems [32], this work can be extended for MIMO configuration with enhanced isolation in the future.

Table 2. Performance comparison of the proposed CP ME dipole antenna with the reported works.

Ref.	Antenna Type	Antenna Volume (λ_0^3)	f_c (GHz)	Polarization	3 dB ARBW (%)	Peak Gain (dBic)
[8]	ME dipole antenna	1.54 × 0.48 × 0.291	29	CP	4.8	5.1
[16]	ME dipole antenna	1.18 × 1.05 × 0.122	24	CP	12.8	7.8
[17]	ME dipole antenna	1.6 × 1.3 × 0.26	2.5	CP	71.5	8
[18]	ME dipole antenna	1.1 × 1.1 × 0.29	2.2	CP	47.7	8.6
[19]	ME dipole antenna	6 × 6 × 0.305	60	CP	23.4	8.6
[20]	ME dipole antenna	1.12 × 3.03 × 0.6	60	CP	11.6	10.4
[21]	ME dipole antenna	1 × 1 × 0.315	60	CP	21.9	7.9
[22]	ME dipole antenna	0.85 × 0.85 × 0.18	2.3	CP	53.2	6.6
[23]	ME dipole antenna array	Not given	28.8	CP	9.7	7
[24]	ME dipole antenna array	Not given	35.2	CP	44	19.2
[25]	ME dipole antenna array	Not given	27	CP	27.8	20.2
This work	ME dipole antenna	0.83 × 0.83 × 0.125	28	CP	18.1	8.5

4. Conclusions

An ME dipole antenna with RHCP radiation for unidirectional radiation characteristics is characterized and achieved for 5G mm-wave communication systems. The proposed antenna comprises two pairs of rotational symmetric electric dipoles and two pairs of metallic vias perpendicular to the ground plane, acting as a magnetic dipole. An L-shape probe is used to excite the antenna. Owing to the shape of the electric dipoles together with the complementary effect, it produces CP radiations. This antenna has impedance bandwidth of 24.6 for $|S_{11}| < -10$ dB and 3 dB AR bandwidth of 18.1%. The antenna has achieved a peak gain of 8.5 dBic with a compact overall size of 0.83 λ_o × 0.83 λ_o × 0.125 λ_o. The wide CP bandwidth, stable radiation characteristics, high efficiency, and low profile of the proposed antenna makes it a suitable candidate for 5G smart devices and sensors.

Author Contributions: Conceptualization, N.H. and H.A.; methodology, H.A. and M.A.S.; validation, S.M.L.; formal analysis, N.H. and H.A.; investigation, H.A. and M.A.S.; writing—original draft preparation, H.A. and N.H.; writing—review and editing, N.H. and H.A.; supervision, N.K.; project administration, N.K. All authors have read and agreed to the published version of the manuscript.

Funding: This work was supported in part by Institute for Information and Communication Technology Promotion (IITP) (A Study on Public Health and Safety in a Complex EMF Environment), under grant 2019-0-00102, and in part by Radio Research Agency (RRA) (Development of Rapid Antenna Measurement Technique for Antennas with New Radio Technology).

Institutional Review Board Statement: Not applicable.

Informed Consent Statement: Not applicable.

Data Availability Statement: All data are included in the manuscript.

Acknowledgments: The author acknowledges the research environment provided by OIP Lab, Chungbuk National University, Korea.

Conflicts of Interest: The authors declare no conflict of interest.

References

1. Dangi, R.; Lalwani, P.; Choudhary, G.; You, I.; Pau, G. Study and investigation on 5G technology: A systematic review. *Sensors* **2022**, *22*, 26. [CrossRef] [PubMed]
2. 5G Frequency Bands. Available online: https://halberdbastion.com/technology/cellular/5g-nr/5g-frequency-bands/n257-28-ghz (accessed on 5 January 2022).
3. Nalanagula, R.; Darimireddy, N.K.; Kumari, R.; Park, C.-W.; Reddy, R.R. Circularly polarized hybrid dielectric resonator antennas: A brief review and perspective analysis. *Sensors* **2021**, *21*, 4100. [CrossRef] [PubMed]
4. Sabban, A. Wearable circular polarized antennas for health care, 5G, energy harvesting, and IoT systems. *Electronics* **2022**, *11*, 427. [CrossRef]
5. Trinh, L.H.; Truong, N.V.; Ferrero, F. Low cost circularly polarized antenna for IoT space applications. *Electronics* **2020**, *9*, 1564. [CrossRef]
6. Nguyen, H.Q.; Le, M.T. Multiband ambient RF energy harvester with high gain wideband circularly polarized antenna toward self-powered wireless sensors. *Sensors* **2021**, *21*, 7411. [CrossRef]

7. Hussain, N.; Jeong, M.; Park, J.; Kim, N. A Broadband circularly polarized Fabry-Perot resonant antenna using a single-layered PRS for 5G MIMO applications. *IEEE Access* **2019**, *7*, 42897–42907. [CrossRef]
8. Zhang, C.; Chen, H. A broadband circularly polarized substrate integrated antenna with dual magnetoelectric dipoles coupled by crossing elliptical slots. In Proceedings of the 2018 IEEE 18th International Conference on Communication Technology (ICCT), Chongqing, China, 8–11 October 2018; pp. 564–567.
9. Nkimbeng, C.H.; Wang, H.; Park, I. Coplanar waveguide-fed bidirectional same-sense circularly polarized metasurface-based antenna. *J. Electromagn. Eng. Sci.* **2021**, *21*, 210–217. [CrossRef]
10. Zambak, M.F.; Yasin, M.N.M.; Adam, I.; Iqbal, J.; Osman, M.N. Higher-order-mode triple band circularly polarized rectangular dielectric resonator antenna. *Appl. Sci.* **2021**, *11*, 3493. [CrossRef]
11. Luk, K.M.; Wong, H. A new wideband unidirectional antenna element. *Int. J. Microw. Opt. Technol.* **2006**, *1*, 35–44.
12. Tan, W.; Xiao, Y.; Li, C.; Zhu, K.; Luo, H.; Sun, H. A wide-band high-efficiency hybrid-feed antenna array for mm-Wave wireless systems. *Electronics* **2021**, *10*, 2383. [CrossRef]
13. Wang, Y.; Tan, W.; Zhu, K.; Luo, H.; Zhao, G.; Sun, H. Design of A Ka-band 3D-printed dual-polarization magnetoelectric dipole antenna array with low sidelobe. *Electronics* **2021**, *10*, 2969. [CrossRef]
14. Mak, K.M.; Luk, K.M. A circularly polarized antenna with wide axial ratio beamwidth. *IEEE Trans. Antennas Propag.* **2009**, *57*, 3309–3312.
15. Ta, S.X.; Park, I. Crossed dipole loaded with magneto-electric dipole for wideband and wide-beam circularly polarized radiation. *IEEE Antennas Wirel. Propag. Lett.* **2014**, *14*, 358–361. [CrossRef]
16. Wu, S.; Zhao, J.; Xu, J. A circularly polarized low-profile magnetoelectric dipole antenna. *Microw. Opt. Technol. Lett.* **2021**, *63*, 2852–2858. [CrossRef]
17. Kang, K.; Shi, Y.; Liang, C.-H. A wideband circularly polarized magnetoelectric dipole antenna. *IEEE Antennas Wirel. Propag. Lett.* **2017**, *16*, 1647–1650. [CrossRef]
18. Li, M.; Luk, K.-M. A wideband circularly polarized antenna for microwave and millimeter-wave applications. *IEEE Trans. Antennas Propag.* **2014**, *62*, 1872–1879. [CrossRef]
19. Ruan, X.; Qu, S.-W.; Zhu, Q.; Ng, K.B.; Chan, C.H. A complementary circularly polarized antenna for 60-GHz applications. *IEEE Antennas Wirel. Propag. Lett.* **2016**, *16*, 1373–1376. [CrossRef]
20. Bai, X.; Qu, S.-W.; Yang, S.; Hu, J.; Nie, Z.-P. Millimeter-wave circularly polarized tapered-elliptical cavity antenna with wide axial-ratio beamwidth. *IEEE Trans. Antennas Propag.* **2015**, *64*, 811–814. [CrossRef]
21. Zhu, Q.; Ng, K.-B.; Chan, C.H. Printed circularly polarized spiral antenna array for millimeter-wave applications. *IEEE Trans. Antennas Propag.* **2016**, *65*, 636–643. [CrossRef]
22. Ding, X.; Wu, S.; Yang, G.; Lan, H. Dual Circularly polarized wideband magneto-electric dipole antenna for wireless applications. *Prog. Electromagn. Res. Lett.* **2022**, *102*, 27–36. [CrossRef]
23. Wang, Y.F.; Wu, B.; Zhang, N.; Zhao, Y.T.; Su, T. Wideband circularly polarized magneto-electric dipole 1x2 antenna array for millimeter-wave applications. *IEEE Access* **2020**, *8*, 27516–27523. [CrossRef]
24. Xiang, L.; Wu, F.; Yu, C.; Jiang, Z.H.; Yao, Y.; Hong, W. A wideband circularly polarized magneto-electric dipole antenna array for millimeter-wave applications. *IEEE Trans. Antennas Propagation.* **2021**. [CrossRef]
25. Feng, B.; Lai, J.; Chung, K.L.; Chen, T.Y.; Liu, Y. A compact wideband circularly polarized magneto-electric dipole antenna array for 5G millimeter-wave application. *IEEE Trans. Antennas Propag.* **2020**, *68*, 6838–6843. [CrossRef]
26. Causse, A.; Rodriguez, K.; Bernard, L.; Sharaiha, A.; Collardey, S. Compact bandwidth enhanced cavity-backed magneto-electric dipole antenna with outer G-shaped probe for GNSS bands. *Sensors* **2021**, *21*, 3599. [CrossRef]
27. Hussain, N.; Jeong, M.; Abbas, A.; Kim, N. Metasurface-based single-layer wideband circularly polarized MIMO antenna for 5G millimeter-wave systems. *IEEE Access* **2020**, *8*, 130293–130304. [CrossRef]
28. Hussain, N.; Jeong, M.; Abbas, A.; Kim, T.; Kim, N. A metasurface-based low-profile wideband circularly polarized patch antenna for 5G millimeter-wave systems. *IEEE Access* **2020**, *8*, 22127–22135. [CrossRef]
29. Li, R.-L.; Fusco, V.F.; Nakano, H. Circularly polarized open-loop antenna. *IEEE Trans. Antennas Propag.* **2003**, *51*, 2475–2477.
30. Askari, H.; Hussain, N.; Domin Choi, M.; Sufian, A.; Abbas, A.; Kim, N. An AMC-based circularly polarized antenna for 5G sub-6 GHz communications. *CMC-Comput. Mater. Contin.* **2021**, *69*, 2997–3013. [CrossRef]
31. Lim, J.H.; Lee, J.W.; Lee, T.K. Cross-pol pattern effects of parabolic reflector antennas on the performance of spaceborne quad-pol SAR systems. *J. Electromagn. Eng. Sci.* **2021**, *21*, 218–227. [CrossRef]
32. Lavdas, S.; Gkonis, P.K.; Zinonos, Z.; Trakadas, P.; Sarakis, L. An adaptive hybrid beamforming approach for 5G-MIMO mmWave wireless cellular networks. *IEEE Access* **2021**, *9*, 127767–127778. [CrossRef]

sensors

Communication

A Conformal Frequency Reconfigurable Antenna with Multiband and Wideband Characteristics

Niamat Hussain [1], Adnan Ghaffar [2], Syeda Iffat Naqvi [3], Adnan Iftikhar [4], Dimitris E. Anagnostou [5] and Huy H. Tran [6,*]

[1] Department of Smart Device Engineering, School of Intelligent Mechatronics Engineering, Sejong University, Seoul 05006, Korea; niamathussain@sejong.ac.kr
[2] Department of Electrical and Electronics Engineering, Auckland University of Technology, Auckland 1010, New Zealand; aghaffar@aut.ac.nz
[3] Telecommunication Engineering Department, University of Engineering & Technology Taxila (UET Taxila), Taxila 47050, Pakistan; iffat.naqvi@uettaxila.edu.pk
[4] Department of Electrical and Computer Engineering, COMSATS University Islamabad, Islamabad 45550, Pakistan; adnaniftikhar@comsats.edu.pk
[5] Institute of Signals, Sensors and Systems, Heriot Watt University, Edinburgh EH14 4AS, UK; d.anagnostou@hw.ac.uk
[6] Faculty of Electrical and Electronic Engineering, PHENIKAA University, Hanoi 12116, Vietnam
* Correspondence: hung.tranhuy@phenikaa-uni.edu.vn

Abstract: A compact flexible multi-frequency antenna for smart portable and flexible devices is presented. The antenna consists of a coplanar waveguide-fed slotted circular patch connected to a rectangular secondary resonator (stub). A thin low-loss substrate is used for flexibility, and a rectangular stub in the feedline is deployed to attain wide operational bandwidth. A rectangular slot is etched in the middle of the circular patch, and a *p-i-n* diode is placed at its center. The frequency reconfigurability is achieved through switching the diode that distributes the current by changing the antenna's electrical length. For the ON state, the antenna operates in the UWB region for −10 dB impedance bandwidth from 2.76 to 8.21 GHz. For the OFF state of the diode, the antenna operates at the ISM band (2.45/5.8 GHz), WLAN band (5.2 GHz), and lower X-band (8 GHz) with a minimum gain of 2.49 dBi and a maximum gain of 5.8 dBi at the 8 GHz band. Moreover, the antenna retains its performance in various bending conditions. The proposed antenna is suitable for modern miniaturized wireless electronic devices such as wearables, health monitoring sensors, mobile Internet devices, and laptops that operate at multiple frequency bands.

Keywords: compact antenna; frequency reconfigurable; multiband; conformal antenna

Citation: Hussain, N.; Ghaffar, A.; Naqvi, S.I.; Iftikhar, A.; Anagnostou, D.E.; Tran, H.H. A Conformal Frequency Reconfigurable Antenna with Multiband and Wideband Characteristics. *Sensors* **2022**, *22*, 2601. https://doi.org/10.3390/s22072601

Academic Editor: Adrian Bekasiewicz

Received: 15 February 2022
Accepted: 23 March 2022
Published: 29 March 2022

1. Introduction

The advancements in wireless technology and electromagnetic spectrum limitations have led to the development of multi-standard and multi-application devices. Considering this, an antenna with the characteristic of adaptability to various practical applications and standards is necessary. Hence, due to the dynamic characteristics and capability to modify properties such as polarization, radiation pattern, and frequency, along with system requirements, reconfigurable antennas have recently received a large amount of attention [1–8]. In particular, a frequency reconfigurable antenna is beneficial for various applications. Frequency reconfigurability can be achieved by using electrical switches [9–11], varactor diodes [12–15], *p-i-n* diodes [16–19], and radio-frequency micro-electromechanical systems (RF-MEMS) [20–22]. The electric switching technique has the advantage of lower voltage requirements, whereas RF-MEMS provide higher switching time. The *p-i-n* diodes have been widely used as reconfigurable techniques due to characteristics such as compactness and good switching time (1 to 100 microseconds) [23].

Moreover, the increasing requirement of compact, conformal devices, and wearable gadgets has drawn researchers' attention towards flexible, low profile, and light-weight antennas. Thus, in addition to reconfigurability, flexibility is also a significant characteristic required by modern-day applications such as e-health monitoring, biosensing, and e-utility [24]. Although frequency reconfigurable antennas have been investigated widely in recent years, most of these designs have used rigid substrates. By comparison, few antenna designs have been recently reported in the literature, in which reconfigurability and flexibility have been integrated [25–29]. The antenna design proposed in [25] is an inkjet-printed flexible, reconfigurable antenna with overall dimensions of 53×31 mm^2. In addition, a *p-i-n* diode is used for reconfigurability purposes. In another work [26], a reconfigurable antenna is reported for wearable applications. For this antenna, reconfigurability is realized using a *p-i-n* diode. In order to improve the performance of the antenna, an artificial magnetic conductor (AMC) surface is assimilated with the antenna. The proposed structure has an overall size of 83×89 mm^2. Similarly, the work in [27] presented an inkjet-printed conformal antenna with geometrical dimensions of 30×40 mm^2, where frequency reconfigurability is achieved by employing two diodes. Another reconfigurable antenna is reported in [28] with a substrate size of 59.8×59.8 mm^2. In addition to reconfigurability, the antenna is flexible, and a conductive fabric is used to design the antenna on a polydimethylsiloxane (PDMS) substrate. Moreover, the antenna design demonstrated in [29] is a conformal antenna, and frequency reconfigurability is obtained using two *p-i-n* diodes. The geometrical size of the reported structure is 50×30 mm^2. The works in [30,31] proposed flexible antennas with frequency reconfigurability. The proposed system presented in [30] obtains reconfigurability by employing two *p-i-n* diodes, and the dimensions of the substrate are 24×19 mm^2 with a thickness of 1.53 mm. It can be observed that, although the reported antennas discussed here are both flexible and reconfigurable, these antennas have relatively larger dimensions, which restrict their usage in wearable and compact devices. In addition, these antennas have a relatively low gain. Thus, it can be deduced from the aforementioned discussion that a compact, high gain, flexible, and reconfigurable frequency antenna having a practical demonstration using diodes is still a challenge for researchers.

In order to overcome the limitations and discrepancies of the earlier reported antenna designs, this work proposes a compact and flexible antenna design for the ISM, WLAN, X-band, and UWB frequency bands. In addition, frequency reconfigurability is achieved by incorporating a *p-i-n* diode. Hence, the three bands can be merged into a single wideband (2.81–8.41 GHz) using this *p-i-n* diode. The omnidirectional radiation pattern and stable performance over a wide range of frequencies make this antenna desirable for a variety of applications.

2. Antenna Design Methodology

This section is divided into subsections for the ease of understanding.

2.1. Geometry of the Proposed Design

A circular planar radiator was chosen as the basis of the antenna design due to its inherent wide bandwidth. A coplanar waveguide (CPW) feed technique is used to excite a circular radiating patch that has a rectangular slot at its center. The circular patch has a radius $R = 11$ mm. The resonant frequency (f) and the corresponding radius of the patch are estimated using [31]:

$$f = \frac{1.8412 \cdot c}{4\pi R_{eff}\sqrt{\varepsilon_{eff}}} \quad (1)$$

where c is the free-space speed of light (3×10^8 m/s), and R_{eff} is the effective radius of the patch whose value is estimated using:

$$R_{eff} = R\sqrt{1 + \frac{2H}{\pi\varepsilon_{eff}R}\left(ln\left(\frac{\pi R}{2H}\right) + 1.7726\right)} \quad (2)$$

In (2), R is the physical radius of the patch, H is the thickness of the substrate and ε_{eff} is the effective dielectric constant, which can be calculated by:

$$\varepsilon_{eff} = \frac{\varepsilon_r + 1}{2} + \frac{\varepsilon_r - 1}{2}\left\{\left(1 + 12\frac{H}{A_x}\right)^{-0.5} + 0.04\left(1 - \frac{H}{A_x}\right)^2\right\} \quad (3)$$

where ε_r and A_x are the dielectric constant and the width of the substrate, respectively.

The schematic of the proposed frequency reconfigurable antenna is shown in Figure 1. A rectangular stub with dimensions $Sx \times Sy$ is also inserted between the patch and feedline. The stub is employed to enhance the relatively narrow bandwidth of the circular patch. A rectangular slot is etched on the circular radiator to add additional capacitive load and allow the *p-i-n* diode to be inserted at the center of the slot to achieve frequency reconfigurability. The rectangular slot and the *p-i-n* diode control the amount of current by electrically connecting and disconnecting the upper part of the radiator with the lower part. With this arrangement, the diode generates the multiband and UWB modes. A detailed discussion on the working principle of the proposed antenna is presented in the subsequent sections.

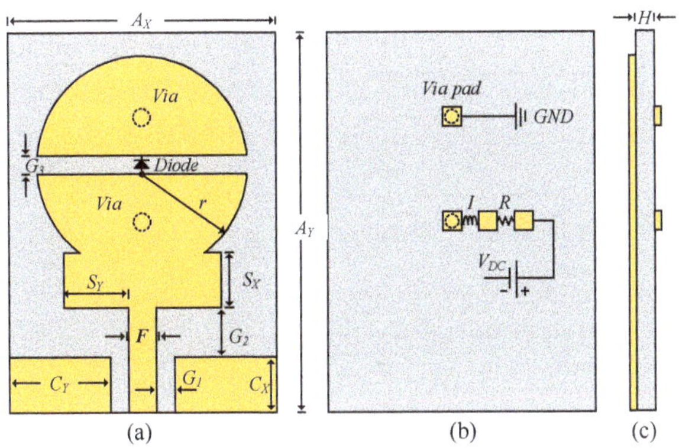

Figure 1. Schematic of proposed frequency reconfigurable antenna: (**a**) top-view, (**b**) bottom-view, and (**c**) side-view.

The antenna is printed on a thin (h = 0.254 mm) Rogers RT5880LZ substrate with relative permittivity ε_r = 2.1 and loss tangent $tan\delta$ = 0.0024 [32] This substrate is made of polytetrafluoroethylene (PTFE) composites and was chosen for its low-loss and flexible nature. These attributes make it a desirable substrate for flexible antennas.

2.2. Simulation Setup

The commercial finite element method-based HFSS [33] was used to simulate the proposed antenna. To avoid unwanted fabrication tolerance errors and effects from the SMA connector on the antenna, a 3D model of the 50-Ω SMA connector was designed and used to excite the proposed antenna in simulations. To model the real *p-i-n* diode switches, the equivalent circuit model of the Infineon model 3 BAR-50C-SC79 [34] was incorporated using lumped element parallel RLC boundary conditions. The equivalent circuit model of the diode in ON and OFF states is shown in Figure 2. In the ON state, a 4.5 Ω resistor is in series with a 0.15 nH inductor (Figure 2a). In the OFF state, a 0.15 nH inductor is in series with the parallel combination of a 5 kΩ resistor and a 0.15 pF capacitor (Figure 2b). Keeping in mind the practical issues related to measurements i.e., limiting the RF and DC current flowing towards the diode, an RF choke comprising a 68 nH inductor and a 1-kΩ resistor were utilized, as shown in Figure 2c. In addition, to accommodate biasing circuitry for the diode during measurements and eliminate any effect of the biasing circuitry on the radiating patch, two small biasing pads were incorporated on the bottom side of the proposed antenna. The optimized design is shown in Figure 1 and has the following dimensions: A_Y = 35 mm, A_x = 25 mm, H = 0.254 mm, Cx = 5 mm, C_y = 11 mm, S_x = 6 mm, S_y = 9.5 mm, r = 11 mm, F = 1 mm, G_3 = 1 mm, G_1 = 1 mm, G_2 = 2.3 mm.

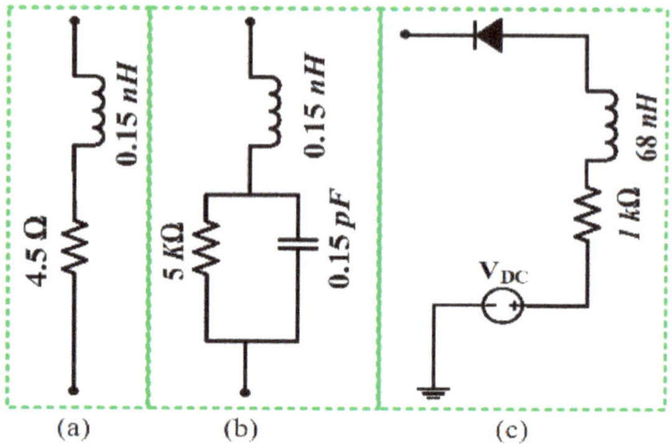

Figure 2. Diode equivalent model: (**a**) ON-state, (**b**) OFF-state, and (**c**) biasing circuit.

2.3. UWB Antenna Design

The fundamental CPW fed circular patch antenna was designed using (1)–(3). The resulting antenna resonates at 2.45 GHz with an impedance bandwidth of 2.25–2.72 GHz. Figure 3 (blue curve) illustrates the reflection coefficient of the elementary circular radiator. To broaden the narrowband operation, various techniques, including metamaterials, complex geometrical structures such as DGS, and etching slots, have been adopted [24–27]. Here, a uniplanar rectangular secondary resonator (stub) is introduced between the circular radiator and the feed. The stub acts as a high-frequency resonator and adds higher resonating frequencies to the circular radiator, resulting in a more wideband antenna. In other words, the field distribution of the conventional monopole antenna is altered due to the insertion of the rectangular stub, which supports multiple higher-order resonances instead of having only the single matched resonance offered by the circular radiator. The added bandwidth of the stub-loaded antenna results in a combined 5.8 GHz (2.4–3.8 GHz and 4.8–9.2 GHz) bandwidth, instead of only the 470 MHz (2.25 GHz–2.72 GHz) bandwidth that was achieved by the circular monopole itself.

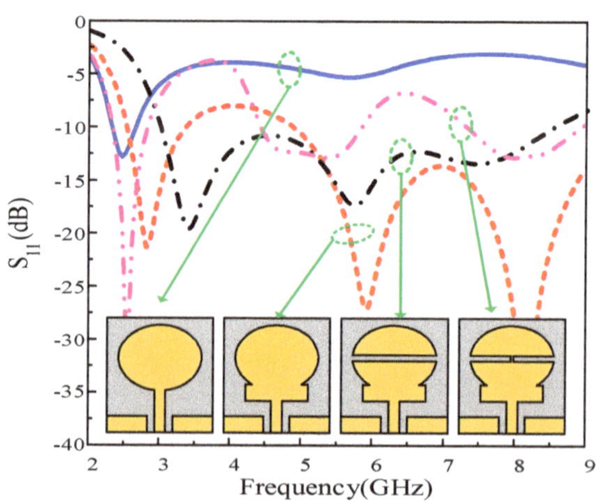

Figure 3. $|S_{11}|$ comparison among various steps included in the antenna design.

Although a wideband antenna was achieved, a small portion of the bandwidth was slightly mismatched. Capacitor loading is a well-known technique to enable lower resonances and improve impedance matching [35]. Instead of loading the antenna with a physical capacitor, here, an electrical method was employed by etching a rectangular slot at the center of the circular radiator. The slot thickness was optimized so that the mismatched band could be matched by controlling the capacitance generated by the slot. The impedance bandwidth of the antenna before and after the insertion of the slot is shown in Figure 3 (black curve). It is observed that the stub-loaded antenna with the slot exhibits an ultrawide impedance bandwidth of 5.6 GHz, ranging from 2.81 GHz to 8.41 GHz.

2.4. Multiband Antenna Design

The designed UWB antenna was further utilized to develop a tri-band antenna. A small path is provided for the current to flow from the lower to the upper part of the radiator through a *p-i-n* diode in the ON state, as shown in Figure 3 (magenta curve). This path alters the capacitance of the slot, and the antenna exhibits a tri-band resonance. The first notch band was expected due to the disturbance in the capacitive load of the antenna. It is observed from Figure 3 that the antenna with the stub has a notched band inside the UWB region. The second notched band is due to the presence of two rectangular slots formed as a result of setting the diode ON. These slots behave like a band stop filter and thus cause higher band mitigation on the radiator. Moreover, a significantly lower current is present around the stub, which results in the suppression of the 3.75 and 6.5 GHz bands, as depicted in Figures 4b and 4d, respectively. The geometric modifications in the resultant antenna exhibit three passbands having resonances at 2.45, 5.5, and 8 GHz, as shown in Figure 3 (magenta curve).

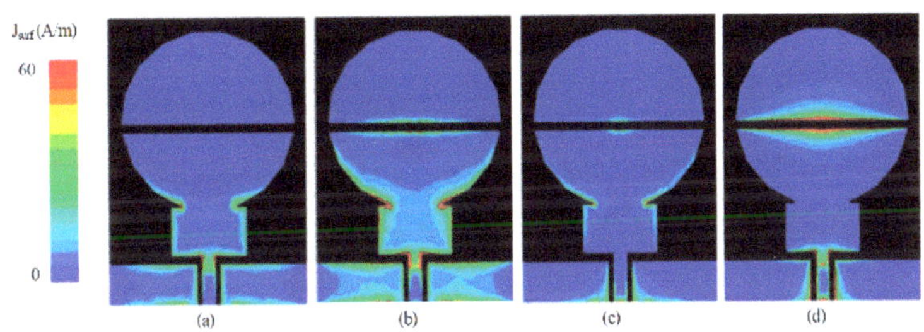

Figure 4. Distribution of current density on the surface of antenna at (**a**) 3.75 GHz [diode-OFF], (**b**) 3.75 GHz [diode-ON], (**c**) 6.5 GHz [diode-OFF], and (**d**) 6.5 GHz [diode-ON]. Comparison among the various steps included in the antenna design.

2.5. Parametric Analysis

A parametric study was performed to analyze the effects of the different antenna parameters on the antenna impedance. For better understanding, the diode OFF state was parametrically analyzed. It was noticed that the length (*Sy*) and width (*Sx*) of the rectangular patch deployed between the CPW and circular patch plays a key role in matching the impedance at different frequencies by controlling the amount of current flow on the antenna geometry. Figure 5a shows the effect of *Sy* on $|S_{11}|$. An increase in the length of Sy from the optimized value of 9.5 mm results in better matching while disturbing the operational bandwidth. Conversely, a decrease in Sy results in comparatively better bandwidth. However, the reflection coefficient increases significantly.

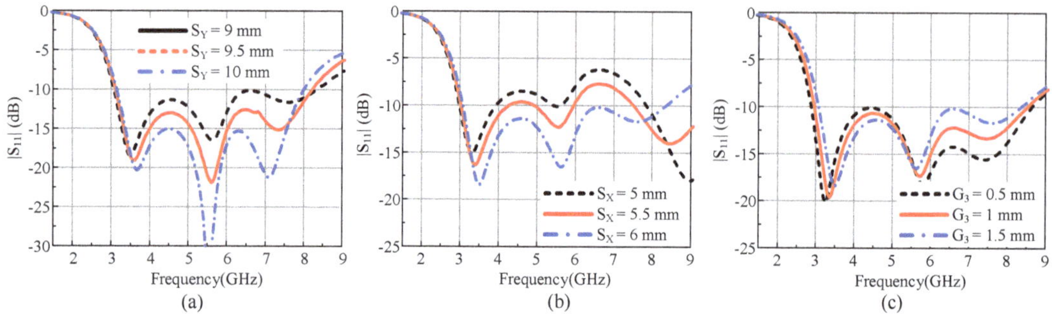

Figure 5. $|S_{11}|$ analysis for different (**a**) length of rectangular patch *Sy*, (**b**) width of rectangular patch *Sx*, and (**c**) width of slot G_3.

Similarly, Figure 5b shows that a reduction in the length of *Sx* from the optimized value of 6 mm results in better matching ($|S_{11}| < -10$ dB) at higher frequencies. Contrarily, by decreasing the width of the slot G_3 from the optimized 1 mm, the reflection coefficient increases at lower frequencies, whereas an increase in G_3 results in a mismatch at higher frequencies (Figure 5c). Considering these parameters, the optimized values were chosen to achieve the maximum bandwidth by considering $|S_{11}| < -10$ dB.

3. Results and Discussion

The simulation results and its comparison with measured results are presented in this discussion.

3.1. Measurement Setup

To validate the working principle of the antenna, the antenna shown in Figure 1 was fabricated, and a photograph of the fabricated prototype is shown in Figure 6. Standard chemical etching was used for the fabrication and the scattering (S) parameters of the antenna were measured using a calibrated HP 8720D Vector Network Analyzer (VNA). To practically verify the reconfigurable operation, an Infineon (#BAR-50C SC79) *p-i-n* diode was soldered to the top side of the antenna, as depicted in Figure 6a. The biasing circuit was defined on the backside of the antenna (see Figure 6b) to prevent degradation of the radiation characteristics of the antenna. Two conducting vias were drilled in the radiating patches to provide bias voltages for the diode operation. A battery of 3 V was connected for the flow of current through the resistor and inductor, to turn on the diode, named ON-state. When the battery was disconnected, no current flowed through the diode and it behaved like an open circuit, referred to as its OFF-state.

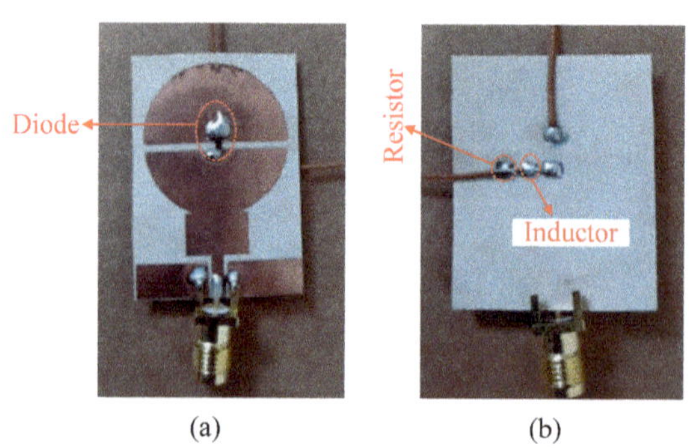

Figure 6. Fabricated prototype: (**a**) top-view and (**b**) bottom view.

3.2. Scattering Parameters

In Figure 7, the measured and simulated S-parameters are compared. When the diode is OFF, the antenna exhibits matching less than −10 dB over the 5.34 GHz band ranging from 2.76 to 8.1 GHz, compared with the simulated 5.6 GHz from 2.81 to 8.41 GHz. When the diode is switched ON, the antenna resonates at three frequencies: 2.47, 5.25, and 8.1 GHz, having an impedance bandwidth of 920 MHz (2.12–3.2 GHz), 2170 MHz (3.95–6.12 GHz), and 1200 MHz (7.71–8.83 GHz), respectively, as depicted in Figure 7b. The respective simulated values show resonances at 2.45, 5.2, and 8 GHz and with similar respective bandwidths.

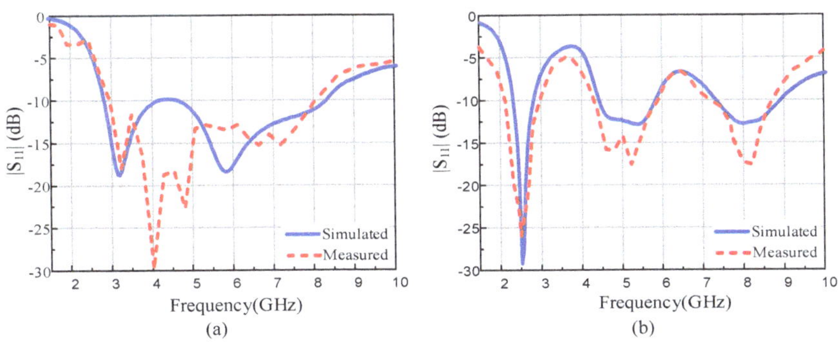

Figure 7. Simulated and measured $|S_{11}|$: (**a**) diode ON and (**b**) diode OFF.

3.3. Conformability Analysis

Flexibility and conformability are key requirements of flexible devices and a key advantage of the proposed antenna. Ideally, the antenna radiation should remain unchanged under both flat and flexed conditions. The conformability analysis was performed by bending the antenna on a cylindrical foam along the *x*- and *y*-axis, as depicted in Figure 8. The radius of the foam cylinder was chosen to be 20 mm as a realistic arm radius. With the diode in the OFF state, the antenna exhibits wide operational bandwidth and good agreement between simulations and measurements under both bending scenarios.

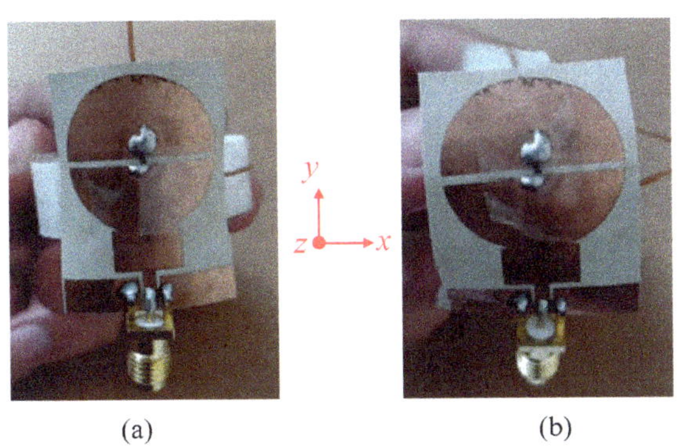

Figure 8. Antenna under conformal condition: (**a**) bending along the *x*-axis and (**b**) bending along the *y*-axis.

Similarly, Figure 9b illustrates that with the diode ON, the antenna exhibits a tri-band mode with almost identical $|S_{11}|$. With the overall flat and flexed conditions having practically similar performance, the application of this antenna for both rigid and flexible/wearable devices was validated.

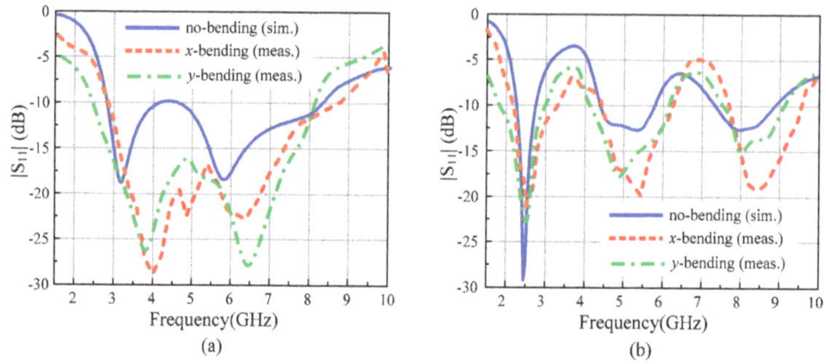

Figure 9. Conformability analysis of the proposed antenna: (**a**) diode ON, (**b**) diode OFF.

3.4. Far-Field Analysis

To observe the far-field radiation, the antenna was measured in a calibrated anechoic chamber as shown in Figure 10. The fabricated prototype was placed on a turntable in front of a broadband double-ridged horn at a far-field distance. Figure 11 shows the measured radiation patterns of the antenna for *p-i-n* diode ON and OFF at both *E*- and *H*-planes. The antenna has near-omnidirectional patterns at 2.45, 5.2, and 8 GHz in the principal *H*-plane, whereas for the *E*-plane, a tilted bi-directional pattern is observed, which is more prominent at higher frequencies (Figure 11a–c). A similar omni-directional *H*-plane pattern is observed for the ON state with a slightly tilted bi-directional *E*-plane at the selected frequencies of 3.2 and 5.8 GHz (Figure 12a,b). Overall, excellent agreement between measurements and simulations is observed at all frequencies for the diode's ON and OFF states.

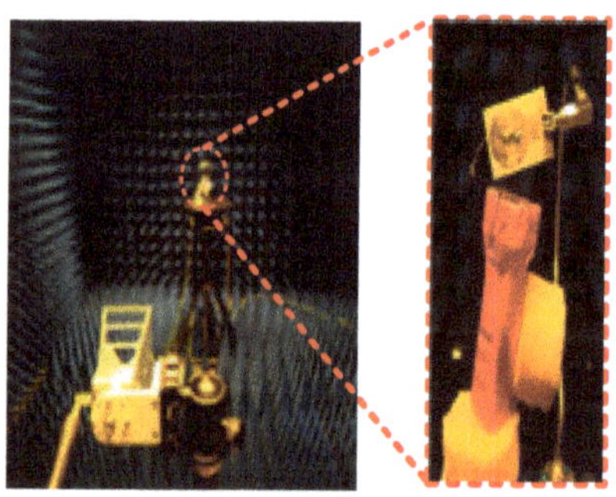

Figure 10. Far-field radiation pattern measurement setup.

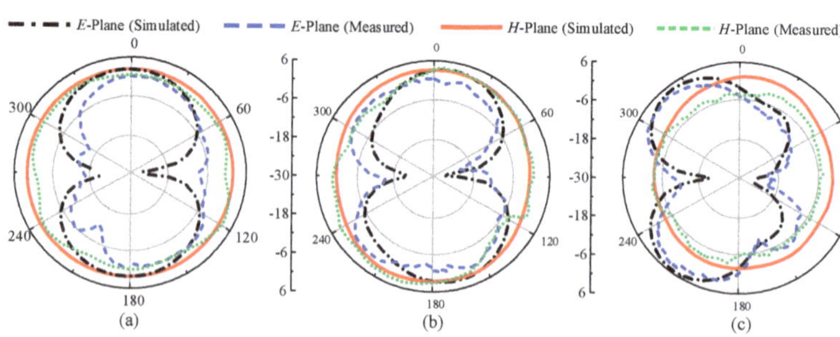

Figure 11. Radiation patterns of the proposed antenna for multiband mode: (**a**) 2.45 GHz, (**b**) 5.2 GHz, and (**c**) 8 GHz.

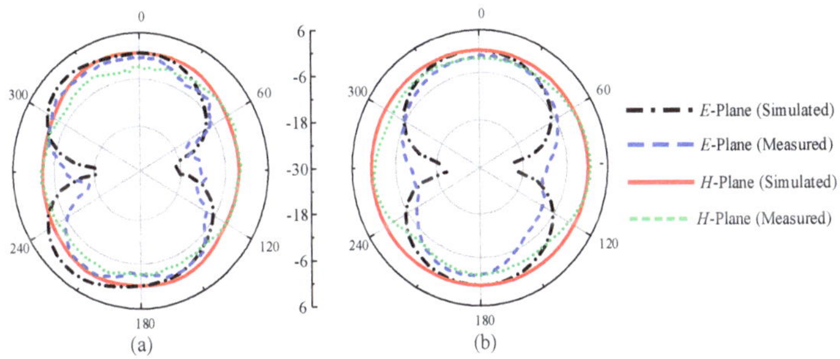

Figure 12. The proposed antenna's radiation patterns for UWB mode: (**a**) 3.2 GHz and (**b**) 5.8 GHz.

Figure 13a compares the simulated and measured gain in the ON and OFF states. The antenna exhibits a minimum gain of 2.49 dBi at 2.45 GHz and a maximum gain of 5.8 dBi at 8 GHz for the passband. Moreover, the gain decreases by up to −3 dBi in the band stop regions, which suffices to reject potential undesired interference in the ON state. Similarly, the simulated efficiency of the antenna has a minimum value of 80% in the operational band, whereas in the band stop region it decreases by up to 22%, as shown in Figure 13b. Thus, the antenna efficiently operates in the UWB and tri-band modes, depending upon the user requirements, by simply switching a single *p-i-n* diode.

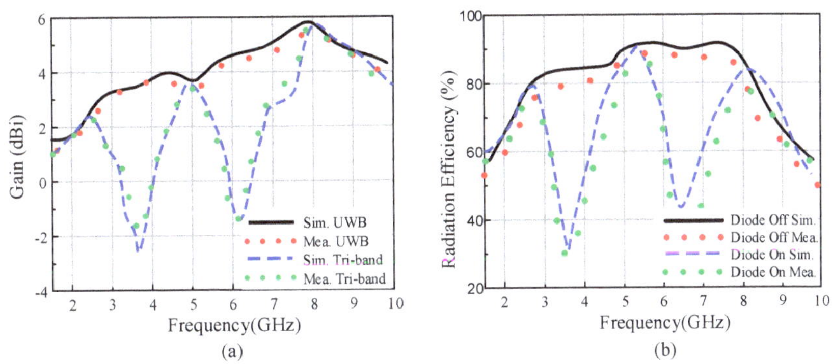

Figure 13. (**a**) Gain and (**b**) radiation efficiency of the proposed antenna.

3.5. Performance Comparision

The presented work was compared with the state-of-the-art works reported in literature, as summarized in Table 1. It can be observed that few works [17,18,27] are reconfigurable with good impedance bandwidth and significant gain; however, these designs have been employed on rigid substrates, which limit their effectiveness for various wireless communication applications. Other works presented in [25,26,28–30], demonstrate the advantages of reconfigurability and flexibility; however, these antenna structures either have larger dimensions or exhibit narrow bandwidth and low gain as compared to the proposed antenna. It is worth noting that this design uses only a single diode to achieve multiple reconfigurable bands. This comparative analysis verifies the usefulness and suitability of the proposed reconfigurable antenna for various modern-day wireless communication systems.

33. *Ansys*, version 2016.2; Ansys Corporation: Pittsburgh, PA, USA, 2017.
34. Infineon Technologies. Available online: www.infineon.com (accessed on 30 March 2020).
35. Ghaffar, A.; Awan, W.A.; Hussain, N.; Ahmad, S.; Li, X.J. A Compact Dual-Band Flexible Antenna for Applications at 900 and 2450 MHz. *Prog. Electromagn. Res. Lett.* **2021**, *99*, 83–91. [CrossRef]

sensors

Communication

Wideband Filtering Slot Antenna Design with Stable Gain Using Characteristic Mode Analysis

Chao Ni [1,2], Biyang Wen [1,*], Weijun Wu [2] and Ping Ren [2]

1 Radar and Signal Processing Laboratory, Electronic Information School, Wuhan University, Wuhan 430064, China; nichao@whu.edu.cn
2 Science and Technology on Electromagnetic Compatibility Laboratory, China Ship Development and Design Centre, Wuhan 430060, China; emc1218wu@126.com (W.W.); 13476061860@163.com (P.R.)
* Correspondence: bywen@whu.edu.cn

Abstract: A filtering slot antenna with a simple structure combination using characteristic mode analysis (CMA) is proposed. To realize filtering characteristics, characteristic magnetic currents of line and ring slots are analyzed and designed. Then, the folding-line slot and double-ring slot are selected to realize radiation null separately and combined to construct the basic slot antenna. By properly exciting the selected characteristic modes, a wide filtering bandwidth and a stable gain are obtained. To validate the design process, a prototype antenna with a finite ground plane of about 1.1 $\lambda \times 1.1\ \lambda$ is designed and fabricated. Simulated and measured results agree well, which both show a sharping roll rate in the lower and higher frequency and a flat gain realization in the pass band. The filtering bandwidth is 32.7%, the out-of-band suppression level at the higher frequency is over 20 dB, and the gain in the working frequency varies from 3.9 to 5.2 dB.

Keywords: filtering antenna; stable gain; characteristic mode analysis; radiation nulls; magnetic current

Citation: Ni, C.; Wen, B.; Wu, W.; Ren, P. Wideband Filtering Slot Antenna Design with Stable Gain Using Characteristic Mode Analysis. *Sensors* **2022**, *22*, 2780. https://doi.org/10.3390/s22072780

Academic Editors: Naser Ojaroudi Parchin, Chan Hwang See and Raed A. Abd-Alhameed

Received: 16 March 2022
Accepted: 3 April 2022
Published: 5 April 2022

1. Introduction

With the rapid development of wireless communication technology, compact size, high integration, and multiple functions are highly demanded in electronic systems including various RF circuits and components. Therefore, due to the advantages of selectivity, out-of-band suppression, antennas with filtering response and compact size attract lots of attention [1–15]. It is common to design filters and antennas separately and cascade them with good impedance matching to achieve filtering characteristics, which inevitably brings insertion loss and introduces extra sizes. Therefore, new co-design methods for filter and antenna integration are proposed. One typical approach for filtering antenna realization is to replace the filter's last stage resonator with the antenna radiator [6–8]. Antennas using this method usually need multiple filter resonators and specific antenna radiator selection, which would introduce extra area and insertion loss. Another approach is to introduce specific element units to realize radiation nulls in the far zone. These units include stacked patch and multiple shorting pins [9], metasurfaces, and parasitic elements [10–12]. As this design introduces additional units as well, the whole profile is still high and the design methods are relatively complex.

Nowadays, due to its direct insight into the antenna radiation principle, CMA becomes popular [16–22]. It has been found that simple structures without complex combinations can be used to achieve wide bandwidth and other outstanding characteristics [16–19]. Therefore, with a special design, it is possible to obtain a wideband filtering antenna if the characteristic current of the antenna can cancel each other. However, after our further survey, we find that there are few published reports on filtering antenna realization using CMA.

In this paper, a simple-structured combined-slot antenna with filtering response is proposed and analyzed using CMA. The antenna employs two simple slot structures including a deformed folding-line slot and a double-ring slot. The deformed folding-line slot is designed from the basic line slot, and the double-ring slot is designed from the single-ring. Using CMA, specific characteristic modes and characteristic magnetic currents are selected to generate radiation nulls. Then, the filtering characteristic is achieved by merging the radiation nulls with a properly designed feeding structure. To demonstrate the design process, a prototype of the antenna is fabricated and measured. Simulated and measured results both indicate that the proposed slot-combined antenna obtains a wide filtering bandwidth and a flat realized gain with sharping roll rate.

This paper is organized as follows. In Section 2, the antenna design stages are given, and two simple slot structures including the folding-line slot and the double-ring slot are analyzed separately using CMA. Then, the filtering antenna with a combination of the above two slots is proposed and its CMs are analyzed to explain the filtering achievement. In Section 3, a prototype antenna is fabricated and measured to verify the total analysis and the design process. Section 4 gives the conclusion of this paper.

2. Antenna Design

2.1. Antenna Design Stages

It has been proven that combined CMs can be introduced by a probe-fed slot antenna, and the characteristic magnetic currents with proper excitation on the slot antenna can be used to broaden the antenna's bandwidth with additional stubs [19]. Inspired by the fact that radiation nulls can be generated if electric or magnetic currents on the antenna flow in opposite directions, we consider that slot antennas using CMA can be easily used to bring radiation nulls in specific frequencies. Furthermore, filtering response can be achieved when two radiation nulls are realized separately. Figure 1 shows the filtering antenna design stages. Firstly, two simple structures including a sing-ring slot and line slot are selected. Based on the basic slot structures in Ant.1, Ant.2 can be created to realize radiation nulls using CMA. Then, a double-ring slot and folding-line slot are combined (Ant.3) to achieve filtering characteristics. To improve the impedance matching, Ant.4 (the proposed design) is designed based on Ant.3 by introducing a pair of additional arc-shaped slots inside the inner ring, which are used for impedance matching.

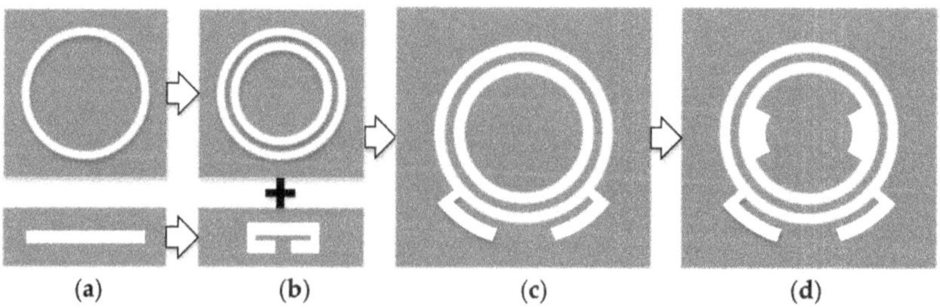

Figure 1. Antenna design stages. (**a**) antenna 1, (**b**) antenna 2, (**c**) antenna 3, (**d**) antenna 4.

In the following section, a detailed analysis is given using CMA, including a folding-line slot antenna, double-ring slot antenna, and combined-slot antenna. All the simulations performed below were carried out with CADFEKO Suite 7.0. All the cases discussed below are presented on the FR4 substrate, its permittivity ε_r is 4.4 and its height is 0.5 mm. For simulation simplicity, the folding-line slot and double-ring slot are fed by a metal probe, whose both ends are connected with the metal ground. In addition, infinite ground planes were used, and planar Green's function aperture was adopted to simulate the slot [23]. The radius of the feeding probe for simulation is 0.2 mm. The white part refers to the slot, the gray part refers to the metal ground, and the red line refers to the feeding probe.

2.2. Folding-Line Slot

It is well known that a line slot antenna has a 1/2-wave characteristic mode (CM), 1-wave CM, and 3/2-wave CM at least. Figure 2a illustrates the CMs distribution. It can be observed that the 1/2-wave CM has characteristic magnetic currents flowing in one direction, which is easy to be excited with probe feeding in the middle of the line [19]. For achieving radiation null using this CM, one intuitive idea is to bend two ends of the slot to change the total flowing direction of characteristic magnetic currents.

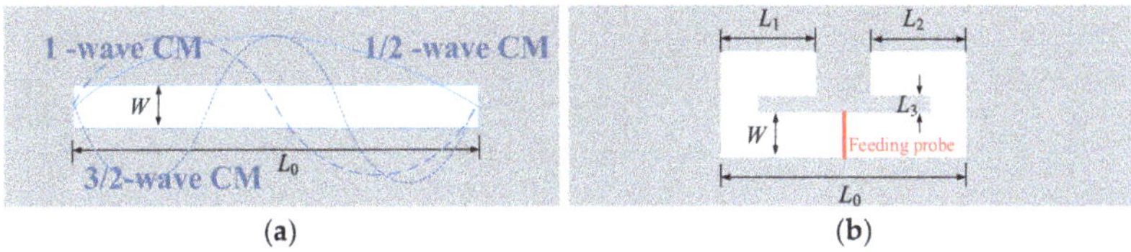

Figure 2. Geometry of the different line-slot antenna. (**a**) line slot antenna, (**b**) folding-line slot antenna.

Figure 2b shows the ends-folding-line slot antenna. Generally, the length-to-width ratio of the slot should be large enough to be regarded as a thin and linear slot antenna. We chose the line slot length L_0 = 80 mm, the bended part $L_1 = L_2$ = 32 mm, the length of the joint slot L_3 = 7 mm, and the width W = 3 mm.

Figure 3 shows the modal significance (MS) curves of the folding-line slot antenna, and three dominant CMs including CM1, CM3, and CM4 can be found. The magnitude of CM2's MS is too low to radiate.

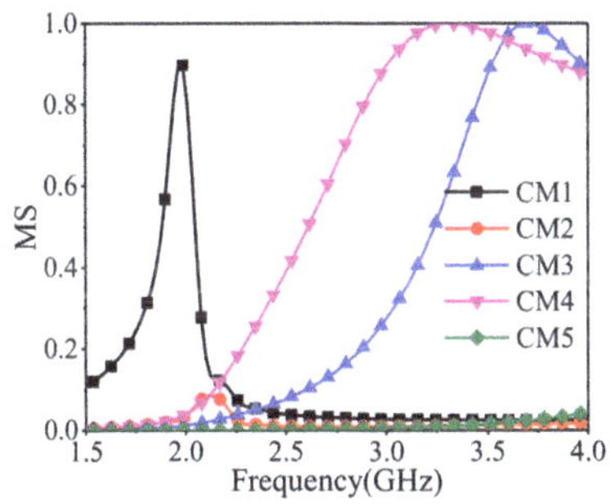

Figure 3. Modal significances of the folding-line slot antenna.

Figure 4 illustrates the characteristic magnetic currents of the above three dominant CMs (CM_1, CM_3, and CM_4) on the folding-line slot antenna. J_n represents the modal magnetic current of mode n in the slot, and the modal currents are shown at the resonant frequencies. It can be seen that CM_1, CM_3, and CM_4 are similar to the 1/2-wave CM, 3/2-wave CM, and 2-wave CM on the line-slot antenna. J_1 and J_4 flow in opposite directions at two ends of the slot, and the total currents cancel each other, which may be used in radiation null achievement.

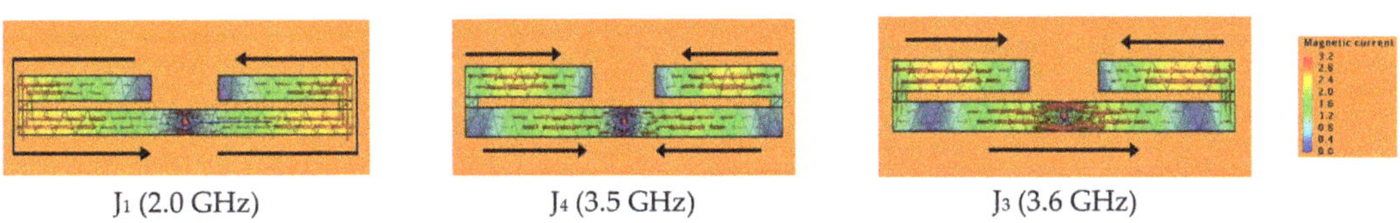

Figure 4. Characteristic magnetic currents of the folding-slot antenna.

As a narrow line slot structure can be seen as a magnetic dipole, its bandwidth is limited. CM_1 and CM_4 cannot be excited to radiate simultaneously within their working band. Due to the simplicity of J_1's distribution, CM_1 can be chosen to achieve radiation null. It is indicated that filtering response on one side of frequency can be realized by adjusting the line slot length and the bent slot length. Together with other filtering structures, the antenna's filtering characteristics can be realized.

2.3. Double-Ring Slot

Figure 5a shows the geometry of a typical single-ring slot antenna and its main CMs' magnetic currents at resonant frequencies. It can be observed that a single-ring slot antenna is hard to generate radiation nulls in the lower frequency, as the corresponding characteristic magnetic currents in the lower frequency flow towards one direction. In addition, it is hard to excite high-order CM as the characteristic magnetic currents are complex. For achieving filtering characteristics, one intuitive idea is to insert one ring-slot into another to form an opposite current flow.

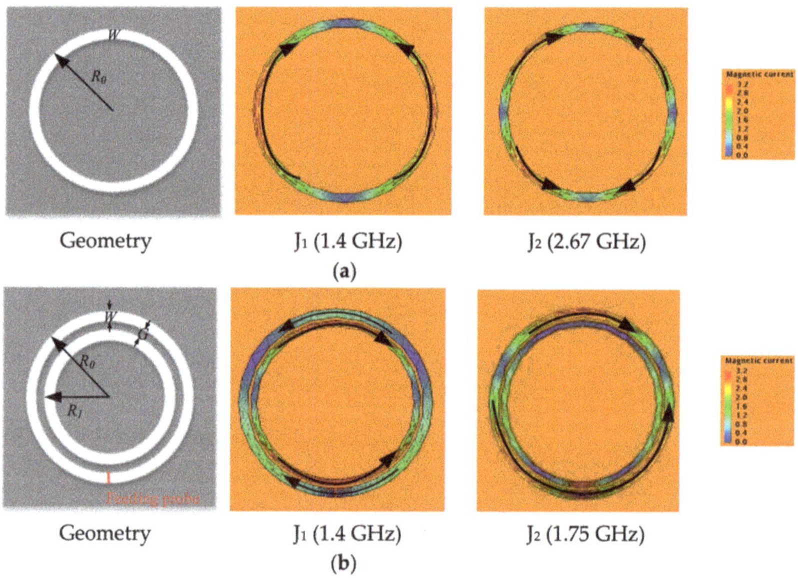

Figure 5. Sing-ring/double-ring slot antenna with R_0 = 32.3 mm, W = 3 mm, R_1 = 28.3 mm, G = 1 mm, and corresponding characteristic magnetic current distribution of two typical CMs.

Figure 5b shows the double-ring slot antenna. It consists of two ring slots with a fixed gap G. The radii of the outer and inner ring slots are R_0 and R_1 respectively, and the width of the slot is W. As the inner ring slot is not easily fed, the feeding probe is placed across the outer ring. It can be observed that the characteristic magnetic current distribution of the double-ring slot changes a lot compared with the single-ring slot. The total eigencurrents of CM_2 flow in one direction, and it is not in the aimed mode.

CM_1's characteristic magnetic currents in the inner and outer ring slots flow in opposite directions. Besides, J_1 is mainly distributed in the inner ring. With the probe feeding across the outer ring, magnetic currents in the outer ring can be improved. Together with the excited CM_1, magnetic currents in the inner and outer ring slot can be canceled, and radiation null can be realized. Therefore, the double-ring slot can be used with other structures to form a better filtering antenna due to the simplicity of the geometry.

2.4. Working Principles of the Proposed Antenna

To achieve a good filtering response, the above structures are combined together. The geometry of the proposed antenna is given in Figure 6. The antenna is printed on an FR4 substrate, its permittivity ε_r is 4.4, and its height h is 0.5 mm. It has a dimension of 100 mm × 100 mm. The yellow, green, and black parts refer to the metal ground, FR4 substrate, and the stepped metal microstrip line used for 50 Ω excitation lying on the back side of the substrate. The end of the feeding-line is connected with the ground plane via a metal shorting pin, which is shown in red color. The proposed feeding structure with the shorting pin is similar to the above probe-fed structure [18], which is easier to be fabricated and ensures the whole slot structure's excitation.

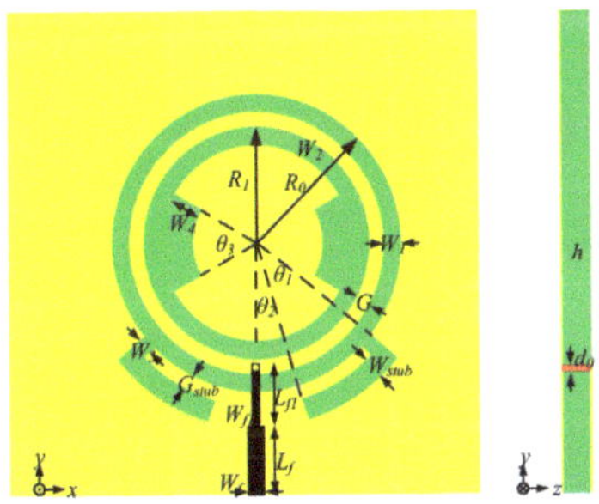

Figure 6. Proposed filtering antenna. R_0 = 21.8 mm, R_1 = 17.5 mm, W_1 = 3 mm, W_2 = 2 mm, W_{stub} = 1.8 mm, W_3 = 6 mm, G_{stub} = 1 mm, G = 1 mm, W_4 = 3 mm, θ_1 = 52°, θ_2 = 10°, θ_3 = 60°, h = 0.5 mm, L_f = 24 mm, L_{f1} = 8.2 mm, W_f = 0.92 mm, W_{f1} = 0.5 mm, h = 0.5 mm.

Figures 7 and 8 show the modal significances of the first five CMs and the corresponding magnetic eigencurrents. CM_3's MS is too low to be radiated. CM_1 and CM_4 are distributed at both ends of the frequency band, the resonant frequency difference is beyond 2 GHz, and they are hard to be excited at the same time. Figure 7 also depicts the modal weight coefficient (MWC) curves of the first five CMs. It can be seen that CM_2 and CM_5 are mainly excited within the aimed frequency range. Although other CMs such as CM_3 are excited, their magnitude of MS is too low to be radiated. Therefore, only CM_2 and CM_5 should be considered.

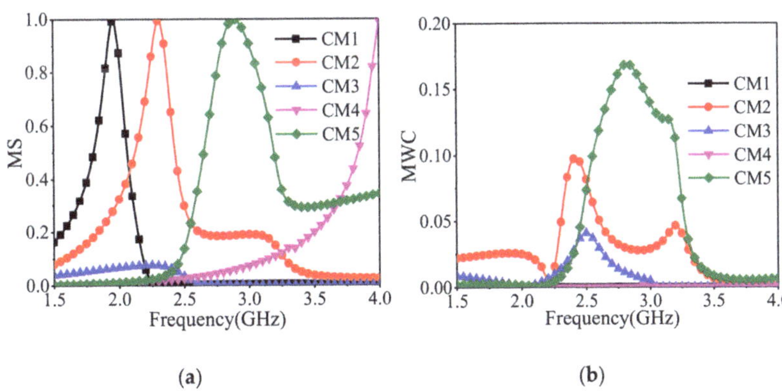

Figure 7. Modal significances and modal weighting coefficients for the first five modes of the proposed antenna. (**a**) modal significances, (**b**) modal weighting coefficients.

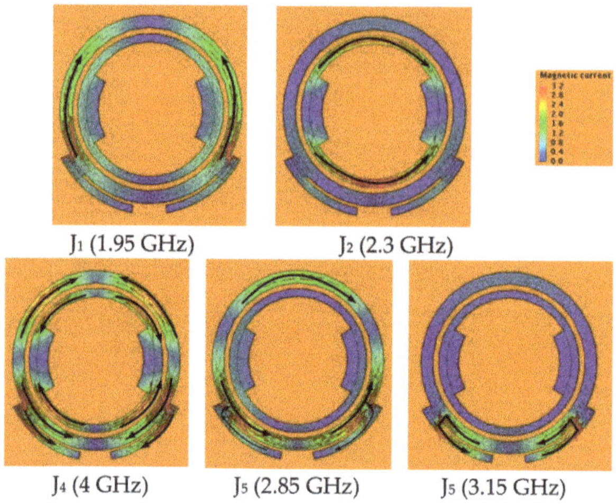

Figure 8. Magnetic eigencurrents of CMs.

It can be seen that magnetic eigencurrents of CM_2 and CM_5 are mainly distributed in the inner-ring and outer-ring slots separately. CM_2 is similar to the corresponding CM_1 of the double-ring slot. CM_5 is a new CM introduced after slots combination compared with the double-ring slot and is similar to the superposition of corresponding CMs of the above slots. The resonant frequency spacing between CM_2 and CM_5 is nearly 500 MHz. With the proposed feeding structure, CM_2 and CM_5 are mainly excited, and wide bandwidth can be achieved. Meanwhile, as magnetic currents in the outer ring and slot stubs are excited and enhanced, a radiation null in the lower frequency can be achieved for CM_2, just like CM_1's behavior in the double-ring slot. As shown in Figure 8, for CM_5 in the upper frequency, magnetic current distribution in the outer-ring and slot stubs changes, and the eigencurrents are mainly distributed in the slot stubs. Similar to CM_1's behavior in the folding-line slot, it brings another radiation null. Therefore, we can adjust the frequency band utilizing CM_2 and CM_5 by changing the proper elements of the proposed antenna.

Besides, characteristic magnetic currents of excited CM_2 and CM_5 flow in one direction, which ensures the antenna has a good linear polarization.

Figure 9 shows the magnetic current distribution at the first depression frequency in the lower and upper bands. It can be easily seen that the filtering response in the lower and upper bands is predominantly realized by the excited magnetic currents in the double-ring and folding-line slot separately. This agrees with the former analysis. Figure 10 gives the simulated S11 and realized gain. It can be seen that the realized gain has a sharp rolling rate at the upper and lower frequencies, which shows a high out-of-band suppression characteristic and a good filtering response within the 10 dB impedance bandwidth.

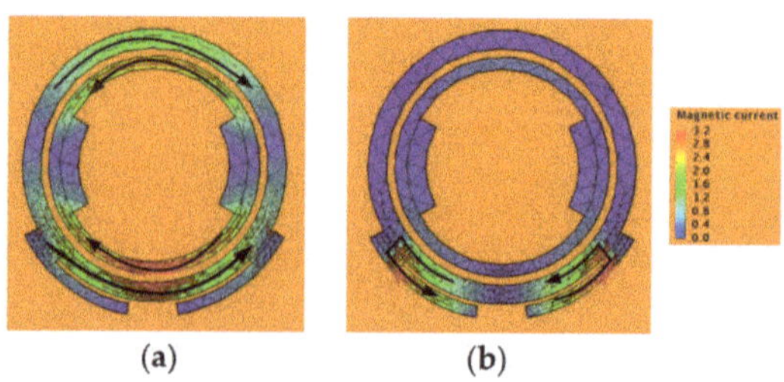

Figure 9. Simulated magnetic current distribution of the proposed antenna at different frequencies. (**a**) 2.08 GHz (**b**) 3.23 GHz.

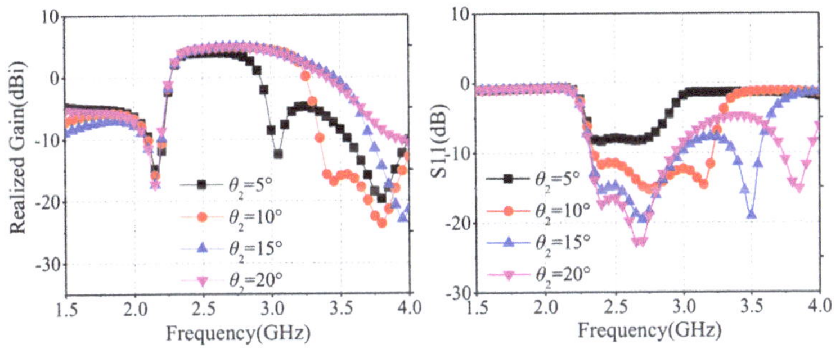

(a) (b)

Figure 10. Simulated results for different parameter θ_2, (**a**) realized gain, (**b**) S11.

To further illustrate the proposed antenna's mechanism, we choose θ_2 and W_4 for parameter sweep analysis. Figure 10 shows that the high out-of-band suppression at the higher frequency varies with θ_2. This is because that θ_2 is associated with the total length of the folding-line slot. As θ_2 increases, the total length of the folding-line slot decreases, and the corresponding resonant frequency increases.

Figure 11 depicts the filtering bandwidth that varies with W_4 around lower frequencies. W_4 is the width of the inner arc-shaped slot stub. As characteristic magnetic currents of CM_2 and CM_5 around the inner arc-shaped slots are rare, W_4 can be chosen to adjust the impedance bandwidth, and the corresponding filtering frequency varies.

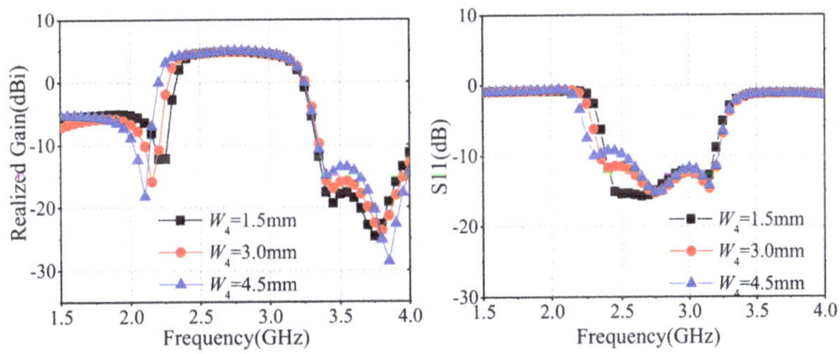

(a) (b)

Figure 11. Simulated results for different parameter W_4, (**a**) realized gain, (**b**) S11.

3. Results and Discussion

According to the analysis results mentioned above, the designed antenna was prototyped and measured.

Figure 12 gives the comparison between the simulated and measured results of S11 and realized gain. It can be seen that the measured bandwidth ranges from 2.32 GHz to 3.13 GHz, and the simulated and measured results agree well. The gain ranges from 3.9 dB to 5.2 dB, and the average suppression around the upper frequencies is above 20 dB, which demonstrates the filtering characteristic. The measured S11 and gain shift a little left around 3.2 GHz. The difference is mainly caused by manufacturing errors and welding deviation. Figure 13 gives the radiation patterns of the measured and simulated results in different frequencies. It shows that the proposed antenna has a good broadside radiation characteristic and a stable radiation pattern across the whole working bandwidth. The cross-polarization suppression levels at different planes are mostly lower than 20 dB, but a little higher at the upper frequencies, which is caused by the imperfect measurement environment and the connected SMA.

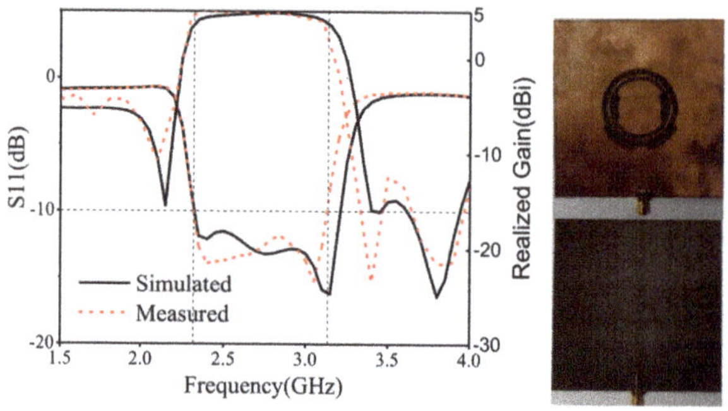

Figure 12. Comparisons of the simulated/measured S11 and realized gain, prototype of the antenna.

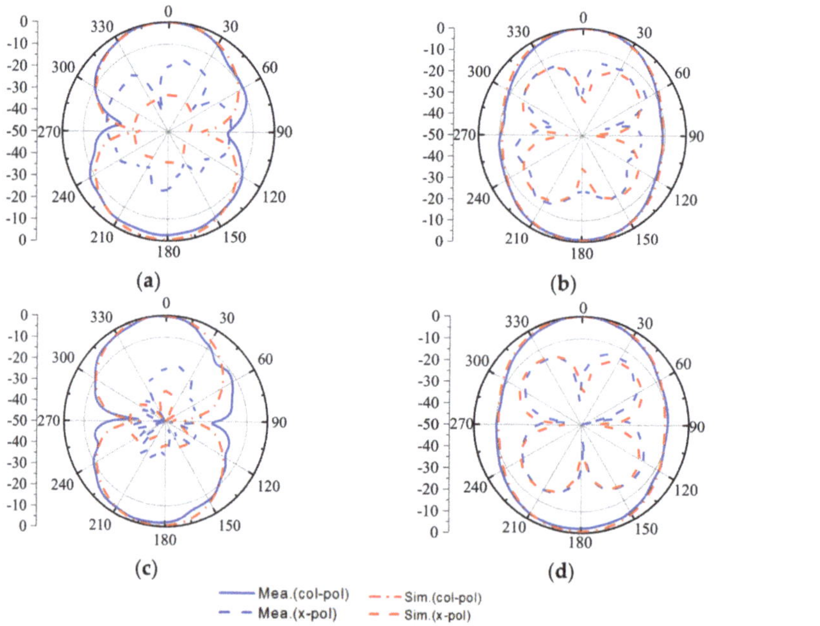

Figure 13. Simulated/measured radiation patterns of the proposed antenna, (**a**) xoz plane@2.5 GHz (**b**) yoz plane@2.5 GHz (**c**) xoz plane@3 GHz (**d**) yoz plane@3 GHz.

Comprehensive comparisons with reported filtering antennas of different types are given in Table 1. In [1,10], complex structures including stacked patches with specific shorting pins and metasurface design are utilized. In [2], a conventional patch antenna with designed etched slots is used to achieve filtering response, but the bandwidth is limited. Due to its 2-order filter design, the filtering antenna in [6] shows a good stable gain, but a limited bandwidth and complex structure. It can be seen that the proposed filtering antenna using CMA has advantages of low profile, simple structure, wide bandwidth, and stable gain.

Table 1. Comparisons of filtering antennas.

Reference	Configuration	Size ($\lambda_0{}^3$)	BW	Gain (dBi)
[1]	Stacked slot patch with shoring pins	1.13 * 0.42 * 0.06	23.5%	4.7~7.5
[2]	Single patch with etched slots	0.7 * 0.59 * 0.03	7.1%	4.6~6.6
[6]	Single slot patch with n-order filter	0.25 * 0.3 * 0.02	18.9%	0.7~2.3
[10]	Metasurface with designed slots	0.77 * 0.77 * 0.04	17.6%	7~9
Prop.	Single combined slot patch	1.1 * 1.1 * 0.01	30.6%	3.9~5.2

4. Conclusions

In this paper, a combined-slot antenna with filtering characteristics and stable gain using CMA is proposed. With CMA, double-ring slot and folding-line slot are analyzed to achieve radiation nulls at different frequencies. By combining the above slot structures, two radiation nulls can be realized. Further, two CMs with proper excitation are selected to obtain wide bandwidth. A fabricated prototype achieves a wide bandwidth of 32.7%, and a stable gain of 3.9~5.2 dB, which demonstrates the design process. Compared with other filtering design antennas, the proposed one exhibits the advantage of wide bandwidth, low profile, and simple structure, which shows the potential to be used in high-integrated microwave systems.

Author Contributions: Conceptualization, C.N.; methodology, B.W.; investigation, W.W., P.R.; writing—original draft preparation, C.N.; writing—review and editing, W.W.; supervision, B.W.; All authors have read and agreed to the published version of the manuscript.

Funding: This research received no external funding.

Institutional Review Board Statement: Not applicable.

Informed Consent Statement: Not applicable.

Data Availability Statement: The data supporting this research article are available upon request to the author.

Conflicts of Interest: The authors declare no conflict of interest.

References

1. Li, J.; Chen, Z.N.; Wu, D.; Zhang, G.; Wu, Y. Dual-Beam Filtering Patch Antennas for Wireless Communication Application. *IEEE Trans. Antennas Propag.* **2018**, *66*, 3730–3734. [CrossRef]
2. Jin, J.Y.; Liao, S.; Xue, Q. Design of filtering-radiating patch antennas with tunable radiation nulls for high selectivity. *IEEE Trans. Antennas Propag.* **2018**, *66*, 2125–2130. [CrossRef]
3. Mallat, N.K.; Amjad, I. Out-of-band suppressed SIW-DRA based filter-antenna subsystem with flexible bandwidth and transmission zeros. *AEU-Int. J. Electron. Commun.* **2021**, *135*, 153735. [CrossRef]
4. Allam, V.K.; Madhav, B.T.; Anilkumar, T.; Maloji, S. A Novel Reconfigurable Bandpass Filtering Antenna for IoT Communication Applications. *Prog. Electromagn. Res. C* **2019**, *96*, 13–26. [CrossRef]
5. Al-Yasir, Y.I.A.; AAlhamadani, H.; Kadhim, A.S.; Ojaroudi Parchin, N.; Saleh, A.L.; Elfergani, I.T.E.; Rodriguez, J.; Abd-Alhameed, R.A. Design of a Wide-Band Microstrip Filtering Antenna with Modified Shaped Slots and SIR Structure. *Inventions* **2020**, *5*, 11. [CrossRef]
6. Chen, X.; Zhao, F.; Yan, L.; Zhang, W. A compact filtering antenna with flat gain response within the passband. *IEEE Antennas Wirel. Propag. Lett.* **2013**, *12*, 857–860. [CrossRef]
7. Lin, C.K.; Chung, S.J. A compact filtering microstrip antenna with quasi-elliptic broadside antenna gain response. *IEEE Antennas Wirel. Propag. Lett.* **2011**, *10*, 381–384. [CrossRef]
8. Wu, W.J.; Yin, Y.Z.; Zuo, S.L.; Zhang, Z.Y.; Xie, J.J. A new compact filter-antenna for modern wireless communication systems. *IEEE Antennas Wirel. Propag. Lett.* **2011**, *10*, 1131–1134. [CrossRef]

9. Zhang, X.Y.; Duan, W.; Pan, Y.-M. High-gain filtering patch antenna without extra circuit. *IEEE Trans. Antennas Propag.* **2015**, *63*, 5883–5888. [CrossRef]
10. Yang, W.; Chen, S.; Xue, Q.; Che, W.; Shen, G.; Feng, W. Novel Filtering Method Based on Metasurface Antenna and Its Application for Wideband High-Gain Filtering Antenna with Low Profile. *IEEE Trans. Antennas Propag.* **2019**, *67*, 1535–1544. [CrossRef]
11. Mao, C.X.; Gao, S.; Wang, Y.; Wang, Z.; Qin, F.; Sanz-Izquierdo, B.; Chu, Q.X. An integrated filtering antenna array with high selectivity and harmonics suppression. *IEEE Trans. Microw. Theory Tech.* **2016**, *64*, 1798–1805. [CrossRef]
12. Park, J.; Jeong, M.; Hussain, N.; Rhee, S.; Park, S.; Kim, N. A low-profile high-gain filtering antenna for fifth generation systems based on nonuniform metasurface. *Microw. Opt. Technol. Lett.* **2019**, *61*, 2513–2519. [CrossRef]
13. Wang, W.; Zheng, Z.; Fang, X.; Zhang, H.; Jin, M.; Lu, J.; Luo, Q.; Gao, S. A waveguide slot filtering antenna with an embedded metamaterial structure. *IEEE Trans. Antennas Propag.* **2019**, *67*, 2953–2960. [CrossRef]
14. Mao, C.X.; Gao, S.; Wang, Y.; Cheng, Z. Filtering antenna with two-octave harmonic suppression. *IEEE Antennas Wirel. Propag. Lett.* **2017**, *16*, 1361–1364. [CrossRef]
15. Naeem, U.; Iqbal, A.; Shafique, M.F.; Bila, S. Efficient design methodology for a complex DRA-SIW filter-antenna subsystem. *Int. J. Antennas Propag.* **2017**, *2017*, 6401810. [CrossRef]
16. Elias, B.B.Q.; Soh, P.J.; Al-Hadi, A.A.; Akkaraekthalin, P.; Vandenbosch, G.A.E. Bandwidth Optimization of a Textile PIFA with DGS Using Characteristic Mode Analysis. *Sensors* **2021**, *21*, 2516. [CrossRef]
17. Jabire, A.H.; Ghaffar, A.; Li, X.J.; Abdu, A.; Saminu, S.; Alibakhshikenari, M.; Falcone, F.; Limiti, E. Metamaterial Based Design of Compact UWB/MIMO Monopoles Antenna with Characteristic Mode Analysis. *Appl. Sci.* **2021**, *11*, 1542. [CrossRef]
18. Liang, P.; Wu, Q. Duality Principle of Characteristic Modes for the Analysis and Design of Aperture Antennas. *IEEE Trans. Antennas Propag.* **2018**, *66*, 2807–2817. [CrossRef]
19. Lin, J.; Chu, Q. Increasing Bandwidth of Slot Antennas with Combined Characteristic Modes. *IEEE Trans. Antennas Propag.* **2018**, *66*, 3148–3153. [CrossRef]
20. Zhao, C.; Wang, C.-F. Characteristic Mode Design of Wide Band Circularly Polarized Patch Antenna Consisting of H-Shaped Unit Cells. *IEEE Access* **2018**, *6*, 25292–25299. [CrossRef]
21. Deng, C.; Feng, Z.; Hum, S.V. MIMO mobile handset antenna merging characteristic modes for increased bandwidth. *IEEE Trans. Antennas Propag.* **2016**, *64*, 2660–2667. [CrossRef]
22. Gao, G.; Zhang, R.-F.; Geng, W.-F.; Meng, H.-J.; Hu, B. Characteristic Mode Analysis of a Nonuniform Metasurface Antenna for Wearable Applications. *IEEE Antennas Wirel. Propag. Lett.* **2020**, *19*, 1355–1359. [CrossRef]
23. Altair. CADFEKO Suite 7.0. Available online: https://help.altair.com/2021/feko/index.htm (accessed on 1 March 2021).

electronics

Article

Design of a Low-Profile Wideband Magnetoelectric Dipole Antenna with Reduced Gain Drop

Zhiyi Li, Xing Chen *, Yuzhu Tang, Liangbing Liao, Linwan Deng and Zhifan Zhao

College of Electronics and Information Engineering, Sichuan University, Chengdu 610065, China; lzy502762410@126.com (Z.L.); tyznature@163.com (Y.T.); liaoliangbing@126.com (L.L.); dlw2335219243@126.com (L.D.); zzf525522988@126.com (Z.Z.)

* Correspondence: xingc@live.cn

Abstract: In this paper, a novel low-profile magnetoelectric (ME) dipole antenna with wideband is presented. The conventional vertical fixing structure is bended four times from the center to the sides. The Γ-shaped feeding structure is bended two times to lower the height of the antenna step by step. The effect of three kinds of vertical wall is discussed to show their influence on boresight gain. Through comparison, only one vertical wall is erected on the left side of the ground to decrease the boresight gain drop at 2.2 GHz. Both simulation and analysis are made to sufficiently explain the working principle. At last, the proposed ME dipole antenna has only $0.095\lambda_0$ (λ_0 is the center operating wavelength in free space) in height, and the wideband property is still maintained. By simulation, the relative bandwidth for VSWR < 2.0 is 47.9% (from 1.35 to 2.2 GHz). The boresight gain ranges from 8.1 to 9.6 dBi in the operating band. The measured relative bandwidth for VSWR < 2.0 is 50.3% (from 1.34 to 2.24 GHz), and the boresight gain ranges from 7.38 to 8.73 dBi. The gain drop on boresight is less than 1.4 dBi. Radiation patterns show a unidirectional characteristic in the whole operating band. Additionally, the cross-polarization level is less than −25 dB on boresight. The simulating and measuring results agree well with each other. Therefore, the proposed antenna is suitable for applications of limited height and wideband.

Keywords: low profile; wideband; ME dipole; multiple bending; unidirectional

Citation: Li, Z.; Chen, X.; Tang, Y.; Liao, L.; Deng, L.; Zhao, Z. Design of a Low-Profile Wideband Magnetoelectric Dipole Antenna with Reduced Gain Drop. *Electronics* **2022**, *11*, 1156. https://doi.org/10.3390/electronics11071156

Academic Editors: Naser Ojaroudi Parchin, Chan Hwang See and Raed A. Abd-Alhameed

Received: 20 February 2022
Accepted: 4 March 2022
Published: 6 April 2022

1. Introduction

To meet the increasing requirement of wireless communication systems, wideband equipment is widely designed and produced by suppliers. Traditional narrow band antennas are no longer suitable for current wideband and ultra-wideband (UWB) wireless applications [1]. Conventionally, two methods are adopted to satisfied the wideband communication: the first one is to use multiple narrow band antennas that worked at a continuous frequency band to realize a wideband operation. The second one is to design a wideband antenna to cover the whole operating band. For the first method, the multiple antennas will occupy much installation space, and lots of cables will be needed to connect to the input port of each antenna element. This inevitably increases the complexity of the whole system. Therefore, to adopt a wideband antenna to realize the wideband communication is feasible and rational.

To save the installation space and decrease the wind drag, a low-profile antenna is widely adopted in lot of scenarios, such as the surface of aircraft, vessel, building, and so on. Up until now, lots of low-profile antennas were designed in the literature. In [2], an ultralow-profile patch antenna was designed with only $0.01\lambda_0$ in height by using the slot-loading technique. In [3–5], three kinds of metamaterial loaded methods were adopted to realize a low profile. In [6], a filter antenna has a profile of $0.026\lambda_0$ but with single-frequency point operation. Recently, a multimode technique was taken to broaden the bandwidth of a low-profile antenna. In [7], with a profile of $0.07\lambda_0$, three modes (TM10, TM12, and antiphase TM22) were excited simultaneously to realize a relative bandwidth of 26.2% for

S11 < −15 dB. In [8], based on multiple mode partial aperture, a microstrip antenna is able to work at the bandwidth of 57.3% with a profile of $0.044\lambda_0$.

In recent years, the ME dipole antenna, which was first proposed by Alvin Chlavin [9] and developed by Luk [10], has attracted lots of research. This kind of antenna has the advantages of wideband, stable unidirectional radiation, low front–back ratio, and equal E and H planes [11–14]. To reduce the profile of the ME dipole, many methods were proposed in the literature [15–19]. In [15], the author utilized an obtuse triangular structure to reduce the antenna thickness to $0.097\lambda_0$, and an impedance bandwidth of 28.2% was achieved. In [16], by using a pair of vertically oriented folded shorted patches, the antenna height is reduced from $0.25\lambda_0$ to $0.116\lambda_0$. In [19], the magnetic dipole mode of ME dipole antenna was formed by the slot-aperture between patches. Therefore, the height of the antenna can be as low as $0.11\lambda_0$, and the impedance bandwidth is 27.6% for S11 < −15 dB.

In this paper, a novel wideband low-profile ME dipole antenna is presented. To decrease the height of the traditional ME dipole, a multi-bending technique is introduced. Both the fixing structure and Γ-shaped feeding structure are bended to lower the profile step by step. Additionally, the effect of three kinds of vertical wall is discussed, and only one vertical wall is erected on the left side of the ground to avoid the gain drop on boresight at last. Finally, the total height of the proposed ME dipole antenna is only $0.095\lambda_0$, and the measured relative bandwidth is up to 50.3%.

2. Technique and Geometry

2.1. Low-Profile Technique

Ahead of this design, four kinds of ME dipole antennas are depicted in Figure 1 to show the current low-profile technique [15]. Antenna 1 is the traditional ME dipole with a profile of $0.25\lambda_0$. For antenna 2, the vertical fixing structure is bended and the height can be lowered to $0.2\lambda_0$. For antenna 3, an obtuse-triangular structure is introduced to lower the antenna height to $0.16\lambda_0$. For antenna 4, the feeding line is moved to the outside of the fixing structure, and the final height is $0.116\lambda_0$.

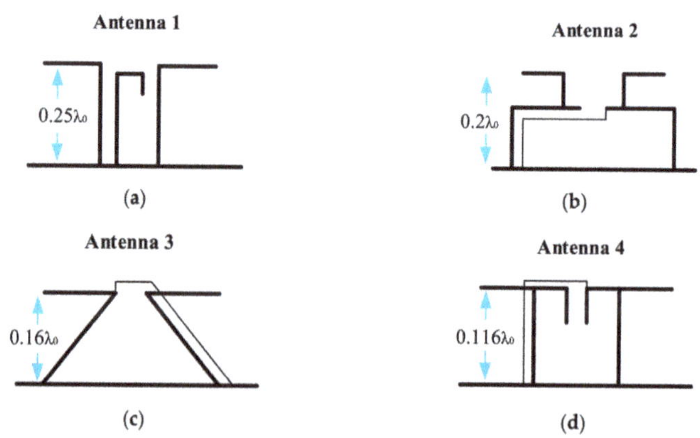

Figure 1. A depiction of the current low-profile techniques for an ME dipole antenna. (**a**) antenna 1; (**b**) antenna 2; (**c**) antenna 3; (**d**) antenna 4.

2.2. Evolution of the Proposed Antenna

To demonstrate the current low-profile technique, an evolution of the proposed ME dipole is depicted in Figure 2. Here, five evolution processes (type 1, type 2, type 3, type 4, and type 5, respectively) are depicted. Type 1 is the normal ME dipole designed by Luk [10]. By bending the vertical portion of type 1, type 2 is obtained on the right-top in Figure 2. Furtherly, type 3 and type 4 are formed by an oblique extension and a meandering process for the bending portion of type 2, as shown on the left-center and right-center in

Figure 2, respectively. Finally, the proposed antenna (type 5) is obtained through bending the Γ-shaped feeding structure of type 4 to the right side.

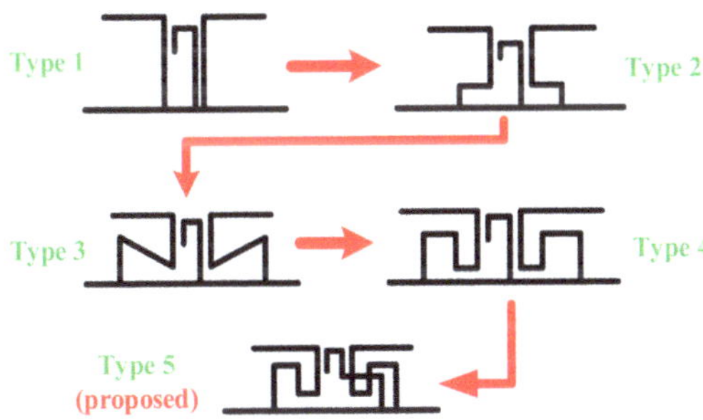

Figure 2. The evolution of the proposed low-profile ME dipole.

2.3. Realization of the Proposed Low-Profile ME Dipole

In this part, the structure of the proposed low-profile ME dipole is presented. The front view and top view are demonstrated in Figure 3a,b. As shown in Figure 3a, the proposed ME dipole takes the multi-bending technique as presented in Figure 2. On the one hand, the fixing structure (in blue) is bended from the center to the side four times and erected on the ground. On the other hand, the traditional Γ-shaped feeding line is bended two times to guarantee a good impedance matching to a 50 ohm coaxial line at a low-frequency band. The feeding line passes through the bended fixing structure on the right side by a rectangular slot, as depicted in Figure 3c. The detail of the feeding line is shown in Figure 3d. Finally, a vertical wall with height H is erected on the left side of the ground. The optimized sizes of the structure are listed in Table 1.

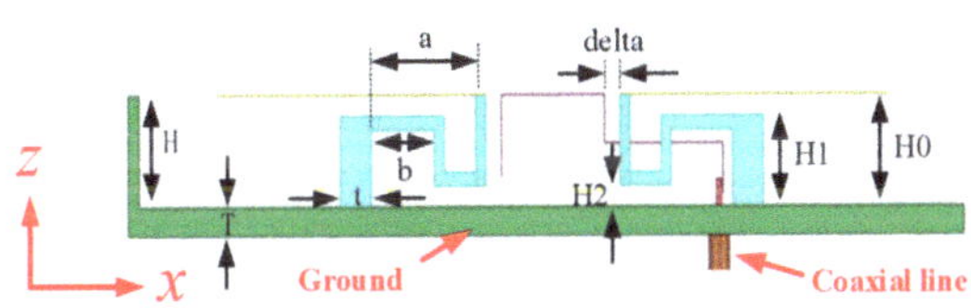

(**a**)

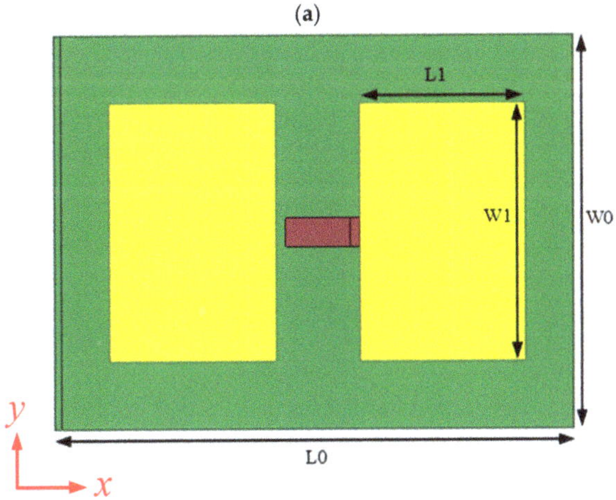

(**b**)

Figure 3. *Cont.*

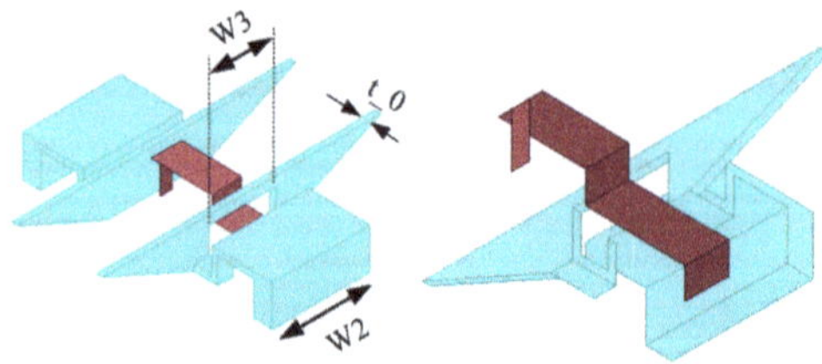

(c)

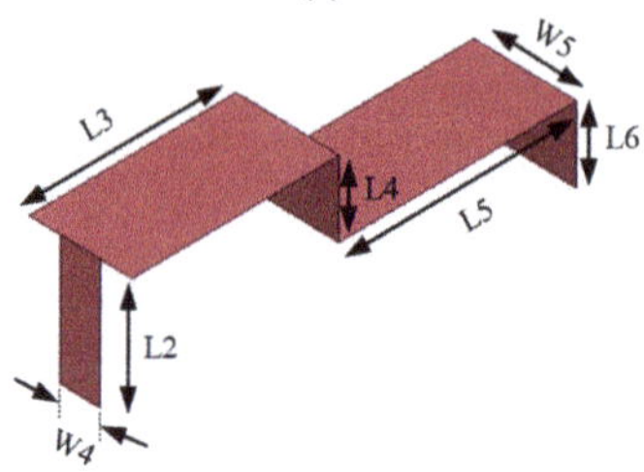

(d)

Figure 3. The structure of the proposed low-profile ME dipole. (**a**) Front view, (**b**) top view, (**c**) bended fixing structure, (**d**) feeding line.

Table 1. Parameter size of the low-profile ME dipole antenna (unit: mm).

L0	**L1**	**L2**	**L3**	**L4**	**L5**	**L6**	**W0**	**W1**	**W2**
162 ($0.97\lambda_0$)	52 ($0.31\lambda_0$)	12 ($0.07\lambda_0$)	20 ($0.12\lambda_0$)	7 ($0.04\lambda_0$)	23 ($0.14\lambda_0$)	7 ($0.04\lambda_0$)	140 ($0.84\lambda_0$)	91 ($0.54\lambda_0$)	32 ($0.19\lambda_0$)
W3	**W4**	**W5**	**T**	**t**	**t_0**	**H**	**H0**	**H1**	**H2**
22 ($0.13\lambda_0$)	4 ($0.02\lambda_0$)	10 ($0.06\lambda_0$)	4 ($0.02\lambda_0$)	6 ($0.04\lambda_0$)	2 ($0.01\lambda_0$)	16 ($0.095\lambda_0$)	16 ($0.095\lambda_0$)	13 ($0.08\lambda_0$)	3 ($0.018\lambda_0$)
a	**b**	**Delta**							
20 ($0.12\lambda_0$)	12 ($0.07\lambda_0$)	3 ($0.018\lambda_0$)							

In this design, the electric dipole and the magnetic dipole is realized by the top horizontal sheet and the space between the bended fixing structure and the ground, respectively. The schematic diagram is demonstrated in Figure 4. The red arrow denotes an electric dipole and the blue arrow denotes a magnetic dipole. The combination of the electric dipole and magnetic dipole forms an ME dipole antenna.

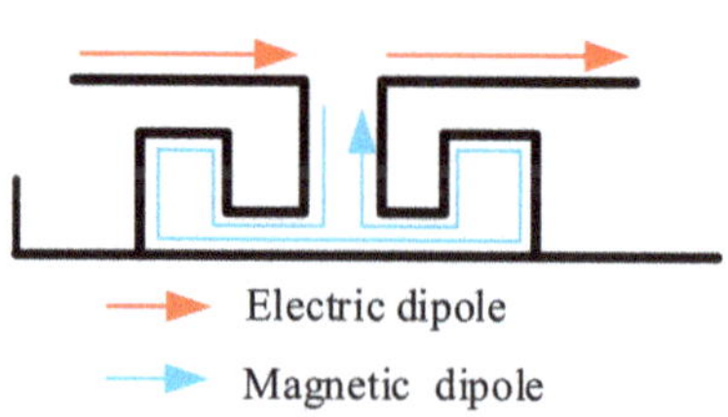

Figure 4. The schematic diagram of electric dipole and magnetic dipole.

3. Simulation, Comparison, and Analysis

3.1. VSWR for Different Evolutions

To show the low-profile property of the proposed ME dipole antenna, three types of evolution antennas (ANT 1, ANT 2, and ANT 3) are simulated and compared in this part and are shown in Figure 5. For the sake of fairness, all the antennas have the same height (H0 = 16 mm) and are well optimized. As shown in Figure 5a, these antennas from top to bottom correspond to type 2, type 4, and type 5 in Figure 2, respectively. The fixing structure for ANT 1 is only bended one time. The fixing structure for ANT 2 is bended four times. Both the fixing structure and Γ-shaped feeding line are bended for ANT 3. The optimized VSWR for the above three antennas are depicted in Figure 5b and marked by blue, black, and red, respectively. For ANT 1, the impedance bandwidth ranges from 2.25 to 3.3 GHz for VSWR < 2. For ANT 2, the impedance bandwidth ranges from 2.05 to 2.95 GHz. However, the optimized impedance bandwidth (VSWR < 2) ranges from 1.35 to 2.2 GHz for ANT 3. Furthermore, the impedance of above three antennas is also plotted in Figure 5c for comparison. Obviously, ANT 1 has a small resistance in the band of 1.35–2.2 GHz. ANT 2 has a small resistance in the band of 1.35–1.8 GHz and has a large resistance and reactance around 1.9 GHz. This leads to ANT 1 and ANT 2 being hard to match a pure 50 Ohm resistance at the input port. Meanwhile, ANT 3 increases the resistance and moderates the reactance by bending the feeding structure. It is seen that the resistance and reactance of ANT 3 is around 50 Ohm and 0 Ohm, respectively, in the band from 1.35 to 2.2 GHz. Therefore, ANT 3 can operate at a lower band and also keep the wideband characteristic when compared with ANT 1 and ANT 2 after the evolution.

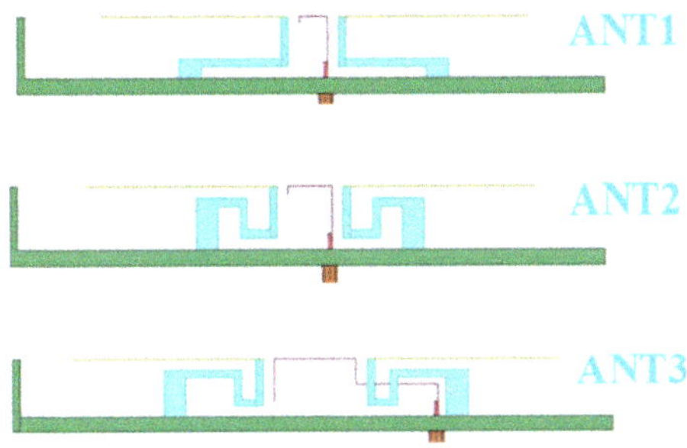

(**a**)

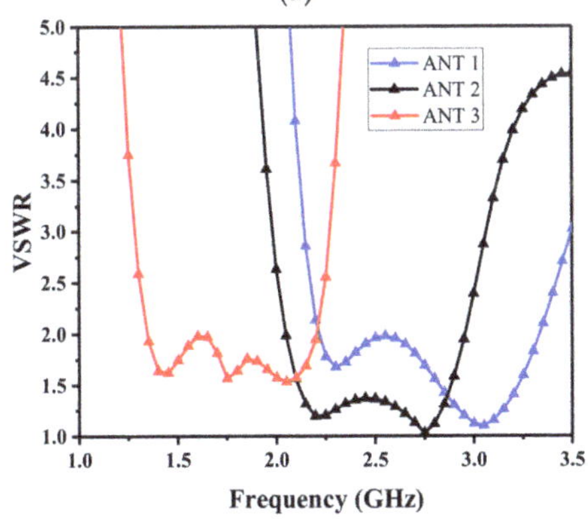

(**b**)

Figure 5. *Cont.*

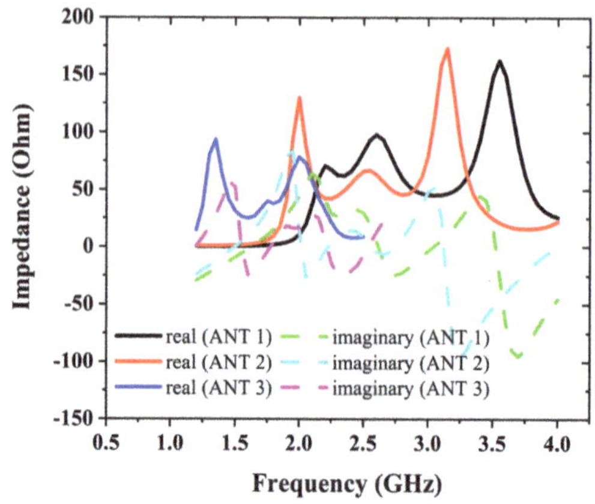

(c)

Figure 5. (**a**) The structure of ANT 1, ANT 2, and ANT3; (**b**) VSWR of ANT 1, ANT 2, and ANT3; (**c**) impedance of ANT 1, ANT 2, and ANT 3.

3.2. *Effect of Different Grounds*

The antennas with three types of ground are discussed in this section. The first type is the one with two symmetric vertical walls, as shown in Figure 6a. The second type is shown in Figure 6b. This antenna has a flat ground. The third one is our proposed antenna, which has only one vertical wall on the left side of the flat ground, as shown in Figure 6c. The above three types of antennas are named antenna 1, antenna 2, and antenna 3, respectively. The simulated VSWR and boresight gain of them are depicted in Figure 6d for comparison, simultaneously. Obviously, the impedance bandwidth (VSWR < 2) of antenna 1 and antenna 2 ranges from 1.35 to 2.25 GHz (50%) and from 1.75 to 2.2 GHz (22%), respectively. The boresight gain ranges from 6.6 to 9.7 dBi and from 9.04 to 9.6 dBi, respectively. From the red curve in Figure 6d, it is seen that a significant drop on boresight gain happens at 2.2 GHz for antenna 1. The frequency is at the upper side of the operating band. Here, we point out that this drop at 2.2 GHz for antenna 1 is due to the maximum radiation direction in E plane deviating from the boresight, as depicted in Figure 6e. The angle of deviation is $\theta = 10°$. However, from the purple curve in Figure 6d, antenna 2 has a good boresight gain without drop but a bad VSWR at the lower band. To maintain the advantages of the impedance bandwidth of antenna 1 and the boresight gain of antenna 2 simultaneously, a method to erect only one vertical wall on the side of the ground is taken. Firstly, a single vertical wall is erected on the right side of the ground. However, the radiation field deviates from the boresight to the right side more seriously, as shown in Figure 6f. At this step, the angle of deviation is up to $\theta = 21°$. Therefore, an idea of only erecting a single vertical wall on the left side of the ground is adopted. As the radiation field shows in Figure 6g, the maximum radiating direction is close to the boresight when the single vertical wall was erected on the left side. The angle of deviation is $\theta = 4°$ at this time. From the black curves in Figure 6d, the impedance bandwidth (VSWR < 2) of antenna 3 ranges from 1.35 to 2.2 GHz (47.9%) and the boresight gain ranges from 8.1 to 9.6 dBi. As mentioned above, antenna 3 indeed keeps the bandwidth and avoids the gain drop of antenna 1.

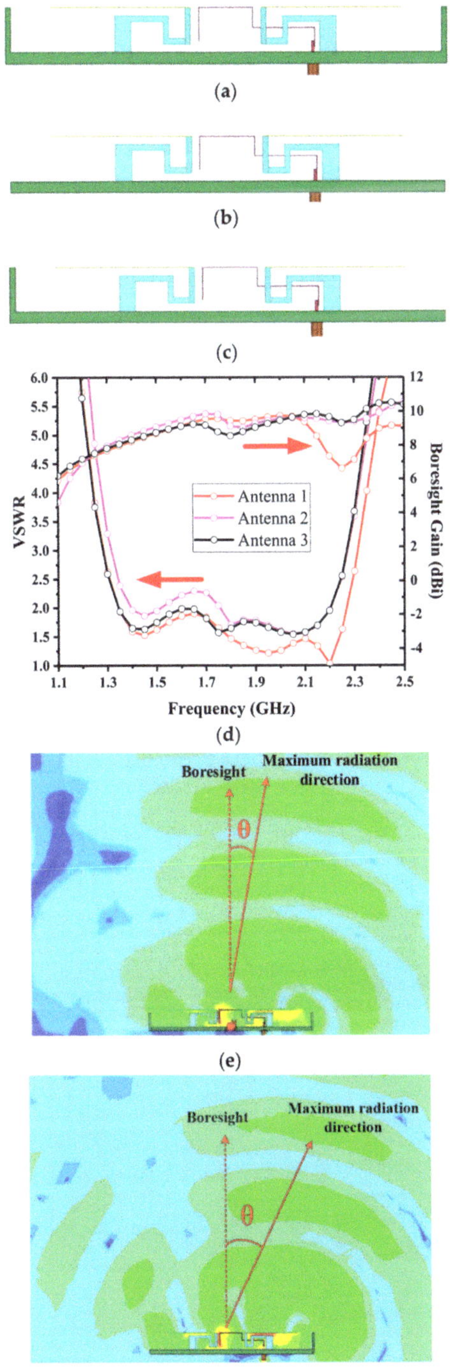

Figure 6. *Cont.*

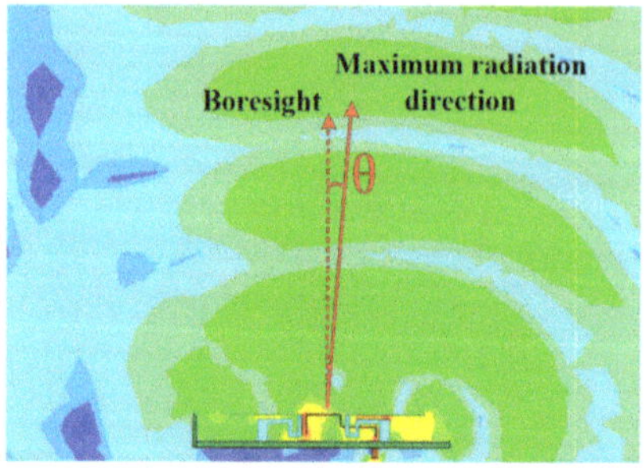

(g)

Figure 6. The structure of the antenna with (**a**) two symmetric vertical walls; (**b**) only flat ground plane; (**c**) one vertical wall; (**d**) the VSWR for different ground; (**e**) the radiating E field at 2.2 GHz for antenna 1; (**f**) the radiating E field at 2.2 GHz for the antenna with the single vertical wall on the left side; (**g**) the radiating E field at 2.2 GHz for antenna 3.

3.3. Working Mechanism

The working mechanism of the proposed antenna is analyzed by plotting the distribution of surface current. Here, it is pointed out that the proposed ME dipole antenna has its ME mode at 1.4 GHz. Therefore, the surface current in one period (1T) is given at 1.4 GHz and is shown in Figure 7. For simplicity, let t = 0 when the antenna is initially excited. As shown in Figure 7a, the surface current on the whole antenna is very weak at t = 0 and large current is just going through the feeding port to the antenna. However, the current on the top sheet, the bended surface, and the ground between these bended surfaces are strong at t = T/4, as shown in Figure 7b. The current flow is marked by a red arrow. At t = T/2, the surface current is similar to the case at t = 0 except for a reverse current flow, as shown in Figure 7c. At t = 3T/4, the current on the top sheet, the bended surface, and the ground between these bended surfaces become strong again, as shown in Figure 7d. From the principle of the ME dipole antenna shown in Figure 4, it is concluded that the electric dipole and magnetic dipole work simultaneously at t = T/4 and 3T/4 while they do not work at t = 0 and T/2.

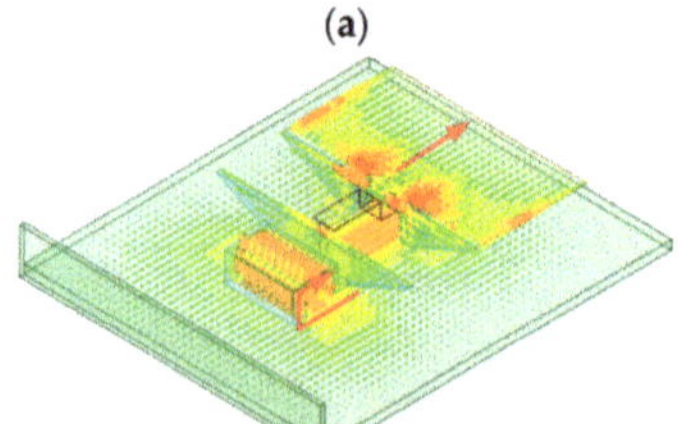

(a)

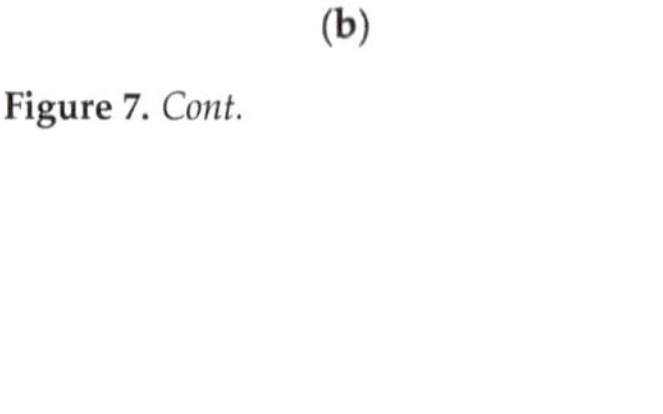

(b)

Figure 7. *Cont.*

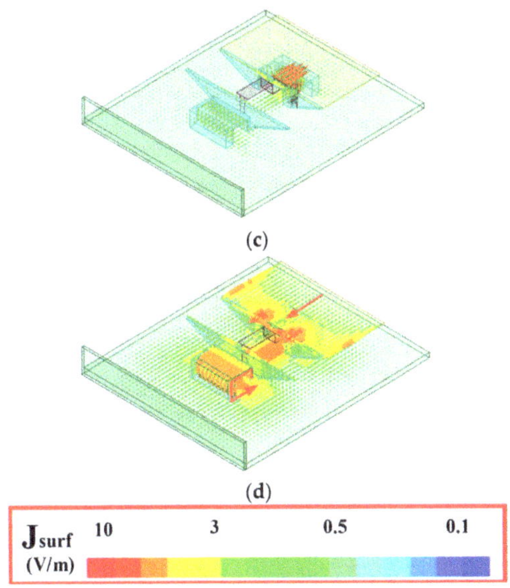

Figure 7. The distribution of electric current in one period. (**a**) at t = 0, (**b**) at t = T/4, (**c**) at t = T/2, (**d**) at t = 3T/4.

3.4. Parameter Analysis

To demonstrate the sensitivity of the structure parameters, four key parameters are discussed in this section. Firstly, W3 is chosen and analyzed. As shown in Figure 8a, with the decrease in W3 (meaning the slot on the bent structure is narrower), the VSWR on the middle and upper band becomes larger, while it becomes better around 1.45 GHz. As shown in Figure 8b, the operating band moves to a lower frequency slightly with the increase in L2. The in-band matching becomes better when L2 ranges from 4 to 12 mm. Therefore, L2 is a key parameter to make a fine tuning to the impedance matching in the operating band. As shown in Figure 8c, the VSWR on the lower band becomes better with the increase in a. However, the middle and upper bands deteriorate. As shown in Figure 8d, with the increase in b, the VSWR on the lower band also becomes better while the middle and upper bands deteriorate again. To compromise the low-profile characteristic and wideband operation, these parameters of W3 = 22 mm, L2 = 12 mm, a = 20 mm, and b = 12 mm are chosen for the proposed antenna.

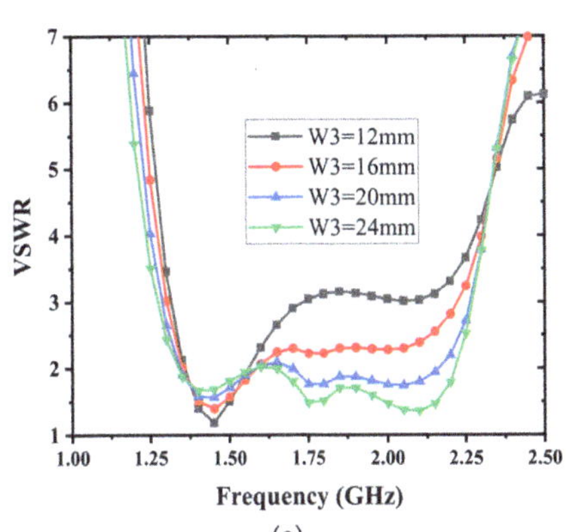

Figure 8. *Cont.*

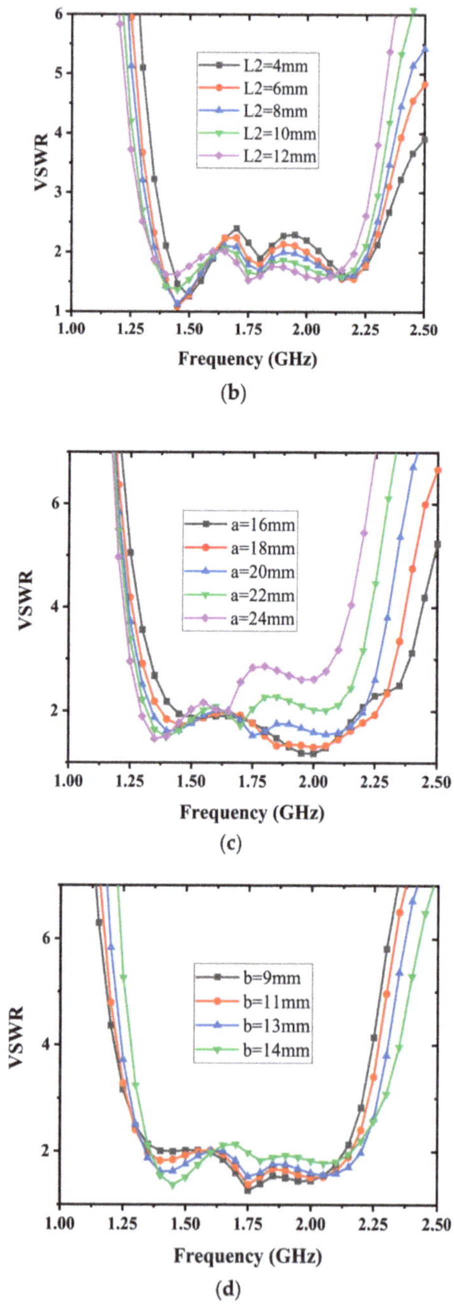

Figure 8. The parameter analysis for (**a**) W3, (**b**) L2, (**c**) a, and (**d**) b.

4. Fabrication and Measurement

To validate the feasibility and correctness of the proposed low-profile ME dipole antenna, a prototype has been fabricated and tested. The bended fixing structure and ground were made of aluminum. The feeding structure and the top horizontal patch were made of a thin copper sheet. To prevent the waggle of the top patches (namely, the electric dipole), two foams are adopted to support them. Finally, the prototype was tested in a microwave chamber as a receiving antenna. An ultra-wideband ridged horn (working from 500 MHz to 18 GHz) acted as a transmitting antenna, which directed to the receiving antenna, as shown in Figure 9.

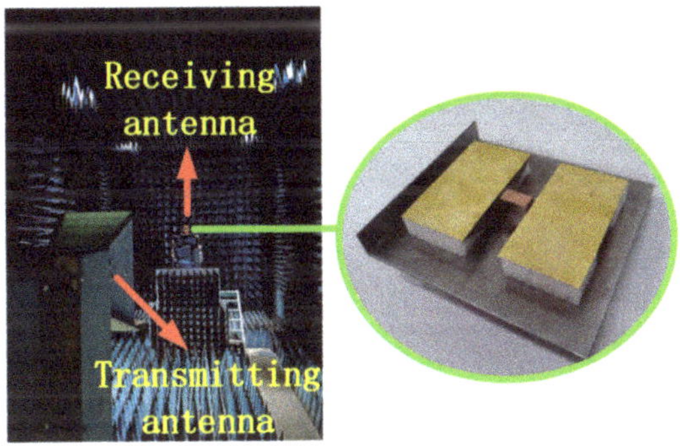

Figure 9. The measurement of proposed antenna.

The VSWR was tested by an Agilent vector network analyzer N5230A. Both simulated and measured results are plotted in Figure 10 for comparison. Obviously, the tested bandwidth for VSWR < 2 ranges from 1.34 to 2.24 GHz with a relative bandwidth of 50.3%. Considering the center operating frequency of 1.79 GHz, the profile of the antenna is only $0.095\lambda_0$. In the impedance bandwidth, the tested boresight gain ranges from 7.38 to 8.73 dBi. The gain drop in the whole band is less than 1.4 dB. Additionally, both simulated and measured normalized radiation patterns are plotted in Figure 11 at 1.4 GHz, 1.8 GHz, and 2.2 GHz for comparison. The E plane and H plane are the xoz plane and yoz plane in Figure 3, respectively. It is obvious that the radiation at the above three frequency points maintains a unidirectional property, and the cross-polarization level on boresight is less than −25 dB. Both tested and simulated patterns agree well with each other.

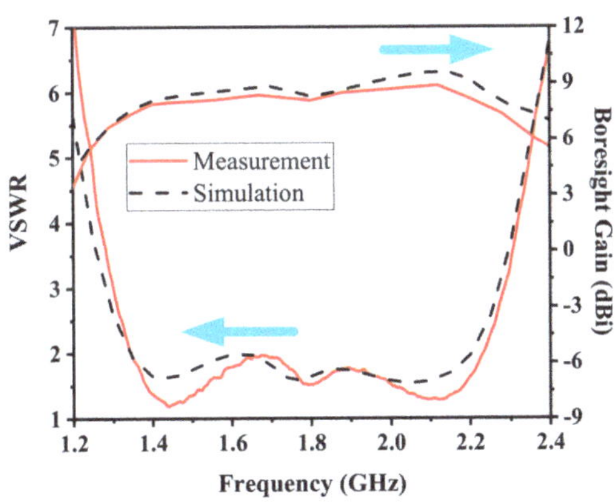

Figure 10. The comparison of simulated VSWR and boresight gain.

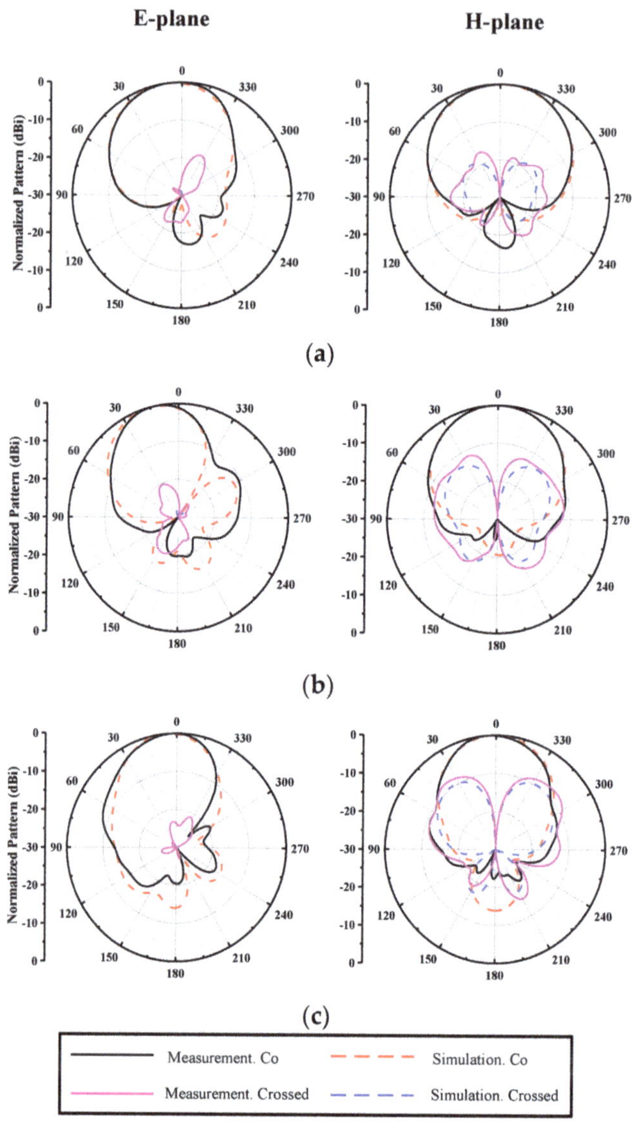

Figure 11. The simulated and measured radiation patterns at (**a**) 1.4 GHz, (**b**) 1.8 GHz, (**c**) 2.2 GHz.

Finally, a performance comparison between the proposed ME dipole and other published ME dipoles is listed in Table 2. For reference [10], which is the classical ME dipole antenna, the profile is up to $0.25\lambda_0$ together with an average gain of 8 dBi. With the development of different low-profile techniques, the profile can be reduced to $0.097\lambda_0$ in reference [15]. However, the relative bandwidth is only 28.2%. In reference [17], the relative bandwidth is up to 54.8%, while the profile is only $0.173\lambda_0$. In reference [19], the antenna has a profile of $0.11\lambda_0$, but the bandwidth is only 27.6%. Among these ME dipole antennas in Table 2, the one proposed in this work is a competitive candidate in low-profile and wideband applications due to the thickness of $0.095\lambda_0$ and relative bandwidth of 50.3%.

Table 2. The performance comparison between the proposed antenna and other ME dipoles.

Reference	Profile	Center Frequency	Bandwidth	Average Gain	Gain Drop	Lateral Dimension
[10]	$0.25\lambda_0$	2.37 GHz	43.8%	8 dBi	0.7 dBi	$0.95\lambda_0 \times 0.95\lambda_0$
[13]	$0.11\lambda_0$	5.495 GHz	18.74%	7 dBi	0.4 dBi	$0.66\lambda_0 \times 0.66\lambda_0$
[15]	$0.097\lambda_0$	1.945 GHz	28.2%	9.2 dBi	2.2 dBi	$0.99\lambda_0 \times 0.99\lambda_0$
[16]	$0.116\lambda_0$	2.315 GHz	43.6%	9 dBi	7.4 dBi	$0.88\lambda_0 \times 0.88\lambda_0$
[17]	$0.173\lambda_0$	2.59 GHz	54.8%	8.6 dBi	2 dBi	$0.97\lambda_0 \times 0.97\lambda_0$
[18]	$0.169\lambda_0$	1.68 GHz	45.6%	8.1 dBi	1.6 dBi	$1.04\lambda_0 \times 1.04\lambda_0$
[19]	$0.11\lambda_0$	3.75 GHz	27.6%	8.2 dBi	2.2 dBi	$0.75\lambda_0 \times 0.75\lambda_0$
Proposed	$0.095\lambda_0$	1.96 GHz	50.3%	8.06 dBi	1.4 dBi	$0.96\lambda_0 \times 0.83\lambda_0$

5. Conclusions

To realize a low profile for a traditional ME dipole antenna, a multi-bending technique is proposed. Both the fixing structure and the feeding structure are bended four and two times, respectively. A profile of only $0.095\lambda_0$ in height is achieved. The effect of three different grounded antennas is analyzed and compared. At last, only one single vertical wall is erected on the left side of the ground to eliminate the gain drop at 2.2 GHz. To validate the feasibility and correctness, a prototype is fabricated and measured. By measurement, the relative bandwidth for VSWR < 2 is 50.3% (from 1.34 to 2.24 GHz). The boresight gain ranges from 7.38 to 8.73 dBi. The difference in gain is less than 1.4 dB in the whole operating band. Both the simulated and measured radiation patterns are compared at 1.4 GHz, 1.8 GHz, and 2.2 GHz to show a stable boresight radiation. The cross-polarization level on boresight is less than −25 dB. Therefore, the proposed ME dipole antenna has the advantages of lower profile and good electrical characteristics and can be used in the future low-profile applications.

Author Contributions: Conceptualization, X.C.; methodology, Z.L.; software, Z.L., Y.T. and L.L.; validation, L.D. and Z.Z.; formal analysis, Z.L.; investigation, Z.L.; resources, Z.L.; data curation, Z.L. and X.C.; writing—original draft preparation, Z.L.; writing—review and editing, X.C.; visualization, L.D. and Z.Z.; supervision, X.C.; project administration, X.C.; funding acquisition, X.C. All authors have read and agreed to the published version of the manuscript.

Funding: This work is supported by the National Natural Science Foundation of China under Grant number U19A2054.

Data Availability Statement: Data is contained within the article.

Conflicts of Interest: The authors declare no conflict of interest.

References

1. Schantz, H. *The Art and Science of Ultrawideband Antennas*, 2nd ed.; Artech House Antennas and Propagation Library: Boston, MA, USA, 2015.
2. Xiao, S.; Wang, B.-Z.; Shao, W.; Zhang, Y. Bandwidth-enhancing ultralow-profile compact patch antenna. *IEEE Trans. Antennas Propag.* **2005**, *53*, 3443–3447. [CrossRef]
3. Liu, W.; Chen, Z.N.; Qing, X. Metamaterial-based low-profile broadband mushroom antenna. *IEEE Trans. Antennas Propag.* **2013**, *62*, 1165–1172. [CrossRef]
4. Liu, W.; Chen, Z.N.; Qing, X. Metamaterial-based low-profile broadband aperture-coupled grid-slotted patch antenna. *IEEE Trans. Antennas Propag.* **2015**, *63*, 3325–3329. [CrossRef]
5. Pan, Y.M.; Hu, P.F.; Zhang, X.Y.; Zheng, S.Y. A low-profile high-gain and wideband filtering antenna with metasurface. *IEEE Trans. Antennas Propag.* **2016**, *64*, 2010–2016. [CrossRef]
6. Wang, X.-Y.; Tang, S.-C.; Shi, X.-F.; Chen, J.-X. A low-profile filtering antenna using slotted dense dielectric patch. *IEEE Antennas Wirel. Propag. Lett.* **2019**, *18*, 502–506. [CrossRef]
7. Cheng, B.; Du, Z.; Huang, D. A broadband low-profile multimode microstrip antenna. *IEEE Antennas Wirel. Propag. Lett.* **2019**, *18*, 1332–1336. [CrossRef]

8. An, W.; Li, S.; Sun, W.; Li, Y. Low-profile wideband microstrip antenna based on multiple modes with partial apertures. *IEEE Antennas Wirel. Propag. Lett.* **2019**, *18*, 1372–1376. [CrossRef]
9. Chlavin, A. A new antenna feed having equal e- and h-plane patterns. *Trans. IRE Prof. Group Antennas Propag.* **1954**, *2*, 113–119. [CrossRef]
10. Luk, K.M.; Wong, H. A new wideband unidirectional antenna element. *Int. J. Microw. Opt. Technol.* **2006**, *1*, 35–44.
11. Chen, S.; Luk, K.M. A dual-mode wideband MIMO cube antenna with magneto-electric dipoles. *IEEE Trans. Antennas Propag.* **2014**, *62*, 5951–5959. [CrossRef]
12. Cui, X.; Yang, F.; Gao, M.; Zhou, L.; Liang, Z.; Yan, F. A Wideband magnetoelectric dipole antenna with microstrip line aperture-coupled excitation. *IEEE Trans. Antennas Propag.* **2017**, *65*, 7350–7354. [CrossRef]
13. Lai, H.W.; Wong, H. Substrate integrated magneto-electric dipole antenna for 5G Wi-Fi. *IEEE Trans. Antennas Propag.* **2014**, *63*, 870–874. [CrossRef]
14. Luk, K.M.; Wu, B. The magnetoelectric dipole—A wideband antenna for base stations in mobile communications. *Proc. IEEE* **2012**, *100*, 2297–2307. [CrossRef]
15. Ding, C.; Luk, K.M. Low-profile magneto-electric dipole antenna. *IEEE Antennas Wirel. Propag. Lett.* **2016**, *15*, 1642–1644. [CrossRef]
16. Ge, L.; Gao, S.; Zhang, D.; Li, M. Magnetoelectric dipole antenna with low profile. *IEEE Antennas Wirel. Propag. Lett.* **2018**, *17*, 1760–1763. [CrossRef]
17. Ge, L.; Luk, K.M. A low-profile magneto-electric dipole antenna. *IEEE Trans. Antennas Propag.* **2012**, *60*, 1684–1689. [CrossRef]
18. Ge, L.; Luk, K.M. A magneto-electric dipole antenna with low-profile and simple structure. *IEEE Antennas Wirel. Propag. Lett.* **2013**, *12*, 140–142. [CrossRef]
19. Yang, S.J.; Pan, Y.M.; Zhang, Y.; Gao, Y.; Zhang, X.Y. Low-profile dual-polarized filtering magneto-electric dipole antenna for 5G applications. *IEEE Trans. Antennas Propag.* **2019**, *67*, 6235–6243. [CrossRef]

sensors

MDPI

Communication

Standing-Wave Feeding for High-Gain Linear Dielectric Resonator Antenna (DRA) Array

Kerlos Atia Abdalmalak [1,2,*], Ayman Abdulhadi Althuwayb [3], Choon Sae Lee [4], Gabriel Santamaría Botello [5], Enderson Falcón-Gómez [1], Luis Emilio García-Castillo [1] and Luis Enrique García-Muñoz [1]

1 Department of Signal Theory and Communications, Carlos III University of Madrid, 28903 Madrid, Spain; efalcon@pa.uc3m.es (E.F.-G.); luise@tsc.uc3m.es (L.E.G.-C.); legarcia@ing.uc3m.es (L.E.G.-M.)
2 Electrical Engineering Department, Aswan University, Aswan 81542, Egypt
3 Department of Electrical Engineering, College of Engineering, Jouf University, Sakaka 72388, Saudi Arabia; aaalthuwayb@ju.edu.sa
4 Electrical and Computer Engineering Department, Southern Methodist University, Dallas, TX 75205, USA; csl@lyle.smu.edu
5 Electrical, Computer and Energy Engineering Department, University of Colorado, Boulder, CO 80309, USA; gasantam@pa.uc3m.es
* Correspondence: kerlos.atia@alumnos.uc3m.es

Abstract: A novel feeding method for linear DRA arrays is presented, illuminating the use of the power divider, transitions, and launchers, and keeping uniform excitation to array elements. This results in a high-gain DRA array with low losses with a design that is simple, compact and inexpensive. The proposed feeding method is based on exciting standing waves using discrete metallic patches in a simple design procedure. Two arrays with two and four DRA elements are presented as a proof of concept, which provide high gains of 12 and 15 dBi, respectively, which are close to the theoretical limit based on array theory. The radiation efficiency for both arrays is about 93%, which is equal to the array element efficiency, confirming that the feeding method does not add losses as in the case of standard methods. To facilitate the fabrication process, the entire array structure is 3D-printed, which significantly decreases the complexity of fabrication and alignment. Compared to state-of-the-art feeding techniques, the proposed method provides higher gain and higher efficiency with a smaller electrical size.

Keywords: antenna array feeds; dielectric resonator antenna (DRA); linear antenna arrays; standing wave; high-gain antennas; high radiation efficiency; 3D printing

Citation: Abdalmalak, K.A.; Althuwayb, A.A.; Lee, C.S.; Botello, G.S.; Falcón-Gómez, E.; García-Castillo, L.E.; García-Muñoz, L.E. Standing-Wave Feeding for High-Gain Linear Dielectric Resonator Antenna (DRA) Array. *Sensors* **2022**, *22*, 3089. https://doi.org/10.3390/s22083089

Academic Editor: Qammer Hussain Abbasi

Received: 24 February 2022
Accepted: 15 April 2022
Published: 18 April 2022

1. Introduction

For decades, dielectric resonators (DRs) have been widely utilized as tuners or amplifiers in microwave-circuit applications [1] due to their high Q-factors. The utilization of antennas had to wait for a long time, until Long et al. presented the first cylindrical dielectric resonator antenna (DRA) in 1983 [2]. Ever since then, dielectric antennas have been intensively studied as potential substitutions of traditional less efficient metal radiators, which have serious problems at high frequencies [3–5]. There are some metal antennas such as lens [6] and slot [7] antennas that can avoid such losses and provide high gain with high efficiency, but they still have the drawbacks of being bulky and complex, respectively. From this perspective, DRAs come with numerous beneficial features such as high radiation efficiency, easy excitation scheme, light weight, and small size [8–11]. One of the main limitations of the DRA is its relatively low gain [12]; hence, dielectric resonator antenna (DRA) arrays are a preferred choice for many applications [13,14].

Several excitation schemes are used to feed DRA arrays [15], such as series or corporate microstrip lines [16,17], standard rectangular waveguide (RWG) [18–21], substrate integrated waveguide (SIW) [22–24], and dielectric image waveguide (DIG) [25–27]. Traditional corporate-feeding networks have a number of power dividers that cause spurious radiation, resulting in high losses [28]. Because of this, new feeding techniques focus on avoiding such losses, for example, in [29], the authors proposed the use of a single microstrip line to feed the whole array. Although this feeding technique is simple and of low cost, the resulting gain is relatively small (15 dBi using 9 elements), and its efficiency is below 80% [30]. Efficiency can be enhanced in the case of feeding only a single element and turning the rest of the array elements into parasitic elements, as in [31]; however, field distribution is not the same for all elements, which results in relatively low gain. For example, in [32], a gain of about 6.6 dBi was obtained using five elements. The RWG feeding scheme has two major drawbacks, high production costs and a bulky structure, which hinder array design [33]. Another feeding method is SIW, which results in significant leakage losses through multiple metallic vias [22] and extra complexity in the design. Lastly, in the DIG feeding scheme, considerable backradiation is a major issue [25]. Additionally, tapered rectangular waveguides are needed to launch the DIG that, in turn, renders the structure bulky and increases the fabrication cost [34].

Although state-of-the-art feeding methods provide some good characteristics, there is still a main research gap to find a novel feeding method that simultaneously illuminates the use of power dividers, transitions, or launchers with keeping uniform excitation to the array elements and low losses in the feeding network. This results in a high-gain DRA array with high efficiency, and renders the design simple, compact, and inexpensive, which are essential parameters for wireless applications such as sensing applications and 5G base station antennas. Hence, this paper presents a novel feeding scheme for linear arrays based on the standing-wave concept [35], compared to our previous standing-wave DRA array presented in [36] where the standing wave was formed by dielectric bridges with metal cover between elements. Here, it is formed within discrete metallic patches printed in the same substrate layer. This ensures more uniform excitation between elements, and could thus achieve performance close enough to the theoretical limit of the array, and achieve a similar gain with a lower number of elements compared to [36], which, in turn, would significantly decrease the size of the solution, as is demonstrated in Section 5.

This study is organized as follows. The detailed design procedure of the array element which acts as the unit cell of the array is presented in Section 2. A two-element array design based on a standing-wave feed is explained in Section 3 with the electric field distribution in the feeding network to confirm the forming of standing waves. Then, the concept is extended to a larger array of four elements in Section 4 with the theoretical calculations of the array factor to validate the feeding concept and confirm the uniform excitation of the array elements using the novel feeding method. Fabrication and measured results are given in Section 5 along with a comparison to DRA arrays fed by state-of-the-art feeding methods from different points of view. Lastly, the main remarks and future work are concluded in Section 6.

2. Design of DRA Array Unit Cell

Figure 1 presents a general high-level block diagram of DRA array fed by the proposed standing wave feeding method. The design consists of discrete metallic patches called feed and center patches transferring the signal from the coaxial probe to the dielectric resonator which acts as the radiating element. The feed patch was designed to form a standing wave under the stripline to feed the antenna element. Then, another smaller patch (called a center patch) is inserted with a gap distance to control the field excitation under the dielectric resonator. Lastly, a dielectric resonator (radiating element) is introduced above the center patch. The total distance between neighboring elements is about one and a half guided wavelengths for maximal array gain. It is clear here that the proposed feeding method excited the radiating element without the need for any power dividers, transitions, or launchers, which would significantly affect DRA array performance, as is demonstrated in the rest of the paper.

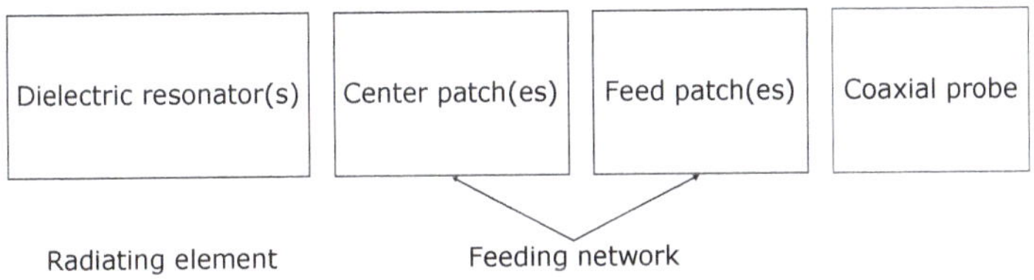

Figure 1. Block diagram of DRA array based on the proposed feeding method.

As an initial step of the array design, the radiating element was designed, which is a rectangular dielectric resonator as shown in Figure 2. Here, a higher-order mode TE_{113} is excited at a frequency of 3.9 GHz to obtain higher gain than that of the fundamental mode of TE_{111} (enhancement of about 2 dB). The optimized dimensions of the resonator were $L_r = 26.2$ mm, $W_r = 26.2$ mm, and $H_r = 45.8$ mm. It was composed of polylactic acid (PLA) with dielectric constant $\varepsilon_r = 3.549$ [37,38]. High-permittivity resonators (ε_r up to 140 [12]) can be used if compact sizes are needed, but at a high cost. The feed-network substrate was Arlon 25N with dielectric constant $\varepsilon_{r(sub)} = 3.38$, loss tangent $\delta_{sub} = 0.0015$, thickness $h_{sub} = 1.5$ mm, and surface area 120×160 mm^2 ($W_{sub} \times L_{sub}$).

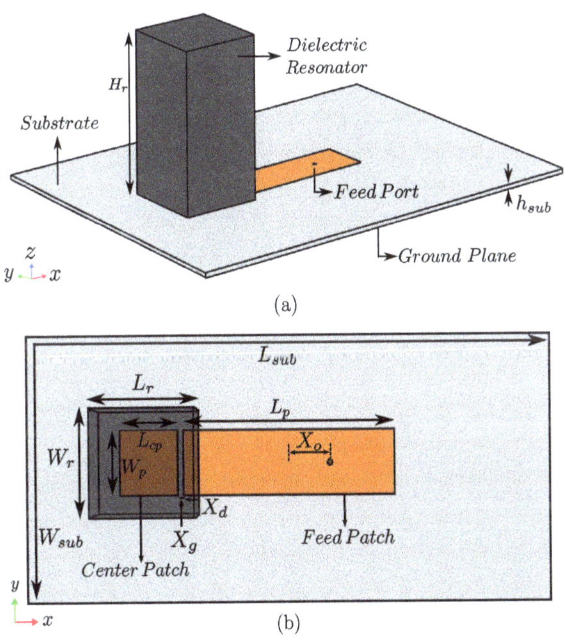

Figure 2. (**a**) 3D view of DRA Unit cell and (**b**) top view with transparent resonator.

The next step is to add the feeding network to have the unit cell for the linear DRA array. To create such a unit cell, feed and center patches with dimensions of 61.5×17 mm ($L_p \times W_p$) and 15×17 mm ($L_{cp} \times W_p$) were introduced with a gap of $X_g = 1.4$ mm between them, as shown in Figure 2. The designed dielectric resonator was placed above the center patch with an overlapping distance ($X_d = 4.2$ mm) between resonator and feed patch, which is needed for sufficient coupling between resonator and feed patch. A 50 Ω coaxial probe was placed at an offset distance of $X_o = 7.6$ mm from the center of the feed patch for proper impedance matching.

The simulated field distribution of TE_{113} mode in the H-plane (y–z plane) inside the DRA at the resonance frequency of 3.9 GHz is shown in Figure 3, along with the reflection coefficient and gain as a function of frequency. Figure 3a shows interested mode TE_{113}

excited at 3.9 GHz and three other neighbor resonant modes at frequencies of 2.6, 4.5, and 5.2 GHz, which correspond to TE_{111}, TE_{113}, and a hybrid mode, respectively. The design and coaxial probe can be easily readjusted to match any of these resonances. The second and third modes were a single TE_{113}, which was split into two resonances due to the impedance mismatching. Hence, by changing the antenna element, matching can be enhanced, and modes come closer which needs a larger bandwidth; however, this is omitted here, as the main contribution for the paper is array feeding and not element design. The antenna produced a peaked gain of 9.3 dBi in the boresight direction ($\theta = 0°$), which outperformed DRA elements operating at TE_{113} [39,40]. This illustrates the efficient feeding scheme of the standing wave, even for a single-element DRA antenna.

The dimensions were optimized for maximal gain using high-frequency structure simulator (HFSS) [41]. The design was simulated under the driven modal on the basis of the finite element method (FEM) and adaptive meshing. The convergence condition was set to achieve target delta S-parameters below 2%, which implied dividing the structure into 20 to 40 thousand tetrahedral meshes for single and four elements, respectively. The used boundary conditions are absorbing boundary condition (ABC) for a surrounding air box that is a quarter wavelength farther away than the antenna edges. Metal feed and center patches were approximated with the perfect electric conductor (PEC) boundary condition during optimization, and were replaced by normal copper units with a thickness of 35 μm at the final simulations.

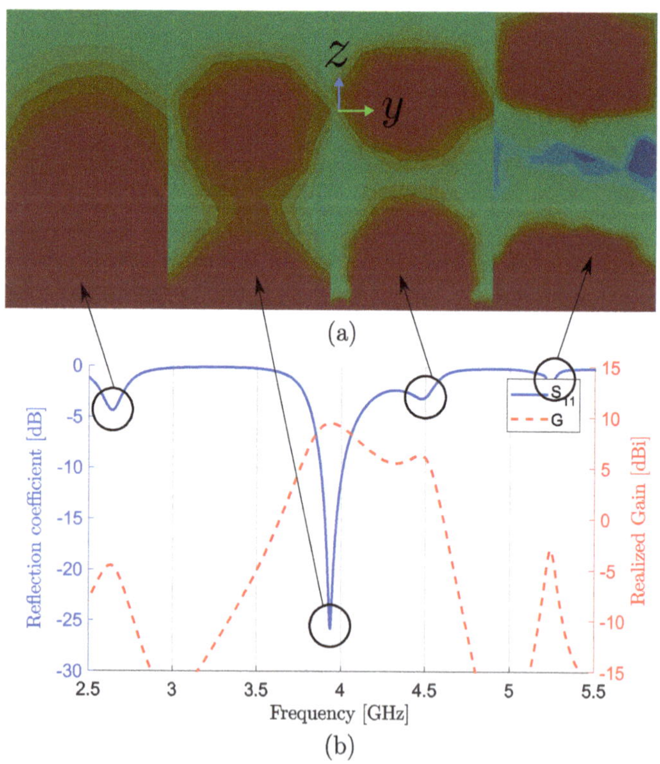

Figure 3. (**a**) Field distribution of TE_{113} mode at 3.9 GHz and neighboring resonant modes inside DRA. (**b**) Simulated S_{11} and gain vs. frequency of antenna element.

3. Two-Element Standing-Wave Linear Array

In this section, a 1 × 2 linear array based on the standing-wave feed method is presented. To form the two-element array, the unit cell in Figure 2 is replicated twice. Since the second cell was the end element of the array, there was no need to add another feed patch at the right of the second element, as the feed patch at the middle would already feed the two elements. Hence, for a general $1 \times N$ element array, we needed N center patches lying underneath each dielectric resonator and $N - 1$ feed patches to uniformly deliver energy to all elements.

A standing wave is formed when two waves travel in opposite directions with an equal magnitude within the feeding network, where null locations do not move. Figure 4 shows the fixed places of null field points in the feeding network. Simulation results are shown in Figure 5, where a maximal realized gain of about 12 dBi was obtained at the resonant frequency. Such a high gain is 2.7 dB over the single-element gain.

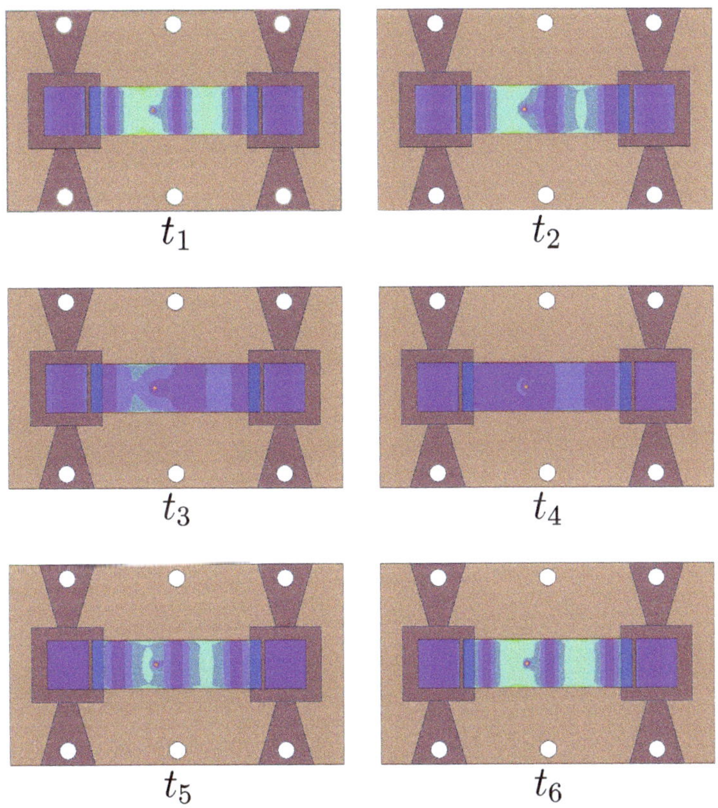

Figure 4. Snapshots of electric field distribution in proposed array demonstrating standing-wave excitation in the feed network.

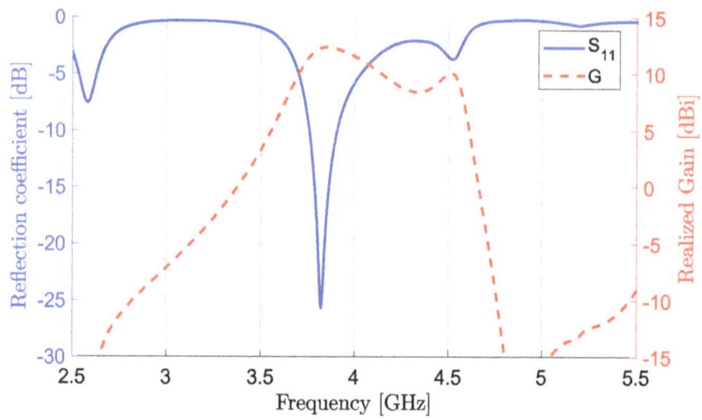

Figure 5. Simulated S_{11} and gain vs. frequency of the proposed two-element standing-wave DRA array.

4. Four-Element Standing-Wave Linear Array

The proposed configuration was extended to create a four-element linear array by adding two more unit cells to the two-element design as shown in Figure 6. Considering that the number of elements (N) was 4, and following our discussion above, the four-element array consists of 4 dielectric resonators lying above 4 center patches with 3 connecting feed patches. The optimized geometrical parameters, as summarized in Table 1, remained almost the same as the number of array elements increased. Such a feature provides a convenient design procedure for various array sizes.

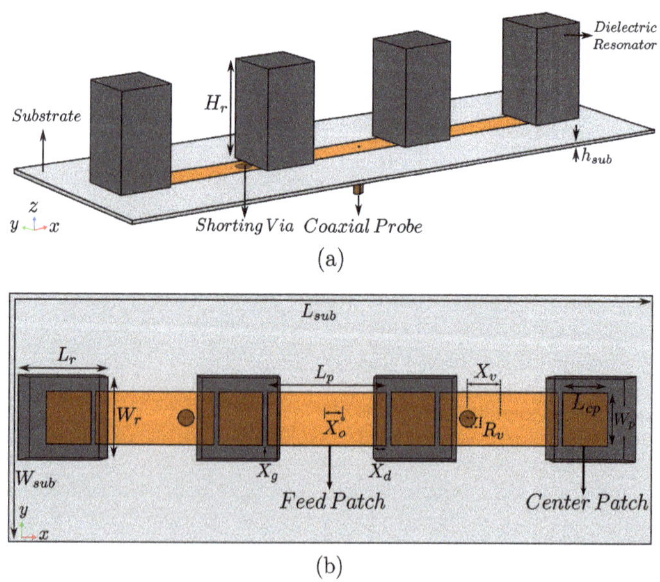

Figure 6. Schematic diagram of proposed four-element standing-wave DRA array: (**a**) 3D view and (**b**) top view with transparent resonators.

Table 1. Geometrical parameters of DRA array (unit: mm).

L_{sub}	W_{sub}	h_{sub}	L_r	W_r	H_r	L_p
280	70	1.52	26.2	26.2	45.8	61.5
L_{cp}	W_p	X_g	X_o	X_d	R_v	X_v
15	17	1.4	7.6	4.2	4	22

In a standing-wave array, the entire array acts as a single large resonator; hence, more than one mode can be strongly excited near the operating frequency, especially in a large array. Unfortunately, some of the excited modes do not produce radiation in the intended direction. Figure 7a presents the field distribution and 3D radiation pattern of the array, being red, yellow, green, and blue colors represent the strength from the max to the minimum. It shows such an example where the fields of the two end elements are in the opposite direction to those of the two inner elements, producing diminished radiation in the boresight direction. To eliminate such unwanted modes, shorting pins are used. Figure 7b shows that the added pins suppress those undesirable modes to give more uniform fields among all array elements for maximal directivity.

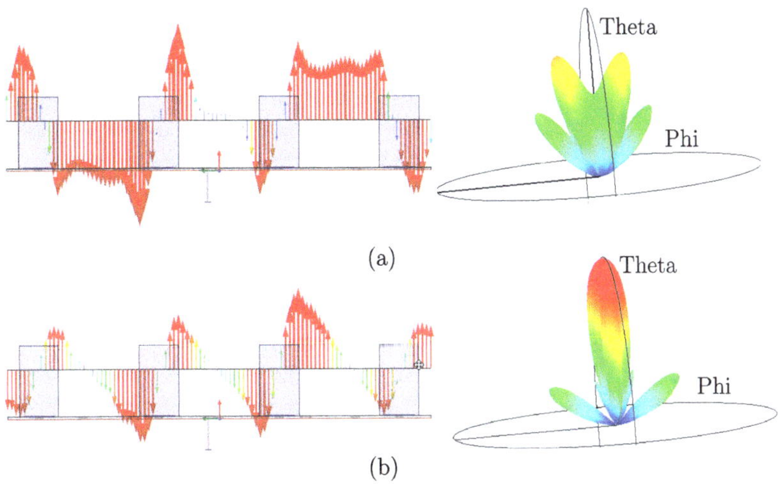

Figure 7. Field distribution and radiation pattern of proposed four-element DRA array (**a**) without and (**b**) with shorting pins.

Figure 8 presents the effect of the shorting-pin position on the array gain. The three pin positions for maximal gain indicate three null electric-field points of the desired mode under the patch because the presence of the pins does not affect the field distribution optimal for gain, but puts down other undesirable modes with finite field strength at the pin location. The radius of the shorting pins was also optimized, so that the undesired modes become eliminated as much as possible, while the selected mode is least affected by the vias.

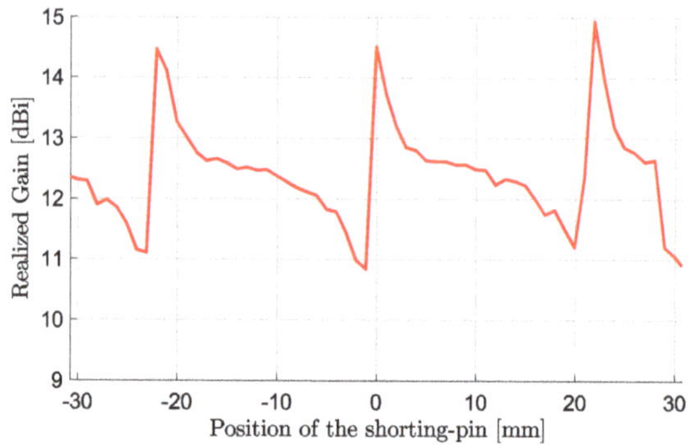

Figure 8. Gain of four-element antenna vs. position of shorting pin along feed patch.

The simulated total efficiency of the four-element design is shown in Figure 9. High efficiency of about 93% was achieved for the four-element array at the resonant frequency, which was only 1% less than that of the single-element design. This demonstrates the benefit of the proposed feeding technique compared to other feeding methods that introduce extra losses in the array structure, resulting in a reduction in antenna efficiency.

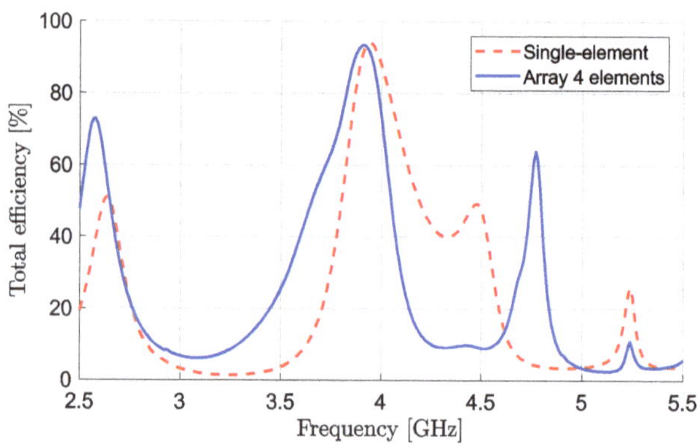

Figure 9. Efficiency of proposed DRA array.

For comparison purposes and to validate the performance of the proposed array feeding method, the array factor for an ideal array was estimated, as is shown in Figure 10a. Hence, four equispaced isotropic elements with the same element spacing as the proposed array and fed with the same amplitude and phase. With this, an upper limit of the gain improvement could be estimated.

Such array factor can be added (in dB) to the actual radiation patterns of the array element (Figure 10b) to compute the estimated total pattern of the array in such an ideal case (uniform feeding to all antenna elements). Figure 10c demonstrates that the proposed feeding method efficiently excites the four elements as the actual simulated radiation patterns of the array are in excellent agreement with the theoretical estimated especially near the boresight. It is clear that the simulated one provided a peak gain of about 15 dBi, which was almost equal to the upper theoretical limit. in other words, the gain enhancement for the four-element array over that of a single element is 5.7 dB, which is approximately equal to the theoretical array gain of a four isotropic-element array with the same element spacing of 1.03 λ_0 [42].

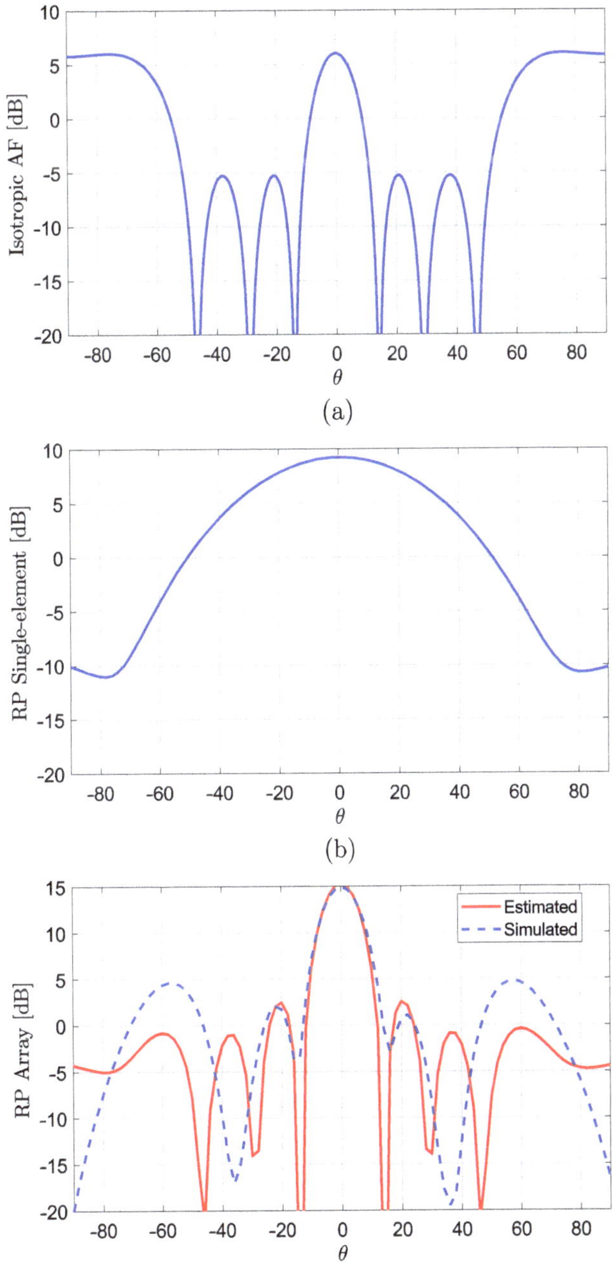

Figure 10. (**a**) Theoretical array factor of four isotropic sources; (**b**) simulated element factor; and (**c**) simulated and estimated radiation pattern of proposed DRA array.

5. Fabrication and Measurements

A four-element standing-wave DRA array was fabricated (Figure 11a) using 3D printing, which is a promising technique for antennas at microwave ranges [43,44]. Two small arms were included in each dielectric element to ensure accurate alignment. Simulated and measured reflection coefficients, and realized gain were in excellent agreement, as shown in Figure 11b.

For gain and radiation pattern measurements, the DRA array was adjusted in an anechoic chamber in the receiving mode with transmitting the signal from the vector network analyzer (VNA) through a standard commercial horn. The spacing between the two antennas was selected to ensure that the tested antenna would be in the far-field region. The signal was amplified using a power amplifier before delivering to the horn to compensate the free space and cables losses, and ensure good receiving power levels above the noise floor of the measurement setup [45]. On the basis of the direct comparison method using a third horn with known gain vs. frequency values, the realized gain as a function of frequency of the proposed DRA array was measured, as shown in Figure 12. Compared to a maximal simulated realized gain of 15 dBi, the measured gain showed a value of 14 dBi at the resonant frequency. The difference in gains was due to uncertainties in material properties, as is explained later in this section. The array had an impedance and 3 dB gain bandwidth product of about 4%, which was equal to those of the single-element antenna. In other words, the proposed array feeding method did not reduce either impedance or gain bandwidths.

The simulated and measured radiation patterns of the dielectric array at the resonance frequency in both E- and H-planes were in relatively good agreement (Figure 13). The simulated sidelobe level was about −10 dB, which was the best achievable value of DRA working in TE_{113} mode [40,46]; however, the measured sidelobe level was lower than that of the simulated. Investigating the possible sources for the mismatching more, and Figures 11 and 12 show that there was a frequency shift up to higher frequency. From this point, it appeared that the reason behind this mismatching was the difference in the dielectric properties of polylactic acid (PLA) due to 3D printing conditions.

Despite the fact that PLA dielectric constant should be about 3.5 at a frequency range around 4 GHz [37,38], the value practically depends on 3D manufacturing properties such as printing temperature [47] and printing resolution [48]. Hence, by considering actual manufacturing conditions (printing temperature of 220 °C and layer thickness of 50 μm) and at the used frequency band, the dielectric constant would be affected to be slightly smaller (around 2.7) [47–50].

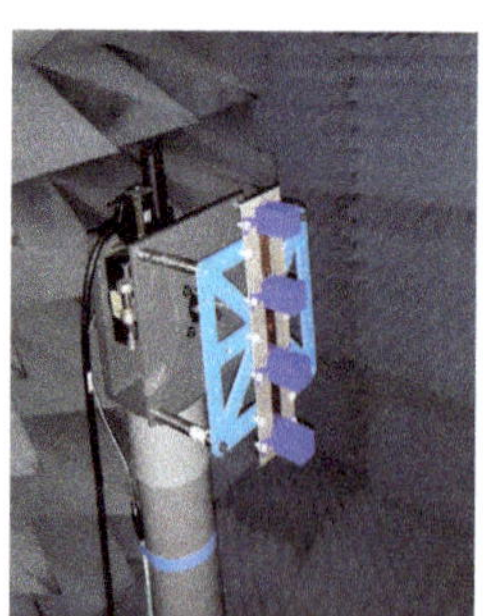

(a)

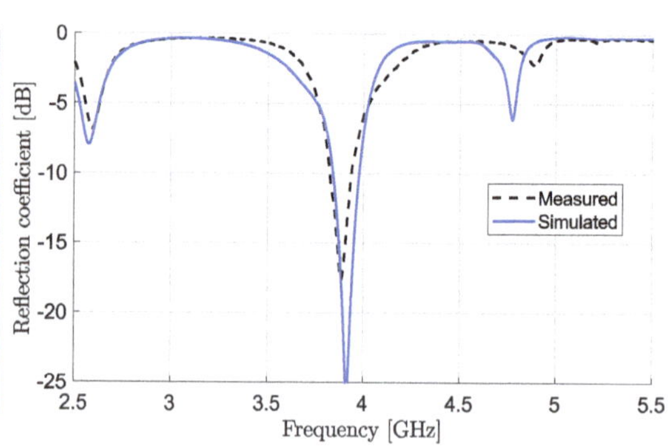

(b)

Figure 11. (**a**) Manufactured prototype of four-element DRA array. (**b**) Measured and simulated S_{11} of manufactured four-element standing-wave DRA array prototype.

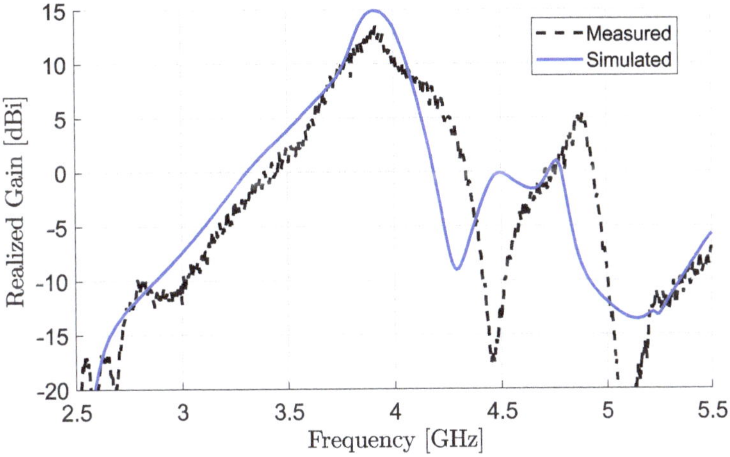

Figure 12. Simulated and measured realized gain of four-element standing-wave DRA array.

To see the effect of such changes in the dielectric constant, the array was resimulated using the new estimated dielectric constant. The resimulated and measured gain and radiation patterns were in excellent agreement, as shown in Figure 14.

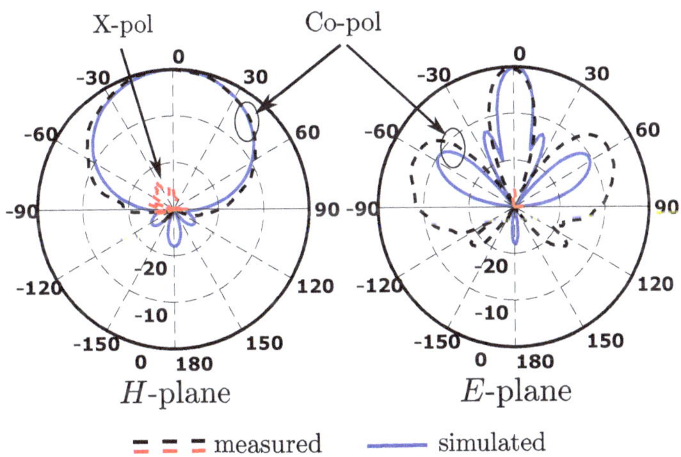

Figure 13. Simulated and measured radiation patterns of four-element standing-wave DRA array.

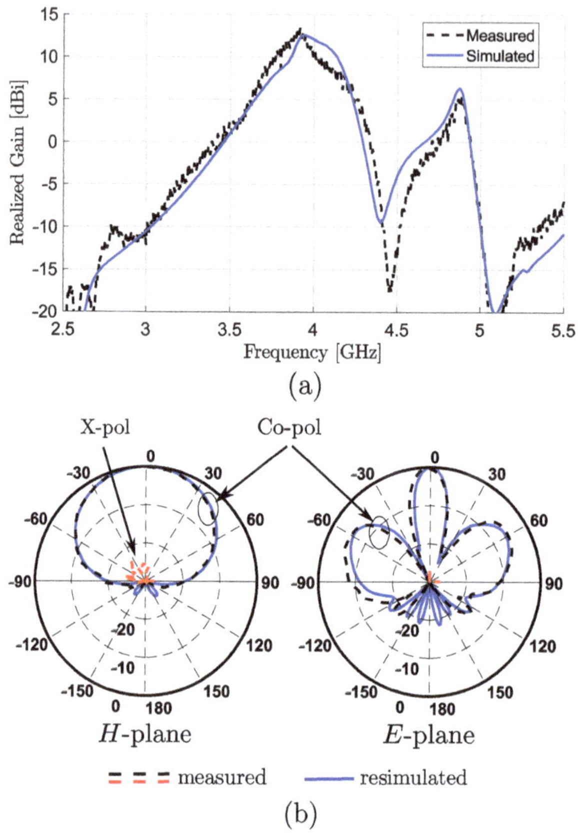

Figure 14. Resimulated and measured: (**a**) realized gain; (**b**) radiation patterns after considering mismatching in dielectric characteristics.

Table 2 shows a comparison of the four-element dielectric array based on the proposed feeding scheme with those of the state-of-the-art feeding techniques for DRA arrays with highlighting their drawbacks in red.The proposed feeding method showed a significant gain improvement of about 3 to 6.5 dB compared with recently reported designs of other feeding schemes. It also remarkably enhanced efficiency compared to standard microstrip lines or DIG feeds. Moreover, in comparison to DIG or SIW, the proposed feeding produced a comparable gain that was provided by the 4 times greater number of elements, which resulted in a very compact design of about 10% of the size of other feeds, in addition to a decrease in cost and complexity. Even for our previous work based on standing-wave feeding [36] which successfully achieved most of the previously mentioned advantages, the feed in this paper is more compact as a similarly high gain of 14 dBi was achieved here by 4 elements ($2\lambda^3$), while 9 elements ($18.7\lambda^3$) are needed for the previous feeding method to achieve a comparable gain of 14 dBi. Regarding the impedance and 3 dB gain bandwidth (BW) product, the proposed feed showed comparable BW to SIW, microstrip lines, and standing-wave feeding methods with lower BW compared to DIG and parasitic methods. However, due to the previously mentioned advantage, the proposed feeding still presented the best overall performance. Additionally, as confirmed in Section 5, this feeding did not degrade array performance, so larger BW could easily be achieved by updating the array element using any standard DRA bandwidth enhancement techniques, as is briefly discussed in Section 6; the proposed array would then have as large a bandwidth as that of the array element.

Table 2. Comparison of proposed standing-wave DRA array with arrays based on state-of-the-art feeding techniques.

Ref	DRA Array Feeding Method	No. of Elements	Gain @f_0 (dBi)	Antenna Size (λ^3)	Impedance/Gain BW Product (%)	Efficiency @f_0 (%)
[31]	Parasitic	3	9.25	$0.89(1.5 \times 1 \times 0.59)$	9	92
[23]	SIW	4	10.6	$1(1.13 \times 3.5 \times 0.26)$	5	93
[16]	Microstrip lines	4	10	$0.9(2.5 \times 2.13 \times 0.17)$	3	85
[25]	DIG	7	7.61	$15.4(4.66 \times 9.77 \times 0.34)$	10	64.6
[24]	SIW	8	11.5	$11(8.4 \times 6.3 \times 0.21)$	2	85
[36]	Standing-wave	9	15	$18.7(4.11 \times 3.73 \times 1.22)$	7	91
[25]	DIG	15	12.46	$25.2(4.66 \times 16 \times 0.34)$	15	63
This work	Standing-wave	2	12	$0.9(0.91 \times 1.58 \times 0.61)$	4	93
This work	Standing-wave	4	14	$2(0.91 \times 3.64 \times 0.61)$	4	93

6. Conclusions and Future Work

In this paper, a high-gain linear DRA array with a standing-wave feed network is presented without the need of using power dividers, transitions, or launchers. The proposed feed technique provided relatively high gain and excellent efficiency compared with those of standard feeding methods. Such attractive features come in addition to a simple and compact design with a straightforward impedance-matching procedure, similar to that of regular microstrip antennas. Arrays based on the proposed feeding method with a different number of elements were presented with manufacturing a four-elements array with low-cost 3D-printing technology. Good agreement was realized between the simulated and measured results with a peak measured gain of 14 dBi. The small deviation between theory and experiments was investigated and was due to changes in the dielectric properties of the resonator material depending on 3D manufacturing properties.

Possible extensions for the work are divided into three main research lines. First, improving the bandwidth, which can be achieved by designing an array on the basis of the same feeding method but using different DRA array element shapes. For example, using an element consisting of multiple dielectric slabs instead of the standard rectangle one to increase the bandwidth (can provide an improvement around 3 times) or using metallic patches as the superstrate. Second, improving the gain even further by using pyramidal horns instead of rectangle resonators. Lastly, scaling the design to work at higher frequencies such as mm wave ranges.

Author Contributions: K.A.A., simulation, manufacturing and measurements, writing, A.A.A.; simulation and writing, C.S.L.; analysis methodology, supervision, funding, G.S.B.; manufacturing, E.F.-G. and L.E.G.-C.; review and editing, and funding, L.E.G.-M.; supervision. All authors have read and agreed to the published version of the manuscript.

Funding: This research was funded by Fundación SENER (REFTA), Comunidad de Madrid MARTINLARA-CM (P2018/NMT-4333), Agencia Estatal de Investigacion (PID2019-109984RB-C41), and RTC2017-6394-7 projects.

Institutional Review Board Statement: Not applicable.

Informed Consent Statement: Not applicable.

Data Availability Statement: Not applicable.

Conflicts of Interest: the authors declare no conflict of interest.

References

1. Kajfez, D.; Guillon, P. *Dielectric Resonators*; Artech House Microwave Library, Artech House: London, UK, 1986.
2. Long, S.; McAllister, M.; Shen, L. The resonant cylindrical dielectric cavity antenna. *IEEE Trans. Antennas Propag.* **1983**, *31*, 406–412. [CrossRef]
3. Brown, E.R.; Abdalmalak, K.A.; Zhang, W. Effect of Metal Resistive Losses on the Gain of a THz Planar Spiral Antenna. In Proceedings of the 2020 14th European Conference on Antennas and Propagation (EuCAP), Copenhagen, Denmark, 15–20 March 2020; pp. 1–4.
4. Kesavan, A.; Al-Hassan, M.; Ben Mabrouk, I.; Denidni, T.A. Wideband Circular Polarized Dielectric Resonator Antenna Array for Millimeter-Wave Applications. *Sensors* **2021**, *21*, 3614. [CrossRef] [PubMed]
5. Rivera-Lavado, A.; García-Muñoz, L.E.; Lioubtchenko, D.; Preu, S.; Abdalmalak, K.A.; Santamaría-Botello, G.; Segovia-Vargas, D.; Räisänen, A.V. Planar Lens–Based Ultra-Wideband Dielectric Rod Waveguide Antenna for Tunable THz and Sub-THz Photomixer Sources. *J. Infrared Millim. Terahertz Waves* **2019**, *40*, 838–855. [CrossRef]
6. García-Muñoz, E.; Abdalmalak, K.A.; Santamaría, G.; Rivera-Lavado, A.; Segovia-Vargas, D.; Castillo-Araníbar, P.; Dijk, F.V.; Nagatsuma, T.; Brown, E.R.; Guzman, R.C.; et al. Photonic-based integrated sources and antenna arrays for broadband wireless links in terahertz communications. *Semicond. Sci. Technol.* **2019**, *34*, 054001. [CrossRef]
7. Warmowska, D.; Abdalmalak, K.A.; Muñoz, L.E.G.; Raida, Z. High-Gain, Circularly-Polarized THz Antenna with Proper Modeling of Structures with Thin Metallic Walls. *IEEE Access* **2020**, *8*, 125223–125233. [CrossRef]
8. Petosa, A. *Dielectric Resonator Antenna Handbook*; Artech House Antennas and Propagation Library, Artech House: London, UK, 2007.
9. Leung, K.W.; Lim, E.H.; Fang, X.S. Dielectric Resonator Antennas: From the Basic to the Aesthetic. *Proc. IEEE* **2012**, *100*, 2181–2193. [CrossRef]
10. Gaya, A.; Jamaluddin, M.H.; Ali, I.; Althuwayb, A.A. Circular Patch Fed Rectangular Dielectric Resonator Antenna with High Gain and High Efficiency for Millimeter Wave 5G Small Cell Applications. *Sensors* **2021**, *21*, 2694. [CrossRef]
11. Petosa, A.; Ittipiboon, A. Dielectric Resonator Antennas: A Historical Review and the Current State of the Art. *IEEE Antennas Propag. Mag.* **2010**, *52*, 91–116. [CrossRef]
12. Luk, K.; Leung, K. *Dielectric Resonator Antennas*; Antennas, S., Ed.; Research Studies Press: New York, NY, USA, 2003.
13. Abdalmalak, K.A.; Santamaría-Botello, G.; Lee, C.S.; Rivera-Lavado, A.; García-Castillo, L.E.; Segovia-Vargas, D.; García-Muñoz, L.E. Microwave Radiation Coupling into a WGM Resonator for a High-Photonic-Efficiency Nonlinear Receiver. In Proceedings of the 2018 48th European Microwave Conference (EuMC), Madrid, Spain, 23–27 September 2018; pp. 781–784. [CrossRef]
14. Abdalmalak, K.A.; Botello, G.S.; Suresh, M.I.; Falcón-Gómez, E.; Lavado, A.R.; García-Muñoz, L.E. An Integrated Millimeter-Wave Satellite Radiometer Working at Room-Temperature with High Photon Conversion Efficiency. *Sensors* **2022**, *22*, 2400. [CrossRef]
15. Petosa, A.; Mongia, R.; Ittipiboon, A.; Wight, J. Investigation of various feed structures for linear arrays of dielectric resonator antennas. In Proceedings of the IEEE Antennas and Propagation Society International Symposium, 1995 Digest, Newport Beach, CA, USA, 18–23 June 1995; Volume 4, pp. 1982–1985. [CrossRef]
16. Mishra, N.K.; Das, S.; Vishwakarma, D.K. Beam steered linear array of Cylindrical Dielectric Resonator Antenna. *AEU Int. J. Electron. Commun.* **2019**, *98*, 106–113. [CrossRef]
17. Petosa, A.; Ittipiboon, A.; Cuhaci, M. Array of circular-polarised cross dielectric resonator antennas. *Electron. Lett.* **1996**, *32*, 1742–1743. [CrossRef]
18. Leung, K.W.; So, K.K. Rectangular waveguide excitation of dielectric resonator antenna. *IEEE Trans. Antennas Propag.* **2003**, *51*, 2477–2481. [CrossRef]
19. Zhang, Y.; Kishk, A.A.; Yakovlev, A.B.; Glisson, A.W. Analysis of Wideband Dielectric Resonator Antenna Arrays for Waveguide-Based Spatial Power Combining. *IEEE Trans. Microw. Theory Tech.* **2007**, *55*, 1332–1340. [CrossRef]
20. Eshrah, I.A.; Kishk, A.A.; Yakovlev, A.B.; Glisson, A.W. Excitation of dielectric resonator antennas by a waveguide probe: modeling technique and wide-band design. *IEEE Trans. Antennas Propag.* **2005**, *53*, 1028–1037. [CrossRef]
21. Leung, K.W.; Lo, H.Y.; So, K.K.; Luk, K.M. High-permittivity dielectric resonator antenna excited by a rectangular waveguide. *Microw. Opt. Technol. Lett.* **2002**, *34*, 157–158. [CrossRef]
22. Xu, F.; Wu, K. Guided-wave and leakage characteristics of substrate integrated waveguide. *IEEE Trans. Microw. Theory Tech.* **2005**, *53*, 66–73. [CrossRef]
23. Abdel-Wahab, W.M.; Busuioc, D.; Safavi-Naeini, S. Millimeter-Wave High Radiation Efficiency Planar Waveguide Series-Fed Dielectric Resonator Antenna (DRA) Array: Analysis, Design, and Measurements. *IEEE Trans. Antennas Propag.* **2011**, *59*, 2834–2843. [CrossRef]
24. Abdallah, M.S.; Wang, Y.; Abdel-Wahab, W.M.; Safavi-Naeini, S. Design and Optimization of SIW Center-Fed Series Rectangular Dielectric Resonator Antenna Array with 45 Linear Polarization. *IEEE Trans. Antennas Propag.* **2018**, *66*, 23–31. [CrossRef]
25. Al-Zoubi, A.S.; Kishk, A.A.; Glisson, A.W. A Linear Rectangular Dielectric Resonator Antenna Array Fed by Dielectric Image Guide with Low Cross Polarization. *IEEE Trans. Antennas Propag.* **2010**, *58*, 697–705. [CrossRef]
26. Al-Zoubi, A.; Kishk, A.; Glisson, A.W. Slot-aperture-coupled linear dielectric resonator array fed by dielectric image line backed by a reflector. In Proceedings of the 2008 IEEE Antennas and Propagation Society International Symposium, San Diego, CA, USA, 5–11 July 2008; 2008; pp. 1–4. [CrossRef]

27. Wyville, M.; Petosa, A.; Wight, J. DIG feed for DRA arrays. In Proceedings of the 2005 IEEE Antennas and Propagation Society International Symposium, Washington, DC, USA, 3–8 July 2005; Volume 2A, pp. 176–179. [CrossRef]
28. Petosa, A.; Ittipiboon, A.; Antar, Y.M.M.; Roscoe, D.; Cuhaci, M. Recent advances in dielectric-resonator antenna technology. *IEEE Antennas Propag. Mag.* **1998**, *40*, 35–48. [CrossRef]
29. Elkarkraoui, T.; Delisle, G.Y.; Hakem, N.; Coulibaly, Y. High gain cross DRA antenna array for underground communications. In Proceedings of the 2014 IEEE Antennas and Propagation Society International Symposium (APSURSI), Memphis, TN, USA, 6–11 July 2014; pp. 1942–1943. [CrossRef]
30. Elkarkraoui, T.; Hakem, N.; Delisle, G.Y.; Coulibaly, Y. A Novel Design Approach for a 60 GHz Circularly Polarized EBG Antenna. *Prog. Electromagn. Res.* **2016**, *69*, 37–51. [CrossRef]
31. Shahadan, N.H.; Jamaluddin, M.H.; Kamarudin, M.R.; Yamada, Y.; Khalily, M.; Jusoh, M.; Dahlan, S.H. Steerable Higher Order Mode Dielectric Resonator Antenna With Parasitic Elements for 5G Applications. *IEEE Access* **2017**, *5*, 22234–22243. [CrossRef]
32. Movahedinia, R.; Sebak, A.R.; Chaharmir, M.R.; Ranjbar Nikkhah, M.; Kishk, A.A. *X* -Band Circularly Polarized Electronically Steerable Parasitic Array Radiator of DRA. *IEEE Trans. Antennas Propag.* **2018**, *66*, 721–728. [CrossRef]
33. Rivera-Lavado, A.; García-Muñoz, L.E.; Generalov, A.; Lioubtchenko, D.; Abdalmalak, K.A.; Llorente-Romano, S.; García-Lampérez, A.; Segovia-Vargas, D.; Räisänen, A.V. Design of a Dielectric Rod Waveguide Antenna Array for Millimeter Waves. *J. Infrared Millim. Terahertz Waves* **2017**, *38*, 33–46. [CrossRef]
34. Bhartia, P.; Bahl, I. *Millimeter Wave Engineering and Applications*; A Wiley-Interscience Publication; Wiley: Seoul, Korea, 1984.
35. Lakshmanan, A.; Lee, C.S. A Standing-Wave Microstrip Array Antenna. *IEEE Trans. Antennas Propag.* **2011**, *59*, 4858–4861. [CrossRef]
36. Althuwayb, A.A.; Abdalmalak, K.A.; Lee, C.S.; Santamaría-Botello, G.; García-Castillo, L.E.; Segovia-Vargas, D.; García-Muñoz, L.E. 3-D-Printed Dielectric Resonator Antenna Arrays Based on Standing-Wave Feeding Approach. *IEEE Antennas Wirel. Propag. Lett.* **2019**, *18*, 2180–2183. [CrossRef]
37. Huber, E.; Mirzaee, M.; Bjorgaard, J.; Hoyack, M.; Noghanian, S.; Chang, I. Dielectric property measurement of PLA. In Proceedings of the 2016 IEEE International Conference on Electro Information Technology (EIT), Grand Forks, ND, USA, 19–21 May 2016; pp. 0788–0792. [CrossRef]
38. Kumar, P.; Dwari, S.; Utkarsh.; Singh, S.; Kumar, J. Investigation and Development of 3D Printed Biodegradable PLA as Compact Antenna for Broadband Applications. *IETE J. Res.* **2020**, *66*, 53–64. [CrossRef]
39. Fang, X.S.; Chow, C.K.; Leung, K.W.; Lim, E.H. New single-/dual-mode design formulas of the rectangular dielectric resonator antenna using covariance matrix adaptation evolutionary strategy. *IEEE Antennas Wirel. Propag. Lett.* **2011**, *10*, 734–737. [CrossRef]
40. Petosa, A.; Thirakoune, S. Rectangular Dielectric Resonator Antennas With Enhanced Gain. *IEEE Trans. Antennas Propag.* **2011**, *59*, 1385–1389. [CrossRef]
41. ANSYS Simulation Driven Product Development, HFSS. Available online: https://www.ansys.com/ (accessed on 1 February 2022).
42. Balanis, C.A. *Antenna Theory: Analysis and Design*; John Wiley & Sons: Hoboken, NJ, USA, 2016.
43. Kremer, H.I.; Leung, K.W.; Wong, W.C.; Lo, K.K.W.; Lee, M.W.K. Design of Dielectric Resonator Antenna Using Dielectric Paste. *Sensors* **2021**, *21*, 4058. [CrossRef] [PubMed]
44. Abdalmalak, K.A.; Botello, G.S.; Llorente-Romano, S.; Rivera-Lavado, A.; Flygare, J.; Fernández, J.A.L.; Puente, J.M.S.; García-Castillo, L.E.; Segovia-Vargas, D.; Pantaleev, M.; et al. Ultrawideband Conical Log-Spiral Circularly Polarized Feed for Radio Astronomy. *IEEE Trans. Antennas Propag.* **2020**, *68*, 1995–2007. [CrossRef]
45. Lamkaddem, A.; El Yousfi, A.; Abdalmalak, K.A.; Posadas, V.G.; Segovia-Vargas, D. Circularly Polarized Miniaturized Implantable Antenna for Leadless Pacemaker Devices. *IEEE Trans. Antennas Propag.* **2022**, 1. [CrossRef]
46. Petosa, A.; Thirakoune, S.; Ittipiboon, A. Higher-order modes in rectangular DRAs for gain enhancement. In Proceedings of the 2009 13th International Symposium on Antenna Technology and Applied Electromagnetics and the Canadian Radio Science Meeting, Banff, AB, Canada, 15–18 February 2009, pp. 1–4. [CrossRef]
47. Dichtl, C.; Sippel, P.; Krohns, S. Dielectric properties of 3D printed polylactic acid. *Adv. Mater. Sci. Eng.* **2017**, *2017*, 6913835. [CrossRef]
48. Veselý, P.; Tichý, T.; Šefl, O.; Horynová, E. Evaluation of dielectric properties of 3D printed objects based on printing resolution. *IOP Conf. Ser. Mater. Sci. Eng.* **2018**, *461*, 012091. [CrossRef]
49. Abdalmalak, K.A. Analysis and Design of Antennas and Radiometers for Radio Astronomy Applications in Microwave, Mm-wave, and THz Bands. Ph.D. Thesis, Universidad Carlos III de Madrid, Madrid, Spain, 2022.
50. Zhang, S.; Njoku, C.C.; Whittow, W.G.; Vardaxoglou, J.C. Novel 3D printed synthetic dielectric substrates. *Microw. Opt. Technol. Lett.* **2015**, *57*, 2344–2346. [CrossRef]

sensors

MDPI

Article

A Novel Meander Line Metamaterial Absorber Operating at 24 GHz and 28 GHz for the 5G Applications

Syed Aftab Naqvi [1], Muhammad Abuzar Baqir [1,*], Grant Gourley [2], Adnan Iftikhar [1], Muhammad Saeed Khan [3] and Dimitris E. Anagnostou [2,*]

[1] Department of Electrical and Computer Engineering, COMSATS University Islamabad, Sahiwal 57000, Pakistan; aftabnaqvi@cuisahiwal.edu.pk (S.A.N.); adnaniftikhar@comsats.edu.pk (A.I.)
[2] Institute of Signals, Sensors and Systems, Heriot Watt University, Edinburgh EH14 4AS, UK; gjg1@hw.ac.uk
[3] Department of Information Engineering, University of Padova, 35122 Padova, Italy; mskj786@hotmail.com
* Correspondence: abuzar@cuisahiwal.edu.pk (M.A.B.); d.anagnostou@hw.ac.uk (D.E.A.)

Abstract: Fifth generation (5G) communication systems deploy a massive MIMO technique to enhance gain and spatial multiplexing in arrays of 16 to 128 antennas. In these arrays, it is critical to isolate the adjacent antennas to prevent unwanted interaction between them. Fifth generation absorbers, in this regard, are the recent interest of many researchers nowadays. The authors present a dual-band novel metamaterial-based 5G absorber. The absorber operates at 24 GHz and 28 GHz and is composed of symmetric meander lines connected through a transmission line. An analytical model used to calculate the total number of required meander lines to design the absorber is delineated. The analytical model is based on the total inductance offered by the meander line structure in an impedance-matched electronic circuit. The proposed absorber works on the principal of resonance and absorbs two 5G bands (24 GHz and 28 GHz). A complete angular stability analysis was carried out prior to experiments for both transverse electric (TE) and transverse magnetic (TM) polarizations. Further, the resonance conditions are altered by changing the substrate thickness and incidence angle of the incident fields to demonstrate the functionality of the absorber. The comparison between simulated and measured results shows that such an absorber would be a strong candidate for the absorption in millimetre-wave array antennas, where elements are placed in proximity within compact 5G devices.

Keywords: meander line; metamaterial absorber; massive MIMO; 5G

Citation: Naqvi, S.A.; Baqir, M.A.; Gourley, G.; Iftikhar, A.; Saeed Khan, M.; Anagnostou, D.E. A Novel Meander Line Metamaterial Absorber Operating at 24 GHz and 28 GHz for the 5G Applications. *Sensors* **2022**, *22*, 3764. https://doi.org/10.3390/s22103764

Academic Editors: Chan Hwang See, Raed A. Abd-Alhameed and Naser Ojaroudi Parchin

Received: 30 March 2022
Accepted: 9 May 2022
Published: 15 May 2022

1. Introduction

The exponential proliferation of frequency-selective devices (including complex wireless electronic systems and internet of things (IoT), etc.) demand large bandwidths, high data rates, and low latencies to operate efficiently and purposefully. The 5G telecommunication network system is the proposed solution to this problem, providing a bandwidth of around 5 GHz, a data rate of up to 5 Gb/s for high mobility and 50 Gb/s for pedestrians, and a low latency of 1 ms [1,2]. The availability of such a wide bandwidth in the mm-wave band was approved in August 2018 by the Federal Communications Commission (FCC) for 5G, during the first 5G spectrum auction for the 24 GHz (24.25–24.45 and 24.75–25.25) and 28 GHz (27.5–28.35 GHz) bands. This allowed researchers to focus on new designs at these frequencies [3].

Novel metamaterial absorbers (MAs), which work on the principal of resonance, have been explored for use with wireless communication devices (emitters, filters, sensors, photodetectors, photovoltaic solar cells, and infrared camouflage) [4–12]. Significant work on MA designs is available in the literature for microwave, terahertz, visible, and ultraviolet frequencies [13–18]. On the contrary, 5G and especially 24 GHz and 28 GHz bands are relatively unattended so far, in this regard [19–22]. Additionally, the limitations of previous works on narrowband, wideband, and ultra-wideband absorbers involve the facts of their complex geometrical structures, large number of layers to trap electromagnetic waves, and costly materials [13–18,23–26].

A meander line is a U-shaped compact transmission line (TL), formed by connecting two parallel TLs with another TL [27,28]. Meander lines achieve compactness as they offer similar impedance Z, inductance L, and other microwave network parameters to straight planar TLs [27–30]. Their reduced overall size also allows them to utilize the given design space more effectively. The total electrical length of the meander line is similar to that of a straight line; hence, the operating frequency remains the same. The ease of integration of a meander line with the rest of the electronic circuitry, as well as the possibility of shifting its frequency by varying the number of meanders, are extremely useful features of meander lines [31]. Meander lines are frequently used in antennas; however, there are very few examples of absorbers employing meander lines.

In this paper, a novel dual-band absorber operating at 24 GHz and 28 GHz for 5G applications is proposed. The material used to design the absorber is cost-effective and the structure is easy to design when compared to the absorbers presented in the literature to date. Novelty lies in the fact that this is the very first attempt to design an absorber for 5G applications at the 24 GHz and 28 GHz frequency bands. Moreover, the derivation of the number of meander lines to devise a structure for the absorber at any frequency by varying the parameters of the meander line is a unique feature of the proposed work in this domain. The utilization of the meander line structures to design an absorber at millimetre-wave frequency is an attempt which is the first of its kind, to the best of the authors' knowledge.

The analytical derivation used to calculate the number of meander lines in order for the absorber to operate at 24 GHz and 28 GHz, as well as the overall design procedure, is presented in Section 2. The simulated and measured results of the absorber are discussed in Section 3. Finally, the conclusion is discussed in Section 4.

2. Design Procedure and Absorption Mechanism

Figure 1b shows the front view of the unit cell of the proposed metamaterial absorber. The absorber consists of two pairs of symmetric meander lines connected to an I-shape TL. The schematic representation of the finite absorber sheet consisting of the proposed metamaterial absorber is depicted in Figure 1a. The procedure for calculating the total number of meander lines needed to design the proposed novel metamaterial absorber is delineated here. This will help the reader and scientific community in designing on-demand metamaterial absorbers for their respective applications in numerous disciplines, according to the requirements of the end user. The procedure for calculating the number of meander lines at the specific frequencies is shown below:

The characteristic impedance of the parallel placed twin TL is [32]

$$Z_0 = \frac{\eta}{\pi} \log \frac{2a}{b} \tag{1}$$

where η is the intrinsic impedance of the free space, a is the distance between the lines, and b is the diameter of a TL. If l is the length of a line terminated with a load impedance Z_L, then the input impedance of the twin TL is given by [28]

$$Z_{in} = Z_0 \frac{Z_L + jZ_0 \tan \beta l}{Z_0 + jZ_L \tan \beta l} \tag{2}$$

where $\beta = \frac{2\pi}{\lambda}$, $\lambda = \frac{c}{f\sqrt{\varepsilon_r}}$, c is the speed of light, f the operating frequency, and ε_r the relative permittivity of the material. Z_{in} is the input impedance of the short-terminated lines which will help to propagate the matched impinged EM waves on the structure. Meander lines are truncated with a short-circuit load (Z_L = 0), so (2) becomes

$$Z_{in} = jZ_0 \tan \beta l \tag{3}$$

Figure 2 illustrates the absorptivity versus the frequency of the MA with a substrate thickness of t = 0.8 mm from 0° to 30° of both TE- and TM-polarized incident waves. Figure 2a shows the absorption for TE polarization corresponding to different angles of incidence from θ = 0° to 30° (with step size of 10°). It is depicted that for θ = 0°, an absorption peak is attained at 26 GHz (as shown in Figure 2a). The results show that with the increase in incidence angle, the absorption peak produces a redshift. Further, it is observed that absorption peaks are attained at 26 GHz, 25.8 GHz, 25.6 GHz, and 25.3 GHz for their respective angles of incidence at θ = 0°, 10°, 20°, and 30°. Notably, for θ = 30°, the proposed MA is most useful in absorbing the 5G band ranging from 24.57 GHz to 25.25 GHz, having a full width half maximum (FWHM) of 0.8 GHz. This shows that the MA covers the entire 5G band. Figure 2b shows the absorption for TM polarization, keeping angular operating conditions like the previously discussed case in Figure 2a. Moreover, it is noticeable that absorption behaviour is changed by altering the angle of incidence. However, for $\theta > 20°$, similar absorptivity is attained as observed for the TE-polarized case depicted in Figure 2a. From this, it can be inferred that for $\theta > 20°$, MA acts as a 5G absorber.

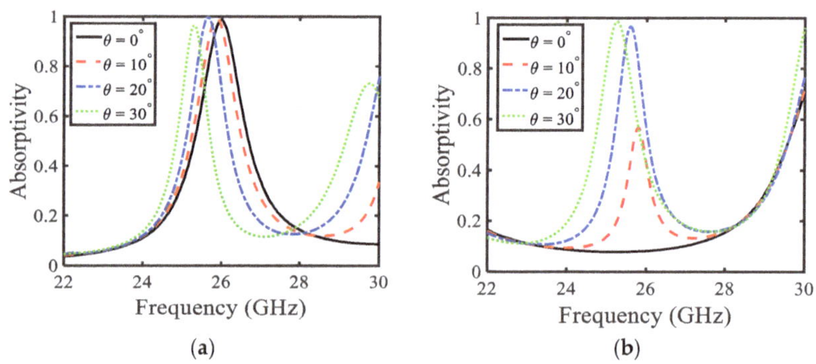

Figure 2. Absorption of the proposed MA keeping substrate thickness t = 0.8 mm (**a**) TE polarization and (**b**) TM polarization.

Figure 3 illustrates absorption when the substrate thickness is increased to t = 1.2 mm. It is noticeable that with the increase in substrate thickness, absorption spectra are altered, compared with the previous case discussed in Figure 2. Figure 3a illustrates the absorptivity of the MA for a TE-polarized wave. Absorption peaks show a redshift with the increase in incidence angle for TE-polarized waves. It is observed that for incidence angles of θ = 20° and 30°, the MA covers the 5G bands (27.5 GHz–28.35 GHz), whereas, for excitation θ = 10°, another 5G band (24.75 GHz–25.25 GHz) is absorbed by the proposed MA. In addition, it is noticed that the highest absorption peak with near-unity absorption is attained for θ = 20°, just before 24 GHz, shown by dash-dotted blue line. In examining the second absorption peaks at higher frequencies, it is observed that these absorption peaks cover another 5G band (27.5 GHz–28.35 GHz) for every incident angle from 0° to 30° (as shown in Figure 3a). The absorption has a large value of FWHM in this case. Moreover, absorptivity increases with the increase in incidence angle. It is noticed that for θ = 0°, the absorption peak attains absorptivity of 75%, and for θ = 30°, the absorption peak has absorptivity of 95%. Additionally, considering Figure 3b, it can be concluded that for θ = 0°, 10°, and 20° in TM-polarized cases, absorptivity lies within the 5G band (24.25 GHz–24.45 GHz and 24.75 GHz–25.24 GHz) with more than 92% absorption.

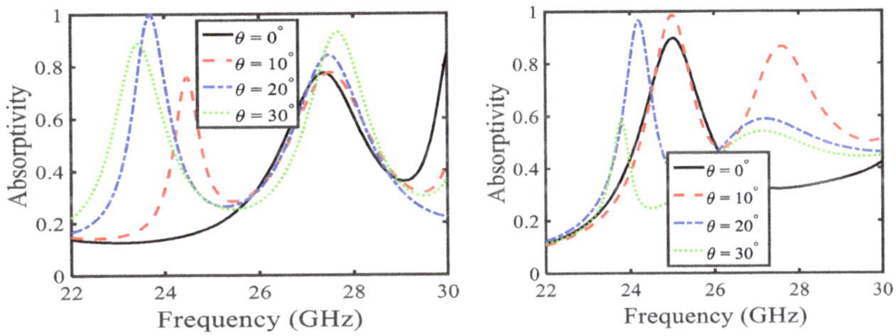

(**a**) (**b**)

Figure 3. Absorption of the proposed MA keeping substrate thickness t = 1.2 mm for (**a**) TE polarization (**b**) TM polarization.

The absorption peaks of the proposed MA are studied for thickness of the substrate at t = 1.6 mm. Figure 4 represents the simulated absorptivity of the chosen MA design. It is observed that with the increase in substrate thickness, absorptivity shows a considerable increase. Further, it can be clearly seen that the number of absorption peaks are increased, as compared to the previous cases discussed in Figures 2 and 3. Figure 4a shows the absorptivity for the TE-polarized wave. Here, it is observed that for $\theta = 0°$, $10°$, $20°$, and $30°$, MA successfully covers the 24 GHz (i.e., 24.25 GHz–24.45 and 24.75 GHz–25.25 GHz) band, with absorption higher than 90% for all values of θ. On the other hand, for the 28 GHz 5G band, it is observed that absorption peaks lie in this band for $\theta = 0°$ and $10°$, as shown by black solid and dashed red lines in Figure 4a. However, the absorption peak shifts to the higher frequency for $\theta = 20°$, as shown by the dashed-dotted blue line in Figure 4a, whereas, for $\theta = 30°$, the absorption peak is attained with an absorptivity of 62% just before 28 GHz (as shown by the dashed green line in Figure 4a). Figure 4b shows the absorptivity for the TM polarization. It is observed that the absorption peaks are found in the 28 GHz frequency range of the 5G—the remaining absorption peaks are not in our interest.

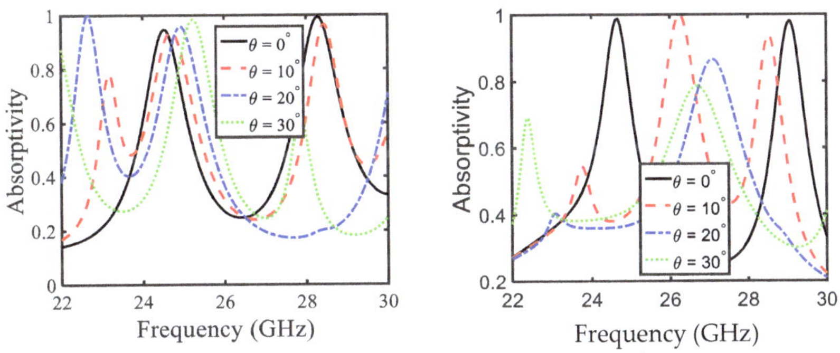

(**a**) (**b**)

Figure 4. Absorption of the proposed PA keeping substrate thickness t = 1.6 mm for (**a**) TE polarization (**b**) TM polarization.

To create a better understanding of the absorption mechanism, the surface current density of the top metasurface and bottom layer is depicted in Figures 5 and 6, corresponding to the absorption peaks at 24.7 and 28.3 GHz, respectively. Figure 5 shows the surface current density for the absorption at 24.7 GHz, and it is noticeable that the maximum surface current remains confined between the top and lower sides of the adjacent mender lines and the middle of the transmission line. Furthermore, the direction of the surface currents is parallel for the top metasurface and bottom conductor in most parts of the unit cell, which leads to electric resonance. However, the surface current on the mender line is antiparallel to the surface current of the bottom layer, which leads to the magnetic dipole. Hence, the absorption is due to the electric and magnetic resonance. Similarly, the surface current density plots at 28.3 GHz are shown in Figure 6, which reveals the formation of electric and magnetic dipoles at the surface of the meander line and the resultant phenomenon of absorption.

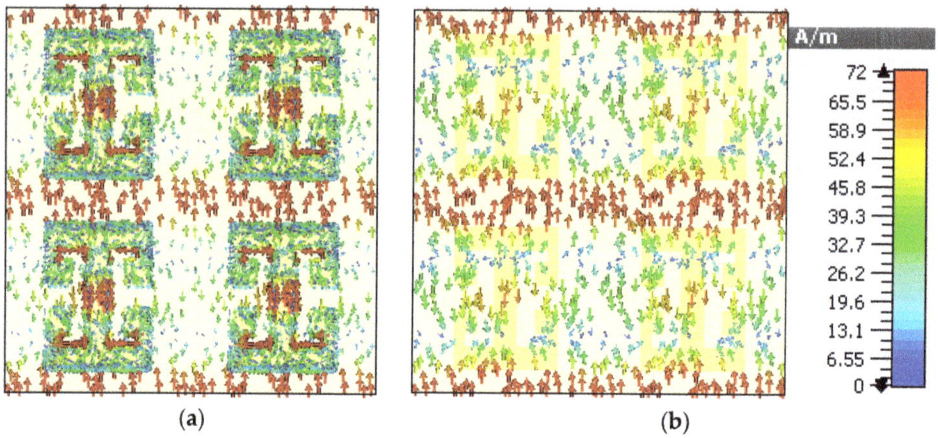

Figure 5. Surface current density for 24.7 GHz (**a**) top metasurface and (**b**) bottom surface.

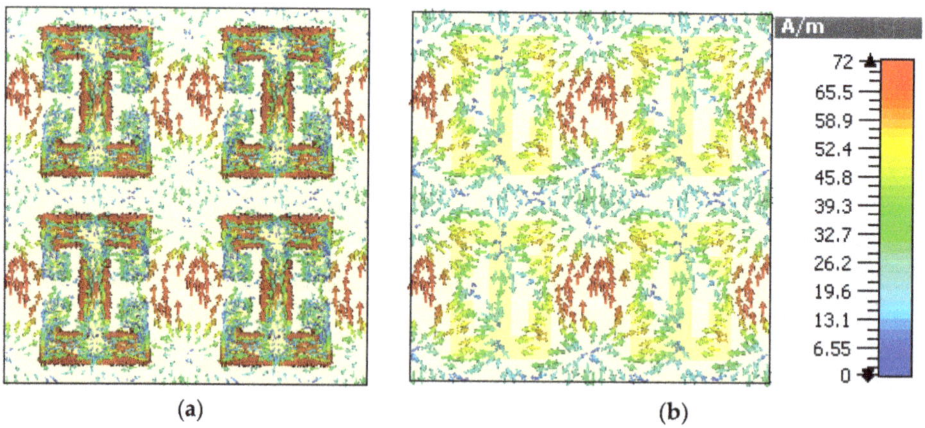

Figure 6. Surface current density for 28.3 GHz (**a**) top metasurface (**b**) bottom surface.

The impedance was also studied to further explore the physical mechanism of the absorption phenomena. Figure 7 illustrates both the real and imaginary effective impedance for $\theta = 0°$, keeping the substrate thickness as $t = 1.6$ mm. Herein, the impedance-matching condition is considered. It is noteworthy that due to the resonance, the free-space impedance matches the top metasurface impedance; therefore, all incidence waves are allowed to enter in the substrate of the metamaterial absorber. The effective impedance of the proposed MA is deduced by employing the impedance formula [18]. The plot depicted in Figure 7 shows the effective impedance for a normal-incidence TE-polarized wave with a dielectric substrate thickness of 1.6 mm. It is noticeable that the real part of the effective impedance shows a sharp dip at ~24.5 GHz and ~28.2 GHz (in the 5G band allocated by FCC), confirming that the proposed MA absorbs frequencies within this band. The imaginary effective impedance attains negative values at the aforesaid frequencies: impedance at ~24.5 GHz and ~28.2 GHz matches with the free-space impedance, allowing the electromagnetic waves to penetrate within the dielectric substrate. Thereby, all the incidence waves remain trapped inside the dielectric substrate around the reported frequencies. Consequently and corresponding to the impedance-matching condition, two absorption peaks are observed in Figure 4a for the normally incident excited wave.

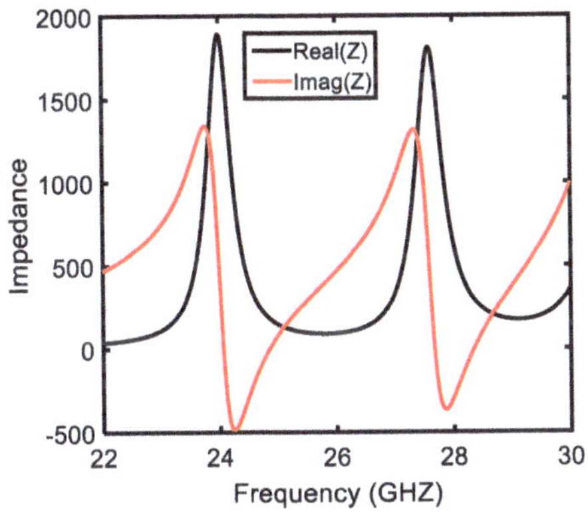

Figure 7. Effective impedance of the metamaterial absorber.

To further investigate the absorption of the proposed metamaterial absorber, the surface electric field is studied with the absorption peak at 24.5 GHz for a normally incident wave. Figure 8a,b illustrate the surface electric field for the TE and TM polarizations, respectively. It is noticeable from the figures that the electric field is maximally concentrated around the edges of the meander line. It can also be observed that the transmission line of the meander line absorber has a lower concentration of electric field, as presented in Figure 8.

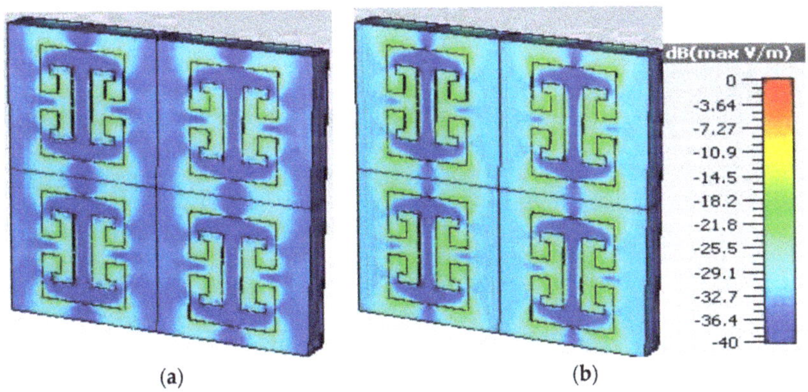

Figure 8. Simulated electric field distribution for 24.5 GHz for (**a**) TE polarized wave and (**b**) TM polarized wave.

To verify the simulation results, an array of 12 × 12 unit cells was fabricated on FR4 dielectric substrate with copper on its bottom side. Although the simulation results of t = 0.8 mm and t = 1.2 mm are encouraging, a substrate thickness of 1.6 mm was chosen for the fabrication of the proposed absorber because of its commercial availability. A photograph of the measurement setup and the finite MA consisting of 12 × 12 unit cells of the proposed meander-line-based absorber printed on 1.6 mm thick FR4 substrate is shown in Figure 9. In the measurement setup, initially, the system was calibrated by placing a metallic sheet in front of the transmitter and receiver horn antennas which were connected to a well-calibrated vector network analyser (VNA). It was also ensured that the horn antennas were placed at a far-field distance from the MA to avoid electromagnetic inference which may have altered the results.

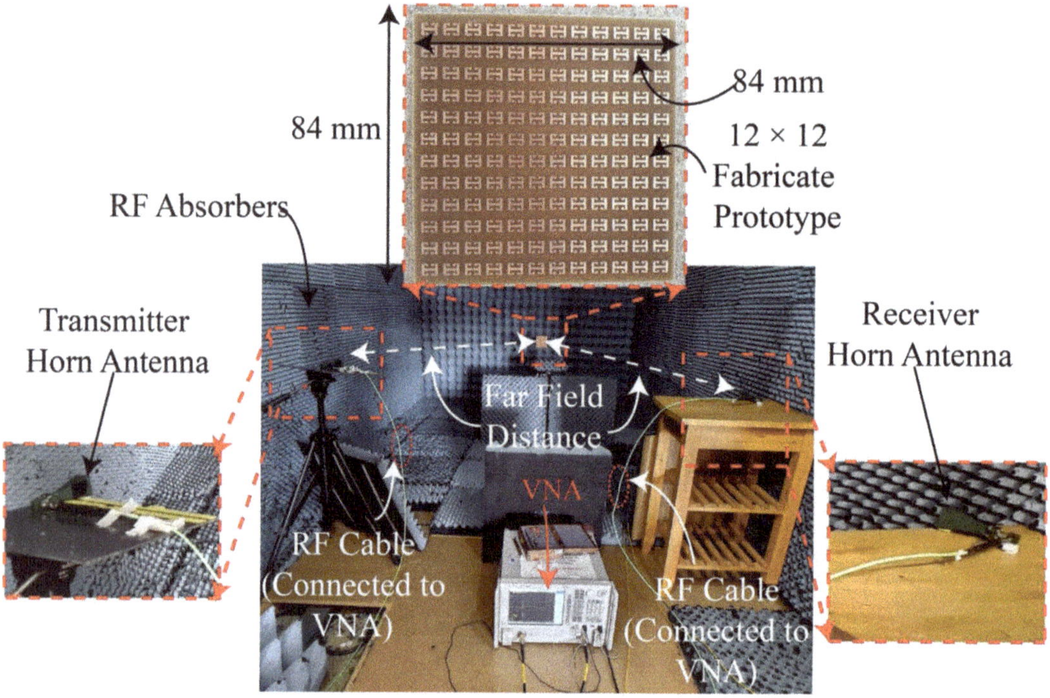

Figure 9. A photograph of the fabricated prototype and experimental setup used for the measurement of the proposed MA.

After the system was calibrated, the fabricated MA was placed in front of the horn antennas, and the magnitude of the transfer coefficient ($|S_{21}|$ (dB) was measured using the VNA for a TE-polarized wave. Then, the MA sheet was rotated horizontally (along the azimuthal plane) from 0° to 30° with a step size of 10°, and the ($|S_{21}|$ (dB) was recorded on the VNA. Similarly, the ($|S_{21}|$ (dB) was recorded by rotating the horn antennas by 90° for a TM polarization and by rotating the sheet from 0° to 30° in azimuth. The recorded values of the ($|S_{21}|$ (dB) were then post processed to calculate the absorption peaks over the frequency range using $1 - |S_{11}|^2 - |S_{21}|^2$ [32]. The $|S_{21}|$ is the wave received by the receiver horn antenna, after impinging on the proposed MA from the transmitting horn. The measured $|S_{11}| \approx 0$, here $|S_{11}|$ is the reflected EM wave by the respective horn antenna, on a linear scale. Figure 10 shows the comparison of the simulated and measured results for the TE wave. For the sake of clarity, the comparison of results for different incidence angles is presented separately.

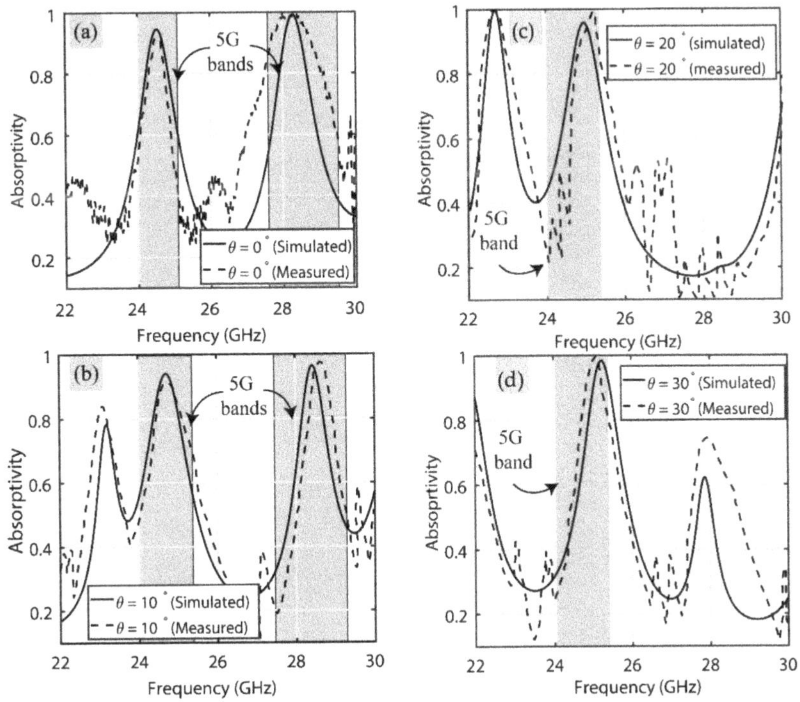

Figure 10. Performance comparison of the proposed MA for TE polarization at (**a**) 0°, (**b**) 10°, (**c**) 20° and (**d**) 30°.

Figure 10a presents the results of when the incident angle is 0°. From this, it is evident that the absorber acts as a dual-band one, operating in the 5G regions of 24 GHz and 28 GHz, i.e., one peak lies in 24.25 GHz–24.45 GHz and the other in the 27.5 GHz–28.35 GHz region. Moreover, the absorptivity is more than 92% over these same frequency bands in both simulated and measured results. Figure 10b shows the comparison for $\theta = 10°$. Again, there are two peaks in the 5G frequency bands, namely 24 GHz (24.75 GHz–25.25 GHz) and 28 GHz (27.5 GHz–28.35 GHz). The simulated and measured results are both in good agreement with each other and, overall, absorptivity is greater than 90% for both.

Figure 10c depicts that at $\theta = 20°$, the proposed absorber absorbs a single 5G frequency band. It is noticeable that at this specific incidence angle for a TE polarization, the proposed absorber exhibits around 95% absorptivity in the 24.75 GHz–25.25 GHz frequency region of 5G. Figure 10d reflects the absorption at $\theta = 30°$, and dual-band absorption behaviour in 5G frequency bands can also be noticed here. One absorption peak is observed in the allocated 5G band ranging from 24.75 GHz to 25.25 GHz with near-unity absorptivity in its measurements, while the other peak is observed in the 28 GHz (27.5 GHz–28.35 GHz) frequency band with around 75% measured absorptivity. Overall, both the simulated and measured absorption results of the proposed absorber in the 5G frequency bands are in good agreement at different obliquity angles for TE polarization.

Figure 11 shows the comparison of simulated and measured results when the TM mode is excited on the proposed absorber. Again, the comparison of results for the different incidence angles is presented separately for clarity. It is noticeable that unlike for the TE-polarized wave case, the proposed absorber here behaves like a single-band absorber when polarization is TM. To explain, it either behaves like an absorber in the 24 GHz or one in the 28 GHz frequency band for different angles of incidence.

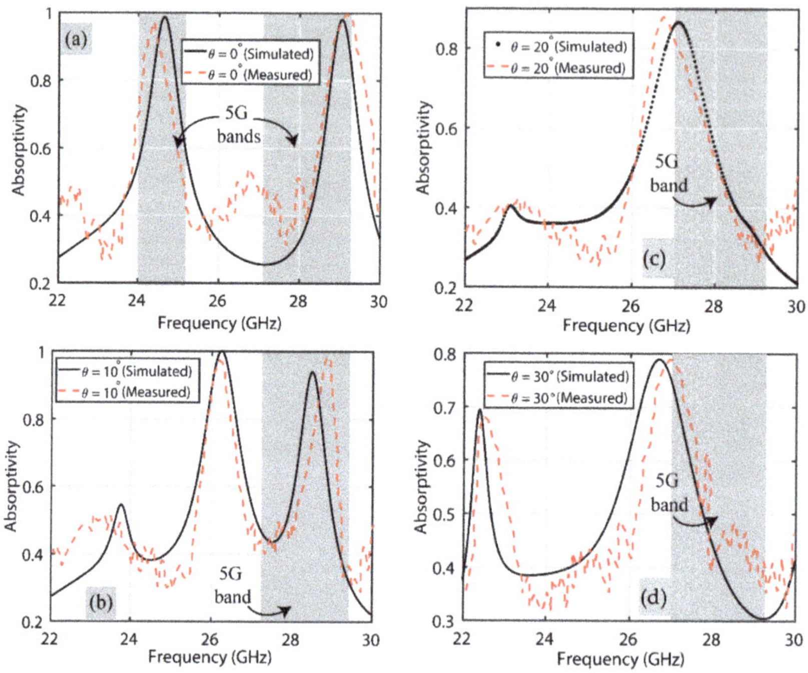

Figure 11. Performance comparison of the proposed MA for TM polarization at (**a**) 0°, (**b**) 10°, (**c**) 20° and (**d**) 30°.

Figure 11a shows the behavior of the absorber at 24 GHz (24.25 GHz–24.45 GHz) with a 98% absorptivity when the incident angle is 0°, whilst more than 95% absorptivity is observed at the 28 GHz (27.5 GHz–28.35 GHz) frequency band for θ = 10°, as depicted in Figure 11b. Both the simulated and measured results for incidence angles of θ = 20° and 30° are shown in Figure 11c,d, respectively, where, again, the results at 28 GHz (27.5 GHz–28.35 GHz) are reported as an absorptivity of around 85% and 80%. It is reported, again, that the overall simulated and measured results are in good agreement (within the respective 5G band) with each other. Ripples at different frequency bands other than the 5G ones are attributed to the imperfect fabrication of the absorber sheet and the use of RF cables within the anechoic chamber.

4. Conclusions

A metamaterial-based absorber operating at 24 GHz and 28 GHz frequency bands and allocated for 5G applications by the FCC was investigated here. The metamaterial structure consists of four meander lines connected with an I-shaped TL. The number of meander lines necessary to design the absorber at the proposed frequency bands was calculated using an analytical formulation, based on the phenomenon of total inductance produced by a meander line structure, under the matched-impedance conditions in the given electronic circuit at the operating frequencies. The meander line structure, created using this analytical formulae, was used to design the resultant structure and was then optimized and simulated in CST to demonstrate the efficiency of the analytical model along with the performance of the absorber. A complete parametric analysis was carried out to demonstrate the design flexibility of the proposed absorber. To explore the practical employability and verify the simulated results of the absorber, a finite sheet of the proposed MA, with 12 × 12 unit cells, was fabricated and tested in an anechoic chamber. In addition, the simulation results of the proposed MA for TE and TM polarizations at different angles of the MA were also verified by measurements, via the rotation of the fabricated prototype at different angles for both the TE- and TM-polarized wave. The performance comparison showed good agreement between simulated and measured results. It is concluded that the desired absorption band can be attained by suitably tailoring the absorber's constituents. Furthermore, these results validated that the proposed absorber would be useful for 5G communication applications, especially for the absorption of frequencies in 5G massive MIMO antenna arrays, to avoid the unwanted near-field interference.

Author Contributions: Conceptualization, S.A.N. and M.A.B.; methodology, S.A.N., G.G., A.I. and D.E.A.; software, S.A.N. and M.A.B.; validation, D.E.A., G.G. and M.S.K.; formal analysis, S.A.N., M.A.B. and A.I.; investigation, S.A.N., D.E.A., M.A.B. and A.I.; data curation, M.A.B. and A.I.; writing—original draft preparation, S.A.N., M.A.B. and G.G.; writing—review and editing, S.A.N., M.A.B., A.I., M.S.K., G.G. and A.I.; visualization, A.I. and M.A.B.; supervision, D.E.A.; project administration, D.E.A.; funding acquisition D.E.A. All authors have read and agreed to the published version of the manuscript.

Funding: This work was supported in part by the EU H2020 Marie Skłodowska-Curie Individual Reintegration Fellowship under grant no. 840854 'VisionRF'.

Institutional Review Board Statement: Not applicable.

Informed Consent Statement: Not applicable.

Data Availability Statement: Not applicable.

Conflicts of Interest: The authors declare no conflict of interest.

References

1. Andrews, J.G.; Buzzi, S.; Choi, W.; Hanly, S.V.; Lozano, A.; Soong, A.C.; Zhang, J.C. What will 5G be? *IEEE J. Sel. Areas Commun.* **2014**, *32*, 1065–1082. [CrossRef]
2. Rappaport, T.S.; Sun, S.; Mayzus, R.; Zhao, H.; Azar, Y.; Wang, K.; Wong, G.N.; Schulz, J.K.; Samimi, M.; Gutierrez, F. Millimetre wave mobile communications for 5G cellular: It will work! *IEEE Access* **2013**, *10*, 335–349. [CrossRef]
3. Federal Communications Commission (FCC). *FCC Establishes Procedures for First 5G Spectrum Auctions*; Federal Communications Commission (FCC): Washington, DC, USA, 2018.
4. Wu, Y.; Wang, J.; Lai, S.; Zhu, X.; Gu, W. Transparent and flexible broadband absorber for the sub-6G band of 5G mobile communication. *Opt. Mater. Express* **2018**, *8*, 3351–3358. [CrossRef]
5. Chen, X.; Chen, X.; Wu, Z.; Zhang, Z.; Wang, Z.; Heng, L.; Wang, S.; Zou, Y.; Tang, Z. An ultra-broadband and lightweight fishnet-like absorber in microwave region. *J. Phys. D Appl. Phys.* **2018**, *51*, 285002. [CrossRef]
6. Li, J.; Zhao, C.; Liu, B.; You, C.; Chu, F.; Tian, N.; Chen, Y.; Li, S.; An, B.; Cui, A.; et al. Metamaterial grating-integrated graphene photodetector with broadband high responsivity. *Appl. Surf. Sci.* **2019**, *473*, 633–673. [CrossRef]
7. Wang, Y.; Sun, T.; Paudel, T.; Zhang, Y.; Ren, Z.; Kempa, K. Metamaterial-Plasmonic Absorber Structure for High Efficiency Amorphous Silicon Solar Cells. *Nano Lett.* **2012**, *12*, 440–445. [CrossRef]
8. Lee, N.; Kim, T.; Lim, J.-S.; Chang, I.; Cho, H.H. Metamaterial-Selective Emitter for Maximizing Infrared Camouflage Performance with Energy Dissipation. *ACS Appl. Mater. Interfaces* **2019**, *11*, 21250–21257. [CrossRef]
9. Kairm, H.; Delfin, D.; Shuvo, M.A.I.; Chavez, L.A.; Garcia, C.R.; Barton, J.H.; Gaytan, S.M.; Cadena, M.A.; Rumpf, R.C.; Wicker, R.B.; et al. Concept and Model of a Metamaterial-Based Passive Wireless Temperature Sensor for Harsh Environment Applications. *IEEE Sens. J.* **2015**, *15*, 1445–1452. [CrossRef]
10. Liu, W.; Tian, J.; Yang, R.; Pei, W. Design of a type of broadband metamaterial absorber based on metal and graphene. *Curr. App. Phys.* **2021**, *31*, 122–131. [CrossRef]
11. Baqir, M.A.; Choudhury, P.K. Hyperbolic Metamaterial-Based UV Absorber. *IEEE Photonic Technol. Lett.* **2017**, *29*, 1548–1551. [CrossRef]
12. Bilal, R.; Baqir, M.; Iftikhar, A.; Ali, M.; Rahim, A.; Akhtar, M.N.; Mughal, M.; Naqvi, S. A novel omega shaped microwave absorber with wideband negative refractive index for C-band applications. *Optik* **2021**, *242*, 167278. [CrossRef]
13. Metamaterial Microwave Absorber. *IEEE Antenn. Wirel. Propag. Lett.* **2019**, *18*, 1016–1020. [CrossRef]
14. Banadaki, M.D.; Heidari, A.A.; Nakhkash, M. A Metamaterial Absorber with a New Compact Unit Cell. *IEEE Antennas Wirel. Propag. Lett.* **2017**, *17*, 205–208. [CrossRef]
15. Wang, B.-X.; Xie, Q.; Dong, G.; Huang, W.Q. Simplified design for broadband and polarisation-insensitive terahertz metamaterial absorber. *IEEE Photonic Technol. Lett.* **2018**, *30*, 1115–1118. [CrossRef]
16. So, S.; Yang, Y.; Lee, T.; Rho, J. On-demand design of spectrally sensitive multiband absorbers using an artificial neural network. *Photonic Res.* **2021**, *9*, B153. [CrossRef]

17. Baqir, M.A. Conductive metal–oxide-based tunable, wideband, and wide-angle metamaterial absorbers operating in the near-infrared and short-wavelength infrared regions. *Appl. Opt.* **2020**, *59*, 10912–10919. [CrossRef]
18. Wang, X.; Sang, T.; Li, G.; Mi, Q.; Pei, Y.; Wang, Y. Ultrabroadband and ultrathin absorber based on an encapsulated T-shaped metasurface. *Opt. Express* **2021**, *29*, 31311. [CrossRef]
19. Bilal, R.; Baqir, M.; Choudhury, P.; Ali, M.; Rahim, A.; Kamal, W. Polarization-insensitive multi-band metamaterial absorber operating in the 5G spectrum. *Optik* **2020**, *216*, 164958. [CrossRef]
20. Qu, S.; Hou, Y.; Sheng, P. Conceptual-based design of an ultrabroadband microwave metamaterial absorber. *Proc. Natl. Acad. Sci. USA* **2021**, *118*, e2110490118. [CrossRef]
21. Amiri, M.; Tofigh, F.; Shariati, N.; Lipman, J.; Abolhasan, M. Ultra Wideband Dual Polarization Metamaterial Absorber for 5G frequency spectrum. In Proceedings of the 2020 14th European Conference on Antennas and Propagation (EuCAP), Copenhagen, Denmark, 15–20 March 2020; pp. 1–5. [CrossRef]
22. Sargun; Verma, S.; Aryan, S.; Jain, P. Design of Metamaterial Absorber for 5G Applications. *Int. J. Adv. Sci. Technol.* **2020**, *29*, 13689–13698.
23. Chen, T.; Li, S.J.; Cao, X.Y.; Gao, J.; Guo, Z.X. Ultra-wideband and polarization-insensitive fractal perfect metamaterial absorber based on a three-dimensional fractal tree microstructure with multi-modes. *Appl. Phys. A* **2019**, *125*, 232. [CrossRef]
24. Li, S.J.; Wu, P.X.; Xu, H.X.; Zhou, Y.L.; Cao, X.Y.; Han, J.F.; Zhang, C.; Yang, H.H.; Zhang, Z. Ultra-wideband and polarization-insensitive perfect absorber using multilayer metamaterials, lumped resistors, and strong coupling effects. *Nanoscale Res. Lett.* **2018**, *13*, 386. [CrossRef] [PubMed]
25. Wen, D.-E.; Yang, H.; Ye, Q.; Li, M.; Guo, L.; Zhang, J. Broadband metamaterial absorber based on a multi-layer structure. *Phys. Scr.* **2013**, *88*, 015402. [CrossRef]
26. Long, C.; Yin, S.; Wang, W.; Li, W.; Zhu, J.; Guan, J. Broadening the absorption bandwidth of metamaterial absorbers by transverse magnetic harmonics of 210 mode. *Sci. Rep.* **2016**, *6*, 21431. [CrossRef]
27. Nakano, H.; Tagami, H.; Yoshizawa, A.; Yamauchi, J. Shortening ratios of modified dipole antennas. *IRE Trans. Antennas Propag.* **1984**, *32*, 385–386. [CrossRef]
28. Rashed, J.; Tai, C.-T. A new class of resonant antennas. *IEEE Trans. Antenn. Propag.* **1991**, *39*, 1428–1430. [CrossRef]
29. Best, S.; Morrow, J. Limitations of inductive circuit model representations of meander line antennas. *IEEE Antenn. Propag. Soc. Int. Symp.* **2004**, *1*, 852–855. [CrossRef]
30. Tuan, C.C.; Chang, C.H.; Chang, Y.J.; Chen, C.H.; Jeong, H.D.; Huang, W.T. A highly reliable platform with a serpentine antenna for IEEE 802.15.4 over a wireless sensor network. *WSEAS Trans. Circuit Sys.* **2012**, *11*, 182–197.
31. Subbaraj, S.; Kanagasabai, M.; Alsath, M.G.N.; Palaniswamy, S.K.; Kingsly, S.; Kulandhaisamy, I.; Shrivastav, A.K.; Natarajan, R.; Meiyalagan, S. A Compact Frequency Reconfigurable Antenna with Independent Tuning for Hand-held Wireless Devices. *IEEE Trans. Antenn. Propag.* **2020**, *68*, 1151–1154. [CrossRef]
32. Balanis, C.A. *Advanced Engineering Electromagnetics*; John Wiley & Sons: Hoboken, NJ, USA, 1999.
33. Endo, T.; Sunahara, Y.; Satoh, S.; Katagi, T. Resonant frequency and radiation efficiency of meander line antennas. *Electron. Commun. Jpn. Part II* **2000**, *83*, 52–58. [CrossRef]

electronics

MDPI

Article

Complementary Multi-Band Dual Polarization Conversion Metasurface and Its RCS Reduction Application

Fengan Li and Baiqiang You *

Department of Electronic Engineering, Xiamen University, Xiamen 361005, China; lifengan1112@foxmail.com
* Correspondence: youzhou@xmu.edu.cn

Abstract: In this paper, we present a metasurface composed of complementary units that can realize orthogonal linear and linear-to-circular polarization conversion in multi-band. Linear polarization conversion has seven high-conversion frequency bands: 9.1–9.7 GHz, 15.6–17.6 GHz, 19.4–19.7 GHz, 21.2–23.1 GHz, 23.5–23.8 GHz, 26.2 GHz, and 27.9 GHz. Linear-to-circular polarization conversion also has seven frequency bands with axial ratios (ARs) less than 3 dB: 8.9–9.0 GHz, 9.9–14.7 GHz, 19.1–19.3 GHz, 23.2–23.35 GHz, 23.4 GHz, 24.1–25.4 GHz, and 27.2–27.8 GHz, with the generation of multiple bands extended by the combination of complementary units. Then, we utilize the combined polarization conversion unit's mirror placement to form a 4 × 4 array to realize the phase difference cancellation of the reflective field, giving the metasurface the radar cross section (RCS) reduction function and the dual-band 10-dB monostatic RCS reduction bandwidth: 8.9–9.7 GHz and 15.5–26.1 GHz. The measured and simulated results were essentially identical. Because the design uses the complementary units to form an array to expand the polarization conversion frequency bands, it provides a novel idea for future designs and can be applied to multiple microwave frequency bands.

Keywords: polarization conversion metasurface (PCM); complementary unit; linear polarization conversion; circular polarization conversion; RCS reduction

Citation: Li, F.; You, B. Complementary Multi-Band Dual Polarization Conversion Metasurface and Its RCS Reduction Application. *Electronics* **2022**, *11*, 1645. https://doi.org/10.3390/electronics11101645

Academic Editors: Naser Ojaroudi Parchin, Chan Hwang See and Raed A. Abd-Alhameed

Received: 30 April 2022
Accepted: 17 May 2022
Published: 21 May 2022

1. Introduction

Polarization is an important characteristic of electromagnetic waves that is used to represent the time-varying trajectory of the electric field strength vector terminal of electromagnetic waves in space. In modern communication, controlling the polarized state of electromagnetic waves is critical in the transmission and reception of wireless signals, and it has also become a frontier direction of discipline exploration. Metasurfaces, as a special non-natural material, have been applied to the precise control of electromagnetic wave polarization with its easy fabrication, good conversion characteristic, small size, and powerful functionality to manipulate the beams, which can effectively avoid the defects of traditional design methods [1–7]. In recent years, various polarization conversion metasurfaces (PCMs) have been proposed to realize different polarization conversions, including incident and reflective orthogonal linear polarization (LP-LP) [8–11], circular-to-circular polarization (CP-CP) [12–15], and linear-to-circular polarization (LP-CP) [16–19]. However, most of these metasurfaces realize a single polarization conversion function, and there are relatively few metasurfaces with multiple polarization conversion functions. Therefore, it is urgent to study the multifunctional PCM for different frequency bands. For this demand, the literature [20] proposed various functions of polarization converters, with the linear polarization conversion bandwidth reaching 95.2%. Gao et al. designed a voltage-controlled reconfigurable broadband PCM for LP-LP and LP-CP conversions [21]. In [22], a four-band, multi-function reflective polarization converter based on linear and circular polarizations was proposed.

The development of metasurfaces has also promoted the improvement of RCS reduction technology. The key to RCS reduction is to suppress reflected waves. Metasurfaces such as the frequency-selective surface [23–25], frequency-absorbing surface [26,27], electromagnetic band-gap structure [28,29], and artificial magnetic conductor [30,31] can adjust different electromagnetic parameters to realize the absorption, diffuse reflection, and abnormal reflection of reflected waves, thus providing new methods for RCS reduction. Specifically, frequency-selective and frequency-absorbing surfaces exhibit total reflection or transmission characteristics around the resonant frequency of the unit, which can be seen as band-pass or band-resistance filters, to control electromagnetic wave propagation in certain frequency bands. The artificial magnetic conductor has different reflection phases at different frequencies, which can be used to achieve phase cancellation. In particular, when mirroring the PCM when a specific polarization wave is incident, the phase difference of the reflected waves by the two mirror-arranged sub-arrays is π, and the two waves are cancelled to achieve overall reflected wave suppression [32,33].

In this work, a novel, complementary, multifunctional PCM is proposed. The designed complementary surface double units have the function of polarization conversion. By combining the two complementary units, a metasurface that realizes orthogonal linear polarization conversion and linear-to-circular polarization conversion is finally designed. Both linear polarization conversion and linear-to-circular polarization conversion have seven frequency bands. Moreover, a mirror-symmetrical RCS reduction surface is designed using this PCM, which realizes double broadband monostatic RCS reduction and improves the performance of related works.

2. Structural Design and Principle

The designed PCM unit is shown in Figure 1. The overall structure is divided into three layers: the PCM unit on the top layer, the FR4 dielectric substrate ($\varepsilon = 4.4$, $tan\delta = 0.02$) with a thickness of H = 1.2 mm in the middle, and the bottom layer being a perfect electric conductor (PEC). Unit 1 is a diagonal structure formed by orthogonal placement of two rectangular patches with equal arm lengths, and unit 2 is formed by stacking three square patches at a certain distance. The two units are complementary, as shown in Figure 1a. The optimized geometric dimensions are L = 8 mm, C = 0.1 mm, and W = 1.95 mm. Figure 1b,c shows that the complementary metal surface units are placed at an angle of 45 degrees along the *X*-axis (*Y*-axis). The two units can be combined into a square patch, and both have polarization conversion functions. Figure 1c explains the principle of polarization conversion from the physical theory of electromagnetic waves. Assuming that the electromagnetic wave is incident along the x polarization direction, and the incident electric field E_i is decomposed into orthogonally polarized waves along the *U*-axis and *V*-axis, we can obtain the following:

$$E_i = \vec{x} E_i e^{j\varphi} = \left(\vec{u} E_{iu} + \vec{v} E_{iv}\right) e^{j\varphi} \tag{1}$$

$$E_r = \vec{u} E_{ru} + \vec{v} E_{rv} = \left(\vec{u} r_u E_{iu} e^{j\varphi_u} + \vec{v} r_v E_{iv} e^{j\varphi_v}\right) e^{j\varphi} \tag{2}$$

where E_{iu} and E_{iv} are the magnitudes of the incident field components on the *U*-axes and *V*-axes, respectively, and φ_u and φ_v are the phases of the corresponding components. Because the bottom is an ideal metal floor structure, it exhibits total reflection characteristics under the incidence of electromagnetic waves, so we find $r_u = r_v = r$. However, under the incidence of U- and V-polarized waves, the resonant structures of the unit in the two directions are different, and the reflective waves have a phase difference $\Delta\varphi = |\varphi_u - \varphi_v|$. When $\Delta\varphi = 180^{\circ}$, Equation (2) becomes

$$
\begin{aligned}
E_r &= r\left(\vec{u}\,E_{iu}e^{j\varphi_u} + \vec{v}\,E_{iv}e^{j(\pi-\varphi_u)}\right)e^{j\varphi} \\
&= r\left(\vec{u}\,E_{iu} - \vec{v}\,E_{iv}\right)e^{j\varphi_u}e^{j\varphi} \\
&= \vec{y}\,E_r e^{j\varphi_u}e^{j\varphi}
\end{aligned}
\tag{3}
$$

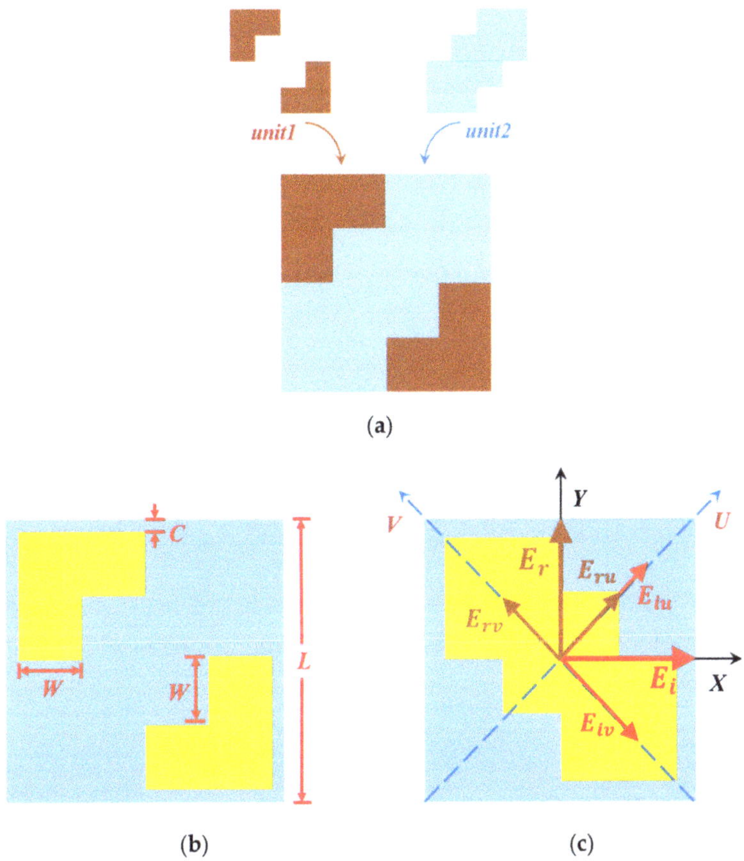

Figure 1. Polarization conversion unit design. (**a**) Complementary unit. (**b**) Structure of unit 1. (**c**) Structure of unit 2.

Through the above derivation, we complete the transformation from the incident wave with x polarization to the reflective wave with y polarization. Because the PCM is placed symmetrically along the diagonal, the incident conditions of the x polarization and y polarization are exactly the same, and the conversion principle is also suitable for the y-polarized incident wave and the x-polarized reflective wave. Similarly, when $\Delta\varphi = 90^\circ$, the synthesized wave is circularly polarized, and the structure realizes the function of linear-to-circular polarization conversion.

3. Simulation Analysis

To better understand the polarization conversion performance of the designed unit, we used the Floquet port in the simulation software with the master-slave boundary to simulate the infinite array condition. Figure 2a,b illustrates the reflection coefficients of unit 1 and unit 2 when an x-polarized wave is incident. The resonance points of unit 1 were at 17.9 GHz and 20.1 GHz, while the resonance points of unit 2 were at 7.2 GHz, 12.3 GHz, and 19.7 GHz. Their cross-polarized reflection coefficients R_{yx} were 0 dB at these frequency points, and the co-polarized reflection coefficients R_{xx} were very small, indicating that polarization conversion occurred. We characterized the polarization conversion by the polarization conversion ratio (PCR), which is defined as follows [34]:

$$\mathrm{PCR}_x = \frac{R_{yx}^2}{R_{yx}^2 + R_{xx}^2}\ ;\ \mathrm{PCR}_y = \frac{R_{xy}^2}{R_{xy}^2 + R_{yy}^2} \tag{4}$$

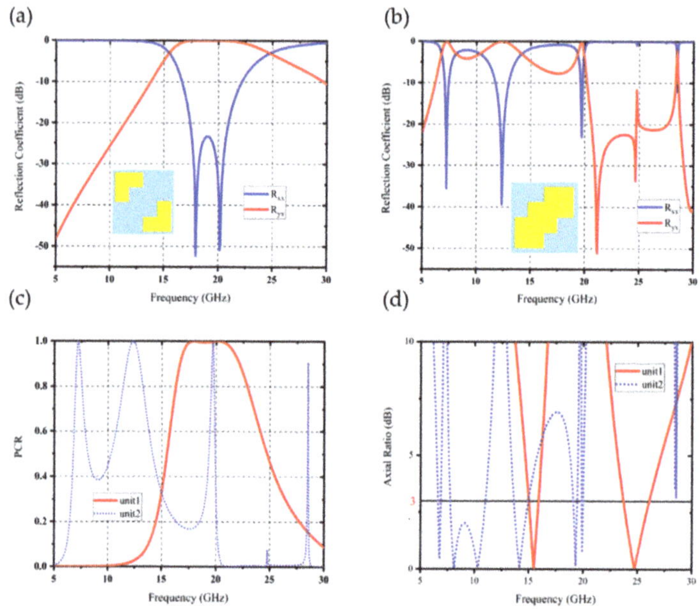

Figure 2. Simulation results of unit. (**a**) Reflection coefficient of unit 1. (**b**) Reflection coefficient of unit 2. (**c**) PCR for units 1 and 2. (**d**) AR for units 1 and 2.

Under the incidence of x-polarized waves, the PCRs of units 1 and 2 are shown in Figure 2c. At the corresponding resonance frequency, the PCR reached 100%, explaining that the conversion of the x-polarized incident wave to the y-polarized reflective wave was completely realized.

In order to realize the functions of multiple polarization conversion on the metasurface, we also explored the circular polarization conversion function of the units. We used the axial ratio (AR) to describe the circular polarization conversion characteristics. When AR < 3 dB, it can be considered that the electromagnetic wave is a circular polarization wave. The AR is expressed by the following formula [35]:

$$\mathrm{AR} = \left(\frac{|R_{xx}|^2 + |R_{yx}|^2 + \sqrt{a}}{|R_{xx}|^2 + |R_{yx}|^2 - \sqrt{a}} \right)^{\frac{1}{2}} \tag{5}$$

$$a = |R_{xx}|^4 + |R_{yx}|^4 + 2|R_{xx}|^2|R_{yx}|^2\cos(2\Delta\varphi) \tag{6}$$

The ARs of units 1 and 2 are shown in Figure 2d, and it can be seen that the 3-dB circular polarization bandwidths of unit 1 were dual-band 15.0–15.9 GHz and 23.72–26.11 GHz, and the 3-dB axial ratio bandwidths of unit 2 were 6.55–6.83 GHz, 7.69–10.96 GHz,

13.61–14.92 GHz, 19.08–19.45 GHz, and 19.83–19.98 GHz. Therefore, both units 1 and 2 realized the polarization conversion function of multi-band linear polarization incident to circular polarization reflection.

Observing Figure 2, we found that there were many polarization conversion frequencies for unit 1 and unit 2, but they were not evenly distributed in the entire solution frequency band, and the conversion bandwidths of the two units did not overlap with each other. Therefore, we could synthesize the conversion characteristics of the two complementary units, which may extend the polarization conversion frequency bands. Based on this idea, we finally designed a combined polarization conversion unit, using the same simulation method to analyze it. The designed structure is shown in Figure 3. Unit 1 and unit 2 are crossed to form a symmetrical structure.

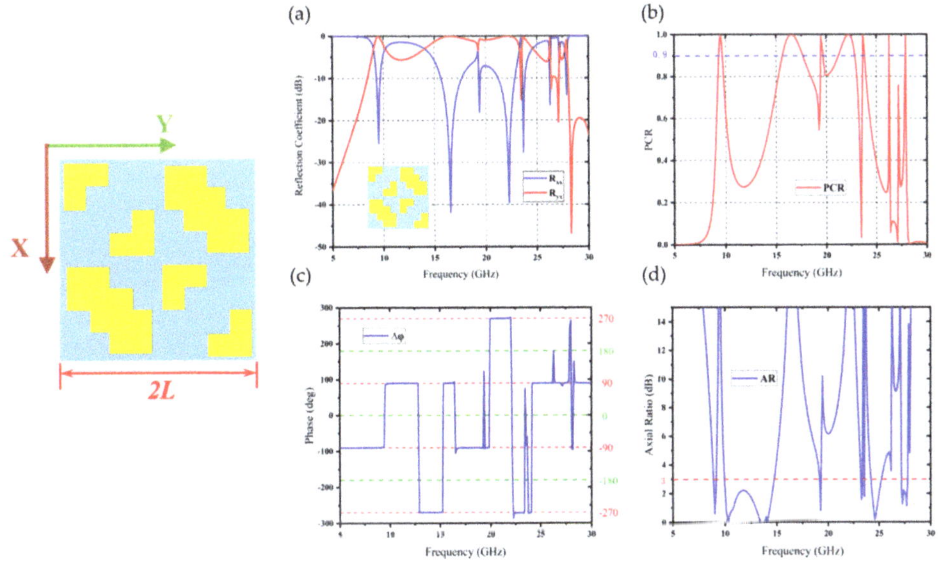

Figure 3. Simulation results of combined unit. (**a**) Reflection coefficient. (**b**) PCR. (**c**) Phase difference. (**d**) AR.

The reflection coefficient and PCR of the structure are shown in Figure 3a,b. The figure shows that the combined unit $R_{xx} < -10$ dB had seven resonant frequency points, the PCR of the corresponding frequency point was 100%, and it had seven PCR bandwidths >90%: 9.1–9.7 GHz, 15.6–17.6 GHz, 19.4–19.7 GHz, 21.2–23.1 GHz, 23.5–23.8 GHz, 26.2 GHz, and 27.9 GHz. In addition, the phase difference of the co-polarized and cross-polarized reflection coefficients are shown in Figure 3c. The phase difference is defined as $\Delta\varphi = \varphi_{xx} - \varphi_{yx}$ when $\Delta\varphi = 0, \pm\pi$. It can be known from the knowledge of the polarization phase that the linear polarization conversion occurred at this time. The phase differences of R_{yx} and R_{xx} at the seven resonance points in the figure all satisfied this condition, suggesting that the complete conversion of x polarization to y polarization was achieved. When comparing the simulation results of the two units, it was found that there was a difference in the offset of the conversion frequency bands, and the new resonance frequency bands were generated. The main reason for this is that after the combined structure, the periodic units changed. The change in the inductance or capacitance generated by the coupling between units 1 and 2 in the overall equivalent circuit resulted in a shift in the resonant frequencies, especially in the relatively high frequency band of 23.5–28 GHz. This change in the equivalent circuit excited two new resonant frequency bands that contributed to the polarization conversion at higher frequencies. As a result, the newly formed unit could expand the frequency and bandwidth of linear polarization conversion.

In addition, the basic conditions of the circular polarization conversion must be satisfied. In the case that the co-polarized reflection coefficient R_{xx} and the cross-polarized reflection coefficient R_{yx} have the same amplitude, the phase difference between them must be satisfied ($\Delta\varphi = \pm n\pi/2$, where n is an odd number). It can be seen from Figure 3a that the possible frequency band for circular polarization conversion occurred near the intersection with the R_{xx} and R_{yx} coefficients, where the amplitudes of them were roughly equal, and the linear-to-circular polarization conversion could be completed as long as the phase difference condition was satisfied. From Figure 3d, it can be seen that there were seven bandwidths with ARs less than 3 dB: 8.9–9.0 GHz, 9.9–14.7 GHz, 19.1–19.3 GHz, 23.2–23.35 GHz, 23.4 GHz, 24.1–25.4 GHz, and 27.2–27.8 GHz. Meanwhile, when observing Figure 3c, it is found that at these frequency bands, the phase differences between co-polarization and cross-polarization were around -270°, -90°, 90°, and 270°, which satisfies the circular polarization conversion phase. Therefore, the combination unit realized the conversion from the incident linear polarization wave to the reflective circular polarization wave in the above frequency bands.

We also analyzed the conversion under oblique incidence conditions, and the incidence angle was set to 0–45 degrees, as shown in Figure 4. What follows is category discussion: (1) for linear polarization conversion, compared with normal incidence, a PCR under 15 or 30 degrees of incidence had six conversion frequency bands, and 45 degrees had only 4 conversion frequency bands, indicating that oblique incidence would reduce the number of frequency bands for linear polarization conversion. When the frequency band was below 18 GHz, the oblique incidence had only a small effect on the frequency band of linear polarization conversion, which is basically consistent with the normal incidence and mainly affects the linear polarization conversion at higher frequencies. Therefore, the linear polarization conversion under oblique incidence conditions still maintained more than 4 conversion bandwidths, especially below 24 GHz. (2) Regarding circular polarization conversion, the oblique incidence had no effect on the circular polarization conversion at 8.9–9.0 GHz but had a great impact on the broadband of 9.9–14.7 GHz, while the oblique incidence in the higher frequency band only had an offset effect on the frequency bands. Therefore, the circular polarization conversion could still maintain the multi-band function under oblique incidence, and the whole was still stable in the higher frequency band.

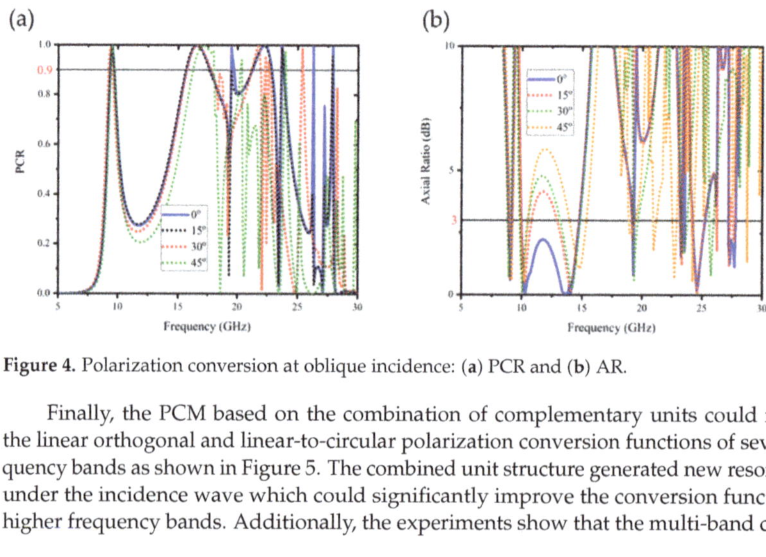

Figure 4. Polarization conversion at oblique incidence: (**a**) PCR and (**b**) AR.

Finally, the PCM based on the combination of complementary units could realize the linear orthogonal and linear-to-circular polarization conversion functions of seven frequency bands as shown in Figure 5. The combined unit structure generated new resonances under the incidence wave which could significantly improve the conversion function of higher frequency bands. Additionally, the experiments show that the multi-band conversion function of the metasurface could still be well maintained under an oblique incidence of 0–45 degrees, and it had angle insensitivity.

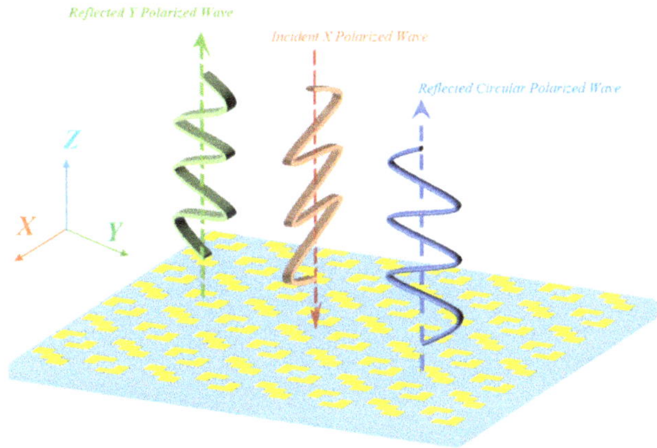

Figure 5. Schematic diagram of PCM.

4. RCS Reduction

The polarization conversion characteristics of the PCM are described above. In the working frequency bands of the metasurface, under a normal incidence, the cross-polarization reflection amplitude of the basic unit was almost the same as that of the mirror image, but the two phases were opposite. Therefore, the PCM unit's mirror arrangement and the RCS reduction in a certain frequency band could be realized by using the phase cancellation method of the scattered field.

The plane of RCS based on the PCM is shown in Figure 6. The whole structure is composed of 4 × 4 units, and the single direction is 2 × 2 units, which are arranged with mirrors up, down, left, and right. The surface was simulated with the radiation boundary, and the simulation results are shown in Figure 7. The control group was a metal plane of equal size. We used RCS reduction to represent the RCS capability. The calculation formula of RCS reduction is expressed in Equation (7) [36], where E_{rx} and E_{ix} represent the reflective field and incident field in the far region, respectively, and r represents the detection distance:

$$\text{RCS reduction(dB)} = 10\,\text{lg}\left[\frac{\lim\limits_{r\to\infty} 4\pi r^2 \frac{|E_{rx}|^2}{|E_{ix}|^2}}{\lim\limits_{r\to\infty} 4\pi r^2 |1|^2}\right] \quad (7)$$

RCS reduction peaked at 9.2 GHz and 16.5 GHz, and the 10-dB dual-band RCS reduction bandwidths were 8.9–9.7 GHz and 15.5–26.1 GHz, while the maximum reduction reached 25.8 dB. Therefore, the metasurface achieved a broadband dual-frequency monostation RCS reduction effect. When observing the functional frequency bands of the PCM and the RCS metasurface, it is shown that the 10-dB reduction bandwidth included the frequency band of the polarization conversion. In particular, the frequency at the reduced peak was roughly coincident with the first two resonant frequencies of polarization conversion, and the polarization conversion at this time was the best, resulting in the generation of the reduced peak, which verifies that our designed PCM can achieve excellent RCS reduction function. The small error of the bandwidth of the two was mainly caused by the finiteness of the array. PCM uses Floquet ports to simulate infinite periodic conditions, while the RCS surface is simulated with finite elements.

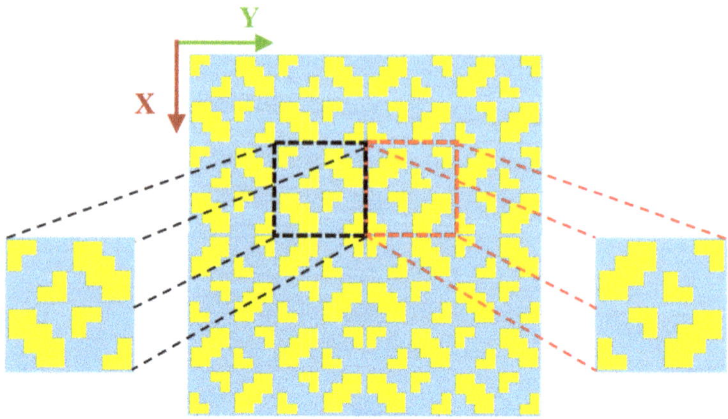

Figure 6. RCS reduction metasurface-based PCM.

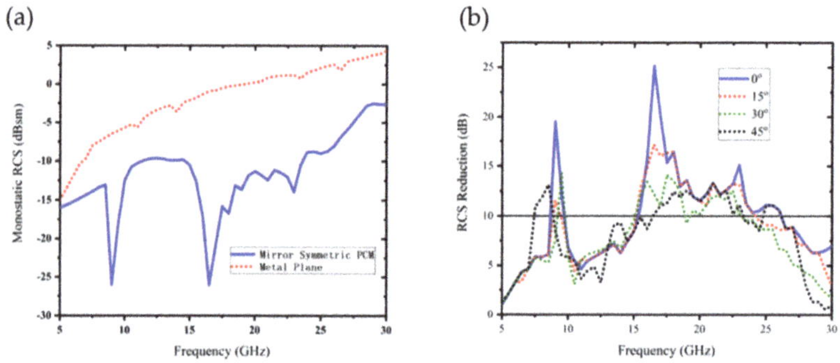

Figure 7. (**a**) Monostatic RCS. (**b**) RCS reduction.

At 15–45 degrees of oblique incidence, the reduced peaks were shifted, but the impact on the reduced peak in the lower frequency band was small and only had a greater impact on the peak in the higher frequency bands. The maximum reduced peak decreased, and with the increase in the incident angle, the reduced peak bandwidth between 15 and 25 GHz was gradually reduced, but the overall 10-dB dual-band reduction function was still maintained. This indicates that the effects of the angle on the conversion ability of the PCM and the reduction ability of the RCS metasurface were similar. Therefore, the RCS reduction metasurface based on PCM also had angle insensitivity under an oblique incidence of 0–45 degrees.

To better understand how the designed metasurface achieved RCS reduction, we present a 3D scattering field of the control metal plane and the designed RCS plane at the peak reduction point. It can be seen from Figure 8 that at 9.2 GHz, compared with the metal plane, the scattering field of the PCM mirror plane had no main lobe, and the main lobe was suppressed by the phase difference of the reflective field, which exhibited a diffuse reflection state. At 16.5 GHz, the designed plane divided the main lobe into four scattered beams, which were reflected in different directions, diminishing the effect of the main lobe and weakening the side lobes to accomplish RCS reduction.

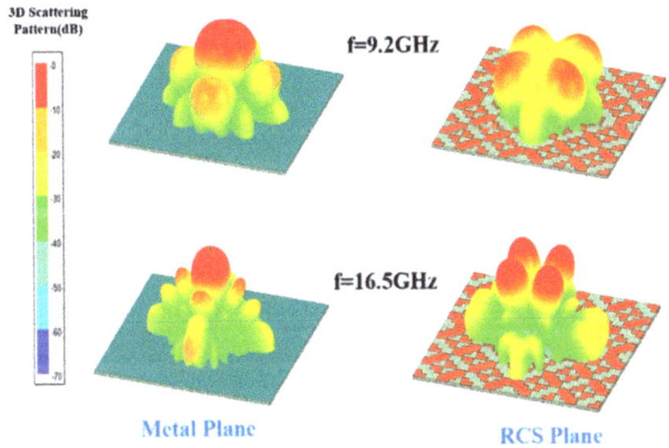

Figure 8. 3D scattering field of metal plane and RCS plane at 9.2 GHz and 16.5 GHz.

Figure 9 shows the bistatic RCS effect at the peak frequency points. Two observation angles are set: $\varphi = 0^{\circ}$ and $\varphi = 90^{\circ}$. The results show that at 9.2 GHz, the RCS plane at two observation angles achieved a good bistation RCS reduction in the interval $\theta \in [-30, 30]$. At 16.5 GHz, the value of the bistation RCS in the entire θ angle range was below -20 dB, showing a perfect bistation RCS reduction function, and the two-angle observation had the largest reduction, reaching 25.8 dB when $\theta = 0^{\circ}$. The results of monostatic and bistatic RCS show that the designed mirror PCM-RCS metasurface can achieve perfect results in a wide frequency band.

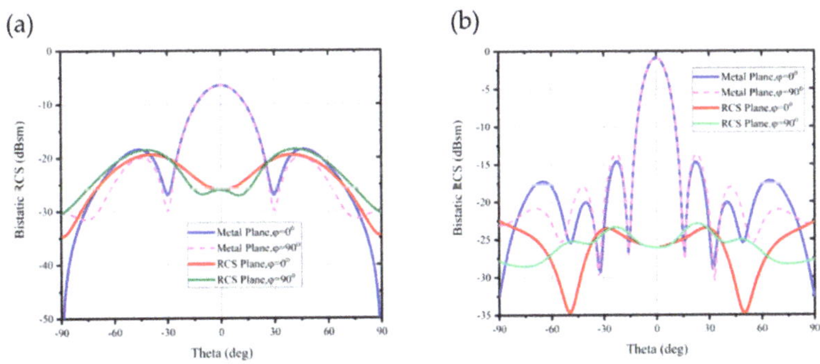

Figure 9. Bistatic RCS reduction: (**a**) f = 9.2 GHz and (**b**) f = 16.5 GHz.

5. Fabrication and Measurement

To verify the correctness of the simulation results, we fabricated and measured the metasurface. The samples are shown in Figure 10, where Figure 10a,b shows the PCM and the RCS reduction metasurfaces, respectively. The same-sized metal plane was used as a reference to obtain the actual RCS reduction. The measured device placement and environmental schematic are shown in Figure 11. The samples were tested in an anechoic chamber with two horn antennas connected to the vector network analyzer (VNA), and the frequency range was 5–30 GHz. In the process of measuring the reflection coefficient, both the co-polarized and cross-polarized reflection coefficients should be measured. When measuring co-polarization, two horn antennas are placed in the same direction. When measuring cross-polarization, the transmitting horn antenna is placed horizontally, and the receiving horn antenna is placed vertically. In monostatic RCS measurement, the center of the sample is at the same height as the two antennas, where the separation angle of the two antennas should be less than 5°, and the distance between the antenna and the sample should be more than 2 m to satisfy the far-field scattering pattern. Figure 12 shows the comparison of the measured and simulated results. It can be seen from the figure that the measured reflection coefficient also had multiple polarization conversion frequency bands, and the frequency range was not much different from the simulation results. The measured wide frequency bands of the monostatic RCS were 9.1–9.8 GHz and 15.3–25.8 GHz. The final reflection coefficient and RCS reduction were roughly consistent with the measured results and the simulation results, which verifies the correctness of our design. The small errors were mainly caused by the fabrication process, the measurement errors, and the finite size of the metasurface.

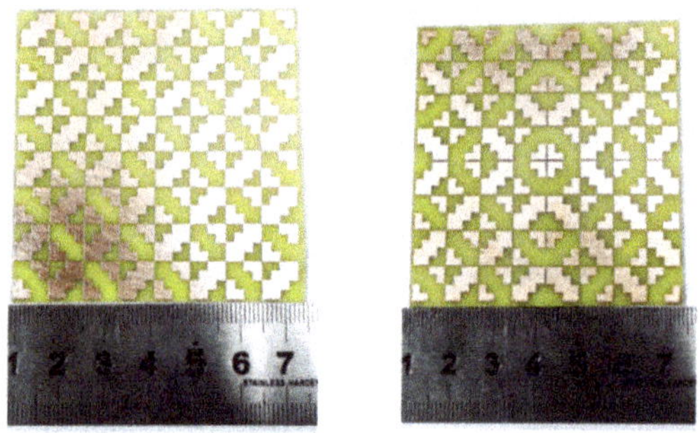

(a) (b)

Figure 10. Physical fabrication: (**a**) PCM and (**b**) RCS plane.

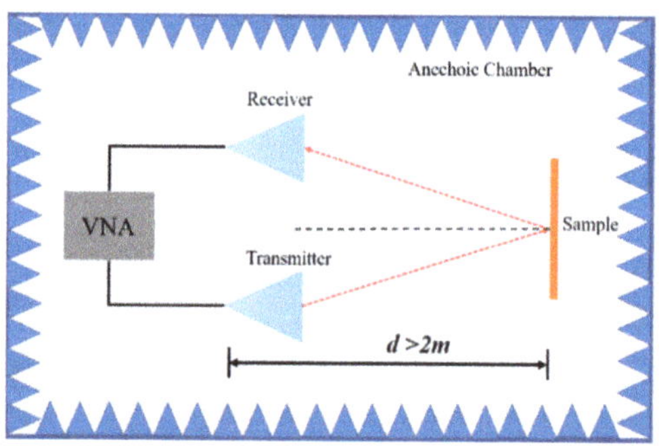

Figure 11. Schematic diagram of measured device and environment.

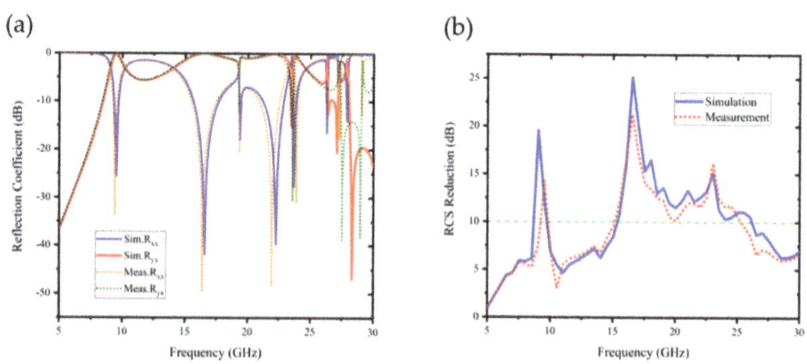

Figure 12. Comparison of measured and simulated results. (**a**) Measured and simulated reflection coefficients. (**b**) Measured and simulated RCS reduction.

A comparison of the performance of this work with the other literature is shown in Table 1, from which it can be seen that the PCM we designed had the most conversion bands as a function of line-to-line orthogonal and line-to-circle polarization conversion and had a wide bandwidth coverage. The RCS reduction metasurface designed using this PCM also had the 2 widest 10-dB reduction bandwidths, further validating the validity of this work. Overall, our work improves the performance and metrics in the field of PCMs, promotes the research of multifunctional PCMs, and provides new ideas and performance designs for electromagnetic stealth design of metasurfaces.

Table 1. Comparisons between this work and other references.

Ref.	Freq. (GHz)		Size (Width × Length × Thickness)	RCS Reduction Band (GHz)
	LP-LP (PCR ≥90%)	LP-CP (AR ≤3 dB)		
[7]	4.63–5.54	6.65–7.62	0.25 $\lambda_\circ$ × 0.25 $\lambda_\circ$ × 0.05 $\lambda_\circ$	3.49–3.62 5.94–6.69
[21]	3.9–7.9	4.9–8.2	0.32 $\lambda_\circ$ × 0.32 $\lambda_\circ$ × 0.12 $\lambda_\circ$	—
[22]	4.19–4.40 6.8–7.64 11.54–13.07 14.98–15.30	3.95–4.14 4.75–5.95 8.35–8.8 14.35–14.6	0.147 $\lambda_\circ$ × 0.147 $\lambda_\circ$ × 0.042 $\lambda_\circ$	—
[34]	9.4–14.0 15.5–20.9	—	0.43 $\lambda_\circ$ × 0.43 $\lambda_\circ$ × 0.13 $\lambda_\circ$	10.2–14.0 15.3–20.7
This work	9.1–9.7 15.6–17.6 19.4–19.7 21.2–23.1 23.5–23.8 26.2 27.9	8.9–9.0 9.9–14.7 19.1–19.3 23.2–23.35 23.4 24.1–25.4 27.2–27.8	0.45 $\lambda_\circ$ × 0.45 $\lambda_\circ$ × 0.07 $\lambda_\circ$	8.9–9.7 15.5–26.1

6. Conclusions

In general, the multi-band PCM can not only realize the effect of an x (y)-polarized wave incident to y (x)-polarized wave reflection but also achieve the characteristics of linear-to-circular polarization, and the unit of PCM is obtained by combining complementary square units, which expands the bandwidth compared with a single unit. Simultaneously, the mirror-combined PCM surface can realize the broadband 10-dB RCS reduction function by using phase cancellation of the reflected waves, and the bistatic RCS reduction effect is also remarkable. Through physical fabrication, the simulation results are consistent with the experimental verification results. Therefore, the proposed PCM can be applied to the design of multifunctional polarization converters for the X, Ku, K, and Ka microwave frequency bands while providing methods for antenna RCS reduction, polarization beam modulation, and electromagnetic stealth design of military equipment.

Author Contributions: Conceptualization, methodology, writing—original draft preparation, and writing—review and editing, F.L.; supervision and project administration, B.Y. All authors have read and agreed to the published version of the manuscript.

Funding: This work was supported by the Natural Science Foundation of Fujian Province of China under Grant 2019J01045 and the Natural Science Foundation of China (NSFC) under Grant No. 62071403.

Acknowledgments: Physical production is supported by the Microwave and Photonics Laboratory of Xiamen University.

Conflicts of Interest: The authors declare no conflict of interest.

References

1. Ling, Y.; Huang, L.; Hong, W.; Liu, T.; Luan, J.; Liu, W.; Lai, J.; Li, H. Polarization-controlled dynamically switchable plasmon-induced transparency in plasmonic metamaterial. *Nanoscale* **2018**, *10*, 19517–19523. [CrossRef] [PubMed]
2. Mercader-Pellicer, S.; Goussetis, G.; Medero, G.M.; Legay, H.; Bresciani, D.; Fonseca, N.J.G. Cross-Polarization Reduction of Linear-to-Circular Polarizing Reflective Surfaces. *IEEE Antennas Wirel. Propag. Lett.* **2019**, *18*, 1527–1531. [CrossRef]
3. Sima, B.; Chen, K.; Luo, X.; Zhao, J.; Feng, Y. Combining Frequency-Selective Scattering and Specular Reflection through Phase-Dispersion Tailoring of a Metasurface. *Phys. Rev. Appl.* **2018**, *10*, 064043. [CrossRef]
4. Li, Z.Y.; Li, S.J.; Han, B.W.; Huang, G.S.; Guo, Z.X.; Cao, X.Y. Quad-Band Transmissive Metasurface with Linear to Dual-Circular Polarization Conversion Simultaneously. *Adv. Theory Simul.* **2021**, *4*, 2100117. [CrossRef]
5. Han, B.; Li, S.; Li, Z.; Huang, G.; Tian, J.; Cao, X. Asymmetric transmission for dual-circularly and linearly polarized waves based on a chiral metasurface. *Opt. Express* **2021**, *29*, 19643–19654. [CrossRef]
6. Li, S.J.; Li, Y.B.; Li, H.; Wang, Z.X.; Zhang, C.; Guo, Z.X.; Li, R.Q.; Cao, X.Y.; Cheng, Q.; Cui, T.J. A Thin Self-Feeding Janus Metasurface for Manipulating Incident Waves and Emitting Radiation Waves Simultaneously. *Ann. Phys.* **2020**, *532*, 2000020. [CrossRef]
7. Li, S.J.; Li, Y.B.; Zhang, L.; Luo, Z.J.; Han, B.W.; Li, R.Q.; Cao, X.Y.; Cheng, Q.; Cui, T.J. Programmable Controls to Scattering Properties of a Radiation Array. *Laser Photonics Rev.* **2021**, *15*, 2170016. [CrossRef]
8. Sun, H.; Gu, C.; Chen, X.; Li, Z.; Liu, L.; Martin, F. Ultra-wideband and broad-angle linear polarization conversion metasurface. *J. Appl. Phys.* **2017**, *121*, 174902. [CrossRef]
9. Mei, Z.L.; Ma, X.M.; Lu, C.; Zhao, Y.D. High-efficiency and wide-bandwidth linear polarization converter based on double U-shaped metasurface. *AIP Adv.* **2017**, *7*, 125323. [CrossRef]
10. Sui, S.; Ma, H.; Wang, J.; Feng, M.; Pang, Y.; Xia, S.; Xu, Z.; Qu, S. Symmetry-based coding method and synthesis topology optimization design of ultra-wideband polarization conversion metasurfaces. *Appl. Phys. Lett.* **2016**, *109*, 014104. [CrossRef]
11. Khan, M.I.; Fraz, Q.; Tahir, F.A. Ultra-wideband cross polarization conversion metasurface insensitive to incidence angle. *J. Appl. Phys.* **2017**, *121*, 045103. [CrossRef]
12. Huang, Y.; Yang, L.; Li, J.; Wang, Y.; Wen, G. Polarization conversion of metasurface for the application of wide band low-profile circular polarization slot antenna. *Appl. Phys. Lett.* **2016**, *109*, 054101. [CrossRef]
13. Fernandez, O.; Gomez, A.; Basterrechea, J.; Vegas, A. Reciprocal Circular Polarization Handedness Conversion Using Chiral Metamaterials. *IEEE Antennas Wirel. Propag. Lett.* **2017**, *16*, 2307–2310. [CrossRef]
14. Wu, L.; Yang, Z.; Cheng, Y.; Gong, R.; Zhao, M.; Zheng, Y.; Duan, J.; Yuan, X. Circular polarization converters based on bi-layered asymmetrical split ring metamaterials. *Appl. Phys. A-Mater. Sci. Process.* **2014**, *116*, 643–648. [CrossRef]
15. Wang, Q.; Plum, E.; Yang, Q.; Zhang, X.; Xu, Q.; Xu, Y.; Han, J.; Zhang, W. Reflective chiral meta-holography: Multiplexing holograms for circularly polarized waves. *Light-Sci. Appl.* **2018**, *7*, 25. [CrossRef]
16. Li, Y.; Zhang, J.; Qu, S.; Wang, J.; Zheng, L.; Pang, Y.; Xu, Z.; Zhang, A. Achieving wide-band linear-to-circular polarization conversion using ultra-thin bi-layered metasurfaces. *J. Appl. Phys.* **2015**, *117*, 044501. [CrossRef]
17. Baena, J.D.; Glybovski, S.B.; del Risco, J.P.; Slobozhanyuk, A.P.; Belov, P.A. Broadband and Thin Linear-to-Circular Polarizers Based on Self-Complementary Zigzag Metasurfaces. *IEEE Trans. Antennas Propag.* **2017**, *65*, 4124–4133. [CrossRef]
18. Lin, B.; Lv, L.; Guo, J.; Liu, Z.; Ji, X.; Wu, J. An Ultra-Wideband Reflective Linear-to-Circular Polarization Converter Based on Anisotropic Metasurface. *IEEE Access* **2020**, *8*, 82732–82740. [CrossRef]
19. Liu, Y.; Luo, Y.; Liu, C.; Song, K.; Zhao, X. Linear polarization to left/right-handed circular polarization conversion using ultrathin planar chiral metamaterials. *Appl. Phys. A—Mater. Sci. Process.* **2017**, *123*, 571. [CrossRef]
20. Lin, B.; Guo, J.; Lv, L.; Wu, J.; Ma, Y.; Liu, B.; Wang, Z. Ultra-wideband and high-efficiency reflective polarization converter for both linear and circular polarized waves. *Appl. Phys. A—Mater. Sci. Process.* **2019**, *125*, 76. [CrossRef]
21. Gao, X.; Yang, W.L.; Ma, H.F.; Cheng, Q.; Yu, X.H.; Cui, T.J. A Reconfigurable Broadband Polarization Converter Based on an Active Metasurface. *IEEE Trans. Antennas Propag.* **2018**, *66*, 6086–6095. [CrossRef]
22. Dutta, R.; Ghosh, J.; Yang, Z.; Zhang, X. Multi-Band Multi-Functional Metasurface-Based Reflective Polarization Converter for Linear and Circular Polarizations. *IEEE Access* **2021**, *9*, 152738–152748. [CrossRef]
23. Liu, Y.; Hao, Y.; Wang, H.; Li, K.; Gong, S. Low RCS Microstrip Patch Antenna Using Frequency-Selective Surface and Microstrip Resonator. *IEEE Antennas Wirel. Propag. Lett.* **2015**, *14*, 1290–1293. [CrossRef]
24. Al-Gburi, A.J.A.; Ibrahim, I.M.; Zakaria, Z.; Abdulhameed, M.K.; Saeidi, T. Enhancing Gain for UWB Antennas Using FSS: A Systematic Review. *Mathematics* **2021**, *9*, 3301. [CrossRef]
25. Decoster, B.; Maes, S.; Cuiñas, I.; García Sánchez, M.; Caldeirinha, R.; Verhaevert, J. Dual-Band Single-Layer Fractal Frequency Selective Surface for 5G Applications. *Electronics* **2021**, *10*, 2880. [CrossRef]
26. Liu, T.; Cao, X.; Gao, J.; Zheng, Q.; Li, W.; Yang, H. RCS Reduction of Waveguide Slot Antenna with Metamaterial Absorber. *IEEE Trans. Antennas Propag.* **2013**, *61*, 1479–1484. [CrossRef]
27. Liu, Y.; Zhao, X. Perfect Absorber Metamaterial for Designing Low-RCS Patch Antenna. *IEEE Antennas Wirel. Propag. Lett.* **2014**, *13*, 1473–1476. [CrossRef]
28. Jiang, H.; Xue, Z.; Zeng, Q.; Li, W.; Ren, W. High-gain low-RCS slot antenna array based on checkerboard surface. *IET Microw. Antennas Propag.* **2018**, *12*, 237–240. [CrossRef]

29. Han, Z.-J.; Song, W.; Sheng, X.-Q. Gain Enhancement and RCS Reduction for Patch Antenna by Using Polarization-Dependent EBG Surface. *IEEE Antennas Wirel. Propag. Lett.* **2017**, *16*, 1631–1634. [CrossRef]
30. Iriarte Galarregui, J.C.; Tellechea Pereda, A.; Luis Martinez de Falcon, J.; Ederra, I.; Gonzalo, R.; de Maagt, P. Broadband Radar Cross-Section Reduction Using AMC Technology. *IEEE Trans. Antennas Propag.* **2013**, *61*, 6136–6143. [CrossRef]
31. Vasanelli, C.; Boegelsack, F.; Waldschmidt, C. Reducing the Radar Cross Section of Microstrip Arrays Using AMC Structures for the Vehicle Integration of Automotive Radars. *IEEE Trans. Antennas Propag.* **2018**, *66*, 1456–1464. [CrossRef]
32. Jia, Y.; Liu, Y.; Guo, Y.J.; Li, K.; Gong, S.-X. Broadband Polarization Rotation Reflective Surfaces and Their Applications to RCS Reduction. *IEEE Trans. Antennas Propag.* **2016**, *64*, 179–188. [CrossRef]
33. Liu, J.; Li, J.-Y.; Zhou, S.-G. Polarization Conversion Metamaterial Surface with Staggered-Arrangement Structure for Broadband Radar Cross Section Reduction. *IEEE Antennas Wirel. Propag. Lett.* **2019**, *18*, 871–875. [CrossRef]
34. Ameri, E.; Esmaeli, S.H.; Sedighy, S.H. Ultra Wideband Radar Cross Section Reduction by Using Polarization Conversion Metasurfaces. *Sci. Rep.* **2019**, *9*, 478. [CrossRef]
35. Mao, C.; Yang, Y.; He, X.; Zheng, J.; Zhou, C. Broadband reflective multi-polarization converter based on single-layer double-L-shaped metasurface. *Appl. Phys. A-Mater. Sci. Process.* **2017**, *123*, 767. [CrossRef]
36. Fu, C.; Han, L.; Liu, C.; Sun, Z.; Lu, X. Dual-Band Polarization Conversion Metasurface for RCS Reduction. *IEEE Trans. Antennas Propag.* **2021**, *69*, 3044–3049. [CrossRef]

usually works in the millimeter-wave band, which is advantageous for reducing the size of the antenna element and the components in the array, making it easier for the entire communication system to be active, higher requirements are proposed for the installation accuracy of the array element position. This is because even minor installation errors are likely to be of the same order as the working wavelength, which will have a significant influence on the antenna's EM performance and channel quality. Therefore, it is necessary to quantitatively study the sensitivity of the 5G/6G communication system performance to the random position error of the antenna elements in the x, y and z directions, respectively.

Therefore, the coupling relationship between the structural factors of base station active phased array antennas and the channel capacity was studied, and the influence of factors including the element position offset, pointing deflection, and feed error caused by thermal power consumption of RF devices on the communication quality was analyzed comprehensively. Moreover, the sensitivity model of the electric field strength and channel capacity of the array antenna to the random position error of the elements is also constructed.

2. Electromechanical Coupling Modeling of Channel Capacity

2.1. Establishment of Electromechanical Coupling Model of Channel Capacity

As shown in Figure 1, a 5G/6G base station phased array antenna is arranged in an equidistant rectangular grid, with a total of $M \times N$ array elements. The spaces between the elements along the x- and y-axes are d_x and d_y, respectively, and the maximum beam direction is (θ_0, ϕ_0). The direction cosine of the target direction (θ, ϕ) at the receiving end can be expressed with respect to the coordinate axis as $(cos\alpha_x, cos\alpha_y, cos\alpha_z)$, shown in Equation (1), so the direction cosine of the maximum base station transmitting beam pointing direction can be expressed as $(u_0, v_0, w_0) = (cos\alpha_{x_0}, cos\alpha_{y_0}, cos\alpha_{z_0})$.

$$\begin{cases} u = cos\alpha_x = sin\theta cos\phi \\ v = cos\alpha_y = sin\theta sin\phi \\ w = cos\alpha_z = cos\theta \end{cases} \tag{1}$$

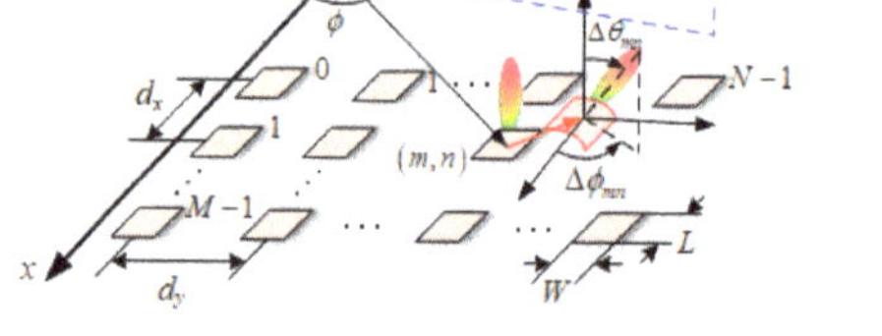

Figure 1. A 5G/6G base station phased array antenna.

During the working process, the element position offset and pointing deflection can be caused by structural distortion, manufacturing, and assembly error of the elements, where the positional offset of the (m, n) $(0 \le m \le M-1, 0 \le n \le N-1)$ element is assumed as $(\Delta x_{mn}, \Delta y_{mn}, \Delta z_{mn})$ and the pointing deflection is $(\Delta\theta_{mn}, \Delta\phi_{mn})$, as shown in Figure 1. In addition, the feeding errors of elements can be generated owing to the thermal power consumption of a large number of electronic devices in T/R modules, where the normalized amplitude error and phase error can be expressed as $\Delta A_{mn}(T)$ and $\Delta\varphi_{mn}(T)$, respectively.

Based on the electromechanical coupling model [22], the pattern function of the 5G/6G base station phased array antenna under the influences of structural distortion, manufacturing, thermal power consumption, and assembly error of the elements is expressed as below when neglecting the mutual coupling between the array elements.

$$F_{BS}(\theta,\phi) = \sum_{m=0}^{M-1}\sum_{n=0}^{N-1} f_{mn}(\theta-\Delta\theta_{mn},\phi-\Delta\phi_{mn})I_{mn}[1+\Delta A_{mn}(T)]\cdot \\ exp\{j[k(md_x u+nd_y v)+\Delta\Phi_{mn}+\Delta\varphi_{mn}(T)]\} \quad (2)$$

where $I_{mn} = A_{mn}exp(j\varphi_{mn})$ is the initial excitation current of the (m,n) element, A_{mn} and φ_{mn} are the amplitude and phase, respectively, $\Delta\Phi_{mn}$ is the far-field spatial phase change caused by the offset between the (m,n) element and the $(0,0)$ element, as expressed in (3), $f_{mn}(\theta-\Delta\theta_{mn},\phi-\Delta\phi_{mn})$ is the pattern of the element itself, as expressed in (4), $k = 2\pi/\lambda$ is the wave propagation constant, and λ is the wavelength of the base station phased array antenna.

$$\Delta\Phi_{mn} = k[(\Delta x_{mn}-\Delta x_{0,0})(u-u_0)+(\Delta y_{mn}-\Delta y_{0,0})(v-v_0)+(\Delta z_{mn}-\Delta z_{0,0})w] \quad (3)$$

$$\begin{cases} f_{mn}(\theta,\phi) = cos\phi f(\theta,\phi) - cos\theta sin\phi f(\theta,\phi) \\ f(\theta,\phi) = sinc\left(\frac{kW}{2}sin\theta sin\phi\right)cos\left(\frac{kL}{2}sin\theta cos\phi\right) \end{cases} \quad (4)$$

The pointing deflection angle $(\Delta\theta_{mn},\Delta\phi_{mn})$ can be found by the following step. First, assume that the surface equation of the entire base station antennas after deformation can be expressed by $z = f(x,y)$. Then, the normal vector of the tangent plane at the point of an element on the surface can be expressed by $\vec{l}_{mn} = \pm\left(-\frac{\partial z}{\partial x},-\frac{\partial z}{\partial y},1\right)$. The relationship between the direction cosine of the normal vector $\vec{l}_{mn}$ and its angle $(\Delta\theta_{mn},\Delta\phi_{mn})$ with respect to the coordinate axis can be obtained as follows.

$$\begin{cases} cos|\Delta\theta_{mn}| = \frac{1}{\sqrt{1+(\partial f(x,y)/\partial x)^2+(\partial f(x,y)/\partial y)^2}} \\ cos|\Delta\phi_{mn}| = \frac{|\partial f(x,y)/\partial x|}{sin|\Delta\theta_{mn}|\sqrt{1+(\partial f(x,y)/\partial x)^2+(\partial f(x,y)/\partial y)^2}} \end{cases} \quad (5)$$

In the communication downlink, a 5G/6G base station phased array antenna is used as the transmitting antenna. Channel noise is inevitable in the transmitting process, usually assumed to be additive white Gaussian noise (AWGN) in communication systems. AWGN is very representative and widely used as it is the most important and common noise and interference model in communication channels. The Shannon capacity formula gives the maximum achievable capacity (transmission bit rate) of a given channel in which the noise characteristics, operating bandwidth, and other indicators are known [26]. Assume that the operating bandwidth of the channel is B (Hz) and the signal-to-noise ratio at the receiving end is SNR, the maximum amount of information C (in bps) that the channel can carry is

$$C = B\times log_2(1+SNR) = B\times log_2\left(1+10lg\frac{P_R}{BN_0}\right) \quad (6)$$

where the SNR is used to evaluate the performance measurement characteristics of the communication system, representing the ratio between the signal of the channel output (meaningful information) and the background noise power, and N_0 is the AWGN power spectral density (W/Hz). Channel environments and transmission distance are different in the analysis of actual problems, but a certain value can always be given in a specific condition. Therefore, N_0 is often treated as a measurable constant when calculating the noise performance of a communication system.

According to the equivalent circuit principle of the receiving antenna, the receiving power of the antenna is

$$P_R = \frac{|E(\theta,\phi)|^2 {F_R}^2(\theta,\phi)}{240\pi k^2} G_R \gamma_R cos^2(\xi) \tag{7}$$

where $E(\theta,\phi)$ is the electric field strength of the incoming wave at the receiving antenna, $F_R(\theta,\phi)$ is the normalized pattern function of the receiving antenna, G_R is the gain of the receiving antenna, γ_R is the matching coefficient at the receiving end, representing the matching degree between the receiving antenna and the load, and $\gamma_R = 1$ when the receiving antenna and the load for the conjugate match, $cos(\xi)$ is the polarization matching factor. $\xi = 0$ and $cos(\xi) = 1$ when it is polarization matching.

According to Poynting vector method, the radiative power flux density of the transmitting antenna in the far field area is

$$S(\theta,\phi) = \frac{1}{2}E(\theta,\phi) \times H(\theta,\phi)^* = \frac{|E(\theta,\phi)|^2}{240\pi} \tag{8}$$

Meanwhile, the radiative power flux density can also be expressed as

$$S(\theta,\phi) = \frac{U(\theta,\phi)}{r^2} = \frac{U_M F_{T,BS}^2(\theta,\phi)}{r^2} = \frac{P_T G_{T,BS} F_{T,BS}^2(\theta,\phi)\gamma_T}{4\pi r^2} \tag{9}$$

where U_M is the radiation intensity in the maximum radiation direction of the transmitting antenna, $G_{T,BS}$ is the gain of the transmitting antenna, $F_{T,BS}(\theta,\phi) = \frac{F_{BS}}{|F_{BS}|_{\max}}$ is the normalized electric field strength pattern function of the transmitting antenna, P_T is the input power of the feeding device at the transmitting end, and γ_T is the efficiency of the feed system at the transmitting end. $\gamma_T = 1$ in an ideal situation.

According to (8) and (9), it can be determined that

$$|E(\theta,\phi)|^2 = \frac{60 P_T G_{T,BS} F_{T,BS}^2(\theta,\phi)\gamma_T}{r^2} \tag{10}$$

Then, the received power of the array antenna can be obtained by substituting (10) into (7), as follows:

$$P_R = \frac{\lambda^2}{16\pi^3 r^2} P_T G_R {F_R}^2(\theta,\phi)\gamma_R G_{T,BS} {F_{T,BS}}^2(\theta,\phi)\gamma_T cos^2(\xi) \tag{11}$$

Finally, substituting (11) into (6), the coupling relationship between the channel capacity and the EM performance of the transmitting phased array antenna can be obtained as follows, which can be used to describe the influence mechanism of the structural and feed errors caused by the thermal power of RF devices on the communication system channel quality in 5G/6G base station phased array antennas. In addition, it is assumed that the transmitting and receiving antennas are both under an ideal matching state and that the receiving antenna works in the ideal condition; therefore, the electromechanical-thermal coupling problem of the receiving antenna is not considered.

$$C = B \times log_2\left(1 + 10lg\frac{\lambda^2\gamma_T\gamma_R P_T G_R {F_R}^2(\theta,\phi)}{16\pi^3 r^2 B N_0} G_{T,BS} F_{T,BS}^2(\theta,\phi)\right) \tag{12}$$

2.2. Sensitivity Model Establishment of Array Element Position

Based on the established electromechanical-thermal coupling model of the 5G/6G base station phased array antenna, the sensitivity calculation model of the electric field strength and channel capacity to the position error of the array element is obtained by separately solving the partial derivatives of the two technical indicators to the array element position. The model can be used to demonstrate the influence of the array element position error on the communication technical indicators. The random position error of the array element is very small. Thus, it has an unobvious influence on the array factor pattern, whereas the array element pattern can be considered unchanged, so the change of the array element pattern is not taken into consideration in the sensitivity calculation model. The electromechanical-thermal coupling model of the 5G/6G base station phased array antenna in Equation (2) can be simplified to

$$\begin{aligned} f_a(\theta,\phi) &= \sum_{m=0}^{M-1}\sum_{n=0}^{N-1} I_{mn}\cdot exp\{jk[(md_x+\Delta x_{mn}-\Delta x_{0,0})(u-u_0)+ \\ &(nd_y+\Delta y_{mn}-\Delta y_{0,0})(v-v_0)+\ (\Delta z_{mn}-\Delta z_{0,0})w]\} \\ &= \sum_{m=0}^{M-1}\sum_{n=0}^{N-1} I_{mn}\cdot exp\{jk[(x'_{mn}-\Delta x_{0,0})(u-u_0)+ \\ &(y'_{mn}-\Delta y_{0,0})(v-v_0)+\ (z'_{mn}-\Delta z_{0,0})w]\} \end{aligned} \tag{13}$$

The partial derivatives of the array factor pattern function $f_a(\theta,\phi)$ to the actual position $(x'_{mn}, y'_{mn}, z'_{mn})$ of the antenna element (m,n) are as follows:

$$\begin{aligned} \frac{\partial f_a(\theta,\phi)}{\partial x'_{mn}} &= |I_{mn}|jk(u-u_0)\cdot exp\{jk[(u-u_0)(x'_{mn}-\Delta x_{0,0})+ \\ &(v-v_0)(y'_{mn}-\Delta y_{0,0})+w(z'_{mn}-\Delta z_{0,0})]\} \end{aligned} \tag{14}$$

$$\begin{aligned} \frac{\partial f_a(\theta,\phi)}{\partial y'_{mn}} &= |I_{mn}|jk(v-v_0)exp\{jk[(u-u_0)(x'_{mn}-\Delta x_{0,0})+ \\ &(v-v_0)(y'_{mn}-\Delta y_{0,0})+w(z'_{mn}-\Delta z_{0,0})]\} \end{aligned} \tag{15}$$

$$\begin{aligned} \frac{\partial f_a(\theta,\phi)}{\partial z'_{mn}} &= |I_{mn}|jkwexp\{jk[(u-u_0)\cdot(x'_{mn}-\Delta x_{0,0})+ \\ &(v-v_0)(y'_{mn}-\Delta y_{0,0})+w(z'_{mn}-\Delta z_{0,0})]\} \end{aligned} \tag{16}$$

According to the above, the sensitivity matrix of the electric field strength of the base station antenna array factor to the (m,n) position error of the array element is

$$S_{mn}^{f_a} = \left[\frac{\partial f_a(\theta,\phi)}{\partial x'_{mn}}, \frac{\partial f_a(\theta,\phi)}{\partial y'_{mn}}, \frac{\partial f_a(\theta,\phi)}{\partial z'_{mn}}\right] \tag{17}$$

Then, by combining the sensitivity values in the x, y, and z directions of all the antenna elements according to the array arrangement, the sensitivity matrix of the array electric field strength factor to all the element positional errors can be obtained.

Similarly, the partial derivative of the channel capacity C to the actual position $(x'_{mn}, y'_{mn}, z'_{mn})$ of the array element can be obtained as follows:

$$\frac{\partial C}{\partial x'_{mn}} = \frac{20B}{[1+10lg\alpha f_a^2(\theta,\phi)]ln2f_a(\theta,\phi)ln10}\cdot\frac{\partial f_a(\theta,\phi)}{\partial x'_{mn}} \tag{18}$$

$$\frac{\partial C}{\partial y'_{mn}} = \frac{20B}{[1+10lg\alpha f_a^2(\theta,\phi)]ln2f_a(\theta,\phi)ln10}\cdot\frac{\partial f_a(\theta,\phi)}{\partial y'_{mn}} \tag{19}$$

$$\frac{\partial C}{\partial z'_{mn}} = \frac{20B}{[1+10lg\alpha f_a^2(\theta,\phi)]ln2f_a(\theta,\phi)ln10}\cdot\frac{\partial f_a(\theta,\phi)}{\partial z'_{mn}} \tag{20}$$

where the parameter α is $\alpha = \frac{\lambda^2\gamma_T\gamma_R P_T G_R F_R{}^2(\theta,\phi)G_{T,BS}}{16\pi^3 r^2 BN_0}$.

The sensitivity matrix of the channel capacity to the position error of the (m,n) array element can be obtained as

$$S_{mn}^{C} = \left[\frac{\partial C}{\partial x'_{mn}}, \frac{\partial C}{\partial y'_{mn}}, \frac{\partial C}{\partial z'_{mn}}\right] \tag{21}$$

The performance of the communication indicators is mainly related to the main-lobe area of the transmitting and receiving beams. Thus, (θ,ϕ) in the above two sensitivity matrices are selected to choose the main-lobe area of the base station antenna.

3. Simulation Analysis and Discussion

3.1. Analysis and Discussion of Electromechanical Coupling Model of Channel Capacity

A 5G/6G array antenna model of the communication base station was built with 256 array elements, where the rectangular microstrip patch antenna is used as the element, and the interval between each element is $\lambda/2$. The frequency of the antenna is 28 GHz. The structural parameters of the array element are shown in Table 1, where the structural parameters of the array element have been optimized at the specified frequency 28 GHz.

Table 1. Structural parameters of the array element.

Antenna Structure	Structural Parameters	Variables	Values (mm)
Microstrip antenna	Length	L_d	3.43
	Width	W_d	3.55
Substrate	Thickness	h_s	0.20
	Length	$L_s = 2L_d$	6.85
	Width	$W_s = 2W_d$	7.10
Feeder position	Distance	L_1	0.52

The base station ambient temperature is 25 °C. The thermal power dissipation of a RF chip on the antenna array is 40mW, and the convective heat transfer coefficient of the array antenna is 1.2 W/(m^2·K). In addition, according to the working conditions of the base station, the four corners of the antenna array are fully constrained, and the thermal simulation is conducted in ANSYS. The temperature fields of the base station array antenna are shown in Figure 2.

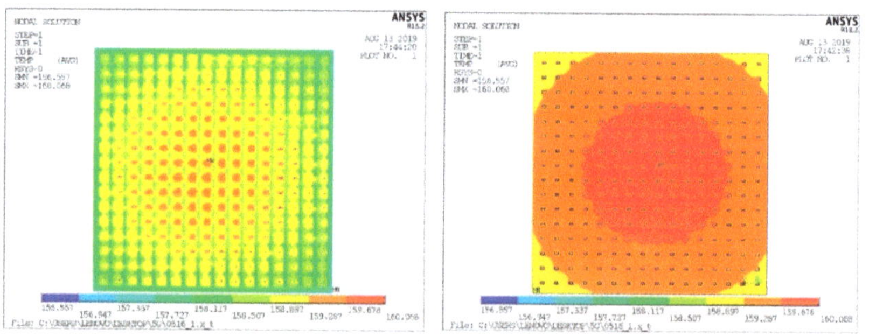

(a) (b)

Figure 2. Temperature field of base station array antenna: (**a**) Antenna element surface; (**b**) RF device surface.

It can be seen that the temperature of the front and back of the array antenna gradually decreases from the center to the surroundings in the temperature field distribution and is vertically and horizontally symmetrical. The overall maximum temperature is at the center of the surface of the RF device. Only linear strain exists during thermal expansion while the shear strain is zero. Hence, thermal deformation can be regarded as the node displacement caused by temperature loads. The temperature field distribution of the array is then used as the load of the structural displacement field analysis to obtain the thermal deformation. Meanwhile, the interpolation algorithm is used to add temperature data to the grid nodes meshed by ANSYS. The ANSYS simulation results of the thermal deformation of the array antenna structure are shown in Figure 3. The deformation of the entire surface is approximately symmetrical about the center, which is consistent with the distribution of the thermal deformation caused by the symmetric temperature distribution.

The maximum displacement occurs in the z-axial direction at the central area of the surface, and the maximum position offset is 3.435 mm.

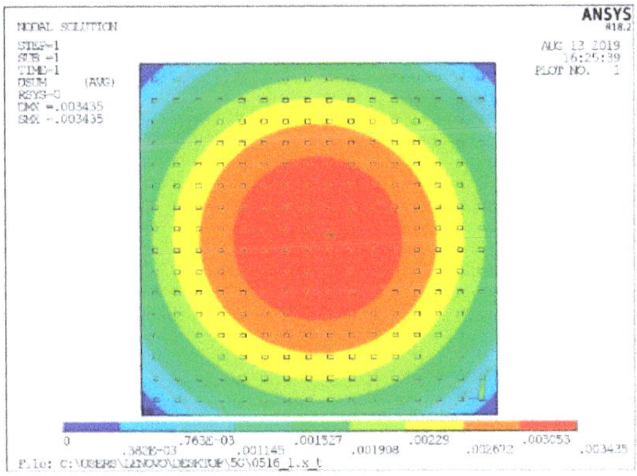

Figure 3. Thermal deformation of base station array antenna.

By extracting the node displacement of the finite element model of the base station array antenna after deformation, the surface fitting is performed in MATLAB, and the surface fitting equation can be expressed as

$$f(x,y) = \begin{array}{l} 3.426 - 0.004323x - 0.004586y \\ -0.001003x^2 + 4.894\text{e} - 06xy - 0.001034y^2 \end{array} \tag{22}$$

According to the electromechanical-thermal coupling model of the base station phased array antenna, the EM performances of the base station antenna before and after thermal deformation are calculated using MATLAB. All of the elements have the same amplitude and phase (when the antenna works in an unscanned state). The E-plane ($\phi = 0°$) and H-plane ($\phi = 90°$) power pattern of the base station array antennas, before and after the thermal deformation during EM performance, are shown in Figure 4. The main EM performance parameters of the base station antenna are shown in Table 2.

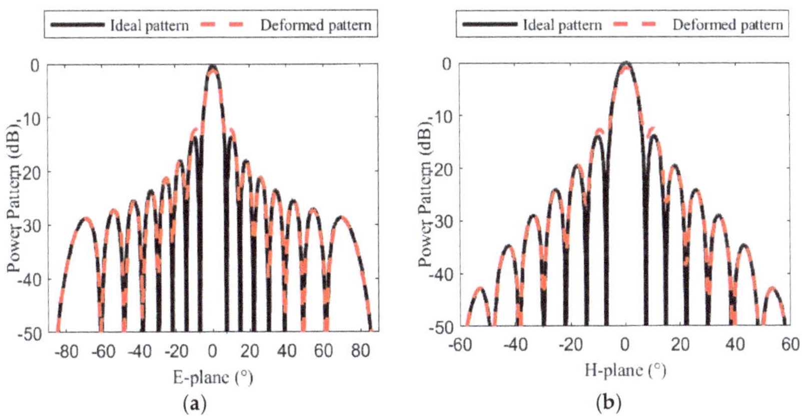

Figure 4. Temperature field of base station array antenna: (**a**) Antenna element surface; (**b**) RF device surface.

Table 2. Parameter variation of base station phased array antenna.

Electrical Performance Parameter	$\phi = 0°$		$\phi = 90°$	
	Ideal Situation	After Deformation	Ideal Situation	After Deformation
Gain loss/dB	0	−0.94	0	−0.94
Maximum pointing direction/°	0	0.02	0	0.03
The first SLL on the left/dB	−13.28	−11.92	−13.78	−12.53
The first SLL change on the left/dB	0	+1.36	0	+1.25
The first SLL on the right/dB	−13.28	−11.92	−13.78	−12.53
The first SLL change on the right/dB	0	+1.36	0	+1.25
The second SLL on the left/dB	−17.91	−17.69	−19.42	−19.24
The second SLL change on the left/dB	0	+0.22	0	+0.18
The second SLL on the right/dB	−17.91	−17.69	−19.42	−19.24
The second SLL change on the right/dB	0	+0.22	0	+0.18
The third SLL on the left/dB	−20.95	−21.06	−24.12	−24.14
The third SLL change on the left/dB	0	−0.11	0	−0.02
The third SLL on the right/dB	−20.95	−21.06	−24.12	−24.14
The third SLL change on the left/dB	0	−0.11	0	−0.02

(PS: "+" means increase, and "−" means decrease).

It can be learned from Figure 4 and Table 2 that:

1. The gain of the base station array antenna decreases because of the thermal deformation, and the gain loss can reach 0.94 dB.
2. The uplift amount of SLL shows a trend of increasing from the far field to the near field in both E-plane ($\phi = 0°$) and H-plane ($\phi = 90°$), reaching a maximum of 1.36 dB.
3. The maximum direction of the base station array antenna on the E-plane and H-plane has an offset of 0.02° and 0.03°, respectively. The reason is that the thermal deformation of the array antenna is approximately symmetrical.

To further evaluate the change in the channel capacity, the *SNR* of the base station antenna system is set to 30 dB under the ideal working situation. The ratio of SNR (Equation (22)) before and after deformation can be obtained by combining Equation (6) and Equation (7). Meanwhile, the ratio of channel capacity (Equation (23)) before and after deformation can be obtained from Equation (6). Then, the ratio of *SNR* and the channel capacity before and after structural deformation and feed error are 80.52 % and 93.92 %, respectively. The channel capacity value is rounded to 3 Gbps under the ideal working situation, for the convenience of subsequent calculations. Thus, the peak rate of the channel is lost by approximately $3 \times 1024 \times (1 - 93.92\%)$=186.8 Mbps when the EM performance of the base station array antenna is degraded.

$$\frac{SNR_{deformed}}{SNR_{ideal}} = \frac{G_{T,BS,deformed}F^2_{T,BS,deformed}(\theta,\phi)}{G_{T,BS,ideal}F^2_{T,BS,ideal}(\theta,\phi)} \tag{23}$$

$$\frac{C_{deformed}}{C_{ideal}} = \frac{B \cdot log_2\left(1 + SNR_{deformed}\right)}{B \cdot log_2(1 + SNR_{ideal})} \tag{24}$$

3.2. Analysis and Discussion of Sensitivity Model of Array Element Position

The 5G/6G base station phased array antenna model mentioned above is used as an example, and the excitation amplitude obeys Taylor's weighted distribution. The main-lobe area of the far field pattern is selected as $\theta \in (-0.1396, 0.1396)$ and $\phi \in (0, 2\pi)$, respectively. The sensitivity numerical distribution of the array factor electric field strength and channel capacity to the random position error of the antenna elements can be obtained, as shown in Figures 5–7.

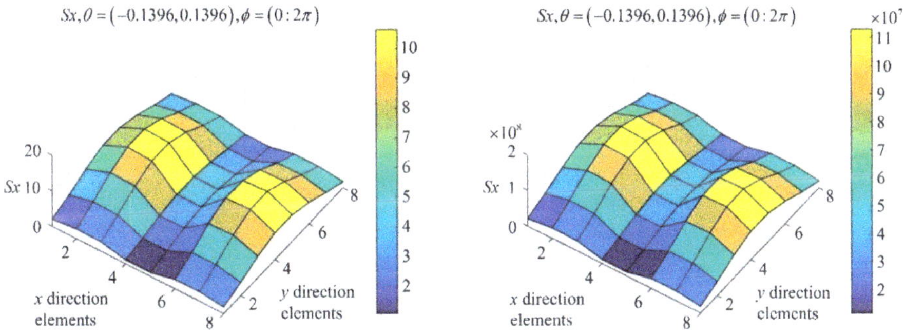

(**a**) (**b**)

Figure 5. Sensitivity distribution in the x direction: (**a**) electric field strength of array factor; (**b**) channel capacity.

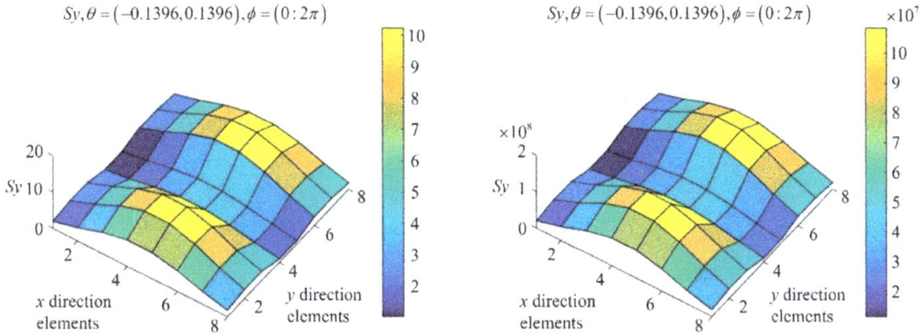

(**a**) (**b**)

Figure 6. Sensitivity distribution in the y direction: (**a**) electric field strength of array factor; (**b**) channel capacity.

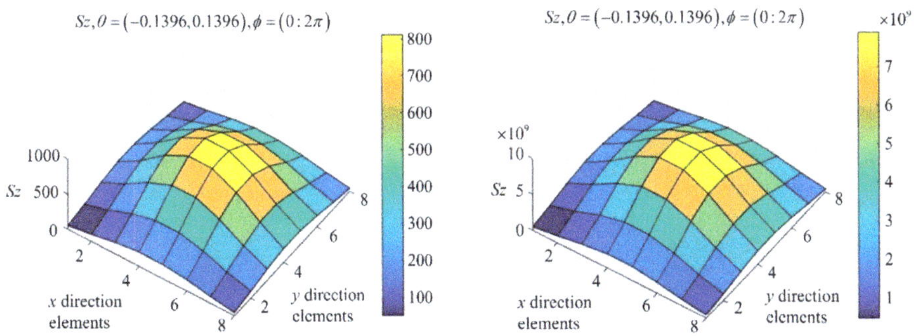

(**a**) (**b**)

Figure 7. Sensitivity distribution in the z direction: (**a**) electric field strength of array factor; (**b**) channel capacity.

It can be determined from Figures 5–7 that:

1. The sensitivity of the electric field strength of the array factor and channel capacity to the random position error in the z-axis direction is much greater than the sensitivity in the x- and y-axis directions, indicating that the tolerance design of the array elements in the z-axis direction should be more strictly controlled during manufacturing and installation.
2. The random position error in the x- and y-axis directions has approximately the same effect on these two technical indicators and the influence on the sensitivity distribution is similar (rotating by almost 90°) and having a certain periodicity along the x- and y-axes, respectively.
3. From the sensitivity of the random position error in the z-axis direction, it can be seen that the sensitivity value shows a decreasing trend from the central to the edge area, indicating that the tolerance design of the central area of the array in the z-axis direction should be stricter than the edge area.
4. For the random position error in the same direction, although the sensitivity distribution of the electric field strength is similar to that of channel capacity, the magnitude of the latter is far above the former, indicating that the influence of the same random position error on the channel capacity is much greater than the influence on the array electric field strength.

In the process of beam matching between the transmitting beam of the actual base station antenna and the receiving beam of the user-end, the most optimal beam direction of the user-end is likely not the direction (0°,0°) of the transmitting beam. Therefore, it is necessary to further analyze the sensitivity of the communication performance indicators to the random position error in x, y, z directions under the situation where the beam is in scanning state. The base station is usually sectored into multiple parts. In the 5G era, a single macro base station usually has 6 sectors, which is superior to the traditional 3-sector honeycomb structure, for it allows the communication system to make full use of spatial multiplexing technology to provide additional system service capacity. It is assumed that the base station is a 6-sector antenna structure. The sensitivity of the array factor electric field strength to the random position error of the array element in the x, y, and z directions in the $\phi = 0°$ plane and $\phi = 90°$ plane when the base station antenna pattern scans $\theta \in (-30°, 30°)$ are shown in Figures 8–11. In addition, it can be seen from Figures 5–7 that the sensitivity distribution of the channel capacity and electric field strength to the random position error are similar. Therefore, only the electric field strength of the array factor is selected for analysis.

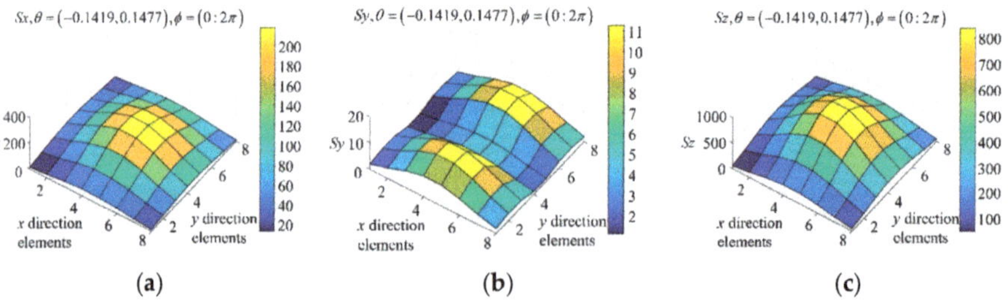

Figure 8. Sensitivity distribution of random position error in $\phi = 0°$ plane with scanning angle $\theta = 15°$: (**a**) x direction; (**b**) y direction; (**c**) z direction.

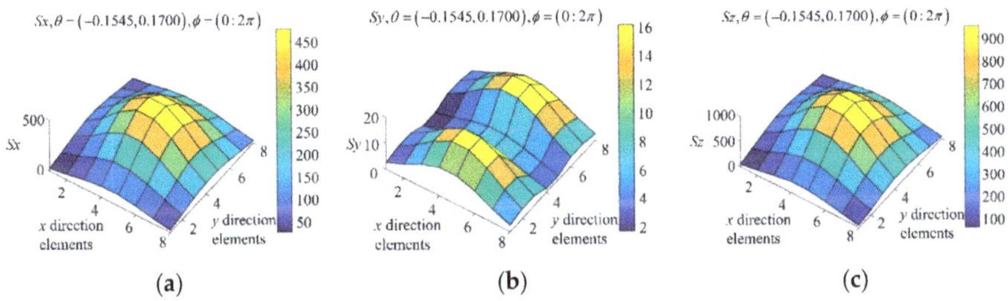

Figure 9. Sensitivity distribution of random position error in $\phi = 0°$ plane with scanning angle $\theta = 30°$: (**a**) x direction; (**b**) y direction; (**c**) z direction.

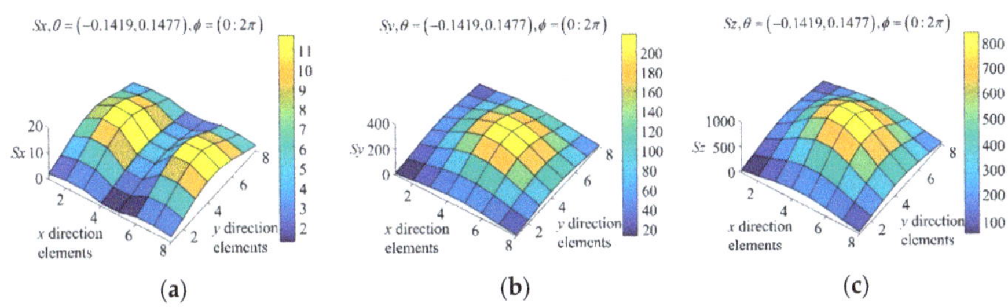

Figure 10. Sensitivity distribution of random position error in $\phi = 90°$ plane with scanning angle $\theta = 15°$: (**a**) x direction; (**b**) y direction; (**c**) z direction.

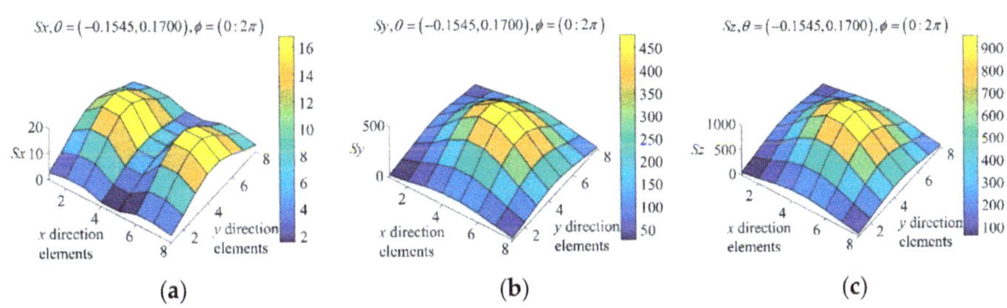

Figure 11. Sensitivity distribution of random position error in $\phi = 90°$ plane with scanning angle $\theta = 30°$: (**a**) x direction; (**b**) y direction; (**c**) z direction.

It can be learned from Figures 8–11 that:

1. The sensitivity of the electric field strength of the array factor and channel capacity to the random position error in the z-axis direction is much greater than the sensitivity in the x- and y-axis directions, indicating that the tolerance design of the array elements in the z-axis direction should be more strictly controlled during manufacturing and installation.
2. The random position error in the x- and y-axis directions has approximately the same effect on these two technical indicators and the influence on the sensitivity distribution is similar (rotating by almost 90°) and having a certain periodicity along the x- and y-axes, respectively.
3. With an increase in the scanning angle, the sensitivity value in the z-direction does not exhibit an obvious change but still increases gradually.
4. With the increasing of the scanning angle, the sensitivity value in the z-direction does not have an obvious change, but still increases gradually.

In summary, from the simulation of the influence of thermal deformation on the electric field strength and channel capacity and the simulation of the sensitivity of these two technical indicators to the random error of elements, it can be demonstrated that:

- The EM performance of the base station array antenna will degrade when the antenna structure undergoes thermal deformation and feed error, thereby causing a loss in the channel capacity of the communication system to a certain degree.
- The random error of the element position has a significant influence on the performance of the 5G/6G communication system. The sensitivity of the two technical indicators of electric field strength and channel capacity are different from the random error of the element position. The channel capacity is much more sensitive to random errors.
- When the beam is in the scanning state, as the scanning angle increases, the sensitivity of the channel capacity and the electric field strength to the random error of the element position in all three directions increases gradually.

4. Conclusions

Owing to the evolution directions of activeness, integration, and miniaturization of the base station phased array antenna system, electromechanical–thermal coupling is gradually becoming a major challenge in the process of commercializing 5G/6G communications. Therefore, this paper effectively coordinates the coupling relationship between the design of various disciplines from the viewpoint of electromechanical coupling and interdisciplinarity, taking into consideration the complex working environment of base stations. The relationship between the field coupling of the structural displacement field, temperature field, and electromagnetic field of the 5G/6G base station phased array antenna is studied, and the sensitivity distribution characteristics of the channel capacity of the antenna electric field strength against element position under different working circumstances are quantitatively evaluated. The work done in this paper could provide a theoretical basis for electromechanical coupling design and control technology to promote the commercialization and development of high-quality communication systems for the next generation.

Author Contributions: Conceptualization, Y.Y., Y.W. (Yan Wang) and C.W.; methodology, Y.Y., Y.W. (Yan Wang) and C.W.; software, Y.Y. and B.H.; validation, Y.Y. and X.H.; formal analysis, B.H. and P.L.; investigation, Z.W., K.Y., M.W. and G.L.; data curation, Y.W. (Yang Wu); writing—original draft preparation, Y.Y. and X.H.; writing—review and editing, Y.Y., Y.W. and B.H.; supervision, Z.W. and Y.W. (Yang Wu); project administration, Y.W. (Yan Wang) and C.W.; funding acquisition, Y.W. (Yan Wang) and C.W. All authors have read and agreed to the published version of the manuscript.

Funding: This research was funded by National Natural Science Foundation of China under grant number 51975447 and 52105272, the National Defense Basic Scientific Research Program of China under grant number JCKY2021210B007, the Scientific Research Program Funded by Shaanxi Provincial Education Department under No. 21JK0721, and Youth Innovation Team of Shaanxi Universities under No. 201926.

Institutional Review Board Statement: Not applicable.

Informed Consent Statement: Not applicable.

Data Availability Statement: Not applicable.

Conflicts of Interest: Authors declare no conflict of interest.

References

1. Iqbal, A.; Smida, A.; Abdulrazak, L.F.; Saraereh, O.A.; Mallat, N.K.; Elfergani, I.; Kim, S. Low-Profile Frequency Reconfigurable Antenna for Heterogeneous Wireless Systems. *Electronics* **2019**, *8*, 976. [CrossRef]
2. Wang, F.; Duan, Z.; Wang, X.; Zhou, Q.; Gong, Y. High Isolation Millimeter-Wave Wideband MIMO Antenna for 5G Communication. *Int. J. Antennas Propag.* **2019**, *2019*, 4283010. [CrossRef]

3. Yang, X.; Matthaiou, M.; Yang, J.; Wen, C.-K.; Gao, F.; Jin, S. Hardware-Constrained Millimeter-Wave Systems for 5G: Challenges, Opportunities, and Solutions. *IEEE Commun. Mag.* **2019**, *57*, 44–50. [CrossRef]
4. Ur-Rehman, M.; Abbasi, Q.H.; Rahman, A.; Khan, I.; Chattha, H.T.; Matin, M.A. Millimetre-Wave Antennas and Systems for the Future 5G. *Int. J. Antennas Propag.* **2017**, *2017*, 6135601. [CrossRef]
5. Agiwal, M.; Roy, A.; Saxena, N. Next Generation 5G Wireless Networks: A Comprehensive Survey. *IEEE Commun. Surv. Tut.* **2016**, *18*, 1617–1655. [CrossRef]
6. Neog, P.; Bera, R. Multi-standard radio for 2G to 5G. In Proceedings of the 2017 2nd International Conference on Telecommunication and Networks (TEL-NET), Noida, India, 10–11 August 2017; pp. 1–5. [CrossRef]
7. Ali, H.; Ren, X.-C.; Bari, I.; Bashir, M.A.; Hashmi, A.M.; Khan, M.A.; Majid, S.I.; Jan, N.; Tareen, W.U.K.; Anjum, M.R. Four-Port MIMO Antenna System for 5G N79 Band RF Devices. *Electronics* **2021**, *11*, 35. [CrossRef]
8. Roh, W.; Seol, J.-Y.; Park, J.; Lee, B.; Lee, J.; Kim, Y.; Cho, J.; Cheun, K.; Aryanfar, F. Millimeter-Wave Beamforming as an Enabling Technology for 5G Cellular Communications: Theoretical Feasibility and Prototype Results. *IEEE Commun. Mag.* **2014**, *52*, 106–113. [CrossRef]
9. Smolders, A.B.; Dubok, A.; Tessema, N.M.; Chen, Z.; Al Rawi, A.; Johannsen, U.; Bressner, T.; Milosevic, D.; Gao, H.; Tangdiongga, E. Building 5G Millimeter-Wave Wireless Infrastructure: Wide-Scan Focal-Plane Arrays with Broadband Optical Beamforming. *IEEE Antennas Propag. Mag.* **2019**, *61*, 53–62. [CrossRef]
10. Khan, J.; Ullah, S.; Tahir, F.A.; Tubbal, F.; Raad, R. A Sub-6 GHz MIMO Antenna Array for 5G Wireless Terminals. *Electronics* **2021**, *10*, 3062. [CrossRef]
11. Kim, K.J.; Choi, K.J.; Lee, S.R.; Kim, K.S. Multi-User Massive MIMO for next-Generation WLAN Systems. *Electron. Lett.* **2015**, *51*, 792–794. [CrossRef]
12. Ojaroudi Parchin, N.; Jahanbakhsh Basherlou, H.; Alibakhshikenari, M.; Ojaroudi Parchin, Y.; Al-Yasir, Y.I.; Abd-Alhameed, R.A.; Limiti, E. Mobile-Phone Antenna Array with Diamond-Ring Slot Elements for 5G Massive MIMO Systems. *Electronics* **2019**, *8*, 521. [CrossRef]
13. Oliveri, G.; Gottardi, G.; Massa, A. A New Meta-Paradigm for the Synthesis of Antenna Arrays for Future Wireless Communications. *IEEE Trans. Antennas Propag.* **2019**, *67*, 3774–3788. [CrossRef]
14. Andrews, J.G.; Buzzi, S.; Choi, W.; Hanly, S.V.; Lozano, A.; Soong, A.C.; Zhang, J.C. What Will 5G Be? *IEEE J. Sel. Areas Commun.* **2014**, *32*, 1065–1082. [CrossRef]
15. Chen, Y.; Jian, R.; Ma, S.; Mohadeskasaei, S.A. A Research for Millimeter Wave Patch Antenna and Array Synthesis. In Proceedings of the 26th Wireless and Optical Communication Conference (WOCC), Newark, NJ, USA, 7–8 April 2017; pp. 1–5. [CrossRef]
16. Przesmycki, R.; Bugaj, M.; Nowosielski, L. Broadband Microstrip Antenna for 5G Wireless Systems Operating at 28 GHz. *Electronics* **2020**, *10*, 1. [CrossRef]
17. Anselmi, N.; Rocca, P.; Feuchtinger, S.; Biscontini, B.; Barrera, A.M.; Massa, A. Optimal Capacity-Driven Design of Aperiodic Clustered Phased Arrays for Multi-User MIMO Communication Systems. *IEEE Trans. Antennas Propag.* **2022**. [CrossRef]
18. Comisso, M.; Palese, G.; Babich, F.; Vatta, F.; Buttazzoni, G. 3D Multi-Beam and Null Synthesis by Phase-Only Control for 5G Antenna Arrays. *Electronics* **2019**, *8*, 656. [CrossRef]
19. Rocca, P.; Anselmi, N.; Benoni, A.; Massa, A. Probabilistic Interval Analysis for the Analytic Prediction of the Pattern Tolerance Distribution in Linear Phased Arrays with Random Excitation Errors. *IEEE Trans. Antennas Propag.* **2020**, *68*, 7866–7878. [CrossRef]
20. Crișan, N.; Cremene, L.C. Rewriting MIMO Channel Capacity for Antenna Configuration Comparison. *Carpathian J. Electron. Comput. Eng.* **2015**, *8*, 31–34.
21. Ando, N.; Fujimoto, M. Effect of Directivity of On-Vehicle Antenna on Spread and Channel Capacity. In Proceedings of the International Symposium on Antennas and Propagation (ISAP), Busan, Korea, 23–26 October 2018; pp. 1–2.
22. Wang, C.S.; Duan, B.Y.; Zhang, F.S.; Zhu, M.B. Coupled Structural-Electromagnetic-Thermal Modelling and Analysis of Active Phased Array Antennas. *IET Microw. Antennas Propag.* **2010**, *4*, 247–257. [CrossRef]
23. Wang, C.; Kang, M.; Wang, W.; Duan, B.; Lin, L.; Ping, L. On the Performance of Array Antennas with Mechanical Distortion Errors Considering Element Numbers. *Int. J. Electron.* **2017**, *104*, 462–484. [CrossRef]
24. Wang, C.; Wang, Y.; Yang, X.; Gao, W.; Jiang, C.; Wang, L.; Zhang, Y.; Wang, M. Effect of Randomness in Element Position on Performance of Communication Array Antennas in Internet of Things. *Wirel. Commun. Mob. Comput.* **2018**, *2018*, 6492143. [CrossRef]
25. Obara, T.; Inoue, Y.; Aoki, Y.; Suyama, S.; Lee, J.; Okumurav, Y. Experiment of 28 GHz Band 5G Super Wideband Transmission Using Beamforming and Beam Tracking in High Mobility Environment. In Proceedings of the 27th Annual International Symposium on Personal, Indoor, and Mobile Radio Communications (PIMRC), Valencia, Spain, 4–8 September 2016; pp. 1–5. [CrossRef]
26. Shannon, C.E. A Mathematical Theory of Communication. *Bell Syst. Tech. J.* **1948**, *27*, 379–423. [CrossRef]

electronics

MDPI

Article

On-Chip Polarization Reconfigurable Microstrip Patch Antennas Using Semiconductor Distributed Doped Areas (ScDDAs)

Rozenn Allanic [1,*], Denis Le Berre [1], Cédric Quendo [1], Douglas Silva De Vasconcellos [2], Virginie Grimal [2], Damien Valente [2] and Jérôme Billoué [2]

[1] Laboratoire des Sciences et Techniques de l'Information, de la Communication et de la Connaissance (Lab-STICC), Université de Brest, 29238 Brest, France; denis.le-berre@univ-brest.fr (D.L.B.); cedric.quendo@univ-brest.fr (C.Q.)

[2] Laboratoire GREMAN, Université de Tours, 37071 Tours, France; douglas.silva@univ-tours.fr (D.S.D.V.); virginie.grimal@univ-tours.fr (V.G.); damien.valente@univ-tours.fr (D.V.); jerome.billoue@univ-tours.fr (J.B.)

* Correspondence: rozenn.allanic@univ-brest.fr

Abstract: This paper presents two polarization reconfigurable patch antennas using semiconductor distributed doped areas (ScDDAs) as active components. One proposed antenna has a switching polarization between two linear ones, while the other one has a polarization able to commute from a linear to a circular one. The antennas are designed on a silicon substrate in order to have the ScDDAs integrated in the substrate, overcoming the needs of classical PIN diodes. Therefore, the proposed co-design method between the antenna and the ScDDAs permits us to optimize the global reconfigurable function, designing both parts in the same process flow. Both demonstrators have a resonant frequency of around 5 GHz. The simulated results fit well with the measured ones.

Keywords: antenna; microstrip; patch antenna; polarization; reconfigurable; ScDDAs; switchable

Citation: Allanic, R.; Le Berre, D.; Quendo, C.; De Vasconcellos, D.S.; Grimal, V.; Valente, D.; Billoué, J. On-Chip Polarization Reconfigurable Microstrip Patch Antennas Using Semiconductor Distributed Doped Areas (ScDDAs). *Electronics* **2022**, *11*, 1905. https://doi.org/10.3390/electronics11121905

Academic Editors: Naser Ojaroudi Parchin, Chan Hwang See and Raed A. Abd-Alhameed

Received: 24 May 2022
Accepted: 14 June 2022
Published: 17 June 2022

1. Introduction

Nowadays, wireless systems are everywhere, and designers have to find new solutions in order to respond to challenges in terms of performance, size and cost. Indeed, the systems have to coexist without creating disturbances for others, all while providing an increase in terms of performance, compactness and a reduction in manufacturing cost. The antennas are one of the main components in these communicating systems and their tunability allows for a reduced size in a multiple standards system. They can be tuned in terms of frequency [1–9] in order to work in several frequency bands of radiation patterns [10,11] in order to modify the orientation of the beam and the polarization [12–21]. Among the various topologies of antennas, microstrip patch antennas are very common antennas used for their compactness, ease of manufacture and low cost.

A novel way to design microwave tunable devices has been developed and consists of a co-design method between active components and passive transmission lines, as in [22,23]. This integration solution with the so-called semiconductor distributed doped areas (ScDDAs) allows us to overcome the needs of classical soldered components. Indeed, the passive devices are designed on a silicon substrate, and thanks to this particular substrate, doped areas can behave electrically via through the substrate thickness, making the device reconfigurable.

Therefore, the passive part is optimized in the same amount of time as a global function. This offers design flexibility, ease of commuting between the working states, a low switching voltage and no parasitic effects between the active and the passive parts. With this monolithic integration of the active elements in a semiconductor substrate, it is also possible to have three working states with a unique DC command [24] and continuous tuning with a triangular-shaped doped area [25].

In this paper, a novel way to co-design on-chip polarization, reconfigurable microstrip patch antennas using ScDDAs are proposed. The flexibility brought by this approach allows for several implementations inducing different kinds of reconfigurability. Hence, the second section presents a switchable patch antenna offering two linear polarizations. A third section proposes a patch antenna with a polarization which commutes from a circular polarization to a linear one.

2. Two Linear Polarizations Reconfigurable Antenna

2.1. First Antenna Design

The idea is based on a patch antenna design with truncated corners. When the limit conditions are modified in both corners positioned diagonal, this modifies the surface current flow and consequently the polarization of the antenna.

Therefore, the patch antenna is designed on a high-resistivity silicon substrate. This semiconductor substrate allows us to have two semiconductor distributed doped areas (ScDDAs) forming N^+PP^+ integrated junctions. These areas act as integrated switches in the substrate thickness depending on their DC bias voltage, modifying the limit conditions onto two corners commuting from open circuits to short circuits. A top view of the antenna design is illustrated in Figure 1a and its side view with the associated technology is shown in Figure 1b. The dimensions of the final design are given in Table 1 with *wa*, which stands for the width of the access line *la*, and *wq* the width of the quarter wavelength line *lq*. *lc* is for the length of the truncated corner, i.e., the smallest dimension of the doped area, while *ldop* is the longest dimension of the doped area and *wdop* is the width of the doped area.

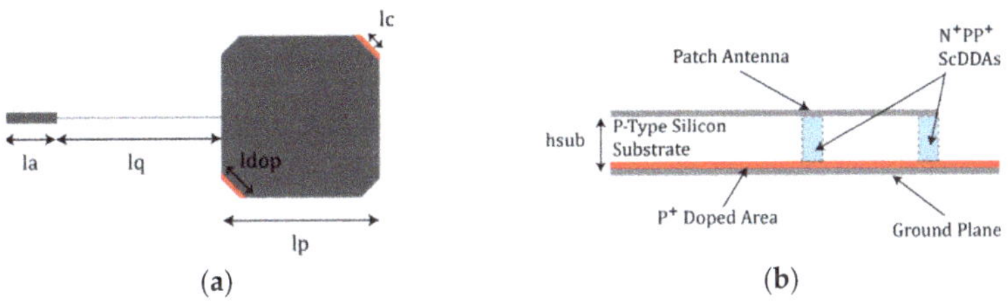

Figure 1. The polarization reconfigurable antenna. (**a**) Top view. (**b**) Side view.

Table 1. Dimensions (in mm) of the two linear polarization reconfigurable antenna.

lp	lq	wq	la	wa	lc	ldop	wdop	hsub
8.7	9.2	0.016	0.3	0.56	1.41	1.84	0.3	0.675

2.2. First Antenna Simulations

The simulations were performed using a full-wave electromagnetic simulator HFSS^{TM} from Ansys. In the OFF state (considering the junctions without bias voltage), the doping areas are neglected and the resistivity of ρ = 2500 Ω.cm is taken into account in the loss tangent calculation thanks to Equation (1) [26].

$$tan\delta = \frac{1}{\rho\omega\varepsilon_0\varepsilon_r} + 0.0018 \quad (1)$$

In the ON state (when a direct bias voltage is applied to the junctions), the junctions are simulated with a homogenous resistivity of ρ = 0.5 Ω.cm in the whole substrate thickness under the surfaces of the doped areas.

Figure 2 presents the simulated reflection coefficient in both states. In the OFF state, the −10 dB bandwidth is between 4.86 GHz and 4.95 GHz, whereas in the ON state, the −10 dB bandwidth is between 4.95 GHz and 5.03 GHz. Figure 3 illustrates the electrical field in both states. In the OFF state, there is no short circuit in the substrate, and the polarization is a linear one, parallel to the slots, so the patch radiates as a classical one. In the ON state, the ScDDAs are some short circuits in the substrate, creating a perturbation in the current flow and modifying the polarization way to a linear at 45° from the original one.

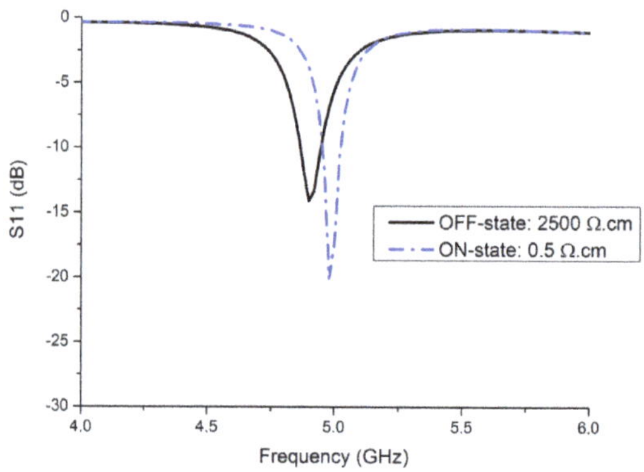

Figure 2. The simulated reflection coefficient in both states of the two linear polarization antennae.

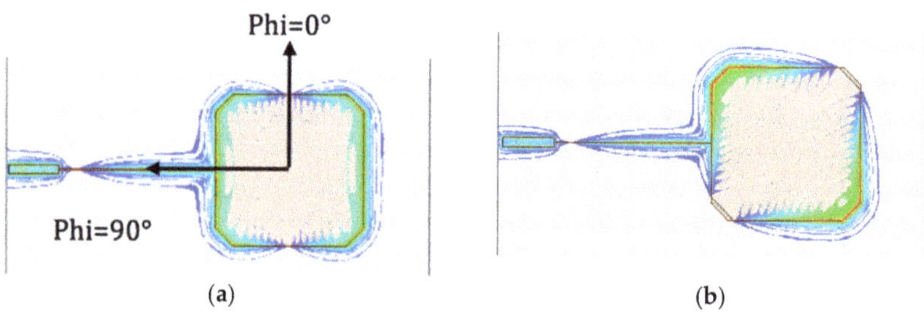

Figure 3. A simulated EM field. (**a**) Horizontal Linear Polarization. (**b**) 45° Linear Polarization.

Figure 4 shows the normalized radiation patterns in both states. The diagram is omnidirectional as a classical patch with a difference between the co- and cross-polarization higher than 45 dB in the OFF state and 12 dB in the ON state.

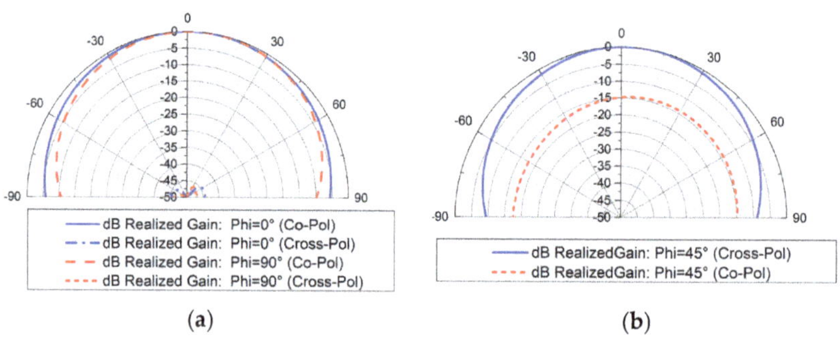

Figure 4. Simulated radiation patterns of the realized gain. (**a**) In the OFF state at Phi = 0° and at Phi = 90°; (**b**) in the ON state at Phi = 45°.

2.3. First Antenna Fabrication

A high-resistivity silicon P-type substrate was selected for manufacturing to minimize the loss tangent and the cost. The P^+ and N^+ areas are doped with a solgel solution and diffusion technique to reach around 10^{19} atoms/cm^3 with Boron and Phosphorus atoms, respectively. The depth of the two junctions is around 3 µm. The process steps are described in [27].

Indeed, only two masks are required for the manufacturing, as illustrated in Figure 5. Figure 6 shows a photograph of the fabricated demonstrator. In fact, the chosen SMA connector allows for measurement of a device with a thickness of 1.2 mm, whereas the silicon substrate thickness used is 0.675 µm, which is why an aluminum plate was put under the antenna to ease the placement of the connector, thus making the measurement possible.

(a)

(b)

Figure 5. (**a**) The metallization mask. (**b**) The doping mask.

Figure 6. A photograph of the first prototype.

2.4. First Antenna Measurement

Firstly, the S_{11} parameter was measured using a vector network analyzer (VNA ZVA 67 from Rohde & Schwarz®). The antenna commutation is provided by applying a DC bias voltage with a DC source connected directly to the analyzer. The DC bias voltage is applied with the RF signal, and due to the kind of junctions, a negative voltage is required to apply direct bias to them. A bias voltage of −6 V is applied at the output of the source. Figure 7 presents the measured results of the reflection coefficient in the OFF and ON states. In the OFF state, the resonant frequency is 4.93 GHz with a S_{11} level lower than −22 dB. In the ON state, the resonant frequency is 4.96 GHz with a S_{11} level lower than −10.2 dB.

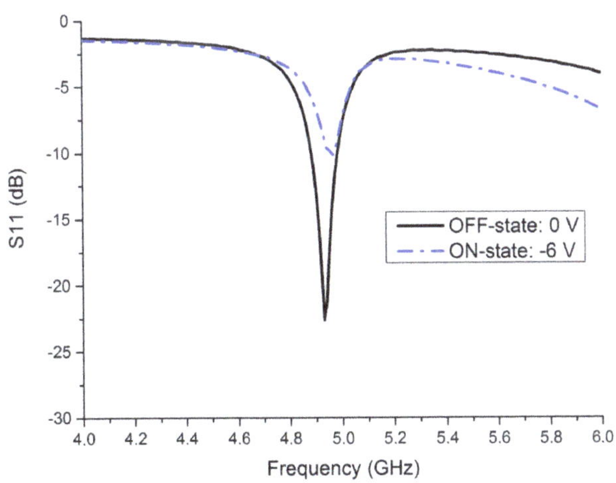

Figure 7. Measured results of the two linear polarization reconfigurable antenna in both states.

Secondly, the antenna was measured in an anechoic chamber, as illustrated in Figure 8. Two supports were made in 3D-printing technology; the white one is for the 0° and 90° radiation patterns measurement and the red one is added for the 45° radiation pattern measurement.

(a) (b)

Figure 8. (**a**) An antenna under measurement in the anechoic chamber. (**b**) A close-up of the positioner.

Figure 9a shows the co-polarization and the cross-polarization in both 0° and 90° planes for the OFF state. The shapes of the measured radiation patterns are identical to those simulated even if in the OFF state, and the cross-polarization level is higher than the simulated one.

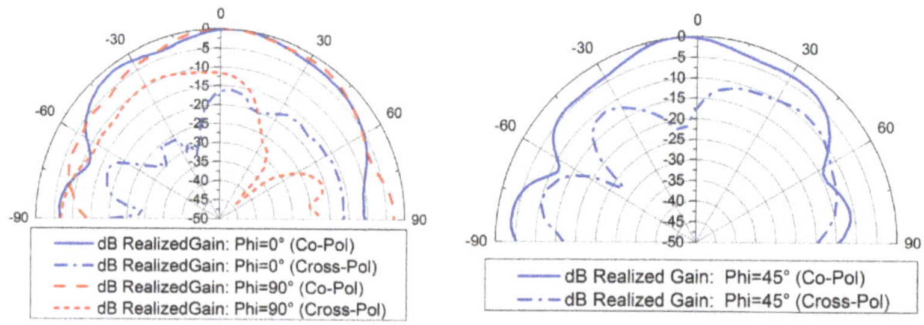

(a) (b)

Figure 9. (**a**) Measured radiation patterns of the realized gain. (**a**) In the OFF state at Phi = 0° and at Phi = 90°; (**b**) in the ON state at Phi = 45°.

This difference is due to the measurement setup. Indeed, the antenna is not soldered to the connector but only inserted between the pins of the connector and maintained by adding an aluminum plate. This induces misalignments between the emitting antenna and the patch antenna, which explains the higher cross-polarization level. Figure 9b shows the co-polarization and the cross-polarization in the 45° plan for the ON state. The shape and the difference between the co- and cross-polarization are almost the same as in the simulated results.

3. Circular to Linear Polarization Reconfigurable Antenna

3.1. Second Antenna Design

Figure 10 presents a second demonstrator able to commute from a circular polarization to a linear one. The patch antenna has two truncated corners to assure the circular polarization in the OFF state, and two doped areas are located in the other two corners to modify the current flows by adding two electrical short circuits in the substrate in the ON state. Table 2 summarizes the antenna and the doped area dimensions with *wa*, which stands for the width of the access *la*, and *wq* the width of the quarter wavelength line *lq*. *lc* is for the length of the truncated corner, *ldop* is for the longest dimension of the doped area and *wdop* is for the width of the doped area.

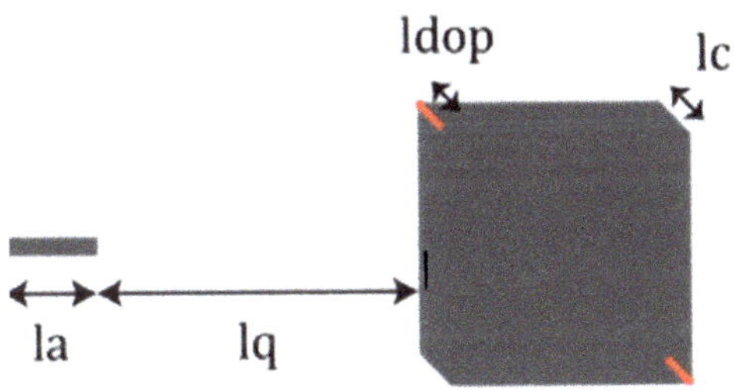

Figure 10. A top view of the circular to linear polarization switchable antenna design.

Table 2. Dimensions (in mm) of the two circular to linear polarization reconfigurable antennae.

lp	lq	wq	la	wa	lc	ldop	wdop
8.7	10.2	0.016	0.3	0.56	1.41	1	0.2

3.2. Second Antenna Simulations

This antenna was designed to have a resonant frequency of around 5 GHz in both states. Figure 11 shows the simulated results of the reflection coefficient. In the OFF state, the −10 dB bandwidth is between 4.77 GHz and 4.96 GHz and in the ON state, with a resistivity in the substrate thickness of 0.5 Ω.cm, the −10 dB bandwidth is between 4.9 GHz and 4.98 GHz.

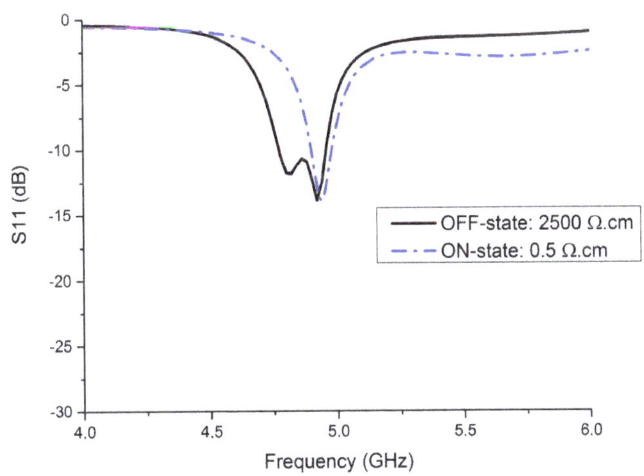

Figure 11. S_{11} simulated results of circular to linear polarization reconfigurable antennae in both states.

Figure 12 presents the simulated radiation patterns in both states. In the OFF state, the co- and cross-polarizations of the phi angle have the same level, which shows that the polarization of the antenna is a circular one. The axial ratio is lower than 2.7 dB. In the

ON state, at phi = 45°, the difference between the co- and cross-polarization is higher than 12 dB, which is a linear polarization.

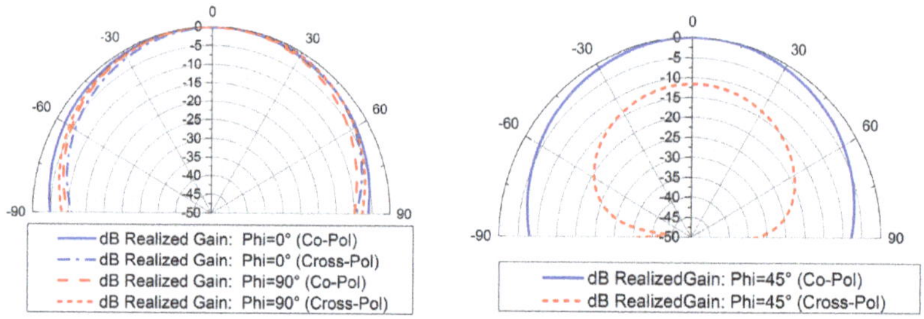

(a) (b)

Figure 12. Simulated radiation patterns of the realized gain. (**a**) In the OFF state in a circular polarization; (**b**) in the ON state in a linear polarization.

3.3. Second Antenna Measurement

A demonstrator was manufactured, and Figure 13 shows the antenna with its SMA connector and its aluminum plate to assure the best maintenance as possible. The measured results are presented in both states in Figure 14. In the OFF state, the -10 dB bandwidth is between 4.75 GHz and 4.86 GHz, with a reflection coefficient of −10 dB at 4.96 GHz. In the ON state, the −10 dB bandwidth is between 4.89 GHz and 4.99 GHz.

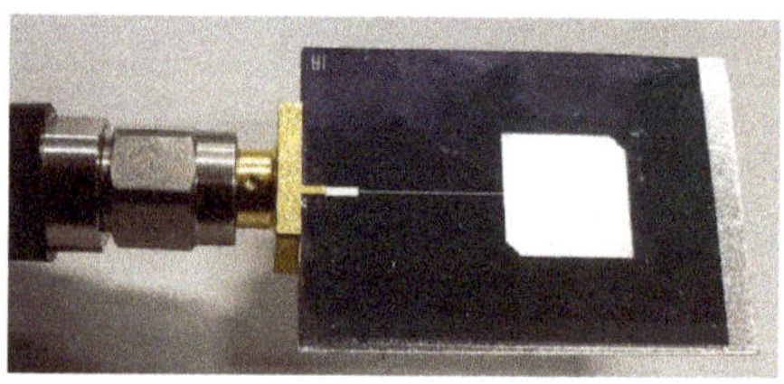

Figure 13. A photograph of the second antenna.

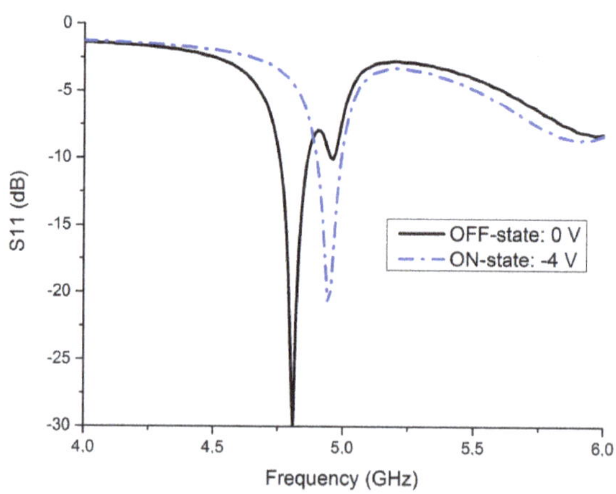

Figure 14. Measured results of the circular to linear polarization reconfigurable antenna.

The radiation patterns of this antenna were measured in both states in an anechoic chamber. The measured results are presented in Figure 16a,b in the OFF state with a circular polarization and in the ON state with a linear polarization, respectively.

Therefore, except for a slight rise in the reflection coefficient in the bandwidth in the OFF state, there is a good agreement between the simulated and measured results, as shown in Figure 15.

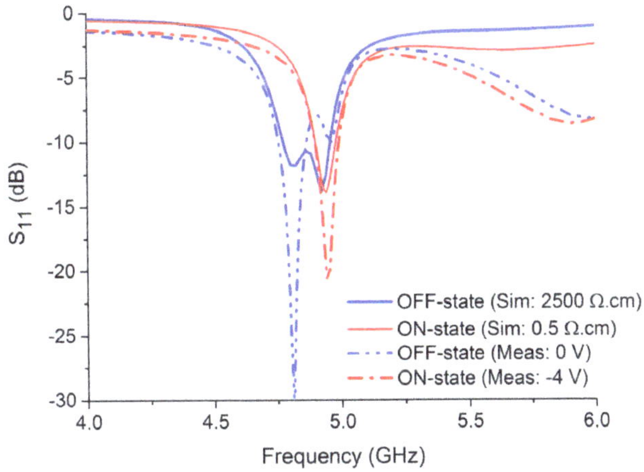

Figure 15. A comparison between simulated and measured results in both states.

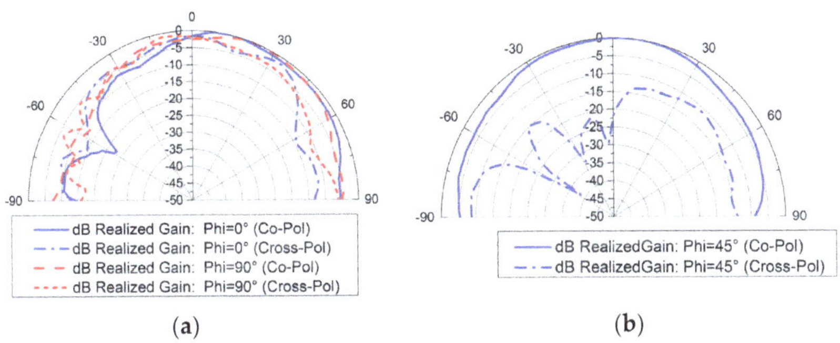

Figure 16. Measured results. (**a**) In the OFF state, in a circular polarization; (**b**) in the ON state, in a linear polarization.

4. Discussion

Table 3 shows a comparison between previous polarization reconfigurable antennas and this work. The antennas are either circularly polarized (CP) in a right-hand circular polarization (RHCP) and/or left-hand circular polarization (LHCP) configuration or linearly polarized (LP) in a horizontal polarization (HP) and/or in a vertical polarization (VP) and/or in a 45° polarization. Few polarization reconfigurable antennae can be found in the literature above 3 GHz. The reconfigurability is achieved thanks to PIN diodes, causing disturbances that increase with the increase in frequency. The proposed patch antennas are designed in a simple manufacture process with only two photolithography masks, allowing small and accurate dimensions of the microstrip lines and particular shapes of the doped areas, permitting an increase in frequency. Moreover, the codesign method given by the choice of a silicon substrate allows for on-chip reconfigurable antennas with integrated junctions in the substrate without requiring additional components.

Table 3. A comparison between previous polarization reconfigurable antennae and this work.

Ref.	Topology of the Antenna	Technology	Freq. (GHz)	Polarization	Active Components
[12]	Patch	Multilayers on PCB	2.4	RHCP/LHCP/LP	4 PIN diodes
[13]	Patch	PCB	1.6	CP/LP	2 PIN diodes
[15]	Circular patch	PCB	2.4	RHCP/LHCP/LP	4 PIN diodes
[18]	Etched ring shape slot	Multilayers on PCB	3.1	RHCP/LHCP/VP/HP	4 PIN diodes
[19]	Patch	PCB	2.4	RHCP/LHCP/LP	2 PIN diodes
[20]	U-slot	PCB	5.7	RHCP/LHCP/VP/HP	2 PIN diodes
[28]	Patch	Multilayers on PCB	2.4	RHCP/LHCP/LP	3 PIN diodes
This study A_1	Patch	Silicon	4.9	LP (90°)/LP (45°)	2 Integrated ScDDAs
This study A_2	Patch	Silicon	4.9	CP/LP	2 Integrated ScDDAs

5. Conclusions

Polarization reconfigurable patch antennas were proposed in this paper as part of a monolithic integration technology. One of the antennae commutes between two different linear polarizations (0° to 45°), while the other one switches from a circular to a linear polarization. The active elements, which are some integrated ScDDAs, overcome the need of classical PIN diodes which limit the parasitic effects. Moreover, the co-design method offers flexibility in the ScDDAs shape (such as the complex shape of the proposed demonstrator), allowing for an optimization of the two commutation states. Two demonstrators were proposed and analyzed to show an overview of the possibilities. The measured results validate the proof of concept. These antennas with polarization reconfigurability could be applied to RADAR applications in a suitable band.

Author Contributions: Conceptualization, R.A.; methodology, R.A.; simulation and characterization, R.A.; validation, R.A., D.L.B. and C.Q.; formal analysis, C.Q.; manufacture, D.S.D.V., V.G., D.V. and J.B.; writing—original draft preparation, R.A.; writing—review and editing, D.L.B. and C.Q.; supervision, D.L.B., C.Q. and J.B. All authors have read and agreed to the published version of the manuscript.

Funding: This work was partly supported by both CERTeM Technological and RENATECH network French Platforms. This publication is also supported by the European Union through the European Regional Development Fund (ERDF), by the Ministry of Higher Education and Research and by Brest Métropole, Brittany, through the CPER Project SOPHIE/STIC & Ondes.

Institutional Review Board Statement: Not applicable.

Informed Consent Statement: Not applicable.

Data Availability Statement: Not applicable.

Acknowledgments: The authors would like to thank the TECHYP platform (the High Performance Computing Cluster of the Lab-STICC) for device simulations. The authors would also like to thank ENSTA-Bretagne engineering school for the use of its anechoic chamber and Fabrice Comblet for his help.

Conflicts of Interest: The authors declare no conflict of interest.

References

1. Singh, G.; Kumar, M. Novel frequency reconfigurable microstrip patch antenna based on a square slot for wireless devices. In Proceedings of the 2012 International Conference on Communication Systems and Network Technologies, Rajkot, India, 11–13 May 2012; pp. 27–30.
2. Bhattacharya, A.; Jyoti, R. Frequency reconfigurable patch antenna using PIN diode at X-band. In Proceedings of the 2015 IEEE 2nd International Conference on Recent Trends in Information Systems (ReTIS), Kolkata, India, 9–11 July 2015; Institute of Electrical and Electronics Engineers Inc.: Piscataway, NJ, USA, 2015; pp. 81–86.
3. Awan, W.A.; Zaidi, A.; Hussain, N.; Khalid, S.; Halima; Baghdad, A. Frequency Reconfigurable patch antenna for millimeter wave applications. In Proceedings of the 2019 2nd International Conference on Computing, Mathematics and Engineering Technologies (iCoMET), Sukkur, Pakistan, 30–31 January 2019; Institute of Electrical and Electronics Engineers Inc.: Piscataway, NJ, USA, 2019.

4. Majid, H.A.; Rahim, M.K.A.; Hamid, M.R.; Murad, N.A.; Ismail, M.F. Frequency reconfigurable microstrip patch antenna. In Proceedings of the IEEE Asia-Pacific Conference on Applied Electromagnetics, (APACE), Melaka, Malaysia, 11–13 December 2012; pp. 342–345.
5. Rawal, P.; Rawat, S. A Bowtie Shaped Frequency Reconfigurable Microstrip Patch Antenna with Wide Coverage Area. In Proceedings of the 2020 5th IEEE International Conference on Recent Advances and Innovations in Engineering (ICRAIE), Jaipur, India, 1–3 December 2020; Institute of Electrical and Electronics Engineers Inc.: Piscataway, NJ, USA, 2020.
6. Sharma, B.L.; Parmar, G.; Kumar, M. Design of frequency reconfigurable microstrip patch antenna for S-Band and C-Band applications. In Proceedings of the 2015 4th International Conference on Reliability, Infocom Technologies and Optimization: Trends and Future Directions (ICRITO), Noida, India, 2–4 September 2015; Institute of Electrical and Electronics Engineers Inc.: Piscataway, NJ, USA, 2015.
7. Majid, H.A.; Rahim, M.K.A.; Hamid, M.R.; Murad, N.A.; Ismail, M.F. Frequency-reconfigurable microstrip patch-slot antenna. *IEEE Antennas Wirel. Propag. Lett.* **2013**, *12*, 218–220. [CrossRef]
8. Lee, C.M.; Jung, C.W. Radiation-pattern-reconfigurable antenna using monopole-loop for fitbit flex wristband. *IEEE Antennas Wirel. Propag. Lett.* **2015**, *14*, 269–272. [CrossRef]
9. Hu, Z.; Liu, S.; An, T.; Yang, Y.; He, W. A Novel Frequency Reconfigurable Slot Antenna Based on S-PINs. In Proceedings of the 2021 International Conference on Microwave and Millimeter Wave Technology (ICMMT), Nanjing, China, 23–26 May 2021; pp. 1–3.
10. An, T.; Liu, S.; Hu, Z.; Yang, Y. Design of a Frequency and Polarization Reconfigurable Slot Antenna with S-PINs. In Proceedings of the 2021 International Applied Computational Electromagnetics Society (ACES-China) Symposium, Chengdu, China, 28–31 July 2021; pp. 1–2.
11. Trad, I.B.; Floc'h, J.M.; Rmili, H.; Drissi, M.; Choubani, F. Design of reconfigurable radiation pattern dipole antenna with director and reflector elements for telecommunication systems. In Proceedings of the 2013 Loughborough Antennas & Propagation Conference (LAPC), Loughborough, UK, 11–12 November 2013; pp. 117–121.
12. Alreshaid, A.T.; Cui, Y.; Bahr, R.; Tentzeris, M.M.; Sharawi, M.S. A Millimeter Wave Tri-Polarized Patch Antenna with a Bandwidth-Enhancing Parasitic Element. In Proceedings of the 2021 IEEE International Symposium on Antennas and Propagation and USNC-URSI Radio Science Meeting (APS/URSI), Singapore, 4–10 December 2021; IEEE: Piscataway, NJ, USA, 2021; pp. 1051–1052.
13. Sung, Y.J. Reconfigurable patch antenna for polarization diversity. *IEEE Trans. Antennas Propag.* **2008**, *56*, 3053–3054. [CrossRef]
14. Tolin, E.; Vipiana, F.; Bruni, S.; Bahr, A. Polarization reconfigurable patch antenna for compact and low cost UHF RFID reader. In Proceedings of the 2019 IEEE International Conference on RFID Technology and Applications (RFID-TA), Pisa, Italy, 25–27 September 2019; Institute of Electrical and Electronics Engineers Inc.: Piscataway, NJ, USA, 2019; pp. 128–130.
15. Osman, M.N.; Rahim, M.K.A.; Yussof, M.F.M.; Hamid, M.R.; Majid, H.A. Polarization reconfigurable cross-slots circular patch antenna. In Proceedings of the 2013 Proceedings of the International Symposium on Antennas & Propagation, Nanjing, China, 23–25 October 2013; pp. 1252–1255.
16. Nguyen-Trong, N.; Mobashsher, A.T.; Abbosh, A.M. Reconfigurable shorted patch antenna with polarization and pattern diversity. In Proceedings of the 2018 Australian Microwave Symposium (AMS), Brisbane, QLD, Australia, 6–7 February 2018; Institute of Electrical and Electronics Engineers Inc.: Piscataway, NJ, USA, 2018; pp. 27–28.
17. Wong, H.; Lin, W.; Wang, X.; Lu, M. LP and CP polarization reconfigurable antennas for modern wireless applications. In Proceedings of the 2017 International Symposium on Antennas and Propagation (ISAP), Phuket, Thailand, 30 October–2 November 2017; Institute of Electrical and Electronics Engineers Inc.: Piscataway, NJ, USA, 2017; pp. 1–2.
18. Song, T.; Lee, Y.; Ga, D.; Choi, J. A polarization reconfigurable microstrip patch antenna using PIN diodes. In Proceedings of the 2012 Asia-Pacific Microwave Conference Proceedings (APMC), Kaohsiung, Taiwan, 4–7 December 2012; pp. 616–618.
19. Yang, X.X.; Shao, B.C.; Yang, F.; Elsherbeni, A.Z.; Gong, B. A polarization reconfigurable patch antenna with loop slots on the ground plane. *IEEE Antennas Wirel. Propag. Lett.* **2012**, *11*, 69–72. [CrossRef]
20. Qin, P.Y.; Weily, A.R.; Guo, Y.J.; Liang, C.H. Polarization reconfigurable U-slot patch antenna. *IEEE Trans. Antennas Propag.* **2010**, *58*, 3383–3388. [CrossRef]
21. Mak, K.M.; Lai, H.W. Polarization reconfigurable slotted circular patch antenna. In Proceedings of the 2016 10th European Conference on Antennas and Propagation (EuCAP), Davos, Switzerland, 10–15 April 2016; Institute of Electrical and Electronics Engineers Inc.: Piscataway, NJ, USA, 2016.
22. Allanic, R.; Le Berre, D.; Quéré, Y.; Quendo, C.; Chouteau, D.; Grimal, V.; Valente, D.; Billoué, J. Impact of the doped areas sizes in the performances of microwave SPST switches integrated in a silicon substrate. In Proceedings of the 2018 IEEE 22nd Workshop on Signal and Power Integrity (SPI), Brest, France, 22–25 May 2018.
23. Allanic, R.; Le Berre, D.; Quere, Y.; Quendo, C.; Chouteau, D.; Grima, V.; Valente, D.; Billoue, J. A Fully Integrated Frequency Reconfigurable Antenna Co-Designed as a Whole Circuit on Silicon. In Proceedings of the 2018 Asia-Pacific Microwave Conference (APMC), Kyoto, Japan, 6–9 November 2018; IEEE: Piscataway, NJ, USA, 2018; pp. 1256–1258.
24. Allanic, R.; Le Berre, D.; Quere, Y.; Quendo, C.; Chouteau, D.; Grimal, V.; Valente, D.; Billoue, J. Three-State Microwave Tunable Resonator Integrating Several Active Elements on Silicon Technology in a Global Design. *IEEE Microw. Wirel. Compon. Lett.* **2018**, *28*, 141–143. [CrossRef]
25. Allanic, R.; Le Berre, D.; Quere, Y.; Quendo, C.; Chouteau, D.; Grimal, V.; Valente, D.; Billoue, J. Continuously tunable resonator using a novel triangular doped area on a silicon substrate. *IEEE Microw. Wirel. Components Lett.* **2018**, *28*, 1095–1097. [CrossRef]

26. Krupka, J.; Breeze, J.; Alford, N.M.; Centeno, A.E.; Jensen, L.; Claussen, T. Measurements of Permittivity and Dielectric Loss Tangent of High Resistivity Float Zone Silicon at Microwave Frequencies. In Proceedings of the 2006 International Conference on Microwaves, Radar & Wireless Communications, Krakow, Poland, 22–24 May 2006; pp. 1097–1100.
27. Allanic, R.; le Berre, D.; Quere, Y.; Quendo, C.; Chouteau, D.; Grimal, V.; Valente, D.; Billoue, J. A Novel Synthesis for Bandwidth Switchable Bandpass Filters Using Semi-Conductor Distributed Doped Areas. *IEEE Access* **2020**, *8*, 122599–122609. [CrossRef]
28. Wang, K.X.; Wong, H. A Reconfigurable CP/LP Antenna With Cross-Probe Feed. *IEEE Antennas Wirel. Propag. Lett.* **2017**, *16*, 669–672. [CrossRef]

electronics

Article

Dual-Band High-Gain Shared-Aperture Antenna Integrating Fabry-Perot and Reflectarray Mechanisms

Xianjin Yi [1], Lin Zhou [2], Shuji Hao [3] and Xing Chen [1,*]

[1] School of Electronics and Information Engineering, Sichuan University, Chengdu 610064, China; yixianjin@stu.scu.edu.cn
[2] School of Electrical and Electronic Engineering, Nanyang Technological University, Singapore 639798, Singapore; tony.zhou@ntu.edu.sg
[3] National Key Laboratory of Electromagnetic Environment, China Research Institute of Radiowave Propagation, Qingdao 266107, China; hsj80586@126.com
* Correspondence: chenxing@scu.edu.cn

Abstract: This work presents a dual-band high-gain shared-aperture antenna. The proposed antenna integrates both the Fabry-Perot and reflectarray mechanisms; the antenna works as a Fabry-Perot cavity antenna (FPCA) in the *S*-band (2.45 GHz) and as a reflectarray antenna (RA) in the *X*-band (10 GHz). The antenna has a simple structure made up of only two printed circuit board layers. The bottom layer acts as a source antenna, a ground plane for the FPCA, and as a reflective surface for the RA. The upper layer contains the source antenna for the RA and serves as a partially reflective superstrate for the FPCA. The FPCA and RA thus share the same physical aperture but function independently. As an example, we design, fabricate, and characterize an antenna that operates at 2.45 and 10 GHz with an aperture size of 300×300 mm^2. The measured results are found to be in good agreement with the simulations. We show that the proposed antenna achieves a gain of 16.21 dBi at 2.45 GHz and 21.6 dBi at 10 GHz with a -10 dB impedance bandwidths of 2.39–2.66 GHz and 9.40–10.28 GHz. The isolation between the two antenna ports is found to be larger than 30 dB.

Keywords: dual-band; Fabry-Perot cavity antenna (FPCA); high gain; reflectarray antenna (RA); shared-aperture antenna

Citation: Yi, X.; Zhou, L.; Hao, S.; Chen, X. Dual-Band High-Gain Shared-Aperture Antenna Integrating Fabry-Perot and Reflectarray Mechanisms. *Electronics* **2022**, *11*, 2017. https://doi.org/10.3390/electronics11132017

Academic Editors: Naser Ojaroudi Parchin, Chan Hwang See and Raed A. Abd-Alhameed

Received: 31 May 2022
Accepted: 25 June 2022
Published: 27 June 2022

1. Introduction

Shared-aperture antennas have attracted considerable attention in recent years. Due to the advantages of dual-band/multiband, high aperture utilization rate, low cost and ease of fabrication, the shared-aperture antenna is appropriate for use in base stations [1,2], wireless communications [3,4], radio-frequency identification readers [5], and aperture radars [6]. Various types of shared-aperture antenna have been proposed, including array antennas [7,8], RAs [9,10] and metasurface-based antennas [11]. There are many recent works that have investigated the design of dual-band shared-aperture FPCAs [12–16].

Ref. [12] designed a dual-band FPCA with different polarizations. By using two identical dielectric slabs as superstrate, the proposed antenna realized left-hand and right-hand circular polarization radiations at 9.65 and 11.75 GHz, respectively. Ref. [13] reported on a dual-band FPCA with two frequency selective surface layers. Each superstrate was active at a particular frequency and transparent at another frequency. Ref. [15] proposed a shared-surface dual-band antenna. By embedding a partially reflective surface unit cell into the metasurface, *S* and *Ka* band radiations were realized. Ref. [16] used a shared aperture for an FPCA and folded transmitarray antenna. A four-layered, metallic, double-ring structure was used as the partially reflective surface and phase-shifting surface. Although the above designs can realize good dual-band working properties and high aperture reuse efficiency, the gain at low frequencies is not high [14–16]. In addition, all these designs are based on multilayered superstrate structures [13,14,16].

An FPCA usually comprises a feed antenna, a partially reflective superstrate and a metal ground, as shown in Figure 1a. Electromagnetic waves radiated from the feed antenna experience multiple reflections within the cavity. The distance between the superstrate and metal ground is nearly half the wavelength of the working frequency [17,18]. An RA usually comprises a feed antenna and a reflective surface. The reflective surface is illuminated by the feed antenna, and a planar phase surface is formed in front of the aperture, as shown in Figure 1b.

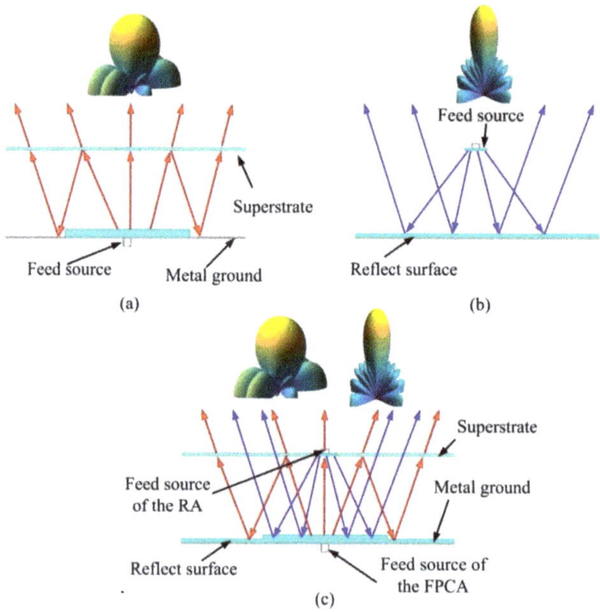

Figure 1. Working principle of a (**a**) Fabry-Perot cavity antenna (FPCA), (**b**)reflectarray antenna (RA), and (**c**) shared-aperture antenna.

This paper explores the design of an FPCA and an RA that share the same physical aperture, as shown in Figure 1c. The feed source of the RA is located at the center of the substrate of the FPCA, and the reflective surface of the RA is combined with the metal ground of the FPCA. The proposed antenna does not need a power division network, and only a single-layer superstrate is used. The FPCA and RA work independently at *S*-band (2.45 GHz) and *X*-band (10 GHz) frequencies, respectively. This study is organized as follows: Section 2 introduces the design principles of the shared-aperture antenna. The simulations and measurements are then presented in Section 3.

2. Shared-Aperture Antenna Design

2.1. FPCA Design

Partially reflective superstrates with a high reflection coefficient should be considered to realize a high gain in FPCAs [19]. The relative gain and resonant cavity height of an FPCA can be calculated as,

$$D = 10\log(\frac{1+R}{1-R}) \tag{1}$$

$$H = \frac{\lambda 1}{4\pi}(\varphi R + \varphi G + 2N\pi) \tag{2}$$

where R is the reflection coefficient of the superstrate, D is the relative gain, λ_1 is the working wavelength, H is the cavity height, φ_R and φ_G represent the reflection phases of the superstrate and ground, respectively. The FPCA designed in this study is intended to work at a frequency of 2.45 GHz. To decrease the complexity of the shared-aperture design, a single-layer uniform superstrate is used. The bandwidth of the FPCA follows,

$$Bandwidth = \frac{\lambda 1}{2\pi H} \cdot \frac{1-R}{\sqrt{R}} \quad (3)$$

Figure 2 shows the structure and simulated model of the superstrate unit cell; this structure comprises a 2-mm-thick dielectric substrate and a 0.035-mm-thick metal layer. The unit cell size of the substrate is 20 × 20 mm^2, and the dielectric constant of the substrate is 2.65. A square patch with a side length of 18 mm is cut away from the middle square patch with a side length of 16 mm to realize a hollow square ring. Figure 3 shows the magnitude of the simulated reflection coefficient, $|S_{11}|$, and magnitude of the transmission coefficient, $|S_{21}|$, of the superstrate unit cell for frequencies from 2 to 12 GHz. The value of $|S_{11}|$ at 2.45 GHz is found to be −0.55 dB. The value of $|S_{21}|$ at 10 GHz is greater than −0.45 dB. The superstrate is almost transparent at 10 GHz.

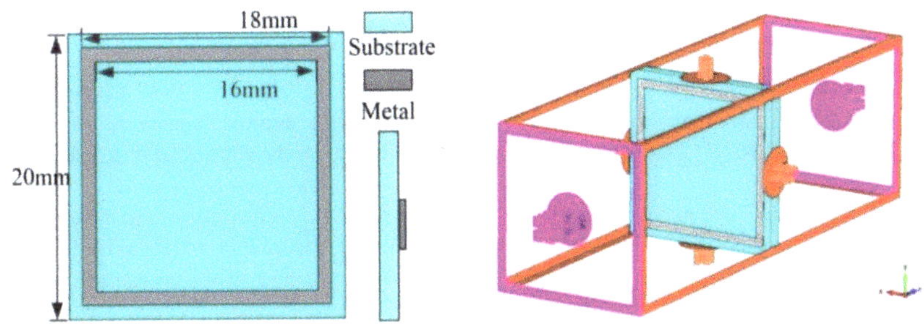

Figure 2. Structure and simulated model of the superstrate unit cell.

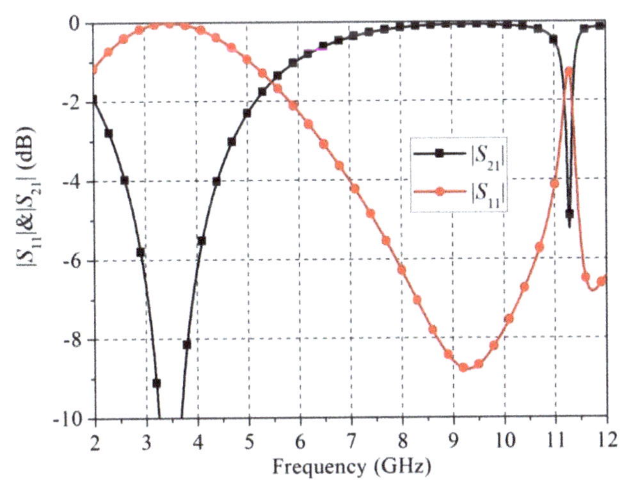

Figure 3. Simulated $|S_{11}|$ and $|S_{21}|$ of a superstrate unit cell.

The whole superstrate is made up of an array of 15 × 15 units, and the aperture size is 300 × 300 mm^2. The simulated values of φ_R and φ_G at 2.45 GHz were found to be −152.68° and 180°, respectively. According to Equation (2), the theoretical resonant cavity height, H, is 65.87 mm, which is nearly half the wavelength at a frequency of 2.45 GHz.

2.2. RA Design

RAs often consist of a feed source and a reflective surface. In this study, a rectangular microstrip antenna is used as the feed source. The phase distribution on the reflective surface should satisfy the parabolic distribution [20],

$$\varphi(x,y) = \frac{2\pi}{\lambda_2}(\sqrt{x^2+y^2+H^2} - H) \pm 2N\pi + \varphi_0 \tag{4}$$

where (x,y) represent the position of the reflective surface unit cell, φ_0 is the reflection phase at the reference point, λ_2 is the working wavelength of the RA, and H is the focal length. In the design used here, H is also the height of the FPCA.

Figure 4 shows the structure and the simulated model of the reflective surface unit cell considered here; this structure comprises a 3-mm-thick dielectric substrate and two metal layers. The unit cell size of the substrate is 10×10 mm^2, and the dielectric constant of the substrate is 2.65. The two metal layers are on either side of the substrate. One of the metal layers is full metal and the other has a modified I-shape structure [21]. The diameter of the I shape is 9 mm, and its width, c, is 1.5 mm. Figure 5 shows the simulated reflection phase distribution of the reflective surface unit cell at a frequency of 10 GHz. By changing the arc angle, a, and the rotated angle, b, the unit cell can realize the required reflection phase range of 0°–360°. It should be noted that the I shape realizes an orthogonal polarization conversion. When the incident wave is of x-polarization, the main reflected radiation wave is of y-polarization. Figure 6 shows the simulated reflection amplitudes of the reflective surface unit cell with co-polarization and cross-polarization. When the angle a is changed from 27° to 87°, the reflection amplitudes corresponding to radiation of cross-polarization are greater than −0.96 dB at 10 GHz. On the other hand, the reflective surface provides a co-polarization reflection at around 2.45 GHz. The reflection amplitudes of the co-polarization radiation maintain the value of −0.001 dB at 2.45 GHz with the angle a changing.

Figure 7 shows the simulated reflection amplitude and phase of a reflect surface unit cell under different incident angles. It is found when the incident angle is less than 60° that the reflection amplitude and phase of the reflective surface unit cell are constant at frequencies of around 10 GHz. The phase difference for b = 45° and b = 135° is independent of the incident angles. The amplitude will decrease when the incident angle is greater than 60°, which may affect the efficiency of the reflectarray. However, in order to realize a high gain (greater than 20 dBi), we retain the use of a reflective surface with a large aperture size.

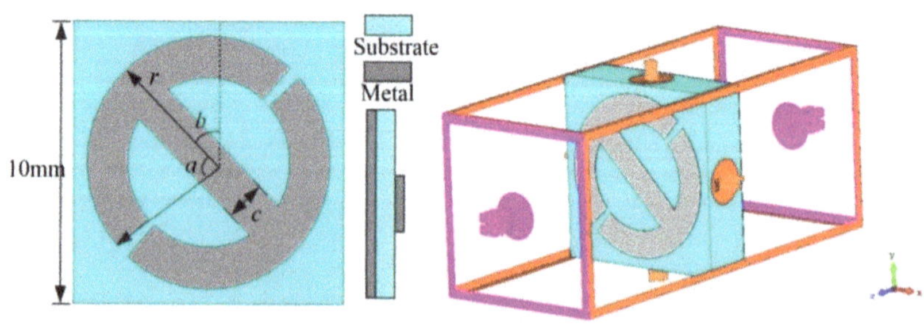

Figure 4. Structure and simulated model of the reflect surface unit cell.

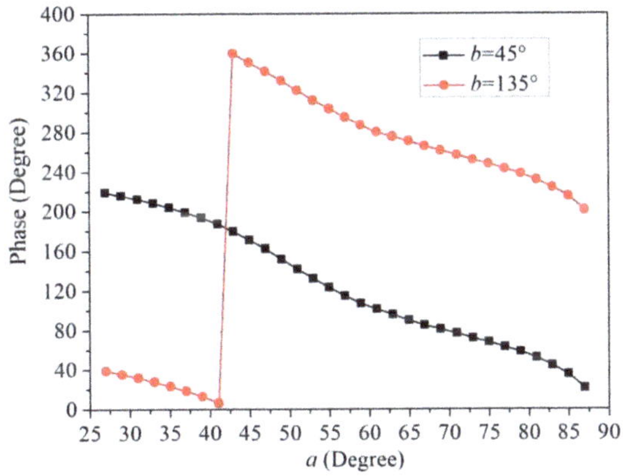

Figure 5. Simulated 10-GHz reflection phase distributions of a reflect surface unit cell.

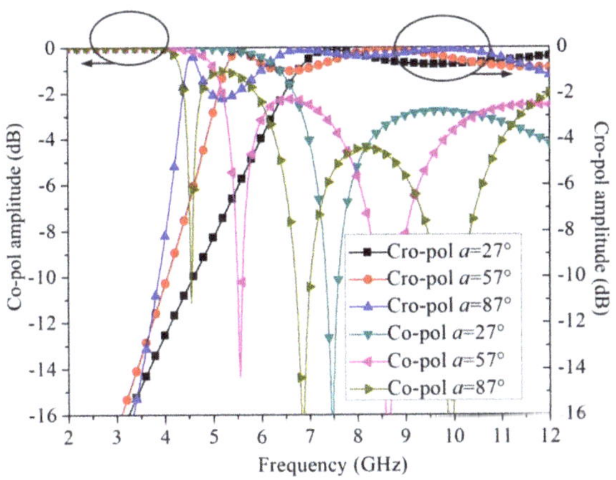

Figure 6. Simulated reflection amplitudes of a reflect surface unit cell with co-polarization and cross-polarization.

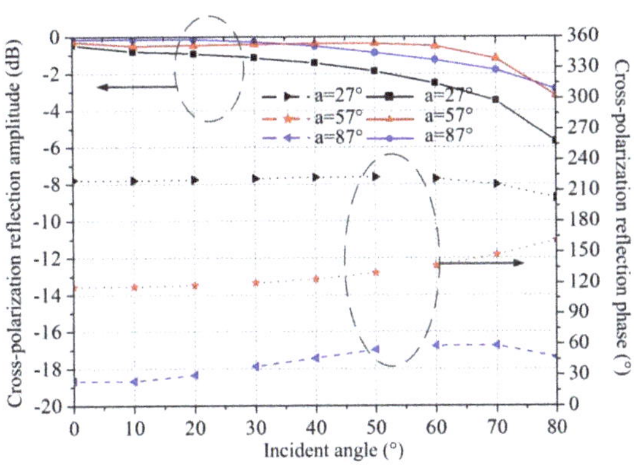

Figure 7. Simulated cross-polarization reflection amplitude and phase of a reflect surface unit cell under different incident angles.

3. Measurement Results and Discussion

The feed source of the FPCA is a microstrip antenna which is located at the center of the lower layer. After optimizing with the method of genetic algorithm, the resonant cavity height, H, is set to be 65 mm. The feed source of the RA is located at the center of the superstrate of the FPCA, and the phase distribution on the reflective surface can be calculated using Equation (3). Figure 8 shows the obtained phase distribution.

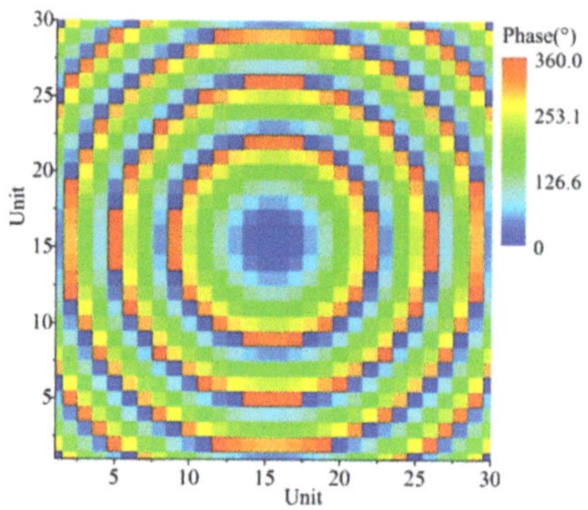

Figure 8. Phase-shift distribution on the reflectarray surface.

The shared-aperture antenna designed here was then fabricated and measured. The fabricated superstrate, reflective surface, and antenna prototype are shown in Figure 9a–c, respectively. Figure 9a,b also shows the properties of the feed antennas. All of the substrates are made of polytetrafluoroethylene ($\varepsilon_r = 2.65$, tan $\delta = 0.001$). Nylon screws were used to control the cavity height and fix the whole antenna. The values of $|S_{11}|$ and $|S_{21}|$ for the antenna were then measured using an Agilent E8362B Network Analyzer. As shown in Figure 10, the measured 10-dB return loss bandwidth of the prototype was found to be 10.82% from 2.385 to 2.658 GHz and 8.94% from 9.4 to 10.28 GHz. The value of $|S_{21}|$ in both the frequency bands considered here was lower than −30 dB, which indicates good isolation between the two antenna ports.

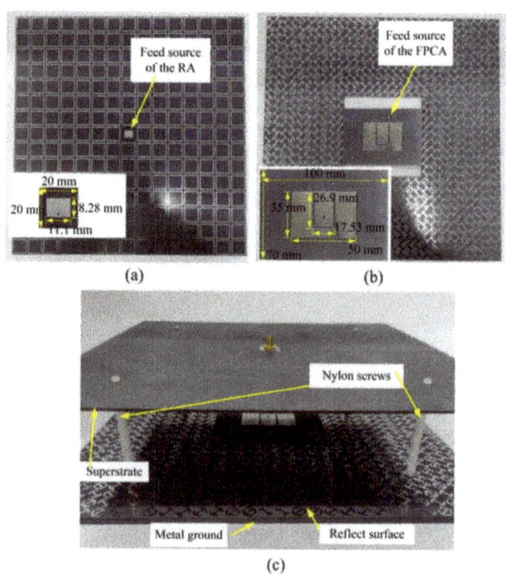

Figure 9. Photographs of (**a**) the fabricated superstrate, (**b**) reflect surface, and (**c**) antenna prototype.

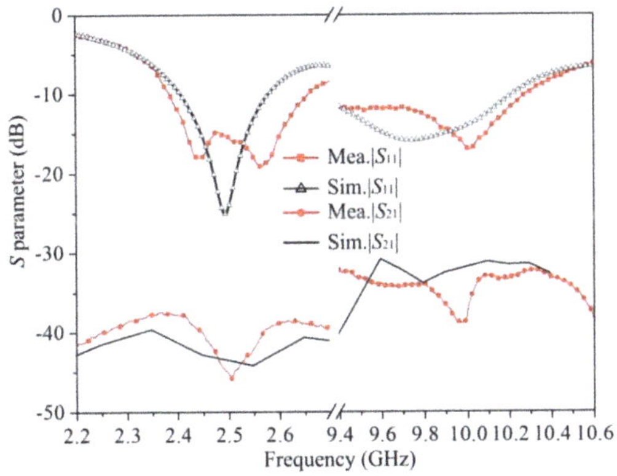

Figure 10. Simulated and measured S parameters.

As shown in Figure 11, the simulated gains at frequencies of 2.45 and 10 GHz were found to be 17.01 and 22.62 dBi, respectively. The measured gains were 16.21 dBi at 2.45 GHz and 21.6 dBi at 10 GHz, and the corresponding radiating efficiencies were 83.2% and 79% [22], respectively. The measured gains were lower than those predicted in the simulations; these reduced values may have been caused by fabrication and measurement errors. The aperture efficiencies at 2.45 and 10 GHz were found to be 55.39% and 11.55% [23], respectively. The feed antenna of the FPCA occupied one-ninth of the reflective surface, which influenced the feed illumination of the RA. According to the simulations, the gain at 10 GHz can be increased from 22.62 to 25.21 dBi by removing the feed structure. The RA was found to have a large subtended (half) angle of 72.9°, which leads to a high spillover efficiency but low taper efficiency, and therefore a low aperture efficiency [24]. The deteriorated performance of the reflective surface unit cell under large incident angles, dielectric loss, and center-feed blocking will also decrease the gain of the RA. The 3-dB gain bandwidths of the two frequency bands were found to be 8.16% and 10.1%. The equivalent 1-dB bandwidths were 1.08% and 6.18%, respectively. When the working frequency deviates from 2.45 GHz, the fixed cavity height will no longer match the optimal resonance height, which will result in a decreased gain and thus a narrow gain bandwidth of the FPCA. In contrast, the gains of the RA depend on the feed source and the reflective surface. The feed source and reflective surface can work well at frequencies of around 10 GHz, thus the RA can exhibit relatively stable gains and a wide gain bandwidth. Figure 12 shows the radiation efficiency of the proposed antenna.

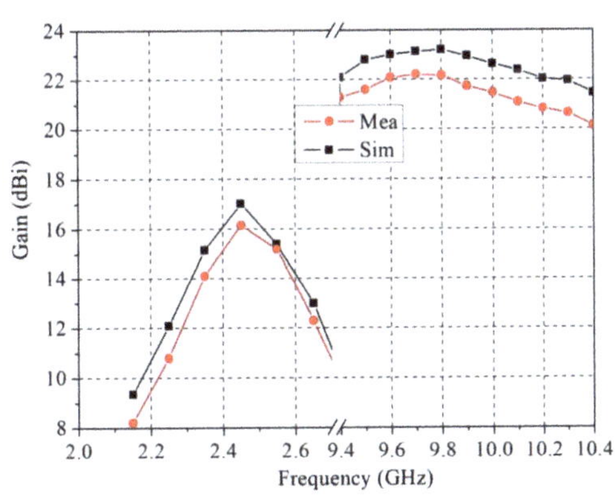

Figure 11. Simulated and measured gains of the proposed antenna at around 2.45 and 10 GHz.

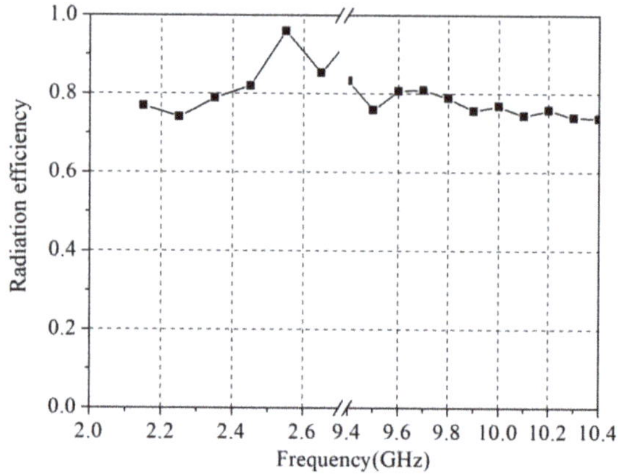

Figure 12. Radiation efficiency of the proposed antenna.

Figures 13 and 14 show the simulated and measured pattern at 2.45 and 10 GHz, respectively. The half-power beamwidth of the E- and H-planes were 22.6° and 21.44° at 2.45 GHz. The beamwidths were 6.31° and 5.46° at 10 GHz. The cross-polarized gains at 2.45 and 10 GHz were 25 dB and 15 dB lower than the co-polarized gains in both planes, respectively. The sidelobe level at 10 GHz was relatively high, which may have also been caused by the feed antenna of the FPCA.

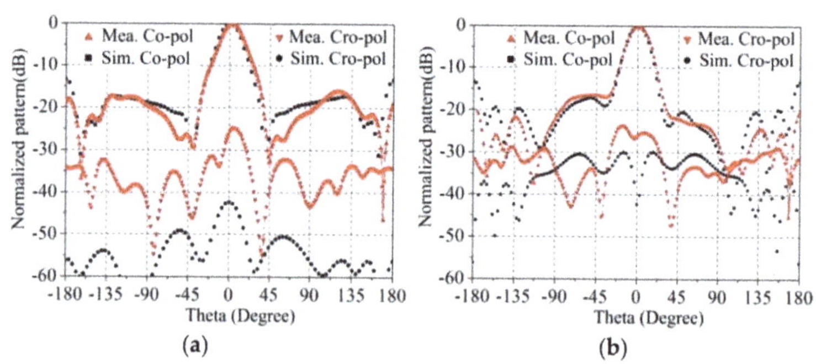

Figure 13. Simulated and measured radiation patterns (dB) of (**a**) E plane and (**b**) H plane at 2.45 GHz.

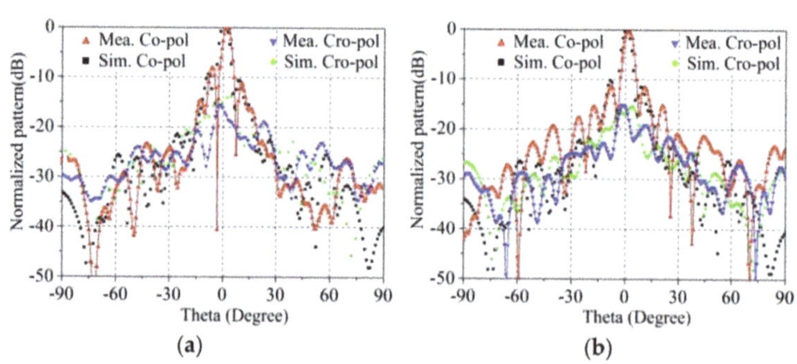

Figure 14. Simulated and measured radiation patterns (dB) of (**a**) E plane and (**b**) H plane at 10 GHz.

Table 1 gives a comparison with some similar works. Compared with [14,16], and [25], our proposed antenna produced high gain in both frequency bands. Although the gain improvement is not high for [13], there is no feeding network in the antenna, and only a single-layer superstrate is used.

Table 1. Comparison with similar works.

Ref.	Frequency (GHz)	Bandwidth	Layer of Superstrate	Gain (dBi)	Size ($\lambda_{low} \times \lambda_{low}$)	Feeding Network
[25]	2.7/9.7	7.7%/4.8%	2	8.4/21.8	1.125 × 1.125	Yes
[14]	3.45/5	1.5%/2.2%	2	13.7/16.8	1.6 × 1.6	No
[16]	10/28	4%/9%	4	13.8/23.6	2.83 × 2.83	No
[13]	5.3/9.6	7.3%/5.7%	2	16.4/20	2.47 × 2.47	Yes
This work	2.45/10	10.8%/8.9%	1	16.2/21.6	2.45 × 2.45	No

4. Conclusions

In this work, a dual-band high-gain shared-aperture antenna integrating the Fabry-Perot and reflectarray mechanisms was proposed. The design hybridizes an FPCA working at 2.45 GHz and an RA working at 10 GHz. The feed source of the RA is located at the center of the substrate of the FPCA, and the reflective surface of the RA is combined with the metal ground of the FPCA. In this work, a shared-aperture antenna with an aperture size of 300 × 300 mm^2 was designed, fabricated, and measured. The proposed antenna was found to have a gain of 16.21 dBi at 2.45 GHz and 21.6 dBi at 10 GHz. The isolation between the two antenna ports was found to be greater than 30 dB. The antenna also has a simple structure with only a single-layer superstrate, which is preferable for practical applications.

Author Contributions: Conceptualization, X.Y. and X.C.; writing—original draft preparation, X.Y., L.Z. and S.H.; writing—review and editing. All authors have read and agreed to the published version of the manuscript.

Funding: This work was supported by Stable-Support Scientific Project of China Research Institute of Radiowave Propagation (Grant No. A132101W09).

Data Availability Statement: Data is contained within the article.

Conflicts of Interest: The authors declare that they have no conflicts of interest.

References

1. He, D.; Yu, Q.; Chen, Y.; Yang, S. Dual-band shared-aperture base station antenna array with electromagnetic transparent antenna elements. *IEEE Trans. Antennas Propag.* **2021**, *69*, 5596–5606. [CrossRef]
2. Chen, Y.; Zhao, J.; Yang, S. A novel stacked antenna configuration and its applications in dual-band shared-aperture base station antenna array designs. *IEEE Trans. Antennas Propag.* **2019**, *67*, 7234–7241. [CrossRef]
3. Zhang, J.F.; Cheng, Y.J.; Ding, Y.R.; Bai, C.X. A dual-band shared-aperture antenna with large frequency ratio, high aperture reuse efficiency, and high channel isolation. *IEEE Trans. Antennas Propag.* **2018**, *67*, 853–860. [CrossRef]
4. Mao, C.X.; Gao, S.; Wang, Y.; Chu, Q.X.; Yang, X.X. Dual-band circularly polarized shared-aperture array for C-/X-band satellite communications. *IEEE Trans. Antennas Propag.* **2017**, *65*, 5171–5178. [CrossRef]
5. Zhang, J.; Shen, Z. Dual-band shared-aperture UHF/UWB RFID reader antenna of circular polarization. *IEEE Trans. Antennas Propag.* **2018**, *66*, 3886–3893. [CrossRef]
6. Chen, Y.; Vaughan, R.G. Dual-Polarized L-Band and Single-Polarized X-Band Shared-Aperture SAR Array. *IEEE Trans. Antennas Propag.* **2018**, *66*, 3391–3400. [CrossRef]
7. Mao, C.X.; Gao, S.; Wang, Y.; Luo, Q.; Chu, Q.X. A shared-aperture dual-band dual-polarized filtering-antenna-array with improved frequency response. *IEEE Trans. Antennas Propag.* **2017**, *65*, 1836–1844. [CrossRef]
8. Kwon, G.; Park, J.; Kim, D.; Hwang, K.C. Optimization of a shared-aperture dual-band transmitting/receiving array antenna for radar applications. *IEEE Trans. Antennas Propag.* **2017**, *65*, 7038–7051. [CrossRef]
9. Smith, T.; Gothelf, U.; Kim, O.S.; Breinbjerg, O. An FSS-backed 20/30 GHz circularly polarized reflectarray for a shared aperture L-and Ka-band satellite communication antenna. *IEEE Trans. Antennas Propag.* **2013**, *62*, 661–668. [CrossRef]
10. Liu, Y.; Cheng, Y.J.; Zhao, M.H.; Fan, Y. Dual-band shared-aperture high-efficiency reflectarray antenna based on structure-reuse technique. *IEEE Antennas Wirel. Propag. Lett.* **2021**, *20*, 366–370. [CrossRef]

11. Li, T.; Chen, Z.N. Metasurface-based shared-aperture 5G S-/K-band antenna using characteristic mode analysis. *IEEE Trans. Antennas Propag.* **2018**, *66*, 6742–6750. [CrossRef]
12. Zeb, B.A.; Nikolic, N.; Esselle, K.P. A high-gain dual-band EBG resonator antenna with circular polarization. *IEEE Antennas Wirel. Propag. Lett.* **2014**, *14*, 108–111. [CrossRef]
13. Qin, F.; Gao, S.S.; Luo, Q.; Mao, C.-X.; Gu, C.; Wei, G.; Xu, J.; Li, J.; Wu, C.; Zheng, K.; et al. A simple low-cost shared-aperture dual-band dual-polarized high-gain antenna for synthetic aperture radars. *IEEE Trans. Antennas Propag.* **2016**, *64*, 2914–2922. [CrossRef]
14. Liu, Z.G.; Yin, R.J.; Ying, Z.N.; Lu, W.B.; Tseng, K.C. Dual-band and shared-aperture Fabry–Pérot cavity antenna. *IEEE Antennas Wirel. Propag. Lett.* **2021**, *20*, 1686–1690.
15. Li, T.; Chen, Z.N. Shared-surface dual-band antenna for 5G applications. *IEEE Trans. Antennas Propag.* **2019**, *68*, 1128–1133. [CrossRef]
16. Mei, P.; Zhang, S.; Pedersen, G.F. A dual-polarized and high-gain X-/Ka-band shared-aperture antenna with high aperture reuse efficiency. *IEEE Trans. Antennas Propag.* **2020**, *69*, 1334–1344. [CrossRef]
17. Decoster, B.; Maes, S.; Cuiñas, I.; Sánchez, M.G.; Caldeirinha, R.; Verhaevert, J. Dual-band single-layer fractal frequency selective surface for 5G applications. *Electronics* **2021**, *10*, 2880. [CrossRef]
18. Al-Gburi, A.J.A.; Ibrahim, I.M.; Zakaria, Z.; Abdulhameed, M.K.; Saeidi, T. Enhancing gain for UWB antennas using FSS: A systematic review. *Mathematics* **2021**, *9*, 3301. [CrossRef]
19. Qin, P.Y.; Ji, L.Y.; Chen, S.L.; Guo, Y.J. Dual-polarized wideband Fabry–Pérot antenna with quad-layer partially reflective surface. *IEEE Antennas Wirel. Propag. Lett.* **2018**, *17*, 551–554. [CrossRef]
20. Yu, S.; Liu, H.; Li, L. Design of near-field focused metasurface for high-efficient wireless power transfer with multifocus characteristics. *IEEE Trans. Ind. Electron.* **2018**, *66*, 3993–4002. [CrossRef]
21. Jia, S.L.; Wan, X.; Su, P.; Zhao, Y.J.; Cui, T.J. Broadband metasurface for independent control of reflected amplitude and phase. *AIP Adv.* **2016**, *6*, 045024. [CrossRef]
22. Lai, Q.; Almpanis, G.; Fumeaux, C.; Benedickter, H.; Vahldieck, R. Comparison of the radiation efficiency for the dielectric resonator antenna and the microstrip antenna at Ka band. *IEEE Trans. Antennas Propag.* **2008**, *56*, 3589–3592.
23. Jiang, M.; Hong, W.; Zhang, Y.; Yu, S.; Zhou, H. A folded reflectarray antenna with a planar SIW slot array antenna as the primary source. *IEEE Trans. Antennas Propag.* **2014**, *62*, 3575–3583. [CrossRef]
24. Pozar, D.M.; Targonski, S.D.; Syrigos, H.D. Design of millimeter wave microstrip reflectarrays. *IEEE Trans. Antennas Propag.* **1997**, *45*, 287–296. [CrossRef]
25. Kothapudi, V.K.; Kumar, V. A multi-layer S/X-band shared aperture antenna array for airborne synthetic aperture radar applications. *Int. J. RF Microw. Comput. Aided Eng.* **2021**, *31*, e22720. [CrossRef]

sensors

MDPI

Article

Performance Analysis of Wearable Dual-Band Patch Antenna Based on EBG and SRR Surfaces

Abdul Wajid [1], Ashfaq Ahmad [2], Sadiq Ullah [1], Dong-you Choi [2,*] and Faiz Ul Islam [3]

1 Department of Telecommunication Engineering, University of Engineering and Technology Mardan, Mardan 23200, Pakistan; wajiduetpeshawar@gmail.com (A.W.); sadiqullah@uetmardan.edu.pk (S.U.)
2 Communication and Wave Propagation Laboratory, Department of Information and Communication Engineering, Chosun University, Gwangju 61452, Korea; ashfaquetb11@gmail.com
3 School of Automation, Nanjing University of Science and Technology, Nanjing 210094, China; uet_faiz@yahoo.com
* Correspondence: dychoi@chosun.ac.kr

Abstract: This paper presents the performance comparison of a dual-band conventional antenna with a split-ring resonator (SRR)- and electromagnetic bandgap (EBG)-based dual-band design operating at 2.4 GHz and 5.4 GHz. The compactness and dual-frequency operation in the legacy Wi-Fi range of this design make it highly favorable for wearable sensor network-based Internet of Things (IoT) applications. Considering the current need for wearable antennas, wash cotton (with a relative permittivity of 1.51) is used as a substrate material for both conventional and metamaterial-based antennas. The radiation characteristics of the conventional antenna are compared with the EBG and SRR ground planes-based antennas in terms of return loss, gain, and efficiency. It is found that the SRR-based antenna is more efficient in terms of gain and surface wave suppression as well as more compact in comparison with its two counterparts. The compared results are found to be based on two distinct frequency ranges, namely, 2.4 GHz and 5.4 GHz. The suggested SRR-based antenna exhibits improved performance at 5.4 GHz, with gains of 7.39 dbi, bandwidths of 374 MHz, total efficiencies of 64.7%, and HPBWs of 43.2 degrees. The measurements made in bent condition are 6.22 db, 313 MHz, 52.45%, and 22.3 degrees, respectively. The three considered antennas (conventional, EBG-based, and SRR-based) are designed with a compact size to be well-suited for biomedical sensors, and specific absorption rate (SAR) analysis is performed to ensure user safety. In addition, the performance of the proposed antenna under bending conditions is also considered to present a realistic approach for a practical antenna design.

Keywords: patch antenna; split-ring resonator (SRR); specific absorption rate (SAR); electromagnetic bandgap (EBG)

Citation: Wajid, A.; Ahmad, A.; Ullah, S.; Choi, D.-y.; Islam, F.U. Performance Analysis of Wearable Dual-Band Patch Antenna Based on EBG and SRR Surfaces. *Sensors* **2022**, *22*, 5208. https://doi.org/10.3390/s22145208

Academic Editors: Raed A. Abd-Alhameed, Chan Hwang See and Naser Ojaroudi Parchin

Received: 23 May 2022
Accepted: 11 July 2022
Published: 12 July 2022

1. Introduction

In the modern communication era, the wearable antenna is a topic of interest, with the recent development of several advanced technologies, such as the Internet of Things (IoT), for communications both off-body and on-body to fixed or mobile local wireless networks [1]. These systems are directly attached to the human body or fixed on clothes; they are of high interest for detecting the motion of a body during exercise and for medical applications, such as monitoring of blood pressure and heartbeat, and they form a local network of several sensors using the local Wi-Fi frequency range [2]. The major challenge for the design of wearable antennas is achieving compactness and the required flexibility while maintaining a high quality of service for a standard application. The microstrip patch antenna is a favorable choice for wearable devices, as it is flexible, conformable, lightweight, inexpensive, easy to fabricate, and low-profile. However, it has certain limitations, such as low gain, reduced efficiency, narrow bandwidth, spurious radiation, and out-of-phase reflections [3]. Such worn antennas with high back-lobe radiations may cause increased

where ε_r is the relative permittivity of the substrate, C is the speed of light, and f_r represents the resonance frequency. The effective dielectric constant (ε_{eff}) can be found as:

$$\varepsilon_{eff} = \frac{\varepsilon_r + 1}{2} + \frac{\varepsilon_r - 1}{2}\left(\frac{1}{\sqrt{1 + \frac{12h}{w}}}\right) \tag{2}$$

where h represents the thickness of the substrate. Therefore, the actual length of the patch antenna is found by:

$$L = l_{eff} - 2\Delta L \tag{3}$$

where ΔL represents the extension length and l_{eff} represent the effective resonance length of the patch.

Table 1. Summary of antenna dimensions.

Parameter	Description	Values (mm)
L_P	Length of patch	52
W_P	Width of patch	54.7
L_S	Length of substrate	98
W_S	Width of substrate	109.4
H	Thickness of substrate	3
mt	Thickness of patch	0.01
W_S	Width of slot	20.35
S_L	Depth of slot	10
W_f	Width of feeder	6
L_f	Length of feeder	30.25

2.2. *Design of Metamaterials EBG and SRR*

A comparison of EBG and SRR structure-based ground planes, as shown in Figures 2 and 3, respectively, is made here by considering 2.4 GHz. Zelt material was used as a conductor, while wash cotton was selected as the dielectric substrate to be compatible with wearable design. The optimized parameters of the proposed EBG- and SRR-based designs, using Sievenpiper's equations [23], are listed in Tables 2 and 3, respectively.

2.2.1. Description of EBG Structure

Initially, a dual-band EBG unit cell was designed and analyzed as shown in Figure 2 (zoom view of EBG unit cell); it was observed that the proposed unit cell had in-phase reflections on the desired bands, i.e., 2.4 and 5.4 GHz, as shown in Figure 3. Then, a 5×5 array was designed from the proposed unit cells. This surface was characterized in terms of surface wave suppression within a specific resonant bandwidth using the two-port arrangement technique of Figure 2 and fixing a 50 Ω transmission line on the EBG surface to transmit the surface waves. The transmission line is excited at both ends using discrete ports, where one port acts as a source and the other acts as a match load. The transmission coefficient (S21) was, at a minimum, below −35 dB within a desired band at a centered frequency of 2.4 GHz. This clearly depicts that, within this band, the surface wave was suppressed and kept to a minimum level. However, at other desired bands at 5.4 GHz, it lacked the ability to suppress the surface wave. The periodicity of the EBG unit cell is Lp + Gp = 31.4 mm. The via radius was chosen as r = 1 mm. The geometrical model of the dual-band EBG structure is shown in Figure 2.

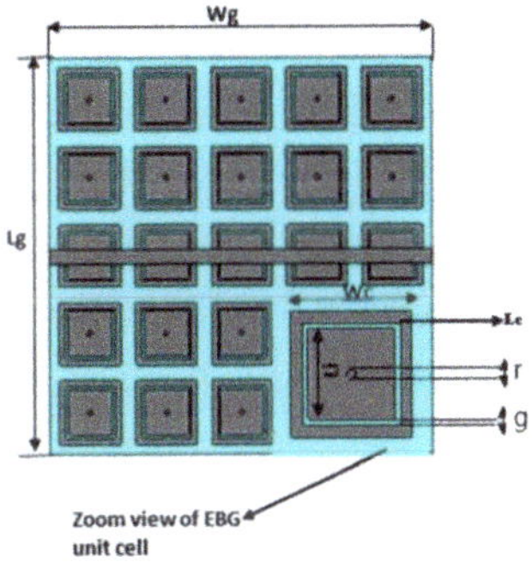

Figure 2. Geometric model of the dual-band EBG.

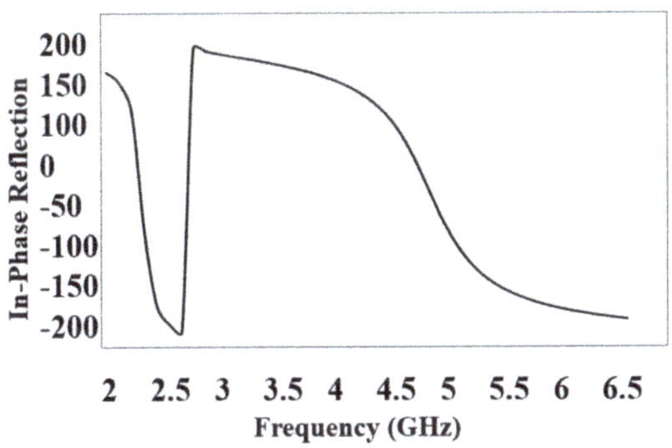

Figure 3. In-phase reflection of EBG unit cell.

Table 2. Optimized parameter values for the array EBG model.

Parameter	Description	Value (mm)
Wc	Width or length of unit cell	29.4
G	Space b/w adjacent cells	2
R	Via radius	1
Lc	Length of outer ring of EBG unit cell	5.525
A	Periodicity	31.4
Li	Length of inner patch	21.875

Table 3. Optimized parameter values for the array SRR model.

Parameters	Descriptions	Value (mm)
g1	Adjacent ring gap	4.5 mm
g3	Inner ring gap	11 mm
g2	Outer split	5.7 mm
A	Periodicity	30.5 mm
L	Width/length of cell	25.5

This surface was characterized in terms of surface wave suppression within a specific resonant bandwidth using the two-port arrangement technique of Figure 4 and fixing a 50 Ω transmission line on the EBG surface to transmit the surface waves. The transmission line is excited at both ends using discrete ports, where one port acts as a source and the other acts as a match load. The transmission coefficient (S_{21}) was, at a minimum, below −35 dB within a desired band at a centered frequency of 2.4 GHz. This clearly depicts that, within this band, the surface wave was suppressed and kept to a minimum level. However, at other desire bands at 5.4 GHz, it lacked the ability to suppress the surface wave. The periodicity of the EBG unit cell is Lp + Gp = 31.4 mm. The via radius was chosen as r = 1 mm. The geometrical model of the dual-band EBG structure is shown in Figure 2.

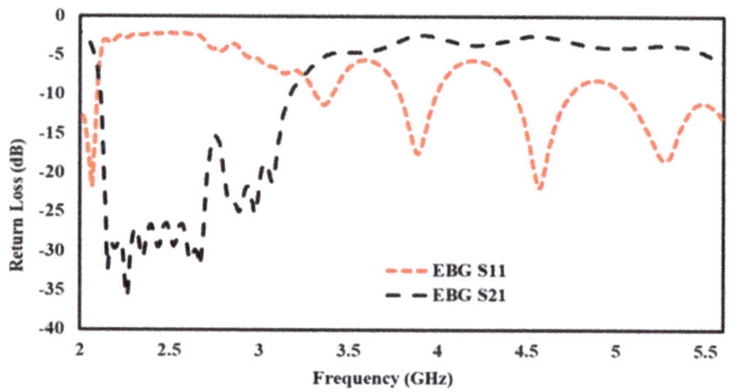

Figure 4. Simulated S11, S21 of EBG array.

2.2.2. Description of SRR Structure

Initially, a dual-band SRR unit cell was design and analyzed as shown in Figure 5 (zoom view of SRR unit cell); it was observed that the proposed unit cell had in-phase reflections on the desired bands, i.e., 2.4 and 5.4 GHz, as shown in Figure 6. Then, a 5 × 5 array was designed from the proposed unit cells. The split-ring resonator (SRR) structure proposed in this paper suppresses the surface wave within a specific band of frequencies. The band of frequencies within which the propagation of the wave is highly restricted is called its bandgap. It consists of two rectangular metallic rings and a split on the opposite sides, as shown in Figure 5. Such a design makes it a *LC* resonator (with distributed *L* and *C*).

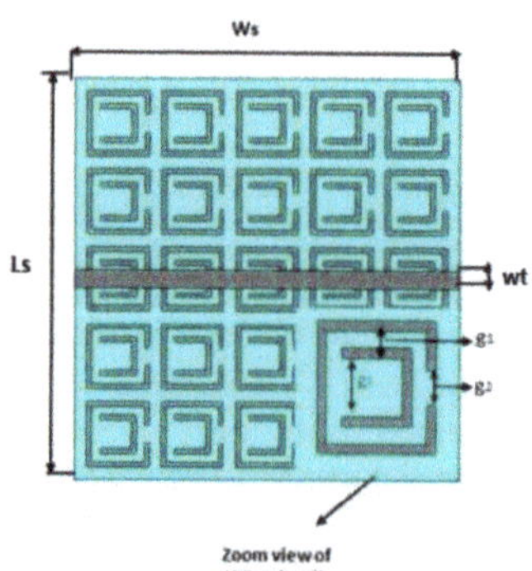

Figure 5. Geometric model of the dual-band SRR.

The geometric model of the SRR array is shown in Figure 5. It is worth mentioning that the SRR suppressed the propagation of the wave at both centered frequencies of 2.4 GHz and 5.4 GHz, as shown in Figure 7.

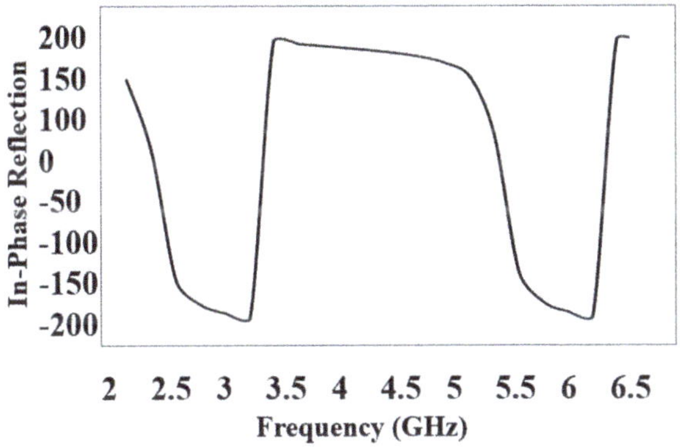

Figure 6. In-phase reflection of SRR unit cell.

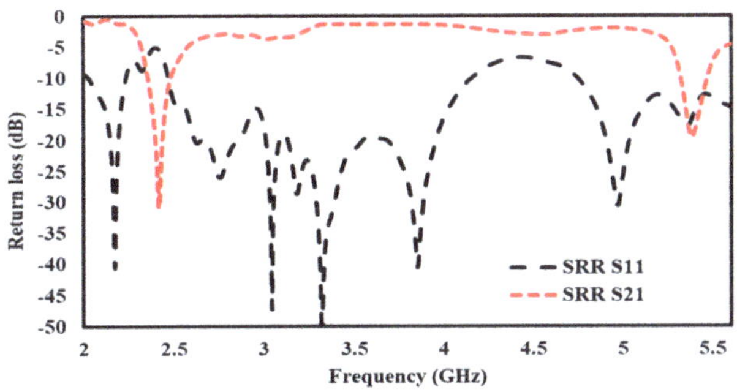

Figure 7. Simulated S11, S21 of 5 × 5 SRR array.

2.3. Mushroom-Like EBG and Split-Ring Resonator (SRR)

A dual-band microstrip patch antenna was placed above the array of EBG cells by removing a few EBG cells and the transmission line; it was excited by a waveguide port using a microstrip feedline and simulated in CST. The resulting EBG-based antenna shown in Figure 8 operated on dual bands, 2.4 GHz and 5.4 GHz.

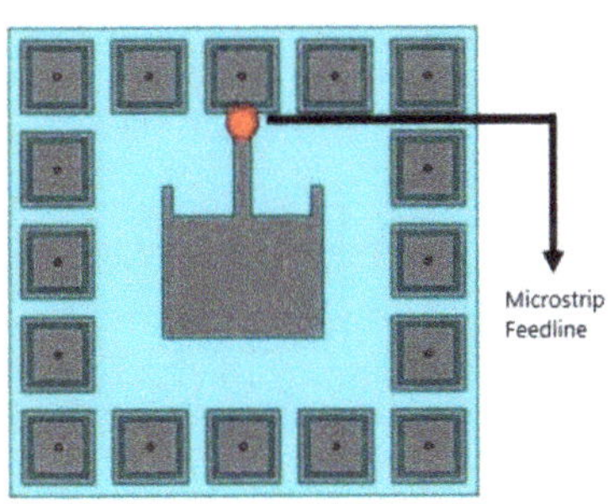

Figure 8. Geometric model of the dual-band EBG-based antenna.

A dual-band microstrip patch antenna was placed above the array of SRR cells by removing a few SRR cells and the transmission line; it was excited by a waveguide port using a microstrip feedline and simulated in CST. In comparison to wired antennas, wearable antennas are typically low-profile, compact, flexible, lightweight, and inexpensive. The resulting SRR-based antenna shown in Figure 9 operated on dual bands, 2.4 GHz and 5.4 GHz.

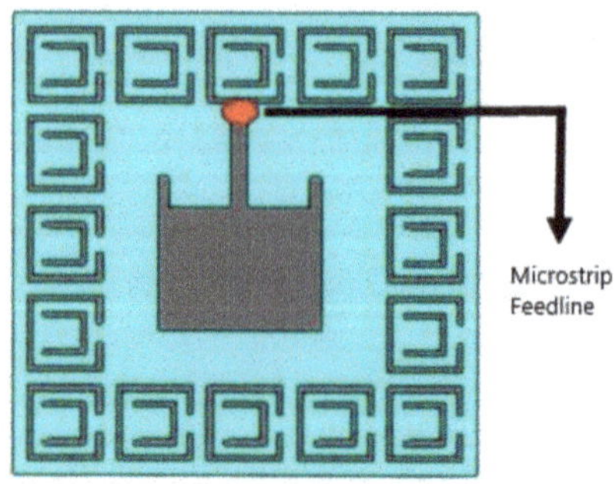

Figure 9. Geometric model (front view) of the dual-band SRR-based antenna.

2.4. *Bending Scenario*

2.4.1. Bending of Conventional Antenna

For on-body analysis other than flat, a microstrip patch antenna was bent on a 120 mm radius cylinder in CST 2011; we then avoided the cylinder to observe the effect of bending on the previous result and simulated it. The figure of the bent microstrip patch antenna is given below in Figure 10a.

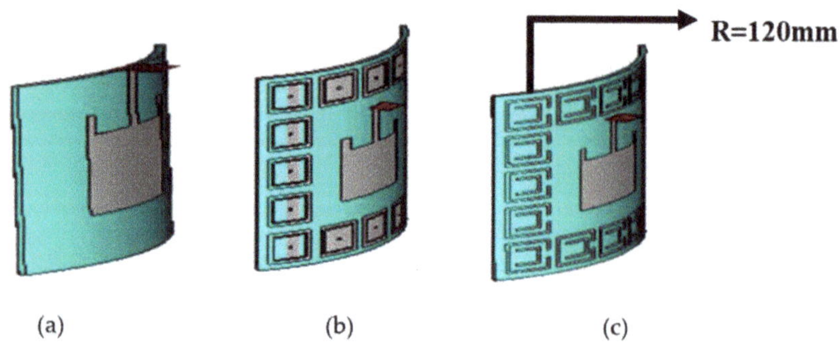

Figure 10. Geometric model of dual-band antennas in free space. (**a**) Conventional antenna. (**b**) EBG-based antenna. (**c**) SRR-based antenna.

2.4.2. Bending of EBG Antenna

For on-body analysis other than flat, an EBG metamaterial-based microstrip patch antenna was bent on a 120 mm radius cylinder; we then avoided the cylinder to observe the effect of bending on the previous result and simulated it. The geometry of the bent EBG-based patch antenna is shown in Figure 10b.

2.4.3. Bending of SRR Antenna

For on-body analysis other than flat, a SRR metamaterial-based microstrip patch antenna was bent on a 120 mm radius cylinder; we then avoided the cylinder to observe the effect of bending on the previous result and simulated it. The geometry of the bent SRR-based patch antenna is shown in Figure 10c.

2.4.4. Bending of Conventional Antenna on Body

To observe the specific absorption rate (SAR) of a conventional patch antenna, a 120 mm radius arm of the body was designed in CST Microwave Studio which consisted of skin as the upper layer and fat, muscle, and bone as the innermost layer of the body. A conventional patch antenna was placed and bent on the arm of the human body and simulated. The geometry is shown in Figure 11a.

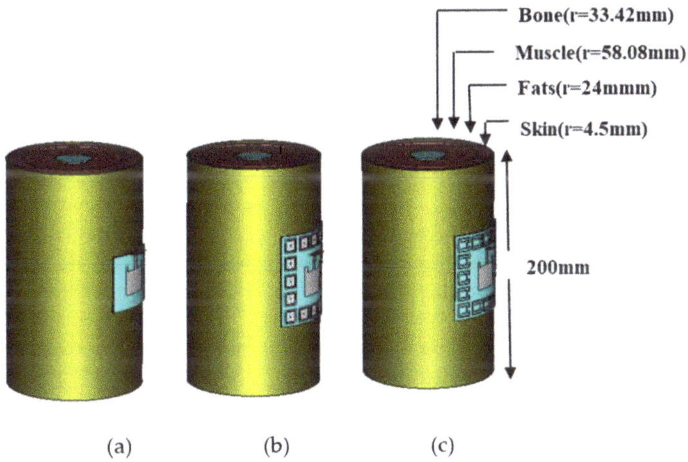

Figure 11. Geometric model of the dual-band antennas bent on human arm. (**a**) Conventional antenna. (**b**) EBG-based antenna. (**c**) SRR-based antenna.

2.4.5. Bending of EBG-Based Antenna on Body

To observe the specific absorption rate (SAR) of an EBG-based patch antenna, a 120 mm radius arm of the body was designed in CST Microwave Studio which consisted of skin as the upper layer and fat, muscle, and bone as the innermost layer of the body. An EBG-based patch antenna was placed and bent on the arm of the human body—the geometry of which is shown in Figure 11b as below—and simulated.

2.4.6. Bending of SRR-Based Antenna on Body

To observe the specific absorption rate (SAR) of a SRR-based patch antenna, a 120 mm radius arm of the body was designed in CST Microwave Studio which consisted of skin as the upper layer and fat, muscle, and bone as the innermost layer of the body. A SRR-based patch antenna was placed and bent on the arm of the human body and simulated. The geometry of the SRR-based patch antenna on the arm is given in Figure 11c. The intrinsic properties of each layer are summarized in Table 4.

Table 4. Properties of human body tissues.

Tissue	Permittivity (ε_r)	Conductivity (S/m)	Density (Kg/m^3)
Skin	35.61	3.2185	1100
Fat	4.6023	0.58521	1100
Muscle	49.278	4.2669	1060
Bone	9.946	1.0101	1850

3. Results and Discussion

The results of the different scenarios (Normal, OFF Body, ON Body) are presented here in terms of their return loss comparison and 2D and 3D antenna radiation pattern.

3.1. Normal Scenario

The results of the normal scenario are discussed in Figures 12 and 13.

Table 5. Comparative results of (conventional vs. EBG-based vs. SRR-based) antennas in normal scenario.

Parameters	Conventional	With EBG	With SRR	Conventional	With EBG	With SRR
Frequency (GHz)	2.4	2.4	2.4	5.4	5.4	5.4
Gain (dB)	6.55	7.01	7.22	6.17	7.13	7.39
Bandwidth (MHz)	82	77	79	308	360	374
Total Efficiency (%)	60.25	64.52	62.28	62.6	59.2	64.77
HPBW (deg)	71.2	66.00	62.2	42.1	36.2	43.2

3.2. *Bent Scenario*

3.2.1. OFF Body Analysis

The bent EBG- and SRR-based antennas are compared here and analyzed in the absence of a human arm with a conventional patch antenna using CST Microwave Studio 2011.

3.2.2. Return Losses

Bending affects the impedance of the conventional patch antenna and the EBG- and SRR-based antennas, and the resonance frequencies in all of the antennas slightly shifted toward the left and right from their basic resonance frequencies of 2.4 GHz and 5.4 GHz, as shown in Figure 15.

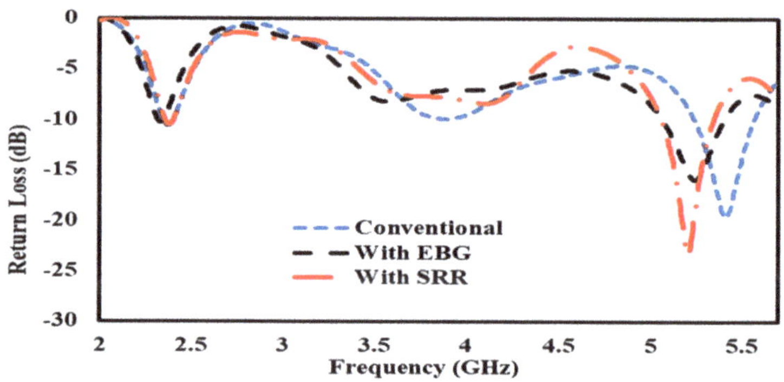

Figure 15. S11 comparison of conventional antenna, antenna on EBG, and SRR/HIS in free space.

3.2.3. 2D and 3D Gain Plots

The off-body E-plane and H-plane simulated gains of the antennas are compared in Figure 16. It can be seen that the main lobe radiation of the SRR-based antenna was maximal, at 4.1 dB, as compared to the EBG-based (3.7 dB) and conventional patch antennas (4.1 dB) at 2.4 GHz. In addition, the main lobe radiation of the SRR-based antenna was highest, at 6.2 dB, as compared to the EBG-based (5.4 dB) and conventional patch antennas (5.5 dB) at 5.4 GHz. Similarly, the −3 dB beamwidth of the SRR-based antenna grew up to (81.2 deg) relative to the EBG-based (73.9 deg) and conventional patch antennas (76.3 deg) at 2.4 GHz. In addition, the −3 dB beamwidth of the conventional patch antenna was highest (47.4 deg) relative to the EBG-based (32.7 deg) and SRR-based antennas (37.5 deg) at 5.4 GHz. The simulated gains comparison of conventional and EBG- and SRR-based antennas at 2.4 GHz and 5.4 GHz is shown in Figure 16a–d.

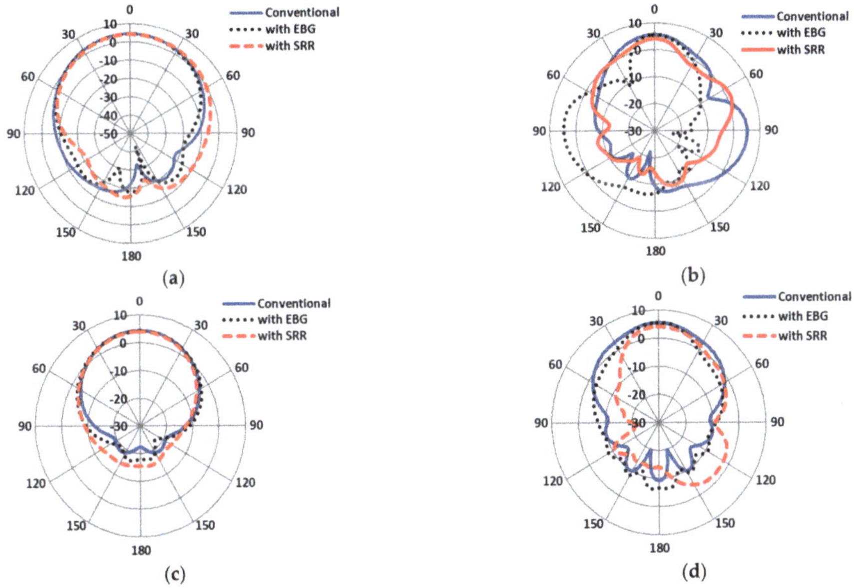

Figure 16. Gain patterns comparisons of conventional antenna under bent condition (free space). (**a**) E-plane at 2.4 GHz. (**b**) E-plane at 5.4 GHz. (**c**) H-plane at 2.4 GHz. (**d**) H-plane at 5.4 GHz.

The 3D gains of normal conventional patch and EBG- and SRR-based antennas are shown in Figure 17. The gains of the conventional patch and the EBG- and SRR-based antennas are 4.31 dB, 4.08 dB, and 4.05 dB, respectively, at 2.4 GHz. Similarly, the gains of the conventional patch and the EBG- and SRR-based antennas are 5.54 dB, 5.5 dB, and 4.04 dB, respectively, at 5.4 GHz. Table 6 depicts that, comparatively, the boresight gain of the antenna does not show a good result, due to mismatching of the impedance owing to free space. The 3D gain patterns of antennas in free space are presented in Figure 17a–f.

Table 6. Summary of results (conventional vs. EBG-based vs. SRR-based) bent antennas in free space.

Parameters	Conventional	With EBG	With SRR	Conventional	With EBG	With SRR
Frequency (GHz)	2.37	2.33	2.38	5.41	5.24	5.44
Gain (dB)	4.31	4.08	4.05	5.54	5.5	4.04
Bandwidth (MHz)	55.5	36.27	58	324	333.8	287
Total Efficiency (%)	36.4	26.9	34.5	58.00	50.38	41.5
HPBW (deg)	76.3	73.9	101.8	47.4	32.37	34.5

3.3. ON Body Analysis

For practical applications, to observe results for conventional and EBG- and SRR-based antennas, a model of the human arm was proposed with a radius of 120 mm. It consisted of four layers: skin, fat, muscle, and bone.

3.3.1. Return Losses

The high-loss nature of the human body lowers the radiating efficiency of a mounted antenna. This section provides the simulation results in terms of return losses and input impedance for a scenario where the proposed EBG- and SRR-based antennas are mounted over a human arm. For comparison, the results from a conventional patch antenna are also presented. The bending of the antenna patch is an important factor to be considered, as it causes a significant reduction in the driving point of the antenna impedance. This results in a small change in the resonant frequency and return loss. Figure 18 shows that the metamaterial-inspired antenna performed very well as compared to the conventional patch antenna in providing low return loss with bending.

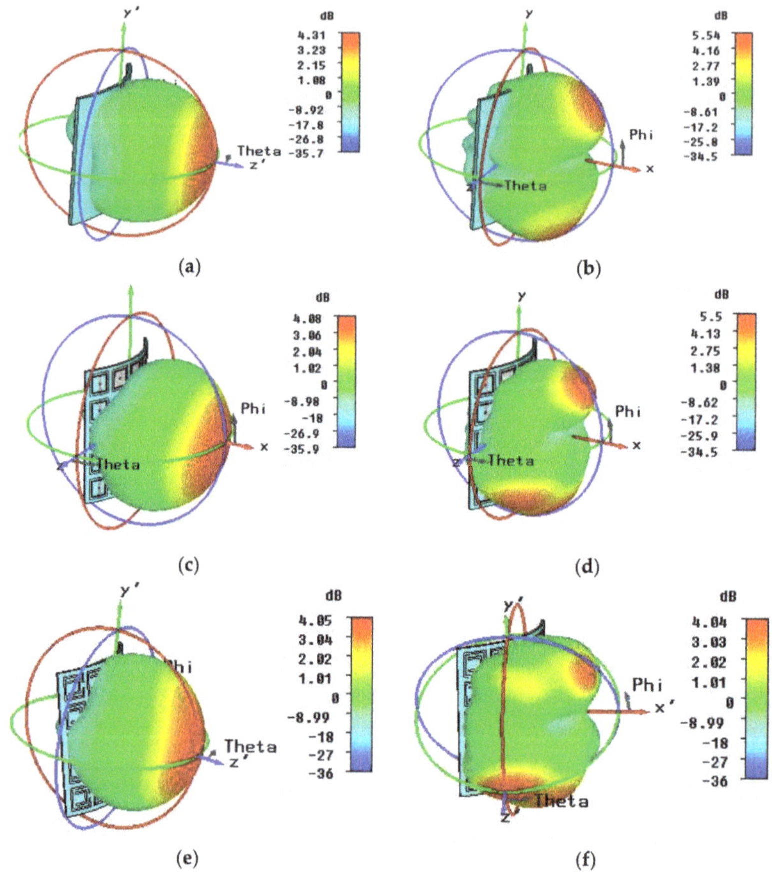

Figure 17. 3D radiation pattern of bent scenario (free space). (**a**) Conventional antenna at 2.4 GHz. (**b**) Conventional antenna at 5.4 GHz. (**c**) EBG-based antenna at 2.4 GHz. (**d**) EBG-based antenna at 5.4 GHz. (**e**) SRR-based antenna at 2.4 GHz. (**f**) SRR-based antenna at 5.4 GHz.

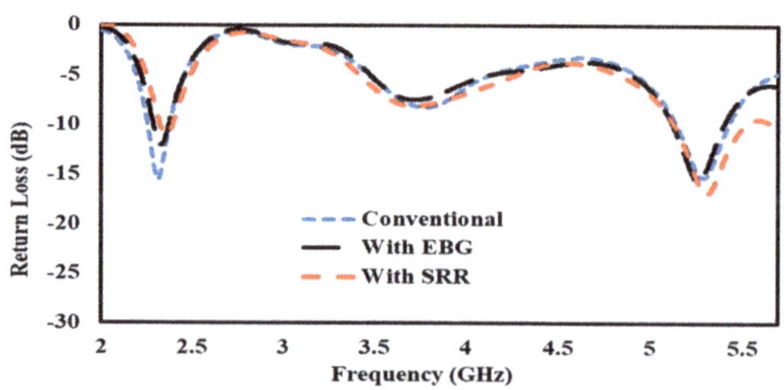

Figure 18. Return loss comparison of conventional antenna, antenna on EBG, and SRR/HIS bent on human arm.

3.3.2. 2D and 3D Gain Plots

The E-plane and H-plane on-body gains of the antennas are compared in Figure 19. The simulation results show that the main lobe radiation of the EBG-based antenna was minimal, at 3.79 dB, as compared to the SRR-based (4.21 dB) and conventional patch antennas (4.03 dB) at 2.4 GHz. In addition, the main lobe radiation of the SRR-based antenna was highest, at 6.22 dB, as compared to the EBG-based (5.39 dB) and conventional patch antennas (5.56 dB) at 5.4 GHz. Similarly, the −3 dB beamwidth of the EBG-based antenna grew up to (89.8 deg) relative to the SRR-based (85.7 deg) and conventional patch antennas (83.0 deg) at 2.4 GHz. In addition, the −3 dB beamwidth of the SRR-based antenna decayed up to (22.3 deg) relative to the EBG-based (39.8 deg) and conventional patch antennas (30.5 deg) at 5.4 GHz. The simulated gains comparison of the conventional and the EBG- and SRR-based antennas at 2.4 GHz and 5.4 GHz is shown in Figure 19a–d.

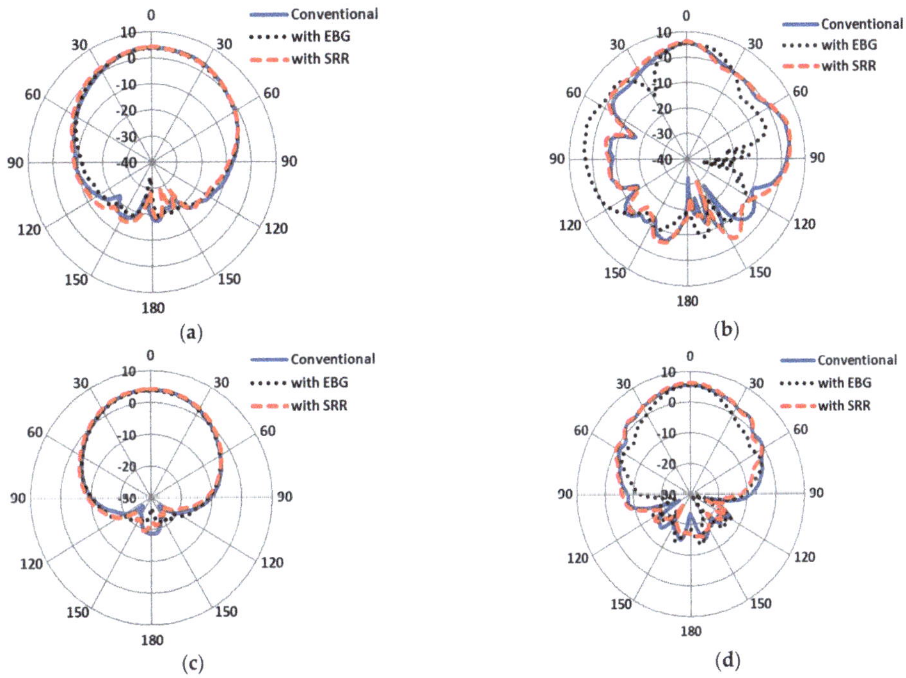

Figure 19. Gain patterns comparisons of conventional antenna under bent condition (on-body). (**a**) E-plane at 2.4 GHz. (**b**) E-plane at 5.4 GHz. (**c**) H-plane at 2.4 GHz. (**d**) H-plane at 5.4 GHz.

The 3D gains of the normal conventional patch and the EBG- and SRR-based antennas are shown in Figure 20. The gains of the conventional patch and EBG- and SRR-based antennas are 4.03 dB, 3.79 dB, and 4.21 dB, respectively, at 2.4 GHz. Similarly, the gains of the conventional patch and the EBG- and SRR-based antennas are 5.56 dB, 5.39 dB, and 6.22 dB, respectively, at 5.4 GHz. Table 7 illustrates that the comparative results of the SRR-based antenna is maximal in terms of gain, bandwidth, and total efficiency at 5.4 GHz. The 3D gain patterns of the conventional and EBG- and SRR-based antennas at 2.4 GHz and 5.4 GHz are presented in Figure 20a–f.

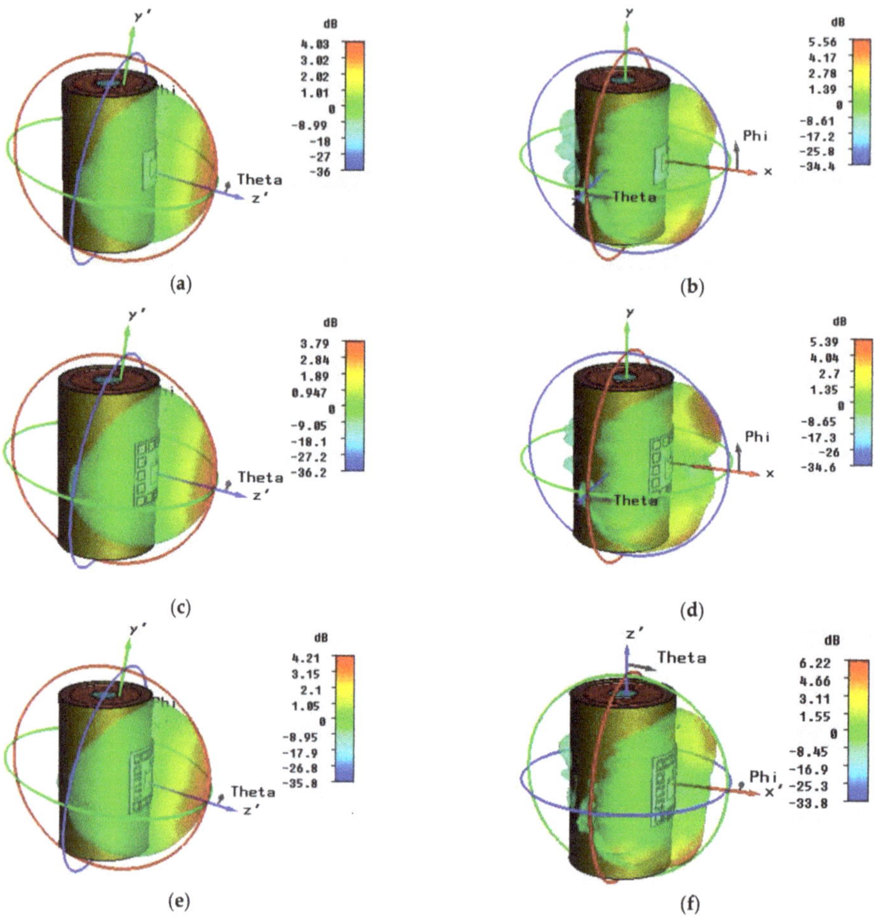

Figure 20. 3D radiation patterns in bent scenarios (on-body). (**a**) Conventional antenna at 2.4 GHz. (**b**) Conventional antenna at 5.4 GHz. (**c**) EBG-based antenna at 2.4 GHz. (**d**) EBG-based antenna at 5.4 GHz. (**e**) SRR-based antenna at 2.4 GHz. (**f**) SRR-based antenna at 5.4 GHz.

Table 7. Summary of results (conventional vs. EBG-based vs. SRR-based) of bent antennas on human arm.

Parameters	Conventional	With EBG	With SRR	Conventional	With EBG	With SRR
Frequency (GHz)	2.4	2.4	2.4	5.4	5.4	5.4
Gain (dB)	4.03	3.79	4.21	5.56	5.39	6.22
Bandwidth (MHz)	223	62.9	128	286.5	267	313
Total Efficiency (%)	23.34	32.58	37.30	41.2	49.8	52.45
HPBW (deg)	83.00	89.8	85.7	30.5	39.8	22.3

4. SAR Analysis

For the proposed design of a dual-band antenna, in order to demonstrate its compatibility for human-wearable applications, specific absorption analysis (SAR) was conducted by considering an arm of 200 mm radius. The bending was considered in E-plane in CST MW Studio using the IEEE C95.3 averaging method. The ICNIRP EU standard of 10 g of tissue volume was considered during this analysis. As the human body consists of several conductive and dielectric substances, by bringing an antenna in proximity to the body with an imperfect impedance matching, the human body detunes and reflects backward the radiated energy and then absorbs some part of it. This results in high SAR values, and the use of a high-impedance metamaterial surface as the ground serves as an isolation sheet between the human body and the radiating surface.

Simulation Setup and SAR Calculation

It is clearly shown in Figure 21 that most of the power was reflected and absorbed in the human body in an unloaded metamaterial antenna; therefore, the SAR value at 2.4 GHz was 11 W/kg and 5.88 W/kg at 5.4 GHz, which is above a safe level. Similarly, in the case of the EBG metamaterial-based antenna, the SAR value was 3.40 W/kg at 2.4 GHz and 1.24 W/kg at 5.4 GHz. In the case of the SRR metamaterial-based antenna, the SAR value was 1.97 w/kg at 2.4 GHz and 1.18 w/kg at 5.4 GHz, which is below a safe level (European standard: 2 W/kg). The complete SAR comparison is listed in Table 8. A significant reduction in SAR values was obtained through the metamaterial-inspired design of the antenna because of the HIS (high-impedance surface) when the metamaterial surface was utilized for the ground plane. With such a configuration, resonance at the desired frequency was obtained due to the surface wave bandgap. In addition, this surface also insulates the antenna from electromagnetic radiations from the body.

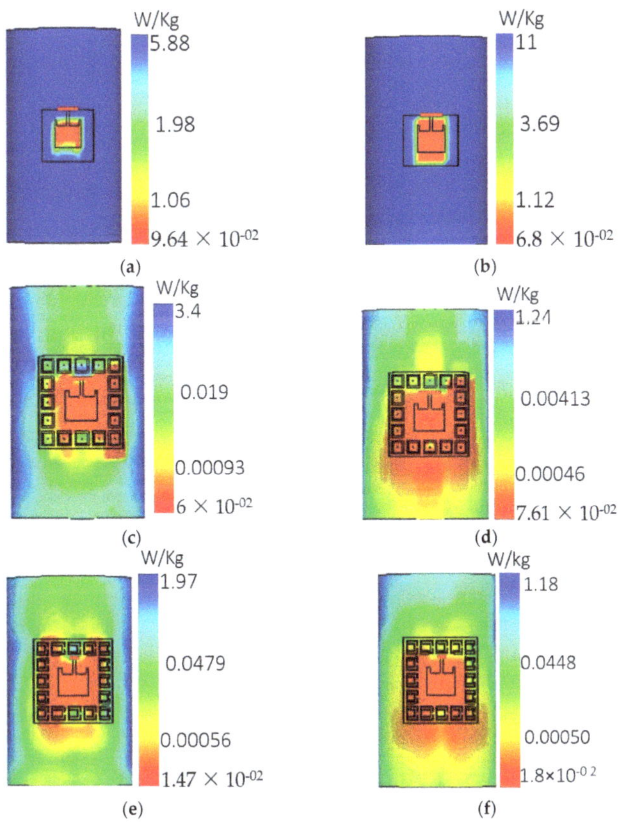

Figure 21. Conventional antenna bent around human arm: (**a**) SAR at 2.4 GHz. (**b**) SAR at 5.4 GHz, EBG-based antenna bent around human arm. (**c**) SAR at 2.4 GHz. (**d**) SAR at 5.4 GHz, SRR-based antenna bent around human arm. (**e**) SAR at 2.4 GHz. (**f**) SAR at 5.4 GHz.

Table 8. Summary of simulated SAR normalized to 0.5 W (r.m.s.) at 2.4 GHz and 5.4 GHz.

Antenna	10 g Peak SAR (W/Kg) at 2.4 GHz	10 g Peak SAR (W/Kg) at 5.4 GHz
Conventional	11	5.88
EBG-based	3.40	1.24
SRR-based	1.97	1.18

Various works have been carried out in the past for the enhancement of parameters such as gain, bandwidth, and efficiency, etc. Some of the concluded data from the past works [35–38] are summarized in the Table 9. It is deduced from the table that the proposed antenna shows supremacy, with reference to the previous work carried out, in terms of gain, bandwidth, and radiation efficiency.

Table 9. Comparison table.

Ref. No	Antenna Size (mm^2)	Operating Frequency	Gain (db/dbi)	Bandwidth	Efficiency
[35]	100 × 100	2.42	2.42 db	298 MHz	40%
[36]	50 × 50	2.4 ISM	4.12 dbi	65 MHz	78.97%
[37]	102 × 102	2.4/5.5	5.2 db	55 MHz	83%
[38]	84 × 162.25	1.85 -3.15	3.38 db	130 MHz	NA
Proposed work	98 × 109.4	2.4/5.4	7.13/7.39 db	374 MHz	64.77

5. Conclusions

In this paper, a conventional and metamaterial (EBG, SRR)-based flexible dual-band (2.4 GHz, 5.4 GHz) wearable patch antenna design is presented and analyzed in terms of radiation characteristics. A signigicant contribution is made in this paper with the proposal of a flexible and conformable antenna design for biomedical wearable applications by using a clothing material (wash cotton) as a dielectric. It was observed that the comparative results were based on two different frequency ranges, i.e., 2.4 GHz and 5.4 GHz. At 5.4 GHz, the proposed SRR-based antenna showed enhanced results in terms of gain, bandwidth, total efficiency, and HPBW, which were 7.39 db, 374 MHz, 64.77%, and 43.2 deg, respectively, while, in bent condition, the obtained results were 6.22 db, 313 MHz, 52.45%, and 22.3 deg, respectively. The metamaterial surfaces (EBG and SRR) are proposed to be used as a high-impedance surface for protection from the risks of electromagnetic radiations, and the results show that the SAR values were successfully reduced to the safe limit by employing the metamaterial surfaces. It was observed that, in terms of gain, the SRR-based antenna performance was more dominant compared to the conventional and EBG-based antennas. The performance of the three proposed antennas was evaluated in a bent condition. It was demonstrated that the proposed design offered a great amount of flexibility, and bending had no prominent effects on the performance of the recommended antennas. However, the proposed design has a limited bandwidth, low power-handling capacity, and high power loss at high frequencies, i.e., >12 GHz. Additionally, due to the unavalibility of the fabrication facilities, measured results are not included in the current version.

Author Contributions: Conceptualization, A.W. and S.U.; formal analysis, A.W. and S.U.; funding acquisition, D.-y.C. and S.U.; methodology, A.W., S.U., A.A. and D.-y.C.; project administration, S.U. and D.-y.C.; resources, A.W. and S.U.; software, A.W., S.U. and A.A.; supervision, S.U., D.-y.C., A.A. and F.U.I.; validation, S.U., D.-y.C. and A.W.; visualization, S.U., F.U.I. and. A.A.; writing—original draft, A.W., S.U., A.A. and F.U.I.; writing—review and editing, D.-y.C., S.U., F.U.I., A.A. and A.W. All authors have read and agreed to the published version of the manuscript.

Funding: This research was supported by the BrainKorea21Four Program through the National Research Foundation of Korea (NRF) funded by the Ministry of Education (4299990114316).

Institutional Review Board Statement: Not applicable.

Informed Consent Statement: Not applicable.

Conflicts of Interest: The authors declare no conflict of interest.

References

1. Ullah, S.; Ahmad, I.; Raheem, Y.; Ullah, S.; Ahmad, T.; Habib, U. Hexagonal shaped CPW Feed based Frequency Reconfigurable Antenna for WLAN and Sub-6 GHz 5G applications. In Proceedings of the 2020 International Conference on Emerging Trends in Smart Technologies (ICETST), Piscataway, NJ, USA, 26 March 2020; pp. 1–4.
2. Ali, U.; Ullah, S.; Khan, J.; Shafi, M.; Kamal, B.; Basir, A.; Seager, R.D. Design and SAR analysis of wearable antenna on various parts of human body, using conventional and artificial ground planes. *J. Electr. Eng. Technol.* **2017**, *12*, 317–328. [CrossRef]
3. Faisal, F.; Ahmad, A.; Ali, U.; Ullah, S.; Ullah, K. Performance analysis of a 2.4 GHz planar antenna using different types of wearable artificial ground planes. In Proceedings of the 12th International Conference on High-Capacity Optical Networks and Enabling/Emerging Technologies (HONET), Islamabad, Pakistan, 21–23 December 2015; pp. 1–5.
4. Soh, P.J.; Giman, F.N.; Jamlos, M.F.; Lago, H.; Al-Hadi, A.A. C-slotted dual band textile antenna for WBAN applications. In Proceedings of the URSI Asia-Pacific Radio Science Conference (URSI AP-RASC), New Delhi, India, 9–15 March 2019; pp. 1621–1624.
5. Faisal, F.; Yoo, H. A Miniaturized Novel-Shape Dual-Band Antenna for Implantable Applications. *IEEE Trans. Antennas Propag.* **2018**, *67*, 774–783. [CrossRef]
6. Abdullah, S.; Xiao, G.; Amaya, R.E. A Review on the History and Current Literature of Metamaterials and Its Applications to Antennas & Radio Frequency Identification (RFID) Devices. *IEEE J. Radio Freq. Identif.* **2021**, *5*, 427–445. [CrossRef]
7. Kumar, R.; Kumar, M.; Chohan, J.S.; Kumar, S. Overview on metamaterial: History, types and applications. *Sci. Direct J. Mater. Today* **2022**, *56*, 3016–3024. [CrossRef]
8. Ahmad, S.; Paracha, K.N.; Sheikh, Y.A.; Ghaffar, A.; Butt, A.D.; Alibakhshikenari, M.; Soh, P.J.; Khan, S.; Falcone, F. A Metasurface-Based Single-Layered Compact AMC-Backed Dual-Band Antenna for Off-Body IoT Devices. *IEEE Access* **2021**, *9*, 159598–159615. [CrossRef]
9. Sheikh, Y.A.; Paracha, K.N.; Ahmad, S.; Bhatti, A.R.; Butt, A.D.; Rahim, S.K.A. Analysis of Compact Dual-Band Metamaterial-Based Patch Antenna Design for Wearable Application. *Arab. J. Sci. Eng.* **2022**, *47*, 3509–3518. [CrossRef]
10. Shahzad, M.A.; Paracha, K.N.; Naseer, S.; Ahmad, S.; Malik, M.; Farhan, M.; Ghaffar, A.; Hussien, M.; Sharif, A.B. An Artificial Magnetic Conductor-Backed Compact Wearable Antenna for Smart Watch IoT Applications. *Electronics* **2021**, *10*, 2908. [CrossRef]
11. Altintas, O.; Aksoy, M.; Akgol, O.; Ünal, E.; Karaaslan, M.; Sabah, C. Fluid, Strain and Rotation Sensing Applications by Using Metamaterial Based Sensor. *J. Electrochem. Soc.* **2017**, *164*, B567–B573. [CrossRef]
12. Garg, B.; Verma, R.D.; Samadhiya, A. Design of rectangular microstrip patch antenna incorporated with innovative metamaterial structure for dual band operation and amelioration in patch antenna parameters with negative μ and ε. *Int. J. Eng. Technol.* **2012**, *1*, 205–216. [CrossRef]
13. Liu, S.; Wang, Z.; Dong, Y. Compact Wideband SRR-Inspired Antennas for 5G Microcell Applications. *IEEE Trans. Antennas Propag.* **2021**, *69*, 5998–6003. [CrossRef]
14. Manoharan, S.; Ramasamy, P.; Singaravelu, R. A quad-band fractal antenna with metamaterial resonator-backed ground for sub-6 GHz, C and X band applications. *Appl. Phys. A* **2021**, *127*, 703. [CrossRef]
15. Desai, A.; Palandoken, M.; Kulkarni, J.; Byun, G.; Nguyen, T.K. Wideband Flexible/Transparent Connected-Ground MIMO Antennas for Sub-6 GHz 5G and WLAN Applications. *IEEE Access* **2021**, *9*, 147003–147015. [CrossRef]
16. Ohashi, R.; Tanaka, T.; Yamamoto, S.; Takikawa, M.; Inasawa, Y. A Design of Multi-band Mushroom-type EBG Structure with Multi-layer Configuration. In Proceedings of the 2020 IEEE International Symposium on Antennas and Propagation (ISAP), Osaka, Japan, 25–28 January 2021; pp. 115–116.
17. Ramachandran, T.; Faruque, M.R.I.; Siddiky, A.M.; Islam, M.T. Reduction of 5G cellular network radiation in wireless mobile phone using an asymmetric square shaped passive metamaterial design. *Sci. Rep.* **2021**, *11*, 1–22. [CrossRef]
18. Sharma, N.; Bhatia, S.S. Metamaterial inspired fidget spinner-shaped antenna based on parasitic split ring resonator for multi-standard wireless applications. *J. Electromagn. Waves Appl.* **2019**, *34*, 1471–1490. [CrossRef]
19. Nagulpelli, A.S.; Varun, D. Hexagonal Shaped Magnetically Coupled EBG for X-Band Antenna Bandwidth Enhancement. In Proceedings of the 3rd IEEE International Conference for Convergence in Technology (I2CT), Pune, India, 6–8 April 2018; pp. 1–6.
20. Mouhouche, F.; Azrar, A.; Dehmas, M.; Djafri, K. Design a Compact UWB Monopole Antenna with Triple Band-Notched Characteristics Using EBG Structures. *Frequenz* **2018**, *72*, 479–487. [CrossRef]
21. Moniruzzaman, M.; Islam, M.T.; Islam, R.; Misran, N.; Samsuzzaman, M. Coupled ring split ring resonator (CR-SRR) based epsilon negative metamaterial for multiband wireless communications with high effective medium ratio. *Results Phys.* **2020**, *18*, 103248. [CrossRef]
22. Alnaiemy, Y.; Nagy, L. Improved Antenna Gain and Efficiency Using Novel EBG Layer. In Proceedings of the IEEE 15th International Conference of System of Systems Engineering (SoSE), Budapest, Hungary, 2–4 June 2020; pp. 271–276.
23. Mohamadzade, B.; Lalbakhsh, A.; Simorangkir, R.B.V.B.; Rezaee, A.; Hashmi, R.M. Mutual Coupling Reduction in Microstrip Array Antenna by Employing Cut Side Patches and Ebg Structures. *Prog. Electromagn. Res. M* **2020**, *89*, 179–187. [CrossRef]

24. Prasad, K.V.; Prasad MV, S.; Kumar, M.S.; Alekhya, B. Surface wave suppression in patch arrays using EBG structures. In Proceedings of the Conference on Signal Processing and Communication Engineering Systems (SPACES), Vijayawada, India, 4–5 January 2018; pp. 99–104.
25. Wagih, M.; Hilton, G.S.; Weddell, A.S.; Beeby, S. Millimeter-Wave Power Transmission for Compact and Large-Area Wearable IoT Devices Based on a Higher Order Mode Wearable Antenna. *IEEE Internet Things J.* **2021**, *9*, 5229–5239. [CrossRef]
26. Abirami, B.S.; Sundarsingh, E. EBG-Backed Flexible Printed Yagi–Uda Antenna for On-Body Communication. *IEEE Trans. Antennas Propag.* **2017**, *65*, 3762–3765. [CrossRef]
27. Kaabal, A.; El Halaoui, M.; El Jaafari, B.; Ahyoud, S.; Asselman, A. Design of EBG antenna with multi-sources excitation for high directivity applications. *Procedia Manuf.* **2018**, *22*, 598–604. [CrossRef]
28. Ashyap, A.Y.I.; Abidin, Z.Z.; Dahlan, S.H.; Majid, H.A.; Shah, S.M.; Kamarudin, M.R.; Alomainy, A. Compact and Low-Profile Textile EBG-Based Antenna for Wearable Medical Applications. *IEEE Antennas Wirel. Propag. Lett.* **2017**, *16*, 2550–2553. [CrossRef]
29. Dhanaraj, P.; Maheswari, S.U. Performance analysis of electrically coupled SRR bowtie antenna for wireless broadband communications. *Wirel. Networks* **2020**, *26*, 5271–5283. [CrossRef]
30. Kadam, A.A.; Deshmukh, A.A. Pentagonal shaped uwb antenna loaded with slot and ebg structure for dual band notched response. *Prog. Electromagn. Res. M* **2020**, *95*, 165–176. [CrossRef]
31. Kiani, S.; Rezaei, P.; Navaei, M. Dual-sensing and dual-frequency microwave SRR sensor for liquid samples permittivity detection. *Measurement* **2020**, *160*, 107805. [CrossRef]
32. Selvi, N.T.; Selvan, P.T.; Babu, S.; Pandeeswari, R. Multiband metamaterial-inspired antenna using split ring resonator. *Comput. Electr. Eng.* **2020**, *84*, 106613. [CrossRef]
33. Balanis, C.A. *Antenna Theory: Analysis and Design*; John Wiley & Sons: Hoboken, NJ, USA, 2016.
34. Ahmad, A.; Faisal, F.; Khan, S.; Ullah, S.; Ali, U. Performance analysis of a wearable and dual band planar antenna using a mushroom-like electromagentic bandgap (EBG) ground plane. In Proceedings of the International Conference on Open-Source Systems & Technologies (ICOSST), Lahore, Pakistan, 17–19 December 2015; pp. 24–29. [CrossRef]
35. Lago, H.; Soh, P.J.; Jamlos, M.F.; Shohaimi, N.; Yan, S.; VandenBosch, G.A.E. Textile antenna integrated with compact AMC and parasitic elements for WLAN/WBAN applications. *Appl. Phys. A* **2016**, *122*, 1059. [CrossRef]
36. Agarwal, K.; Guo, Y.-X.; Salam, B. Wearable AMC Backed Near-Endfire Antenna for On-Body Communications on Latex Substrate. *IEEE Trans. Compon. Packag. Manuf. Technol.* **2016**, *6*, 346–358. [CrossRef]
37. Chahat, N.; Zhadobov, M.; Sauleau, R.; Mahdjoubi, K. Improvement of the on-body performance of a dual-band textile antenna using an EBG structure. In Proceedings of the 2010 Loughborough Antennas & Propagation Conference, Loughborough, UK, 8–9 November 2010; pp. 465–468. [CrossRef]
38. Hussin, E.F.N.M.; Soh, P.J.; Jamlos, M.F.; Lago, H.; Al-Hadi, A.A.; Rahiman, M.H.F. A wideband textile antenna with a ring-slotted AMC plane. *Appl. Phys. A* **2016**, *123*, 46. [CrossRef]

electronics

MDPI

Article

A Flower-Shaped Miniaturized UWB-MIMO Antenna with High Isolation

Weidong Mu [1], Han Lin [1,*], Zhonggen Wang [1], Chenlu Li [2], Ming Yang [3], Wenyan Nie [4] and Juan Wu [1]

1 School of Electrical and Information Engineering, Anhui University of Science and Technology, Huainan 232001, China; mwd18755802702@163.com (W.M.); zgwang@ahu.edu.cn (Z.W.); yj165@mail.ustc.edu.cn (J.W.)
2 School of Electrical and Information Engineering, Hefei Normal University, Hefei 230061, China; chenluli@hfnu.edu.cn
3 Department of Electrical and Communications Engineering, West Anhui University, Lu'an 237012, China; myang@ahu.edu.cn
4 School of Mechanical and Electrical Engineering, Huainan Normal University, Huainan 232001, China; wynie5240@163.com
* Correspondence: hanlin@aust.edu.cn

Abstract: An ultra-wideband (UWB) multiple-input, multiple-output (MIMO) antenna with a reasonably compact size of 30 × 18 × 1.6 mm^3 is presented in this paper. The proposed antenna contains two radiating components, each of which is made up of three elliptically shaped patches situated 60 degrees apart, and resembles the shape of a flower. Moreover, the proposed antenna design incorporates a T-like ground branch that functions as a decoupling structure, and is composed of two modified inverted-L branches and an I-shaped stub, offering an isolation of more than 20 dB over the whole operation band (4.3–15.63 GHz). Furthermore, the proposed antenna system was fabricated and tested, and the envelope correlation coefficient (ECC), diversity gain (DG), and total active reflection coefficient (TARC), as well as the radiation characteristics and MIMO performance, were analyzed. The proposed UWB-MIMO antenna may be a suitable candidate for diverse UWB applications, based on the simulated and measured results of this study.

Keywords: UWB; MIMO antenna; decoupling; diversity performance

Citation: Mu, W.; Lin, H.; Wang, Z.; Li, C.; Yang, M.; Nie, W.; Wu, J. A Flower-Shaped Miniaturized UWB-MIMO Antenna with High Isolation. *Electronics* **2022**, *11*, 2190. https://doi.org/10.3390/electronics11142190

Academic Editors: Naser Ojaroudi Parchin, Chan Hwang See and Raed A. Abd-Alhameed

Received: 25 June 2022
Accepted: 11 July 2022
Published: 13 July 2022

1. Introduction

Ultra-wideband (UWB) technology has been extensively applied in areas including short-range communications, radar, location, and tracking due to its extraordinarily low transmission power and high data speed [1]. The employment of multiple-input, multiple-output (MIMO) technology in a UWB wireless communication system enhances the signal-to-noise ratio and data capacity of the communication system by enabling multiplexing, hence improving the overall performance of the system [2]. The performance of a UWB MIMO system is influenced by factors such as bandwidth, isolation, and other functional characteristics. In recent years, researchers have used a variety of ways to improve the performance of UWB antennas for MIMO systems.

In [3], an antenna designed on a 34 × 34 × 1.6 mm^3 FR4 substrate was proposed, with L-shaped and C-shaped slots embedded in each radiator and an electromagnetic bandgap (EBG) structure loaded near the microstrip feeding line, thereby allowing the antenna to exhibit triple-band slot characteristics, resulting in a bandwidth of 2.5–12 GHz. Here, in addition to the EBG structure being used to enhance the gain of the UWB antennas, the frequency selection surface (FSS) method will also have a significant gain-enhancing effect. Reference [4] presented a method for enhancing UWB antennas with FSS, which minimized power loss in the undesired transmission area of the antenna and blocked possible interference from undesirable and wasteful radiation, in order to obtain constant gain.

Additionally, ref. [5] utilized FSS to separate and effectively isolate the antenna elements. A compact UWB antenna printed on a Rogers RO4003 substrate with a size of 30 $\times$ 31 mm^2 for personal communication and with Bluetooth capability was proposed in [6], where the UWB characteristics were achieved by employing a conventional cylindrical radiating patch and an improved partial ground plane. Through the operation of a small resonator with capacitors, the antenna could also work in the Bluetooth band. In addition, a slit resonator was integrated in the radiating bulk to prevent interference in the WLAN band, resulting in a band-notched characteristic. In [7], sound isolation between antenna components was available throughout the UWB by adding a vertical stub and an H-slot in the ground plane. In [8], the combination of the ground stub on the bottom layer and the EBG structure between the two rectangular patches on the top layer led to a remarkably low mutual coupling between the two radiating patches. This antenna design possessed a compact size of 26 $\times$ 31 mm^2 and displayed a frequency range of 3.1–11 GHz. To increase the VSWR bandwidth, a bending and defective ground plane for the basic radiator was proposed in [9]. A longer ground stub was also installed to increase the bandwidth to meet current automotive needs. However, it had a larger size of 42 $\times$ 24 mm^2. Progressively, as described in [10], two homogeneous, semi-circular radiating elements with a synchronous stepped elliptical structure and an I-shaped ground structure were developed to generate strong isolation and a broad bandwidth between 1.99 GHz and 10.02 GHz. A fence-style structure and an L-shaped parasitic branch were placed on the ground to enhance the impedance bandwidth and isolation at low frequencies, as prescribed in the literature [11]. In [12], a flower-shaped radiator was utilized to boost the isolation of the MIMO elements, and the isolation was further improved by placing a swastika-shaped stub on the ground to achieve a return loss of S11 (<-10 dB) and isolation of S12 (<-18 dB) on an FR4 substrate of 40 $\times$ 40 mm^2, capable of covering the whole UWB spectrum (3.1–14 GHz). Furthermore, the authors of [13] proposed a UWB-MIMO antenna with four suppression bands and a T-shaped stepped stub on the back ground for achieving 3–11 GHz impedance bandwidth and -15 dB isolation. In [14], a triple bandgap CSRR-loaded EBG structure was inserted near the UWB antenna feedline, encompassing 2.5–12 GHz. The overall size of the proposed MIMO/diversity antenna was 30 $\times$ 44 mm^2. In addition, a hexagonal slot and a mirrored pair of F-shaped stubs were employed to decrease the mutual coupling. A four-port and overt-leaf-shaped MIMO antenna with coplanar waveguide feeding was proposed in [15] to achieve wideband (12.75–16.05 GHz) by optimizing the ground plane and radiating elements. In addition, a fan-shaped decoupler was inserted in the middle of the back surface of the substrate in sequence, with further low coupling between components to provide more than 20 dB of isolation. However, the above-mentioned design approaches have fundamental flaws, such as complex structure or excessive size.

In this work, we demonstrate a compact and unique dual-port UWB-MIMO antenna with an incredibly simple construction. Each radiating element has three flower-like elliptically shaped patches situated 60 degrees apart. On one hand, the branches of the modified ground structure are used to generate multiple frequencies in order to broaden the frequency band through resonance, and on the other hand, these branches are utilized to achieve a high level of isolation by effectively absorbing the current and reducing the mutual coupling between the two radiating patches. The ECC, DG, and TARC of the proposed system are all within an acceptable range.

The structural layout of this paper is as follows: Section 2 discusses the proposed UWB-MIMO system's structure, design evolution, parameter analysis, and current distribution. Section 3 presents the proposed MIMO system's simulated and measured performance, including S-parameters, far-field characteristics, and MIMO features. A comparison with literature is provided in Section 4 to emphasize the benefits of the proposed design. The conclusions are detailed in Section 5.

2. The Proposed Antenna System

2.1. Antenna Geometry

Figure 1 depicts the topology of the proposed dual-port, flower-shaped UWB-MIMO antenna system, and Figure 2 shows the fabricated prototype. Compared with the antennas reported in [3,6,8,9,11,12,14], the MIMO antenna system proposed in this study has a smaller size of 30 × 18 mm^2 (0.84λ × 0.50λ), and was designed on an FR4 substrate with 1.6 mm thickness (tanδ = 0.02 and εr = 4.4). Two similar flower-shaped radiating elements and a metal ground make up the overall antenna model. Each flower-shaped radiating element directly supplied by a microstrip line is made up of three elliptically shaped patches set above the substrate, each at an angle of 60 degrees from the others. Next, improved and inverted L-shaped branches with mirror symmetry and an I-shaped stub above L-shaped branches are added to create a T-like branch at the bottom of the substrate and above the rectangular floor, thereby establishing the proposed ground structure with a rectangular floor. The role of the T-like branching in this design is comparable to that of the ladder resonator proposed in [16], which will effectively block or absorb the surface current between the patch antenna elements at the operating frequency, thereby reducing the mutual influence. The specific design process and principle are described below. The parameters of the proposed dual-port UWB-MIMO antenna are listed in Table 1.

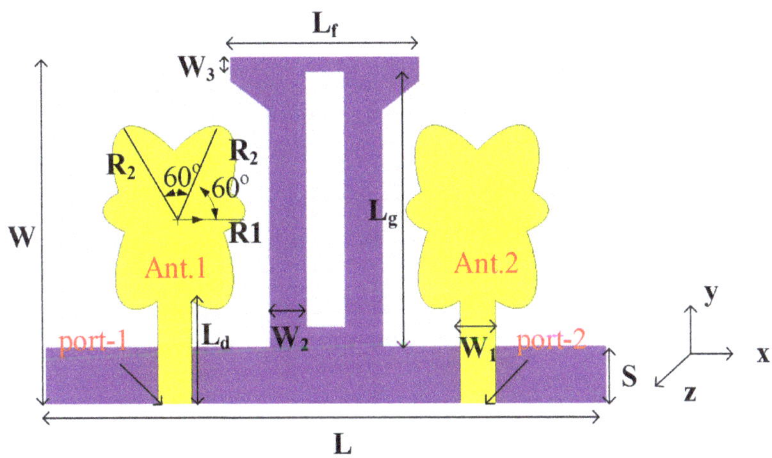

Figure 1. The proposed dual-port, flower-shaped UWB-MIMO antenna system structure.

(a) (b)

Figure 2. Fabricated prototype of the proposed dual-port, flower-shaped UWB-MIMO antenna: (**a**) front view, (**b**) back view.

Table 1. Dimensions of the proposed MIMO antenna structure (f = 4 GHz).

Parameter	Dimension	Value (mm)
L	Length of MIMO antenna	30 (0.84λ)
W	Width of MIMO antenna	18 (0.50λ)
L_d	Length of microstrip feed line	5.416 (0.15λ)
W_1	Width of microstrip feed line	1.8 (0.05λ)
R_1	Radius of ellipse 1	3.75 (0.10λ)
R_2	Radius of ellipse 2	5.25 (0.15λ)
S	Width of rectangular metallic ground	3 (0.084λ)
L_g	Vertical length of modified L-shaped ground branch	14 (0.39λ)
W_2	Width of modified L-shaped ground branch	2 (0.056λ)
L_f	Length of I-shaped ground stub	10 (0.28λ)
W_3	Width of I-shaped ground stub	1 (0.028λ)

2.2. Design Evolution Stages of the MIMO Antenna

To examine the implications of different MIMO antenna configurations, the 50 Ω transmission line feed is utilized in combination with the fractional ground plane. The overall design procedure for the proposed UWB-MIMO antenna system is elaborated in Figure 3, and the MIMO system's working principle, using its reflection coefficient and transmission coefficient curves, is presented in Figures 4 and 5.

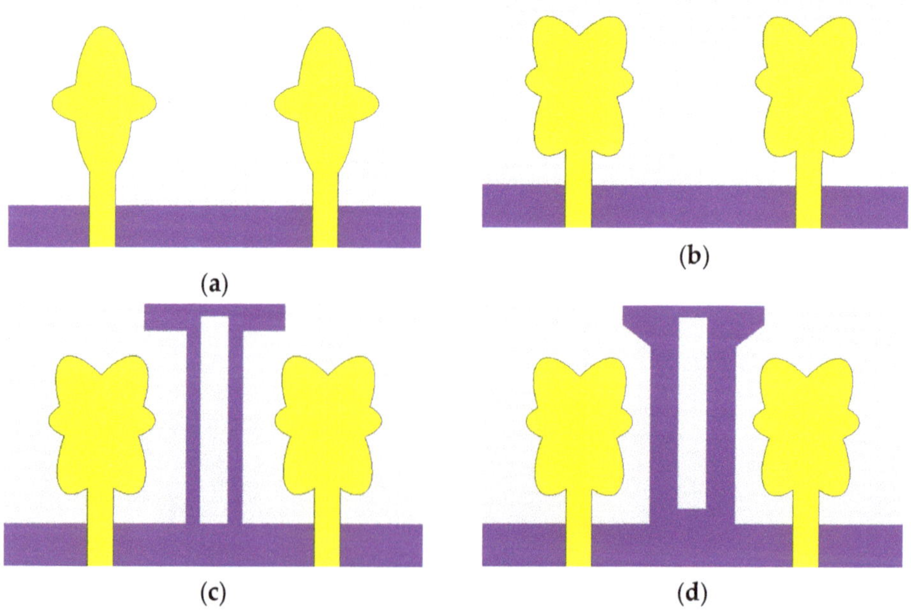

Figure 3. Evolution of the design process of UWB-MIMO system: (**a**) step 1, (**b**) step 2, (**c**) step 3, (**d**) step 4 (proposed MIMO system).

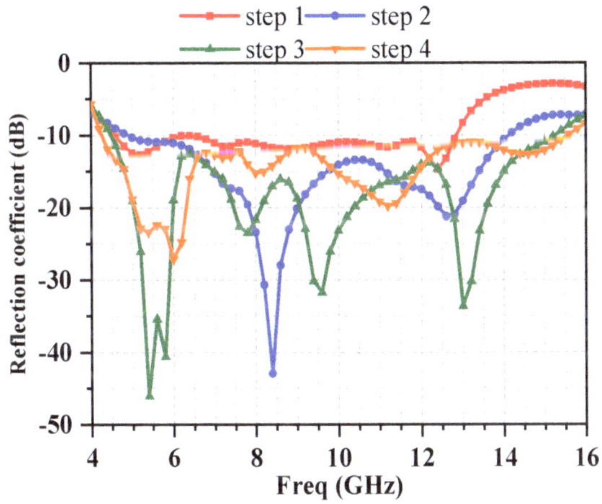

Figure 4. Comparison of reflection coefficients of 4 proposed UWB-MIMO antenna designs.

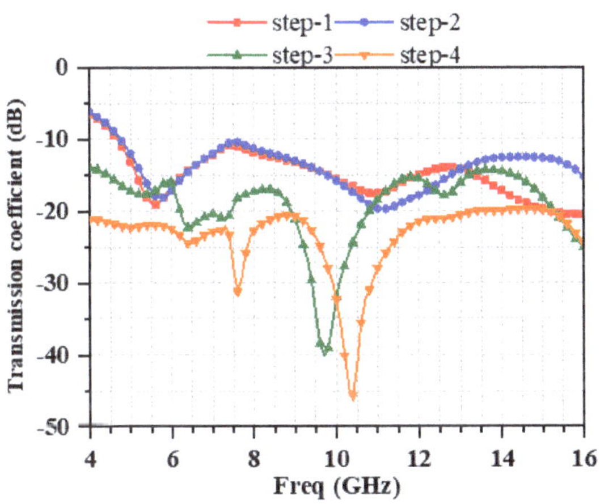

Figure 5. Comparison of transmission coefficients of 4 proposed UWB-MIMO antenna designs.

The radiating element in step 1 (Figure 3a) is made up of two mutually perpendicular elliptical patches and a microstrip line, along with a full rectangular ground at the bottom. Notably, the radiating element in this case is identical to the one proposed in [17], which is utilized for 5G communication. Although the antenna designed in step1 can cover 4.58–12.85 GHz, its reflection coefficient is poor, and the best value for S11 is only −15.2 dB, as shown in Figure 4. Meanwhile, the isolation between the antennas in the covered frequency range is less than 19 dB, since there is no decoupling structure involved, as illustrated in Figure 5.

Therefore, the radiating element is modified by merging three elliptically shaped patches in step 2, resembling a flower, but the ground structure is left unchanged. This antenna structure generates two resonant modes at 8.4 GHz and 12.8 GHz from its reflection coefficient, and the impedance-matching performance is improved. In the resonant modes, reflection coefficients are −43 and −21 dB, respectively, and the impedance bandwidth reaches 4.89–14.13 GHz. The transmission coefficient between the antennas, on the other hand, has not improved.

Progressively, in step 3, a horizontal I-shaped stub is placed above the inverted L-shaped branch to produce a T-like branch on ground, which improves the impedance matching and isolation compared with step 2. The antenna is stimulated into five resonant modes (5.38, 5.8, 7.8, 9.6, and 13 GHz), as shown by its reflection coefficient results, thus suggesting that the new T-like branch functions as a resonator, hence extending the bandwidth so that it spans between 4.48 and 15.26 GHz. From these results, we can see that the antenna's bandwidth has been increased. Furthermore, the total reflection coefficient is lowered, indicating that the impedance-matching ability is improved. Due to the separation impact of the T-like branch on the antenna components, isolation is enhanced, reaching more than 14.2 dB. These findings suggest that the adoption of the T-like branch is important for boosting the bandwidth and isolation.

In the MIMO antenna system, increasing the independence between the antenna components has long been a desired aim. However, as a result, additional reductions in the correlation and improved isolation between the antenna components are required. Accordingly, we enhanced the shape of the metallic ground in step 4, which completed the design of the proposed UWB-MIMO system. To construct the final ground structure, the shape of the inverted L-shaped branch was slightly modified, i.e., the vertical width was extended and the triangular patches were proportionately cut out at the edges of the inverted L-shaped branches. The resonance frequencies stimulated in step 4 are shifted to the right compared with those of step 3, which are now 5.4, 6, 8, 11.2, and 14.6 GHz, respectively, as displayed in Figure 4. Although the reflection coefficient performance is worse than that of step 3, the design of step 4 still covers the frequency range of 4.3–15.63 GHz. More importantly, Figure 5 shows that the isolation in step 4 has been enhanced compared with that in step 3, reaching more than 20 dB, which suggests that the mutual coupling has been decreased. Essentially, these results imply that the improved ground structure makes a significant contribution toward improving the isolation.

2.3. Parameter Analysis

The lengths of the modified T-like branches L_g and L_f have a dramatic impact on the UWB-MIMO system's impedance-matching and isolation performance. Only the impact of these particular factors on system performance is examined, while other parameters are kept constant. The S-parameters for tuning L_g from 12 mm to 14 mm are illustrated in Figure 6. The impedance bandwidth is further improved when L_g changes from 12 mm to 14 mm, as shown in Figure 6a. Neither UWB features can be achieved when L_g has a value of 12 mm or 13 mm. Different values of L_g also exhibit various isolation effects in terms of their influence on isolation. Although the isolation effect is optimal overall when L_g is at 14 mm, as shown in Figure 6b, the transmission coefficient must be enhanced in the 6–8 GHz band region when L_g is at 14 mm. These results demonstrate that the S-parameters of the system are significantly influenced by the vertical length of the modified L-shaped ground branch (L_g), and it is best to select a value of 14 mm for L_g when taking into account the size of the system.

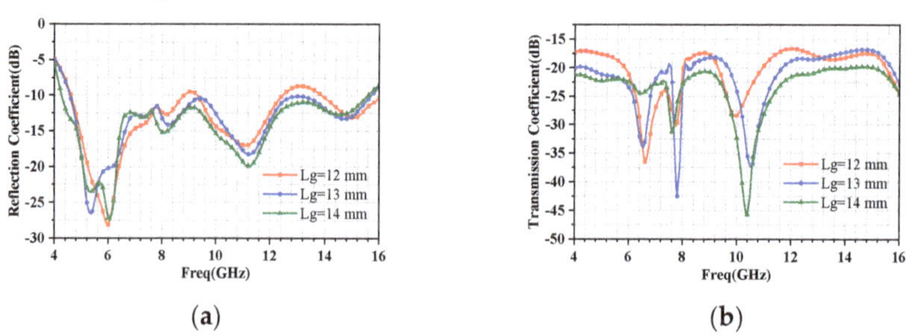

Figure 6. Simulated S-parameters for tuning L_g: (**a**) reflection coefficient, (**b**) transmission coefficient.

The simulated variation in characteristics of L_f from 9 mm to 11 mm are presented in Figure 7. As can be seen from Figure 7a, the effect of L_f on return loss is not very significant at these values, and all can achieve acceptable ultra-wideband properties. However, the effect of L_f on isolation is even more pronounced. As can be observed in Figure 7b, the optimal isolation effect is obtained when L_f is 10 mm, that is, when the I-shaped ground stub and modified L-shaped ground branch are exactly combined to form a T-like branch, which also confirms the authenticity of the proposed decoupling structure.

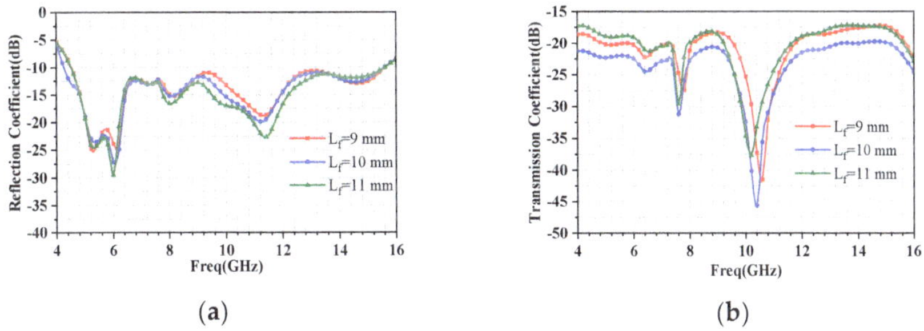

Figure 7. Simulated S-parameters resulting from the tuning of L_f: (**a**) reflection coefficient, (**b**) transmission coefficient.

2.4. Current Distribution

Figure 8 presents the surface current distribution in the resonance modes, to visually emphasize the decoupling effect of ground plane geometry. In the proposed MIMO system, port 1 is stimulated, while port 2 is terminated with a 50 Ω matched load. When just port 1 is stimulated, the current is largely distributed on antenna 1 and its modified T-like branch on the same side, as shown in Figure 8, whereas the current distribution on the surface of antenna 2 is relatively weak. It can be deduced that current-absorbing effect of the improved ground branch successfully improves the port isolation between the two monopole antennas.

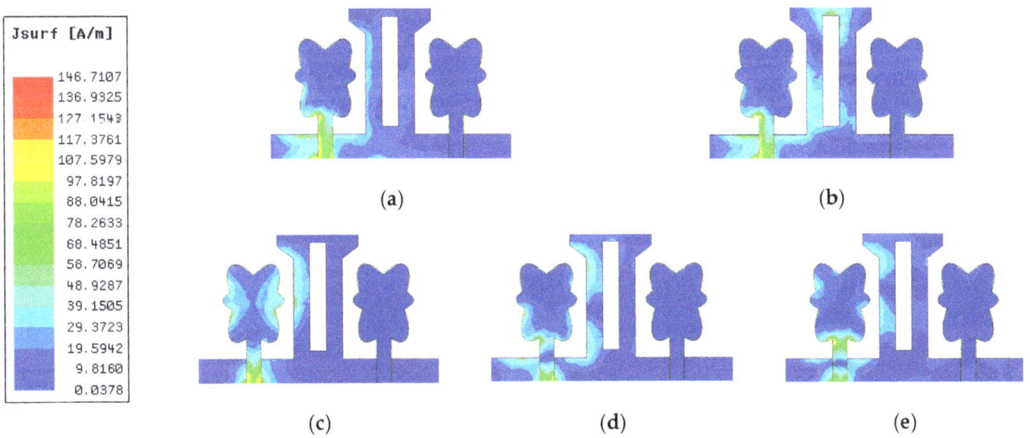

Figure 8. Surface current distribution when port 1 is stimulated at (**a**) 5.4 GHz, (**b**) 6 GHz, (**c**) 8 GHz, (**d**) 11.2 GHz, (**e**) 14.6 GHz.

3. Results and Discussion

3.1. S-Parameter Results

HFSS and an Agilent N5247A vector network analyzer were used to simulate and measure the proposed UWB-MIMO antenna, and the results are provided in Figure 9. The measured S11 can cover 4.51–15.1 GHz, while S12 is below −15 dB, as shown in Figure 9. It is worthwhile to note that there are significant differences between the measured and simulated results of S12, especially in 6.5–7.5 GHz and 10–11 GHz bands, where some frequencies with a difference of more than 20 dB can be seen. Manufacturing and measurement errors might be responsible for the discrepancy between the simulated and measured S12 results, which does not affect the overall high-isolation performance of the system. From these findings, the proposed antenna offers a wide operation band with high isolation.

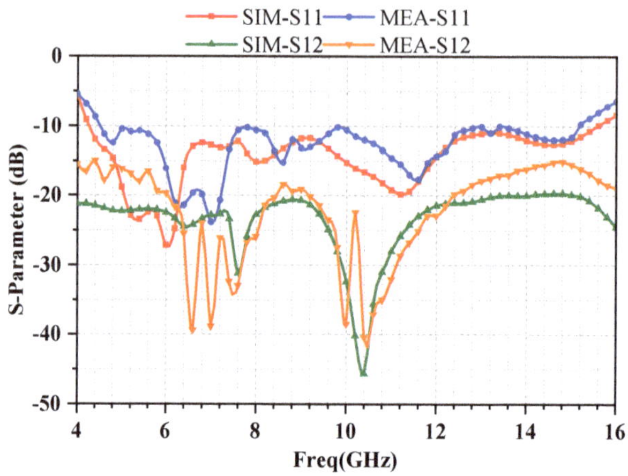

Figure 9. Simulated and measured S-parameters.

3.2. Far-Field Properties

The UWB-MIMO antenna was tested in an anechoic chamber for its radiation patterns, and the results are presented in Figure 10. By stimulating port 1 and terminating port 2 with a matched load, the corresponding radiation patterns were measured. It can be seen from Figure 10a,b that the antenna almost achieved omnidirectional radiation on both the XOZ plane and the YOZ plane at low frequencies (5.4 GHz and 6 GHz). At 8 GHz from (c), the radiation direction of the antenna on the XOZ surface was mainly distributed between 180° and 360°, and the maximum radiation gain reached 4.4 dB at approximately 255°, while on the YOZ surface, the maximum radiation direction was approximately 0° and 180°. At 11.2 GHz, the XOZ surface achieved almost omnidirectional radiation characteristics, while on the YOZ surface, the radiation gain obtained its maximum value at approximately 90°. However, at high frequency, it can be seen from Figure 10e that the radiation pattern of the XOZ plane was not as good as that of the YOZ plane. According to these results, the antenna had acceptable radiation characteristics on the XOZ (E-plane) and YOZ (H-plane) at frequencies of 5.4, 6, 8, 11.2, and 14.6 GHz, and the measured results are consistent with the simulated results.

The proposed MIMO antenna's radiation efficiency and peak gain are shown on Figure 11. As the radiators of the proposed MIMO antenna were structured symmetrically, the following values are shown for only one radiator. The radiation efficiency varied from 85 to 93 percent, implying that the majority of the energy was radiated away. The maximum gain of a single antenna at 5.4 GHz was 5.35 dBi, whereas the peak gain values at 6.4–7.6 GHz were relatively low, in the range of 2.5–4 dBi. Throughout the whole operational frequency range, the proposed MIMO antenna showed a positive gain value. Based on the results described above, the proposed UWB-MIMO antenna system offers promising radiation features.

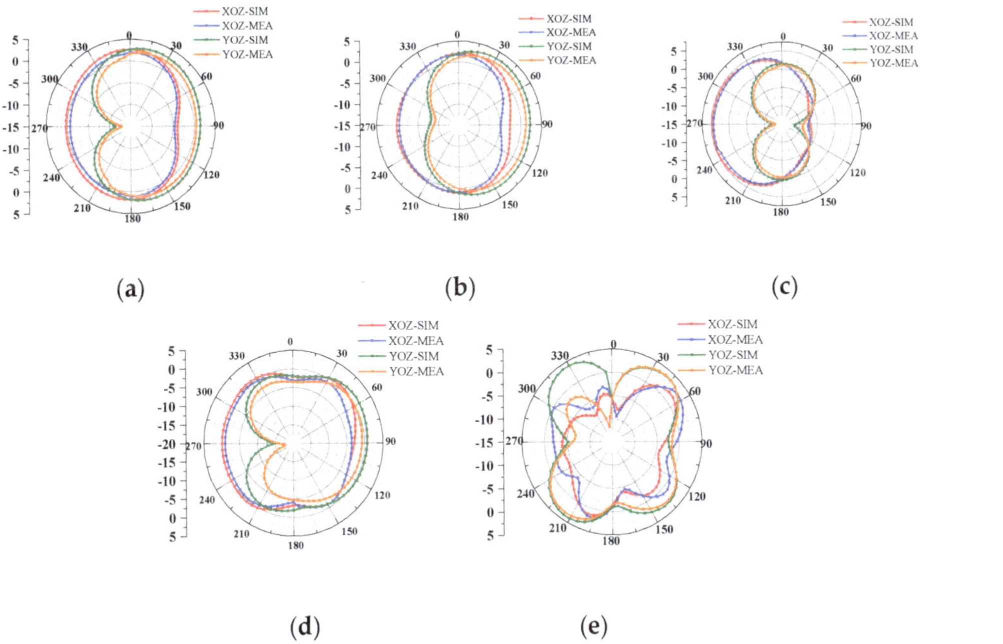

Figure 10. Simulated and measured far-field patterns on XOZ and YOZ planes at (**a**) 5.4 GHz, (**b**) 6 GHz, (**c**) 8 GHz, (**d**) 11.2 GHz, and (**e**) 14.6 GHz.

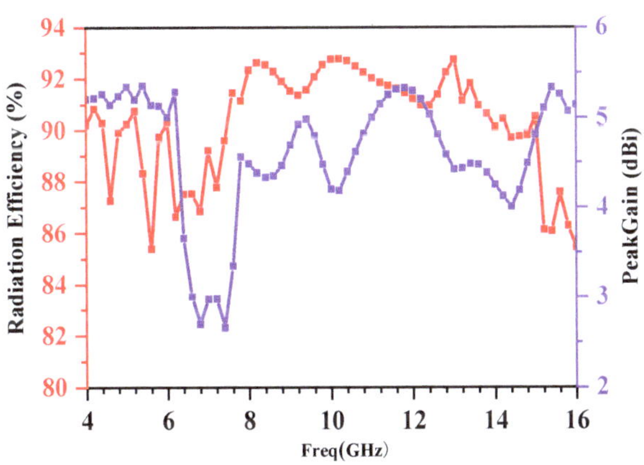

Figure 11. Calculated radiation efficiency and peak gain.

3.3. MIMO Performance

Different performance metrics of MIMO antennas, such as ECC (envelope correlation coefficient), DG (diversity gain), and TARC (total active reflection coefficient), should be examined to ensure their efficient operation.

The ECC value is a critical parameter for evaluating the performance of the radiation patterns of MIMO radiators. A lower ECC value indicates a lower effect on other antennas while working alone, and greater efficiency. To guarantee effective operation of each antenna, the specified ECC value of a MIMO system in wireless communication networks is generally less than 0.5 [18]. The ECC can be calculated from the S-parameters using

Equation (1). The results of simulated and measured ECC are displayed in Figure 12, which shows an acceptable isolation performance.

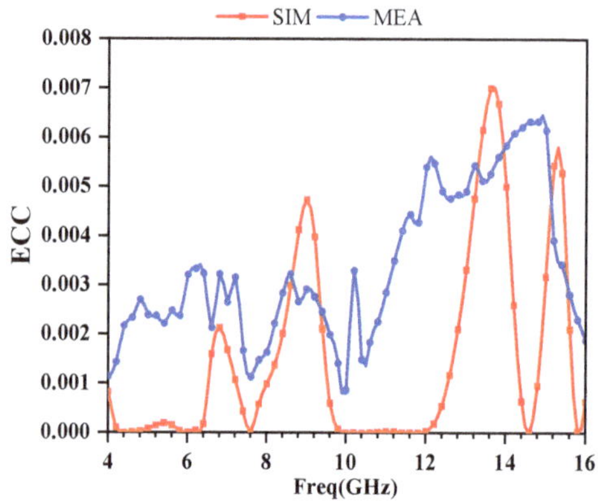

Figure 12. Simulated and measured ECC.

ECC can be computed from the S-parameters using the following equation.

$$\text{ECC} = \frac{|S_{11}^{*}S_{12} + S_{21}^{*}S_{22}|^2}{(1 - |S_{11}|^2 - |S_{21}|^2)(1 - |S_{22}|^2 - |S_{12}|^2)}, \tag{1}$$

Another key MIMO performance metric is DG, which describes how effective the diversity is. ECC is used to represent its value, which can be computed by Equation (2).

$$\text{DG} = 10 \times \sqrt{1 - |\text{ECC}|}, \tag{2}$$

The value of DG, calculated from the S-parameters of the proposed MIMO antenna system, is shown in Figure 13. In the working frequency band, DG is larger than 9.96 dB, thereby indicating a strong MIMO diversity performance.

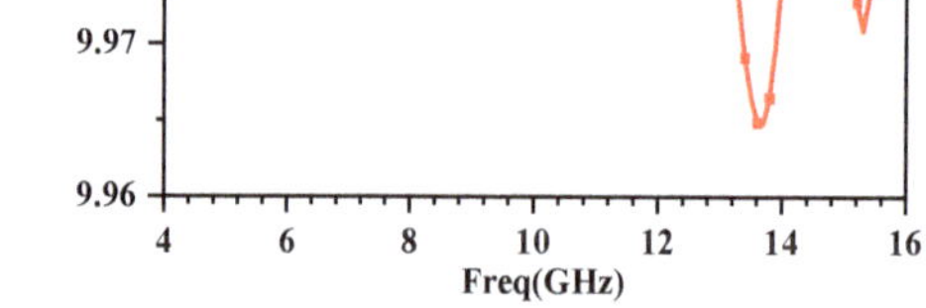

Figure 13. Calculated DG.

Furthermore, the statistic TARC is related to the total reflected power and total incident power, and it is used to assess the MIMO system's effectiveness. TARC should ideally be zero, which indicates that the antenna receives all of the incident power [19]. TARC can be computed using Equation (3) for a two-port system. Figure 14 displays a comparison between simulated and measured TARC values, which reveals that TARC is less than −20 dB in operating frequency range.

$$\mathrm{TARC} = -\sqrt{\frac{(S_{11}+S_{12})^2+(S_{21}+S_{22})^2}{2}}, \quad (3)$$

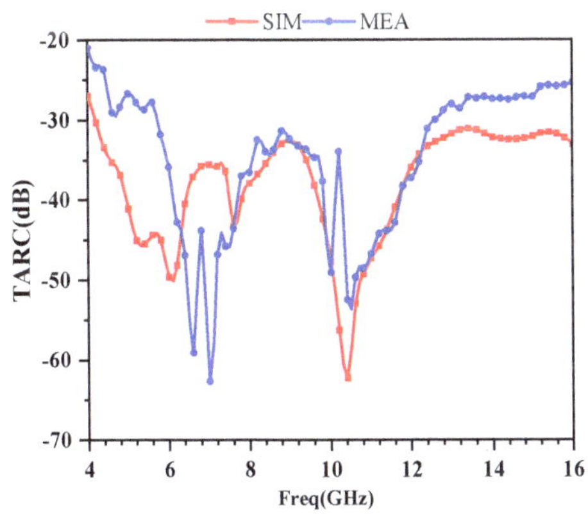

Figure 14. The comparison of simulated and measured TARC.

4. Comparison with Existing Models

Table 2 compares the performance of the proposed structure with the performance of various previously published UWB-MIMO antenna configurations. In contrast with the other reported designs, the proposed UWB-MIMO antenna structure provides obvious benefits. As can be observed from the table, the proposed UWB-MIMO antenna structure outperforms all the other designs. The proposed candidate is ideally suited for numerous UWB wireless applications owing to its small size, moderate impedance bandwidth, strong isolation capability, low ECC, high gain, and relatively steady radiation efficiency.

Table 2. Comparison of several reported UWB-MIMO antennas.

Ref	Size (mm^2)	Ports Number	Bandwidth (GHz)	Isolation (dB)	Decoupling Technique	ECC	Radiation Efficiency (%)	Gain (dBi)
[3]	34 × 34	4	2.5–12	15	Perpendicular Placement and a Parasitic Strip	<0.05	>75	2.5–5.5
[8]	26 × 31	2	3.1–11	25	Ground Stub and EBG	<0.01	>70	0–5.5
[9]	42 × 24	2	3.1–10.9	15	Vertical Placement	<0.2	>75	0–3.5
[12]	40 × 40	4	3.1–14	18	Swastika-shaped Stub	<0.012	>89	5.5
[20]	42 × 27	2	3.1–11.5	15	Defected Ground Structure (DGS)	<0.005	>75	0–2
[21]	29 × 23	2	3.0–12.0	15	Inverted L-shaped Stub and CSRR	<0.15	>82	4.7
[22]	26 × 28	2	2.9–10.8	15	T-shape Stub	<0.08	Not Given	1.6–4
[23]	32 × 32	2	2.9–12	15	Placed Perpendicularly	<0.04	>60	1.7–4.2
[24]	35 × 35	4	3.8–15	15	Ground Stubs	<0.07	>70	3–5
[25]	28 × 22	2	2.9–11.8	20	H-shape Slot	<0.03	Not Given	1.4–3.7
This work	30 × 18	2	4.3–15.63	20	Ground Branch	<0.0075	85–93	2.5–5.35

5. Conclusions

A compact-sized, dual-port, flower-shaped UWB-MIMO antenna with high isolation is presented in this paper. Three elliptically shaped metal patches, located at 60 degrees from each other, were combined to form a flower-shaped radiating element. The adopted ground structure improvements, including two modified inverted L-shaped branches and an I-shaped stub, expand the impedance bandwidth by generating multiple resonance modes to cover the 4.3–15.63 GHz (relative bandwidth 113.4%) range, which is typically used for multi-standard wireless applications, such as 5G N79 (4.4–5 GHz), WLAN (5.15–5.35 GHz/5.72–5.825 GHz), 5G spectrum band (5.9–6.4 GHz), X-band for satellite communication (8–12 GHz), FSS (11.45–11.7 GHz/12.5–12.75 GHz), and Ku band (12–18 GHz). These improvements also effectively reduce the mutual coupling between the antennas by absorbing the current, thereby enhancing the isolation to more than 20 dB. In addition, the measured results suggest that the proposed antenna displays favorable radiation patterns, where the radiation efficiency is between 85% and 93%, while the peak gain ranges from 2.5–5.35 dBi. Moreover, the antenna also possesses acceptable values for ECC (<0.0075), DG (>9.96 dB), and TARC (<−20 dB), demonstrating that the proposed MIMO antenna is highly compatible with the UWB communication systems.

Author Contributions: Conceptualization, Z.W.; methodology, H.L. and M.Y.; software, W.M.; validation, W.M. and C.L.; investigation, W.N.; writing—original draft preparation, W.M.; writing—review and editing, H.L. and Z.W.; supervision, J.W. All authors have read and agreed to the published version of the manuscript.

Funding: This research was funded in part by the Anhui Provincial Natural Science Foundation under grant 2108085MF200, the Natural Science Foundation of Anhui Provincial Education Department under grant KJ2020A0307 and KJ2020A0768, the Academic Funding Project for Distinguished Top Talents of Colleges and Universities in Anhui Province under grant gxbjZD2021088, and the Graduate Innovation Fund of Anhui University of Science and Technology under grant 2021CX2070.

Institutional Review Board Statement: Not applicable.

Informed Consent Statement: Not applicable.

Data Availability Statement: The simulated and measured data used to support the findings of this study are included within the article.

Conflicts of Interest: The authors declare no conflict of interest.

References

1. Mohamadzade, B.; Simorangkir, R.B.V.B.; Hashmi, R.M.; Lalbakhsh, A. A Conformal Ultrawideband Antenna With Monopole-Like Radiation Patterns. *IEEE Trans. Antennas Propag.* **2020**, *68*, 6383–6388. [CrossRef]
2. Chandel, R.; Gautam, A.K.; Rambabu, K. Tapered Fed Compact UWB MIMO-Diversity Antenna with Dual Band-Notched Characteristics. *IEEE Trans. Antennas Propag.* **2018**, *66*, 1677–1684. [CrossRef]
3. Chen, Z.J.; Zhou, W.S.; Hong, J.S. A Miniaturized MIMO Antenna With Triple Band-Notched Characteristics for UWB Applications. *IEEE Access* **2021**, *9*, 63646–63655. [CrossRef]
4. Al-Gburi, A.J.A.; Ibrahim, I.M.; Zakaria, Z.; Abdulhameed, M.K.; Saeidi, T. Enhancing Gain for UWB Antennas Using FSS: A Systematic Review. *Mathematics* **2021**, *9*, 3301. [CrossRef]
5. Bilal, M.; Shahid, S.; Khan, Y.; Rauf, Z.; Wagan, R.A.; Butt, M.A.; Khonina, S.N.; Kazanskiy, N.L. A Miniaturized FSS-Based Eight-Element MIMO Antenna Array for Off/On-Body WBAN Telemetry Applications. *Electronics* **2021**, *11*, 522. [CrossRef]
6. Rahman, M.; NagshvarianJahromi, M.; Mirjavadi, S.S.; Hamouda, A.M. Compact UWB Band-Notched Antenna with Integrated Bluetooth for Personal Wireless Communication and UWB Applications. *Electronics* **2019**, *8*, 158. [CrossRef]
7. Masoodi, I.S.; Ishteyaq, I.; Muzaffar, K.; Magray, M.I. A compact band-notched antenna with high isolation for UWB MIMO applications. *Int. J. Microw. Wirel. Technol.* **2020**, *13*, 634–640. [CrossRef]
8. Khan, A.; Bashir, S.; Ghafoor, S.; Qureshi, K.K. Mutual Coupling Reduction Using Ground Stub and EBG in a Compact Wideband MIMO-Antenna. *IEEE Access* **2021**, *9*, 40972–40979. [CrossRef]
9. Alsath, M.G.N.; Kanagasabai, M. Compact UWB Monopole Antenna for Automotive Communications. *IEEE Trans. Antennas Propag.* **2015**, *9*, 4204–4208. [CrossRef]

10. Chithradevi, R.; Sreeja, B.S. A compact UWB MIMO antenna with high isolation and low correlation for wireless applications. In Proceedings of the 2017 IEEE International Conference on Antenna Innovations & Modern Technologies for Ground, Aircraft and Satellite Applications (iAIM), Bangalore, India, 24–26 November 2017; pp. 24–26.
11. Wang, L.L.; Du, Z.H.; Yang, H.L.; Ma, R.Y.; Zhao, Y.C.; Cui, X.Q.; Xi, X.L. Compact UWB MIMO Antenna With High Isolation Using Fence-Type Decoupling Structure. *IEEE Antennas Wirel. Propag. Lett.* **2019**, *18*, 1641–1645. [CrossRef]
12. Suresh, A.C.; Reddy, T. A Flower Shaped Miniaturized 4 × 4 MIMO Antenna for UWB Applications Using Characteristic Mode Analysis. *Prog. Electromagn. Res. C* **2022**, *119*, 219–233. [CrossRef]
13. Wu, L.; Cao, X.; Yang, B. Design and Analysis of a Compact UWB-MIMO Antenna with Four Notched Bands. *Prog. Electromagn. Res. M* **2022**, *108*, 127–137. [CrossRef]
14. Dalal, P.; Dhull, S.K. Design of triple band-notched UWB MIMO/diversity antenna using triple bandgap EBG structure. *Prog. Electromagn. Res. C* **2021**, *113*, 197–209. [CrossRef]
15. Agarwal, S.; Rafique, U.; Ullah, R.; Ullah, S.; Khan, S.; Donelli, M. Double Overt-Leaf Shaped CPW-Fed Four Port UWB MIMO Antenna. *Electronics* **2021**, *10*, 3140. [CrossRef]
16. Roshani, S.; Shahveisi, H. Mutual Coupling Reduction in Microstrip Patch Antenna Arrays Using Simple Microstrip Resonator. *Wirel. Pers. Commun.* **2022**, 1–13. [CrossRef]
17. Bait-Suwailam, M.M.; Almoneef, T.; Saeed, S.M. Wideband MIMO Antenna with Compact Decoupling Structure for 5G Wireless Communication Applications. *Prog. Electromagn. Res. Lett.* **2021**, *100*, 117–125. [CrossRef]
18. Li, Y.X.; Sim, C.Y.D.; Li, Y.; Yang, G.L. High-Isolation 3.5 GHz Eight-Antenna MIMO Array Using Balanced Open-Slot Antenna Element for 5G Smartphones. *IEEE Trans. Antennas Propag.* **2019**, *67*, 3820–3830. [CrossRef]
19. Bhatia, S.S.; Sharma, N. Modified Spokes Wheel Shaped MIMO Antenna System for Multiband and Future 5G Applications: Design and Measurement. *Prog. Electromagn. Res. C* **2021**, *117*, 261–276. [CrossRef]
20. Banerjee, J.; Karmakar, A.; Ghatak, R.; Poddar, D.R. Compact CPW-fed UWB MIMO antenna with a novel modified Minkowski fractal defected ground structure (DGS) for high isolation and triple bandnotch characteristic. *J. Electromagn. Waves Appl.* **2017**, *31*, 1550–1565. [CrossRef]
21. Khan, M.S.; Capobianco, A.D.; Asif, S.M.; Anagnostou, D.E.; Shubair, R.M.; Braaten, B.D. A compact CSRR enabled UWB diversity antenna. *IEEE Antennas Wirel. Propag. Lett.* **2016**, *16*, 808–812. [CrossRef]
22. Zhao, Y.; Zhang, F.S.; Cao, L.X.; Li, D.H. A compact dual band-notched MIMO diversity antenna for UWB wireless applications. *Prog. Electromagn. Res. C* **2019**, *89*, 161–169. [CrossRef]
23. Ren, J.; Hu, W.; Yin, Y.; Fan, R. Compact printed MIMO antenna for UWB applications. *IEEE Antennas Wirel. Propag. Lett.* **2014**, *13*, 1517–1520.
24. Zamir, W.; Kumar, D. A compact 4 × 4 MIMO antenna for UWB applications. *Microw. Opt. Technol. Lett.* **2016**, *58*, 1433–1436.
25. Kang, L.; Li, H.; Wang, X.H.; Shi, X.W. Miniaturized band-notched UWB MIMO antenna with high isolation. *Microw. Opt. Technol. Lett.* **2016**, *58*, 878–881. [CrossRef]

sensors

MDPI

Article

Portable Wideband Directional Antenna Scheme with Semicircular Corrugated Reflector for Digital Television Reception

Bancha Luadang [1], Rerkchai Pukraksa [2], Pisit Janpangngern [2], Khanet Pookkapund [3], Sitthichai Dentri [4], Sompol Kosulvit [2] and Chuwong Phongcharoenpanich [2,*]

1 Faculty of Engineering, Rajamangala University of Technology Rattanakosin, Nakhon Pathom 73170, Thailand; bancha.lua@rmutr.ac.th

2 School of Engineering, King Mongkut's Institute of Technology Ladkrabang, Bangkok 10520, Thailand; 62601099@kmitl.ac.th (R.P.); pisit.janpangngern@gmail.com (P.J.); sompol.antenna@gmail.com (S.K.)

3 Faculty of Science and Technology, Phranakhon Si Ayutthaya Rajabhat University, Phranakhon Si Ayutthaya 13000, Thailand; pkhanet@aru.ac.th

4 College of Industrial Technology, King Mongkut's University of Technology North Bangkok, Bangkok 10800, Thailand; sitthichaid@kmutnb.ac.th

* Correspondence: chuwong.ph@kmitl.ac.th

Abstract: This research proposed a portable wideband horizontally-polarized directional antenna scheme with a radome for digital terrestrial television reception. The operating frequency band of the proposed antenna scheme is 470–890 MHz. The portable antenna scheme was an adaptation of the Yagi-Uda antenna, consisting of a folded bowtie radiator, a semicircular corrugated reflector, and a V-shaped director. Simulations were carried out, and an antenna prototype was fabricated. To validate, experiments were undertaken to assess the antenna performance, including the impedance bandwidth ($|S_{11}| \leq -10$ dB), gain, and unidirectionality. The measured impedance bandwidth was 75.93%, covering 424–943 MHz, with a measured antenna gain of 2.69–4.84 dBi. The radiation pattern was of unidirectionality for the entire operating frequency band. The measured xz- and yz-plane half-power beamwidths were 150°, 159°, 160° and 102°, 78°, 102° at 470, 680, and 890 MHz, with the corresponding cross-polarization below −20 dB and −40 dB. The radome had a negligible impact on the impedance bandwidth, gain, and radiation pattern. The power obtained for the outdoor test, at 514 MHz, was 38.4 dBμV (−70.4 dBm) with a carrier-to-noise ratio (C/N) of 11.6 dB. In addition, the power obtained for the indoor test was 26.6 dBμV (−82.2 dBm) with a C/N of 10.9 dB. The novelty of this research lies in the concurrent use of the Yagi-Uda and bowtie antenna technologies to improve the impedance bandwidth and directionality of the antenna for digital terrestrial television reception.

Keywords: corrugated reflector; digital television; portable antenna; radome; wideband antenna

Citation: Luadang, B.; Pukraksa, R.; Janpangngern, P.; Pookkapund, K.; Dentri, S.; Kosulvit, S.; Phongcharoenpanich, C. Portable Wideband Directional Antenna Scheme with Semicircular Corrugated Reflector for Digital Television Reception. *Sensors* **2022**, *22*, 5338. https://doi.org/10.3390/s22145338

Academic Editors: Raed A. Abd-Alhameed, Chan Hwang See and Naser Ojaroudi

Received: 31 May 2022
Accepted: 13 July 2022
Published: 17 July 2022

1. Introduction

In recent years, attempts have been made to synthesize diverse antenna technologies to improve the impedance bandwidth of antennas for digital television reception. Specifically, in [1], a wideband dipole antenna for digital television reception was proposed using a multi-loop radiator and a coplanar waveguide feed. In [2], an antenna with seven resonance pads on the ground plane was proposed to improve the impedance bandwidth.

In [3], an internal antenna that combined a planar inverted F antenna (PIFA) and a loop antenna for digital television reception achieved an impedance bandwidth of 48%, covering 470–771 MHz. In [4], a miniaturized wideband meander-line antenna using a magneto dielectric material was developed for the DVB-H/LTE13/GSM850/900 applications. Despite wide impedance bandwidth [1–4], these antennas suffer from low gain.

As a result, further attempts were made to develop antennas for digital television reception with wide impedance bandwidth, high gain, and stable radiation pattern throughout the operating frequency band (470–890 MHz). Specifically, the bowtie antenna technology was adopted in the development of antennas for digital television reception due to impedance bandwidth enhancement [5]. In addition, the Yagi-Uda antenna technology was deployed to achieve high gain and directionality [6].

For the bowtie antenna technology, a U-shaped bowtie magneto-electric dipole unidirectional antenna with a dual-layered horn reflector was proposed for ultra-wideband applications [7]. In [8], multiple bowtie patch antenna configurations with broadband operation, adjustable beamwidth, and multiple-beam capability were proposed for broadband indoor wireless communication. In [9], an adaptive ground penetrating radar (GPR) antenna based on bowtie antenna technology was proposed for the classification of soil types.

In [10], an antenna with a bowtie dipole element above the ground plane and a metallic bridge for wide beamwidth was proposed for base station applications. A tapered bowtie slot antenna [11] and a cavity-backed bowtie antenna [12] achieved the unidirectional radiation pattern. In [13], a substrate integrated waveguide (SIW) cavity backed bowtie slot antenna was proposed to improve the gain and unidirectionality. In addition, a unidirectional bowtie array antenna with an incision gap for digital video broadcasting-T2 base station applications was proposed in [14].

The Yagi-Uda antenna is conventionally used for television signal reception, and it is normally installed on the rooftop. The advantages of the Yagi-Uda antenna are its simple design, low cost, and reasonably high gain. The principal elements of the Yagi-Uda antenna include the radiator, the reflector, and the director [15]. In [16,17], the broadband Yagi-Uda antennas were proposed for wireless communication applications.

Currently, the antennas commonly used for digital television signal reception include loop antennas, omnidirectional dipole antennas, monopoles, log-periodic dipole arrays (LPDA), and quasi-Yagi antennas [18]. The loop antennas, omnidirectional dipole antennas, and monopoles can receive signals from every direction but achieve a low gain [19–23]. Meanwhile, the LPDA and quasi-Yagi antennas realize a high gain by using multiple dipoles. However, the unidirectionality of the LPDA and quasi-Yagi antennas gives rise to several installation challenges [24–26]. To overcome such shortcomings, this research thus combined the Yagi-Uda and bowtie antenna technologies in the development of the portable wideband directional antenna scheme with a radome for digital television reception.

By definition, a radome is a structural enclosure that protects an antenna and is constructed of materials that minimally attenuate the electromagnetic signal transmitted and received by the antenna [27,28]. The materials used in the radome construction include fiber reinforced plastics (FRP), glass reinforced plastics (GRP), polypropylene (PP), acrylonitrile butadiene styrene (ABS), polystyrene unplasticized polyvinyl chloride (UPVC), and polycarbonate [29,30].

Specifically, this research proposed a portable wideband directional antenna scheme with a radome for digital terrestrial television reception (470–890 MHz). The portable antenna scheme was an adaptation of the Yagi-Uda antenna, consisting of a folded bowtie radiator, a semicircular corrugated reflector, and a V-shaped director. The radome was of ABS material. In the study, simulations were carried out, and an antenna prototype was fabricated. In addition, experiments were undertaken to determine the antenna performance, including the impedance bandwidth ($|S_{11}| \leq -10$ dB), gain, and unidirectionality. The outdoor and indoor testing was also carried out to assess the outdoor and indoor reception performance of the proposed antenna scheme.

The organization of this research is as follows: Section 1 is the introduction. Section 2 details the antenna structure, and Section 3 deals with the parametric study. Section 4 is concerned with the fabrication of the prototype antenna and experimental results. Section 5 discusses the indoor and outdoor application testing. The concluding remarks are provided in Section 6.

2. Antenna Structure

Figure 1 shows the geometry of the portable wideband directional antenna scheme for digital terrestrial television reception, and Figure 2 illustrates the proposed antenna scheme of 75 Ω input impedance with radome. The proposed antenna scheme was an adaptation of the Yagi-Uda antenna, consisting of the folded bowtie radiator, semicircular corrugated reflector, and V-shaped director.

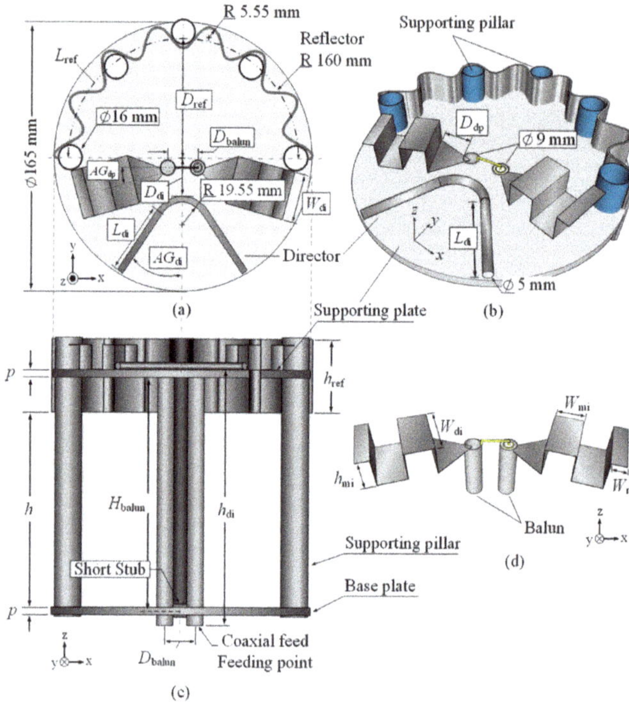

Figure 1. Geometry of the portable wideband directional antenna: (**a**) top view, (**b**) perspective view, (**c**) side view, and (**d**) folded bowtie radiator.

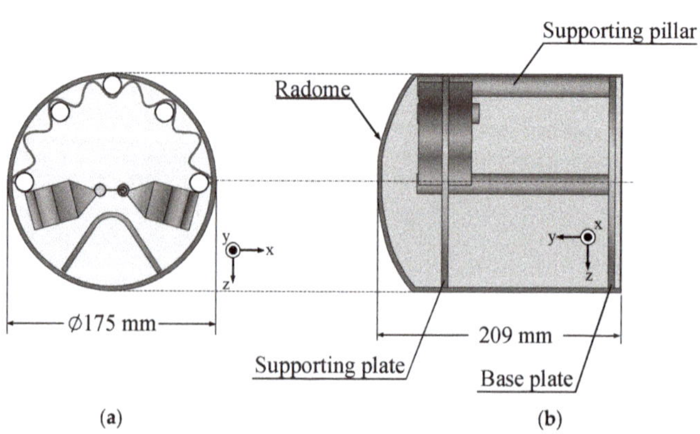

Figure 2. Geometry of the portable wideband directional antenna with radome: (**a**) top view, (**b**) side view.

The folded bowtie radiator and the corrugated reflector were made of a thin stainless-steel sheet (0.2 mm in thickness). The semicircular reflector was corrugated to reduce the overall physical size of the antenna while increasing its electrical size. In addition, the V-shaped director was made of a hollow aluminum tube 8 mm in diameter. The director was incorporated to improve the directivity and gain of the antenna.

The three elements (i.e., the folded bowtie radiator, corrugated reflector, and V-shaped director) were mounted on a circular plastic supporting plate 165 mm in diameter. The antenna was fed by a coaxial cable connected to a 75 Ω F-type connector, and a balun ($\lambda/4$ coaxial balun) was utilized to convert unbalanced to balanced output signal. The portable wideband directional antenna was also covered with a 0.5 mm-thick cylindrical acrylonitrile butadiene styrene (ABS) radome 175 and 209 mm in diameter and height.

Simulations were carried out to optimize the antenna parameters using CST Microwave Studio Suite [31]. Table 1 tabulates the optimal physical dimensions of the portable wideband directional antenna scheme for the 470–890 MHz frequency band. (Note: The experimental results revealed that the ABS radome had a negligible effect on the antenna performance, including the impedance bandwidth, gain, and radiation pattern).

Table 1. The optimal dimensions of the portable wideband directional antenna for digital terrestrial television reception.

Symbol	Description	Physical Size, mm	Electrical Size at 680 MHz
h	Spacing between the lower base plate and the reflector	139.00	0.315 λ_C
p	Thickness of the supporting plate and the base plate	5.00	0.011 λ_C
H_{balun}	Balun height	140.32	0.318 λ_C
h_{di}	Antenna height	156.00	0.353 λ_C
D_{balun}	Distance between center to center of the balun	19.24	0.043 λ_C
h_{ref}	Reflector height	45.00	0.102 λ_C
L_{ref}	Reflector length	226.42	0.513 λ_C
D_{ref}	Distance from center of the supporting plate to the reflector	78.61	0.178 λ_C
AG_{dp}	Angle of the folded bowtie radiator (degree)	15.00	-
W_{di}	Width of the folded bowtie radiator	30.00	0.068 λ_C
D_{di}	Distance between director and center of the supporting plate	25.00	0.056 λ_C
L_{di}	Director's arm length	43.34	0.098 λ_C
AG_{di}	Angle of the director's arm (degree)	34.60	-
D_{dp}	Length of triangular section of the folded bowtie radiator	18.70	0.042 λ_C
h_{mi}	Height of the folded bowtie radiator	15.00	0.034 λ_C
W_{mi}	Width of the 1st and 2nd sections of the folded bowtie radiator	15.20	0.034 λ_C
W_r	Width of the final section of the folded bowtie radiator	6.90	0.015 λ_C

λ_C is the wavelength at the center frequency of the target operation band (470–890 MHz).

3. Parametric Study

3.1. Conceptualization of the Portable Directional Antenna Scheme

Figure 3a–c illustrates the conceptualization of the proposed portable wideband directional antenna scheme for digital terrestrial television reception, starting with the Yagi-Uda antenna (model I), the modified antenna with the U-shaped reflector and director (model II), and the proposed antenna scheme (model III).

Model I is the Yagi-Uda antenna of a wire-type structure. The main advantages of the Yagi-Uda antenna are unidirectionality and high gain. However, model I suffered from a narrow impedance bandwidth. Given the center frequency of 680 MHz, the length of the radiator element was $\lambda_C/2$ (220.58 mm), where λ_C was the wavelength of the center frequency.

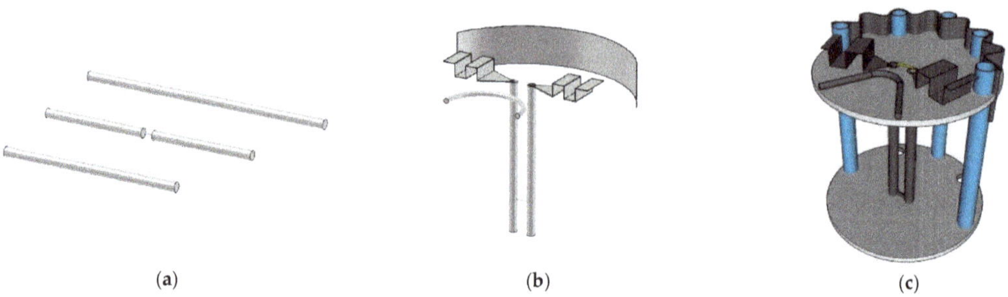

Figure 3. Conceptualization of the portable directional antenna scheme: (**a**) the Yagi-Uda antenna (model I), (**b**) the modified antenna with the U-shaped reflector and director (model II), and (**c**) the proposed portable antenna scheme (model III).

In model II, to improve the impedance bandwidth, the radiator element of model I was replaced with the folded bowtie radiator. The wire reflector was replaced with the stainless-steel U-shaped reflector, and the director was reshaped into the U shape. In model III, the U-shaped reflector was corrugated to reduce the overall physical size of the antenna while increasing the electrical size. Moreover, the corrugated reflector improved the electromagnetic reflection of the proposed antenna scheme [32,33]. The U-shaped director was reshaped into the V shape for the antenna compactness without compromising the antenna performance.

Figure 4 compares the simulated impedance bandwidth ($|S_{11}| \leq -10$ dB) of the antenna models I, II, and III. Model I achieved a narrow impedance bandwidth, covering 620–700 MHz. The narrow impedance bandwidth rendered model I operationally unsuitable for digital television reception. Model II experienced an impedance mismatch between 590–780 MHz, rendering it operationally inapplicable. Model III achieved the simulated impedance bandwidth of 63.85%, covering 468–907 MHz, which encompassed the target frequency band (470–890 MHz). Figure 5 compares the antenna gain of models I, II, and III.

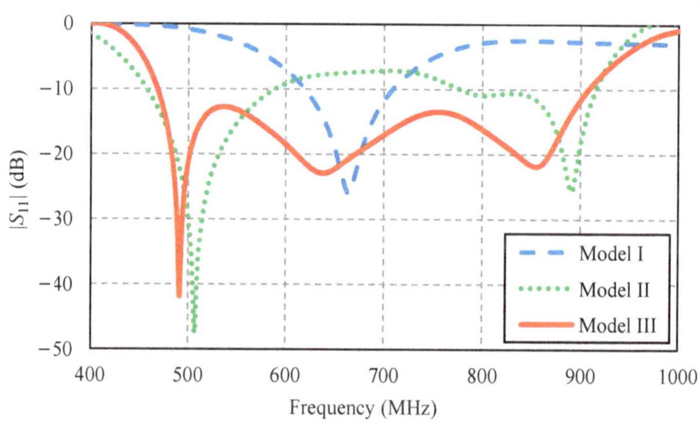

Figure 4. Simulated impedance bandwidth ($|S_{11}| \leq -10$ dB) of the antenna models I, II, and III.

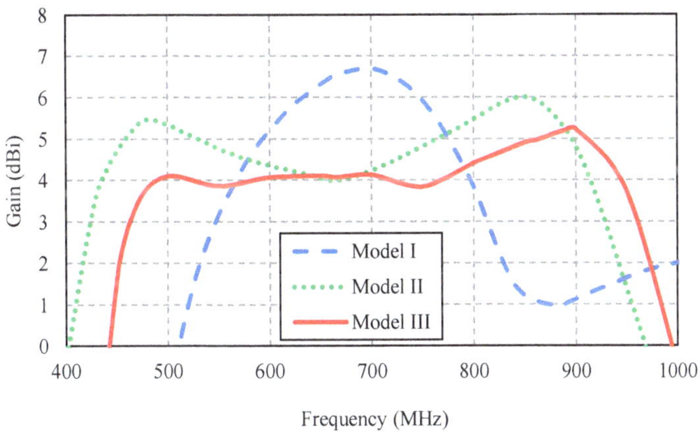

Figure 5. The simulated antenna gains of models I, II, and III.

Figure 6 compares the simulated xz- and yz-plane radiation patterns of the antenna models I, II, and III. The three antenna models exhibited similar radiation patterns. However, models I and II failed to cover the target frequency band (Figure 4). The xz- and yz-plane cross-polarizations of model III were below −40 dB and −20 dB, with the corresponding half-power beam width (HPBW) of 149°, 194°, and 158° and 99°, 92.8°, and 84.2° at 470, 680, and 890 MHz, respectively. The xz- and yz-plane back lobe levels were −10 dB.

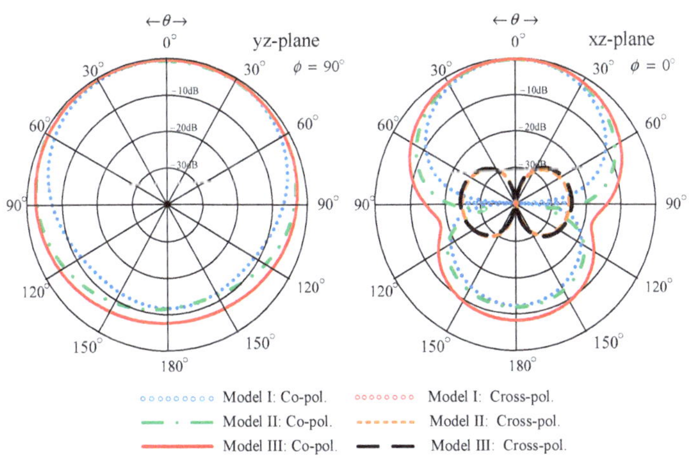

Figure 6. Simulated xz- and yz-plane radiation patterns of the antenna models I, II, and III.

3.2. Evolution of the Portable Wideband Antenna Scheme

This section discusses the evolutionary stages of the proposed portable wideband directional antenna scheme (model III), consisting of three antenna generations: first, second, and third generations. As shown in Figure 7, the first generation was the antenna scheme without a reflector and director, and the second generation was the antenna scheme with the corrugated reflector but without the director. The third generation was the proposed portable wideband directional antenna scheme with the corrugated reflector and the V-shaped director.

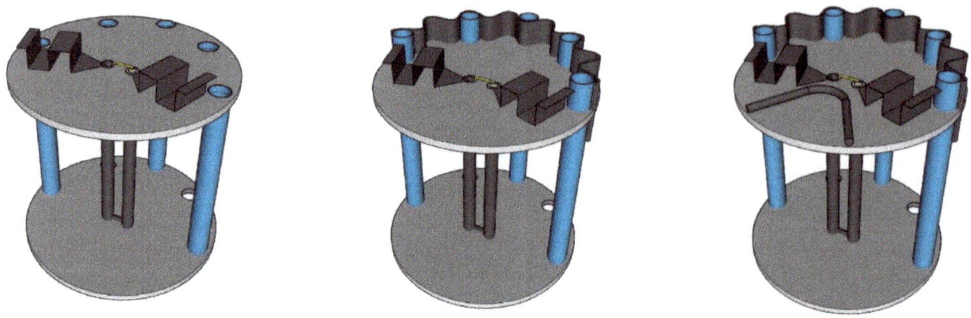

(**a**) (**b**) (**c**)

Figure 7. Evolutionary stages of the portable antenna scheme: (**a**) first generation, (**b**) second generation, and (**c**) third generation.

Figure 8 compares the simulated impedance bandwidth ($|S_{11}| \leq -10$ dB) of the first-, second-, and third-generation antenna schemes. The first-generation antenna scheme achieved a narrow impedance bandwidth, covering 830–925 MHz, rendering the scheme operationally unsuitable for digital terrestrial television reception. The impedance bandwidth of the second-generation antenna scheme closely resembled that of the third-generation scheme, covering a 468–907 MHz frequency band. Both the second- and third-generation antenna schemes covered the entire target operating frequency band of 470–890 MHz. However, the overall gain of the third-generation antenna scheme was higher than that of the second-generation scheme, as shown in Figure 9.

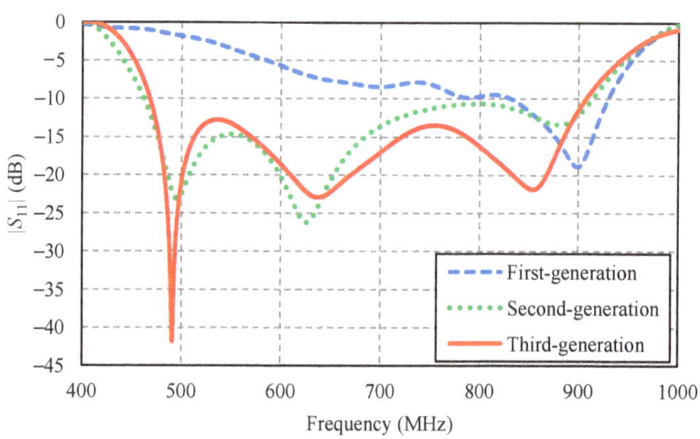

Figure 8. Simulated impedance bandwidth ($|S_{11}| \leq -10$ dB) of the first-, second-, and third-generation antenna schemes.

Figure 10 compares the simulated xz- and yz-plane radiation patterns of the first-, second-, and third-generation antenna schemes. The radiation pattern of the first-generation antenna scheme was of omnidirectionality due to the absence of the corrugated reflector (Figure 7a). The first-generation scheme failed to cover the entire target operating frequency band (Figure 8). The radiation patterns of the second- and third-generation antenna schemes were closely similar. However, the gain of the third-generation scheme was higher than that of the second-generation scheme.

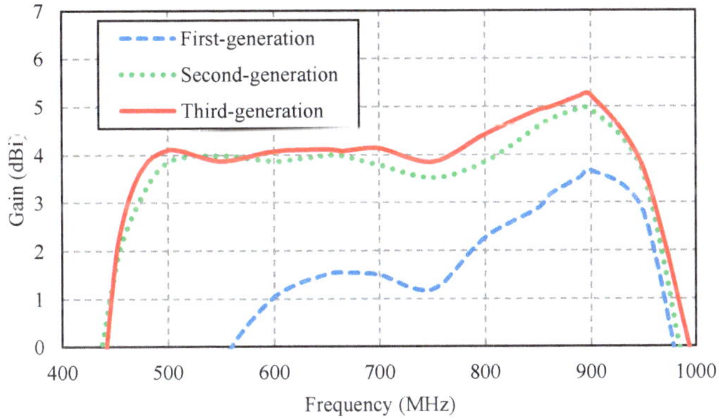

Figure 9. Simulated antenna gains of the first-, second-, and third-generation antenna schemes.

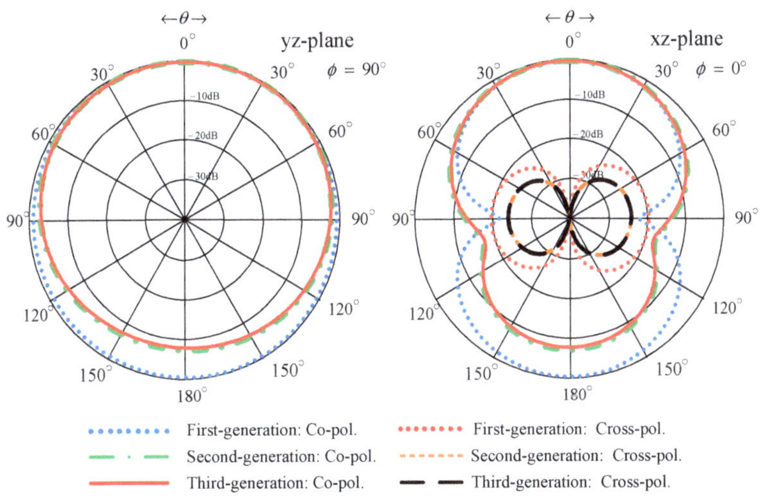

Figure 10. Simulated xz- and yz-plane radiation patterns of the first-, second-, and third-generation antenna schemes.

3.3. Surface Current Distribution

This section focuses on the third-generation antenna scheme (i.e., the proposed portable wideband directional antenna scheme) due to its wide impedance bandwidth (Figure 8), high gain (Figure 9), and unidirectional radiation pattern (Figure 10).

Figure 11a–c illustrates the simulated surface current distribution of the proposed portable wideband directional antenna scheme at 470, 680, and 890 MHz, respectively. At 470 MHz (the first resonant frequency), the currents are concentrated around the feeding point and along the corrugated reflector, as shown in Figure 11a. In Figure 11b, at 680 MHz (the second resonant frequency), the currents are concentrated around the feeding point, the lower part of the balun, and the upper edge of the corrugated reflector. At 890 MHz (the third resonant frequency), the currents are concentrated around the feeding point, the folded bowtie radiator, and the balun, as shown in Figure 11c. Essentially, the third-generation antenna scheme (i.e., the proposed portable wideband directional antenna scheme) covered the target frequency band for the digital terrestrial television reception of 470–890 MHz, as indicated in Figure 8.

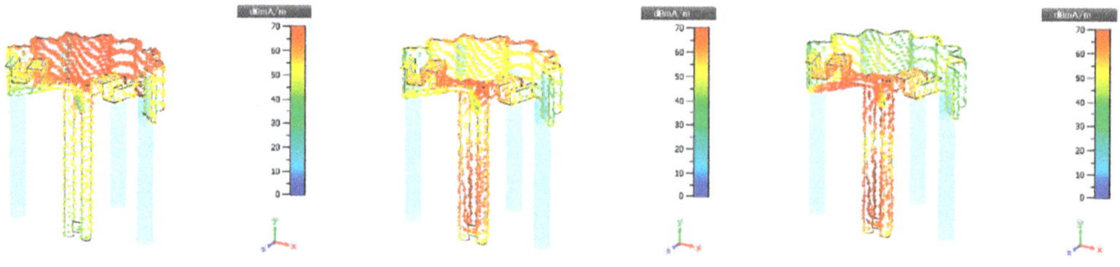

(a) (b) (c)

Figure 11. Simulated surface current distribution of the proposed portable wideband directional antenna scheme at: (**a**) 470 MHz, (**b**) 680 MHz, and (**c**) 890 MHz.

3.4. Parametric Sweep of the Antenna Scheme

This section investigates the effect of variable antenna parameters on the impedance bandwidth of the proposed portable wideband directional antenna scheme. The results are graphically presented in Figures 12–15.

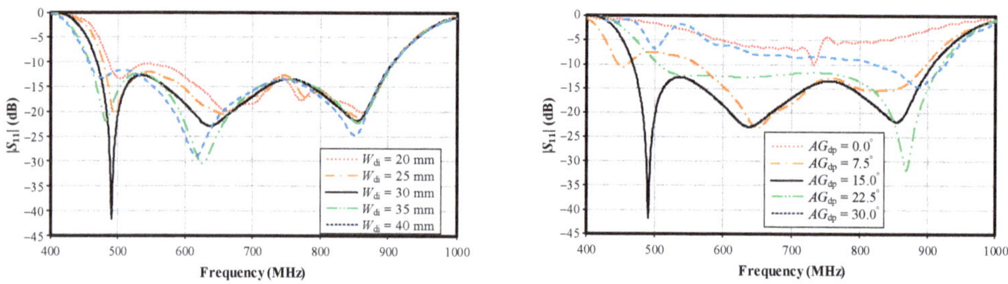

(a) (b)

Figure 12. Simulated impedance bandwidth ($|S_{11}| \leq -10$ dB) under variable parameters: (**a**) width of folded bowtie radiator (W_{di}), (**b**) angle of the folded bowtie radiator (AG_{dp}).

Figure 12a,b shows the simulated impedance bandwidth ($|S_{11}| \leq -10$ dB) under various widths of the folded bowtie radiator (W_{di}; 20, 25, 30, 35, and 40 mm) and angles of the folded bowtie radiator (AG_{dp}; 0, 7.5, 15, 22.5, and 30°). As shown in Figure 12a,b, the optimal W_{di} and AG_{dp} were 30 mm and 15°, respectively.

In addition, the optimal length of the triangular section of the folded bowtie radiator (D_{dp}) was 18.7 mm, and the optimal height of the folded bowtie radiator (h_{mi}) was 15 mm. The optimal width of the first and second sections of the folded bowtie radiator (W_{mi}) were 15.2 and 15.2 mm, while the optimal width of the final section of the folded bowtie radiator (W_r) was 6.9 mm. The optimal length of one arm of the folded bowtie radiator was thus 101 mm. In other words, the optimal overall length of the folded bowtie radiator was 202 mm.

Figure 13a,b shows the simulated impedance bandwidth ($|S_{11}| \leq -10$ dB) under variable distance from the center of the supporting plate to the reflector (D_{ref}) and reflector length (L_{ref}). As shown in Figure 13a,b, the optimal D_{ref} and L_{ref} were 78.61 mm and 226.42 mm, respectively, covering the target operating frequency band of 470–890 MHz. Figure 13c shows the simulated impedance bandwidth under variable reflector height (h_{ref}) (25, 45, 65, and 85 mm), and Figure 13d illustrates the corresponding simulated radiation pattern at 890 MHz. The larger h_{ref} improved the back lobe level (i.e., front-to-back ratio), especially in the upper frequency band. In Figure 13d, despite the high back lobe level of −21.3 dB under h_{ref} = 85 mm, the final antenna was bulky. As a result, the reflector height (h_{ref}) of 45 mm, with a back lobe level of −14.14 dB, was selected for compactness.

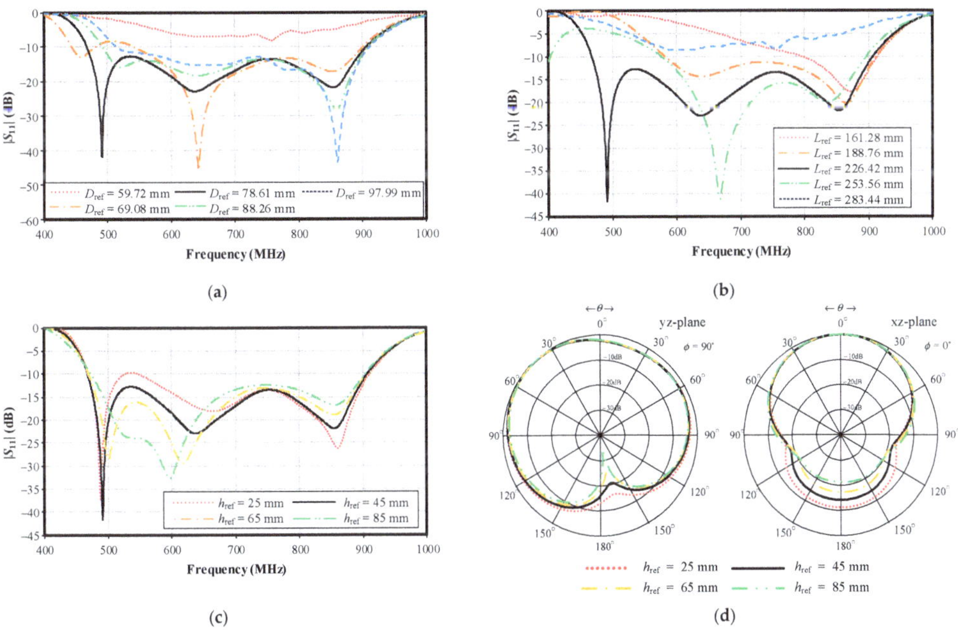

Figure 13. Simulated impedance bandwidth under variable parameters: (**a**) distance from center of the supporting plate to the reflector (D_{ref}), (**b**) reflector length (L_{ref}), (**c**) reflector height (h_{ref}), and (**d**) radiation pattern at 890 MHz under variable h_{ref}.

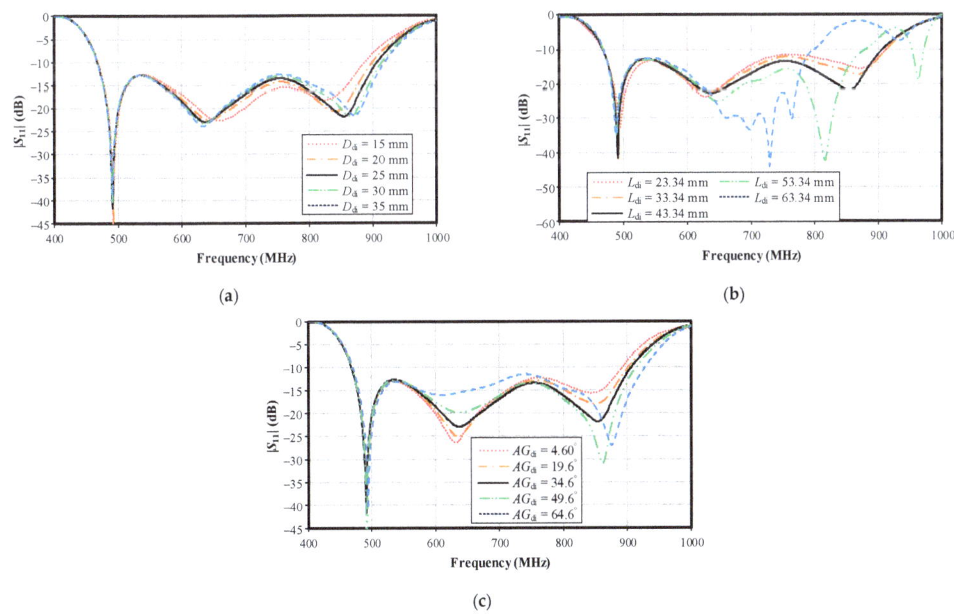

Figure 14. Simulated impedance bandwidth ($|S_{11}| \leq -10$ dB) under variable parameters: (**a**) distance between the director and center of the supporting plate (D_{di}), (**b**) director's arm length (L_{di}), and (**c**) angle of the director's arm (AG_{di}).

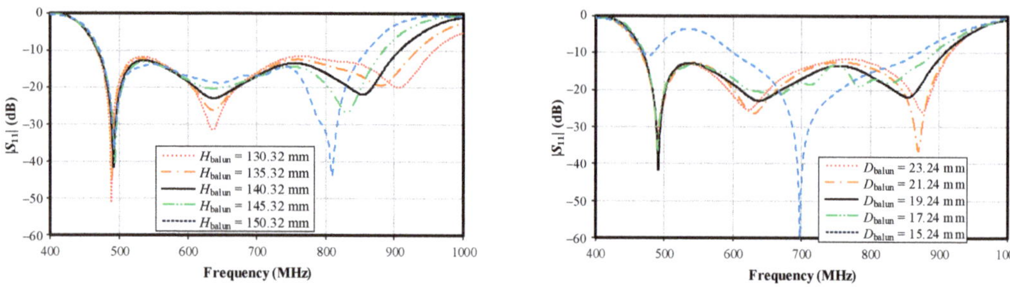

(**a**) (**b**)

Figure 15. Simulated impedance bandwidth ($|S_{11}| \leq -10$ dB) under variable parameters: (**a**) balun height (H_{balun}), (**b**) distance from center to center of the balun (D_{balun}).

Figure 14a–c shows the simulated impedance bandwidth ($|S_{11}| \leq -10$ dB) under variable distance between the director and the center of the supporting plate (D_{di}), the director's arm length (L_{di}), and the angle of the director's arm (AG_{di}). The optimal D_{di}, L_{di}, and AG_{di} were 25 mm, 43.34 mm, and 34.6 mm, respectively, covering the target operating frequency band of 470–890 MHz.

Figure 15a,b shows the simulated impedance bandwidth ($|S_{11}| \leq -10$ dB) under variable balun height (H_{balun}) and distance from the center to the center of the balun (D_{balun}). The optimal H_{balun} and D_{balun} were 140.32 mm and 19.24 mm, respectively. Meanwhile, the antenna height (h_{di}; 156 mm) and the spacing between the lower base plate and the reflector (h; 139 mm) were proportional to the balun height (H_{balun}) (Table 1).

4. Prototype Fabrication and Measured Results

Figure 16a,b depicts the prototype of the portable wideband directional antenna scheme for digital terrestrial television reception without and with ABS radome. Experiments with the antenna prototype (75 Ω input impedance) were carried out in an anechoic chamber using a 50 Ω vector network analyzer (Agilent E5061B) [34]. The impedance transformer (TME ZT-205) was used to match a 75 Ω antenna with a 50 Ω measurement system.

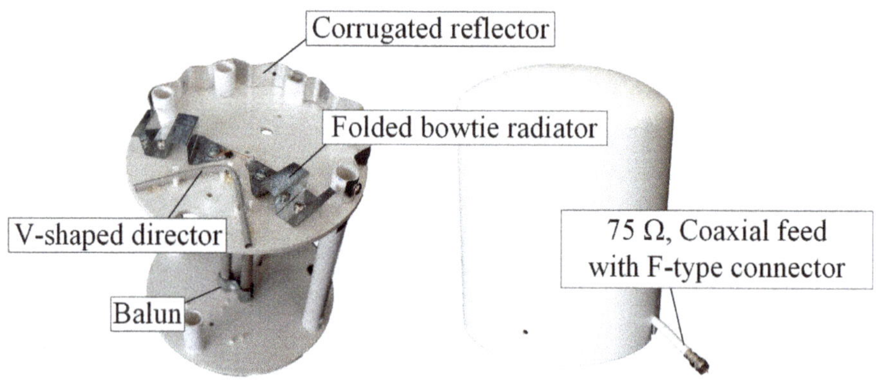

(**a**) (**b**)

Figure 16. Prototype of the portable wideband directional antenna scheme: (**a**) without radome, (**b**) with radome.

Figure 17 compares the simulated and measured impedance bandwidth ($|S_{11}| \leq -10$ dB) of the portable wideband directional antenna scheme. The measured impedance bandwidth was 75.93%, covering the 424–943 MHz frequency band, with the measured antenna gain of 2.69–4.84 dBi (Figure 18).

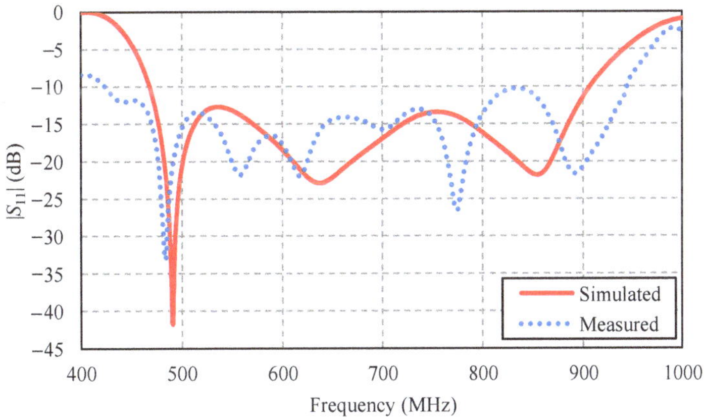

Figure 17. Simulated and measured impedance bandwidths of the portable wideband directional antenna scheme.

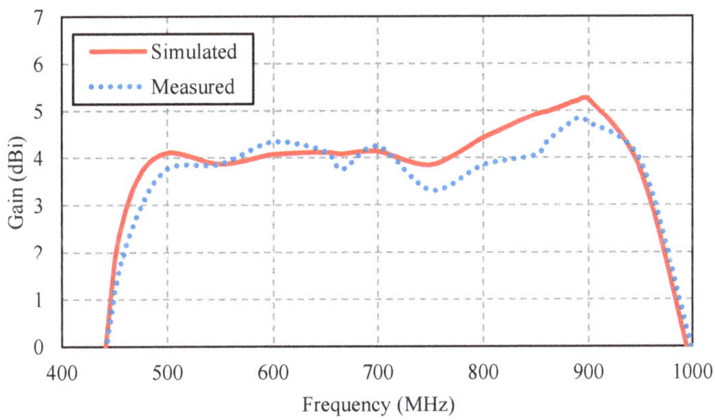

Figure 18. Simulated and measured gains of the portable wideband directional antenna scheme.

As shown in Figure 19, the radiation pattern of the proposed portable wideband horizontally polarized directional antenna scheme was of unidirectionality. The measured xz- and yz-plane cross-polarizations were below −20 dB and −40 dB, respectively, with the corresponding HPBW of 150°, 159°, and 160° and 102°, 78°, and 102° at 470 MHz, 680 MHz, and 890 MHz, respectively. The discrepancy between the simulated and measured cross-polarization could be attributed to the use of self-trapping screws in the antenna prototype fabrication, as shown in Figure 16a. The measured xz- and yz-plane back lobe levels were below −9 dB. The experiments also revealed that the ABS radome had negligible impacts on the impedance bandwidth and the antenna gain, vis-à-vis the antenna scheme without the ABS radome. Table 2 tabulates the simulated and measured performance of the proposed portable wideband directional antenna scheme. Table 3 shows a comparison between previous works and the proposed portable wideband directional antenna scheme.

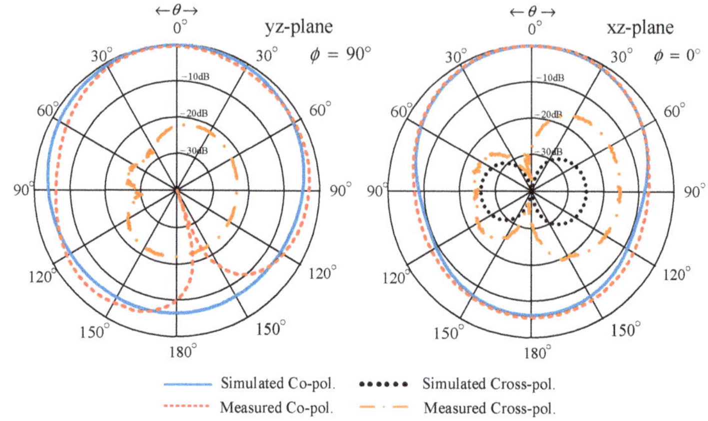

(**a**)

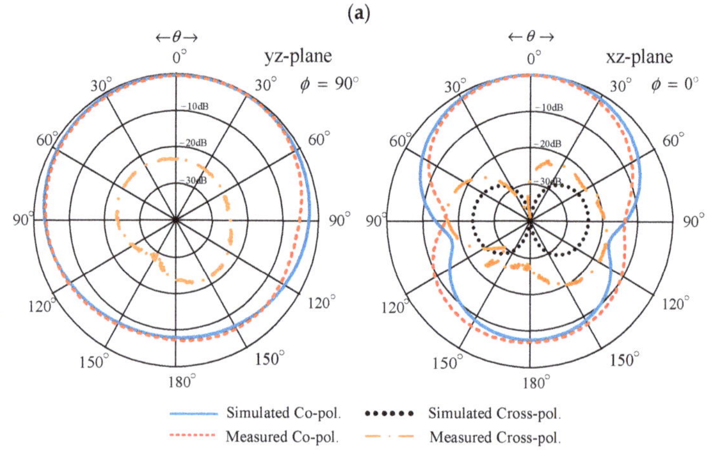

(**b**)

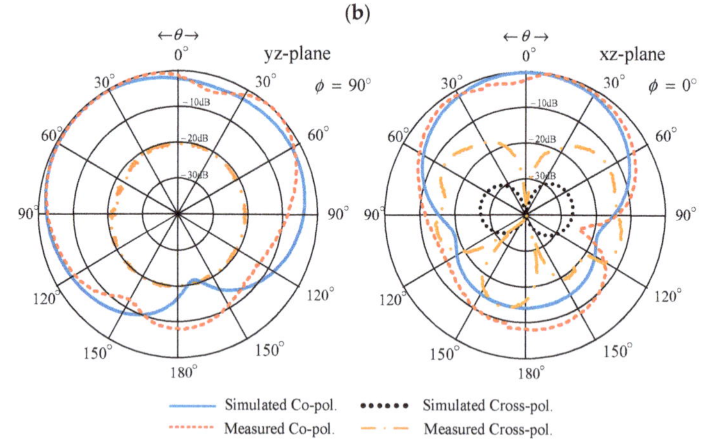

(**c**)

Figure 19. Simulated and measured xz- and yz-plane radiation patterns of the portable wideband directional antenna scheme: (**a**) 470 MHz, (**b**) 680 MHz, and (**c**) 890 MHz.

Table 2. Simulated and measured performances of the portable wideband directional antenna scheme.

Specifics	Simulation			Measurement		
	470 MHz	680 MHz	890 MHz	470 MHz	680 MHz	890 MHz
$\lvert S_{11} \rvert < -10$ dB Bandwidth, %		(468–907), 63.85%			(424–943), 75.93%	
Gain (dBi)	3.45	4.07	5.21	2.69	3.75	4.84
HPBW in xz-plane, (deg.)	149	194	185.1	150	159	160
HPBW in yz-plane, (deg.)	99	92.8	84.2	102	78	102
Cross-pol. in xz-plane (dB)		≤ -40			≤ -20	
Cross-pol. in yz-plane (dB)		≤ -40			≤ -40	
Back lobe level (dB)		≤ -10			≤ -9	
Radiation pattern		Unidirectional			Unidirectional	

Table 3. Comparison between previous works and the proposed portable wideband directional antenna scheme.

Reference	Antenna Type	Bandwidth (MHz), (%)	Gain (dBi)	Dimension (mm) (Electrical Size at 470 MHz)
[1]	Printed dipole with multiple loops antenna	430–1180 (93.16%)	1.78–2.83	58.5 × 241 × 1.6 (0.091λ × 0.377λ × 0.0025λ)
[2]	Printed monopole with spiral loops antenna	470–862 (58.85%)	−19–(−12)	20 × 30 × 0.4 (0.031λ × 0.047λ × 0.0006λ)
[3]	PIFA-loop antenna	470–771 (48.50%)	2.2–4.8	17 × 375 (0.026λ × 0.587λ)
[4]	Printed meander line antenna with magneto dielectric	467–1012 (73.69%)	0.86–1.65	10 × 25 × 1 (0.015λ × 0.039λ × 0.0015λ)
[19]	Printed dipole antenna	455–1070 (80.65%)	−0.57 1.15	20 × 200 × 0.8 (0.031λ × 0.313λ × 0.001λ)
[20]	Printed dipole antenna	452–897 (65.97%)	2.09–3.85	45 × 250 × 1.6 (0.070λ × 0.391λ × 0.002λ)
[21]	Printed dipole antenna	441–890 (67.46%)	4.65 (peak)	35 × 243 × 1.6 (0.054λ × 0.380λ × 0.002λ)
[23]	Printed loop antenna	461–806 (54.45%)	1.9–2.5	165 × 165 × 0.8 (0.258λ × 0.258λ × 0.001λ)
[24]	Printed quasi-Yagi antenna	450–848 (61.32%)	3.5–4.6	200 × 240 × 1.6 (0.313λ × 0.376λ× 0.002λ)
[26]	Log-periodic antenna	470–790 (50.79%)	8 (peak)	302.6 × 356 × 35 (0.474λ × 0.557λ 0.054λ)
Proposed antenna	Yagi-Uda with bowtie antenna	424–943 (75.93%)	2.69–4.84	175 × 175 × 209 (0.274λ × 0.274λ × 0.327λ)

5. Indoor and Outdoor Application Testing

The proposed portable wideband directional antenna scheme was tested for the indoor and outdoor reception of digital terrestrial television signals in the 470–862 MHz frequency range. This research was conducted in Thailand, which adopted the DVB-T2 standard for digital terrestrial television broadcasting. As a result, the indoor and outdoor tests were carried out within the 470–862 MHz frequency band.

The transmitting station is located in Thailand's capital Bangkok at the coordinates 13°45′16″ N 100°32′24″ E, with an effective isotropic radiated power (EIRP) of 50 kW. The broadcast range encompasses the capital and neighboring provinces in the central region, as indicated by the area in pink in Figure 20. The height of the transmission antenna was 328.4 m. The receiving antenna (i.e., antenna under test (AUT)) for both the indoor and outdoor testing was located at the coordinates 14°20′52″ N 100°33′55″ E, which is 65.73 km from the transmitting station. In the indoor and outdoor testing, the AUT was placed at a height of at least 10 m above ground. The transmitting and receiving (AUT) antennas were in the line of sight.

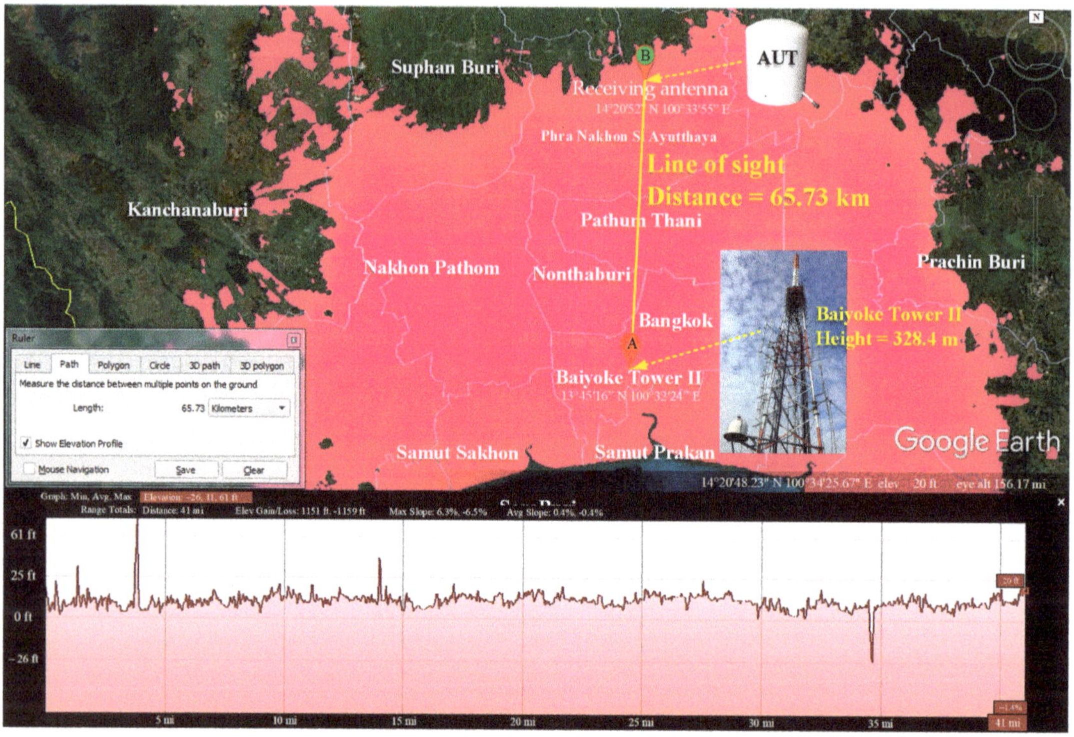

Figure 20. Distance between the transmitting and receiving antennas without obstructions (line of sight).

Figure 21 shows the experimental setup for the outdoor and indoor testing of the proposed portable wideband directional antenna scheme with a digital video broadcasting (DVB) signal receiver (Promax Ranger Neo$^+$). The measurements were taken in a concrete building. Figure 22 depicts the outdoor reception performance of the proposed antenna scheme at the first multiplexer (MUX) of 514 MHz in Thailand for the DVB-T2 system [35]. The power obtained was 38.4 dBμV (−70.4 dBm), with a carrier to noise ratio (C/N) of 11.6 dB. The signal spectrum was of a Rician fading channel, rendering the proposed antenna scheme suitable for outdoor digital terrestrial television reception [34].

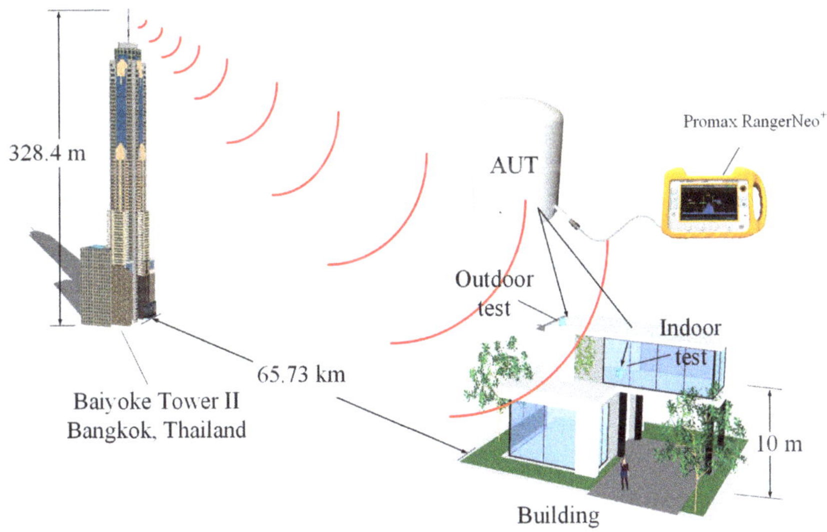

Figure 21. The portable wideband directional antenna scheme with a DVB signal receiver.

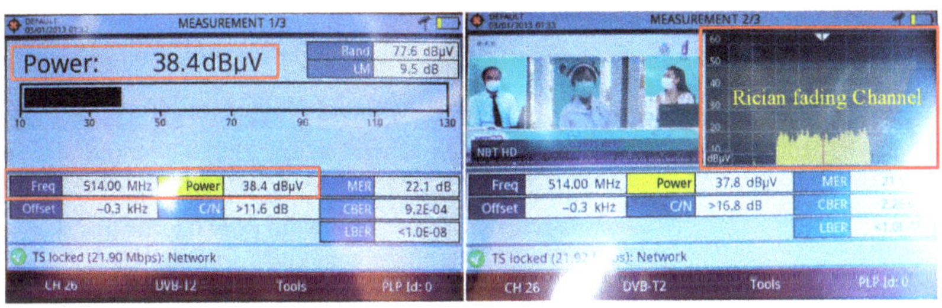

Figure 22. The outdoor reception performance of the portable wideband directional antenna scheme.

Figure 23 demonstrates the indoor reception performance of the proposed antenna scheme at 514 MHz. The power obtained was 26.6 dBμV (−82.2 dBm) with a C/N of 10.9 dB. The signal spectrum was of a Rayleigh fading channel, rendering the proposed antenna scheme suitable for indoor digital terrestrial television reception [36].

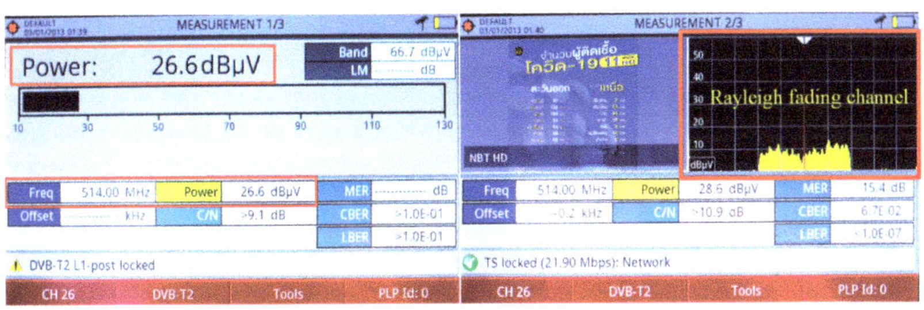

Figure 23. The indoor reception performance of the portable wideband directional antenna scheme.

Figure 24 shows the experimental setup used to assess the reception performance of the proposed antenna scheme in actual use, whereby the AUT was connected by a 75 Ω RG6 coaxial cable to the set top box and a television set. The results indicate that the

23. Yeo, J.; Lee, J.I. Miniaturized wideband loop antenna using a multiple half-circular-ring-based loop structure and horizontal slits for terrestrial DTV and UHD TV applications. *Sensors* **2021**, *21*, 2916. [CrossRef]
24. Lee, J.I.; Yeo, J.; Cho, Y.K. Broadband compact quasi-Yagi antenna for indoor digital TV. *Microw. Opt. Technol. Lett.* **2013**, *55*, 2859–2863. [CrossRef]
25. Rachchompoo, S.; Intarawiset, N.; Akatimagool, S.; Chaiyawong, K. Development of learning innovation of DTV antennas for telecommunication education. In Proceedings of the 7th International Conference on Technical Education (ICTechEd7), Bangkok, Thailand, 25–26 March 2020; pp. 41–44.
26. Mistry, K.K.; Lazaridis, P.I.; Zaharis, Z.D.; Chochliouros, I.P.; Loh, T.H.; Gravas, I.P.; Cheadle, D. Optimization of log-periodic TV reception antenna with UHF mobile communications band rejection. *Electronics* **2020**, *9*, 1830. [CrossRef]
27. Audone, B.; Delogu, A.; Moriondo, P. Radome Design and Measurements. *IEEE Trans. Instrum. Meas.* **1998**, *37*, 292–295. [CrossRef]
28. Moreno, J.; Fernandez, M.; Somolinos, A.; Lozano, L.; Catedra, F.; Gonzalez, I. Analysis and Design of Antenna Radomes. In Proceedings of the Communications, Antennas and Electronic Systems (COMCAS 2013), Tel Aviv, Israel, 21–23 October 2013.
29. What is Radome. Available online: http://www.radomeservices.com/what-is-a-radome/ (accessed on 8 January 2022).
30. Chen, Z.N.; Luk, K.-M. *Radome materials In Antennas for Base Stations in Wireless Communication*; McGraw-Hill: New York, NY, USA, 2009; pp. 56, 246.
31. Studio, C.P. *CST Studio Suite 3D EM Simulation and Analysis Software*; CST Studio Suite; Simulia: Darmstadt, Germany, 2018.
32. Altınoklu, A.; Karaosmanoğlu, B.; Ergül, Ö. Electromagnetic optimization of corrugatedmetallic sheets for maximum power focusing. *J. Electromagn. Waves Appl.* **2017**, *31*, 1–13. [CrossRef]
33. Karaova, G.; Altınoklu, A.; Ergül, Ö. Full-wave electromagnetic optimisation of corrugated metallic reflectors using a multigrid approach. *Sci. Rep.* **2018**, *8*, 1267. [CrossRef] [PubMed]
34. Pumpoung, T.; Wongsiritorn, P. Accurate measurement techniques for 75 ohms antenna with 50 ohms vector network analyzer. *Przegląd Elektrotechniczny* **2020**, *96*, 39–43. [CrossRef]
35. Suwansukho, N.; Promwong, S.; Chamchoy, M. Evaluation of DVB-T2 transmission channel in Bangkok, Thailand. In Proceedings of the 2nd International Conference on Digital Arts, Media and Technology Digital Economy for sustainable growth (ICDAMT), Chiang Mai, Thailand, 1–4 March 2017; pp. 1–6.
36. Antennas and Digital Television. Available online: https://www.fcc.gov/sites/default/files/antennas_and_digital_television.pdf (accessed on 12 May 2022).

sensors

MDPI

Article

Advanced Marine Predator Algorithm for Circular Antenna Array Pattern Synthesis

Eunice Oluwabunmi Owoola [1], Kewen Xia [1,*], Samuel Ogunjo [2], Sandrine Mukase [1] and Aadel Mohamed [3,4]

1 School of Electronics and Information Engineering, Hebei University of Technology, Tianjin 300401, China; bunmso@gmail.com (E.O.O.); ssandrinem@gmail.com (S.M.)
2 Department of Physics, Federal University of Technology, Akure 340271, Ondo State, Nigeria; stogunjo@futa.edu.ng
3 School of Mechanical Engineering, Hebei University of Technology, Tianjin 300401, China; aadelm380@gmail.com
4 Department of Mechanical Engineering, University of Nyala, Nyala 11111, Sudan
* Correspondence: kwxia@hebut.edu.cn

Abstract: The pattern synthesis of antenna arrays is a substantial factor that can enhance the effectiveness and validity of a wireless communication system. This work proposes an advanced marine predator algorithm (AMPA) to synthesize the beam patterns of a non-uniform circular antenna array (CAA). The AMPA utilizes an adaptive velocity update mechanism with a chaotic sequence parameter to improve the exploration and exploitation capability of the algorithm. The MPA structure is simplified and upgraded to overcome being stuck in the local optimum. The AMPA is employed for the joint optimization of amplitude current and inter-element spacing to suppress the peak sidelobe level (SLL) of 8-element, 10-element, 12-element, and 18-element CAAs, taking into consideration the mutual coupling effects. The results show that it attains better performances in relation to SLL suppression and convergence rate, in comparison with some other algorithms for the optimization case.

Keywords: advanced marine predator algorithm; beam pattern; circular antenna array; sidelobe level

Citation: Owoola, E.O.; Xia, K.; Ogunjo, S.; Mukase, S.; Mohamed, A. Advanced Marine Predator Algorithm for Circular Antenna Array Pattern Synthesis. *Sensors* **2022**, *22*, 5779. https://doi.org/10.3390/s22155779

Academic Editors: Naser Ojaroudi Parchin, Chan Hwang See and Raed A. Abd-Alhameed

Received: 12 July 2022
Accepted: 1 August 2022
Published: 2 August 2022

1. Introduction

Antenna arrays have many applications in communications systems, radars, medical imaging, remote sensing, Internet of Things (IoT), and signal processing because of their ability to provide high directive characteristics [1], spatial diversity [2], adaptive beamforming, and beam steering [3]. The effectiveness of antenna arrays is highly dependent on their geometry, the number of antennas, element spacing, beam pattern, and sidelobe level (SLL). To meet the necessities of long-distance communication, antenna arrays with very directional qualities are required in many applications [4]. To deal with the high free-space route loss problem of millimeter-wave (mmWave) transmissions and fulfill the increasing capacity demand, the fifth-generation (5G) cellular communication technology requires large-scale antenna array systems [5]. The urgency to meet the demand for qualitative and quantitative energy has necessitated vast research into improving the efficiency of antenna arrays, in the aspect of SLL reduction, interference nulling, narrow directivity, and high gains.

Many research works have been dedicated to developing techniques for the pattern synthesis of antenna arrays with a narrow main lobe and reduced sidelobe level [2,6]. Pattern synthesis of different types of antenna such as linear array, circular array, hexagonal array, concentric circular array, time modulated arrays, and many more using some heuristic algorithms has been carried out by several researchers [7–11]. To solve the antenna array optimization, the choice of the optimization algorithm is essential, as is the objective function capturing the designated relevance of the antenna array. Mutual coupling is likewise a relevant factor to consider in the design of the antenna array, for it has a great effect on the field performance of the antennas. Thus, recent research work carefully put into consideration the mutual coupling effect during the antenna array synthesis [7,9,12].

Over the years, the addition of refined or optimal parameters and hybridization of multiple metaheuristic algorithms has been an effective way of improving the efficiency of the algorithms combined for antenna array synthesis. Ref. [13] applied an invasive weed optimization algorithm to the synthesis of a rectangular array, circular arrays, and the bee-hive. The modification of the IWO algorithm for the optimization of time-modulated LAA was proposed by Basak et al. [14]. The IWO was improved by including two parallel populations and a new routine of changing the mutation step-size with iterations, which made the algorithm achieve better performance. Jang et al. performed the circular lattice array structure optimization in aperture synthesis radiometers using particle swarm optimization (PSO) [15]. Hosseini and Jafarian [16] came up with the concept of hybridizing invasive weed optimization and particle swarm optimization (HIWOPSO) so that IWO will benefit from the swarm intelligence of PSO and avoid being trapped in local solutions. A hybrid invasive weed optimization (IWO) and wind driven optimization (WDO) was also proposed and adopted for the synthesis of uniformly spaced LAA and non-uniform CAA by Mahto and Choubey [17].

The optimization of a circular array using efficient biogeography-based optimization (EBBO) was proposed by ref. [18]. The authors improved the standard BBO and applied it to optimize jointly the amplitude current and element position of the CAA. A convex improved genetic algorithm (CIGA) was proposed by Li et al. [19] for the beam pattern synthesis of sparse arrays. This hybrid algorithm was able to adjust the excitation and positions of the arrays to suppress the peak SLL and obtain better results as compared to some other existing algorithms. The least mean square (LMS) algorithm and sample matrix inversion (SMI) were also combined to optimize the weight of the adaptive antenna array at different arrival angles [20]. Ref. [21] introduced the use of a hybrid cuckoo search algorithm with convex programming (CS–CP) to solve the sparse LAA based on a subarray scheme synthesis problem. In each subarray, the number, excitation, and spacing are optimized with the minimum or maximum spacing constraints. The optimized array was able to achieve an improved peak SLL and reduced excitation control number. Wei et al. [22] proposed a hybrid artificial neural network (ANN) and convex optimization for linear and planar antenna array design, taking into consideration the coupling effect between the antenna elements. The ANN was applied to design and optimize the radiation pattern and element positions of the array antenna, whereas the convex optimization was employed for the synthesis of the main lobe direction and SLL of the array.

An improved differential evolution (DE) algorithm combined with the successful-parent-selecting (SPS) framework, named SPS-JADE, was proposed and adopted for the pattern synthesis of LAA by Zhang et al. [23]. In this algorithm, the SPS framework was used to mitigate the stagnation problem of DE algorithms, and hence improved the overall performance of the algorithm in terms of the global searchability, convergence rate, and robustness. To the synthesis of LAA using GWO [24], Lakhlef et al. [25] proposed a modified GWO by introducing the competitive exclusion selection inspired from the genetic algorithm in the proposed algorithm with the inclusion of Gaussian function. However, a Jaya-grey wolf optimizer was suggested for the design of the antenna array for 5G communications systems [26]. Both GWO and the Jaya algorithm have the advantage of only two parameters setting; therefore, the features of both algorithms are aggregated to form a new hybrid, which proved to be effective.

The marine predator algorithm (MPA) [27] has recently been proposed for solving optimization problems of various kinds. The MPA optimization method has been applied to solving some engineering problems such as fault diagnosis of rolling bearing [28] and the PID tuning problem for DC motor speed control [29], but it is yet to be adopted in solving antenna synthesis problems. Therefore, this work introduces the MPA to the electromagnetic world both in its standard form and also in its improved form. Research has proven the standard MPA to exhibit deficiency in terms of the inability to produce a diverse initial population with high productivity, slow escape from the local optimum, and a little exploration of the search space [29,30]. We propose a novel algorithm called the advanced marine predator algorithm (AMPA) for better optimization performance in solving engineering problems, most especially in the electromagnetic field. AMPA's velocity update strategy aids MPA in overcoming the problem of slow escape from the local optimum and increases the algorithm's convergence rate.

The major contributions of the work are as follows:

1. Introduction of an improved velocity (stepsize) update strategy and an adaptive parameter to control the velocity of the prey, thus improving the exploration capability and mitigating the stagnancy problem of MPA.
2. A chaotic sequence called the Chebyshev map is added to the prey update to aid the stepsize and thus enhance the exploitation ability of the algorithm. These improvements effectively help MPA to overcome the problem of being stuck easily in local optimum, as well as improve the convergence rate of the overall algorithm.
3. We formulated the peak SLL suppression optimization problem for the synthesis of CAA and applied the novel AMPA to solve the objective functions. The excitation current and the inter-element spacing of four examples of CAA elements are jointly optimized by using AMPA and five other existing algorithms.

The rest of the work is arranged as follows. Section 2 explains the CAA model and optimization objective. Section 3 describes the AMPA technique and Section 4 presents the simulation results. Lastly, Section 5 gives the conclusion.

2. Antenna Array Model and Problem Formulation

2.1. The CAA Model

As shown in Figure 1, we consider an asymmetric circular antenna array of N isotropic elements with equal excitation and element spacing. The array factor is expressed as [18]:

$$AF(\theta) = \sum_{m=1}^{M} I_m \times e^{i(kr.\cos(\theta-\phi_m)+\alpha_m)}. \tag{1}$$

$$kr = \frac{2\pi r}{\lambda}\sum_{j=1}^{M} d_j. \tag{2}$$

$$\phi_m = \left(\frac{2\pi r}{\lambda}\right).\sum_{j=1}^{m} d_j. \tag{3}$$

where AF, I_m, d_m, and α_m denote the array factor, the excitation, the position of the mth element, and the phase of the antenna array.

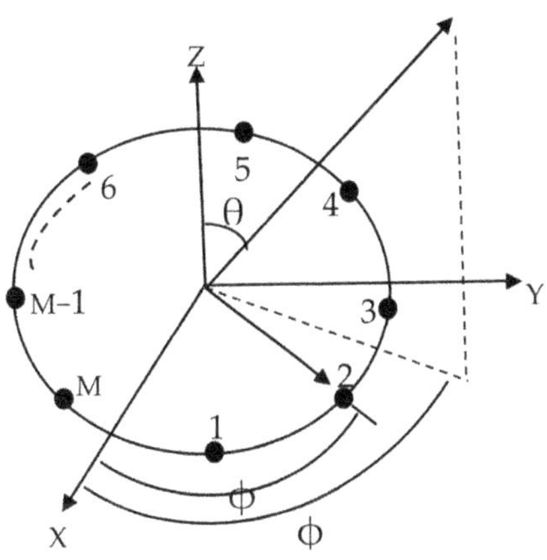

Figure 1. Asymmetric N isotropic circular antenna array (CAA).

2.2. Problem Formulation

This work aims at minimizing the SLL of the CAA by jointly optimizing its excitation amplitude and inter-element spacing. Hence, the objective function is given as:

$$f = w_1 \times \underbrace{\left(\max(AF(\theta_{PSL}^{R_1})) + \max(AF(\theta_{PSL}^{R_2}))\right)}_{\text{SLL supression}}. \quad (4)$$

where w_1 is the weight, and $\theta_{PSL}^{R_1}$ and $\theta_{PSL}^{R_2}$ denote the peak SLL suppression angles. The equation suppresses the SLL region of the array pattern; thus, the optimization problem is formulated as:

$$\min f(I_1, I_2, I_3, \ldots\ldots, I_n, d_1, d_2, d_3, \ldots\ldots, d_n). \quad (5)$$

$$s.t.?\ \theta_{PSL}^{R} = \in \max([-\pi, \theta_{R1}] \cup [\theta_{R2}, \pi]). \quad (6)$$

$$0 \leq I_m < 1, m = 1, 2, 3, \ldots\ldots, M \quad (7)$$

$$d_{\text{mi}} - d_{\text{m}} \geq d_{\text{min}} < d_{\text{max}}, m = 1, 2, 3, \ldots\ldots, M \quad (8)$$

where $[-\pi, \theta_{R1}]$ and $[\theta_{R2}, \pi]$ are the regions of the SLL suppression. d_{min} and d_{max} are the minimum and maximum inter-element distances.

3. Proposed Algorithm

This section discusses the MPA, and the novel proposed advanced marine predator algorithm's (AMPA) operation mechanisms.

3.1. Marine Predators Algorithm (MPA)

In MPA, the initial solution is randomly generated and uniformly distributed over the search space, keeping in view the upper and lower range of values of the design variables. The fittest solution is nominated as the top predator to construct a matrix called Elite. In this algorithm, both the predator and prey are considered search agents because as the predator is in search of its prey, the prey is also in search of its food.

The optimization phase of the standard MPA is divided into three stages based on the velocity ratio and the entire activity of the predator and prey. The first phase is the exploration stage, and it occurs in the first third of iterations. The mathematical expression is written as:

$$\begin{aligned}
&\textit{While Iteration} < \tfrac{1}{3}\ \textit{Max_Iteration} \\
&\overrightarrow{Stepsize_j} = \vec{r}_b \otimes (\overrightarrow{Elite} - \vec{r}_b \otimes \overrightarrow{\text{Prey}_j})\ j = 1, 2, \ldots n \\
&\overrightarrow{\text{Prey}_j} = \overrightarrow{\text{Prey}_j} + C \cdot \vec{r} \otimes \overrightarrow{Stepsize_j}
\end{aligned} \quad (9)$$

where r_b represents the Brownian motion, which is a random number based on normal distribution, and r is a uniform random number between 0 and 1. C is a constant number that equals 0.5. The second phase of the algorithm commences at the two-third stage of the iteration, while the predator and the prey are moving at the same pace. This can be expressed as:

$$\begin{aligned}
&\textit{While } \tfrac{1}{3}\ \textit{Max_Iteration} < \textit{iteration} < \tfrac{2}{3}\ \textit{Max_Iteration} \\
&\text{For the first half of the population} \\
&\overrightarrow{Stepsize_j} = \vec{r}_l \otimes (\overrightarrow{Elite_j} - \vec{r}_l \otimes \overrightarrow{\text{Prey}_j})\ j = 1, 2, \ldots n/2 \\
&\overrightarrow{\text{Prey}_j} = \overrightarrow{\text{Prey}_j} + C \cdot \vec{r} \otimes \overrightarrow{Stepsize_j} \\
&\text{For the sec ond half of the population} \\
&\overrightarrow{Stepsize_j} = \vec{r}_b \otimes (\vec{r}_b \otimes \overrightarrow{Elite_j} - \vec{r}_l \otimes \overrightarrow{\text{Prey}_j})\ j = n/2, \ldots n \\
&\overrightarrow{\text{Prey}_j} = \overrightarrow{Elite_j} + C \cdot AP \otimes \overrightarrow{Stepsize_j}
\end{aligned} \quad (10)$$

where $\vec{r}_l$ represents the Lévy movement, which is a random number based on the Lévy distribution. *AP* is the adaptive parameter assigned to control the step size of the predator movement and it is expressed as:

$$AP = (1 - \frac{Iteration}{Max\_Iteration})^{(2\frac{Iteration}{Max\_Iteration})} \quad (11)$$

The third phase occurs when the iteration is greater than two-thirds of the maximum iteration. This phase is presented as:

$$\begin{aligned} \overrightarrow{Stepsize}_j &= \vec{r}_l \otimes (\vec{r}_l \otimes \overrightarrow{Elite} - \overrightarrow{\mathrm{Prey}_j})\ j = 1,2,\ldots n \\ \overrightarrow{\mathrm{Prey}_j} &= \overrightarrow{Elite_j} + C \cdot AP \otimes \overrightarrow{Stepsize}_j \end{aligned} \quad (12)$$

Environmental factors, such as the fish aggregating devices (FADs) effects or the eddy formation effect, also cause behavioral changes in the predator. These FADs serve as local optima where the predators can be easily trapped. To escape the FADs and avoid stagnation, the predator or prey needs to make a longer jump, else it will become trapped. The FADs effect is expressed as:

$$\overrightarrow{\mathrm{Prey}_j} = \begin{cases} \overrightarrow{\mathrm{Prey}_j} + AP\left[\vec{x}_{\min} + \vec{r} \otimes (\vec{x}_{\max} - \vec{x}_{\min})\right] \otimes \vec{u} & if\ r \le FADs \\ \overrightarrow{\mathrm{Prey}_j} + \left[FADs(1 - r^1) + r^1\right](\overrightarrow{\mathrm{Prey}_{r1}} - \overrightarrow{\mathrm{Prey}_{r2}}) & if\ r > FADs \end{cases} \quad (13)$$

where $\vec{x}_{\min}$ and $\vec{x}_{\max}$ represent the vectors containing the lower and upper bound of the dimensions. r^1 is a random number, whereas $r1$ and $r2$ subscripts are random indexes of the prey matrix. $\vec{u}$ is a binary vector that has arrays that include 0 and 1. The probability of the FADs effect equals 0.2.

The steps of MPA can be found in [27].

3.2. *Advanced Marine Predators Algorithm (AMPA)*

The operation mechanisms of the AMPA are as follows:

3.2.1. Initialization

The predator and prey are a matrix formed from the best or fittest randomly generated solution, which are called Elite and Prey, respectively, as conformed to the standard MPA. The format of the Elite and Prey for the AMPA is expressed as:

$$Elite/\mathrm{Prey} = \begin{bmatrix} X_{1,1} & X_{1,2} & \cdots\cdots & X_{1,d} \\ X_{2,1} & X_{2,2} & \cdots\cdots & X_{2,d} \\ \vdots & \vdots & \vdots & \vdots \\ X_{n,1} & X_{n,2} & \cdots\cdots & X_{n,d} \end{bmatrix}_{n \times d} \quad (14)$$

where n is the population size, d is the dimension, and X represents the randomly generated solution.

3.2.2. Optimization Process

The AMPA optimization process is divided into two phases, unlike the standard MPA which has three phases. This algorithm is also subjected to the same environmental factor effects as the standard MPA.

(1) Phase 1: In the first phase, an improved velocity update and position update are introduced to the algorithm. This velocity update takes into consideration both the personal best fitness of individual prey or Elite, and also their global best fitness. This gives room for the expansion of the exploration capacity of the algorithm, unlike the standard MPA

that uses only the best fitness. The velocity update used in this technique is expressed mathematically as:

$$\overrightarrow{Stepsize_{i,j}} = AP \times \overrightarrow{Stepsize_{i,j}} + b_1 e^{-\beta r_p^2}(pbest_{i,j} - \text{Prey}_{i,j}) + b_2 e^{-\beta r_g^2}(gbest_{i,j} - \text{Prey}_{i,j}) \quad (15)$$

where AP, $stepsize$, and β are the adaptive weight, velocity, and distance sight coefficient, respectively. b_1 and b_2 are learning coefficients. $gbest$ and $pbest$ symbolize the global and personal best positions, respectively. r_p represents the Euclidean distance between the personal best positions and the current positions of the prey, whereas r_g represents the Euclidean distance between the global best position and the current positions of the prey. The updated Prey is thus expressed as:

$$\overrightarrow{\text{Prey}_{i,j}} = \overrightarrow{\text{Prey}_{i,j}} + C \times RC \times \overrightarrow{Stepsize_{i,j}} \quad (16)$$

where C is a constant number that equals 0.4, and RC is the chaos called the Chebyshev map.

Chebyshev map: This is a one-dimensional chaotic map with an easy and simple to implement iterative equation. It exhibits pseudo-randomness of output sequences, as well as high ergodic property and sensitivity to parameter and initial value [31]. It is mathematically expressed as [31]:

$$x_{s+1} = \cos(k \cos^{-1}(x_s)) \quad (17)$$

In this work, the Chebyshev map is used to generate chaotic sequences to improve the stepsize, and hence enhance the exploitation and convergence of the algorithm. The chaos introduces randomness generated by a simple deterministic dynamic system called the chaotic system into the algorithm to help the algorithm explore the search space more dynamically and globally without being trapped into local optima. The starting value and the number of iterations used for the Chebyshev map are 0.7 and 100, respectively.

(2) Phase 2: The second phase of this algorithm is made to conform to the last phase of the standard MPA algorithm. At this phase, the predator is assumed to move at a faster pace than the prey, with high exploitation capability, and a low-velocity ratio of 0.1 (v = 0.1). Here, the predator's movement is Lévy, and the phase is presented as shown in Equation (12).

The mathematical expression for the Lévy movement is given as:

$$\vec{r}_l = 0.05 * \left(\frac{N_{randn} \times \delta}{[abs(N_{randn})]^{\frac{1}{\beta}}}\right) \quad (18)$$

$$\delta = \left\{\frac{\Gamma(1+\beta) \times \sin(\frac{\pi}{2} \times \beta)}{\left[\Gamma(\frac{1+\beta}{2}) \times \beta \times 2^{(\frac{\beta-1}{2})}\right]}\right\}^{\frac{1}{\beta}} \quad (19)$$

where N_{randn}, β, Γ, and δ represent the random number, spatial exponent, gamma function, and random variable, respectively. This optimization process is subjected to the same FAD effect as that of the standard MPA expressed in Equation (13).

The basic steps of the AMPA are summarized in the pseudo-code shown in Algorithm 1.

Algorithm 1 Pseudo-code of advanced marine predator algorithm (AMPA) (AMPA code)

Initialize parameters: m, FADs, C
Initialize search agents (Prey) populations $x_i = (x_i, x_i, \ldots, x_i)$ and step size
While maximum iteration is not met
Evaluate the fitness
Construct the Elite and Prey matrix using Equation (14)
Accomplish memory saving

If rand < 0.6
Update the stepsize and prey using Equations (15) and (16)
Else
Update prey using Equation (12)
End (if)
Evaluate fitness of the updated prey
Save and update the Elite
Apply the FADs effect and update using Equation (13)
End **while**

4. Simulation Results

The performance of the proposed algorithm is evaluated by applying it to the circular antenna array pattern synthesis. The simulation results obtained by the AMPA are compared with five other state-of-art algorithms, which are MPA, arithmetic optimization algorithm (AOA), moth flame optimization (MFO) algorithm, grey wolf optimization (GWO), and IWO. The parameter settings of all algorithms are given in Table 1. The simulations are performed on an AMD A10-5750M APU with a Radeon (tm) HD Graphics personal computer with Windows 10 operating system, 2.5 GHz processor, and 8 GB ram, using MATLAB 2020a. The tuning of the parameter values of the proposed AMPA for CAA pattern synthesis is also performed.

Table 1. Parameter settings of the algorithms.

Algorithm	Parameters	Values
AMPA	b_1 and b_2 FADs and C	1.7 [0.2, 0.4]
MPA	FADs and C	[0.2, 0.5]
AOA	α and μ	[5, 0.499]
MFO	Spiral constant Convergence constant Random factor	1 [−1, −2] [−1, 1]
IWO	Exponent	2
	Minimum and maximum number of seeds Initial and final SD	[0, 5] [0.01, 0.1]
GWO	Control parameter	[2, 0]

4.1. Parameter Value Tuning

The parameters of AMPA are tuned for antenna optimization. The learning coefficient values 'b_1 and b_2' are very sensitive parameters that play a great role in improving the exploration capability of the algorithm; therefore, it is being tuned. We vary the values of b_1 and b_2 from 1.0 to 2.0, at a step size of 0.1, while using the AMPA to solve the formulated optimization problem. The test was repeated 50 times to eliminate random bias, and the best results are presented in Figure 2. We observed that the 8-CAA and 18-CAA elements obtained an optimal peak SLL at a value of 1.7. Therefore, we choose to set the values of the learning coefficients to 1.7 for the CAA synthesis in this research work.

4.2. Simulation Result for the CAA Pattern Synthesis

In this subsection, the proposed AMPA and other algorithms stated in Table 1 are applied to solve the formulated CAA optimization problem in Section 2. The maximum number of iterations, population size, and number of runs are set to 100, 20, and 50, respectively, for all the algorithms. The simulation is performed on four different examples of CAA, which are 8-element, 10-element, 12-element, and 18-element CAA. To cater for the mutual coupling between the elements and avoid possible grating lobes, the inter-element spacing was constrained to range between 0.252 λ and 0.999 λ. The excitation amplitude ranged from 0 to 1. The best optimization results for each algorithm are recorded in this work.

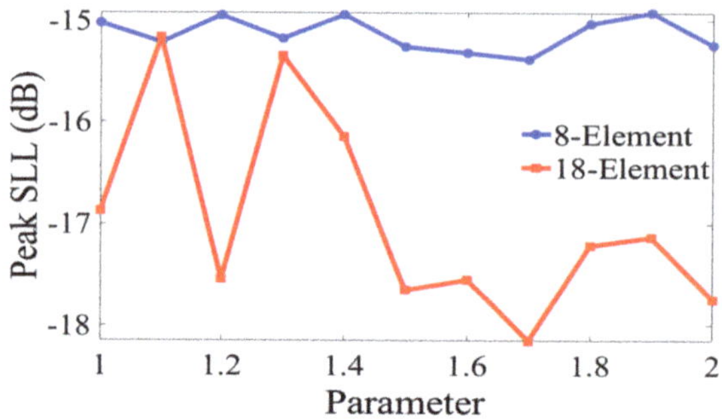

Figure 2. Selection of b1 and b2 for both 8–element and 16–element CAA.

4.2.1. Beam Pattern Synthesis of 8-Element CAA

Table 2 shows the peak SLL, FNBW, circumference, and total computational time obtained by different optimization algorithms for the synthesis of 8-element CAA. AMPA obtained the lowest SLL of −15.3811 dB, which is 0.9659 dB, 5.3435 dB, 2.2027 dB, 5.3183 dB, 2.2158 dB, and 11.2109 dB lower than MPA, AOA, MFO, GWO, and IWO, respectively. The optimized excitation amplitude and element spacing obtained by AMPA is displayed in Table 3. Though AMPA achieved the best SLL, it was also able to acquire an FNBW of 81.00°, which is narrower than that of AOA and GWO. The radiation pattern and convergence rate of this analysis are shown in Figure 3a,b. The convergence curve clearly shows how AMPA is able to overcome being stuck in local optimum, and thus achieve the best SLL suppression results.

Table 2. Results of 8-element CAA obtained by different algorithms for PSLL minimization.

Algorithm	Peak SLL (dB)	FNBW (°)	Circumference (λ)	CPU Time (s).
Uniform	−4.1702	70.00	3.75	0.00
IWO	−13.1653	80.00	4.49	0.34
GWO	−10.0628	80.00	4.45	0.39
MFO	−13.1784	81.00	4.41	0.38
AOA	−10.0376	86.00	4.40	0.38
MPA	−14.4152	80.00	4.52	0.66
AMPA	−15.3811	80.00	4.55	0.77

Table 3. Amplitude (I) and element spacing (d) obtained by using AMPA for 8-element CAA PSLL minimization.

Element	1	2	3	4	5	6	7	8
I	0.8111	0.4236	0.9577	0.9793	0.0435	0.3654	0.8533	0.0962
d (λ)	0.3137	0.8028	0.8627	0.6000	0.3684	0.4822	0.7883	0.3272

4.2.2. Beam Pattern Synthesis of 10-Element CAA

The peak SLL values of −14.4185 dB, −12.4175 dB, −9.2885 dB, −12.2772 dB, −8.8086 dB, and −12.7904 dB are achieved by IWO, GWO, MFO, AOA, MPA, and AMPA, respectively, as shown in Table 4. The proposed AMPA obtained the best SLL suppression with an FNBW of 64° and an aperture length of approximately 6.00 λ. MPA obtained the narrowest FNBW, followed by the AOA and other algorithms, whereas IWO had the largest FNBW. With the aid of the velocity update mechanism, it became easy for the AMPA to escape local optimum, while the chaos improves its ability to exploit a larger search space. This is obvious from the radiation pattern and convergence plot depicted in Figure 4a,b. Table 5 shows the optimized parameters used to plot the radiation pattern of AMPA for the 10-element CAA. The 3D beam pattern reflecting the level of improvement between the uniform and AMPA-improved CAA for 8 , and 10-element CAA is shown in Figure 5a–d.

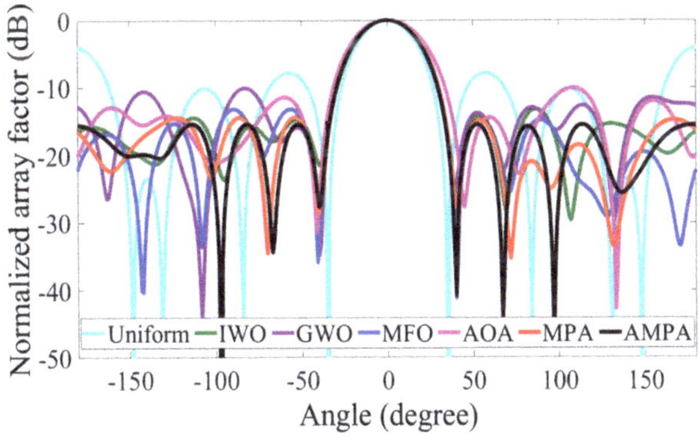

(**a**)

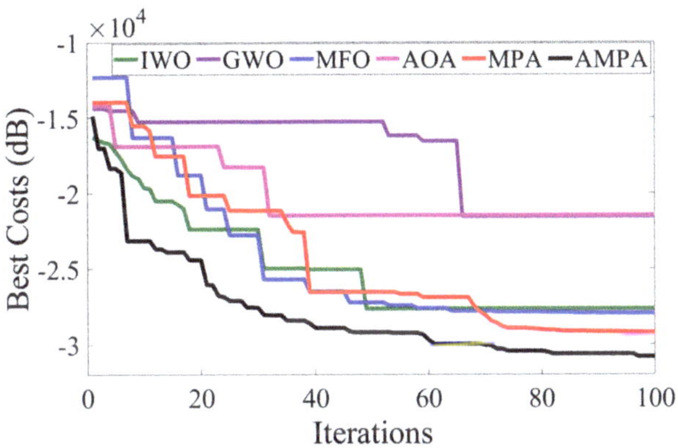

(**b**)

Figure 3. 8−element CAA. (**a**) Radiation patterns obtained by different algorithms for reducing the PSLL. (**b**) Convergence rates of different algorithms for reducing the PSLL.

Table 4. Optimization results of 10-element CAA obtained by different algorithms for peak sidelobe level (PSLL) minimization.

Algorithm	Peak SLL (dB)	FNBW (°)	Circumference (λ)	CPU Time (s)
Uniform	−3.5975	56.00	4.75	0.00
IWO	−12.7904	67.00	5.87	0.59
GWO	−8.8086	63.00	5.97	0.47
MFO	−12.2772	65.00	5.83	0.45
AOA	−9.2885	62.00	5.75	0.49
MPA	−12.4175	60.00	6.00	0.85
AMPA	−14.4185	64.00	6.00	0.87

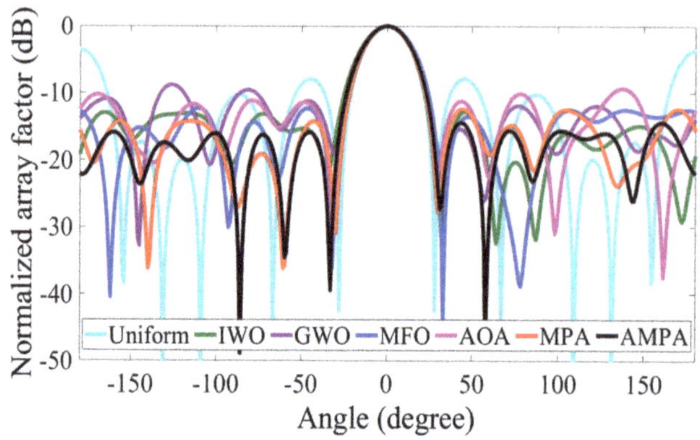

(a)

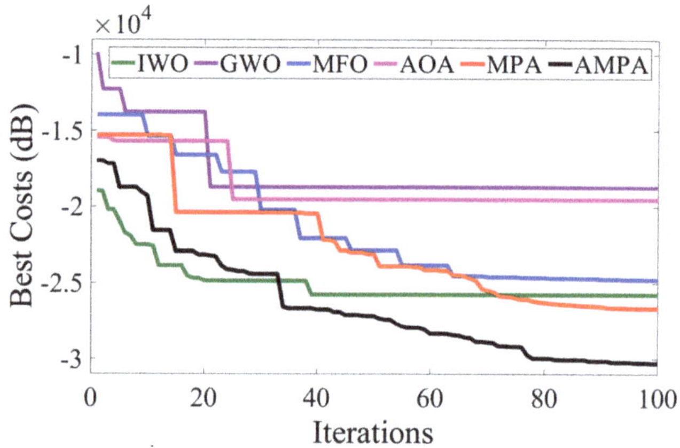

(b)

Figure 4. 10−element CAA. (**a**) Radiation patterns obtained by different algorithms for reducing the PSLL. (**b**) Convergence rates of different algorithms for reducing the PSLL.

Table 5. Amplitude (I) and element spacing (d) obtained by using AMPA for 10-element CAA PSLL minimization.

Element	1	2	3	4	5	6	7	8	9	10
I	0.9540	0.4040	0.3468	0.9940	0.9864	0.3329	0.5148	0.1317	0.9996	0.3978
d (λ)	0.2920	0.9990	0.4042	0.9990	0.5748	0.9509	0.5501	0.4142	0.4803	0.3313

4.2.3. Beam Pattern Synthesis of 12-Element CAA

The numerical results in terms of the PSLL, FNBW, circumference, and total computational time are recorded in Table 6. The solutions I and d, obtained by AMPA for this optimization goal, are recorded in Table 7. It can be seen from Table 6 and Figure 6a that the proposed AMPA not only attained the lowest peak SLL as compared with the other benchmark algorithms, but also had an FNBW of 41.00°, which is narrower than that of the uniform array by 5°. This proves that the AMPA has an excellent performance. MFO achieved the highest aperture length of 9.35 λ and the narrowest FNBW. GWO had the highest peak SLL, followed by AOA and MFO. The computational time of both AMPA and MPA compares favorably. The convergence curve of the algorithms is shown in Figure 6b.

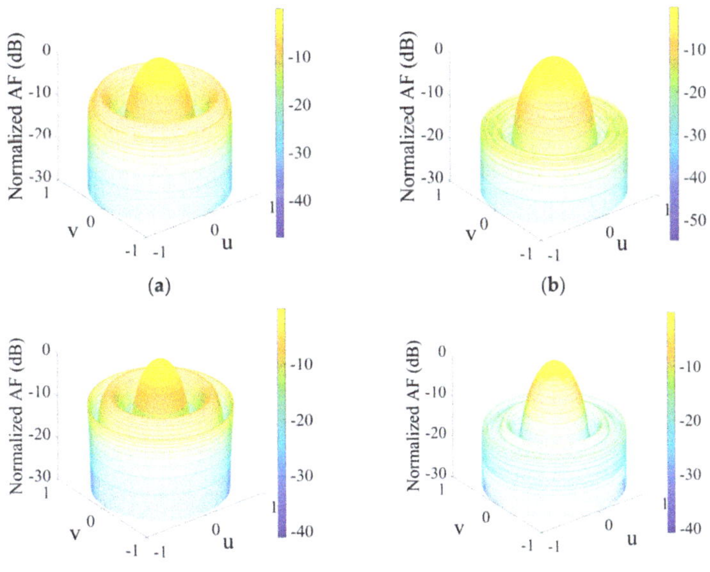

Figure 5. 3D radiation patterns of CAA. (**a**) 8−element CAA uniform array; (**b**) 8−element CAA for AMPA; (**c**) 10−element CAA uniform array; (**d**) 10−element CAA for AMPA.

Table 6. Optimization results of 12-element CAA obtained by different algorithms for PSLL minimization.

Algorithm	Peak SLL (dB)	FNBW (°)	Circumference (λ)	CPU Time (s)
Uniform	−7.165	46.00	5.75	0.00
IWO	−13.0541	48.00	7.23	0.64
GWO	−9.34022	44.00	7.66	0.53
MFO	−11.6163	35.00	9.35	0.47
AOA	−10.3039	57.00	5.97	0.53
MPA	−13.5153	47.00	7.34	1.00
AMPA	−14.9518	41.00	9.15	1.07

Table 7. Amplitude (I) and element spacing (d) obtained by using AMPA for 12-element CAA PSLL minimization.

Element	**1**	**2**	**3**	**4**	**5**	**6**
I	0.9993	0.7689	0.0865	0.6451	0.9850	0.9998
d (λ)	0.6822	0.9854	0.9781	0.9981	0.6357	0.4636
Element	7	8	9	10	11	12
I	0.8172	0.8431	0.0100	0.8646	0.5589	0.9988
d (λ)	0.4424	0.9990	0.3813	0.9463	0.9515	0.6832

4.2.4. Beam Pattern Synthesis of 18-Element CAA

The optimization result comparison of the 18-element CAA is made in Table 8, and the array pattern is shown in Figure 7a. According to the result, it is observed that AMPA attained a peak SLL of −18.1481 dB, which obviously supersedes MPA, AOA, MFO, GWO, IWO, and uniform array by −3.9114 dB, −7.4804 dB, −5.3536 dB, −8.5261 dB, 4.5531 dB, and −10.2312 dB, respectively. The solutions obtained by AMPA are recorded in Table 9. The capability of the AMPA is reflected in the radiation pattern and convergence curve (Figure 7). Though AMPA covered the largest circumference, it still attains excellent SLL reduction with an FNBW of 37.00°. With the low SLL obtained by AMPA, it is easy to avoid the interference with the other system operating in the same frequency band. Figure 8a–d depict the 3D beam pattern that compares the 12- and 18-element CAA's uniform and AMPA-improved beam patterns. The computational time obtained by each algorithm for optimizing the 8-, 10-, 12-, and 18-element CAA is presented in Figure 9. It can be seen that the AMPA has the highest computational time throughout the entire simulation. This is owed to the algorithm's computational complexity, as it requires more mathematical steps to achieve the given optimal results than others. Figure 9 also shows that each algorithm's computational time increases as the CAA's elements increases.

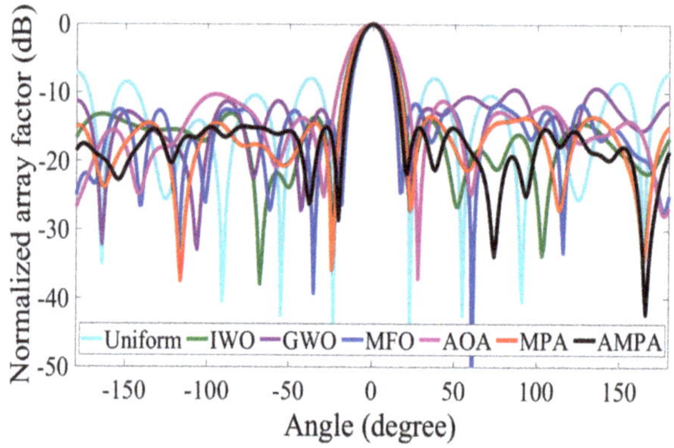

(**a**)

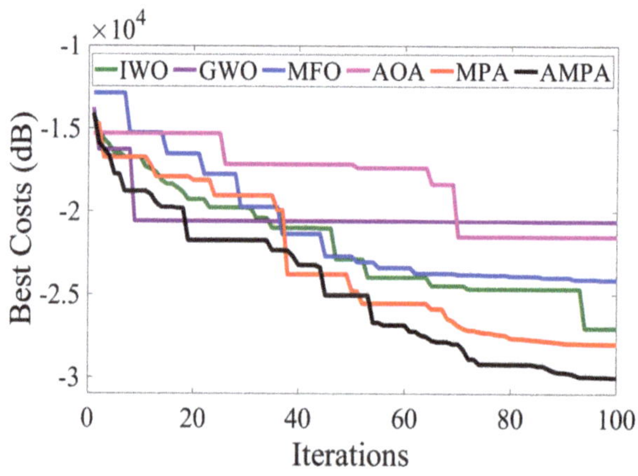

(**b**)

Figure 6. 12−element CAA. (**a**) Radiation patterns obtained by different algorithms for reducing the PSLL. (**b**) Convergence rates of different algorithms for reducing the PSLL.

Table 8. Optimization results of 18-element CAA obtained by different algorithms for PSLL minimization.

Algorithm	Peak SLL (dB)	FNBW (°)	Circumference (λ)	CPU Time (s)
Uniform	−7.9169	30.00	8.75	0.00
IWO	−13.5950	40.00	9.15	0.68
GWO	−9.6220	38.00	9.17	0.70
MFO	−12.7945	37.00	9.07	0.85
AOA	−10.6677	36.00	9.02	0.66
MPA	−14.2367	39.00	9.16	1.41
AMPA	−18.1481	37.00	10.69	1.46

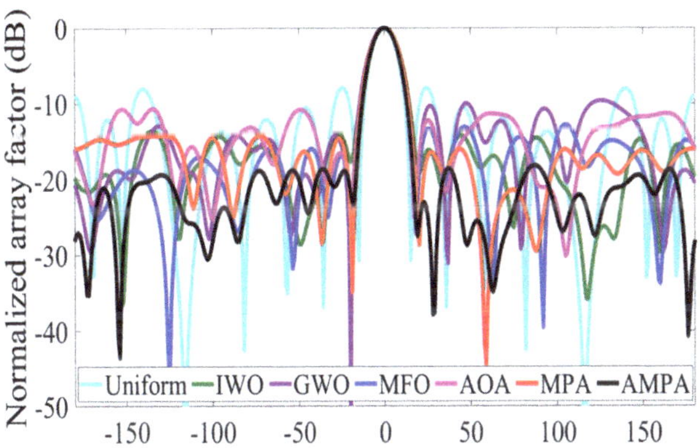

(a)

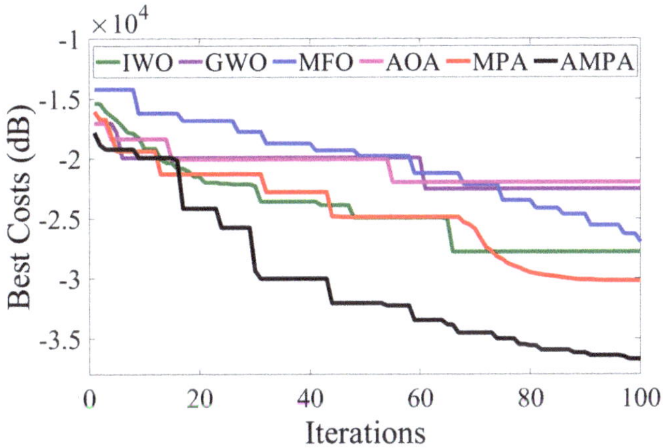

(b)

Figure 7. 18−element CAA. (**a**) Radiation patterns obtained by different algorithms for reducing the PSLL. (**b**) Convergence rates of different algorithms for reducing the PSLL.

Table 9. Amplitude (I) and element spacing (d) obtained by using AMPA for 18-element CAA PSLL minimization.

Element	1	2	3	4	5	6	7	8	9
I	0.9215	0.6189	0.5579	0.3879	0.0850	0.8766	0.8956	0.6880	0.9204
d (λ)	0.3028	0.4996	0.9128	0.6433	0.7766	0.5094	0.9317	0.4235	0.2744
Element	10	11	12	13	14	15	16	17	18
I	0.8215	0.7992	0.6829	0.7112	0.3970	0.4189	0.3077	0.8632	0.7655
d (λ)	0.3560	0.4245	0.9459	0.6335	0.7798	0.5515	0.9937	0.4096	0.3166

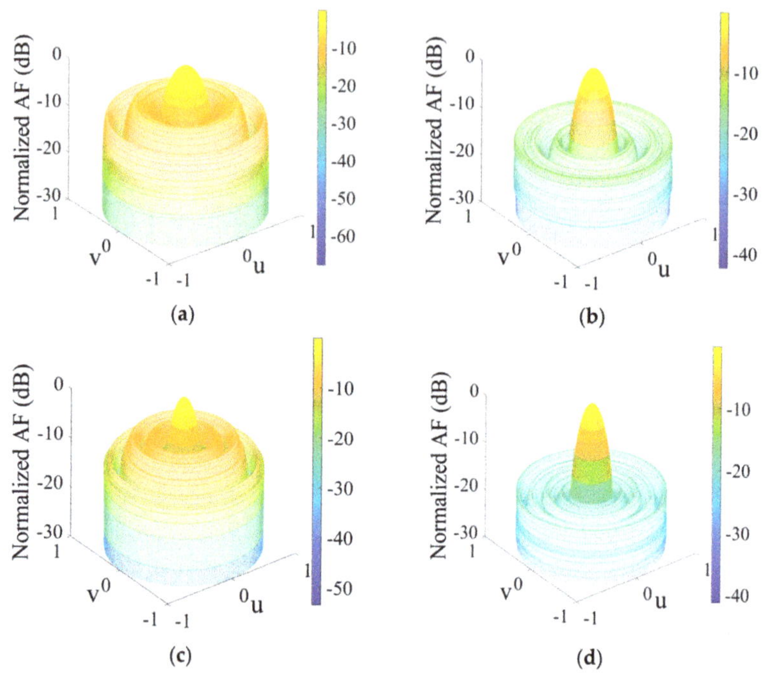

Figure 8. 3D radiation patterns of CAA. (**a**) 12−element CAA uniform array; (**b**) 12−element CAA for AMPA; (**c**) 18−element CAA uniform array; (**d**) 18−element CAA for AMPA.

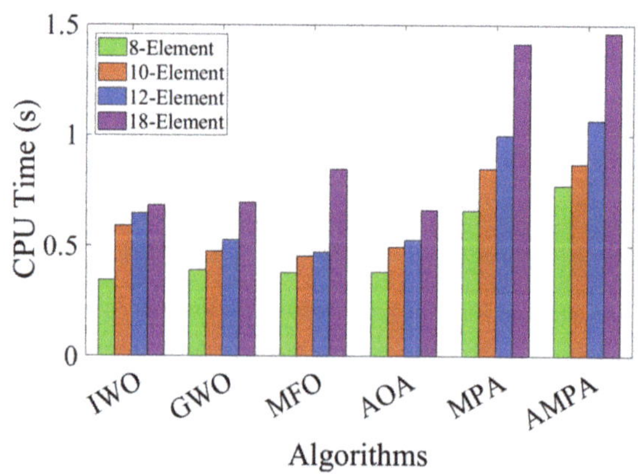

Figure 9. Bar graph for the computational time obtained by the algorithms for each CAA example.

5. Conclusions

This study proposes a new method called advanced marine predator algorithm (AMPA) for the synthesis of CAA. The proposed method is easy to implement, and it effectively optimizes both excitation current and element-spacing simultaneously. It is computationally fast and has a better convergence rate than MPA, AOA, MFO, GWO, and IWO, which it was compared with; thus, it proves to be more efficient in the CAA optimization. The optimal peak SLL values of −15.3811 dB, −14.4185 dB, −14.9518 dB, and −18.1481 dB, obtained by the AMPA for 8-, 10-, 12-, and 18- element CAA, respectively, verified the efficacy of the algorithm to easily escape local optima and obtain better optimization solutions. With the information provided in this research, a practical antenna array design can be made to meet the needs of the optimal communication system. Further improvement can be made to this algorithm by exploring other chaotic sequences to enhance its optimization potential.

Author Contributions: Conceptualization, E.O.O.; formal analysis, E.O.O., S.O. and S.M.; funding acquisition, K.X.; methodology, E.O.O.; resources, A.M.; software, S.M.; supervision, K.X.; validation, K.X.; writing—original draft, E.O.O.; writing—review and editing, S.O. and A.M. All authors have read and agreed to the published version of the manuscript.

Funding: This work was supported by the National Natural Science Foundation of China (No. 42075129), the Hebei Province Natural Science Foundation (No. E2021202179), and the Key Research and Development Project from Hebei Province (No. 19210404D, No. 20351802D, No. 21351803D).

Informed Consent Statement: Not applicable.

Conflicts of Interest: The authors declare no conflict of interest.

References

1. Kazema, T.; Michael, K. Investigation and analysis of the effects of geometry orientation of array antenna on directivity for wire-less communication. *Cogent Eng.* **2016**, *3*, 1232330. [CrossRef]
2. Rahman, S.U.; Cao, Q.; Ahmed, M.M.; Khalil, H. Analysis of linear antenna array for minimum side lobe level, half power beamwidth, and nulls control using PSO. *J. Microw. Optoelectron. Electromagn. Appl.* **2017**, *16*, 577–591. [CrossRef]
3. Cheng, Y.F.; Ding, X.; Shao, W.; Liao, C. A High-Gain Sparse Phased Array with Wide-Angle Scanning Performance and Low Sidelobe Levels. *IEEE Access* **2019**, *7*, 31151–31158. [CrossRef]
4. Shihab, M.; Najjar, Y.; Dib, N.; Khodier, M. Design of non-uniform circular antenna arrays using particle swarm optimization. *J. Electr. Eng.* **2008**, *59*, 216–220.
5. Almagboul, M.A.; Shu, F.; Qian, Y.; Zhou, X.; Wang, J.; Hu, J. Atom search optimization algorithm based hybrid antenna array receive beamforming to control sidelobe level and steering the null. *AEU-Int. J. Electron. Commun.* **2019**, *111*, 152854. [CrossRef]
6. Taser, A.E.; Guney, K.; Kurt, E. Circular antenna array synthesis using multiverse optimizer. *Int. J. Antennas Propag.* **2020**, *2020*, 3149826. [CrossRef]
7. Owoola, E.O.; Xia, K.; Wang, T.; Umar, A.; Akindele, R.G. Pattern Synthesis of Uniform and Sparse Linear Antenna Array Using Mayfly Algorithm. *IEEE Access* **2021**, *9*, 77954–77975. [CrossRef]
8. Zheng, T.; Liu, Y.; Sun, G.; Liang, S.; Han, J.; Ju, Q.; Li, S. Joint sidelobe suppression and nulls control of large-scale linear antenna array using particle swarm optimization with global search and population mutation. *Int. J. Numer. Model. Electron. Netw. Devices Fields* **2020**, *33*, e2710. [CrossRef]
9. Zheng, T.; Liu, Y.; Sun, G.; Zhang, L.; Liang, S.; Wang, A.; Zhou, X. IWORMLF: Improved Invasive Weed Optimization With Random Mutation and Lévy Flight for Beam Pattern Optimizations of Linear and Circular Antenna Arrays. *IEEE Access* **2020**, *8*, 19460–19478. [CrossRef]
10. Shen, H.O.; Wang, B.H.; Li, L.J. Effective approach for pattern synthesis of sparse reconfigurable antenna arrays with exact pattern matching. *IET Microw. Antennas Propag.* **2016**, *10*, 748–755. [CrossRef]
11. Liang, L.; Sun, J.; Li, H.; Liu, J.; Jiang, Y.; Zhou, J. Research on Side Lobe Suppression of Time-Modulated Sparse Linear Array Based on Particle Swarm Optimization. *Int. J. Antennas Propag.* **2019**, *2019*, 7130106. [CrossRef]
12. Das, A.; Mandal, D.; Kar, R. An optimal circular antenna array design considering the mutual coupling employing ant lion optimization. *Int. J. Microw. Wirel. Technol.* **2021**, *13*, 164–172. [CrossRef]
13. Oraizi, H.; Bahreini, B. A comparison among circular, rectangular and bee-hived array geometries using the invasive weed optimization algorithm. In Proceedings of the 2016 16th Mediterranean Microwave Symposium (MMS), Abu Dhabi, United Arab Emirates, 14–16 November 2016; pp. 16–19.
14. Basak, A.; Pal, S.; Das, S.; Abraham, A.; Snasel, V. A modified invasive weed optimization algorithm for time-modulated linear antenna array synthesis. In *Proceedings of the 2010 IEEE World Congress on Computational Intelligence, WCCI 2010-2010 IEEE Congress on Evolutionary Computation, Barcelona, Spain, 18–23 July 2010*; IEEE: Piscataway, NJ, USA, 2010.
15. Jang, C.H.; Hu, F.; He, F.; Wu, L. A Novel Circular Array Structure and Particle Swarm Optimization in Aperture Synthesis Radiometers. *IEEE Antennas Wirel. Propag. Lett.* **2015**, *14*, 1758–1761. [CrossRef]
16. Hosseini, Z.; Jafarian, A. A Hybrid Algorithm based on Invasive Weed Optimization and Particle Swarm Optimization for Global Optimization. *Int. J. Adv. Comput. Sci. Appl.* **2016**, *7*, 295–303. [CrossRef]
17. Mahto, S.K.; Choubey, A. A novel hybrid IWO/WDO algorithm for nulling pattern synthesis of uniformly spaced linear and non-uniform circular array antenna. *AEU-Int. J. Electron. Commun.* **2016**, *70*, 750–756. [CrossRef]
18. Sun, G.; Liu, Y.; Liang, S.; Wang, A.; Zhang, Y. Beam pattern design of circular antenna array via efficient biogeography-based optimization. *AEU-Int. J. Electron. Commun.* **2017**, *79*, 275–285. [CrossRef]

19. Li, J.; Ren, S.; Guo, C. Synthesis of Sparse Arrays Based On CIGA (Convex Improved Genetic Algorithm). *J. Microw. Optoelectron. Electromagn. Appl.* **2020**, *19*, 444–456. [CrossRef]
20. Al-Sadoon, M.; Abd-Alhameed, R.A.; Elfergani, I.T.E.; Noras, J.M.; Rodriguez, J.; Jones, S.M.R. Weight Optimization for Adaptive Antenna Arrays Using LMS and SMI Algorithms. *WSEAS Trans. Commun.* **2016**, *15*, 206–214.
21. Wang, R.Q.; Jiao, Y.C. Synthesis of Sparse Linear Arrays With Reduced Excitation Control Numbers Using a Hybrid Cuckoo Search Algorithm With Convex Programming. *IEEE Antennas Wirel. Propag. Lett.* **2020**, *19*, 428–432. [CrossRef]
22. Wei, W.Y.; Shi, Y.; Yang, J.X.; Meng, H.X. Artificial neural network and convex optimization enable antenna array design. *Int. J. RF Microw. Comput.-Aided Eng.* **2021**, *31*, e22593. [CrossRef]
23. Zhang, R.; Zhang, Y.; Sun, J.; Li, Q. Pattern synthesis of linear antenna array using improved differential evolution algorithm with sps framework. *Sensors* **2020**, *20*, 5158. [CrossRef]
24. Saxena, P.; Kothari, A. Optimal Pattern Synthesis of Linear Antenna Array Using Grey Wolf Optimization Algorithm. *Int. J. Antennas Propag.* **2016**, *2016*, 1205970. [CrossRef]
25. Lakhlef, N.; Oudira, H.; Dumond, C. Optimal pattern synthesis of linear antenna arrays using modified grey wolf optimization algorithm. *Instrum. Mes. Metrol.* **2020**, *19*, 255–261. [CrossRef]
26. Goudos, S.K.; Yioultsis, T.V.; Boursianis, A.D.; Psannis, K.E.; Siakavara, K. Application of New hybrid jaya grey Wolf optimizer to antenna design for 5G communications systems. *IEEE Access* **2019**, *7*, 71061–71071. [CrossRef]
27. Faramarzi, A.; Heidarinejad, M.; Mirjalili, S.; Gandomi, A.H. Marine Predators Algorithm: A nature-inspired metaheuristic. *Expert Syst. Appl.* **2020**, *152*, 113377. [CrossRef]
28. Chen, X.; Qi, X.; Wang, Z.; Cui, C.; Wu, B.; Yang, Y. Fault diagnosis of rolling bearing using marine predators algorithm-based support vector machine and topology learning and out-of-sample embedding. *Measurement* **2021**, *176*, 109116. [CrossRef]
29. Ramezani, M.; Bahmanyar, D.; Razmjooy, N. A New Improved Model of Marine Predator Algorithm for Optimization Problems. *Arab. J. Sci. Eng.* **2021**, *46*, 8803–8826. [CrossRef]
30. Chen, G.; Xiao, Y.; Long, F.; Hu, X.; Long, H. An Improved Marine Predators Algorithm for Short-term Hydrothermal Scheduling. *IAENG Int. J. Appl. Math.* **2021**, *51*, 936–949.
31. Bouchekara, H. Solution of the optimal power flow problem considering security constraints using an improved chaotic electromagnetic field optimization algorithm. *Neural Comput. Appl.* **2020**, *32*, 2683–2703. [CrossRef]

sensors

MDPI

Communication

A Compact Four-Port MIMO Antenna for UWB Applications

Aiting Wu, Mingyang Zhao, Pengquan Zhang * and Zhonghai Zhang

School of Electronics and Information, Hangzhou Dianzi University, Hangzhou 310018, China; wuaiting@hdu.edu.cn (A.W.); my1227@hdu.edu.cn (M.Z.); zhangzhonghai@hdu.edu.cn (Z.Z.)

* Correspondence: zpq@hdu.edu.cn

Abstract: A compact four-port multiple-input multiple-output (MIMO) antenna for ultrawideband (UWB) applications is presented in this paper. The proposed antenna has four unit cell antennas. Each unit cell is placed orthogonal to its adjacent elements. The radiation element of each unit cell is composed of a cut semicircular patch and a stepped microstrip feed line. The whole ground on the back side consists of four parts of defective ground and their extended branches, which are connected through a "卍" structure. The main decoupling technology used in the MIMO antenna is polarization diversity. In addition, protruded ground and parasitic elements are added to achieve a higher isolation. This compact antenna has a small area of 45 mm × 45 mm and is printed on a single layer substrate (FR4) with an $\varepsilon_r = 4.4$ and a thickness of 1.6 mm. This antenna has an impedance bandwidth (S11 < −10 dB) of 3.1–13.1 GHz (123%) and an isolation of less than −17 dB. The envelope correction coefficient (ECC) is less than 0.02 and the average gain is 4 dBi. The ultrawide bandwidth and compact size of the proposed antenna make it a promising candidate for UWB applications.

Keywords: UWB antenna; MIMO antenna; compact antenna; polarization diversity; high isolation

Citation: Wu, A.; Zhao, M.; Zhang, P.; Zhang, Z. A Compact Four-Port MIMO Antenna for UWB Applications. *Sensors* **2022**, *22*, 5788. https://doi.org/10.3390/s22155788

Academic Editors: Naser Ojaroudi Parchin, Chan Hwang See and Raed A. Abd-Alhameed

Received: 2 July 2022
Accepted: 1 August 2022
Published: 3 August 2022

1. Introduction

UWB technology has become a hot topic in recent research. It is a communication method implemented by a series of pulses with very short period; it is also called pulse communication technology. Compared with traditional narrowband technology, it has many advantages, such as high data rate wireless transmission, rich multipath diversity, and very low power consumption. Due to the overlap between UWB signal frequency range and the existing narrowband signal frequency range, the FCC limits the transmission power of UWB (3.1–10.6 GHz). In order to obtain better performance with the limited transmission power, extensive research has been conducted and many solutions proposed. A promising method is to combine MIMO and UWB technology.

Multiple antennas at the transmitter and receiver are used in MIMO technology. This manages to suppress channel fading due to its multipath characteristics and can significantly improve the spectrum utilization. In order to integrate multiple antenna elements on a small substrate, appropriate decoupling is introduced between antenna elements to increase the isolation.

Many decoupling methods have been proposed. In [1], a neutralization line is added between two radiation elements, and it connects to the elements. The neutralization line contains two metal strips connected via rhombus plate. The line effectively reduces the coupling current at ground and achieves a wideband decoupling current. A high isolation is achieved by exploiting the polarization of the multiple elements [2]. A UWB MIMO antenna with a high isolation (less than −22 dB) by loading a parasitic unit decoupling structure on the floor was proposed in [3]. The parasitic element consists of a T-junction and a pair of symmetrical bending lines. Other decoupling methods include defective ground structure (DGS) [4], protruded ground [5], self-decoupling, electromagnetic band gap (EBG) structures [6], decoupling network, and metamaterial [7]. The frequency selective surfaces (FSSs) [8,9] could enhance the antenna's gain for UWB frequencies.

In [2], a four-element MIMO antenna with polarization diversity technology has a high isolation and a small size. However, the substrate of this antenna is Rogers TMM4. The material is very expensive. The size of the four-element MIMO antenna mentioned in [10] is small, but the isolation is not good, only less than −16 dB. For the antennas described in [11,12], due to their single decoupling structure, their isolation is not ideal; it is about −15 dB for both.

A compact four-port MIMO antenna for UWB applications with a high isolation and a wide bandwidth is proposed in this paper. The antenna achieves a high isolation, less than −17 dB, by using the polarization diversity method, protruded ground structures, and parasitic elements. Besides, this antenna provides a good impedance matching from 3.1 to 13.1 GHz and has a good average gain of about 4 dBi. In addition, the ECC result is less than 0.02. It indicates that the antenna meets the polarization diversity requirements. In summary, the highlights of the proposed antenna are its ultrawide bandwidth and compact size. At the same time, the other properties are also good, including the isolation, the gain, and the ECC.

The rest of the paper is organized in the following sections: Section 2 presents the geometry of the antenna. To improve the isolation, three configurations (polarization diversity, protruded ground, and parasitic elements) are analyzed in detail. The wide bandwidth is developed by cutting the current path of the patch. Section 3 presents the measured and the simulated results, including return loss, isolation, ECC, diversity gain (DG), radiation pattern, and gain, with detailed analysis. Section 4 is the conclusion.

2. Antenna Design

The design of the proposed antenna starts from a circular monopole antenna called ant1, as shown in Figure 1a. The return loss (S11) of ant1 is shown in Figure 1b. The monopole antenna is cut into two identical parts to obtain two semicircular radiation elements for miniaturizing called ant2, as shown in Figure 1a. In order to reduce the mutual coupling between these two elements, protruded ground structure is added. It can be seen from Figure 1b that the ant2′s return loss is improved compared with ant1.

Based on ant2, a four-element MIMO antenna is constructed, named ant3. It consists of an orthogonal structure that aims to improve the isolation between elements, as shown in Figure 2a. It also retains the protruded ground structure in ant2. The return loss and isolation (S21, S31) of ant3 are shown in Figure 2b. The S11 of ant3 is poor in high-frequency band, and the isolation of this antenna is above −15 dB in the low-frequency band.

Figure 3 depicts the surface current distribution at 6.5 GHz of ant3. In order to improve the bandwidth, the geometry of the antenna patch is changed to adjust the current distribution, trying to improve impedance matching. As shown in Figure 4a, the result after cutting the patch of ant3 is ant4. Figure 4b shows the return loss and isolation of ant4. Both the bandwidth and isolation have been improved. However, the isolation still does not meet the requirement in some frequency bands.

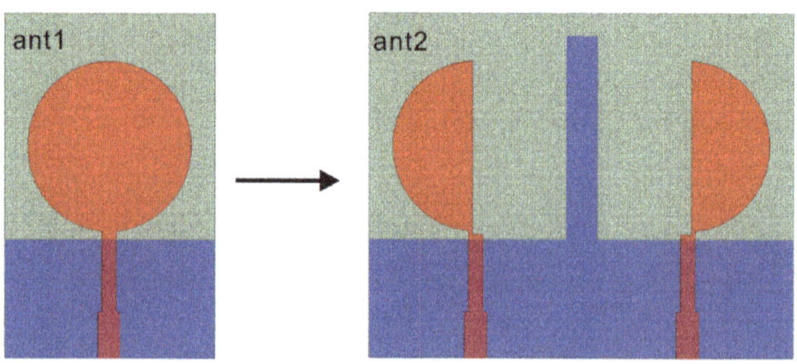

(a)

Figure 1. *Cont.*

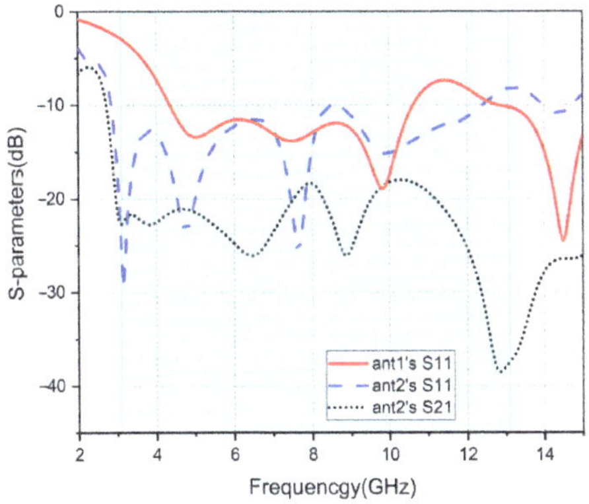

(**b**)

Figure 1. ant1 and ant2: (**a**) geometry; (**b**) S-parameters.

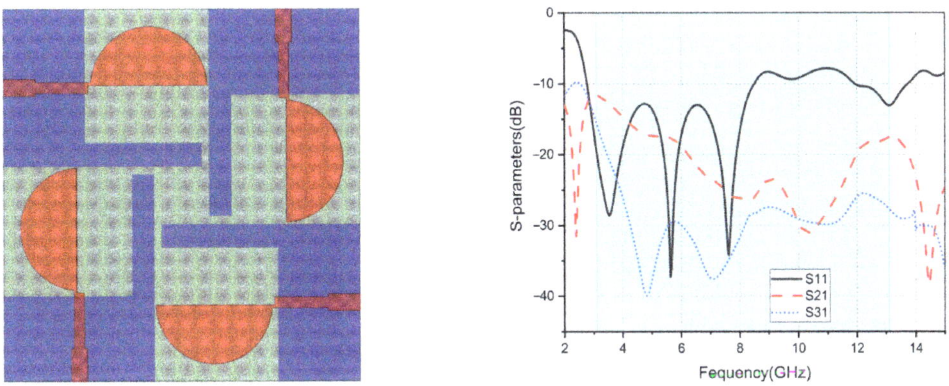

(**a**) (**b**)

Figure 2. ant3: (**a**) geometry; (**b**) S-parameters.

Figure 3. Surface current distribution at 6.5 GHz for ant3.

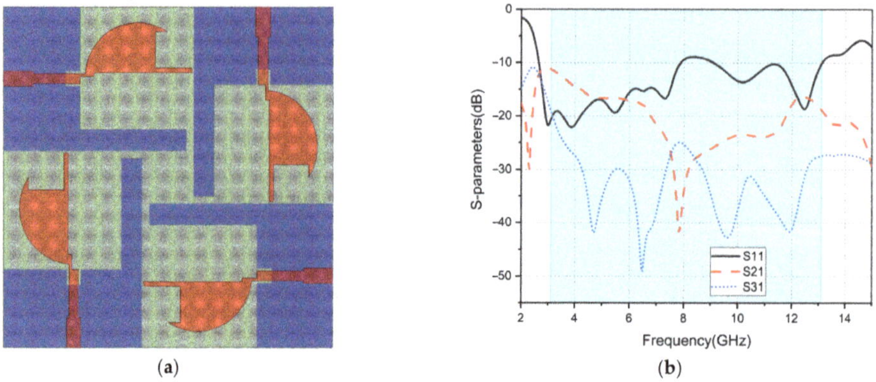

Figure 4. ant4: (**a**) geometry; (**b**) S-parameters.

To further improve the isolation, a "卍" structure is introduced on the metal ground side, as shown in Figure 5a. This generates ant5. It can be seen in Figure 5b, after adding this structure, that the isolation between adjacent elements is significantly improved. This is because the magnitude of mutual coupling currents generated between adjacent elements is the same, while their directions are opposite. They offset each other after passing through the "卍" structure. Besides, the structure also improves the impedance matching of the antenna, and produces a better return loss.

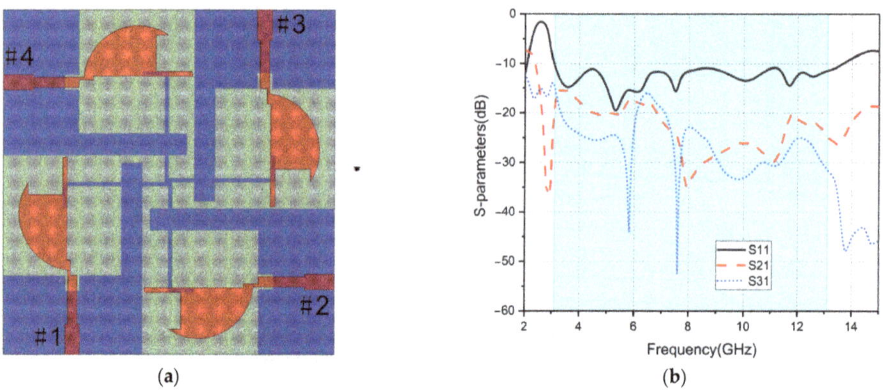

Figure 5. ant5: (**a**) geometry; (**b**) S-parameters.

The schematic diagram of the final proposed antenna (ant5) is shown in Figure 6. The antenna's geometric parameters are listed in Table 1.

Table 1. Geometric parameters of the proposed MIMO antenna.

Dimension	Value (mm)	Dimension	Value (mm)	Dimension	Value (mm)
W	45	Wg4	14.4	W2	1.4
L	45	Lg	10.3	W3	4.8
H	1.6	Lg1	2	Lf	11
Wg	19	Lg2	22.2	Lf1	4
Wg1	2	Lg3	2.5	Wf1	2
Wg2	0.5	Lg4	3.5	Wf2	1.3
Wg3	2.8	W1	9.5	a1	0.95
b1	15.5	a2	1.45	b2	13.5
R	7.4				

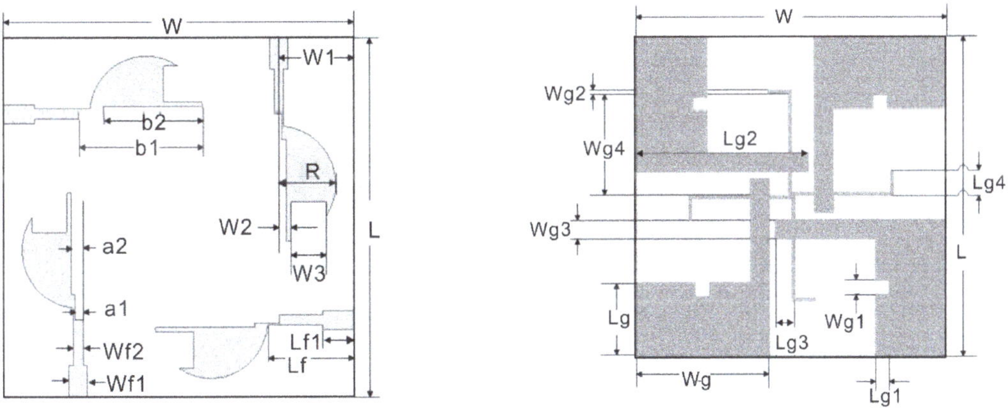

(**a**) (**b**)

Figure 6. The schematic diagram of the proposed antenna: (**a**) top; (**b**) bottom.

3. Results and Discussion

The proposed antenna is fabricated on a FR4 substrate with a thickness of 1.6 mm, as shown in Figure 7. The measured results are presented and discussed in the following subsections.

(**a**) (**b**)

Figure 7. Photograph of the fabricated antenna: (**a**) top; (**b**) bottom.

3.1. S-Parameters

The simulated and measured return loss results of port 1 are provided in Figure 8. The simulated bandwidth (S11 < −10 dB) is 10.5 GHz (3.1–13.6 GHz) and the measured bandwidth is 10 GHz (3.1–13.1 GHz). Due to the manufacturing process, welding problems, and substrate quality, there are slight discrepancies between the simulated and the measured results.

The simulated and measured isolation results between port 1 and port 2, and between port 1 and port 3, are given in Figure 9. The isolation of adjacent elements is less than −17 dB, and that of diagonal elements is less than −20 dB. Because the antenna is completely symmetric, these two parameters only are enough to characterize the antenna's isolation.

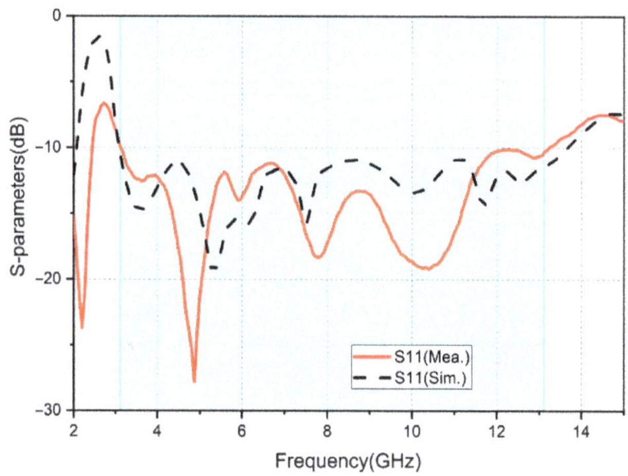

Figure 8. Simulated and measured return loss (S11).

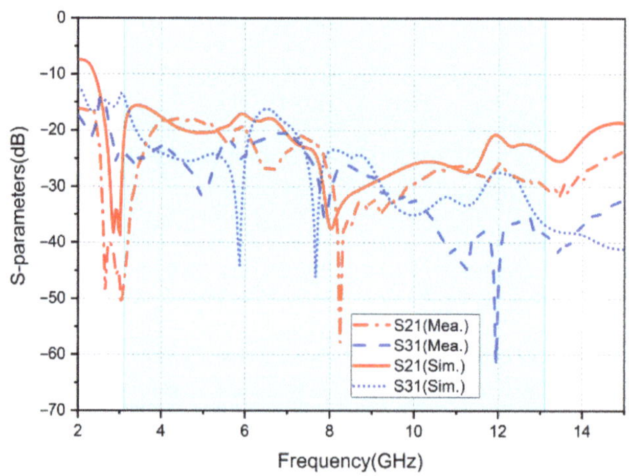

Figure 9. Simulated and measured isolation (S21, S31).

The surface current distribution of the antenna is shown in Figure 10. All the ports of the prototype are terminated to 50 Ω broadband matched loads, except the excited one. The solution frequencies are 3.5 GHz, 7 GHz, and 10 GHz, respectively. As shown in Figure 10, the currents are confined within the area of the active element, protruded ground, and parasitic elements. In the adjacent ports, the currents are very low. The surface current distribution exhibits good isolation.

3.2. Diversity Performance

The diversity performance of the UWB-MIMO antenna system is evaluated by ECC, diversity gain (DG), and total active reflection coefficient (TARC).

ECC parameters reflect the degree of correlation between the adjacent elements of MIMO antennas. For a MIMO system, the ECC parameters are expected to be less than

0.5. When the radiation efficiency is high, ECC could be calculated from the following equation [13,14]:

$$ECC = \frac{\left|S_{ii}^{*}S_{ij} + S_{ji}^{*}S_{jj}\right|^{2}}{\left(1 - |S_{ii}|^{2} - |S_{ji}|^{2}\right)\left(1 - |S_{jj}|^{2} - |S_{ij}|^{2}\right)} \quad (1)$$

The ECC parameters between the antenna's ports are shown in Figure 11; these are less than 0.02 in the operating frequency band. This indicates good isolation between ports.

(a)

(b)

Figure 10. *Cont.*

(c)

Figure 10. The surface current distribution at: (**a**) 3.5 GHz; (**b**) 7 GHz; (**c**) 10 GHz.

DG evaluates the quantified improvement in signal-to-noise ratio when the antennas in the MIMO system receive the RF signal. Equation (2) provides the diversity gain [15].

$$DG = 10\sqrt{1 - |ECC|^2} \tag{2}$$

The diversity gain of the proposed antenna can be calculated by the ECC. The DG parameters between different ports are given in Figure 12. A larger value of DG indicates better diversity characteristics. Figure 12 shows that over the whole operating frequency band (3.1–13.1 GHz), the DG values are above 9.9985.

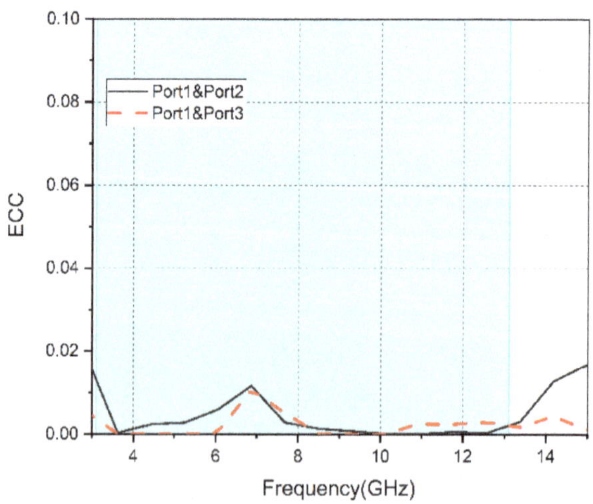

Figure 11. The ECC parameters between different ports.

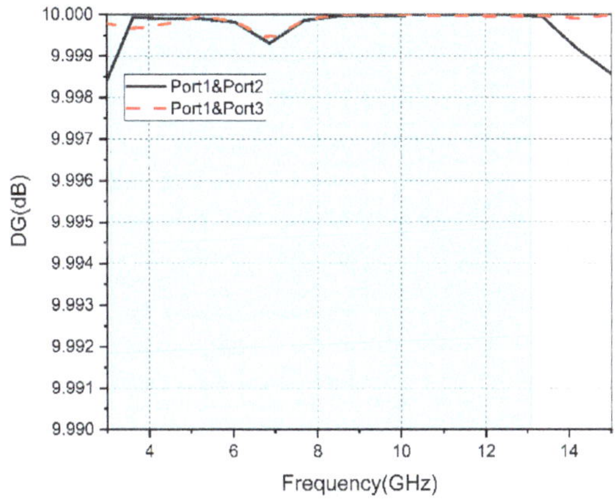

Figure 12. Diversity gain between different ports.

TARC is defined as the ratio of the square root of the total reflected power to the square root of total incident power. It could accurately characterize the effective operating bandwidth of the whole MIMO antenna system and the effect of the change in the phase of signal on the bandwidth of the MIMO antenna. Equation (3) can be used to calculate TARC of the four-port MIMO antenna [13].

$$TARC = \frac{\sqrt{\begin{array}{l}\left|\left(S_{11}+S_{12}e^{j\theta}+S_{13}e^{j\theta'}+S_{14}e^{j\theta''}\right)\right|^2+\left|\left(S_{21}+S_{22}e^{j\theta}+S_{23}e^{j\theta'}+S_{24}e^{j\theta''}\right)\right|^2\\+\left|\left(S_{31}+S_{32}e^{j\theta}+S_{33}e^{j\theta'}+S_{34}e^{j\theta''}\right)\right|^2+\left|\left(S_{41}+S_{42}e^{j\theta}+S_{43}e^{j\theta'}+S_{44}e^{j\theta''}\right)\right|^2\end{array}}}{2} \quad (3)$$

We obtain the average TARC curve by selecting 10 sets of random phases (θ, θ′, θ″) as shown in Figure 13. TARC typically takes less than 0 dB to characterize the antenna's performance. The measured TARC obtained is less than −25 dB. It shows that the proposed antenna has a good performance in the MIMO system.

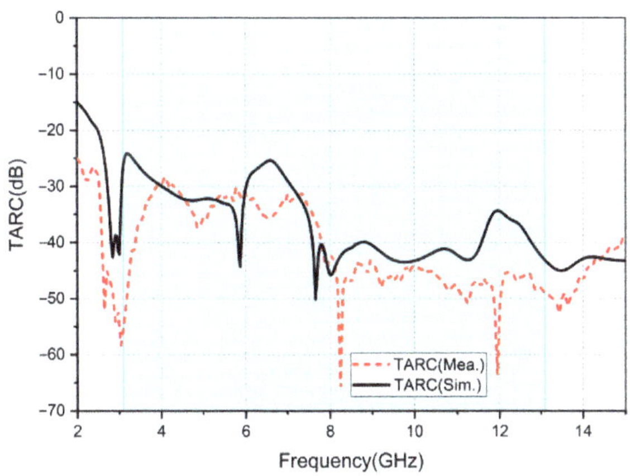

Figure 13. Simulated and measured TARC.

3.3. Radiation Pattern, Gain, and Efficiency

The far-field results are obtained in a microwave anechoic chamber; the excited port is connected to the testing cable and the other ports terminated to 50Ω matched loads. Figure 14 shows the normalized measured and simulated radiation patterns of one element of the antenna at 3.5 GHz, 7 GHz, and 10 GHz, respectively. The radiation pattern of the antenna is distorted with the increase in the frequency. This is mainly because the electrical size of the antenna becomes larger as the frequency increases. It is no longer an electrically small antenna, so the radiation pattern becomes distorted.

The radiation pattern is almost omni-directional in the H-plane. The 3 dB bandwidth at 3.5 GHz is 40° in the E-plane, at 7 GHz is 120° and at 10 GHz is 30°.

The measured gain and simulated gain are shown in Figure 15. From 3.1 GHz to 13.1 GHz, the gain value ranges from 1.6 dBi to 7.3 dBi and the average gain is 4 dBi. There are some discrepancies between the simulated and measured gain. The reason is that the SMA welded at the antenna's ports and three 50 Ω matched loads are equivalent to metal conductors, which will reflect electromagnetic waves. Therefore, these metal conductors will affect the radiation and the gain of the antenna.

The simulated radiation efficiency is shown in Figure 15. It is between 73% and 95% over the entire UWB frequency band. The radiation efficiency of the antenna is relatively stable.

The comparisons between the proposed antenna and existing ones reported in the recent publications are summarized in Table 2. With regard to the characteristics, including size, bandwidth, ECC, gain, and isolation, the proposed antenna has its own advantages compared with others.

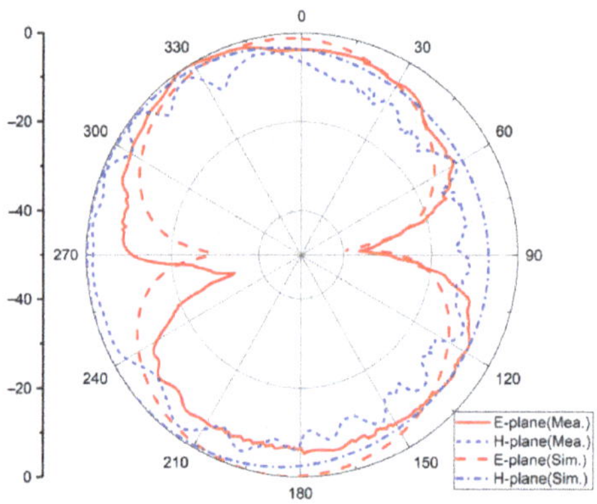

(a)

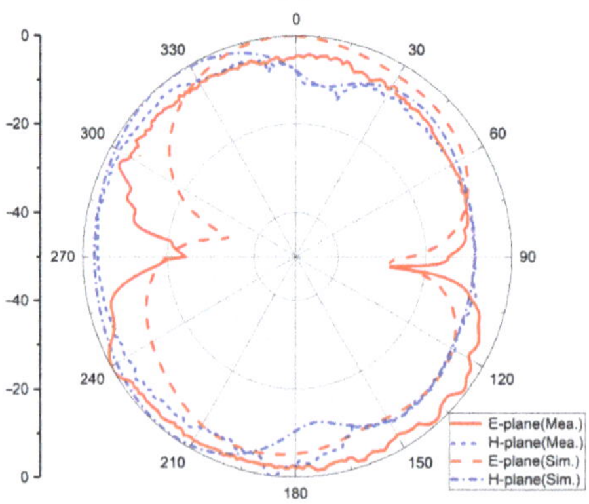

(b)

Figure 14. *Cont.*

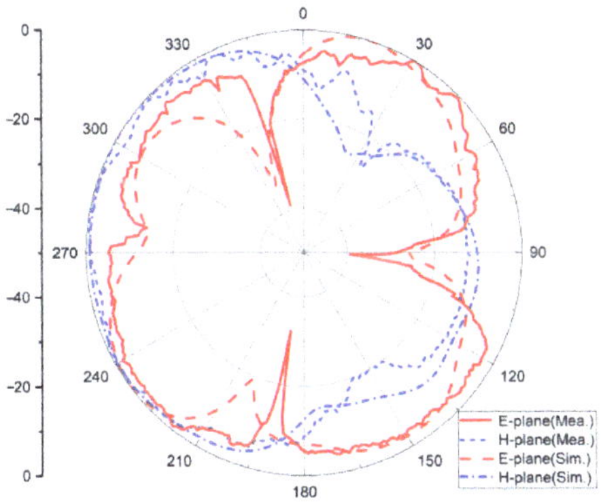

(c)

Figure 14. Radiation pattern at: (**a**) 3.5 GHz; (**b**) 7 GHz; (**c**) 10 GHz.

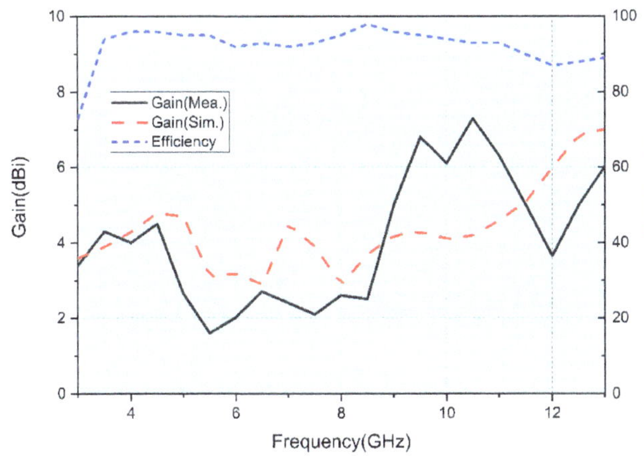

Figure 15. Antenna gain and radiation efficiency.

Table 2. Comparison of the stated UWB MIMO antennas within recent studies.

Ref.	Operating Frequency (GHz)	Bandwidth (%)	Size (λ_0) [1]	Gain (dBi)	ECC	Iso. (dB)	Radiation Efficiency	Substrate	No. of Elements
[2]	2.5~12	131%	$0.42\lambda_0 \times 0.33\lambda_0$ (50 mm × 39.8 mm)	N/A	<0.03	>17	N/A	TMM4	4
[12]	3~11	114%	$0.42\lambda_0 \times 0.42\lambda_0$ (42 mm × 42 mm)	3.5	<0.05	>15	>70%	FR4	4
[15]	3~11	114%	$0.6\lambda_0 \times 0.6\lambda_0$ (60 mm × 60 mm)	3.4	<0.02	>20	>68%	FR4	4
[16]	3.1~10.6	109%	$0.49\lambda_0 \times 0.97\lambda_0$ (47 mm × 93 mm)	3.5	<0.2	>31	>70%	FR4	2
[17]	3.1~17.3	139%	$0.78\lambda_0 \times 0.78\lambda_0$ (75.10 mm × 75.19 mm)	5.5	<0.1	>13	N/A	FR4	4
[18]	2.7~10.6	119%	$0.54\lambda_0 \times 0.54\lambda_0$ (60 mm × 60 mm)	3.5	<0.063	>15	N/A	FR4	4

Table 2. *Cont.*

Ref.	Operating Frequency (GHz)	Bandwidth (%)	Size (λ_0) [1]	Gain (dBi)	ECC	Iso. (dB)	Radiation Efficiency	Substrate	No. of Elements
[19]	3~13.2	126%	$0.383\lambda_0 \times 0.383\lambda_0$ (38.3 mm × 38.3 mm)	4.1	<0.02	>17	>72%	Taconic RF-45	4
[20]	3~13.5	127%	$0.58\lambda_0 \times 0.58\lambda_0$ (58 mm × 58 mm)	2.9	<0.008	>22	N/A	FR4	4
[21]	3.1~11	112%	$0.46\lambda_0 \times 0.46\lambda_0$ (45 mm × 45 mm)	4.3	<0.015	>16	N/A	FR4	4
Pro.	3.1~13.1	123%	$0.46\lambda_0 \times 0.46\lambda_0$ (45 mm × 45 mm)	4.0	<0.02	>17	>73%	FR4	4

[1] λ_0 represents the wavelength in air at the lowest frequency.

4. Conclusions

A compact four-port UWB MIMO antenna with overall dimensions of 45 mm × 45 mm × 1.6 mm is described in this paper. The proposed antenna can operate in the whole UWB band (3.1–10.6 GHz) with a high isolation (−17 dB). Polarization diversity, protruded ground, and parasitic elements are used in the MIMO antenna to achieve the higher isolation. The measured return loss, isolation, and radiation pattern agree with the simulated ones. The antenna has an average gain of 4 dBi. The diversity performance of the proposed antenna is good, with an ECC less than 0.02, DG larger than 9.9985 dB, and TARC less than −25 dB. In summary, the proposed antenna is a promising candidate for UWB-MIMO wireless applications.

Author Contributions: Conceptualization, A.W. and M.Z.; methodology, A.W. and M.Z.; validation, A.W. and M.Z.; formal analysis, A.W. and M.Z.; investigation, M.Z. and Z.Z.; writing—original draft preparation, M.Z.; writing—review and editing, A.W., M.Z. and Z.Z.; supervision, P.Z.; project administration, P.Z. All authors have read and agreed to the published version of the manuscript.

Funding: This research was funded by the National Natural Science Foundation of China under Grant 61501153.

Institutional Review Board Statement: Not applicable.

Informed Consent Statement: Not applicable.

Data Availability Statement: Not applicable.

Conflicts of Interest: The authors declare no conflict of interest.

References

1. Srivastava, N.; Rao, P.K.; Mishra, R. Decoupling Function for UWB MIMO Antenna to Enhance Bandwidth with Neutralization Line. In Proceedings of the 2019 IEEE 5th International Conference for Convergence in Technology (I2CT), Bombay, India, 29–31 March 2019; pp. 1–3. [CrossRef]
2. Khan, M.S.; Capobianco, A.D.; Asif, S.; Iftikhar, A.; Braaten, B.D. A 4 element compact Ultra-Wideband MIMO antenna array. In Proceedings of the IEEE International Symposium on Antennas & Propagation & USNC/URSI National Radio Science Meeting, Vancouver, BC, Canada, 19–24 July 2015.
3. Zhang, K.; Li, J.; Zhai, H.; Wang, Z.; Zeng, Y. UWB-MIMO antenna decoupling based on a wideband parasitic unit structure. In Proceedings of the 2020 International Conference on Microwave and Millimeter Wave Technology (ICMMT), Shanghai, China, 20–23 September 2020; pp. 1–3. [CrossRef]
4. Pandit, S.; Mohan, A.; Ray, P.; Rana, B. Compact Four-Element MIMO Antenna Using DGS for WLAN Applications. In Proceedings of the 2018 International Symposium on Antennas and Propagation (ISAP), Busan, Korea, 23–26 October 2018; pp. 1–2.
5. Zhang, S.; Ying, Z.; Xiong, J.; He, S. Ultrawideband MIMO/Diversity Antennas with a Tree-Like Structure to Enhance Wideband Isolation. *IEEE Antennas Wirel. Propag. Lett.* **2009**, *8*, 1279–1282. [CrossRef]
6. Khan, A.; Bashir, S.; Ghafoor, S.; Qureshi, K.K. Mutual Coupling Reduction Using Ground Stub and EBG in a Compact Wideband MIMO-Antenna. *IEEE Access* **2021**, *9*, 40972–40979. [CrossRef]

7. Zhou, W.; Li, Y. A Taichi-Bagua-like Metamaterial Covering High Isolation MIMO for 5G Application. In Proceedings of the 2021 IEEE 4th International Conference on Electronic Information and Communication Technology (ICEICT), Xi'an, China, 18–20 August 2021; pp. 948–949. [CrossRef]
8. Al-Gburi, A.J.A.; Ibrahim, I.M.; Zakaria, Z.; Abdulhameed, M.K.; Saeidi, T. Enhancing Gain for UWB Antennas Using FSS: A Systematic Review. *Mathematics* **2021**, *9*, 3301. [CrossRef]
9. De Sabata, A.; Matekovits, L.; Buta, A.; Dassano, G.; Silaghi, A. Frequency Selective Surface for Ultra-Wide Band Filtering and Shielding. *Sensors* **2022**, *22*, 1896. [CrossRef] [PubMed]
10. Gurjar, R.; Upadhyay, D.K.; Kanaujia, B.K. Compact four-element 8-shaped self-affine fractal UWB MIMO antenna. In Proceedings of the 2018 3rd International Conference on Microwave and Photonics (ICMAP), Dhanbad, India, 9–11 February 2018.
11. Ellatif, W.A.; Aziz, D.; Mahmoud, R. A 4-elements performance analysis of compact UWB antenna for MIMO-OFDM systems. In Proceedings of the IEEE International Conference on Wireless for Space and Extreme Environments (WiSEE) 2016, Aachen, Germany, 26–28 September 2016.
12. Mathur, R.; Dwari, S. A compact 4-port UWB-MIMO/diversity antenna for WPAN application. In Proceedings of the 3rd International Conference on Microwave and Photonics (ICMAP 2018), Dhanbad, India, 9–11 February 2018.
13. Chen, Z.; Zhou, W.; Hong, J. A Miniaturized MIMO Antenna with Triple Band-Notched Characteristics for UWB Applications. *IEEE Access* **2021**, *9*, 63646–63655. [CrossRef]
14. Tripathi, S.; Mohan, A.; Yadav, S. A Compact Koch Fractal UWB MIMO Antenna with WLAN Band-Rejection. *IEEE Antennas Wirel. Propag. Lett.* **2015**, *14*, 1565–1568. [CrossRef]
15. Ahmad, S.; Khan, S.; Manzoor, B.; Soruri, M.; Alibakhshikenari, M.; Dalarsson, M.; Falcone, F. A Compact CPW-Fed Ultra-Wideband Multi-Input-Multi-Output (MIMO) Antenna for Wireless Communication Networks. *IEEE Access* **2022**, *10*, 25278–25289. [CrossRef]
16. Radhi, A.H.; Nilavalan, R.; Wang, Y.; Al-Raweshidy, H.S.; Eltokhy, A.A.; AbAziz, N. Mutual coupling reduction with a wideband planar decoupling structure for UWB-MIMO antennas. *Int. J. Microw. Wirel. Technol.* **2018**, *10*, 1143–1154. [CrossRef]
17. Kayabasi, A.; Toktas, A.; Yigit, E.; Sabanci, K. Triangular quad-port multi-polarized UWB MIMO antenna with enhanced isolation using neutralization ring. *AEU Int. J. Electron. Commun.* **2018**, *85*, 47–53. [CrossRef]
18. Khac, K.N.; Bao, P.; Ngoc, C.D. Design of Compact 4 × 4 UWB-MIMO Antenna with WLAN Band Rejection. *Int. J. Antennas Propag.* **2014**, *2014*, 539094.
19. Gómez-Villanueva, R.; Jardón-Aguilar, H. Compact UWB Uniplanar Four-Port MIMO Antenna Array with Rejecting Band. *IEEE Antennas Wirel. Propag. Lett.* **2019**, *18*, 2543–2547. [CrossRef]
20. Raheja, D.K.; Kanaujia, B.K.; Kumar, S. Compact four-port MIMO antenna on slotted-edge substrate with dual-band rejection characteristics. *Int. J. RF Microw. Comput.-Aided Eng.* **2019**, *29*, e21756. [CrossRef]
21. Kumar, S.; Lee, G.; Kim, D.; Mohyuddin, W.; Choi, H.; Kim, K. A compact four-port UWB MIMO antenna with connected ground and wide axial ratio bandwidth. *Int. J. Microw. Wirel. Technol.* **2020**, *12*, 75–85. [CrossRef]

sensors

MDPI

Communication

Coplanar Meta-Surface-Based Substrate-Integrated Waveguide Antennas with Broadband and Low Reflections for K-Band Beam Scanning

Chunli Wang, Dongxing Gao *, Likai Liang and Yanling Wang

School of Mechanical, Electrical and Information Engineering, Shandong University, Weihai 264209, China
* Correspondence: dxgao@sdu.edu.cn

Abstract: Four novel substrate-integrated waveguide (SIW) antennas are proposed, in order to obtain K-band beam scanning through the coplanar meta-surfaces of properly devised complementary splitting resonators. More specifically, coplanar rhombus- and hexagon-shaped meta-surfaces replace the metallized via holes in the traditional SIW structure, achieving low reflection and wide bandwidth, respectively. Another trapezoid-shaped meta-surface is introduced, in order to realize good leaky-wave radiation performance with high-gain beam scanning in both rhombus- and hexagon-shaped SIW components. These designs are further extended to two different mixed types of two-row meta-surfaces, with the rhombus and hexagon structures combined in different orders to enhance the complex SIW transmission lines and antennas, which can simultaneously obtain good reflection and bandwidth with different priority, depending on the arrangement. We explain the performance differences with rhombus and hexagon meta-surfaces through the analysis of relevant equivalent circuit models and extracting the effective medium parameters, and we verify the bandwidths and radiations of four SIW antennas both numerically and experimentally. The maximum gains of the four antennas are 18.1 dBi, 17.0 dBi, 18.8 dBi and 17.1 dBi, where the corresponding relative bandwidths are 10.74%, 19.42%, 14.13% and 18.38%. The maximum simulated radiation efficiency and aperture efficiency of the proposed antennas are 91.20% and 61.12%, respectively. Our approach for generating flexible and selectable tuned electromagnetic fields from SIWs is applicable for the development of mm-Wave antennas or sensors on PCB-integrated platforms for highly directive scanning radiation.

Keywords: substrate-integrated waveguide (SIW); mm-Wave antennas; meta-surface; beam scanning; broadband

Citation: Wang, C.; Gao, D.; Liang, L.; Wang, Y. Coplanar Meta-Surface-Based Substrate-Integrated Waveguide Antennas with Broadband and Low Reflections for K-Band Beam Scanning. *Sensors* **2022**, 22, 6353. https://doi.org/10.3390/s22176353

Academic Editors: Naser Ojaroudi Parchin, Chan Hwang See and Raed A. Abd-Alhameed

Received: 8 August 2022
Accepted: 23 August 2022
Published: 24 August 2022

1. Introduction

Substrate-integrated waveguides (SIWs) have attracted the attention of various researchers, due to their great capacity for easy integration and good propagation characteristics on printed circuit board (PCB) substrates, where the double-coated copper and two rows of metallized via holes forming the low-profile and closed structure naturally adapt to the PCB board [1]. The integration advantage of SIWs is due to their similarity to the micro-strip transmission line, while avoiding the large radiation loss from the micro-strip structure as the frequency increases [2,3]. On the other hand, the propagating modes of an SIW imitate those of rectangular waveguides, with the desirable features of low return loss, high Q, and large power capacity, while solving the miniaturization compatibility problem of conventional waveguides [4,5]. As mobile communication technology shifts to 5G and 6G bands, PCB-integrated SIWs are expected to become more widely applied in antennas and sensors [6–8]. Drilling and metallization techniques become more difficult in mm-Wave bands, due to the reduced volume of the electromagnetic devices. The thinner PCB board is also easily deformed under mechanical pressure, resulting in a deterioration in electromagnetic performance. Meanwhile, the flexible adjustment of bandwidth and radiation is urgently required in antennas or sensors for various complex platforms.

The use of meta-surfaces [9–12] has been shown to be an efficient way to build novel electromagnetic devices, including high-gain beam redirection [13–15] and multi-beam control with enhanced radiations [16,17], where the abrupt phase discontinuities provided by meta-surfaces could offer a straightforward method to manipulate electromagnetic waves. Such a method is also similar to frequency selective surfaces (FSSs), which is one of the periodic interface structures, where the sub-wavelength units enable us to tune the electrical properties, such as amplitude, phase and polarization, of the incoming waves, thus fulfilling the high-gain radiations through reshaping the wavefront simply at surfaces [18,19]. Moreover, the capacity of meta-surfaces of perfectly integrating with PCB platforms simply by using copper cladding has also been demonstrated [20–23]. On this basis, specifically designed classical complementary split-ring resonators have been proposed to imitate a virtual electric wall in the SIW design and remove conventional metallized via holes [24]. However, in order to design mm-Wave SIW devices with stronger performance and more flexible parameters, it is necessary to further couple the meta-surface and SIW structure. We can achieve the free manipulation of electromagnetic fields and specific adjustment of the bandwidth, reflection coefficient and pattern performance through the introduction of different types of sub-wavelength meta-atom arrays into the desired SIWs [25–29], thus initiating the quest for tangible applications in the SIW transmission line and antenna scenarios.

In this paper, we propose four packs of SIW antennas to obtain beam scanning, low reflection and wide bandwidth performance through arranging series of properly devised coplanar meta-surfaces. Rhombus- and hexagon-shaped meta-units are independently built to simulate virtual electric walls and replace metal via holes, where the rhombic structure has a steeper *S*-curve and numerically better electric wall performance, while the hexagonal structure focuses on realizing a wider frequency band. These different performances are well explained through analysis of relevant circuit models and calculating the normalized impedance. Rhombus and hexagon meta-surface inspired SIW transmission lines are thus established, achieving lower reflection and wider bandwidths, respectively, as expected. By bringing trapezoid-shaped meta-surfaces to the rhombus and hexagon SIW lines, the coplanar meta-surface-based SIW antennas are further established, demonstrating better leaky-wave performance as well as preservation of the original bandwidth parameters. Finally, we combine the rhombus and hexagon structures in different orders to simultaneously obtain good reflection and bandwidth, with different emphasis due to the relative SIW transmission lines. Trapezoid-shaped meta-surfaces are also introduced to these hybrid SIWs, forming two other new coplanar meta-surface-based SIW antennas, which achieve the expected radiations. All four of the proposed antennas have the advantages of more flexibly controlling the reflection coefficient and bandwidth, while also demonstrating good leaky-wave and beam-scanning performances. Our work aims to guide the selection of a specific combination of meta-surfaces to meet the various demands of electromagnetic devices, thus broadening the application scope of mm-Wave PCB-integrated antennas or sensors.

2. Meta-Surface-Based SIW Component Design and Analysis

2.1. Unit Cell Design

Let us start with the design of unit cells for the rhombus- and hexagon-shaped meta-surfaces around 20 GHz, as shown in Figure 1. The two unit elements share a similar complementary split-ring resonator template, but have different geometric structures and sizes, leading to different *S*-parameter simulations and circuit parameters. Considering a traditional SIW, the metallized via holes are equivalent to small dipoles and, thus, act as an electric wall, fulfilling the surface current loop driven by the TE_{m0} mode flowing along each via [1]. Note that the classical complementary split-ring resonator is capable of imitating the virtual electric wall and keeping the horizontal incoming waves fixed in the closed cavity at specific narrow bands [30,31]. Hence, we modify the geometric structures of classical resonators and rebuild the rhombus- and hexagon-shaped unit cells with different wide-bands around 20 GHz. The unit length a is set to 2.5 mm that is $\lambda_0/6$ at 20 GHz, which meets the requirements that the periodic composite structure could be applied as sub-wavelength atoms of meta-surfaces at such volume scale and specific frequency [32–36]. Further, the length l_1 and l_2 of the rhombus- and hexagon-shaped cells could be determined by their specific geometry and unit length a, following the traditional design method of meta-surface units. The rhombus (or hexagon)-shaped slots are etched on the top and bottom PEC planes, where the two slots have the same geometric sizes as listed in the figure caption but opposite opening directions of the gaps, as shown in Figure 1a (or Figure 1b). To verify the blocking performance generated from the coupling effect between the complementary rings of the unit, the S-parameter performances are simulated based on the enhanced Floquet mode with a vertically polarized electric field feeding in port 1 and receiving in port 2, where the ports are attached to the front and back of the substrates, which are vertical to the y-axis. The symmetrical spaces above and below the substrates are filled with air, with the top and bottom boundaries as a perfectly matched layer (PML), while the remaining boundaries are set as perfect magnetic conductors (PMCs). Periodic boundary conditions are, thus, indirectly enforced, repeating the modeled structure periodically along the x- and y-axes. The simulated results are shown in Figure 1c,d. We can observe that both the rhombus and hexagon units present sufficiently wide bands, where the S_{11} results higher than -10 dB range from 18.53 to 26.72 GHz and 15.86 to 26.66 GHz, respectively. The -1 dB bandwidth of S_{11} results used for evaluating shielding effects are 19.45 to 25.05 GHz for the rhombus-shaped unit cell and 16.98 to 24.10 GHz for the hexagon-shaped unit cell, as shown in the gray areas. Moreover, the phase component of the S_{11} value is also simulated for further analysis, plotted as black dotted lines in Figure 1c,d. Both the phases of the rhombus- and hexagon-shaped units decrease approximately and linearly within the transmission interval and the shielding interval, while experiencing abrupt value changing near the resonance frequency points. A high S_{11} level corresponds to a low S_{21} value, which means that the unit is equivalent to an electric wall at these frequency ranges, and that the propagating mode through the cell is prohibited. We can also observe that the rhombus unit has a steeper slope near the minimum transmission regions at 19.97 GHz and 24.17 GHz, while the hexagon structure has 31.87% more bandwidth, based on the S_{11} values.

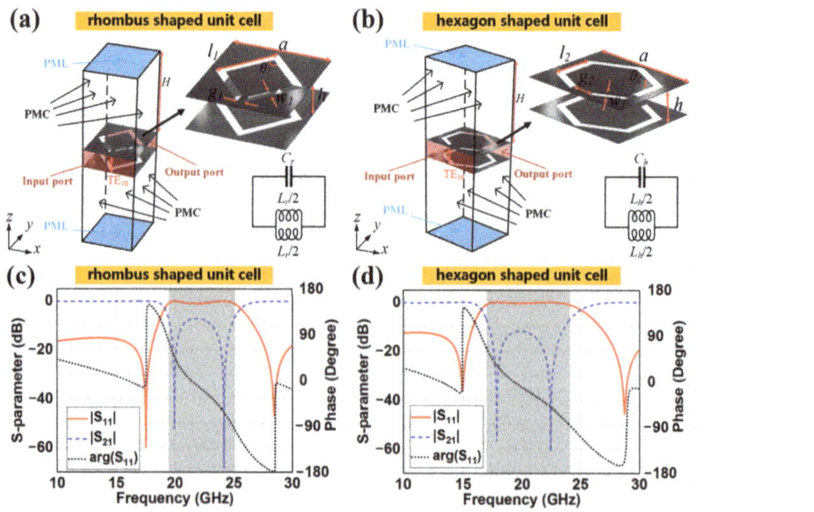

Figure 1. The Floquet model and corresponding equivalent circuits of the (**a**) rhombus- and (**b**) hexagon-shaped unit cells. Each unit has the dimensions of 2.5 mm × 2.5 mm × 0.787 mm and

the substrate is chosen as RT5880 material, with relative permittivity of 2.2 and loss tangent of 0.001. Moreover, a = 2.5 mm, h = 0.787 mm, H = 3.75 mm, l_1 = 1.58 mm, l_2 = 1.2 mm, θ_1 = 90°, θ_2 = 120°, g_1 = 0.16 mm, g_2 = 0.5 mm, w_1 = 0.2 mm, and w_2 = 0.2 mm, based on the optimization results. The S-parameter of the (**c**) rhombus- and (**d**) hexagon-shaped unit cells is also provided.

To explain these different resonant properties, we propose the equivalent circuit models of rhombus and hexagon cells, as shown in Figure 1a,b, based on the basic equivalent circuits of classical split-ring resonators [37,38], as well as the duality principle [39–42]. In particular, the equivalent circuit of the classic double-layer split ring resonator is an LC parallel resonance circuit [24,43,44]. We start with such a classical resonance circuit extracted from a circular ring and change the circuit form to adapt to rhombus and hexagon unit cells using the duality principle and analyzing geometry changes. For the physical structure of the classical circular ring and the proposed rhombus and hexagon rings, the corresponding circuit parameters could be calculated through analyzing the detailed micro-structure of the rings, where the intervals between rings mainly represent capacitance, and the ring with a specific length mainly represents inductance. Therefore, the corresponding capacitance and inductance could be updated through comparing the geometric differences between the classical circular ring and proposed rhombus and hexagon rings. The total capacitance is obtained by calculating the series capacitances $C_{SRR}/2$ of two semi-circular ring loops. As the ring is a double-layer structure, the unit length capacitance of the microstrip line can also be calculated as a series capacitance of two microstrip lines on both layers. That is, the unit length capacitance C_m of each side can be expressed as $C_m = \sqrt{\varepsilon_{\text{eff}}}/c_0 Z_0$, with its substrate height equal to half of the total substrate thickness, where ε_{eff}, c_0 and Z_0 are the effective dielectric constant, speed of light and microstrip line characteristic impedance, respectively [45]. The total unit length capacitance is equal to $C_m/2$, when considering the series effect of the two layers. Thus, the capacitance value, C_{SRR}, can be expressed as $C_{\text{SRR}} = C_m L_{\text{eff}}/2$, where L_{eff} is the effective perimeter of the ring. The total inductance of the standard circular ring loop can be expressed as $L_{SRR} = u_0 u_r (D_{\text{eff}}/2)(\ln(8D_{\text{eff}}/w) - 2)$, based on [46], where D_{eff} and w represent the effective diameter and wire width of the ring, respectively. On this basis, the complementary split-ring resonators of rhombus and hexagon cells are equivalent to a duality circuit of the LC parallel resonance circuit mentioned above, as the units used here are complementary rings [44,47]. Thus, the converted capacitance C_r (C_h) is calculated by multiplying the original inductance L_{SRR} by a factor of $4\varepsilon_0/\mu_0$, while the converted inductance L_r (L_h) is equal to the quotient of the original capacitance C_{SRR} and the same factor [38]. The derived formulas are as follows:

$$C_r = 2\varepsilon_0 \mu_r \frac{l_1}{\sqrt{\pi}} \left(\ln \frac{8l_1}{\sqrt{\pi}\omega_1} - 2 \right), \tag{1}$$

$$C_h = 2\varepsilon_0 \mu_r \frac{l_2}{\sqrt{\frac{3\sqrt{3}}{2\pi}}} \left(\ln \frac{8l_2}{\sqrt{\frac{3\sqrt{3}}{2\pi}}\omega_2} - 2 \right), \tag{2}$$

$$L_r = \frac{\mu_0}{8\varepsilon_0} \frac{\sqrt{\varepsilon_{\text{eff}}}}{c_0 z_0} \left[4(l_1 - \omega_1) - \sqrt{2} g_1 \right], \tag{3}$$

$$L_h = \frac{\mu_0}{8\varepsilon_0} \frac{\sqrt{\varepsilon_{\text{eff}}}}{c_0 z_0} \left[6\left(l_2 - \frac{\omega_2}{\sqrt{3}} \right) - \frac{2}{\sqrt{3}} g_2 \right], \tag{4}$$

where l_1, l_2, w_1, w_2, g_1 and g_2 are labeled in Figure 1 and their values are listed in the corresponding figure caption. The effective diameter and perimeter of the two meta-units, expressed as D_{eff} and L_{eff} in the original formulas, can be represented by the specific structural parameters, according to the geometric properties of the rhombus and hexagon unit cells. By comparing the values of the geometric parameters for the rhombus- and hexagon-shaped units shown in Figure 1, we obtained the following expressions: $C_h > C_r$ and $L_h > L_r$. Therefore, the hexagon unit design has lower resonant frequencies than the rhombus unit. Considering the electromagnetic response of the periodic composite material at different volume scales and frequency ranges, the analytical model of the electromagnetic structure with constant size may change from the classical medium to the meta-surface, then move to a frequency-selective surface as the frequency increases, where the wavelength gradually becomes comparable to the geometry of electromagnetic devices [32–35]. This means that the media properties of the rhombus and hexagon unit cells gradually move to the transition region of the meta-surface and frequency-selective surface when the second resonant frequency (i.e., the upper stop-band frequency point) appears as the frequency continues to increase. Compared to the lower-side resonant frequency, the upper-side frequency is less sensitive to the geometric microstructure of the unit cell, leading to the fact that the upper-side frequencies of the rhombus and hexagon meta-units are closer to each other. This phenomenon and Equations (1)–(4) work well together to explain that both of the resonant frequencies of the hexagon unit (17.85 GHz and 22.41 GHz) are lower than those of the rhombus unit (19.97 GHz and 24.17 GHz), and the frequency difference of the lower-side frequencies of the two unit cells is larger than that for the upper-side frequencies. In summary, the hexagon-shaped design has a wider bandwidth than that of the rhombus design.

Another method for analyzing the electromagnetic response of a meta-surface is effective medium theory [48,49], which can be extracted and calculated from the S-parameter results. We used the MATLAB software to calculate the normalized transfer impedance Z_T of the lower- and upper-side resonant frequencies of the rhombus and hexagon units, which can be applied to quantify the shielding effectiveness of the virtual electric walls. The normalized transfer impedances of the rhombus unit were 7.79×10^{-7} and 4.63×10^{-7} at 19.97 GHz and 24.17 GHz, respectively, while the normalized transfer impedances of the hexagon unit were 1.05×10^{-5} and 7.00×10^{-5} at 17.85 GHz and 22.41 GHz, respectively. A lower Z_T value means that the incoming waves experience a stronger impedance mismatch and amplitude decay, resulting in steeper slopes and narrower bands in the resonant curve. We can conclude that the rhombus design has stronger shielding effectiveness and better virtual electric wall performance near the minimum transmission regions, despite the narrower bands, while the hexagon design has a significantly broader band with acceptably raised S_{21} performance.

2.2. SIW Topology

First of all, we would like to demonstrate the design method and logical relationship of all proposed coplanar SIW transmission lines and antennas that are constructed with meta-surfaces of sub-wavelength meta-atom arrays mentioned in Figure 1, which are summarized and listed in Table 1. One row of rhombus (or hexagon) units could construct one side of the virtual electric walls of basic meta-surface-based SIW transmission lines, as shown from Figures 2–4. For the extended version of SIW lines, another row of meta-units is introduced to combine into rhombus–hexagon (or hexagon–rhombus) virtual electric walls, in order to obtain different bandwidth performances, as shown in Figure 5. Furthermore, all four proposed SIW transmission lines could be evolved into four SIW antennas through enabling trapezoid-shaped meta-surfaces, respectively. We now practically construct the basic version of coplanar SIW transmission lines and antennas with meta-surfaces, as shown in Figure 2. The double-layer rhombus (or hexagon)-shaped unit slots are etched on the top metal surface and bottom grounded plane instead of drilling the substrate, substituting for the metallized via holes of the standard SIW structure near the same locations and assembling the coplanar meta-surface-based SIW line, as shown in Figure 2a (or Figure 2b). The gaps of the double-layer ring slots are face-to-face on the top and in opposite directions on the bottom. The micro-strip feeding lines and transitions are attached to the SIW structure on the both ends. As the transmission lines or antennas exhibit periodic structures along the direction of wave propagation, we cut off the central parts of the SIWs, in order to better illustrate the detailed structure, in Figure 2. The width of the virtual electric walls from meta-surface-based SIW lines, expressed as w_r (or w_h), is the key geometric dimension in the proposed design, where most of other parameters are related to it. Since the meta-surfaces would substitute the metallized via holes shown in the top areas of Figure 2a, the geometry sizes of the novel SIW structure, including w_r or w_h, could be calculated and optimized from the classical method of the traditional SIW design. The equivalent width of the traditional SIW expressed as w_{eff} could be calculated as [50]

$$w_{\text{eff}} = w - 1.08\frac{d^2}{s} + 0.1\frac{d^2}{w} \tag{5}$$

where w refers to the physical width of the two rows of via holes, d is the diameter of a single hole, and s is the distance between neighboring holes. Thus, the parameter w_{eff} would determine the transmission characteristics and operating frequency [50], which would help us to determine the initial dimensions of w_r and w_h, and we could further tune and optimize these values to fulfill the actual working frequency around 20 GHz. On this basis, the other parameter values could be determined within specific ranges. The W value is almost the same as the entire width of the traditional SIW, if choosing the same frame structure, material and working frequency. The L value depends on the actual requirements of the PCB board. Finally, the size of the transition structures, which are sensitive to the S-parameter, should be initialized and further optimized to satisfy the impedance matching requirements when we introduce the 50 Ω micro-strip lines at both ends [51–53], and such values should also be limited in the ranges smaller than w_r and w_h.

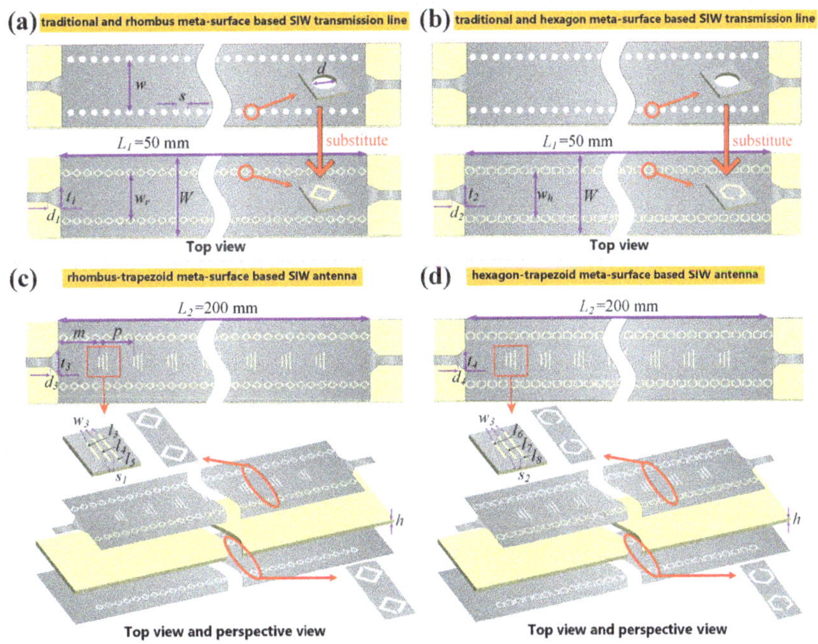

Figure 2. The coplanar (**a**) rhombus and (**b**) hexagon meta-surface-based SIW transmission lines compared with the traditional SIW structure, as well as the coplanar (**c**) rhombus–trapezoid and (**d**) hexagon–trapezoid meta-surface-based SIW leaky-wave antennas. The structural parameters are L_1 = 50 mm, W = 22 mm, t_1 = 6 mm, d_1 = 4 mm, w_r = 13 mm, t_2 = 5 mm, d_2 = 2.5 mm, w_h = 13 mm, L_2 = 200 mm, m = 9.4 mm, p = 8 mm, t_3 = 7 mm, d_3 = 1.8 mm, t_4 = 5.7 mm, d_4 = 1.7 mm, and h = 0.787 mm. For the trapezoid unit, the geometric sizes are set as: l_3 = 2.9 mm, l_4 = 3.9 mm, l_5 = 4.9 mm, s_1 = 0.9 mm, w_3 = 0.2 mm, l_6 = 3 mm, l_7 = 4 mm, l_8 = 5 mm, and s_2 = 1 mm.

Table 1. The design method and logical relationship of all proposed coplanar meta-surface-based SIW transmission lines and antennas.

	Basic Meta-Surface-Based Design Figures 2–4		Extended Meta-Surface-Based Design Figure 5	
Transmission line	Rhombus row	Hexagon row	Rhombus–hexagon rows	Hexagon-rhombus rows
Antenna	Rhombus row (Trapezoid enabled)	Hexagon row (Trapezoid enabled)	Rhombus–hexagon rows (Trapezoid enabled)	Hexagon–rhombus rows (Trapezoid enabled)

Figure 2c,d further illustrate the rhombus and hexagon meta-surface-based SIW slot-antenna designs, respectively. The transverse multiple-slot meta-surfaces are introduced and periodically etched on the top plane of the SIW transmission lines, in order to construct the coplanar leaky-wave antennas with slow-wave structure. Such periodic profile modulation would introduce infinite space harmonics [54–57], and the phase constant β_n of the nth space harmonic is calculated as

$$\beta_n = \beta_0 + \frac{2n\pi}{p} \quad ,n = 0, \pm1, \pm2 \ldots \tag{6}$$

where β_0 refers to the phase constant of the dominant mode, and p is the periodic length of the slot-units shown in Figure 2c. The space harmonic with phase constant β_{-1} is usually chosen to be a fast mode, capable of converting transmission wave to radiation wave with backward beam angle as $\theta_{-1} = \sin^{-1}(\beta_{-1}/k_0)$, where k_0 is the wavenumber in free space, while the other harmonics remain in slow-wave mode and do not generate radiation. Therefore, these conditions could be quantified as $\beta_0 > k_0$, $|\beta_{-1}| < k_0$ and $|\beta_n| > k_0$, where $n < -1$. Further, we can derive the range of p expressed as $2\pi/(\beta_0 + k_0) < p < 4\pi/(\beta_0 + k_0)$ from these formulas to satisfy the condition that only $n = -1$ space harmonic can generate radiated fields. In addition, the minimum length of the slot can be calculated as $l = \lambda_0/(4\sqrt{\varepsilon_r})$ [58]. Taking 20.0 GHz as an example, we could obtain β_0 = 601.57 rad/m, k_0 = 418.88 rad/m, β_{-1} = −183.83 rad/m, β_{-2} = −969.23 rad/m, l = 2.53 mm and 6.16 mm < p < 12.32 mm, which exactly meets the requirements of $n = -1$ mode radiation. At the beginning, the main single slot contributing the radiation is thus built using these calculated parameters, with dimensions of l_5 and w_3 in Figure 2c, as well as l_8 and w_3 in Figure 2d. Based on the single-slot structure, the updated unit consists of three transverse slots arranged in a trapezoid shape, which is proposed to construct the multiple-slot meta-surface and etch on the top plane of the SIW component. Inside each unit cell, the slot located in the end side with the longest length acts as the original main slot, and the extra two shorter slots that are linearly trapezoidal are designed to reduce undesired reflection, where the width gradient increases along the direction of the inside transmission wave. Since the parameter values or ranges, especially l and p, are calculated by classical theory, fulfilling the $n = -1$ space harmonic that adapts to a standard SIW structure with metallized via holes, we take them as the initial values or ranges and continue to optimize the geometric sizes of the main slot to satisfy the specific meta-surface-based designs, where the rest of the dimensions of the trapezoid unit would undergo synchronized optimization to hold the low reflections. All the corresponding geometric sizes in Figure 2 are labeled and listed in the figure and caption, and the sizes of the trapezoid units have a minor difference between the rhombus and hexagon meta-surface-based designs, in order to achieve their respective optimum performance. Compared to the conventional leaky-wave antenna with identical single slots, each unit of the trapezoid meta-surface in our proposed antennas enables local impedance transitions to suppress extra reflections, where all the slots are distributed with uniform aperture. There are also two common impedance gradient methods in the traditional SIW leaky-wave antenna designs, such as locating tapered increased single slots at the beginning [59,60] and periodically changing the slot length throughout the entire line [61,62]. Both of these design methods can achieve a significantly optimized S_{11} level, while the overall length of the SIW structure cannot be arbitrarily set, due to the demands involved with adapting to a gradually changing geometry. Meanwhile, our proposed trapezoid meta-surface could also obtain low reflections, and the minimum changing step size is the small distance p of the trapezoid unit. As a result, the rhombus-, hexagon- and trapezoid-shaped meta-surfaces assembled in the SIWs all have a coplanar design and can be etched by copper cladding technology, which avoids the drilling of metallized via holes while realizing easy fabrication and significantly increasing device reliability.

2.3. Full-Wave Simulations

Full-wave simulations (CST Microwave Studio with Frequency Domain Solver) were performed to verify the coplanar SIW transmission lines and SIW antennas, as shown in Figures 3 and 4, respectively. Figure 3a (or Figure 3b) demonstrates the two-dimensional E-field distributions on the cutting plane of $h/2$ for the rhombus (or hexagon) meta-surface-based SIW transmission line at 19.99 GHz (or 19.27 GHz). We can observe that the TE_{10} mode, the same as the rectangular waveguide, was successfully established, with minor leakage of the E-field outside the meta-surface rows. The E-field results prove that the two proposed meta-surface-based SIW lines can well imitate virtual electric walls and replace metallized via holes. The rhombus design demonstrates better electric wall performance with lower sidewall leakage of E-fields than the hexagon design. Figure 3c,d illustrate the corresponding S-parameters of the rhombus and hexagon SIW lines, respectively. We can observe that the rhombus design has S_{11} values less than -10 dB ranging from 18.59 GHz to 20.37 GHz, with a minimum value of -38.37 dB at 19.99 GHz and relative bandwidth of 9.14%, while the hexagon design has S_{11} values less than -10 dB ranging from 16.78 GHz to 20.68 GHz, with a minimum value of -27.47 dB at 19.27 GHz and relative bandwidth of 20.82%. We can also observe that there were some frequency shifts in the SIW lines and the original units, which can mainly be attributed to the fact that the unit cells were stimulated using ideal periodic Floquet models, while the meta-surfaces etched in practical structures would be affected by truncated boundary conditions. However, both the rhombus and hexagon SIW lines still achieved satisfactory power transfer with a lower reflection coefficient and wider bandwidth, respectively.

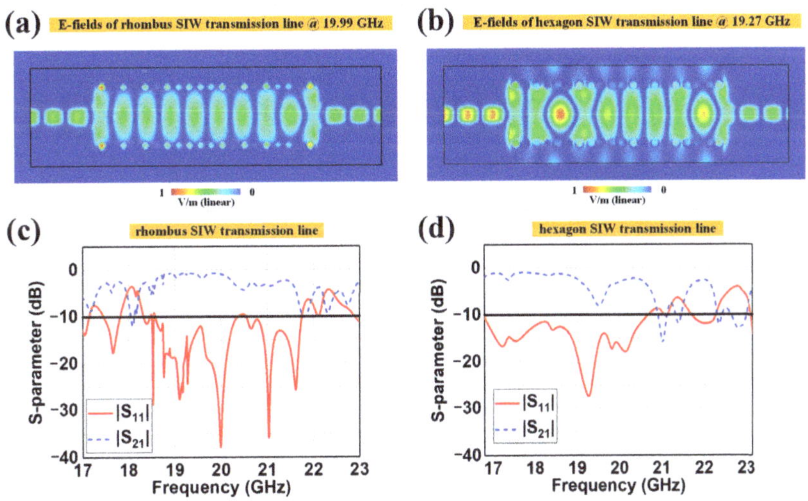

Figure 3. The E-field distributions of the (**a**) rhombus and (**b**) hexagon meta-surface-based SIW transmission lines, as well as the corresponding S-parameters of the (**c**) rhombus and (**d**) hexagon meta-surface-based SIW transmission lines.

Figure 4a,b demonstrate the corresponding E-field distributions on the top plane of the rhombus and hexagon meta-surface-based SIW antennas at 19.60 GHz and 18.00 GHz, respectively. We can observe that the significant E-fields were distributed over the trapezoid-based meta-surfaces, which would excite high-gain radiations in the far-field. The field leakage at the sides still remained at a low level, as with the previous SIW lines. Figure 4c,d illustrate the corresponding *S*-parameters of the rhombus and hexagon SIW antennas with their comparison groups, respectively. In order to verify the effect of suppressing extra reflections from the trapezoid meta-surfaces, the traditional single-slot SIW antennas with the same external rhombus and hexagon meta-surface-based structures were also included in the S_{11} simulations, for comparison purposes. Note that the proposed designs in Figure 4c,d and their comparison groups of single-slot structures should optimize the respective sizes of the microstrip transitions, achieving their best *S*-parameter performances. The -10 dB bandwidths of the trapezoid meta-surface designs are marked as red areas, while those of single-slot designs are marked as blue areas, and the intersections are marked as purple areas. We can observe that the trapezoid meta-surface designs had better reflection coefficients than the single-slot design, which proves that our trapezoid design enables better impedance matching through the transitions of the gradient slots. The rhombus–trapezoid design has S_{11} values less than -10 dB ranging from 18.94 GHz to 21.09 GHz, with a minimum value of -50.04 dB at 19.31 GHz and a relative bandwidth of 10.74%, while the hexagon–trapezoid design has S_{11} values less than -10 dB ranging from 17.20 GHz to 20.90 GHz, with a minimum value of -20.65 dB at 20.29 GHz and a relative bandwidth of 19.42%. We can also observe that the trapezoid meta-surface SIW antennas and the corresponding transmission lines showed similar trends in S_{11}, indicating that the introduction of a trapezoidal meta-surface only affects the reflection coefficient, to a certain extent.

Figure 4e–h continue to illustrate the two- and three-dimensional far-field radiation patterns of the coplanar trapezoid meta-surface-based SIW antennas, as well as the maximum gains over frequencies. We can observe that both the proposed designs generate highly directive fan beams and achieve satisfactory beam scanning within their operating frequency bands. For example, at the frequencies of 19.6 GHz and 18.0 GHz, the maximum gains of the rhombus and hexagon SIW antennas were 18.10 dBi and 17.00 dBi, respectively. The radiation efficiency and aperture efficiency thus could be obtained through comparing the gains to the directivity or aperture of the antenna. To be more specific, the radiation efficiency is calculated from the ratio of gain to directivity, expressed as $\eta_{ra} = G/D$, where the directivity can be obtained from the simulations [63–65]. Meanwhile, the aperture efficiency of the meta-surface-based SIW antenna can be calculated by $\eta_{ap} = (G\lambda^2)/(4\pi A)$, where A is the physical aperture of the antenna [66–68]. Therefore, the radiation efficiencies of the rhombus–trapezoid and hexagon–trapezoid SIW antennas are 87.10% and 81.28%, respectively, and the corresponding aperture efficiencies are 52.02% and 47.88%, respectively. The rhombus–trapezoid design has a 3-dB beamwidth of 4.7° in the E-plane and 58.9° in the H-plane, while the hexagon–trapezoid design has a 3-dB beamwidth of 5.4° in the E-plane and 46.1° in the H-plane. We can also observe that the beam-scanning performances of the two proposed antennas are well demonstrated when the operating frequency varies within certain ranges with high-gain performance. The rhombus–trapezoid meta-surface-based SIW antenna has scanning beams in $-24°$, $-22°$, $-20°$, $-18°$ and $-16°$ at 19.2 GHz, 19.4 GHz, 19.6 GHz, 19.8 GHz and 20.0 GHz, respectively, while the hexagon–trapezoid design has scanning beams in $-42°$, $-37°$, $-32°$, $-28°$ and $-23°$ at 17.2 GHz, 17.6 GHz, 18.0 GHz, 18.4 GHz and 18.8 GHz, respectively. Every 0.4 GHz frequency offset would result in 4° beam steering for the rhombus antenna and 5° beam steering for the hexagon antenna. Some of the three-dimensional radiation patterns are shown in Figure 4g,h. Both the rhombus–trapezoid- and hexagon–trapezoid-shaped antennas maintain high-gain radiations over the examined bandwidth, from 19.0 GHz to 20 GHz and from 17.0 GHz to 18.8 GHz, respectively. Although there are some fluctuations in the maximum gains, this is reasonable, as the electromagnetic responses from the meta-surfaces vary with frequency.

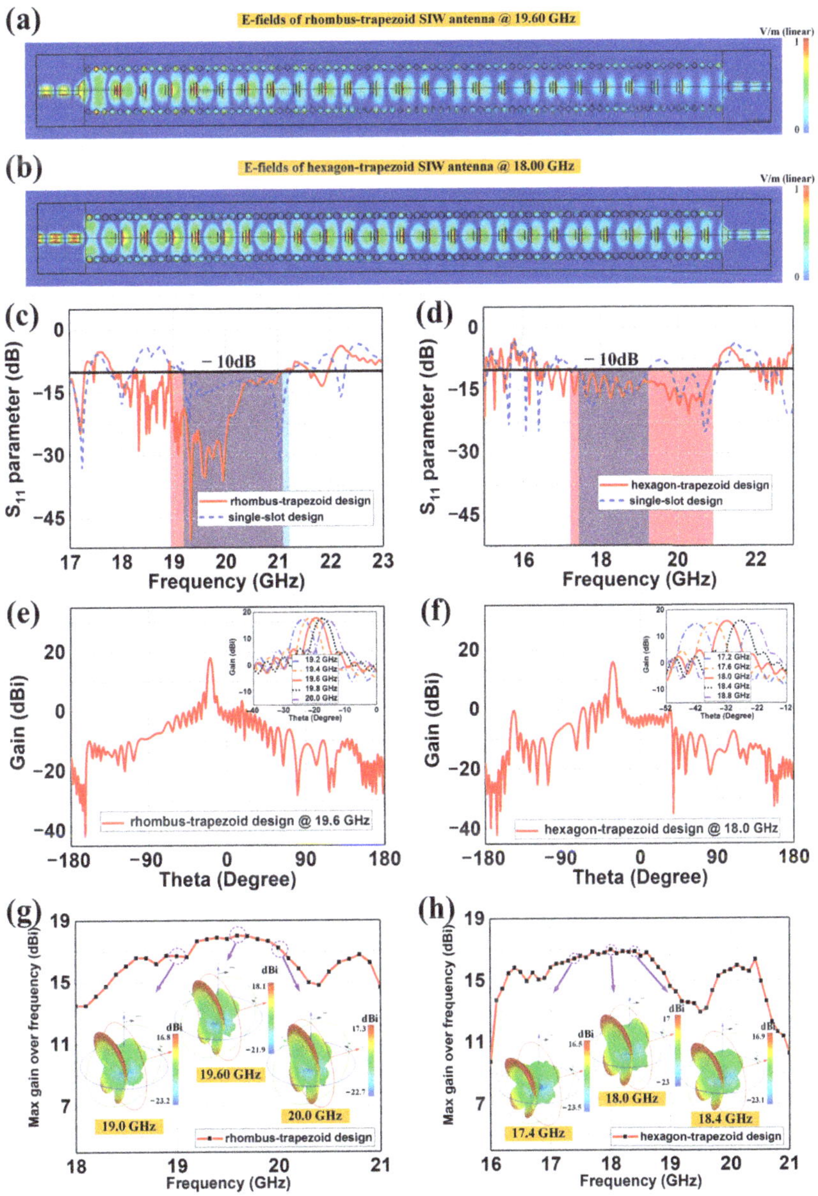

Figure 4. The E-field of the (**a**) rhombus and (**b**) hexagon meta-surface-based SIW leaky-wave antennas, and the corresponding reflection coefficients of the (**c**) rhombus- and (**d**) hexagon-shaped SIW antenna designs with comparison groups. The 2D radiation patterns of the (**e**) rhombus and (**f**) hexagon meta-surface-based SIW leaky-wave antennas, and the corresponding maximum gains and 3D radiations of the (**g**) rhombus- and (**h**) hexagon-shaped designs.

2.4. Extended SIW Design and Simulations

We extend our design to coplanar two-row based meta-surfaces by combining rhombus and hexagon structures in different orders, based on the original transmission lines and antennas in Figure 2. Figure 5a demonstrates the rhombus–hexagon meta-surface-based SIW transmission lines, where the rows of hexagon unit cells are etched at one cell spacing outside the rhombus cell rows of the original rhombus meta-surface-based SIW line. Similarly, Figure 5b demonstrates the opposite structure to Figure 5a, with hexagon rows inside and rhombus rows outside. Moreover, the same trapezoid-shaped meta-surfaces are also introduced to the two mixed SIW transmission lines, thus constructing the complex coplanar meta-surface combined SIW antennas shown in Figure 5c,d. Based on the optimization process of the original meta-surface-based SIWs, the four proposed designs in Figure 5 could use the optimized results from Figure 2 as initial values and further simulate and adjust specific geometric sizes such as transition sizes to obtain low reflections. The corresponding E-field distributions and S_{11} parameters of the four proposed designs are illustrated in Figure 5e–h. We can observe, from the E-fields, that good TE_{10} mode transmission is achieved in each of the transmission lines and antennas, with less edge leakage compared to that shown in Figures 3 and 4. The -10 dB bandwidths of the meta-surface-based antennas are marked as red areas, while those of transmission lines are marked as blue areas, and the intersections are marked as purple areas. The rhombus–hexagon transmission line in Figure 5a has S_{11} values less than -10 dB ranging from 17.62 GHz to 20.50 GHz, with a minimum value of -46.22 dB at 17.71 GHz and a relative bandwidth of 15.11%, while the hexagon–rhombus transmission line in Figure 5b has S_{11} values less than -10 dB ranging from 16.63 GHz to 20.21 GHz, with a minimum value of -49.39 dB at 16.95 GHz and a relative bandwidth of 19.44%. For the complex SIW antennas in Figure 5c,d, the corresponding ranges are from 17.89 GHz to 20.61 GHz with a minimum of -35.61 dB at 20.05 GHz and relative bandwidth of 14.13%, and from 17.24 GHz to 20.73 GHz with a minimum of -46.99 dB at 20.02 GHz and relative bandwidth of 18.38%. The original rhombus structures in Figure 2a,c concentrate on obtaining lower reflections, while the new rhombus–hexagon designs in Figure 5a,c increase the bandwidth on this basis. Similarly, the original hexagon structures in Figure 2b,d have significantly wider bands with acceptable reflections, while the hexagon–rhombus designs in Figure 5b,d retain these broad bands and achieve greatly reduced reflections at specific frequencies. These different results are mainly attributed to the primary shielding effect of the inside meta-surfaces and the secondary shielding effect of the outside meta-surfaces, where the primary meta-unit rows determine the overall transmission performance and the secondary cells fine-tune the reflections on this basis. As a result, all of the proposed devices obtain good reflection and bandwidth, but with different priority.

For the complex coplanar meta-surface combined SIW antennas, the far-field radiation patterns, including the beam-scanning performance, are demonstrated in Figure 5i–l. We can observe that both the proposed antennas in Figure 5 obtained similar high-gain fan beams and satisfactory beam scanning, compared to the originals in Figure 4. For example, at 19.6 GHz and 18.2 GHz, the maximum gains of the two coplanar SIW antennas were 18.8 dBi and 17.1 dBi, respectively. Based on the same calculation method for the rhombus and hexagon antennas from Figure 3, the radiation efficiencies of the rhombus–hexagon and hexagon–rhombus meta-surface-based SIW antennas are 91.20% and 83.18%, respectively. Moreover, the corresponding aperture efficiencies are 61.12% and 47.92%, respectively. The other radiation results of the SIW antennas from Figure 5c,d, including the 3-dB beamwidth on the E- and H-plane, beam scanning angles at specific frequencies and beam scanning rate, are all summarized and listed in Table 2. We can conclude that the extended SIW transmission lines and antennas in Figure 5 are still capable of propagating the electromagnetic fields well and generating high-gain beam-scanning fan beams. Moreover, every 0.4 GHz frequency offset resulted in approximately 4° and 5° beam steering for the proposed meta-surface-based SIW antennas. Moreover, the S-parameter results appear more flexible and tunable, compared to those of the original devices. The complex coplanar meta-surface-based designs provide promising candidates for building novel electromagnetic devices on PCB-based systems. The four SIWs can flexibly handle different bandwidth requirements, where the length value L of the two antennas could be freely adjusted with a minimum step value of p, instead of a larger length, for impedance matching with traditional single-slot designs.

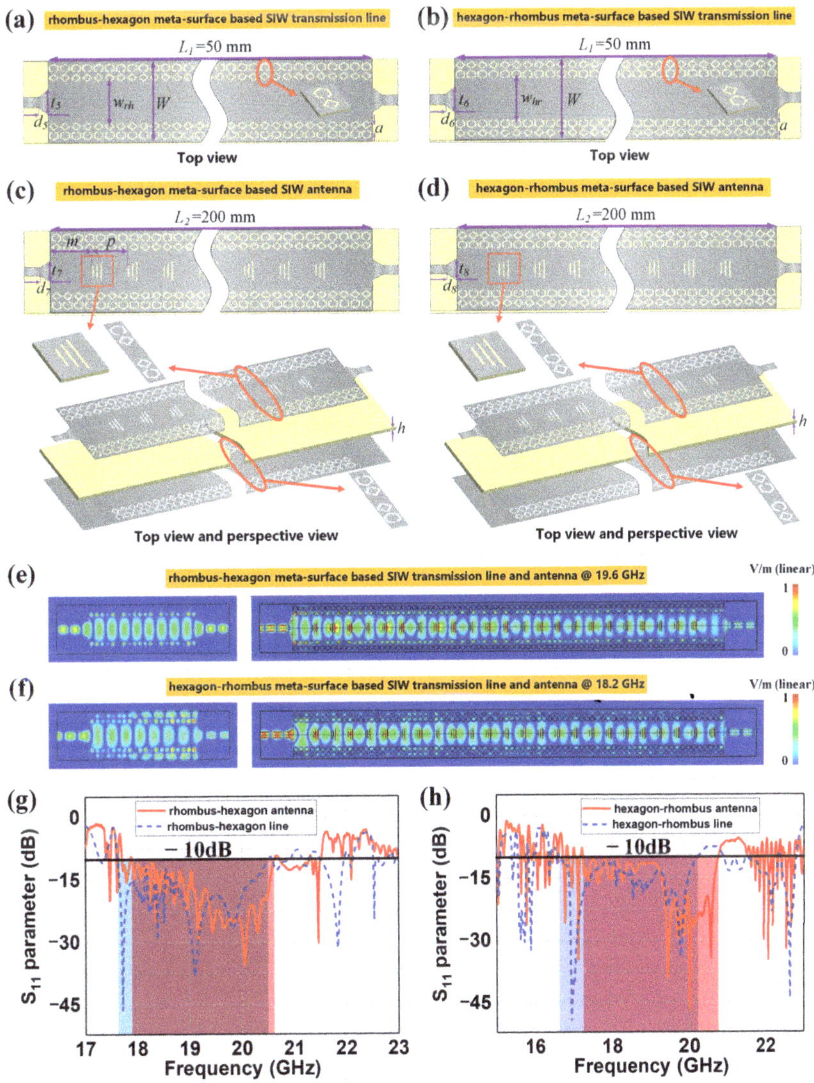

Figure 5. *Cont.*

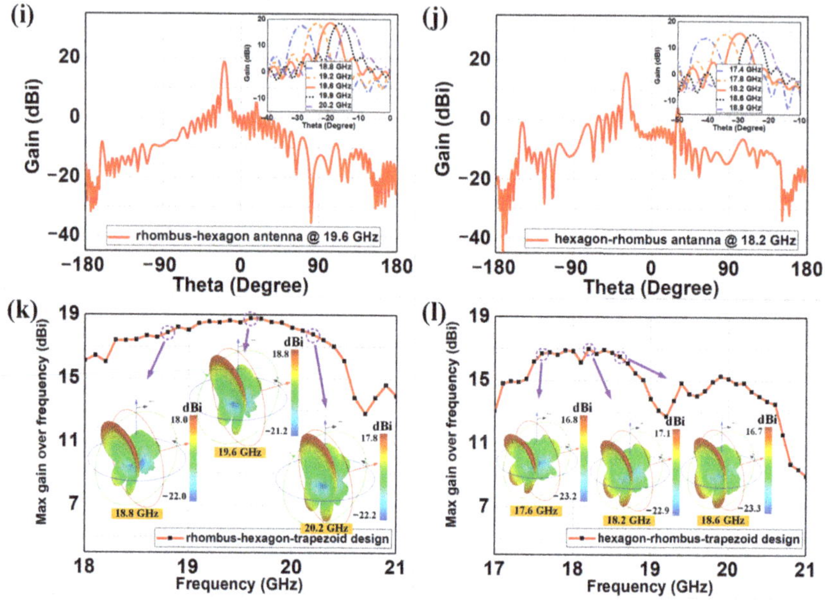

Figure 5. The coplanar (**a**) rhombus–hexagon and (**b**) hexagon–rhombus meta-surface-based SIW transmission lines, and the corresponding trapezoid-enabled meta-surface-based SIW leaky-wave antennas, as shown in (**c**,**d**). The geometric sizes are L_1 = 50 mm, W = 22 mm, t_5 = 6 mm, d_5 = 4 mm, w_{rh} = 13 mm, t_6 = 6 mm, d_6 = 2.5 mm, w_{hr} = 13.1 mm, L_2 = 200 mm, m = 9.4 mm, p = 8 mm, t_7 = 7 mm, d_7 = 1.8 mm, t_8 = 5 mm and d_8 = 2 mm. E-field distributions of the meta-surface-based SIW designs: (**e**) rhombus–hexagon and (**f**) hexagon–rhombus transmission lines and antennas. The corresponding reflection coefficients of the four SIW designs: (**g**) rhombus–hexagon and (**h**) hexagon–rhombus transmission lines and antennas. The far-field radiations of the meta-surface-based SIW leaky-wave antennas are demonstrated as follows: 2D patterns of (**i**) rhombus–hexagon and (**j**) hexagon–rhombus, and maximum gains of (**k**) rhombus–hexagon and (**l**) hexagon–rhombus with 3D patterns.

Table 2. Comparison of radiation performance between rhombus–hexagon and hexagon–rhombus meta-surface-based SIW antennas.

Radiation Performance	Rhombus–Hexagon Meta-Surface-Based SIW Antenna	Hexagon–Rhombus Meta-Surface-Based SIW Antenna
3-dB beamwidth on E-plane	4.7°	5.2°
3-dB beamwidth on H-plane	60.1°	49.3°
Beam-scanning angles @ frequency	−29° @ 18.8 GHz −24° @ 19.2 GHz −20° @ 19.6 GHz −16° @ 19.9 GHz −13° @ 20.2 GHz	−41° @ 17.4 GHz −35° @ 17.8 GHz −30° @ 18.2 GHz −25° @ 18.6 GHz −23° @ 18.9 GHz
Beam-scanning rate	4°/0.4 GHz	5°/0.4 GHz

3. Discussion of Experimental Results

Finally, we fabricated the coplanar rhombus, hexagon, rhombus–hexagon and hexagon–rhombus meta-surface-based SIW leaky-wave antennas and carried out corresponding experiments in a microwave chamber to verify the proposed designs, as shown in Figure 6a,b. Through simple copper cladding and etching technology, the meta-surfaces and microstrip structures were printed on both sides of the substrate slabs. In addition, standard SMA connectors were soldered on both ends of the substrates for feeding excitation. The four meta-surface-based SIW antennas under test with equal height to the receiving antenna were set as transmit antennas and properly assembled on the center position of the turntable. A vector network analyzer was applied for measuring the reflection coefficients of the proposed antennas, and a standard diagonal horn antenna with the distance of 3.72 m to the center of the turntable was utilized to receive the electromagnetic fields and measure the radiation performance. The experimental far-field radiation results of four proposed SIW antennas are demonstrated in Figure 6c–f, with a comparison to the corresponding simulations. We can observe that all proposed antennas demonstrated highly directive radiation and beam-scanning performance, as designed. The measured gain values of the center frequencies were 17.06 dBi, 15.79 dBi, 17.9 dBi and 15.97 dBi at 19.6 GHz, 18 GHz, 19.6 GHz and 18.2 GHz, respectively, as shown in Figure 6c–f. Based on these gain results, we could calculate the measured radiation efficiencies of the rhombus, hexagon, rhombus–hexagon and hexagon–rhombus meta-surface-based SIW antennas as 68.55%, 61.52%, 74.13% and 64.12%, and the measured aperture efficiencies were 40.94%, 36.23%, 49.68% and 36.94%, using the same methods as those for calculating the simulated radiation and aperture efficiencies. The measured efficiencies are approximately 10–18% lower than the simulated efficiencies. The corresponding beam-scanning measurements are also illustrated, with nearly the same frequency ranges as in Figures 4 and 5. The corresponding S_{11} results are shown in Figure 6g,h with a comparison to the corresponding simulations, where the trends of the S-parameter curves are similar to those in the original simulations, achieving their different bandwidth design aims. For Figure 6g, the rhombus SIW antenna has S_{11} values less than -10 dB ranging from 18.05 GHz to 21.1 GHz, with a minimum value of -49.35 dB at 18.55 GHz and relative bandwidth of 15.58%. Meanwhile, the hexagon SIW antenna has S_{11} values less than -10 dB ranging from 16.20 GHz to 21.40 GHz, with a minimum value of -30.33 dB at 17.45 GHz and relative bandwidth of 27.66%. For Figure 6h, the rhombus–hexagon SIW antenna has S_{11} values less than -10 dB ranging from 17.35 GHz to 21.00 GHz, with a minimum value of -34.57 dB at 17.85 GHz and relative bandwidth of 19.04%. Meanwhile, the hexagon–rhombus SIW antenna has S_{11} values less than -10 dB ranging from 16.80 GHz to 20.95 GHz, with a minimum value of -23.04 dB at 20.65 GHz and relative bandwidth of 21.99%. We can observe that the measured gains and S-parameters were mostly in agreement with the simulations, with only a few degradations or frequency shifts. As a result, there are some discrepancies in the simulated and measured results from the proposed antennas. These degradations were mainly attributed to the fabrication tolerance of the meta-surfaces, the welding quality of the SMA connectors and the measurement deviations. The fabrication tolerance could affect the geometric sizes and substrate electromagnetic properties of the unit cells, which would result in discrepancies in the electromagnetic response from the meta-surfaces. The welding quality would affect the transmission characteristics of the SMAs, which in turn affects the radiation performance of the antennas. Moreover, the potential antenna deformation during the measurement and the effects of fixed brackets on electromagnetic waves could also lead to measurement deviations. However, all four of the proposed meta-surface-based SIW antennas demonstrated their beam-scanning ability, with flexible reflection coefficients and bandwidths.

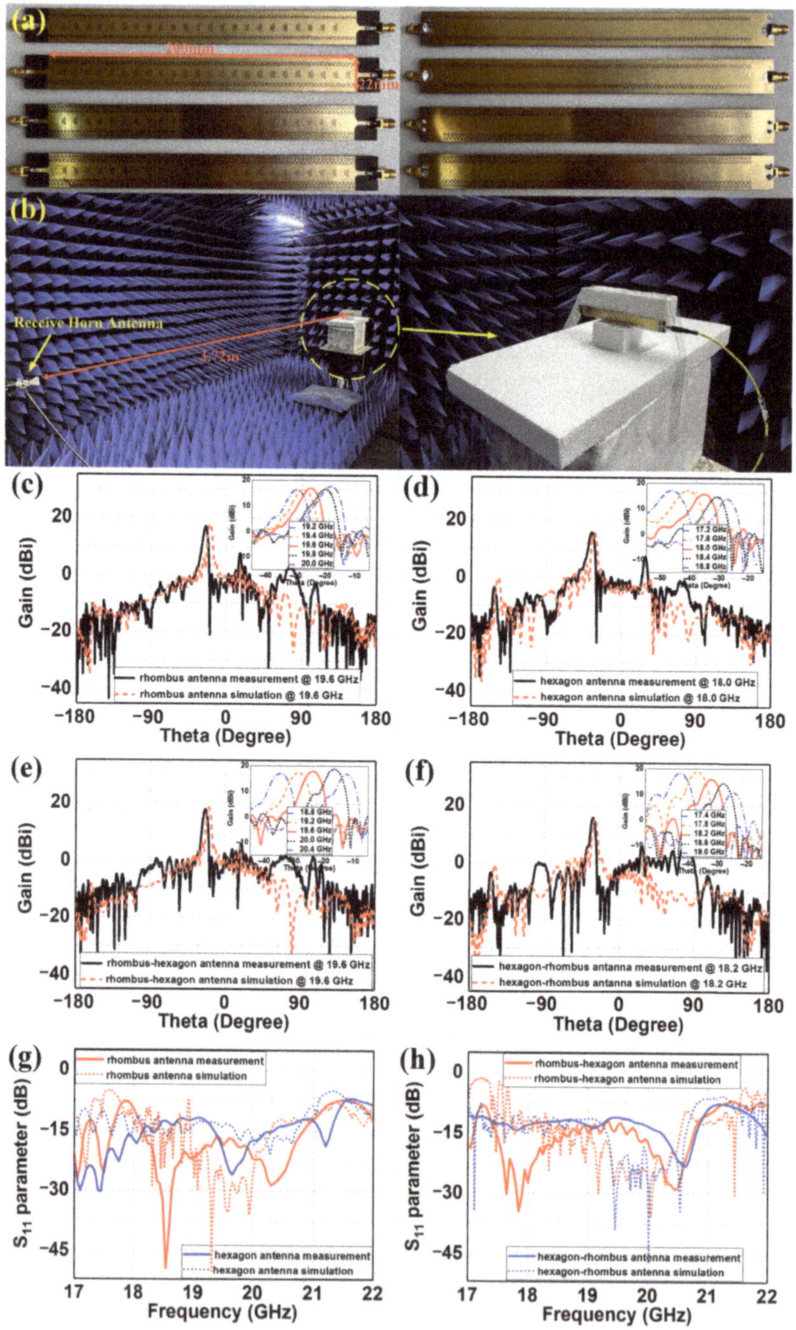

Figure 6. Photos and experimental results of the manufactured meta-surface-based SIW leaky-wave antennas with comparison to the corresponding simulations. (**a**) Photos of the four SIW antennas. (**b**) The corresponding experiments in a microwave chamber. The measured radiation patterns of the (**c**) rhombus, (**d**) hexagon, (**e**) rhombus–hexagon and (**f**) hexagon–rhombus meta-surface-based SIW leaky-wave antennas, The measured reflection coefficients of the (**g**) rhombus and hexagon designs, and (**h**) rhombus–hexagon and hexagon–rhombus designs.

4. Conclusions

In conclusion, we introduced coplanar rhombus- and hexagon-shaped meta-surfaces into the design of SIW transmission lines, and constructed four associated coplanar meta-surface-based SIW antennas through the application of trapezoid-shaped meta-surfaces. The four antennas obtain maximum gains of 18.1 dBi, 17.0 dBi, 18.8 dBi and 17.1 dBi, where the simulated radiation efficiencies are 87.10%, 81.28%, 91.20% and 83.18%, and the simulated aperture efficiencies are 52.02%, 47.88%, 61.12% and 47.92%. By replacing metallized via holes with specifically modified meta-surfaces, all of the proposed designs were shown to possess more flexible and tunable reflection coefficients and bandwidth performances, where the relative bandwidths thus are 10.74%, 19.42%, 14.13% and 18.38%. Further, the four coplanar meta-surface-based SIW antennas presented highly directive beam-scanning fan beams at K-bands with beam-scanning rate 4° (rhombus and rhombus–hexagon designs) and 5° (hexagon and hexagon–rhombus designs) per 0.4 GHz, paving the way for the development of mm-Wave PCB-integrated antennas or sensors with freely customizable electrical performance.

Author Contributions: Conceptualization, C.W. and D.G.; methodology, C.W.; software, C.W.; validation, L.L. and Y.W.; formal analysis, D.G.; investigation, L.L. and Y.W.; data curation, D.G., L.L. and Y.W.; writing—original draft preparation, C.W.; writing—review and editing, D.G.; visualization, C.W. and D.G.; supervision, D.G.; project administration, D.G.; and funding acquisition, D.G. All authors have read and agreed to the published version of the manuscript.

Funding: This research was funded by the National Natural Science Foundation of China (grant number 42104167) and Shandong Provincial Natural Science Foundation (grant number ZR2021QD045).

Institutional Review Board Statement: Not applicable.

Informed Consent Statement: Not applicable.

Data Availability Statement: Not applicable.

Conflicts of Interest: The authors declare no conflict of interest.

References

1. Wu, K. Integration and Interconnect Techniques of Planar and Non-Planar Structures for Microwave and Millimeter-Wave Circuits-Current Status and Future Trend. In Proceedings of the 2001 Asia-Pacific Microwave Conference, Taipei, Taiwan, 3–6 December 2001; pp. 411–416.
2. Kuznetcov, M.V.; Buendía, V.G.-G.; Shafiq, Z.; Matekovits, L.; Anagnostou, D.E.; Podilchak, S.K. Printed Leaky-Wave Antenna with Aperture Control Using Width-Modulated Microstrip Lines and Tm Surface-Wave Feeding by Siw Technology. *IEEE Antennas Wirel. Propag. Lett.* **2019**, *18*, 1809–1813. [CrossRef]
3. Zerfaine, A.; Djerafi, T. Ultrabroadband Circularly Polarized Antenna Array Based on Microstrip to Siw Junction. *IEEE Trans. Antennas Propag.* **2021**, *70*, 2346–2351. [CrossRef]
4. Ballesteros, J.A.; Belenguer, A.; Fernandez, M.D.; González, H.E.; Boria, V.E. Improved Microstrip-to-Esiw Transition with Elliptical Dielectric Taper in Ku-and Ka-Bands. *IEEE Access* **2022**, *10*, 51412–51418. [CrossRef]
5. Gheisanab, H.N.; Pourmohammadi, P.; Iqbal, A.; Kishk, A.; Denidni, T.A. Siw-Based Self-Quadruplexing Antenna for Microwave and Mm-Wave Frequencies. *IEEE Antennas Wirel. Propag. Lett.* **2022**, *21*, 1482–1486.
6. Guo, Y.J.; Ziolkowski, R.W. *Advanced Antenna Array Engineering for 6G and Beyond Wireless Communications*; John Wiley & Sons: Hoboken, NJ, USA, 2021.
7. Spurek, J.; Raida, Z. Siw-Based Circularly Polarized Antenna Array for 60 GHz 5g Band: Feasibility Study. *Sensors* **2022**, *22*, 2945. [CrossRef]
8. Dam, T.H.L.; Houzet, G.; Lacrevaz, T.; Vuong, T.P. High Gain Vivaldi Antenna from 26 up to 40 GHz for 5g Applications. *Microw. Opt. Technol. Lett.* **2022**, *64*, 1267–1271. [CrossRef]
9. Yu, N.; Genevet, P.; Kats, M.A.; Aieta, F.; Tetienne, J.-P.; Capasso, F.; Gaburro, Z. Light Propagation with Phase Discontinuities: Generalized Laws of Reflection and Refraction. *Science* **2011**, *334*, 333–337. [CrossRef]
10. Imani, M.F.; Gollub, J.N.; Yurduseven, O.; Diebold, A.V.; Boyarsky, M.; Fromenteze, T.; Pulido-Mancera, L.; Sleasman, T.; Smith, D.R. Review of Metasurface Antennas for Computational Microwave Imaging. *IEEE Trans. Antennas Propag.* **2020**, *68*, 1860–1875. [CrossRef]
11. Iyer, A.K.; Alu, A.; Epstein, A. Metamaterials and Metasurfaces—Historical Context, Recent Advances, and Future Directions. *IEEE Trans. Antennas Propag.* **2020**, *68*, 1223–1231. [CrossRef]

12. Abadal, S.; Cui, T.-J.; Low, T.; Georgiou, J. Programmable Metamaterials for Software-Defined Electromagnetic Control: Circuits, Systems, and Architectures. *IEEE J. Emerg. Sel. Top. Circuits Syst.* **2020**, *10*, 6–19. [CrossRef]
13. Asadchy, V.S.; Albooyeh, M.; Tcvetkova, S.N.; Díaz-Rubio, A.; Ra'di, Y.; Tretyakov, S. Perfect Control of Reflection and Refraction Using Spatially Dispersive Metasurfaces. *Phys. Rev. B* **2016**, *94*, 075142. [CrossRef]
14. Díaz-Rubio, A.; Asadchy, V.S.; Elsakka, A.; Tretyakov, S.A. From the Generalized Reflection Law to the Realization of Perfect Anomalous Reflectors. *Sci. Adv.* **2017**, *3*, e1602714. [CrossRef] [PubMed]
15. Arbabi, A.; Arbabi, E.; Horie, Y.; Kamali, S.M.; Faraon, A. Planar Metasurface Retroreflector. *Nat. Photonics* **2017**, *11*, 415–420. [CrossRef]
16. Kan, Y.; Yang, R.; Zhang, A.; Lei, Z.; Jiao, Y.; Li, J. Meta-Surface Cavity-Based Waveguide Slot Array for Dual-Circularly Polarized Dual Beam. *IEEE Trans. Antennas Propag.* **2021**, *70*, 3894–3898. [CrossRef]
17. Williams, R.J.; Gjonbalaj, A.M.; Green, K.D.; Wells, B.M. Generation of Arbitrarily Directed Split Beams with a Reflective Metasurface. *Opt. Express* **2022**, *30*, 25318–25325. [CrossRef]
18. Oh, T.; Cho, C.; Ahn, W.; Yook, J.-G.; Lee, J.; You, S.; Yim, J.; Ha, J.; Bae, G.; You, H.-C. Plasma Generator with Dielectric Rim and Fss Electrode for Enhanced Rcs Reduction Effect. *Sensors* **2021**, *21*, 8486. [CrossRef]
19. Alwareth, H.; Ibrahim, I.M.; Zakaria, Z.; Al-Gburi, A.J.A.; Ahmed, S.; Nasser, Z.A. A Wideband High-Gain Microstrip Array Antenna Integrated with Frequency-Selective Surface for Sub-6 GHz 5g Applications. *Micromachines* **2022**, *13*, 1215. [CrossRef]
20. Chen, M.; Abdo-Sánchez, E.; Epstein, A.; Eleftheriades, G.V. Theory, Design, and Experimental Verification of a Reflectionless Bianisotropic Huygens' Metasurface for Wide-Angle Refraction. *Phys. Rev. B* **2018**, *97*, 125433. [CrossRef]
21. Cerveny, M.; Ford, K.L.; Tennant, A. Reflective Switchable Polarization Rotator Based on Metasurface with Pin Diodes. *IEEE Trans. Antennas Propag.* **2020**, *69*, 1483–1492. [CrossRef]
22. Pitilakis, A.; Tsilipakos, O.; Liu, F.; Kossifos, K.M.; Tasolamprou, A.C.; Kwon, D.-H.; Mirmoosa, M.S.; Manessis, D.; Kantartzis, N.V.; Liaskos, C. A Multi-Functional Reconfigurable Metasurface: Electromagnetic Design Accounting for Fabrication Aspects. *IEEE Trans. Antennas Propag.* **2020**, *69*, 1440–1454. [CrossRef]
23. Zetterstrom, O.; Hamarneh, R.; Quevedo-Teruel, O. Experimental Validation of a Metasurface Luneburg Lens Antenna Implemented with Glide-Symmetric Substrate-Integrated Holes. *IEEE Antennas Wirel. Propag. Lett.* **2021**, *20*, 698–702. [CrossRef]
24. Nitas, M.; Passia, M.T.; Yioultsis, T.V. Fully Planar Slow-Wave Substrate Integrated Waveguide Based on Broadside-Coupled Complementary Split Ring Resonators for Mmwave and 5g Components. *IET Microwaves Antennas Propag.* **2020**, *14*, 1096–1107. [CrossRef]
25. Martinez-Ros, A.J.; Gómez-Tornero, J.L.; Goussetis, G. Holographic Pattern Synthesis with Modulated Substrate Integrated Waveguide Line-Source Leaky-Wave Antennas. *IEEE Trans. Antennas Propag.* **2013**, *61*, 3466–3474. [CrossRef]
26. Cameron, T.R.; Eleftheriades, G.V. Experimental Validation of a Wideband Metasurface for Wide-Angle Scanning Leaky-Wave Antennas. *IEEE Trans. Antennas Propag.* **2017**, *65*, 5245–5256. [CrossRef]
27. Li, T.; Chen, Z.N. Wideband Substrate-Integrated Waveguide-Fed Endfire Metasurface Antenna Array. *IEEE Trans. Antennas Propag.* **2018**, *66*, 7032–7040. [CrossRef]
28. Alibakhshikenari, M.; Virdee, B.S.; Althuwayb, A.A.; Aïssa, S.; See, C.H.; Abd-Alhameed, R.A.; Falcone, F.; Limiti, E. Study on on-Chip Antenna Design Based on Metamaterial-Inspired and Substrate-Integrated Waveguide Properties for Millimetre-Wave and Thz Integrated-Circuit Applications. *J. Infrared Millim. Terahertz Waves* **2021**, *42*, 17–28. [CrossRef]
29. Boyarsky, M.; Sleasman, T.; Imani, M.F.; Gollub, J.N.; Smith, D.R. Electronically Steered Metasurface Antenna. *Sci. Rep.* **2021**, *11*, 4693. [CrossRef]
30. Marques, R.; Martel, J.; Mesa, F.; Medina, F. A New 2d Isotropic Left-Handed Metamaterial Design: Theory and Experiment. *Microw. Opt. Technol. Lett.* **2002**, *35*, 405–408. [CrossRef]
31. Solymar, L.; Shamonina, E. *Waves in Metamaterials;* Oxford University Press: New York, NY, USA, 2009.
32. Brillouin, L. *Wave Propagation in Periodic Structures;* Dover: New York, NY, USA, 1953.
33. Chan, C. *Frequency Selective Surface and Grid Array;* John Wiley & Sons: New York, NY, USA, 1995.
34. Tretyakov, S. *Analytical Modeling in Applied Electromagnetics;* Artech House: Norwood, MA, USA, 2003.
35. Holloway, C.L.; Kuester, E.F.; Gordon, J.A.; O'Hara, J.; Booth, J.; Smith, D.R. An Overview of the Theory and Applications of Metasurfaces: The Two-Dimensional Equivalents of Metamaterials. *IEEE Antennas Propag. Mag.* **2012**, *54*, 10–35. [CrossRef]
36. Pendry, J.B.; Holden, A.J.; Robbins, D.J.; Stewart, W. Magnetism from Conductors and Enhanced Nonlinear Phenomena. *IEEE Trans. Microw. Theory Technol.* **1999**, *47*, 2075–2084. [CrossRef]
37. Marqués, R.; Medina, F.; Rafii-El-Idrissi, R. Role of Bianisotropy in Negative Permeability and Left-Handed Metamaterials. *Phys. Rev. B* **2002**, *65*, 144440. [CrossRef]
38. Baena, J.D.; Bonache, J.; Martín, F.; Sillero, R.M.; Falcone, F.; Lopetegi, T.; Laso, M.A.; Garcia-Garcia, J.; Gil, I.; Portillo, M.F. Equivalent-Circuit Models for Split-Ring Resonators and Complementary Split-Ring Resonators Coupled to Planar Transmission Lines. *IEEE Trans. Microw. Theory Technol.* **2005**, *53*, 1451–1461. [CrossRef]
39. Falcone, F.; Lopetegi, T.; Baena, J.D.; Marqués, R.; Martín, F.; Sorolla, M. Effective Negative-ε Stopband Microstrip Lines Based on Complementary Split Ring Resonators. *IEEE Microw. Wirel. Compon. Lett.* **2004**, *14*, 280–282. [CrossRef]
40. Bonache, J.; Gil, M.; Gil, I.; García-García, J.; Martín, F. On the Electrical Characteristics of Complementary Metamaterial Resonators. *IEEE Microw. Wirel. Compon. Lett.* **2006**, *16*, 543–545. [CrossRef]

41. Bonache, J.; Gil, I.; Garcia-Garcia, J.; Martin, F. Complementary Split Rings Resonators (Csrrs): Towards the Miniaturization of Microwave Device Design. *J. Comput. Electron.* **2006**, *5*, 193–197. [CrossRef]
42. Aznar, F.; Vélez, A.; Durán-Sindreu, M.; Bonache, J.; Martín, F. Open Complementary Split Ring Resonators: Physics, Modelling, and Analysis. *Microw. Opt. Technol. Lett* **2010**, *52*, 1520–1526. [CrossRef]
43. Marqués, R.; Mesa, F.; Martel, J.; Medina, F. Comparative Analysis of Edge-and Broadside-Coupled Split Ring Resonators for Metamaterial Design-Theory and Experiments. *IEEE Trans. Antennas Propag.* **2003**, *51*, 2572–2581. [CrossRef]
44. Wang, J.; Qu, S.; Zhang, J.; Ma, H.; Yang, Y.; Gu, C.; Wu, X.; Xu, Z. A Tunable Left-Handed Metamaterial Based on Modified Broadside-Coupled Split-Ring Resonators. *Prog. Electromagn. Res. Lett.* **2009**, *6*, 35–45. [CrossRef]
45. Bahl, I.; Bhartia, P. *Microwave Solid State Circuit Design*; John Wiley & Sons: New York, NY, USA, 2003.
46. Wheeler, H.A. Formulas for the Skin Effect. *Proc. IRE* **1942**, *30*, 412–424. [CrossRef]
47. Nitas, M.; Yioultsis, T.V. Characterization of Edge-Coupled Broadside-Coupled and Complementary Split-Ring Resonator Periodic Media Based on Numerical Solutions of Eigenvalue Problems. *IEEE Trans. Microw. Theory Technol.* **2021**, *69*, 5259–5269. [CrossRef]
48. Smith, D.; Schultz, S.; Markoš, P.; Soukoulis, C. Determination of Effective Permittivity and Permeability of Metamaterials from Reflection and Transmission Coefficients. *Phys. Rev. B* **2002**, *65*, 195104. [CrossRef]
49. Smith, D.; Vier, D.; Koschny, T.; Soukoulis, C. Electromagnetic Parameter Retrieval from Inhomogeneous Metamaterials. *Phys. Rev. E* **2005**, *71*, 036617. [CrossRef]
50. Xu, F.; Wu, K. Guided-Wave and Leakage Characteristics of Substrate Integrated Waveguide. *IEEE Trans. Microw. Theory Technol.* **2005**, *53*, 66–73.
51. Hao, Z.-C.; Hong, W.; Chen, J.X.; Chen, X.P.; Wu, K. Planar Diplexer for Microwave Integrated Circuits. *IEE Proc.-Microw. Antennas Propag.* **2005**, *152*, 455–459. [CrossRef]
52. Deslandes, D. Design Equations for Tapered Microstrip-to-Substrate Integrated Waveguide Transitions. In Proceedings of the 2010 IEEE MTT-S International Microwave Symposium, Anaheim, CA, USA, 23–28 May 2010; pp. 704–707.
53. Pozar, D.M. *Microwave Engineering*; John Wiley & Sons: New York, NY, USA, 2011.
54. Hessel, A. *General Characteristics of Traveling-Wave Antennas, in Antenna Theory, Part 2*; McGraw-Hill: New York, NY, USA, 1969.
55. Podilchak, S.K.; Matekovits, L.; Freundorfer, A.P.; Antar, Y.M.; Orefice, M. Controlled Leaky-Wave Radiation from a Planar Configuration of Width-Modulated Microstrip Lines. *IEEE Trans. Antennas Propag.* **2013**, *61*, 4957–4972. [CrossRef]
56. Martinez-Ros, A.J.; Gómez-Tornero, J.L.; Losada, V.; Mesa, F.; Medina, F. Non-Uniform Sinusoidally Modulated Half-Mode Leaky-Wave Lines for near-Field Focusing Pattern Synthesis. *IEEE Trans. Antennas Propag.* **2014**, *63*, 1022–1031. [CrossRef]
57. Guan, D.-F.; Zhang, Q.; You, P.; Yang, Z.-B.; Zhou, Y.; Yong, S.-W. Scanning Rate Enhancement of Leaky-Wave Antennas Using Slow-Wave Substrate Integrated Waveguide Structure. *IEEE Trans. Antennas Propag.* **2018**, *66*, 3747–3751. [CrossRef]
58. Mohtashami, Y.; Rashed-Mohassel, J. A Butterfly Substrate Integrated Waveguide Leaky-Wave Antenna. *IEEE Trans. Antennas Propag.* **2014**, *62*, 3384–3388. [CrossRef]
59. Jackson, D.R.; Caloz, C.; Itoh, T. Leaky-Wave Antennas. *Proc. IEEE* **2012**, *100*, 2194–2206. [CrossRef]
60. Ghiotto, A.; Parment, F.; Vuong, T.-P.; Wu, K. Millimeter-Wave Air-Filled Siw Antipodal Linearly Tapered Slot Antenna. *IEEE Antennas Wirel. Propag. Lett.* **2016**, *16*, 768–771. [CrossRef]
61. Liu, J.; Jackson, D.R.; Long, Y. Substrate Integrated Waveguide (Siw) Leaky-Wave Antenna with Transverse Slots. *IEEE Trans. Antennas Propag.* **2011**, *60*, 20–29. [CrossRef]
62. Araghi, A.; Khalily, M.; Xiao, P.; Tafazolli, R.; Jackson, D.R. Long Slot Mmwave Low-Sll Periodic-Modulated Leaky-Wave Antenna Based on Empty Siw. *IEEE Trans. Antennas Propag.* **2021**, *70*, 1857–1868. [CrossRef]
63. Zarifi, D.; Farahbakhsh, A.; Zaman, A.U.; Kildal, P.-S. Design and Fabrication of a High-Gain 60-Ghz Corrugated Slot Antenna Array with Ridge Gap Waveguide Distribution Layer. *IEEE Trans. Antennas Propag.* **2016**, *64*, 2905–2913. [CrossRef]
64. Zhu, Q.; Ng, K.B.; Chan, C.H.; Luk, K.-M. Substrate-Integrated-Waveguide-Fed Array Antenna Covering 57–71 GHz Band for 5g Applications. *IEEE Trans. Antennas Propag.* **2017**, *65*, 6298–6306. [CrossRef]
65. Park, J.; Seong, H.; Whang, Y.N.; Hong, W. Energy-Efficient 5g Phased Arrays Incorporating Vertically Polarized Endfire Planar Folded Slot Antenna for Mmwave Mobile Terminals. *IEEE Trans. Antennas Propag.* **2019**, *68*, 230–241. [CrossRef]
66. Li, Y.; Luk, K.-M. Low-Cost High-Gain and Broadband Substrate-Integrated-Waveguide-Fed Patch Antenna Array for 60-Ghz Band. *IEEE Trans. Antennas Propag.* **2014**, *62*, 5531–5538. [CrossRef]
67. Jiang, M.; Hong, W.; Zhang, Y.; Yu, S.; Zhou, H. A Folded Reflectarray Antenna with a Planar Siw Slot Array Antenna as the Primary Source. *IEEE Trans. Antennas Propag.* **2014**, *62*, 3575–3583. [CrossRef]
68. Xu, H.-X.; Cai, T.; Zhuang, Y.-Q.; Peng, Q.; Wang, G.-M.; Liang, J.-G. Dual-Mode Transmissive Metasurface and Its Applications in Multibeam Transmitarray. *IEEE Trans. Antennas Propag.* **2017**, *65*, 1797–1806. [CrossRef]

sensors

MDPI

Article

Beamforming with 1 × *N* Conformal Arrays

Irfan Ullah [1], Benjamin D. Braaten [2], Adnan Iftikhar [3], Symeon Nikolaou [4,5] and Dimitris E. Anagnostou [6,*]

1 Department of Electrical and Computer Engineering, COMSATS University Islamabad, Abbottabad Campus, Abbottabad 22060, Pakistan
2 Electrical and Computer Engineering Department, North Dakota State University, Fargo, ND 58102, USA
3 Department of Electrical and Computer Engineering, COMSATS University Islamabad, Islamabad 45550, Pakistan
4 Frederick Research Center, Nicosia 1303, Cyprus
5 Department of Electrical Computer Engineering and Informatics, Frederick University, Nicosia 1036, Cyprus
6 Institute of Signals, Sensors and Systems, Heriot Watt University, Edinburgh EH14 4AS, UK
* Correspondence: d.anagnostou@hw.ac.uk

Abstract: The rapid growth of wireless spectrum access through cellular and IoT devices, for example, requires antennas with more capabilities such as being conformal and self-adapting beamforming. In this paper, the adaptive beamforming patterns of microstrip patch antenna arrays on changing flexible (or conformal) curved surfaces are developed by deriving array coefficients based on the projection method that includes the mutual coupling between elements. A linear four-element microstrip patch antenna array is then embedded on two deformed conformal surfaces to investigate the projection method for desired beamforming patterns. The generated beamforming radiation patterns using the computed weighting coefficients are validated with theoretical equations evaluated in MATLAB, full-wave simulations in HFSS and measurement results. The measured results of the fabricated system agree with the simulated results. Furthermore, new guidelines are provided on the effects of mutual coupling and changing conformal surfaces for various beam-forming patterns. Such demonstrations pave the way to an efficient and robust conformal phased-array antenna with multiple beam forming and adaptive nulling capabilities.

Keywords: beam-forming antenna array; conformal surface; self-adapting

Citation: Ullah, I.; Braaten, B.D.; Iftikhar, A.; Nikolaou, S.; Anagnostou, D.E. Beamforming with 1 × *N* Conformal Arrays. *Sensors* **2022**, *22*, 6616. https://doi.org/10.3390/s22176616

Academic Editors: Naser Ojaroudi Parchin, Chan Hwang See and Raed A. Abd-Alhameed

Received: 28 April 2022
Accepted: 25 August 2022
Published: 1 September 2022

1. Introduction

More sophisticated wireless technology will certainly be required to meet the demand of an increasing number of wireless users and IoT devices throughout the world accessing the spectrum. One major component of a wireless system that will fulfill this need of spectrum access is a smart antenna [1–10]. This is because a smart antenna can measure the surrounding electromagnetic environment and steer the main lobe of the pattern towards the desired users and move the nulls of the antenna toward unwanted noise sources; thus, making the antenna adaptive and increasing the capacity of the overall system in real time. Furthermore, to increase the coverage and capacity of multifunctional wireless platforms, conformal smart antennas are being placed in unique locations such as buildings, vehicles and aircraft [9,10]. In many of these locations, however, the conformal smart antennas are subjected to forces that may deform the topology of the antenna and thus degrade the performance.

In the area of conformal antennas, researchers have studied how the radiation pattern of a conformal antenna changes as it is deformed in various ways [11–16]. It has been shown that the overall gain of an antenna array can be reduced by as much as 25 dBi in a particular direction without appropriate phase and amplitude compensation [16]. To mitigate some of the unwanted effects of deformation on the overall gain of the array, several compensation techniques have been proposed [17–26]. The aforementioned literature study shows that the main focus is on the broadside radiation pattern correction of a conformal antenna array attached to a changing surface (i.e., vibrating surface) using composite amplitude and phase errors measurements, mechanical calibration methods, hardware sensor monitoring and the null points adjusting method. Then again, the previous work (with the exceptions of [27–30]) was developed under the assumption that the main lobe direction of the radiation pattern was fixed and the location of the nulls for beamforming was not defined.

Thus, the objective of this work is to study the theory of beamforming for conformal antenna arrays attached to the changing wedge- and cylindrical-shaped surfaces in Figure 1a,b for a specific main lobe direction and null locations for the first time. Moreover, the changing surface effects on the radiation properties of conformal beamforming antennas will be explored. In particular, this work will focus on the beamforming for the one dimensional $1 \times N$ arrays on the wedge- and cylindrical-shaped surfaces in Figure 1a,b, respectively. For both Figure 1a,b, the location of each antenna element on the conformal surface is shown as a black dot and denoted as $A_{\pm n}$ where N is assumed to be even and $n = -N/2, \ldots, -1, 1, \ldots, N/2$. It should be mentioned that similar work could be carried out for an odd value of N. Moreover, the wavefront is denoted as a grey dotted line and the direction of propagation is shown with a black arrow. The wavefront is shown away from the origin for illustration purposes only.

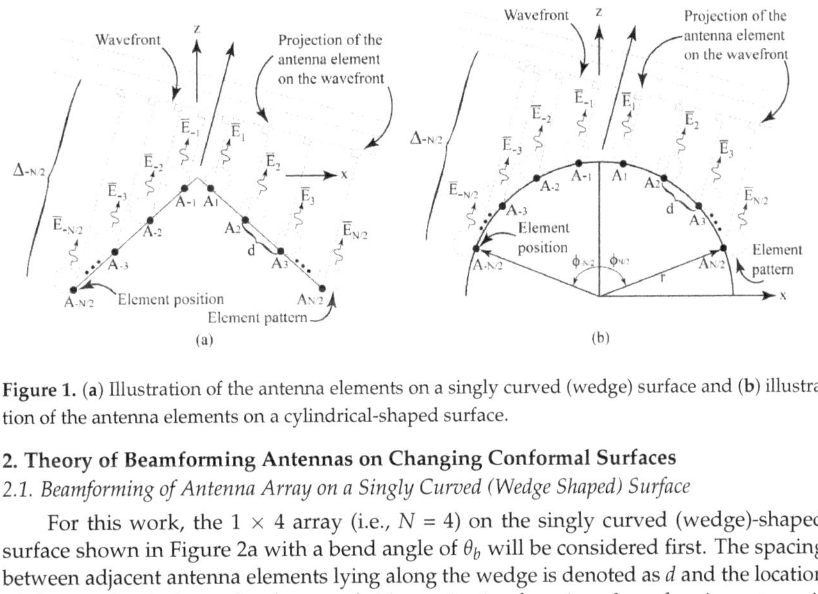

Figure 1. (**a**) Illustration of the antenna elements on a singly curved (wedge) surface and (**b**) illustration of the antenna elements on a cylindrical-shaped surface.

2. Theory of Beamforming Antennas on Changing Conformal Surfaces

2.1. Beamforming of Antenna Array on a Singly Curved (Wedge Shaped) Surface

For this work, the 1×4 array (i.e., $N = 4$) on the singly curved (wedge)-shaped surface shown in Figure 2a with a bend angle of θ_b will be considered first. The spacing between adjacent antenna elements lying along the wedge is denoted as d and the location of element $A_{\pm n}$ is denoted as $(x_{\pm n}, z_{\pm n})$. The motivation for using a beamforming antenna is the feature of being able to generate a main beam in a desired direction while simultaneously defining nulls in the radiation pattern in other directions. The main lobe can then be used to communicate with a desired user in a particular direction and the nulls can be used to reduce the incoming signal from users not of interest. For this work, the angle of the signal of interest will be denoted as θ_{SOI} and one user of interest will be assumed while the angle of the nth user not of interest will be denoted as θ_{SNOI_n}. Because a four-element array is being considered here, three nulls at θ_{SNOI_1}, θ_{SNOI_2} and θ_{SNOI_3} are defined. Now, the goal is to define appropriate array weights that can be used to drive the four-element array in Figure 1a,b to give a main lobe at θ_{SOI} and three nulls at θ_{SNOI_1}, θ_{SNOI_2} and θ_{SNOI_3}. Furthermore, to determine how the shape of the wedge changes the weights, they are to be determined for various bend angles θ_b of the wedge.

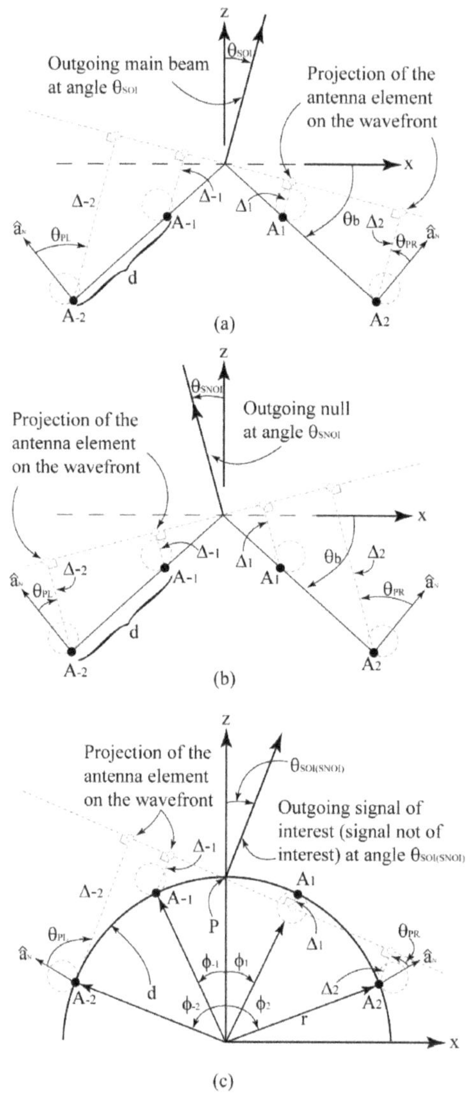

Figure 2. Geometrical illustration of the proposed projection method for (**a**) antenna array on a wedged-shaped surface with desired signal at angle θ_{SOI}; (**b**) antenna array on a wedged-shaped surface with undesired signal at angle θ_{SNOI} and (**c**) antenna array on a cylindrical curvature with desired (undesired) signal at angle $\theta_{SOI(SNOI)}$.

To compute the surface-dependent array weights, the projection method [31] will be used along with the matrix method for computing antenna weights defined in [1]. Since a maximum radiation in the direction of θ_{SOI} and nulls in the directions of θ_{SNOI_1}, θ_{SNOI_2} and θ_{SNOI_3} are desired, the array will be considered as a transmitter. First, suppose the antenna array is radiating in the direction of θ_{SOI}, as shown in Figure 2a. One method used to provide a field in the direction of θ_{SOI} is to ensure that the fields radiated from elements A_{-2}, A_{-1}, A_1 and A_2 reach at the reference wavefront with equal phases, thereby resulting in a broad-side pattern to the wavefront. To ensure that these fields arrive with the same phase, the voltage phase driving each individual element can be adjusted with a phase shifter. The amount of phase required is equal and opposite to the sign of the phase introduced by the free-space propagation from the antenna element to the wavefront. To compute this distance, the values for Δ_{-2}, Δ_{-1}, Δ_1 and Δ_2 in Figure 2a, which denote the distance from the elements A_{-2}, A_{-1}, A_1 and A_2, respectively, to the wavefront (or the projected elements on the wavefront), need to be computed. Furthermore, the expressions for A_{-2}, A_{-1}, A_1 and A_2 should be written in a general manner including surface dependent position of each antenna element and the bend angle θ_b. A similar argument can be made for the case when the array is transmitting a null in the direction of θ_{SNOI_n}, which is shown in Figure 2b. For this case, the values of $\Delta_{\pm n}$ still need to be computed.

2.2. Computing the Distance to the Projected Elements on the Wave Front

For this work, the values of θ_{SOI} and θ_{SNOI_n} are between $-\pi/2$ and $\pi/2$. Because of this, the problem will be broken down into two cases. For the first case, $-\theta_b \leq \theta_{SOI(SNOI_n)} \leq \theta_b$. For these angles of $\theta_{SOI(SNOI_n)}$, the projected elements on the wavefront are all outside of the wedge-shaped surface. For the second case, $\theta_b \leq \theta_{SOI(SNOI_n)} \leq \pi/2$ or $-\pi/2 \leq \theta_{SOI(SNOI_n)} \leq -\theta_b$. For these angles of $\theta_{SOI(SNOI_n)}$, half of the projected elements are outside of the wedge-shaped surface and half are inside of the surface. This is because when $\theta_{SOI(SNOI_n)} \geq \theta_b$ or $\theta_{SOI(SNOI_n)} \leq -\theta_b$, the projected elements for A_1 and A_2 or A_{-1} and A_{-2}, respectively, are inside of the wedge-shaped surface in Figure 2a,b.

2.2.1. Case 1: Computing $\Delta_{\pm n}$ for $-\theta_b \leq \theta_{SOI(SNOI_n)} \leq \theta_b$

Using the notation in Figure 2a,b and several identities, the projected distance from antenna elements on the wedge to the reference wavefront is given by:

$$\Delta_{\pm n} = d\frac{|n|}{2n}(2|n| - 1)\sin\left(\theta_b \mp \theta_{SOI(SNOI_n)}\right) \tag{1}$$

for $0 \leq \theta_{SOI(SNOI_n)} \leq \theta_b$ and

$$\Delta_{\pm n} = d\frac{|n|}{2n}(2|n| - 1)\sin\left(\theta_b \pm \left|\theta_{SOI(SNOI_n)}\right|\right) \tag{2}$$

for $-\theta_b \leq \theta_{SOI(SNOI_n)} \leq 0$. Notice that the expressions in (1) and (2) are written as a function of the flexing (bend) angle θ_b of the wedge and the location of each antenna element, indicating the generality of the expressions.

2.2.2. Case 2: Computing $\Delta_{\pm n}$ for $\theta_b \leq \theta_{SOI(SNOI_n)} \leq \pi/2$ or $-\pi/2 \leq \theta_{SOI(SNOI_n)} \leq -\theta_b$

Next, for the remaining values of $\theta_{SOI(SNOI_n)}$, the projected distance from each antenna element in the array to the reference wavefront can be calculated as:

$$\Delta_{\pm n} = d\frac{|n|}{2n}(2|n| - 1)\sin\left(\theta_{SOI(SNOI_n)} \mp \theta_b\right) \tag{3}$$

for $\theta_b \leq \theta_{SOI(SNOI_n)} \leq \pi/2$ and

$$\Delta_{\pm n} = d\frac{|n|}{2n}(2|n| - 1)\sin\left(\left|\theta_{SOI(SNOI_n)}\right| \pm \theta_b\right) \tag{4}$$

for $-\pi/2 \leq \theta_{SOI(SNOI_n)} \leq \theta_b$. Notice that (3) and (4) are also written in terms of θ_b of the wedge and the position of each antenna element in a general fashion.

2.3. Computing the Radiation in the Direction of θ_{SOI}, θ_{SNOI_1}, θ_{SNOI_2} and θ_{SNOI_3}

Next, the field from the array in the direction of θ_{SOI} (signal of interest), and θ_{SNOI_1}, θ_{SNOI_2} and θ_{SNOI_3}(signals of no interest) on the singly curved surface (wedge) is computed with the Array Factor (AF_w) expression in [15]:

$$AF_w(\theta,\phi) = \sum_{n=-2}^{-1} F_{n,L} e^{jk[ux_n+vy_n+z_n \cos\theta]} + \sum_{n=1}^{2} F_{n,R} e^{jk[ux_n+vy_n+z_n \cos\theta]} \quad (5)$$

where

$$F_{n,L} = w_n \cos\theta_{PL} e^{\pm jk\Delta_{\pm n}} \quad (6)$$

and

$$F_{n,R} = w_n \cos\theta_{PR} e^{\pm jk\Delta_{\pm n}} \quad (7)$$

A spherical coordinates system is assumed in (5) with $u = \sin\theta\cos\phi$, $v = \sin\theta\sin\phi$, $\Delta_{\pm n}$ is defined in (1)–(4), (x_n, y_n, z_n) is the location of the nth array element and w_n are the array excitation weights (amplitudes and phases) to drive the individual antenna elements. Furthermore, the element patterns for A_1 and A_2 are denoted as $e_R(\theta) = \cos\theta_{PR}$ and the element patterns for A_{-1} and A_{-2} are denoted as $e_L(\theta) = \cos\theta_{PL}$ where θ_{PR} and θ_{PL} are defined in Figure 2.

2.4. Computing the Array Weights for N = 4 Elements

The complex weighting functions (i.e., array weights) in (5) are of interest in this work because these are the array weights that need to be computed for beamforming of the array. To compute these weights, the matrix method for smart antennas presented in [1] will be used. Because the array studied here has four elements, four array weights are needed. This then requires a set of four equations with four unknowns that can be written as a square matrix. Then, the array weights can be computed using a matrix solver. The sum in (5) has four terms for N = 4 and to obtain four equations with four unknowns, (5) will be evaluated at each value of θ_{SOI}, θ_{SNOI_1}, θ_{SNOI_2} and θ_{SNOI_3}. This then gives the following array factor matrix for antenna array on wedge-shaped conformal surface AF_w:

$$\boldsymbol{AF_w} = \begin{bmatrix} AF_w(\theta_{SOI}) \\ AF_w(\theta_{SNOI_1}) \\ AF_w(\theta_{SNOI_2}) \\ AF_w(\theta_{SNOI_3}) \end{bmatrix} \quad (8)$$

where $\theta = \theta_{SOI}, \theta_{SNOI_1}, \theta_{SNOI_2}$ or θ_{SNOI_3} and $\phi = 0$. Then, factoring out the array weights gives,

$$\boldsymbol{AF_w} = \boldsymbol{AW} \quad (9)$$

where $\boldsymbol{A}$ is the array factor in (5) for antenna array on wedge-shaped conformal surface with factoring out the complex weights and

$$\boldsymbol{W} = \begin{bmatrix} w_{-2} \\ w_{-1} \\ w_1 \\ w_2 \end{bmatrix} \quad (10)$$

Next, to ensure that the conformal antenna will have a main beam at the scan angle θ_{SOI}, $AF_w(\theta_{SOI})$ must equal 1 in (8), or $AF_w(\theta_{SOI}) = 1$. Then, in order to provide nulls in the direction of θ_{SNOI_1}, θ_{SNOI_2} and θ_{SNOI_3}, $AF_w(\theta_{SNOI_1}) = AF_w(\theta_{SNOI_2}) = AF_w(\theta_{SNOI_3}) = 0$ in (8). This can be written in matrix form in the following manner:

$$\boldsymbol{AW} = \boldsymbol{C} \quad (11)$$

where

$$C = \begin{bmatrix} 1 \\ 0 \\ 0 \\ 0 \end{bmatrix} \tag{12}$$

and AW is defined in (9). Finally, solving for the weights in (11) gives

$$W = A^{-1}C \tag{13}$$

By setting the first element in (12) to be 1 and the rest of the elements to be 0, the array factor is forced to give nulls (or a zero field) analytically at the angles of θ_{SNOI_1}, θ_{SNOI_2} and θ_{SNOI_3}. Thus, the solution of (13) are the array weights required to give these pattern null features. Furthermore, since the values of $\Delta_{\pm n}$ in the array factor expression in (8) are written in terms of the wedge angle θ_b and the location of each antenna element, the weights computed with (13) are determined in a setting where the antenna, and hence the wedge, can change shape.

2.5. Computing the Array Weights for N Elements

The previous expressions were determined for N = 4 and can be generalized for more elements. More, specifically, (8)–(13) can be generalized in the following manner:

$$\boldsymbol{AF_w} = \begin{bmatrix} AF_w(\theta_{SOI}) \\ \cdot \\ \cdot \\ \cdot \\ AF_w(\theta_{SNOI_1}) \\ \cdot \\ \cdot \\ \cdot \\ AF_w(\theta_{SNOI_n}) \end{bmatrix} \tag{14}$$

where

$$AF_{wn}(\theta,\phi) = \sum_{n=-\frac{N}{2}}^{-1} F_{n,L} e^{jk[ux_n+vy_n+z_n\cos\theta]} + \sum_{n=1}^{\frac{N}{2}} F_{n,R} e^{jk[ux_n+vy_n+z_n\cos\theta]} \tag{15}$$

and $F_{n,L}$ and $F_{n,R}$ are defined in (6) and (7), respectively. Then, factoring out the array weights gives

$$\boldsymbol{AF_{wn}} = \boldsymbol{A_n W_n} \tag{16}$$

where $\boldsymbol{A_n}$ is the array factor in (15) for antenna array on wedge-shaped conformal surface with factoring out the complex weights and

$$\boldsymbol{W_n} = \left[w_{-\frac{N}{2}} \ldots w_{-1}\; w_1 \ldots w_{\frac{N}{2}} \right]^T \tag{17}$$

Finally, the coefficients in (12) can be generalized as

$$\boldsymbol{C_n} = [c_1\; c_2 \ldots c_n]^T \tag{18}$$

and the weights can be computed using

$$\boldsymbol{W_n} = \boldsymbol{A_n}^{-1}\boldsymbol{C_n} \tag{19}$$

Equation (19) can then be used to solve for the array weights that will result in a main radiation pattern in the direction of θ_{SOI} and nulls in the directions of $\theta_{SNOI_1}, \theta_{SNOI_2}, \cdots \theta_{SNOI_{(N-1)}}$ for an N-element array, on a changing wedge-shaped surface.

2.6. Beamforming of a 1 × 4 Array on a Cylindrical-Shaped Surface

Next, the complex weights for antenna array on the cylindrical curvature shown in Figure 2c are computed. Again, for illustration, the derivations will be carried out for $N = 4$ elements. As with the wedge-shaped conformal surface, the projected distance from the antenna elements on the cylindrical surface to the transmitted wavefront at angle $\theta_{SOI(SNOI_n)}$ must be computed. Again, these distances are denoted as $\Delta_{\pm n}$ in Figure 2c. First, the distance from the point where the cylinder intersects the z-axis (denoted as point P in Figure 2c) to each antenna element is computed using the following equation

$$h_{\pm n} = \sqrt{(0 - x_{\pm n})^2 + (r - z_{\pm n})^2}, \tag{20}$$

where again $(x_{\pm n}, z_{\pm n})$ is the location of the nth element on the cylindrical surface, $x_{\pm n} = r\cos\phi_{\pm n}$, $z_{\pm n} = r\sin\phi_{\pm n}$ and $\phi_{\pm n}$ is defined in Figure 2c. As with the wedge-shaped surface, the problem will be considered as two different cases.

The first case is for $\theta_{SOI(SNOI_n)} \geq 0$ and the second case is for $\theta_{SOI(SNOI_n)} \leq 0$. The first case is shown in Figure 2c. The projected elements of A_{-2}, A_{-1} and A_2 in the direction of $\theta_{SOI(SNOI_n)}$ are outside of the cylindrical surface and the projected element of A_1 in the direction of $\theta_{SOI(SNOI_n)}$ is inside of the cylindrical surface. Because some projected elements are outside of the cylinder and some are inside, each case will be broken into two parts. The first part will compute the distance from the elements A_{-1} and $A_{\pm 2}$ to the projected elements on the wavefront and the second part will compute the distance from the element A_1 to the projected element on the wavefront (as shown in Figure 2c).

2.6.1. Case 1: $\theta_{SOI(SNOI_n)} \geq 0$

Next, using (20) and several trigonometric identities, the projected distance from the elements A_{-1} and $A_{\pm 2}$ to the transmitted wavefront is calculated as:

$$\Delta_{\pm n} = h_{\pm n}\sin(\mp\theta_{SOI(SNOI_n)} + \theta_{\pm n}) \tag{21}$$

where $\theta_{\pm n} = \cos^{-1}|x_{\pm n}/h_{\pm n}|$. Then, for the element A_1 the distance to the projected element can be computed as:

$$\Delta_{+n} = h_{+n}\sin\left(\theta_{SOI(SNOI_n)} - \theta_{+n}\right) \tag{22}$$

2.6.2. Case 2: $\theta_{SOI(SNOI_n)} \leq 0$

Again, using (20) and several trigonometric identities, the distance from the elements A_1 and $A_{\pm 2}$ to the projected elements can be computed as:

$$\Delta_{\pm n} = h_{\pm n}\sin\left(\pm\left|\theta_{SOI(SNOI_n)}\right| + \theta_{\pm n}\right) \tag{23}$$

where $\theta_{\pm n} = \cos^{-1}|x_{\pm n}/h_{\pm n}|$. Then, for the element A_{-1}, the distance to the projected element can be computed as:

$$\Delta_{-n} = h_{-n}\sin\left(\left|\theta_{SOI(SNOI_n)}\right| - \theta_{-n}\right) \tag{24}$$

Several comments can be made about (21)–(24). As with the wedge-shaped surface, these expressions have been written in a general manner in terms of antenna position on the cylindrical surface and the radius of curvature. Moreover, special care should be taken when implementing these equations if $\left|\theta_{SOI(SNOI_n)}\right| \leq \theta_{\pm n}$. This is because for these projected elements, angles on the wavefront will be outside of the cylindrical surface. In this case, the distance to the wavefront can be computed using (21) or (23) (depending on whether $\theta_{SOI(SNOI_n)} \leq 0$ or $\theta_{SOI(SNOI_n)} \geq 0$ and the computations of (22) and (24) are not required.

2.7. Computing the Array Weights for N = 4 Elements

Next, as with the antenna array on the wedge-shaped conformal surface, the array weights can be computed using (13). For the computations of these array weights on the cylinder, the distance from the elements to the wavefront should be computed using (21)–(24).

2.8. Computing the Array Weights for N Elements

Finally, (19) can be used (with the updated distances computed using (21)–(24)) to compute the array weights for an N-dimensional array on a cylindrical surface with radius r. As with the wedge-shaped surface, the expressions to compute the array weights have been written in a general manner that includes element spacing and the radius of the cylinder. This makes the technique presented here useful for an array attached to a changing cylindrical surface.

2.9. Computing the Weighting Coefficients with Mutual Coupling

For the previous derivations of (13) and (19), it was assumed that there was no mutual coupling between the elements. To model the mutual coupling, the work presented in [32] was considered. More specifically, the work in [32] proposed a model for the mutual coupling in adaptive arrays and demonstrated that the mutual coupling between the elements of an adaptive array can cause a significant degradation in the signal-to-interference-plus-noise ratio (SINR). The methods in [32] will be adopted here to model the mutual coupling effects on the radiation pattern of the conformal beamforming array. The coupling between the antenna elements can be modelled as an $N + 1$ port network, as shown in Figure 3. The antenna elements in the array are all terminated with Z_L and denoted as ports $1, 2, \ldots N$. The antenna port driven by a voltage source V_s is denoted as $N + 1$. The port driven with V_s is the representation of the transmitted (or incoming) signals at angles θ_{SOI} or θ_{SNOI_n}. Then, using the Kirchhoff relations for the $N + 1$ terminal network, the voltage at the terminated port n can be written as:

$$V_{Tn} = I_1 Z_{n,1} + I_2 Z_{n,2} + \ldots + I_n Z_{n,m} + \ldots + I_s Z_{n,s} \tag{25}$$

where $Z_{n,m}$ is the mutual impedance between the nth and m^{th} port, I_n is the current going through the terminating load on the nth port, $Z_{n,n}$ is the self-impedance of the nth port and $Z_{n,s}$ represents the mutual coupling term between the driven element with V_s and the nth antenna element. Furthermore, the current at the nth port can be written in terms of the terminal voltage and load impedance in the following manner:

$$I_n = \frac{-V_{Tn}}{Z_L} \tag{26}$$

Then, making use of the open circuit condition and removing the terminating impedances results in $I_1 = I_2 = \ldots I_N = 0$. This simplifies (25) to $V_{Tn} = I_s Z_{n,s}$. Under these conditions, (25) represents the open circuit voltages at the nth port caused by the mutual coupling between the driven element and the nth port, and can be computed as $V_{Tn} = I_s Z_{n,s} V_{OCn}$. Next, substituting (26) into (25), making use of the open-circuit condition and writing (25) in matrix form results in the following:

$$\mathbf{Z_c V_T} = \mathbf{V_{OC}} \tag{27}$$

where

$$\mathbf{Z_c} = \begin{bmatrix} 1+\frac{Z_{1,1}}{Z_L} & \frac{Z_{1,2}}{Z_L} & \cdots & \frac{Z_{1,N}}{Z_L} \\ \frac{Z_{2,1}}{Z_L} & 1+\frac{Z_{2,2}}{Z_L} & \cdots & \frac{Z_{2,N}}{Z_L} \\ \vdots & \vdots & \ddots & \vdots \\ \frac{Z_{N,1}}{Z_L} & \frac{Z_{N,2}}{Z_L} & \cdots & 1+\frac{Z_{N,N}}{Z_L} \end{bmatrix} \tag{28}$$

$$\mathbf{V_T} = [V_{T1}\ V_{T2} \ldots V_{Tn}]^T \tag{29}$$

and

$$\mathbf{V_{OC}} = [V_{OC1}\ V_{OC2} \ldots V_{OCn}]^T \tag{30}$$

The normalized impedance matrix $\mathbf{Z_c}$ includes self- and mutual-terms, and can be determined from a 3D full wave simulator such as HFSS [33]. The open circuit voltage column matrix $\mathbf{V_{OC}}$ represents the array weights (i.e., the complex voltages used to drive the beamforming conformal antenna) without including mutual coupling. This then results in $\mathbf{V_{OC}} = \mathbf{W}_n$ where $\mathbf{W}_n$ is given in (19). The terminal voltage column matrix $\mathbf{V_T}$ represents the array weights that include the mutual coupling and can be computed from (27) as follows:

$$\mathbf{V_T} = \mathbf{Z_c}^{-1}\mathbf{V_{OC}} \tag{31}$$

Next, to write the array weights with mutual coupling in terms of $\mathbf{Z_c}$, substitute $\mathbf{V_{OC}} = \mathbf{W}_n$ into (19) and rearranging gives the following expression:

$$\mathbf{A_n}\mathbf{V_{OC}} = \mathbf{C_n} \tag{32}$$

Then, solving for $\mathbf{V_{OC}}$ in (32) and substituting into (31) gives:

$$\mathbf{V_T} = \mathbf{Z_c}^{-1}\mathbf{A_n}^{-1}\mathbf{C_n} = \mathbf{W}_n^c \tag{33}$$

where $\mathbf{W}_n^c$ represents the new array weights with the coupling included in the computations. In the next Section, validation of $\mathbf{W}_n$ using (19) and $\mathbf{W}_n^c$ using (33) for various values of θ_{SOI} and θ_{SNOI_n} will be presented followed by the characteristics of the array weights for various conformal surfaces.

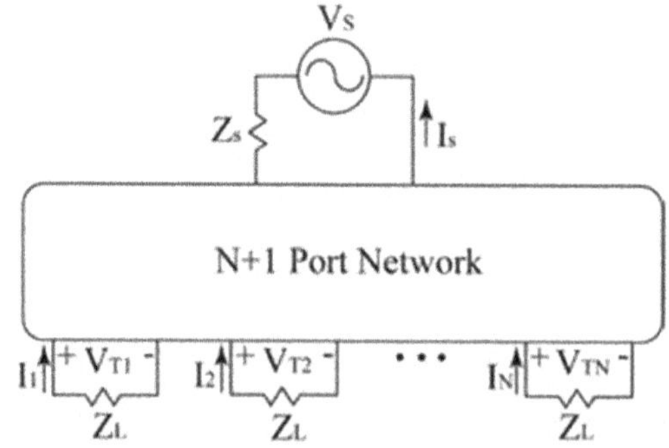

Figure 3. *N*-port network illustration of conformal array with a signal of interest at angle θ_{SOI} or θ_{SNOI_n}.

3. Validation with Analytical, Simulations and Measurement Results

In this Section, a beamforming array is used to validate the previously derived array weight expressions using simulations and measurements. More specifically the array weights are computed using (19) and (33) for the 1 × 4 element array on the singly curved wedge and cylindrical curvature surfaces, shown in Figure 2. Two different beam-formation patterns were considered, and each pattern was evaluated on three different conformal surfaces. The characteristics of each beam-formation pattern, denoted as pattern 1 and pattern 2, are mentioned in Table 1. Furthermore, the three conformal surfaces considered were the singly curved wedge surface with θ_b = 30° and θ_b = 45° and a cylindrical surface with r = 10 cm.

Table 1. Specifications of the beamforming patterns.

Variable	Pattern 1	Pattern 2
θ_{SOI}	0°	40°
θ_{SNOI_1}	−30°	−45°
θ_{SNOI_2}	30°	−25°
θ_{SNOI_3}	40°	10°

3.1. Four-Element Beamforming Array

For measurement purposes, the four-element beamforming array shown in Figure 4a was manufactured. The array consisted of connectorized voltage-controlled phase shifters, voltage-controlled attenuators, a four-way power divider, an amplifier and four microstrip patches designed to operate at 2.47 GHz. A picture of the attenuators, phase shifters and power divider of the manufactured array is shown in Figure 4b, and a picture of the four microstrip patches is shown in Figure 4c. Four individual microstrip patches were used for the convenience of placing the array on the various conformal surfaces. The phase shifters were manufactured by Hittite Microwave Corporation [34] (PN: HMC928LP5E) and the power divider, attenuators and amplifiers were manufactured by Mini-Circuits [35] (PNs: ZN8PD1-53-S+, ZX73-2500-S+ and ZX60-33LN-S+, respectively). Identical SMA cables were used to connect each patch to a port on the power divider.

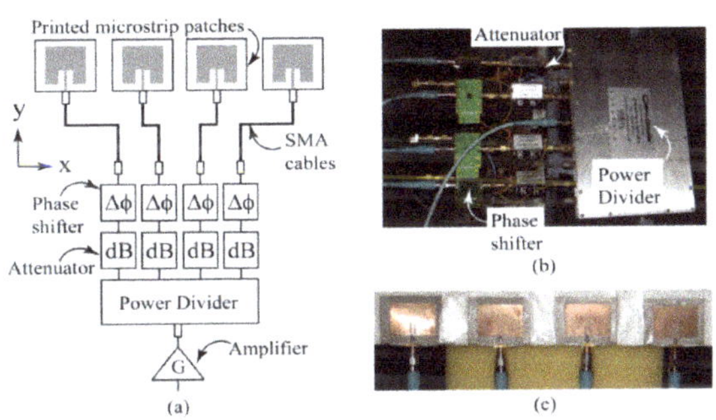

Figure 4. (**a**) Topology of the four-element beamforming array, (**b**) a picture of the power divider, voltage-controlled phase shifters and voltage-controlled attenuators used for measurements and (**c**) a picture of the microstrip patch elements used for attachment to conformal surfaces.

3.2. Beamforming Results on the Conformal Wedge Antenna Array with θ_b = 30°

First measurements taken were for the 1 × 4 element beamforming array on the singly curved flexing wedge in Figure 2a for θ_b = 30°, and the array weights were computed using (33). The inter-element spacing was 0.5 λ and a picture of the array being measured on the surface with a 2-port network analyzer in a fully calibrated anechoic chamber is shown in Figure 5a. The results from these measurements for both patterns summarized in Table 1 are shown in Figure 6a,b. Next, the four-element beamforming array was modelled in HFSS and the weights computed using (33) were used to drive the array. More specifically, the separate substrates, conducting layers and SMA connectors were modelling in HFSS to provide an accurate representation of the measurement setup in Figure 5. The radiation pattern from these simulations can also be seen in Figure 6a,b at 2.47 GHz. Then, for a third comparison both patterns were computed analytically using the array factor expression in (5) with the weights determined using (19). These results are also shown in Figure 6a,b. Finally, new weights that include the mutual coupling were computed using (33) and used in (5) to compute the radiation pattern analytically. These results are shown in Figure 6a,b. Overall, agreement between measurements, simulations and analytical computations (with both sets of array weights) is shown. The array weights for the results in Figure 6a,b are shown in Tables A1 and A2 in Appendix A.

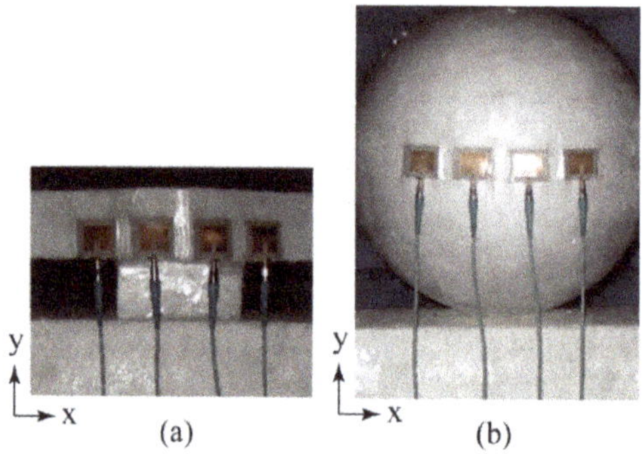

Figure 5. (**a**) Photograph of the four-element beamforming array being measured on the singly curved non-conducting wedge surface with $\theta_b = 30°$ and (**b**) photograph of the four-element beamforming array being measured on the cylindrical-shaped surface with $r = 10$ cm.

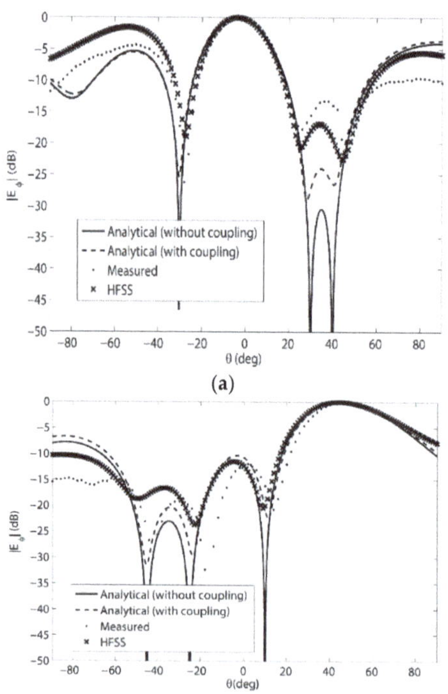

Figure 6. (**a**) Pattern 1 beamforming results for the 1 × 4 microstrip patch array on the singly curved (wedge-shaped) surface with $\theta_b = 30°$; (**b**) Pattern 2 beamforming results for the 1 × 4 microstrip patch array on the singly curved (wedge-shaped) surface with $\theta_b = 30°$.

3.3. Beamforming Results on the Conformal Wedge Antenna Array with $\theta_b = 45°$

Next, measurements were taken for the four-element beam-forming array on the singly curved flexing wedge in Figure 2a for $\theta_b = 45°$. The adjacent element spacing was again 0.5 λ and (33) was used to compute the new array weights. The results from these measurements for both patterns summarized in Table 1 are shown in Figure 7a,b. Next, the four-element beamforming array was simulated in HFSS with the array weights computed using (33) for $\theta_b = 45°$ and the radiation pattern can also be seen in Figure 7a,b at 2.47 GHz. Then, for a third comparison, the expression in (19) was used to compute the analytical results shown in Figure 7a,b. Next, the four-element beamforming array was simulated in HFSS with the array weights computed using (33) for $\theta_b = 45°$ and the radiation pattern can also be seen in Figure 7a,b at 2.47 GHz. Then, for a third comparison the expression in (19) was used to compute the analytical results shown in Figure 7a,b. New weights that include the mutual coupling were also computed using (33) for the new value of $\theta_b = 45°$. These weights were then used in (5) to compute the radiation pattern and these results are shown in Figure 7a,b. As with the $\theta_b = 30°$ results, agreement between measurements, simulations and analytical computations is shown. The array weights for the results in Figure 7a,b are shown in Tables A3 and A4 in Appendix A.

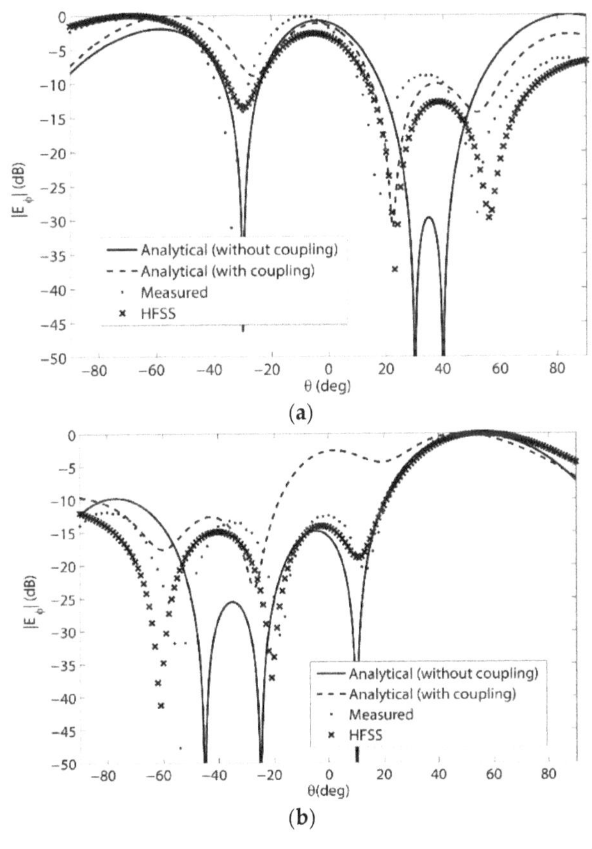

Figure 7. (**a**) Pattern 1 beamforming results for the 1 × 4 microstrip patch array on the singly curved (wedge-shaped) surface with $\theta_b = 45°$; and (**b**) Pattern 2 beamforming results for the 1 × 4 microstrip patch array on the singly curved (wedge-shaped) surface with $\theta_b = 45°$.

3.4. Beamforming Results on the Conformal Cylindrical Antenna Array with r = 10 cm

Finally, the same comparison between measurements, simulations and analytical computations was conducted for the four-element array on the cylindrical-shaped surface with $r = 10$ cm. The array being measured in the full anechoic chamber is shown in Figure 5b and the results are shown in Figure 8. It should be mentioned that in order to measure the array on a cylindrical surface, a sphere was used and the antenna elements were placed along the equator. This then resulted in an antenna shape similar to a cylindrical surface. Simulation results and analytical computations with and without the coupling are also shown in Figure 8 and overall agreement is shown. The array weights for the results in Figure 8 are shown in Tables A5 and A6 in Appendix A.

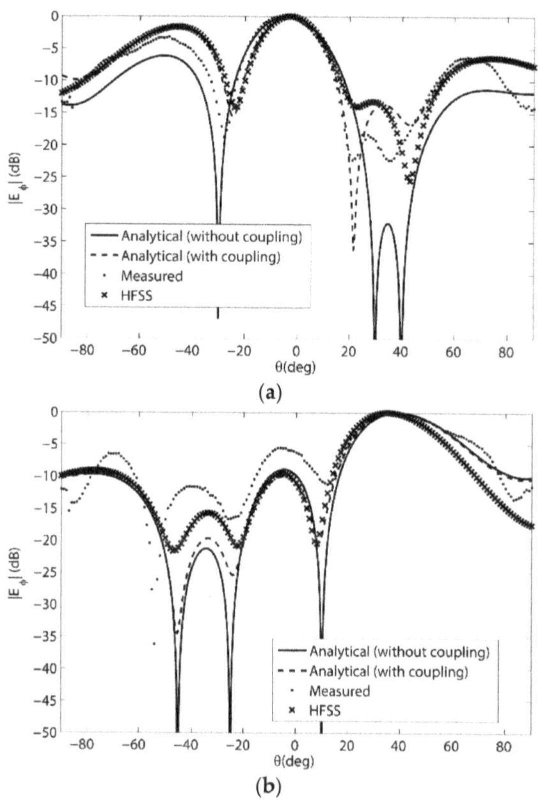

Figure 8. (**a**) Pattern 1 beamforming results for the 1 × 4 microstrip patch array on the cylindrical curvature surface with $r = 10$ cm; (**b**) Pattern 2 beamforming results for the 1 × 4 microstrip patch array on the cylindrical curvature surface with $r = 10$ cm.

4. Discussions

For the results in Figures 6 and 7, the most agreement is between the measured results, the HFSS simulations and the analytical computations with the array weights including the mutual coupling. This illustrates the improved accuracy of the weights computed using (33).

Moreover, when comparing the pattern 2 results in Figure 6b and b, it is shown that more agreement between the results around $-60°$ exist for the $\theta_b = 45°$ surface than for the $\theta_b = 30°$ surface. This is thought to be due to the more severe surface deformation that exists for the $\theta_b = 45°$. Overall though, the array weights computed using (19) and (33) have been shown to be accurate and the effects due to the mutual coupling and surface deformations on the radiation pattern have been demonstrated. Thus again, the weights computed using (33) can be used to model the coupling, and with proper optimization, an improved beam-formation could be achieved in a general setting that includes the mutual coupling between elements on a changing conformal surface.

The approach to beamforming presented in this paper has many applications ranging from IoT devices, 5G and mobile satellite communications. As devices in these spaces become smaller, phased array antennas are required to operate in the vicinity of scattering surfaces and other array elements, hence requiring knowledge of mutual coupling.

5. Dependence of Complex Weights on the Geometry of the Conformal Surface

Next, the array weights for a three-element array were computed analytically using (19). This was undertaken to show how the array weights are dependent on bend angle θ_b and the radius of curvature r. It should be mentioned that = 3 was chosen because the problem becomes much more complex for $N > 3$ and requires a numerical analysis of the matrices in (19). After several algebraic steps it can be shown that the array weights can be written as:

$$w_{+1} = \frac{1}{F_{1,R}{}^{\Delta} - DF_{-1,L}{}^{\Delta}} \tag{34}$$

$$w_0 = 1 - w_{+1}[F_{1,R}(\theta_{SOI}) - DF_{-1,L}(\theta_{SOI})] \tag{35}$$

and

$$w_{-1} = -w_{+1}D \tag{36}$$

where

$$D = \frac{F_{1,R}(\theta_{SNOI_2}) - F_{1,R}(\theta_{SNOI_1})}{F_{-1,L}(\theta_{SNOI_2}) - F_{-1,L}(\theta_{SNOI_1})} \tag{37}$$

$F_{1,R}{}^{\Delta} = F_{1,R}(\theta_{SOI}) - F_{1,R}(\theta_{SNOI_1})$, $F_{-1,L}{}^{\Delta} = F_{-1,L}(\theta_{SOI}) - F_{-1,L}(\theta_{SNOI_1})$ and $F_{n,L}$ and $F_{n,R}$ are defined in (6) and (7), respectively, with the weights factored out. For a small value of N the results in (34)–(36) show that the expression for the array weights can be quite complicated. Then again, it does also show the dependency of each weight on the angle $\theta_{SOI(SNOI_n)}$ and the location of each antenna element in the array.

To illustrate the behavior of the weights using (34)–(36) for various surfaces, the three-element array was considered on a wedge-shaped surface for various values of θ_b. For these computations, the values of θ_{SOI} and θ_{SNOI_n} were $\theta_{SOI} = 40^{\circ}$, $\theta_{SNOI_1} = -45^{\circ}$ and $\theta_{SNOI_2} = 10^{\circ}$. The amplitude and phase of the array weights computed using (34)–(36) are shown in Figure 9. The results show that for a fixed beam-formation there is a strong relationship between the array weights and the conformal surface. A similar derivation could be conducted to compute (33) for the array weights that include coupling, and the same relationship between the array weights and the surface deformation was shown to exist for the cylindrical surface.

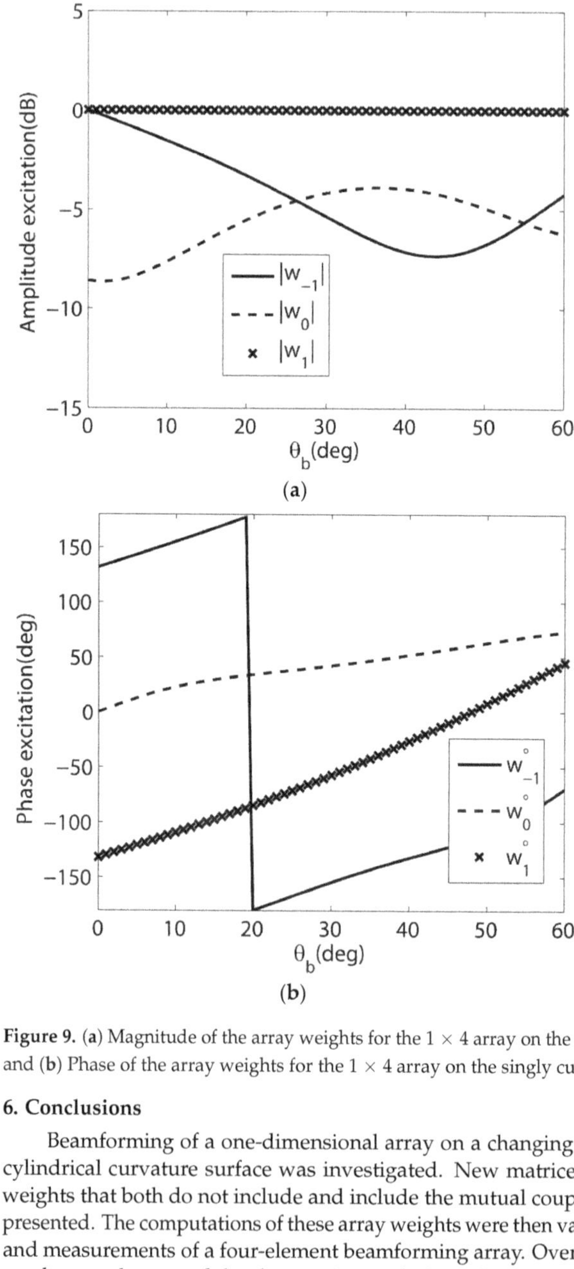

Figure 9. (**a**) Magnitude of the array weights for the 1 × 4 array on the singly curved (wedge) surface and (**b**) Phase of the array weights for the 1 × 4 array on the singly curved (wedge) surface.

6. Conclusions

Beamforming of a one-dimensional array on a changing singly curved wedge and cylindrical curvature surface was investigated. New matrices for computing the array weights that both do not include and include the mutual coupling between elements are presented. The computations of these array weights were then validated with the simulation and measurements of a four-element beamforming array. Overall agreement between the results was shown and the characteristics of a beamforming array on a conformal surface were discussed.

Author Contributions: Conceptualization, I.U. and B.D.B.; methodology, I.U. and B.D.B.; software, validation, and analysis, I.U. and B.D.B.; writing—original draft preparation, I.U. and B.D.B.; writing—review and editing, A.I., S.N. and D.E.A.; supervision, D.E.A.; project administration, S.N. and D.E.A.; funding acquisition S.N. and D.E.A. All authors have read and agreed to the published version of the manuscript.

Funding: This research is partially funded by the project ICARUS—EXCELLENCE/0918/0365, which is co-funded by the Republic of Cyprus and the European Regional Development Fund (through the "EXCELLENCE" RESTART 2016–2020 Program for Research, Technological Development and Innovation). This research is also partially funded by the European H2020 Marie Skłodowska-Curie Individual Fellowship (MSCA-IF) grant #840854 (VisionRF)

Institutional Review Board Statement: Not applicable.

Informed Consent Statement: Not applicable.

Data Availability Statement: Not applicable.

Conflicts of Interest: The authors declare no conflict of interest.

Appendix A

The following tables summarize the array weights for the results in Figures 6–8.

Table A1. Array weights of the conformal antenna on the wedge-shaped surface with θ_b = 30° without coupling using (19).

Pattern 1 Weights	Pattern 2 Weights
$w_{-2} = 0.98\angle + 128.76^{\circ}$	$w_{-2} = 0.31\angle - 72.72^{\circ}$
$w_{-1} = 0.21\angle - 75.85^{\circ}$	$w_{-1} = 0.56\angle + 135.57^{\circ}$
$w_1 = 1.00\angle + 35.24^{\circ}$	$w_1 = 1.00\angle - 2.34^{\circ}$
$w_2 = 0.59\angle - 174.73^{\circ}$	$w_2 = 0.84\angle - 78.45^{\circ}$

Table A2. Array weights of the conformal antenna on the wedge-shaped surface with θ_b = 30° with coupling using (33).

Pattern 1 Weights	Pattern 2 Weights
$w_{-2} = 0.96\angle + 88.34^{\circ}$	$w_{-2} = 0.35\angle - 112.26^{\circ}$
$w_{-1} = 0.26\angle - 108.76^{\circ}$	$w_{-1} = 0.61\angle + 91.65^{\circ}$
$w_1 = 1.00\angle - 8.1^{\circ}$	$w_1 = 1.00\angle - 39.52^{\circ}$
$w_2 = 0.65\angle + 146.62^{\circ}$	$w_2 = 0.84\angle - 124.1^{\circ}$

Table A3. Array weights of the conformal antenna on the wedge-shaped surface with θ_b = 45° without coupling using (19).

Pattern 1 Weights	Pattern 2 Weights
$w_{-2} = 1.00\angle - 154^{\circ}$	$w_{-2} = 0.26\angle - 14.85^{\circ}$
$w_{-1} = 0.74\angle - 31.9^{\circ}$	$w_{-1} = 0.26\angle + 165.14^{\circ}$
$w_1 = 0.62\angle + 65.6^{\circ}$	$w_1 = 1.00\angle + 39.57^{\circ}$
$w_2 = 0.52\angle - 107.1^{\circ}$	$w_2 = 0.88\angle - 16.86^{\circ}$

Table A4. Array weights of the conformal antenna on the wedge-shaped surface with θ_b = 45° with coupling using (33).

Pattern 1 Weights	Pattern 2 Weights
$w_{-2} = 1.00\angle - 155.1^{\circ}$	$w_{-2} = 0.34\angle - 15.8^{\circ}$
$w_{-1} = 0.41\angle - 56.29^{\circ}$	$w_{-1} = 0.14\angle + 121.26^{\circ}$
$w_1 = 0.24\angle + 55.72^{\circ}$	$w_1 = 0.17\angle + 21.0^{\circ}$
$w_2 = 0.53\angle - 116.44^{\circ}$	$w_2 = 1.00\angle - 28.94^{\circ}$

Table A5. Array weights of the conformal antenna on the cylindrical-shaped surface with *r* = 10 cm without coupling using (19).

Pattern 1 Weights	Pattern 2 Weights
$w_{-2} = 0.95\angle + 48.27^{\circ}$	$w_{-2} = 0.53\angle - 133.55^{\circ}$
$w_{-1} = 0.22\angle + 38.35^{\circ}$	$w_{-1} = 0.86\angle + 83.9^{\circ}$
$w_{1} = 1.00\angle - 7.36^{\circ}$	$w_{1} = 0.91\angle - 51.41^{\circ}$
$w_{2} = 0.60\angle + 110^{\circ}$	$w_{2} = 1.00\angle - 136.1^{\circ}$

Table A6. Array weights of the conformal antenna on the cylindrical-shaped surface with *r* = 10 cm with coupling using (33).

Pattern 1 Weights	Pattern 2 Weights
$w_{-2} = 0.97\angle + 5.39^{\circ}$	$w_{-2} = 0.59\angle + 168.29^{\circ}$
$w_{-1} = 0.26\angle + 102.15^{\circ}$	$w_{-1} = 0.94\angle + 29.34^{\circ}$
$w_{1} = 1.00\angle - 75.79^{\circ}$	$w_{1} = 1.00\angle - 102.68^{\circ}$
$w_{2} = 0.93\angle + 61.92^{\circ}$	$w_{2} = 0.88\angle + 169.59^{\circ}$

References

1. Balanis, C.A. *Antenna Theory: Analysis and Design*, 3rd ed.; John Wiley and Sons: Hoboken, NJ, USA, 2005.
2. Lockyer, A.J.; Alt, K.H.; Coughlin, D.P.; Durham, M.D.; Kudva, J.N.; Goetz, A.C.; Tuss, J. Design and developement of a conformal load bearing smart skin antenna: Overiew of the AFRL smart skin structure technology demonstration (S3TD). In Proceedings of the 1999 Symposium on Smart Structures and Materials, Newport Beach, CA, USA, 1 March 1999; Volume 3674, p. 410.
3. Bellofiore, S.; Balanis, C.; Foutz, J.; Spanias, A. Smart-antenna systems for mobile communication networks. Part 1. Overview and antenna design. *IEEE Antennas Propag. Mag.* **2003**, *44*, 145–154. [CrossRef]
4. Balanis, C.A. Smart antennas for future reconfigurable wireless communication networks. In Proceedings of the 2003 IEEE Topical Conference on Wireless Communication Technology, Honolulu, HI, USA, 15–17 October 2003; pp. 181–182.
5. Lockyer, A.J.; Alt, K.H.; Kinslow, R.W.; Kan, H.-P.; Goetz, A.C. Development of a structurally integrated Conformal load-bearing multifunction antenna: Overview of the Air Force Smart Skin Structures Technology Demonstration Program. *Proc. SPIE* **1996**, *2722*, 55–64.
6. Varadan, V.K.; Varadan, V.V. Design and development of conformal smart spiral antenna. In Proceedings of the 1996 Symposium on Smart Structures and Materials, San Diego, CA, USA, 25–29 February 1996; Volume 2722, p. 46.
7. Benedetti, M.; Azaro, R.; Massa, A. Experimental validation of full-adaptive smart antenna prototype. *Elec. Lett.* **2008**, *44*, 661–662. [CrossRef]
8. Ohira, T. Electronically steerable passive array radiator antennas for low-cost analog adaptive beamforming. In Proceedings of the 2000 IEEE International Conference on Phased Array Systems and Technology, Dana Point, CA, USA, 21–25 May 2000; pp. 101–104.
9. Choi, S.; Shim, D. A novel adaptive beamforming algorithm for a smart antenna system in a CDMA mobile communication environment. *IEEE Trans. Veh. Technol.* **2002**, *49*, 1793–1806. [CrossRef]
10. Schippers, H.; Verpoorte, J.; Jorna, P.; Hulzinga, A.; Meijerink, A.; Roeloffzen, C.; Heideman, R.G.; Leinse, A.; Wintels, M. Conformal phased array with beamforming on airborne satellite communication. In Proceedings of the International ITG Workshop on Smart Antennas, Hamburg, Germany, 26–27 February 2008; pp. 343–350.
11. Schippers, H.; Spalluto, G.; Vos, G. Radiation analysis of conformal phased array antennas on distorted structures. In Proceedings of the 12th International Conference on Antennas and Propagation (ICAP 2003), Exeter, UK, 31 March–3 April 2003; pp. 160–163.
12. Schippers, H.; Knott, P.; Deloues, T. Vibrating antennas and compensation techniques research in NATO/RTO/SET 087/RTG 50. In Proceedings of the IEEE Aerospace Conference, New York, NY, USA, 3–10 March 2007; pp. 1–13.
13. Knott, P. Deformation and vibration of conformal antenna arrays and compensation techniques. In Proceedings of the Meeting Proceedings RTO-MP-AVT-141, Neuilly-sur-Seine, France, October 2006; Multifunctional Structures/Integration of Sensors and Antennas. pp. 19-1–19-12.
14. Wincza, K.; Gruszczynski, S. Influence of Curvature Radius on Radiation Patterns in Multibeam Conformal Antennas. In Proceedings of the 36th European Microwave Conference, Manchester, UK, 10–15 September 2006; pp. 1410–1413.
15. Haupt, R.L. *Antenna Arrays: A Computational Approach*; John Wiley and Sons, Ltd.: Hoboken, NJ, USA, 2010; pp. 287–315.
16. Braaten, B.D.; Roy, S.; Nariyal, S.; Al Aziz, M.; Chamberlain, N.F.; Ullah, I.; Reich, M.T.; Anagnostou, D. A Self-Adapting Flexible (SELFLEX) Antenna Array for Changing Conformal Surface Applications. *IEEE Trans. Antennas Propag.* **2013**, *61*, 655–665. [CrossRef]

17. O'Donovan, P.L.; Rudge, A.W. Adaptive control of a flexible linear array. *Electron. Lett.* **1973**, *9*, 121–122. [CrossRef]
18. Lier, E.; Purdy, D.; Kautz, G. Calibration and Integrated Beam Control/Conditioning System for Phased-Array Antennas. US Patent 6163296, 19 December 2000.
19. Lier, E.; Purdy, D.; Ashe, J.; Kautz, G. An On-Board Integrated Beam Conditioning System for Active Phased Array Satellite Antennas. In Proceedings of the 2000 IEEE International Conference on Phased Array Systems and Technology, Dana Point, CA, USA, 21–25 May 2000; pp. 509–512.
20. Lier, E.; Zemlyansky, M.; Purdy, D.; Farina, D. Phased array calibration and characterization based on orthogonal coding: Theory and experimental validation. In Proceedings of the 2010 IEEE International Symposium on Phased Array Systems and Technology (ARRAY), Greater Boston, MA, USA, 12–15 October 2010; pp. 271–278.
21. Guo, J.-L.; Li, J.-Y. Pattern synthesis of conformal array antenna in the presence of platform using differential evolution algorithm. *IEEE Trans. Antennas Propag.* **2009**, *57*, 2615–2621.
22. Wu, Y.; Warnick, K.F.; Jin, C. Design study of an L-band phaed array feed for wide-field surveys and vibration compensation on FAST. *IEEE Trans. Antennas Propag.* **2013**, *61*, 3026–3033. [CrossRef]
23. Purdy, D.; Ashe, J.; Lier, E. System and Method for Efficiently Characterizing the Elements in an Array Antenna. U.S. Patent 2002/0171583 A1, 21 November 2002.
24. Yang, P.; Yang, F.; Nie, Z.-P.; Li, B.; Tang, X. Robust adaptive beamformer using interpolation technique for conformal antenna array. *Prog. Electromagn. Res. B* **2010**, *23*, 215–228. [CrossRef]
25. Seidel, T.J.; Rowe, W.S.T.; Ghorbani, K. Passive compensation of beam shift in a bending array. *Prog. Electrom. Res.* **2012**, *29*, 41–53. [CrossRef]
26. Salonen, P.; Rahmat-Samii, Y.; Schaffrath, M.; Kivikoski, M. Effect of textile materials on wearable antenna performance: A case study of GPS antennas. in Proc. *IEEE Antennas Propag. Soc. Int. Symp.* **2004**, *1*, 459–462.
27. Irfan, U.; Braaten, B.D. Nulls of a conformal beamforming array on an arbitrary wedge-shaped surface. In Proceedings of the 2014 IEEE Antennas and Propagation Society International Symposium (APSURSI). Memphis, TN, USA, 6–11 July 2014; pp. 1788–1789.
28. Irfan, U.; Braaten, B.D. On the effects of a changing cylindrical surface on the nulls of a conformal beamforming array. In Proceedings of the 2014 IEEE Antennas and Propagation Society International Symposium (APSURSI). Memphis, TN, USA, 6–11 July 2014; pp. 1790–1791.
29. Ullah, I.; Munsif, H.; Razzaq, S.; Najam, A.I. Cylindrical phased array with adaptive nulling using eigen-correlation technique. *Int. J. RF Microw. Comput.-Aided Eng.* **2021**, *32*, e22969. [CrossRef]
30. Munsif, H.; Najam, A.I.; Irfanullah. Eigenbeam analysis of singly curved conformal antenna array. *Prog. Electromagn. Res. M* **2021**, *103*, 27–36.
31. Chiba, I.; Hariu, K.; Sato, S.; Mano, S. A projection method providing low sidelobe pattern in conformal array antennas. In Proceedings of the Digest on Antennas and Propagation Society International Symposium, San Jose, CA, USA, 26–30 June 1989.
32. Gupta, I.J.; Ksienski, A.A. Effect of mutual coupling on the performance of adaptive arrays. *IEEE Trans. Antennas Propag.* **1983**, *31*, 785–791. [CrossRef]
33. Ansys Inc. Ansoft HFSS, Version 13.0.1. Available online: http://www.ansoft.com (accessed on 24 August 2022).
34. Hittite Microwave Corporation. Available online: http://www.hittite.com (accessed on 24 August 2022).
35. Mini Circuits. Available online: http://www.minicircuits.com (accessed on 24 August 2022).

electronics

MDPI

Article

A Simple and Effective Approach for Scattering Suppression in Multiband Base Station Antennas

Madiha Farasat [1,*], Dushmantha Thalakotuna [1], Zhonghao Hu [2] and Yang Yang [1]

1 School of Electrical and Data Engineering, University of Technology, Sydney 2000, Australia
2 Wireless Business Unit, Rosenberg Technology Australia, Northmead 2152, Australia
* Correspondence: madiha.farasat@student.uts.edu.au; Tel.: +61-4-1306-6996

Abstract: The high band pattern distortions in an 1810–2690 MHz frequency band, introduced due to low band radiators working in 690–960 MHz, are mitigated by a simple yet effective change to the low band-radiating elements. A novel horizontal and vertical radiating element is designed instead of a conventional slant polarized low band-radiating element to reduce the scattering. The slant polarization is achieved from the horizontal and vertical dipoles, using a 180° hybrid coupler. The vertical dipole length is optimized to improve the high band patterns. The experimental results verified that the proposed horizontal and vertical low band dipole result in the reduction of high band pattern distortions. The low band-radiating elements provide >12 dB return loss over the entire frequency band 690–960 MHz and provide comparable pattern performance to a conventional slant low band dipole.

Keywords: base station antenna; hybrid-matching circuits; pattern distortion; crossband scattering suppression; wideband matching

Citation: Farasat, M.; Thalakotuna, D.; Hu, Z.; Yang, Y. A Simple and Effective Approach for Scattering Suppression in Multiband Base Station Antennas. *Electronics* **2022**, *11*, 3423. https://doi.org/10.3390/electronics11213423

Academic Editors: Naser Ojaroudi Parchin, Raed A. Abd-Alhameed and Chan Hwang See

Received: 29 September 2022
Accepted: 17 October 2022
Published: 22 October 2022

1. Introduction

The growth in connected devices requires high connection density in 5G networks. In fact, requirements specify more than 1 million devices per square kilometer for 5G massive machine-type communication (mMTC), a 10-fold increase compared to 4G [1]. This is catered through small cell densification and improved spectrum utilization, which both require compact multiband antenna systems. The compactness and need for multiband antenna systems present conflicting demands, since the need for multiple bands requires multiple arrays and thus more space [2,3]. The only way to realize a compact antenna is to interleave these multiband arrays. The main challenge is to then ensure the radiation patterns are not distorted due to scattering from other radiating elements in the vicinity. In such multiband interleaved array, the high-frequency element patterns are impacted due to the scattering from the currents induced in low-frequency radiating elements and vice versa. The impact can be on multiple parameters such as beamwidth, squint, and cross-polarization discrimination (XPD). Often, it is attempted to ensure that each radiating element is transparent to each other in their operating frequency bands, which is a challenging task.

The common approach to reducing the crossband scattering is to adopt different sizes and shapes of metal cavities or walls [4–7]. Cavity-backed antenna is used in [8] to improve the isolation of antenna elements in an array. In [9], the high-frequency band (HB) pattern distortions caused by the lower frequency band (LB) radiating elements are minimized by introducing chokes into the LB element. These chokes are quarter-wavelength open-circuit segments at high band frequencies, which minimizes scattering. In [10], the printed dipole is segmented into smaller segments that are not resonant in the higher-frequency band region. Each segment is connected to the other by inductive thin lines. This makes the lower band elements transparent to the high band-radiating elements. In [11], a cloaked antenna system is realized to minimize the scattering of closely located antennas. Filtering techniques with different configurations are also employed to obtain better isolation between radiating elements. As in [12], the filtering antenna as array elements has been proposed. The radiating element realized the filtering response by adding the shorting pins and E-slot to reduce the crossband scattering. A coupled resonator-based decoupling network is deployed in [13] to improve crossband scattering. The currents induced by the coupled resonator help to suppress the strong coupling between the antennas operated in two frequency bands. The addition of a frequency-selective surface between the high band and low band elements has been deployed in [14] to reduce the crossband scattering. Although these techniques suppress the crossband scattering, they do require significant modifications to LB or HB elements which can increase the cost of manufacturing due to added complexity of products.

The novelty in this work is in the simple and effective approach proposed to reduce the crossband scattering at HB by orienting the LB radiating element in horizontal (H) and vertical (V) directions, as shown in Figure 1b, in contrast to the conventional slant LB dipole configuration shown in Figure 1a. The main challenge then is to generate the slant polarization (±45°) from an H and V oriented dipoles. This has been overcome by utilizing a 180° hybrid coupler. Below is a list of abbreviations in Table 1 to provide a better understanding of acronyms used in the proposed technique.

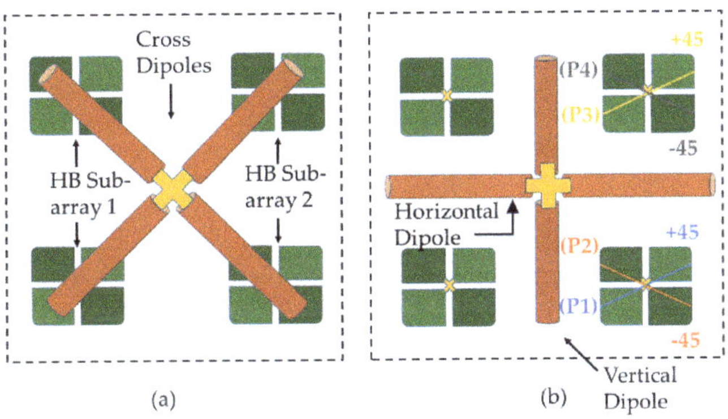

Figure 1. (**a**) Slant dipole configuration used in traditional interleaved scheme for dual-band dual-polarized BSA. (**b**) The proposed horizontal and vertical LB dipole configuration (LBHV) with HB subarrays. P1 and P3 refer to +45 polarizations; P2 and P4 refer to −45 polarization.

Table 1. List of Acronyms.

No.	Acronym	Description
1	HB	High band
2	LB	Low band
3	BSA	Base station antenna
4	LBHV	Low band horizontal–vertical
5	HPBW	Half power beamwidth
6	HD	Horizontal dipole
7	VD	Vertical dipole
8	TL	Transmission line

2. Low Band Horizontal and Vertical Configuration

The HB and LB elements are interleaved in the current base station antenna (BSA) for space savings. Due to the use of ±45 polarized elements, the dipoles are oriented in a slant configuration, as shown in Figure 1a. It should be noted that a full array has multiple such elements arranged vertically, and that only a segment is depicted in this figure for clarity. The presence of the LB radiator on top of the HB radiators can induce HB currents on the LB dipoles, acting as resonating structures and distorting the patterns. In addition, being directly in front of the HB radiators disrupts the nearfield distribution and scatters the desired HB patterns. This results in undesirable HB patterns, as shown in Figure 2.

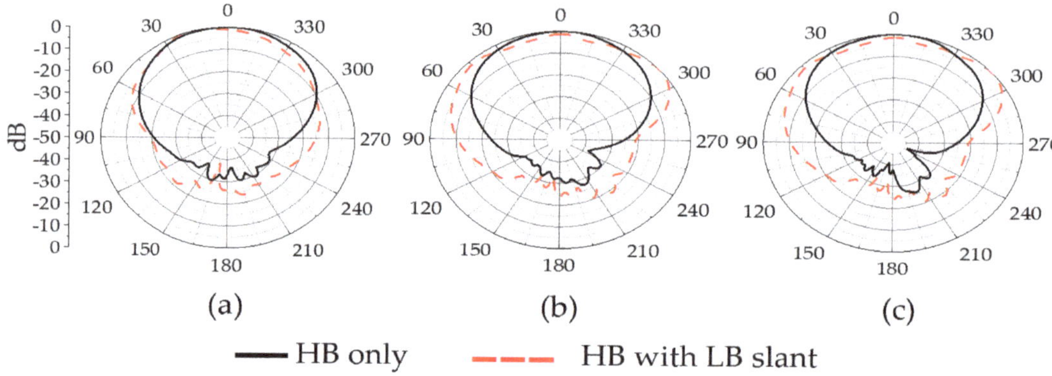

Figure 2. High band-only and high band with slant LB antenna-measured azimuth patterns at (**a**) 1.8 GHz, (**b**) 2.4 GHz and (**c**) 2.6 GHz.

The aim of this work is to mitigate such HB pattern effects due to LB antenna element. The location of the dipoles of the LB antennas on top of the HB antenna elements was investigated. It was found that pattern distortions were significant as long as the LB dipole was directly above the HB antenna element, irrespective of the type of the HB radiator. This study uses a square-shaped cross-dipole antenna as the HB element [4]. The HB array itself, without any LB radiators, has cleaner patterns with half-power beamwidth (HPBW) around $65 \pm 5^{\circ}$ and almost no squint, as shown in Figure 2.

A careful observation of the current distribution at HB frequencies revealed that the induced HB currents reduce in magnitude as the LB dipoles are moved away from a slant position to horizontal and vertical positions. As a result, HB patterns are cleaner, as shown in Figure 3. This LB configuration is referred to as Low Band Horizontal–Vertical (LBHV) from here onwards. The measured results show that the pattern performance of HB arrays with LBHV elements is similar to HB-only radiation patterns.

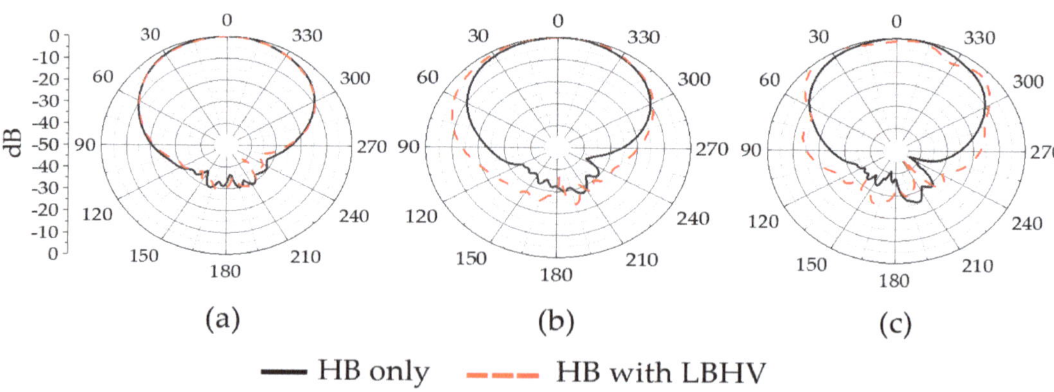

Figure 3. High-band measured azimuth radiation patterns with a LBHV element at (**a**) 1.8 GHz, (**b**) 2.4 GHz, and (**c**) 2.6 GHz.

It is observed that the main beam shows a significant 3 dB and 10 dB beam squint with the slant dipoles compared to the LBHV, as shown in Figure 4a–f. The proposed work is also compared with the recent state-of-the-art works in Table 2, which shows isolation performances and the addition of structure type to radiating elements.

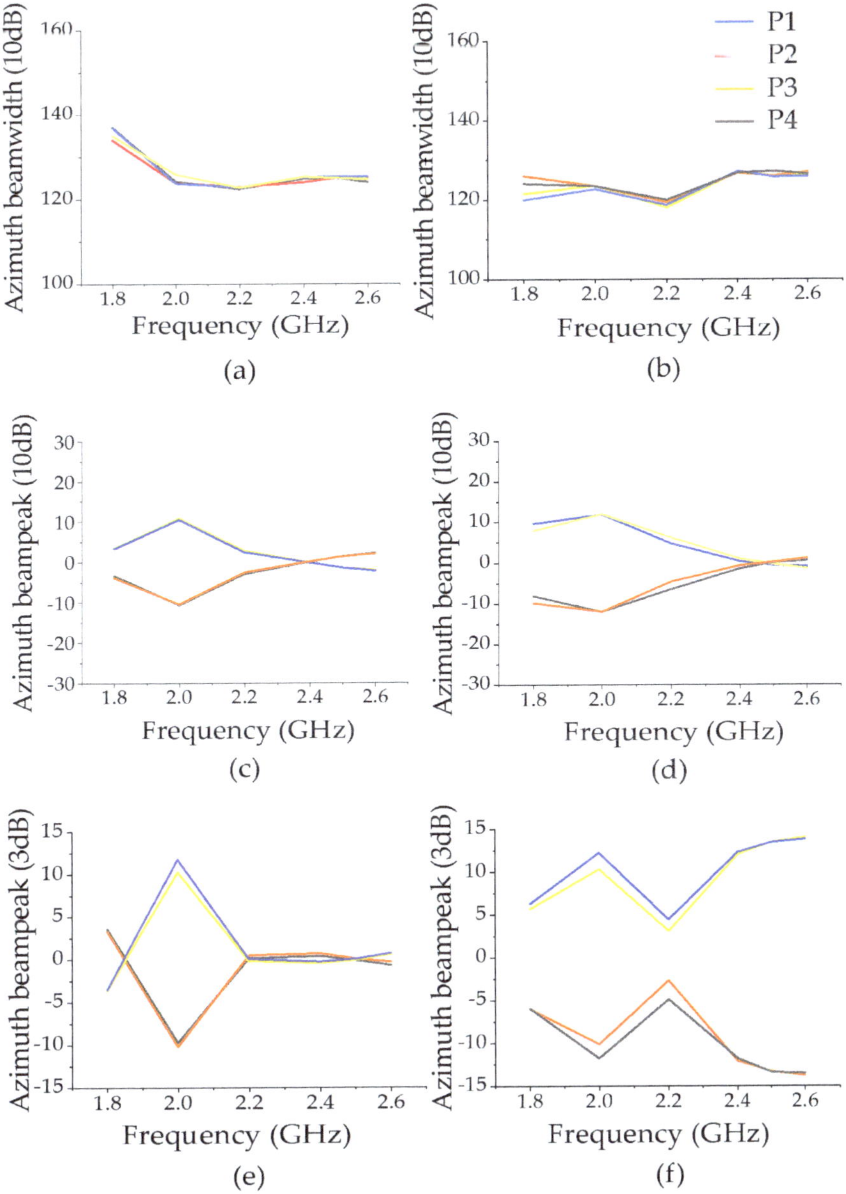

Figure 4. Measured 10 dB azimuth beamwidth of HB arrays with (**a**) LBHV element (**b**) LB slant element, 10 dB azimuth beam peak of HB arrays with (**c**) LBHV element (**d**) LB slant element, 3 dB azimuth beam peak of HB arrays with (**e**) LBHV element (**f**) LB slant element. The P1, P2, P3, and P4 refer to polarizations indicated in Figure 1.

Table 2. Comparison of recent state-of-the-art works with proposed work.

References	Modification Type to Improve Isolation	Frequency Band (GHz)	Isolation (dB)	HPBW (Measured)
[4]	Metal baffles	0.77–0.98 1.65–2.9	>23 17.5	64.5 ± 57.1° 84.4 ± 74.1°
[5]	Arc-shaped baffle plates	0.74–0.96 1.7–2.6	>27.5	61.5° 90°
[9]	Chokes in LB element	0.82–1 1.71–2.28	NG	69.5 ± 4° 65 ± 5°
[12]	Filtering antenna elements	1.71–1.88 1.9–2.17	>30	65 ± 5°
Proposed Antenna	No added structure	0.69–0.96 1.8–2.6	>36 >20	60 ± 5° 65 ± 11°

3. Virtual Slant Polarizations

The base station antennas require ±45 polarized antenna elements, and the proposed horizontal and vertical dipoles provide only horizontal and vertical polarization. Therefore, a novel technique creates a ±45 polarization virtually from H- and V-oriented dipoles. The 180° hybrid coupler is a common microwave component which can be realized either as a rat-race, ring or magic T-hybrid. The 180° hybrid coupler is commonly used to get in and out of phase signals. Different design approaches have been followed in the literature, just as in [15] where a specialized TL is designed for a compact rat race coupler.

As shown in Figure 5a, a 180° hybrid coupler is used between the inputs and the LBHV element. Port 2 feeds the horizontal polarized element, and port 3 of the hybrid provides the vertically polarized dipole.

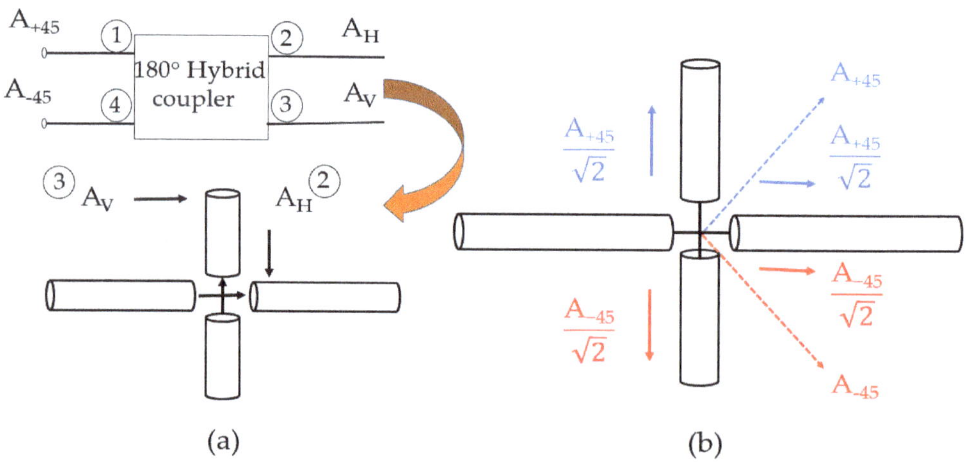

Figure 5. (**a**) Conversion of slant-polarized inputs to H and V polarized inputs, (**b**) creation of virtual slant polarizations from H and V feed signals.

The A_H and A_V can be expressed in terms of the +45 and −45 input signals A_{+45} and A_{-45} as:

$$\begin{bmatrix} A_H \\ A_V \end{bmatrix} = [S_{180^\circ}] \begin{bmatrix} A_{+45} \angle & \theta_{+45} \\ A_{-45} \angle & \theta_{-45} \end{bmatrix}$$

$$A_H = \frac{1}{\sqrt{2}} [A_{+45} \angle \theta_{+45} + A_{-45} \angle \theta_{-45}]$$

$$A_v = \frac{1}{\sqrt{2}}[A_{+45} \angle \theta_{+45} + A_{-45} \angle \theta_{-45} - 180°]$$

The output at port 2, fed to the horizontal-polarized dipole, contains the in-phase components of both the +45 and −45 polarizations, while the output port 3 contains signals of +45 in-phase and a −45 signal with a 180° phase shift. Since the +45 signal is fed in phase to both the H and V polarizations, it virtually creates a sum vector in the +45 direction, as shown in Figure 5b. The −45 component fed to the vertically polarized element has a 180° phase difference compared to the horizontally polarized dipole. Therefore, it creates a signal polarized in the −45 direction, as shown in Figure 5b. By the theory of reciprocity, the received signals at H and V dipoles provide the sum and difference signals, which are +45 and −45 signals, back to the baseband unit.

4. Low Band Element

The LBHV element should have a comparable performance in terms of patterns to a slant dipole, while providing minimum distortions to the HB patterns. The length of the dipoles in the H and V polarizations are optimized to meet these requirements. The optimized values were obtained through a series of parametric studies. It was found that the vertical dipole length presents more resonant behavior at the high band and needs to be shorter to reduce HB pattern distortions. This can result in a wider 3 dB beamwidth for LB patterns. A longer dipole close to a full wavelength is required in the H orientation to compensate for this.

The LB dipoles need to be impedance-matched for the 690–960 MHz band. The impedance-matching network is based on a feed network design proposed in [16]. A ladder-type band-pass filter using a serial and shunt resonant LC circuit is used for wideband matching. As shown in Figure 6, a circuit model contains three tuners, i.e., a series resonator and a shunt resonator used to control/reduce the changing reactance, followed by a quasi-quarter wavelength transmission line (TL2). The optimization and physical realization were conducted in a CST studio to obtain the best match over the required band.

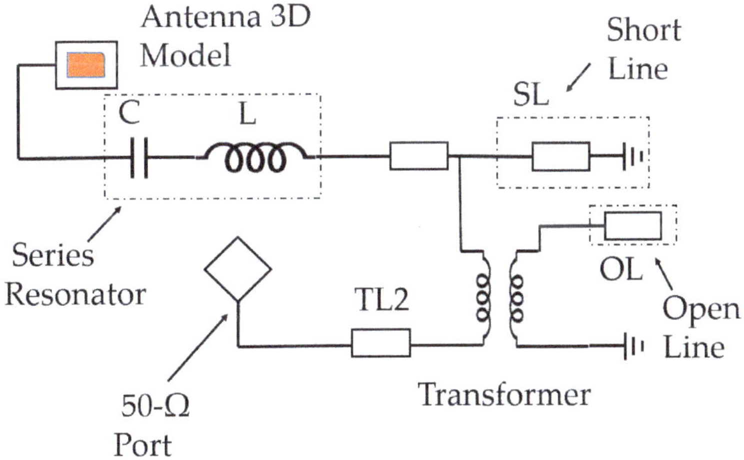

Figure 6. Circuit theory model of matching circuit.

The matching circuit was implemented using microstrip technology, as shown in Figure 7a. The lines TL1 (transmission line) and SL (short line) are realized as coupled lines that are printed on the back-conducting layer, while TL2 and OL (open line) are conventional microstrip layers.

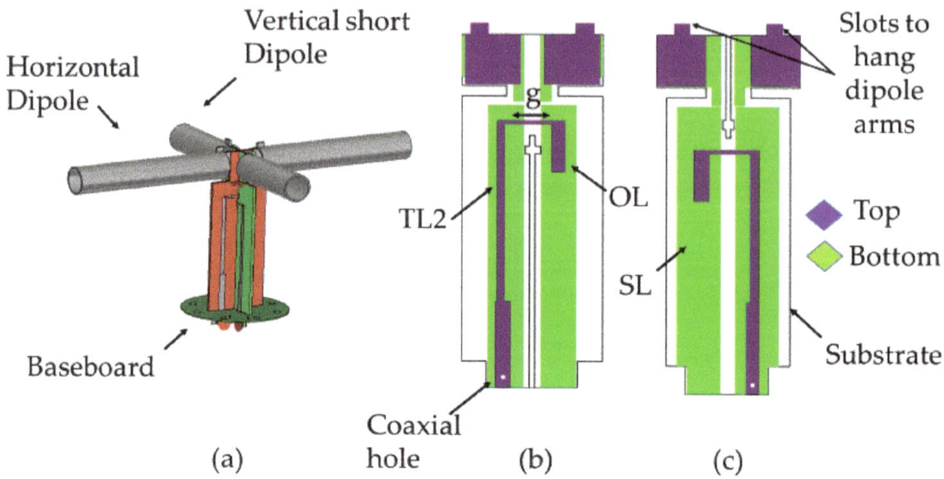

Figure 7. (**a**) Perspective view of the LBHV. The horizontal dipole is 135 mm long, and the vertical dipole is 60 mm long, (**b**) details of the LB horizontal dipole (LBHD) feed, (**c**) details of the LB vertical dipole (LBVD) feed.

The configuration of horizontal dipole balun is shown in Figure 7b, while short dipole balun is depicted in Figure 7c. They are orthogonally arranged to feed the pair of horizontal and vertical LB arms. Although matching the short dipole is quite challenging, the antenna is well-matched with a reflection coefficient of better than 12 dB for horizontal polarization, and better than 10 dB for vertical polarization from 690 MHz to 960 MHz, as shown in Figure 8a. To have an insight into the working mechanism of the proposed antenna, VSWRs are studied and shown in Figure 8b. The optimized dimensions of the structure are those listed in Table 3. A fabricated prototype of the antenna is shown in Figure 9.

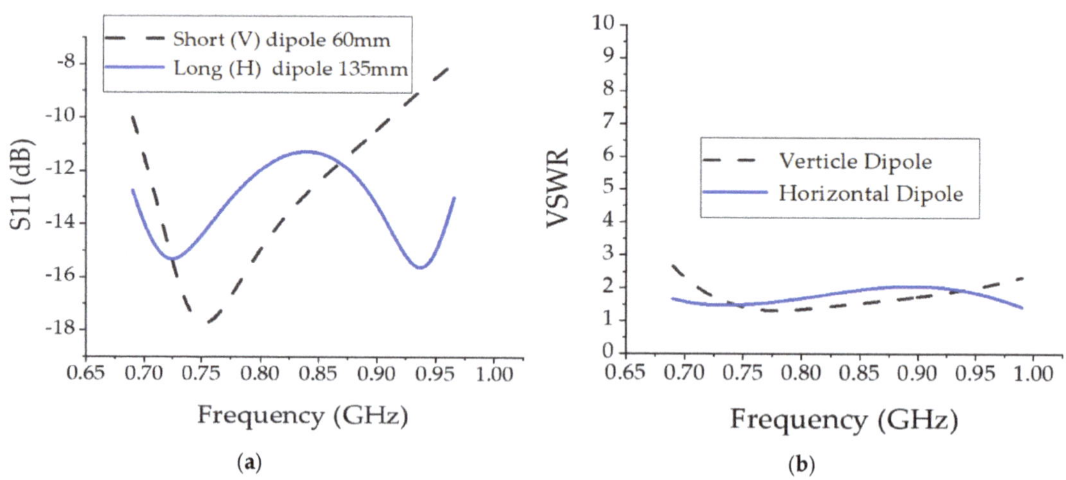

Figure 8. (**a**) The measured return loss of the short V-dipole and long H-dipole. (**b**) VSWR of proposed antenna LBHV.

Table 3. Optimized parameters of the proposed antenna.

Parameters.	Values LB-HD (mm)	Values LB-VD (mm)	Description
W-SL	6.5	10	Width of SL
L-SL	95	95	Length of SL
W-TL1	8.5	8.5	Width of TL1
L-TL1	17	17	Length of TL1
W-TL2	2	3	Width of TL2
L-TL2	60	50	Length of TL2
W-OL	4	5	Width of OL
L-OL	18	17	Length of OL
g	19	13	Gap between SL

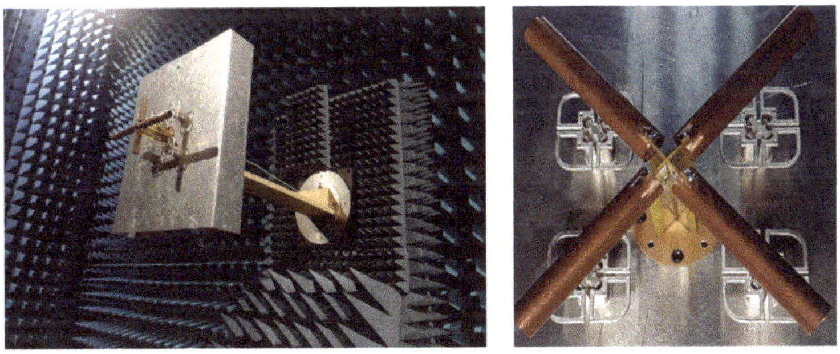

(a) (b)

Figure 9. Fabricated prototype with (**a**) LBHV antenna element (**b**) LB slant element.

Despite this, the achieved HPBW with LBHV configuration is relatively stable, with a reflection coefficient larger than 12 dB. The port-to-port-measured worst-case isolation of LBHV configuration in hybrid mode is 36 dB at 960 MHz and 20 dB at 2 GHz, as shown in Figure 10, in the desired whole-band 690–960 MHz and 1810–2690 MHz, respectively.

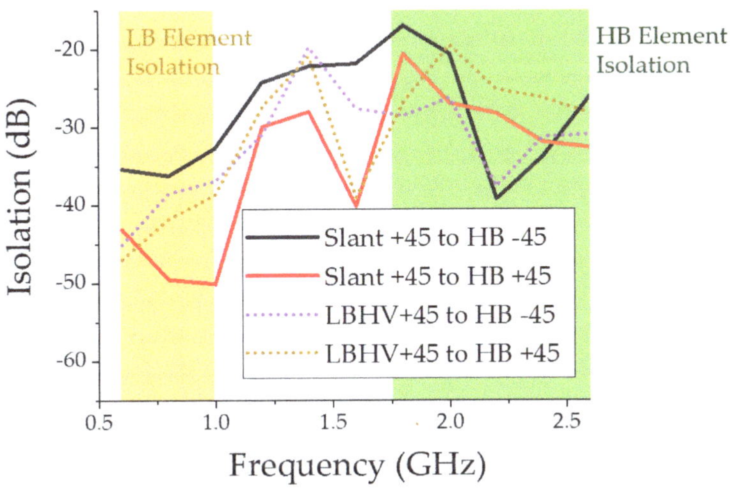

Figure 10. Measured isolation of proposed antenna LBHV with slant.

5. Conclusions

The high band pattern distortions can be reduced by using a horizontal and vertical low band element instead of a conventional slant low band element. The scattering from this low band element is minimized since dipoles are moved away from the high band-radiating elements. This method is simpler compared to other techniques available in the literature that require significant modifications to radiating element design, resulting in an increased cost. The construction of the horizontal low band element is almost similar to a conventional low band element design, and therefore, construction and assembly are simple. The only addition is a 180° hybrid coupler, which is used to generate the slant polarizations. The new low band element is matched over the entire low band frequency range with >12 dB return loss and the low band patterns are identical to conventional slant low band element pattern with 6.2 dB gain, $60 \pm 5°$ beam width for a single element.

Author Contributions: All authors conceived and proposed the idea; D.T. contribute towards structure of the paper. Wrote the paper under the supervision of D.T. All authors have read and agreed to the published version of the manuscript.

Funding: This research received no external funding.

Conflicts of Interest: The authors declare no conflict of interest.

References

1. Osseiran, A.; Parkvall, S.; Persson, P.; Zaidi, A.; Magnusson, S.; Balachandran, K. *5G Wireless Access: An Overview*; 1/28423-FGB1010937; Ericsson: Stockholm, Sweden, 2020.
2. Kelly, I.; Zimmerman, M.; Butler, R.; Zheng, Y. Base Station Antenna Selection for LTE Networks. 2017. Available online: https://www.commscope.com/globalassets/digizuite/2096-bsa-selection-for-lte-networks-wp-108976.pdf?r=1 (accessed on 1 September 2022).
3. Mandhyan, A. 4G and 5G Capacity Solutions-Comparative Study. 2019. Available online: https://telecoms.com/intelligence/4g-and-5g-capacity-solutions-comparative-study/ (accessed on 1 September 2022).
4. Huang, H.; Liu, Y.; Gong, S. A Novel Dual-Broadband and Dual-Polarized Antenna for 2G/3G/LTE Base Stations. *IEEE Trans. Antennas Propag.* **2016**, *64*, 4113–4118. [CrossRef]
5. Huang, H.; Liu, Y.; Gong, S. A Dual-Broadband, Dual-Polarized Base Station Antenna for 2G/3G/4G Applications. *IEEE Antennas Wirel. Propag. Lett.* **2017**, *16*, 1111–1114. [CrossRef]
6. He, Y.; Pan, Z.; Cheng, X.; He, Y.; Qiao, J.; Tentzeris, M.M. A Novel Dual-Band, Dual-Polarized, Miniaturized and Low-Profile Base Station Antenna. *IEEE Trans. Antennas Propag.* **2015**, *63*, 5399–5408. [CrossRef]
7. He, Y.; Tian, W.; Zhang, L. A novel dual-broadband dual-polarized electrical downtilt base station antenna for 2G/3G applications. *IEEE Access* **2017**, *5*, 15241–15249. [CrossRef]
8. Elsherbini, A.; Wu, J.; Sarabandi, K. Dual Polarized Wideband Directional Coupled Sectorial Loop Antennas for Radar and Mobile Base-Station Applications. *IEEE Trans. Antennas Propag.* **2015**, *63*, 1505–1513. [CrossRef]
9. Sun, H.H.; Ding, C.; Zhu, H.; Jones, B.; Guo, Y.J. Suppression of Cross-Band Scattering in Multiband Antenna Arrays. *IEEE Trans. Antennas Propag.* **2019**, *67*, 2379–2389. [CrossRef]
10. Isik, O.; Gripo, P.; Thalakotuna, D.; Liversidge, P. Cloaked Low Band Elements for Multiband Radiating Arrays. US Patent 10 439 285 B2, 16 October 2019.
11. Soric, J.C.; Monti, A.; Toscano, A.; Bilotti, F.; Alu, A. Dual-Polarized Reduction of Dipole Antenna Blockage Using Mantle Cloaks. *IEEE Trans. Antennas Propag.* **2015**, *63*, 4827–4834. [CrossRef]
12. Zhang, Y.; Zhang, X.Y.; Ye, L.; Pan, Y. Dual-Band Base Station Array Using Filtering Antenna Elements for Mutual Coupling Suppression. *IEEE Trans. Antennas Propag.* **2016**, *64*, 3423–3430. [CrossRef]
13. Zhao, L.; Wu, K. A Dual-Band Coupled Resonator Decoupling Network for Two Coupled Antennas. *IEEE Trans. Antennas Propag.* **2015**, *63*, 2843–2850. [CrossRef]
14. Zhu, Y.; Chen, Y.; Yang, S. Decoupling and Low-Profile Design of Dual-Band Dual-Polarized Base Station Antennas Using Frequency-Selective Surface. *IEEE Trans. Antennas Propag.* **2019**, *67*, 5272–5281. [CrossRef]
15. Lalbakhsh, A.; Mohamadpour, G.; Roshani, S.; Ami, M.; Roshani, S.; Sayem, A.S.M.; Alibakhshikenari, M.; Koziel, S. Design of a Compact Planar Transmission Line for Miniaturized Rat-Race Coupler With Harmonics Suppression. *IEEE Access* **2021**, *9*, 129207–129217. [CrossRef]
16. Ding, C.; Sun, H.; Zhu, H.; Guo, Y. Achieving Wider Bandwidth with Full-Wavelength Dipoles for 5G Base Stations. *IEEE Trans. Antenna Propag.* **2020**, *68*, 1119–1127. [CrossRef]

sensors

MDPI

Article

Radio Astronomical Antennas in the Central African Region to Improve the Sampling Function of the VLBI Network in the SKA Era?

Marcellin Atemkeng [1,*], Patrice Okouma [1,2], Eric Maina [3], Roger Ianjamasimanana [4] and Serges Zambou [5]

1 Department of Mathematics, Rhodes University, Artillery Rd, Grahamstown 6139, South Africa
2 Gabon Space Agency, Libreville P.O. Box 3850, Gabon
3 Department of Physics and Electronics, Rhodes University, Grahamstown 6139, South Africa
4 Instituto de Astrofísica de Andalucía (CSIC), Glorieta de la Astronomía, E-18008 Granada, Spain
5 Semiconductor Development Division, Beneq Oy, Olarinluoma 9, 02200 Espoo, Finland
* Correspondence: m.atemkeng@ru.ac.za

Abstract: On the African continent, South Africa has world-class astronomical facilities for advanced radio astronomy research. With the advent of the Square Kilometre Array project in South Africa (SA SKA), six countries in Africa (SA SKA partner countries) have joined South Africa to contribute towards the African Very Long Baseline Interferometry (VLBI) Network (AVN). Each of the AVN countries aims to construct a single-dish radio telescope that will be part of the AVN, the European VLBI Network, and the global VLBI network. The SKA and the AVN will enable very high sensitivity VLBI in the southern hemisphere. In the current AVN, there is a gap in the coverage in the central African region. This work analyses the increased scientific impact of having additional antennas in each of the six countries in central Africa, i.e., Cameroon, Gabon, Congo, Equatorial Guinea, Chad, and the Central African Republic. A number of economic human capital impacts of having a radio interferometer in central Africa are also discussed. This work also discusses the recent progress on the AVN project and shares a few lessons from some past successes in ground stations retrofitting.

Keywords: radio astronomy; square kilometre array; very long baseline interferometry; radio interferometer

Citation: Atemkeng, M.; Okouma, P.; Maina, E.; Ianjamasimanana, R.; Zambou, S. Radio Astronomical Antennas in the Central African Region to Improve the Sampling Function of the VLBI Network in the SKA Era? *Sensors* **2022**, *22*, 8466. https://doi.org/10.3390/s22218466

Academic Editors: Raed A. Abd-Alhameed, Chan Hwang See and Naser Ojaroudi Parchin

Received: 20 September 2022
Accepted: 30 October 2022
Published: 3 November 2022

1. Introduction

The spatial resolution of a particular telescope determines how well one can see the details of cosmic objects. This resolution depends on the size of the telescope and the wavelength of the astronomical sources. To have a better resolution, one solution is to build a telescope with a large diameter. However, there is a practical limitation on the size of a telescope, and this led to the development of interferometry. That is, instead of having one large telescope, one can cross-correlate signals from individual antennas, and the resolution of the combined array of the antennas (the so-called interferometer) is determined by the largest separation (baseline) between the individual antennas. The combined array is therefore equivalent to a large single-dish telescope with a diameter equal to the longest baseline. Many radio telescopes were built after the development of the interferometry techniques. However, by the mid-1960s [1], it was realised that some radio sources could not be resolved even with radio telescopes of a few hundred km baselines. The quest for higher resolution led to the development of the Very Long Baseline Interferometry (VLBI) [2]. The VLBI is a technique of cross-correlating signals recorded by different antennas (and/or an array of antennas) separated by a large distance of up to the diameter of the earth. With this technique, detailed images of astronomical objects at milliarcsecond resolution have been obtained. In addition, high-precision astrometry has also been achieved.

South Africa and Australia are currently leading one of the most powerful and modern radio telescopes, the Square Kilometre Array (SKA) [3]. The SKA will be split in a mid-frequency (350 MHz–14 GHz) part build in South Africa, which will incorporate the MeerKAT telescope [4], and a low-frequency (50–350 MHz) part in Australia. To enable high-resolution interferometry through the VLBI, the SKA South Africa currently leads an effort to convert existing unused telecommunication dishes in partner countries (Botswana, Ghana, Kenya, Madagascar, Mauritius, Mozambique, Namibia, and Zambia) to radio telescopes. The converted antennas will then become part of a network of antennas distributed throughout Africa to form the African VLBI Network (AVN) [5]. Ghana has already successfully converted its old telecommunication dish to a working radio telescope [6]. Efforts to do the same in other African partner countries are underway. The AVN will significantly improve the science capabilities of the global VLBI community [7]. The AVN combined with the existing international VLBI facilities will produce huge quantities of data, presenting new challenges in data processing and storage. New techniques to manage the data must be developed, including: storage systems and data compression techniques; machine learning methods; software design techniques, control, and monitoring systems that parallel the internet of things; data flow architecture; and systems dealing with massive-scale computing. All of these challenges will strengthen the scientific collaborations between South Africa and its partner countries. In addition, Africa will become an international science and technology focus.

The central African states are currently not part of the AVN. This paper investigates the technical impact of this in terms of the AVN image quality and science capabilities. We will demonstrate by means of simulations how the AVN image signal-to-noise ratio (S/N) improves if antennas were added in these countries. This paper also highlights the economical and technological benefits for these countries should they join the AVN project. The main objective of this work is to stimulate a discussion within the global VLBI community on the need for additional antennas in the central African region.

2. Motivations

The Central African States (ECCAS—Economic and Community of Central African States) is a group of countries located in the centre of the African continent. These countries are located along the equator, have an ideal position, with a tropical climate, constant duration between days and nights, and no harsh climate, hence being ideal for observation. The ECCAS countries are mostly French speaking, with little use of English, and their integration to world scientific projects has been significantly slower compared to that of other African countries. Except for Cameroon, the countries in that space rank at the bottom of all continental and worldwide scientific metrics. In Africa, most scientific projects are split/shared between the North and the South. Northern African states are very close to Europe, hence they enjoy the benefit of that proximity in terms of scientific collaboration and industrial outsourcing. Southern African states are mostly English speaking and slightly wealthier than other parts of the continent, hence they have been very successful in attracting world-class scientific projects to their area. A similar trend is observed in the East and West African countries which have a large number of English-speaking countries. The ECCAS states are lagging, and the low scientific output, in terms of projects and journal publications in the region, correlates these facts [8]. To date, the ECCAS have not joined the AVN, even though significant advantage could open up for the region by becoming part of the AVN. Moreover, due to their geographical locations, the existing AVN community will significantly benefit from the participation of that region. Below are some non-exhaustive benefits of joining the AVN.

2.1. Education and Research Impacts

Joining the AVN will boost international cooperation in the field of astronomy and engineering and enable participation in international scientific research. Running a radio telescope requires skilled engineers, scientists, and technicians who will manage and run the facility. These personnel need to be trained in various disciplines, from radio astronomy and astrophysics to computer science and engineering. The decision to join the AVN network will trigger the development of critical skills and the institutional capacity necessary to optimise the ECCAS participation in the SKA. The high-end technologies and high-performance computing facilities needed to operate and maintain the telescopes are being developed in South Africa. The Centre for High-Performance Computing (CHPC) is already in place and running.

Ghana has already successfully converted its old telecommunication dish into a functioning radio telescope [6]. This is a demonstration that Africa can participate in high-level scientific research. The skills and experience from the development of these facilities will be transferred to the ECCAS if they join the AVN project. In addition, the project will bring new scientific opportunities to the ECCAS countries in a relatively short timescale. Currently, students in the AVN partner countries are benefiting from a number of scholarships to pursue further studies, train, and acquire more skills. Some of those scholarships include but are not limited to: the South African Radio Astronomy observatory (SARAO) scholarship; the Development for Africa with Radio Astronomy (DARA) project in the United Kingdom; and a number of South African Research Chairs Initiative (SARChI) of the Department of Science and Technology and the National Research Foundation. Another benefit in joining the project includes the training of African scientists, engineers, and technicians, with a view to ensuring that partner countries benefit from the second phase of the SKA. Along this issue of human capacity consolidation, a number of surveys have already highlighted the relatively weak hands-on exposure to core engineering and artificial intelligence skills among graduates trained in mathematical sciences in Africa. Towards possibly mitigating this diffuse pathology in the African Higher Education ecosystem, it is worth noting that a growing number of open-source hardware (and software) platforms offer, today, compelling tools for early exposure to what we could naively term the applied component of mathematical sciences. Due to its cross-discipline nature, radio interferometry has long been recognised as a suitable operational framework for impacting both conceptual and practical aptitudes to graduates. This can be illustrated by a pathfinding teaching and learning approach jointly adopted by Otago University (New Zealand), Rhodes University, and Stellenbosch University (South Africa) in deploying three mature versions of the Transient Array Radio Telescope (TART). TART (https://en.wikipedia.org/wiki/Transient_Array_Radio_Telescope accessed on 15 June 2022, https://www.womeninscience.africa/newly-launched-radio-telescope-opens-up-stem-learning-opportunities/ accessed on 15 June 2022) is an inexpensive open-source 24-elements radio telescope specifically designed to be a platform for experimental imaging algorithms for undergraduate and graduate students as well as researchers. It has the unique feature to allow the filling of multi-scale gaps in a value chain ranging from a basic knowledge of a diode to the most advanced mathematics embedded in artificial intelligence and related fields [9]. TART holds the range of versatility and affordability suitable for various intensities and targets for training African graduates in aperture synthesis as a way to better integrate a number of them into the AVN and SKA era.

2.2. Economic Impacts

The AVN will trigger foreign investment and expenditure (including local contracts and suppliers). The skills development that will be promoted by the project should enhance ECCAS' engineering and scientific capabilities, promoting science and engineering breakthroughs for other sectors, such as medicine, remote sensing, and telecommunication, thereby enhancing innovation and competitiveness among industries. The construction or conversion of the telescopes will pave the way for a major boost to the local businesses, e.g., tourism industries, construction industries, and hospitality industries, thus creating new job opportunities and enhancing local revenue.

2.3. Scientific Impacts and Capabilities

Each AVN antenna will be designed to work as a single-dish radio telescope and as part of a larger VLBI network. The Ghana Radio Astronomy Observatory (GRAO) can serve as a reference case of what would be the science capabilities of the AVN antenna.

The GRAO currently operates at 6.7 and 5 GHz frequencies. Future bands potentially include the L-band and the K-band receivers. Thus, as a single dish, it can be used to detect methanol masers, water masers, and pulsars; perform spectral line imaging; and observe continuum sources from AGNs and star-forming regions. The first detection of the methanol maser, G9.621+0.196E, was successfully performed by the GRAO on 21 November 2016. In addition, the GRAO detected the Vela pulsar (PSR J0835-451) at 5 GHz on 30 April 2017 [10]. In addition, the GRAO is currently being used to promote astronomy outreach [11] and was also used to detect the planet Mercury as it transited in November 2019 [12]. As part of a VLBI network, the AVN has a range of applications. This includes astrometry, geodesy, spacecraft tracking, and high-resolution imaging of distant radio sources. When combined with the global VLBI network worldwide, the AVN will significantly improve the imaging capability of the current VLBI network by providing more arrays and longer baselines.

2.4. Recent Progress on the AVN Project

The latest developments in the AVN project in each of the partner countries were reported during the 6th SKA Africa Partner Countries Meeting from 14 to 17 October 2019 in Pretoria, South Africa. As reported by South Africa's Department of Science and Technology (DST), capacity building was one of the top priorities for each of the AVN partner countries. For example, through the two-dish interferometer concepts, universities in partner countries will be equipped with small radio telescopes aimed at radio astronomy training and astronomy outreach activities. The Botswana International University for Science and Technology already secured its two-aerial dish interferometer, UK Newton Funding, and a 10-PC laboratory to train students in radio astronomy. In this regard, Kenya also reported during the meeting that it plans to set up its two-dish interferometer in the near future. Training in the field of high-performance computing and big data is also underway, as presented in Section 2.5.

2.5. Scientific and Technical Training

The DARA project is intended to develop technical and scientific skills in a number of African countries, targeted to form the AVN. This program consisted of four training modules that were provided to 10 students from three countries, namely Zambia, Kenya, and Namibia, and four more AVN partners, which are Botswana, Ghana, Madagascar, and Mozambique. The JUMPING JIVE WP9 collaborated with the DARA project in a number of ways. From its pool of scientific and technical expertise, the JUMPING JIVE WP9 incorporated necessary skills into the already existing DARA network. More specifically, the JUMPING JIVE WP9 mobilised and supported European Union experts and scientists to tutor, advertise, and develop an interest in astronomy within the AVN partners. The JUMPING JIVE WP9 also allowed for African personnel to visit or have placements at European institutes as well as setting up an AVN technical support forum. Some of the organisations involved in providing personnel from the UK are the universities of Leeds, Oxford, Manchester, Central Lancashire, and Bristol. South African institutes and universities, including the SARAO, also contribute to the DARA project. In order to develop the technical and scientific skills necessary for the ECCAS to join the AVN network, it would be necessary for the DARA to extend its network into the ECCAS and initiate training programs in these countries.

3. Background and Past Successes on Ground Stations Retrofitting

In this section, we discuss some currently abandoned telecommunication antennas in the ECCAS region and present the requirements for a station to join the AVN, then discuss lessons learned from past successes on ground stations retrofitting.

3.1. Existing Abandoned Telecommunication Facilities

The potential benefit of converting ground stations for radio astronomy has been recognised worldwide. The SKA Africa partner countries that currently host such redundant large antennas are South Africa (3 antennas), Ghana, Kenya, Madagascar, and Zambia. Similar facilities have been located in Algeria (2), Benin, Cameroon (2), the Congo Peoples Republic, Egypt (2), Ethiopia, Malawi, Morocco, Niger, Nigeria (3), Senegal, Tunisia, Uganda, Gabon (2), and Zimbabwe. As illustrated in Figure 1, Gabon is a country in the ECCAS community currently hosting such an idling ground station. This station is typical of others in the ECCAS region. It was of the type commissioned between 1970 and 1985, as a node of access into the global network of the Intelsat Satellite Earth Stations. The radio bands allocated for these satellite communications were typically 5.925–6.425 GHz for the up-link and 3.700–4.200 GHz for the down-link. For optimal operations, these antennas had to be at least 30 m in diameter [5]. Today, while remaining the property of the state, the idling dish in Figure 1 is part of the infrastructure currently leased to a private operator in the mobile telephony sector. It is located in an area known as Nkoltang, in the Northern part of Libreville. The dish is adjacent to the telemetry ground station used by the French CNES to perform follow-up of the ARIANE launch from French Guyana. If refurbished, idling satellite earth station antennas such as the one in Gabon will operate from existing facilities relatively close to cities in a relatively populated neighbourhood. The issue of a relatively high level of pre-existing Radio Frequency Interference (RFI) will therefore need to be dealt with appropriately. Standard procedures currently in place in Kuntunse easily serve as a template. Kuntunse, in Ghana, is currently the site hosting the first ever idling ground station in Africa that has been successfully refurbished to become an operational radio telescope. In the absence of dedicated protocols, the dominant approach there consists of performing standard flagging during the data analysis with standard packages. Such an approach has the advantage of providing further opportunities for building capacity in standard procedures, using standard resources often very well supported by the community. Coupled with the advent of the MOOC (Massive open online course), the initial demand on expert human capital as trainers is therefore minimised.

Given their relative availability on the African continent, the case for ground stations retrofitting raises two points of interest that we explore in the next sections. Namely, (1) what is required for a typical station to join the AVN and (2) what are some of the lessons learned from previous success stories of ground stations retrofitting.

3.2. Minimal Requirements for a Typical AVN Station

We use the GRAO as a benchmark for a typical AVN station. The GRAO is a former telecommunications satellite earth station whose initial conversion phase was concluded in August 2017. It was in part funded by South Africa and the African SKA partnership agreement. The GRAO is the first station of the intended network of radio telescopes, on the African continent, to support the existing VLBI. In the African context (excluding South Africa), and given the effort already spent on it, we conservatively base indications for minimal specifications for joining the AVN on the GRAO ones listed below. We base most of this section on [6,10,13].

With a primary 32 m diameter for its reflector and an estimated aperture efficiency of 0.5, the GRAO can achieve a System Equivalent Flux Density (SEFD) of 860 Jy using an ambient (uncooled) receiver at a temperature of 125 K. The switch to a cryogenically cooled receiver was assessed to be able to provide a decrease by a factor of 5 in the GRAO's SEFD [13]. The GRAO has a fully steerable, Cassegrain beam-waveguide design dish. It is connected to a room temperature receiver and is currently capable of observing at 5 and 6.7 GHz (C band) with a bandwidth of up to 400 MHz. Although still short of funding, the current intentions are for the future acquisition of a 1.4–1.7 GHz (L-band) receiver. With further funding, subsequent plans are to replace the original C-band feed horn with a wider band design covering more VLBI bands and introducing cryogenic receivers for improved sensitivity [6].

Figure 1. Decommissioned 32-m large satellite earth station antennas in Gabon.

As the VLBI observations require coherent per-baseline sampling while on target, the timing and frequency reference considerations suggested the choice of a hydrogen maser frequency reference, on-site, for the GRAO. Although the choice of such a frequency reference comes with issues of sensitivity to vibrations, thermal and magnetic field changes, this provides a benchmark that could be systematised for other facilities of a similar nature on the continent. At the GRAO, the VLBI backend is supported by the standard EVN hardware comprised of a Digital-Baseband Convertor (DBBC) and Mark-5b data recorder. A data archiving system using an HDF5 container and data-visualisation software was adopted as a standard for the GRAO [13,14].

The e-VLBI mode of operation consists of transferring data over a suitable link directly from the station to the correlator. It is a modern departure from the standard VLBI data transfer via physical disks shipping. In the e-VLBI mode, the required data rate (in Megabits per second) is given by $\Delta\nu \times n_{\text{pol}} \times n_{\text{bit}} \times 2$, where $\Delta\nu$ is the bandwidth in Megahertz, n_{pol} is the number of observed polarisation, and n_{bit} is the number of bits per sample. A 128 MHz observing frequency would require 1024 Mbps connectivity to support a dual-polarisation e-VLBI mode observation using 2-bit sampling [13]. Presently, at the GRAO, internet connectivity to support the e-VLBI mode is not available yet.

3.3. A Few Lessons from Some Past Successes in Ground Stations Retrofitting

Ghana is uniquely located at 5 deg north of the equator. This allows viewing the entire plane of the Milky Way galaxy and nearly the whole sky, thus suggesting potentially compelling science cases. Already in 2017, the very first VLBI fringe tests were performed. These tests were satisfactorily conducted with the C-band (3.8–6.4 GHz) communication receiver. The outputs were clear VLBI fringes obtained at the JIVE on baselines between the GRAO and stations at Medicina (Italy), Yebes (Spain), Zelenchukskaya (Russia), and Hartebeesthoek (South Africa) [15]. The Jodrell Bank Centre for Astrophysics, supported by the UK's STFC/Newton Fund, has developed a new pulsar timing system (Hebe) for the GRAO, with the first pulsar detection made in the C band on 30 April 2017 [10]. The targeted radio continuum flux measurements and emission lines spectroscopy make up most of the single-dish science cases [6]. The GRAO science cases above should be contrasted with those at similar retrofitted facilities elsewhere in the world. Below, we list a few of other such facilities while highlighting the congruence and potential paths for the GRAO and future AVN stations.

In 2001, the National Astronomical Observatory of Japan (NAOJ) was given a 32 m antenna from the KDDI corporation. The NAOJ successfully converted the latter into the Yamaguchi 32 m class radio telescope, in collaboration with the Yamaguchi University and other institutes. The facility was dedicated to (1) use as an element of the Japanese global VLBI network, (2) methanol (6.7 GHz) and water (22 GHz) maser monitoring campaigns, and (3) the education and popularisation of radio astronomy among surrounding communities [16]. Most of these goals are congruent with what is at stake at the GRAO and future AVN stations.

In 2008, an idling telecommunications station with a 32 m parabolic antenna was donated to the Geophysical Institute of Peru. With major support from the National Astronomical Observatory of Japan, the facility was converted into the Sicaya radio telescope. Initially, in the single-dish mode, it was aimed at methanol maser and young stellar objects observations, using an uncooled 6.7 GHz (C band). Joining the VLBI observations remains in sight [17].

Around 2011, a consortium of universities was set up with the goal of using the decommissioned telecommunications infrastructure at the Goonhilly Earth Stations in Cornwall (UK) for astronomical purposes at the L, C, and K bands. The simulations showed that the inclusion of the 30 m class antennas at the Goonhilly Earth Station into e-MERLIN would double the resolving power of the array by increasing the maximum baseline from 217 to 441 km. Resolving the molecular gas at high-redshift with a resolution exceeding that of all but the highest ALMA bands was recognised as a major strength of the intended facility [18]. This illustrates the potential at stake when proper human and financial capital are assembled. One notes with strong interest the increasingly tight working relationship between key members of the consortium above and AVN partners through such initiatives as the DARA.

The 3-year-long conversion, in New Zealand, of the Warkworth 30 m antenna into a radio telescope is probably the most recent example that bears the most relevance to the phased process at the GRAO while also providing foresight on what can be routed beyond retrofitting for future AVN stations. The Warkworth 30 m radio telescope had its "official" First Light on 4 July 2014. By that time, it was already a fully steerable radio telescope operating at the C band for commissioning observations using an uncooled receiver, much like the GRAO today. Already a valuable instrument in its own right for single-dish observations, on 11 December 2014, the first geodetic and astrophysical scientific VLBI experiment was conducted. Although the VLBI fringes checks have been performed at the GRAO, both the full-scale geodetic and astrophysical scientific VLBI remain to be tested at the GRAO. The observatory has access to a 10 Gbit/s connection to the Research and Education Advanced Network New Zealand, providing high-speed data transfers for the e-VLBI. The plans underway include equipping the facility with a cryogenically cooled C-band receiver, an X-band receiver, and a 2 Gbps recording system. For the science case, both the VLBI and single-dish spectral observations in the C band can potentially include spectral line observations of OH maser lines at 6.03 and 6.035 GHz and methanol masers at 6.7 GHz. The addition of S- and X-band receivers would enable the use of the facility for geodetic research [19,20]. The human capital impact has already proven valuable with the radio telescope providing a facility for unparalleled laboratory work and students' research projects, some of it geared towards design work for the SKA, from New Zealand. Unambiguously, the Warkworth 30 m radio telescope provides a compelling example for what a typical AVN station can be and do, once established.

4. Radio Interferometer, uv-Coverage, and VLBI

Following the van Cittert–Zernike theorem [21], the output from a two-elements interferometer is a complex measurement; the so-called "visibility', under specific conditions (e.g., assuming no sampling and other corruption effects), is given by:

$$\mathcal{V}(u,v,w) = \int\int \frac{\mathcal{I}(l,m)}{n} e^{-2i\pi(ul+vm+w(n-1))} \mathrm{dldm}, \tag{1}$$

where $n^2 = 1 - l^2 - m^2$ and $\mathcal{I}$ is the apparent sky with coordinates (l,m). The signals received by each of the two-elements interferometer are cross-correlated, either in real time or offline, these cross-correlation products are accumulated during a defined period (the integration time) and at each channel. If the number of antennas is n, then the instantaneous number of correlation during the integration time is $n(n-1)/2$. Because the relative orientation of the antennas and the sources change as the earth rotates, one can take advantage of the earth's rotation to measure more samples. For an in-depth discussion, we refer the reader to [21,22].

The differential of the spatial frequencies u, v, and w measured in wavelength as a function of the baseline vector with components L_x, L_y, L_z along the axes of the International Terrestrial Reference Frame (ITRF, a terrestrial coordinate system that provides the coordinates of points on the earth's surface [23]) is given by:

$$\frac{\partial u}{\partial t} = \frac{\omega_e}{\lambda}(L_x \cos h - L_y \sin h) \tag{2}$$

$$\frac{\partial v}{\partial t} = \frac{\omega_e}{\lambda}(L_x \sin\delta \sin h + L_y \sin\delta \cos h) \tag{3}$$

$$\frac{\partial w}{\partial t} = \frac{-\omega_e}{\lambda}(L_z \cos\delta \sin h + L_y \cos\delta \cos h), \tag{4}$$

where the baseline is tracking a source at declination δ and hour angle h, $\omega_e = 7.2925 \times 10^{-5}$ rad·s^{-1} is the angular velocity of the earth. The uv-coverage is the set of all the projected baseline vectors, (u,v,w) in the Fourier plane or uv-plane. An efficient way to fill the uv-coverage is to add many antenna telescopes together while making use of the earth's rotation, the frequency coverage, and antennas layout of the interferometer. The more complete the uv-coverage, the better the response of the instrument, and therefore the image quality. In order to achieve a milliarcsecond resolution, the network of antennas requires baselines longer than 10^4 km. Achieving such a high-resolution observation requires a VLBI technique.

Why the Missing Samples in the EVN + MeerKAT + Kuntunse uv-Coverage?

In this section, we discuss the performances of the uv-coverage density of the combined Kuntunse antenna in Ghana with the MeerKAT telescope in South Africa, correlated to the full EVN. The full EVN consists of 12 stations across the globe, i.e., Badary, Effelsberg, Hartebeesthoek, Jodrell Bank, Medicina, Noto, Onsala, Shanghai, Svetloe, Torun, Westerbork, and Zelenchukskaya. Figure 2 shows an African map where the green points are the Kuntunse antenna in Ghana, the MeerKAT telescope in South Africa, and the EVN. There are some stations of the EVN that do not appear on the map (e.g., Shanghai); these are stations that are on the other side of the globe. The 64 antennas of the MeerKAT telescope do not appear all on the map; this is because the antennas are very close to each other, making it difficult to visualise them on a bigger scale. The points in red are some of the locations of the abandoned old telecommunications satellites in the ECCAS countries or locations suitable to build new radio antennas.

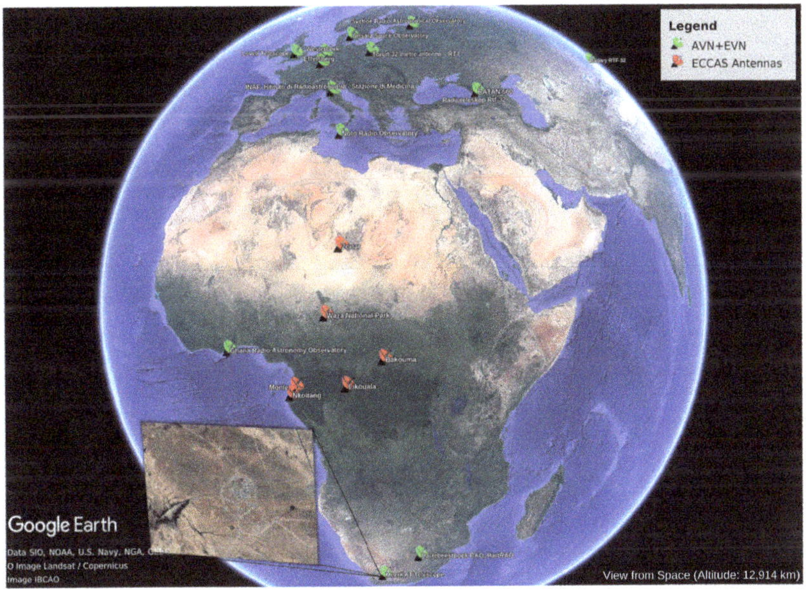

Figure 2. Green points: locations of the Kuntunse antenna in Ghana, the MeerKAT stations in South Africa, and the EVN. Red points: locations of abandoned old telecommunication satellite facilities in the ECCAS region and/or possible sites to build new radio telescopes.

Figure 3 shows the *uv*-coverage of the MeerKAT telescope correlated to the EVN (left panels) and the Kuntunse combined with the MeerKAT telescope and both correlated to the EVN (right panels). These *uv*-coverages are obtained by a simulation at a frequency of 16 GHz, during a total period of 10 h with 1 s integration time and 16 MHz bandwidth divided into 64 channels. Using the casacore (https://pypi.org/project/python-casacore/, accessed on 15 June 2022), we generate the casa table from the antenna positions specified in the ITRF coordinate system [23] and then we create an empty measurement set using simms (https://github.com/ratt-ru/simms, accessed on 15 June 2022). From the empty measurement set, the *uv*-coverage is extracted. The *uv*-coverages are tracking a source at the declination of −20, +20, and +60 deg. It is clearly seen from the correlation between the MeerKAT telescope and the EVN (left panels in Figure 3) that there are missing samples in the middle area. The latter correlation is then combined with the Kuntunse antenna in Ghana which still does not optimally sample these missing areas (see the right panels in Figure 3). In these *uv*-coverages, the samples from the core are from shorter baselines; these shorter baselines are the internal baselines of the MeerKAT telescope and the EVN. The samples at the outer core are from the longer baselines; these baselines relate the AVN antennas in the northern hemisphere to the MeerKAT telescope and Hartebeesthoek in the southern hemisphere. There are few medium-length and long baselines coming from the correlation between the antennas in the northern hemisphere or southern hemisphere to the Kuntunse antenna in Ghana. The areas sampled by these medium-length and long baselines linked to the Kuntunse antennas are seen in blue from Figure 3, right panels. To fill these missing samples, we need more medium-length baselines. These medium-length baselines can only be obtained if some of the antennas are placed around the equatorial line in Africa. Most of the countries in Africa around the equator are the French-speaking countries or the ECCAS countries. Using simulations, we show in this work that if one were to build radio telescopes and/or convert old abandoned telecommunication satellite antennas to radio telescopes in the ECCAS countries, no matter where these antennas are to be located in each of these countries, this should significantly improve the *uv*-coverage of the current EVN + MeerKAT + Kuntunse.

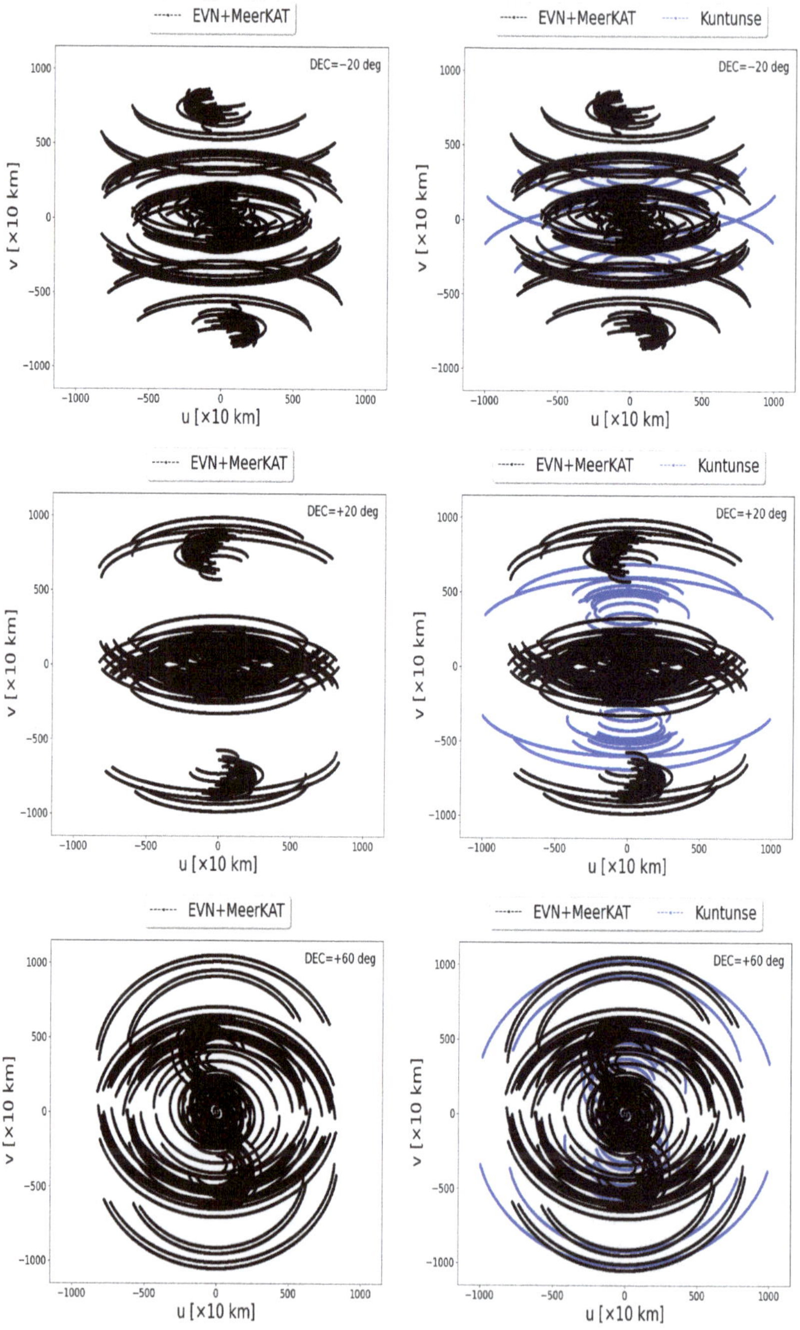

Figure 3. The EVN combined with the MeerKAT telescope (left panels) and the EVN combined with the MeerKAT telescope and the Kuntunse antenna in Ghana (right panels).

5. Simulations and Discussion

Two simulations are performed using antennas, as shown in Figure 2. Firstly, to evaluate the performance of the single interferometer array represented by the six antennas of the ECCAS countries (i.e., the antennas with their position marked in red in Figure 2), we generate a measurement set using the procedure described in Section 4 and then fill the measurement set with simulated visibilities using MeqTrees [24]. Secondly, using the same simulation tools, we correlate the six ECCAS antennas with the AVN and the EVN and then we simulate the visibilities. To demonstrate the scientific advantages of adding these six antennas to the current VLBI network, for each of the two simulations, we made an image and measure its S/N.

5.1. Performance Assessment of the uv-Coverage of the ECCAS Antennas

Figure 4 shows the *uv*-coverage of the six antennas in the ECCAS countries at 1.4 GHz and for four declinations. This *uv*-coverage is simulated during 10 h total time with 1 s integration time and a total bandwidth of 16 MHz divided into 64 channels. The positions of the antennas are shown in Figure 2, red points. As expected, the *uv*-coverage is very poor as the six antennas are spread over a large distance. Each of these antennas can function as a single-dish radio telescope and can perform high-level science, e.g., observing pulsars, masers, and hot gas from the Milky Way or distant galaxies.

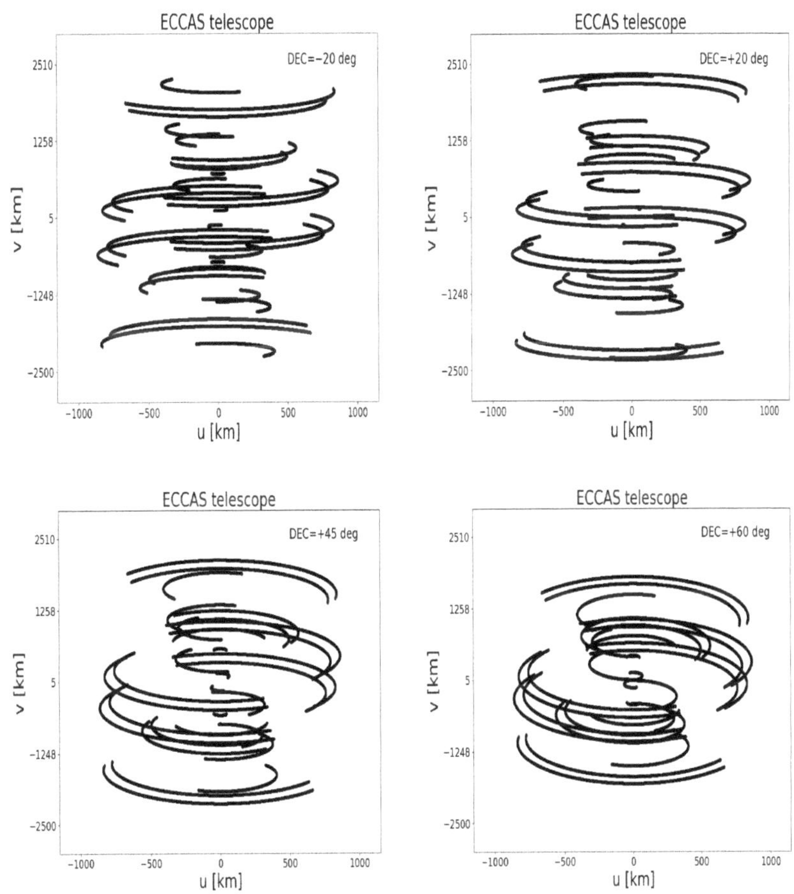

Figure 4. ECCAS antennas *uv*-coverage at 1.4 GHz at four declinations (−20, +20, +45, and +60 deg), 10 h observation, and 16 MHz total bandwidth showing a lot of holes or gaps.

5.2. Filled *uv*-Coverage for ECCAS + MeerKAT + EVN + Kuntunse

This time, the MeerKAT telescope, EVN, and Kuntunse are correlated to the ECCAS antennas. The full antennas used in the simulation are shown in Figure 2. The black points in Figure 5 are the data from the EVN, MeerKAT telescope, and Kuntunse antenna while the red points are the data coming from the ECCAS antennas and their correlation to the EVN, MeerKAT telescope, and Kuntunse. We note that while the six ECCAS antennas on their own give poor *uv*-coverage, as shown in Figure 4, they significantly improve the current *uv*-coverage of the full VLBI network; the *uv*-coverage is now well filled because of the extra medium-length baselines that relate the EVN and MeerKAT telescope to the ECCAS antennas. These *uv*-coverages can even be optimised further if more antennas are added at the equatorial line. Below, using simulated images, we demonstrate that adding these antennas at the equatorial line will improve the S/N and therefore the image quality of the combined radio interferometers.

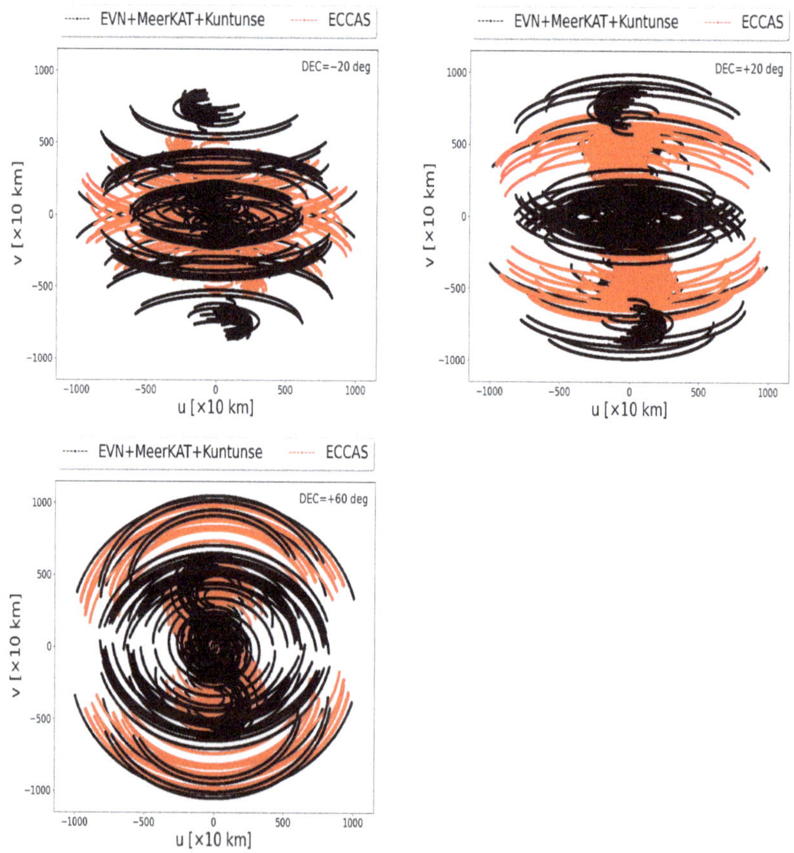

Figure 5. Performance of the global VLBI *uv*-coverage. The ECCAS antennas are correlated with the Kuntunse antenna, the MeerKAT telescope, and the EVN. The *uv*-coverage is well-filled because of the extra medium-length baselines that relate the EVN and MeerKAt telescope to the ECCAS antennas.

5.3. An Estimate of the Image S/N

In this section, we describe the procedure used to estimate the S/N in our images. As discussed in [25], an approximate of the S/N is given as:

$$S/N \sim \frac{\mathcal{A}_{\rm flux}}{\mathcal{C} + \sigma}, \tag{5}$$

where $\mathcal{A}_{\text{flux}}$ is the brightness distribution of a point source, $\mathcal{C}$ is the confusion noise, i.e., the unwanted signals that contaminate the field of the source of interest. The root mean square noise σ describes the sensitivity of the instrument and is approximated as:

$$\sigma \sim \frac{\sigma_{\text{vis}}}{\sqrt{n_{\text{ch}} n_{\text{int}} n_{\text{pol}} n_{\text{bal}}}}, \tag{6}$$

where σ_{vis} and n_{ch} are the noise per visibility and the number of channels in the bandwidth, respectively, n_{int} and n_{pol} are the number of correlations in time and polarisations, respectively, and n_{bal} is the total number of baselines. The product $n_{\text{ch}} n_{\text{int}} n_{\text{pol}} n_{\text{bal}}$ is the total number of visibilities in the uv-plane. For an in-depth discussion on how to derive Equation (6), we refer the reader to [25]. During an observation, the confusion noise decreases with the increasing sampled Fourier coefficients in the uv-coverage. Moreover, adding antennas increases the number of baselines which then decreases the root mean square noise. In other words, by adding the ECCAS antennas to the current AVN, we optimise for the unsampled Fourier coefficients in the AVN uv-coverage which then limits the far-field contamination while reducing the dynamic range required to image the source and relaxes the requirements for deconvolving the source. Thus, assuming that $\mathcal{A}_{\text{flux}}$ remains constant for both the EVN + MeerKAT + Kuntunse and EVN + MeerKAT + Kuntunse + ECCAS, we see that the S/N will increase for the EVN + MeerKAT + Kuntunse + ECCAS because, by definition, as shown in Equation (6), $\mathcal{C}$ and σ are smaller when observing with the EVN + MeerKAT + Kuntunse + ECCAS compared to the EVN + MeerKAT + Kuntunse. To demonstrate the latter, the measurement sets described above are used to simulate the EVN + MeerKAT + Kuntunse and the EVN + MeerKAT + Kuntunse + ECCAS telescopes.

To populate the measurement sets with visibilities, the MeqTrees [24] software is used to simulate a sky model with two point sources: a 1 Jy point source at the phase centre and a far-field point source with 5 Jy brightness located at 6 arcmin from the phase centre. The 1 Jy point source at the phase centre represents our source of interest for which we want to measure its S/N. This source is simulated at the phase centre to limit the amplitude attenuation or the effect of smearing so that one can estimate the $\mathcal{A}_{\text{flux}}$ effectively from the simulation. In the VLBI regime, a source at 6 arcmin is already a far-field source and, therefore, can be used to measure the contamination, $\mathcal{C}$, at the phase centre. MeqTrees is used to corrupt the simulated visibility data with 1 Jy Gaussian noise per visibility. The WSclean imager [26] is then used to translate the visibilities to the image using uniform weighting, as shown in Figure 6, where a few pixels are imaged to visualise only the single source at the phase centre. The left panel and the right panel of Figure 6 show the 1 Jy point source as seen by the EVN + MeerKAT + Kuntunse and EVN + MeerKAT + Kuntunse + ECCAS telescopes, respectively. It is easy to see from this result that the point source sidelobes as seen by the EVN + MeerKAT + Kuntunse + ECCAS telescope are lower compared to the sidelobes of the same source as seen by the EVN + MeerKAT + Kuntunse telescope: this is a straightforward result from sampling more Fourier coefficients in the uv-plane. We also note that the resolution of both images in Figure 6 does not vary (i.e., the size or shape of the source main lobe does not change): this is due to adding more antenna telescopes that do not increase the longest Fourier modes and, therefore, preserve the same resolution of the instrument. To estimate the S/N of the source at the phase centre, we adopt the following procedures using uniform weighting:

- The 1 Jy point source at the phase centre is simulated separately (as a single source) for each of the radio interferometers and its amplitude $\mathcal{A}_{\text{flux}}$ is measured for each image after cleaning.
- The 5 Jy point source at 6 arcmin from the phase centre is also simulated separately as a single point source. Note that one would expect no signal at the phase centre except the contamination $\mathcal{C}$ from this far-field source.
- An empty measurement set (without a source) is simulated for each of the instruments, and then a 1 Jy Gaussian noise per visibility is used to corrupt the measurement sets. The root mean square of the image around the phase centre provides an approximate of σ.

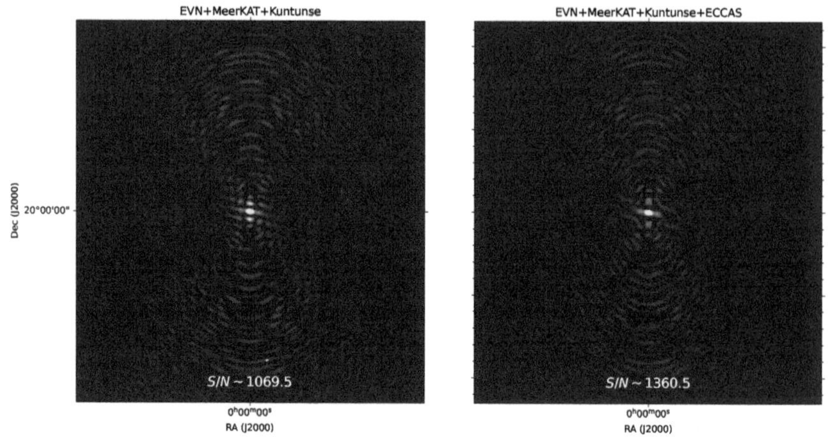

Figure 6. Simulated sky model with two point sources: a 1 Jy point source at the phase centre and a far-field point source with 5 Jy brightness located at 6 arcmin from the phase centre as seen by the EVN + MeerKAT + Kuntunse telescope (left) and by the EVN + MeerKAT + Kuntunse + ECCAS telescope (right) for a simulated observation at 16 GHz. To corrupt the simulation, 1 Jy Gaussian noise per visibility is used. The data are sampled during a total period of 10 h with 1 s integration time and using a total bandwidth of 16 MHz divided into 64 channels. Then, a few pixels are imaged to visualise only the source at the phase centre.

At the end of these simulations, we then estimate the S/N from the measured parameters in Table 1. The EVN + MeerKAT + Kuntunese station gave an S/N of ~1069.5. By adding the ECCAS station, i.e., the EVN + MeerKAT + Kuntunese + ECCAS combination, the S/N improved significantly, and we measure an S/N of ~1360.5. This clearly demonstrates the strength of the ECCAS station when combined with the AVN network.

Table 1. Measured parameters used to calculate S/N.

Telescope	$\mathcal{A}_{\text{flux}}$	$\mathcal{C}$	σ	S/N
EVN + MeerKAT + Kuntunse	1	0.000767	0.000168	1069.5
EVN + MeerKAT + Kuntunse + ECCAS	1	0.000589	0.000146	1360.5

6. Conclusions

We conclude that the *uv*-coverage for a full VLBI observation will improve if a few antennas were to be added in the ECCAS region. The resulting Fourier transform of the *uv*-coverage compactness will only lead to a low confusion noise limit which is suitable for the high signal-to-noise or dynamic range images requirement. Building or converting old, abandoned satellite telecommunication facilities in the ECCAS region is a guarantee that the science results from these antennas will expand the universities' international visibility. As part of the VLBI, the scientific community of the ECCAS region will be fully prepared for strong scientific involvement with the SKA.

Author Contributions: M.A. conceived the experiments; M.A. and P.O. conducted the experiments; M.A., P.O., E.M., R.I. and S.Z. analysed the results. All authors reviewed the manuscript. All authors have read and agreed to the published version of the manuscript.

Funding: This work is based upon research supported by the South African Research Chairs Initiative of the Department of Science and Technology and National Research Foundation. The MeerKAT telescope is operated by the South African Radio Astronomy Observatory, which is a facility of the National Research Foundation, an agency of the Department of Science and Innovation.R.I. receives financial support from grant RTI2018-096228-B-C31 (MCIU/AEI/FEDER,UE), and from the State Agency for Research of the Spanish Ministry of Science, Innovation and Universities through the

"Center of Excellence Severo Ochoa" awarded to the Instituto de Astrofísica de Andalucía (SEV-2017-0709), from the grant IAA4SKA (Ref. R18-RT-3082) from the Economic Transformation, Industry, Knowledge and Universities Council of the Regional Government of Andalusia and the European Regional Development Fund from the European Union.

Institutional Review Board Statement: Not applicable.

Informed Consent Statement: Not applicable.

Data Availability Statement: Not applicable.

Acknowledgments: We thank our colleagues Aard Keimpema and Griffin Foster for their insights and comments on early drafts of this work.We would like to thank the anonymous reviewers, for their comments that improved the paper.

Conflicts of Interest: The authors declare no conflict of interest.

References

1. Clark, B. A review of the history of VLBI. *Radio Astron. Fringe* **2003**, *300*, 1.
2. Kellermann, K.; Moran, J. The development of high-resolution imaging in radio astronomy. *Annu. Rev. Astron. Astrophys.* **2001**, *39*, 457–509. [CrossRef]
3. Wild, S. Giant radio telescope scaled back to contain costs. *Nature* **2017**, *547*, 391–392. [CrossRef]
4. Booth, R.; Jonas, J. An overview of the MeerKAT project. *Afr. Skies* **2012**, *16*, 101.
5. Gaylard, M.; Bietenholz, M.; Combrinck, L.; Booth, R.; Buchner, S.; Fanaroff, B.; MacLeod, G.; Nicolson, G.; Quick, J.; Stronkhorst, P.; et al. An African VLBI Network of radio telescopes. *arXiv* **2014**, arXiv:1405.7214.
6. Asabere, B.D.; Gaylard, M.; Horellou, C.; Winkler, H.; Jarrett, T. Radio astronomy in Africa: The case of Ghana. *arXiv* **2015**, arXiv:1503.08850.
7. Venturi, T.; Paragi, Z.; Lindqvist, M.; Bartkiewicz, A.; Beswick, R.; Bogdanović, T.; Brisken, W.; Charlot, P.; Colomer, F.; Conway, J.; et al. VLBI20-30: A scientific roadmap for the next decade—The future of the European VLBI Network. *arXiv* **2020**, arXiv:2007.02347.
8. Adams, J.; Gurney, K.; Hook, D.; Leydesdorff, L. International collaboration clusters in Africa. *Scientometrics* **2014**, *98*, 547–556. [CrossRef]
9. Scheel, M. Instrumentation and Calibration of the Transient Array Radio Telescope. Ph.D. Thesis, University of Otago, Dunedin, New Zealand, 2018.
10. Scragg, T.; Stappers, B.; Breton, R.; Smith, J.; Adomako, D.; Asabere, B.D.; Chibueze, J.; Cloete, K. Pulsar Observations at the Ghana Radio Astronomy Observatory. *Proc. Int. Astron. Union* **2017**, *13*, 410–411. [CrossRef]
11. Aworka, R.; Proven-Adzri, E.; Ansah-Narh, T.; Koranteng-Acquah, J.; Aggrey, E. Using Ghana's 32-m radio telescope to promote astronomy outreach. *Nat. Astron.* **2021**, *5*, 1199–1202. [CrossRef]
12. Proven-Adzri, E.; Ansah-Narh, T.; Aworka, R.; Fosuhene, S.; Sottie, S.; Gyasi, G. Mercury transit observed in Ghana. *Phys. Educ.* **2022**, *57*, 045007. [CrossRef]
13. Copley, C.; Thondikulam, V.; Loots, A.; Bangani, S.; Cloete, K.; Combrinck, L.; Gioio, S.; Ludick, J.; Nicolson, G.; Pollak, A.; et al. The african very long baseline interferometry network: The Ghana antenna conversion. *arXiv* **2016**, arXiv:1608.02187.
14. Tuccari, G.; Buttaccio, S.; Nicotra, G.; Alef, W.; Keller, R.; Nalbach, M.; Wunderlich, M. DBBC-A Flexible Environment for VLBI and Space Research: Digital Receiver and Back-end Systems. In Proceedings of the 18th European VLBI for Geodesy and Astrometry Work Meeting, Vienna, Austria, 12–13 April 2007; pp. 45–49.
15. Gurvits, L.I.; Beswick, R.; Hoare, M.; Njeri, A.; Blanchard, J.; Sharpe, C.; Tiplady, A.; De Witt, A. High-resolution radio astronomy: An outlook for Africa. *Proc. Int. Astron. Union* **2019**, *15*, 137–142. [CrossRef]
16. Fujisawa, K.; Mashiyama, H.; Shimoikura, T.; Kawaguchi, N. The yamaguchi 32-m radio telescope. In Proceedings of the 8th Asian-Pacific Regional Meeting, Tokyo, Japan, 2–5 July 2002; Volume II, pp. 3–4.
17. Ishitsuka, J.; Kobayashi, H.; Miyoshi, M. Capabilities and Present Status of The Sicaya Radio Telescope in Peru. *Rev. Mex. Astron. Astrofis. Conf. Ser.* **2017**, *49*, 59.
18. Heywood, I.; Kloeckner, H.; Beswick, R.; Garrington, S.; Hatchell, J.; Hoare, M.; Jarvis, M.; Jones, I.; Muxlow, T.; Rawlings, S. Expanding e-MERLIN with the Goonhilly Earth Station. *arXiv* **2011**, arXiv:1103.1214.
19. Woodburn, L.; Natusch, T.; Weston, S.; Thomasson, P.; Godwin, M.; Granet, C.; Gulyaev, S. Conversion of a New Zealand 30-metre telecommunications antenna into a radio telescope. *Publ. Astron. Soc. Aust.* **2015**, *32*, e017. [CrossRef]
20. Petrov, L.; Natusch, T.; Weston, S.; McCallum, J.; Ellingsen, S.; Gulyaev, S. First scientific VLBI observations using New Zealand 30 meter radio telescope WARK30M. *Publ. Astron. Soc. Pac.* **2015**, *127*, 516. [CrossRef]
21. Thompson, A.R. Fundamentals of radio interferometry. In *Synthesis Imaging in Radio Astronomy II*; Astronomical Society of the Pacific: San Francisco, CA, USA, 1999; Volume 180, p. 11.
22. Fomalont, E.B. Earth-rotation aperture synthesis. *Proc. IEEE* **1973**, *61*, 1211–1218. [CrossRef]
23. Taylor, G.B.; Carilli, C.L.; Perley, R.A. *Synthesis Imaging in Radio Astronomy II*; Astronomical Society of the Pacific: San Francisco, CA, USA, 1999; Volume 180.

24. Noordam, J.E.; Smirnov, O.M. The MeqTrees software system and its use for third-generation calibration of radio interferometers. *Astron. Astrophys.* **2010**, *524*, A61. [CrossRef]
25. Atemkeng, M.; Smirnov, O.; Tasse, C.; Foster, G.; Keimpema, A.; Paragi, Z.; Jonas, J. Baseline-dependent sampling and windowing for radio interferometry: Data compression, field-of-interest shaping, and outer field suppression. *Mon. Not. R. Astron. Soc.* **2018**, *477*, 4511–4523. [CrossRef]
26. Offringa, A.; McKinley, B.; Hurley-Walker, N.; Briggs, F.; Wayth, R.; Kaplan, D.; Bell, M.; Feng, L.; Neben, A.; Hughes, J.; et al. WSCLEAN: An implementation of a fast, generic wide-field imager for radio astronomy. *Mon. Not. R. Astron. Soc.* **2014**, *444*, 606–619. [CrossRef]

electronics

MDPI

Article

Design of a Dual-Polarization Dipole Antenna for a Cylindrical Phased Array in Ku-Band

Ning Zhang [1], Zhenghui Xue [2,*], Pei Zheng [3], Lu Gao [3] and Jia Qi Liu [3]

[1] EMC and Microwave System Laboratory, Beijing Institute of Technology, Beijing 100081, China
[2] School of Integrated Circuits and Electronics, Beijing Institute of Technology, Beijing 100081, China
[3] National Key Laboratory of Science and Technology on Test Physics and Numerical Mathematics, Beijing 100076, China
* Correspondence: zhxue@bit.edu.cn

Abstract: This paper proposes a dual-polarization dipole antenna for a cylindrical phased array working in Ku-band. The dipole antenna is double-layer structured and is composed of two orthogonal butterfly shaped dipole radiators, two ground co-planar waveguide (GCPW) feeding structures and vias. Each dipole is in the shape of a butterfly. The dipole patch is grooved triangularly and one side of it is bent into an N shape, which effectively expands the working frequency band of the antenna. The double-layer structure improves the isolation between the antenna ports. The antenna works between 15 GHz to 16.2 GHz and the isolation between the antenna's two feeding ports in this band is better than 20 dB. The proposed dipole antenna is applied in a 32-element cylinder array. The simulation and measured results show that the array can scan between −60° to +60° in the azimuth plane with a gain fluctuation less than 2.5 dB. Therefore, the proposed design is an attractive candidate for conformal devices at Ku-band frequencies, and it also has a great potential for application in larger antenna arrays.

Keywords: cylindrical phased array; dual-polarization; ground co-planar waveguide (GCPW); Ku-band

Citation: Zhang, N.; Xue, Z.; Zheng, P.; Gao, L.; Liu, J.Q. Design of a Dual-Polarization Dipole Antenna for a Cylindrical Phased Array in Ku-Band. *Electronics* **2022**, *11*, 3796. https://doi.org/10.3390/electronics11223796

Academic Editors: Naser Ojaroudi Parchin, Chan Hwang See and Raed A. Abd-Alhameed

Received: 20 October 2022
Accepted: 14 November 2022
Published: 18 November 2022

1. Introduction

Compared with traditional planar arrays, conformal arrays have unparalleled advantages in reducing aerodynamic drag, wide-angle coverage and low radar cross section (RCS) on aircraft and missiles [1]. Therefore, the conformal phased array antenna has great potential in radar system.

Dual-polarization antennas radiate and receive electromagnetic waves with the same frequency and orthogonal polarizations, and are widely applied to different fields. In wireless communications, the application of dual-polarized antennas can enlarge channel capacity and enhance spectral efficiency [2,3]. When used in wireless energy transmission systems, they can enhance the efficiency of energy transfer [4]. Particularly, it is an efficient approach for radar systems to solve the anti-jamming and multipath immunity problems. In radar detecting, they can further obtain other information except amplitude and phase from the scattered wave of the target and lead to a higher detecting level [5–8].

In recent years, many researchers focus on the study of conformal arrays and dual-polarization antennas. P. Wang et al. [9] introduce a broadband conformal antenna array installed on a large metal cylinder. By using several Vivaldi antenna elements, the antenna array achieves ultra-wideband and low cross polarization level. However, the antennas cannot fit the cylinder completely, and reduce the aerodynamic performance of the carrier. Microstrip antenna has always been one of the main candidates for conformal arrays for its good conformal ability, though it often suffers from the problem of narrow bandwidth. Ref. [10] proposes a cylindrical conformal array with microstrip antenna elements. It introduces the defective ground structures (DGS) to solve the problem of narrow bandwidth caused by the deformation of the array antenna when conformal to the cylindrical surface.

In [11], a full solid angle scanning cylindrical-and-conical conformal phased array antenna is proposed. As a part of it, a cylindrical array with microstrip antennas as the elements is designed. It shows a good scanning performance, but the bandwidth is only 1.74%. In [12], the differential feeding technique is applied to dual-polarized microstrip antenna to achieve a good cross polarization level. Based on a dual-polarized patch antenna element with vertical orthogonal baluns, a non-planar conformal dual-polarized phased array antenna is proposed in [13,14]. The antenna has a port isolation of better than 40 dB, but its profile is relatively high. Due to the virtues of low profile, various forms, easy processing etc., printed dipoles are also widely used in conformal antennas. In [15–17], tightly coupled dipole are applied to realize dual-polarization cylindrical conformal arrays. Ultra-wide bandwidth and large scanning range can be achieved. Ref. [18] introduces a novel conformal dipole antenna with polarization diversity. The polarization diversity is achieved by placing two dipole antennas orthogonally on the outside of a capsule shell. In [19], two curved patch dipoles are arranged orthogonally in the cross contour of a hemispherical shell to realize a circularly polarized antenna for the ground terminal facilities of the global satellite navigation system. Based on the polyimide film, J. J. Peng et al. [20] propose a dipole conformal array working in P-band for the UAV radar. Y. Gao et al. [21] propose a dual-polarized monopulse conical conformal antenna with Yagi antennas as the radiation structures. The antenna achieves wide bandwidth by taking double diamond dipoles as excited elements [22]. In [23], a crossed dipole antenna element is employed to a cylindrical phased array for weather surveillance radars. A port isolation over 50 dB is achieved at the cost of antenna profile.

As stated above, it is difficult for conformal dual polarized phased arrays to simultaneously have wide bandwidth, high isolation, low profile and large scanning angle. In this article, a dual-polarization dipole antenna, which is used to a cylindrical phased array antenna in Ku-band, is proposed. The antenna consists of dual pairs of dipoles placed orthogonally. Different from the conventional printed dipole antenna, this paper proposes a dipole with the shape of a butterfly wing. With the dipole shape modification, the impedance bandwidth and the port isolation of the antenna is greatly improved. Taking it as the antenna element, a linear conformal phased array which has 32 elements is designed, fabricated, and measured. The measured results show that the proposed dual-polarization conformal phased array antenna can realize beam scanning in the range of -60° to 60° with the gain fluctuation less than 2.42 dB.

2. Design of the Dipole Antenna

As a simple planar antenna, the printed dipole antenna is not only simple in forms but also has many variants to improve its performance. It is a promising antenna candidate for conformal antennas. In this paper, a dual polarization printed dipole is designed, which has the characteristics of certain frequency band, high isolation, simple structure, and low profile.

The model of the proposed antenna is shown in Figure 1. It is composed of metal radiators, two dielectric substrates, isolation ground, vias, and grounded coplanar waveguide (GCPW) feed structures. The butterfly shaped dipole is printed on the top surface of the upper substrate. The GCPW feeding structure is printed on bottom surface of the lower substrate. The isolation ground is between the two substrates, which improves the port isolation by reducing the effect of the CPW slot to the antenna's radiation. The substrates are both F4B, which has better flexibility and makes the antenna array easy to install on the cylindrical surface. The dielectric constant is 2.2 and the loss tangent is 0.007. The thicknesses of the upper and lower substrates are both 1 mm.

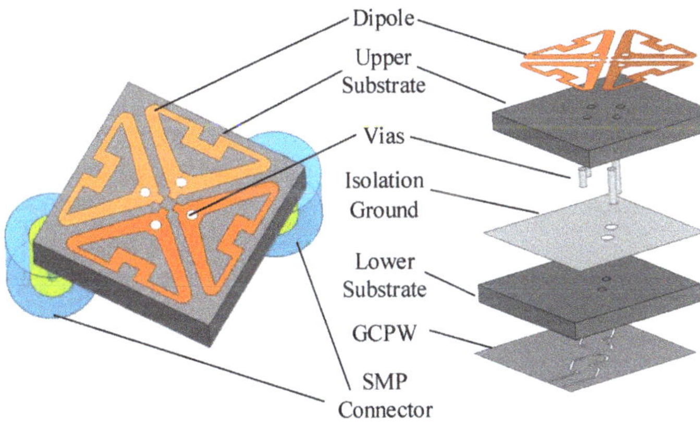

Figure 1. Dipole Antenna model.

The radiator consists of two pairs of rotationally symmetrical dipoles. The length of one dipole arm is $0.2\lambda_0$ (L = 3.8 mm), and the length of a pair of dipoles is $0.437\lambda_0$, where λ_0 is the free space wavelength at the center frequency. The dipoles are designed in the shape of a butterfly by grooving the redating patch and bending the side into an N shape. Through the bending structure, an additional capacitance is introduced, and the antenna bandwidth is widened. The antenna is simulated with Ansys HFSS 2019. Figure 2 shows the impedance performance and the port isolation of two equal sized antennas with and without bending. Because the two pair of dipoles are rotationally symmetrical, only |S11| is shown in the figure. Figure 2a shows that the frequency band of |S11| less than −10 dB is 15.6–16.0 GHz before side bending, while after side bending it becomes 15.0–16.2 GHz. An improvement of the relative bandwidth from 1.9% to 8% is achieved with the addition of the bending structure. Figure 2b shows the port isolation of the antennas. It can be observed that the isolation is better than 20 dB only in the frequency range of 16.2–16.5 GHz before side bending. After side bending, the bandwidth of 20 dB isolation increases to 14.9–16.56 GHz, with the −10 dB impedance bandwidth contained.

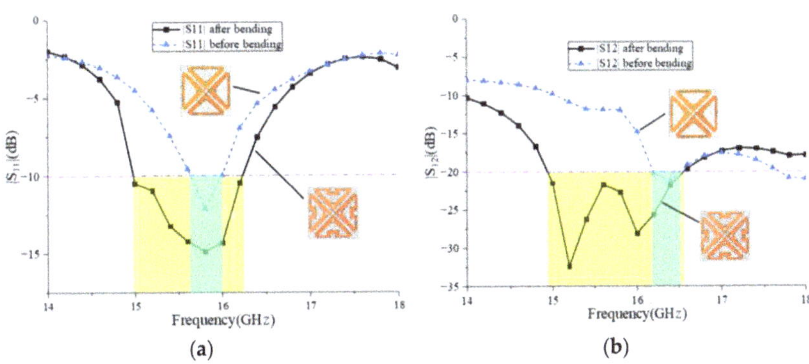

Figure 2. Effect of side bending on antenna bandwidth. (**a**) |S11| and (**b**) |S12|.

Since the distance between the two dipole feeding points is only 1.7 mm, directly connecting the SMP connectors with the feeding points will lead to the contradiction. Therefore, a grounded coplanar waveguide (GCPW) structure is introduced. It combines the advantages of microstrip line and coplanar waveguide structure [24,25]. The GCPW structure not only solves the problem of feeding contradiction, but also leads to a better impedance match. Four vias are drilled through the substrates, with two for grounding, and the others as part of the feed to connect the dipole with the GCPW.

The detailed antenna structure and parameters are shown in Figure 3. The parameters are analyzed and optimized to obtain good antenna performance. The optimal parameters of the proposed antenna are listed in Table 1.

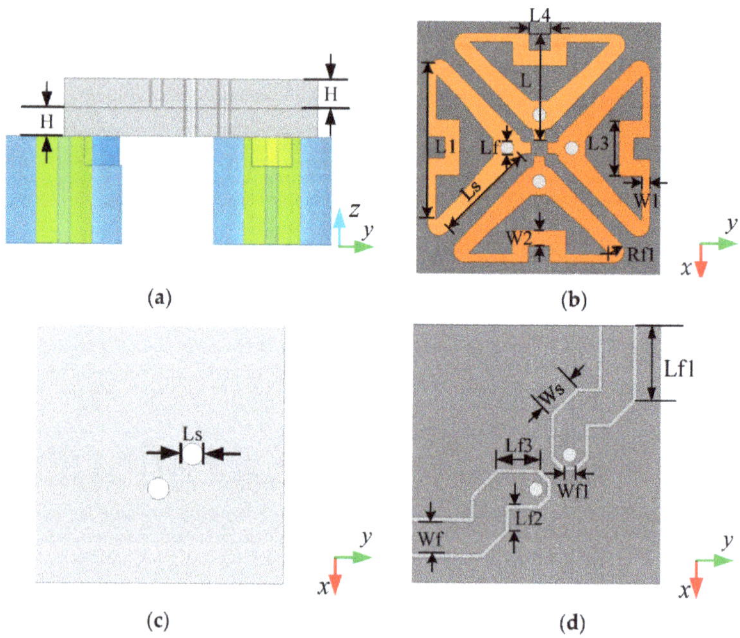

Figure 3. Detailed antenna structure: (**a**) side view; (**b**) top view; (**c**) isolation ground; and (**d**) bottom view.

Table 1. Dimensions for the Proposed Antenna (unit: mm).

Parameter	Size	Parameter	Size
L	3.8	Rs	1.9
L1	6.25	Lf1	2.54
L3	2	Lf2	0.84
L4	0.8	Lf3	1.39
L5	3	Wf	1.1
Ls	3	Wf1	0.4
W1	3	Ws	1.21
W2	2	H	1
Lf	0.75		

Figure 4 shows the simulated and measured results of $|S_{11}|$, $|S_{12}|$ and radiation patterns of the proposed dipole antenna. It can be observed that the measured results agree well with the simulated results. The measured bandwidth of $|S_{11}|$ and $|S_{12}|$ is just as the simulated ones described above. Figure 4c shows the co-polarization and cross-polarization patterns of the antenna when one port is fed. Because of the rotational symmetry, the patterns are the same when the other port is fed. As can be seen from the figure, the antenna has a gain of about 6.2 dBi, and the cross-polarization level is lower than −16.5 dB.

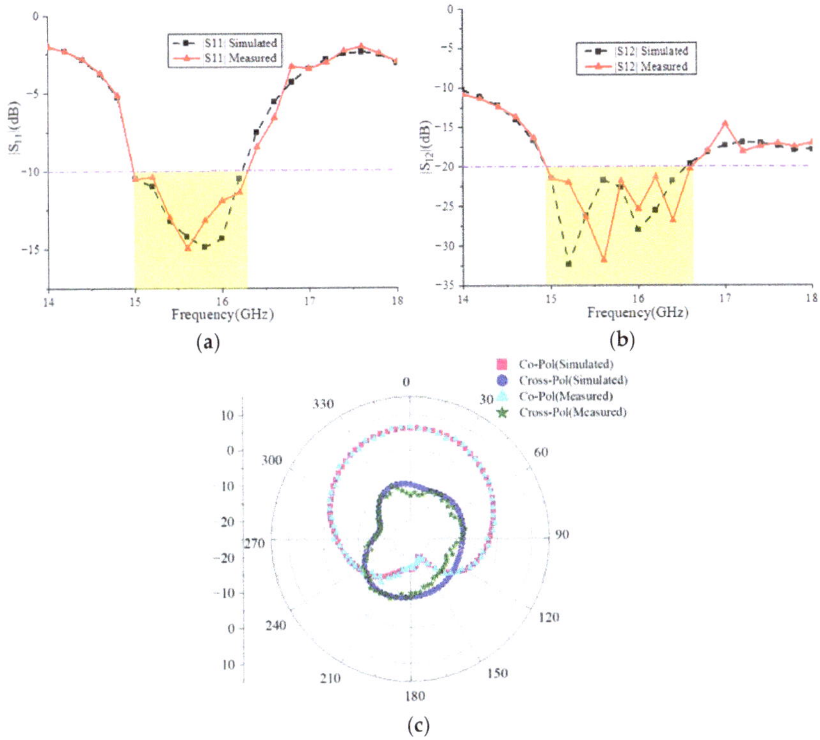

Figure 4. Simulation and measurement results of the dipole antenna: (**a**) |S11|; (**b**) |S12|; and (**c**) co/cross-pol radiation patterns.

The surface current distributions of the antenna at 15.6 GHz are shown in Figure 5. It can be seen that the synthesizing surface current on the dipole is mainly along the y-axis when Port 1 is excited, while along the x-axis with Port 2 fed. Therefore, two orthogonal polarizations are achieved with a relatively good purity of polarization.

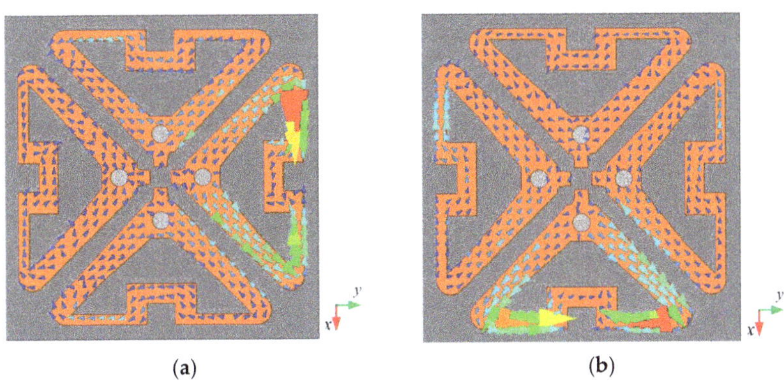

Figure 5. Current distributions of the proposed antenna fed by (**a**) Port 1 and (**b**) Port 2.

3. Design of the Cylindrical Array

With the proposed dipole antenna, a dual-polarization cylindrical phased array antenna is designed for Ku-band radar applications. The geometry and prototype of the antenna array is depicted in Figure 6. By comprehensive consideration of the size of the cylindrical surface and the grating lobe suppression of the antenna array, the element spacing is set to be 9 mm (approximately $0.47\lambda_0$). The array is conformal to a cylindrical aluminum surface The radius of the cylindrical surface is 185 mm, the height is 30 mm, the overall center angle of the metal cylindrical surface is 96°, and the array spanning angle is 88°. A distance of 1 mm is reserved between the antenna and the cylindrical surface to reduce the influence of the metal surface on the feed. There are holes in the metal cylindrical surface through which the SMP connectors are connected to the antennas.

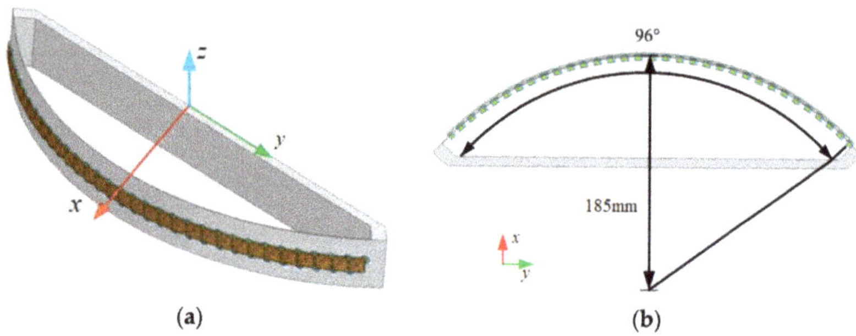

Figure 6. Array antenna model of (**a**) overall view (**b**) top view.

The antenna elements conformal to the cylindrical surface are excited with equal amplitude and proper phase to direct the array beam. The feeding phase of the ith element is

$$phase_i = -\frac{2\pi}{\lambda}(x_i \sin\theta_0 \cos\varphi_0 + y_i \sin\theta_0 \sin\varphi_0 + z_i \cos\theta_0) \tag{1}$$

where (θ_0, φ_0) is the scanning angle, and (x_i, y_i, z_i) represents the coordinate of the ith element. With the coordinate system shown in Figure 6, the scanning angles in the azimuth plane have θ_0 of 90°. Figure 7 shows the simulation patterns for different scanning angles at the center frequency 15.6 GHz. Since the scanning performances of the array for horizontal polarization and vertical polarization are similar, only the patterns for the horizontal polarization are given.

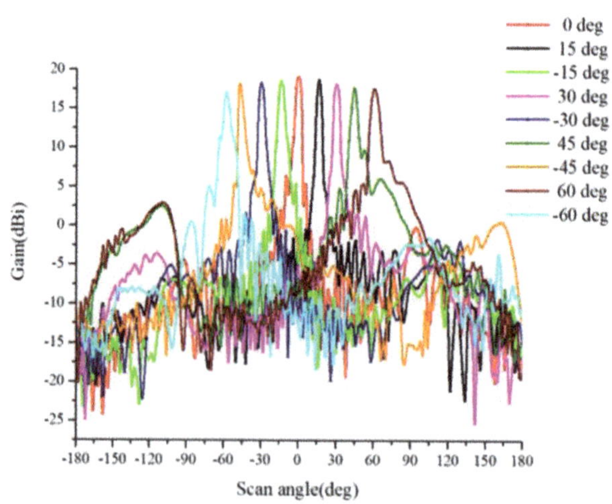

Figure 7. Simulation results of the array in the azimuthal plane.

It can be seen from the simulation results that when the scanning angle is 0°, the maximum gain is 19.06 dBi, and when the scanning angle is −60°, the gain is17.05 dBi. The gain fluctuation is approximately 2 dB in the scanning range of −60° to +60°. In order to verify the simulation results of the proposed cylindrical array, it is fabricated and measured in a microwave anechoic chamber. The array photograph and the measurement scene are as shown in Figure 8. The size of the microwave anechoic chamber is 6.4 m × 5 m × 3.2 m. Its equipment includes a 2 m × 2 m scanning shelf, a rotating platform, a vector network analyzer, and a control cabinet. Making use of a standard Ku-band horn as the receiving antenna, the radiation pattern of the array is measured by far-field measurement.

(a)

(b)

Figure 8. The prototype of the phased array. (**a**) Array photograph (**b**) the measurement scene.

Figure 9 shows the simulated and measured results of 0°, ±30°, ±60° for the horizontal-polarization patterns. It can be seen that the measured results are in good agreement with the simulated ones. Table 2 lists the gains at these scan angles. It is observed that the measured gains are slightly lower than the simulation ones. This is caused by the losses of the connectors and cables. At broadside, all the 32 elements contribute to the main lobe and the gain is the maximum. When the beam is scanned, some of the elements on the opposite side of the scanning angle have less contribution to the main lobe, resulting in the decreasing gains. However, the scan loss is only 2.42 dB from 0° to +60°, which shows the excellent wide-angle scanning performance for a cylindrical conformal array.

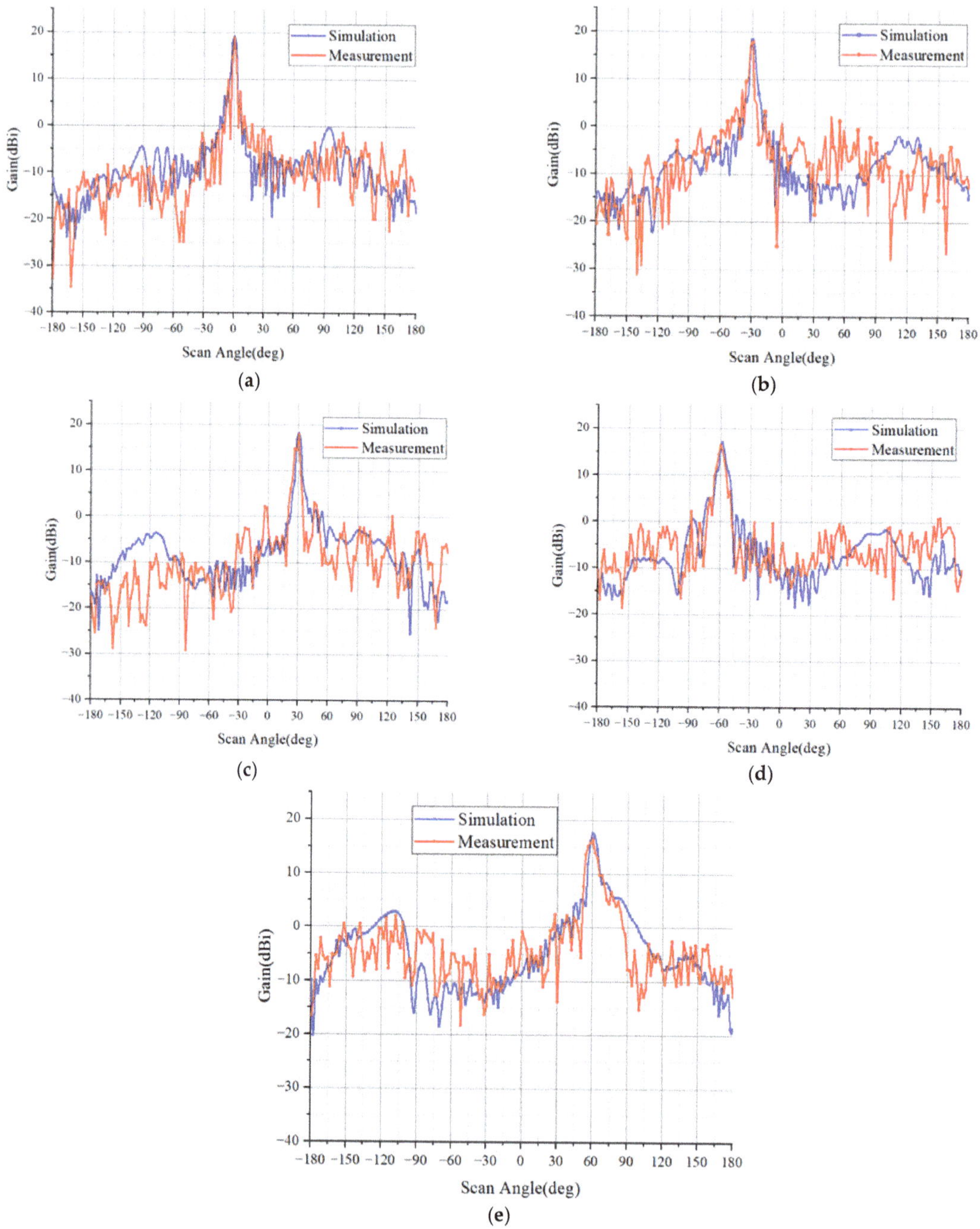

Figure 9. Simulated and measured results for different scan angles. (**a**) φ = 0°. (**b**) φ = −30°. (**c**) φ = 30°. (**d**) φ = −60°. (**e**) φ = 60°.

Table 2. Gain of the array at different scan angles.

	−60°	−30°	0°	30°	60°
Simulation (dBi)	17.32	18.29	19.06	18.12	17.56
Measurement (dBi)	16.66	17.83	18.90	17.96	16.48

Figure 10 shows the simulated radiation efficiency of the array at broadside. As can be seen, the array has a radiation efficiency better than 96% in the working frequency band.

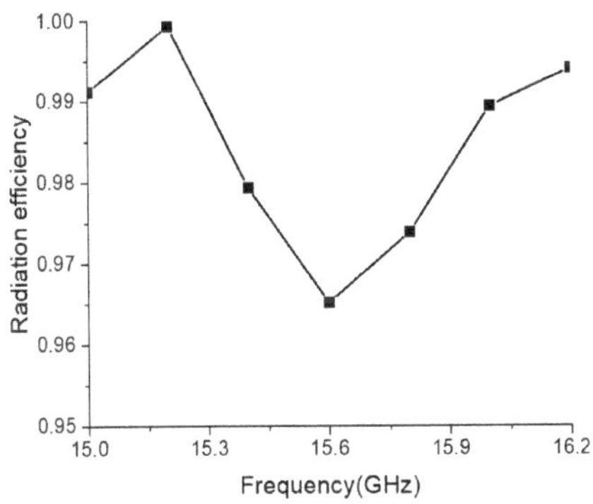

Figure 10. Simulated radiation efficiency of the array at broadside.

A comparison between this study and other cylindrical arrays in existing literatures is carried out in Table 3. The proposed dual polarization conformal array has the widest scan angle with small scan loss and almost the lowest profile. The bandwidth and port isolation are medium, which is sufficient for many applications.

Table 3. Comparison of cylindrical arrays.

Reference	Bandwidth	Port Isolation	Thickness	Scan Angle	Scan Loss	Gain (Broadside)
[11]	1.74%	Sigle-polarized	$0.052l_0$	−30°~+30°	1.25 dB	20.8 dBi
[13]	20.98%	>45 dB	$0.211l_0$	−45°~+45°	1.88 dB	14.92 dBi
[15]	100%	>17 dB	$0.281l_0$	−60°~+60°	2.5 dB	Not mentioned
[23]	7.14%	>50 dB	$1.051l_0$	Switching beam	-	Not mentioned
This work	8.0%	>21 dB	$0.111l_0$	−60°~+60°	2.42 dB	19.06 dBi

4. Conclusions

This article proposes a dual-polarization dipole antenna in Ku-band. By grooving the patch and bending the dipole side, the antenna bandwidth is effectively widened, and the port isolation is greatly improved. Additionally, the proposed double-layered design, which introduces an isolation ground between two substrates also plays an important role in the port isolation improvement. The designed dipole antenna is applied to a cylindrical phased array, which realize a scan range of −60° to 60° with the gain fluctuation less than 2.42 dB. The measured results verify the simulation results, which provides a guidance for the design of large-scale arrays and dual polarization arrays.

This study only discusses the performance of a one-dimensional linear array. An array on a two-dimensional surface will be further studied in the future.

Author Contributions: Conceptualization, N.Z. and Z.X.; methodology, N.Z. and L.G.; software, N.Z., P.Z. and L.G.; validation, N.Z. and P.Z.; formal analysis, N.Z. and P.Z.; investigation, L.G.; resources, J.Q.L.; data curation, N.Z. and J.Q.L.; writing—original draft preparation, N.Z.; writing—review and editing, N.Z. and Z.X.; visualization, P.Z. and L.G.; supervision, J.Q.L.; project administration, J.Q.L.; funding acquisition, Z.X. All authors have read and agreed to the published version of the manuscript.

Funding: This research received no external funding.

Data Availability Statement: The raw/processed data required to reproduce these findings cannot be shared at this time as the data also forms part of an ongoing study.

Conflicts of Interest: The authors declare no conflict of interest.

References

1. Mailloux, R.J. *Phased Array Antenna Handbook*; Artech House: Boston, MA, USA, 2005; p. 185.
2. Li, H.; Kang, L.; Wei, F.; Cai, Y.M.; Yin, Y.Z. A Low-Profile Dual-Polarized Microstrip Antenna Array for Dual-Mode OAM Applications. *IEEE Antennas Wirel. Propag. Lett.* **2017**, *16*, 3022–3025. [CrossRef]
3. Yang, B.; Huang, H.; Feng, B.; Deng, L. A Dual-Polarized Zhe-Shaped Conformal Patch Antenna for 5G Millimeter-Wave Applications. In Proceedings of the IEEE 4th International Conference on Electronic Information and Communication Technology (ICEICT), Xi'an, China, 18–20 August 2021; pp. 866–868.
4. Liao, L.; Li, Z.; Tang, Y.; Chen, X. Dual-Polarized Dipole Antenna for Wireless Data and Microwave Power Transfer. *Electronics* **2022**, *11*, 778. [CrossRef]
5. Benny, S.; Sahoo, S. Analysis of Optimized Subarray Configuration for Cross Polarization Reduction for Phased Array Antennas used in Weather Radar. In Proceedings of the IEEE International Symposium on Antennas and Propagation and USNC-URSI Radio Science Meeting (APS/URSI), Singapore, 4–10 December 2021; pp. 519–520.
6. Wang, X.; Zhou, S.; Liu, H.; Yan, J.; Su, H.; Li, H.; Yin, K.; Sun, W.; Zhou, L. Polarization Parameter Estimation of Conformal MIMO Radar Targets. In Proceedings of the IEEE Radar Conference (RadarConf20), Florence, Italy, 21–25 September 2020; pp. 1–5.
7. Huai, W.; Yanyan, Z.; Xinghui, P.; Xinying, B. An S/C Dual-Band Dual-Polarization Array for Synthetic Aperture Radar. In Proceedings of the IEEE 9th International Symposium on Microwave, Antenna, Propagation and EMC Technologies for Wireless Communications (MAPE), Chengdu, China, 26–29 August 2022; pp. 184–187.
8. Kobayashi, T.; Ko, K.; Choi, S.J.; Choi, J.H. Orthogonal Dual Polarization GPR Measurement for Detection of Buried Vertical Fault. *IEEE Geosci. Remote Sens. Lett.* **2022**, *19*, 1–5. [CrossRef]
9. Wang, P.; Wen, G.; Zhang, H.; Sun, Y. A Wideband Conformal End-Fire Antenna Array Mounted on a Large Conducting Cylinder. *IEEE Trans. Antennas Propag.* **2013**, *61*, 4857–4861. [CrossRef]
10. Liu, Y.; Wang, T.; Zhong, P.; Wang, S. Cylindrical conformal broadband array antenna based on defective ground structure. In Proceedings of the IEEE 4th Advanced Information Technology, Electronic and Automation Control Conference (IAEAC), Chengdu, China, 20–22 December 2019; pp. 382–385.
11. Xia, Y.; Muneer, B.; Zhu, Q. Design of a Full Solid Angle Scanning Cylindrical-and-Conical Phased Array Antennas. *IEEE Trans. Antennas Propag.* **2017**, *65*, 4645–4655. [CrossRef]
12. Zhang, J.; Chai, S.; Xiao, K.; Ding, L.; Zhao, F. Differential feeding technique for full-polarization conformal phased array. In Proceedings of the International Workshop on Electromagnetics: Applications and Student Innovation Competition, London, UK, 30 May–1 June 2017; pp. 87–88.
13. Lang, L.; Mei, L.; Zhang, N. Design and Simulation of Dual-polarized Non-planar Conformal Phased Array Antenna. In Proceedings of the Photonics & Electromagnetics Research Symposium—Fall (PIERS—Fall), Xiamen, China, 17–20 December 2019; pp. 2385–2392.
14. Li, J.; Yang, S.; Gou, Y.; Hu, J.; Nie, Z. Wideband dual-polarized magnetically coupled patch antenna array with high port isolation. *IEEE Trans. Antennas Propag.* **2016**, *64*, 117–125. [CrossRef]
15. Xiao, S.; Yang, S.; Zhang, H.; Xiao, Q.; Chen, Y.; Qu, S.-W. Practical Implementation of Wideband and Wide-Scanning Cylindrically Conformal Phased Array. *IEEE Trans. Antennas Propag.* **2019**, *67*, 5729–5733. [CrossRef]
16. Long, X.-P.; Zhou, S.-Z.; Zong, Z.-Y.; Wu, W. Design of Broadband Dual-Polarized Conformal Phased Antenna. In Proceedings of the IEEE International Symposium on Antennas and Propagation and USNC-URSI Radio Science Meeting (APS/URSI), Singapore, 4–10 December 2021; pp. 989–990.
17. Xiao, Q.; Yang, S.; Bao, H.; Chen, Y.; Qu, S. A wideband dual-polarized conformal phased array based on tightly coupled dipoles. In Proceedings of the Sixth Asia-Pacific Conference on Antennas and Propagation (APCAP), Xi'an, China, 16–19 October 2017; pp. 1–3.
18. Wang, Y.; Yan, S. Novel Conformal Dipole Antenna with Polarization Diversity for Biomedical Applications. In Proceedings of the International Conference on Microwave and Millimeter Wave Technology (ICMMT), Shanghai, China, 20–23 September 2020; pp. 1–3.

19. Yan, Y.-D.; Jiao, Y.-C.; Zhang, C.; Zhang, Y.-X.; Chen, G.-T. Hemispheric Conformal Wide Beamwidth Circularly Polarized Antenna Based on Two Pairs of Curved Orthogonal Dipoles in Space. *IEEE Trans. Antennas Propag.* **2021**, *69*, 7900–7905. [CrossRef]
20. Peng, J.-J.; Qu, S.-W.; Xia, M.; Yang, S. Wide-Scanning Conformal Phased Array Antenna for UAV Radar Based on Polyimide Film. *IEEE Antennas Wirel. Propag. Lett.* **2020**, *19*, 1581–1585. [CrossRef]
21. Gao, Y.; Jiang, W.; Hu, W.; Wang, Q.; Zhang, W.; Gong, S. A Dual-Polarized 2-D Monopulse Antenna Array for Conical Conformal Applications. *IEEE Trans. Antennas Propag.* **2021**, *69*, 5479–5488. [CrossRef]
22. Gao, Y.; Jiang, W.; Hong, T.; Gong, S. A High-gain Conical Conformal Antenna with Circularly Polarization and Axial Radiation in X-band. In Proceedings of the 13th European Conference on Antennas and Propagation (EuCAP), Krakow, Poland, 31 March–5 April 2019; pp. 1–4.
23. Golbon-Haghighi, M.-H.; Mirmozafari, M.; Saeidi-Manesh, H.; Zhang, G. Design of a Cylindrical Crossed Dipole Phased Array Antenna for Weather Surveillance Radars. *IEEE Open J. Antennas Propag.* **2021**, *2*, 402–411. [CrossRef]
24. Hu, J.; Sligar, A.; Chang, C.H.; Lu, S.L.; Settaluri, R.K. A grounded coplanar waveguide technique for microwave measurement of complex permittivity and permeability. *IEEE Trans. Magn.* **2006**, *42*, 1929–1931.
25. Chen, J.S. A CPWG-fed dual-frequency rectangular patch antenna. *Microw. Opt. Technol. Lett.* **2002**, *34*, 397–398. [CrossRef]

future internet

MDPI

Article

Evolutionary Computation for Sparse Synthesis Optimization of CCAAs: An Enhanced Whale Optimization Algorithm Method

Bohao Tang, Lihua Cai, Shuai Yang, Jiaxing Xu and Yi Yu *

Changchun Institute of Optics, Fine Mechanics and Physics, Chinese Academy of Sciences, Changchun 130033, China
* Correspondence: yuyi@ciomp.ac.cn

Abstract: Concentric circular antenna arrays (CCAAs) can obtain better performance than other antenna arrays. However, high overhead and excessive sidelobes still make its application difficult. In this paper, we consider the sparse synthesis optimization of CCAAs. Specifically, we aim to turn off a specific number of antennas while reducing the sidelobe of CCAAs. First, we formulate an optimization problem and present the solution space. Then, we propose a novel evolutionary method for solving the optimization problem. Our proposed method introduces hybrid solution initialization, hybrid crossover method, and hybrid update methods. Simulation results show the effectiveness of the proposed algorithm and the proposed improvement factors.

Keywords: concentric circular antenna arrays; sparse synthesis; optimization

Citation: Tang, B.; Cai, L.; Yang, S.; Xu, J.; Yu, Y. Evolutionary Computation for Sparse Synthesis Optimization of CCAAs: An Enhanced Whale Optimization Algorithm Method. *Future Internet* **2021**, *14*, 347. https://doi.org/10.3390/fi14120347

Academic Editors: Naser Ojaroudi Parchin, Chan Hwang See and Raed A. Abd-Alhameed

Received: 17 October 2022
Accepted: 17 November 2022
Published: 22 November 2022

1. Introduction

With the rapid development of computer technology and wireless communication technology, increasing numbers of users are being connected to the network. However, due to the large number of users accessing 5G and 6G networks, a single common antenna often cannot provide sufficient network support for users. Increasing the transmission power of the antenna alone often cannot improve the network quality, and creates severe interference problems. In this case, using multiple antennas to form an antenna array and conduct beamforming can be seen as an effective technique to enhance wireless communication performance [1].

Specifically, multiple antennas can be controlled by a single controller and then emit the same electromagnetic wave towards the target. These electromagnetic waves are superimposed and canceled in free space to generate a beam pattern with a distribution of strengths and weaknesses [2]. We note that the direction with the strongest gain is called the mainlobe and the rest are called the sidelobes. By pointing the mainlobe to the target user, the power received by the target user can be greatly enhanced, thus strengthening the performance of the antenna system in long-range and high-speed scenarios.

The antennas can be distributed in different manners to form different antenna arrays. Common antenna arrays include linear antenna arrays, square antenna arrays, Yagi–Uda antenna arrays, etc. In recent years, concentric circular antenna arrays (CCAAs) [3] have received a lot of attention from researchers due to their unique advantages and inherent merits [4]. As shown in Figure 1, CCAAs consist of some concentric circular rings, and each ring contains a various number of antennas. Compared with other antenna arrays, CCAAs can perform 360-degree beam scanning to achieve more comprehensive coverage [5]. On the other hand, CCAAs can obtain relatively small sidelobes, thus reducing the interference signals.

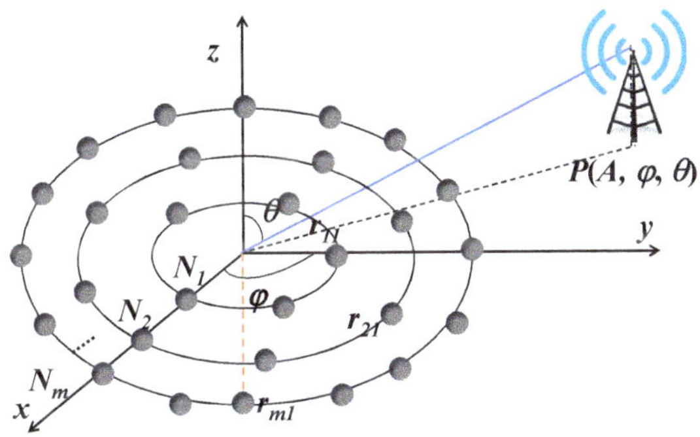

Figure 1. CCAA model.

In existing CCAAs, the excitation current weights of array elements are often set as uniform, as this scheme requires no further design and is easy to implement. However, these unoptimized and simple uniform CCAAs have some drawbacks that lead to their limited application in the real world. Specifically, CCAAs often contain a huge number of array elements which indeed effectively improve the transmission performance of CCAAs but cause a large overhead. Actually, in most scenarios, not all array elements need to be turned on. On the other hand, the sidelobes of uniform CCAAs are also difficult to suppress, which may generate a great deal of interference to the surrounding devices [6]. Therefore, some of the array elements of CCAAs can be turned off and some array elements can be assigned low excitation current weights, thereby completing the sparse synthesis optimization of CCAAs.

The existing methods for performing sparse synthesis are based on the following types of methods. Some work has utilized deterministic optimization methods to perform sparse synthesis of various antenna arrays, such as convex optimization methods to determine which antennas need to be turned off and optimize the excitation current accordingly. However, these methods often require the relaxation of certain key constraints, leading to a degradation of the optimization performance. Several works have employed evolutionary computation to perform sparse synthesis tasks. This is due to the fact that evolutionary computation is able to solve many NP-hard problems and does not require information about the gradient of the problem. Therefore, evolutionary computation can be seen as an effective method to perform the sparse synthesis optimization of CCAAs. For instance, references [7–9] considers using moth flame optimization, a teaching–learning-based optimization algorithm, and a cuckoo search algorithm to solve the sparse synthesis optimization of CCAAs and obtain suitable results.

Most of these methods only consider reducing the sidelobes or decreasing the number of active antennas as much as possible. However, aiming to keep the transmission power steady, there also is a requirement to reduce the maximum sidelobe level while switching off a fixed number of antennas. This requirement has been overlooked in the literature. Note that turning off a specific number of array elements of CCAAs while optimizing sidelobes is challenging because the structure of the decision variables will be changed. Different from these methods, we aim to turn off the specific number of antennas while reducing the sidelobe of CCAAs via a novel evolutionary algorithm in this work. Our contributions are summarized as follows.

- We formulate a novel optimization problem to turn off a specific number of antennas while reducing the sidelobes of CCAAs. The problem is challenging because it involves both binary and continuous decision variables and is an NP-hard problem.
- We propose an enhanced whale optimization algorithm (EWOA) to solve our formulated problem. EWOA introduces chaos theory and thereby proposes a novel hybrid solution initialization method. Moreover, we propose a hybrid solution crossover approach to balance the exploitation and exploration abilities of the EWOA. Finally, we

propose a hybrid solution update method to handle the specific structure of the problem. After being enhanced, the proposed EWOA is more suitable and efficient in solving our considered problem and has a main difference from the conventional EWOA.
- We conduct extensive simulations to verify the performance of the proposed EWOA. The simulation results show that the proposed EWOA is effective and outperforms other peer algorithms. Moreover, the proposed improved factors are also evaluated and demonstrated.

The rest of the paper is organized as follows. In Section 2, we briefly present some existing works in terms of deterministic methods and evolutionary approaches. Moreover, Section 3 introduces the considered models and formulates our problem. In addition, we propose the EWOA and the corresponding improved factors in Section 4. Furthermore, Section 5 conducts the simulations. Finally, the paper is concluded in Section 6.

2. Related Works

In this work, we consider using an evolutionary algorithm to solve the antenna optimization problem. Thus, in this part, we briefly present some existing works in the fields of deterministic methods and evolutionary approaches.

2.1. Deterministic Methods

Some works use deterministic methods to determine the excitation current weights of array elements. For instance, Buttazzoni et al. [10] proposed an iterative algorithm for sparse antenna arrays which introduces a compressive sensing approach for transforming the original non-convex optimization problem into a convex problem. Moreover, the authors in [11] proposed a deterministic approach to the optimal power synthesis of mask-constrained shaped beams through the CCAAs. Additionally, Fan et al. [12] formulated a unified sparse array design framework for beam pattern synthesis based on this design metric. The authors in [13] discussed a novel technique for sparse array synthesis in which the synthesis strategy was based on a sparse forcing algorithm using an improvement of reweighted minimization. In [14], a novel analytical approach to the synthesis of linear sparse arrays with non-uniform amplitude excitation was presented and thoroughly discussed. Comisso et al. [15] presented an iterative algorithm for the synthesis of the 3D radiation pattern generated by an antenna array of arbitrary geometry. However, these methods often require a relaxation of certain key constraints, leading to a degradation of the optimization performance.

2.2. Evolutionary Computation Approaches

Some approaches have adopted various evolutionary computation approaches to perform sparse synthesis tasks. For example, the authors in [4] used three evolutionary optimization algorithms, quantum particle swarm optimization (QPSO), teaching–learning-based optimization (TLBO), and symbiotic organism search (SOS), to thin the CCAAs. Moreover, Dutta et al. [16] proposed an improved version of QPSO to minimize the SLLs of CCAAs, in which the array elements have been considered to be uniformly excited with unit excitation amplitude. In [7], the authors presented an accurate approach based on moth flame optimization (MFO) to solve the CCAA synthesis problem for achieving lower SLL. In addition, Ismaiel et al. [17] proposed an optimization method based on a comprehensive learning particle swarm optimizer (CLPSO) to reduce the SLL of CCAAs, in which the excitation current weights are considered. In [18], a method based on a combination of ant lion optimizer and sequential quadratic programming was proposed for concentric circular antenna array (CCAA) synthesis, in which excitation amplitudes of array elements were optimized for CCAAs with low maximum sidelobe level (MSL), narrow first null beamwidth (FNBW), and low dynamic range ratio (DRR). Challa et al. [19] presented an optimized CCAA of antennas based on biogeography-based optimization (BBO). In [20], the authors used a political optimizer (PO), which is a novel evolutionary algorithm, to reduce the SLL of the CCAA. However, most of these methods only consider reducing the sidelobes or decreasing the number of the activated antennas as much as possible and do not consider turning off the specific number of array elements of CCAAs while decreasing the sidelobes.

Different from these works, we consider turning off a specific number of antennas while reducing the sidelobe of CCAAs via a novel evolutionary algorithm in this work. The details of our model and problem are presented as follows.

3. Model and Problem

In this section, we first present our considered CCAA model. Then, we formulate our considered sparse synthesis optimization problem in CCAAs.

3.1. CCAA Models

We show CCAA models in Figure 1. As can be seen, CCAAs consist of some concentric circular rings, and each ring contains a various number of antennas. Moreover, we assume that each antenna element of CCAA is radiation isotropic. In addition, all the antenna elements of CCAA are controlled by a centralized process and these antennas are synchronized in terms of time, frequency, and initial phase. Furthermore, all the antenna elements are fixed and the relative positions cannot be changed.

In this case, the electromagnetic waves generated by all the array elements will be precisely superimposed at the receiver and offset in some directions. Intuitively, these antennas will generate a beam pattern that contains a mainlobe and some sidelobes [21]. Mathematically, we can use the array factor as the index to show the signal strength of the CCAA system. In CCAAs, the array factor is a complex function of the excitation current weights of array elements of CCAAs, which is shown as follows [22].

$$F(\theta,\phi) = \sum_{m=1}^{M} \sum_{n=1}^{N_m} I_{mn} \cdot e^{j[kr_m \sin(\theta)\cos(\phi-\phi_{mn})]+\gamma_{mn}}, \tag{1}$$

where M is the number of concentric rings. Moreover, N_m is the number of antennas of the mth concentric rings, where $m = 1, 2, \ldots, M$. Note that we use N_T to denote the total number of antennas in the CCAA, which can be calculated as $N_L = \sum_{m=1}^{M} N_m$. In addition, $k = 2\pi/\lambda$m, where λ is the wavelength. Additionally, r_m is the radius of the mth circle with the value of $N_m d_m/2\pi$, wherein d_m is the spacing between the elements on the mth concentric rings. Furthermore, θ and ϕ are the zenith and azimuth angles, θ_0 and φ_0 are the directions of the mainlobe, and ϕ_{mn} is the central angle of the nth antenna element of the mth concentric rings. The most important parameters of Equation (1) are I_{mn} and γ_{mn}, where I_{mn} is the excitation current weight of nth antenna element of the mth circle and γ_{mn} is the initial phase of nth antenna element of the mth circle. We assume that the initial phase of all the antennas are synchronized, thus γ_{mn} can be calculated as follows.

$$\gamma_{mn} = -k\gamma_m \sin(\theta_0)\cos(\varphi_0 - \varphi_{mn}). \tag{2}$$

3.2. Inactivated or Activated Antenna Model

In this work, we aim to turn off a specific number of array elements in CCAAs while achieving some other optimization objective. Thus, we introduce our inactivated or activated antenna model in this section.

Specifically, we consider the use of a binary vector $\boldsymbol{B} = \{b_1, b_2, \ldots, b_{N_T}\}$ to denote the antenna's inactivated or activated condition. As shown in Figure 2, we use 0 to indicate that the antenna is not turned on, and 1 to indicate that the antenna is turned on. Mathematically, this model can be expressed as follows.

$$b_l = \begin{cases} 0, & \text{The } l\text{th antenna is turned off,} \\ 1, & \text{The } l\text{th antenna is turned on.} \end{cases} \tag{3}$$

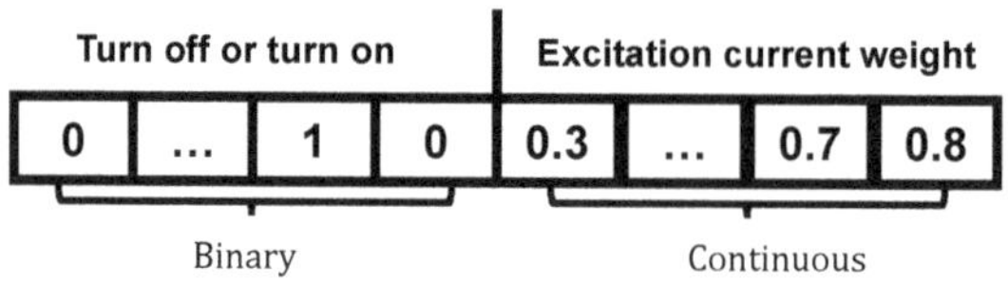

Figure 2. Inactivated or activated antenna model.

3.3. Sidelobe Suppression Model

As previously mentioned, the portion of the signal that is not the mainlobe belongs to the sidelobes. Therefore, it is difficult for us to evaluate the suppression of the sidelobes in a beam pattern. Generally speaking, we use the maximum sidelobe level (SLL) as a performance index to evaluate the suppression of the total sidelobes. This is because once the maximum SLL is suppressed, all the sidelobes will be suppressed. Mathematically, the calculation of the maximum SLL is as follows:

$$f_{sll}(\boldsymbol{I}) = \frac{\boldsymbol{B}|AF(\theta_{SLL})|}{|AF(\theta_{ML})|}, \tag{4}$$

where $\boldsymbol{I} = \{I_1, I_2, ..., I_{N_T}\}$ is a vector denote the excitation current weights of the total antennas.

3.4. Optimization Problem

In this work, we aim to turn off a specific number of antennas while reducing the sidelobe of CCAAs. Based on the models and analyses above, we can carefully control $\boldsymbol{B}$ and $\boldsymbol{I}$ to optimize the maximum SLL. Specifically, our considered optimization problem can be expressed as follows:

$$\min_{\{B,I\}} \quad f = f_{sll} \tag{5a}$$

$$\text{s.t.} \quad 0 \leqslant I_l \leqslant 1, \forall l \in N_T \tag{5b}$$

$$b_l = [0,1], \forall l \in N_T \tag{5c}$$

We mention that our considered problem is a very complex optimization problem. Specifically, controlling the excitation current weights of antennas to minimize the maximum SLL is already an NP-hard problem. Our problem is more complicated than that, which means that our problem is also a complex NP-hard problem. Moreover, most existing works only consider optimizing the excitation current weights, which only have continuous decision variables. Likewise, there are also some works that only consider determining whether the antennas should be switched on or switched off, which only have binary decision variables. Different from these works, we consider both excitation current weight and antenna switched-on optimizations. Thus, our formulated problem involves both continuous and binary decision variables, which is difficult to solve with the existing methods. In this case, a powerful high-performance algorithm is urgently needed to solve our problem.

4. Proposed Algorithm

In this section, we aim to propose a novel and powerful algorithm to solve our considered problem. To this end, we introduce an evolutionary computation method and then enhance the algorithm, thereby achieving better objective values by controlling the decision variables of $\boldsymbol{B}$ and $\boldsymbol{I}$.

4.1. Overview of Evolutionary Computation

Evolutionary computation is a very popular advanced optimization method in recent years. Evolutionary computation refers to the rules and guidelines of nature, physics, and chemistry, so as to abstract important mathematical models with solution significance, thereby achieving the purpose of searching the solution space [23].

Specifically, most evolutionary computation algorithms generate a population in which each individual in the population is a solution to the problem to be solved. These individuals can be evaluated by an objective function, and the corresponding individuals with high objectives will have higher values. Individuals in the population are constantly updated in iteration, and individuals with high values can influence individuals with low values to update. In this way, through continuous iteration, the algorithm can continuously search for better solutions in the solution space.

Mathematically, we denote the population as P, in which $P = \{X_1, X_2, ..., X_{N_{pop}}\}$, wherein X_i is the ith solution of the population. Moreover, X_i must be a feasible solution to the optimization problem shown in Equation (5), which can be denoted as $X_i = \{x_1, x_2, ..., x_{N_{dim}}\} = \{\boldsymbol{B}_i, \boldsymbol{I}_i\}$.

4.2. Conventional Whale Optimization Algorithm

Among the various evolutionary computation algorithms, the whale optimization algorithm (WOA) is a recently proposed algorithm with high performance. In some engineering problems, WOA often achieves better performance than other algorithms. WOA considers th prey-encircling spiral bubble-net feeding maneuver, and the search for prey of the whale population. These models can be expressed as follows.

4.2.1. Encircling Prey

Whales can find prey and surround them. The WOA algorithm allows individuals other than other individuals to update their positions to approach the optimal position, which can be expressed as follows.

$$\begin{aligned} \vec{D} &= \left|\vec{C} \cdot X^*(t) - X(t)\right|, \\ X(t+1) &= X^*(t) - \vec{A} \cdot \vec{D}, \end{aligned} \tag{6}$$

where t indicates the iteration, X^* is the position vector of the best solution of current population, and $||$ is the absolute value operator. The coefficient vectors A and C can be calculated as follows.

$$\begin{aligned} \vec{A} &= 2\vec{a} \cdot r - \vec{a}, \\ \vec{C} &= 2r, \end{aligned} \tag{7}$$

where a is a parameter that is linearly decreased from 2 to 0 through the iteration and r is a random number [0, 1].

4.2.2. Bubble-Net Attacking Method

The bubble-net attacking method contains two methods which are detailed as follows.

(i) Shrinking encircling mechanism. This behavior is achieved by reducing the value of parameter a, and the details can be found in [24].

(ii) Spiral updating position. WOA uses a helix-shaped movement of humpback whales to guide the update of the population, which can be shown as follows.

$$\begin{aligned} \overrightarrow{D'} &= |X^*(t) - X(t)|, \\ X(t+1) &= \overrightarrow{D'} \cdot e^{bl} \cdot \cos(2\pi l) + X^*(t), \end{aligned} \tag{8}$$

where $\overrightarrow{D'}$ is the distance between the current whale and prey, b is a constant parameter, and l is random in $[-1, 1]$.

4.2.3. Search for Prey

WOA develops a random selection method to enhance the exploration capabilities of the algorithm, as follows.

$$\vec{D} = \left|\vec{C} \cdot X_{\text{rand}} - X\right|,$$
$$X(t+1) = X_{\text{rand}} - \vec{A} \cdot \vec{D}, \tag{9}$$

where X_{rand} is a random position vector.

4.2.4. Shortcomings

Although WOA has many inherent advantages, there are still some flaws and urgent problems in WOA that are relevant to solving our problem. For example, the problem requires that the partial decision variables be binary, and the partial decision variables are continuous variables from 0 to 1. This makes our problem a very intractable mixed solution optimization problem, which poses a challenge to the solution of WOA. Second, the development ability and exploration ability required by WOA to solve different problems are not the same. These flaws prompted us to improve it and propose an improved version of WOA.

4.3. Proposed Enhanced Algorithm

In order to make WOA more able to solve our optimization problem, we propose an enhanced WOA (EWOA) in this part. The details and improved factors are shown as follows.

4.3.1. Hybrid Solution Initialization Method

The conventional WOA uses a stochastic number generator to generate the initial population. A generated initial population via this type has the following issues. First, this method will lead to a decrease in the performance of the initial solution, thereby reducing the convergence performance of the algorithm. Second, our problem is a mixed solution problem, and a stochastic number generator can only produce continuous solutions. Due to these reasons, we propose a hybrid solution update method in this part.

First, we generate the continuous part of the solution via chaos maps. Note that chaos maps are special mathematical structures that include periodic and stochastic continuous sequences. Because of their characteristics, they are often used in the improvement of initialization in evolutionary computation algorithms. Specifically, there are about 10 common chaotic maps, namely, Chebyshev, circle, Gauss/mouse, iterative, logistic, piecewise, sine, singer, sinusoidal, and tent. In this work, we select an iterative map to generate the continuous part of the solutions, which can be detailed as follows.

$$p_{q+1} = \sin\left(\frac{c\pi}{p_q}\right), c = 0.7, \tag{10}$$

where p_q is the qth sequence of the iterative map. Then, the continuous part of a solution can be initialized as follows.

$$x_k = lb_k + p_q(ub_k - lb_k), \tag{11}$$

By using this method, the initial solution generated by the algorithm will have a more balanced performance, so the algorithm's performance is better.

Second, we generate the binary part of the solution, which can be detailed as follows.

$$\boldsymbol{B} = \text{Rand}_{1,0}(N_{\text{sel}}, N_S - N_{\text{sel}}), \tag{12}$$

where $\text{Rand}_{1,0}(a, b)$ can generate a vector combined with a ones and b zeros.

4.3.2. Hybrid Solution Crossover

We consider a crossover method to hybridize solutions to enhance the exploration ability of our proposed algorithm. Specifically, the best solution for the population often contains the most information and can guide all individuals to explore in a better direction.

In each iteration, we select some individuals and use the crossover with the best solution to improve them. The details of this method are as follows. As with the proposed hybrid solution initialization method, this crossover method operates on different classes of solutions separately.

For the discrete part, we first find the discrete dimension which is different from the best solution and exchange information between the two. Subsequently, we judge whether the discrete part of the current individual satisfies the selected number of antennas. If the number is too large or too small, we invert the random selection solution from the individual until the number of antennas selected by the individual is appropriate.

For the continuous part, we make individuals exchange information with the best solution with half the probability. The pseudocode for this method is shown in Algorithm 1.

Algorithm 1: Hybrid Solution Crossover.

Define the related parameters: a solution X_i, best solution $X' = [x'_1, x'_2, ..., x'_{N_{dim}}]$, dimension number N_{dim}, etc.;
for $n = 1$ *to* $N_{dim}/2$ **do**
    if $N_{rand} < 0.5$ **then**
        if $x_n != x'_n$ **then**
            $x_n = x'_n$;
        end
        $x_{n+N_{dim}/2} = x'_{n+N_{dim}/2}$;
    end
end
while *sum* $([x_1, x_2, ..., x_{N_{dim}/2}]) == N_{sel}$ **do**
    if *sum* $([x_1, x_2, ..., x_{N_{dim}/2}]) > N_{sel}$ **then**
        Select a dimension with value 1 and assign it to 0;
    end
    if *sum* $([x_1, x_2, ..., x_{N_{dim}/2}]) < N_{sel}$ **then**
        Select a dimension with value 0 and assign it to 1;
    end
end
Return X_i;

4.3.3. Hybrid Solution Update

In this section, we update the two parts of the hybrid solution separately. Specifically, the continuous solution part is still updated using the original WOA update method. For discrete solutions, we consider mutating and crossing them first. The method is specifically shown in Algorithm 2.

Algorithm 2: Hybrid Solution Update.

Define the related parameters: a solution X_i, best solution $X' = [x'_1, x'_2, ..., x'_{N_{dim}}]$, dimension number N_{dim}, etc.;
if $N_{rand} < \frac{1}{3}$ **then**
  $[x_1, ..., x_{r1}, ..., x_{r2}, ..., x_{N_{dim}}] = [x_1, ..., x_{r2}, ..., x_{r1}, ..., x_{N_{dim}}]$;
end
else if $N_{rand} < \frac{2}{3}$ **then**
  for $n = 1$ *to* $N_{dim}/2$ **do**
    if $N_{rand} < 0.5$ **then**
      if $x_n != x'_n$ **then**
        $x_n = x'_n$;
      end
      $x_{n+N_{dim}/2} = x'_{n+N_{dim}/2}$;
    end
  end
end
while $sum\ ([x_1, x_2, ..., x_{N_{dim}/2}]) == N_{sel}$ **do**
  if $sum\ ([x_1, x_2, ..., x_{N_{dim}/2}]) > N_{sel}$ **then**
    Select a dimension with value 1 and assign it to 0;
  end
  if $sum\ ([x_1, x_2, ..., x_{N_{dim}/2}]) < N_{sel}$ **then**
    Select a dimension with value 0 and assign it to 1;
  end
end
Return X_i;

4.3.4. Main Steps and Complexity Analysis of EWOA

The main steps of the proposed EWOA can be shown in Algorithm 3. Moreover, the complexity of EWOA is examined. The calculation of the objective function value, which is considerably more difficult than other steps in our optimization problem, takes the longest. Thus, other steps can be skipped in this situation. As a result, when the maximum number of iterations and population size are denoted as t_{max} and N_{pop}, respectively, the complexity of EWOA is $\mathcal{O}(t_{max} \cdot N_{pop})$.

Algorithm 3: EWOA.

Define the related parameters: population size N_{pop}, bat dimension N_{dim}, maximum iteration t_{max}, and objective function, etc.;
Initialize the population by using Equations (10)–(12);
for $t = 1$ *to* t_{max} **do**
  Calculate the objective values of all individuals;
  Select the individual with the highest objective value as the best solution X^*;
  Update the individuals via conventional WOA method (e.g., Equations (6)–(9));
  Perform crossover operation via Algorithm 1;
  Perform solution update via Algorithm 2;
end
Return X^*; //X^* **is the best solution obtained by the algorithm**

5. Simulations

In this section, we perform simulations to evaluate the performance of EWOA. First, we present our experimental setup, comparison algorithms, etc. Second, we present our optimization results. Third, we illustrate the effectiveness of the improvement factors.

5.1. Setups

In this work, we use a CCAA with 10 rings for the simulation. In this case, the total number of antenna elements in the CCAA is 440, and the number of elements per ring is 8× m, where m is the number of turns. Furthermore, we aim to turn off one half of the antenna elements. Moreover, our used CPU was an 11th Gen Intel(R) Core(TM) i7-11700 @ 2.50 GHz with 16 GB of RAM.

In addition, we introduced some comparison algorithms for comparison, specifically, the dragonfly algorithm (DA) [25], conventional WOA [24], gray wolf optimization (GWO) [26], sine cosine algorithm (SCA) [27], and salp swarm algorithm (SSA) [28]. In this case, all algorithm populations were set to 50, and the number of iterations was set to 200. Note that these algorithms are both evolutionary algorithms, in which their computational complexities correspond to the times of fitness calculations. Thus, in this work, these algorithms have the same complexity which is $t_{max} \times N_{pop}$, i.e., 50 × 200.

5.2. Simulation Results

We first evaluated the objective function values of different algorithms to solve our problem and give the average, maximum, and minimum values in Table 1. It can be seen that our proposed method is superior to all other comparison algorithms. The reason is that we have a better balance between exploration ability and exploitation ability. Moreover, as shown in Figure 3, we give the convergence curves of different algorithms. It can be seen that our algorithm's convergence is the fastest. It can be seen that our proposed initial solution method can effectively speed up convergence. In addition, as shown in Table 2, we give the running time of different algorithms. It can be seen that the running time of all algorithms is similar, and our algorithm does not increase the running complexity of the algorithm. Furthermore, as shown in Figure 4, we provide the beam patterns of the CCAAs obtained by different algorithm. As can be seen, our proposed EWOA achieved the best beam pattern performance. For a more intuitive demonstration, we plot the structure of the optimized CCAAs in Figure 5. Specifically, the entire antenna array is the structure of the original CCAA. After being optimized by our algorithm, some antennas are switched off while the rest are also switched on.

Table 1. Statistical results of the objective values obtained by different approaches.

(DB)	AVE.	MAX.	MIN.
DA	−23.2598	−22.6900	−23.7409
GWO	−23.2644	−22.6404	−24.0011
SSA	−23.2434	−22.6982	−23.8245
WOA	−23.441	−22.8497	−24.0907
SCA	−21.2753	−20.1031	−22.3683
EWOA	−23.7883	−23.2147	−24.1995

Table 2. Statistical results of the running time obtained by different approaches.

(DB)	AVE.	MAX.	MIN.
DA	409.3217	520.8348	286.9482
GWO	271.8296	298.7387	232.12
SSA	273.4268	302.3688	222.1255
WOA	297.5742	312.5098	290.1445
SCA	254.5393	290.0058	189.7762
EWOA	313.075	250.7853	331.1821

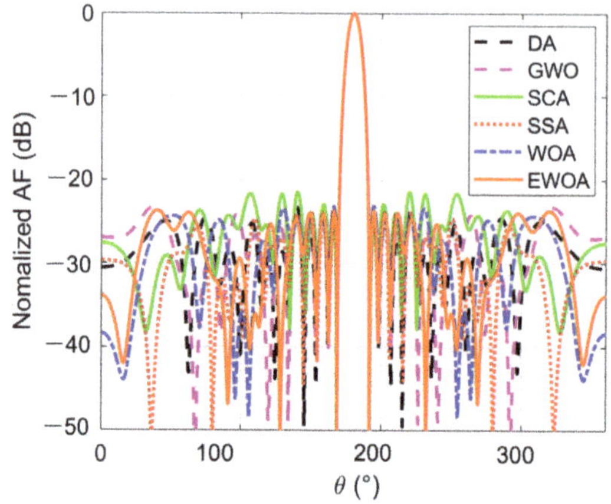

Figure 3. Beam patterns obtained by different algorithms.

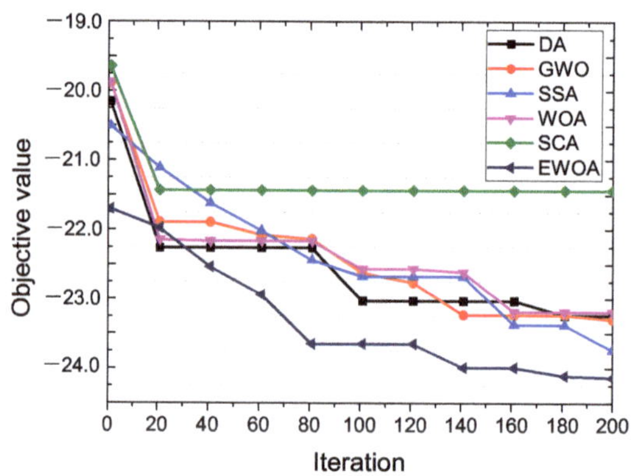

Figure 4. Convergence rates obtained by different algorithms.

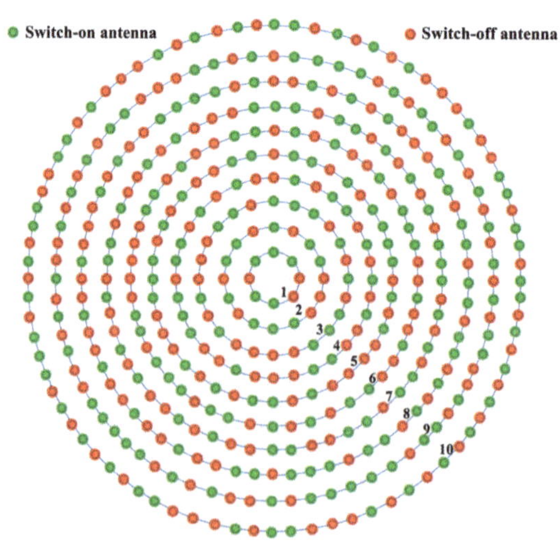

Figure 5. State of the CCAA.

On the other hand, we evaluated the impact of different improved factors on the algorithm. We considered different chaotic map-based initialization methods and proposed crossover methods. The optimization results and running time are shown in Tables 3 and 4, respectively. It can be seen that our selected chaotic map has the best performance, and the proposed crossover method can also effectively improve the performance of the algorithm. In addition, these algorithms are all on the same running time, which means that there is no additional increase in the running time of the algorithm.

Table 3. Statistical results of the objective values obtained by different improved factors.

(DB)	AVE.	MAX.	MIN.
Chebyshev	−23.5074	−23.0789	−23.934
Circle	−23.4647	−23.0118	−23.8412
Guass/mouse	−23.4788	−22.9087	−24.0222
Iterative	−23.5758	−23.0417	−23.9682
Logistic	−23.4941	−22.5767	−23.9068
Piecewise	−23.4257	−22.7275	−23.8546
Sine	−23.4498	−22.8004	−24.0941
Singer	−23.4784	−22.9801	−23.9801
Sinusoidal	−23.508	−23.0097	−23.9226
Tent	−23.4395	−23.0049	−23.9017
Crossover	−23.7153	−23.1617	−24.2604
EWOA	−23.7883	−23.2147	−24.1995

Table 4. Statistical results of the running time obtained by different improved factors.

(s)	AVE.	MAX.	MIN.
Chebyshev	300.2103	311.8449	273.2599
Circle	300.4639	308.2679	264.3575
Guass/mouse	305.5025	321.2203	285.5261
Iterative	311.7976	337.1776	182.6852
Logistic	306.6116	323.0361	290.9548
Piecewise	306.8332	328.9165	279.5588
Sine	331.7841	340.0234	322.1776
Singer	333.9409	341.0207	314.4988
Sinusoidal	337.2349	350.9823	277.0756
Tent	336.8633	354.8735	274.9848
Crossover	302.2373	316.8397	243.6888
EWOA	313.075	331.1821	250.7853

Except for the considered scenario in our original submission, we further added examples: one involving a very low population within 440 antenna elements and another one involving smaller arrays of 100 elements with large and small populations. The corresponding simulation results are presented as follows. Specifically, Table 5 shows the simulation results obtained by different algorithms in the case with 10 populations and 440 elements. Moreover, Table 6 shows the simulation results obtained by different algorithms in the case with 50 populations and 120 elements. In addition, Table 7 shows the simulation results obtained by different algorithms in the case with 10 populations and 120 elements. As can be seen, our proposed algorithm also outperforms other compared algorithms in the three cases. The reason may be that we balance the exploration and exploitation abilities and can make our proposed algorithm more suitable for solving our optimization problem.

Table 5. Statistical results of the objective values obtained by different approach in the case with 10 populations and 440 elements.

(DB)	AVE.	MAX.	MIN.
DA	−22.492	−21.281	−23.5388
GWO	−22.7027	−21.8035	−23.5196
SSA	−22.7403	−22.0197	−23.8178
WOA	−22.7205	−21.647	−23.4629
SCA	−20.1068	−18.9223	−22.2253
EWOA	−23.0437	−22.5879	−23.6858

Table 6. Statistical results of the objective values obtained by different approach in the case with 50 populations and 120 elements.

(DB)	AVE.	MAX.	MIN.
DA	−8.73719	−8.39645	−9.46009
GWO	−9.05977	−8.16812	−9.57519
SSA	−8.99624	−8.71156	−9.36257
WOA	−9.09705	−8.32748	−9.47902
SCA	−8.91788	−8.3524	−9.36407
EWOA	−9.41423	−9.15111	−9.6363

Table 7. Statistical results of the objective values obtained by different approach in the case with 10 populations and 120 elements.

(DB)	AVE.	MAX.	MIN.
DA	−8.27269	−7.7332	−9.08744
GWO	−8.68996	−7.24128	−9.26944
SSA	−8.59402	−7.72822	−9.31861
WOA	−8.67264	−7.46379	−9.38475
SCA	−8.56307	−7.73576	−9.14937
EWOA	−9.1513	−8.67405	−9.45595

6. Conclusions

This paper considered the sparse synthesis optimization of CCAAs. We aimed to turn off a specific number of antennas while reducing the sidelobe of CCAAs. First, we formulated an optimization problem and presented the solution space. Then, we proposed an EWOA method for solving the optimization problem. EWOA introduced hybrid solution initialization, crossover, and update methods. Finally, we conducted simulations to show the effectiveness of the proposed EWOA and the proposed improvement factors.

Author Contributions: Data curation, B.T. and J.X.; Supervision, L.C.; Writing—original draft, S.Y. and Y.Y. All authors have read and agreed to the published version of the manuscript.

Funding: This work was supported by the National Natural Science Foundation of China [grant number 51675506], and the Foundation for Excellent Young Scholars of Jilin Province, China [grant number 20190103015JH].

Institutional Review Board Statement: Not applicable.

Informed Consent Statement: Not applicable.

Data Availability Statement: Not applicable.

Conflicts of Interest: The authors declare no conflict of interest.

References

1. Kala, D.D.; Sundari, D.T. A review on optimization of antenna array by evolutionary optimization techniques. *Int. J. Intell. Unmanned Syst.* **2021.**
2. Geyi, W. The method of maximum power transmission efficiency for the design of antenna arrays. *IEEE Open J. Antennas Propag.* **2021**, *2*, 412–430. [CrossRef]
3. Ioannides, P.; Balanis, C.A. Uniform circular and rectangular arrays for adaptive beamforming applications. *IEEE Antennas Wirel. Propag. Lett.* **2005**, *4*, 351–354. [CrossRef]
4. Dutta, K.P.; Mahanti, G.K. Evolutionary algorithms for optimal synthesis of thinned multiple concentric circular array antenna with constraints. *Int. J. Electron.* **2020**, *107*, 1649–1662. [CrossRef]
5. Dessouky, M.I.; Sharshar, H.; Albagory, Y. Efficient sidelobe reduction technique for small-sized concentric circular arrays. *Prog. Electromagn. Res.* **2006**, *65*, 187–200. [CrossRef]
6. Sun, G.; Liu, Y.; Chen, Z.; Liang, S.; Wang, A.; Zhang, Y. Radiation beam pattern synthesis of concentric circular antenna arrays using hybrid approach based on cuckoo search. *IEEE Trans. Antennas Propag.* **2018**, *66*, 4563–4576. [CrossRef]
7. Das, A.; Mandal, D.; Ghoshal, S.; Kar, R. Concentric circular antenna array synthesis for side lobe suppression using moth flame optimization. *AEU-Int. J. Electron. Commun.* **2018**, *86*, 177–184. [CrossRef]
8. Dutta, K.P.; Mahanti, G.K. Synthesizing uniform concentric circular array antenna for minimized sidelobes using teaching–learning-based optimization algorithm. In *Emerging Technologies in Data Mining and Information Security*; Springer: Berlin/Heidelberg, Germany, 2021; pp. 243–251.
9. Sun, G.; Liu, Y.; Chen, Z.; Zhang, Y.; Wang, A.; Liang, S. Thinning of concentric circular antenna arrays using improved discrete cuckoo search algorithm. In Proceedings of the 2017 IEEE Wireless Communications and Networking Conference (WCNC), San Francisco, CA, USA, 19–22 March 2017; pp. 1–6.
10. Buttazzoni, G.; Babich, F.; Vatta, F.; Comisso, M. Geometrical synthesis of sparse antenna arrays using compressive sensing for 5G IoT applications. *Sensors* **2020**, *20*, 350. [CrossRef]
11. Morabito, A.F.; Nicolaci, P.G. Optimal synthesis of shaped beams through concentric ring isophoric sparse arrays. *IEEE Antennas Wirel. Propag. Lett.* **2016**, *16*, 979–982. [CrossRef]
12. Fan, W.; Liang, J.; Fan, X.; So, H.C. A unified sparse array design framework for beampattern synthesis. *Signal Process.* **2021**, *182*, 107930. [CrossRef]
13. Pinchera, D.; Migliore, M.D.; Schettino, F.; Lucido, M.; Panariello, G. An effective compressed-sensing inspired deterministic algorithm for sparse array synthesis. *IEEE Trans. Antennas Propag.* **2017**, *66*, 149–159. [CrossRef]
14. Caratelli, D.; Vigano, M.C. A novel deterministic synthesis technique for constrained sparse array design problems. *IEEE Trans. Antennas Propag.* **2011**, *59*, 4085–4093. [CrossRef]
15. Comisso, M.; Palese, G.; Babich, F.; Vatta, F.; Buttazzoni, G. 3D multi-beam and null synthesis by phase-only control for 5G antenna arrays. *Electronics* **2019**, *8*, 656. [CrossRef]
16. Dutta, K.P.; Sundi, S.; Khalkho, S.S.; Tirkey, J.; Hembrom, K.K. Minimization of sidelobe level of uniformly excited concentric circular array antenna using evolutionary technique. In *Innovative Data Communication Technologies and Application*; Springer: Berlin/Heidelberg, Germany, 2021; pp. 185–194.
17. Ismaiel, A.M.; Elsaidy, E.; Albagory, Y.; Atallah, H.A.; Abdel-Rahman, A.B.; Sallam, T. Performance improvement of high altitude platform using concentric circular antenna array based on particle swarm optimization. *AEU-Int. J. Electron. Commun.* **2018**, *91*, 85–90. [CrossRef]
18. Taser, A.E.; Guney, K.; Kurt, E. Synthesizing of concentric circular antenna arrays by using a combination of ant lion optimizer and sequential quadratic programming. *Int. J. Numer. Model. Electron. Netw. Devices Fields* **2022**, e3071. [CrossRef]
19. Challa Ram, G.; Girish Kumar, D.; Venkata Subbarao, M. A Novel Optimization for Synthesis of Concentric Circular Array Antenna. In *Intelligent Data Engineering and Analytics*; Springer: Berlin/Heidelberg, Germany, 2022; pp. 343–349.
20. Durmus, A.; Kurban, R. Optimal synthesis of concentric circular antenna arrays using political optimizer. *IETE J. Res.* **2022**, *68*, 768–777. [CrossRef]
21. Li, J.; Kang, H.; Sun, G.; Liang, S.; Liu, Y.; Zhang, Y. Physical Layer Secure Communications Based on Collaborative Beamforming for UAV Networks: A Multi-objective Optimization Approach. In Proceedings of the IEEE INFOCOM 2021-IEEE Conference on Computer Communications, Vancouver, BC, Canada, 10–13 May 2021; pp. 1–10.
22. Balanis, C.A. *Antenna Theory: Analysis and Design*; John Wiley & Sons: Hoboken, NJ, USA, 2015.
23. Parpinelli, R.S.; Plichoski, G.F.; Silva, R.S.D.; Narloch, P.H. A review of techniques for online control of parameters in swarm intelligence and evolutionary computation algorithms. *Int. J. -Bio-Inspired Comput.* **2019**, *13*, 1–20. [CrossRef]
24. Mirjalili, S.; Lewis, A. The whale optimization algorithm. *Adv. Eng. Softw.* **2016**, *95*, 51–67. [CrossRef]
25. Mirjalili, S. Dragonfly algorithm: A new meta-heuristic optimization technique for solving single-objective, discrete, and multi-objective problems. *Neural Comput. Appl.* **2016**, *27*, 1053–1073. [CrossRef]
26. Nadimi-Shahraki, M.H.; Taghian, S.; Mirjalili, S. An improved grey wolf optimizer for solving engineering problems. *Expert Syst. Appl.* **2021**, *166*, 113917. [CrossRef]

27. Mirjalili, S.M.; Mirjalili, S.Z.; Saremi, S.; Mirjalili, S. Sine cosine algorithm: Theory, literature review, and application in designing bend photonic crystal waveguides. *Nat.-Inspired Optim.* **2020**, *811*, 201–217.
28. Faris, H.; Mirjalili, S.; Aljarah, I.; Mafarja, M.; Heidari, A.A. Salp swarm algorithm: Theory, literature review, and application in extreme learning machines. *Nat.-Inspired Optim.* **2020**, *811*, 185–199.

electronics

MDPI

Article

Design of 1 × 2 MIMO Palm Tree Coplanar Vivaldi Antenna in the E-Plane with Different Patch Structure

Nurhayati Nurhayati [1,*], Eko Setijadi [2], Alexandre Maniçoba de Oliveira [3], Dayat Kurniawan [4] and Mohd Najib Mohd Yasin [5]

1 Department of Electrical Engineering, Universitas Negeri Surabaya, Surabaya 60231, Indonesia
2 Departement of Electrical Engineering, Institut Tenologi Sepuluh Nopember, Surabaya 60111, Indonesia
3 Laboratório James Clerk Maxwell de Micro-ondas e Eletromagnetismo Aplicado (LABMAX), Instituto Federal de Educacao, Ciencia e Tecnologia de Sao Paulo, Cubatão 11533-160, Brazil
4 Research Center for Telecommunication, National Research and Innovation Agency (BRIN), Bandung 40135, Indonesia
5 Advanced Communication Engineering (ACE) Centre of Excellence, Faculty Electronic Engineering Technology, Universiti Malaysia Perlis (UniMAP), Kangar 01000, Perlis, Malaysia
* Correspondence: nurhayati@unesa.ac.id

Abstract: In this paper, 1 × 2 MIMO of Palm Tree Coplanar Vivaldi Antenna is presented that simulated at 0.5–4.5 GHz. Some GPR applications require wideband antennas starting from a frequency below 1 GHz to overcome high material loss and achieve deeper penetration. However, to boost the gain, antennas are set up in MIMO and this is costly due to the large size of the antenna. When configuring MIMO antenna in the E-plane, there is occasionally uncertainty over which antenna model may provide the optimum performance in terms of return loss, mutual coupling, directivity, beam squint, beam width, and surface current using a given substrate size. However, the configuration of E-plane antenna in MIMO has an issue of mutual coupling if the distance between elements is less than 0.5λ. Furthermore, it produces grating lobes at high frequencies.We implement several types of patch structures by incorporating the truncated, tilt shape, Hlbert and Koch Fractal, Exponential slot, Wave slot, the lens with elips, and metamaterial slot to the radiator by keeping the width of the substrate and the shape of the feeder. The return loss, mutual coupling, directivity, beam squint, beamwidth, and surface current of the antenna are compared for 1 × 2 MIMO CVA. A continuous patch MIMO has a spacing of 0.458λ at 0.5 GHz, which is equivalent to its element width. From the simulation, we found that Back Cut Palm Tree (BCPT) and Horizontale Wave Structure Palm Tree (HWSPT) got the best performance of return loss and mutual scattering at low-end frequency respectively. The improvement of directivity got for Metamaterial Lens Palm Tree (MLPT) of 4.453 dBi if compared with Regular Palm Tree-Coplanar Vivaldi Antena (RPT) at 4 GHz. Elips Lens Palm Tree (ELPT) has the best beam squint performance across all frequencies of 0°. It also gots the best beamwidth at 4.5 GHz of 3.320. In addition, we incorporate the MLPT into the radar application.

Keywords: bandwith; Coplanar Vivaldi Antenna; mutual coupling; radiation pattern

Citation: Nurhayati, N.; Setijadi, E.; de Oliveira, A.M.; Kurniawan, D.; Yasin, M.N.M. Design of 1 × 2 MIMO Palm Tree Coplanar Vivaldi Antenna in the E-Plane with Different Patch Structure. *Electronics* **2023**, *12*, 177. https://doi.org/10.3390/electronics12010177

Academic Editors: Naser Ojaroudi Parchin, Chan Hwang See and Raed A. Abd-Alhameed

Received: 5 December 2022
Revised: 21 December 2022
Accepted: 23 December 2022
Published: 30 December 2022

1. Introduction

The antenna is an important component of the telecommunications system because it is the final part of the process of sending and receiving data. Presently, there is a continuous increase in data transmission hence research for improving antenna performance continues to rise. Many antenna studies have been carried out for microwave imaging [1–3] military or electronic warfare [4,5], Wifi [6,7], 5G telecommunications [8–11], vehicle communication [12–14], Maritim, airborne [15,16], and RADAR applications [17–19]. The studies conducted on RADAR applications including GPR applications usually work in the low frequency between 500 MHz to 3 GHz [20,21]. Antenna research that discusses its usage in an extensive frequency has been conducted [22–24], however, research on planar antennas that perform at low frequencies below 1 GHz is still quite restricted.

An antenna that has higher bandwidth would contribute immensely to the high imaging resolution of the radar system. UWB imaging system is designed to detect and see the object of interest in a structure that can be classified as through-the-wall imaging biomedical imaging and ground-penetrating radar. The higher the bandwidth the better the image resolution. The lower the frequency the larger the wavelength, so a radar with a very large wavelength will not be able to locate anything very precisely. The lower frequency the poorer the resolution. On the other hand, the lower frequency has its advantage, particularly when using the radar for GPR or through-the-wall imaging applications as this would ease the high material losses that will be encountered when using a higher frequency range and also guarantee deeper penetration of the electromagnetic wave. Furthermore, radar and telecommunications applications require antennas that work with wide bandwidth and high gain.

There have also been studies on how to boost its bandwidth and gain by changing its feeding, ground plane, and radiator, including the inclusion of corrugated slots. Vivaldi is one of the planar antennas that can work in a wide bandwidth. This antenna produces high gain coverage by providing corrugated [25], metamaterial [26,27], lens [28,29], fractal structures [30], and others. In addition to modifying the antenna elements, the increasing gain can be done by arranging the antenna in the form of an array and MIMO [31] as discussed in [32–36]. MIMO (Multiple Input Multiple Output) uses multiple antennas in the transmitter and receiver to improve its gain and data output. The MIMO antenna can be applied for radio astronomy applications, GPR, through-wall imaging applications [37], microwave tomography [38], and 5G applications [39]. The antenna arrangement in the form of MIMO on wideband antennas must pay attention to the mutual coupling between neighboring components in the low-end frequency because it affects the scattering characteristics.

Research on the method of reducing mutual antenna coupling, especially for Vivaldi, has been discussed by [40] using the corrugated slot technique, in [41] by multiple notch structure in the ground plane, and [42] by a triangular director. Mutual coupling is a significant problem in the E-plane because surface current flows in the neighboring element by the continuous patch, causing coupling between elements, but in the H-plane, the coupling is caused by an electric field flowing through the air. Mutual coupling antenna with very wide bandwidth will be risky at low-end frequencies while for high frequencies there is usually no problem. This is at low frequencies, the antenna has a longer wavelength. Therefore, to reduce mutual coupling between elements, the antenna must be more than half the wavelength long. If the antenna is arranged in MIMO for wideband antennas, there will be a trade-off between the mutual coupling performance and the radiation pattern performance at low and high frequencies. At low frequencies, The Mutual coupling performance of the antenna is good if it has the mutual scattering parameter $S_{21} < -20$ dB or the isolation of more than 20 dB. When the distance between elements is too large relative to the wavelength at high frequencies, a free lobe will emerge, thereby deteriorating the performance of the radiation pattern. Many different types of antennas may be used in radar and communications applications, including patch antennas [3], monopole antennas [6,8], and 3D antennas [5]. However, some of those antennas have an omnidirectional radiation pattern, or if they do have a directional radiation pattern, the directivity is minimal. Vivaldi antenna has advantages such as a planar antenna, wide bandwidth, and directional radiation pattern. There are several types of Vivaldi antennas including Coplanar Vivaldi Antenna [43,44], Antipodal Vivaldi Antenna (AVA) [45], and Balanced Antipodal Vivaldi Antenna (BAVA) [46]. The AVA is frequently mentioned in contrast to the CVA, even though the CVA also offers benefits in its gain performance. Additionally, the Vivaldi antenna's elements are explained in greater detail than those of the MIMO antenna. As a consequence, a discussion about the performance of the CVA antenna in MIMO is also suggested.

According to the preceding explanation, many radar applications require antennas that can operate at low frequencies, and the size of antennas that operate at low frequencies is often large. Meanwhile, the production of huge antennas and MIMO setups is prohibitively expensive. Aside from that, other researchers have done studies on Vivaldi antennas with several models, but very few have compared multiple models with the same substrate size. And antenna designers are occasionally perplexed about which antenna model to utilize to optimize certain performances. As a result, a discussion about the performance of the Vivaldi antenna is required, including return loss, mutual coupling, beam squint, and a beamwidth of the antenna of the same size, so that the antenna performance results can be used as a recommendation for which antenna design accommodates the required performance.

As a result, our contribution is to create 15 different slot structures to the Palm Tree CVA to investigate the effect of slot shape on mutual coupling reduction as well as the effect of different slot structure shapes on the performance of the radiation pattern using the same substrate width. We only discuss mutual coupling in the E-field in this discussion because mutual coupling in the E-field is a major problem, particularly at low frequencies. After all, surface currents can flow directly to adjacent elements, whereas mutual coupling in the H field can occur due to electric field induction through the air. We discovered the position and shape of the truncated slot that provides the best performance in terms of return loss performance, the wavy slot shape that provides the best performance in terms of mutual coupling at low frequencies, and the size of the elliptical structure that can provide 0° beam squint performance at all entire frequencies (0.5–4.5 GHz) while also providing the best beamwidth. Furthermore, we developed a metamaterial in the form of a split ring resonator that can boost gain by 4.451 dBi when compared to the RPT-CVA without giving a slot structure, and we evaluate it with measurement data, apply it to radar, and compare it to relevant research.

In this study, we compared the performance of the return loss, mutual coupling, side lobe level, beam squint, and a beamwidth of 15 types of 1 × 2 MIMO Palm Tree Coplanar Vivaldi Antenna using the same feeding shape and the substrate width in the E-plane. The antennas compared are Regular Palm Tree-Coplanar Vivaldi Antena (RPT-CVA), Front Cut Palm Tree (FCPT-CVA), Middle Cut Palm Tree (MCPT-CVA), Back Cut Palm Tree (BCPT-CVA), Complete Cut Palm Tree (CCPT-CVA), Left Tilt Palm Tree (LTPT-CVA), Right Tilt Palm Tree (RTPT-CVA), Hilbert Fractal Structure Palm Tree (HFSPT-CVA), Koch Fractal Structure Palm Tree (KFSPT CVA), Exponential Slot Edge Palm Tree (ESEPT-CVA), Vertical Wave Structure Palm Tree (VWPT-CVA), Horizontale Wave Structure Palm Tree (HWPT-CVA), Regular Lens Palm Tree (RLPT-CVA), Elips Lens Palm Tree (ELPT-CVA) and Metamaterial Lens Palm Tree (MLPT-CVA).

This paper is organized as follows: Section 2 discusses antenna design, Section 3 discussesResults and Discussion, Section 4 discusses Measurement and comparison of related antenna, and Section 5 conclusion.

2. Antenna Configuration

This study designed 1 × 2 antenna Palm Tree Coplanar Vivaldi antenna in the E-plane, as shown in Figures 1 and 2, using the same type and size of the substrate. The antenna is designed with an FR-4 substrate with a thickness of 1.6 mm, while the radiator and feeding shape is made of copper with a thickness of 0.035 mm. The dimensions of the patch element are 250 mm × 250 mm × 0.035 mm but it has an additional substrate width of 12.5 mm on both sides of the antenna width to reduce the mutual coupling. The antenna was simulated in the frequency range between 500 MHz–4.5 GHz with the dimension as illustrated in Table 1. Equation (1) is used to determine the tapered side on the bottom of PT CVA, the flared region, and the ESE structure. The C_1 and C_2 are constants, and R is the exponential growth rate with the value of R_1, R_2, $R_{3,}$ and R_4 shown in Table 1 (based on Figure 1a. The beginning and ending points of an exponential curve are x_1, y_1, x_2, and y_2 [44].

$$y = C_1 e^{Rx} + C_2, \quad C_1 = \frac{y_2 - y_1}{e^{Rx_2} - e^{Rx_1}}, \quad C_2 = \frac{y_1 e^{Rx_2} - y_2 e^{Rx_1}}{e^{Rx_2} - e^{Rx_1}} \tag{1}$$

Figure 1h delivers the Hilbert curve structure in the third iteration with the total length of the slot (l_h), the line segment (*dd*), and iteration (*in*) follows Equation (2) [47]. In this case, we use 3rd iteration.

$$dd = \frac{l_h}{2^{in} - 1} \tag{2}$$

Figure 2c,d show the vertical and horizontal wave slots. The Constant of the wavy slot in Equation (3) is $B_1 = 2$, $B_2 = 1$, $B_3 = 1$, $B_4 = 5$, and $B_5 = 36$. The depth of the wave, the number of waves, and the length of the wave slot can all be modified by changing the value of B_n [48].

$$A(t) = B_1 \left(B_2 + B_3 \cos \left(\frac{B_4 \pi t}{B_5} \right) \right) \tag{3}$$

Table 1. Parameter dimension of the antenna.

Dimension (mm)									
Par	**Dim**	**Par**	**Dim**	**Par**	**Dim**	**Par**	**Dim**	**Par**	**Dim**
a	550	k	100	u	25	E	10	O	12
b	250	l	50	v	93.53	F	188	P	120°
c	135	m	99.78	w	6	G	30	Q	12
d	55	n	38	x	17.9	H	150	R	10
e	0.5	o	30°	y	162	I	50	S	16
f	90	p	30°	z	25.11	J	51	R_1	0.03
g	14	q	35	A	25	K	1	R_2	0.05
h	1005	r	25	B	62.39	L	6	R_3	−0.2
i	50	s	41	C	5	M	0.7	R_4	−02
j	25	t	5	D	25	N	0.4		

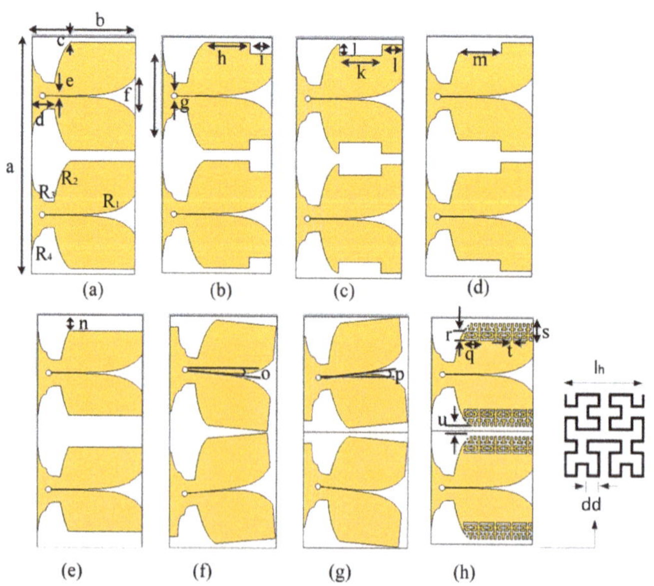

Figure 1. The 1 × 2 Coplanar Vivaldi Antenna of (**a**) Regular Palm Tree (RPT-CVA), (**b**) Front Cut Palm Tree (FCPT-CVA), (**c**) Middle Cut Palm Tree (MCPT-CVA), (**d**) Back Cut Palm Tree (BCPT-CVA), (**e**) Complete Cut Palm Tree (CCPT-CVA), (**f**) Left Tilt Palm Tree (LTPT-CVA), (**g**) Right Tilt Palm Tree (RTPT-CVA), (**h**) Hilbert Fractal Structure Palm Tree (HFSPT-CVA).

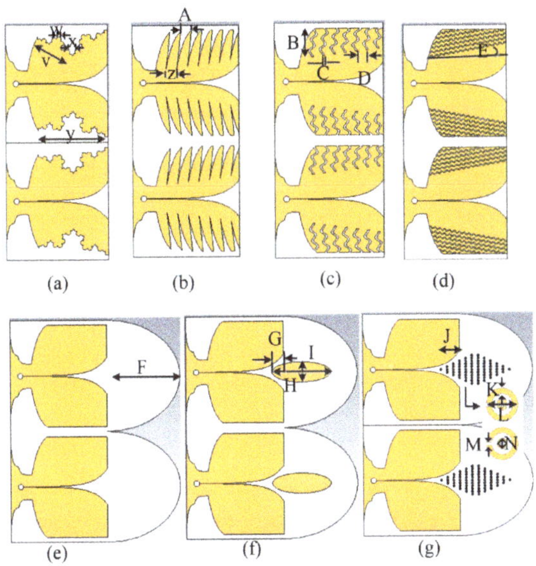

Figure 2. The 1 × 2 MIMO Coplanar Vivaldi Antenna of (**a**) Koch Fractal Structure Palm Tree (KFSPT-CVA), (**b**) Exponential Slot Edge Palm Tree (ESEPT-CVA), (**c**) Vertical Wave Structure Palm Tree (VWPT-CVA), (**d**) Horizontale Wave Structure Palm Tree (HWPT-CVA), (**e**) Regular Lens Palm Tree (RLPT-CVA), (**f**) Elips Lens Palm Tree (ELPT-CVA), (**g**) Metamaterial Lens Palm Tree (MLPT-CVA).

3. Results and Discussion

3.1. Scattering Parameter Performance

The S_{11} and S_{21} performance of RPT, FCPT, and MCPT-CVA are displayed in Figure 3. The best S_{11} is obtained from MCPT, at the low-end frequency. According to the simulation results displayed in Figure 3a, the low-end frequencies for RPT, FCPT, and MCPT-CVA that got S_{11} −10 dB are 0.52 GHz, 0.53 GHz, and 0.5 GHz. FCPT experienced poor return loss due to S_{11} exceeding −10 dB for frequencies 0.62 GHz to 0.81 GHz. The right side of Figure 1 shows the S_{21} performance of the RPT, FCPT, and MCPT. As can be observed, in the first low-end frequency at −20 dB the mutual scattering parameters for RPT, FCPT, and MCPT are 0.769 GHz, 0.848 GHz, and 0.58 GHz respectively. The best S_{21} is achieved for MCPT at low-end frequencies. However, the best S_{21} in overall frequency is acquired for FCPT as shown in Figure 3. The mutual scattering parameter has a lesser value at higher frequencies. In this study, an antenna with a substrate element's width of 275 mm was created, with frequencies of 0.5 GHz and 4.5 GHz and equal to the wavelengths of 600 mm and 66.67 mm, respectively. If the antenna has spacing between elements 275 mm so electrically, the antenna has a size of 0.458λ at a frequency of 0.5 GHz and 4.125λ at a frequency of 4.5 GHz. The mutual coupling of the antenna will be large if it has a size of fewer than 0.5λ (at the low-end frequency). Wideband and Ultra Wideband and also Super wideband antennas will experience mutual coupling issues at low frequencies. The greater the frequency, the less mutual coupling. However, it yields the problem of the grating lobe at high frequency. The performance of return loss and mutual scattering characteristics to the RPPT, BCPT, and CCPT in 1 × 2 MIMO antenna are assessed in Figure 3b. The best return loss performance is attained for BCPT. At 0.5 GHz, the antenna's S_{11} is −10.736 dB. At a frequency of 0.79 GHz, it has an S_{11} of −51.094 dB. The worst return loss for CCPT appears at low frequency. It demonstrates that at 0.699 GHz, it exhibits S_{11} −5.985 dB.

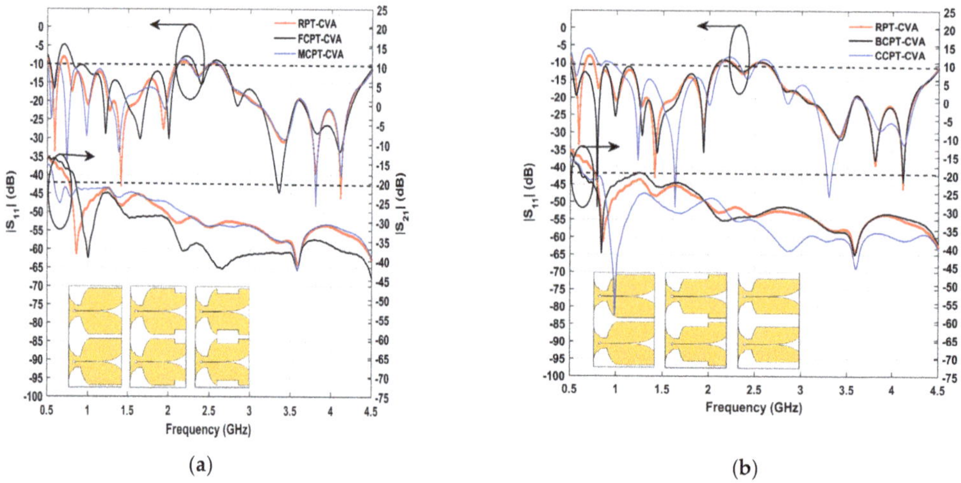

Figure 3. S_{11} and S_{21} performance of 1 × 2 MIMO (**a**). Regular Palm Tree Coplanar Vivaldi Antena (RPT-CVA), Front Cut Palm Tree (FCPT-CVA), Middle Cut Palm Tree (MCPT-CVA), and (**b**). S_{11} and S_{21} of Regular Palm Tree Coplanar Vivaldi Antena (RPT-CVA), Back Cut Palm Tree (BCPT-CVA), Complete Cut Palm Tree (CCPT-CVA).

However, CCPT has an S_{11} of −51.359 dB at 1.628 GHz. The S_{21} of RPT, BCPT, and CCPT might be visible on Figure 3b's bottom side. The best S_{21} performance is in BCPT, CCPT, and RPT CVA. Meanwhile, the CCPT CVA received S_{21} at 0.989 GHz of −55 dB. The CCPT performs well in S_{21} because it has the largest truncated area without copper on both edges of the patch.

Figure 4a shows the comparison of return loss performance and mutual scattering performance of RPT, LTPT, and RTPT CVA. From Figure 4a, it could be seen that LTPT has the best performance, due to covering S_{11} below −10 dB almost in the all-frequency range from 0.5–4.5 GHz. However, the bottom of Figure 4a shows that LTPT has the worst S_{21} performance. At 0.5 GHz LTPT has S_{21} −11.491 dB. LTPT has a distance between the feeding point is 0.495λ and RTPT of 0.422λ The distance between the two feeds for LTPT is getting farther away from the distance between the two centers of the tapered slot. RTPT has the best S_{21} at 0.885 GHz of −47.506. The electric field between two tapered slots will propagate away for RTPT while for LTPT. Figure 4b shows that the HFSPT and KFSPT have better performance of return loss than RPT at low-end frequency. At 0.5 GHz the S_{11} performance of RPT, HFSPT, and KFSPT are −8.311 dB, −12.442 dB, and −14.33 dB respectively. At 4.1 GHz KFSPT result S_{11} of −51.093 dB. S_{21} performance of RPT, HFSPT, and KFSPT is declared at the bottom of Figure 4b. At 0.5 GHz the S_{21} of RPT, HFSPT, and KFSPT are −8.311 dB, −14.502 dB, and −15.291 dB. It can be concluded that the KFSPT has the best S_{21} performance at low-end frequency.

The performance of S_{11} and S_{21} for RPT, ESEPT, VWSPT, and HWSPT is shown in Figure 5a. We can see that the best S_{11} performance is found for RPT. However, ESEPT, VWSPT, and HWSPT have S_{11} of more than −10 dB in some low frequencies. At 0.5 GHz ESEPT has S_{11} −16.859 dB but at 0.517 GHz to 0.628 GHz. it has an S_{11} of more than −10 dB as well as the structure of VWSPT and HWSPT.

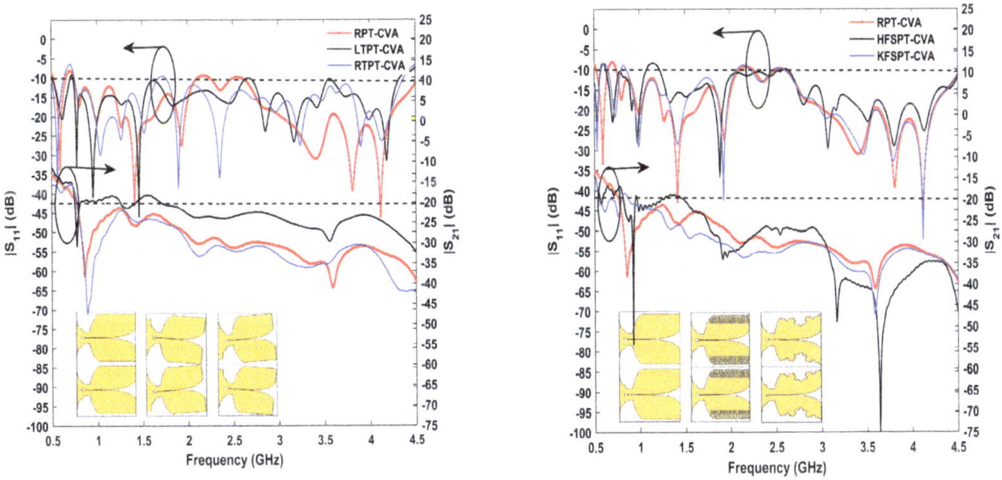

(**a**) (**b**)

Figure 4. S_{11} and S_{21} performance of 1 × 2 MIMO (**a**). Regular Palm Tree-Coplanar Vivaldi Antena (RPT-CVA), Left Tilt Palm Tree (LTPT-CVA), Right Tilt Palm Tree (RTPT-CVA) and (**b**). Regular Palm Tree-Coplanar Vivaldi Antena (RPT-CVA), Hilbert Fractal Structure Palm Tree (HFSPT-CVA), Koch Fractal Structure Palm Tree (KFSPT-CVA).

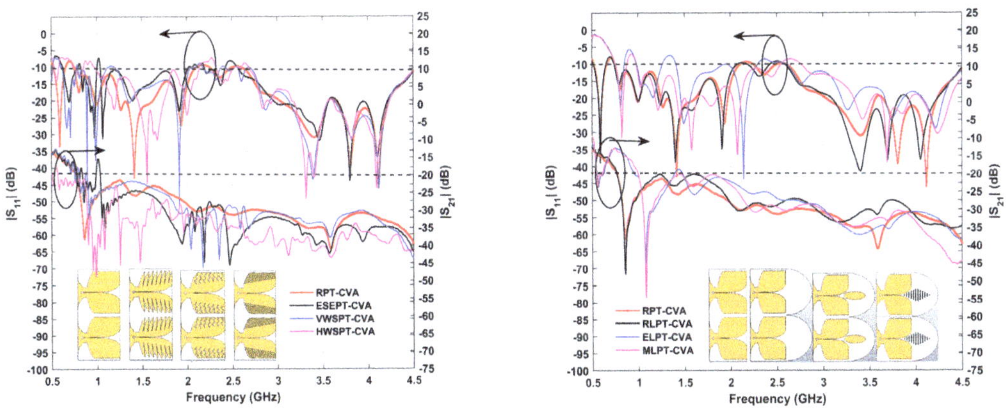

(**a**) (**b**)

Figure 5. S_{11} and S_{21} performance of 1 × 2 MIMO (**a**). Regular Palm Tree-Coplanar Vivaldi Antena (RPT-CVA), Exponential Slot Edge Palm Tree (ESEPT-CVA), Vertical Wave Structure Palm Tree (VWPT-CVA), and Horizontale Wave Structure Palm Tree (HWPTCVA) and (**b**) S_{11} and S_{21} of Regular Palm Tree-Coplanar Vivaldi Antena (RPT-CVA), Elips Lens Palm Tree (ELPT-CVA), and Metamaterial Lens Palm Tree (MLPT-CVA).

The return loss performance of the low frequency will be impacted by the corrugated structure or wave structure. The electric field will be trapped between the corrugated and wave structure. However, the Mutual scattering of HWSPT got the best performance than others. The S_{21} of RPT, ESEPT, and VWSPT is almost the same. Adding corrugated and wave structures in the vertical direction will affect the S_{21} performance at low frequencies. The return loss performance of RPT and RLPT is almost the same, likewise the performance of ELPT and MLPT-CVA. It could be shown in Figure 5b that by adding the Ellips and metamaterial structure in the mouth flared of the two tapered slots, the return loss performance in the low-end frequency becomes worst in the low-end frequency than without adding structure. From Figure 5b, it can also be seen that RPT and RLPT have almost the same return loss at frequencies below 3 GHz, but above 3 GHz there are differences. S_{21} performance of RPT, RLPT, ELPT, and MLPT can be seen at the bottom of Figure 5b. Even though ELPT and MLPT appear to have S_{21} below −20 dB at low-end frequency but only a few frequencies are covered. All four antennas have poor mutual catering parameters at low frequencies, even if at 0.895 GHz MLPT has S_{21} of −54.432 dB. According to the overall scattering parameter simulation findings for the 15 antenna types, BCPT has the best S_{11} performance at low frequencies, while HWSPT has the most effective S_{21} performance at low-end frequencies. Adding the structure in the patch will affect the electric field so that it interferes with the scattering parameters.

3.2. Radiation Pattern Performance

3.2.1. Directivity Performance

The directivity of the element and 1 × 2 RPT, 1 × 2 of FCPT, MCPT, BCPT, and CCPT is displayed in Figure 6a. At 2 GHz, by arranging the antenna into a MIMO, there is an improvement in RPT directivity of 3.2 dBi. At 2 GHz, the directivity of 1 × 2 RPT and 1 × 2 FCPT is 8.894 dBi and 11.587 dBi. This means there is an improvement in a gain of 2.693 dBi. However, at a frequency above 2 GHz, the gain decreases. The highest directivity is obtained for 1 × 2 BCPT at 4 GHz of 12 dBi.

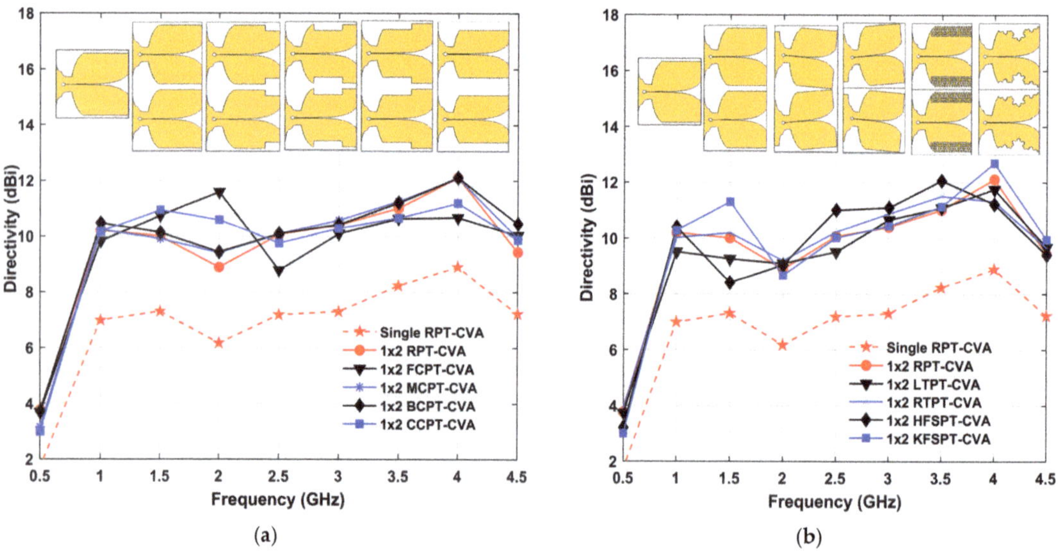

Figure 6. Directivity of: (**a**). Single and 1 × 2 Regular Palm Tree (RPT-CVA), 1 × 2Front Cut Palm Tree (FCPT-CVA), 1 × 2 Middle Cut Palm Tree (MCPT-CVA), 1 × 2 Back Cut Palm Tree (BCPT-CVA), 1 × 2 Complete Cut Palm Tree (CCPT-CVA) and (**b**). Single and 1 × 2 of Regular Palm Tree (RPT-CVA), 1 × 2 Left Tilt Palm Tree (LTPT-CVA), 1 × 2 Right Tilt Palm Tree (RTPT-CVA), 1 × 2 Hilbert Fractal Structure Palm Tree (HFSPT-CVA), 1 × 2 Koch Fractal Structure Palm Tree (KFSPT-CVA).

Figure 6b shows the directivity comparison between single and 1 × 2 RPT, 1 × 2 of LTPT, RTPT, HFSPT, and KFSPT. It demonstrates that the directivity changes slightly by tilting the antenna. It means there is an improvement in directivity by making the antenna position slightly tilted left and right, in this case, tilted 5 degrees, then the directivity does not change much. But at a frequency below 3.5 GHz, the 1 × 2 RTPT has a better performance of directivity than 1 × 2 LTPT and 1 × 2 RPT. At low frequencies, it can be a consideration for arranging the antenna with a larger tilt angle outward position so that the electric field coming out of the two tapered slots does not affect each other. Figure 6b shows the directivity of 1 × 2 KFSPT of 11.299 dBi while 1 × 2 RPT of 8.894 dBi. At Frequency 4 GHz the 1 × 2 KFSPT has the best directivity of 12.108 dBi.

The directivity performance of single and 1 × 2 RPT, 1 × 2 of ESEPT, VWSPT, and HWSPT can be observed a Figure 7a. 1 × 2 ESEPT has better performance of directivity than 1 × 2 RPT and 1 × 2 VWSPT almost at all frequency ranges. The best directivity is obtained at 4 GHz of 12.877 dBi. Figure 7b shows the directivity performance of single and 1 × 2 RPT, 1 × 2 RLPT, 1 × 2 ELPT, and 1 × 2 MLPT. At frequencies, less than 2.5 GHz the 1 × 2 ELPT got the best performance of directivity. However, at a frequency of more than 2.5 GHz, 1 × 2 MLPT has the best directivity. At 4 GHz, the directivity of 1 × 2 RPT, RLPT, ELPT, and MLPT is 12.108 dBi, 14.329 dBi, 15.248 dBi, and 16.561 dBi. It means there is an improvement of directivity of 4.453 dBi. Although the MLPT has a larger substrate size than regular ones, adding a lens and metamaterial structure can trap the electric field in the lens causing the gain increase.

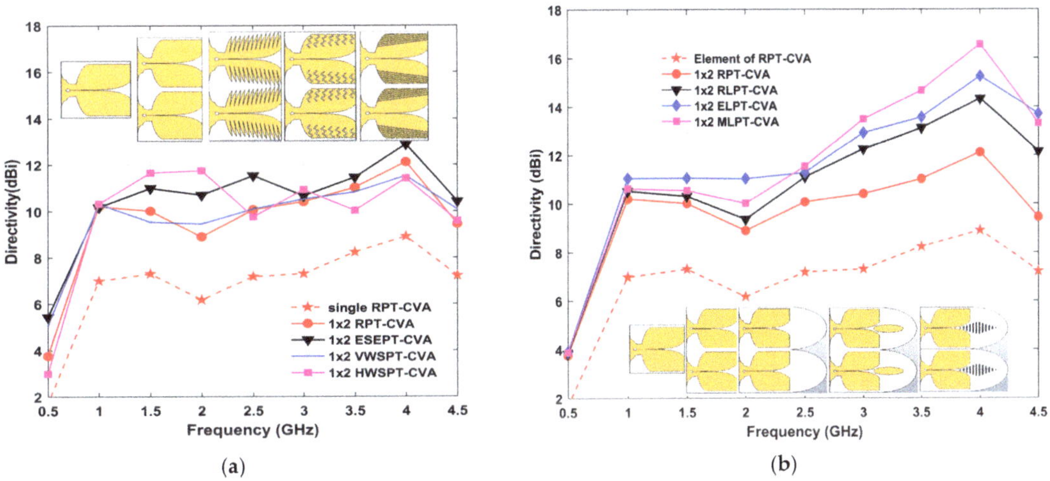

Figure 7. Directivity of (**a**). Element and 1 × 2 Regular Palm Tree- (RPT-CVA), 1 × 2 Exponential Slot Edge Palm Tree (ESE-CVA), 1 × 2 Vertical Wave Structure Palm Tree (VWPT-CVA), 1 × 2 Horizontale Wave Structure Palm Tree (HWPT-CVA), and (**b**). Element and 1 × 2 Regular Palm Tree-Coplanar Vivaldi Antena (RPT-CVA), 1 × 2 Regular Lens Palm Tree (RLPT-CVA), 1 × 2 Elips Lens Palm Tree (ELPT-CVA), And 1 × 2 Metamaterial Lens Palm Tree (MLPT-CVA).

However, in a wideband antenna, the wider the antenna bandwidth, the greater distance between elements relative to their wavelength (especially at high-end frequencies), and this cause a grating lobe which will reduce the directivity of the antenna. The grating lobe is a side lobe that is enlarged and resembles the main lobe, this thing caused by the effect of changing the distance between the antenna elements further apart. In this case the 1 × 2 MIMO has spacing between element is 275 mm and it means that at a frequency of 0.5 GHz, the distance between elements is 0.458λ while at a frequency of 4.5 GHz the distance between elements is 4.125λ. The distance between elements rises at 4.5 GHz, causing the grating lobe and antenna directivity to diminish.

3.2.2. Side Lobe Level Performance

Figure 8a shows the Side Lobe Level (SLL) of element RPT, 1 × 2 of RPT, 1 × 2 of FCPT, MCPT, BCPT, CCPT, LTPT, RTPT, and HFSP. Meanwhile, Figure 8b presents the SLL of element RPT, SLL 1 × 2 of RPT, 1 × 2 of KSPT, ESEPT, VWSPT, HWSPT, RLPT, ELPT, and

MLPT-CVA. From the simulation result, It shows that the best SLL performance is reached for RPT of 11.94 dB at 1 GHz, followed by 1 × 2 HFSPT of 10.357 dB at 3 GHz. Furthermore, most antennas have an SLL greater than −5 dB at frequencies above 1.5 GHz. Although mutual scaring parameters of the antenna at high frequencies are good, it produces a high-level sidelobe. This is because the higher the frequency, the greater the distance between the elements relative to the wavelength, therefore, the grating lobe will occur, enlarging the SLL However the SLL of a single element of RPT is better than others at frequency 2, 2.5, 3.5 and 4 GHz. But at frequency 1.5 GHz shows that 1 × 2 RPT has better performance of SLL than a single element. For wideband antennas, by arranging the antennas into MIMO, the sidelobe level performance can increase at low frequencies but at high frequencies, the SLL performance decreases due to the presence of grating lobes because the distance between elements becomes greater. Figure 8b shows that the 1 × 2 ELPT got the best SLL performance at 1 GHz of −13.757 dB. At 2 GHz, the 1 × 2 KFSPT results in the best performance of SLL of −9.426 dB. Meanwhile, at 2 GHz, the worst SLL of −0.677 dB was obtained for RPT. This means an SLL improvement of 8.749 dB between 1 × 2 KFSPT and 1 × 2 RPT. The best SLL of −6.89 dB was obtained at 2.5 GHz for 1 × 2 MLPT, while at 4 GHz, the 1 × 2 ELPT has the best performance of −4.688 dB. SLL performance can also be improved by adding a structure and lens at both antenna heights.

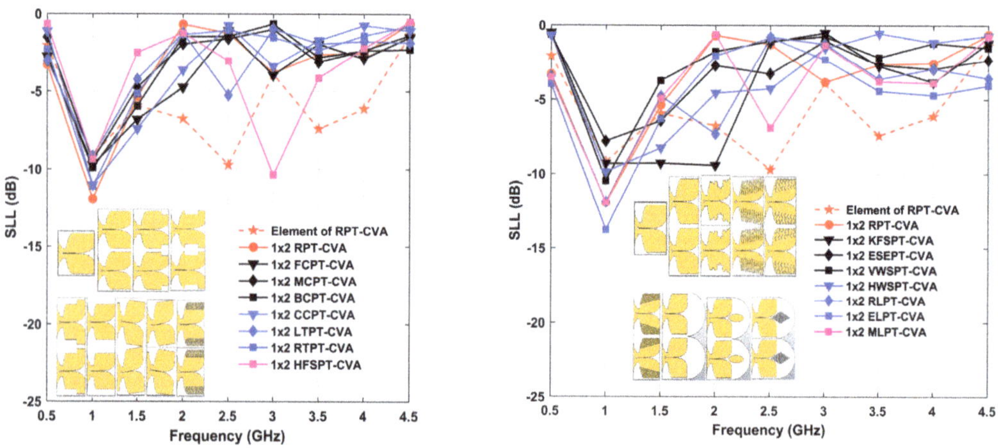

(**a**) (**b**)

Figure 8. Side Lobe Level of (**a**). Element and 1 × 2 Regular Palm Tree (RPT-CVA), 1 × 2 Front Cut Palm Tree (FCPT-CVA), 1 × 2 Middle Cut Palm Tree (MCPT-CVA), 1 × 2 Back Cut Palm Tree (BCPT-CVA), 1 × 2 Complete Cut Palm Tree (CCPT-CVA), 1 × 2 Left Tilt Palm Tree (LTPT-CVA), 1 × 2 Right Tilt Palm Tree (RTPT-CVA), 1 × 2 Hilbert Fractal Structure Palm Tree (HFSPT-CVA), and (**b**). Element and 1 × 2 Regular Palm Tree (RPT-CVA), 1 × 2 Koch Fractal Structure Palm Tree (KFSPT-CVA), 1 × 2 Exponential Slot Edge Palm Tree (ESEPT-CVA), 1 × 2 Vertical Wave Structure Palm Tree (VWSPT-CVA), 1 × 2 Horizontale Wave Structure Palm Tree (HWSPT-CVA), 1 × 2 Regular Lens Palm Tree (RLPT-CVA), 1 × 2 Elips Lens Palm Tree (ELPT-CVA), and 1 × 2 Metamaterial Lens Palm Tree (MLPT-CVA).

3.2.3. Beam Squint and Beamwidth Performance

The boresight to 10 dB beamwidth ratio is known as beam squint. The boresight should be symmetrical if the beam squint is zero. The feeding network settings, the design of the patch or radiator, and the design of the antenna substrate can all determine the polarization of the antenna, which can influence the beam squint. Ideally, the beam squint is zero which means the boresight is symmetrical. Changes in the beam squint can be induced by changes in the polarization of the antenna, which can be caused by the feeding network settings, the geometry of the patch/radiator, and the shape of the antenna substrate. The squint beam will reduce the link budget due to the misalignment of the main beam [42].

Figure 9a shows the beam squint of the element and 1 × 2 RPT, 1 × 2 of FCPT, MCPT, BCPT, CCPT, LTPT, RTPT, and HFSPT-CVA. The simulation results that HFSPT has the worst performance at 0.5 GHz and 2 GHz. At 2 GHz the result of beam squint is 32,425° and −30.613° for 1 × 2 HFSPT and 1 × 2 RTPT respectively. Figure 9b describes the beam squint of the element and 1 × 2 of RPT, 1 × 2 of KFSPT, ESEPT, VWSPT, HWSPT, RLPT, ELPT, and MLPT-CVA. At 0.5 GHz 1 × 2 KFSPT has the worst beam squint as shown in Figure 9b. Adding a corrugated slot can affect the beam squint as well as the return loss performance in the low-end frequency. The 1 × 2 KFSPT also results in the beam squint of 32.87°, 25.44°, and 21.048° at 2 GHz, 2.5 GHz, and 3 GHz respectively. Figure 9b, found that ELPT results in the best beam squint performance, in all frequencies between 0.5 GHz to 4.5. However, the 1 × 2 RPT element maximum Beamsquint is 3.28° at frequencies 0.5 and 4 GHz. Adding structure on. both edges of the patch can increase the gain but also affect the beam squint of the antenna.

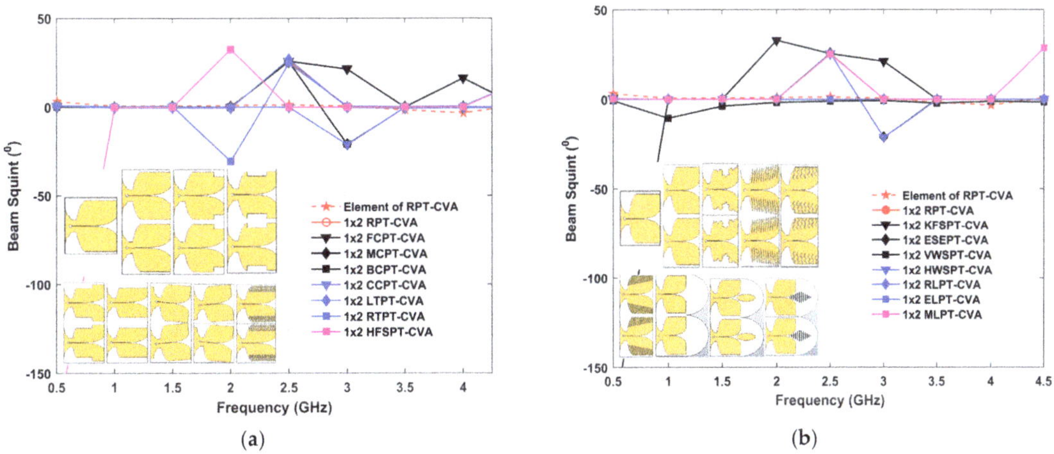

Figure 9. Beam Squint Performance of (**a**) Element and 1 × 2 Regular Palm Tree (RPT-CVA), 1 × 2 Front Cut Palm Tree (FCPT-CVA), 1 × 2 Middle Cut Palm Tree (MCPT-CVA), 1 × 2 Back Cut Palm Tree (BCPT-CVA), 1 × 2 Complete Cut Palm Tree (CCPT-CVA), 1 × 2 Left Tilt Palm Tree (LTPT-CVA), Right Tilt Palm Tree (RTPT-CVA), 1 × 2 Hilbert Fractal Structure Palm Tree (HFSPT-CVA), and (**b**). Element and 1 × 2 Regular Palm Tree (RPT-CVA), 1 × 2 Koch Fractal Structure Palm Tree (KFSPT-CVA), 1 × 2 Exponential Slot Edge Palm Tree (ESEPT-CVA), 1 × 2 Vertical Wave Structure Palm Tree (VWSPT-CVA), 1 × 2 Horizontale Wave Structure Palm Tree (HWSPT-CVA), 1 × 2 Regular Lens Palm Tree (RLPT-CVA), 1 × 2 Elips Lens Palm Tree (ELPT-CVA), and 1 × 2 Metamaterial Lens Palm Tree (MLPT-CVA).

Figure 10 shows that the antenna has a large beamwidth at a frequency of 0.5 GHz, which decreases with increasing frequency. The largest beamwidth obtained by the ELPT-CVA structure at 0.5 was 78.57°, while the smallest was by VWSPT-CVA at 41.41°. The antenna beamwidth at 4 GHz frequency shows the smallest beamwidth of 3.32° for the ELPT-CVA structure. Although at a frequency of 0.5 GHz, the ELPT-CVA antenna has the largest beamwidth, the addition of an elliptical structure, with increasing frequency, makes the beamwidth smaller. The beam width variation across the 15 simulated antenna in 1 × 2 MIMO configurations is modest at the 1–4.5 GHz frequency. Figure 10 illustrates that the single-element beamwidth is greater than the antenna in MIMO at all frequencies. Antennas in MIMO can increase beamwidth performance.

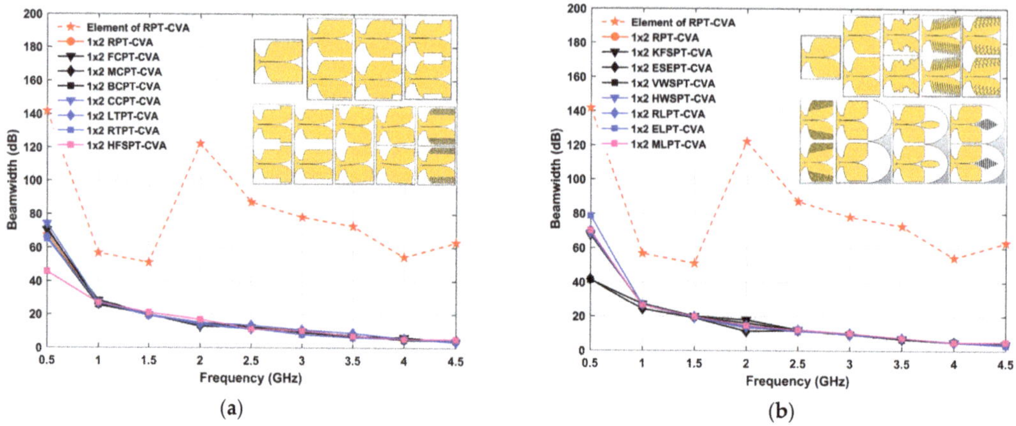

Figure 10. Beamwidth of (**a**). Element and 1 × 2 Regular Palm Tree (RPT-CVA), 1 × 2 Front Cut Palm Tree (FCPT-CVA), 1 × 2 Middle Cut Palm Tree (MCPT-CVA), 1 × 2 Back Cut Palm Tree (BCPT-CVA), 1 × 2 Complete Cut Palm Tree (CCPT-CVA), 1 × 2 Left Tilt Palm Tree (LTPT-CVA), 1 × 2 Right Tilt Palm Tree (RTPT-CVA), 1 × 2 Hilbert Fractal Structure Palm Tree (HFSPT-CVA), and (**b**). Element and 1 × 2 Regular Palm Tree (RPT-CVA), 1 × 2 Koch Fractal Structure Palm Tree (KFSPT-CVA), 1 × 2 Exponential Slot Edge Palm Tree (ESEPT-CVA), 1 × 2 Vertical Wave Structure Palm Tree (VWSPT-CVA), 1 × 2 Horizontale Wave Structure Palm Tree (HWSPT-CVA), 1 × 2 Regular Lens Palm Tree (RLPT-CVA), 1 × 2 Elips Lens Palm Tree (ELPT-CVA), and 1 × 2 Metamaterial Lens Palm Tree (MLPT-CVA).

3.2.4. Rectangular Radiation Characteristic

Figure 11 displays some of the outcomes of the RPT antenna's radiation pattern with a different slot structure. Figure 11a,b show the 2 GHz radiation patterns between the RPT, FCPT, and HWSPT antennas in the E-plane. Figure 11c,d show the radiation patterns comparison of the RPT to the ESEPT, and MLPT antennas at 4 GHz, respectively. At a frequency of 2 GHz, the main lobe RPT is 8.89 dBi, the side lobe level (SLL) is −0.7 dB, the Angular width (3 dB) is 15.2° and the main lobe direction is 0°. The FCPT structure results of 11.6 dBi main lobe, −4.7 dB sidelobe level, 14.1° angular widths, and 0° main lobe direction. While the HWSPT structure generates a major lobe of 11.7 dBi, the main lobe with an angular width of 14.6°, and the main lobe direction of 0°, as well as a sidelobe level of −4.6 dB. The antenna performance of the FCPT and HWSPT structures at 2 GHz is superior to RPT in the main lobe, SLL, and Angular width (3 dB). For instance, At the frequency of 4 GHz, ESEPT generates the main lobe of 12.9 dBi, the main lobe direction of 0°, an angular width of 7.4°, and a side lobe level of −2.5 dB. At the frequency of 4 GHz, the performance of the RPT and ESEPT antenna radiation is almost the same, with only an increase in the directivity of 0.8 dBi. At 4 GHz, the MLPT produces a main lobe of 16.6 dBi, a Mainlobe direction of 0°, an angular width of 23.1°, and a side lobe level of −11.1 dB. The directivity of MLPT increased by 4.5 dBi due to the addition of the lens structure and metamaterial so that electromagnetic waves are embedded in the metamaterial structure.

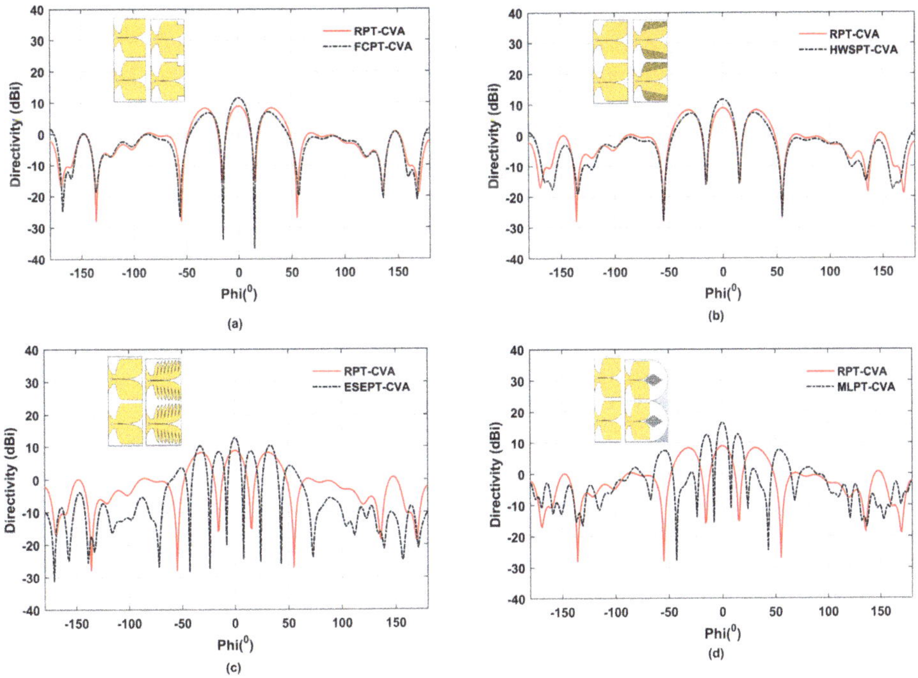

Figure 11. Radiation Pattern in the E-Plane of (**a**). RPT-CVA vs FCPT-CVA at 2 GHz, (**b**). RPT-CVA vs HFSPT-CVA at 2 GHz, (**c**) RPT-CVA vs ESE-CVA at 4 GHz, and (**d**). RPT-CVA vs MLPT-CVA at 4 GHz.

3.3. Surface Current Performance

Surface current is an electric current induced by an electromagnetic field. The surface current distribution varies with frequency. Figure 12 depicts the surface current of several antennas at a frequency of 0.5 GHz while Figure 13 shows the surface current of 15 antennas at 2 GHz.

In this case, we set the maximum surface current of all antennas at 0.5 A/m. Surface current varies for all antenna types, as seen in Figures 12 and 13. We designed the antenna by providing a distance between adjacent patches (copper) of 25 mm (as shown in Figure 1b) so that the surface current does not flow directly to adjacent patch elements (copper radiators) when the antennas are placed closely together. At 0.5 GHz, even though the copper radiators are separated by 25 mm, there is still a dispersion of surface currents with high intensity in nearby elements (indicated in red in the yellow circle with dotted lines) as demonstrated in Figure 12a,b. RPT and FCPT had larger surface currents in both neighboring patches than BCPT. This demonstrates that BCPT outperforms the rest in terms of S11 performance. There are various places with high surface current intensity in slots such as HFSPT, ESEPT, VWSPT, and HWSPT that are highlighted in red. Figure 12 shows that at a frequency of 0.5 GHz, the surface current in the elliptical structure and metamaterial is not excessive. Antennas with a lens structure that has been given an elliptical slot structure or metamaterial show increased surface current concentration at 4 GHz. A significant electric field is described by the existence of a high surface current.

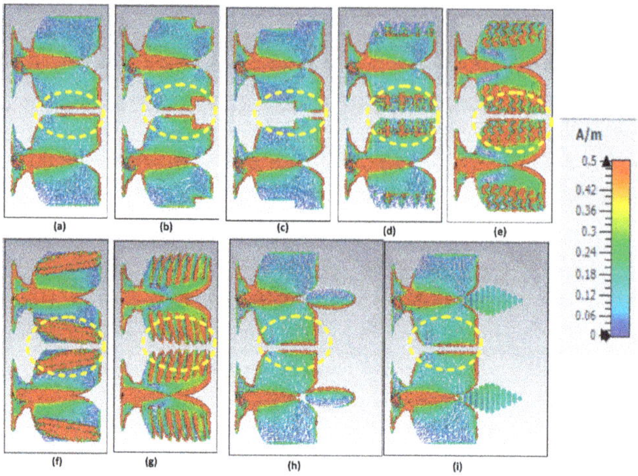

Figure 12. Surface current performance of (**a**) Regular Palm Tree (RPT-CVA), (**b**) Front Cut Palm Tree (FCPT-CVA), (**c**) Back Cut Palm Tree (BCPT-CVA), (**d**) Hilbert Fractal Structure Palm Tree (HFSPT-CVA), (**e**) Vertical Wave Structure Palm Tree (VWSPT-CVA), (**f**) Horizontale Wave Structure Palm Tree (HWSPT-CVA), (**g**) Exponential Slot Edge Palm Tree (ESEPT-CVA), (**h**) Elips Lens Palm Tree (ELPT-CVA), and (**i**) Metamaterial Lens Palm Tree (MLPT-CVA).

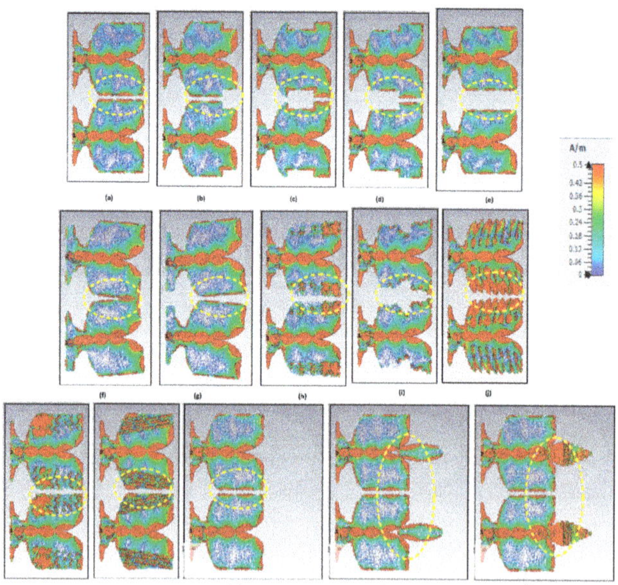

Figure 13. Surface current performance of (**a**) Regular Palm Tree (RPT-CVA), (**b**) Front Cut Palm Tree (FCPT-CVA), (**c**) Middle Cut Palm Tree (MCPT-CVA), (**d**) Back Cut Palm Tree (BCPT-CVA), (**e**) Complete Cut Palm Tree (CCPT-CVA), (**f**) Left Tilt Palm Tree (LTPT-CVA), (**g**) Right Tilt Palm Tree (RTPT-CVA), (**h**) Hilbert Fractal Structure Palm Tree (HFSPT-CVA), (**i**) Koch Fractal Structure Palm Tree (KFSPT-CVA), (**j**) Exponential Slot Edge Palm Tree (ESEPT-CVA), (**k**) Vertical Wave Structure Palm Tree (VWSPT-CVA), (**l**) Horizontale Wave Structure Palm Tree (HWSPT-CVA), (**m**) Regular Lens Palm Tree (RLPT-CVA), (**n**) Elips Lens Palm Tree (ELPT-CVA), and (**o**) Metamaterial Lens Palm Tree (MLPT-CVA).

4. Measurement and Comparison of Related Antenna

Figure 14 depicts the fabrication and comparison of measurement and simulation results for ESEPT-CVA and MLPT-CVA. At 4 GHz, both antennas have a greater gain than the RPT-CVA, as seen in Figure 7. Antenna measurements are carried out by taking S_{11} antenna data using a brand VNA Siglent which works from a frequency of 100 kHz to 3.2 GHz. From the measurement results, it is found that the S_{11} antenna measured by VNA produces better results at low frequencies. It is known that the measurement results are in agreement with the simulation results where most of the antennas have S_{11} below −10 dB. High-gain antennas can be applied to radar applications.

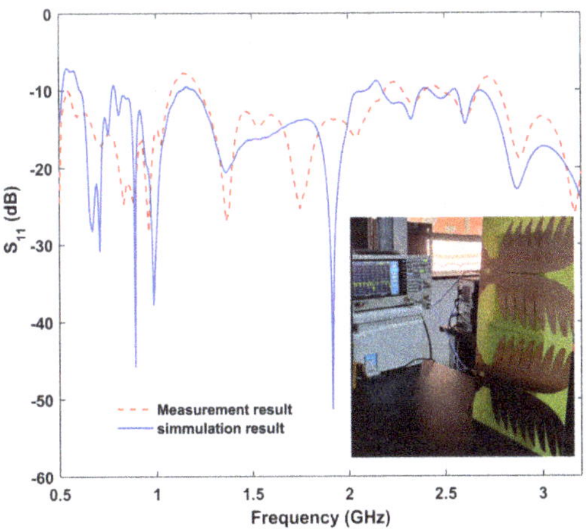

(a)

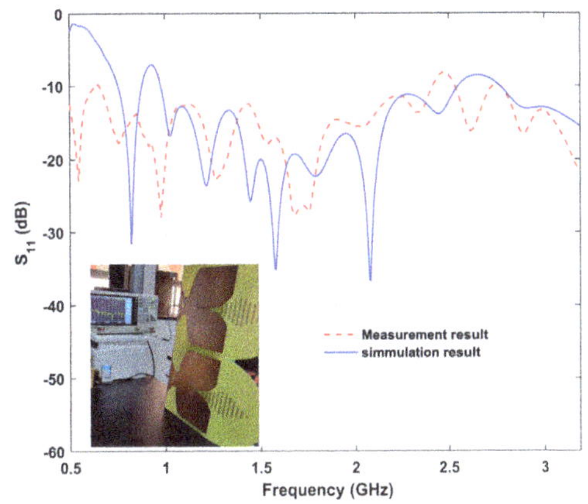

(b)

Figure 14. Simulation and measurement result of (**a**) ESEPT-CVA and (**b**) MLPT-CVA.

The Vector Network Analyzer is a tool that can use for measuring radio frequency scattering properties in radar applications. When two ports of the VNA are linked to the antenna, the S_{21} data can explain the transfer function of the signals emitted and received as shown in Equations (4) and (5). $S_x(t)$ represents the chirp signal with a certain period and bandwidth, whereas $S_y(t)$ represents the received signal chirp [49].

$$S_y(f) = S_{21}S_x(f) \tag{4}$$

$$S_r(t) = F^{-1}S_y(f) \tag{5}$$

In this study, we used an antenna to detect objects behind the wall. The antenna used is MLPT by connecting it to a portable nano VNA that works at a frequency of 0.5–3 GHz. The antenna and the detected object are varied in distance to the wall. The detected object is a laptop varying the object's distance from the wall and the detection process is carried out by placing the antenna on the E-plane. The antenna is connected to the VNA and the VNA is connected to the laptop. The Scattering signal received will be seen on the laptop display as shown in Figure 15.

The detected wall has an area of 60 × 150 cm^2 by dividing the area into several segments area in the x and y direction. The S_{21} data received on the laptop will be processed and a signal reconstruction process will be carried out so that object detection results are produced as shown in Figure 16. The yellow circle shows the position of the detected target in the x-y plane.

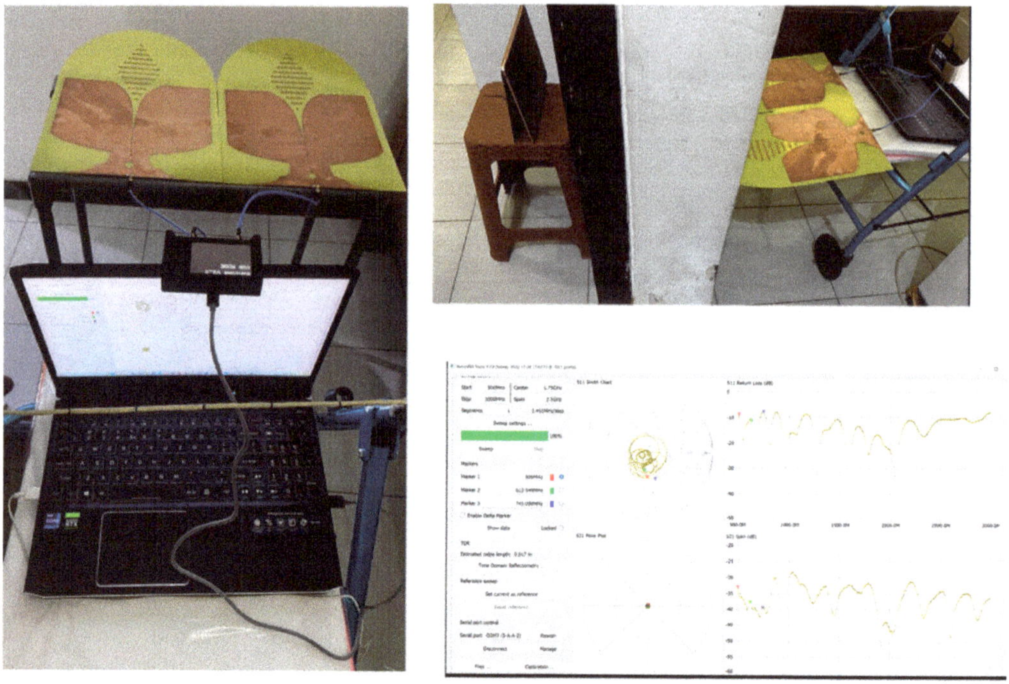

Figure 15. Radar target measurement with MLPT-CVA prototype.

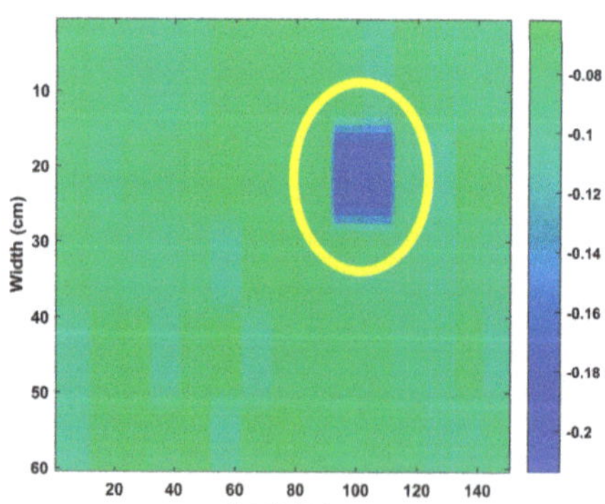

Figure 16. Radar target detection in the xy-planes.

Table 2 reveals that there has been some research on Vivaldi antennas that operate at low frequencies, but they have not taken into account mutual coupling. In this case, we also compare the antenna element. As shown in Table 2, Our antenna has a smaller width than the antenna at [50]. It employs ceramic materials to operate at a frequency of 0.5 to 3 GHz. The antenna has a wide bandwidth of 0.5–6 GHz has been discussed in [51]. That antenna achieves a gain of 8 dBi at a frequency of 2 GHz, but as frequency increased, the gain declined. A big CVA antenna, which operated at a very low frequency has been evaluated in [52]. However, the maximum frequency used in research [53,54] is 2.1 GHz. The AVA antenna for GPR application with metamaterial and DGS structure has been discussed in [55,56] with a larger size than our purposes study. Aspects of our research operate between 0.5 and 4.5 GHz. The ESEPT-CVA and MLPT-CVA antennas, with respective elemental gains of 9.7 dBi and 13.4 dBi at 4 GHz, are used in this comparison. In this study, ESEPT-CVA and MLPT-CVA were arranged in MIMO 1 × 2 to produce 12.87 dBi and 16.561 dBi gains at 4 GHz frequency.

Table 2. Comparison of the proposed antenna and related research.

Ref	Element Dimension (mm)	Ant. Type	Sub. Type	Freq. (GHz)	Gain (dBi)
[50]	300 × 360	AVA	T-ceramic	0.5–3	-
[51]	258 × 150	CVA	FR-4	0.5–6	8
[52]	950 × 780	CVA	-	0.02–0.12	-
[53]	260 × 185	AVA	Taconic	0.5–2	-
[54]	286 × 300 (with metamaterial)	CVA	FR4	0.7–2.1	10.5
[55]	450 × 600	AVA	Rogers 4350	0.3–2	4.4–11.5
[56]	750 × 525	CVA	FR4	0.26–0.34	4.2
ESEPT	275 × 275	CVA	FR4	0.5–4.5	9.7
MLPT	275 × 438	CVA	FR4	0.5–4.5	13.4

The comparative performance of 15 Vivaldi antennas is displayed in Table 3 to get the performance in 1 × 2 Palm Tree MIMO antenna. Table 3 shows that the best S_{11} at the 0.5 GHz frequency was obtained by ESEPT which is −16.86 dB, but for the overall working frequency included in S_{11} less than −10 dB is BCPT as shown in Figure 3b. Table 3 also shows that the best $S21$ at the 0.5 GHz frequency was obtained by HWSPT which was −18.57 dB. Table 3 shows that at 0.5 GHz, practically all models still have S21 > −20 dB because the distance between components is less than 0.5, implying that another strategy for mutual coupling reduction is required. Table 3 further shows that the maximum directivity attained by MLPT is 16.56 dBi, whereas ELPT produced the lowest SLL, best beam squint, and lowest beam width.

Table 3. 1 × 2 MIMO palm tree Coplanar Vivaldi Antena.

Ant. Type	S11 (dB) At 0.5 GHz	S21 (dB) At 0.5 GHz	Max Dir (dB)	Min SLL (dB)	Max Beamsquit (°) (0.5–4.5 GHz)	Min Beamwidth (°)
RPT	−8.33	−13.45	12.11 (4 GHz)	−11.94 (1 GHz)	25.46 (2.5 GHz)	3.98 (4.5 GHz)
FCPT	−7.29	−13.73	11.58 (2 GHz)	−9.73 (1 GHz)	25.84 (2.5 GHz)	3.78 (4.5 GHz)
MCPT	−10.04	−15.19	12.09 (4 GHz)	−9.43 (1 GHz)	25.50 (2.5 GHz)	4.41 (4.45 GHz)
BCPT	−10.48	−14.06	12.10 (4 GHz)	−9.9 (11 GHz)	25.46 (2.5 GHz)	4.27 (4.5 GHz)
CCPT	−6.57	−16.4	11.19 (4 GHz)	−11.02 (1 GHz)	−21.17 (3 GHz)	3.99 (4.5 GHz)
LTPT	−9.36	−11.49	11.75 (4 GHz)	−9.16 (1 GHz)	26.69 (2.5 GHz)	3.64 (4.5 GHz)
RTPT	−10.37	−15.38	11.50 (3.5 GHz)	−11.11 (1 GHz)	−30.63 (2 GHz)	4.4 (4.5 GHz)
HFSP	−12.21	−14.5	12.06 (3.5 GHz)	−10.35 (1 GHz)	−179.62 (0.5 GHz)	5.22 (4.5 GHz)
KFSPT	−14.33	−15.29	12.68 (4 GHz)	−9.43 (2 GHz)	−9.43 (2 GHz)	4.09 (4.5 GHz)
ESEPT	−16.86	−13.43	12.87 (4 GHz)	−7.79 (1 GHz)	−179.81 (0.5 GHz)	4.49 (4.5 GHz)
VWSPT	−15.63	−13.73	11.51 (4 GHz)	−10.46 (1 GHz)	−179.82 (0.5 GHz)	3.69 (4.5 GHz)
HWSPT	−6.53	−18.57	11.74 (2 GHz)	−9.57 (1 GHz)	25.53 (2.5 GHz)	4.09 (4.5 GHz)
RLPT	−8.71	−12.57	14.43 (4 GHz)	−11.89 (1 GHz)	25.63 (2 GHz)	3.79 (4.5 GHz)
ELPT	−8.92	−12.53	15.25 (4 GHz)	−13.77 (1 GHz)	0	3.32 (4.5 GHz)
MLPT	−8.87	−12.74	16.56 (4 GHz)	−11.91 (1 GHz)	29.27 (4.5 GHz)	5.22 (4.5 GHz)

5. Conclusions

We have simulated 15 kinds of palm tree antennas in the 0.5–4.5 GHz frequency with several structures, namely Regular Palm Tree-Coplanar Vivaldi Antenna (RPT-CVA), Front Cut Palm Tree (FCPT-CVA), Middle Cut Palm Tree (MCPT-CVA), Back Cut Palm Tree (BCPT-CVA), Complete Cut Palm Tree (CCPT-CVA), Left Tilt Palm Tree (LTPT-CVA), Right Tilt Palm Tree (RTPT-CVA), Hilbert Fractal Structure Palm Tree (HFSPT- CVA), Koch Fractal Structure Palm Tree (KFSPT-CVA), Exponential Slot Edge Palm Tree (ESEPT-CVA), Vertical

Wave Structure Palm Tree (VWPT-CVA), Horizontal Wave Structure Palm Tree (HWPT-CVA), Regular Lens Palm Tree (RLPT-CVA), Elliptical Lens Palm Tree (ELPT-CVA) and Metamaterial Lens Palm Tree (MLPT-CVA). By giving a different structure to the antenna radiator while maintaining the same substrate width and feed parameters and the similarity of the slope of the two tapered slots, the performance of return loss, mutual scaring, beam squint, and beamwidth produce different performances. From the simulation results, it is found that HWSPT gets the best mutual scaring performance at low frequencies because it has less than -10 dB mutual scaring. The maximum directivity of the RPT-CVA is 12.108 dBi, while the MLPT-CVA has a gain of 16.561 dBi at the 4 GHz frequency. The beam squint at all frequencies is 0° for ELPT-CVA and the lowest beamwidth is also obtained by ELPT-CVA at 4.5 GHz. This comparative analysis can be used as a reference for the selection of MIMO antenna design in considering the performance requirements of return loss, mutual coupling, directivity, beam squint, and beam width. From the results of the return loss measurements, there is also a match between the simulation results and antenna measurements where the antenna can work at a frequency of 0.5–4.5 GHz so that this antenna can be recommended for radar applications.

Author Contributions: Conceptualization, N.N. and A.M.d.O.; methodology, E.S.; software, N.N. and D.K.; validation, N.N. and E.S.; formal analysis, N.N. and M.N.M.Y.; investigation, D.K.; resources, E.S.; data curation, N.N. and D.K.; writing—original draft preparation, N.N.; writing—review and editing, E.S. and M.N.M.Y.; supervision, A.M.d.O.; project administration, N.N.; funding acquisition, N.N. and M.N.M.Y. All authors have read and agreed to the published version of the manuscript.

Funding: We appreciate the funds granted by the Directorate of Research, Technology, and Community Service under Decree No. 313/UN38/HK/PP/2022 and Agreement/Contract No. 039/E5/PG.02.00.PT/2022 and the international grant UNESA-UNIMAP (INTERES).

Institutional Review Board Statement: Not applicable.

Data Availability Statement: Not applicable.

Conflicts of Interest: The authors declare no conflict of interest.

References

1. Awan, W.A.; Soruri, M.; Alibakhshikenari, M.; Limiti, A.E. On-Demand Frequency Switchable Antenna Array Operating at 24.8 and 28 GHz for 5G High-Gain Sensors Applications. *Prog. Electromagn. Res. M* **2022**, *108*, 163–173. [CrossRef]
2. Mohamadi, C.T.; Asefi, M.; Thakur, S.; Paliwal, J.; Gilmore, C. Improved Metallic Enclosure Electromagnetic Imaging Using Ferrite Loaded Antennas. *Electronics* **2022**, *11*, 3804. [CrossRef]
3. Aguilar, S.M.; Al-Joumayly, M.A.; Burfeindt, M.J.; Behdad, N.; Hagness, S.C. Multiband Miniaturized Patch Antennas for a Compact, Shielded Microwave Breast Imaging Array. *IEEE Trans. Antennas Propag.* **2014**, *62*, 1221–1231. [CrossRef]
4. Guo, R.; Ni, Y.; Liu, H.; Wang, F.; He, L. Signal Diverse Array Radar for Electronic Warfare. *IEEE Antennas Wirel. Propag. Lett.* **2017**, *16*, 2906–2910. [CrossRef]
5. Lee, S.; Kim, S.; Park, Y.; Choi, J. A 3D-Printed Tapered Cavity-Backed Flush-Mountable Ultra-Wideband Antenna for UAV. *IEEE Access* **2019**, *7*, 156612–156619. [CrossRef]
6. Ghouz, H.H.M.; Sree, M.F.A.; Ibrahim, M.A. Novel Wideband Microstrip Monopole Antenna Designs for WiFi/LTE/WiMax Devices. *IEEE Access* **2020**, *8*, 9532–9539. [CrossRef]
7. Navarro-Mendez, D.V.; Carrera-Suarez, L.F.; Antonino-Daviu, E.; Ferrando-Bataller, M.; Baquero-Escudero, M.; Gallo, M.; Zamberlan, D. Compact Wideband Vivaldi Monopole for LTE Mobile Communications. *IEEE Antennas Wirel. Propag. Lett.* **2015**, *14*, 1068–1071. [CrossRef]
8. El-Bacha, A.; Sarkis, R. Design of tilted taper slot antenna for 5G base station antenna circular array. In Proceedings of the 2016 IEEE Middle East Conf. Antennas Propagation, MECAP 2016, Beirut, Lebanon, 20–22 September 2016; pp. 5–8. [CrossRef]
9. Ikram, M.; Nguyen-Trong, N.; Abbosh, A. Hybrid Antenna Using Open-Ended Slot for Integrated 4G/5G Mobile Application. *IEEE Antennas Wirel. Propag. Lett.* **2020**, *19*, 710–714. [CrossRef]
10. Fernandez-Martinez, P.; Martin-Anton, S.; Segovia-Vargas, D. Design of a wideband vivaldi antenna for 5G base stations. In Proceedings of the 2019 IEEE International Symposium on Antennas and Propagation and USNC-URSI Radio Science Meeting, Atlanta, GA, USA, 7–12 July 2019; pp. 149–150. [CrossRef]
11. Chu, H.L.; Mishra, G.; Sharma, S.K. Dual Polarized Wideband Vivaldi 4x4 Subarray Antenna Aperture for 5G Massive MIMO Panels with Simultaneous Multiple Beams. In Proceedings of the 2018 18th International Symposium on Antenna Technology and Applied Electromagnetics (ANTEM), Waterloo, ON, Canada, 19–22 August 2018; Volume 2018, pp. 32–33. [CrossRef]

12. Dzagbletey, P.A.; Shim, J.; Chung, J. Communication V2X Communication Measurement. *IEEE Trans. Antennas Propag.* **2019**, *67*, 1957–1962. [CrossRef]
13. He, S.H.; Shan, W.; Fan, C.; Mo, Z.C.; Yang, F.H.; Chen, J.H. An Improved Vivaldi Antenna for Vehicular Wireless Communication Systems. *IEEE Antennas Wirel. Propag. Lett.* **2014**, *13*, 1505–1508. [CrossRef]
14. Li, X.; Ji, Y.; Lu, W.; Fang, G. Analysis of GPR Antenna System Mounted on a Vehicle. *IEEE Antennas Wirel. Propag. Lett.* **2013**, *12*, 575–578. [CrossRef]
15. Kirscht, M.; Mietzner, J.; Bickert, B.; Hippler, J.; Zahn, R.; Boukamp, J. An Airborne Radar Sensor for Maritime & Ground Surveillance and Reconnaissance. In Proceedings of the EUSAR 2014, 10th European Conference on Synthetic Aperture Radar, Berlin, Germany, 3–5 June 2014; pp. 1077–1080.
16. García-Fernández, M.; López, Y.A.; Andrés, F.L.-H. Airborne multi-channel ground penetrating radar for improvised explosive devices and landmine detection. *IEEE Access* **2020**, *8*, 165927–165943. [CrossRef]
17. Kibria, S.; Samsuzzaman, M.; Islam, T.; Mahmud, Z.; Misran, N.; Islam, M.T. Breast Phantom Imaging Using Iteratively Corrected Coherence Factor Delay and Sum. *IEEE Access* **2019**, *7*, 40822–40832. [CrossRef]
18. Liu, H.; Zhao, J.; Sato, M. A Hybrid Dual-Polarization GPR System for Detection of Linear Objects. *IEEE Antennas Wirel. Propag. Lett.* **2015**, *14*, 317–320. [CrossRef]
19. Latha, T.; Ram, G.; Kumar, G.A.; Chakravarthy, M. Review on Ultra-Wideband Phased Array Antennas. *IEEE Access* **2021**, *9*, 129742–129755. [CrossRef]
20. Guan, B.; Ihamouten, A.; Derobert, X.; Guilbert, D.; Lambot, S.; Villain, G. Near-Field Full-Waveform Inversion of Ground-Penetrating Radar Data to Monitor the Water Front in Limestone. *IEEE J. Sel. Top. Appl. Earth Obs. Remote Sens.* **2017**, *10*, 4328–4336. [CrossRef]
21. Karim, M.N.A.; Malek, M.F.A.; Jamlos, M.F.; Seng, L.Y.; Saudin, N. Design of Ground Penetrating Radar antenna for buried object detection. In Proceedings of the 2013 IEEE International RF and Microwave Conference (RFM), Penang, Malaysia, 9–11 December 2013; pp. 253–257. [CrossRef]
22. Chareonsiri, Y.; Thaiwirot, W.; Akkaraekthalin, P. Design of Ultra-Wideband Tapered Slot Antenna by Using Binomial Transformer with Corrugation. *Frequenz* **2017**, *71*, 251–260. [CrossRef]
23. Yan, J.-B.; Gogineni, S.; Camps-Raga, B.; Brozena, J. A Dual-Polarized 2–18-GHz Vivaldi Array for Airborne Radar Measurements of Snow. *IEEE Trans. Antennas Propag.* **2016**, *64*, 781–785. [CrossRef]
24. Dixit, A.S.; Kumar, S. A Survey of Performance Enhancement Techniques of Antipodal Vivaldi Antenna. *IEEE Access* **2020**, *8*, 45774–45796. [CrossRef]
25. Liu, Y.-Q.; Liang, J.-G.; Wang, Y.-W. Gain-improved double-slot LTSA with conformal corrugated edges. *Int. J. RF Microw. Comput. Eng.* **2017**, *27*, e21133. [CrossRef]
26. Sun, M.; Chen, Z.N.; Qing, X. Gain Enhancement of 60-GHz Antipodal Tapered Slot Antenna Using Zero-Index Metamaterial. *IEEE Trans. Antennas Propag.* **2013**, *61*, 1741–1746. [CrossRef]
27. Shi, X.; Cao, Y.; Hu, Y.; Luo, X.; Yang, H.; Ye, L.H. A High-Gain Antipodal Vivaldi Antenna with Director and Metamaterial at 1–28 GHz. *IEEE Antennas Wirel. Propag. Lett.* **2021**, *20*, 2432–2436. [CrossRef]
28. Amiri, M.; Tofigh, F.; Ghafoorzadeh-Yazdi, A.; Abolhasan, M. Exponential Antipodal Vivaldi Antenna With Exponential Dielectric Lens. *IEEE Antennas Wirel. Propag. Lett.* **2017**, *16*, 1792–1795. [CrossRef]
29. Huang, M. Modified Balanced Antipodal Vivaldi antennas with Substrate-Integrated Lenses for 2-18 GHz Application. In Proceedings of the 2018 IEEE International Symposium on Antennas and Propagation & USNC/URSI National Radio Science Meeting, Boston, MA, USA, 8–13 July 2018; pp. 1761–1762. [CrossRef]
30. Biswas, B.; Ghatak, R.; Poddar, D.R. A Fern Fractal Leaf Inspired Wideband Antipodal Vivaldi Antenna for Microwave Imaging System. *IEEE Trans. Antennas Propag.* **2017**, *65*, 6126–6129. [CrossRef]
31. Karmakar, A.; Subhash, N.; College, E. A Wideband Vivaldi Antenna with Fractal Dielectric Lens for Imaging Applications. In Proceedings of the International Conference for Convergence in Technology (I2CT), Pune, India, 6–8 April 2018; pp. 9–13.
32. Hussain, M.; Awan, W.A.; Ali, E.M.; Alzaidi, M.S.; Alsharef, M.; Elkamchouchi, D.H.; Alzahrani, A.; Sree, M.F.A. Isolation Improvement of Parasitic Element-Loaded Dual-Band MIMO Antenna for Mm-Wave Applications. *Micromachines* **2022**, *13*, 1918. [CrossRef] [PubMed]
33. Jiang, P.; Jiang, W.; Gong, S. Vivaldi Array Antenna with Low In-band RCS and Low Cross-polarization Properties by Loading Spoof Surface Plasmon Polariton Absorber. In Proceedings of the 2021 IEEE International Symposium on Antennas and Propagation and USNC-URSI Radio Science Meeting (APS/URSI), Singapore, 4–10 December 2021; pp. 1323–1324. [CrossRef]
34. Prachi, V.G.; Vijay, S. A Novel Design of Compact 28 GHz Printed Wideband Antenna for 5G Applications. *Int. J. Innov. Technol. Explor. Eng.* **2020**, *9*, 3696–3700. [CrossRef]
35. Virone, G.; Sarkis, R.; Craeye, C.; Addamo, G.; Peverini, O.A. Gridded Vivaldi Antenna Feed System for the Northern Cross Radio Telescope. *IEEE Trans. Antennas Propag.* **2011**, *59*, 1963–1971. [CrossRef]
36. Reid, E.W.; Ortiz-Balbuena, L.; Ghadiri, A.; Moez, K. A 324-Element Vivaldi Antenna Array for Radio Astronomy Instrumentation. *IEEE Trans. Instrum. Meas.* **2012**, *61*, 241–250. [CrossRef]
37. Buzdar, A.R.; Buzdar, A.; Bin Tila, H.; Sun, L.; Khan, M.; Khan, U.; Ferozv, W. Low cost Vivaldi array antenna for mobile through wall sensing platforms. In Proceedings of the 2016 IEEE MTT-S International Wireless Symposium (IWS), Shanghai, China, 14–16 March 2016; pp. 9–12. [CrossRef]

38. Ahsan, S.; Kosmas, P.; Sotiriou, I.; Palikaras, G.; Kallos, E. Balanced antipodal Vivaldi antenna array for microwave tomography. In Proceedings of the 2014 IEEE Conference on Antenna Measurements & Applications (CAMA), Antibes Juan-les-Pins, France, 16–19 November 2014; pp. 4–6. [CrossRef]
39. Fernandez-Martinez, P.; Martin-Anton, S.; Segovia-Vargas, D. Dual-Band Array of Cross-Polarized Vivaldi Antennas for 5G Applications. In Proceedings of the 2020 14th European Conference on Antennas and Propagation (EuCAP), Copenhagen, Denmark, 15–20 March 2020; pp. 2–6. [CrossRef]
40. Nurhayati; Hendrantoro, G.; Fukusako, T.; Setijadi, E. Mutual Coupling Reduction for a UWB Coplanar Vivaldi Array by a Truncated and Corrugated Slot. *IEEE Antennas Wirel. Propag. Lett.* **2018**, *17*, 2284–2288. [CrossRef]
41. Zhu, S.; Liu, H.; Chen, Z.; Wen, P. A Compact Gain-Enhanced Vivaldi Antenna Array with Suppressed Mutual Coupling for 5G mmWave Application. *IEEE Antennas Wirel. Propag. Lett.* **2018**, *17*, 776–779. [CrossRef]
42. Herzi, R.; Zairi, H.; Gharsallah, A. Antipodal Vivaldi antenna array with high gain and reduced mutual coupling for UWB applications. In Proceedings of the 2015 16th International Conference on Sciences and Techniques of Automatic Control and Computer Engineering (STA), Monastir, Tunisia, 21–23 December 2015; pp. 789–792. [CrossRef]
43. Yang, Y.; Wang, Y.; Fathy, A.E. Design of compact Vivaldi antenna arrays for UWB see through wall applications. *Prog. Electromagn. Res.* **2008**, *82*, 401–418. [CrossRef]
44. Nurhayati, N.; Setijadi, E.; Hendrantoro, G. Radiation Pattern Analysis and Modelling of Coplanar Vivaldi Antenna Element for Linear Array Pattern Evaluation. *Prog. Electromagn. Res. B* **2019**, *84*, 79–96. [CrossRef]
45. Hasim, N.S.B.; Ping, K.A.H.; Islam, M.T.; Mahmud, M.Z.; Sahrani, S.; Mat, D.A.A.; Zaidel, D.N.A. A slotted uwb antipodal vivaldi antenna for microwave imaging applications. *Prog. Electromagn. Res. M* **2019**, *80*, 35–43. [CrossRef]
46. Jolani, F.; Dadashzadeh, G.; Naser-Moghadasi, M.; Dadgarpour, A. Design and optimization of compact balanced antipodal vivaldi antenna. *Prog. Electromagn. Res. C* **2009**, *9*, 183–192. [CrossRef]
47. Murad, N.A.; Esa, M.; Yusoff, M.F.M.; Ali Ammah, S.H. Hilbert curve fractal antenna for RFID application. In Proceedings of the 2006 International RF and Microwave Conference, Putra Jaya, Malaysia, 12–14 September 2006; pp. 182–186. [CrossRef]
48. Nurhayati, N.; De-Oliveira, A.M.; Chaihongsa, W.; Sukoco, B.E.; Saleh, A.K. A comparative study of some novel wideband tulip flower monopole antennas with modified patch and ground plane. *Prog. Electromagn. Res. C* **2021**, *112*, 239–250. [CrossRef]
49. Pramudita, A.A.; Praktika, T.O.; Jannah, S. Radar modeling experiment using vector network analyzer. In Proceedings of the 2020 International Symposium on Antennas and Propagation (ISAP), Osaka, Japan, 25–28 January 2021; pp. 99–100. [CrossRef]
50. Molaei, A.; Dagheyan, A.G.; Juesas, J.H.; Martinez-Lorenzo, J. Miniaturized UWB Antipodal Vivaldi Antenna for a mechatronic breast cancer imaging system. In Proceedings of the 2015 IEEE International Symposium on Antennas and Propagation & USNC/URSI National Radio Science Meeting, Vancouver, BC, Canada, 19–24 July 2015; pp. 352–353. [CrossRef]
51. Liu, Y.; Zhou, W.; Yang, S.; Li, W.; Li, P.; Yang, S. A Novel Miniaturized Vivaldi Antenna Using Tapered Slot Edge with Resonant Cavity Structure for Ultrawideband Applications. *IEEE Antennas Wirel. Propag. Lett.* **2016**, *15*, 1881–1884. [CrossRef]
52. Elsheakh, D.M.; Abdallah, E.A. Novel shapes of Vivaldi antenna for ground pentrating radar (GPR). In Proceedings of the 2013 7th European Conference on Antennas and Propagation (EuCAP), Gothenburg, Sweden, 8–12 April 2013; pp. 2886–2889.
53. Fioranelli, F.; Salous, S.; Ndip, I.; Raimundo, X. Through-The-Wall Detection with Gated FMCW Signals Using Optimized Patch-Like and Vivaldi Antennas. *IEEE Trans. Antennas Propag.* **2015**, *63*, 1106–1117. [CrossRef]
54. Cheng, H.; Yang, H.; Li, Y.; Chen, Y. A Compact Vivaldi Antenna with Artificial Material Lens and Sidelobe Suppressor for GPR Applications. *IEEE Access* **2020**, *8*, 64056–64063. [CrossRef]
55. Guo, J.; Tong, J.; Zhao, Q.; Jiao, J.; Huo, J.; Ma, C. An Ultrawide Band Antipodal Vivaldi Antenna for Airborne GPR Application. *IEEE Geosci. Remote Sens. Lett.* **2019**, *16*, 1560–1564. [CrossRef]
56. Penaloza-Aponte, D.; Alvarez-Montoya, J.; Clemente-Arenas, M. GPR vivaldi antenna with DGS for archeological prospection. In Proceedings of the 2017 IEEE XXIV International Conference on Electronics, Electrical Engineering and Computing (INTERCON), Cusco, Peru, 15–18 August 2017. [CrossRef]

electronics

MDPI

Article

A Composite Right/Left-Handed Phase Shifter-Based Cylindrical Phased Array with Reinforced Particles Responsive to Magneto-Static Fields

Muhammad Ayaz [1], Adnan Iftikhar [2], Benjamin D. Braaten [3], Wesam Khalil [4] and Irfan Ullah [1,*]

1 Department of Electrical and Computer Engineering, Abbottabad Campus, COMSATS University Islamabad, Abbottabad 22060, Pakistan
2 Department of Electrical and Computer Engineering, COMSATS University Islamabad, Islamabad 45550, Pakistan
3 Department of Electrical and Computer Engineering, North Dakota State University, Fargo, ND 58102, USA
4 Server Security Integration Group, Intel Corporation, 2111 NE 25th Ave, Hillsboro, OR 97124, USA
* Correspondence: eengr@cuiatd.edu.pk

Abstract: A conformal cylindrical phased array antenna excited with composite right/left-handed (CRLH) phase shifters is proposed. The phase tuning of the CRLH phase shifter is achieved by embedding novel magneto-static field-responsive micron-sized particles in its structure. It is shown that through the tiny magnet activation of these novel magneto-static particles at appropriate locations along the length of CRLH stub and inter-digital fingers, variable phase shifts are obtained. The proposed particle-based CRLH phase shifter operates in C-band (5–6) GHz with a low insertion loss and phase error. The 1 × 4 cylindrical phased array is excited with the four unit cells of the proposed particle-embedded CRLH transmission line phase shifters to scan the main beam at desired scan angles. A prototype of a 1 × 4 cylindrical phased array excited with the particle-based CRLH phase shifters was fabricated, and the results show that the simulated results are in close agreement with the measured results. The conformal cylindrical array with the proposed particle-based CRLH phase shifters has great potential for use in printed and flexible electronics design where commercially available phase shifters have a definite drawback.

Keywords: magneto-static particles; conformal antenna; phased array; metamaterial transmission line; composite right/left-handed (CRLH) transmission line

Citation: Ayaz, M.; Iftikhar, A.; Braaten, B.D.; Khalil, W.; Ullah, I. A Composite Right/Left-Handed Phase Shifter-Based Cylindrical Phased Array with Reinforced Particles Responsive to Magneto-Static Fields. *Electronics* **2023**, *12*, 306. https://doi.org/10.3390/electronics12020306

Academic Editors: Naser Ojaroudi Parchin, Chan Hwang See and Raed A. Abd-Alhameed

Received: 13 December 2022
Revised: 4 January 2023
Accepted: 4 January 2023
Published: 6 January 2023

1. Introduction

The metamaterial-based phase shifters that use composite right/left-handed (CRLH) transmission lines have been shown to have practical applications in many wireless communication systems such as radar systems, phased array antenna systems, and other compact microwave devices [1]. Conventional phase shifters uses right-handed transmission line (RH-TL) structures to achieve the phase shift response, by changing the length of the transmission line structure, so a longer length is required to achieve the larger phase shift. However, the CRLH-TL structure has unique phase characteristics, which can introduce positive, zero, and negative phase shifts [1]. In the literature, the phase shift of the CRLH transmission line is varied using conventional PIN/varactor diodes and a microelectromechanical system (MEMS) as switching elements in the fingers and stub of the structure to change its capacitance and inductance and therefore, the phase response. Another more recently introduced approach to obtain the variable phase shift is the use of ceramic material in the conventional CRLH structure, where the effective dielectric constant of the substrate is made variable to obtain the variable phase shift. Here, a brief overview of the most relevant literature studies is presented where these two common approaches have been applied to the CRLH-TL structure to achieve the variable phase response.

The work in [2,3] describes the CRLH-based phase shifter design using lumped capacitors and inductors as left-handed components. The requirements of unique discrete component values and switching between a reference line and CRLH transmission line (CRLH-TL) to calculate the phase shift complicated the design. In [4–6], an n-bit phase shifter based on CRLH-TL was designed. PIN diodes along the length of fingers of CRLH-TL were used as switches to obtain various phase shifts. A phased array antenna using BST/CRLH-TL composite phase shifters was presented in [7]. In [8], the fingers and stub of the CRLH-TL were loaded with varactor diodes to obtain variable phase constants. In [9], bulky MEMS switches were incorporated in the CRLH-TL structure to achieve the desired phase shifts. In [10,11], radiation patterns of a linear antenna array using a tunable phase shifter based on CRLH-TL with mechanically variable metal–insulator–metal (MIM) capacitors were presented. The requirements of external dc biasing circuitry for these diodes/mechanical switches not only complicated the design but could also be a drawback in fully printing electronics design applications and for large antenna array systems comprising thousands of phase shifters. A tunable phase shifter based on Inkjet-printed BST material in between the finger spacings of CRLH-TL was widely studied in [12–15]. In [16], the design of fully printable conformal antennas with polymer/CRLH-TL composite phase shifters was investigated. More details about the physical properties of reinforced polymer composites were presented in [17]. The relative permittivity of BST and the corresponding phase shift were achieved by applying an electrostatic field across it. The major drawback was the requirement of the tuning voltage in hundreds of volts to achieve the required phase shift. In [18,19], metamaterial-based phase shifters using thick-film ceramic material for phased array antennas were investigated. The complex process of integrating ceramic/polymer material in the artificial transmission line and the complex design of the DC biasing circuitry limited its applications in compact electronics component placement for fully printable and flexible applications. In a nutshell, these two common approaches for phase tuning of CRLH-TL have shown promising applications in phased array antenna systems. However, the need of external biasing circuitry, the complex fabrication process, and the requirements of high external voltage (in hundreds of volts) put limitations on these approaches for their use in large antenna systems (requiring thousands of phase shifters) and where space constraint is a major concern.

In the proposed work, a new particle-based CRLH (which we call a P-CRLH) phase shifter is presented to overcome these potential design obstacles. Then, a 1×4 cylindrical phased array is excited with the proposed P-CRLH phase shifters for main beam scanning at the desired scan angles. The use of magneto-static-responsive conductive micron-sized particles has been shown to have practical applications for RF switches [20,21], EBG antennas [22], reconfigurable antennas [23,24], leaky-wave antennas [25,26], metamaterials [27], the phase shift of a microstrip transmission line [28], and comparison with a PIN diode and electromagnetic propagation through a dielectric medium [29]. The main advantages of the proposed design are (a) the use of a single substrate for particle integration unlike the requirements of the two separate substrates in [22,23], (b) no requirement of discontinuous transmission line as in [24], and (c) no need for a direct connection of a biasing circuitry as required in conventional RF switches. A description of the abbreviations used in the study is given in Table A1 in Appendix A.

The rest of the paper is organized as: Section 2 discusses the design and analysis of a unit cell P-CRLH phase shifter; Section 3 presents a detailed simulation analysis of the unit cell P-CRLH phase shifter. Section 4 discusses the P-CRLH excited 1×4 conformal phased array, Section 5 provides detail of the measurement results, and Section 6 concludes the proposed work.

2. The P-CRLH Unit Cell Design and Analysis

A baseline CRLH-TL unit cell is designed first for a 0° phase shift at the design frequency $f_0 = 5.2$ GHz. The unit cell architecture is based on the well-known composite right/left-handed (CRLH) metamaterial transmission line [1] with slight modification of introducing symmetrical stub lines as shown in Figure 1. An inductor stub line is used for left-handed inductance L_L and inter-digital fingers represent the left-handed capacitance C_L. The right-handed components L_R and C_R come from the inherent right-handed parasitic elements present in the microstrip structure. This baseline CRLH unit cell is designed on a TMM4 substrate ($\varepsilon_r = 4.7$, $tan\delta = 0.002$ and thickness = 0.635 mm). After the baseline 0° unit cell has been designed, the variable phase shifts can be achieved by tuning this unit cell. The tuning of the baseline CRLH unit cell is obtained by embedding novel magneto-static micron-sized particles (SM25P20 SILVER FERRITE in [30]) in the substrate along the length of stub (or at appropriate locations in the inter-digital fingers). We call this new CRLH structure with embedded particles, a 'Particle-CRLH (P-CRLH)'. The embedding procedure of these micron-sized conductive particles in the TMM4 substrate is reported in detail in [31] and is briefly mentioned here. A small cylindrical cavity of diameter d and height equal to the thickness of the substrate in Figure 1 along the length of stub length of CRLH-TL is made through the substrate. Then, the cavity is partially filled with the silver-coated magnetic particles. The top and bottom of the cavity are the stub transmission lines and ground plane. This small structure of the particle-filled cavity with top and bottom conducting planes is denoted as the Magneto-static field Responsive Structure (MRS). It is worth mentioning that the work in [22–24] requires separate MRSs to be manufactured before their final integration in the transmission line. That is, the MRSs and transmission line require two separate substrates. Another requirement in these designs is the need for a discontinuous transmission line for the integration of MRSs. In the proposed P-CRLH phase shifter, both of these requirements are overcome by embedding the particles in a single TMM4 substrate. The particles in the cavity are aligned in the direction of the applied static magnetic field as shown in Figure 1b, thereby changing the electrical length of the shunted stub in the P-CRLH structure. The novel P-CRLH structure in Figure 1 consists of six inter-digital fingers and two symmetrical stubs. Sixteen MRS structures numbered from 1 to 16 are designed in the two stubs. A tiny magnet beneath any MRS structure will activate the particles in the cavity, which causes a change in the length of the current path along the stub and thereby changes the value of the left-handed inductance L_L. Similarly, activating these particles in MRS structures at appropriate locations along the inter-digital fingers of CRLH will cause C_L to change. The changes in L_L and C_L will change the phase response of the P-CRLH unit cell.

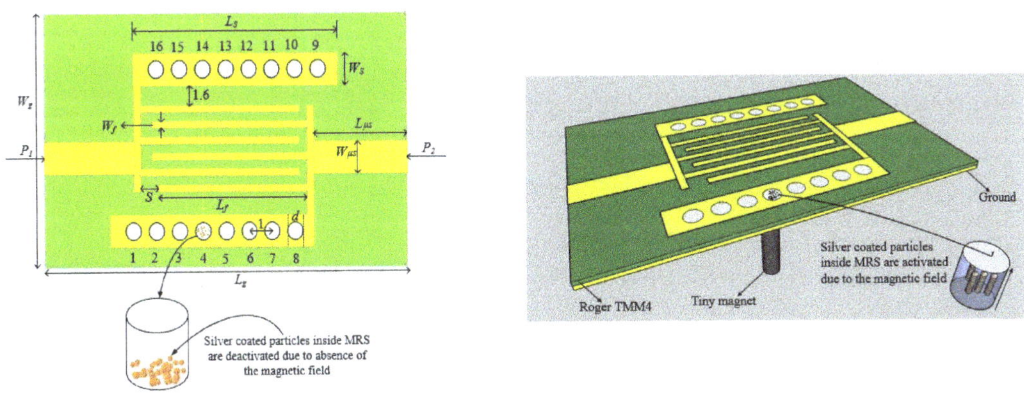

(a) (b)

Figure 1. (a) P-CRLH unit cell top view (*All dimensions are in mm* : $L_S = 8$, $W_S = 1$, $W_f = 0.3$, $L_f = 6$, $L_{\mu s} = 7.6$, $W_{\mu s} = 1.8$, $d = 0.6$, $W_g = 9$, $L_g = 22$, $S = 0.2$); (b) P-CRLH unit cell 3D view.

The equivalent circuit of the proposed P-CRLH unit cell structure is shown in Figure 2, and its phase shift analysis is based on the guidelines in [1]. ${L_s}^{IDC}$ and ${L_s}^{SI}$ represent the inductances of the Inter-Digital Capacitor (IDC) and Stub Inductor (SI) due to their physical lengths. $L_P{}^{SI}$ is the shunted inductance of the current path between the stub inductor and the ground plane controlled by activating and deactivating the embedded micron-sized particles inside the MRS cavity. $C_P{}^{SI}$ and $C_P{}^{IDC}$ are the inherent parallel plate capacitances between the stub inductor (or IDC) and the ground plane. $C_s{}^{IDC}$ is the capacitance due to the spacing among the fingers of IDC. $L_{\mu s}$ and $C_{\mu s}$ are the inductance and capacitance of the microstrip transmission lines on both ends of the P-CRLH unit cell. Finally, the four P-CRLH parameters are calculated as follows [1]:

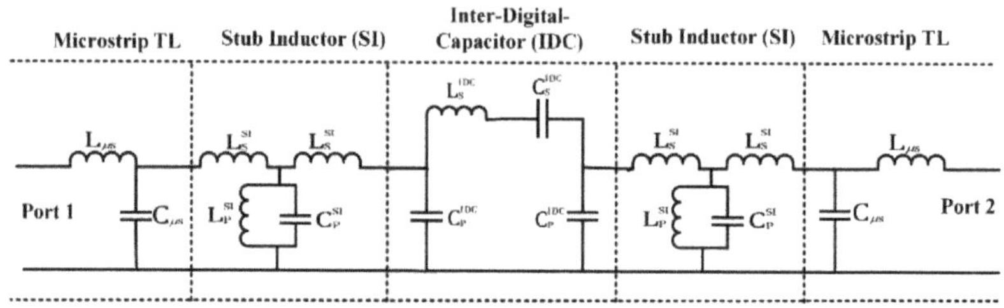

Figure 2. Equivalent circuit model of the P-CRLH unit cell.

$$\left.\begin{array}{l} L_R = L_s{}^{IDC} + \frac{L_s{}^{SI}}{2} \\ C_R = 2C_P{}^{IDC} + C_P{}^{SI} \\ L_L = L_P{}^{SI} \\ C_L = C_s{}^{IDC} \end{array}\right\} \tag{1}$$

Next, with the use of the ABCD parameter approach for the proposed P-CRLH unit cell structure, the phase response is calculated as follows: The cascaded ABCD matrices of the equivalent cicuit model in Figure 2 of the proposed P-CRLH unit cell design are given by

$$\begin{pmatrix} A & B \\ C & D \end{pmatrix}_{PCRLH} = \begin{pmatrix} A & B \\ C & D \end{pmatrix}_{\mu s} \begin{pmatrix} A & B \\ C & D \end{pmatrix}_{SI} \begin{pmatrix} A & B \\ C & D \end{pmatrix}_{IDC} \begin{pmatrix} A & B \\ C & D \end{pmatrix}_{SI} \begin{pmatrix} A & B \\ C & D \end{pmatrix}_{\mu s} \tag{2}$$

Then, the S-parameters of the P-CRLH unit cell design are calculated as [32]

$$\begin{pmatrix} S_{11} & S_{12} \\ S_{21} & S_{22} \end{pmatrix} = \begin{pmatrix} \frac{A + \frac{B}{Z_O} - CZ_O - D}{A + \frac{B}{Z_O} + CZ_O + D} & \frac{2(AD - DC)}{A + \frac{B}{Z_O} + CZ_O + D} \\ \frac{2}{A + \frac{B}{Z_O} + CZ_O + D} & \frac{-A + \frac{B}{Z_O} - CZ_O + D}{A + \frac{B}{Z_O} + CZ_O + D} \end{pmatrix} \tag{3}$$

Finally, the phase response and insertion loss (IL) of the proposed P-CRLH unit cell structure from the S-parameters marix in (3) are calculated as

$$\phi_{S_{21}} = phase\ of\ S_{21} \tag{4}$$

$$\text{IL(dB)} = 20 \times log10|S_{21}| \tag{5}$$

The expressions in Equations (1)–(5) indicate that various phase shifts from the P-CRLH unit cell can be obtained by changing the values of left-handed parameters L_L and C_L. The values of L_L and C_L are controlled through the activation of the silver-coated magnetic particles embedded in the structure. Practically, the value of L_L is varied by changing the physical lengths (L_s) of the shunted stub transmission lines in Figure 1a. The physical lengths of the stub transmission lines are changed by activating the MRSs along their lengths numbered from 1 to 16. Similarly, by varying the spacing among the fingers (s), different values of C_L are obtained. The phase response of the P-CRLH unit cell is a complex function of mainly four parameters (L_R, C_R, L_L, C_L) that is calculated using the expressions in Equations (1)–(5) once the physical values of these parameters are determined.

3. Simulation Results of the P-CRLH Phase Shifter

Initially, particles in MRSs at positions 1 and 9 in Figure 1 are assumed to be activated. This shows that the entire lengths of both stub inductors are in place. Next, the equivalent circuit model parameters in Equation (1) are extracted using the full-wave simulation at the design frequency $f_0 = 5.2$ GHz using the procedure outlined in [1], and the extracted values are $L_s{}^{IDC} = 0.65$ nH, $C_P{}^{IDC} = 1.97$ pF, $C_s{}^{IDC} = 3$ pF, $L_s{}^{SI} = 0.85$ nH, $L_P{}^{SI} = 2.91$ nH, $C_P{}^{SI} = 1.94$ pF, $L_{\mu s} = 0.5$ nH, and $C_{\mu s} = 0.36$ pF. Then, by implementing Equations (1)–(5) in MATLAB, the reflection coefficient, Inserion Loss (IL), and phase response of the P-CRLH phase shifter are shown in Figure 3. The phase response of the P-CRLH phase shifter using the equivalent circuit model in Figure 2 with the extracted circuit values from the ADS circuit simulator are also shown in Figure 3. Next, for full-wave results, the microstrip implementation of the P-CRLH phase shifter in Figure 1 is simulated in the CST microwave studio suite. For MRS structure implementation in the full-wave simulator, the guidelines in [22] are followed, and the corresponding CST simulation results are shown in Figure 3. The analytical (MATLAB) and ADS circuit simulator results in Figure 3 are on top of each other, validating the analytical modeling in (1)–(5). The -10 dB impedance-matching characteristics (S_{11}) and Insertion Loss (S_{12}) in the full-wave simulation for the proposed P-CRLH phase shifter are in good agreement with the analytical and circuit simulation results shown in Figure 3a. The phase deviation between the analytical and full-wave simulation results in Figure 3b is around 4°–5°. Next, the parametric study of the silver-coated magnetic particle activation in MRSs in the symmetrical stub transmission lines of the P-CRLH phase shifter structure in Figure 1 for different phase shift responses is carried out, and the full-wave CST results are shown in Figure 4. The results in Figure 4 shows that almost equal phase slope phase responses are obtained over the C-band by activating MRSs at different locations along the lengths of the stub transmission lines. The maximum shift obtained with the unit cell P-CRLH design is approximtely 45° by activating particles in MRSs 8 and 16 simultaneosuly. The impedance matching characteristics (S_{11}) of the structure deteriorates in the case of MRSs 8 & 16 activation as compared to MRSs 1 & 9 activation as shown in Figure 4. The reason is that for MRSs 1 & 9 activation, the entire lengths of the two shunted stub lines are electrically in place and therefore support the propagation of the electromagnetic wave from port 1 to port 2. In the case of MRSs 8 & 16 activation, almost the entire lengths of the two stubs are effectively out of the circuit, which negatively affects the impedance matching characteristics. Therefore, the insertion loss (S_{12}) for MRSs 8 & 16 activation is approximately 0.48 dB greater than the insertion loss for the MRSs 1 & 9 activation.

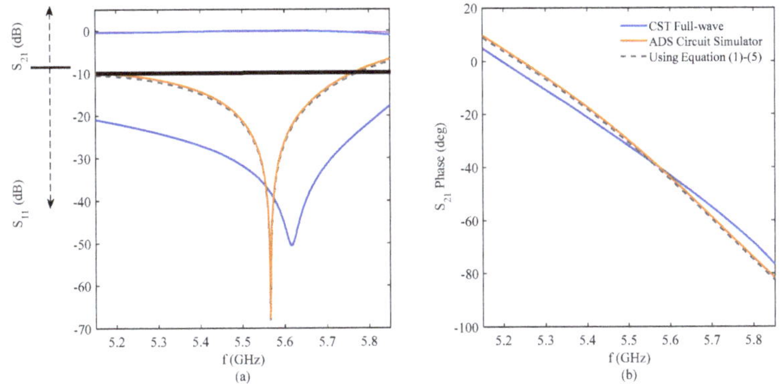

Figure 3. (**a**) Reflection coefficient, insertion loss, and (**b**) phase response of the unit cell P-CRLH phase shifter with MRSs 1 & 9 activation.

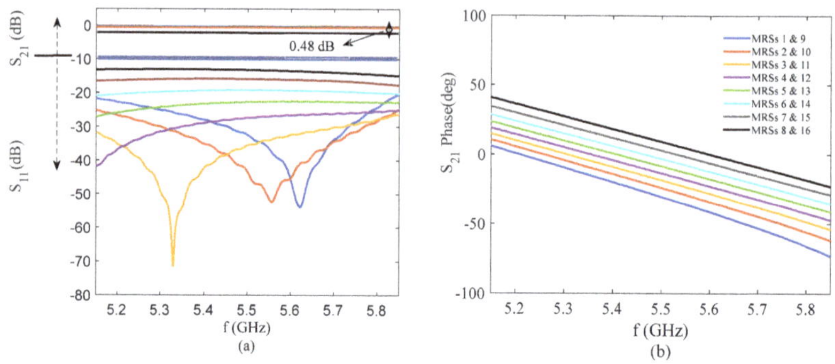

Figure 4. Parametric study of particle activation in MRSs along the symmetrical stub transmission lines. (**a**) Reflection coefficient, insertion loss, and (**b**) phase response of the unit cell P-CRLH phase shifter.

4. Simulated Results of a Cylindrical Phased Array with P-CRLH Phase Shifters

To test the performance of the proposed P-CRLH phase shifter, a 1×4 broadband antenna array placed along the cylindrical arc of radius 30 cm operating in the C-band (5–6 GHz) with inter-element spacing of $\frac{\lambda}{2}$ at $f_0 = 5.2$ GHz was considered a test case as shown in Figure 5. The required phase shifts for broadside, 15°, and 30° main beam scanning are given in Table 1 and were calculated using the technique in [33]. Since the P-CRLH phase shifter in Section 2 can give a maximum 45° phase shift, to achieve the phase shifts in Table 1, four unit cells of P-CRLH were cascaded. The phase shifts achieved with the cascaded P-CRLH phase shifters and the corresponding MRSs activation in the structure are given in Table 2. The activations of specific MRSs for certain required phase shifts are determined using the phase response graphs in Figure 4b. After achieving the required phase shifts, the 1×4 cylindrical antenna array was excited with cascaded P-CRLH phase shifters as shown in Figure 5. Initially in the CST simulator, the 1×4 cylindrical antenna array was directly excited with the phase shifts given in Table 1.

Then, the array was exited with four cascaded P-CRLH phase shifters to achieve the desired radiation patterns. The CST simulated results for directly excited and the P-CRLH phase shifter-excited 1×4 cylindrical antenna array at 5.2 GHz, 5.5 GHz, and 5.8 GHz in the C-band are shown in Figure 6. The results indicate that the proposed P-CRLH phase shifters can achieve the desired broadside radiation pattern and 15°, 30° main beam scanning patterns in the C-band. The broadside and scanning results are shown to agree well with the results reported in [33].

Table 1. Phases (degrees) for broadside, 15,° and 30° main-beam scanning of a cylinder with $r = 30$ cm.

Antenna	θ_s=0°		θ_s=15°		θ_s=30°	
	Calculated Phases From [33]	**Achieved P-CRLH Phases in This Work**	**Calculated Phases from [33]**	**Achieved P-CRLH Phases in This Work**	**Calculated Phases from [33]**	**Achieved P-CRLH Phases in This Work**
A_1	19.4374	19.081	17.4326	19.081	17.4326	19.081
A_2	2.1630	2.309	−44.6481	−49.3426	−88.0606	−88.345
A_3	2.1630	2.309	−91.2355	−93.649	−178.0606	−173.57
A_4	19.4374	19.081	−122.3297	−119.68	−252.5674	−255.342

Table 2. Achieved phases (degrees) with cascaded four-unit-cell P-CRLH phase shifters.

Scan Angle (θ_s)	P-CRLH Phases	MRSs Activations (See Figure 1a for Numbering of Particle Activation in the Cavities			
		Unit Cell-1	Unit Cell-2	Unit Cell-3	Unit Cell-4
0°	19.081	1, 9	1, 9	1, 9	1, 2, 9, 10
	2.309	1, 9	1, 9	1, 9	1, 8, 9, 16
	2.309	1, 9	1, 9	1, 9	1, 8, 9, 16
	19.081	1, 9	1, 9	1, 9	1, 2, 9, 10
15°	19.081	1, 9	1, 9	1, 9	1, 2, 9, 10
	−49.3426	1–7, 9–15	1–7, 9–15	1–7, 9–15	1–7, 9–15
	−93.649	1–5, 9–13	1–5, 9–13	1–5, 9–13	1–7, 9–15
	−119.68	1–4, 9–12	1–4, 9–12	1–4, 9–12	1–7, 9–15
30°	19.081	1, 9	1, 9	1, 9	1, 2, 9, 10
	−88.345	1, 5, 9, 13	1, 5, 9, 13	1,5, 9,13	1, 8, 9, 16
	−173.57	1–3, 9–11	1–3, 9–11	1–3, 9–11	1–3, 9–11
	−255.342	1, 9	1, 9	1, 9	1, 9

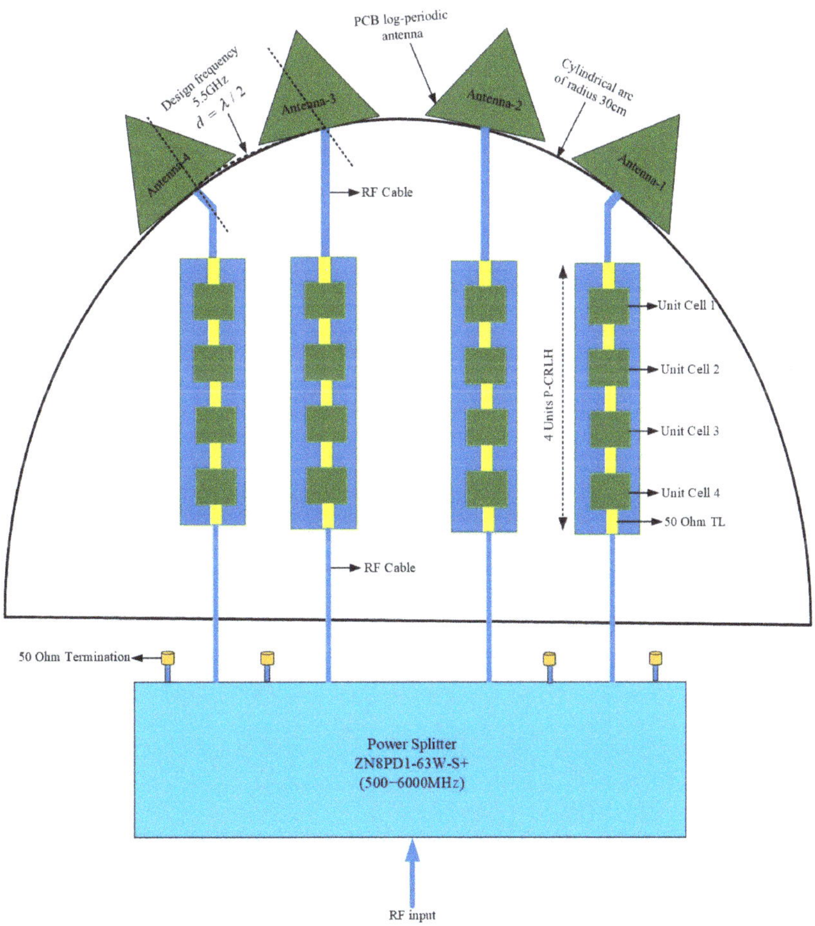

Figure 5. Schematic of the 1 × 4 P-CRLH excited cylindrical phased array.

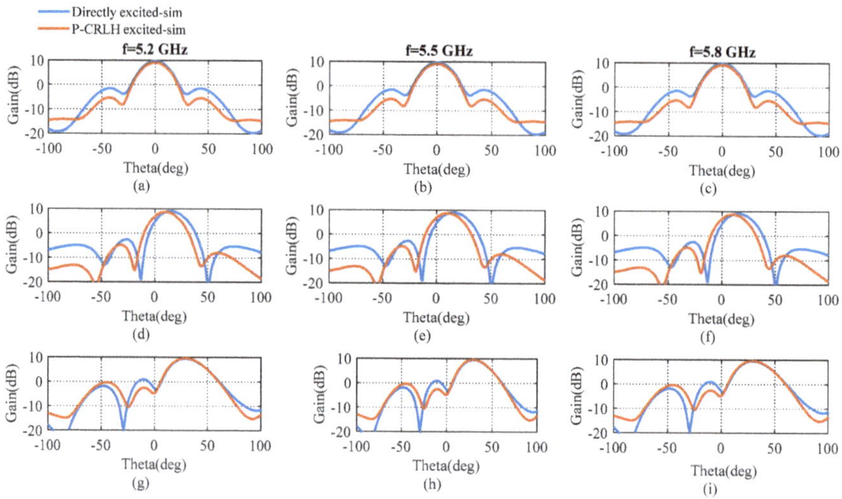

Figure 6. Simulation results of a 1 × 4 P-CRLH excited cylindrical phased array operating in C-band: (**a–c**) broadside patterns, (**d–f**) scanned patterns at 15° and (**g–i**) scanned patterns at 30°.

5. Measurement Results of a Cylindrical Phased Array with P-CRLH Phase Shifters

Next, to validate the simulation results, four cascaded unit cells of the P-CRLH phase shifter were manufactured on a TMM4 substrate ($\varepsilon_r = 4.7$, $tan\delta = 0.002$, and thickness = 0.635 mm). The silver-coated micron-sized magnetic particles used throughout the measurements were manufactured by Potters Industries (Malvern, PA, USA) (CONDUCT-O-Fil, part number SM25P20) [30]. MRSs cavities were drilled through the substrate along the length of stub transmission lines at the locations indicated in Figure 1. Then, the MRSs cavities were partially filled with these particles, the top of the MRS cavity is the stub transmission line and the bottom is a ground plane. The activation of particles inside any MRS cavity was achieved by a tiny magnet beneath the cavity as illustrated in Figure 7a. For measurement purposes, the MRSs cavities given in Table 2 were filled with the particles and were activated with tiny magnets. Next, the C-band 1 × 4 antenna array was placed along the circular arc of a cylindrical surface with radius of 30 cm. The model of the cylindrical-parabolic conformal antenna array in [34] was adopted. Each individual antenna in the array was excited with the cascaded four unit cells of the P-CRLH phase shifter. The entire setup including RF cables in a fully calibrated in-house anechoic chamber (300 KHz–20 GHz) measurement facility is shown in Figure 7b. The Keysight VNA (E5071C) and Diamond Engineering desktop antenna measurement system were used for the radiation patterns synthesis. The measured results of the radiation patterns for broadside, 15°, and 30° main beam scanning using the P-CRLH phase shifter-excited conformal phased array at the central frequency of the C-band (5.5 GHz) are shown in Figure 8. For comparison, the simulated results for the radiation patterns of the P-CRLH-excited conformal phased array are also shown in Figure 8. The differences in the peak gain between the simulated and measured results for the 0°, 15°, and 30° scan angles are 0.55 dB, 0.22 dB, and 0.19 dB, respectively. The corresponding differences in the first sidelobe levels for the three scan angles are 2.8 dB, 1 dB, and 2.3 dB, respectively. Overall, there is a good agreement between the simulated and measured peak gain results except for the sidelobe levels as shown in Table 3. This is thought to be due to imperfect P-CRLH phase shifter fabrication, differences in the manufactured and simulated antenna array modeling, and a lack of perfect symmetry in the anechoic chamber.

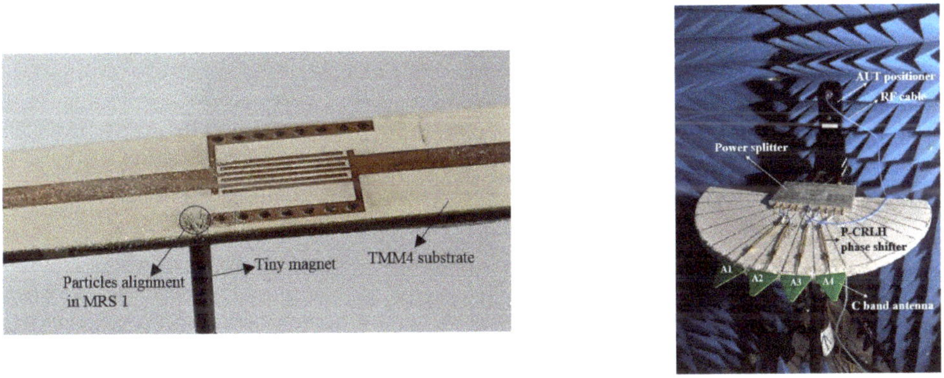

(a) (b)

Figure 7. Photographs of (**a**) A fabricated unit cell P-CRLH phase shifter; (**b**) A P-CRLH phase shifter-based 1 × 4 cylindrical antenna array in an anechoic chamber.

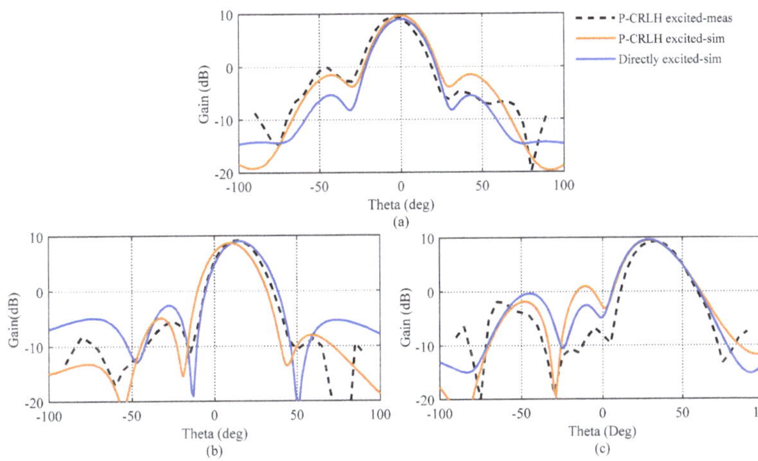

Figure 8. Measured radiation patterns of a 1 × 4 P-CRLH-excited cylindrical phased array: (**a**) broadside pattern, (**b**) main beam scanning at 15°, and (**c**) main beam scanning at 30°.

Table 3. Comparison between simulated and measured results of a 1 × 4 P-CRLH-excited cylindrical phased array.

Scan Angle (θ_s)	Peak Gain (dB)		First Sidelobe Level (dB)	
	Simulated	Measured	Simulated	Measured
0°	9.48	8.93	13.1	10.3
15°	8.87	8.65	14.1	15.1
30°	8.68	8.49	12.9	15.2

6. Conclusions

A novel micron-sized particle-based composite right/left-handed (CRLH) phase shifter is investigated for conformal phased array antennas. These particles are embedded inside the substrate along the stub and inter-digital fingers of a CRLH transmission line. Desired phase shifts are obtained by magnetically aligning these particles with a tiny magnet, which in turn changes the left-handed inductance and capacitance of the CRLH transmission line. A 1 × 4 conformal phased array is then excited with these particle-based CRLH phase shifters for steering the main beam at desired scan angles. The proposed conformal phased array with integrated particle CRLH phase shifters is an excellent choice for flexing electronics applications where commercial phase shifters have drawbacks in fully printing requirements. One such example is the cylindrical array studied in [33], where a separate DC biasing circuitry is needed to operate the phase shifters, which can cause electromagnetic coupling with the phase shifters' functionality. The proposed P-CRLH phase shifter needs a magnetic field to be applied from the ground plane to the top layer of the phase shifter, which effectively isolates the phase shifter from unintended coupling. The phase shifters in [5–7] need the mounting of additional components (resistors, inductors, capacitors) with off-chip ICs, and as such cannot be made flexible and fully printable. The proposed P-CRLH phase shifter does not need the discrete components and therefore can be easily integrated with conformal arrays and other flexible and printing electronics circuitry.

Author Contributions: B.D.B. provided the idea. A.I. and W.K. assisted in simulations. M.A. performed complete simulations, fabrications, and measurements. I.U. assisted in measurements, overall paper management, idea development, and manuscript writing. All authors have read and agreed to the published version of the manuscript.

Funding: This work is supported by the HEC-NRPU, Pakistan, via Project No. 20-14696/NRPU/R&D/HEC/2021. The APC was paid by co-author Wesam Khalil.

Data Availability Statement: All data have been included in study.

Conflicts of Interest: The authors declare no conflict of interest.

Appendix A

Table A1. Abbreviations used in the paper.

Notation	Description
TL	Transmission Line
CRLH	Composite Right/Left-Handed
CRLH-TL	Composite Right/Left-Handed Transmission Line
P-CRLH	Particles-CRLH
MRS	Magneto-static field Responsive Structure
RH-TL	Right-Handed Transmission Line
PIN	p-type intrinsic n-type
MEMS	Microelectromechanical systems
BST	Barium Strontium Titanate
MIM	Metal-Insulator-Metal
RF	Radio Frequency
EBG	Electromagnetic Band-Gap
SI	Stub Inductor
IDC	Inter Digital Capacitor
IL	Insertion Loss

References

1. Caloz, C.; Itoh, T. *Electromagnetic Metamaterials: Transmission Line Theory and Microwave Applications*, 1st ed.; Wiley-IEEE Press; John Wiley and Sons, Ltd.: Hoboken, NJ, USA, 2005; p. 376.
2. Vivos, J.; Crépin, T.; Foulon, M.F.; Sokoloff, J. Unbalanced metamaterials applied to phase shifter: Dedicated design method and application in C-band. *Prog. Electromagn. Res. C* **2019**, *93*, 1–17. [CrossRef]

3. Maassel, M.; Braaten, B.D.; Rogers, D.A. A metamaterial-based multiband phase shifter. In Proceedings of the IEEE International Conference on Electro/Information Technology, Milwaukee, WI, USA, 5–7 June 2014.
4. Zhang, J.; Cheung, S.W.; Yuk, T.I. A compact 6-bit phase shifter with high-power capacity based on composite right/left-handed transmission line. In Proceedings of the 40th European Microwave Conference, Paris, France, 28–30 September 2010.
5. Zhang, J.; Cheung, S.W.; Yuk, T.I. Design of n-bit digital phase shifter using single CRLH TL unit cell. *Electron. Lett.* **2010**, *46*, 506–508. [CrossRef]
6. Zhang, J.; Cheung, S.W.; Yuk, T.I. Design of n-bit phase shifters with high power-handling capability inspired by composite right/left-handed transmission line unit cells. *IET Microw. Antennas Propag.* **2010**, *4*, 991–999. [CrossRef]
7. Li, P.; Niu, J.; Fu, J.; Chen, W.; Lv, B.; Wang, Z.; Li, A. A novel 360° continuously tunable phase shifter based on varactor-loaded CRLH transmission line at exact 2.4 GHz. In Proceedings of the 2016 IEEE International Conference on Electronic Information and Communication Technology (ICEICT), Harbin, China, 20–22 August 2016.
8. Wu, Y.S.; Lin, X.Q.; Zhang, J.; Jiang, Y.; Cheng, F.; Fan, Y. Broadband and wide range tunable phase shifter based on composite right/left handed transmission line. *J. Electromagn. Waves Appl.* **2012**, *26*, 1308–1314. [CrossRef]
9. Hwang, S.H.; Jang, T.; Kim, J.M.; Kim, Y.K.; Lim, S.; Baek, C.W. MEMS-tunable composite right/left-handed (CRLH) transmission line and its application to a phase shifter. *J. Micromech. Microeng.* **2011**, *21*, 125022. [CrossRef]
10. Michishita, N.; Kitahara, H.; Yamada, Y.; Cho, K. Tunable phase shifter using composite right/left-handed transmission line with mechanically variable MIM capacitors. *IEEE Antennas Wirel. Propag. Lett.* **2011**, *10*, 1579–1581. [CrossRef]
11. Michishita, N.; Yamada, Y.; Cho, K. Design of composite right/left-handed transmission line for phase shifter of multi-band base station antenna. *EPJ Appl. Metamateri.* **2011**, *5*, 12. [CrossRef]
12. Nikfalazar, M.; Sazegar, M.; Zheng, Y.; Wiens, A.; Jakoby, R.; Friederich, A.; Kohler, C.; Binder, J.R. Compact tunable phase shifter based on inkjet printed BST thick-films for phased-array application. In Proceedings of the 2013 European Microwave Conference, Nuremberg, Germany, 6–10 October 2013.
13. Nikfalazar, M.; Sazegar, M.; Friederich, A.; Kohler, C.; Zheng, Y.; Wiens, A.; Binder, J.R.; Jakoby, R. Inkjet printed BST thick-films for x-band phase shifter and phased array applications. In Proceedings of the 2013 International Workshop on Antenna Technology (iWAT), Karlsruhe, Germany, 4–6 March 2013.
14. Nikfalazar, M.; Sazegar, M.; Mehmood, A.; Wiens, A.; Friederich, A.; Maune, H.; Binder, J.R.; Jakoby, R. Two-dimensional beam-steering phased-array antenna with compact tunable phase shifter based on BST thick films. *IEEE Antennas Wirel. Propag. Lett.* **2016**, *16*, 585–588. [CrossRef]
15. Sazegar, M.; Zheng, Y.; Maune, H.; Zhou, X.; Damm, C.; Jakoby, R. Compact left handed coplanar strip line phase shifter on screen printed BST. In Proceedings of the 2011 IEEE MTT-S International Microwave Symposium, Baltimore, MD, USA, 5–10 June 2011.
16. Haghzadeh, M.; Jaradat, H.M.; Armiento, C.; Akyurtlu, A. Design and simulation of fully printable conformal antennas with BST/polymer composite based phase shifters. *Prog. Electromagn. Res. C* **2016**, *62*, 167–178. [CrossRef]
17. Ebrahimi, F.; Qaderi, S. Stability analysis of embedded graphene platelets reinforced composite plates in thermal environment. *Eur. Phys. J. Plus.* **2019**, *134*, 349. [CrossRef]
18. Sazegar, M.; Zheng, Y.; Maune, H.; Damm, C.; Zhou, X.; Binder, J.; Jakoby, R. Low-cost phased-array antenna using compact tunable phase shifters based on ferroelectric ceramics. *IEEE Trans. Microw. Theory Tech.* **2011**, *59*, 1265–1273. [CrossRef]
19. Sazegar, M.; Zheng, Y.; Maune, H.; Zhou, X.; Damm, C.; Jakoby, R. Compact artificial line phase shifter on ferroelectric thick-film ceramics. In Proceedings of the 2010 IEEE MTT-S International Microwave Symposium, Anaheim, CA, USA, 23–28 May 2010.
20. Iftikhar, A.; Parrow, J.; Asif, S.; Braaten, B.D.; Allen, J.; Allen, M.; Wenner, B. On using magneto-static responsive particles as switching elements to reconfigure microwave filters. In Proceedings of the 2016 IEEE International Conference on Electro Information Technology (EIT), Grand Forks, ND, USA, 19–21 May 2016.
21. Soufizadeh-Balaneji, N.; Rogers, D.A.; Braaten, B.D. New embodiments of static field micro-particle components for reconfigurable RF applications. In Proceedings of the 2019 IEEE MTT-S International Microwave Symposium (IMS) 2019, Boston, MA, USA, 2–7 June 2019.
22. Iftikhar, A.; Asif, S.M.; Parrow, J.M.; Allen, J.W.; Allen, M.S.; Fida, A.; Braaten, B.D. Changing the operation of small geometrically complex EBG-based antennas with micron-sized particles that respond to magneto-static fields. *IEEE Access* **2020**, *8*, 78956–78964. [CrossRef]
23. Iftikhar, A.; Parrow, J.M.; Asif, S.M.; Fida, A.; Allen, J.; Allen, M.; Braaten, B.D.; Anagnostou, D.E. Characterization of novel structures consisting of micron-sized conductive particles that respond to static magnetic field lines for 4G/5G (Sub-6 GHz) reconfigurable antennas. *Electronics.* **2020**, *9*, 903. [CrossRef]
24. Iftikhar, A.; Parrow, J.; Asif, S.; Allen, J.; Allen, M.; Braaten, B.D. Improving the efficiency of a reconfigurable microstrip patch using magneto-static field responsive structures. *Electron. Lett.* **2016**, *52*, 1194–1196. [CrossRef]
25. Jerika, C.; Mitra, D.; Lewis, J.; Braaten, B.D.; Allen, J.; Allen, M. Devolvement of a sub-unit cell consisting of capacitive gaps and magneto- static particles. In Proceedings of the 2019 IEEE Research and Applications of Photonics in Defense Conference (RAPID), Miramar Beach, FL, USA, 19–21 August 2019.
26. Cleveland, J.; Lewis, J.; Mitra, D.; Braaten, B.D.; Allen, J.; Allen, M. On the image analysis of conducting magneto-responsive micro-particles for applications in leaky wave antenna beam steering. In Proceedings of the 2020 IEEE International Symposium on Antennas and Propagation and North American Radio Science Meeting, Montreal, QC, Canada, 5–10 July 2020.

27. Cleveland, J.; Braaten, B.D.; Allen, M.; Allen, J.; Wenner, B. On using micron-sized silver coated particles to control the electromagnetic response of a metamaterial with complementary split ring resonators and wires in a host dielectric. In Proceedings of the 2018 IEEE Research and Applications of Photonics In Defense Conference (RAPID), Miramar Beach, FL, USA, 22–24 August 2018.
28. Cleveland, J.; Lewis, J.; Mitra, D.; Braaten, B.D.; Allen, J.; Allen, M. On the phase shift of a microstrip transmission line loaded with magneto-responsive conducting micro-particles. In Proceedings of the 2020 IEEE International Symposium on Antennas and Propagation and North American Radio Science Meeting, Montreal, QC, Canada, 5–10 July 2020.
29. Parrow, J.M.; Iftikhar, A.; Asif, S.M.; Allen, J.W.; Allen, M.S.; Wenner, B.R.; Braaten, B.D. On the bandwidth of a microparticle-based component responsive to magnetostatic fields. *IEEE Trans. Electromagn. Compat.* **2017**, *59*, 1053–1059. [CrossRef]
30. Potters Industries LLC. Available online: http://www.pottersbeads.com/ (accessed on 12 December 2022).
31. Parrow, J.M. Equivalent circuit modeling and signal integrity analysis of magneto-static responsive structures, and their applications in changing the effective permittivity of microstrip transmission lines. Ph.D. Thesis, Electrical and Computer Engineering Department, North Dakota State University, Fargo, ND, USA, 2016.
32. Pozar, D.M. *Microwave Engineering*, 4th ed.; John Wiley and Sons, Ltd.: Hoboken, NJ, USA, 2011; p. 752.
33. Braaten, B.D.; Roy, S.; Nariyal, S.; Aziz, M.A.; Chamberlain, N.F.; Irfanullah, I.; Reich, M.T.; Anagnostou, D.E. A self-adapting flexible (SELFLEX) antenna array for changing conformal surface applications. *IEEE Trans. Antennas Propag.* **2013**, *61*, 655–665. [CrossRef]
34. Yang, K.; Zhao, Z.; Nie, Z.; Ouyang, J.; Liu, Q.H. Synthesis of conformal phased arrays with embedded element pattern decomposition. *IEEE Trans. Antennas Propag.* **2011**, *59*, 2882–2888. [CrossRef]

electronics

MDPI

Article

A Novel Integrated UWB Sensing and 8-Element MIMO Communication Cognitive Radio Antenna System

D Srikar [1], Anveshkumar Nella [2], Ranjith Mamidi [3], Ashok Babu [4], Sudipta Das [5,*], Sunil Lavadiya [6], Abeer D. Algarni [7] and Walid El-Shafai [8,9]

1 Chaitanya Bharathi Institute of Technology, Hyderabad 500075, India
2 School of EEE, VIT Bhopal University, Bhopal 466114, India
3 Vaagdevi College of Engineering, Warangal 506005, India
4 Institute of Aeronautical Engineering, Hyderabad 500043, India
5 Department of Electronics & Communication Engineering, IMPS College of Engineering and Technology, Malda 732103, India
6 Department of Information and Communication Technology, Marwadi University, Rajkot 360003, India
7 Department of Information Technology, College of Computer and Information Sciences, Princess Nourah bint Abdulrahman University, P.O. Box 84428, Riyadh 11671, Saudi Arabia
8 Security Engineering Lab, Computer Science Department, Prince Sultan University, Riyadh 11586, Saudi Arabia
9 Department of Electronics and Electrical Communications Engineering, Faculty of Electronic Engineering, Menoufia University, Menouf 32952, Egypt
* Correspondence: sudipta.das1985@gmail.com

Abstract: In this article, a cognitive radio (CR) integrated antenna system, which has 1 sensing and 24 communication antennas, is proposed for better spectrum utilization efficiency. In the 24 communication antennas, 3 different operating band antennas are realized with an 8-element MIMO configuration. The sensing antenna linked to port 1 is able to sense the spectrum that ranges from 2 to 12 GHz, whereas the communication 8-element MIMO antennas linked with ports 2 to 9, ports 10 to 17 and ports 18 to 25 perform operations in the 2.17–4.74 GHz, 4.57–8.62 GHz and 8.62–12 GHz bands, respectively. Mutual coupling is found to be less than −12 dB between the antenna elements. Peak gain and radiation efficiency of the sensing antenna are better than 2.25 dBi and 82%, respectively, whereas the peak gains and radiation efficiencies of all communication antennas are more than 2.5 dBi and 90%, respectively. Moreover, diversity characteristics of the MIMO antenna are assessed by parameters such as DG, ECC and CCL. It is found that ECC and CCL are less than 0.42 and 0.46 bits/s/Hz, respectively, and also DG is more than 9.1 dB.

Keywords: CCL; cognitive radio; DGG; ECC; MIMO; spectrum sensing; spectrum utilization efficiency

Citation: Srikar, D.; Nella, A.; Mamidi, R.; Babu, A.; Das, S.; Lavadiya, S.; Algarni, A.D.; El-Shafai, W. A Novel Integrated UWB Sensing and 8-Element MIMO Communication Cognitive Radio Antenna System. *Electronics* **2023**, *12*, 330. https://doi.org/10.3390/electronics12020330

Academic Editors: Naser Ojaroudi Parchin, Chan Hwang See and Raed A. Abd-Alhameed

Received: 22 November 2022
Revised: 12 December 2022
Accepted: 13 December 2022
Published: 8 January 2023

1. Introduction

In the licensed spectrum, channels are unutilized most of the time, thus leading to inefficient spectrum utilization. Hence, spectrum utilization efficiency deteriorates. The unutilized channels (i.e., licensed) can be used effectively for other applications to reduce the wastage of spectrum issue. CR technology mainly uses the concept of using the unutilized channels in the spectrum overlay approach. It is believed that the primary users in the spectrum overlay approach are the owners of the licensed spectrum and do not utilize their channels in the licensed spectrum completely. So, there exists a continuous monitoring in the radio environment to find the white spaces (i.e., spectrum holes). When a white space is detected at any moment, secondary users can use the channel that consists of the detected white space for other applications until the primary users want to use that channel. When primary users use that channel, secondary users should shift to any other unutilized licensed channels.

A new frequency-agile CR MIMO patch antenna was proposed in [1]. It comprises two patch antenna elements, which are hexagonal shaped. To achieve compactness and increase isolation, hexagonal-shaped defected ground structures are incorporated in the ground plane. The compactness is further attained by utilizing reactive loading. Varactor diodes are employed in the microstrip feed line to achieve frequency reconfigurability. Consequently, frequency tuning is achieved at 1.42–2.27 GHz. However, the proposed CR MIMO antenna cannot sense the spectrum as it does not have a sensing antenna. A frequency-reconfigurable CR MIMO antenna system for interweave scenario was presented in [2]. It comprises four reconfigurable antenna elements that are pentagonal slot-based. Varactor diodes are employed to alter the capacitance of the slot. As a result, a wide tuning range, which ranges from 3.2 GHz to 3.9 GHz, is achieved with a minimum of 100 MHz impedance bandwidth in each band. However, the proposed MIMO antenna does not have a sensing antenna. Additionally, the mutual coupling in the proposed CR MIMO antenna system is just less than −10 dB. A four-port CR MIMO antenna that works for overlay and underlay approaches for 5G applications in the 2.5–4.20 GHz range was proposed in [3]. This type of feature of the proposed antenna is obtained by controlling the operating modes of the multifunctional reconfigurable filter since the multifunctional reconfigurable filter, which works in three operational modes such as tunable bandpass filter, tunable band reject filter, and all pass filter, is integrated with UWB sensing antenna. Four filtennas are integrated on FR-4 epoxy substrate to make a MIMO system. Additionally, they are well isolated with the help of reflectors. A two element MIMO antenna, which comprises sensing antenna operating at 2.2–7 GHz and two similar reconfigurable antennas, was proposed in [4]. By varying the capacitance of varactor diodes, reconfigurable antennas can be tuned to any frequency from 2.3 to 6.3 GHz. A four-port MIMO antenna in which two antennas are dedicated for sensing the spectrum 2.35–5.9 GHz and two other antennas are dedicated for communication was proposed. The reconfigurable narrowband antennas for communication can be tuned to any frequency from 2.6 to 3.6 GHz.

The undesired radio frequency interference can be avoided when sensing antenna and communication antennas are well isolated. Consequently, good performance is guaranteed for a CR device. When spectrum holes change rapidly in the radio environment and a single reconfigurable antenna is used for sensing the spectrum and performing the communication tasks, it becomes very hard for the antenna to switch from sensing mode to communication mode and vice versa. Moreover, in case of reconfigurable antennas, loses become high due to the presence of lumped elements, diodes, etc. Consequently, the performance of the antenna deteriorates to some extent. Additionally, it is noticed that the existing CR MIMO antennas in the literature [5–19] have a small tuning range since it is quite hard to tune the antenna to a desired frequency in a wide bandwidth. The drawbacks, which are associated with the available reconfigurable CR antennas in the literature, are power consumption, use of extra hardware, nonlinear effects of switches and biasing line effects. Moreover, the reconfigurable mechanisms may require the presence of motors and some additional biasing circuitry at times. Nevertheless, they have been extensively used by many researchers to make the antenna compact and tunable. Additionally, they are very difficult to implement in real time. In addition, the available reconfigurable CR antennas in the literature are able to perform only communication operations despite multiple spectrum holes being identified. Since it is well known that low profile planar antennas are advantageous and reconfigurable antennas have some unavoidable drawbacks [20,21], integrated sensing antennas and multiple communication antenna systems [22–26] are treated as the best alternative to reconfigurable antenna systems for CR applications. These antenna systems have a striking feature of performing multiple communication operations simultaneously due to the presence of multiple communication antennas. Since the spectrum is utilized in an efficient manner by performing multiple communication operations simultaneously, spectrum utilization efficiency increases significantly with these antenna systems. Whenever a white space (spectrum hole) identified by the sensing antenna matches with the operating frequency of the communication antenna, the respective communication antenna is given access to the secondary users. Otherwise, it is terminated by a 50-ohm load. In integrated sensing antennas and multiple communication antenna systems, communication antennas are selected by an excitation switching reconfigurable mechanism.

In this article, a cognitive radio (CR) integrated antenna system, which has 1 sensing and 24 communication antennas, is proposed for better spectrum utilization efficiency. The sensing antenna linked to port 1 is able to sense the spectrum that ranges from 2 to 12 GHz, whereas the communication 8-element MIMO antennas linked with ports 2 to 9, ports 10 to 17 and ports 18 to 25 perform operations in the 2.17–4.74 GHz, 4.57–8.62 GHz and 8.62–12 GHz bands, respectively. Mutual coupling is less than −12 dB. The HFSS EM simulation tool is used for the proposed structure design. The fabricated prototype is verified and a good agreement is noted between the simulated and measured results. To the best of our knowledge, this is the only sensing and 8-element MIMO communication integrated antenna system that has the notable feature of performing a maximum of three communication operations concurrently in the CR environment. Moreover, this antenna system is the best alternative for all reconfigurable CR MIMO antenna systems. Additionally, it is less complex compared to all the other existing CR MIMO antennas in the literature.

The rest of the paper is organized as follows. The design of the proposed structure, UWB and narrow band antenna design steps, along with performance characteristics, are reported. Later, results and discussion on each and every performance parameter are provided. Thereafter, MIMO diversity characteristics are discussed. Finally, the conclusion of the work is given.

2. Twenty-Five Port CR Integrated Antenna System

The proposed CR integrated antenna system schematic is shown in Figure 1. The performance and 10 dB return loss bandwidth of each antenna are given in Table 1. The dimensions of the proposed CR integrated antenna system are given in Table 2. Sensing antenna is linked to port 1, whereas the communication 8-element MIMO antennas are linked with ports 2 to 9, ports 10 to 17 and ports 18 to 25. FR4 epoxy substrate of a thickness of 1.6 mm is used in the present design. Additionally, the design procedures for sensing and communication antennas are explained in this section.

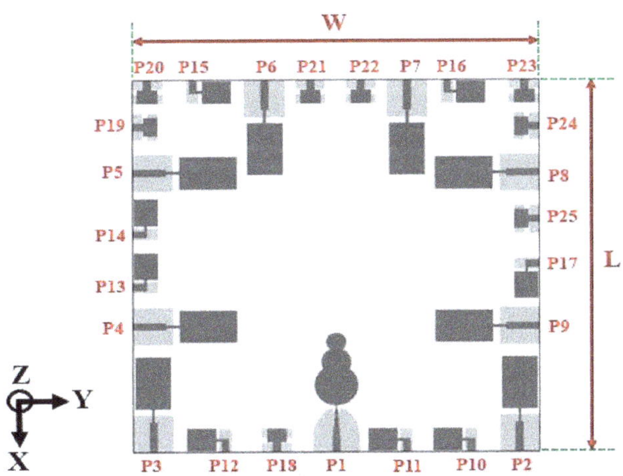

(a)

Figure 1. *Cont.*

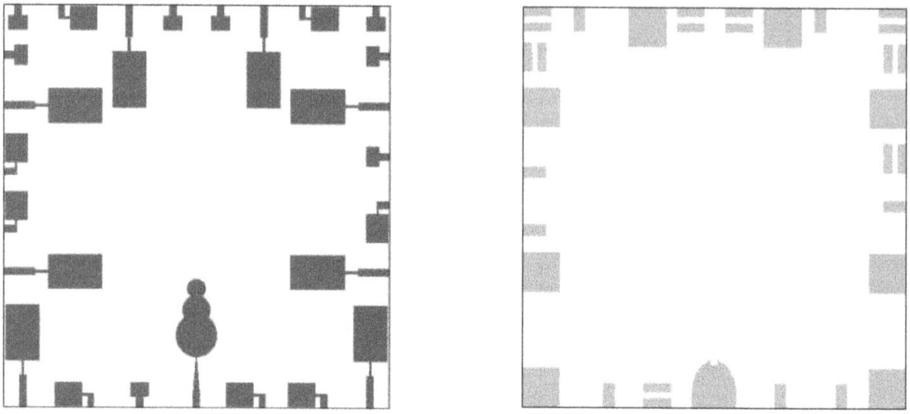

(b) (c)

Figure 1. Geometry of the 25-port CR MIMO antenna. (**a**) Structure of the antenna. (**b**) Top view. (**c**) Back view.

Table 1. Specifications of the 25-port CR MIMO antenna.

Antenna	Usage	10 dB Return Loss Bandwidth
Ant (P1)	Sensing	2–12 GHz
Ant (P2) to Ant (P9)	Communication	2.17–4.74 GHz
Ant (P10) to Ant (P17)	Communication	4.57–8.62 GHz
Ant (P18) to Ant (P25)	Communication	8.62–12 GHz

Table 2. Dimensions of the 25-port CR MIMO antenna.

Parameter	Dimension (mm)	Parameter	Dimension (mm)
R1	8.78	l1	22.5
R2	5.28	w1	14
R3	4	lg1	16
p1	1.16	wg1	16
lg	5	ln1	5.5
wg	18.98	wn1	1
wf	3	p2	2.5
ln	1.57	lw1	13
wn	4.5	wf1	3
p4	0.2	l2	11.5
wf3	3.2	w2	9.5
W	160	L	160
lg2	5	w3	8
wg2	10	lm	4
wf2	3	wm	12
wn2	1.1	lg3	4
p3	0.95	wg3	12
l3	5.8	g1	1.16

3. Design Process of the Sensing Antenna Linked with Port 1

The structure of the sensing antenna that operates at 2–12 GHz is shown in Figure 2. As depicted in Figure 3, the design process is finished in five steps. A traditional circular shaped monopole antenna (i.e., Ant I in Figure 3) is designed in the first step, and its impedance bandwidth ranges from 2.6 GHz to 10.6 GHz, as illustrated in Figure 4. The electrical lengthening is performed in the second step by combining a circular patch of radius R2 with the circular radiator of radius R1, as depicted in Ant II in Figure 3. As a result, the lower band edge (LBE) frequency of Ant II in Figure 1 is less than the LBE frequency of Ant I in Figure 4. However, the impedance matching of Ant II in Figure 4 at 3–5.75 GHz is poor. The electrical lengthening is further performed in the third step by combining a circular patch of radius R3 with the radiator of Ant II. Consequently, the LBE frequency of Ant III in Figure 4 becomes 2.07 GHz, which is less than the LBE frequency of Ant II in Figure 4, but the reflection coefficient curve of Ant III in Figure 4 is below −10 dB at 2.55–4.3 GHz and 8.1–8.8 GHz. So, to attain impedance matching throughout a wideband, i.e., 2–12 GHz, the ground plane of Ant IV in Figure 3 is made semi-elliptical shaped and some portion of the 50-ohm feed line is tapered towards the radiator. Thus, better impedance matching is achieved compared to Ant III in Figure 4, but the reflection coefficient of Ant IV is a bit better than −10 dB at 2.82–3.57 GHz and 5.56–6.34 GHz, as shown in Figure 4. In the last stage, good impedance matching is achieved over a wideband (i.e., 2–12 GHz) by etching a rectangle-shaped notch in the ground plane of Ant V.

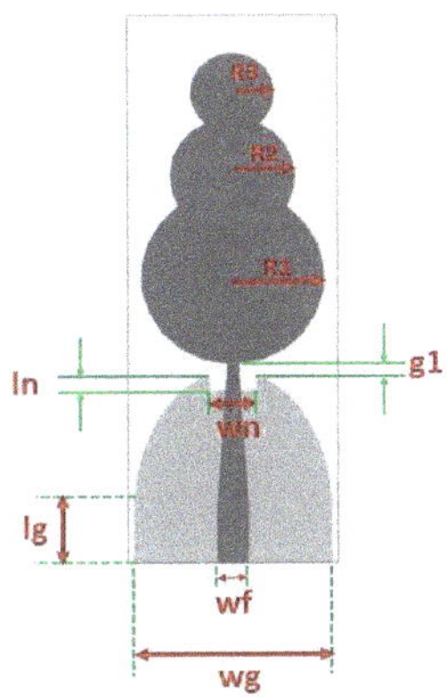

Figure 2. Schematic of the antenna linked with port 1.

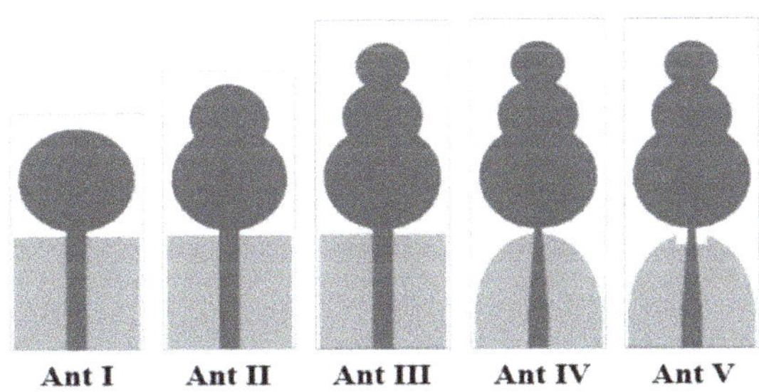

Figure 3. Evolution of antenna linked with port 1.

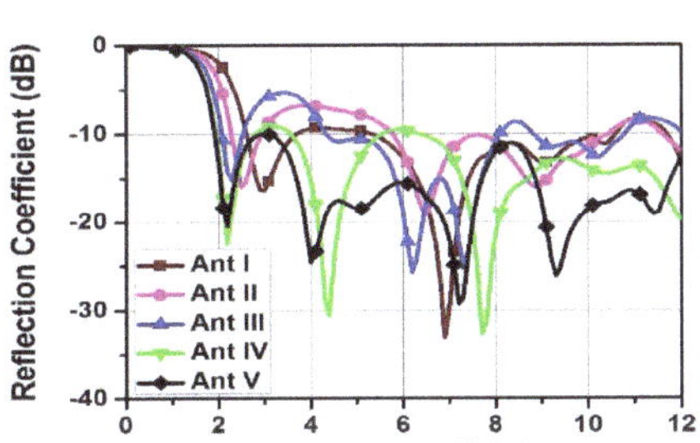

Figure 4. Evolution of the antenna linked with port 1.

As a result, the lower band edge (LBE) frequency of Ant II in Figure 1 is less than the LBE frequency of Ant I in Figure 4. However, the impedance matching of Ant II in Figure 4 at 3–5.75 GHz is poor. The electrical lengthening is further performed in the third step by combining a circular patch of radius R3 with the radiator of Ant II. Consequently, the LBE frequency of Ant III in Figure 4 becomes 2.07 GHz, which is less than the LBE frequency of Ant II in Figure 4, but the reflection coefficient curve of Ant III in Figure 4 is below −10 dB at 2.55–4.3 GHz and 8.1–8.8 GHz. So, to attain impedance matching throughout a wideband, i.e., 2–12 GHz, the ground plane of Ant IV in Figure 3 is made semi-elliptical shaped and some portion of the 50-ohm feed line is tapered towards the radiator. Thus, better impedance matching is achieved compared to Ant III in Figure 4, but the reflection coefficient of Ant IV is a bit better than −10 dB at 2.82–3.57 GHz and 5.56–6.34 GHz, as shown in Figure 4. In the last stage, good impedance matching is achieved over a wideband (i.e., 2–12 GHz) by etching a rectangle-shaped notch in the ground plane of Ant V.

4. Design Process of the Communication Antennas Linked with P2 to P9

The structure of the antenna linked with port 2 is depicted in Figure 5. The antenna linked with port 2 is targeted to operate at low frequencies and its design process ends in two stages. In the first stage, a normal monopole antenna with rectangle-shaped patch is designed, as illustrated in Figure 6. Furthermore, this antenna is realized with an 8-element MIMO configuration at ports P2 to P9. Dimensions of the rectangular radiator are selected in such a way that the LBE frequency of the antenna linked with port 2 is about 2.3 GHz. The mathematical formula for calculating the LBE of Ant I in Figure 6 is given below.

$$f_{\mathrm{LAntI(P2)}} = \frac{7.2}{(\mathrm{l1} + \mathrm{r1} + \mathrm{p2}) \times \mathrm{k}}\ \mathrm{GHz} \quad (1)$$

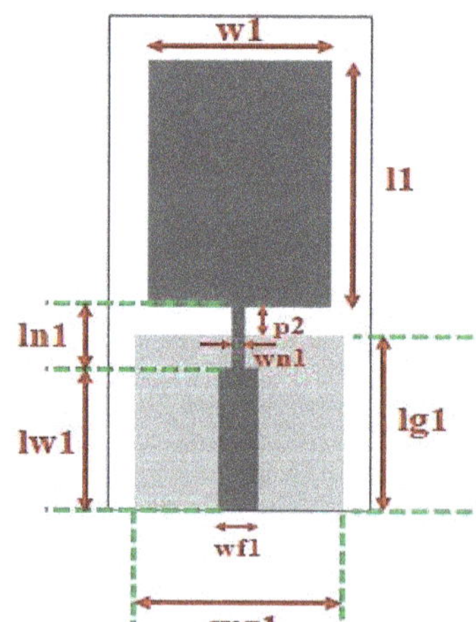

Figure 5. Schematic of the antennas linked with P2 to P9.

In Equation (1), l1 and p2 are the length of the rectangular patch and feed gap in centimeters, respectively. The term 'r1′ in centimeters can be calculated from the width of rectangular patch since $\mathrm{r1} = \frac{w1}{2\pi}$. The term 'k' is constant and is equal to 1.15.

After finding out the values of l1, r1, p2 and k, the LBE frequency of Ant I in Figure 1 is found as 2.3 GHz, whereas it is found as 2.34 GHz in simulation. It is evident from Figure 7 that Ant I operates at 2.34- 2.9 GHz and 4.9–7.75 GHz. As the antenna linked with port 2 is aimed to operate at low frequencies, a 1 mm width narrow strip line is placed between rectangular patch and the 50-ohm feed line of 3 mm width in the second stage, as depicted in Ant II in Figure 6. A parametric study is performed by altering the length of the

1 mm strip line, as depicted in Figure 8. It is obvious from Figure 8 that lower band edge frequency does not change much when the 1 mm strip line's length changes. Additionally, bandwidth slightly decreases when the 1 mm strip line's length increases. Finally, the 1 mm strip line's length is selected as 5.5 mm as good impedance matching and large −10 dB reflection coefficient bandwidth (i.e., 2.17–4.75 GHz) are achieved.

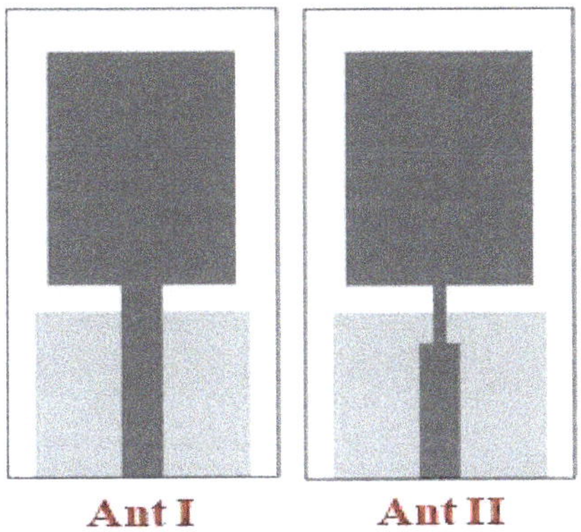

Figure 6. Evolution of the antennas linked with P2 to P9.

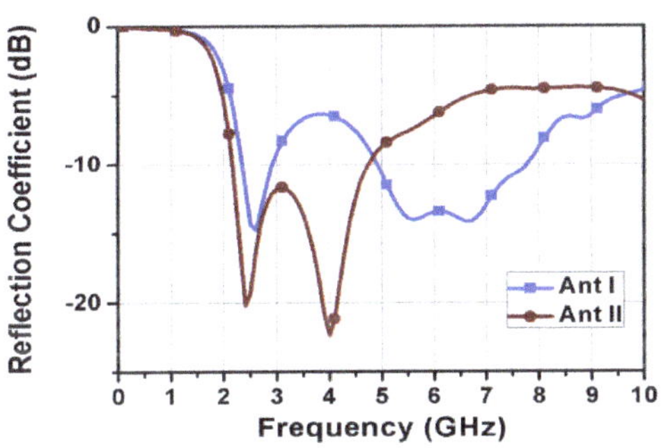

Figure 7. Reflection coefficients of the antennas in intermediate steps of the antenna linked with port 2.

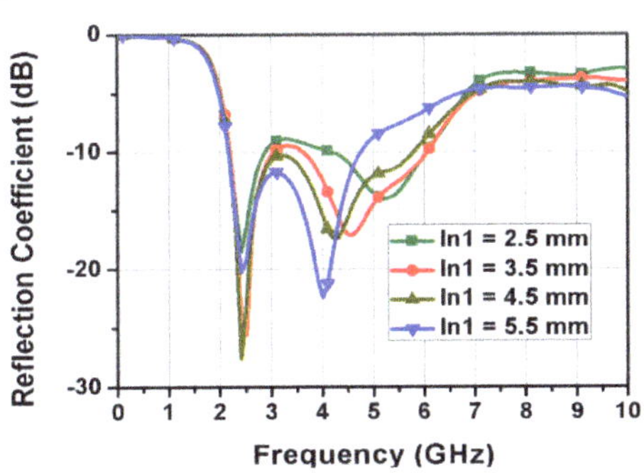

Figure 8. Reflection coefficients of the antenna linked with port 2 for different values of ln1.

5. Design Process of the Communication Antennas Linked with P10 to P17

The structure of the communication antennas linked with port 10 is depicted in Figure 9. A bent microstrip line feed is used to achieve good impedance matching in the appropriate band.

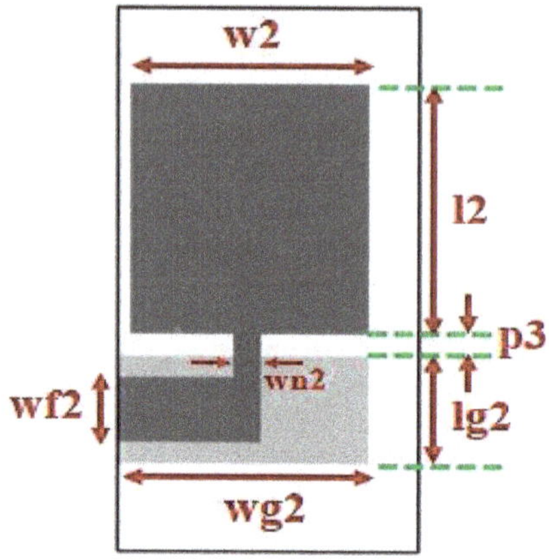

Figure 9. Schematic of the antennas linked with P10 to P17.

The design process of the communication antenna linked with port 10 ends in two stages. In the first stage, a rectangular radiator of length 11.5 mm and width 9.5 mm is fed by a bent microstrip line feed, as depicted in Ant I in Figure 10. Furthermore, this antenna is realized with an 8-element MIMO configuration at ports P10 to P17. The mathematical formula for calculating the LBE of Ant I in Figure 10 is given below.

$$f_{LAntI(P10)} = \frac{7.2}{(l2 + r2 + p3) \times k} GHz \quad (2)$$

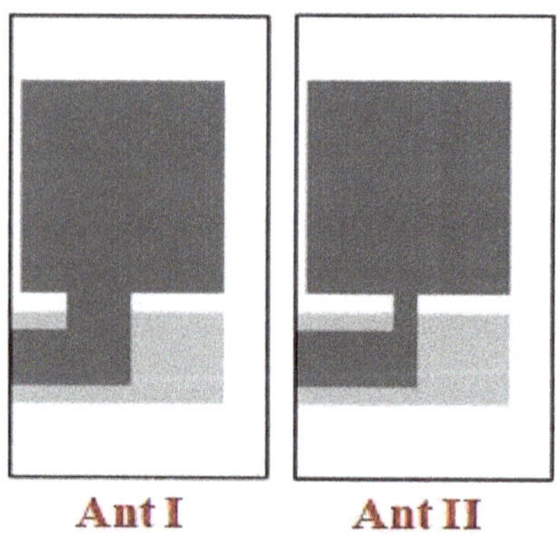

Figure 10. Evolution of the antennas linked with P10 to P17.

In Equation (2), l2 and p3 are the length of the rectangular patch and feed gap in centimeters, respectively. The term 'r2' in centimeters can be calculated from the width of the rectangular patch since r2 = $\frac{w2}{2\pi}$. The term 'k' is constant and is equal to 1.15. After finding out the values of l2, r2, p3 and k, the LBE frequency of Ant I in Figure 10 is found as 5.15 GHz, whereas it is found as 4.95 GHz in simulation, as shown in Ant I in Figure 11. In the next stage, the reflection coefficient performance is analyzed by altering the width of the strip line of length ln. As illustrated in Figure 12, the strip line of a width of 1.1 mm offers best impedance matching among different widths of strip line of length ln and the return loss is better than 10 dB in 4.6–8.7 GHz.

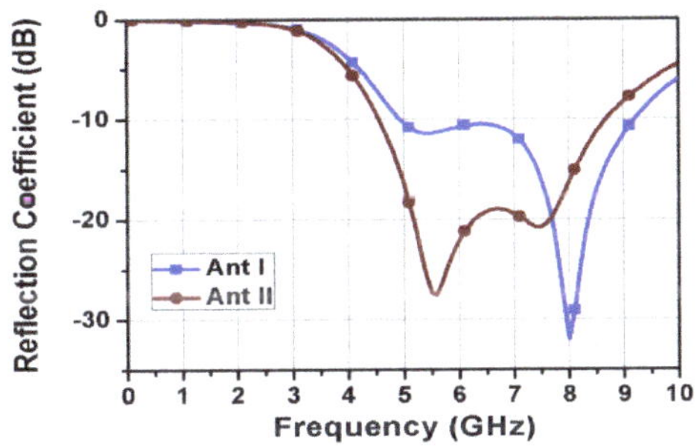

Figure 11. Reflection coefficients of the antennas in intermediate steps of the antenna linked with port 10.

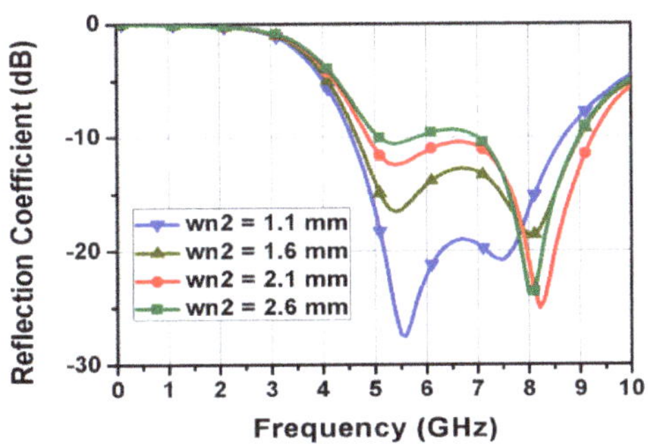

Figure 12. Reflection coefficients of the antenna linked with port 10 for different values of wn2.

6. Design Process of the Communication Antennas Linked with P18 to P24

The structure of the communication antennas linked with P18 to P24 is depicted in Figure 13. Since the communication antennas linked with ports 2 and 10 are designed to operate at 2.17–4.74 GHz and 4.57–8.62 GHz, respectively, it is aimed to design a communication antenna that operates at frequencies which are greater than 8.62 GHz to cover the complete FCC UWB of 3.1–10.6 GHz. Furthermore, this antenna is realized with an 8-element MIMO configuration at ports P18 to P24. As depicted in Figure 14, the design process of the communication antenna linked with port 18 ends in two stages. In the first stage, a normal monopole antenna with a rectangle-shaped patch is designed, as illustrated in Figure 14. The length and width of the rectangular patch are selected in such a way that the LBE frequency is approximately 8.6 GHz.

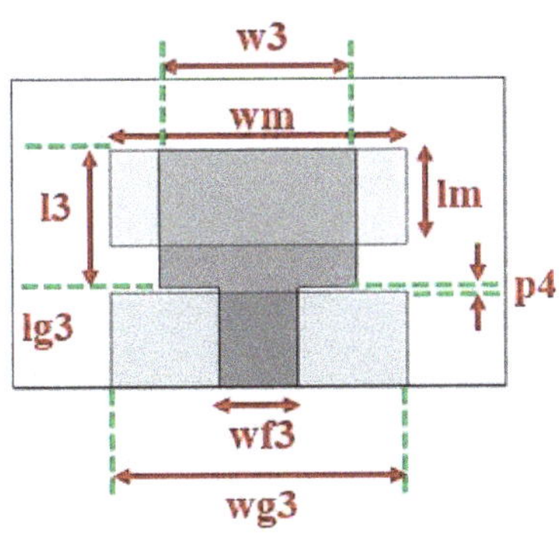

Figure 13. Schematic of the antennas linked with P18 to P24.

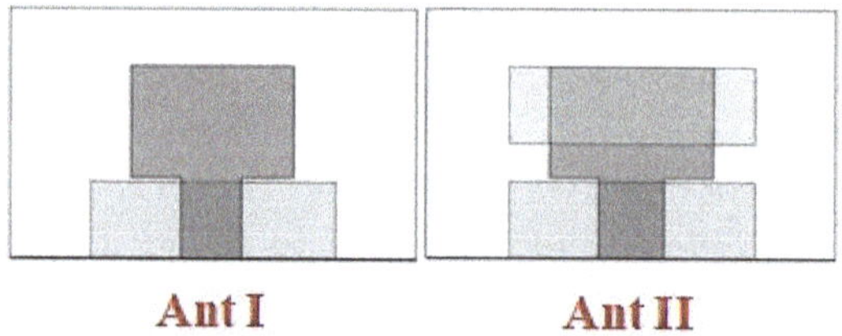

Figure 14. Evolution of the antennas linked with P18 to P24.

The mathematical formula for calculating the LBE frequency of Ant I in Figure 14 is given as:

$$f_{LAntI(P18)} = \frac{7.2}{(l3 + r3 + p4) \times k}\,GHz \tag{3}$$

where l3 and p4 are length of the rectangular patch and feed gap in centimeters, respectively. The term 'r3′ in centimeters is expressed in terms of width of the rectangular patch (i.e., $r3 = \frac{w3}{2\pi}$). The term 'k′ is constant and is equal to 1.15. After substituting the values of l3, r3, p4 and k in Equation (3), the LBE frequency of Ant I in Figure 14 is found as 8.6 GHz, whereas it is found as 8.5 GHz in simulation, as shown in Ant I in Figure 15.

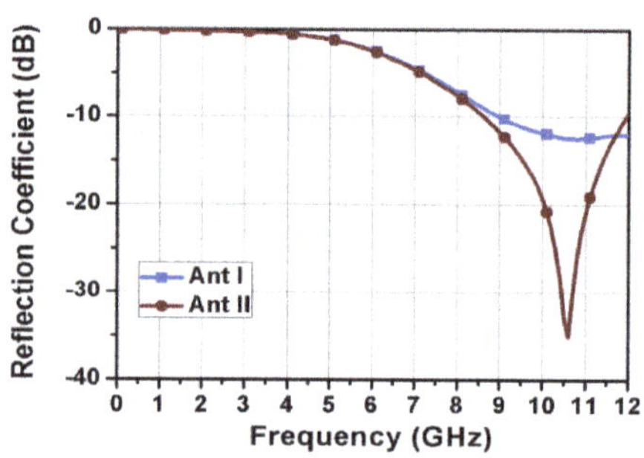

Figure 15. Reflection coefficients of the antennas in intermediate steps of the antenna linked with port 18.

It is seen from Figure 15 that the reflection coefficient curve of Ant I in Figure 1 is below −10 dB from 8.5 GHz and impedance matching is just decent. To make the lower band edge frequency 8.6 GHz and improve impedance matching, a rectangular-shaped patch is printed on the back portion of the substrate and the width of the 50-ohm feed line is chosen as 3.2 mm in the next stage. Figure 16 depicts a parametric study performed by altering the feed line's width. It is concluded from Figure 16 that the desired performance is achieved with a feed line of a width of 3.2 mm.

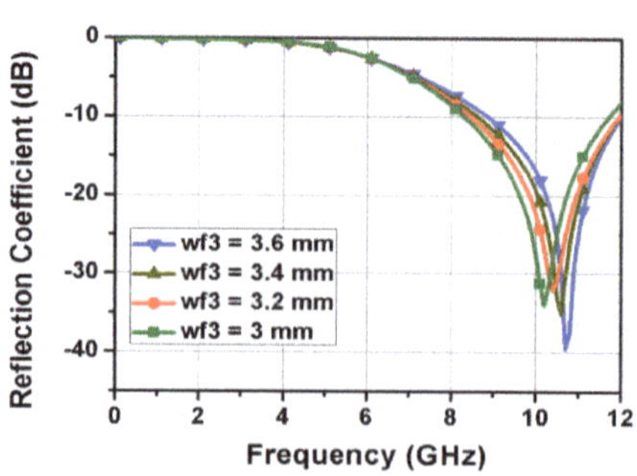

Figure 16. Reflection coefficient of the antenna linked with port 10 for different values of wf3.

7. Results and Discussions

After designing a sensing antenna and three communication antennas, the sensing antenna and eight copies of each communication antenna are placed as depicted in Figure 17.

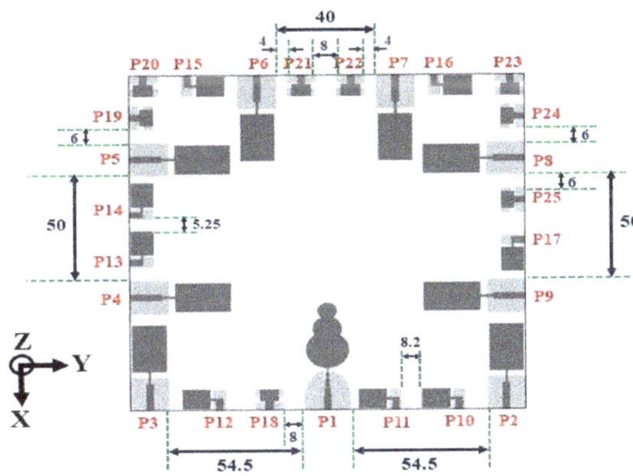

Figure 17. Inter-elemental spacing of the 25-port CR MIMO antenna.

The reflection coefficients of all antennas in the proposed 25-port CR MIMO antenna are depicted in Figures 18–20. An Agilent N5232A PNA-L network analyzer was used to check the correctness of the antenna. Good isolation between two antennas, without using any decoupling mechanism, can be attained in four possible cases. In the first case, low mutual coupling can be attained by placing two antennas in orthogonal fashion. In the second case, good isolation can be attained between two adjacent antennas that are similar to each other when the separation between those two antennas is greater than the one-fourth of the wavelength corresponding to their LBE frequency. In the third case, good isolation can be attained between two different antennas when the distance between them is greater than one-fourth of the wavelengths corresponding to their LBE frequencies. In the fourth case, if two similar antennas are even separated vertically at a distance that is slightly less than one-fourth of the wavelength corresponding to their LBE frequency, low mutual coupling can be attained between them. It can be seen from Figures 21–23 that an isolation of better than 12 dB is attained without using any decoupling mechanisms.

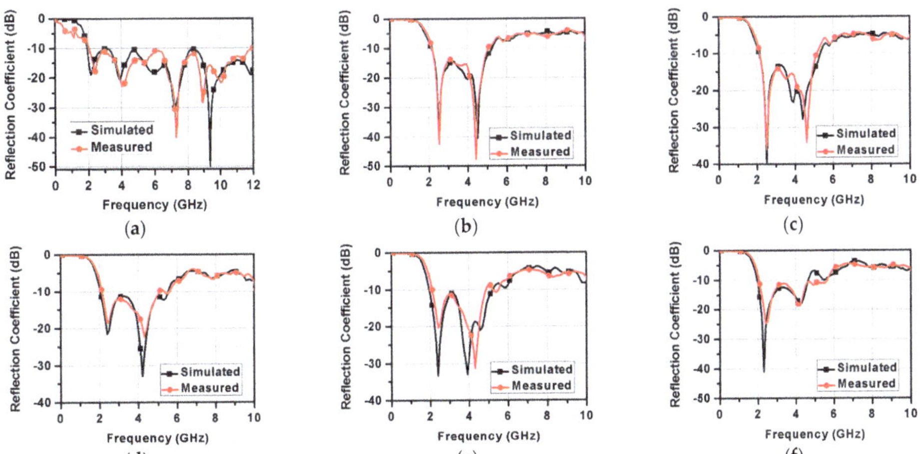

Figure 18. *Cont.*

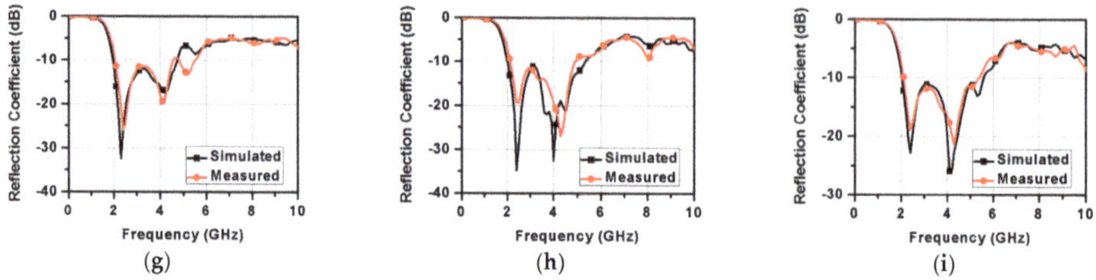

Figure 18. Plot of (**a**) S_{11}, (**b**) S_{22}, (**c**) S_{33}, (**d**) S_{44}, (**e**) S_{55}, (**f**) S_{66}, (**g**) S_{77}, (**h**) S_{88}, (**i**) S_{99}.

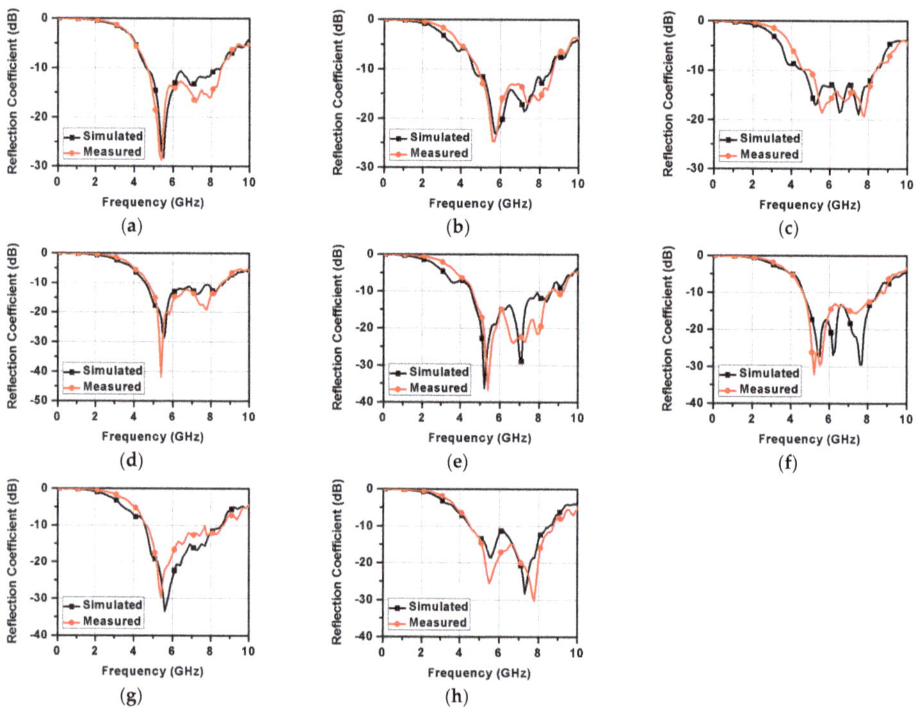

Figure 19. Plot of (**a**) $S_{10\,10}$, (**b**) $S_{11\,11}$, (**c**) $S_{12\,12}$, (**d**) $S_{13\,13}$, (**e**) $S_{14\,14}$, (**f**) $S_{15\,15}$, (**g**) $S_{16\,16}$, (**h**) $S_{17\,17}$.

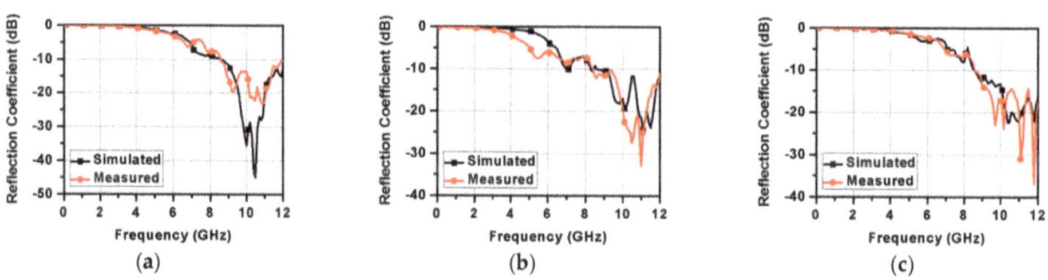

Figure 20. *Cont.*

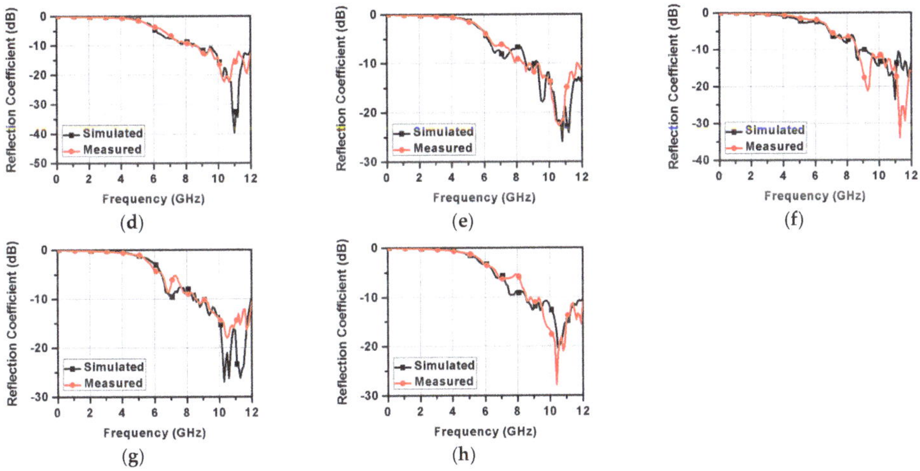

Figure 20. Plot of (**a**) $S_{18\,18}$, (**b**) $S_{19\,19}$, (**c**) $S_{20\,20}$, (**d**) $S_{21\,21}$, (**e**) $S_{22\,22}$, (**f**) $S_{23\,23}$, (**g**) $S_{24\,24}$, (**h**) $S_{25\,25}$.

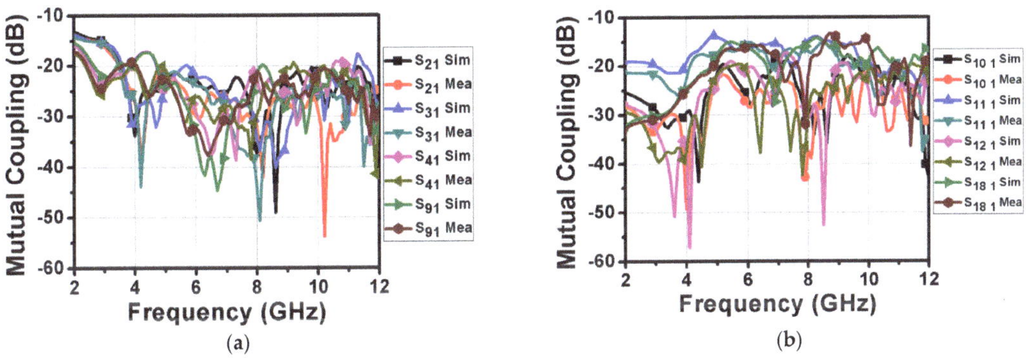

Figure 21. Mutual coupling between sensing and communication antennas (**a**) S_{21}, S_{31}, and S_{41} (**b**) $S_{10\,1}$, $S_{11\,1}$, $S_{12\,1}$, and $S_{18\,1}$.

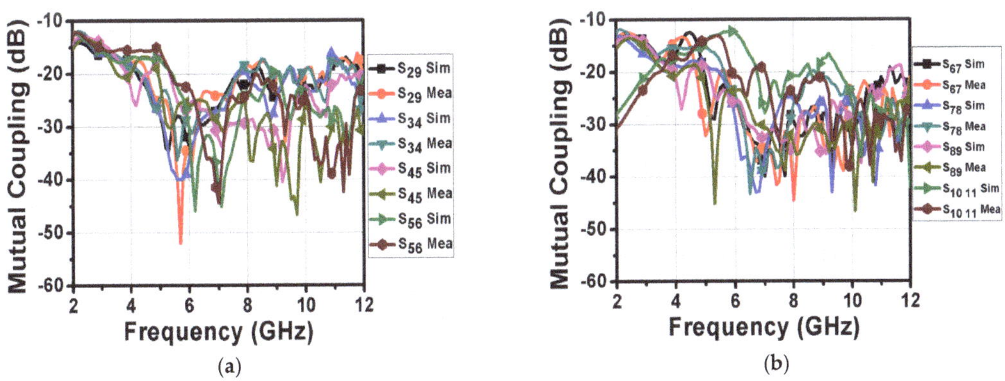

Figure 22. *Cont.*

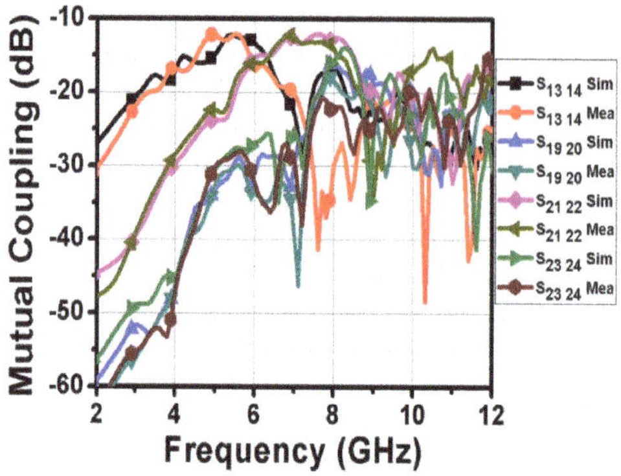

(c)

Figure 22. Mutual coupling between identical communication antennas (**a**) S_{29}, S_{34}, S_{45}, and S_{56} (**b**) S_{67}, S_{78}, S_{89}, and $S_{10\,11}$ (**c**) $S_{13\,14}$, $S_{19\,20}$, $S_{21\,22}$, and $S_{23\,24}$.

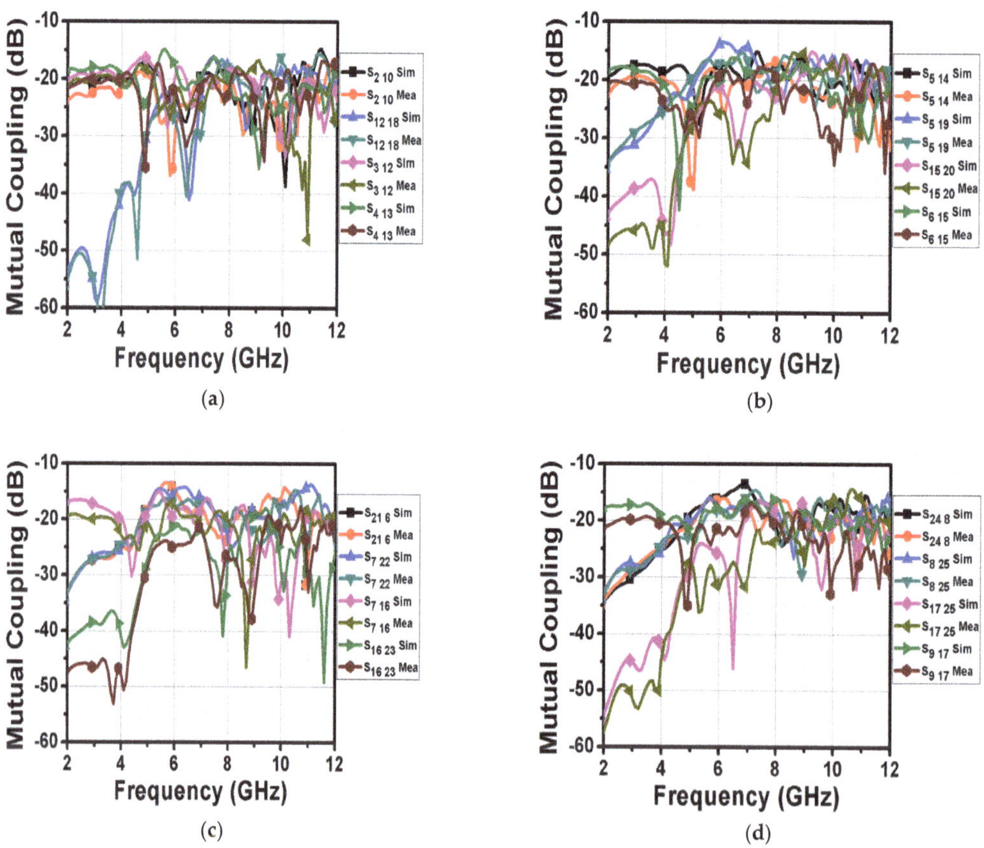

Figure 23. Mutual coupling between non-identical communication antennas. (**a**) $S_{2\,10}$, $S_{12\,18}$, $S_{3\,12}$, and $S_{4\,13}$ (**b**) $S_{5\,14}$, $S_{5\,19}$, $S_{15\,20}$, and $S_{6\,15}$ (**c**) $S_{21\,6}$, $S_{7\,22}$, $S_{7\,16}$, and $S_{16\,23}$ (**d**) $S_{24\,8}$, $S_{8\,25}$, $S_{17\,25}$, and $S_{9\,17}$.

In the proposed 25-port MIMO antenna system, if any two antennas are considered, they come under any one of the above four discussed cases. Hence, all the antennas in the proposed 25-port CR MIMO antenna system are well isolated. The mutual coupling of other possible combinations is not shown in Figures 21–23 since two antennas in those combinations are separated by large distances. So, mutual coupling of better than −15 dB is guaranteed. Pyramidal horn antenna and the proposed CR MIMO antenna are used as transmitting and receiving antennas, respectively, in an anechoic chamber while measuring radiation patterns. A microwave analog signal generator is used to connect the transmitting antenna. In the far-field region, the designed antenna connected to a coaxial detector is kept in receiving mode and used as receiving antenna. The far-field patterns of the sensing antenna are depicted in Figure 24. The communication antennas linked with ports 2, 10 and 18 are illustrated in Figures 25–27, respectively. Dipole-natured patterns and nearly omnidirectional patterns are attained at low (2.5 and 5 GHz) and high frequencies (7.5 and 10 GHz), respectively, whereas dipole-natured patterns are attained in both the planes in case of communication antennas. The far-field patterns of the communication antennas accessed at ports 2, 10 and 18 are similar to that of a dipole antenna at 3 GHz, 6 GHz and 9 GHz, respectively, as depicted in Figures 25–27.

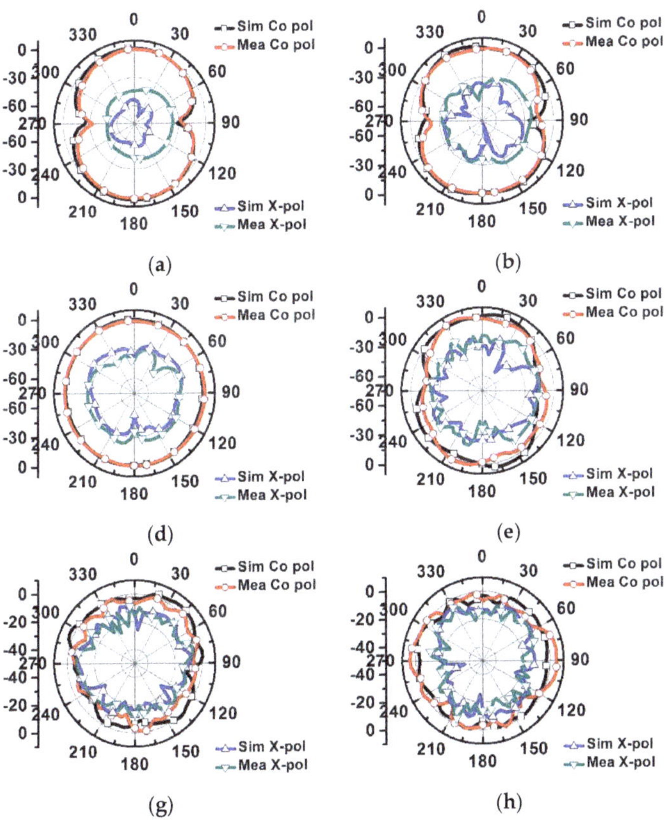

Figure 24. Far-field patterns of the sensing antenna in orthogonal planes at (**a**) 2.5 GHz (XZ), (**b**) 2.5 GHz (YZ), (**c**) 5 GHz (XZ), (**d**) 5 GHz (YZ), (**e**) 7.5 GHz (XZ), (**f**) 7.5 GHz (YZ), (**g**) 10 GHz (XZ), (**h**) 10 GHz (YZ).

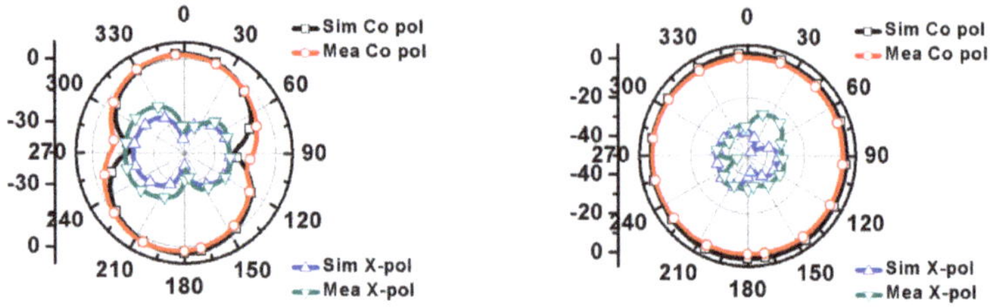

(a) (b)

Figure 25. Far-field patterns of the communication antenna linked with port 2 at (**a**) 3 GHz (XZ) and (**b**) 3 GHz (YZ).

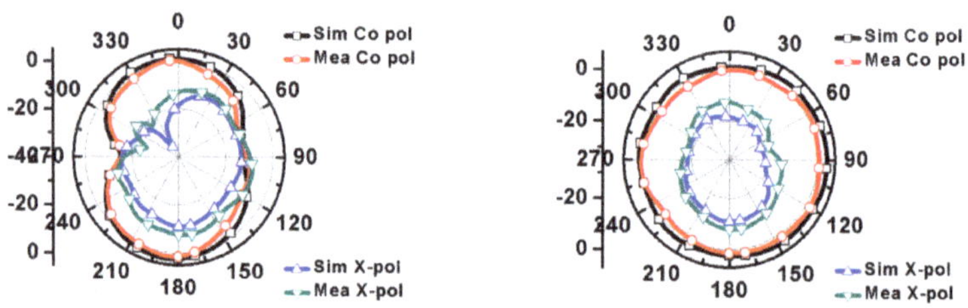

(a) (b)

Figure 26. Patterns of the communication antenna linked with port 10 at 6 GHz in (**a**) XZ and (**b**) YZ.

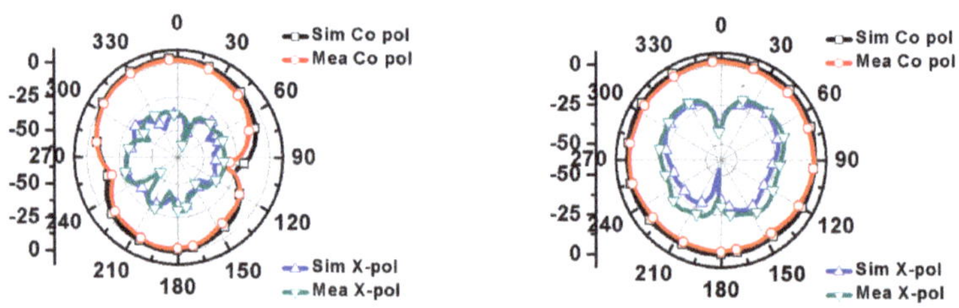

(a) (b)

Figure 27. Far-field patterns of the communication antenna linked with port 18 at (**a**) 9 GHz (XZ) and (**b**) 9 GHz (YZ).

The slight shifts and little deviations in the resonances of the reflection coefficient curves may be because of the faults in fabrication of the prototype, material impurities, and imperfections in soldering and connector losses. However, the measured reflection coefficients of all the communication antennas in the 25-port MIMO antenna are below −10 dB in their impedance bandwidths, which indicates that the simulated reflection coefficient results of the 25-port CR MIMO antenna are in good agreement with the measured reflection coefficient results. The cross polarization levels of all the antennas in the proposed antenna are less than −20 dB, as shown in Figures 24–27. It indicates that the vertically polarized surface currents are more dominant than the horizontally polarized surface currents in all the antennas in the proposed 25-port CR MIMO antenna. Moreover, the obtained omnidirectional radiation patterns of all the antennas are well suitable for CR MIMO applications.

The radiation efficiency and peak gain of the sensing antenna are illustrated in Figures 28 and 29, respectively. It is seen that the simulated peak gain and radiation efficiency of the sensing antenna are better than 82% and 2.25 dBi, respectively. The radiation efficiencies and peak gains of the communication antennas are illustrated in

Figures 30 and 31, respectively. It is obvious from Figures 30 and 31 that the simulated peak gains and radiation efficiencies of the sensing antenna are better than 90% and 2.5 dBi, respectively. The measured peak gains are slightly less than the simulated peak gains because of substrate losses, magnetic field of the earth and errors in measurement. The fabricated prototype of the proposed 25-port CR MIMO antenna is provided in Figure 32.

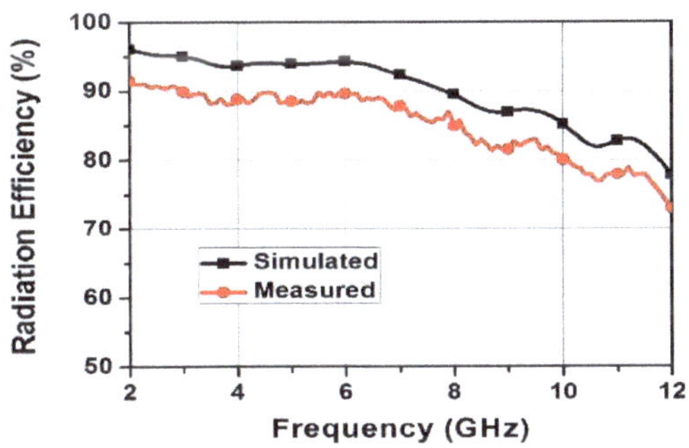

Figure 28. Radiation efficiency of the sensing antenna.

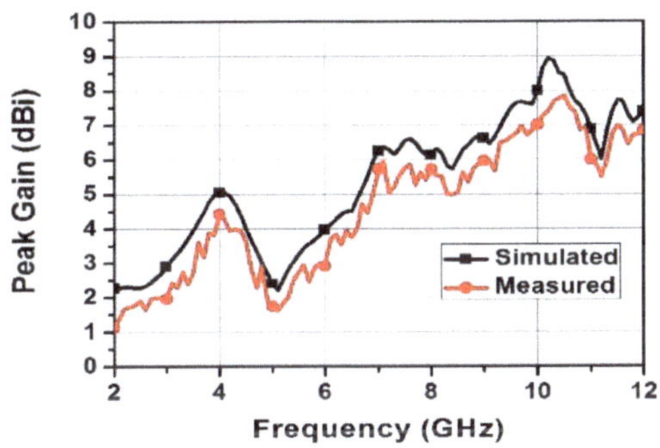

Figure 29. Peak gain of the sensing antenna.

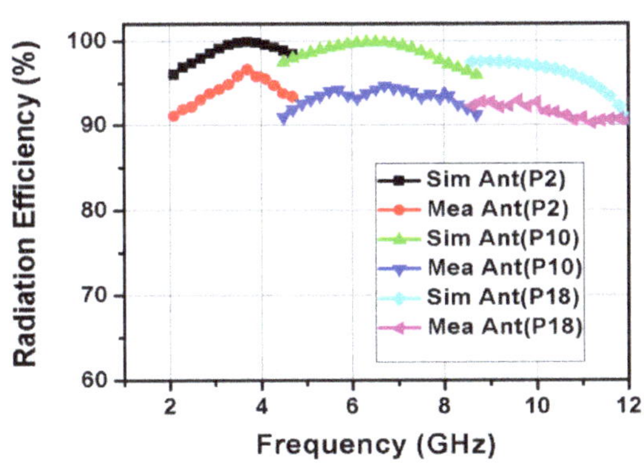

Figure 30. Radiation efficiencies of the communication antennas.

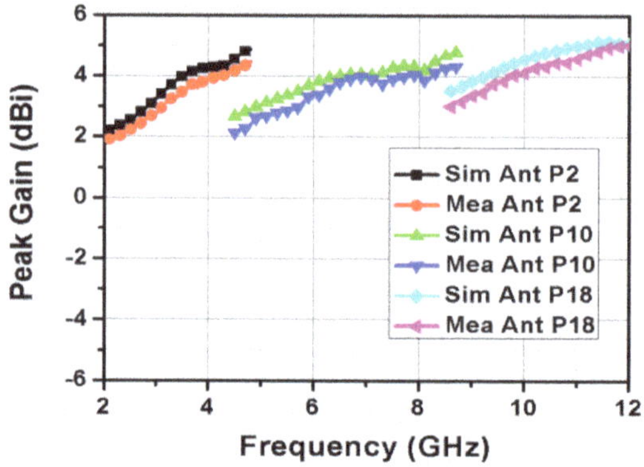

Figure 31. Peak gains of the communication antennas.

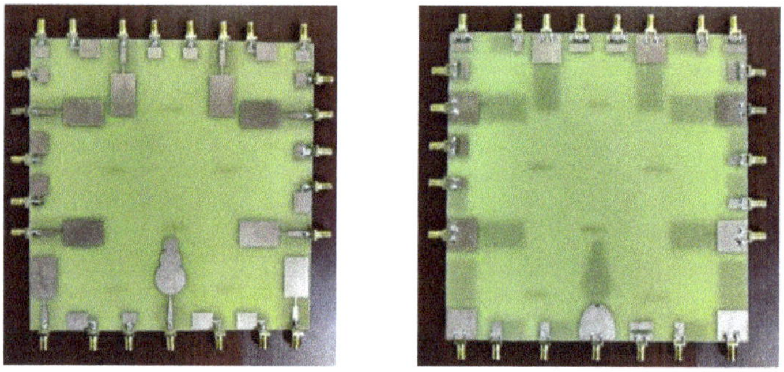

(**a**) (**b**)

Figure 32. Fabricated prototype of the 25-port CR MIMO antenna. (**a**) Top view. (**b**) Bottom view.

It can be observed from Table 3 that no CR 8-element MIMO antennas appear in the existing literature. Additionally, the proposed 8-element MIMO antenna achieves polarization diversity and covers all communication bands from 2.17 to 12 GHz. Moreover, it does not have any switching elements such as PIN diodes, MEMS, varactor diodes, or other factors.

Table 3. Comparison of the 25-port CR MIMO antenna with other antennas in the literature.

Ref.	Size (mm^2)	Range of the Sensing Antenna (GHz)	Range Covered by Communication Antennas (GHz)	Min. Isolation (dB)	n-Element MIMO Antenna	Reconfiguration Mechanism
[1]	50 × 100	-	1.42–2.27	12	2	Varactor diodes
[2]	60 × 120	-	3.2–3.9	10	4	Varactor diodes
[3]	109 × 109	2–5.7	2.5–4.2	15	4	PIN and Varactor diodes
[4]	60 × 40	2.2–7	2.3–6.3	18	2	Varactor diodes
[5]	100 × 120	2.3–5.5	2.5–4.2	15	4	PIN and Varactor diodes
[6]	60 × 120	1–4.5	0.9–2.6	12.5	2	PIN and Varactor diodes
[7]	65 × 120	0.72–3.44	A few frequencies in 0.72–3.44	15.5	2	PIN and Varactor diodes
[8]	52.2 × 35	1.7–10.6	5.1–5.5, 6.6–7.2 and 9.7–10.2	15	2	-
[9]	50 × 110	-	1.73–2.28 and 2.45	10	2	Varactor diodes
[10]	110 × 70	2.45–5.3	2.5–3.6	12	4	Varactor diodes
[11]	152 × 126	2.4–6	5.8–6.3	20	4	PIN diodes
[12]	24 × 25	-	2.4, 3.5, 5.25	25	2	PIN diodes
This work	160 × 160	2–12	All frequencies in 2.17–12	12	8	-

8. Performance Analysis of the 25-port CR MIMO Antenna

The diversity characteristics of the proposed MIMO antenna can be assessed by the two parameters diversity gain (DG) and the envelope correlation coefficient (ECC) [27–30]. ECC can be calculated mathematically by using the S-parameters, as shown in Equation (4). An ideal MIMO antenna system has an ECC of 0. ECC of the proposed 8-element MIMO antenna is illustrated in Figure 1. DG can be determined mathematically once ECC is known using the equation given in Equation (1), and its ideal value for a good MIMO antenna system is 10 dB.

$$\mathrm{ECC}(\rho)=\frac{\left|S_{ii}{}^{*}S_{ii}+S_{ji}{}^{*}S_{jj}\right|^{2}}{\left(1-\left|S_{ii}\right|^{2}-\left|S_{ij}\right|^{2}\right)\left(1-\left|S_{ji}\right|^{2}-\left|S_{jj}\right|^{2}\right)} \tag{4}$$

$$\mathrm{ECC}(\rho)=\frac{\left|\iint_{4\pi}\left[F_i(\theta,f)*F_j(\theta,f)\right]d\Omega\right|^{2}}{\iint_{4\pi}\left|F_i(\theta,f)\right|^{2}d\Omega\iint_{4\pi}\left|F_j(\theta,f)\right|^{2}d\Omega} \tag{5}$$

where $i, j \in N$, $i < j$, $j \leq 4$, $F_i(\theta, \varnothing)$ is nothing but the far-field pattern of the antenna when port i and port j are excited, and * denotes the Hermitian product. However, calculation of ECC using far-field patterns is very tedious. After ECC is calculated, DG in dB can be determined using Equation (6).

$$\text{Diversity Gain (dB)} = 10\times\sqrt{1-|0.99\rho|^{2}} \tag{6}$$

The quality of the proposed 25-port CR MIMO antenna system can be assessed by another crucial parameter called channel capacity loss (CCL). When the number of antennas increases, channel capacity increases under certain conditions without increasing transmitted power or bandwidth. However, channel capacity decreases when a correlation between the links exists. Additionally, as the correlation increases, CCL increases. CCL can be calculated from the equation [27–30] given below.

$$C_{loss} = -\log_2 \det\left(\Psi^R\right) \tag{7}$$

where Ψ^R represents correlation matrix of the receiving antenna. Its matrix representation is given below.

$$\Psi^R = \begin{pmatrix} \rho_{11} & \rho_{12} \\ \rho_{21} & \rho_{22} \end{pmatrix} \tag{8}$$

$$\rho_{ii}=1-\left|S_{ii}\right|^{2}-\left|S_{ij}\right|^{2} \text{ and } \rho_{ij}=-\left(\left|S_{ii}{}^{*}S_{ij}+S_{ji}{}^{*}S_{jj}\right|\right) \tag{9}$$

It is evident from Figure 33 that the ECC of the proposed 25-port CR MIMO antenna is less than 0.42 and DG of the proposed 25-port CR MIMO antenna is more than 9.1 dB. It can also be clearly seen that the CCL of the proposed 25-port CR MIMO antenna is less than 0.46 bits/s/Hz, as illustrated in Figure 34. Since the acceptable level of ECC and CCL of a good MIMO are 0.5 and 0.5 bits/s/Hz, the proposed 25-port CR MIMO antenna is well suitable for CR MIMO applications.

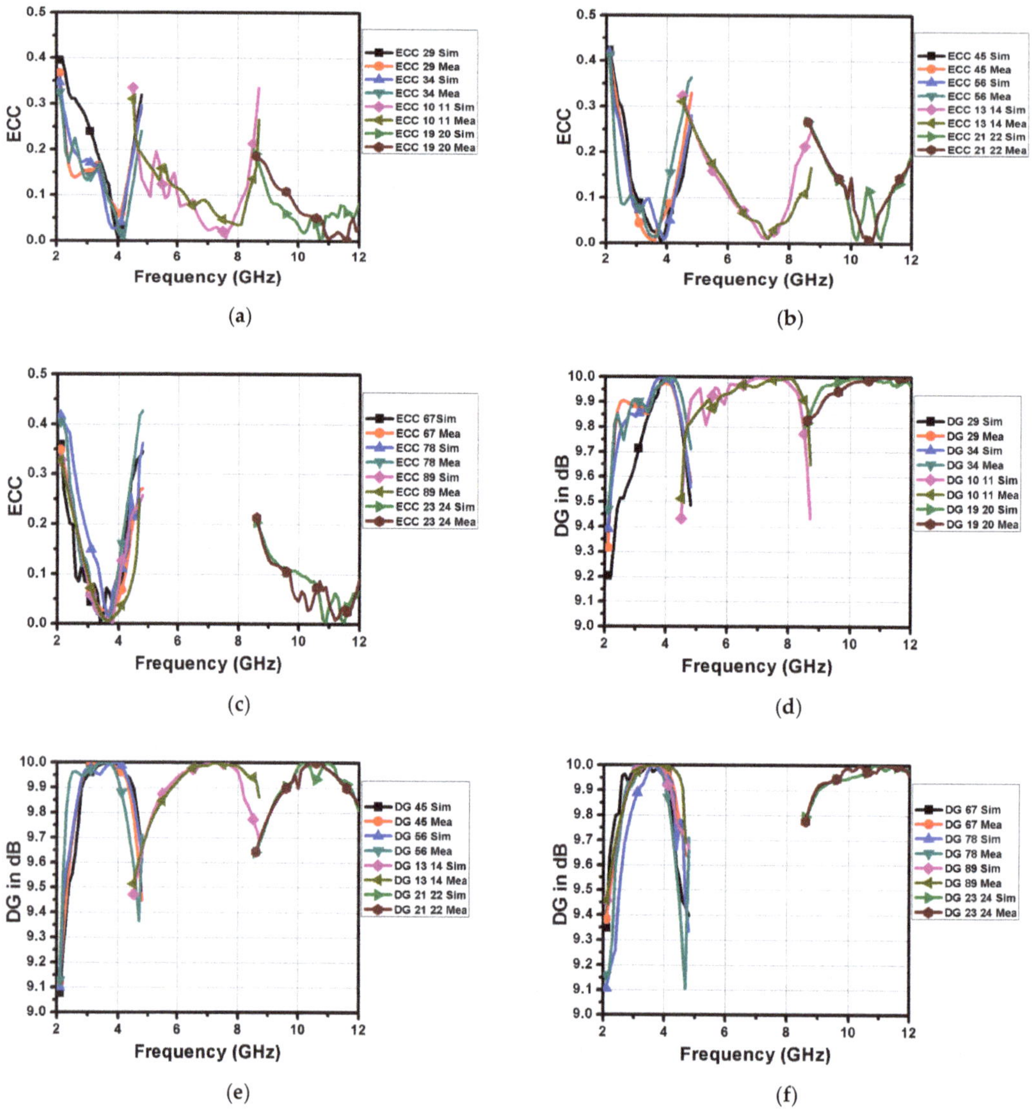

Figure 33. ECC and DG of the 25-port CR MIMO antenna. (**a**) ECC 29, ECC 34, ECC 10 11 and ECC 19 20(**b**) ECC 45, ECC 56, ECC 13 14 and ECC 21 22 (**c**) ECC 67, ECC 78, ECC 89 and ECC 23 24 (**d**) DG 29, DG 34, DG 10 11 and DG 19 20 (**e**) DG 45, DG 56, DG 13 14 and DG 21 22 (**f**) DG 67, DG 78, DG 89 and DG 23 24.

The total active reflection coefficient (TARC) is one of the crucial parameters to evaluate the diversity performance of the antenna. It is nothing but the ratio of total incident power to the total power that is outgoing when a multiport antenna system is present. It can be calculated using the Equation (10) given in [27].

$$\text{TARC} = \frac{\sqrt{\sum_{k=1}^{N} |b_k|^2}}{\sqrt{\sum_{k=1}^{N} |a_k|^2}} \quad (10)$$

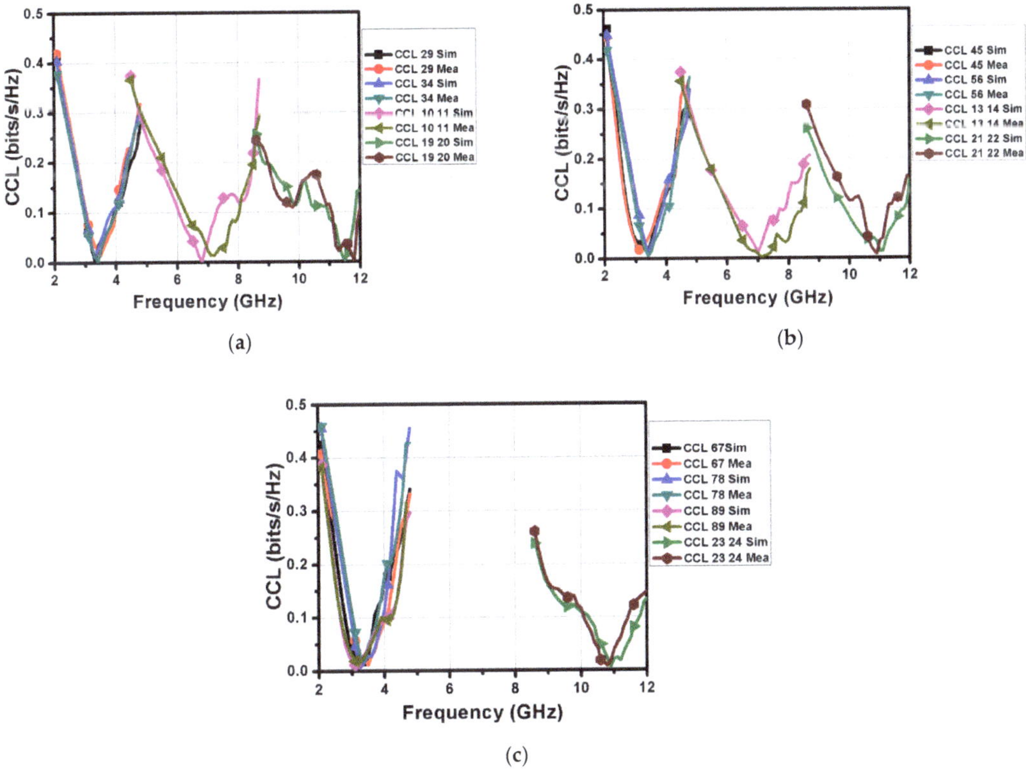

Figure 34. CCL of the 25-port CR MIMO antenna. (**a**) CCL 29, CCL 34, CCL 10 11 and CCL 19 20 (**b**) CCL 45, CC 56, CCL 13 14 and CCL 21 22 (**c**) CCL 67, CCL 78, CCL 89 and CCL 23 24.

where $|a|$ is the excitation parameter and $|b|$ is the scattering parameter in Equation (10). In order to check the effect of TARC on the 10 dB return loss bandwidth of the communication antennas, the proposed MIMO antenna is integrated with an ideal phase shifter in which scan angle is changed from 45° to 180°. It is obvious from Figure 35 that TARC for all 8-element communication antennas is less than −10 dB. So, it is confirmed that all the power that is delivered is accepted by the other antenna element without affecting the 10 dB return loss bandwidth of 8-element MIMO communication antennas.

Mean effective gain (MEG) is another important parameter to assess the diversity performance of the antenna in wireless channels. It ascertains the antenna element's ability to accept electromagnetic signals in the presence of rich fading channels. Practically, it ranges from −3 dB to −12 dB for a MIMO antenna with good diversity performance. It is evident from Figure 36 that the MEG for all MIMO communication antennas is less than −3 dB. So, it can be concluded that the 8-element MIMO communication antennas is a promising candidate for CR MIMO applications.

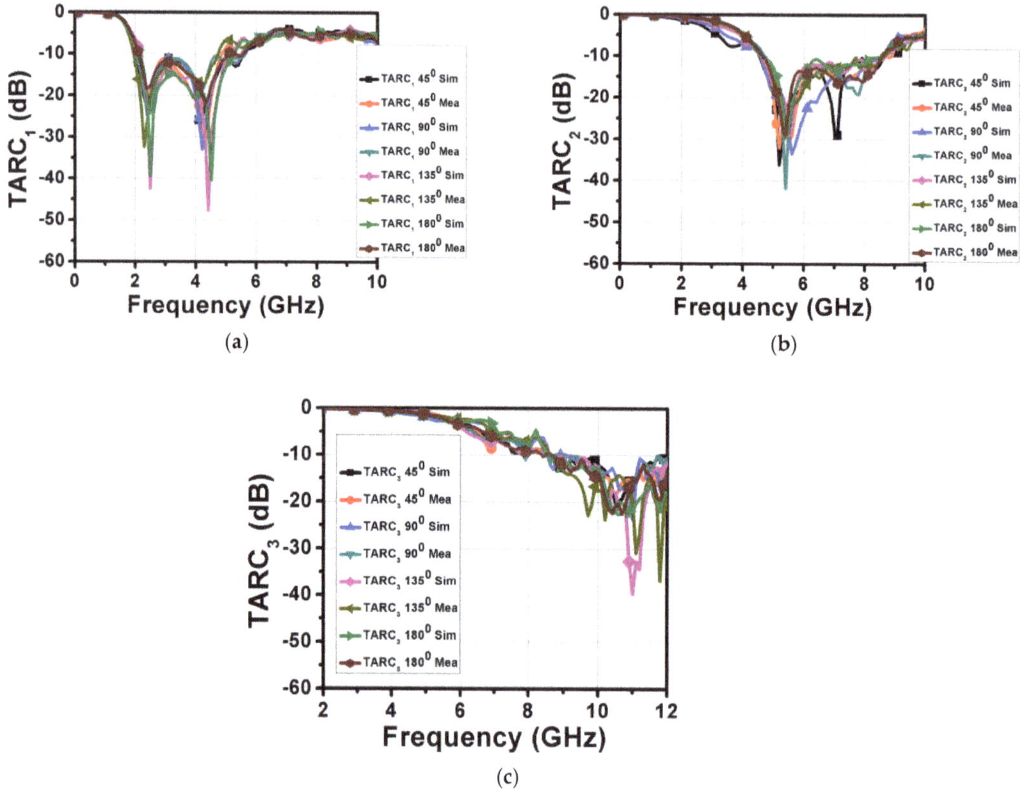

Figure 35. TARC of the 8-element communication antennas associated with (**a**) ports 2 to 9, (**b**) ports 10 to 17, and (**c**) ports 18 to 25.

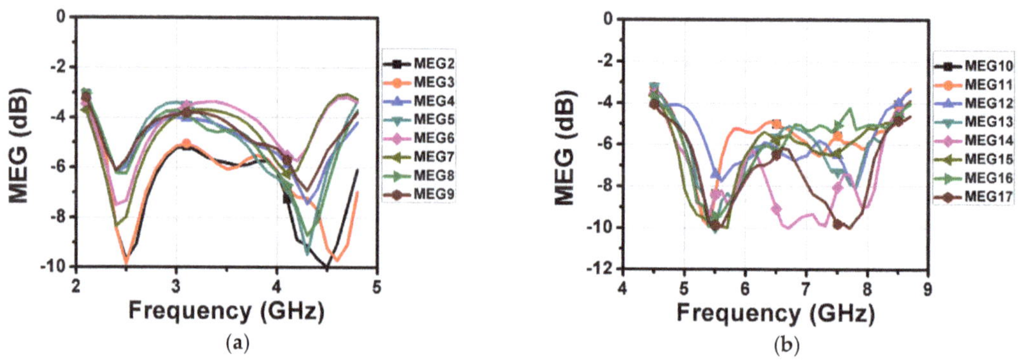

Figure 36. *Cont.*

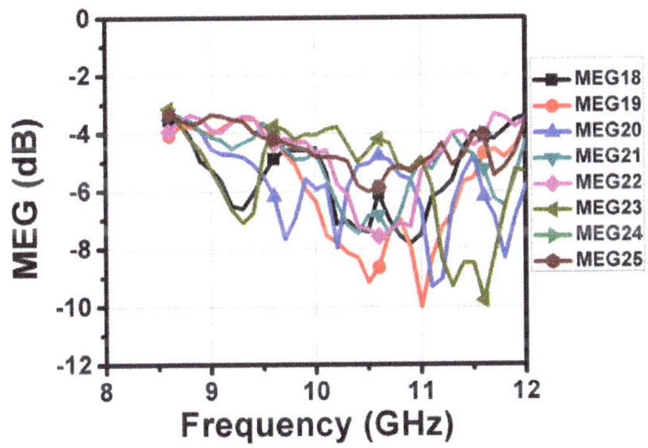

(c)

Figure 36. MEG of the 8-element communication antennas associated with (**a**) ports 2 to 9, (**b**) ports 10 to 17, and (**c**) ports 18 to 25.

9. Conclusions

In this article, a CR-integrated antenna system, which can perform maximum of three communication operations, has been presented to improve spectrum utilization efficiency. The sensing antenna linked to port 1 was able to sense the spectrum that ranges from 2 to 12 GHz, whereas the communication MIMO antennas linked with ports 2 to 9, ports 10 to 17 and ports 18 to 25 perform operations in the 2.17–4.74 GHz, 4.57–8.62 GHz and 8.62–12 GHz bands, respectively. Mutual coupling in the proposed CR MIMO antenna was less than −12 dB. Peak gain and radiation efficiency of the sensing antenna are found to be better than 2.25 dBi and 82%, respectively, whereas the peak gains and radiation efficiencies of all communication antennas were more than 2.5 dBi and 90%, respectively. It is inexpensive, easily implementable and has less complexity compared to the traditional reconfigurable CR MIMO antennas. Additionally, it can overcome all the drawbacks that are associated with reconfigurable CR MIMO antennas. Its performance has also been assessed by evaluating ECC, DG, CCL, TARC and MEG. The simulated and measured ECC, DG, CCL, TARC and MEG are within their acceptable limits. Hence, the proposed CR MIMO antenna is a promising candidate for CR MIMO applications.

Author Contributions: Conceptualization: D.S., A.N., R.M., A.B. and S.D.; methodology: D.S., A.N., S.D., S.L. and A.B.; software: D.S., A.B., A.D.A. and W.E.-S.; validation: A.D.A., W.E.-S. and S.L.; writing—original draft preparation: D.S., A.N. and S.D.; writing—review, and editing: S.L., W.E.-S. and R.M.; supervision: S.D., A.N., R.M. and A.B.; project administration: W.E.-S.; funding acquisition A.D.A. All authors have read and agreed to the published version of the manuscript.

Funding: This work is supported by Princess Nourah bint Abdulrahman University Researchers Supporting Project number (PNURSP2023R51), Princess Nourah bint Abdulrahman University, Riyadh, Saudi Arabia.

Institutional Review Board Statement: Not applicable.

Informed Consent Statement: Not applicable.

Data Availability Statement: The data presented in this research are available on request from the corresponding author.

Acknowledgments: The authors would like to acknowledge the Princess Nourah bint Abdulrahman University Researchers Supporting Project number (PNURSP2023R51), Princess Nourah bint Abdulrahman University, Riyadh, Saudi Arabia.

Conflicts of Interest: The authors declare no conflict of interest.

References

1. Riaz, S.; Zhao, X.; Geng, S. A Frequency Reconfigurable MIMO Antenna with Agile Feedline for Cognitive Radio Applications. *Int. J. RF Microw. Comput. Aided Eng.* **2019**, *30*, e22100.
2. Hussain, R.; Raza, A.; Khan, M.U.; Shammim, A.; Sharawi, M.S. Miniaturized Frequency Reconfigurable Pentagonal MIMO Slot Antenna for Interweave CR Applications. *Int. J. RF Microw. Comput. Aided Eng.* **2019**, *29*, e21811. [CrossRef]
3. Alam, T.; Thummaluru, S.R.; Chaudhary, R.K. Integration of MIMO and Cognitive Radio for Sub-6 GHz 5G Applications. *IEEE Antennas Wirel. Propag. Lett.* **2019**, *18*, 2021–2025. [CrossRef]
4. Cheng, S.; Lin, K. A Reconfigurable Monopole MIMO Antenna with Wideband Sensing Capability for Cognitive Radio using Varactor Diodes. In Proceedings of the IEEE International Symposium on Antennas and Propagation & USNC/URSI National Radio Science Meeting, Vancouver, BC, Canada, 19–24 July 2015; pp. 2233–2234.
5. Thummaluru, S.R.; Ameen, M.; Chaudhary, R.K. Four-Port MIMO Cognitive Radio System for Midband 5G Applications. *IEEE Trans. Antennas Propag.* **2019**, *67*, 5634–5645. [CrossRef]
6. Zhao, X.; Riaz, S.; Geng, S. A Reconfigurable MIMO/UWB MIMO Antenna for Cognitive Radio Applications. *IEEE Access* **2019**, *7*, 46739–46747. [CrossRef]
7. Hussain, R.; Sharawi, M.S. A Cognitive Radio Reconfigurable MIMO and Sensing Antenna System. *IEEE Antennas Wirel. Propag. Lett.* **2015**, *14*, 257–260. [CrossRef]
8. Pahadsingh, S.; Sahu, S. Four Port MIMO Integrated Antenna System with DRA for Cognitive Radio Platforms. *AEU-Int. J. Electron. Commun.* **2018**, *92*, 98–110. [CrossRef]
9. Hussain, R.; Khan, M.U.; Sharawi, M.S. An Integrated Dual MIMO Antenna System with Dual-Function GND-Plane Frequency Agile Antenna. *IEEE Antennas Wirel. Propag. Lett.* **2018**, *17*, 142–145. [CrossRef]
10. Alam, T.; Thummaluru, S.R.; Chaudhary, R.K. Improved Multifunctional MIMO Cognitive Radio System for Integrated Interweave-Underlay Operations. *IEEE Trans. Microw. Theory Tech.* **2022**, *70*, 631–640. [CrossRef]
11. Jain, P.; Saptarshi, G. A Reconfigurable Four-Port MIMO Antenna for Sub-6 GHz Applications. In Proceedings of the IEEE Wireless Antenna and Microwave Symposium (WAMS), Rourkela, India, 5–8 June 2022; pp. 1–5.
12. Mansoul, A.; Mourad, N. Compact and Reconfigurable Multiband 2-Element MIMO Slot Antenna for Advanced Communication Systems. In Proceedings of the IEEE International Symposium on Antennas and Propagation and North American Radio Science Meeting, Montreal, QC, Canada, 5–10 July 2020; pp. 575–576.
13. Hussain, R.; Sharawi, M.S. 4-Element Planar MIMO Reconfigurable Antenna System for Cognitive Radio Applications. In Proceedings of the IEEE International Symposium on Antennas and Propagation & USNC/URSI National Radio Science Meeting, Vancouver, BC, Canada, 19–24 July 2015; pp. 717–718.
14. Hussain, R.; Sharawi, M.S.; Shamim, A. 4-Element Concentric Pentagonal Slot-Line-Based Ultra-Wide Tuning Frequency Reconfigurable MIMO Antenna System. *IEEE Trans. Antennas Propag.* **2018**, *66*, 4282–4287. [CrossRef]
15. Chacko, B.P.; Augustin, G.; Denidni, A. Electronically Reconfigurable Uniplanar Antenna with Polarization Diversity for Cognitive Radio Applications. *IEEE Antennas Wirel. Propag. Lett.* **2015**, *14*, 213–216. [CrossRef]
16. Hussain, R.; Sharawi, M.S.; Shamim, A. An Integrated Four-Element Slot-Based MIMO and a UWB Sensing Antenna System for CR Platforms. *IEEE Trans. Antennas Propag.* **2018**, *66*, 978–983. [CrossRef]
17. Tawk, Y.; Constantine, J.; Christodoulou, C.G. Reconfigurable Filtennas and MIMO in Cognitive Radio Applications. *IEEE Trans. Antennas Propag.* **2014**, *62*, 1074–1083. [CrossRef]
18. Islam, H.; Das, S.; Ali, T.; Bose, T.; Prakash, O.; Kumar, P. A Frequency Reconfigurable MIMO Antenna with Bandstop Filter Decoupling Network for Cognitive Communication. *Sensors* **2022**, *22*, 6937. [CrossRef]
19. Hussain, R.; Sharawi, M.S. An Integrated Slot-Based Frequency-Agile and UWB Multifunction MIMO Antenna System. *IEEE Antennas Wirel. Propag. Lett.* **2019**, *18*, 2150–2154. [CrossRef]
20. Nella, A.; Abhay, S.G. A survey on planar antenna designs for cognitive radio applications. *Wirel. Pers. Commun.* **2018**, *98*, 541–569. [CrossRef]
21. Nella, A.; Abhay, S.G.; Vigneswaran, D. Cognitive radio paradigm and recent trends of antenna systems in the UWB 3.1–10.6 GHz. *Wirel. Netw.* **2020**, *26*, 3257–3274.
22. Srikar, D.; Anuradha, S. Twelve Port MIMO Antenna with Polarisation Diversity for Cognitive Radio Applications. *Electron. Lett.* **2019**, *55*, 1165–1168. [CrossRef]
23. Srikar, D.; Anuradha, S. A Compact Six Port Antenna for Better Spectrum Utilization Efficiency in Cognitive Radio Applications. *Int. J. RF Microw. Comput. Aided Eng.* **2020**, *30*, e22383. [CrossRef]
24. Srikar, D.; Anuradha, S. A New Two-Element MIMO Antenna System for Cognitive Radio Applications. *Circuit World* **2022**, *48*, 111–125. [CrossRef]
25. Nella, A.; Gandhi, A.S. A Five-Port Integrated UWB and Narrowband Antennas System Design for CR Applications. *IEEE Trans. Antennas Propag.* **2018**, *66*, 1669–1676. [CrossRef]
26. Anvesh Kumar, N.; Gandhi, A.S. A Compact Novel Three-Port Integrated Wide and Narrow Band Antennas System for Cognitive Radio Applications. *Int. J. Antennas Propag.* **2016**, *2016*, 2829357.
27. Nandi, S.; Mohan, A. CRLH Unit Cell Loaded Triband Compact MIMO Antenna for WLAN/WiMAX Applications. *IEEE Antennas Wirel. Propag. Lett.* **2017**, *16*, 1816–1819.

28. Balaji, V.R.; Tathababu, A.; Arpan, D.; Anveshkumar, N.; Truong, K.N. An Inverted L-Strip Loaded Ground with Hollow Semi-Hexagonal Four-Element Polarization Diversity UWB-MIMO Antenna. *Trans. Emerg. Telecommun. Technol.* **2022**, *33*, e4381. [CrossRef]
29. Sharma, P.; Tiwari, R.N.; Singh, P.; Kumar, P.; Kanaujia, B.K. MIMO Antennas: Design Approaches, Techniques and Applications. *Sensors* **2022**, *22*, 7813. [CrossRef]
30. Addepalli, T.; Desai, A.; Elfergani, I.; Anveshkumar, N.; Kulkarni, J.; Zebiri, C.; Rodriguez, J.; Abd-Alhameed, R. 8-Port Semi-Circular Arc MIMO Antenna with an Inverted L-Strip Loaded Connected Ground for UWB Applications. *Electronics* **2021**, *10*, 1476. [CrossRef]

sensors

MDPI

Article

A Single-Fed Wideband Circularly Polarized Cross-Fed Cavity-Less Magneto-Electric Dipole Antenna

Linyu Cai and Kin-Fai Tong *

Department of Electronic and Electrical Engineering, University College London, Torrington Place, London WC1E 7JE, UK
* Correspondence: k.tong@ucl.ac.uk

Abstract: In this paper, we proposed a new wideband circularly polarized cross-fed magneto-electric dipole antenna. Different from conventional cross-dipole or magneto-electric dipole antennas, the proposed simple geometry realizes a pair of complementary magnetic dipole modes by utilizing the two open slots formed between the four cross-fed microstrip patches for achieving circular polarization and high stable gain across a wide frequency band. No parasitic elements are required for extending the bandwidths; therefore, both the radiation patterns and in-band gain are stable. The simulated field distributions demonstrated the phase complementarity of the two pairs of magnetic and electric dipole modes. A parametric study was also performed to demonstrate the radiation mechanism between the electric and magnetic dipole modes. The radiating elements are realized on a piece of double-sided dielectric substrate fed and mechanically supported by a low-cost commercial semirigid cable. The overall thickness of the antenna is about $0.22\lambda_0$ at the center frequency of axial ratio bandwidth. The measured results show a wide impedance bandwidth ($|S_{11}| < -10$ dB) of 70.2% from 2.45 to 5.10 GHz. The in-band 3-dB axial ratio bandwidth is 51.5% from 3.0 to 5.08 GHz. More importantly, the gain of the antenna is 9.25 ± 0.56 dBic across the 3-dB axial ratio bandwidth.

Keywords: circular polarization; wideband; stable high gain; magneto-electric dipoles; cross-fed

Citation: Cai, L.; Tong, K.-F. A Single-Fed Wideband Circularly Polarized Cross-Fed Cavity-Less Magneto-Electric Dipole Antenna. *Sensors* **2023**, *23*, 1067. https://doi.org/10.3390/s23031067

Academic Editors: Raed A. Abd-Alhameed, Antonio Lázaro, Chan Hwang See and Naser Ojaroudi Parchin

Received: 3 December 2022
Revised: 31 December 2022
Accepted: 12 January 2023
Published: 17 January 2023

1. Introduction

Antennas with wide bandwidths and stable high gain features are key devices for achieving the high data rate and low latency in future wireless communications and sensing systems [1,2]. These features can compensate the free-space path loss and improve the signal stability and reliability for ensuring the quality of systems. Magneto-electric (ME) dipole antennas, which have a pair of slightly separated resonances between their magnetic dipole (M-dipole) and electric dipole (E-dipole) modes, are well-known for providing such advantages [3–5]. In addition to the gain improvement contributed by the reflection from the ground plane, the complementary radiation of the M- and E-dipole modes of the antennas can further increase the gain and maintain stable radiation patterns across the wide passband [6,7]. Various wideband linearly polarized (LP) ME dipole antennas have been reported [8–10]; however, circularly polarized (CP) designs are rare.

In a multipath-rich environment, the polarization of the signal will be altered at reflections. Polarization loss will happen when the polarization of the signal and antennas are not aligned. CP antennas have been widely used to solve such problems [11,12]. As no polarization alignment is required, the wireless channels are more stable and robust. However, for single-fed CP antennas, the axial ratio (AR) bandwidth is usually narrow, typically less than 10% [13,14]. Wideband CP ME dipole antennas have recently been drawing attention [15–19]. An aperture coupled CP ME dipole antenna array fed by a low-loss gap waveguide has achieved 14.5% of AR bandwidth [16]. A dielectric-based ME dipole antenna achieves a wide impedance bandwidth of 56.7% (VSWR $\leq$ 2) and 3-dB AR bandwidth of 41%. However, the gain fluctuation is about $\pm$2.6 dBic within the operating frequency band [20]. A dual-fed bowtie dipole utilizing a wideband 90° hybrid Wilkinson power divider feed network was proposed in [21]. By introducing two metallic strips connecting the bowtie dipoles, the gain is improved to 9 dBic with the help of a relatively large ground plane. However, wideband CP ME dipole antennas with stable (less than ±1 dB variation) and high gain (larger than 9 dBic) with reasonable footprint (about one square wavelength) are rare.

Recently, wideband crossed-dipole CP antennas with parasitic elements have also been proposed for improving the 3-dB AR bandwidth [22–25]. In [22], a similar geometry of four cross-fed rectangular patches was reported. As the open slots are not aligned in a straight line, the cross-dipole design solely relies on the addition of curved-delay line to provide the required phase difference in order to enhance the AR bandwidth. Large ground plane, cavity reflectors, and the addition of parasitic elements are also utilized to boost the gain and bandwidths. However, without the presence of the complementary M-dipole modes as proposed, these designs either suffer from low gain and/or large gain fluctuation within the passband. Some crossed-dipole antennas utilize complex cavity reflectors to mitigate the fluctuating gain problems [26,27], but the improvement is limited, and the antenna geometries become complicated. A comparison of the measured results of recently reported wideband CP antennas and the proposed antenna is presented in Table 1. It can be observed that the proposed antenna demonstrates wide impedance and 3-dB AR bandwidths without suffering from low and/or high gain fluctuation within the bandwidth. The simple single-fed geometry can also reduce the fabrication cost and potential errors, and additionally, no cavity reflector is required for achieving the stable high gain.

Table 1. Comparison of measured results of recently reported wideband circular polarized antennas.

	BW (%)	AR BW (%)	Inline BW (%)	Gain (dBic)	Feed Pt. No.	Footprint (λ_o^2)	Height (λ_o)	Cavity Reflector
[21]	57.6	51.9	51.9	9.50 ± 0.30	2	1.43	0.18	nil
[22]	57.6	38.9	38.9	9.35 ± 1.35	2	0.83	0.26	nil
[23]	59.8	26.8	26.8	8.0 ± 0.50	1	0.41	0.16	single
[24]	57.3	50.9	43.5	9.56 ± 1.19	1	1.02	0.27	single
[25]	69.0	58.6	58.6	8.51 ± 0.89	1	0.63	0.27	single
[26]	93.1	87.6	83.2	3.25 ± 5.38	1	1.24	0.42	nil
[27]	61.8	51.6	51.6	3.70 ± 0.65	1	0.18	0.07	nil
[28]	77.7	68.1	68.1	3.63 ± 4.63	1	0.83	0.22	nil
[29]	79.4	66.7	66.7	8.51 ± 1.29	1	0.74	0.36	complex
[30]	56.7	38.9	38.9	9.35 ± 1.35	2	0.83	0.26	complex
[31]	50	56.5	50	6.7 ± 1.75	1	1.58	0.16	nil
[32]	65.1	71.5	65.1	8.5 ± 1.0	1	1.98	0.26	nil
[33]	22.6	6.8	6.8	10.1 ± 0.1	1	6.67	0.34	nil
This work	70.2	51.5	51.5	9.31 ± 0.56	1	1.16	0.24	nil

On the other front, as described in [5], the M-dipole mode of a general ME dipole antenna is usually realized by a pair of vertical cavity walls. A linearly polarized ME dipole antenna without the conventional quarter-wavelength vertical wall was reported [34]; it demonstrated that the gap between the wide E-dipole patches, operating as an open slot, can replace the quarter-wavelength cavity walls for generating the M-dipole mode. More importantly, the wide bandwidth and stable high gain features can be retained without the vertical walls. The profile of the cavity-less ME dipole antenna can also be reduced to 0.16 λ_o at the center frequency. With the facilitation of a pair of cross-dipoles, the cavity-less concept in [34] is extended to circular polarization in this paper. The proposed wideband

CP cavity-less ME dipole antenna is designed to operate at 5G mid-band. CST Studio Suite 2020 [35] was used to perform the simulations.

In Section 2, the antenna geometry and operating principle will be presented. The parametric study of the critical parameters of the antenna in Section 3 will confirm the operating principle and serve as a reference for designing the proposed antennas for different applications. The simulated and measured results of the final design are provided in Section 4. The work is concluded in Section 5.

2. Geometry and Operating Principle

2.1. Antenna Geometry

Figure 1 shows the isometric view of the proposed antenna; the detailed dimensions are provided under the caption. The proposed antenna consists of three components: (i) The radiating layer accommodating Patches A, B, C, and D is a piece of Rogers RT5880LZ substrate (ε_r = 1.96, h_{sub} = 1.27 mm and tan δ = 0.0019@10 GHz). Patches A and B are on the top side, whereas Patches C and D are on the bottom side of the radiating layer. (ii) A semirigid coaxial cable (RG402, 50 Ω, ∅ 3.58 mm), which is connected to the corresponding patches through the two crossed-feeding networks. (iii) A square aluminium planar ground plane is located about a quarter-wavelength under the radiating layer. The 1 mm thick ground plane also serves as a reflector of the backward radiation; in addition to the complimentary of electric and magnetic dipole modes, it provides a unidirectional radiation pattern and higher front-to-back ratio.

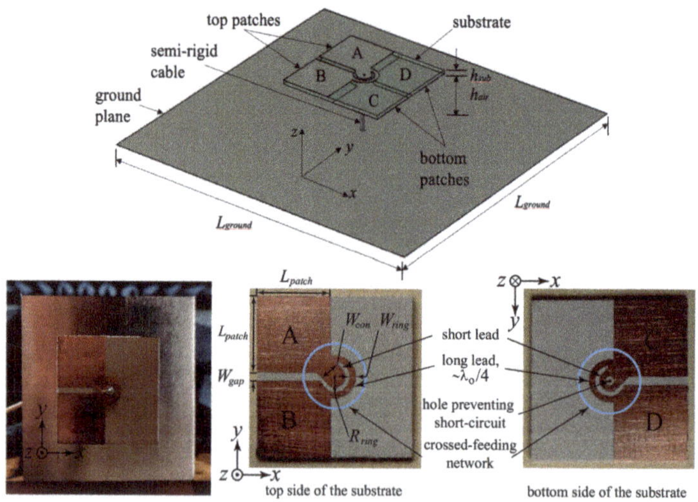

Figure 1. Geometry and fabricated prototype of the wideband CP cavity-less ME dipole antenna. L_{ground} = 80.0, h_{sub} = 1.27, h_{air} = 17.7 (0.24 λ_0), L_{patch} = 18.5, W_{gap} = 2.1, R_{ring} = 5.2, W_{ring} = 1.5, W_{con} = 5.2. All dimensions are in millimeters.

Figure 1 also shows the detailed metallic patterns of the radiating elements and crossed-feeding network on the top and bottom sides of the radiating layer of the realized prototype. Patches A, B, C, and D are basically four square patches and have the same dimensions with length L_{patch}. Through the cross-feeding network, Patches A and B are connected to the inner conductor of the semirigid coaxial cable, and Patches C and D are connected to the outer conductor of the cable and the ground plane of the antenna. Avoiding the short-circuit to the center conductor, a hole on the short lead of the bottom cross-feeding network is etched.

The four patches are designed to operate as half-wave electric printed dipoles; therefore, the length of patch (L_{patch}) was first assigned as roughly a quarter-wavelength at the center frequency in the simulations. Extending the concept reported in [27], which showed that the gap between the wide electric dipole can replace the vertical walls for creating the magnetic dipole, the two gaps between the four patches will form a pair of crossed magnetic dipoles, in the form of an open slot, for generating the circular polarization in this configuration. Together with the two crossed electric dipoles formed by the four patches and proper phase arrangement, four resonances contributing to the wideband CP radiation can be generated. The crossed-feeding network, which has two short and two long leads, is located at the center of the radiating layer. It plays a critical role in distributing equal power and proper phase to each patch to achieve circular polarization. The length of long leads is roughly a quarter wavelength. Such arrangement will introduce a 90° phase difference between Patches A and B and Patches C and D; therefore, it always keeps a 90° phase difference between adjacent patches for CP radiation. The length of the long leads is determined by its radius R_{ring}. The widths of the long lead W_{ring} and the short lead W_{con} are adjusted for best impedance matching.

Finally, the square ground plane is set at about a quarter-wavelength below the radiating layer, so that the gain will be increased. Furthermore, the size of the ground plane is set at about one wavelength to further improve the gain of the ME dipole antenna. The finalized values of the parameters will be further discussed in Section 3.

2.2. Operating Principle

Conventionally, circular polarization is generated by two orthogonal resonances; they should have equal magnitude and a phase difference of 90°. The two resonances operate at slightly different frequency points for wider 3-dB AR bandwidth. In contrast, the proposed antenna has two pairs of 45° angularly separated modes, as shown in Figure 2a. The two electric dipoles operate at a lower frequency band as a result of their longer effective electrical length, whereas the two magnetic dipoles, in the form of open-end slots between the patches, operate at the higher frequency band, as shown in Figure 2a; therefore, two minimum AR frequency points and wide AR bandwidth were achieved, as shown in the AR plots.

In the proposed wideband CP cavity-less ME dipole antenna, the two pairs of rhombic-shaped electric dipoles are located on the diagonals of the square radiating layer. Patches A and C are connected to the short leads of the crossed-feeding network, as shown in Figure 1. On the other diagonal lie Patches B and D; they are the two arms of the Electric Dipole 2 and are fed by the two long leads. As the length of the long leads is roughly a quarter-wavelength longer than the short leads, the phase of signal traveling to Electric Dipole 2 is 90° delayed when compared to that of Electric Dipole 1. Through such an arrangement, CP radiation in the lower frequency band could be achieved.

The two open-end slots, which are formed by the gaps between the patches (highlighted in light blue in Figure 2a), radiate as magnetic dipole antennas. As shown in Figure 2a, Magnetic Dipole 1 is located along the x-axis, whereas Magnetic Dipole 2 is in the y-direction. Through the same crossed-feeding network, a 90° phase difference can be achieved between the two magnetic dipole modes. In considering the shorter length of the open slots, CP radiation at the high-frequency band is generated. The phase relations between the electric dipoles and magnetic dipoles were also illustrated in Figure 2a by assuming the phase of Patch A is 0°, which explains how the electric field can be produced in the open-end slots. The simulated E-field at a different time of the period (T) shown in Figure 2b verified the above explanations.

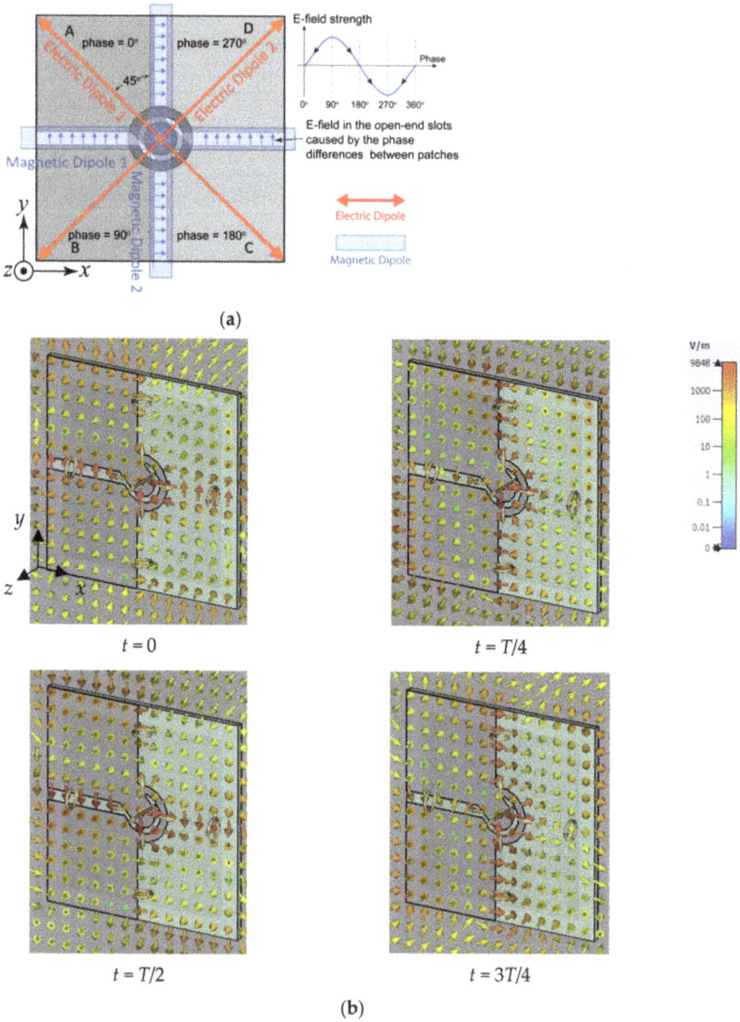

Figure 2. (**a**) The four fundamental modes of the proposed CP cavity-less ME dipole antenna, (**b**) *E*-field of the magnetic dipole modes at $t = 0$, $T/4$, $T/2$, and $t = 3T/4$ at 4 GHz.

To further illustrate the four modes contributing to the wide impedance and 3-dB AR bandwidth of the proposed antenna, Figure 3 shows the simulated real and imaginary parts of the impedance of the proposed antenna. It can be observed that the four modes operate at around 2.8 GHz, 3.6 GHz, 4.4 GHz, and 4.9 GHz, respectively. As the two electric dipoles are operating at the diagonal direction on the *xy*-plane, in addition to the fringing field that exists between the edges of the electric dipoles and the ground plane, their effective lengths are longer than that of the magnetic dipoles. Therefore, the two electric dipoles operate at lower frequency (2.8 GHz and 3.6 GHz), and magnetic dipoles operate at the high-frequency band (4.4 GHz and 4.9 GHz). Depending on the frequency separation between the modes, the imaginary part of the 2nd E-dipole and 2nd M-dipole modes are only close to, but not equal to, zero. Further verifications will be provided in the Parametric Study in Section 3. By combining the resonances with the proper physical parameters, wide impedance bandwidth and 3-dB AR bandwidth can be achieved.

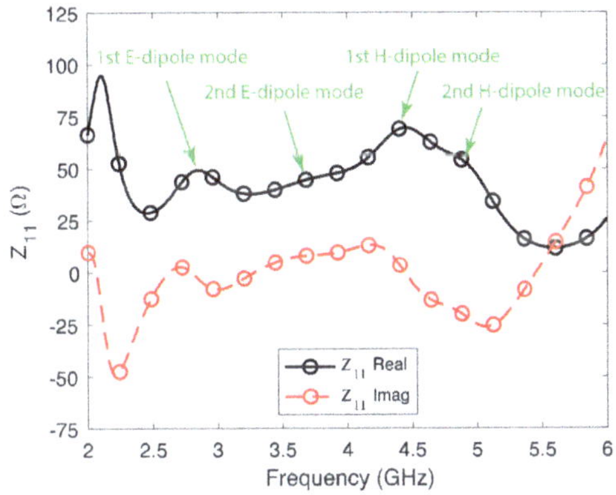

Figure 3. Simulated Z_{11} of the proposed wideband CP cavity-less ME dipole antenna.

Figure 4 illustrates the electric fields at a different time (t) in one period of oscillation (T) of the proposed wideband CP antenna at the center frequency point of the 3-dB AR bandwidth, i.e., 4 GHz. It can be observed that the E-field rotates in an anticlockwise direction, with the wave propagating in the positive z-direction, i.e., a right-hand circular polarization. Left-hand circular polarization could be achieved by mirroring the radiating layer against the y-axis.

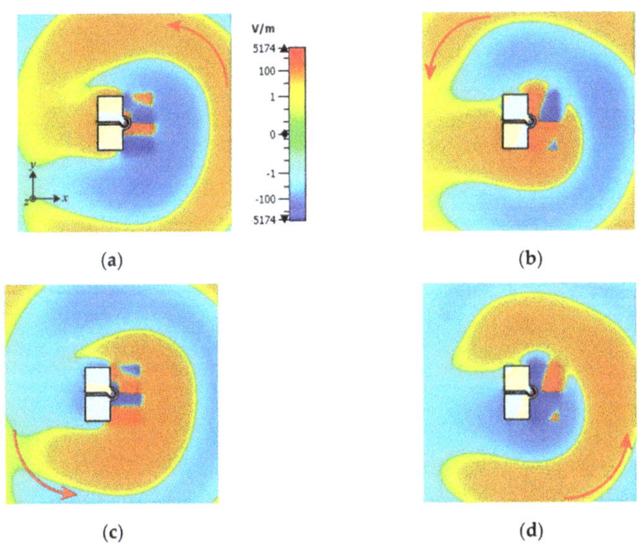

Figure 4. The z-components of E-field of the proposed antenna (**a**) $t = T/8$; (**b**) $t = 3T/8$; in xy-plane; (**c**) $t = 5T/8$, (**d**) $t = 7T/8$, at 4 GHz.

3. Parametric Study

A parametric study is presented in this section to further demonstrate the operating principle of the proposed wideband CP ME dipole antenna. Firstly, the four patches on the radiating layer play a key role in generating the fundamental resonances which provide the stable high gain across the wide operating frequency band, so the length of patch (L_{patch}) is studied first. The crossed-feeding network, enabling the circular polarization and wide 3-dB AR bandwidth, is essential for wideband impedance matching; the width of the short leads (W_{con}), width of the long leads (W_{ring}), and radius of the arc (R_{ring}) are then investigated. Thirdly, the height between radiating layer and the ground plane (h_{air}) and the length of square ground plane (L_{ground}) were also investigated to study their influence on gain and bandwidths.

3.1. Length of the Patches (L_{patch})

The four patches on the radiating layer are the main radiating elements. Together with the two narrow gaps between the patches, four resonant modes are produced, so the size of patches determines the frequencies of four resonances. Figure 5a depicts the real and imaginary parts of the impedance with different L_{patch}. Since the resonant frequency of the half-wavelength electric dipole and the magnetic dipole are determined by this parameter, when L_{patch} is longer, all four modes move to a lower frequency band. Although the resonant frequency changes with L_{patch} varying from 15.5 mm to 20 mm (19.5 λ_g to 25.2 λ_g), the value of impedance is generally not affected significantly, except for in the fourth mode, so the bandwidth is still wide generally, while the high-frequency band is affected more, as shown in Figure 5b,c. By tuning L_{patch}, the operating frequency of the proposed antenna shifts accordingly, but the impedance bandwidth remains wide at about 60%.

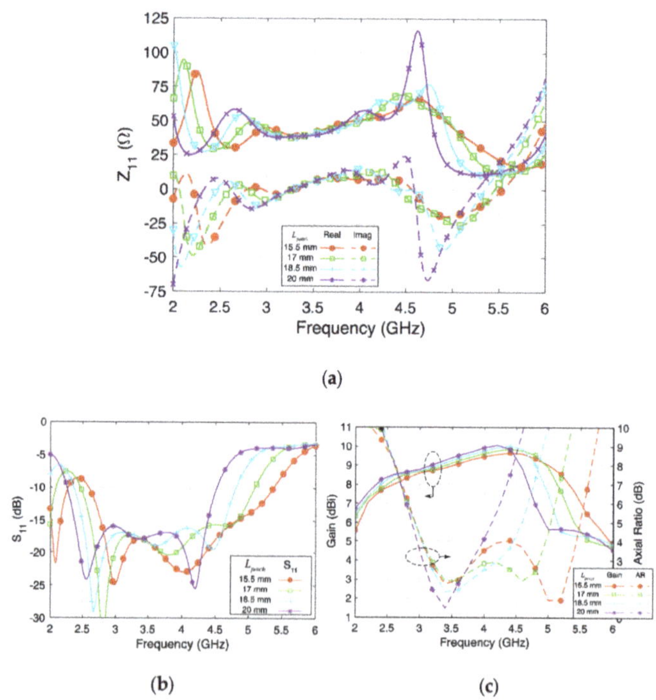

Figure 5. (**a**) Impedance (Z_{11}), (**b**) reflection coefficient (S_{11}), and (**c**) gain and axial ratio of the proposed antenna at different L_{patch}.

3.2. Width of the Short Lead (W_{con})

The cross-feeding network is the critical part that transfers the properly split signal from the semirigid coaxial cable to each patch and the open-end slots. It plays an essential role in creating the 90° phase difference and allocates equal power to the four patches to generate circular polarization. The signal is spilt into four paths and travels to each patch through the feed network. Thus, the width of each lead becomes a critical value that affects the impedance matching.

Figure 6 shows the influence of W_{con} on Z_{11} and S_{11}. Generally, the wider W_{con} is, the wider the S_{11} bandwidth will be. As can be observed from Figure 6a, larger values of W_{con} will cause both the real and the imaginary part of the impedance of the antenna that is balanced and close to 50 + j0 Ω. In this design, 5.2 mm (0.065 λ_g) is almost the maximum width that can be achieved, as it is limited by the space available. Figure 6c shows that by tuning W_{con} from 3.7 mm (0.046 λ_g) to 5.2 mm (0.066 λ_g), the gain and AR are not significantly affected, although the gain at the high-frequency band slightly increased. The AR remains low in the investigated range of W_{con}, and the AR profile shows that W_{con} is related to the balance of AR between the low-frequency and high-frequency band. To sum up the analysis state so far, keep the following in mind when setting up the value of short lead W_{con}: Firstly, the width should be as wide as possible if a wide impedance bandwidth is desired. Secondly, based on the required AR bandwidth, choose the value of W_{con} for the antenna for a balanced circular polarization between the low- and high-frequency bands.

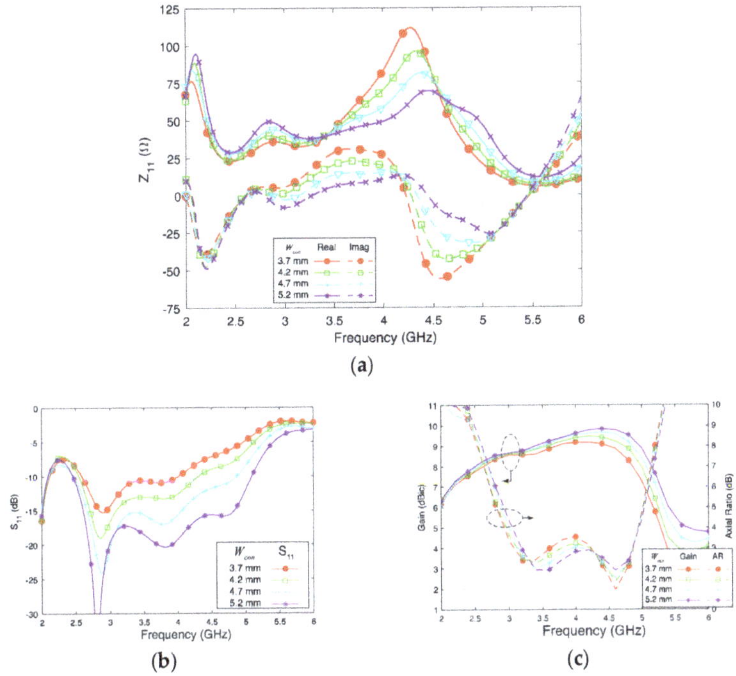

Figure 6. (**a**) Impedance (Z_{11}), (**b**) reflection coefficient (S_{11}), and (**c**) gain and axial ratio of the proposed antenna at different W_{con}.

3.3. Width of the Long Lead (W_{ring})

Figure 7a,b show the influence of W_{ring} on Z_{11} and S_{11}, respectively. W_{ring} affects the real part of the second and fourth modes and the imaginary part of the first and the third modes of the antenna. Generally, a wider W_{ring} means wider impedance bandwidth in the investigated range, and 1.5 mm is almost the maximum width that can be achieved in this design.

In Figure 7c, W_{ring} is tuned from 0.6 mm (0.008 λ_g) to 1.5 mm (0.019 λ_g); the gain remains high except in the high frequency since the impedance matching worsens when W_{ring} is narrow. Because W_{ring} affects the second and fourth modes of the proposed antenna, both circular polarizations could be affected. Relatively narrow W_{ring} separates, while wide W_{ring} brings the two AR center points closer.

When deciding the values of W_{con} and W_{ring}, it is always good to keep a large value so that the antenna will have a wide impedance bandwidth. More importantly, the W_{con} and

W_{ring} affect the AR more significantly than the impedance bandwidth. The final optimized value should be carefully selected according to the AR, as wide AR bandwidth is the priority of this wideband CP cavity-less ME dipole antenna.

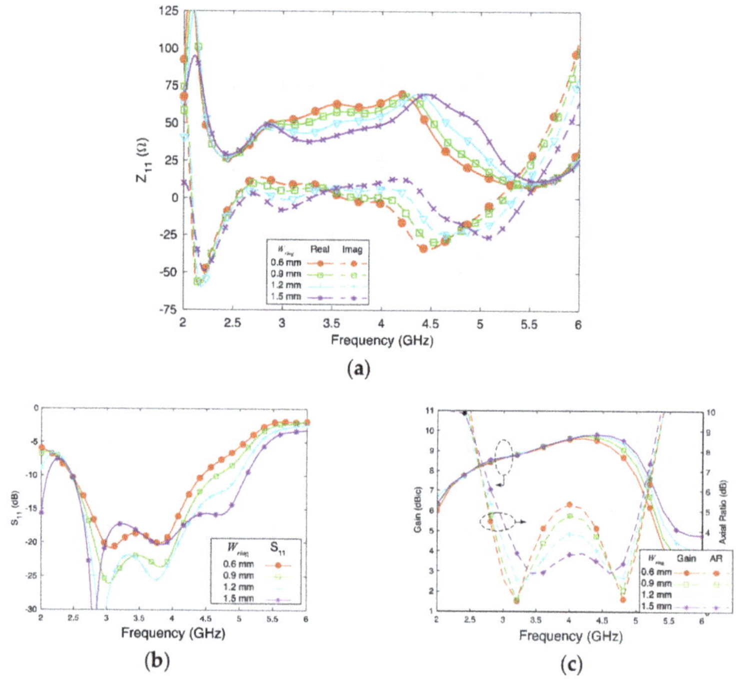

Figure 7. (**a**) Impedance (Z_{11}), (**b**) reflection coefficient (S_{11}), and (**c**) gain and axial ratio of the proposed antenna at different W_{ring}.

3.4. Radius of the Quarter-Wavelength Arc (R_{ring})

R_{ring} represents the radius of the quarter-wavelength arc from center to the middle of the arc. The signal from the feed network travels along a longer path to Patches B and D due to the extra length of the long lead. Thus, the length of the long lead is the key for creating the 90° phase difference. Figure 8a,b show the influence of R_{ring} on Z_{11} and S_{11}, respectively. It can be observed that the changes of R_{ring} affect all four modes of the antenna. Shorter R_{ring} makes the impedance match better at the high-frequency band, and wide bandwidth will be achieved.

Figure 8c shows the gain and axial ratio of the antenna when R_{ring} varies from 4.9 mm to 5.5 mm. The gain remains high, and the gain in the high frequency decreased slightly since the impedance does not match well. R_{ring} determines the phase difference between the orthogonal resonances, so the two circularly polarized frequency points could be affected. A shorter R_{ring} brings the two AR center frequency points closer and results in a narrower 3-dB AR bandwidth. Simultaneously, a desirable value of R_{ring} will arrange the two AR center points to be properly separated so that a wide AR bandwidth can be achieved.

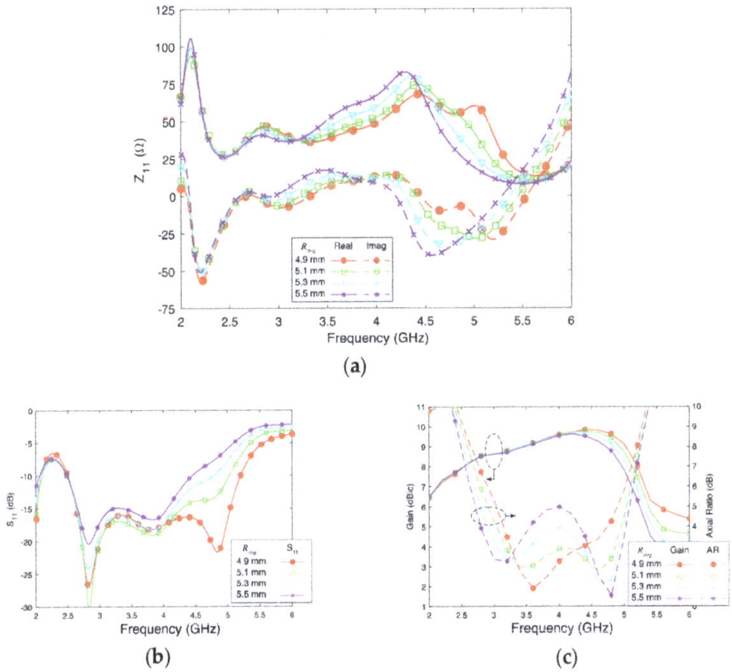

Figure 8. (**a**) Impedance (Z_{11}), (**b**) reflection coefficient (S_{11}), and (**c**) gain and axial ratio of the proposed antenna at different R_{ring}.

3.5. Radiating Layer to Ground Plane Separation (h_{air})

h_{air} represents the height between the radiating layer to the ground plane. Figure 9 shows the influence of h_{air} on Z_{11} and S_{11} of the proposed antenna. By tuning h_{air} from 14 mm (0.177 λ_g) to 23 mm (0.290 λ_g), the two lower electric dipole modes are affected more, as the fringing field at the edges is reduced at shorter h_{air}, while the two magnetic dipole modes are less subjected by the change of h_{air}. Therefore, the lower 3-dB AR bandwidth decreased with h_{air}. In contrast, since the value of Z_{11} remains relatively stable and close to 50 + $j0$ Ω, the impedance bandwidth remains wide.

3.6. Length of Ground Plane (L_{ground})

Figure 10 depicts the influence of L_{ground} on the proposed antenna. A square ground plane with a size equal or larger size than λ_0^2 will usually give better gain enhancement [6]. The value of L_{ground} was set from 40 mm (0.504 λ_g), i.e., same size of the radiating layer, to 100 mm (1.25 λ_0) for investigation. Figure 10a,b show that Z_{11} and S_{11} are barely affected by the length of the square ground, except in the case of 40 mm since it is only about a half-wavelength of the low-frequency band. Figure 10c shows the gain and AR with different values of L_{ground}. The 3-dB AR in the lower band is slightly improved when L_{ground} varies from 60 mm to 100 mm, as more fringing field can reach the ground plane. However, the gain of the antenna is affected more by the size of the ground plane size. In the investigation range, the gain increased with a larger size of the ground plane. When the size increased to 100 mm, the peak gain was about 9.9 dBic at 4 GHz. It is worth mentioning that when the size of the ground plane is extended to 1.3 λ_0, a slightly wider simulated 3-dB AR bandwidth and high peak gain can be achieved. In this paper, L_{ground} = 80 mm was selected to strike the right balance between the size and performance.

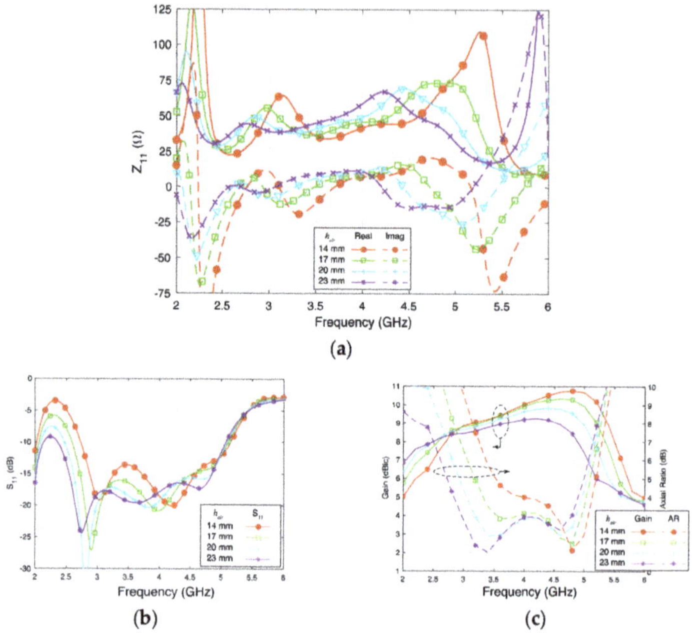

Figure 9. (**a**) Impedance (Z_{11}), (**b**) reflection coefficient (S_{11}), and (**c**) gain and axial ratio of the proposed antenna at different h_{air}.

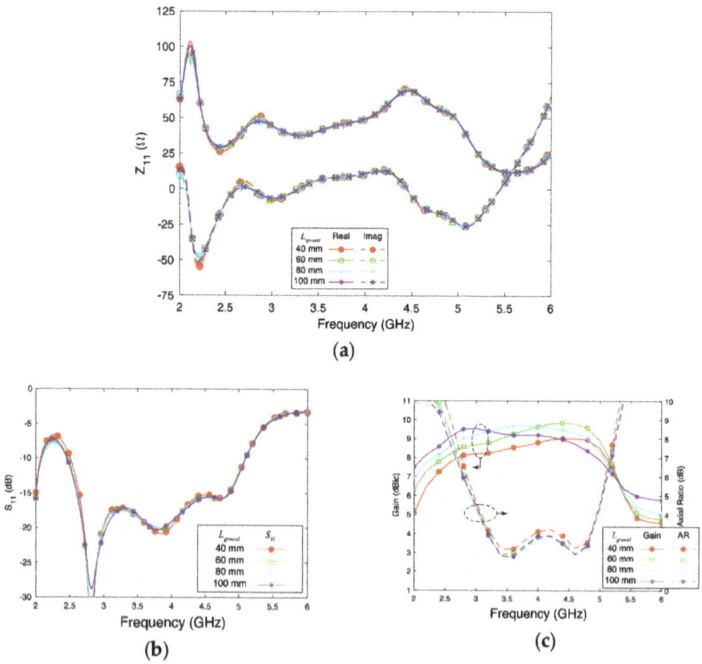

Figure 10. (**a**) Impedance (Z_{11}), (**b**) reflection coefficient (S_{11}), and (**c**) gain and axial ratio of the proposed antenna at different L_{ground}.

4. Measurement Results and Discussions

Antenna prototypes, whose physical parameters are referenced to the results of the parametric study, were fabricated and measured to verify the performance of the proposed antenna. Figure 1 shows one of the fabricated CP cavity-less ME dipole antenna prototypes mounted inside the anechoic chamber during the measurement. Three prototypes were fabricated for the measurements, and the average values were taken to improve the accuracy.

The simulated and measured S_{11} results present a reasonable consistency, as shown in Figure 11. The measured impedance bandwidth ($|S_{11}| < -10$ dB) is 70.2% from 2.45 to 5.10 GHz. Figure 12 shows the gain and 3-dB AR of the proposed antenna; the gains of the three antenna prototypes were measured by using the well-known three-antenna method, i.e., eliminating the duplicated transmitted power and antenna separation in the Friis equation in three sets of measurements. It can be observed that the measured average gain is about 9.31 dBic, with stable gain ± 0.56 dBic across the 3-dB AR bandwidth. The maximum measured gain of the antenna is about 9.9 dBic at 4.2 GHz. The axial ratio of the three antenna prototypes were measured by rotating the wideband linearly polarized dual-ridge horn antenna to different axial angles. The measured 3-dB AR bandwidth is 51.5% from 3.0 to 5.08 GHz, which is wider than the simulated results. In addition to the discrepancy in fabrication, the difference may also be caused by the misalignment when rotating the antenna from $\phi = 0°$ to 45°, 90°, and 135° for AR measurement and the averaging effect between the results of the three prototypes built. However, the trend and shape of the results are reasonably close to each other.

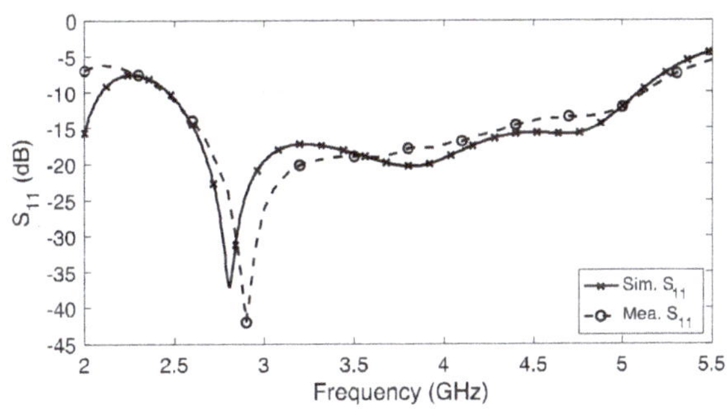

Figure 11. Simulated and measured S_{11} of the proposed antenna.

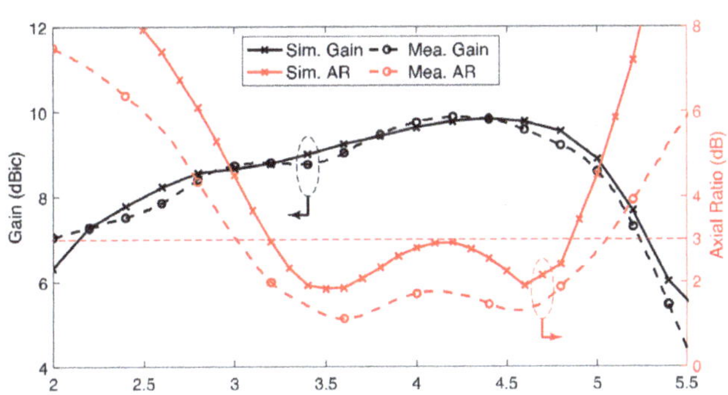

Figure 12. Simulated and measured gain and axial ratio of the proposed antenna.

To ensure of the stability of a wideband communication system, the radiation pattern of a wideband antenna at different frequency points within the operating bandwidth is expected to be consistent across the bandwidth. Figure 13 shows the simulated and measured radiation patterns at low- (3.2 GHz), mid- (4.0 GHz), and high- (4.8 GHz) frequency points of the 3-dB AR bandwidth. The co- and cross-polarization radiation patterns were measured at the four major planes, i.e., $\phi = 0°, 45°, 90°$, and $135°$. The measured results have the acceptable consistency with the simulated results. From Figure 13, high symmetry can be observed, the front-to-back ratio is about 20 dB in different planes over the passband, and the cross-polarization is also acceptable in general.

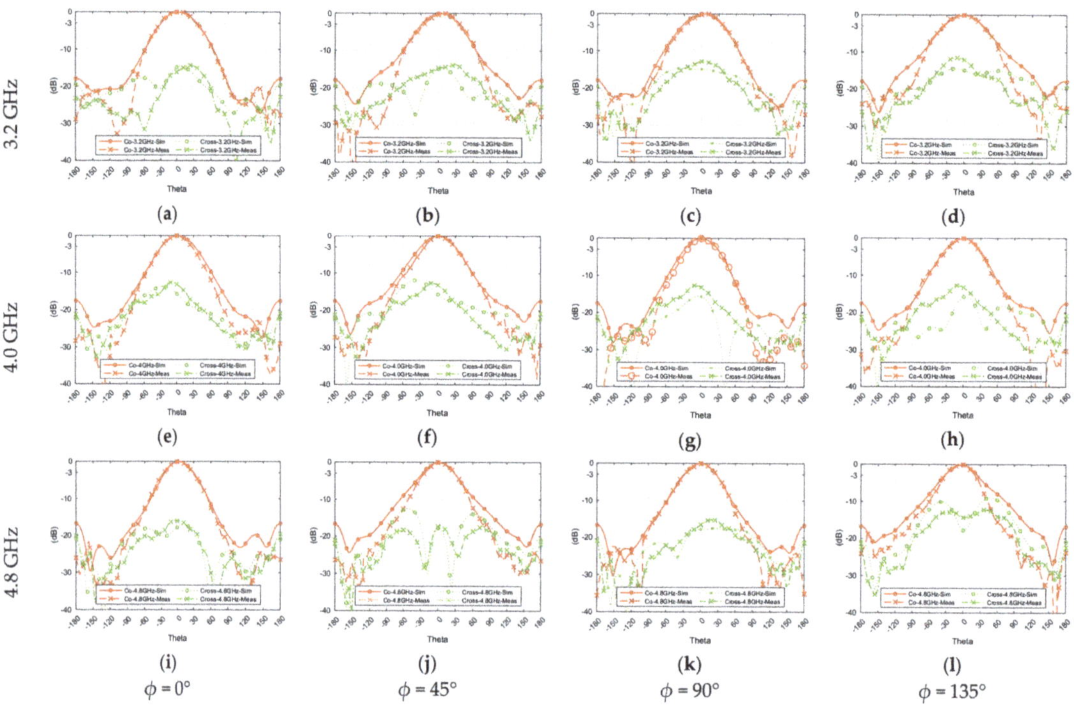

Figure 13. Radiation patterns of the proposed antenna, (**a–d**) at 3.2 GHz, (**e–h**) at 4 GHz, and (**i–l**) at 4.8 GHz, in the four principal planes: $\phi = 0°, 45°, 90°$, and $135°$.

Table 2 compares the simulated and measured results of 3-dB beamwidth at the low-, mid-, and high-frequency points. Two parameters, $|\Delta_1|$ and $|\Delta_2|$, were introduced for comparing the radiation patterns at different frequency points and principal planes. $|\Delta_1|$ represents the 3-dB beamwidth variations across the frequency in the same plane, whereas $|\Delta_2|$ is the variations in different planes at the same frequency. $|\Delta_1|$ shows good consistency at each plane, with a small variation between 9.2° to 12.1°, whereas the angular variation at different planes $|\Delta_2|$ ranges from 1.8° to 4.3°. This means the proposed CP cavity-less ME dipole antenna has a similar 3-dB beamwidth regarding the four principal planes over the frequency band. As explained, the low-frequency band mainly is contributed by the E-dipoles, while the high-frequency band is the result of the M-dipoles. Therefore, circular polarization should have the most average distribution at the center frequency (around 4 GHz), and that explains that $|\Delta_2|$ has the minimum value of 1.8° at the center frequency.

Table 2. Measured and simulated 3-db beamwidth at low-, mid-, and high-frequency points.

3-dB Beam-Width (°)	3.2 GHz		4.0 GHz		4.8 GHz		$\mid\Delta_1\mid$ (°)	
	mea.	sim.	mea.	sim.	mea.	sim.	mea.	sim.
$\phi = 0°$	65.5	59.1	70.3	67.9	74.7	67.3	9.2	8.8
$\phi = 45°$	63.7	59.9	68.5	65.1	75.8	67.8	12.1	7.9
$\phi = 90°$	66.0	63.2	69.9	64.3	76.8	69.0	10.8	5.8
$\phi = 135°$	68.0	63.0	69.0	62.9	77.5	76.3	9.5	13.3
$\mid\Delta_2\mid$ (°)	4.3	4.1	1.8	5.0	2.8	9.0		

It is worth mentioning that usually the 3-dB beamwidth of a wideband antenna decreases at the higher frequency band due to the electrically larger radiation aperture when frequency increases. However, in this wideband CP cavity-less ME dipole antenna, the 3-dB beamwidth increases with frequency. The reason explaining this feature is that the M-dipole modes operate in the high-frequency band, and the aperture for the M-dipole is determined by the parameter $2 \times L_{patch} + W_{gap}$, not the diagonal of the radiating layer. Therefore, the aperture of the antenna responsibility to the radiation at the high-frequency band is not electrically increased accordingly, so the 3-dB beamwidth at the high-frequency band does not decrease.

5. Conclusions

In this paper, a novel wideband circularly polarized cavity-less magneto-electric dipole antenna with stable high gain was presented. By utilizing the four square-patches and the corresponding gaps between the patches to realize the two pairs of properly power-fed and phase-separated electric dipole and magnetic dipole modes, the proposed antenna has a wide impedance bandwidth of 70.2% from 2.45 GHz to 5.1 GHz. The measured AR bandwidth achieved is 51.5% from 3.0 GHz to 5.1 GHz, and its height equals about 0.24 λ_o at the center frequency of 3-dB axial ratio bandwidth. Inheriting the advantages from the cavity-less ME dipole antenna concept, the average gain is 9.31 dBic, and the in-band variation is only ±0.56 dBic. The proposed antenna has a simple geometry, which has no parasitic elements or cavity ground plane, for low-cost and easy fabrication.

Author Contributions: Conceptualization, L.C. and K.-F.T.; methodology, L.C.; software, L.C.; validation, L.C.; formal analysis, L.C.; investigation, L.C.; resources, K.-F.T.; data curation, L.C.; writing—original draft preparation, L.C.; writing—K.-F.T.; visualization, L.C. and K.-F.T.; supervision, K.-F.T.; project administration, K.-F.T.; funding acquisition, K.-F.T. All authors have read and agreed to the published version of the manuscript.

Funding: This work was partly supported by the UK Research and Innovation (UKRI) of United Kingdom Project 49034—Ref.134042.

Institutional Review Board Statement: Not applicable.

Informed Consent Statement: Not applicable.

Data Availability Statement: Please contact Prof. Kin-Fai Tong at k.tong@ucl.ac.uk.

Acknowledgments: The authors would like to express their appreciations to K. M. Luk, City University of Hong Kong, for the inspirational discussions and constructive advice on the development of different wideband ME dipole antennas.

Conflicts of Interest: The authors declare no conflict of interest.

References

1. Pi, Z.; Khan, F. An introduction to millimeter-wave mobile broadband systems. *IEEE Commun. Mag.* **2011**, *49*, 101–107. [CrossRef]
2. Rappaport, T.S.; Sun, S.; Mayzus, R.; Zhao, H.; Azar, Y.; Wang, K.; Wong, G.N.; Schulz, J.K.; Samimi, M.; Gutierrez, F. Millimeter Wave Mobile Communications for 5G Cellular: It Will Work! *IEEE Access* **2013**, *1*, 335–349. [CrossRef]

3. Chlavin, A. A new antenna feed having equal E -and H-plane patterns. *Trans. IRE Prof. Group Antennas Propag.* **1954**, *2*, 113–119. [CrossRef]
4. Wong, H.; Mak, K.-M.; Luk, K.-M. Wideband Shorted Bowtie Patch Antenna With Electric Dipole. *IEEE Trans. Antennas Propag.* **2008**, *56*, 2098–2101. [CrossRef]
5. Luk, K.-M.; Wu, B. The Magnetoelectric Dipole—A Wideband Antenna for Base Stations in Mobile Communications. *Proc. IEEE* **2012**, *100*, 2297–2307. [CrossRef]
6. Tong, K.F.; Lee, K.F.; Luk, K.M. On the effect of ground plane size to wideband shorting-wall probe-fed patch antennas. In Proceedings of the 2011 IEEE-APS Topical Conference on Antennas and Propagation in Wireless Communications, Turin, Italy, 12–16 September 2011; Volume 147, p. 486.
7. Ge, L.; Luk, K.M. Beamwidth Reconfigurable Magneto-Electric Dipole Antenna Based on Tunable Strip Grating Reflector. *IEEE Access* **2016**, *4*, 7039–7045. [CrossRef]
8. Luk, K.; Wong, H. A new wideband unidirectional antenna element. *Int. J. Microw. Opt. Technol.* **2006**, *1*, 35–44.
9. Ng, K.B.; Wong, H.; So, K.K.; Chan, C.H.; Luk, K.M. 60 GHz Plated Through Hole Printed Magneto-Electric Dipole Antenna. *IEEE Trans. Antennas Propag.* **2012**, *60*, 3129–3136. [CrossRef]
10. Ge, L.; Luk, K.M. A Low-Profile Magneto-Electric Dipole Antenna. *IEEE Trans. Antennas Propag.* **2012**, *60*, 1684–1689. [CrossRef]
11. Tong, K.-F.; Lacotte, G.; Huang, J. Wideband single-fed proximity coupled circularly polarised annular slot antenna. *IET Microw. Antennas Propag.* **2010**, *4*, 1451–1455. [CrossRef]
12. So, K.K.; Wong, H.; Luk, K.M.; Chan, C.H. Miniaturized Circularly Polarized Patch Antenna With Low Back Radiation for GPS Satellite Communications. *IEEE Trans. Antennas Propag.* **2015**, *63*, 5934–5938. [CrossRef]
13. Tong, K.-F.; Huang, J. New Proximity Coupled Feeding Method for Reconfigurable Circularly Polarized Microstrip Ring Antennas. *IEEE Trans. Antennas Propag.* **2008**, *56*, 1860–1866. [CrossRef]
14. Sze, J.-Y.; Chen, W.-H. Axial-Ratio-Bandwidth Enhancement of a Microstrip-Line-Fed Circularly Polarized Annular-Ring Slot Antenna. *IEEE Trans. Antennas Propag.* **2011**, *59*, 2450–2456. [CrossRef]
15. EnCheng, W.; LuYao, S. An Improved Wideband Dipole Antenna for Global Navigation Satellite System. *IEEE Antennas Wirel. Propag. Lett.* **2014**, *13*, 1305–1308. [CrossRef]
16. Cao, J.; Wang, H.; Mou, S.; Quan, S.; Ye, Z. W-Band High-Gain Circularly Polarized Aperture-Coupled Magneto-Electric Dipole Antenna Array With Gap Waveguide Feed Network. *IEEE Antennas Wirel. Propag. Lett.* **2017**, *16*, 2155–2158. [CrossRef]
17. Cao, J.; Wang, H.; Mou, S.; Soothar, P.; Zhou, J. An Air Cavity-Fed Circularly Polarized Magneto-Electric Dipole Antenna Array With Gap Waveguide Technology for mm-Wave Applications. *IEEE Trans. Antennas Propag.* **2019**, *67*, 6211–6216. [CrossRef]
18. Scalise, G.; Boccia, L.; Amendola, G.; Rousstia, M.; Shamsafar, A. Magneto-Electric Dipole antenna for 5-G applications. In Proceedings of the 2020 14th European Conference on Antennas and Propagation (EuCAP), Copenhagen, Denmark, 15–20 March 2020; pp. 1–3. [CrossRef]
19. Mustacchio, C.; Boccia, L.; Maggiora, R.; Arnieri, E.; Amendola, G. E-Band Magneto-Electric Dipole Antenna for 5G Backhauling Applications. In Proceedings of the 2022 52nd European Microwave Conference (EuMC), Milan, Italy, 27–29 September 2022. [CrossRef]
20. Li, M.; Luk, K.-M. A Wideband Circularly Polarized Antenna for Microwave and Millimeter-Wave Applications. *IEEE Trans. Antennas Propag.* **2014**, *62*, 1872–1879. [CrossRef]
21. Zhang, Z.-Y.; Liu, N.-W.; Zhao, J.-Y.; Fu, G. Wideband Circularly Polarized Antenna With Gain Improvement. *IEEE Antennas Wirel. Propag. Lett.* **2013**, *12*, 456–459. [CrossRef]
22. He, Y.; He, W.; Wong, H. A Wideband Circularly Polarized Cross-Dipole Antenna. *IEEE Antennas Wirel. Propag. Lett.* **2014**, *13*, 67–70. [CrossRef]
23. Ta, S.X.; Park, I. Crossed Dipole Loaded With Magneto-Electric Dipole for Wideband and Wide-Beam Circularly Polarized Radiation. *IEEE Antennas Wirel. Propag. Lett.* **2014**, *14*, 358–361. [CrossRef]
24. Tran, H.H.; Park, I. Wideband Circularly Polarized Cavity-Backed Asymmetric Crossed Bowtie Dipole Antenna. *IEEE Antennas Wirel. Propag. Lett.* **2015**, *15*, 358–361. [CrossRef]
25. Tran, H.H.; Park, I.; Nguyen, T.K. Circularly Polarized Bandwidth-Enhanced Crossed Dipole Antenna with a Simple Single Parasitic Element. *IEEE Antennas Wirel. Propag. Lett.* **2017**, *16*, 1. [CrossRef]
26. Feng, G.; Chen, L.; Wang, X.; Xue, X.; Shi, X. Broadband Circularly Polarized Crossed Bowtie Dipole Antenna Loaded With Parasitic Elements. *IEEE Antennas Wirel. Propag. Lett.* **2017**, *17*, 114–117. [CrossRef]
27. Yang, W.J.; Pan, Y.M.; Zheng, S.Y. A Compact Broadband Circularly Polarized Crossed-Dipole Antenna With a Very Low Profile. *IEEE Antennas Wirel. Propag. Lett.* **2019**, *18*, 2130–2134. [CrossRef]
28. Wang, L.; Fang, W.X.; En, Y.F.; Huang, Y.; Shao, W.H.; Yao, B. Wideband Circularly Polarized Cross-Dipole Antenna With Parasitic Elements. *IEEE Access* **2019**, *7*, 35097–35102. [CrossRef]
29. Nguyen, T.K.; Tran, H.H.; Nguyen-Trong, N. A Wideband Dual-Cavity-Backed Circularly Polarized Crossed Dipole Antenna. *IEEE Antennas Wirel. Propag. Lett.* **2017**, *16*, 3135–3138. [CrossRef]
30. Qu, S.-W.; Chan, C.H.; Xue, Q. Wideband and High-Gain Composite Cavity-Backed Crossed Triangular Bowtie Dipoles for Circularly Polarized Radiation. *IEEE Trans. Antennas Propag.* **2010**, *58*, 3157–3164. [CrossRef]
31. Trinh-Van, S.; Van Trinh, T.; Yang, Y.; Lee, K.-Y.; Hwang, K.C. A broadband circularly polarized magneto-electric dipole array antenna for 5G millimeter-wave applications. *Appl. Phys. Lett.* **2021**, *119*, 023503. [CrossRef]

32. Kang, K.; Shi, Y.; Liang, C.-H. A Wideband Circularly Polarized Magnetoelectric Dipole Antenna. *IEEE Antennas Wirel. Propag. Lett.* **2017**, *16*, 1647–1650. [CrossRef]
33. Dadgarpour, A.; Sorkherizi, M.S.; Kishk, A.A. High-Efficient Circularly Polarized Magnetoelectric Dipole Antenna for 5G Applications Using Dual-Polarized Split-Ring Resonator Lens. *IEEE Trans. Antennas Propag.* **2017**, *65*, 4263–4267. [CrossRef]
34. Cai, L., Wong, H.; Tong, K. F. A Simple Low-Profile Coaxially-Fed Magneto-Electric Dipole Antenna Without Slot-Cavity. *IEEE Open J. Antennas Propag.* **2020**, *1*, 233–238. [CrossRef]
35. CST. CST Microwave Studio. Available online: https://www.3ds.com/products-services/simulia/products/cst-studio-suite/ (accessed on 11 January 2023).

electronics

MDPI

Article

Antipodal Linearly Tapered Slot Antenna with Quasi-Hemispherical Pattern Using Subwavelength Elements

Rui Wang [1], Dashuang Liao [2] and Feng Yang [1,*]

[1] School of Electronic Science and Engineering, University of Electronic Science and Technology of China, Chengdu 611731, China
[2] International Joint Innovation Center, Key Laboratory of Advanced Micro/Nano Electronic Devices & Smart Systems of Zhejiang, Zhejiang University, Haining 314499, China
* Correspondence: yangf@uestc.edu.cn

Abstract: Antennas with quasi-hemispherical radiation patterns are preferred in many wide−area wireless communication systems which require the signals to uniformly cover a wide two−dimensional region. In this work, a simple but effective beamwidth broadening technique based on an antipodal linearly tapered slot antenna (ALTSA) is first proposed and then experimentally verified. Compared with most of the reported designs, the proposed antenna can significantly widen beamwidth and achieve a quasi-hemispherical radiation pattern without increasing the overall size and structural complexity. Only two rows of subwavelength metallic elements (eight elements in total) are simply and skillfully printed at specified positions on the dielectric substrate (relative permittivity ε_r = 2.94 and thickness h = 1.5 mm) of a general ALTSA whose peak gain is 11.7 dBi, approximately 200% half-power beamwidth (HPBW) enlargement can be obtained in all cut-planes containing the end-fire direction at the central frequency of 15 GHz, and the HPBW extensions in different cut-planes have good consistency. Thus, a quasi-hemispherical beam pattern can be acquired. Thanks to the simplicity of this method, the antenna size and structural complexity do not increase, resulting in the characteristics of easy fabrication and integration, being lightweight, and high reliability. This proposed method provides a good choice for wide−beam antenna design and will have a positive effect on the potential applications of wide-area wireless communication systems.

Keywords: antipodal linearly tapered slot antenna; wide beam; subwavelength elements; HPBW

Citation: Wang, R.; Liao, D.; Yang, F. Antipodal Linearly Tapered Slot Antenna with Quasi-Hemispherical Pattern Using Subwavelength Elements. *Electronics* **2023**, *12*, 628. https://doi.org/10.3390/electronics12030628

Academic Editors: Naser Ojaroudi Parchin, Chan Hwang See and Raed A. Abd-Alhameed

Received: 11 January 2023
Revised: 22 January 2023
Accepted: 25 January 2023
Published: 27 January 2023

1. Introduction

Nowadays, wide-beam antennas have a range of applications requiring wide coverage [1–7], such as satellite systems [8], radar detection [9,10], and 5G wireless communication [11–13]. Although wide beamwidth can be obtained by reducing antenna gain, the directivity of the antenna will also be reduced and cause a large loss of radiation energy. To solve this contradiction between high directivity and wide beamwidth, beam broadening technology is worth studying. In most of the reported studies, beamwidth widening is only aimed at either one or both of the E- and H-planes, while few pieces of research focus on all cut-planes to achieve a quasi-hemispherical radiation pattern, especially for end-fire antennas. Meanwhile, to achieve the goal of beam broadening, most of the reported studies have paid the cost of overall size and structural complexity increase. All of these insufficiencies limit the applications of wide-beam antennas.

In recent years, a large number of techniques have been investigated to broaden beamwidth. In [14], a binary coding technique was proposed to determine the complex metal patterns of a patch antenna. In this way, the beamwidth in the H−plane for port 1 was effectively broadened. However, it also significantly increased the design complexity. The authors of [15] employed four separated microstrip patches to load on a dielectric substrate as the radiators. By adjusting the distance between these patches, the beamwidth in the

E-plane could be widened. Unfortunately, the antenna suffered from design complexity, and beam broadening in all cut-planes could not be achieved. In [16], a new aperture-coupled cylindrical dielectric resonator antenna was designed. By means of exciting the basic and high order modes in the dielectric resonator, a widened beam could be realized in the E- and H-planes. Nonetheless, the overall size of the antenna was enlarged due to the dielectric resonator loading, and the stacking of multi-layered cylindrical disks also increased the manufacturing difficulty. Another dielectric resonator antenna was also proposed in [17]. Owing to the loading of the engraved groove and comb-like metal wall, the HPBW in the E- and H- planes could be widened, but at the cost of the size and design complexity. In [18], a magnetoelectric dipole could achieve a stable wide beam in the H-plane; regrettably, however, the HPBW extension was not large enough and the structure was complex. The authors of [19] combined a U-shaped reflector with L-shaped metal arms as a magnetic dipole and added a parasitic patch to the L-shaped metal arms as an electric dipole. Thereby, the impedance bandwidth and the antenna HPBW were broadened. Nevertheless, a larger antenna size was needed and fabrication at high frequencies was not easy. In [20], to achieve a broad beamwidth, the authors combined a parasitic strip with a rectangular microstrip magnetic dipole antenna (MMDA). The MMDA was equivalent to a magnetic dipole toward the y-axis direction, and the parasitic strip worked as an electric dipole toward the x-axis direction. The beamwidth in the *yoz* cut-plane could be widened from about 70° to 180° due to the superposition of multiple complementary sources. However, the beamwidth was basically widened in a single cut-plane. In [21], a subarray, including a main radiant patch and two parasitic patches, was created to widen the beamwidth in the E-plane. By adjusting the spacing between the patches, the amplitude and phase of the coupling energy from the main patch to the parasitic patches could be controlled and the superposed radiation patterns could be affected. Nevertheless, this also only widened beamwidth in a single cut-plane. In [22], the mushroom structure and a pair of equal-amplitude and out-of-phase Huygens sources were introduced into the dielectric resonator antenna. The beamwidth was widened from 90° to 194°. However, the beamwidth could only be widened within one cut-plane.

The antipodal linearly tapered slot antenna (ALTSA), developed from the Vivaldi aerial [23], has numerous excellent qualities, such as good radiation orientation, low cost, being lightweight, easy fabrication, and easy integration, etc. The combination of the ALTSA and wide-beam antenna has the potential to achieve broader applications. However, research on this subject is scarce. In [24,25], artificial material units were loaded on the surface of an antipodal tapered slot antenna (ATSA) to improve directivity. Nevertheless, these methods further narrowed the beamwidth and required the printing of a large number of units, which increased the overall size. In [26], a smart configuration was proposed. Two antipodal tapered patches were designed as a fan-like shape for widening the beam, and good results were realized. However, the HPBW amplification was limited, and the design was also complex.

It can be observed that none of the above designs focus on widening the beams uniformly in all cut-planes without sacrificing antenna size and complexity, especially for end-fire antennas. This limits the applications of wide-beam antennas.

In this paper, a beam-widening design based on an ALTSA is first proposed. By loading two rows of subwavelength metallic strips on the blank area between the two tapers of the basic ALTSA, the HPBW can be widened by about 200% in entire cut-planes that include the boresight at the central frequency of 15 GHz, and the difference between each HPBW amplification is small. Thus, a quasi-hemispherical pattern can be acquired. In this way, there is almost no increase in overall size and structural complexity. Along with the above characteristics, the proposed design also has the advantages of being low cost, lightweight, and having strong reliability, which makes it a good candidate for a wide-beam antenna.

2. Antenna Configuration and Performance Analysis

2.1. Antenna Configuration

The configurations of the proposed wide-beam antenna are presented in Figure 1, which comprises a basic ALTSA and a pair of subwavelength elements. For this basic ALTSA, as shown in Figure 1a, two symmetrical metal tapers with slots are printed on the top and bottom side of the dielectric plate (relative permittivity ε_r = 2.94 and thickness h = 1.5 mm) as radiators. The microstrip line serves as a feed for both tapers, and the substrate-integrated waveguide (SIW) structure is used to optimize the impedance matching between them. The slots in the metal tapers are employed to reduce the antenna size and optimize the antenna pattern, where their width, depth, and spacing are indicated.

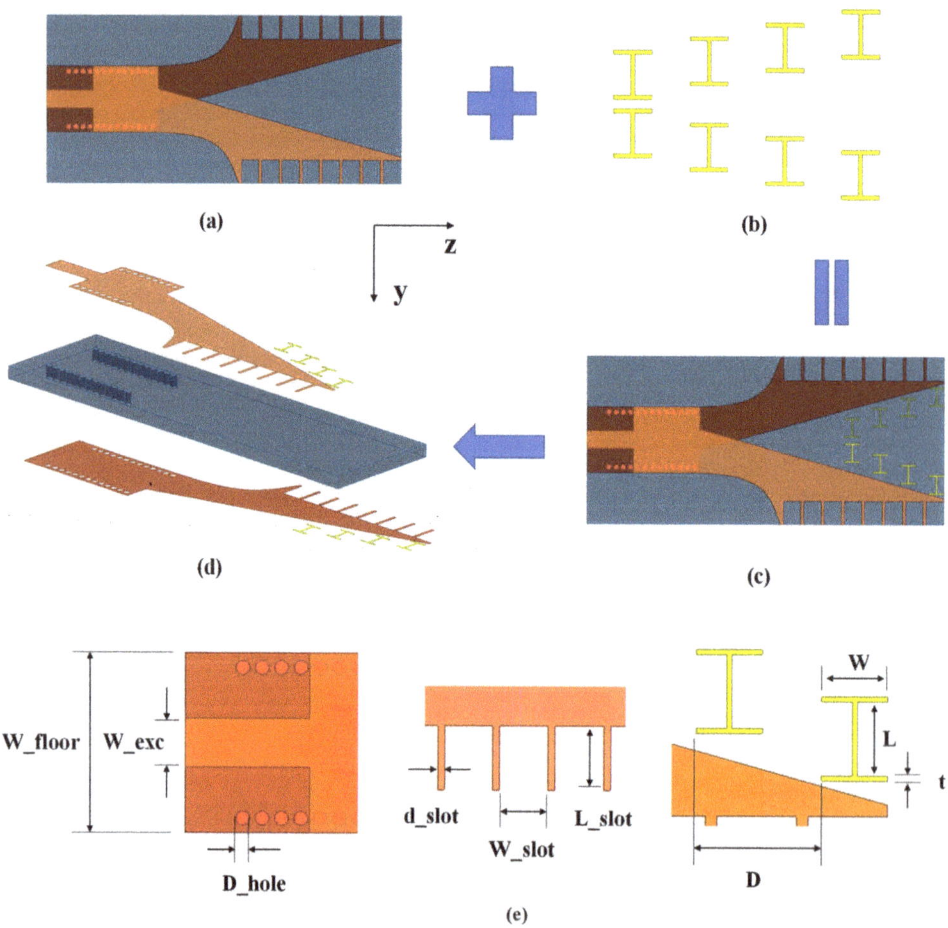

Figure 1. Configurations of the proposed wide-beam antenna. (**a**) Basic ALTSA. (**b**) Subwavelength units. (**c**) Proposed wide-beam antenna. (**d**) Exploded view. (**e**) Local geometries.

A pencil beam toward the boresight is radiated by the basic ALTSA. To widen the beamwidth, two oblique rows of subwavelength metal units, as shown in Figure 1b, are loaded on the dielectric substrate of the basic ALTSA. The length of the subwavelength metal unit is about $0.3\lambda_0$, where λ_0 is the free-space wavelength at the central frequency of 15 GHz. Each row has four units, and the subwavelength structures are designed as simple

I-shapes. Since there are only eight subwavelength elements and they are etched on the surfaces of the dielectric substrate, the overall size and structural complexity of the antenna are hardly increased.

After the combination of the basic ALTSA and the subwavelength metal units, the proposed antenna in our article is formed, as shown in Figure 1c, and the wide−beam effect can be realized. Figure 1d shows the exploded view. From these two Figures, it can be seen that the subwavelength elements are attached to the blank area between the two tapers near the end of the antenna. It is notable that the two groups of subwavelength units are not integrated into the same plane, but are printed separately on the top and bottom sides of the dielectric substrate together with one of the tapers. Additionally, the closer to the end of the antenna, the closer the subwavelength units are to the metal tapers, but they do not touch. Figure 1e shows the local geometries of the proposed wide−beam antenna, and the optimized parameter values are shown in Table 1.

Table 1. Optimized geometrical parameters (unit: mm).

Parameter	W_floor	W_exc	D_hole	d_slot	W_slot
Value	12.5	3.4	1	0.5	3.5
Parameter	L_slot	D	W	L	t
Value	4.5	5.5	2.9	3.43	0.06

2.2. Simulated Results and Analysis

The simulations and optimizations of the proposed antenna were carried out in Ansoft High-Frequency Structure Simulator (HFSS). The proposed antenna can effectively broaden the beamwidth in the frequency band of 14.5–15.5 GHz. Figures 2 and 3 show the simulation results of $|S_{11}|$ and the gain patterns of the proposed antenna. Meanwhile, the corresponding simulation results of the basic ALTSA are also presented for comparison. As shown in Figure 2, the $|S_{11}|$ of these two antennas were both less than −10 dB in the band of 14.5–15.5 GHz, indicating a good impedance matching. Considering the characteristic of the narrow band, the far-field radiation performances of the wide-beam antenna were demonstrated and analyzed only at the center frequency f_c = 15 GHz. Figure 3 shows the gain results of these two antennas in each cut-plane, including the boresight (+*z*-axis direction). We present patterns in cut-plane increments of 15 degrees (phi) to illustrate the performance. Phi represents the azimuth angle, phi = 0° represents the *xoz*-plane, phi = 90° represents the *yoz*-plane, and the coordinate system is given in Figure 1. For observing the amplification of HPBW in each cut-plane more directly, the relationship between the amplification of HPBW and phi is shown in Figure 4.

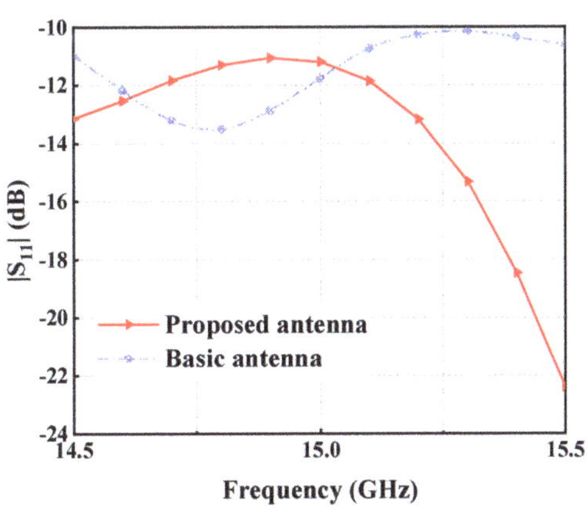

Figure 2. Simulated reflection coefficients of the proposed wide−beam antenna and basic ALTSA.

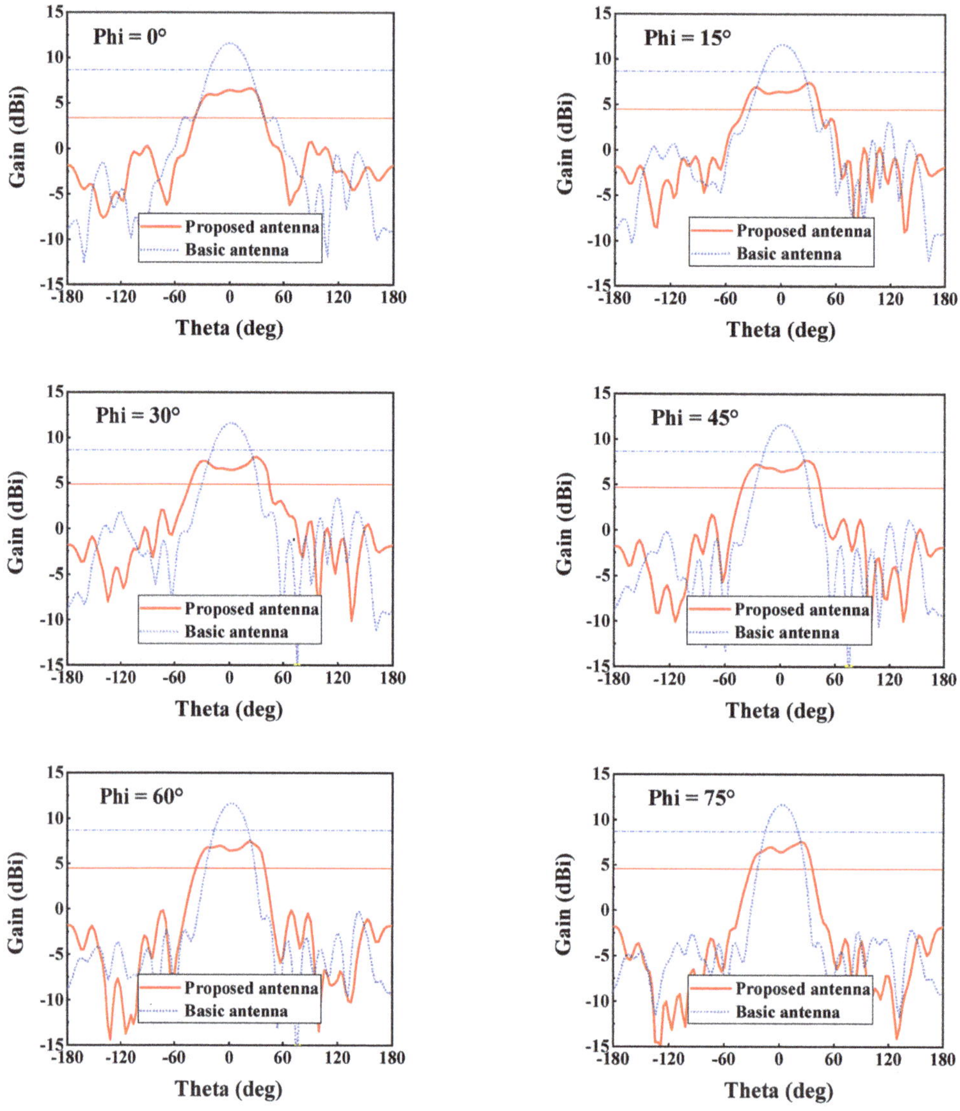

Figure 3. *Cont.*

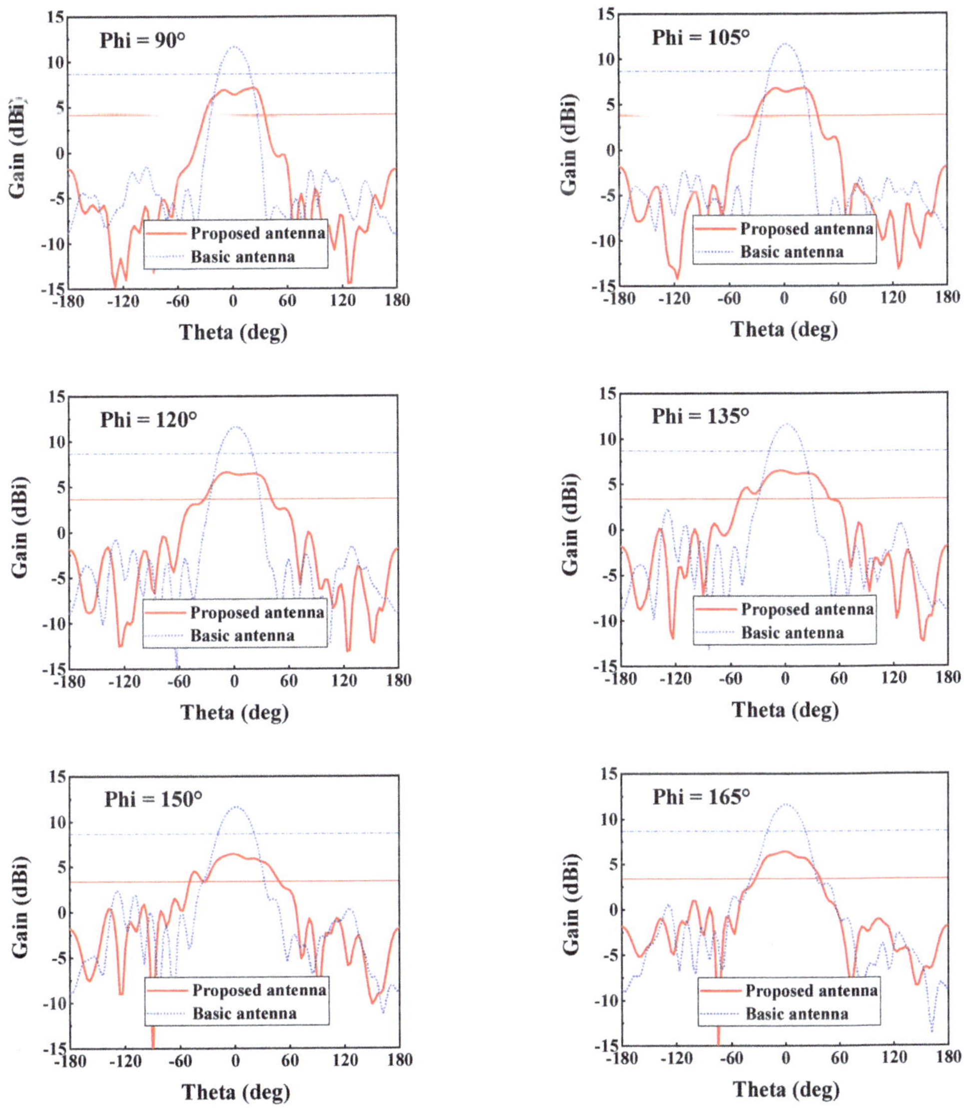

Figure 3. Simulated gain results of the proposed wide−beam antenna and basic ALTSA in different cut-planes.

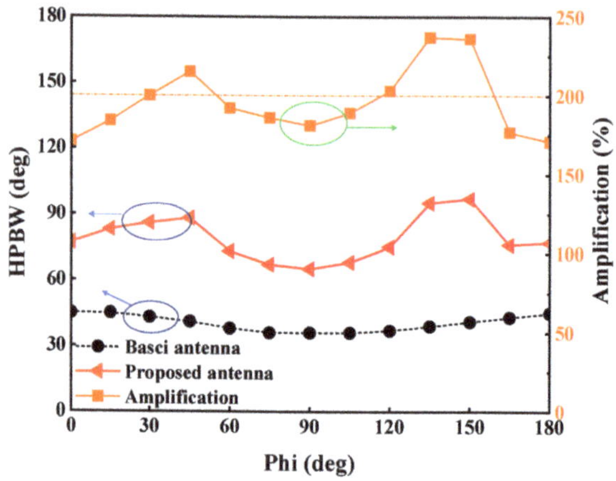

Figure 4. HPBW values of two antennas and their ratios in different cut-planes.

It can be seen from Figures 3 and 4 that the HPBW in each cut-plane of the basic ALTSA was 36–45° with a peak gain of 11.7 dBi, while the HPBW of the proposed antenna was 65–97°. The red and blue horizontal lines represent a gain drop of 3 dB for the proposed antenna and the basic antenna, respectively. The HPBW in each cut-plane was widened, and all widening ranges were about 200%, showing good consistency. The beamwidths of the basic ALTSA were relatively small due to their high gain, which also limits the absolute values of the broadened beamwidth. Proper selection of a basic antenna with a lower gain is beneficial to obtain a larger absolute value of beamwidth.

It should be pointed out that this research aims to demonstrate a new beam−broadening idea based on an ALTSA. Depending on the practical beamwidth requirements, other basic ALTSAs which have different gain and beamwidth can be flexibly selected. Moreover, the basic antenna and subwavelength elements in this paper were designed with simple structures and ordinary performances. The achievements of beam broadening can be further enhanced if basic ALTSA and subwavelength units with more complicated structures and better performance are chosen.

Considering that the number of subwavelength elements loaded has a great impact on the performances of the antenna, parameter analysis was carried out. Figure 5 shows the performances of antennas with different numbers of subwavelength elements in each row. As shown in Figure 5a, when there were three or four units in each row, the $|S_{11}|$ of the antenna were almost lower than −10 dB in the whole frequency band, indicating good impedance matching. Additionally, as shown in Figure 5b, when there were two to four units in each row, the amplification of HPBW in all cut-planes was similar, but the amplification of HPBW with four units in each row was the largest. Hence, four units in each row were selected.

To qualitatively analyze the reason why HPBW was broadened, the E-field amplitude and phase results of the basic ALTSA and the proposed wide−beam antenna are given, as shown in Table 2. At the same time, considering that the performances in each cut-plane were relatively close, only the corresponding phase diagrams in the phi = 0°, 45°, 90°, and 135° cut-planes at 15 GHz are presented here. As can be seen from Table 2, before loading subwavelength units, the E-field energy distribution between the two tapers of basic ALTSAs was relatively uniform, and the beam energy was mainly shot toward the boresight. However, after adding subwavelength units, the E−field energy distribution in this area significantly changed, and the energy was mainly concentrated near the subwavelength units, deviating from the boresight and transmitting to both sides. Similarly, the subwavelength elements change the equivalent dielectric constant of the relevant region, thus affecting the transmission phase of the electromagnetic wave.

Additionally, the shape of the wavefront then changed, making the beam radiate to both sides. These are beneficial to broadening the beamwidth. The current distribution of the proposed antenna at 15 GHz is also given in Figure 6. From Figure 6, we can observe that the subwavelength units had a strong current response, which indicates that the subwavelength units play an important role in widening beamwidth.

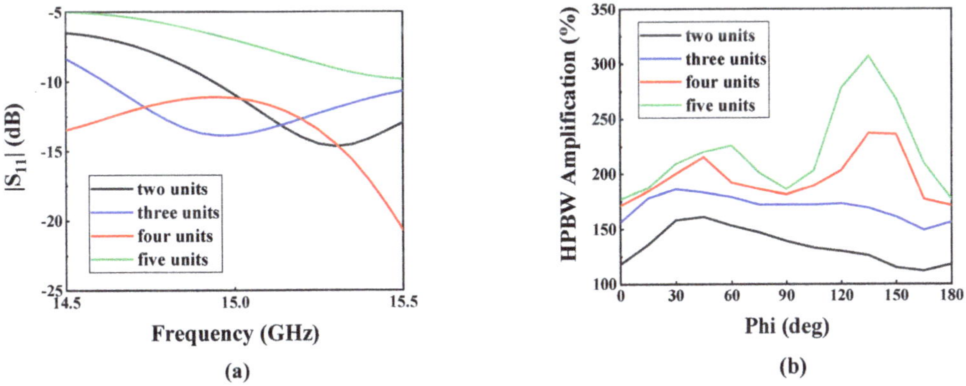

Figure 5. The performances of antennas with different numbers of subwavelength elements in each row. (**a**) $|S_{11}|$. (**b**) The amplification of HPBW in all cut-planes.

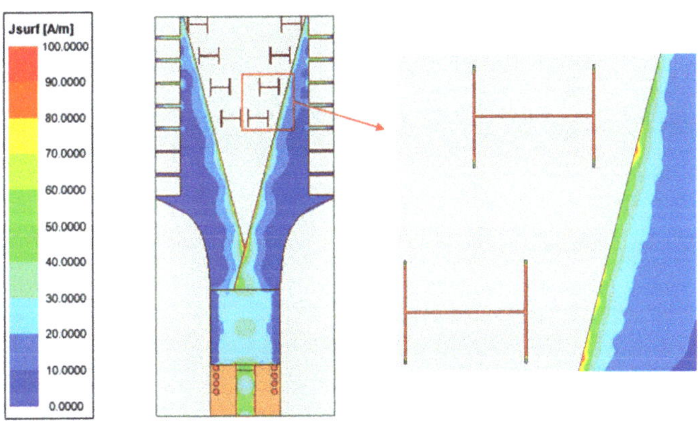

Figure 6. The current distribution of the proposed antenna.

2.3. Performance Comparison

The performance of the proposed antenna in this work were compared with other wide−beam antennas in publications, as shown in Table 3. It can be seen that there are few instances of research into broadening HPBW in all cut-planes, especially with end-fire characteristics. Moreover, most of the reported articles could not achieve beam broadening without sacrificing antenna size and complexity. For example, in [14,17,19], some interesting designs were presented that could effectively achieve wide beams, but all experienced the disadvantages of increased size, complexity, or both. In this work, based on an ALTSA with an end-fire pattern, two rows of quasi-intersecting subwavelength elements were loaded, realizing a wide beam in all cut−planes without increasing the antenna size and complexity, and the extensions in different cut-planes were similar. As mentioned above, the proposed antenna meets the demands of wide beam, low cost, and easy fabrication in many applications.

Table 2. E−field amplitude and phase distributions of the proposed and basic antennas.

	Proposed Antenna	Basic Antenna	
Amplitude	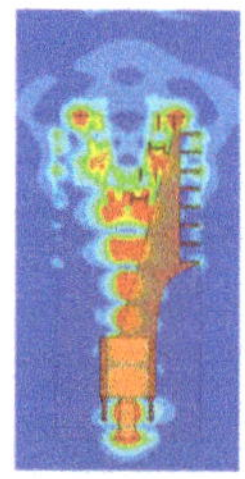	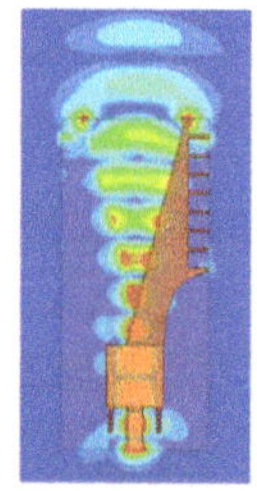	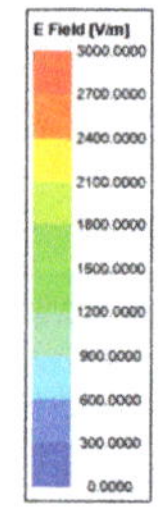
Phase (Phi = 0)	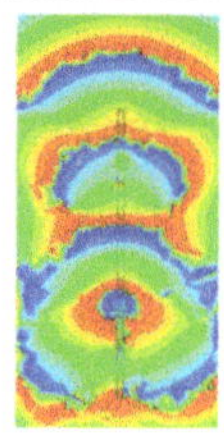	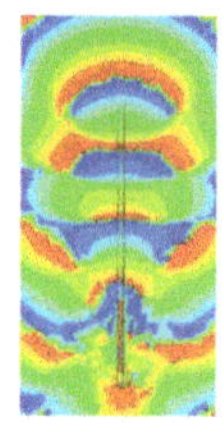	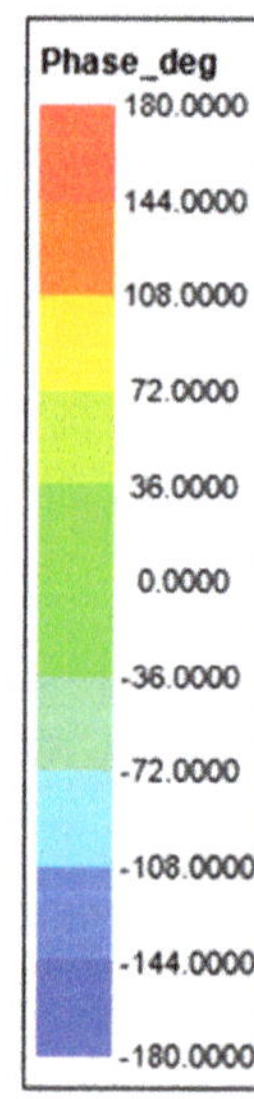
Phase (Phi = 45)	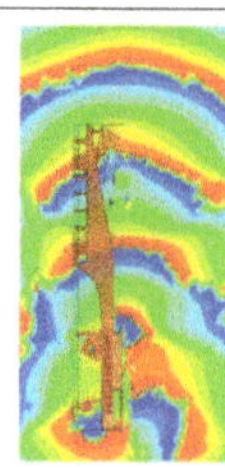	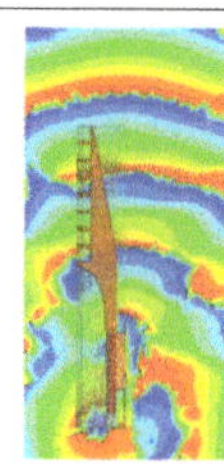	
Phase (Phi = 90)	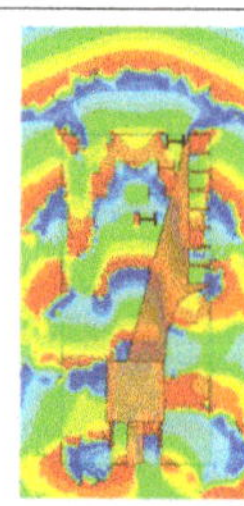	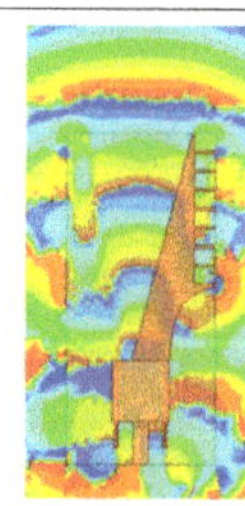	
Phase (Phi = 135)	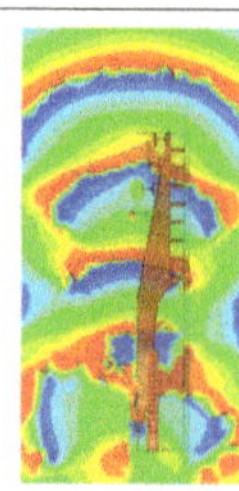	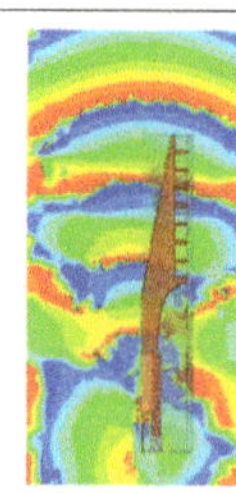	

Table 3. Comparison between wide-beam antennas.

Ref.	Type	Radiation Direction	Aim at All Cut-Planes	# HPBW	* Increased Size	* Increased Complexity
[2]	Dielectric resonator antenna	Broadside	No	189%	Yes	Yes
[14]	Patches antenna	Broadside	No	~200%	No	Yes
[17]	Dielectric resonator antenna	Broadside	No	~350%	Yes	Yes
[19]	Dipole antenna	Broadside	No	~230%	Yes	Yes
[26]	Tapered slot antenna	End-fire	No	<200%	No	Yes
[27]	Microstrip antenna	Broadside	No	~230%	Yes	Yes
[28]	Dielectric resonator antenna	Broadside	No	~330%	Yes	Yes
[29]	Microstrip antenna	Broadside	No	241%	No	Yes
This work	ALTSA	End-fire	Yes	237%	No	No

#: Maximum amplification of the HPBW in the given cut-planes mentioned in the literature. *: Increased size or complexity compared to basic antennas, and little addition treated as no increase. ~: Not mentioned directly but estimated according to the figures in the reference.

3. Fabrication and Measurement

To prove the properties of the proposed wide-beam antenna, a prototypical antenna was fabricated using the printed-circuit-board (PCB) process, and measured in the microwave anechoic chamber, as shown in Figure 7. The top and bottom view of the fabricated wide-beam antenna are presented in Figure 7a,b, respectively, and Figure 7c shows the measurement scene. In this section, we present the measured reflection coefficient in Figure 8. It can be observed that the $|S_{11}|$ were less than -10 dB in the whole frequency band. Additionally, we also display the measured gain results in different cut-planes at the central frequency and compare them with the simulated ones. Considering that Section 2 has given the simulated gain results, where phi was chosen from 0° to 180° at an interval of 15°, the HPBW amplifications in different cut-planes were relatively consistent. Therefore, the gain consequences only in the phi = 0°, 45°, 90°, and 135° cut-planes are demonstrated here, as shown in Figure 9. It can be seen that the measured and simulated results had a general consistency, with slight deviations caused by the dielectric loss and fabrication tolerances. Taking the phi = 0° cut-plane results as an example, the measured HPBW was about 69°, covering a range from $-32°$ to 37°, with a peak gain of 6.1 dBi. Meanwhile, the simulated HPBW in this cut-plane was approximately 78°, overriding a scale from $-38°$ to 40°, with a peak gain of 6.7 dBi. The difference between the measured and simulated HPBW was about 9°. On the whole, the measured results are in good agreement with the simulated results.

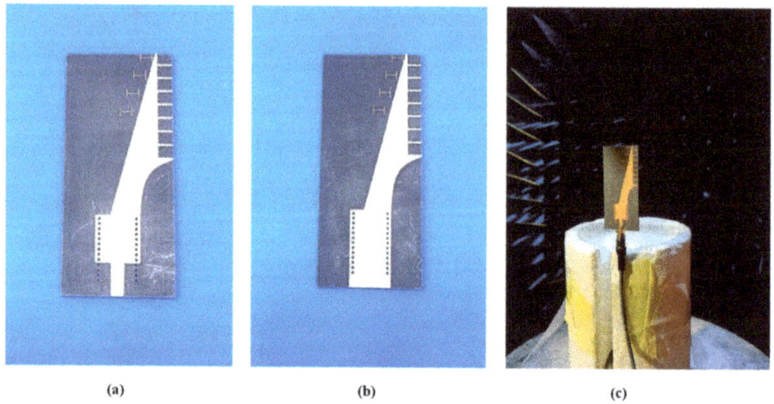

Figure 7. (**a**) Top view of the fabricated antenna. (**b**) Bottom view of the fabricated antenna. (**c**) Photograph of the measurement scene.

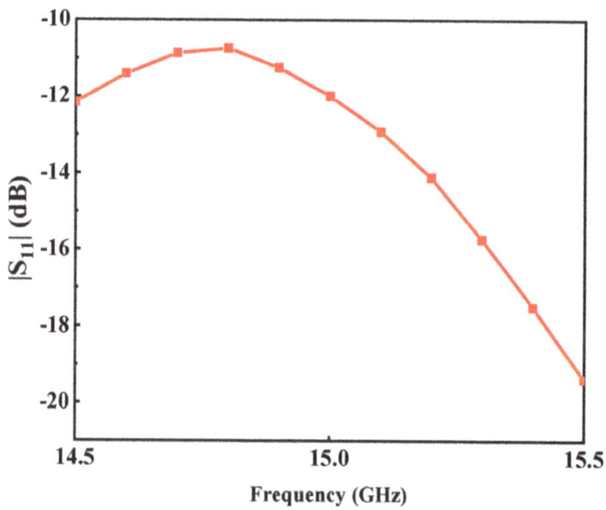

Figure 8. Measured reflection coefficients of the proposed antenna.

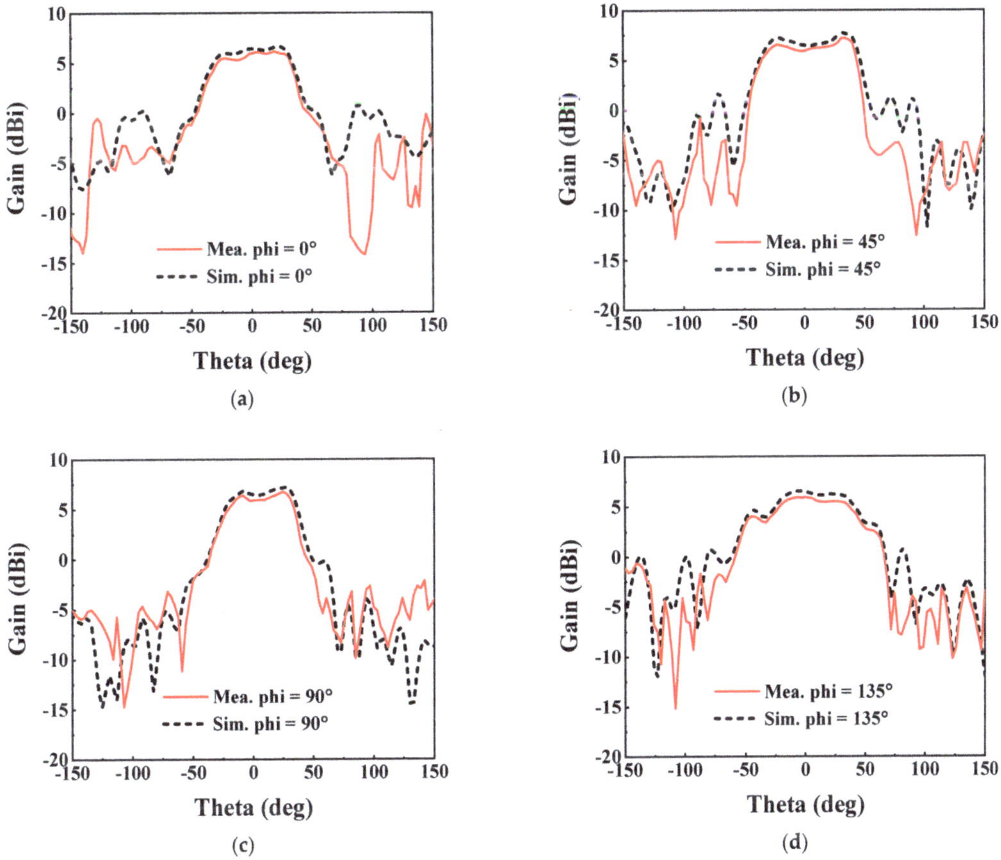

Figure 9. Measured and simulated gain results of the proposed antenna in different cut−planes of (**a**) phi = 0°, (**b**) phi = 45°, (**c**) phi = 90°, and (**d**) phi = 135°.

4. Conclusions

In most of the reported studies, beam-widening could only be realized in one or both of the E- and H-planes, and came at the cost of increased antenna size and complexity. These obviously limit the applications of wide-beam antennas. To address these problems, in this paper, a novel and simple method to widen the antenna beamwidth was put forward. By simply printing two quasi-intersecting rows of subwavelength cells on the dielectric substrate of a basic ALTSA, the beamwidth can be widened in all cut-planes that include the boresight direction with a tiny increase in antenna size and complexity. Additionally, the widening amplitudes at the center frequency are stable at around 200%, which produces a quasi-hemispherical pattern. Moreover, the main goal of this paper is to show the feasibility of this beam-broadening method. It is completely possible to use the basic ALSTA and subwavelength units with more complex structures and excellent performances to obtain better results. The designed antenna also has the characteristics of being low-cost and lightweight, easy to fabricate, and easy to integrate, making it a good candidate for satellite communication antennas.

Author Contributions: Conceptualization, R.W.; investigation, R.W. and D.L.; methodology, R.W. and F.Y.; validation, R.W. and F.Y.; software, R.W. and D.L.; fabrication, R.W. and D.L.; data curation, R.W. and D.L.; writing—original draft, R.W.; supervision, F.Y.; funding acquisition, D.L. and F.Y. All authors have read and agreed to the published version of the manuscript.

Funding: This research was funded in part by the National Natural Science Foundation of China under Grant 61871101, Grant 61721001, and Grant 61631006, in part by the Joint Fund of Equipment Pre-Research of Aerospace Science and Technology under Grant 6141B061008, and in part by the National Key Research and Development Program of China under Grant 2016YFC0303501.

Data Availability Statement: Not available.

Conflicts of Interest: The authors declare no conflict of interest.

References

1. Wang, T.; Huang, K.; Zhang, B.; Liu, W.; Zhang, Z.; Hou, J.; Rasool, N. A Novel Lens Antenna with Wide Beamwidth Using Three Dimensional Printing. *Microw. Opt. Technol. Lett.* **2021**, *63*, 2192–2198. [CrossRef]
2. Yang, N.; Weng, Z.; Wang, L.; Liao, T. A Hybrid Dual-Mode Dielectric Resonator Antenna with wide Beamwidth. *Int. J. RF Microw. Comput. Aided Eng.* **2020**, *30*, e22337. [CrossRef]
3. Wang, J.Y.; Lu, W.J.; Zhang, W.L.; Zhu, L. Balanced Low-Profile Wide Beamwidth Circularly Polarized Stacked Loop Antenna. *Int. J. RF Microw. Comput. Aided Eng.* **2021**, *31*, e22848. [CrossRef]
4. Zhang, W.-H.; Xue, Q.; Liao, S.; Che, W.; Yang, W. Low-Profile Compact Microstrip Magnetic Dipole Antenna with Large Beamwidth and Broad Bandwidth for Vehicular Applications. *IEEE Trans. Veh. Technol.* **2021**, *70*, 5445–5456. [CrossRef]
5. Pradhan, S.; Gupta, B. Radome Enclosed Circularly Polarized Antenna System with Enhanced Beamwidth. *Radioengineering* **2022**, *31*, 54–61. [CrossRef]
6. Mashhadi, S.H.H.; Jiao, Y.-C.; Chen, J. Wide 3-dB Beamwidth Step-Walled Rectangular Dielectric Resonator Antenna. *J. Electromagn. Waves Appl.* **2020**, *34*, 349–361. [CrossRef]
7. Zhao, Z.-B.; Lu, W.-J.; Zhu, L.; Yu, J. Wideband Wide Beamwidth Full-Wavelength Sectorial Dipole Antenna under Dual-Mode Resonance. *IEEE Trans. Antennas Propag.* **2021**, *69*, 14–24. [CrossRef]
8. Zhang, Y.-Q.; Qin, S.-T.; Li, X.; Guo, L.-X. Novel Wide-Beam Cross-Dipole CP Antenna for GNSS Applications. *Int. J. RF Microw. Comput. Aided Eng.* **2018**, *28*, e21272. [CrossRef]
9. Yang, X.; Liu, X. Design of a Wide-Beam Microstrip Array Antenna for Automotive Radar Application. *IEEE Access* **2021**, *9*, 142340–142347. [CrossRef]
10. Li, Z.; Zhang, G.; Golbon-Haghighi, M.-H.; Saeidi-Manesh, H.; Herndon, M.; Pan, H. Initial Observations with Electronic and Mechanical Scans Using a Cylindrical Polarimetric Phased Array Radar. *IEEE Geosci. Remote Sens. Lett.* **2021**, *18*, 271–275. [CrossRef]
11. Wang, Z.; Liang, T.; Dong, Y. Metamaterial-Based, Compact, Wide Beam-Width Circularly Polarized Antenna for 5G Indoor Application. *Microw. Opt. Technol. Lett.* **2021**, *63*, 2171–2178. [CrossRef]
12. Park, J.-S.; Ko, J.-B.; Kwon, H.-K.; Kang, B.-S.; Park, B.; Kim, D. A Tilted Combined Beam Antenna for 5G Communications Using a 28-GHz Band. *IEEE Antennas Wirel. Propag. Lett.* **2016**, *15*, 1685–1688. [CrossRef]
13. Mittra, R.; Nasri, A.; Arya, R.K. Wide-Angle Scanning Antennas for Millimeter-Wave 5G Applications. *Engineering* **2022**, *11*, 60–71. [CrossRef]
14. Wang, L.; Geng, J.; Wang, K.; Zhou, H.; Ren, C.; Wu, H.; Zhao, X.; He, C.; Liang, X.; Zhu, W.; et al. Wideband Dual-Polarized Binary Coding Antenna with Wide Beamwidth and Its Array for Millimeter-Wave Applications. *IEEE Antennas Wirel. Propag. Lett.* **2020**, *19*, 636–640. [CrossRef]
15. Chen, X.; Qin, P.-Y.; Guo, Y.J.; Fu, G. Low-Profile and Wide-Beamwidth Dual-Polarized Distributed Microstrip Antenna. *IEEE Access* **2017**, *5*, 2272–2280. [CrossRef]
16. Mashhadi, S.H.H.; Jiao, Y.-C.; Chen, J. Broadbeam Cylindrical Dielectric Resonator Antenna. *IEEE Access* **2019**, *7*, 112653–112661. [CrossRef]
17. Li, R.Y.; Jiao, Y.C.; Zhang, Y.X.; Zhang, L.; Wang, H.Y. A DRA with Engraved Groove and Comb-like Metal Wall for Beamwidth Enhancement in Both E- and H-Planes. *IEEE Antennas Wirel. Propag. Lett.* **2022**, *20*, 543–547. [CrossRef]
18. Li, Y.; Luk, K.-M. A Linearly Polarized Magnetoelectric Dipole with Wide H-Plane Beamwidth. *IEEE Trans. Antennas Propag.* **2014**, *62*, 1830–1836. [CrossRef]
19. Chang, L.; Chen, L.-L.; Zhang, J.-Q.; Chen, Z.-Z. A Compact Wideband Dipole Antenna with Wide Beamwidth. *IEEE Antennas Wirel. Propag. Lett.* **2021**, *20*, 1701–1705. [CrossRef]
20. Zhang, W.-H.; Xue, Q.; Liao, S.; Yang, W.; Che, W.; Chen, H. A Linearly Polarized Low-Profile Complementary Antenna with Enhanced Bandwidth and Wide Broadside Beamwidth. *IEEE Antennas Wirel. Propag. Lett.* **2021**, *20*, 1332–1336. [CrossRef]
21. Su, G.-R.; Li, E.S.; Jin, H.; Hsu, P.-H.; Sun, J.-S.; Chin, K.-S. 79-GHz SIW slot-coupled patch antenna array with low cross polarization and wide beamwidth. *J. Electromagn. Waves Appl.* **2022**, *37*, 38–52. [CrossRef]

22. Wang, Z.; Zhao, S.; Dong, Y. Metamaterial-Based Wide-Beam Dielectric Resonator Antenna for Broadband Wide-Angle Beam-Scanning Phased Array Applications. *IEEE Trans. Antennas Propag.* **2022**, *70*, 9061–9072. [CrossRef]
23. Gibson, P.J. The Vivaldi Aerial. In Proceedings of the 1979 9th European Microwave Conference, Brighton, UK, 17–20 September 1979; pp. 101–105. [CrossRef]
24. Chen, L.; Lei, Z.; Yang, R.; Fan, J.; Shi, X. A Broadband Artificial Material for Gain Enhancement of Antipodal Tapered Slot Antenna. *IEEE Trans. Antennas Propag.* **2015**, *63*, 395–400. [CrossRef]
25. Zhou, B.; Cui, T.J. Directivity Enhancement to Vivaldi Antennas Using Compactly Anisotropic Zero-Index Metamaterials. *IEEE Antennas Wirel. Propag. Lett.* **2011**, *10*, 326–329. [CrossRef]
26. Sun, M.; Qing, X.; Chen, Z.N. 60-GHz End-Fire Fan-Like Antennas with Wide Beamwidth. *IEEE Trans. Antennas Propag.* **2013**, *61*, 1616–1622. [CrossRef]
27. Zhao, C.; Li, X.; Sun, C.; Huang, H. Single-Feed Microstrip Antenna for Wide Angle Circular Polarizations with Six Metal Columns. *IET Microw. Antennas Propag.* **2019**, *13*, 2569–2574. [CrossRef]
28. Boyuan, M.; Huang, S.; Pan, J.; Liu, Y.-T.; Yang, D.; Guo, Y.-X. Higher-Order Characteristic Modes-Based Broad-Beam Dielectric Resonator Antenna. *IEEE Antennas Wirel. Propag. Lett.* **2022**, *21*, 818–822. [CrossRef]
29. He, Y.; Li, Y. Dual-Polarized Microstrip Antennas with Capacitive via Fence for Wide Beamwidth and High Isolation. *IEEE Trans. Antennas Propag.* **2020**, *68*, 5095–5103. [CrossRef]

sensors MDPI

Communication

Compact UWB MIMO Antenna for 5G Millimeter-Wave Applications

Mohamed Atef Abbas [1,2,*], Abdelmegid Allam [3], Abdelhamid Gaafar [2], Hadia M. Elhennawy [1] and Mohamed Fathy Abo Sree [2,*]

1 Department of Electronics and Communications Engineering, Ain Shams University, Cairo 11566, Egypt
2 Department of Electronics and Communications Engineering, Arab Academy for Science, Technology and Maritime Transport, Cairo 11799, Egypt
3 Head of Information Technology, GUC, Cairo 101516, Egypt
* Correspondence: m.atef.abbas@aast.edu (M.A.A.); mohamed.fathy@aast.edu (M.F.A.S.)

Abstract: This paper presents a printed multiple-input multiple-output (MIMO) antenna with the advantages of compact size, good MIMO diversity performance and simple geometry for fifth-generation (5G) millimeter-wave (mm-Wave) applications. The antenna offers a novel Ultra-Wide Band (UWB) operation from 25 to 50 GHz, using a Defective Ground Structure (DGS) technology. Firstly, its compact size makes it suitable for integrating different telecommunication devices for various applications, with a prototype fabricated having a total size of 33 mm × 33 mm × 0.233 mm. Second, the mutual coupling between the individual elements severely impacts the diversity properties of the MIMO antenna system. An effective technique of orthogonally positioning the antenna elements to each other increased their isolation; thus, the MIMO system provides the best diversity performance. The performance of the proposed MIMO antenna was investigated in terms of S-parameters and MIMO diversity parameters to ensure its suitability for future 5G mm-Wave applications. Finally, the proposed work was verified by measurements and exhibited a good match between simulated and measured results. It achieves UWB, high isolation, low mutual coupling, and good MIMO diversity performance, making it a good candidate and seamlessly housed in 5G mm-Wave applications.

Keywords: MIMO antenna; UWB; 5G; DGS; ECC; CCL; FDTD

Citation: Abbas, M.A.; Allam, A.; Gaafar, A.; Elhennawy, H.M.; Sree, M.F.A. Compact UWB MIMO Antenna for 5G Millimeter-Wave Applications. *Sensors* **2023**, *23*, 2702. https://doi.org/10.3390/s23052702

Academic Editors: Antonio Lázaro, Christian Vollaire, Naser Ojaroudi Parchin, Chan Hwang See and Raed A. Abd-Alhameed

Received: 27 December 2022
Revised: 16 February 2023
Accepted: 23 February 2023
Published: 1 March 2023

1. Introduction

There is no doubt that this era is one of wireless communication technology that provides its contributed to the development of very high data rates, higher capacity, and reliability in communication applications used daily by a wide range of people [1,2]. Recently, Ultra-Wide Band (UWB) technology has been addressed in wireless communication systems due to its high data rates and low fabrication cost. Unfortunately, reflection and diffraction of an electromagnetic wave are the main obstacles in the dense medium, which cause multipath fading problems in UWB technology. Multiple-Input Multiple-Output (MIMO) technology has been proposed to alleviate multipath fading and increase transmission quality in wireless communication systems [3,4]. MIMO antenna systems associated with UWB technology provide an increase in the capacity linearly with the number of antennas without the need to add an additional frequency spectrum or increase the power resources. However, the design is affected by the mutual coupling between the antenna elements and the spatial correlation between them [5,6].

Concerning the separation between the antenna elements, the mutual coupling is inversely proportional to it. However, this distance should be chosen to be at a minimum between $\lambda/4$ and $\lambda/2$ to have a mutual coupling of less than −15 dB and therefore have an efficient working MIMO system. Additionally, a larger MIMO system size results from the increased separation between antenna units. As a result, while designing MIMO antennas, trade-offs between the total size of MIMO systems and the minimizing of mutual coupling should be taken into account [7].

Many researchers aim to enhance the performance of different parameters such as bandwidth, compact size, gain, efficiency, mutual coupling and diversity properties in the Fifth Generation (5G) Millimeter-Wave (mm-Wave) antenna design. So, they use several enhancement techniques to obtain the best performance, like substrate choice, corrugation, multi-element, dielectric lens and mutual coupling reduction techniques [8,9]. First and foremost, the substrate used in antenna fabrication is critical. For antenna fabrication, various substrates with varying permittivity and loss tangents are available. As a result, a substrate with a lower relative permittivity and a lower loss tangent will increase gain while decreasing power loss [10]. Second, the corrugation structure, which removes the metal part of the radiator's edge, can help improve the bandwidth and front-to-back ratio [11]. In terms of the multi-element technique, it increases an antenna's gain while also increasing its bandwidth and efficiency. As a result, structures can meet requirements such as high gain and wide bandwidth that a single antenna cannot. Furthermore, the dielectric lens can transmit electrostatic radiation in only one direction, increasing an antenna's directivity and gain [12,13]. Dielectric lenses of various shapes are created using the same or different substrate material. Finally, mutual coupling reduction techniques mitigate the effect of multiple elements on antenna performance. This technique, also known as the isolation technique, is critical for achieving the best diversity performance from MIMO antennas.

A variety of approaches have been used to reduce mutual coupling between MIMO antenna elements. Neutralization techniques, simultaneous matching of orthogonal feeding or elements, pattern diversity, and significant polarisation are all included [11–13]. In addition, many methods have been employed in an attempt to reduce coupling, including Defective Ground Structures (DGS), Electromagnetic Band Gap (EBG) substrates, complementary split ring resonators (CSRR), metamaterials, and parasitic resonators [14–17]. These methods, however, call for a sizable amount of circuit board space. Etching the isolating slots between the antenna elements is a different strategy. Numerous slot configurations, such as an F-shaped stub, a T-shaped slot, an L-shaped slot, or a vertical slot, have been tested [18–21]. Although the mutual coupling between the radiating antennas is reduced by these slots, the overall gain remains unchanged. In a 4-element MIMO antenna, linear slots were inserted in between the radiating slots to lessen mutual coupling [22]. In [23], etching linear slots between the radiators of a 4-element MIMO antenna system reduced the mutual coupling by an amount of 20 dB. A 4-element MIMO antenna with four T-shaped radiating elements was proposed by M. Alibakhshikenari [24]. This small, wideband antenna had a maximum gain of 5 dBi and a minimum isolation of 15 dB. This resulted from stubs on the bottom ground and a shortened ground plane. It was suggested in [25] to use a 2-port textile MIMO antenna with two half-ring-shaped antenna elements. In order to obtain a minimum isolation of 15.5 dB, the ground plane is defective. Dual patched wideband MIMO antennas with an inverted pair of L-shaped stubs in the ground were another design [26]. In comparison to other proposed designs, the isolation was roughly 20 dB, and the highest gain was 5.32 dBi. The main drawbacks of this system are that it is more complex and has low gain.

On the other hand, Diversity properties of MIMO systems like envelope correlation coefficient (ECC), Channel capacity loss (CCL), Diversity Gain (DG), Multiplexing Efficiency (ME) and Total Active Reflection Coefficient (TARC) are strongly affected by the mutual coupling between antenna elements of MIMO system [27–30]. Therefore, it is evident that a MIMO antenna system should have more diversity parameters to be evaluated in addition to the antenna's parameters. These metrics are not required for single-antenna elements but are required for multi-antenna devices.

Although significant work and research were made in the last half-decade to introduce reliable MIMO antennas with acceptable diversity parameters, In this work, the DGS and antenna element orientation are developed to achieve acceptable isolation along the operating band. In this study, a 4-port MIMO antenna system covering the frequency range of 25 to 50 GHz is developed, constructed, and tested. The ground plane's carved slots between the antenna elements and the orthogonal orientation of the antenna elements form the design's foundation. A DGS is then used to deploy for mutual coupling reduction and gain enhancement after a reference antenna is tuned for bandwidth enhancement. The work is organised as follows; Section 2 presents the antenna design and optimization of a single antenna. Section 3, which is concerned with simulation and measurable outcomes, presents a brief summary of the theoretical basis, including diversity parameters. Section 5 serves as the paper's conclusion.

2. Antenna Design

Since the antenna is the pivot element of MIMO antenna systems, many researchers present different antennas adapted for this technology. Concerning mm-Wave applications, printed antennas are the best candidates for this technology because of their low cost, low profile, compact size, and good choice to achieve a functional element with a proper balance of performance and manufacturing complexity for millimeter wave applications.

Figure 1 depicts the suggested four-element MIMO antenna design taken into consideration in this study. In (a), the top view is shown, while (b) shows the bottom perspective (b). The constructed model of the antenna structure is shown in Figure 2. The modelling and simulation of this design are done using the Finite-Difference Time-Domain (FDTD) (CST Microwave Studio). The choice of the substrate material, RO4003C, for electronic circuit boards that provide high-frequency performance and low-cost circuit manufacturing was made since it is one of the most widely used industry-wide standard substrate material formats [31,32]. Figure 3 illustrates the measurement setup of the constructed antenna using the ROHDE & SCHWARZ ZVA67 Vector Network Analyzer [33]. The dimensions of the MIMO are $33 \times 33 \times 0.203$ mm^3.

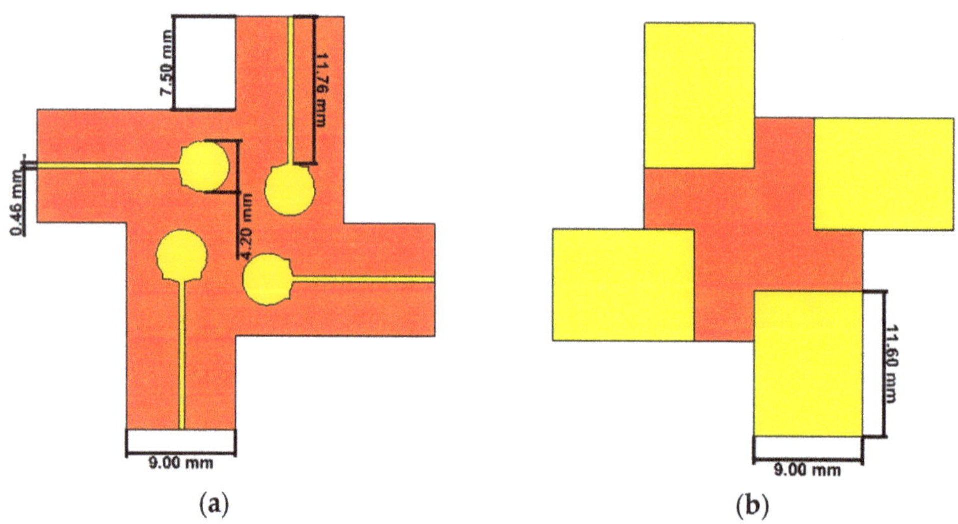

Figure 1. The proposed antenna (**a**) Top view. (**b**) Bottom view.

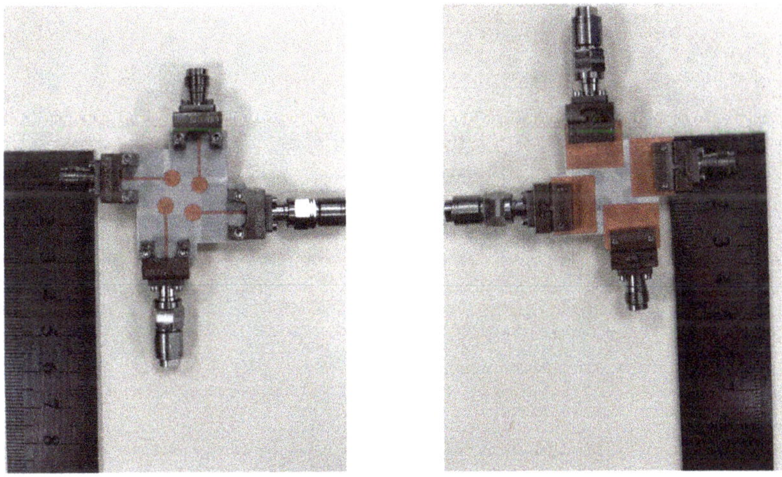

(**a**) (**b**)

Figure 2. The fabricated prototype of 4 elements MIMO antenna (**a**) Top View (**b**) Bottom view.

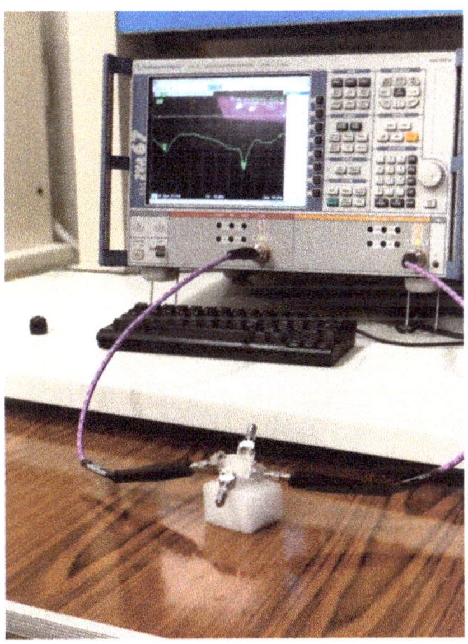

Figure 3. Measurement setup of proposed MIMO using ROHDE & SCHWARZ ZVA67 Vector Network Analyzer.

An overview of the sequential design evolution is shown in Figure 4. Figure 4 shows how the modelling begins with a single antenna element (a). as an initial step. Using well-known mathematical formulas, a rectangular patch antenna with inset feed and the full ground plane is produced it resonates at 30 GHz. The preliminary values of length L and width W can be determined from [34]:

$$W = \frac{c}{2f_r}\sqrt{\frac{2}{\varepsilon_r + 1}} \quad (1)$$

$$L = L_{eff} - 2\Delta L \quad (2)$$

$$\Delta L = 0.412h \frac{\left(\varepsilon_{reff} + 0.3\right)\left(\frac{W}{h} + 0.264\right)}{\left(\varepsilon_{reff} - 0.258\right)\left(\frac{w}{h} + 0.8\right)} \quad (3)$$

$$\varepsilon_{reff} = \frac{\varepsilon_r + 1}{2} + \frac{\varepsilon_r - 1}{2}\left(1 + 12\frac{h}{W}\right)^{-0.5}, \frac{W}{h} > 1 \quad (4)$$

$$L_{eff} = \frac{c}{2f_r\sqrt{\varepsilon_{reff}}} \quad (5)$$

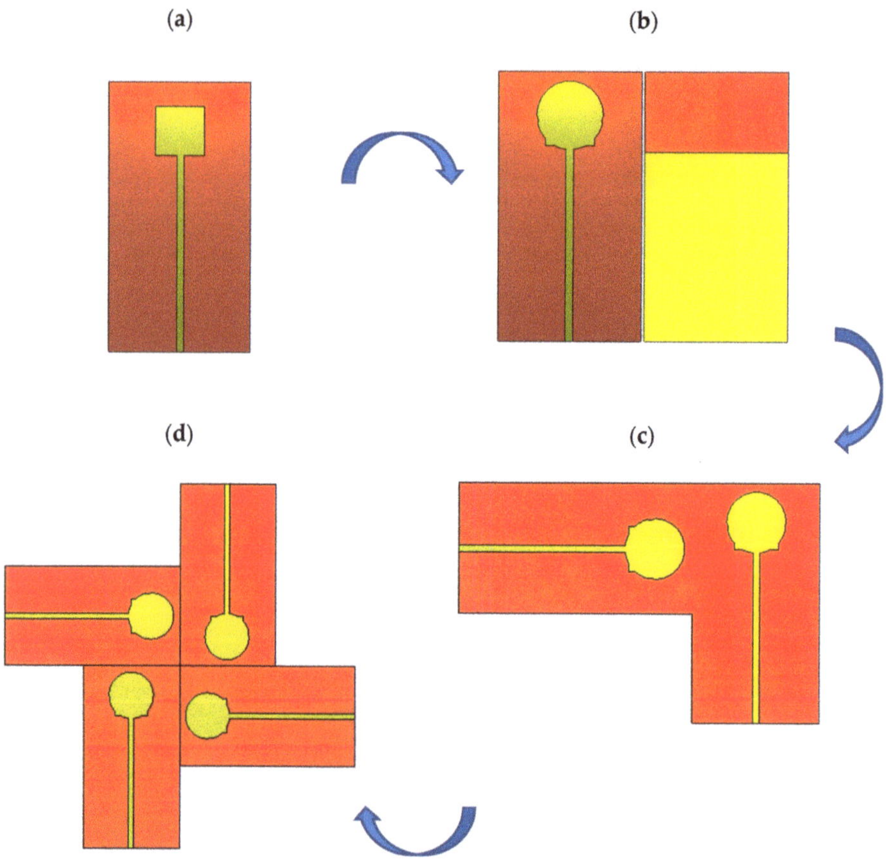

Figure 4. Steps of design (**a**) rectangular patch antenna with inset feed and full ground (**b**) single proposed antenna (top and bottom) (**c**) 2-port MIMO antenna with orthogonal orientation (**d**) Final step for 4-port MIMO antenna.

The rectangular patch antenna is improved for a larger band in the second step by etching the ground (DGS) and adding vertically and horizontally oriented circular slots to the three sides, as shown in Figure 4b. To reduce the mutual coupling, the MIMO structure is created for two ports utilizing an orthogonal orientation, as shown in Figure 4c. Based on this setup, Figure 4d shows a MIMO antenna structure with four ports to span the frequency range of 25 to 50 GHz.

The long microstrip feed in Figure 4b is a typical choice for measurements in the millimetric band. To accommodate the special connector needed for the antenna element, additional length and width would need to be added to the grounded substrate. Figure 5 demonstrates the fixation of the connector of each individual antenna element. In addition, a separation between the fixation nuts of the connector and the radiator of MIMO antenna should attain a long length to separate them as shaded by black color.

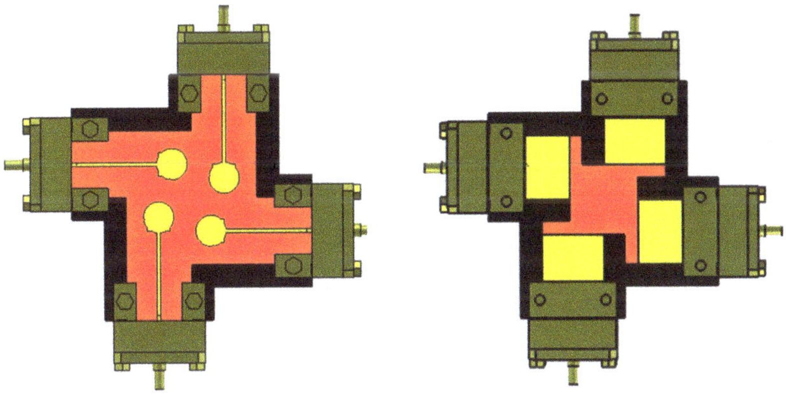

(a) (b)

Figure 5. The prototype of 4 elements MIMO antenna with connector: (**a**) Top View (**b**) Bottom view.

The fabrication process includes many steps. Based on the available dielectric substrate, we select the appropriate one suitable for the required frequency band. The antenna is modelled on CST microwave studio simulator. The optimization process is developed. Then, the Gerber file and ".Sab" file are exported to the manufacturer. At this step the duty of the academic researchers is finished. On the fabrication side the mask of design is developed. The copper is etched from the top and bottom surfaces using chemical process.

3. MIMO Antenna Parameters for Performance Evaluations

When presenting a MIMO antenna along with other fundamental characteristics, Diversity parameters analysis is a must in addition to the fundamental antenna performance evaluations and parameters, such as bandwidth, resonance frequency, radiation patterns, gain, and efficiency. These characteristics are necessary for multi-antenna devices but not for single-antenna elements. The parameters used for this include the TARC, CCL, DG, and ECC The next subsections go over these variables in relation to the suggested antenna design.

The antenna is fabricated as shown in Figure 3 and measured using ROHDE & SCHWARZ ZVA67 vector Network Analyzer. A demonstration of the S-parameters coefficients values can be seen in Figure 6. It is clear that the reflection coefficient level is below −10 dB over the whole band from 25 to 50 GHz, as shown in Figure 6a. The difference between measurement and simulated results is noticeable in the higher frequency band due to the connectors in the millimetric band and in the other environments. The measurement setup is not in a completely shielded room. On the other hand, the coupling parameters are below −20 dB on average. It reaches −40 dB to −50 dB in individual sub-bands, as shown in Figure 6b.

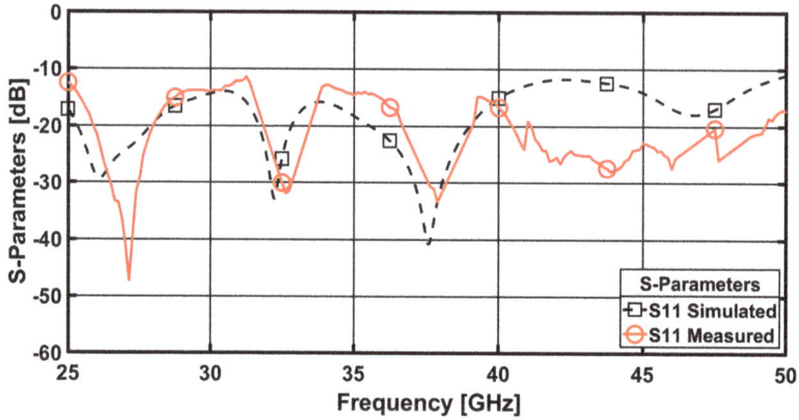

(**a**)

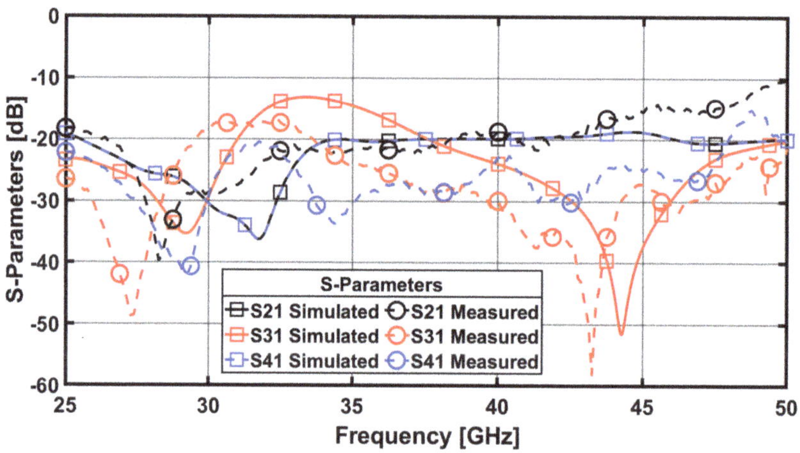

(**b**)

Figure 6. S-parameters coefficient (**a**) Reflection coefficient (**b**) Transmission coefficient.

3.1. Envelope Correlation Coefficient (ECC)

The level of correlation between various antenna components in a MIMO configuration is measured by ECC. Lower ECC results in less interdependence between the components and, as a result, better MIMO diversity performance. Equations (6) and (7), respectively, show how to compute the ECC from both the far field radiation pattern and the scattering parameters [28,35].

$$ECC = \rho_e = |\rho_{ij}| = \frac{|\int \int 4\pi[F_i(\theta,\phi)] \bullet [F_j(\theta,\phi)]d\Omega|^2}{\int \int 4\pi|F_i(\theta,\phi)|^2 d\Omega \int \int 4\pi|F_j(\theta,\phi)|^2 d\Omega} \quad (6)$$

where $i, j = 1, 2, 3, 4$, ρ_{ij} is the ECC between i^{th} and j^{th} antenna elements, while $Fi\ (\theta,\ \varphi)$ is the 3D radiation pattern field with excitation at port "i" and "•" denotes the Hermitian product and "Ω" is the solid angle.

$$ECC = \rho_e = |\rho_{ij}| = \frac{\left|S_{ii}^{*}S_{ij} + S_{jl}^{*}S_{jj}\right|^2}{\left(1 - \left(|S_{ii}|^2 + |S_{ji}|^2\right)\right)\left(1 - \left(|S_{jj}|^2 + |S_{ij}|^2\right)\right)} \quad (7)$$

where "∗" is the complex conjugate of the S-parameter. When assessing any lossy antenna, it is crucial to note that obtaining the ECC values using S-parameters is inaccurate and significantly underestimates its values. Therefore, because they are more accurate, the results acquired from far-field radiation patterns are more useful [36]. However, this process is complex and requires advanced calculations. If the ECC is lower than 0.5, then it is within the acceptable limit and the MIMO diversity is considered good as shown in Figure 7. Which indicates that no correlation between elements.

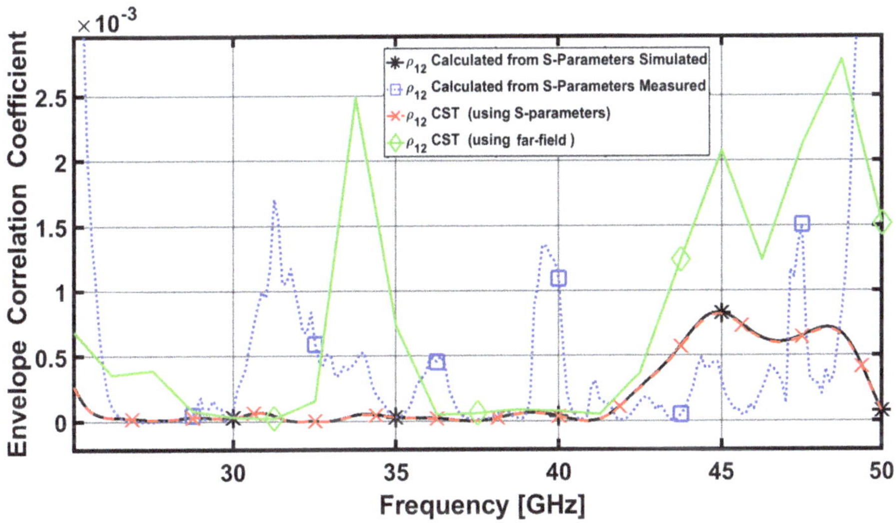

Figure 7. ECC of the proposed four-element MIMO antenna between port 1 and 2.

3.2. Diversity Gain (DG)

DG value is obtained from ECC, and it can be easily calculated using the following equation:

$$DG = 10 * \sqrt{1 - \rho_{ECC}} \quad (8)$$

It is clear that ECC and DG are inversely proportional to one another. A high DG value therefore indicates superior performance. The DG may be determined using both the far field radiation pattern and the scattering parameters, much like the ECC can [27]. Better isolation between the patch elements of the MIMO antenna is correlated with higher diversity gain values. More than 9.95 dB should be present in the DG. The estimated calculated values of the DG are higher than 9.99 dB across the antenna's operating band, as shown in Figure 8. Which indicates the good diversity performance of the proposed MIMO antenna system.

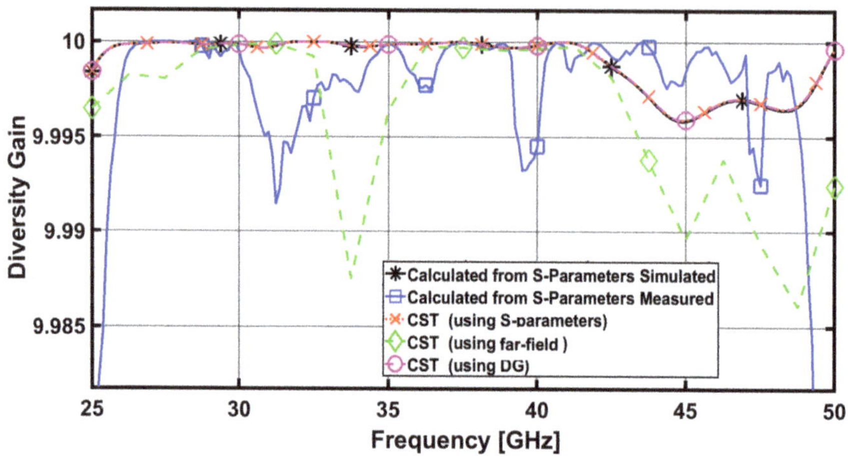

Figure 8. DG between ports 1 and 2 for the proposed four-element MIMO antenna.

3.3. *Channel Capacity Loss (CCL)*

The CCL parameter, which specifies the highest achievable limit of the information transmission rate, can be used to describe the channel capacity of MIMO systems. It calculates the ideal information transfer rate, in other words. An Equations from [37] can be used to derive the CCL parameter.

$$C(Loss) = -\log_2 \det(\psi^R) \tag{9}$$

where ψ^R indicates the correlation matrix at the receiving antenna.

$$\psi^R = \begin{bmatrix} \rho 11 & \rho 12 \\ \rho 21 & \rho 22 \end{bmatrix} \tag{10}$$

$$\rho_{ii} = 1 - \left(|S_{ii}|^2 + |S_{ij}|^2\right), for\ \ i,j = 1or2 \tag{11}$$

$$\rho_{ij} = -\left(S_{ii}{}^* S_{ij} + S_{ji}{}^* S_{ij}\right), for\ \ i,j = 1or2 \tag{12}$$

Figure 9 depicts the calculated and measured CCL. The CCL value is clearly less than 0.3 bps/Hz across the entire band. So, using the available CCL, it is possible to confirm that the proposed antenna provides a higher transmission data rate in any scattering environment.

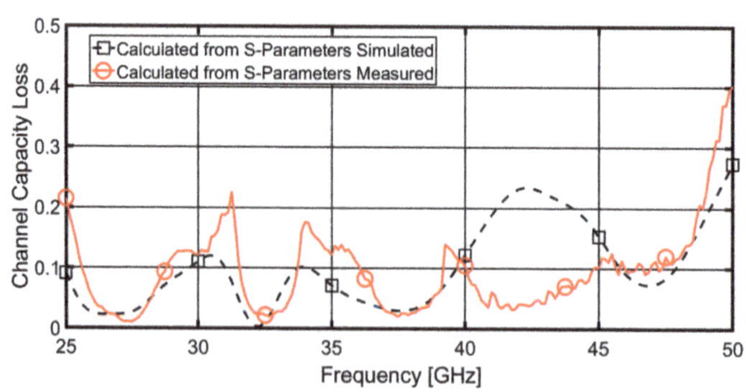

Figure 9. CCL of the proposed four-element MIMO antenna between port 1 and 2.

3.4. Multiplexing Efficiency (ME)

ME is defined as the ratio between the power of the real antenna and that of the ideal one. The maximum ME has been calculated as [38]:

$$\eta_{max} = \sqrt{\eta_i \eta_j \left(1 - |\rho_{eij}|^2\right)} \quad (13)$$

where η_i *and* η_j is the total efficiency of the ith and j^{th} antenna elements and ρ_{eij} is the magnitude of complex correlation coefficients between i^{th} and j^{th} antenna elements. The simulated and the measured ME are around −1 dB within the band from 25 GHz to 50 GHz, as illustrated in Figure 10. Which indicate that the MIMO antenna system are considered as ideal from 80% to 92% all over the band.

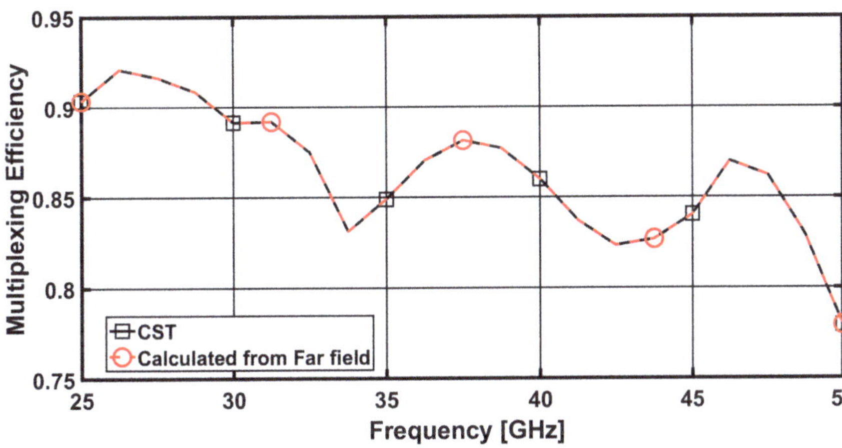

Figure 10. ME of the proposed four-element MIMO antenna between port 1 and 2.

3.5. Total Active Reflection Coefficient (TARC)

The ratio between the square roots of the total power reflected and the total power incident, is known as the total active reflection coefficient (TARC). It may be estimated using the S-parameters and depicts the random signal combinations and mutual couplings between the ports. TARC is the sole MIMO parameter that takes into account the unpredictable phases of incoming signals, which can have a significant impact on MIMO array behavior in certain circumstances. TARC of 4-port MIMO antenna can be calculated as [39,40]:

$$TARC = N^{-0.5} * \sqrt{\sum_{i=1}^{N} \left| \sum_{k=1}^{N} S_{ik} e^{j\theta_{k-1}} \right|^2} \quad (14)$$

where S_{ik} is the scattering parameter between the ith and k^{th} ports, and θ_{k-1} is the excitation phase difference between the ith and k^{th} ports.

The value of TARC for a MIMO communication system shouldn't exceed 0 dB. Figure 11 displays the measured and simulated values of TARC. The value of the proposed antenna's simulated and measured TARC is clearly under 10 dB for nearly the whole frequency range between 25 and 50 GHz where it is near to the reflection coefficients (the reference band), as can be seen from the Figure. The fabrication tolerance, test connectors, and connecting cables, as well as the surrounding test environment may all contribute to a slight difference between the measured and simulation findings.

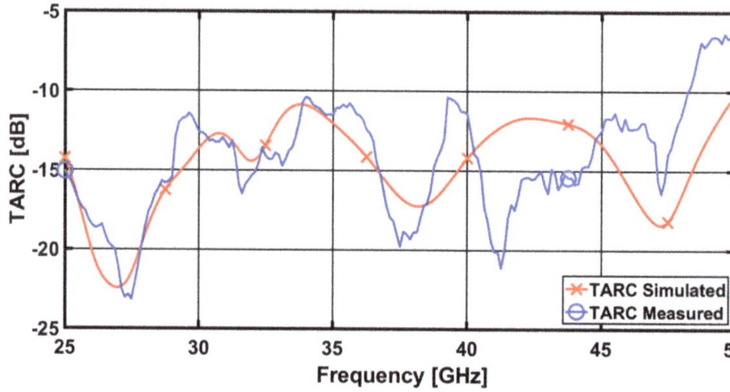

Figure 11. TARC simulated and measured results of the proposed four-element MIMO antenna.

4. Performance Comparison

In Table 1, a comparison between the proposed MIMO antenna system and the most recent MIMO work working in the 25–50 GHz region is made. The analysis takes several MIMO antenna characteristics into account. The proposed antenna with the same number of ports is more compact than [41–43]. In terms of realized gain, radiation efficiency, and isolation between antenna elements, it delivers superior MIMO performance than other small designs in [44–46]. Additionally, it has a high fractional bandwidth and a decreased mutual coupling. Its size is not the smallest, but compared to other literature work, it exhibits a considerable improvement in high isolation and superior MIMO diversity performance. This makes it possible to use the suggested antenna for 5G mm-wave applications. It's also crucial to note that the proposed design is straightforward and has symmetric geometry, making it simple to include into larger MIMO antennas with more radiating elements.

Table 1. Comparisons between the proposed MIMO antenna system and the published literature operating in 25–50 GHz band.

Reference	Physical Dimension (mm^2), Number of Ports	Substrate ε_r Thickness	Frequency, Band (B.W), Fractional B.W (GHz%)	MIMO Parameters					Mutual Coupling Reduction Technique
				ECC	DG	CCL	ME	TARC	
[41]	80 × 80 4 ports	Roger 5880, $\varepsilon_r = 2.2$ 0.508	23–40, 17, 53%	<0.0014	NA	NA	NA	NA	DGS
[42]	23.75 × 42.5 4 Ports	Roger 5880, $\varepsilon_r = 2.2$ 0.508	24–28, 4, 14.2%	<0.002	9.9	<0.001	N/A	N/A	DGS
[43]	45 × 45 4 Ports	FR-4 $\varepsilon_r = 4.2$, 1.6	4.98–5.98, 1, 16.7%	<0.0022	9.95	<0.4	N/A	CALC.	Floating Parasitic Element
[44]	30 × 30 4 ports	Roger 5880, $\varepsilon_r = 2.2$ 1.575	25–31, 6, 21.1%	<0.0005	NA	0.15	NA	NA	DGS
[45]	17 × 17 4 Ports	FR-4 $\varepsilon_r = 4.2$, 0.8	3.4–3.6, 0.2, 5.5%	0.2	NA	57	CALC.	NA	Orthogonal Polarization
[46]	5 × 4 Single	Roger 5880, $\varepsilon_r = 2.2$ 1.575	5.5–9.5, 4, 40%	N/A	N/A	N/A	N/A	N/A	N/A
Proposed Antenna	33 × 33 4 Ports	RO4003C, $\varepsilon_r = 3.55$ 0.203	25–50, 25, 66%	<0.005	≈10	<0.25	>0.85	<−10	Orientation and DGS

5. Conclusions

The proposed MIMO antenna system has been designed to provide high isolation and UWB capability, with a perfect alignment orientation between the four elements and etching the ground using DGS. The proposed antenna system has achieved excellent performance parameters, such as low mutual coupling of less than −10 dB, low ECC of less than 0.005, low CCL of 0.25 bits/s/Hz, high DG of 9.999 dBi and low TARC of less than −10 dB, which shows a superior diversity performance. In addition, the wide operational bandwidth of 25 GHz, from 25 to 50 GHz, enables the antenna to be used in multiple applications. The compact design and low-cost design features make it suitable for mm-wave 5G applications and can be easily integrated into telecommunication devices. Furthermore, the suggested MIMO antenna system offers great performance and portability, making it an ideal choice for a wide range of 5G mm-wave applications.

Author Contributions: Conceptualization, A.A. and M.A.A.; methodology, A.A.; software, M.F.A.S. and M.A.A.; validation, M.F.A.S. and M.A.A.; writing—original draft preparation, M.A.A.; writing—review and editing, A.A., M.F.A.S. and M.A.A.; supervision, A.A., A.G. and H.M.E.; All authors have read and agreed to the published version of the manuscript.

Funding: This research received no external funding.

Institutional Review Board Statement: Not applicable.

Informed Consent Statement: Not applicable.

Data Availability Statement: Not all data is applicable. However, the MATLAB portion of the data is useful and can be shared. To access this data, please reach out to the first author via email.

Conflicts of Interest: The authors declare no conflict of interest.

References

1. Rappaport, T.S.; Sun, S.; Mayzus, R.; Zhao, H.; Azar, Y.; Wang, K.; Wong, G.N.; Schulz, J.K.; Samimi, M.; Gutierrez, F. Millimeter Wave Mobile Communications for 5G Cellular: It Will Work! *IEEE Access* **2013**, *1*, 335–349. [CrossRef]
2. Hasan, M.; Faruque, M.R.I.; Islam, M.T. Dual Band Metamaterial Antenna For LTE/Bluetooth/WiMAX System. *Sci. Rep.* **2018**, *8*, 1240. [CrossRef]
3. Katalinic, A.; Nagy, R.; Zentner, R. Benefits of MIMO Systems in Practice: Increased Capacity, Reliability and Spectrum Efficiency. In Proceedings of the Proceedings ELMAR 2006, Zadar, Croatia, 7–10 June 2006; pp. 263–266. [CrossRef]
4. Molisch, A.; Win, M. MIMO systems with antenna selection. *IEEE Microw. Mag.* **2004**, *5*, 46–56. [CrossRef]
5. Paulraj, A.; Gore, D.; Nabar, R.; Bolcskei, H. An Overview of MIMO Communications—A Key to Gigabit Wireless. *Proc. IEEE* **2004**, *92*, 198–218. [CrossRef]
6. Pi, Z.; Khan, F. An introduction to millimeter-wave mobile broadband systems. *IEEE Commun. Mag.* **2011**, *49*, 101–107. [CrossRef]
7. Nadeem, I.; Choi, D.-Y. Study on Mutual Coupling Reduction Technique for MIMO Antennas. *IEEE Access* **2018**, *7*, 563–586. [CrossRef]
8. Kumar, S.; Dixit, A.S.; Malekar, R.R.; Raut, H.D.; Shevada, L.K. Fifth Generation Antennas: A Comprehensive Review of Design and Performance Enhancement Techniques. *IEEE Access* **2020**, *8*, 163568–163593. [CrossRef]
9. Dixit, A.S.; Kumar, S. A Survey of Performance Enhancement Techniques of Antipodal Vivaldi Antenna. *IEEE Access* **2020**, *8*, 45774–45796. [CrossRef]
10. Farida, S.F.; Hadalgi, P.; Hunagund, P.; Ara, S.R. Effect of substrate thickness and permittivity on the characteristics of rectangular microstrip antenna. In Proceedings of the 1998 Conference on Precision Electromagnetic Measurements Digest (Cat. No. 98CH36254), Washington, DC, USA, 6–10 July 1998. [CrossRef]
11. Honari, M.M.; Ghaffarian, M.S.; Mousavi, P.; Sarabandi, K. A Wideband High-Gain Planar Corrugated Antenna. In Proceedings of the 2020 IEEE International Symposium on Antennas and Propagation and North American Radio Science Meeting, Montreal, QC, Canada, 5–10 July 2020; pp. 7–8. [CrossRef]
12. Wang, Z.; Dou, W. Dielectric lens antennas designed for millimeter wave application. In Proceedings of the 2006 Joint 31st International Conference on Infrared Millimeter Waves and 14th International Conference on Teraherz Electronics, Shanghai, China, 18–22 September 2006; p. 376. [CrossRef]
13. Zhang, Z.; Li, X.; Sun, C.; Liu, Y.; Han, G. Dual-Band Focused Transmitarray Antenna for Microwave Measurements. *IEEE Access* **2020**, *8*, 100337–100345. [CrossRef]
14. Dey, S.; Dey, S.; Koul, S.K. Isolation Improvement of MIMO Antenna Using Novel EBG and Hair-Pin Shaped DGS at 5G Millimeter Wave Band. *IEEE Access* **2021**, *9*, 162820–162834. [CrossRef]

15. Goswami, C.; Roy, B.; Bhattacharya, P.; Ray, S.; Ghosh, S.; Ghosh, D.; Saha, D.; Jana, T.; Chatterjee, S. Complementary Split Rings Resonator Based Circular Microstrip Antenna for Wireless Applications. In Proceedings of the 2019 IEEE International Electromagnetics and Antenna Conference (IEMANTENNA), Vancouver, BC, Canada, 17–19 October 2019; pp. 065–067. [CrossRef]
16. Ziolkowski, R.W. Metamaterials and Antennas. In *Handbook of Antenna Technologies*; Chen, Z., Liu, D., Nakano, H., Qing, X., Zwick, T., Eds.; Springer: Singapore. [CrossRef]
17. Chen, Z.; Tang, M.-C.; Wang, Y.; Li, M.; Li, D. Mutual Coupling Reduction Using Planar Parasitic Resonators for Wideband, Dual-Polarized, High-Density Patch Arrays. In Proceedings of the 2019 IEEE MTT-S International Wireless Symposium (IWS), Guangzhou, China, 19–22 May 2019; pp. 1–3. [CrossRef]
18. Iqbal, A.; Saraereh, O.A.; Ahmad, A.W.; Bashir, S. Mutual Coupling Reduction Using F-Shaped Stubs in UWB-MIMO Antenna. *IEEE Access* **2017**, *6*, 2755–2759. [CrossRef]
19. Chandel, R.; Gautam, A.K.; Rambabu, K. Tapered Fed Compact UWB MIMO-Diversity Antenna with Dual Band-Notched Characteristics. *IEEE Trans. Antennas Propag.* **2018**, *66*, 1677–1684. [CrossRef]
20. Karaboikis; Soras; Tsachtsiris; Makios Compact dual-printed inverted-F antenna diversity systems for portable wireless devices. *IEEE Antennas Wirel. Propag. Lett.* **2004**, *3*, 9–14. [CrossRef]
21. Li, Z.; Yin, C.; Zhu, X. Compact UWB MIMO Vivaldi Antenna With Dual Band-Notched Characteristics. *IEEE Access* **2019**, *7*, 38696–38701. [CrossRef]
22. El-Din, M.S.H.S.; Shams, S.I.; Allam, A.M.M.A.; Gaafar, A.; Elhennawy, H.M.; Sree, M.F.A. SIGW Based MIMO Antenna for Satellite Down-Link Applications. *IEEE Access* **2022**, *10*, 35965–35976. [CrossRef]
23. Alibakhshikenari, M.; Virdee, B.S.; See, C.H.; Abd-Alhameed, R.; Falcone, F.; Limiti, E. A new waveguide slot array antenna with high isolation and high antenna bandwidth operation on Ku- and K-bands for radar and MIMO systems. In Proceedings of the 2018 15th European Radar Conference (EuRAD), Madrid, Spain, 26–28 September 2018; pp. 401–404.
24. Alibakhshikenari, M.; Virdee, B.S.; See, I.C.H.; Abd-Alhameed, R.; Falcone, F.; Andújar, A.; Anguera, J.; Limiti, E. Study on antenna mutual coupling suppression using integrated metasurface isolator for SAR and MIMO applications. In Proceedings of the 2018 48th European Microwave Conference (EuMC), Madrid, Spain, 23–27 September 2018; pp. 1425–1428.
25. Biswas, A.K.; Chakraborty, U. Textile Multiple Input Multiple Output Antenna for X-Band and Ku-Band Uplink-downlink Applications. In Proceedings of the 2020 National Conference on Emerging Trends on Sustainable Technology and Engineering Applications (NCETSTEA), Durgapur, India, 7–8 February 2020; pp. 1–4. [CrossRef]
26. Kumar, N.; Kumar, P.; Sharma, M. Superwideband Dual Notched Band Square Monopole MIMO Antenna for UWB/X/Ku Band Wireless Applications. In Proceedings of the 2019 IEEE Indian Conference on Antennas and Propogation (InCAP), Ahmedabad, India, 19–22 December 2019; pp. 1–5. [CrossRef]
27. Sharawi, M.S. Printed Multi-Band MIMO Antenna Systems and Their Performance Metrics [Wireless Corner]. *IEEE Antennas Propag. Mag.* **2013**, *55*, 218–232. [CrossRef]
28. Sharawi, M.S. *Printed MIMO Antenna Engineering*; Artech House: Norwood, MA, USA, 2014.
29. Manteghi, M.; Sami, Y.R. Broadband Characterization of the Total Active Reflection Coefficient of Multiport Antennas. In Proceedings of the IEEE Antennas and Propagation Society International Symposium (APS2003), Columbus, OH, USA, 22–27 June 2003; Volume 3, pp. 20–23.
30. Abdelraheem, A.; Elregaily, H.; Mitkees, A.A.; Abdalla, M. A Hybrid Isolation in Two-Element Directive UWB MIMO Antenna. *IETE J. Res.* **2020**, *69*, 499–508. [CrossRef]
31. Available online: https://rogerscorp.com/advanced-electronics-solutions/ro4000-series-laminates/ro4003c-laminates. (accessed on 20 December 2021).
32. Available online: https://rogerscorp.com/-/media/project/rogerscorp/documents/advanced-electronics-solutions/english/data-sheets/ro4000-laminates-ro4003c-and-ro4350b---data-sheet.pdf. (accessed on 20 December 2021).
33. Available online: https://scdn.rohde-schwarz.com/ur/pws/dl_downloads/dl_common_library/dl_brochures_and_datasheets/pdf_1/ZVA_dat-sw_en_5213-5680-22_v1400.pdf. (accessed on 20 December 2021).
34. Balanis, C. *Antenna Theory: Analysis and Design*; Wiley: New York, NY, USA, 1997.
35. Sarkar, D.; Mikki, S.; Srivastava, K.V.; Antar, Y. Far-field Envelope Correlation Coefficient and Near-field Reactive Energy of MIMO Antennas: An FDTD-IDM-CGF Approach. In Proceedings of the 2019 URSI Asia-Pacific Radio Science Conference (AP-RASC), New Delhi, India, 9–15 March 2019. [CrossRef]
36. Mei, X.; Wu, K.-L. Envelope Correlation Coefficient for Multiple MIMO Antennas of Mobile Terminals. In Proceedings of the 2020 IEEE International Symposium on Antennas and Propagation and North American Radio Science Meeting, Montreal, QC, Canada, 5–10 July 2020; pp. 1597–1598. [CrossRef]
37. Tariq, S.; Naqvi, S.I.; Hussain, N.; Amin, Y. A Metasurface-Based MIMO Antenna for 5G Millimeter-Wave Applications. *IEEE Access* **2021**, *9*, 51805–51817. [CrossRef]
38. Tian, R.; Lau, B.K.; Ying, Z. Multiplexing Efficiency of MIMO Antennas. *IEEE Antennas Wirel. Propag. Lett.* **2011**, *10*, 183–186. [CrossRef]
39. Chae, S.H.; Kawk, W.I.; Park, S.-O.; Lee, K. Analysis of mutual coupling in MIMO antenna array by TARC calculation. In Proceedings of the 2006 Asia-Pacific Microwave Conference, Yokohama, Japan, 12–15 December 2006; pp. 2090–2093. [CrossRef]
40. Wu, A.; Zhao, M.; Zhang, P.; Zhang, Z. A Compact Four-Port MIMO Antenna for UWB Applications. *Sensors* **2022**, *22*, 5788. [CrossRef]

41. Bilal, M.; Naqvi, S.I.; Hussain, N.; Amin, Y.; Kim, N. High-Isolation MIMO Antenna for 5G Millimeter-Wave Communication Systems. *Electronics* **2022**, *11*, 962. [CrossRef]
42. Fat, S.Y.A.; Hamad, E.K.I.; Swelam, W.; Allam, A.M.; Mohamed, H.A.E. Design of compact 4-port MIMO antenna based on Minkowski fractal shape DGS for 5G applications. *Prog. Electromagn. Res. C* **2021**, *113*, 123–136. [CrossRef]
43. Kumar, A. Compact 4 × 4 CPW-Fed MIMO Antenna with Wi-Fi and WLAN Notch for UWB Applications. *Radioelectron. Commun. Syst.* **2021**, *64*, 92–98. [CrossRef]
44. Hussain, M.; Ali, E.M.; Jarchavi, S.M.R.; Zaidi, A.; Najam, A.I.; Alotaibi, A.A.; Althobaiti, A.; Ghoneim, S.S.M. Design and Characterization of Compact Broadband Antenna and Its MIMO Configuration for 28 GHz 5G Applications. *Electronics* **2022**, *11*, 523. [CrossRef]
45. Li, M.-Y.; Ban, Y.-L.; Xu, Z.-Q.; Guo, J.; Yu, Z.-F. Tri-Polarized 12-Antenna MIMO Array for Future 5G Smartphone Applications. *IEEE Access* **2017**, *6*, 6160–6170. [CrossRef]
46. Fatah, S.Y.A.; Hamad, E.K.I.K.I.; Swelam, W.; Allam, A.M.M.A.; Sree, M.F.A.; Mohamed, H.A. Design and Implementation of UWB Slot-Loaded Printed Antenna for Microwave and Millimeter Wave Applications. *IEEE Access* **2021**, *9*, 29555–29564. [CrossRef]

electronics

MDPI

Article

Design of Broadband Low-RCS Array Antennas Based on Characteristic Mode Cancellation

Jialiang Han [1], Dan Jia [1,*], Biao Du [1], Guodong Han [1], Yongtao Jia [2] and Zekang Zhao [1]

1 The 54th Research Institute of China Electronics Technology Group Corporation, Shijiazhuang 050081, China
2 Science and Technology on Antenna and Microwave Laboratory, Xidian University, Xi'an 710071, China
* Correspondence: jiad87@163.com

Abstract: In this letter, a design method for low radar cross section (RCS) array antennas based on characteristic mode cancellation (CMC) is presented. Based on the characteristic mode theory (CMT), two novel microstrip elements are designed by introducing rectangular slots and cross slots, which produce 180° scattering phase difference by adjusting the size of slots. The dominant characteristic modes of the two elements achieve broadband dual-linear polarization CMC and similar radiation performances. The 4 × 4 array antenna consisting of these two antenna elements is designed. The operating band is from 4.55 GHz to 5.49 GHz (relative bandwidth 18.7%). The gain loss of the proposed array is about 0.1 dB compared to the reference array. The monostatic RCS is reduced for dual−linear polarized waves, and the 6 dB radar cross section reduction (RCSR) bandwidths are 62.3% and 35.7%, respectively. The prototype is fabricated and measured. The measured results of radiation pattern and RCS are in good agreement with the simulated results.

Keywords: antenna array; broadband radar cross section reduction; characteristic mode theory; characteristic mode cancellation

Citation: Han, J.; Jia, D.; Du, B.; Han, G.; Jia, Y.; Zhao, Z. Design of Broadband Low-RCS Array Antennas Based on Characteristic Mode Cancellation. *Electronics* **2023**, *12*, 1536. https://doi.org/10.3390/electronics12071536

Academic Editors: Naser Ojaroudi Parchin, Chan Hwang See and Raed A. Abd-Alhameed

Received: 26 February 2023
Revised: 18 March 2023
Accepted: 23 March 2023
Published: 24 March 2023

1. Introduction

The stealth technology of antennas has attracted much attention, and many efforts have been made to reduce the radar cross section (RCS) of array antennas in recent years. Traditional methods of antenna radar cross section reduction (RCSR) include shape modification [1], frequency selective surface (FSS) [2–4], radar absorbing material (RAM) [5–8], and scattering cancellation method [9–18].

Shape modification, such as loading slots on the antenna patch, can realize broadband and wide−angle RCSR. Shape modification usually requires abundant design experience, coupled with lots of optimization processes. Shape modification is mostly used in the design of element antennas and is difficult to directly apply to RCSR of array antennas. As a kind of spatial filter, FSS is transparent to in-band electromagnetic waves and has a cutoff effect on out-of-band electromagnetic waves. Therefore, a stealth radome based on FSS can reduce the out−of−band RCS of array antennas, but then it is difficult to reduce the in-band RCS. RAM can significantly reduce the RCS of array antennas over broadband, but it usually results in gain loss. It is also difficult to reduce the in-band RCS of antenna. The scattering cancellation method is based on the scattering phase difference of two objects, which can realize broadband RCSR with little gain loss. When the scattering phase difference is 180, scattering cancellation can be realized. Some researchers have focused on metasurface technology to realize scattering cancellation with antennas. In 2017, the quasi-fractal AMC structure was proposed, with a value of reflection phase difference between AMC and PEC of 180° ± 37° from 6.4 GHz to 21.7 GHz [15]. Compared to the reference antenna, the RCS of the proposed antenna was reduced from 6.0 GHz to 30.0 GHz. In 2021, three different AMC structures were proposed as the ground of a 1-D phased array [14]. Compared to the reference array antenna, the RCS was reduced from 12.0 GHz to 16.0 GHz for TE polarization and from 10.8 GHz to 15.2 GHz for TM polarization, respectively. Antenna and metasurface scattering cancellation is mainly used for RCSR of element antennas, and would increase the transverse size of the antenna element. When these structures are applied to array antennas, it will result in a scanning range that is limited due to the increase in element spacing. Some researchers propose to realize scattering cancellation by two kinds of element antennas. The element scattering cancellation of an array antenna can reduce its RCS without changing the array size, which has great application value. In 2019, a novel element with a U−shaped slot was designed [17]. The novel element had opposite scattering phases with a traditional microstrip element. The RCS of the proposed array was reduced from 4 GHz to 8 GHz, with a maximum reduction value of 23 dB. In 2022, two antenna elements were proposed by introducing L-feeding patches along the orthogonal direction to the metasurface [16]. The 6 dB RCSR bandwidth of the proposed array was from 5.1 GHz to 6.9 GHz. However, due to a lack of theoretical analysis methods, the design of the scattering cancellation elements relied more on engineering experience and optimization.

Characteristic mode theory (CMT) defines a series of orthogonal characteristic currents for electromagnetic objects of arbitrary shape. These characteristic currents are their inherent properties and can essentially explain radiation characteristics and scattering characteristics. Therefore, CMT is widely used in antenna design [19–21]. Recently, CMT has been used to distinguish between radiation modes and scattering modes, which are applied to realize radar cross section reduction (RCSR) of antennas [22–28]. In 2018, CMT was proposed to reduce the RCS of slot microstrip antennas, and RCS reduction of element antennas was realized from 2 GHz to 4 GHz [29]. The RCSR methods based on CMT have been proved to be useful for antenna element design. In 2019, metasurface design based on CMT was studied, and characteristic mode cancellation (CMC) method was proposed and applied to metasurface design. The metasurface was applied among the element antenna and the RCS reduction of the element antennas was realized from 6 GHz to 18 GHz [30]. In the above work, CMT was applied to the design of the microstrip element antenna or metasurface, and the RCS reduction of element antennas was realized. It is understood that the application of CMT to RCSR of array antennas has not been reported in public.

In this work, the scattering theory of array antennas based on CMT is studied. The RCSR method of array antennas based on CMC is proposed. For array antennas composed of two kinds of element, theoretical derivation shows that the RCS of the array antenna can be reduced by using CMC. The radiation and scattering characteristic modes of the antenna elements with rectangular slots and cross slots are proposed and analyzed. The scattering mode phase can be controlled by adjusting the size of the slots, and the broadband effective phase difference is realized. By using CMC, the array antenna is designed to realize RCSR over a wide frequency band in the case of small gain loss and broadband radiation performance. More clearly, a theory guide for scattering cancellation array antenna design is provided by CMT, which can be used for wide-band low-RCS array antenna design.

Key contributions of this paper are shown as follows:

1. A method of low-RCS array antenna design based on CMC is presented.
2. A dual-linear polarization low−RCS microstrip array antenna with a 6 dB RCSR bandwidth of 62.3% and 35.7% is obtained.

2. Characteristic Mode Theory and its Application to Array Antennas

Both the radiation and scattering performances of an antenna can be investigated by CMT. Therefore, CMT is not only widely applied in the radiation performance design of antenna, but is also suitable for antenna RCSR design. According to CMT, the scattering current of an object can be expressed as a superposition of a series of orthogonal and complete characteristic currents [31]:

$$\vec{J} = \sum_{n=1}^{N} \alpha_n \vec{J}_n = \alpha_1 \vec{J}_1 + \alpha_2 \vec{J}_2 + \ldots + \alpha_N \vec{J}_N \tag{1}$$

where α_n is the mode weighting coefficient (MWC) and $\vec{J}_n$ is the characteristic current of the nth mode. Furthermore, α_n denotes $\alpha_n = |\alpha_n|e^{j\varphi_n}$ with the corresponding magnitude of α_n and phase of φ_n.

The scattering field is determined by superposition of the characteristic fields. The scattering field of an object can be calculated as [32]:

$$\vec{E}_S = \sum_{n=1}^{N} \alpha_n \vec{E}_n = \alpha_1 \vec{E}_1 + \alpha_2 \vec{E}_2 + \ldots + \alpha_N \vec{E}_N \tag{2}$$

where $\vec{E}_S$ is the total scattering field of the object. The scattering field for the nth mode $\vec{E}_n$ denotes $\vec{E}_n = |E_n|e^{j\gamma_n}$ with the corresponding magnitude of $|E_n|$ and phase of γ_n. Therefore, $\vec{E}_S$ can be written as:

$$\vec{E}_S = \sum_{n=1}^{N} |\alpha_n|\cdot|E_n|e^{j(\gamma_n+\varphi_n)} = \sum_{n=1}^{N} A_n \cdot e^{j\Phi_n} \tag{3}$$

where A_n $(A_n = |\alpha_n|\cdot|E_n|)$ and Φ_n $(\Phi_n = \gamma_n + \varphi_n)$ are the amplitude and phase of the nth characteristic mode field.

The array antenna composed of two types of elements is shown in Figure 1. The scattering field of an array antenna can be approximately expressed as the product of the element scattering field and the scattering array factors when the mutual influence among the antenna elements is ignored. The scattering fields of an antenna element can be expressed as in Equations (2) and (3). The scattering fields of the two subarrays could then be respectively written as:

$$\vec{E}_S^{1} = \left(\sum_{n=1}^{N} \alpha_n \cdot \vec{E}_n^{1}\right) \cdot \sigma^1 \tag{4}$$

$$\vec{E}_S^{2} = \left(\sum_{m=1}^{M} \beta_m \cdot \vec{E}_m^{2}\right) \cdot \sigma^2 \tag{5}$$

where α_n and β_m are the MWC of the two elements, and σ^1 and σ^2 are the scattering array factors of the two subarrays. When these two kinds of element have the same quantity and same array form, the scattering array factors are the same ($\sigma^1 = \sigma^2 = \sigma_0$). Therefore, the total scattering field of the array antenna composed of these two subarrays with element 1 and element 2 could be given as:

$$\begin{aligned} \vec{E}_S = \vec{E}_S^{1} + \vec{E}_S^{2} &= \left(\sum_{n=1}^{N} \alpha_n \cdot \vec{E}_n^{1} + \sum_{m=1}^{M} \beta_m \cdot \vec{E}_m^{2}\right) \cdot \sigma_0 \\ &= \left(\sum_{n=1}^{N} A_n \cdot e^{j\Phi_n} + \sum_{m=1}^{M} B_m \cdot e^{j\psi_m}\right) \cdot \sigma_0 \end{aligned} \tag{6}$$

where A_n and Φ_n are the amplitude and phase of the characteristic mode field of element 1, and B_m and ψ_m are the amplitude and phase of the characteristic mode field of element 2. It can be concluded from Equation (6) that the scattering field of the array antenna can be reduced by decreasing the superposition of the scattering characteristic fields of these two elements. When the scattering amplitude of the two elements has equal value ($A_n = B_m = A_0$), Equation (6) can be written as:

$$\vec{E}_S = A_0 \cdot \left(\sum_{n=1}^{N} e^{j\Phi_n} + \sum_{m=1}^{M} e^{j\psi_m}\right) \cdot \sigma_0 \tag{7}$$

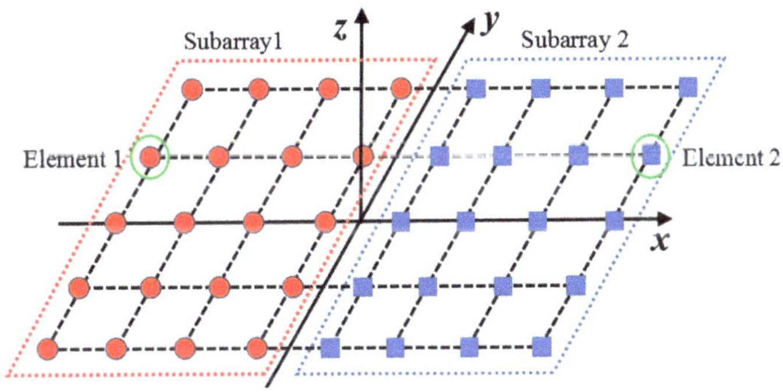

Figure 1. Illustration of the array antenna composed of two kinds of elements.

For example, when the two element antennas have two dominant modes, the amplitude of the scattering field of the array can be calculated as:

$$\begin{aligned}\vec{E}_S &= A_0 \cdot \left[\left(e^{j\phi_1} + e^{j\phi_2}\right) + \left(e^{j\psi_1} + e^{j\psi_2}\right)\right] \cdot \sigma_0 \\ &= A_0 \cdot \left[\left(e^{j\phi_1} + e^{j\psi_1}\right) + \left(e^{j\phi_2} + e^{j\psi_2}\right)\right] \cdot \sigma_0\end{aligned} \tag{8}$$

In this paper, the phase of the characteristic mode is regulated by the modification technology. When $\phi_1 - \psi_1 = 180^\circ$ or $\phi_2 - \psi_2 = 180^\circ$, the CMC can be realized and the RCS is reduced. When the scattering phase difference of the characteristic mode field of two elements is 180°, $\vec{E}_S = 0$ will be obtained to achieve minimal RCS. Therefore, the RCS of an array antenna can be reduced by introducing CMC technology.

3. Design of the Antenna Element Based on CMT

In this paper, two novel antenna elements are proposed based on CMT. The phase of the characteristic mode of an element is regulated by the shape modification technology to realize CMC. The ground plane would be connected when the patch elements are arranged in an array. Therefore, in order to maintain the mode phase characteristics of elements in the array, the modification is carried out on the radiation patch. Firstly, a novel antenna element with rectangular slots (called antenna A) is designed to achieve CMC, and the reference antenna is a traditional square patch antenna element. The radiation characteristic currents of the reference antenna and antenna A are similar, and the scattering characteristic currents are opposite. Next, in order to expand the CMC bandwidth, another novel antenna element (called antenna B) etched with cross slots is designed. Antenna A and antenna B can realize broadband dual-polarization CMC, which lays a foundation for the wide-band RCS reduction for the array antenna.

3.1. Characteristic Mode Analysis of the Reference Antenna and Antenna A

The geometries of reference antenna and antenna A are shown in Figure 2 and their main dimension parameters are shown in Table 1. The reference antenna and antenna A have similar structures and the dielectric material is polytetrafluoron ($\varepsilon_r = 2.2$, $\tan\delta_r = 0.009$). The lower layer is a ground plane structure with rectangular slots which is used to adjust the antenna resonant frequency. The middle layer adopts an L-shaped coupling feed structure to widen the operating bandwidth. For the reference antenna, the upper layer is a square patch, as shown in Figure 2a. For antenna A, the upper patch is a square patch with two rectangular slots through which the characteristic mode can be regulated, as shown in Figure 2b. For a y-polarized wave, the equivalent electrical length of the scattering current can be increased, which impacts the phase of the characteristic mode.

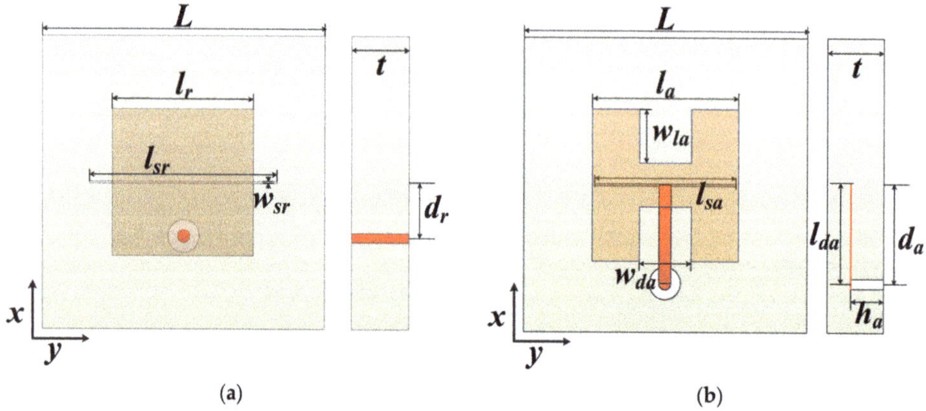

Figure 2. Geometry of the reference antenna and antenna A. (**a**) Reference antenna; (**b**) antenna A.

Table 1. Parameters of the design elements (Unit: mm).

Symbol	Value	Symbol	Value	Symbol	Value	Symbol	Value
L	30	w_{sb}	0.2	l_{sa}	18	d_a	4
t	6	w_{db}	1	w_{da}	5.5	l_r	14.5
l_b	19	l_{db}	9	l_a	15.5	h_a	3.5
l_{sb}	20	d_b	8	l_{da}	9	l_{sr}	21

The simulation results of the reflection coefficient (S_{11}) and far-field patterns of the two antenna elements are shown in Figure 3. The two elements have the same resonant frequency and similar far−field patterns in both the *xoz* plane and *yoz* plane at 5.1 GHz. The operating frequency with the reflection coefficient less than −10 dB is from 4.64 GHz to 5.52 GHz.

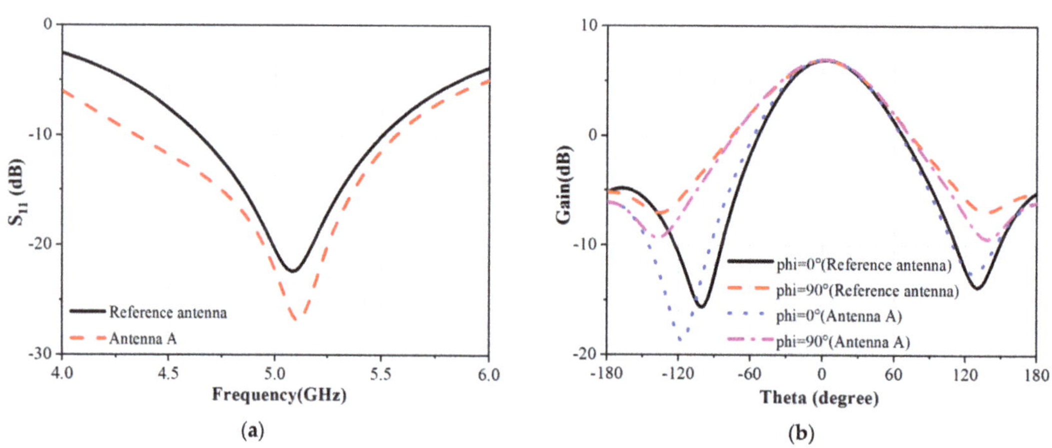

Figure 3. Radiation performance simulation results of the reference antenna and antenna A. (**a**) S_{11}; (**b**) far−field patterns.

To maintain stability while effectively reducing RCS, the radiation and scattering characteristic modes analyses need to be undertaken.

The radiation characteristic modes analysis is firstly carried out. The MWC (α_n) and characteristic electric field ($\vec{E}_n$) are analyzed. The value of A_n can be calculated, as shown in Figure 4. Mode 1 is the dominant mode for the two antenna elements. The characteristic current distribution of the two dominant modes is similar, as shown in Figure 5. Therefore, the radiation properties of these two antennas are similar.

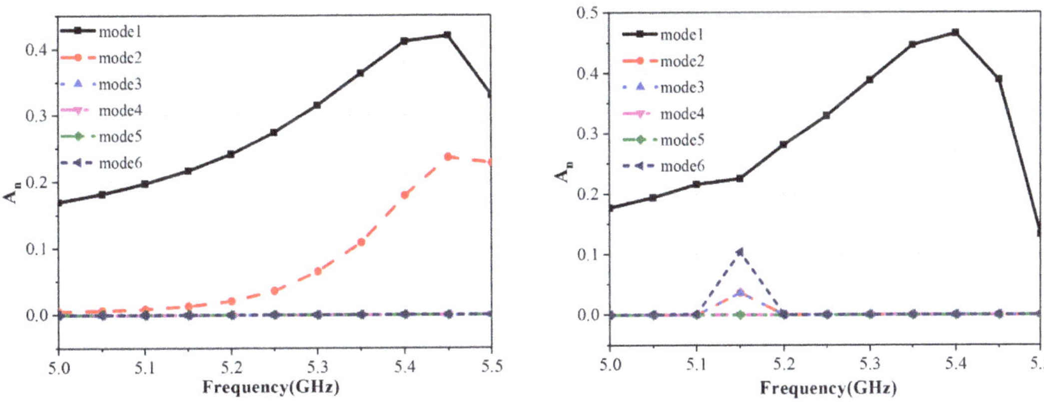

(a) (b)

Figure 4. A_n of the radiation characteristic modes of the reference antenna and antenna A. (**a**) Reference antenna; (**b**) antenna A.

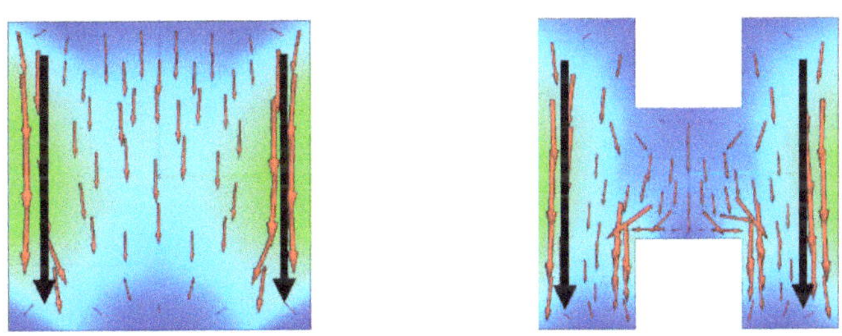

(a) (b)

Figure 5. Radiation characteristic current distribution on the radiation patch at 5.1 GHz. (**a**) Reference antenna; (**b**) antenna A.

The scattering characteristic mode analyses of the reference antenna and antenna A are performed from 4.0 GHz to 8.0 GHz. For the $y-$polarized incident wave, the MWC (α_n) and characteristic electric field ($\vec{E}_n$) of the two elements are simulated. The amplitudes (A_n) of the first six modes are calculated in Figure 6 and the phases (ϕ_n) of the two dominant modes are given Figure 7. It can be seen from Figure 6 that modes 1 and 2 are the dominant modes for the reference antenna and antenna A. The rectangular slots on antenna A lengthen the current path of mode 1 and mode 2, so A_n of the modes moved to lower frequency. Figure 7a shows ϕ_n of the reference antenna and antenna A. The value of the phase difference is shown in Figure 7b, in which $\Delta\phi_1$ is the phase difference between mode 1 of the reference antenna and antenna A and $\Delta\phi_2$ is the phase difference between mode 2 of the reference antenna and antenna A. For out-band 4.4 GHz and in-band 5.0 GHz, the values of A_n of the cancelled modes (mode 1 and mode 2) are similar. The characteristic current and far−field of the two cancelled modes are shown in Figures 8 and 9. The characteristic current of the reference antenna is opposite to that of antenna A, and the characteristic far−field patterns of the two modes are similar. Therefore, mode 1 of the reference antenna and antenna A form cancellation at 4.4 GHz. Mode 2 of the reference antenna and antenna A form cancellation at 5.0 GHz.

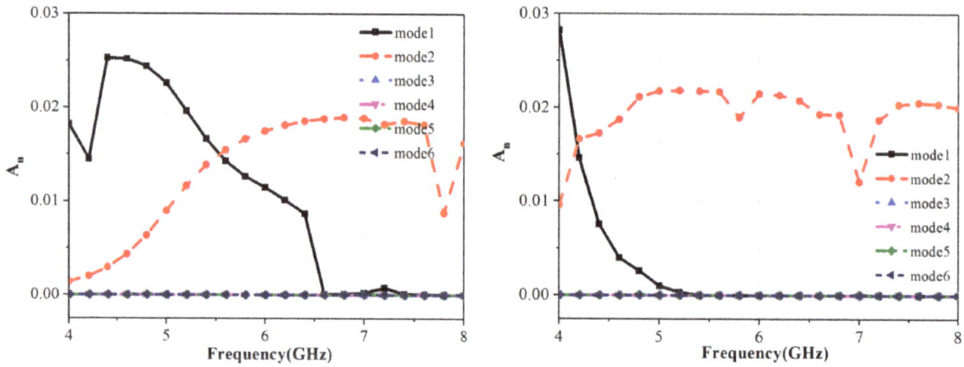

(a) (b)

Figure 6. A_n of scattering characteristic modes of the reference antenna and antenna A. (**a**) Reference antenna; (**b**) antenna A.

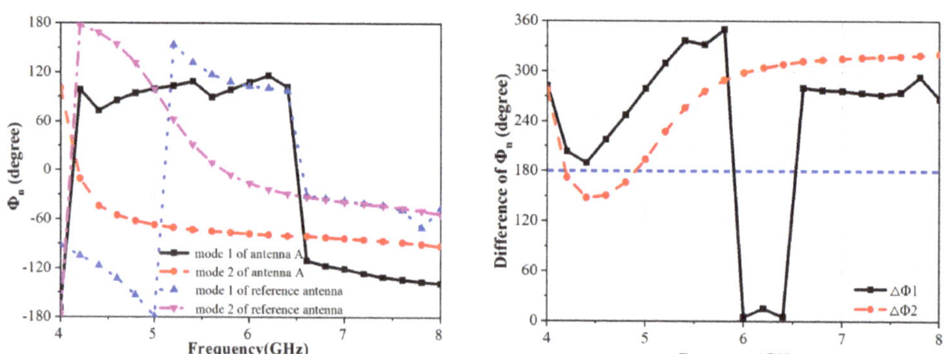

(a) (b)

Figure 7. The ϕ_n and the difference of ϕ_n of the dominant scattering characteristic modes of the reference antenna and antenna A. (**a**) The ϕ_n of the reference antenna and antenna A; (**b**) difference of ϕ_n of the dominant scattering characteristic modes of the reference antenna and antenna A.

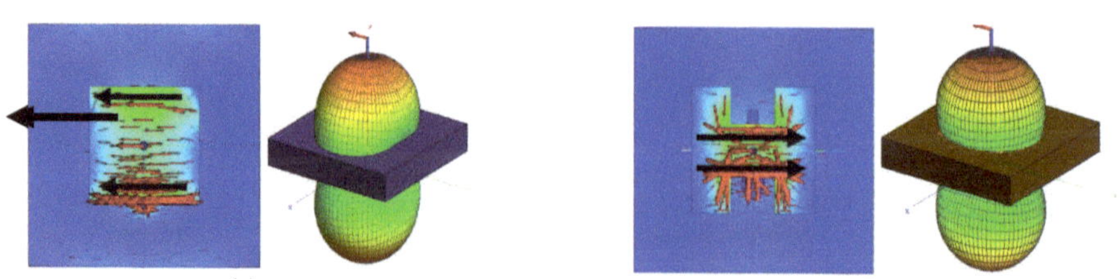

(a) (b)

Figure 8. Scattering characteristic current distribution and far−field patterns comparison at 4.4 GHz. (**a**) Mode 1 of the reference antenna; (**b**) mode 1 of antenna A.

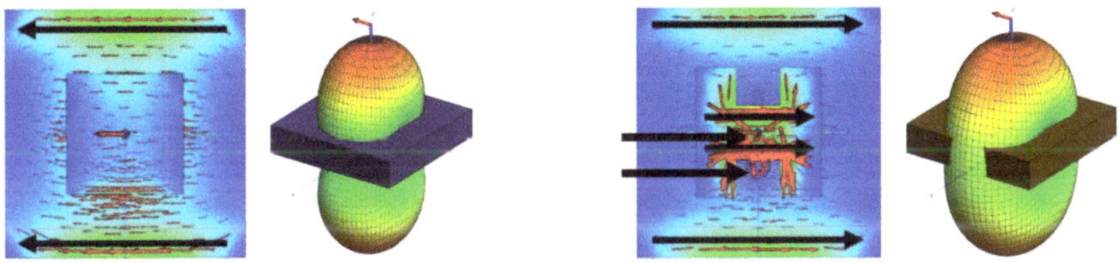

(**a**) (**b**)

Figure 9. Scattering characteristic currents distribution and far−field patterns comparison for at 5.0 GHz. (**a**) Mode 2 of the reference antenna; (**b**) mode 2 of antenna A.

3.2. Characteristic Mode Analysis of Antenna B

Though the CMC between the reference antenna and antenna A is realized at 4.4 GHz and 5.0 GHz, the mode phases no longer satisfy CMC conditions at higher frequencies. To achieve wideband RCSR, the characteristic modes of the reference antenna should be changed. Therefore, a pair of cross slots is etched on the radiation patch (called antenna B) to control the scattering phase, as shown in Figure 10. The antenna structure is similar to antenna A and the main dimension parameters are shown in Table 1. The cross slots can expand the CMC bandwidth and realize dual−polarization scattering cancellation.

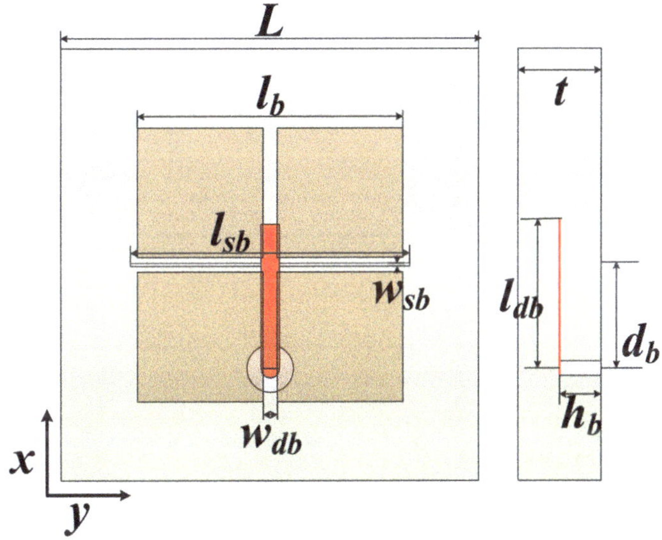

Figure 10. Geometry of antenna B.

The radiation characteristic modes of antenna B are first analyzed. Figure 11 shows A_n and the characteristic current distribution. It can be seen from Figure 11a that mode 1 is the dominant mode of antenna B. The characteristic current distribution of mode 1 of antenna B is similar to antenna A, as shown in Figure 11b.

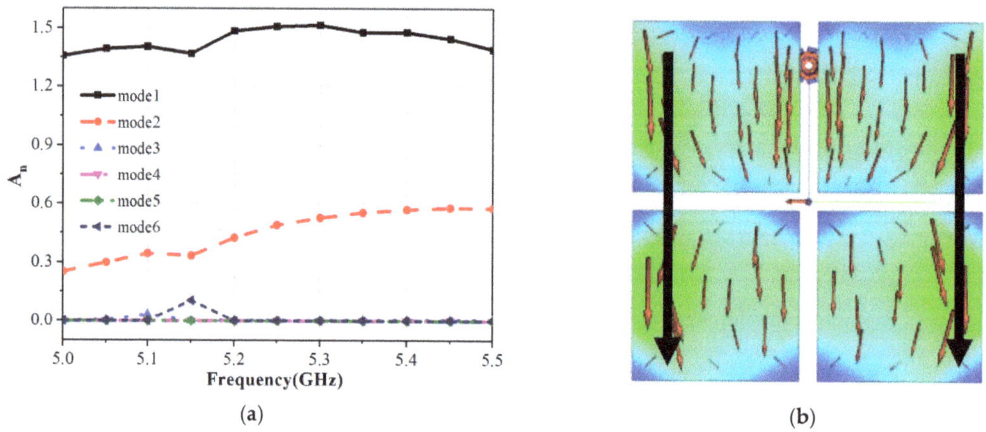

Figure 11. Radiation characteristic modes analysis results of antenna B. (**a**) A_n; (**b**) characteristic current distribution at 5.1 GHz.

The simulation results of the reflection coefficient (S11) and far-field patterns of the two antenna elements are shown in Figure 12. It can be seen that antenna A and antenna B have the same resonant frequency and similar far-field patterns in both the *xoz* plane and *yoz* plane at 5.1 GHz. The operating frequency with a reflection coefficient less than −10 dB is from 4.55 GHz to 5.49 GHz.

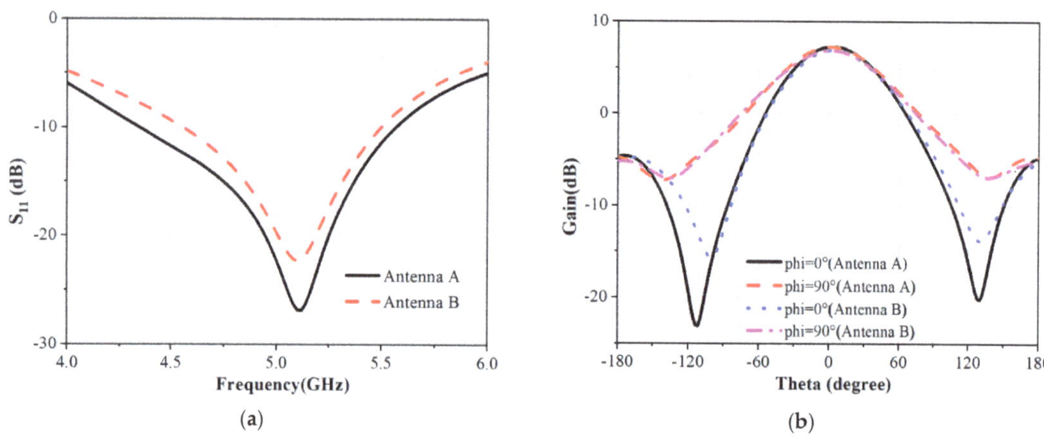

Figure 12. Radiation performances of antenna A and antenna B. (**a**) S_{11}; (**b**) far−field patterns.

The scattering characteristic mode analysis of antenna B is performed from 4.0 GHz to 8.0 GHz. For the y−polarized incident wave, the MWC (α_n) and characteristic electric field ($\vec{E}_n$) of antenna B are analyzed. The key parameters (A_n and ϕ_n) of the characteristic modes are calculated as shown in Figure 13. It can be seen from Figure 13a that modes 1 and 2 are the dominant modes. The phase (ϕ_n) of the dominant scattering modes is calculated, as shown in Figure 13b. The phase difference between antenna A and antenna B is shown in Figure 13c. There, $\Delta\phi_3$ is the phase difference between mode 1 of antenna A and antenna B and $\Delta\phi_4$ is the difference between mode 2 of antenna A and antenna B, and $\Delta\phi_4$ is about 180° from 5.6 GHz to 7.6 GHz. The values of A_n of the cancelled modes are similar. Therefore, broadband scattering cancellation is realized. The characteristic currents and far-field patterns of antenna A and antenna B at 7.0 GHz are shown in Figure 14. The current direction of mode 2 of antenna A is opposite to that of mode 2 of antenna B, and the far-field patterns of the modes are similar. Therefore, a cancellation effect of antenna A and antenna B is obvious, and the RCSR is expected to achieve between 4.0 GHz and 8.0 GHz for the *y*-polarized wave.

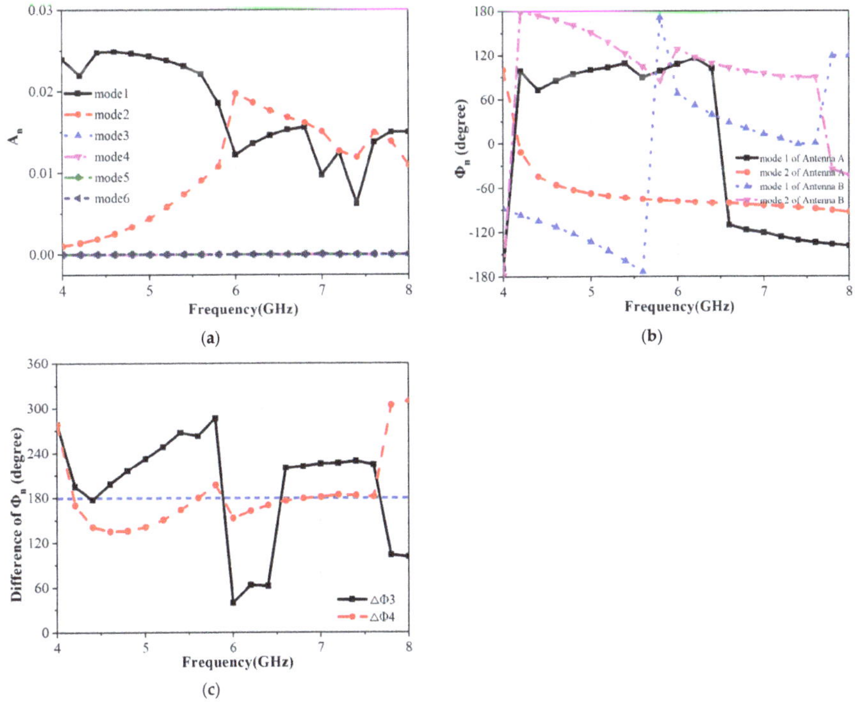

Figure 13. Scattering characteristic mode analysis results of antenna B. (**a**) A_n of antenna B; (**b**) ϕ_n of antenna A and antenna B; (**c**) phase difference of antenna A and antenna B.

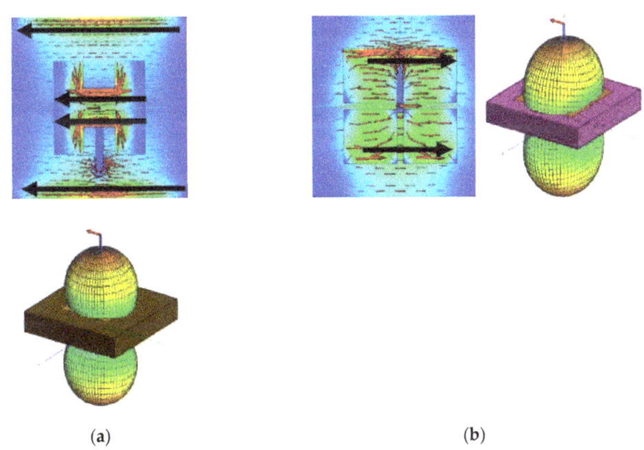

Figure 14. Scattering characteristic currents distribution and far−field patterns comparison for at 7.0 GHz. (**a**) Mode 2 of antenna A; (**b**) mode 2 of antenna B.

For the *x*-polarized incident wave, the scattering characteristic mode analysis of antenna A and antenna B is performed from 4.0 GHz to 8.0 GHz. Figures 15 and 16 show the calculations of A_n and ϕ_n of the scattering characteristic modes. It can be seen from

Figure 15 that modes 1 and 2 are the dominant modes for antenna A and antenna B. The mode amplitude (A_n) is small in the antenna operating band, which originates from co-polarization absorption due to antenna impedance matching. The scattering phase (ϕ_n) is shown in Figure 16a and the phase difference between antenna A and antenna B is shown in Figure 16b, where $\Delta\phi_5$ is the phase difference between mode 1 of antenna A and antenna B, and $\Delta\phi_6$ is the phase difference between mode 2 of antenna A and antenna B. There, $\Delta\phi_6$ is 175° at 6.0 GHz and the corresponding values of A_n are about 0.015. Therefore, a cancellation effect of antenna A and antenna B is produced outside of operating band. It can be concluded that the RCSR of an array antenna composed of antenna A and antenna B can be obtained from 4.2 GHz to 6.6 GHz for the x-polarized incident wave.

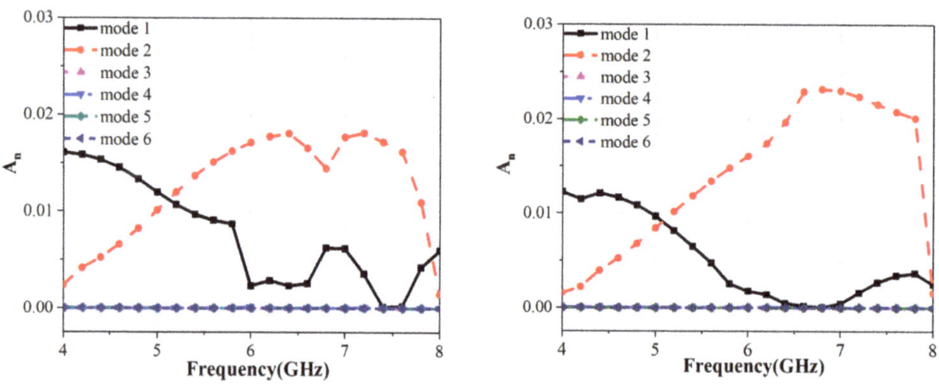

(a) (b)

Figure 15. A_n of scattering characteristic modes of the reference antenna and antenna A. (**a**) Antenna A; (**b**) antenna B.

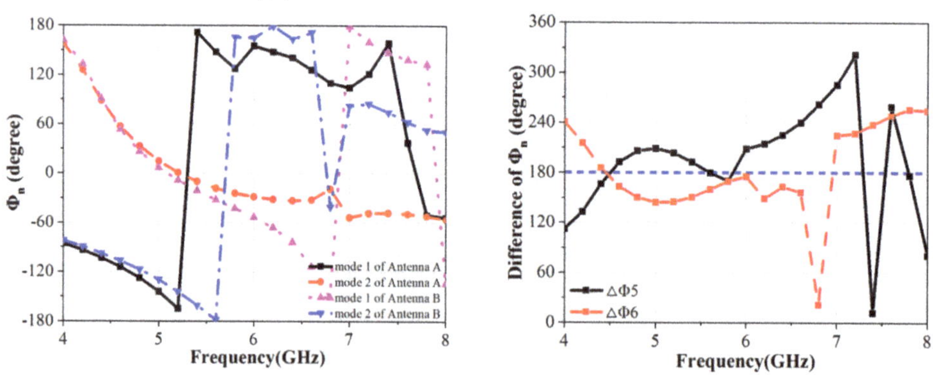

(a) (b)

Figure 16. ϕ_n and difference of ϕ_n of the dominant scattering characteristic modes of the reference antenna and antenna A. (**a**) ϕ_n of antenna A and antenna B; (**b**) difference of ϕ_n of the dominant scattering characteristic modes of the reference antenna and antenna A.

4. Design of the Array Antennas

The 4 × 4 array antenna consisting of antenna A and antenna B is designed and the number of antenna As and antenna Bs is equal. Considering the influence of scattering coupling on RCS of array antennas, the RCS of different array forms is studied. The array antennas are in forms of ABAB, ABBA, AABB, and checkerboard respectively, as shown in Figure 17. The size of the four arrays is 120 mm × 120 mm. The RCS simulation results of the four arrays are shown in Figure 18. As seen from the Figure 18, the RCS reduction of different array forms is all realized in the 4 GHz–8 GHz band, but the RCS of the ABAB array forms is the lowest. Therefore, the form ABAB is used to conform to CMC array antennas. The element structures are the same in the x−direction and are arranged in the form of ABAB in the y direction.

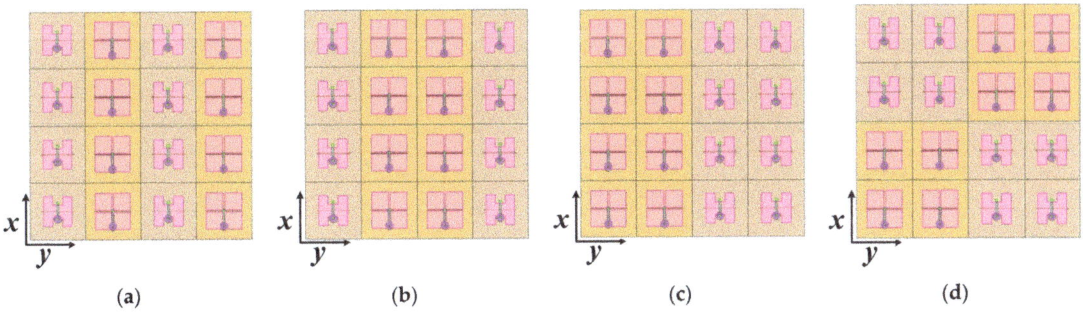

Figure 17. Four forms of proposed arrays. (**a**) ABAB; (**b**) ABBA; (**c**) AABB; (**d**) checkerboard.

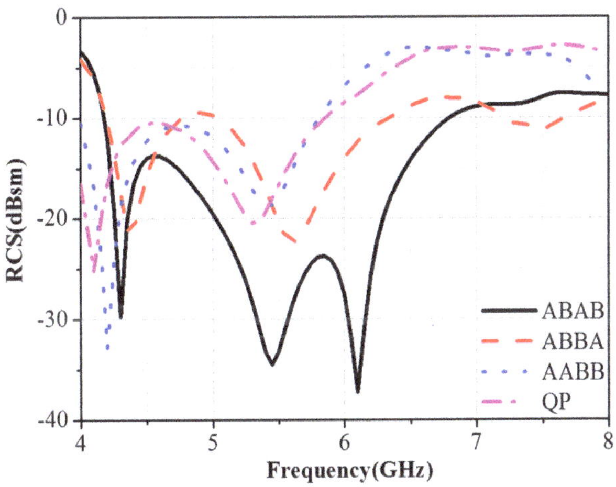

Figure 18. RCS simulation results of different array forms.

A conventional 4 × 4 patch array antennas consisting of the reference antenna is shown in Figure 19. The size of the reference array is 120 mm × 120 mm. The radiation far−field pattern simulation results of the array antennas are shown in Figure 20. The far−field patterns of the reference array and proposed array are basically the same and the gain difference is about 0.1 dB.

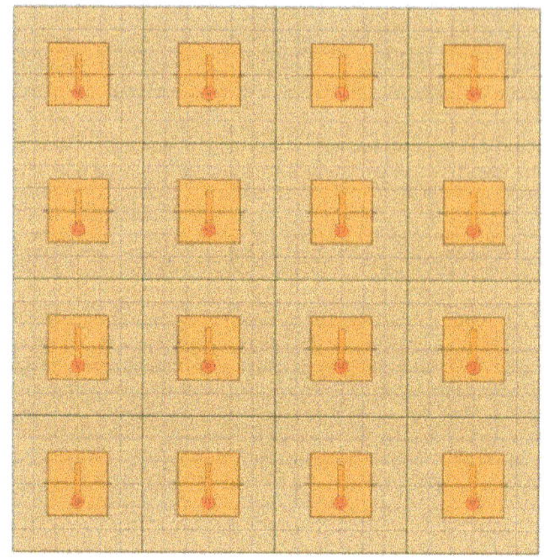

Figure 19. Geometry of reference array.

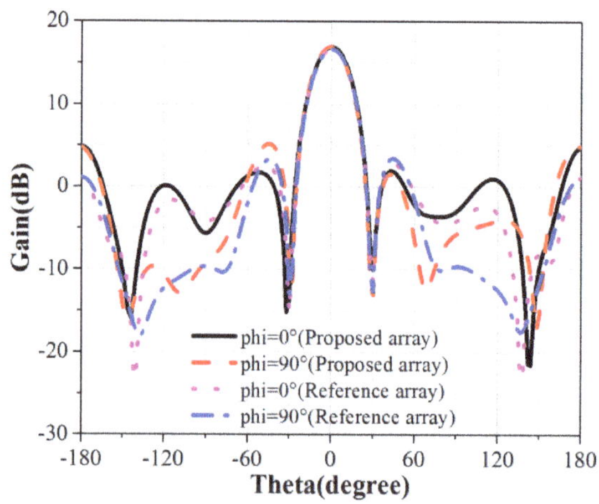

Figure 20. Radiation far−field patterns of the reference and proposed array antenna.

The simulation results comparison of the monostatic RCS between the two arrays is shown in Figure 21. For the x-polarized incident wave, the RCS of the proposed array is reduced from 4.0 GHz to 8.0 GHz compared with the reference array. The maximum value of RCSR is 32.3 dB at 6 GHz and the average value of in−band RCSR is 9.5 dB. The 6 dB RCSR bandwidth of the proposed array antenna is from 4.6 GHz to 6.6 GHz. Combined with the scattering characteristic mode analysis for the x-polarized incident wave, it can be concluded that the in-band RCSR of the x−polarized incident wave comes from the absorption and the out-of-band RCSR comes from CMC of the two antenna elements. For the y-polarized incident wave, the RCS of the proposed array is reduced from 4.0 GHz to 8.0 GHz compared with the reference array. The maximum reduction is 24 dB at 5.6 GHz, and the average RCSR value is 13.1 dB. The average value of in−band RCSR is 16 dB. The 6 dB RCSR bandwidth of the proposed antenna array is from 4.2 GHz to 8.0 GHz. Combining the scattering characteristic mode analysis for the x-polarized incident wave, it can be concluded that the RCSR of the y-polarized incident wave comes from CMC of the two antenna elements.

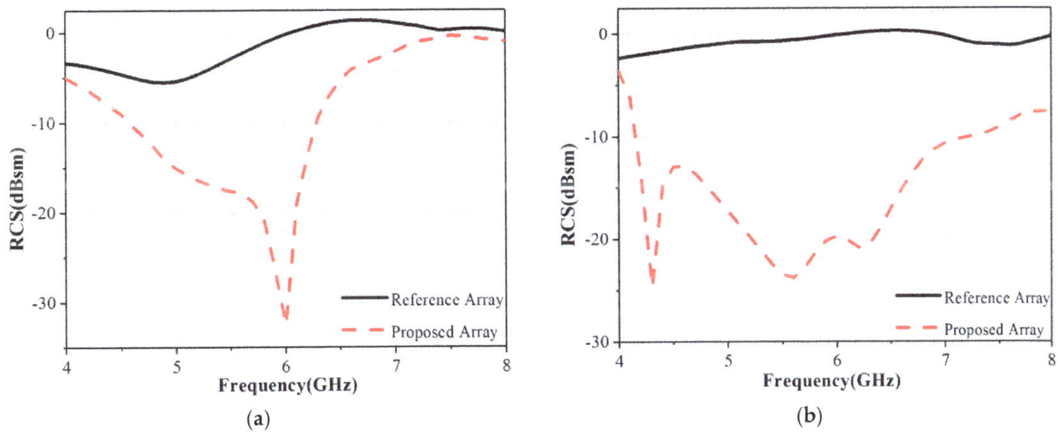

Figure 21. Monostatic RCS simulated results of the reference and proposed array antennas. (**a**) $x-$polarized; (**b**) $y-$polarized.

The comparison of the radiation and RCS characteristics among typical low-RCS array antennas and the proposed array antenna is shown in Table 2. The methods of FSS, absorbing material loading technology, and cancellation technology are listed. For antenna size, the transverse size and profile height of the antenna array proposed in this work are nearly the same as the size of the array antenna in Refs. [16,17], and the profile height is reduced compared with array antenna in Refs. [4,8]. For radiation performance, the operating bandwidth of the array antenna in this work achieves significantly wider operating bandwidth and it is only slightly lower than the array design based on FSS in Ref. [8]. The aperture efficiency of the proposed antenna is higher. For scattering performance, the cross−polarization RCSR bandwidth is nearly the same as other array antennas. The mean value of RCSR in operating bandwidth is improved for both co-polarization and cross−polarization. Compared with the low−RCS array antennas designed by antenna element cancellation (A.E.C.), the operating bandwidth is wider and the RCSR mean value in the operating bandwidth is bigger.

Table 2. Comparison with typical low-RCS arrays.

Ref.	Electric at Size (λ^3)	OB (GHz)	Gain (dB)/Aperture Efficiency	Co/Cross−Polarization RCSR Mean Value in OB (dB)	Co/Cross−Polarization RCSR B.W. (GHz)		Method
Ref. [4]	2.2 × 2.2 × 0.8	-	15.8/62.5%	11.8/19.8	-	6–18 (100%)	FSS
Ref. [8]	4.7 × 4.7 × 0.7	26.7–34.2 (24.6%)	20/36.7%	12/12	-	28–48 (52%)	Absorptive
Ref. [16]	1.8 × 1.8 × 0.05	5.05–5.42 (7.1%)	-	-/16.5	-	4–8 (66.7%)	A.E.C.
Ref. [17]	2 × 2 × 0.1	4.75–5.25 (10%)	-	13/13	4–8 (66.7%)	4–8 (66.7%)	A.E.C.
This work	2 × 2 × 0.1	4.55–5.49 (18.7%)	16.6/89%	14.7/22.8	4–8 (66.7%)	4–8 (66.7%)	CMC

OB: operating bandwidth. B.W.: bandwidth. A.E.C.: antenna element cancellation.

5. Experimental Results and Discussion

The prototype consists of an antenna radiation array and power dividers. The beam synthesis for the 16 units of the array is realized by power dividers. The antenna radiation array is composed of two layers of PCB structure. These two layers are fabricated by printed board (PCB) technology. The upper PCB is printed with the radiation patch of the antenna. Both sides of the lower PCB are printed with a different structure. The upper surface is printed with the antenna feeder, and the lower surface is printed with rectangular slots. Finally, the two layers of PCB structure are spliced together. The prototype of the proposed array antenna is shown in Figure 22a.

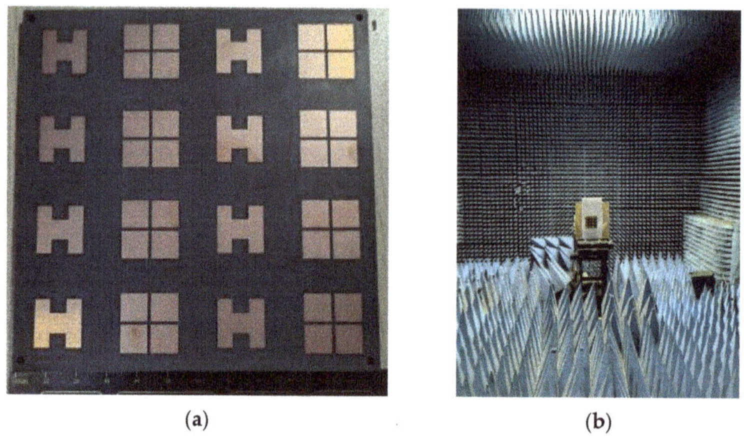

Figure 22. Fabricated prototype of the proposed array antenna. (**a**) Structural diagram; (**b**) radiation far-field patterns test scene.

The S_{11} of the antenna prototype is measured by a vector network analyzer. Antenna elements are fed by coaxial probes. The measured and simulated S_{11} of antenna A and antenna B are shown in Figure 23a. The measured reflection coefficient less than −10 dB is from 4.54 GHz to 5.46 GHz, which is almost the same as the simulated results.

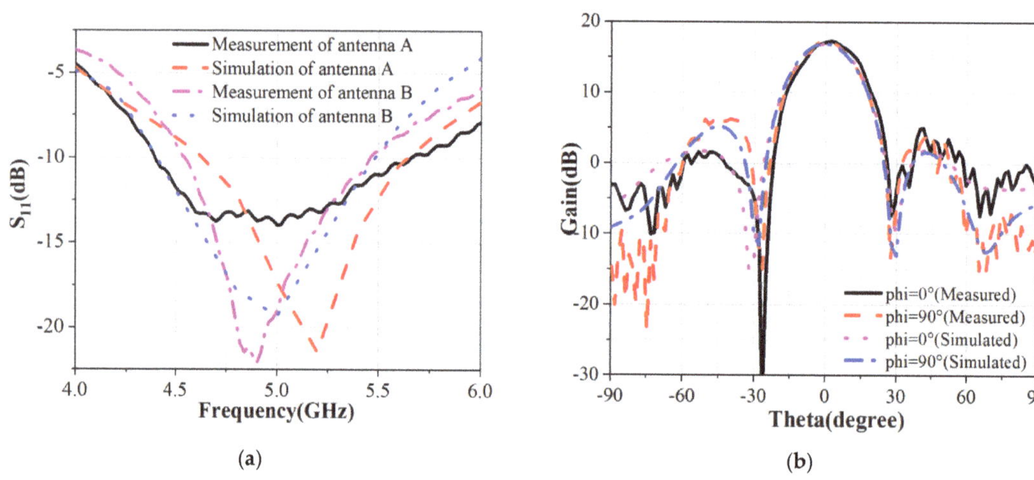

Figure 23. Measured and simulated radiation performances of the proposed patch array antenna. (**a**) S_{11}; (**b**) far−field patterns.

The realized gain and radiation patterns measurement of the antenna prototype are carried out in the microwave anechoic chamber, as shown in Figure 22b. The far−field condition is satisfied between the antenna prototype and the standard horn antenna. The array antenna is fed by a power divider when measuring its far-field patterns. The far−field patterns of the fabricated array antenna are tested, as shown in Figure 23b. Because some energy is lost in the SMA connector and coaxial line, the measured gain is 0.5 dB lower than the simulated result.

The scattering performance of the array antenna prototype is tested in the compact range anechoic chamber as shown in Figure 24. Array antenna elements are all matched

50 Ω loads, and the prototype is placed on a low RCS foam support. The plane wave of the test field is vertically incident on the antenna radiation surface, and the frequency sweep test is carried out. The measured and simulated monostatic RCS results of the proposed antenna for vertical incident plane waves with $x-$polarization and $y-$polarization are shown in Figure 25. The measured results are in good agreement with the simulated results. The difference between the measured results and simulated results is mainly due to assembly error and measurement error.

(**a**) (**b**)

Figure 24. The RCS test scenario of the array antenna prototype. (**a**) Array antenna prototype; (**b**) the RCS test environment.

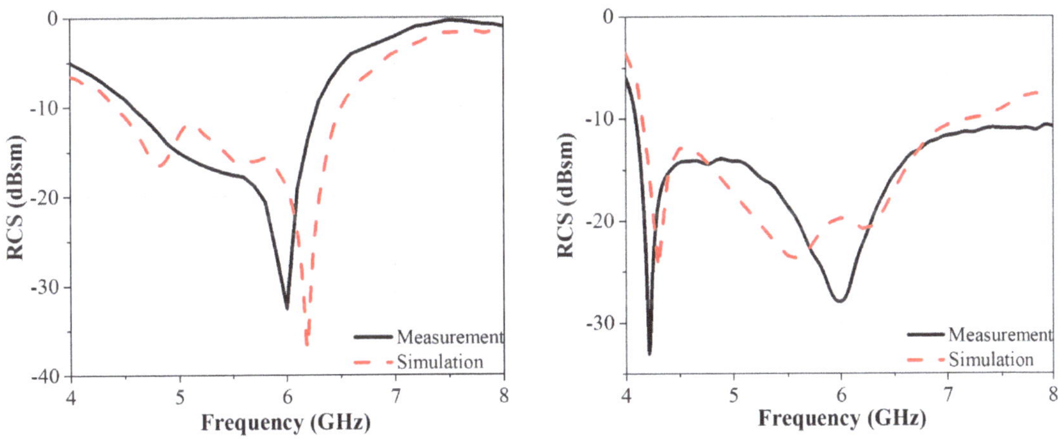

(**a**) (**b**)

Figure 25. Measured and simulated monostatic RCS of the proposed array antenna for vertical incident plane waves. (**a**) $x-$polarized; (**b**) $y-$polarized.

In order to illustrate the measured results, the impact of the manufacturing error is analyzed. The dielectric constant of the PCB printed board is more accurate. The pattern-etching machining technology is mature and the etching precision is higher. Therefore, the machining error introduced by the PCB can be ignored. The gap and the displacement between the double PCB layers play a major role in manufacturing error of the prototype. For the $y-$polarized incident wave, we calculated RCS in conditions of the space gap widths of 0 mm, 0.3 mm, and 0.5 mm respectively, as shown in Figure 26a. We also calculated RCS in conditions of the displacement sizes of 1 mm, 2 mm, and 3 mm respectively, as shown in Figure 26b. It can be found that the RCS simulation results with the error show the same trend of change as the measured results. Therefore, the gap and the displacement tolerances should be strictly controlled in the manufacturing process.

28. Liu, Y.; Du, B.; Jia, D.; Jia, Y. Ultra-Wideband Radar Cross-section Reduction for Ring-shaped Microstrip Antenna based on Characteristic Mode Analysis. *Microw. Opt. Technol. Lett.* **2021**, *63*, 1538–1546. [CrossRef]
29. Zhao, Z.; Chen, Y.; Yang, S. In-Band Radar Cross-Section Reduction of Slot Antenna Using Characteristic Modes. *IEEE Antennas Wirel. Propag. Lett.* **2018**, *17*, 1166–1170. [CrossRef]
30. Shi, Y.; Meng, Z.; Wei, W.; Zheng, W.; Li, L. Characteristic mode cancellation method and its application for antenna RCS reduction. *IEEE Antennas Wirel. Propag. Lett.* **2019**, *18*, 1784–1788. [CrossRef]
31. Harrington, R.; Mautz, J. Theory of characteristic modes for conducting bodies. *IEEE Trans. Antennas Propag.* **1971**, *19*, 62271. [CrossRef]
32. Bhavani, K.D.; Madhav, B.T.P.; Das, S.; Hussain, N.; Ali, S.S.; Babu, K.V. Development of Metamaterial Inspired Non-Uniform Circular Array Superstate Antenna Using Characteristic Mode Analysis. *Electronics* **2022**, *11*, 2517. [CrossRef]

electronics

MDPI

Communication

Experimental Demonstration of Beam Scanning of Dual-Metasurface Antenna

Lucia Teodorani [1,*], Francesco Vernì [1,†], Giorgio Giordanengo [2], Rossella Gaffoglio [2] and Giuseppe Vecchi [1]

1 Antenna and EMC Lab (LACE), Department of Electronics and Telecommunication (DET), Politecnico di Torino, 10129 Torino, Italy
2 Advanced Computing, Photonics and Electromagnetics Area, Fondazione LINKS, 10138 Torino, Italy
* Correspondence: lucia.teodorani@polito.it
† Current address: Huawei, Milan Research Center, 20054 Segrate, Italy.

Abstract: Beam-scanning antennas are employed in a wide range of applications, such as in satellite communications and 5G networks. Current commercial solutions rely mostly on electronically reconfigurable phased arrays, which require complex feeding networks and are affected by high losses, high costs, and are often power-hungry. In this paper, a novel beam scanning architecture employing a pair of planar metasurfaces, for use in thin reconfigurable antennas, is presented and experimentally demonstrated. The structure consisted of a radiative passive (non-reconfigurable) modulated metasurface, and a second metasurface that controls beam pointing, operating as a variable-impedance ground plane. Unlike other existing approaches, surface impedance variation was obtained by on-plane varactor diodes, no vias and a single voltage bias. This paper presents a design procedure based on an approximate theoretical model and simulation verification; a prototype of the designed antenna was fabricated for operation in X band, and a good agreement between measured results and simulations was observed. In the presented simple embodiment of the concept, the angular scanning range was limited to 10°; this limitation is discussed in view of future applications.

Keywords: beam steering; leaky-wave antenna; reconfigurable metasurface; varactors

Citation: Teodorani, L.; Vernì, F.; Giordanengo, G.; Gaffoglio, R.; Vecchi, G. Experimental Demonstration of Beam Scanning of Dual-Metasurface Antenna. *Electronics* **2023**, *12*, 1833. https://doi.org/10.3390/electronics12081833

Academic Editors: Naser Ojaroudi Parchin, Raed A. Abd-Alhameed and Chan Hwang See

Received: 13 March 2023
Revised: 9 April 2023
Accepted: 11 April 2023
Published: 12 April 2023

1. Introduction

In recent years, reconfigurable antennas have been the subject of constantly growing interest, since beam steering is required in a wide range of present, emerging, and future applications, as in satellite communications, radars, 5G, and beyond-5G networks [1–3].

Commercial solutions, current and under development, rely mostly on electronically reconfigurable phased arrays, but the inherent complex feeding networks and high losses make these antennas less than optimal [4]; this has prompted research into alternative architectures. Among these, solutions are favored in which the radiating part and the power distribution structure coexist; the interplay between wave guiding (spatial power distribution) and radiation is well-captured by the leaky-wave paradigm [5–7], which allows for approximate designs. Beam steering at a fixed frequency has been achieved using materials with tunable electric properties [8–11], performed electromechanically [12] or by employing a multitude of (lumped) active components such as varactors [4,13–18].

In [18], varactors were used to individually load the grooves of a corrugated microstrip line, with operation below 6 GHz; the resulting antenna was thin and simple but the use of (TEM) microstrip guiding likely affected operation at higher frequencies because of the intrinsic losses. In [14], varactors were employed as tuning elements for a high-impedance surface in a 1D Fabry–Perot leaky-waveguide; the inherent transverse resonance mechanism required a thickness of about 3/4 wavelength. In [4], varactors were inserted into tunable-impedance phase-shifting side walls in a waveguide antenna (a modified WR90) derived from a standard waveguide slot array (in leaky-wave operation mode).

One of the most recent uses of metasurfaces is as refracting or reflective intelligent surfaces [19–21], also beamformers [22] for satellite-terrestrial networks have been revamped.

In this paper, a novel fixed-frequency beam scanning mechanism is introduced and demonstrated. The main goal of the present study was to investigate the potential of beam steering via distributed varactor diodes with a *single* DC control, i.e., on-plane varactors and no vias: this can be realized with existing standard low-cost technologies (e.g., pick-and-place). Using the metasurface paradigm, varactors can act as variable (reactive) loads in a suitably designed texture, yielding a variable impedance surface, with a surface impedance value controlled by the DC bias.

By combining a modulated upper metasurface (responsible for radiation) with a reconfigurable impedance plane (responsible for beam scanning), we effectively separated RF wave propagation from DC bias, while maintaining a very small form factor.

In this paper, Section 2 illustrates the adopted design techniques and describes the geometry of the designed antenna, while simulation results and experimental measurements of the fabricated prototype are reported in Section 3. In Section 4, these results are discussed and future work is outlined. Finally, conclusions are drawn in Section 5.

2. Materials and Methods

2.1. Principle of Operation

The beam scanning mechanism was conceived as a *dual* metasurface structure (see Figure 1a). Radiation is effected by a (static) transparent metasurface with metal texture on a dielectric (inspired by [23]); in a fixed-beam configuration, the radiating metasurface would be backed by a metal plane providing wave guiding. Here, the ground plane was replaced by a *tunable-impedance metasurface*. The principle of operation was based on the radiation of a guided wave, whose phase velocity was controlled by the variable-impedance surface.

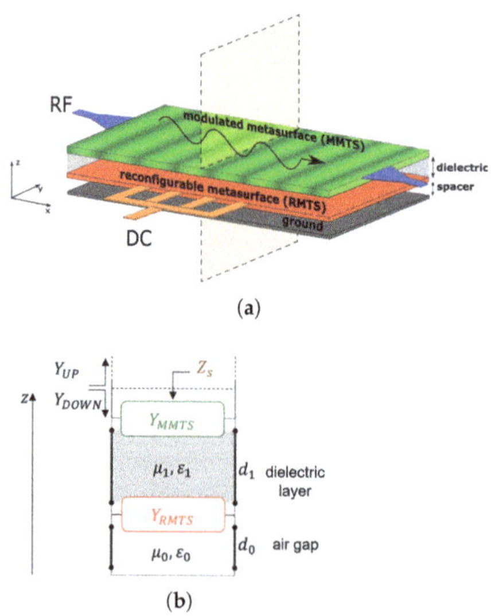

Figure 1. Structure of the proposed antenna. (**a**) Schematic architecture. (**b**) Transverse equivalent network: Y_{MMTS} is the admittance of the upper metasurface, Y_{RMTS} is the admittance of the lower (reconfigurable) metasurface, d_1 is the thickness of the dielectric layer, d_0 indicates the air gap, Y_{UP} and Y_{DOWN} are the admittances looking up and down from the interface between the antenna and free space, Z_s is the equivalent impenetrable surface impedance that approximates the whole multilayer structure.

The present study considered an antenna with scan in the vertical plane; the transverse size can range from small to multi-wavelength, depending on the type of feeding structure.

For the sake of simplicity, here the intermediate size of one wavelength was considered, which would allow for star-type 2D scanning [24].

Radiation of a guided mode is conveniently framed in the leaky-wave paradigm; hence, radiation happens via the spatially-modulated upper metasurface, exploiting both well-known results [5] and recent advances in metasurface antennas [23]. The operation principle, and thus the first-pass design, can be conveniently understood for the simplest modulation of the radiating metasurface, i.e., for the sinusoidal one [23]:

$$Z_s(x) = jX_{s,ave}\left[1 + M\cos\left(\frac{2\pi}{p}x\right)\right], \quad (1)$$

where Z_s is the equivalent impenetrable surface impedance that approximates the whole multilayer structure and takes into account the effects of both metasurfaces (Figure 1b), x is the coordinate along the direction of wave propagation, M is the modulation index, p is the modulation period, and $X_{s,ave}$ is the average surface reactance.

In this case, leaky-wave radiation happens for the $n = -1$ Floquet harmonic of the traveling wave [5,23]; for TM mode propagation, the radiation angle θ_{-1} is linked to the average surface reactance $X_{s,ave}$ by the approximate expression [18,23]:

$$\theta_{-1} = \arcsin\left(\sqrt{1 + \left(\frac{X_{s,ave}}{\eta_0}\right)^2} - \frac{2\pi}{k_0 p}\right), \quad (2)$$

where η_0, k_0 are the free-space impedance and wavenumber, respectively.

2.2. Design Method

The design was carried out in terms of standard guided-mode analysis [25] via the Transverse-Resonance Equation (TRE). As depicted in Figure 1b, the total admittance from the antenna-free space boundary was controlled by the variable impedance layer, resulting in a controllable wavenumber of the guided wave.

The use of the usual "adiabatic" approximation for slow variations of the top impedance allowed us to approximate the antenna with a transverse equivalent network (TEN) (see Figure 1b) at any given sample point of the modulated structure. From this transmission-line representation of the structure, the *local* value of the surface impedance $Z_s(x)$ can be retrieved by solving the corresponding TRE:

$$Y_{UP}(x) + Y_{DOWN}(x) = 0, \quad (3)$$

where Y_{UP} and Y_{DOWN} are the admittances looking up and down from a reference plane located at the interface between the antenna and free space. Indicating with d_1 the thickness of the dielectric layer between the two metasurfaces and with d_0 the air gap between reconfigurable metasurface and ground plane, standard transmission-line theory yields [26]:

$$Y_{UP} = Y_0^{TM} \quad (4)$$

$$Y_{DOWN} = Y_{MMTS} + Y_1^{TM}\frac{Y_{EGP} + jY_1^{TM}\tan(k_{z1}d_1)}{Y_1^{TM} + jY_{EGP}\tan(k_{z1}d_1)} \quad (5)$$

$$Y_{EGP} = Y_{RMTS} - jY_0^{TM}\cot(k_{z0}d_0), \quad (6)$$

where Y_{MMTS} is the admittance of the upper metasurface, Y_{RMTS} is the admittance of the lower (reconfigurable) metasurface, and Y_{EGP} is the admittance of the equivalent ground plane looking down from just above the tunable impedance plane (Figure 1b). In (4)–(6), Y_0^{TM} and Y_1^{TM} are the TM wave admittances in free space and dielectric substrate:

$$Y_0^{\mathrm{TM}} = \frac{\omega\varepsilon_0}{k_{z0}} \tag{7}$$

$$Y_1^{\mathrm{TM}} = \frac{\omega\varepsilon_1}{k_{z1}}, \tag{8}$$

where k_{z0} and k_{z1} are the wavenumbers in the transverse direction:

$$k_{z0} = \sqrt{k_0^2 - k_x^2} \tag{9}$$

$$k_{z1} = \sqrt{k_1^2 - k_x^2}, \tag{10}$$

and k_x is the longitudinal wavenumber. From (5) and (6), it can be seen that the solution of the TRE depends on the sheet admittances of both metasurfaces (Y_{MMTS} and Y_{RMTS}). Since all quantities in (4)–(6) can be expressed as a function of k_x, it is possible to solve (3) for k_x and then obtain the (local) surface impedance as:

$$Z_s = \mathrm{j}X_s = \frac{1}{Y_{\mathrm{DOWN}}}. \tag{11}$$

It is worth noting that the longitudinal wavenumber k_x considered above is used only to compute the local surface impedance based on the adiabatic approximation, and does not correspond to the actual wavenumber of the leaky wave that arises from the modulation of such impedance.

The average value of X_s is linked to the radiation angle through (2). This means that tuning Y_{RMTS} directly affects the beam direction, thus validating the concept of this architecture. Once a desired radiation angle for the $n = -1$ harmonic is chosen, the modulation period p is retrieved from (2) after selecting a proper value of $X_{s,\mathrm{ave}}$ among those physically attainable with the considered structure.

The design process now requires the choice of the geometry of both metasurfaces and the computation of their sheet admittances Y_{MMTS} and Y_{RMTS}, in order to solve (3). For the upper layer, a unit cell was chosen such that its admittance can be easily modulated by varying only one geometric feature, e.g., the gap width between two adjacent metal strips, as in [23]. For the implementation of the tunable metasurface, the unit cell shown in Figure 2 was used. This shape was inspired by the geometry described in [27]. Each unit cell contained two MAVR-011020-1411 varactors that were approximated as RC series elements for simulation purposes; the values of resistance and voltage-dependent capacitance were taken from the manufacturer's datasheet [28]. Varying the bias voltage changes the varactors' capacitance, which in turn affects the sheet admittance of this layer.

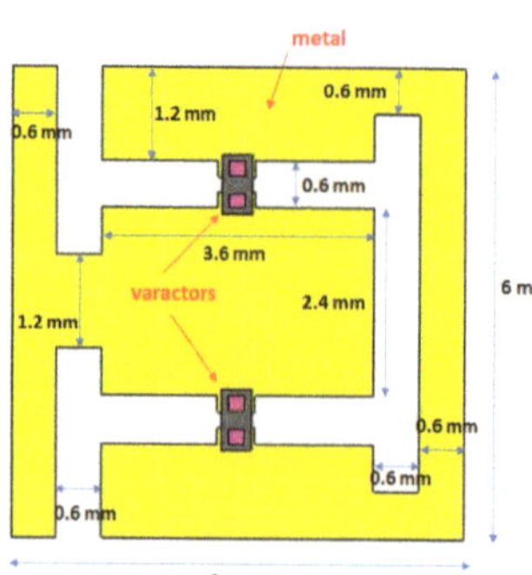

Figure 2. Unit cell of the reconfigurable metasurface.

The computation of Y_{MMTS} must be carried out for every possible value of the gap width in the constitutive unit cell, while Y_{RMTS} must be determined for every possible biasing state of the varactor diodes. This allowed us to obtain the equivalent impenetrable surface reactance X_s by solving the TRE for every combination of gap width in the upper metasurface and capacitance value in the reconfigurable plane. Moreover, at any given frequency, both admittances are also dependent on the longitudinal wavenumber k_x [14,29]: namely, $Y_{MMTS} = Y_{MMTS}(g, k_x)$ and $Y_{RMTS} = Y_{RMTS}(C, k_x)$, where g and C are the gap width and the varactors' capacitance, respectively. Therefore, to compute such admittances, the technique described in [14,30] was adopted, where the dependence on the longitudinal wavenumber was retrieved by performing scattering simulations of the constitutive unit cells for different angles of incidence. In particular, similarly to what was carried out in [30], the rational functions used to express Y_{MMTS} and Y_{RMTS} are:

$$Y_{MMTS}(g, k_x) = \mathrm{j}\frac{[g - g_{z1}(k_x)][g - g_{z2}(k_x)]}{g[g - g_{p1}(k_x)]\,[g - g_{p2}(k_x)]} \tag{12}$$

$$Y_{RMTS}(C, k_x) = \mathrm{j}\frac{[C - C_{z1}(k_x)][C - C_{z2}(k_x)][C - C_{z3}(k_x)]}{C - C_p(k_x)}. \tag{13}$$

In (12) and (13), subscripts z and p indicate zeros and poles of the rational functions. These quantities are expressed as third degree polynomials of k_x. In the frequency range of interest ([10.55–10.75] GHz), for the considered geometries, Y_{MMTS} is capacitive and Y_{RMTS} is always inductive.

Figure 3 shows how the surface reactance of the whole antenna is affected by a change in the varactors' bias voltage. The curve relating X_s to the gap width is shifted when the bias voltage goes from 0 V to 8.2 V. This means that, for a fixed spatial modulation in the upper layer, the sinusoidal profile of the surface reactance (and its average value $X_{s,ave}$) seen by the traveling wave is dependent on the varactors' basing state. Since $X_{s,ave}$ is linked to the radiation angle through (2) and the period p is fixed by the upper metasurface, beam steering occurs.

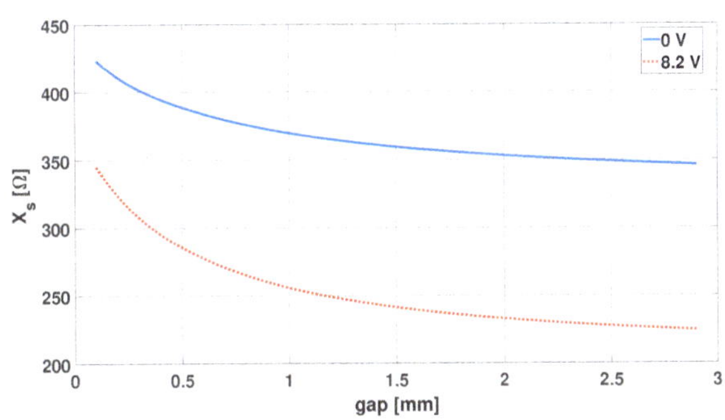

Figure 3. Surface reactance of the whole structure vs gap width in the array of strips, for two different values of the varactors' bias voltage.

2.3. Implementation

A center-band frequency of 10.65 GHz was considered. A Rogers RT5880 ($\varepsilon_r = 2.2$, $\tan\delta = 0.0009$) 3.175 mm-thick substrate was placed between the upper radiating layer and the lower variable-impedance plane. The unit cell of the modulated metasurface was chosen to be about 1/10 of the wavelength, i.e., 3 mm. The air gap between the reconfigurable plane and the ground was set to 1.5 mm in order to allow enough space for the varactors.

The modulation period p of the top impedance surface was computed from (2), setting a radiation angle of 20° for a 0 V bias voltage of the varactors, and this resulted in $p = 27$ mm. Since the unit cell of the modulated metasurface was 3 mm wide, the cosinusoidal variation of the reactance in one period was sampled in nine points, i.e., with sufficient sampling. For the radiating part, eight modulation periods were chosen for a total length of 216 mm; this was a good compromise between beam width and (leaky-wave) tapering and, at any rate, yielding low S_{21} towards the matched load.

The bias scheme of the varactors in the variable-impedance plane is depicted in Figure 4, which shows an alternate-potential scheme. The necessary single DC bias was conveyed by two 0.15 mm-wide buses placed at the opposite sides of the metasurface; the RF reactance of the very thin (high-impedance) lines was considered enough to decouple RF and DC without the need for a filter, and that was confirmed by simulations.

The resulting complete structure of the designed antenna is shown in Figure 5. The upper modulated metasurface is depicted in Figure 5a, while Figure 5c shows the lower reconfigurable impedance plane; the vertical stackup of the antenna can be seen in Figure 5b. The overall thickness of the structure was less than $\lambda_0/6$ at the working frequency of 10.65 GHz, resulting in a very small form factor. It should be noted from Figure 6 that the unit cell of the reconfigurable metasurface was twice as large as that of the upper layer, in order to reduce the total number of varactors needed in the structure, which amounted to 360.

The excitation was provided with a 50 Ohm coaxial cable, which was matched to the antenna by a properly designed tapered section (Figure 5a).

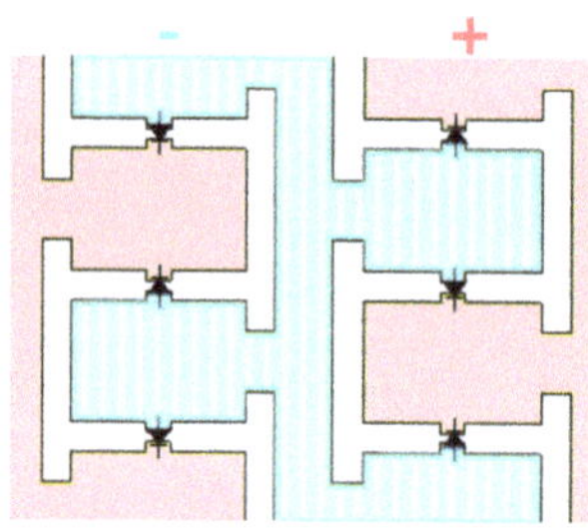

Figure 4. DC voltage distribution in the reconfigurable metasurface.

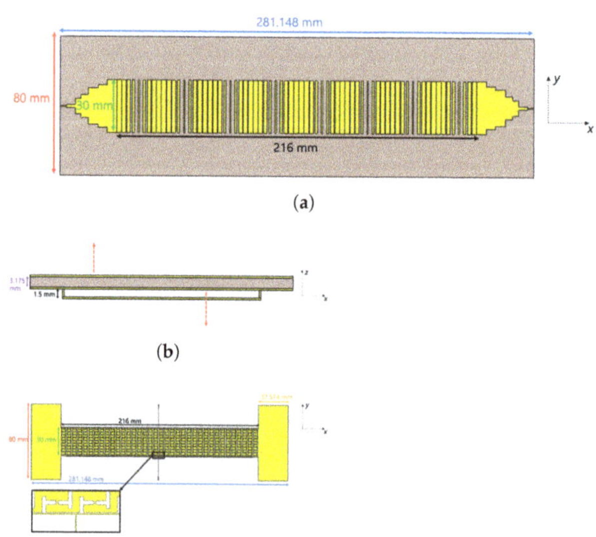

Figure 5. Complete model of the designed antenna. (**a**) Upper sinusoidally modulated reactance surface. (**b**) Lateral view of the antenna. (**c**) Reconfigurable plane; the inset shows one of the two DC buses.

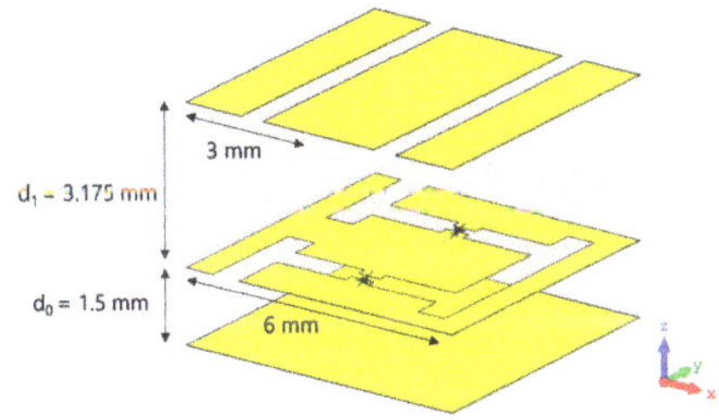

Figure 6. Unit cell of the fully-stacked structure.

3. Results

Full-wave simulations of the designed antenna were performed with the commercial software CST Studio Suite [31]. The varactor diodes were used in a 0–8.2 V biasing range and their capacitance went from 0.233 pF to 0.0548 pF, according to the manufacturer's datasheet [28]. The beam pointing angles obtained from these simulations at the working frequency of 10.65 GHz are listed in Table 1, showing excellent agreement with the main beam directions computed using the approximate model described in Section 2.2.

Table 1. Calculated, simulated, and measured radiation angles at 10.65 GHz.

Voltage [V]	Calculated	Simulated	Measured
0	22.7°	20°	19.5°
8.2	11.9°	13°	12.5°

A prototype was fabricated and tested (Figure 7). The modulated metasurface and the reconfigurable plane are pictured in Figure 7a,b, while Figure 7c,d show the tapered feeding section on the upper layer and the varactor diodes soldered in the lower metasurface, respectively. Radiation measurements were performed in an NF-FF spherical range; the full radiation pattern was obtained, allowing evaluation of directivity and radiation efficiency.

Measured far-field patterns at 10.65 GHz for different bias voltages are shown in Figure 8, together with the simulation results, which are in excellent agreement.

The antenna gain bandwidth was 200 MHz around the working frequency of 10.65 GHz; the impedance bandwidth was larger than the gain one. The scattering parameters S_{11} and S_{21} of the fabricated prototype are shown in Figure 9: S_{11} was always below −10 dB in the working band for all varactors' bias voltages, while S_{21} remained below −20 dB, which is very important for the efficiency of a traveling-wave antenna.

In the considered frequency range, the gain spanned from 11.6 dB to 13.4 dB with varying bias voltage, while the radiation efficiency was between 67% and 76%. The measured aperture efficiency went from 20% to 32%.

In Figure 8, a side-lobe is noticeable in the backward radiation direction; this lobe can be ascribed to radiation of the $n = -2$ harmonic, which falls within the fast wave region at the working frequency [23]. These results are consistent with what can be predicted using (2), i.e., the $n = -2$ harmonic radiating at about −41° and −57° for a bias voltage equal to 0 V and 8.2 V, respectively. The radiation pattern also presents an irregular shape and high side-lobes; this can be ascribed to the anisotropy of the impedance planes, in particular of the lower reconfigurable metasurface. In fact, its unit cell geometry and the placement of the varactor diodes inside it are such that a change in the bias voltage alters not only the component of the sheet impedance in the direction of wave propagation, but also the transverse one. This may cause disturbances in the wave propagation.

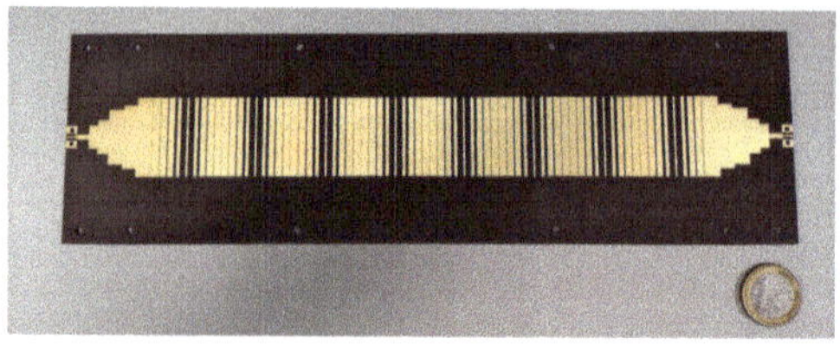

(**a**)

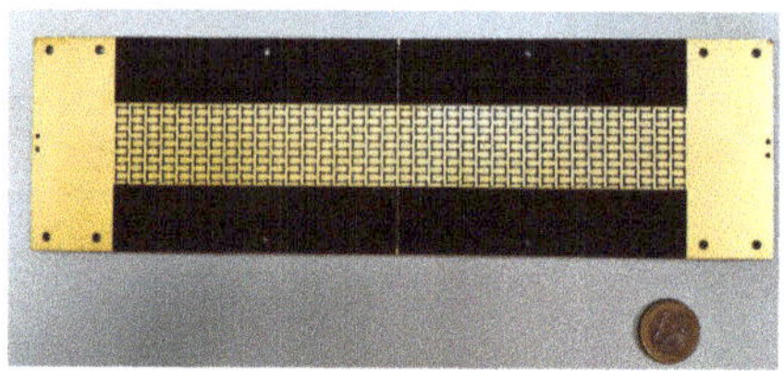

(**b**)

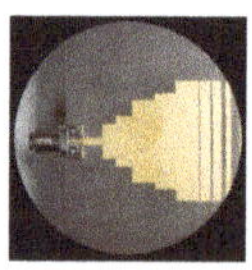

(**c**)

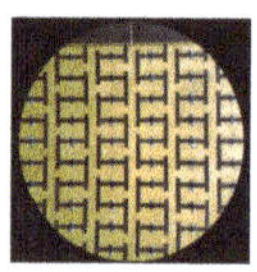

(**d**)

Figure 7. Fabricated antenna prototype. (**a**) Top modulated metasurface. (**b**) Reconfigurable impedance plane. (**c**) Coaxial cable and matched tapered input section at the left end of the upper metasurface (same tapered section is present at the opposite end). (**d**) Zoom showing the soldered varactors in the lower reconfigurable metasurface.

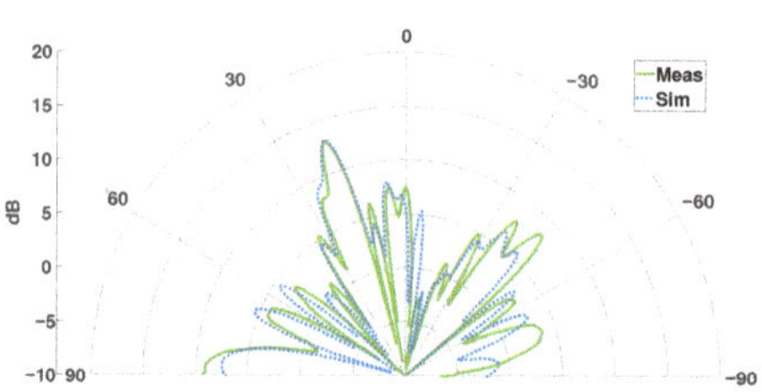

(**a**) 0 V

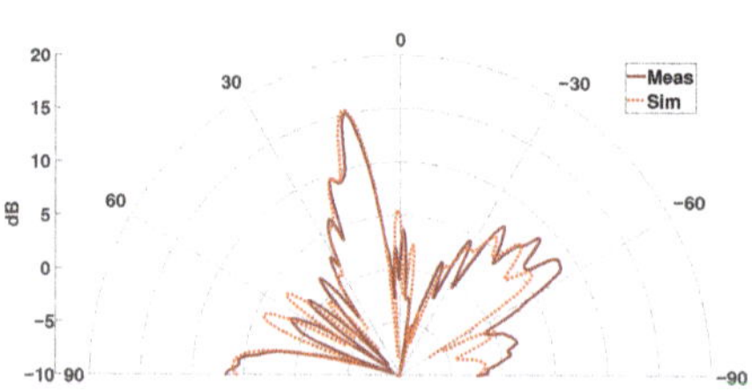

(**b**) 8.2 V

Figure 8. Measured and simulated radiation pattern in the E-plane at 10.65 GHz for different values of bias voltage.

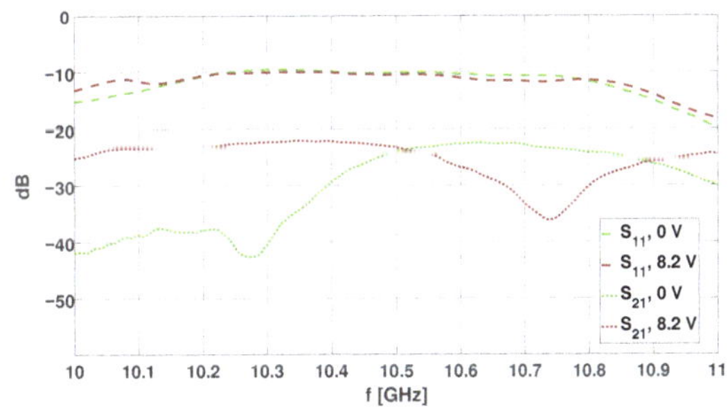

Figure 9. Measured S_{11} and S_{21} for two different values of bias voltage.

4. Discussion

In Table 2, the performances of the antenna described in this paper are compared to those of other solutions found in the literature. In the table, "efficiency" stands for radiation efficiency.

Table 2. Comparison with other LWAs found in the literature.

Reference	Frequency [GHz]	Scanning Range	Antenna Length	Thickness	Efficiency	Gain [dB]
[14]	5.6	21°	$5\lambda_0$	$3\lambda_0/4$	60–75%	5–13
[18] Band 2	5.75	22°	$7.15\lambda_0$	$\lambda_0/17$	40–50%	9.5–12
[4]	9.3	43°	$13\lambda_0$	$\lambda_0/3$	38–46%	8–11.8
This work	10.65	7°	$7.7\lambda_0$	$\lambda_0/6$	67–76%	11.6–13.4

Radiation efficiency and gain values are comparable to those reported in the other works (above the average). The comparison is conservative because all reported references work at a lower frequency, which inherently entails lower losses. The small form factor, simple biasing and high efficiency make the proposed architecture interesting for several low-profile applications. The angular scanning range achieved with this first prototype is, on the other hand, limited: improvement of this is thus important and the subject of current work. Here, some of the reasons for this limitation are listed and ways to improve it are proposed.

A possible degree of improvement is represented by the choice of the dielectric substrate and its thickness, especially for the side lobes. In particular, a higher relative permittivity results in a wider range of surface reactance values, X_s, which allows for a larger modulation factor M and, consequently, a higher leakage constant [23]. This in turn translates into the reduction of side-lobes.

Figure 10 shows the simulated radiation pattern at 10.75 GHz for an alternative design of the proposed antenna, featuring a Rogers RO3006 substrate ($\varepsilon_r = 6.5$) of a thickness equal to 2.56 mm: the radiation angle steers from 15° to 5° when the bias voltage goes from 0 V to 8.2 V. In this case, no other harmonic beside the $n = -1$ radiates, and the very small backward-directed side-lobes are due to the spurious radiation of a second, higher-order TM mode. Directivity values are the same as in the previous configuration. A further reduction of side-lobes could be obtained by tapering the modulation index M along the length of the antenna [32], which was not pursued here.

Another margin of improvement involves the geometry of the reconfigurable metasurface's unit cell. The tunable unit cell used in this first prototype was inspired by the literature and had a different original use [27]; a new geometry could be devised to obtain a wider impedance range for the same diodes.

Moreover, the design presented in this paper was based on the combination of a capacitive impedance (top) and an inductive one (bottom), and the propagation of the TM mode. However, different combinations of impedances can be employed (e.g., capacitive-

capacitive) that may increase the scanning range if properly designed, especially if a dielectric material with higher electric permittivity is used. In particular, analyses carried out with the approximate analytical model show that use of the TE mode (instead of the TM) yields a much larger steering range. This possibility is currently being studied and will be the subject of future work (excitation of the TE mode is less natural than that of the TM).

Finally, the radiating surface employs the standard sinusoidal modulation scheme; the scan range can be improved (for the same impedance variation of the bottom plane) by more advanced impedance modulation schemes, and especially by ad-hoc numerical optimization of the profile [33,34]; this has shown the potential of increasing sensitivity to phase velocity, thus enhancing the scanning range.

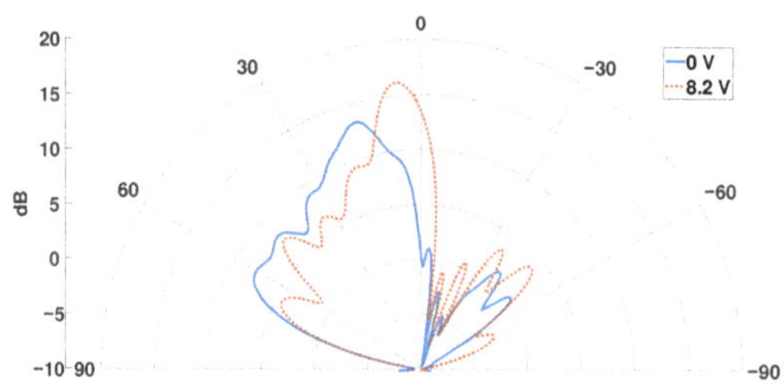

Figure 10. Simulated radiation pattern at 10.75 GHz for different bias voltages for an alternative design with Rogers RO3006 substrate ($\varepsilon_r = 6.5$).

5. Conclusions

In this work, a novel beam steering configuration with varactor diodes, a single voltage bias, and no bias has been demonstrated. The small form factor, simple biasing, and high efficiency make the proposed architecture interesting for low-profile applications. The angular scanning range achieved with this first prototype is limited, but several ways to improve it were discussed.

Author Contributions: Conceptualization, G.V.; methodology, L.T. and F.V.; software, L.T.; validation, L.T., F.V., G.G. and R.G.; formal analysis, L.T. and F.V.; investigation, L.T., F.V. and G.G.; resources, G.G. and G.V.; data curation, L.T. and R.G.; writing—original draft preparation, L.T.; writing—review and editing, G.G., R.G. and G.V.; visualization, L.T., G.G. and R.G.; supervision, G.V.; project administration, G.V.; funding acquisition, G.V. All authors have read and agreed to the published version of the manuscript.

Funding: This work was supported by the Italian Ministry of Research PRIN 2017S29ZLA "Metasurface Antennas for Space Applications".

Institutional Review Board Statement: Not applicable.

Informed Consent Statement: Not applicable.

Data Availability Statement: Not applicable.

Acknowledgments: The authors would like to acknowledge the invaluable help and the technological skill provided by Giuseppe Franco (Fondazione LINKS) in the realization of the prototype.

Conflicts of Interest: The authors declare no conflict of interest.

References

1. Christodoulou, C.G.; Tawk, Y.; Lane, S.A.; Erwin, S.R. Reconfigurable Antennas for Wireless and Space Applications. *Proc. IEEE* **2012**, *100*, 2250–2261. [CrossRef]
2. Schoebel, J.; Buck, T.; Reimann, M.; Ulm, M.; Schneider, M.; Jourdain, A.; Carchon, G.; Tilmans, H. Design Considerations and Technology Assessment of Phased-Array Antenna Systems with RF MEMS for Automotive Radar Applications. *IEEE Trans. Microw. Theory Tech.* **2005**, *53*, 1968–1975. [CrossRef]

3. Huang, K.C.; Wang, Z. Millimeter-Wave Circular Polarized Beam-Steering Antenna Array for Gigabit Wireless Communications. *IEEE Trans. Antennas Propag.* **2006**, *54*, 743–746. [CrossRef]
4. Ohadi, A.; Eleftheriades, G.V. Fixed-Frequency Beam-Steering Using Slotted Waveguide With Tunable Impedance Walls. *IEEE Open J. Antennas Propag.* **2021**, *2*, 978–990. [CrossRef]
5. Oliner, A.; Hessel, A. Guided Waves on Sinusoidally-Modulated Reactance Surfaces. *IRE Trans. Antennas Propag.* **1959**, *7*, 201–208. [CrossRef]
6. Jackson, D.R.; Caloz, C.; Itoh, T. Leaky-Wave Antennas. *Proc. IEEE* **2012**, *100*, 2194–2206. [CrossRef]
7. Monticone, F.; Alu, A. Leaky-Wave Theory, Techniques, and Applications: From Microwaves to Visible Frequencies. *Proc. IEEE* **2015**, *103*, 793–821. [CrossRef]
8. Esquius-Morote, M.; Gómez-Díaz, J.S.; Perruisseau-Carrier, J. Sinusoidally Modulated Graphene Leaky-Wave Antenna for Electronic Beamscanning at THz. *IEEE Trans. Terahertz Sci. Technol.* **2014**, *4*, 116–122. [CrossRef]
9. Torabi, E.; Rozhkova, A.; Chen, P.Y.; Erricolo, D. Compact and Reconfigurable Leaky Wave Antenna Based on a Tunable Substrate Integrated Embedded Metasurface. In Proceedings of the 2020 IEEE International Symposium on Antennas and Propagation and North American Radio Science Meeting, Montréal, QC, Canada, 5–10 July 2020; pp. 163–164. [CrossRef]
10. Martini, E.; Pavone, S.; Albani, M.; Maci, S.; Martorelli, V.; Giodanengo, G.; Ferraro, A.; Beccherelli, R.; Toso, G.; Vecchi, G. Reconfigurable Antenna Based on Liquid Crystals for Continuous Beam Scanning with a Single Control. In Proceedings of the 2019 IEEE International Symposium on Antennas and Propagation and USNC-URSI Radio Science Meeting, Atlanta, GA, USA, 7–12 July 2019; pp. 449–450. [CrossRef]
11. Pavone, S.C.; Martini, E.; Caminita, F.; Albani, M.; Maci, S. Surface Wave Dispersion for a Tunable Grounded Liquid Crystal Substrate Without and With Metasurface on Top. *IEEE Trans. Antennas Propag.* **2017**, *65*, 3540–3548. [CrossRef]
12. Rabbani, M.S.; Churm, J.; Feresidis, A.P. Continuous Beam-Steering Low-Loss Millimeter-Wave Antenna Based on a Piezo-Electrically Actuated Metasurface. *IEEE Trans. Antennas Propag.* **2022**, *70*, 2439–2449. [CrossRef]
13. Sievenpiper, D. Forward and Backward Leaky Wave Radiation with Large Effective Aperture from an Electronically Tunable Textured Surface. *IEEE Trans. Antennas Propag.* **2005**, *53*, 236–247. [CrossRef]
14. Guzman-Quiros, R.; Gomez-Tornero, J.L.; Weily, A.R.; Guo, Y.J. Electronically Steerable 1-D Fabry-Perot Leaky-Wave Antenna Employing a Tunable High Impedance Surface. *IEEE Trans. Antennas Propag.* **2012**, *60*, 5046–5055. [CrossRef]
15. Gregoire, D.J.; Patel, A.; Quarfoth, R. A Design for an Electronically-Steerable Holographic Antenna with Polarization Control. In Proceedings of the 2015 IEEE International Symposium on Antennas and Propagation & USNC/URSI National Radio Science Meeting, Vancouver, BC, Canada, 19–25 July 2015; pp. 2203–2204. [CrossRef]
16. Quarfoth, R.G.; Patel, A.M.; Gregoire, D.J. Ka-Band Electronically Scanned Artificial Impedance Surface Antenna. In Proceedings of the 2016 IEEE International Symposium on Antennas and Propagation (APSURSI), Fajardo, PR, USA, 26 June–1 July 2016; pp. 651–652. [CrossRef]
17. Shlezinger, N.; Alexandropoulos, G.C.; Imani, M.F.; Eldar, Y.C.; Smith, D.R. Dynamic Metasurface Antennas for 6G Extreme Massive MIMO Communications. *IEEE Wirel. Commun.* **2021**, *28*, 106–113. [CrossRef]
18. Wang, M.; Ma, H.F.; Tang, W.; Zhang, H.C.; Jiang, W.; Cui, T.J. A Dual-Band Electronic-Scanning Leaky-Wave Antenna Based on a Corrugated Microstrip Line. *IEEE Trans. Antennas Propag.* **2019**, *67*, 3433–3438. [CrossRef]
19. Lin, Z.; Niu, H.; An, K.; Wang, Y.; Zheng, G.; Chatzinotas, S.; Hu, Y. Refracting RIS-Aided Hybrid Satellite-Terrestrial Relay Networks: Joint Beamforming Design and Optimization. *IEEE Trans. Aerosp. Electron. Syst.* **2022**, *58*, 3717–3724. [CrossRef]
20. Niu, H.; Lin, Z.; An, K.; Wang, J.; Zheng, G.; Al-Dhahir, N.; Wong, K.K. Active RIS Assisted Rate-Splitting Multiple Access Network: Spectral and Energy Efficiency Tradeoff. *IEEE J. Sel. Areas Commun.* **2023**. [CrossRef]
21. Niu, H.; Lin, Z.; Chu, Z.; Zhu, Z.; Xiao, P.; Nguyen, H.X.; Lee, I.; Al-Dhahir, N. Joint Beamforming Design for Secure RIS-Assisted IoT Networks. *IEEE Internet Things J.* **2023**, *10*, 1628–1641. [CrossRef]
22. Lin, Z.; Lin, M.; Champagne, B.; Zhu, W.P.; Al-Dhahir, N. Secrecy-Energy Efficient Hybrid Beamforming for Satellite-Terrestrial Integrated Networks. *IEEE Trans. Commun.* **2021**, *69*, 6345–6360. [CrossRef]
23. Patel, A.M.; Grbic, A. A Printed Leaky-Wave Antenna Based on a Sinusoidally-Modulated Reactance Surface. *IEEE Trans. Antennas Propag.* **2011**, *59*, 2087–2096. [CrossRef]
24. Martinez-Ros, A.J.; Gomez-Tornero, J.L.; Goussetis, G. Broadside Radiation from Radial Arrays of Substrate Integrated Leaky-Wave Antennas. In Proceedings of the 2012 6th European Conference on Antennas and Propagation (EUCAP), Prague, Czech Republic, 26–30 March 2012; pp. 252–254. [CrossRef]
25. Walter, C. *Traveling Wave Antennas*; McGraw-Hill: New York, NY, USA, 1965.
26. Pozar, D.M. *Microwave Engineering*, 4th ed.; John Wiley & Sons: Hoboken, NJ, USA, 2011.
27. Wu, Z.; Grbic, A. Serrodyne Frequency Translation Using Time-Modulated Metasurfaces. *IEEE Trans. Antennas Propag.* **2020**, *68*, 1599–1606. [CrossRef]
28. MACOM Technology Solutions. Available online: https://www.macom.com/products/product-detail/MAVR-011020-1411 (accessed on 12 March 2021).
29. Maci, S.; Caiazzo, M.; Cucini, A.; Casaletti, M. A Pole-Zero Matching Method for EBG Surfaces Composed of a Dipole FSS Printed on a Grounded Dielectric Slab. *IEEE Trans. Antennas Propag.* **2005**, *53*, 70–81. [CrossRef]
30. Guzmán Quirós, R. Analysis and Design of New Electronically Reconfigurable Periodic Leaky-Wave Antennas. Ph.D. Dissertation, Universidad Politécnica de Cartagena, Cartagena, Spain, 2014.

31. Dassault Systèmes Simulia. CST Studio Suite 2019. Available online: www.cst.com (accessed on 12 March 2021).
32. Yang, D.; Nam, S. Tapered Unit Cell Control of a Sinusoidally Modulated Reactance Surface Antenna. *IEEE Antennas Wirel. Propag. Lett.* **2018**, *17*, 2479–2483. [CrossRef]
33. Vernì, F. Advanced Computational Electromagnetics for Metasurfaces. Ph.D. Dissertation, Politecnico di Torino, Turin, Italy, 2020.
34. Caminita, F.; Martini, E.; Minatti, G.; Sabbadini, M.; Maci, S. Low-Profile Dual-Polarized Isoflux Antennas for Space Applications. *IEEE Trans. Antennas Propag.* **2021**, *69*, 3204–3213. [CrossRef]

sensors

Article

Multi-Layer Beam Scanning Leaky Wave Antenna for Remote Vital Signs Detection at 60 GHz

Solomon Mingle *, Despoina Kampouridou and Alexandros Feresidis

Department of Electronic, Electrical and Systems Engineering, University of Birmingham, Birmingham B15 2TT, UK
* Correspondence: sxm1274@student.bham.ac.uk

Abstract: A multi-layer beam-scanning leaky wave antenna (LWA) for remote vital sign monitoring (RVSM) at 60 GHz using a single-tone continuous-wave (CW) Doppler radar has been developed in a typical dynamic environment. The antenna's components are: a partially reflecting surface (PRS), high-impedance surfaces (HISs), and a plain dielectric slab. A dipole antenna works as a source together with these elements to produce a gain of 24 dBi, a frequency beam scanning range of 30°, and precise remote vital sign monitoring (RVSM) up to 4 m across the operating frequency range (58–66 GHz). The antenna requirements for the DR are summarised in a typical dynamic scenario where a patient is to have continuous monitoring remotely, while sleeping. During the continuous health monitoring process, the patient has the freedom to move up to one meter away from the fixed sensor position.The proposed multi-layer LWA system was placed at a distance of 2 m and 4 m from the test subject to confirm the suitability of the developed antenna for dynamic RVSM applications. A proper setting of the operating frequency range (58 to 66 GHz) enabled the detection of both heart beats and respiration rates of the subject within a 30° angular range.

Keywords: 60 GHz band doppler radar sensors; beam scanning; contact monitoring device; leaky wave antenna; remote vital sign monitoring

Citation: Mingle, S.; Kampouridou, D.; Feresidis, A. Multi-Layer Beam Scanning Leaky Wave Antenna for Remote Vital Signs Detection at 60 GHz. *Sensors* **2023**, *23*, 4059. https://doi.org/10.3390/s23084059

Academic Editors: Marta Cavagnaro

Received: 23 February 2023
Revised: 21 March 2023
Accepted: 7 April 2023
Published: 17 April 2023

1. Introduction

Conventional methods of monitoring vital signs, such as an electrocardiogram (ECG), pulse oximetry, and capnography, require sensors to be attached directly to the patient's body, which is either uncomfortable for the patient or not possible under certain circumstances [1–8]. Due to the fact that sensors do not require direct contact with the body, remote monitoring of vital signs with a Doppler radar is more convenient than conventional methods [8]. This makes remote monitoring of vital signs with a Doppler radar an attractive option. A Doppler radar is a non-contact technology that can detect small movements of the body caused by the heartbeat and breathing. This technology has the potential to revolutionize healthcare by providing continuous and non-invasive monitoring of vital signs, particularly for patients who require frequent monitoring or who are in critical condition. Remote vital sign monitoring (RVSM) with a Doppler radar has the potential to be used in a variety of disciplines, including general and specialised healthcare, emergency services, security, and defence. However, for RVSM to be useful in practice, further research needs to be conducted on the accuracy and reliability of the results in different settings and populations [1–10].

The main advantages of using millimetre-wave frequencies for RVSM are (i) higher detection sensitivity due to shorter signal wavelengths, (ii) smaller form factors for more compact devices, and (iii) the ability to transmit and receive signals more precisely without interference from the environment [4,9]. Conventional methods such as electrocardiography (ECG) and photoplethysmography (PPG) require physical contact with the subject, which can be uncomfortable and can interfere with normal activity. Additionally, ECG and PPG provide measurements at a single point in time, while millimetre-wave technology can provide continuous monitoring over an extended period. Furthermore, ECG and PPG are often limited in their ability to penetrate through clothing and tissue, which can limit their accuracy in certain situations.

The 60 GHz frequency band (57–66 GHz) has attracted RVSM interest due to its licence-free nature and its widespread use for several different wireless services. Antennas that operate in this band are particularly challenging to design and fabricate. RVSM measurements require the focus of the antenna beams, as well as enough gain to compensate for path loss in order to achieve high precision. Moreover, antenna beam steering capability is necessary to continuously monitor health with a fixed position sensor over a long period of time, especially when the person may move in practical scenarios, such as during sleep, inside a room, during transport in an ambulance, and at work [11–13]. Some examples of beam-steered antenna applications for RVSM are depicted in Figure 1.

Recent publications [4,8,9] discuss the design of mm-wave antennas for health monitoring with a Doppler radar. However, these are fixed beam antennas, and their gain does not exceed 20 dBi. Recent reports have shown that digital beamforming antenna systems operating in the mm-waveband can be used in a variety of industrial and automotive applications. In RVSM sensors, spatial arrays, compressive sensing, route sharing, and MIMO antennas have high power requirements, which limits their application. Furthermore, their hardware is complicated and expensive. The development of beam steering antennas for use in health monitoring on mm-waves has received a limited number of papers to the best of our knowledge. According to [11,13,14], this is likely the result of the limited frequency of electronic components available today.

Leaky wave antennas (LWAs) are a special family of antennas that are able to steer their beam with frequency [15]. They provide a number of advantages that makes them very suitable for RVSM application. These advantages include beam steering with frequency, which allows the antenna beam to be directed towards the subject of interest without any mechanical rotation; high directivity and bandwidth, which improves the signal-to-noise ratio and increases the range of the system; low profile, which makes such antennas suitable for use in compact and portable monitoring systems.

Our study presents a multi-layer LWA used for remote vital signs detection in a typical situation in which a patient is lying on the bed and has some random movements during sleep. The non-contact detection of vital signs can be achieved using antenna beam characteristics that provide high gain (58–66 GHz) and wide bandwidth performance. A measured antenna bandwidth of 8 GHz and a maximum gain of 24 dBi are measured across the operating band. The proposed antennas are tested experimentally to validate their predicted detection coverage. This antenna design measures the respiratory rate (BR) and the heart rate (HR) from a distance of up to 4 m from the body of the person at five different radiation angles between 6° and 36°. A very good agreement between the measured results and the predicted and simulated results has been achieved.

This work is an improvement from the antennas already published in [16,17]. The reported antenna bandwidth, maximum gain, and beam scanning range in [16] are 3.78 GHz, 20.35 dBi, and 12°, respectively, compared to 16 GHz, 24 dBi, and 30° in this work. The dielectric image line antenna of [17] has a maximum gain of 19 dBi as compared to 24dBi gain of the proposed multi-layer LWA. Due to this high gain, the multiplayer LWA in this work has more advantages in terms of distance applications, such as radar, and many IoT applications.

This paper is organised as follows: The antenna design is described in Section 2, antenna measurements and discussions are presented in Section 3, health monitoring measurements are discussed in Section 4, and this work is brought to a conclusion in Section 5.

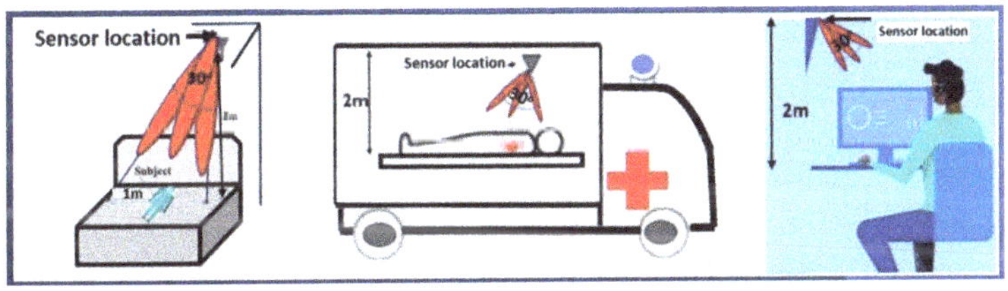

Figure 1. Examples of RVSM applications.

2. Antenna Design Methods

2.1. Unit Cell Design

To analyse the proposed design, we begin by applying ray optics in order to measure multiple reflections between partial reflective surfaces (PRSs) and high impedance surfaces (HISs) on the ground plane [18,19]. The PRS and HIS reflection characteristics are related to antenna gain, bandwidth, and radiation patterns. The same periodicity P is selected for the HIS and PRS unit cells. The reflectance properties of PRS and HIS unit cells are determined at normal incidence. More details about this analysis technique and the relevant equations can be found in [18–22].

Both PRSs and HISs are metasurface arrays made up of square patch elements and square ring elements, respectively, with a ground plane etched on a planar printed circuit board. Figure 2 depicts the structures of both unit cells. The metasurface elements are printed on a Rogers RT/Duroid5880 substrate with a dielectric constant of $\epsilon_r = 2.2$ and a tangent loss of 0.0019.

The design process of the PRS and HIS unit cells is depicted in Figure 3 with respect to important parameters. For both unit cells, the copper cladding thickness is 0.035 mm. The substrate thickness for PRS is selected at $N = 0.508$ mm so that a highly reflective response is achieved, which is necessary for a high gain antenna (Figure 3a). The top patch is square in order to maximize the reflection coefficient magnitude and symmetry [19,22,23]. The HIS unit cell of $M = 1.57$ mm (Figure 3b) uses a square metallic ring element with an inner loop, optimised to achieve a phase change with frequency [18]. The parametric study of Figure 3b for a selected dimension $L_{out} = 1.5$ mm shows that a rapid phase change is achieved with a selection of $L_{in} = 0.6$ mm.

Periodic boundary conditions were utilised in CST Microwave Studio^{TM} along the x and y axis for both unit cell simulations. Open boundaries were established along the incident wave's $+z$-axis in accordance with the design process first detailed in [21] to extract the reflection coefficients and optimize the dimensions.

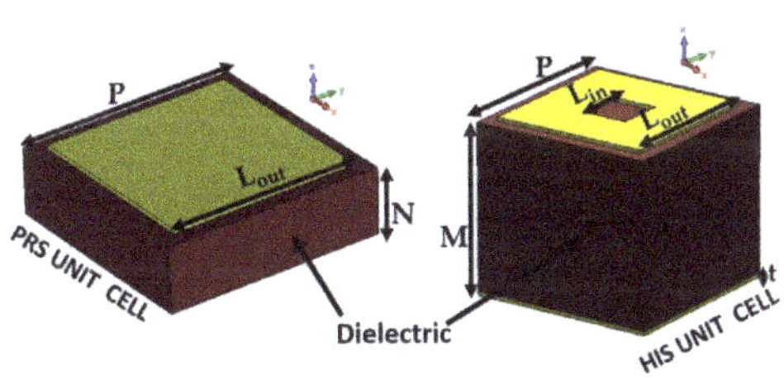

Figure 2. PRS and HIS unit cell structure where $L_{out} = 1.50$ mm, $L_{in} = 0.60$ mm and periodicity $P = 1.75$ mm.

Figure 3a shows the simulated reflection coefficient results of the PRS unit cell, exhibiting an absolute magnitude $|\Gamma| = 0.9$ and reflection phase ϕ_{PRS} response =145° at the frequency of 62.5 GHz, respectively, with slow variation in frequency. Figure 3b depicts the simulated reflection coefficient of S_{11} results of the HIS unit cell, which demonstrate a phase response of $-95°$ at 62.5 GHz. This conforms to the pattern of reflection that is expected from free-standing frequency selective surfaces (FSSs). From the simulated values at 62.5 GHz, the ray tracing method predicts an antenna bandwidth (BW) and a gain of approximately 8 GHz and 24 dBi, respectively, for PRS layers with an finite thickness when fed through by a dipole.

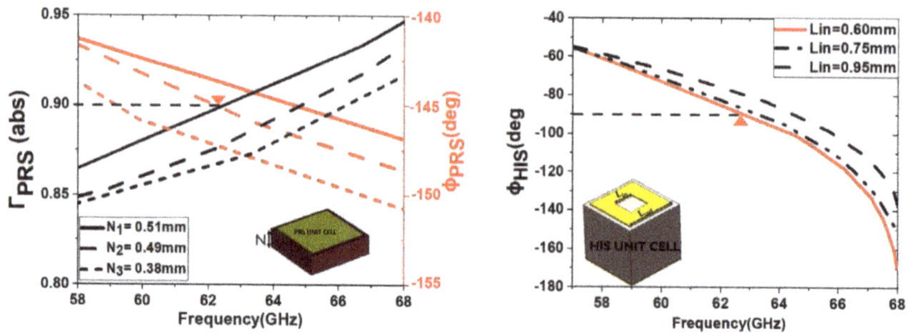

Figure 3. (**a**) Magnitude and phase reflection coefficient of the PRS unit cell. (**b**) Phase reflection coefficient of the HIS unit cell with respect to L_{in} with L_{out} = 1.50 mm.

2.2. Printed Dipole Antenna Design

In this section, a printed dipole antenna (PDA) is designed, which will serve as a primary feeding for the complete leaky wave antenna structure. A design model of this feeding dipole antenna can be seen in Figure 4a. Its dimensions are optimised for best performance in the 60 GHz frequency range. The printed dipole antenna (PDA) is etched on RT/Duroid 5880 with a dielectric constant of $\epsilon_r = 2.2$ and a thickness of 0.13 mm. A reflector is placed on the backside of the printed dipole antenna (PDA) at an optimised distance of 0.41 mm to improve the functionality of the LWA in the xy plane [24]. Figure 4b shows that the simulation and measured results of S_{11} and realised gain are in good agreement.

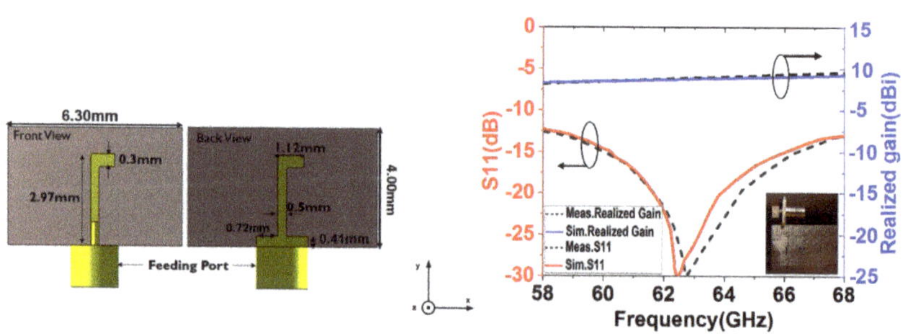

Figure 4. The feeding PDA structure. (**a**) dimensions in front view and back view; (**b**) simulated and measured S_{11} and realised gain.

2.3. Multi-Layer LWA Design

The full proposed multi-layer LWA design is shown in Figure 5b. Both the partially reflective surface (PRS) and the high impedance surfaces (HISs) are formed by arrays of 50 × 15 elements. This size of the antenna was selected as a compromise between a relatively compact size, a satisfactory high gain, and the infinite size assumption of leaky wave theory. The rectangular shape of the antenna was selected in order to achieve a highly directive beam at the E-plane. A fed dipole (Figure 5a) is positioned in the cavity between PRS and HIS towards one end of the antenna. On the same side, there is a vertical metallic wall where the antenna is terminated (Figure 5b). A 3D view of the antenna elements is depicted in Figure 5c.

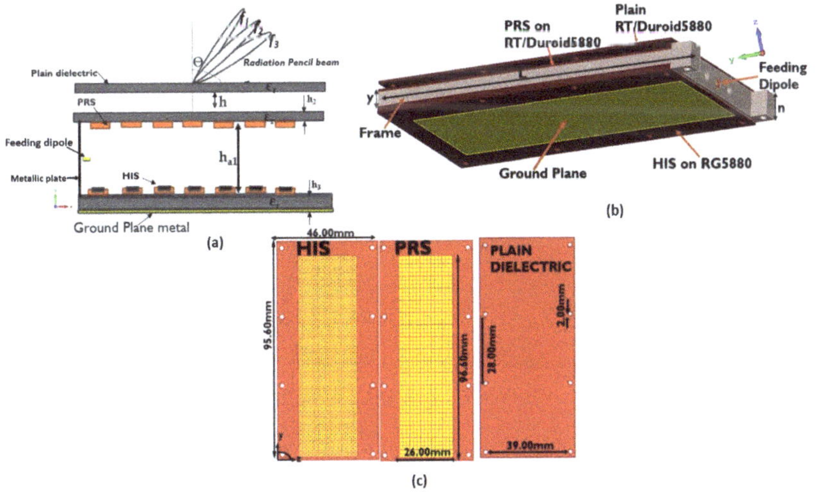

Figure 5. Proposed antenna structure: (**a**) the geometry of LWA; (**b**) view of the LWA along the xy plane; (**c**) 3D view of the LWA with a frame where y = 5 mm, n = 5.2 mm, and L = 102 mm.

The height of the cavity between the PRS and the HIS ground plane, which is represented by the symbol h_{a1} is calculated to be 2.90 mm. Following the ray optic method introduced in [25], the pointing angle of the radiation pattern "P" can be predicted as:

$$P = \frac{1 - \Gamma^2}{1 + \Gamma^2 - cos(\phi_{PRS} - \phi_{HIS} + \frac{4\pi}{\lambda_0} h_{a1} cos\theta)} F(\theta)^2 \tag{1}$$

where $F(\theta)$ equal to 1, Γ and ϕ_{PRS} represent the magnitude and phase of reflection coefficients of PRS, and λ is the operating wavelength.

The phase reflection coefficients of PRS and HIS as a function of frequencies are used to demonstrate LWA beam scanning. These coefficients are derived using Equation (1), which produces a wide scanning angle of 30°, as shown in Figure 6a. It was found that when HIS is replaced by a ground plane, the beam scanning reduces drastically to 18° with few frequencies, as shown in Figure 6b. This is because the ground plane is not sensitive enough to produce high phase reflection coefficients compared to HIS arrays. The cavity height h_{a1} and maximum gain calculation can be found in [20]. The far-field radiation patterns (FRPs) of the proposed antenna design are shown in Figure 7a for a cavity height of 2.90 mm. In particular, Figure 7a depicts the 2-D FRP in the E-plane at various frequencies within the operational range, and Figure 7b depicts the 3-D FRP at 62.5 GHz.

To prove the design concept, three additional variations of our proposed multi-layer LWA were considered, which have been demonstrated through simulations in CST Microwave Studio. Our proposed multi-layer LWA consists of a HIS-PRS cavity and a dielectric on top (as in Figure 5). Three variations of this model are compared: the proposed HIS-PRS LWA of Figure 5 without the dielectric layer on top, the LWA of Figure 5 with a mere ground plane instead of a HIS, and the proposed LWA with a mere ground plane and without the top dielectric layer.

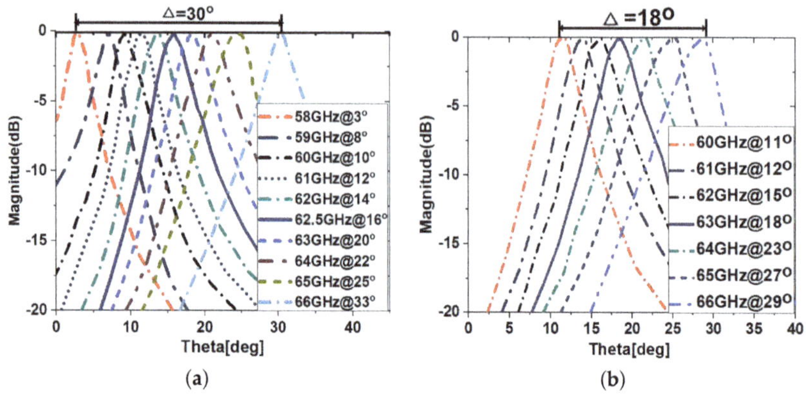

Figure 6. Theoretically estimated beam scanning range of the LWA from 58 to 66 GHz for a fixed cavity height of (**a**) h_{a1} = 2.90 mm with HIS and (**b**) h_{a1} = 3.32 mm without HIS.

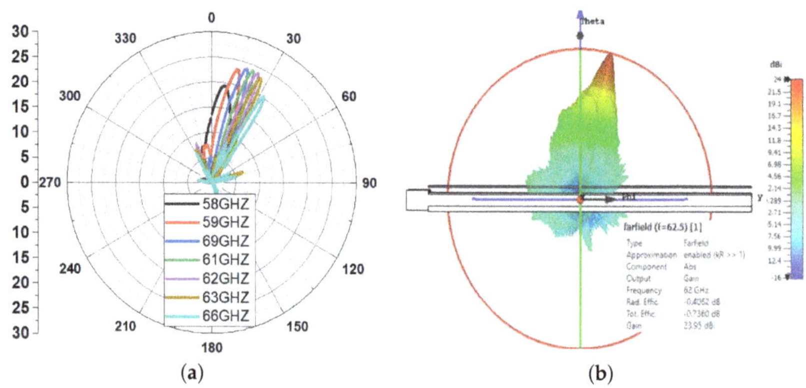

Figure 7. LWA simulation results: (**a**) 2-D FRPs in *E*-plane at various frequencies within the operating band; (**b**) 3-D FRP of the proposed LWA at 62.5 GHz.

The S_{11} parameters for the gain comparison for these four cases are depicted in Figure 8. The S_{11} (Figure 8a) is well below −10 dB for all four LWA models for cavity heights h_{a1} = 2.90 mm (antenna with HIS) and h_{a1} = 3.32 mm (antenna without HIS), and the realised gain remains above 19 dBi for the frequency range of 58 to 67 GHz. It is furthermore evident from Figure 8 that the proposed multi-layer LWA impedance matching, gain, and S_{11} bandwidth are significantly improved (24.3 dBi and 8 dB) by adding an extra plain dielectric slab over the PRS array with a beam scanning range of about 30° and a scanning loss of 3 dB. The proposed multi-layer LWA with HIS and dielectric exhibits a narrower half-power beamwidth (HPBW) compared to the other configurations, spanning from 9° at 58 GHz to 6° at 66.7 GHz. Additionally, the sidelobe level of the proposed multi-layer LWA remains below −10 dB in all the frequency bands. The antenna's simulated results are closely related to their previously calculated expected results. The aforementioned results are summarised in Table 1.

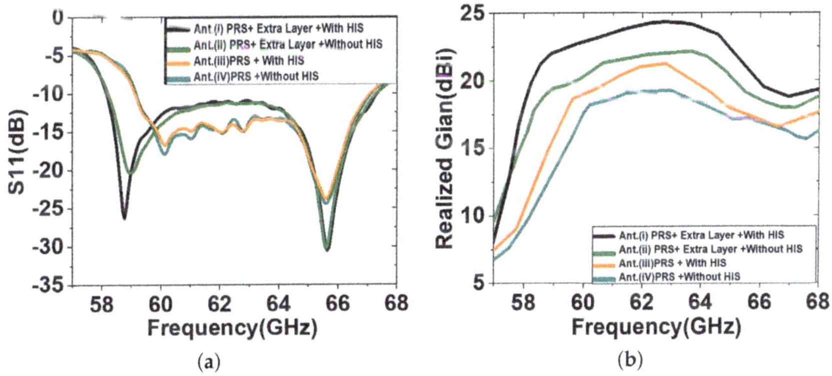

Figure 8. Simulated LWA models (**a**) S_{11} and (**b**) realised gain for four different LWA configurations.

Table 1. Comparison of simulated LWA results at a resonant frequency of 62.5 GHz.

Antenna Type	Operating Band $S_{11} < -10$ dB	Realised Gain (dBi)	Beam angles (deg)	HPBW (deg)
with HIS	58–67 GHz	22	10°–33°	12°–5°
without HIS	60–66.6 GHz	19.8	13°–33°	14°–6°
with HIS and dielectric	*58–66.7 GHz*	*24.3*	*4°–34°*	*9°–6°*
without HIS and dielectric	60–66.7 GHz	18	12°–33°	16°–7°

3. Multi-Layer LWA Measured Results and Discussion

The entire antenna design, including the coaxial feed, is encased in a low-loss polytetrafluoroethylene (PTFE) so that it can be placed in the most realistic environment possible and measurements can be made. An external connection is made using a 1.85 mm flanged launcher and a GB185 glass bead. The antenna layers were fabricated using a typical low-cost PCB fabrication method. The measurement configuration for the multi-layer LWA's S_11 prototype photo is shown in Figure 9, along with two prototypes of the proposed LWA (ant. A and B) for RVSM application.

The measurement of the antenna was carried out in an anechoic chamber. The gain-comparison method was used to determine the realised gain of our proposed multi-layer, leaky wave antenna. This method calls for two antennas: one that serves as a receive antenna and has a known gain (in our case, a horn antenna) and another that serves as a transmitter antenna whose gain is unknown (our proposed multi-layer LWA). The antennas were placed at a far-field distance of 1 m. The antenna gain was next calculated using Friis' equation.

A rotatable base connected to the computer for data acquisition was used to measure the patterns of antenna radiation. The antenna is rotated through 0°, 10°, 20°, 30°, up to 360° and then returned to the starting position after calibrating the VNA to 0 dB insertion loss. The information is then gathered and kept on file to forecast the signal strength. Figure 10 compares the simulated and measured results of the S_{11} and the gain of the proposed multi-layer LWA for RVSM. The measured values for the S_{11} bandwidth and the maximum gain of ant. A is 7.89 GHz and 23.95 dBi, respectively, which are slightly higher than the simulation values (6.8 GHz and 23.5 dBi). Over the key frequencies of interest ranging from 58 to 66 GHz, the S-parameter stays lower than -10 dB. In addition, the beam scanning ranges of each of the tested antennas are the same, with a scanning range of 30°. The small discrepancies between simulations and measurements are attributed to errors during the fabrication process, which affect the performance of the antenna at mm-wave frequencies.

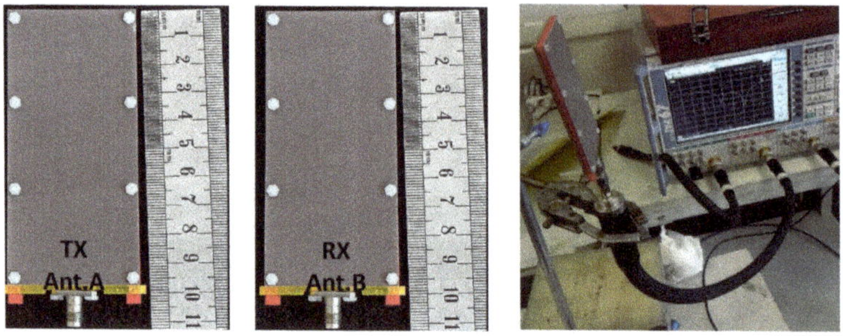

(a) (b)

Figure 9. (**a**) Fabricated LWA prototype with external connections (ant. and ant. B) (**b**) S_{11} test setup.

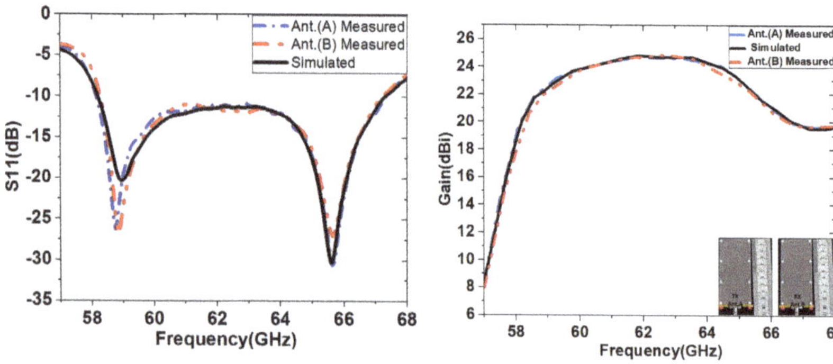

(a) (b)

Figure 10. Measured LWA results compared to simulated in (**a**) S_{11} and (**b**) Realised gain.

To validate the design concept, the measured far-field radiation patterns of the co-polar and cross-polar in the *E*-plane and *H*-plane radiation patterns at 62.5 GHz are presented in Figure 11. Manufacturing and radiation measurement tolerances, such as link losses, lateral reflections, and antenna misalignment, may be to blame for any differences between the simulated and observed S_{11} and FRPs. A complete comparison between simulated and measured *E*-plane radiation patterns at different frequencies is given in Figure 12a. They show excellent agreement between simulations and measurements.

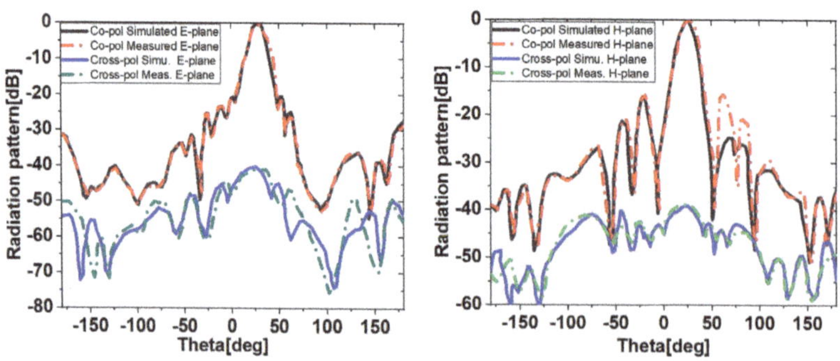

(a) (b)

Figure 11. Measured and simulated copolar and crosspolar (**a**) *E*-plane and (**b**) *H*-plane radiation patterns at 62.5 GHz.

Both the measured and simulated efficiency of antennas A and B remained at approximately 95% for the frequency range above 60 GHz (Figure 12b). For this measurement, first, a current was run through the antenna terminals, and then the strength of the electromagnetic field that went out into space was measured. In the next section, the measured antenna characteristics will be used for the RVSM experiment

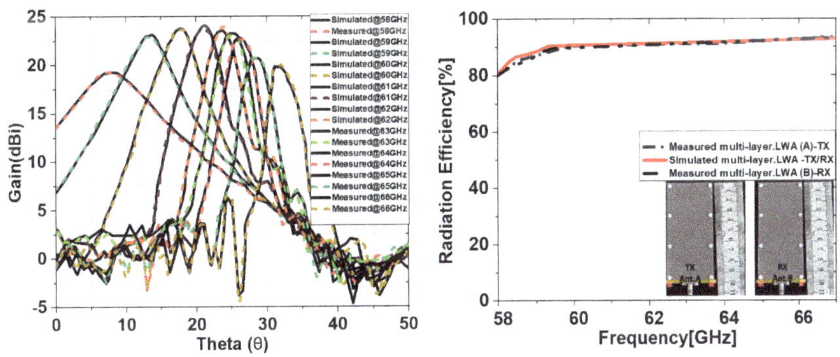

(**a**) (**b**)

Figure 12. The measured and simulated (**a**) *E*-plane radiation patterns of the multi-layer LWA and (**b**) radiation efficiency.

4. RVSM Measurements with Doppler Radar

In this experiment, we focus only on heart rate and respiration rate measured on the chest. Our proposed leaky wave antenna with a beam scanning range of 30° is sufficient for this type of experiment. However, it should be noted that to scan other parts of the body with this method, more complex systems should be used that reach beyond the scope of this paper.

The proposed RVSM process is shown in a block diagram in Figure 13a. The transmitting antenna (Tx) transmits a one-tone electromagnetic (EM), continuous wave (CW signal) in RVSM using a DR technique for a predetermined period of time. After the electromagnetic signal is scattered on the chest of a person standing at a determined distance in front of the antennas, it is simultaneously picked up by a receiving antenna (Rx). The quasi-periodic oscillations in the chest caused by breathing and heartbeats during the designated time phase modulate the received signal. The received phase-modulated signal and the transmitted signal are correlated for demodulation. The time domain (TD) information for the roughly recorded baseband signal at the Rx is provided below [4]:

$$R(t) = cos[\theta(t) + \frac{4\pi x_b(t)}{\lambda} + \frac{4\pi x_h(t)}{\lambda}] \quad (2)$$

where $\theta(t)$ represents the total phase shift brought on by the signal path (d), reflected signals from the environment and the subject, and residual phase noise. The $x_b(t)$ and $x_h(t)$ are the vibrational shifts of the chest that are, respectively, represented by the breathing and heartbeat, where λ is the operating wavelength. These shifts occur when the chest is subjected to vibration. Due to the periodic nature of $x_b(t)$ and $x_h(t)$, an approximation of them may be made as follows: $x_b(t) = m_b sin(2\pi f_b t)$ and $x_h(t) = m_h sin(2\pi f_h t)$, where m_b and m_h are the displacement amplitudes of the chest vibration. The recorded demodulated received signal can be expanded in a Fourier series in the manner described below and thus translated into the Frequency Domain (FD) [26]:

$$R(t) = \sum_{i=-\infty}^{\infty} \sum_{j=-\infty}^{\infty} J_j[\frac{4\pi m_b}{\lambda}] J_i[\frac{4\pi m_h}{\lambda}] \times \\ cos(j2\pi f_b t + i2\pi f_h t + \theta) \quad (3)$$

where $J(X)$ is the argument X representing a Bessel function of the first order. Equation (3) above contains the required BR and HR details, in addition to noise distortions and vibrations from the surroundings. After that, the appropriate digital filters are applied to Equation (3) in order to separate the essential BR and HR signals from the background noise as well as the harmonics that are not required. Figure 13b depicts the whole of digital signal post-processing. Figure 13c shows the experimental setup for the RVSM. The vector network analyser (VNA) with a maximum frequency limit of 67 GHz was used as the transceiver (TRx), which has the proposed Tx and Rx antenna modules connected to it for our RVSM measurements. An extensive series of experiments was carried out to verify the region that the beam-scanning antennas predicted for detecting vital signs while stationary.

To accomplish this, the VNA is first calibrated to send and receive electromagnetic waves with a transmit power of 0 dBm and 201 sampling points in the frequency range between 58 and 66 GHz. As shown in Figure 13c, the test subject sits at a certain radial distance and angular position in front of the antennas. A CW sweep with a tone for 60 s is performed. In a short time, the Tx antenna on port 1 of the VNA sends out a signal, and the Rx antenna on port 2 picks up the reflected signal that the subject sends back. The phase of the received signal is demodulated and recorded by the VNA for 60 s in an S_{21}-phase format. The next step is to extract the needed BR and HR data from the recorded S_{21} phase data through a signal processing programme in MATLAB. The main parts of the signal processing program are bandpass filters that allow BR and HR frequencies, discrete fast Fourier transform (DFFT), and notch filters (10th order Butterworth digital filters) that reduce noise, read out data, and display the heart rate and breathing rate. The values of BR and HR was compared with a hospital blood-pressure monitor for heart rate and manually counted average values of BR to make sure the measurements are accurate. For this reason, the experiment recorded the pulse values of BR and HR five times each and compared their average values with the values of BR and HR obtained by the DR system. For this reason, electrocardiograms (ECGs) are not used to compare waveforms. The reason for this is that, in the presented beam scanning DR to support the possibility of RVSM detection in connection with the suggested antenna beam steering, we are only interested in the average BR and HR (1/min) results in FD (and not in the BR and HR waveforms in time).

Fourier transform-based DR signalling has a RVSM acquisition time and BR/HR resolution trade-off, which must be strictly balanced. For instance, a signal recorded for 60 s will result in a frequency resolution of 1 pulse/s. Likewise, a recorded signal time of 30 s will reduce the BR/HR resolution to 0.5 pulses/s. Signal processing uses time windows. The DR signal is continuously captured in time windowing, but the signal processing is applied to a predetermined shorter time period (e.g., 60 s for a pulse resolution of 1/min), and the findings are updated after a shorter time interval (e.g., every 5 s) [27]. High-resolution RVSM results may be averaged across time using Fourier transform-based signal processing. Other complex signal-processing techniques can be seen in [14].

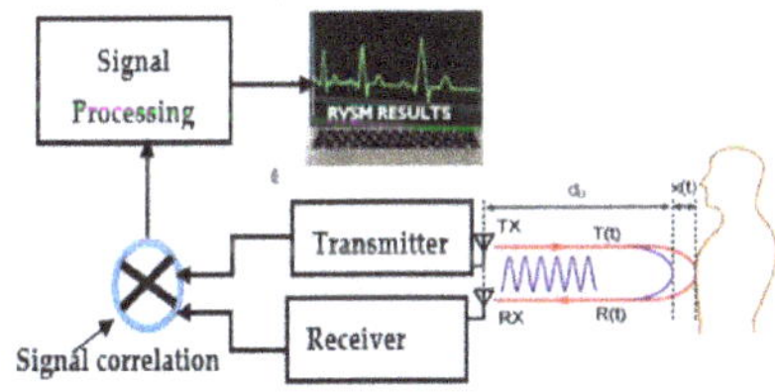

(a)

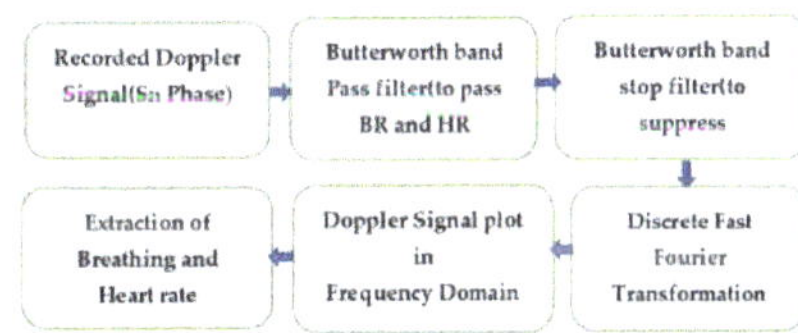

(b)

(c)

Figure 13. Demonstration of the RVSM experiment setup measurements.

In this experiment, the target is 2 m away from the transmitter and the receiver antennas at 5°, 10°, 15°, 20°, and 30°. For each angular point, the demodulated Doppler signal in TD, which contains the S_21 phase data, is recorded at 58 GHz. The results are replicated for frequencies of 60 GHz, 62.5 GHz, 64 GHz, and 66 GHz. As a result, we are able to collect five sets of S_21 phase data at the five specified angular locations for 58 GHz, 60 GHz, 62.5 GHz, 64 GHz, and 65 GHz, as shown in Figure 14 and Table 2. On the left of each subplot is the Doppler signal that was recorded in TD, and on the right is the corresponding processed signal in FD. It can be seen that the Doppler signal with the biggest amplitude compared to the level of background noise was picked up at a wider angle when the operating frequencies were higher. A good illustration of this is the modest peak HR, which is more easily detected by noise.

The estimated BR signal amplitude is seen in the first peak at a frequency of about 19 (1/min), while the predicted HR signal amplitude is seen in the second peak at a frequency of about 74 (1/min). At angles of 5° and 15°, the observed BR and HR signals exhibited amplitudes that were considerably larger than the noise values for the locations under examination. The results confirm predictions that the antenna radiation beam at 62.5 GHz is centred at roughly 25° and has an HPBW of greater than 12° (see Figure 11a and Table 2). It is also evident that the largest and clearest HR peak at 58 GHz is at 5° and 15°, whereas at 62.5 GHz and 64 GHz, it is at 5°, 15°, 25°, and 30°, respectively.

The HR peak is still visible at 66 GHz, and it is greater than the surrounding noise at 30° (See Figure 14e). It can be concluded that the RVSM detection response of the proposed antenna is suitable for a distance of 2 m at an angular range of 5° to 30°, which corresponds to an angular separation of 0.9 m when the operating frequency oscillates between 58 GHz and 66 GHz. Due to the antenna gain and transmit power, the angular range for RVSM detection can be increased beyond 0.9 m if the antennas are placed far away from the object [28].

A hospital blood pressure monitor that also records patients' heart rates on a daily basis was used to validate the results on the same individual, as shown in Figure 15. The BR and HR results from the experiment are compared with the results from the contact device, and it can be seen that both sets of results are in good agreement. This experiment, therefore, demonstrates that the proposed antenna can be used to measure both heart rate and breathing for longer distances (2–4 m) from the subject to provide more than 2 m of angular coverage.

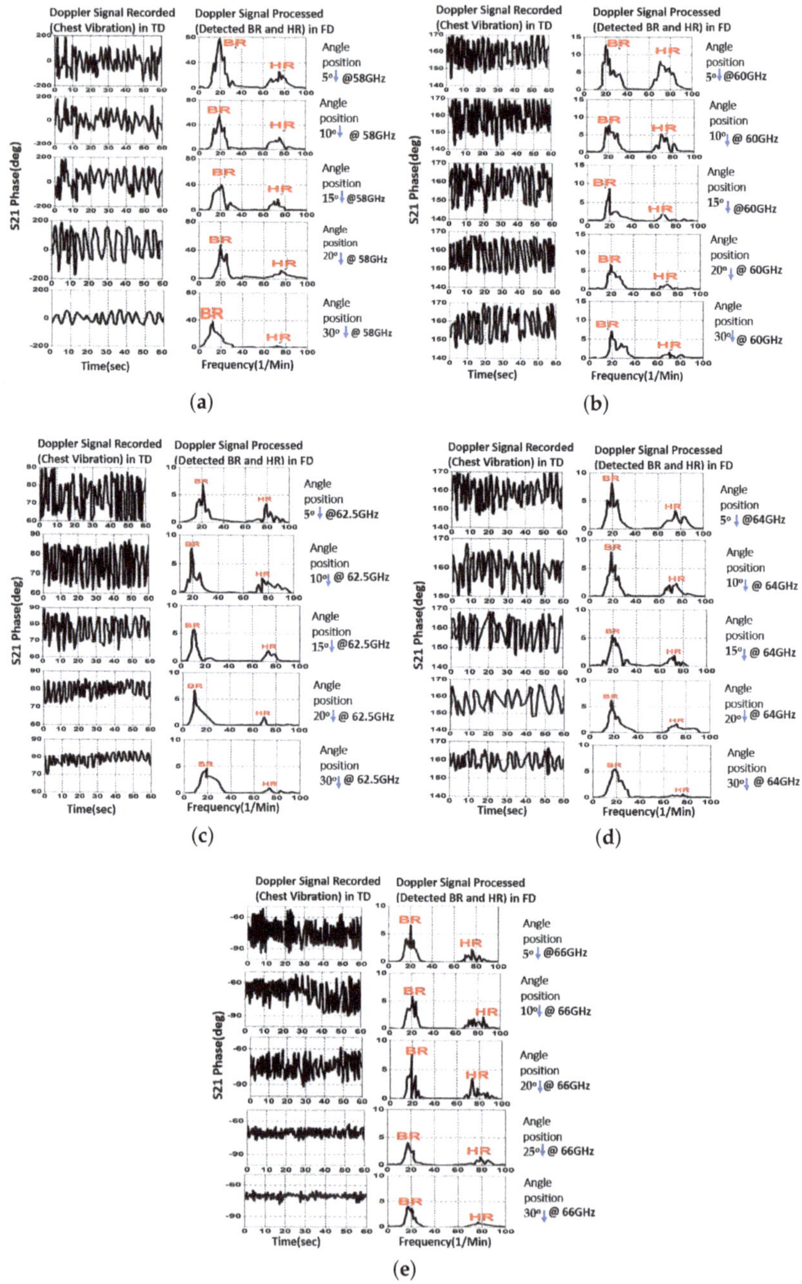

Figure 14. Measured RVSM results from 2 m distance at (**a**) 58 GHz, (**b**) 60 GHz, (**c**) 62.5 GHz, (**d**) 64 GHz, and (**e**) 66 GHz.

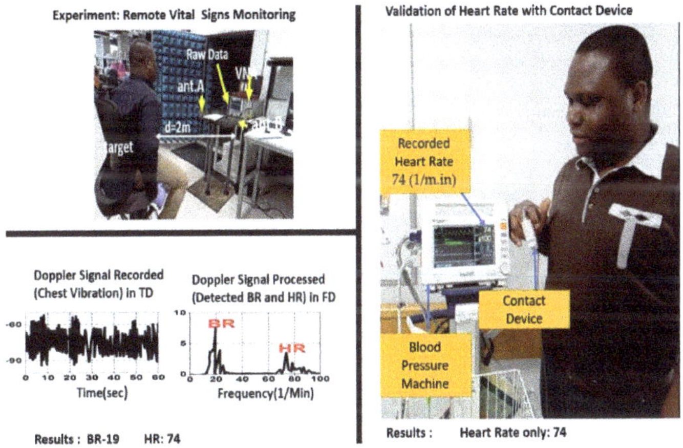

Figure 15. Validation of RVSM using (LHS) non-contact device (RHS) traditional blood pressure machine.

In order to demonstrate the antennas' range at a radial distance of 4 m, the experiment was repeated multiple times with the same subject, who sat at varying angles away from the antennas each time and had their vital signs recorded. The purpose of this was to validate the results and compare them to those obtained from a range of 2 m. The other experimental parameters are kept as previously described. Three RVSM measurements were taken at the best frequencies and beam angles for maximum detect ability, according to Figure 16: 58 GHz for 5°, 62.5 GHz for 25°, and 66 GHz for 30°.

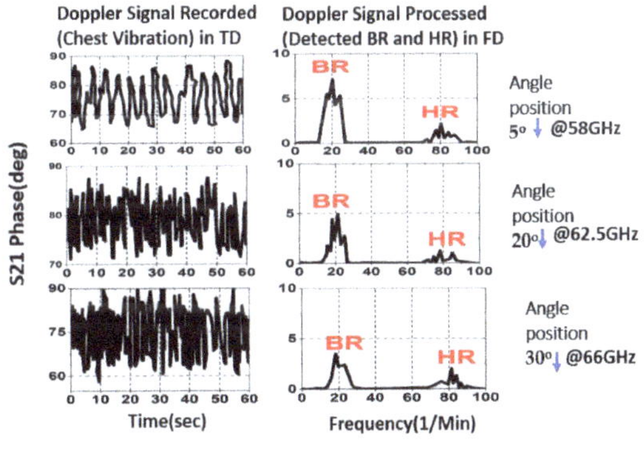

Figure 16. Measured RVSM results from a 4 m distance at 58 GHz, 62.5 GHz, and 66 GHz.

Table 2. Measured BR and HR with HIS-PRS LWA operation results at angles of 5°, 10°, 15°, 20°, and 30° for each frequency.

58 GHz	60 GHz	62.5 GHz	64 GHz	66 GHz	Manually Counted BR and HR Range with Hospital Machine		Max. Error
BR-HR (1/min)	BR-HR (1/min)	BR-HR (1/min)	BR-HR (1/min)	BR-HR (1/min)	BR (1/min)	HR (1/min)	BR-HR (1/min)
20–79	19–78	20–78	20–78	21–80	19–21 (avg. 20)	78–80 (avg. 79)	5–2%

The proposed multi-layer LWA design with Doppler radar technology is well-suited for use as a non-contact wireless sensor for health monitoring. As the experiment has shown, the sensor can detect both breathing rate (BR) and heart rate (HR) accurately and reliably at a radial distance of up to 4 metres. The non-contact nature of the sensor means that it can be used in a variety of scenarios, including when a patient is travelling in an ambulance or when they are sleeping or in occupational settings, where workers' health can be monitored without the need for physical contact (see Figure 1). Additionally, the sensor's ability to measure breathing rate and heart rate parameters within an angular range of 5° to 30° makes it a versatile tool which can be used in different settings. The proposed LWA design's potential applications are not limited to healthcare; however, the sensor's ability to detect and monitor physiological parameters from a distance could also be useful in sports, where athletes' performance and recovery can be monitored remotely.

Table 3 compares this work with others from relative recent literature with RVSM applications. Overall, the major contribution of our proposed LWA is the scanning distance of up to 4 metres from the human body with an improved accuracy of 99.3%, which is a significant advancement compared to other antennas and systems that have been used for RVSM. Additionally, our proposed antenna operates with an enhanced realised gain of 24.3 dBi and radiation efficiency of 95.5% compared to relevant work from the literature. The achieved bandwidth of our work is more limited compared to [17], which is due to the well-known trade-off between gain and beam aperture, which explains the more limited range of our antenna compared to [17]. However, the obtained realised gain of our proposed design makes it suitable not only for RVSM but for a number of other applications as well, such as 5G communications and IoT.

Table 3. Comparison of the proposed LWA with literature.

	[10]	[14]	[16]	[17]	This Work
Frequency (GHz)	60	24	62–65	50–66	**58–66**
Radiation efficiency (%)	94	-	95	83.2	**95.5**
Scanning range (°)	0	90	12	70	**30**
Gain (dBi)	24	-	20.1	19.3	**24.3**
Bandwidth (%)	12	-	3.38	16	**8**
RVSM distance (m)	1	3.4	3	1.5	**4**
RVSM accuracy (%)	98.6	99.12	90.8	95	**99.3**

5. Conclusions

The proposed multi-layer leaky wave antenna is a highly promising technology due to its numerous advantages. It operates in the 60 GHz band, which is ideal for applications requiring high data rates, low interference, and short-high-range communication. Additionally, the antenna has a high gain and enough beam-steering range for RVSM (Remote Vital Sign Monitoring) applications, which is essential for accurately tracking the movements of a person's body. It is manufactured using low-cost PCB (Printed Circuit Board) fabrication methods, which makes it a highly attractive option for widespread adoption. Additionally, the design has been tested and characterised through two prototypes, and good agreement has been observed between the simulations and measurements. The experiment conducted to test the antennas for respiration and heartbeat detection showed that the proposed antenna is highly effective. The antenna was tested at radial distances of 2–4 m and angular positions of 5° to 30° on the subject's body, with successful results. This indicates that the antenna is capable of accurately detecting BR and HR parameters from a distance, making it ideal for use in healthcare and other fields. The design method could be used to make low-cost antennas with high gain that can also work at higher mm-wave frequencies and have a wider range of radiation.

Author Contributions: Design and Conceptualisation, S.M.; methodology, S.M. and D.K.; software, S.M.; validation, S.M., D.K. and A.F.; investigation, S.M.; resources, D.K. and A.F.; data analysis, S.M.; writing—original draft preparation, S.M.; writing—review and editing, S.M., D.K. and A.F.; supervision, D.K. and A.F. All authors have read and agreed to the published version of the manuscript.

Funding: This research received no external funding.

Institutional Review Board Statement: Ethical review and approval were waived for this study due to an exception criterion of the AUT Ethics Committee where the investigators are the subject of their research and where no physically or psychologically hazardous procedure is involved.

Informed Consent Statement: Informed consent was obtained from all subjects involved in this study.

Data Availability Statement: The data that support the findings of this study are available from the corresponding author upon reasonable request.

Acknowledgments: We thank Roger Corporation, USA, for contributing their substrate material (RT5880) for this research.

Conflicts of Interest: The authors declare no conflict of interest.

References

1. Obeid, D.; Zaharia, G.; Sadek, S.; El Zein, G. Microwave Doppler radar for heartbeat detection vs electrocardiogram. *Microw. Opt. Technol. Lett.* **2012**, *54*, 2610–2617. [CrossRef]
2. Rabbani, M.S.; Ghafouri-Shiraz, H. Accurate remote vital sign monitoring with 10 GHz ultra-wide patch antenna array. *AEU Int. J. Electron. Commun.* **2017**, *77*, 36–42. [CrossRef]
3. Rabbani, M.S.; Ghafouri-Shiraz, H. Frequency selective surface antenna for remote vital sign monitoring with ultra-wide band doppler radar. *Microw. Opt. Technol. Lett.* **2017**, *59*, 818–823. [CrossRef]
4. Rabbani, M.S.; Ghafouri-Shiraz, H. Ultra-Wide Patch Antenna Array Design at 60 GHz Band for Remote Vital Sign Monitoring with Doppler Radar Principle. *J. Infrared Millim. Terahertz Waves* **2017**, *38*, 548–566. [CrossRef]
5. Droitcour, A.D.; Seto, T.B.; Park, B.K.; Yamada, S.; Vergara, A.; El Hourani, C.; Shing, T.; Yuen, A.; Lubecke, V.M.; Boric-Lubecke, O. Non-contact respiratory rate measurement validation for hospitalised patients. In Proceedings of the 31st Annual International Conference of the IEEE Engineering in Medicine and Biology Society: Engineering the Future of Biomedicine, EMBC 2009, Minneapolis, MN, USA, 3–6 September 2009. [CrossRef]
6. Villarroel, M.; Guazzi, A.; Jorge, J.; Davis, S.; Watkinson, P.; Green, G.; Shenvi, A.; McCormick, K.; Tarassenko, L. Continuous non-contact vital sign monitoring in neonatal intensive care unit. *Healthc. Technol. Lett.* **2014**, *1*, 87–91. [CrossRef] [PubMed]
7. Suzuki, S.; Matsui, T.; Kagawa, M.; Asao, T.; Kotani, K. An approach to a non-contact vital sign monitoring using dual-frequency microwave radars for elderly care. *J. Biomed. Sci. Eng.* **2013**, *6*, 34091. [CrossRef]
8. Gu, C. Short-range noncontact sensors for healthcare and other emerging applications: A review. *Sensors* **2016**, *16*, 1169. [CrossRef]
9. Li, C.; Lubecke, V.M.; Boric-Lubecke, O.; Lin, J. A review on recent advances in doppler radar sensors for noncontact healthcare monitoring. *IEEE Trans. Microw. Theory Tech.* **2013**, *61*, 2046–2060. [CrossRef]
10. Yang, Z.; Pathak, P.H.; Zeng, Y.; Liran, X.; Mohapatra, P. Vital sign and sleep monitoring using millimeter wave. *ACM Trans. Sens. Netw.* **2017**, *13*, 14. [CrossRef]
11. Shen, T.M.; Kao, T.Y.J.; Huang, T.Y.; Tu, J.; Lin, J.; Wu, R.B. Antenna design of 60-GHz micro-radar system-in-package for noncontact vital sign detection. *IEEE Antennas Wirel. Propag. Lett.* **2012**, *11*, 1702–1705. [CrossRef]
12. Hall, T.; Nukala, B.T.; Stout, C.; Brewer, N.; Tsay, J.; Lopez, J.; Banister, R.E.; Nguyen, T.; Lie, D.Y. A phased array non-contact vital signs sensor with automatic beam steering. In Proceedings of the 2015 IEEE MTT-S International Microwave Symposium, IMS 2015, Phoenix, AZ, USA, 17–22 May 2015. [CrossRef]
13. Hall, T.; Lie, D.Y.; Nguyen, T.Q.; Mayeda, J.C.; Lie, P.E.; Lopez, J.; Banister, R.E. Non-contact sensor for long-term continuous vital signs monitoring: A review on intelligent phased-array doppler sensor design. *Sensors* **2017**, *17*, 2632. [CrossRef] [PubMed]
14. Peng, Z.; Li, C. Portable Microwave Radar Systems for Short-Range Localization and Life Tracking: A Review. *Sensors* **2019**, *19*, 1136. [CrossRef] [PubMed]
15. Jackson, D.R.; Oliner, A.A. Leaky-Wave Antennas. In *Modern Antenna Handbook*; Wiley: Hoboken, NJ, USA, 2007. [CrossRef]
16. Rabbani, M.S.; Churm, J.; Feresidis, A.P. Fabry-Perot Beam Scanning Antenna for Remote Vital Sign Detection at 60 GHz. *IEEE Trans. Antennas Propag.* **2021**, *69*, 3115–3124. [CrossRef]
17. Mingle, S.; Kampouridou, D.; Feresidis, A. Bidirectional Leaky-Wave Antenna Based on Dielectric Image Line for Remote Vital Sign Detection at mm-Wave Frequencies. *IEEE Open J. Antennas Propag.* **2022**, *3*, 1003–1012. [CrossRef]
18. Feresidis, A.P.; Vardaxoglou, J.C. High gain planar antenna using optimised partially reflective surfaces. *IEE Proc. Microw. Antennas Propag.* **2001**, *148*, 345–350. [CrossRef]
19. Goussetis, G.; Feresidis, A.P.; Vardaxoglou, J.C. Tailoring the AMC and EBG characteristics of periodic metallic arrays printed on grounded dielectric substrate. *IEEE Trans. Antennas Propag.* **2006**, *54*, 82–89. [CrossRef]

20. Mingle, S. Frequency Scannable Leaky-Wave Antenna Design for Remote Vital Signs Radar Application at 60GHz Band. In Proceedings of the 3rd International Conference on Electrical, Communication and Computer Engineering, ICECCE 2021, Kuala Lumpur, Malaysia, 12–13 June 2021. [CrossRef]
21. Bancroft, R. *Microstrip and Printed Antenna Design*; IET Digital Library: London, UK, 2019. [CrossRef]
22. Feresidis, A.P.; Goussetis, G.; Wang, S.; Vardaxoglou, J.C. Artificial magnetic conductor surfaces and their application to low-profile high-gain planar antennas. *IEEE Trans. Antennas Propag.* **2005**, *53*, 209–215. [CrossRef]
23. Zhang, Y.; Von Hagen, J.; Younis, M.; Fischer, C.; Wiesbeck, W. Planar Artificial Magnetic Conductors and Patch Antennas. *IEEE Trans. Antennas Propag.* **2003**, *51*, 2704–2712. [CrossRef]
24. Saqib Rabbani, M.; Churm, J.; Feresidis, A. 60GHz-Band Leaky-Wave Antenna for Remote Health Monitoring. In Proceedings of the 13th European Conference on Antennas and Propagation, EuCAP 2019, Krakow, Poland, 31 March–5 April 2019.
25. Von Trentini, G. Partially Reflecting Sheet Arrays. *IRE Trans. Antennas Propag.* **1956**, *4*, 666–671. [CrossRef]
26. Li, C.; Lin, J. Optimal carrier frequency of non-contact vital sign detectors. In Proceedings of the 2007 IEEE Radio and Wireless Symposium, RWS, Long Beach, CA, USA, 9–11 January 2007. [CrossRef]
27. Rabbani, M.S.; Ghafouri-Shiraz, H. Dual-layer partially reflective surface antennas based on extended size unit cells for 60 GHz band WLAN/WPAN. *IET Microw. Antennas Propag.* **2018**, *12*, 789–795. [CrossRef]
28. Biglarbegian, B.; Fakharzadeh, M.; Busuioc, D.; Nezhad-Ahmadi, M.R.; Safavi-Naeini, S. Optimised microstrip antenna arrays for emerging millimeter-wave wireless applications. *IEEE Trans. Antennas Propag.* **2011**, *59*, 1742–1747. [CrossRef]

sensors

MDPI

Communication

Maximizing Antenna Array Aperture Efficiency for Footprint Patterns

Cibrán López-Álvarez [1], María Elena López-Martín [2], Juan Antonio Rodríguez-González [3] and Francisco José Ares-Pena [3,*]

[1] Center for Research in NanoEngineering, Campus Diagonal-Besòs, Polytechnic University of Catalonia, 08019 Barcelona, Spain; cibran.lopez@upc.edu
[2] Department of Morphological Sciences, University of Santiago de Compostela, 15782 Santiago de Compostela, Spain; melena.lopez.martin@usc.es
[3] Department of Applied Physics, University of Santiago de Compostela, 15782 Santiago de Compostela, Spain; ja.rodriguez@usc.es
* Correspondence: francisco.ares@usc.es

Abstract: Despite playing a central role in antenna design, aperture efficiency is often disregarded. Consequently, the present study shows that maximizing the aperture efficiency reduces the required number of radiating elements, which leads to cheaper antennas with more directivity. For this, it is considered that the boundary of the antenna aperture has to be inversely proportional to the half-power beamwidth of the desired footprint for each ϕ-cut. As an example of application, it has been considered the rectangular footprint, for which a mathematical expression was deduced to calculate the aperture efficiency in terms of the beamwidth, synthesizing a rectangular footprint of a 2:1 aspect ratio by starting from a pure real flat-topped beam pattern. In addition, a more realistic pattern was studied, the asymmetric coverage defined by the European Telecommunications Satellite Organization, including the numerical computation of the contour of the resulting antenna and its aperture efficiency.

Keywords: antenna array apertures; pattern synthesis; footprint patterns

Citation: López-Álvarez, C.; López-Martín, M.E.; Rodríguez-González, J.A.; Ares-Pena, F.J. Maximizing Antenna Array Aperture Efficiency for Footprint Patterns. *Sensors* **2023**, *23*, 4982. https://doi.org/10.3390/s23104982

Academic Editor: Raed A. Abd-Alhameed, Hendrik Rogier, Chan Hwang See and Naser Ojaroudi Parchin

Received: 28 March 2023
Revised: 12 May 2023
Accepted: 19 May 2023
Published: 22 May 2023

1. Introduction

In IEEE Standard for Definitions of Terms for Antennas [1], the antenna illumination efficiency is defined as the ratio of the maximum directivity of an antenna aperture to its standard directivity, and the antenna aperture efficiency as the ratio of the maximum effective area of the antenna to the aperture area. In some cases, illumination and aperture efficiencies might coincide. It is remarkable that those antenna arrays synthesized with small aperture efficiency require more radiating elements than necessary.

Kim and Elliott [2] proved that the extensions of Tseng–Cheng distributions, which give a flat-topped beam in every ϕ-cut, are inefficient because they use rectangular boundary arrays, and, as a consequence, the obtained shaped patterns are almost rotationally symmetric, and they present ring side lobes that are not circular ($\theta_{peak} \neq constant$). Obviously, the optimal boundary for this distribution should be circular and not rectangular.

The vast majority of works regarding footprint pattern synthesis use rectangular, circular, or elliptical antennas independently of the shape of the desired pattern ([3–10] being some representative examples). Even some works related to reflectarrays suffer from this issue [11,12].

In [13,14], a synthesis was implemented that tried to slightly optimize antenna aperture efficiency, but without analyzing the problem in depth.

Elliott and Stern [15] have suggested that, in order to obtain a highly efficient antenna, its contour has to be inversely proportional to its half-power beamwidth (HPBW) in every ϕ-cut. This technique was developed by Ares et al. [16] in order to synthesize square footprints. Afterwards, Fondevila et al. [17] numerically optimized the contour of an antenna to obtain rectangular footprints.

More recently, López-Álvarez et al. [18] presented an efficient iterative method that, starting from a circular aperture and removing those elements with low-amplitude excitations, generates footprint patterns.

In this work, a study is presented which tries to maximize aperture efficiency for rectangular footprints as well as for the case of the asymmetric coverage defined by the European Telecommunications Satellite Organization (EuTELSAT) for the first time to the best of our knowledge. The role of aperture efficiency in the synthesis of high-performance antennas is highlighted, a topic that is usually disregarded in modern studies, given that, as previously stated, high aperture efficiencies guarantee not using more radiating elements than needed. The use of conventional methods, which do not maximize aperture efficiency, would increase the price and even diminish the directivity for those antennas in which illumination and aperture efficiencies coincide, requiring larger antenna areas.

In order to synthesize optimal antenna patterns with specific ripple and side-lobe levels, it is necessary to optimize the disposition of radiating elements within the antenna. If this is not achieved, the antenna contour has to be optimized. This work proposes antenna contours that are very close to the optimal solution, which allows obtaining this with global or even local optimization methods.

2. Materials and Methods

A unidirectional planar, circular aperture of radius a and continuous aperture distribution $K_0(\rho)$ (that is, a linearly polarized planar aperture distribution), with the notation expressed in Figure 1, produces the ϕ-symmetric pattern [19,20] from Equation (1).

$$F(\theta) = 2\pi \int_0^a K_0(\rho) J_0(k\rho \sin(\theta))\rho d\rho \tag{1}$$

where λ is the wavelength, J_0 is the zeroth Bessel function of the first kind, and θ is the azimuthal angle, that is, the angle measured from the zenith of this aperture. Consider the following relations:

$$u = \frac{2a}{\lambda}\sin(\theta); \quad p = \frac{\pi}{a}\rho; \quad g_0 = \frac{2a^2}{\pi}K_0(\rho) \tag{2}$$

where the parameter u defines the pointing direction in real space, consisting of an equation that relates the angle and the wavelength, this angle defining the HPBW of the flat-topped beam, and ρ is the radial coordinate of the aperture. These substitutions transform the last equation into

$$F(u) = \int_0^\pi p g_0(p) J_0(up) dp \tag{3}$$

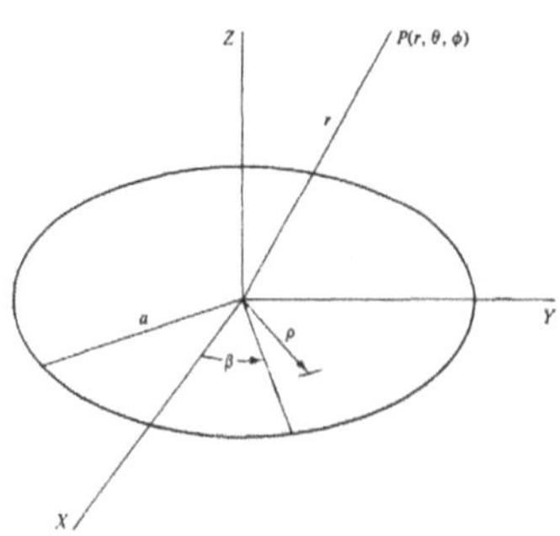

Figure 1. Circular aperture of radius a. The cylindrical coordinates ρ, β refer to the antenna aperture, while the spherical coordinates r, θ, ϕ are used for defining the field point P.

For instance, the case of a constant, uniform aperture $g_0(p) = 1$ yields the well-known pattern

$$F_0(u) = \frac{J_1(\pi u)}{\pi u} \tag{4}$$

consisting of a main beam surrounded by a family of ring side lobes (given the existing axial symmetry), where J_1 is the first-order Bessel function of the first kind.

Then, by representing the aperture distribution in terms of the roots of J_1, such that $J_1(\pi\gamma_{1n}) = 0$, $n = 0, 1, 2\ldots$, we obtain [19]:

$$g_0(p) = \sum_{n=0}^{\infty} B_n J_0(\gamma_{1n} p) \tag{5}$$

Integrating the initial ϕ-symmetric pattern (Equation (3)) with the previous aperture distribution (Equation (5)), evaluated at the roots γ_{1n}, $n = 0, 1, 2\ldots$, we obtain the aperture distribution

$$g_0(p) = \frac{2}{\pi^2} \sum_{n=0}^{\infty} \frac{F(\gamma_{1n}) J_0(\gamma_{1n} p)}{J_0^2(\gamma_{1n}\pi)} \tag{6}$$

Thus, if the roots $u_n = \gamma_{1n}$ for $n \geq \bar{n}$ are kept ($\bar{n}$ being the transition parameter), but the inner roots are displaced for $n = 1, 2, \ldots, \bar{n} - 1$ to the new complex positions $u_n + jv_n \neq \gamma_{1n}$, the pattern becomes a rotationally symmetric field that can be radiated by a circular aperture, with properly filled nulls in the shaped region and controlled side lobe levels in the unshaped region. With appropriate values of u_n and v_n, it is possible to synthesize both real and complex flat-topped beam patterns using [15,17,19]. This is accomplished by dividing Equation (4) by its $(1+\epsilon)M + s$ first zeros and multiplying by the new, displaced ones:

$$F(u) = \frac{J_1(\pi u)}{\pi u} \frac{\prod_{n=1}^{M}\left(1 - \frac{u^2}{(u_n + jv_n)^2}\right)\left(1 - \frac{u^2}{(u_n - jv_n)^2}\right)^{\epsilon} \prod_{n=M+1}^{M+s}\left(1 - \frac{u^2}{u_n^2}\right)}{\prod_{n=1}^{(1+\epsilon)M+s}\left(1 - \frac{u^2}{\gamma_{1n}^2}\right)} \tag{7}$$

where $\epsilon = \{0, 1\}$ (the pattern is real if $\epsilon = 1$ and complex if $\epsilon = 0$). As a result, $F(u_n) \neq 0$ unless $v_n = 0$; thus, there exist new complex roots positions $(u_n + jv_n)$ for which the pattern has properly filled roots in the shaped region and controlled side lobe levels in the rest. For $n \in (M+1, M+s)$, the peak levels of the inner s side lobes depend on the values of u_n, with a decay of $u^{-3/2}$ for distant side lobes. The flat-topped beam is composed of a central beam surrounded by M annular ripples of the same height, with the depth of the troughs between these components depending on the u_n and v_n for $n \in [1, M]$. The corresponding aperture distribution given by Equation (6) truncates at $\bar{n} = (1+\epsilon)M + s + 1$.

Therefore, a flat-topped beam extended about $u_0 = \frac{2a}{\lambda}\sin(\theta_0)$ such that $F(u_0) = -3$ dB will give a θ_0 value that is smaller if $\frac{a}{\lambda}$ is larger. The achieved flat-topped beam for real patterns ($\epsilon = 1$) is broader than those corresponding to the complex patterns ($\epsilon = 0$), and the u_0 value is also bigger in real patterns. The angle θ_0 will define the HPBW of the flat-topped beam. Consequently, the flat-topped HPBW is inversely proportional to $\rho_{max}(\beta)$, which is the distance along the β line in the XY plane out of the periphery.

$$\frac{2a}{\lambda}\sin(\theta_0) = \frac{2\rho_{max}(\beta)}{\lambda}\sin(\theta(\beta)) \tag{8}$$

Thus, the product of the antenna size by the HPBW ($a \cdot \beta\omega_0$) is conserved:

$$a \cdot \beta\omega_0 = \rho_{max}(\beta) \cdot \beta\omega(\beta) \tag{9}$$

As a particular case, we can now consider a rectangular footprint, with quadrant symmetry. The maximum radius for the rectangular footprint is

$$\rho_{max}(\beta) = \begin{cases} \frac{\beta\omega_0}{\beta\omega_a} a \cdot \cos(\beta) \text{ if } 0 \leq \beta < \alpha \\ \frac{\beta\omega_0}{\beta\omega_b} a \cdot \sin(\beta) \text{ if } \alpha \leq \beta < \pi/2 \end{cases} \tag{10}$$

Given this, for some angle α, $\rho_{max}(\beta)$ must be the same for both cases, with $\beta\omega_a, \beta\omega_b$ being the HPBW in each axial direction:

$$\tan(\alpha) = \frac{\beta\omega_b}{\beta\omega_a} \tag{11}$$

By considering $\beta\omega_a \geq \beta\omega_b$, and given that the required antenna for synthesizing this specific pattern (from Equation (10)) cannot exceed the available one (that is, the antenna defined initially, a circular aperture of radius a), it can be shown that there is the following constraint regarding the ratios of the HPBW in each axial direction:

$$\frac{\beta\omega_0}{\beta\omega_b} a \sin\beta \cos\beta = \frac{\beta\omega_0}{\beta\omega_b} a \frac{\sin 2\beta}{2} = f(\beta) \leq a \tag{12}$$

with $\alpha \leq \beta \leq \frac{\pi}{2}$. It is straightforward to obtain the maximum value of the function $f(\beta)$:

$$\frac{df}{d\beta} = 0 \Rightarrow \frac{\beta\omega_0}{\beta\omega_b} a \cos 2\beta = 0 \Rightarrow \beta_m = \frac{\pi}{4} \tag{13}$$

For that angle, the function takes the value

$$f(\beta_m) = a \frac{\beta\omega_0}{\beta\omega_b} \frac{1}{2} \tag{14}$$

Considering that the HPBW associated with the aperture of radius a is such that $\beta\omega_0 = \beta\omega_a$, the constraint regarding the ratios of the HPBW results in

$$f(\beta_m) = a \frac{\beta\omega_a}{\beta\omega_b} \frac{1}{2} \Rightarrow \beta\omega_a \leq 2\beta\omega_b \tag{15}$$

therefore:

$$1 \leq \frac{\beta\omega_a}{\beta\omega_b} \leq 2 \tag{16}$$

or, otherwise, the shape of the antenna could not verify the aspect ratio for the desired footprint.

The effective area of the antenna, in radial coordinates, is computed as the sum of two terms, depending on the value of $\rho_{max}(\beta)$:

$$A_e = \frac{1}{2} \int_0^{\frac{\pi}{2}} \rho^2{}_{max}(\beta) d\beta \tag{17}$$

Therefore, the effective area (for a quadrant) is

$$A_e = \frac{(a \cdot \beta\omega_0)^2}{8} \left(\frac{2\alpha + \sin(2\alpha)}{(\beta\omega_a)^2} + \frac{\pi - 2\alpha + \sin(2\alpha)}{(\beta\omega_b)^2} \right) \tag{18}$$

On the other hand, the antenna aperture area for such a pattern (with the shape of a rectangle) is

$$A = \frac{(a \cdot \beta\omega_0)^2}{(\beta\omega_a)(\beta\omega_b)} \tag{19}$$

The antenna efficiency, as previously defined, is

$$\eta_a = \frac{A_e}{A} = \frac{1}{8}\left(\frac{\beta\omega_b}{\beta\omega_a}(2\alpha + \sin(2\alpha)) + \frac{\beta\omega_a}{\beta\omega_b}(\pi - 2\alpha + \sin(2\alpha))\right) \tag{20}$$

The effective area of the antenna can be expressed in terms of the directivity [20] as

$$A_e = \eta_a A = D_{max}\frac{\lambda^2}{4\pi} \tag{21}$$

where A is the area of the antenna, and $\eta_a \leq 1$ is the antenna efficiency of an aperture-type antenna. On the other hand, the standard directivity [21], the directivity that can be obtained with an aperture A, is

$$D_{std} \leq A\frac{4\pi}{\lambda^2} \tag{22}$$

As a result, considering the maximum value for D_{std}, η_i being the illumination efficiency and taking into account Equation (21), we have

$$\eta_i = \frac{D_{max}}{D_{std}} = \frac{\lambda^2}{4\pi}\frac{D_{max}}{A} = \frac{A_e}{D_{max}}\frac{D_{max}}{A} = \frac{A_e}{A} = \eta_a \tag{23}$$

that is, the illumination efficiency equals the antenna aperture efficiency. Thus, in this case, a good antenna aperture efficiency implies a good illumination efficiency.

3. Discussion

3.1. Application to Footprints with Quadrant Symmetry

For this set of applications, Equation (20) is implemented, taking into account quadrantal symmetry. We might check the case of a clover generating a square pattern (Figure 2). Considering $\alpha = \frac{\pi}{4}$ and $\beta\omega_a = \beta\omega_b$, the aperture efficiency would be

$$\eta_{1:1} = 0.64 \tag{24}$$

As an example of application of the synthesis of a square footprint pattern of approximately $20^\circ \times 20^\circ$ using an aperture of the shape of Figure 2, in [16], a continuous aperture distribution $g(p)$ (that is, pure real $g_0(p)$ from Equation (6) truncated at $\bar{n} = (1+\epsilon)M + s + 1$ with $\epsilon = 1$, $M = 2$, and $s = 2$) was stretched into a distribution within its boundary and afterwards sampled to be applied to a rectangular grid array; 36% of the elements would be saved (thus, 52 elements of a total of 144 for each quadrant). This coincides with the fact that the effective area is 64% of the total area (from Equation (24)).

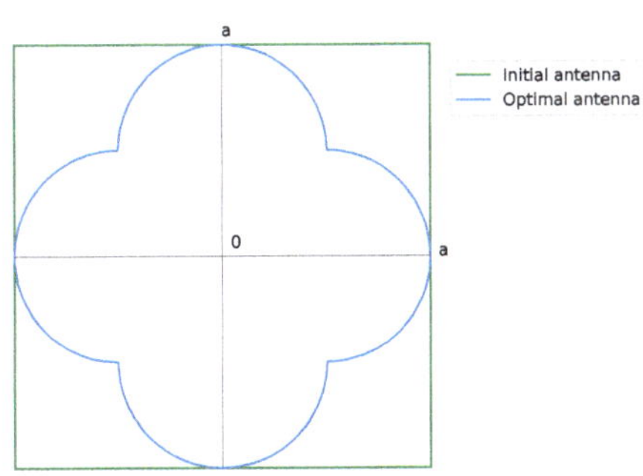

Figure 2. Shape of the required antenna for a rectangular footprint pattern of 1:1 aspect ratio. The required antenna matches the available one. The contour is obtained from Equation (10). The solid green line represents the initial antenna, while the solid blue line is the optimal antenna.

Then, a rectangular footprint pattern of a 2:1 aspect ratio (Figure 3), with $\tan(\alpha) = \frac{1}{2}$, leads to

$$\eta_{2:1} = 0.86 \tag{25}$$

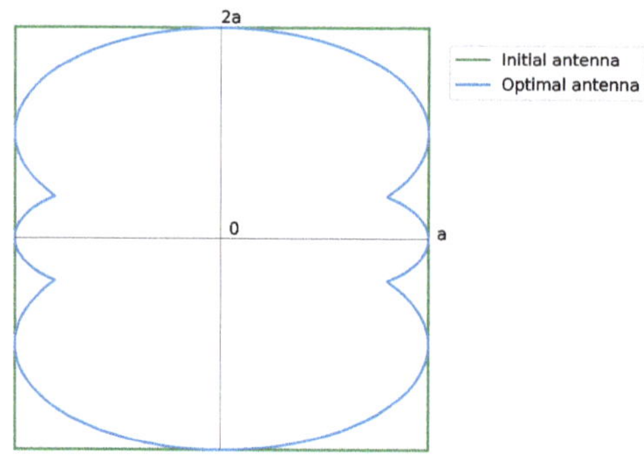

Figure 3. Shape of the required antenna for a rectangular footprint pattern of 2:1 aspect ratio. The required antenna matches the available one. The contour is obtained from Equation (10). The solid green line represents the initial antenna, while the solid blue line is the optimal antenna.

We consider a rectangular footprint pattern of a 3:1 aspect ratio (Figure 4), with $\tan(\alpha) = \frac{1}{3}$. As can be seen, the required antenna exceeds the available one, indicated with solid lines in Figure 4. Thus, the real aperture efficiency η^r has to consider the complete required antenna, with dashed lines Figure 4, which are

$$\begin{aligned} \eta_{3:1} &= 1.21 \\ \eta^r_{3:1} &= 0.80 \end{aligned} \tag{26}$$

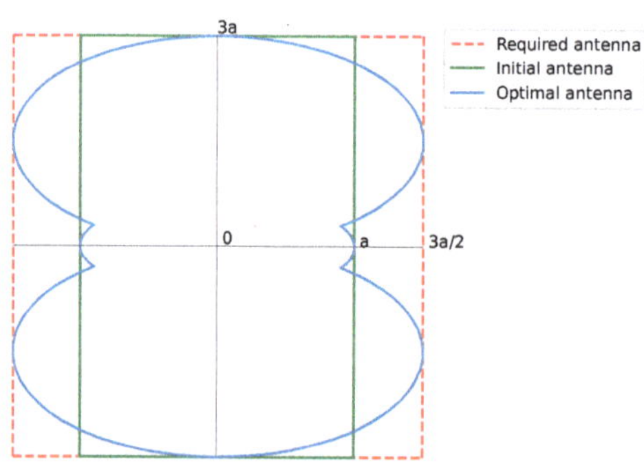

Figure 4. Shape of the required antenna for a rectangular footprint pattern of 3:1 aspect ratio. The required antenna exceeds the available one. The contour is obtained from Equation (10). The solid green line represents the initial antenna, while the dashed red line is the required antenna, and the solid blue line is the optimal antenna.

Furthermore, this effect is accentuated for a rectangular footprint pattern of a 4:1 aspect ratio (Figure 5), with $\tan(\alpha) = \frac{1}{4}$, which leads to

$$\begin{aligned} \eta_{4:1} &= 1.59 \\ \eta^r_{4:1} &= 0.80 \end{aligned} \tag{27}$$

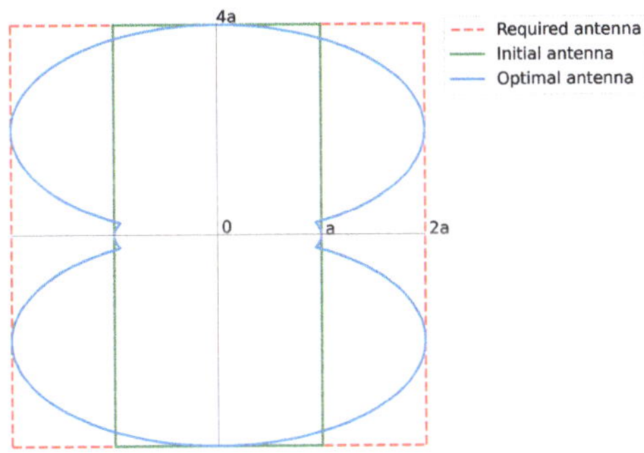

Figure 5. Shape of the required antenna for a rectangular footprint pattern of 4:1 aspect ratio. The required antenna exceeds the available one. The contour is obtained from Equation (10). The solid green line represents the initial antenna, while the dashed red line is the required antenna, and the solid blue line is the optimal antenna.

The synthesis of a rectangular footprint of a 2:1 aspect ratio is exemplified, by starting from a pure real flat-topped beam pattern ($\epsilon = 1$) with a side-lobe level $SLL = -25$ dB, $\bar{n} = 6$, $M = 2$ filled nulls, and a ripple level of ± 0.5 dB; the method depicted in [18] synthesized a pattern with $SLL = -25$ dB and a ripple level of ± 0.8 dB. The resulting array has 1044 elements 0.5λ spaced. Figure 6 shows the normalized aperture distribution as well as the final pattern.

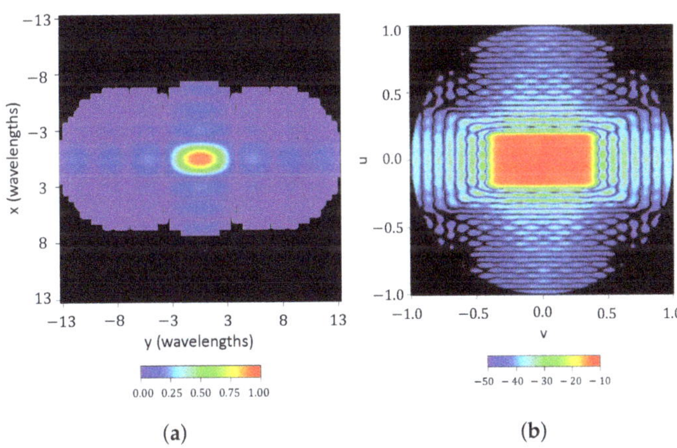

Figure 6. Normalized aperture distribution and interpolated image of the reconstructed pattern with a threshold level set at −50 dB from the antenna contour shown in Figure 3. (**a**) Normalized aperture distribution. (**b**) Reconstructed pattern.

3.2. Application to Asymmetric Footprints

As an initial pattern to compose all the footprints in this section, both real and complex (obtained with the methods described in [15,19] and shown in Figures 7 and 8, respectively) flat-topped beam pattern boundaries are considered. Both sets of roots are implemented with two filled zeros (Figure 9) and one only filled zero (Figure 10). The latter requires much smaller antennas than the former.

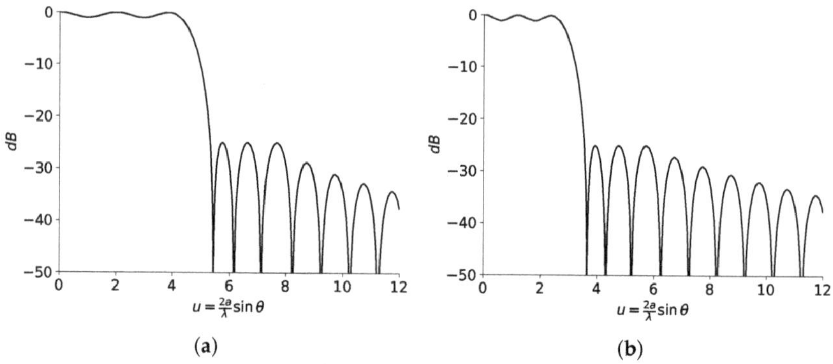

Figure 7. Patterns from (**a**) real and (**b**) complex roots. The HPBW for each set of roots is obtained at (**a**) $u_0 = 4.54$ and (**b**) $u_0 = 2.86$. Produced with a side-lobe level $SLL = -25$ dB, $\bar{n} = 6$ inner roots, $M = 2$ ripple cycles, and a ripple level of ± 0.5 dB.

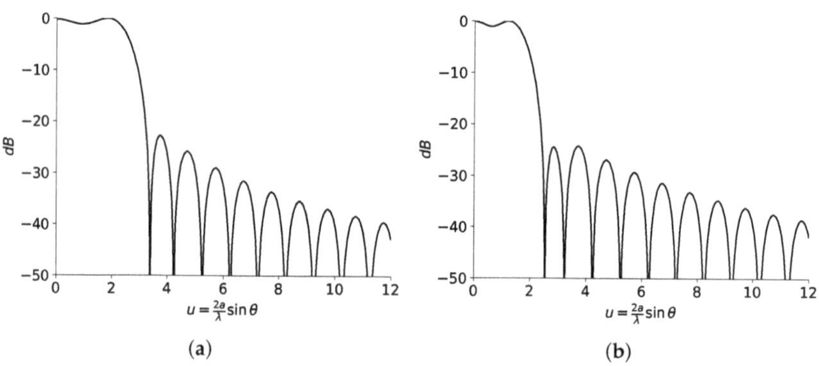

Figure 8. Patterns from (**a**) real and (**b**) complex roots. The HPBW for each set of roots is obtained at (**a**) $u_0 = 2.52$ and (**b**) $u_0 = 1.75$. Produced with a side-lobe level $SLL = -25$ dB, $\bar{n} = 5$ inner roots, $M = 1$ ripple cycles, and a ripple level of ± 0.5 dB.

For this case, the contour of the antenna has to be numerically computed with Equation (9), the values of u_0 from the diagrams of Figures 7 and 8, and the area of the antenna (A_e).

The European coverage defined by the EuTELSAT yields interesting applications from geostationary satellites, with the contour from Figure 11. This configuration leads to the same aperture efficiency for the real and complex cases ($\eta^{real}_{EuTELSAT} = \eta^{complex}_{EuTELSAT} = 0.778$).

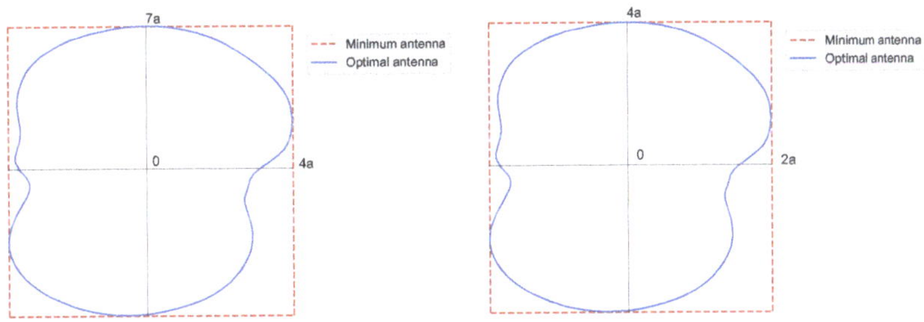

(**a**) (**b**)

Figure 9. Shape of the required antenna for the case of the EuTELSAT antenna contour. The contour is obtained from Equation (9), using as initial pattern both (**a**) real and (**b**) complex flat-topped beam pattern boundaries with two filled zeros. The dashed red line represents the minimum antenna, while the solid blue line is the optimal antenna.

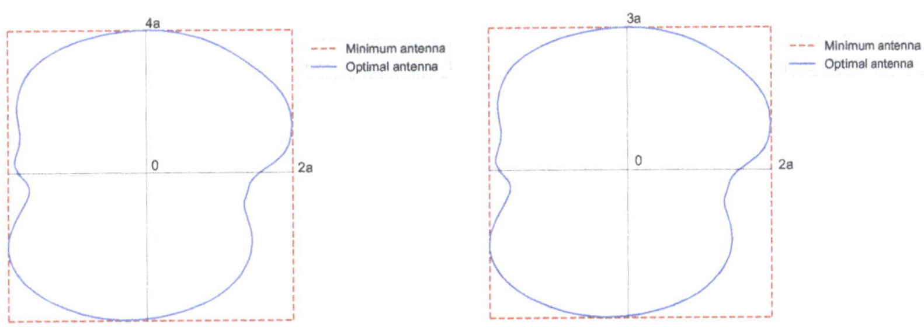

(**a**) (**b**)

Figure 10. Shape of the required antenna for the case of the EuTELSAT antenna contour with one only zero. The contour is obtained from Equation (9), using as initial pattern both (**a**) real and (**b**) complex flat-topped beam pattern boundaries with one filled zero. The dashed red line represents the minimum antenna, while the solid blue line is the optimal antenna.

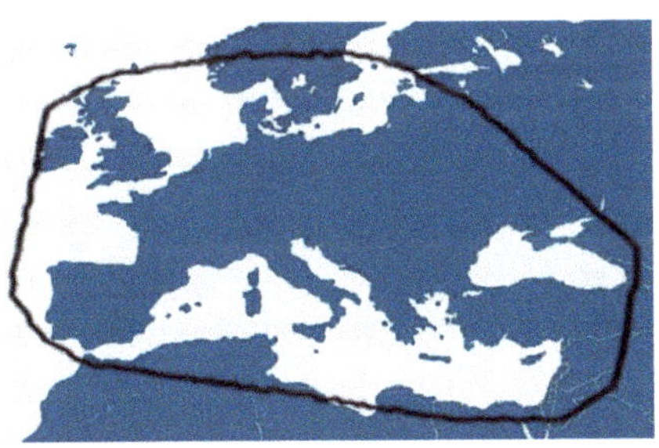

Figure 11. Contour of continental Europe covered by the EuTELSAT satellite.

4. Conclusions

As a consequence of our study, it has been proved that, for rectangular footprints, the HPBW within the principal planes must verify $1 \leq \frac{\beta\omega_a}{\beta\omega_b} \leq 2$, i.e., a maximum aspect ratio of 2:1, in order to be able to obtain an aperture size fitting the aspect ratio of the footprint.

For the case of the EuTELSAT antenna contour, which has been synthesized using both real and complex diagrams, it has been found that the antenna shape from the real pattern is always bigger than from the complex one, but maintaining the same aperture efficiency. If the number of ripple levels is reduced to one, the shape of the antenna also decreases for both pure real and complex cases, verifying the former result. The implementation in this case would lead to the use of fewer elements at the expense of also reducing the antenna directivity in comparison with the case of two ripple levels. For both real and complex patterns, reducing the ripple level implies a small shrinkage of the radiation pattern in the shaped region, which will be greater as the SLL increases.

In the examples shown, we have considered the most favorable case, as we have always used the minimum aperture area that adjusts the effective area as much as possible. Nevertheless, in real cases, the antenna aperture area is expected to be bigger.

This procedure is directly applicable to equispaced linear arrays, where the product of the number of elements and HPBW is approximately constant.

It is recommended that these possible improvements in the aperture efficiency be incorporated in all future syntheses of linear and planar arrays.

Author Contributions: Conceptualization, F.J.A.-P.; methodology, C.L.-Á., J.A.R.-G., F.J.A.-P. and J.A.R.-G.; validation, C.L.-Á. and F.J.A.-P.; investigation, C.L.-Á.; resources, F.J.A.-P. and J.A.R.-G.; writing—original draft preparation, C.L.-Á.; writing—review and editing, C.L.-Á., J.A.R.-G., M.E.L.-M. and F.J.A.-P.; visualization, C.L.-Á.; supervision, J.A.R.-G., M.E.L.-M. and F.J.A.-P.; project administration, F.J.A.-P. and M.E.L.-M.; funding acquisition, F.J.A.-P. and M.E.L.-M. All authors have read and agreed to the published version of the manuscript.

Funding: This work was supported in part by the FEDER/Ministerio de Ciencia e Innovación-Agencia Estatal de Investigación under Project PID2020-119788RB-I00.

Institutional Review Board Statement: Not applicable.

Informed Consent Statement: Not applicable.

Data Availability Statement: Data is contained within the article.

Conflicts of Interest: The authors declare no conflicts of interest.

References

1. *IEEE Std 145-2013*; IEEE Standard for Definitions of Terms for Antennas. IEEE: New York, NY, USA, 2014; pp. 1–92.
2. Kim, Y.U.; Elliott, R.S. Extensions of the Tseng-Cheng Pattern Synthesis Technique. *J. Electromagn. Waves Appl.* **1988**, *2*, 255–268. [CrossRef]
3. Botha, E.; McNamara, D.A. A Contoured Beam Synthesis Technique for Planar Antenna Arrays with Quadrantal and Centro-Symmetry. *IEEE Trans. Antennas Propag.* **1993**, *41*, 1222–1231. [CrossRef]
4. Bucci, O.M.; D'Elia, G.; Mazzarella, G.; Panariello, G. Antenna Pattern Synthesis: A New General Approach. *Proc. IEEE* **1994**, *82*, 358–371. [CrossRef]
5. Chou, H.T.; Hsaio, Y.T.; Pathak; Nepa; Janpugdee. A Fast DFT Planar Array Synthesis Tool for Generating Contoured Beams. *IEEE Antennas Wirel. Propag. Lett.* **2004**, *3*, 287–290. [CrossRef]
6. Villegas, F. Parallel Genetic-Algorithm Optimization of Shaped Beam Coverage Areas Using Planar 2-D Phased Arrays. *IEEE Trans. Antennas Propag.* **2007**, *55*, 1745–1753. [CrossRef]
7. Federico, G.; Caratelli, D.; Theis, G.; Smolders, A.B. A Review of Antenna Array Technologies for Point-to-Point and Point-to-Multipoint Wireless Communications at Millimeter-Wave Frequencies. *Int. J. Antennas Propag.* **2021**, 2021. [CrossRef]
8. Ogurtsov, S.; Caratelli, D.; Song, Z. A Review of Synthesis Techniques for Phased Antenna Arrays in Wireless Communications and Remote Sensing. *Int. J. Antennas Propag.* **2021**, *2021*, 5514972. [CrossRef]
9. Angeletti, P.; Berretti, L.; Maddio, S.; Pelosi, G.; Selleri, S.; Toso, G. Phase-Only Synthesis for Large Planar Arrays via Zernike Polynomials and Invasive Weed Optimization. *IEEE Trans. Antennas Propag.* **2020**, *70*, 1954–1964. [CrossRef]
10. Alijani, M.G.H.; Neshati, M.H. Development of a New Method for Pattern Synthesizing of Linear and Planar Arrays Using Legendre Transform With Minimum Number of Elements. *IEEE Trans. Antennas Propag.* **2022**, *70*, 2779–2789. [CrossRef]
11. Dahri, M.H.; Abbasi, M.I.; Jamaluddin, M.H.; Kamarudin, M.D. A Review of High Gain and High Efficiency Reflectarrays for 5G Communications. *IEEE Access* **2018**, *6*, 5973–5985. [CrossRef]
12. Salucci, M.; Oliveri, G.; Massa, A. An Innovative Inverse Source Approach for the Feasibility-Driven Design of Reflectarrays. *IEEE Trans. Antennas Propag.* **2022**, *70*, 5468–5480. [CrossRef]
13. Savenko, P.O.; Anokhin, V.J. Synthesis of Amplitude-Phase Distribution and Shape of a Planar Antenna Aperture for a Given Power Pattern. *IEEE Trans. Antennas Propag.* **1997**, *45*, 744–747. [CrossRef]

14. Aghasi, A.; Amindavar, H.; Miller, E.L.; Rashed-Mohassel, J. Flat-Top Footprint Pattern Synthesis Through the Design of Arbitrary Planar-Shaped Apertures. *IEEE Trans. Antennas Propag.* **2010**, *58*, 2539–2552. [CrossRef]
15. Elliott, R.S.; Stern, G.J. Footprint Patterns Obtained by Planar Arrays. *IEE Proc. H* **1990**, *137*, 108–112. [CrossRef]
16. Ares, F.; Elliott, R.S.; Moreno, E. Design of Planar Arrays to Obtain Efficient Footprint Patterns With an Arbitrary Footprint Boundary. *IEEE Trans. Antennas Propag.* **1944**, *42*, 1509–1514. [CrossRef]
17. Fondevila-Gomez, J.; Rodriguez-Gonzalez, J.A.; Trastoy, A.; Ares-Pena, F. Optimization of Array Boundaries for Arbitrary Footprint Patterns. *IEEE Trans. Antennas Propag.* **2004**, *52*, 635–637. [CrossRef]
18. López-Álvarez, C.; Rodríguez-González, J.A.; López-Martín, M.E.; Ares-Pena, F.J. An Improved Pattern Synthesis Iterative Method in Planar Arrays for Obtaining Efficient Footprints with Arbitrary Boundaries. *Sensors* **2021**, *21*, 2358.
19. Elliott, R.S.; Stern, G.J. Shaped Patterns from a Continuous Planar Aperture Distribution. *IEE Proc. H* **1988**, *135*, 366–370. [CrossRef]
20. Elliott, R.S. *Antenna Theory and Design, Revised Edition*; IEEE Press: Piscataway, NJ, USA, 2003.
21. Silver, S. *Microwave Antenna Theory and Design*; MIT Rad. Lab. Series; MIT Radiation Laboratory: Cambridge, MA, USA, 1939.

electronics

MDPI

Article

A Patch Antenna with Enhanced Gain and Bandwidth for Sub-6 GHz and Sub-7 GHz 5G Wireless Applications

Shehab Khan Noor [1], Muzammil Jusoh [1,*], Thennarasan Sabapathy [1], Ali Hanafiah Rambe [2], Hamsakutty Vettikalladi [3], Ali M. Albishi [3] and Mohamed Himdi [4]

[1] Advanced Communication Engineering (ACE), Centre of Excellence, Faculty of Electronic Engineering Technology, Universiti Malaysia Perlis, Kangar 01000, Malaysia; shehabrabid97@gmail.com (S.K.N.); thenna84@gmail.com (T.S.)
[2] Department of Electrical Engineering, Universitas Sumatera Utara, Medan 20155, Indonesia; ali3@usu.ac.id
[3] Electrical Engineering Department, College of Engineering, King Saud University, Riyadh 11421, Saudi Arabia; aalbishi@ksu.edu.sa (A.M.A.)
[4] Institute of Electronic Technologies Research (IETR), University of Renne, 35042 Renne, France
* Correspondence: muzammil@unimap.edu.my

Abstract: This paper presents a novel microstrip patch antenna design using slots and parasitic strips to operate at the n77 (3.3–4.2 GHz)/n78 (3.3–3.8 GHz) band of sub-6 GHz and n96 (5.9–7.1 GHz) band of sub-7 GHz under 5G New Radio. The proposed antenna is simulated and fabricated using an FR-4 substrate with a relative permittivity of 4.3 and copper of 0.035 mm thickness for the ground and radiating planes. A conventional patch antenna with a slot is also designed and fabricated for comparison. A comprehensive analysis of both designs is carried out to prove the superiority of the proposed antenna over conventional dual-band patch antennas. The proposed antenna achieves a wider bandwidth of 160 MHz at 3.45 GHz and 220 MHz at 5.9 GHz, with gains of 3.83 dBi and 0.576 dBi, respectively, compared to the conventional patch antenna with gains of 2.83 dBi and 0.1 dBi at the two frequencies. Parametric studies are conducted to investigate the effect of the parasitic strip's width and length on antenna performance. The results of this study have significant implications for the deployment of high-gain compact patch antennas for sub-6 GHz and sub-7 GHz 5G wireless communications and demonstrate the potential of the proposed design to enhance performance and efficiency in these frequency bands.

Keywords: patch antenna; parasitic strips; high gain; 5G; compact antennas; dual-band antennas

Citation: Noor, S.K.; Jusoh, M.; Sabapathy, T.; Rambe, A.H.; Vettikalladi, H.; M. Albishi, A.; Himdi, M. A Patch Antenna with Enhanced Gain and Bandwidth for Sub-6 GHz and Sub-7 GHz 5G Wireless Applications. *Electronics* **2023**, *12*, 2555. https://doi.org/10.3390/electronics12122555

Academic Editors: Naser Ojaroudi Parchin, Chan Hwang See and Raed A. Abd-Alhameed

Received: 8 May 2023
Revised: 2 June 2023
Accepted: 4 June 2023
Published: 6 June 2023

1. Introduction

5G mobile communication systems have significantly improved over the last few decades and are in high demand due to their significant advantages, including low latency, high data rate, and high data capacity [1]. The usage of high data rates, wide bandwidth, and stable gain, which are already in use in many regions, will be greatly impacted by 5G. The 5G New Radio includes frequency bands such as n77 (3.3 GHz–4.2 GHz), n78 (3.3 GHz–3.8 GHz), and n79 (4.4 GHz–5.0 GHz) for sub-6 GHz 5G applications [2]. For applications above 5 GHz, the n96 band falls under sub-7 GHz and ranges from 5.9 to 7.1 GHz [3].

Among many types of antennae in the market and industry, the microstrip patch antenna is a mature design to deploy 5G wireless communications due to its advantages such as its light weight, small volume, low cost, low profile, smaller dimension, and ease of fabrication [4]. However, the main drawbacks of conventional microstrip antennae are their narrow bandwidth and low gain [5]. Nonetheless, the patch antenna can be easily customized to operate at different resonance frequencies, enabling it to be used for many wireless applications with just one antenna.

In recent years, researchers have developed new designs to improve the performance of patch antennas, enabling them to operate in multiple bands, achieve greater gain, wider bandwidth, and be more compact [6–12]. In reference [6], a Teflon substrate-based patch antenna that operates in three different bands under sub-6 GHz 5G, with overall dimensions of 50 mm × 80 mm was reported. The antenna achieves gains of 2.52 dBi, 3.04 dBi, and 4.31 dBi at 2.55 GHz, 3.5 GHz, and 4.75 GHz, respectively. A slotted plus shape patch antenna using Rogers RT5880 substrate operating at 3.12 GHz of the sub-6 GHz band 5G was proposed in reference [7], utilizing the Defected Ground Structure (DGS) method to obtain the desired frequency coverage. The antenna has an overall bandwidth of 2.56 GHz, but the gain is relatively low at 2.44 dBi. Similarly, in reference [8], the DGS method to enable the antenna to operate at two different frequencies, 4.53 GHz and 4.97 GHz, using the RO5880 substrate with a relative permittivity of 2.2 was implemented. The antenna achieves gains of 5 dBi at 4.53 GHz and 4.57 dBi at 4.97 GHz, with overall dimensions of 77 mm × 70.11 mm. However, the bandwidth of the antenna is not reported in [8].

Recent advancements in microstrip patch antenna design have led to significant improvements in the performance of sub-6 GHz 5G communications systems. A high-gain, single-band antenna using Arlon AD300C substrate and operating at 5.65 GHz was proposed in reference [9], with an overall gain of 7.15 dBi and a bandwidth of 135 MHz. To enhance the gain of microstrip patch antennas at 2.4 GHz, the airgap method was reported in [10], which involves inserting an airgap between the substrate and ground plane. By inserting a 3 mm airgap, the gain increased from 7.1 dBi to 7.91 dBi, while the bandwidth reduced from 110.7 MHz to 72.873 MHz. However, the study concluded that there is a trade-off between antenna gain and bandwidth associated with the implementation of this technique. Another approach to increase antenna gain is the use of a reflecting layer, as proposed in [11], which employs four spacers across four different corners of the patch antenna. The antenna operates at 2.392 GHz with an overall gain of 5.2 dBi and a narrow bandwidth of 44.7 MHz, with dimensions of 60 mm × 55 mm × 8 mm. Additionally, a wideband printed antenna using FR-4 substrate to operate in n77 and n78 band of the 5G New Radio (NR) was developed in [12], with overall dimensions of 28 mm × 20 mm. The antenna achieved a maximum gain of 2.5 dBi with a bandwidth of 700 MHz.

A wideband coplanar waveguide (CPW)-fed monopole antenna with overall dimensions of 20 × 8.7 mm^2 where the substrate used was FR-4 with a relative permittivity of 4.3 and thickness of 0.4 mm was proposed by the authors in [13]. The antenna operated in a single band with measured bandwidth coverage for Wi-Fi 5 and Wi-Fi 6 (5.15 GHz to 7.29 GHz). However, the simulated gain at 6.2 GHz was 2.6 dBi while the measured gain was 2.25 dBi. A quad-band circularly polarized antenna was designed to operate at four different bands including WIFI-6E (5925–7125 MHz), 5G n77 (3300–4200 MHz), n78 (3300–3800 MHz), and the GNSS band by the authors [14]. The antenna was fabricated using a FR-4 substrate and the overall dimensions of the antenna were 80 mm × 80 mm. For all the four bands, the antenna obtained wider bandwidth with moderate gain. A flexible antenna fabricated with graphene to operate in dual bands was designed by Zelong Hu et al. [15]. The antenna primarily focused on to generate at Wi-Fi 6e standards (2.4–2.45 GHz and 5.15–7.1 GHz). To make sure that the antenna works well under the conformal conditions, the authors used Polydimethylsiloxane (PDMS) for substrate. The measured realized gain within 5.15 to 7.1 GHz was in the range of 5.25–6.85 dBi and 2.62–2.91 dBi within 2.3–2.5 GHz. The antenna covered the desired frequency bands even when it was bent to 120° and 180°. In [16], a vehicular cellular antenna was designed using FR4 substrate to cover the frequency range 0.617 GHz to 5 GHz which falls under 5G sub-6 GHz. The antenna was designed to achieve an omni-directional radiation pattern which is a major requirement for vehicular systems. To maintain omni-directional radiation patterns, the antenna is intended to be symmetric around the centerline. A high antenna (28.03 × 23.45 mm^2) with defected ground structure (DGS) method to operate at Sub-6 GHz of the 5G communications was proposed in [17]. The authors implemented DGS method to achieve wider bandwidth while keeping the antenna small in shape. The antenna operated from 4.921 to 5.784 GHz with overall gain of 6.2 dBi. A 2 × 2 element patch array antenna on a FR-4 substrate to operate at Wi-Fi 6/6E was designed by the authors in [18]. At 7.1 GHz, the maximum gain of 12.05 dBi was achieved with bandwidth coverage from 5 to 7.5 GHz. The authors used nylon pillars to create some gap between the substrate and a metal plate to keep the antenna simple without degrading its gain and bandwidth. To operate at three different frequency bands, the authors designed a microstrip patch antenna with I-shaped slots and two shorted metallic pins to enhance the bandwidth [19]. The substrate used was Rogers RT-Duroid-5880-DK-2.2 with single patch radiating element with dimensions of 34.6 × 33.05 mm^2. The antenna resonated at 5.2, 5.5, and 5.8 GHz which fall under the 5G Wi-Fi spectrum. The maximum value of gain obtained was 7.2 dBi at 5.5 GHz. A dual band monopole compact antenna to operate at 2.4 GHz and 5.2/5.8 GHz is designed and fabricated in [20]. To reduce the manufacturing cost and complexity of the proposed design, the authors comb-shaped element into the open ring structure. For 2.4 GHz band, the measured gain varied between 3.34 dBi and 3.81 dBi with efficiency ranged between 52% and 61%. Moreover, with regards to the 5.2 GHz band, the measured gain varied between 5.19 dBi and 6.62 dBi with efficiency ranged between 83% and 87%. To improve antenna gain and bandwidth, the use of parasitic elements is a mature and advantageous method [21]. By using dummy elements which can be also called as "parasitic elements" along with the main element known as "driven element", the antenna gain and bandwidth can be enhanced. When the parasitic elements are excited through the radiating edges of the driven element and when all the elements (parasitic and driven) are well-coupled with one another, the resonance of the elements match together and the overall bandwidth increases and it also leads to an increase in antenna gain.

This work proposes a novel method for improving the gain, bandwidth, and efficiency of dual-band antennas without modifying the overall antenna dimension of a conventional patch antenna with slots, by implementing a parasitic patch antenna with a slots-based technique. The proposed antenna is simulated and fabricated using an FR-4 substrate with a relative permittivity of 4.3 and copper with a thickness of 0.035 mm for the ground and radiating planes. To evaluate its performance, the proposed antenna is compared with previous works in terms of size, operating frequencies, bandwidth, and gain. Section 2 summarizes the configuration of the antenna design. The simulated and measured results are discussed in Section 3, along with parametric analysis. Finally, the study's conclusions are presented in Section 4. The proposed technique represents a significant advancement in dual-band antenna design, with the potential to improve the performance and efficiency of 5G wireless communication systems, especially in applications where compact, wideband, and high-gain antennas are essential.

2. Antenna Design Configuration

This study presents a comprehensive comparison of two patch antennas to validate the efficacy of the proposed approach. Antenna-1 is a conventional patch antenna with slots, while Antenna-2 is a novel patch antenna with parasitic elements and slots. The introduction of parasitic elements in the conventional patch antenna can lead to an increase in antenna gain and bandwidth. Figure 1 shows the simulated and fabricated front views of both antennas. The antennas are designed using the concept of a microstrip patch antenna, with FR-4 substrate having a relative permittivity of 4.3 and a thickness of 1.6 mm, and copper with a thickness of 0.035 mm for the radiating and ground planes. Slots are implemented on the radiating patch to enable the antenna to operate at two different bands, and the concept of the inset-fed technique is used for better impedance matching. To design the proposed antenna, Equation (1) to Equation (7) from [22] were used. The width, Wp and the length, Lp of the patch are calculated using the equations as given below:

$$Wp = \frac{c}{2fo\sqrt{\frac{(\varepsilon r+1)}{2}}} \tag{1}$$

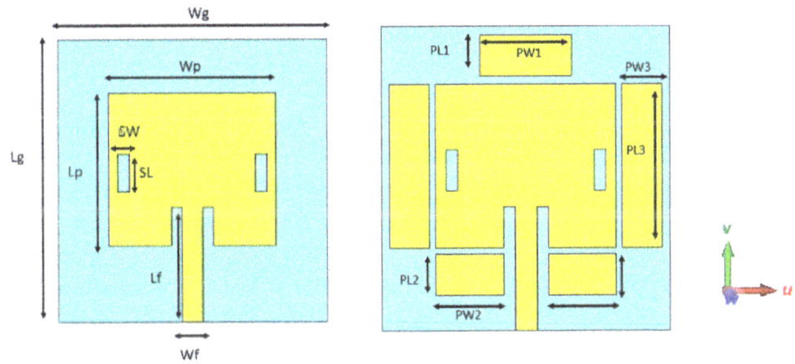

(**a**) (**b**)

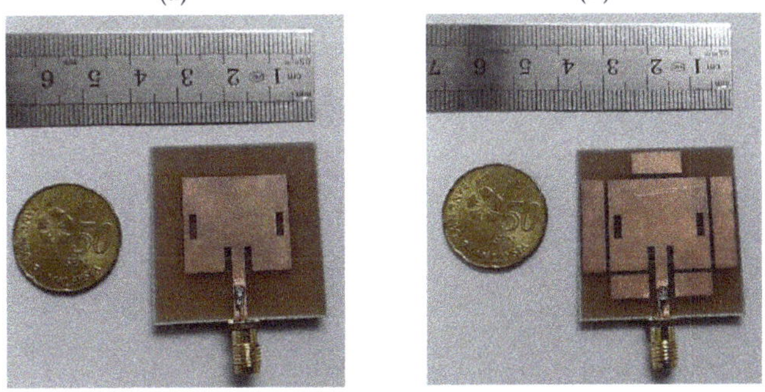

(**c**) (**d**)

Figure 1. Simulated and fabricated designs of the proposed patch antennas: (**a**) Simulated Antenna-1, (**b**) Simulated Antenna-2, (**c**) Fabricated Antenna-1, (**d**) Fabricated Antenna-2.

Here, C = 3×10^8 m/s (light speed), εr = permittivity of substrate and *fo* = resonant frequency (GHz).

$$\text{Lp} = \text{L}_{\text{reff}} - 2\Delta\text{L} \tag{2}$$

where L_{reff} can be found using:

$$\text{L}_{\text{reff}} = \frac{c}{2fo\sqrt{\varepsilon\text{reff}}} \tag{3}$$

Next, ground width (Wg) and ground length (lg) were calculated using:

$$\text{Wg} = 6\text{h} + \text{Wp} \tag{4}$$

$$\text{Lg} = 6\text{h} + \text{Lp} \tag{5}$$

here h = height of the substrate

The width of feed line, Wf:

$$\text{Zo} = \left[87/\sqrt{(\varepsilon\text{r} + 1.141)}\right)\right] \ln(5.98\text{h}/0.8\text{Wf}) \tag{6}$$

The length of feedline, Lf:

$$\text{Lf} = \frac{\text{Lg} - \text{Lp}}{2} \tag{7}$$

The overall antenna dimensions (unit in mm) of both antennas are identical, with Wg = 36, Lg = 35, Wp = 22.5, Lp = 20, SW = 1.5, SL = 5, Lf = 15, Wf = 2.85, PL1 = 5, PW1 = 11.5, PW3 = 5, PL3 = 20, PL2 = 5 and PW2 = 8.4. However, in Antenna-1, there are no parasitic strips. The simulated and fabricated designs of Antenna-1 and Antenna-2 are illustrated in Figure 1a,c, and Figure 1b,d, respectively.

3. Results and Discussion

3.1. Reflection Coefficient (|S_{11}|) and Bandwidth

The reflection coefficient of the fabricated antenna was simulated using Computer Simulation Tool (CST) Microwave Studio (MWS) software version 2022 and measured using a vector network analyzer (VNA) (Agilent E5071C) under five different conditions. The measurement setup for Antenna-1 and Antenna-2 is illustrated in Figure 2a,b, respectively. Figure 2c shows the reflection coefficient plot for the simulated and obtained measurement results. For Antenna-1, the |S_{11}| value at the desired frequency ranges was below −10 dB while resonating at 3.51 GHz with a bandwidth of 140 MHz (3.45 to 3.59 GHz). Additionally, the antenna obtained bandwidth of 150 MHz (5.85 to 6 GHz) while resonating at 5.97 GHz. In contrast, the measured results for the n77 (3.3–4.2 GHz) and n78 (3.3–3.8 GHz) bands showed the antenna operating at 3.62 GHz with a bandwidth of 100 MHz (3.59 to 3.69 GHz), and at 6.14 GHz, covering 6 GHz to 6.21 GHz (210 MHz) for the n96 band. For the proposed Antenna-2, both the simulated and measured |S_{11}| values at the 3.3–4.2 GHz, 3.3–3.8 GHz, and 5.9–7.1 GHz bands were below −10 dB. The simulated bandwidth at 3.45 GHz was 160 MHz (3.34 to 3.49 GHz), while at 5.9 GHz, the antenna resonated at 5.87 GHz, covering a bandwidth from 5.75 GHz to 5.97 GHz (220 MHz). In measurements, the proposed antenna resonated at 3.52 GHz, covering 3.45 GHz to 3.55 GHz (100 MHz), and the antenna covered 5.97 GHz to 6.145 GHz with a bandwidth of 175 MHz, resonating at 6.0 GHz. The simulated and measured results matched well, with a small deviation attributed to the fabrication tolerance and soldering loss. Additionally, the use of SubMiniature version A (SMA) ports to excite the antenna in fabrication resulted in some insertion loss, unlike the waveguide port used in the simulation.

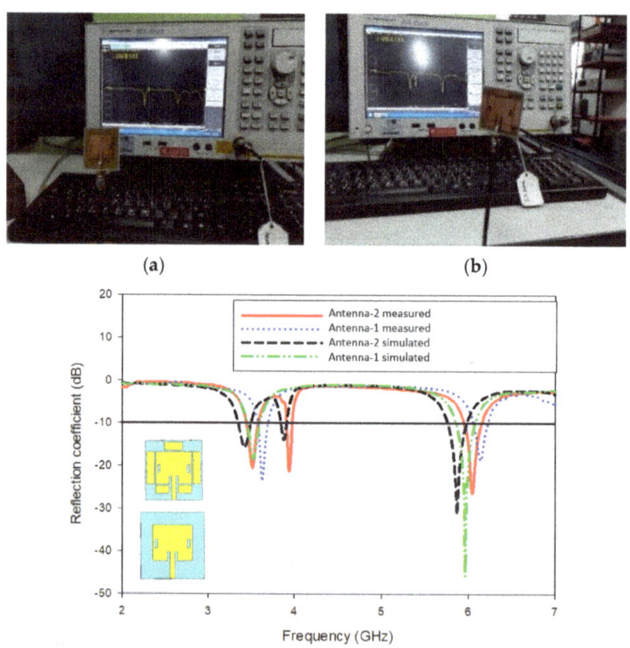

(c)

Figure 2. Simulated and Measured |S_{11}| and bandwidth results (**a**) Antenna-1 setup (**b**) Antenna-2 setup (**c**) Comparison between Antenna-1 and Antenna-2 obtained results.

3.2. Radiation Pattern

To evaluate the performance of the proposed antenna, the radiation pattern was measured inside an anechoic chamber, as shown in Figure 3. The proposed antenna acted as the transmitting antenna, while a standard horn antenna (INFOMW LB-20200-SF) at the other end acted as the receiving antenna. The measurement was carried out by rotating (360°) azimuthally across the horizontal direction (x-direction) as shown in Figure 4. The measurement was conducted at the Advanced Communication Engineering (ACE) research lab at Universiti Malaysia Perlis (UniMAP).

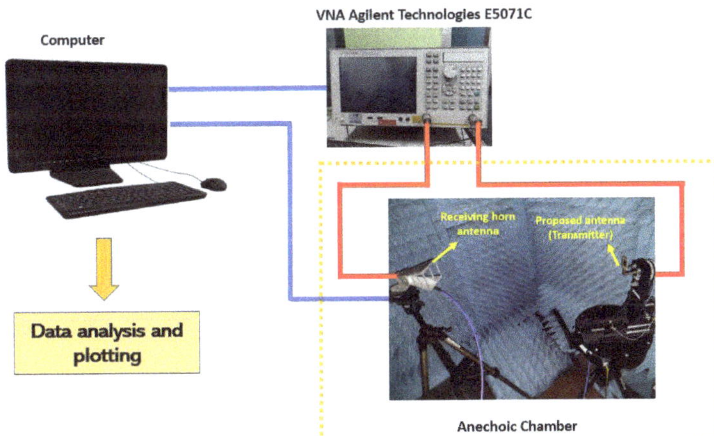

Figure 3. Radiation pattern measurement of the proposed antenna.

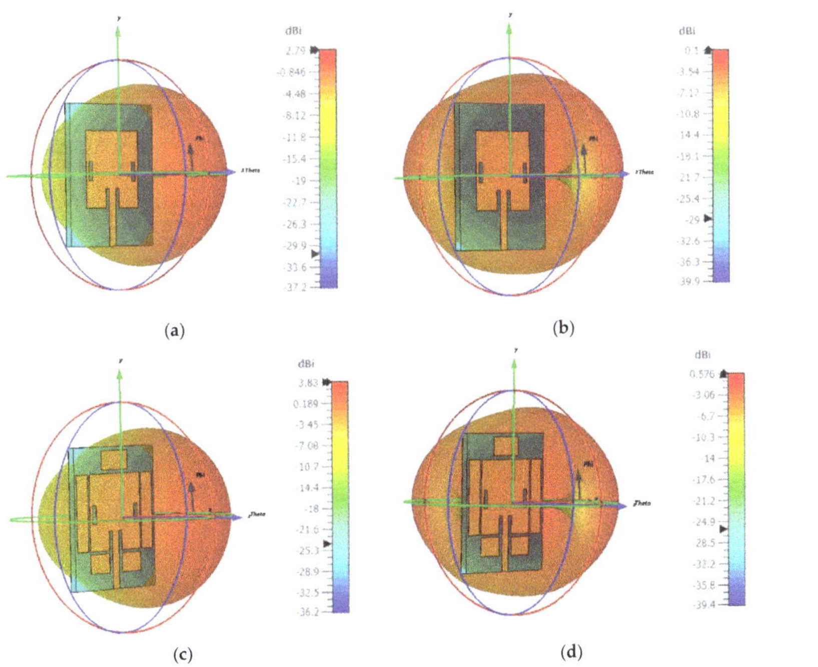

Figure 4. Simulated antenna gains (**a**) Antenna-1 gain at 3.45 GHz (**b**) Antenna-1 gain at 5.9 GHz (**c**) Antenna-2 gain at 3.45 GHz (**d**) Antenna-2 gain at 5.9 GHz.

The gain of Antenna-1 obtained through simulation, are shown in Figure 4a,b, respectively. It can be observed that Antenna-1 has a gain value of 2.79 dBi and 0.1 dBi at 3.45 GHz and 5.9 GHz, respectively. In contrast, Antenna-2 has a gain of 3.83 dBi and 0.576 dBi at 3.45 GHz and 5.9 GHz, respectively, as shown in Figure 4c,d. A comparison of the gain results for Antenna-1 and Antenna-2 reveals that Antenna-2 outperformed Antenna-1 significantly at 3.45 GHz and 5.9 GHz. At 3.45 GHz, the calculated enhanced gain was 40% higher for Antenna-2 than for Antenna-1, while at 5.9 GHz, the calculated enhanced gain for Antenna-2 was almost five times the gain obtained by Antenna-1. These results demonstrate the efficacy of the proposed approach in improving the gain of the proposed antenna.

Figure 5 shows the 2D polar radiation patterns of the proposed antenna. At 3.45 GHz, both the simulated and measured radiation patterns show that the antenna exhibits a strong radiation pattern towards the boresight direction, indicating a high level of directivity and gain, as illustrated in Figure 5a,b. These results are promising for the design of high-performance wireless communication systems that require directional antennas with high gain and efficiency. At 5.9 GHz, however, the simulated and measured radiation pattern is close to omnidirectional, as it is spread over a wider area, as shown in Figure 5c,d. This explains why the simulated gain at 3.45 GHz is higher than the gain at 5.9 GHz. These findings provide useful insights into the performance and suitability of the proposed antenna for 5G wireless communication systems, particularly in applications where directional antennas with high gain and efficiency are required.

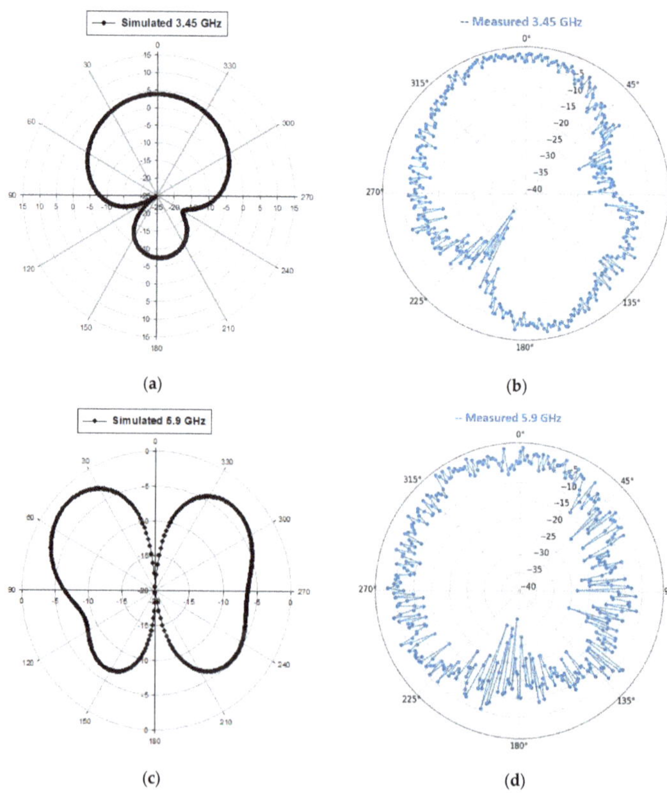

Figure 5. Proposed antenna (Antenna-2) 2D polar radiation pattern (**a**) Simulated 2D polar pattern at 3.45 GHz (**b**) Measured 2D polar pattern at 3.45 GHz (**c**) Simulated 2D polar pattern at 5.9 GHz (**d**) Measured 2D polar pattern at 5.9 GHz.

Figure 6a shows the gain over frequency for Antenna-1 and Antenna-2. It can be observed that the gain at 3.45 GHz for Antenna-2 was higher than the gain for Antenna-1. However, the gain dropped drastically for Antenna-2 while the gain dropped linearly for Antenna-1. From 3.5 GHz, the gain again started to increase for Antenna −2 and it kept increasing till 5.7 GHz approximately. In the case of Antenna-1, the gain was dropping until almost 4.8 GHz and then it started increasing again until 6 GHz. At 3.6 GHz, the gain for Antenna-1 was −3.8 dBi while the gain was 3.18 dBi for Antenna-2. Subsequently, at 4.8 GHz, the gain for Antenna-1 was 0.060 dBi and the gain for Antenna-2 was −1.89 dBi. However, at 5.9 GHz, Antenna-2 achieved higher gain value of 0.576 dBi while Antenna-1 achieved 0.1 dBi. The efficiency over frequency for Antenna-1 and Antenna-2 is illustrated in Figure 6b. The efficiency of Antenna-2 at 3.45 GHz was 59% while the efficiency was 47% for Antenna-1. The efficiency dropped dramatically for Antenna-2 until 4.8 GHz approximately while the efficiency dropped constantly for Antenna-1. At 3.87 GHz, the efficiency for Antenna-1 and Antenna-2 obtained was 46% and 15.9%, respectively. However, at higher frequency onwards, the efficiency of Antenna-2 was more than the efficiency of Antenna-1 because at 5.7 GHz, the gain obtained for Antenna-2 was 36.7% while it was 22% for Antenna-1. Additionally, at 5.9 GHz, the efficiency achieved by Antenna-2 was 45% and Antenna-1 achieved 40%.

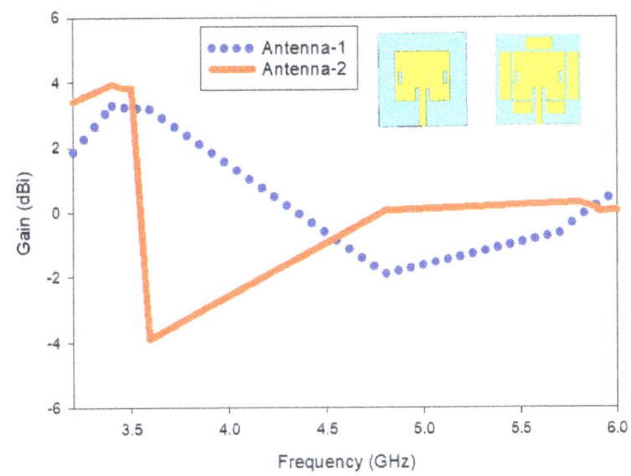

(a)

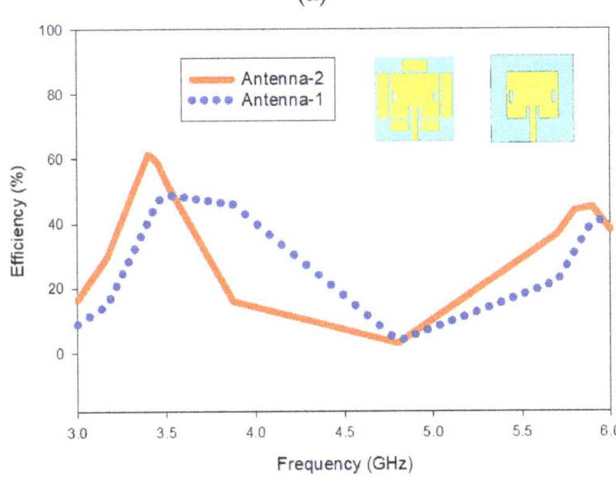

(b)

Figure 6. Simulated results of Antenna-1 and Antenna-2 (**a**) Gain over frequency (**b**) Efficiency over frequency.

3.3. Surface Current Density

Figure 7 shows the surface current distributions of the proposed antenna for 3.45 GHz and 5.9 GHz. The surface current distribution analysis is crucial to identify the components or regions of the antenna responsible for the multi-band frequencies. At 3.45 GHz, the current intensity is more concentrated on the edges of the antenna length and the two-sided parasitic strips, as shown in Figure 7a. These regions are responsible for the antenna to resonate at the desired frequency bands. Meanwhile, for 5.9 GHz, the current intensity is less concentrated across the edges of the antenna length but more concentrated across the antenna width and upper parasitic strips, as demonstrated in Figure 7b. Therefore, these regions are responsible for the antenna to resonate at the n96 band. These results provide valuable insights into the design and optimization of multi-band antennas.

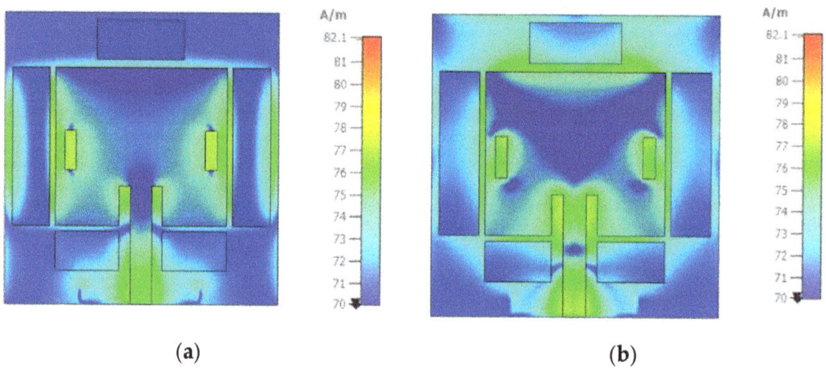

Figure 7. Surface current density of Antenna-2 (**a**) 3.45 GHz (**b**) 5.9 GHz band.

3.4. Parametric Study on Width and Length Parasitic Patch Effect

To investigate the impact of parasitic strip dimensions on the performance of the proposed dual-band microstrip patch antenna, we conducted two parametric studies. In Case-1, we studied the effect of the length of the parasitic strip, as shown in Figure 8a, on the resonating frequency, bandwidth, and gain of the antenna. In Case-2, we investigated the effect of the width of the parasitic strip, as shown in Figure 8b, on the same performance metrics. These studies provide valuable insights into the design and optimization of dual-band microstrip patch antennas.

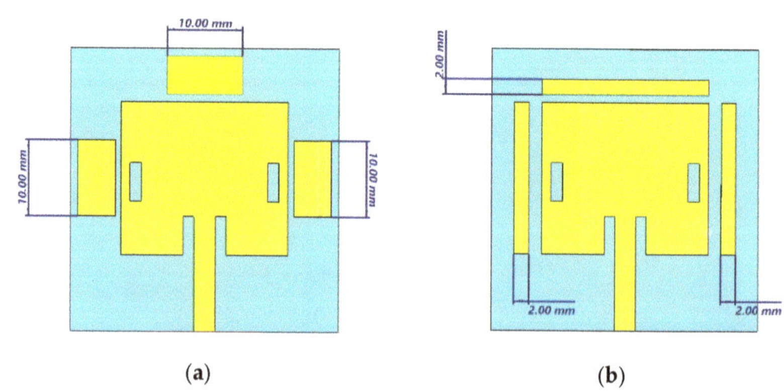

Figure 8. Parametric studies (**a**) Case-1 with reduced length (**b**) Case-2 with reduced width.

3.4.1. $|S_{11}|$ and Bandwidth

Figure 9 illustrates the simulated results of varying the length and width of the parasitic strip in comparison to the finalized design of the proposed dual-band microstrip patch antenna. In Case-1, where the length of the parasitic strip was varied, the antenna resonated at 3.52 GHz with an $|S_{11}|$ value of −19.37 dB, with an overall bandwidth coverage from 3.45 to 3.60 GHz, which is still within the n77/n78 band. For the n96 (5.9–7.1 GHz) band, the antenna resonated at 5.92 GHz, with bandwidth coverage from 5.81 to 6.0 GHz. When the length of the parasitic strip was varied, the operating frequency of the antenna slightly shifted to lower bands in the n77/n78 band, while it shifted slightly to a higher frequency band in the n96 band. In Case-2, where the width of the parasitic strip was varied, the antenna resonated at 3.45 GHz with an $|S_{11}|$ value of −19.36 dB and a bandwidth value of 160 MHz (3.39 to 3.55 GHz). The resonating frequency of the antenna shifted slightly to a higher band, where it resonated at 5.93 GHz with an $|S_{11}|$ value of -22.9 dB and bandwidth coverage from 5.82 to 6.0 GHz.

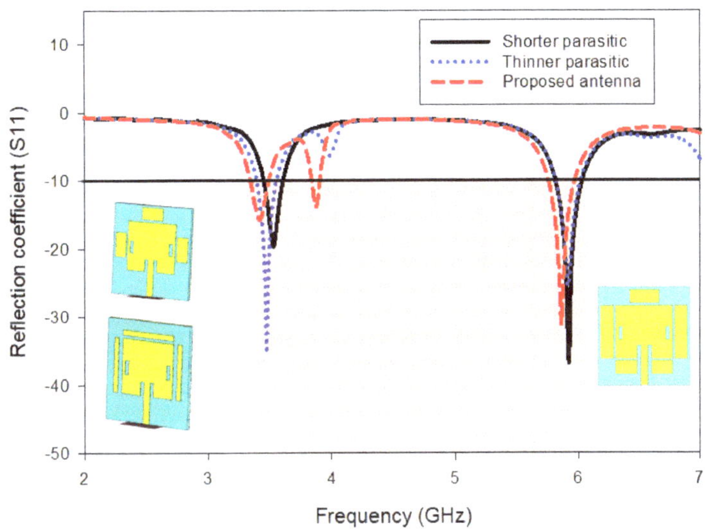

Figure 9. Simulated reflection coefficient and bandwidth for Case-1 and Case-2.

3.4.2. Antenna Gain

We have conducted parametric studies on the proposed dual-band parasitic strip antenna to investigate the effect of reducing its length and width. For Case-1, where the length was reduced, the antenna gain decreased for both 3.45 GHz and 5.9 GHz. The gain at 3.45 GHz was 2.93 dBi, while the gain was 0.464 dBi at 5.9 GHz, as illustrated in Figure 10a,b, respectively. In contrast, for Case-2, where the width was reduced, the gain of the antenna improved. The gain values were 3.73 dBi and 0.602 dBi at 3.45 GHz and 5.9 GHz, respectively, as shown in Figure 10c,d, respectively. These results suggest that by reducing the width of the parasitic strip, the gain of the antenna can be increased for both 3.45 GHz and 5.9 GHz.

The performance comparison between Antenna-1 and the proposed antenna, Antenna-2, is summarized in Table 1, demonstrating that the proposed antenna outperforms Antenna-1 in terms of key performance metrics. Furthermore, compared to previously published works as summarized in Table 2, the proposed dual-band parasitic strip antenna is much smaller in size and exhibits a wider bandwidth and moderate gain for multiple frequency bands.

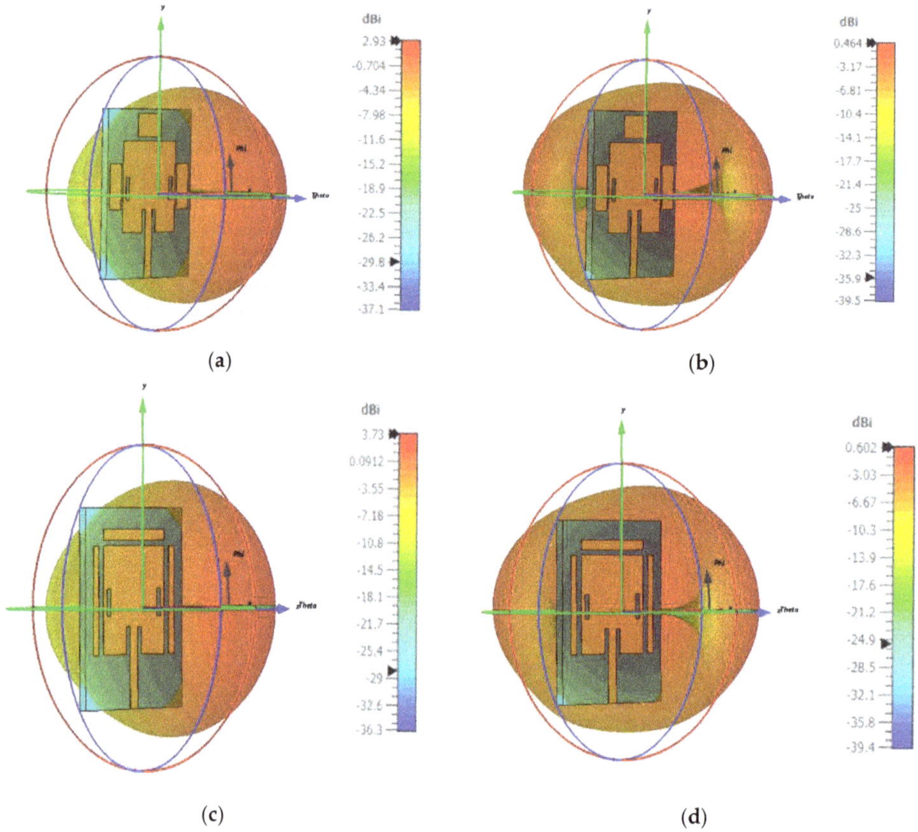

Figure 10. Effect of parasitic strip length and width on gain (**a**) Case-1 at 3.45 GHz (**b**) Case-1 at 5.9 GHz (**c**) Case-2 at 3.45 GHz (**d**) Case-2 at 5.9 GHz.

Table 1. Comparison between Antenna-1 and Antenna-2.

Parameters	Antenna-1		Antenna-2 (Proposed)	
	3.45 GHz	5.9 GHz	3.45 GHz	5.9 GHz
Bandwidth	140 MHz	150 MHz	160 MHz	220 MHz
Gain	2.79 dBi	0.1 dBi	3.83 dBi	0.576 dBi
Directivity	6.08 dBi	1.79 dBi	5.93 dBi	2.75 dBi
Efficiency	47%	40%	59%	45%
Antenna dimension	36 mm × 37 mm		36 mm × 37 mm	

Table 2. Comparison of the proposed antenna (Antenna-2) with some of the previous works.

Ref.	Frequency (GHz)	Antenna Size (mm^2)	Bandwidth (MHz)	Gain (dBi)	Remark
[6]	2.55, 3.5 and 4.75	50 × 80	2920	2.52, 3.05 and 4.31	Antenna is bigger and does not cover n77/n78 band
[7]	3.12	20 × 35	2560	2.44	Single band and low gain
[8]	4.53 and 4.97	77 × 70.11	Not reported	5 and 4.57	Antenna is bigger and did not cover n77/n78 and n96 band

Table 2. *Cont.*

Ref.	Frequency (GHz)	Antenna Size (mm^2)	Bandwidth (MHz)	Gain (dBi)	Remark
[9]	5.65	52.92 × 55.56	135	7.15	Single-band only, the antenna is bigger, and the bandwidth is narrow
[10]	2.4	80 × 80	72.837	7.91	Single band, the antenna is bigger, and the bandwidth is narrow
[11]	2.392	60 × 55	44.7	5.2	Single band, the antenna is bigger and has a very narrow bandwidth
[12]	3.65	28 × 20	700	2.5	Single band and low gain
This work	3.45 and 5.9	36 × 37	160 and 220	3.83 and 0.537	Smaller in size, dual-band, wider bandwidth, and moderate gain for sub-6 GHz band but has low gain for sub-7 GHz band

4. Conclusions

In this work, we have designed and fabricated two FR-4-based dual-band rectangular microstrip patch antennas with an inset-fed technique, operating at 3.45 GHz which falls under sub-6 GHz band and at 5.9 GHz which falls under sub-7 GHz band. One antenna, Antenna-1, is a conventional dual-band patch antenna with slots, while the other, Antenna-2, is a parasitic strip-based antenna proposed in this study. By comparing Antenna-1 and Antenna-2 in terms of bandwidth, gain, directivity, and efficiency, we have demonstrated that Antenna-2 outperforms conventional dual-band patch antennas. The implementation of parasitic strips is the key factor for the enhancement of antenna performance. Moreover, we have carried out measurements to validate the performance of the proposed parasitic strip-based dual-band microstrip patch antenna, which has a compact size of 36 × 37 mm^2 and is easy to design. The proposed antenna exhibits a wider bandwidth above 150 MHz for both sub-6 GHz and sub-7 GHz bands, with a gain value of 3.83 dBi at 3.45 GHz and 0.583 dBi at 5.9 GHz. These measured and tested results have been validated by CST-MWS 2022 software-based simulations. Furthermore, parametric studies have shown that by reducing the width of the parasitic strip, the gain of the antenna can be further increased. Based on our investigations and extensive parametric study, we conclude that the proposed antenna is an excellent competitor for applications below sub-6 GHz of 5G. These findings provide valuable insights into the design and optimization of dual-band microstrip patch antennas for 5G wireless communication systems, particularly in applications where a wide range of frequency bands and high-performance characteristics are required.

Author Contributions: Conceptualization, S.K.N.; Software, M.J.; Validation, T.S. and M.H.; Resources, A.H.R.; Data curation, H.V.; Writing—review & editing, A.M.A. All authors have read and agreed to the published version of the manuscript.

Funding: This research was funded by King Saud University grant number (RSP2023R482).

Acknowledgments: Researchers Supporting Project number (RSP2023R482), King Saud University, Riyadh, Saudi Arabia.

Conflicts of Interest: The authors declare no conflict of interest.

References

1. Azim, R.; Alam, T.; Paul, L.C.; Aktar, R.; Meaze, A.K.M.M.H.; Islam, M.T. Low Profile Multi-slotted Patch Antenna for Lower 5G Application. In Proceedings of the 2020 IEEE Region 10 Symposium (TENSYMP), Dhaka, Bangladesh, 5–7 June 2020; pp. 366–369. [CrossRef]
2. Noor, S.K.; Ismail, A.M.; Yasin, M.N.M.; Osman, M.N.; Ramli, N. Orbital Angular Momentum Vortex Waves Generation Using Textile Antenna Array for 5G Wearable Applications. In Proceedings of the 2022 IEEE Symposium on Wireless Technology & Applications (ISWTA), Kuala Lumpur, Malaysia, 17–18 August 2022; pp. 7–12. [CrossRef]
3. Qualcomm. *Global Update on Spectrum for 4G & 5G*; White Pap.; Qualcomm Inc.: San Diego, CA, USA, 2020; pp. 1–21. Available online: https://www.qualcomm.com/media/documents/files/spectrum-for-4g-and-5g.pdf (accessed on 7 May 2023).

4. Ramli, N.; Ali, M.T.; Yusof, A.L.; Muhamud-Kayat, S.; Aziz, A.A.A. Frequency-reconfigurable stacked patch microstrip antenna using aperture-coupled technique. *Int. J. Microw. Opt. Technol.* **2014**, *9*, 199–205.
5. Pragati; Tripathi, S.L.; Patre, S.R.; Singh, S.; Singh, S.P. Triple-band microstrip patch antenna with improved gain. In Proceedings of the 2016 International Conference on Emerging Trends in Electrical Electronics & Sustainable Energy Systems (ICETEESES), Sultanpur, India, 11–12 March 2016; pp. 106–110. [CrossRef]
6. Tang, X.; Jiao, Y.; Li, H.; Zong, W.; Yao, Z.; Shan, F.; Li, Y.; Yue, W.; Gao, S. Ultra-Wideband Patch Antenna for Sub-6 GHz 5G Communications. In Proceedings of the 2019 International Workshop on Electromagnetics: Applications and Student Innovation Competition (iWEM), Qingdao, China, 18–20 September 2019; pp. 5–7. [CrossRef]
7. Paul, L.C.; Das, S.C.; Rani, T.; Muyeen, S.M.; Shezan, S.A.; Ishraque, M.F. A slotted plus-shaped antenna with a DGS for 5G Sub-6 GHz/WiMAX applications. *Heliyon* **2022**, *8*, e12040. [CrossRef] [PubMed]
8. Sree, M.F.A.; Elazeem, M.H.A.; Swelam, W. Dual Band Patch Antenna Based on Letter Slotted DGS for 5G Sub-6GHz Application. *J. Phys. Conf. Ser.* **2021**, *2128*, 012008. [CrossRef]
9. Tütüncü, B.; Kösem, M. Substrate Analysis on the Design of Wide-Band Antenna for Sub-6 GHz 5G Communication. *Wirel. Pers. Commun.* **2022**, *125*, 1523–1535. [CrossRef] [PubMed]
10. Al Kharusi, K.W.S.; Ramli, N.; Khan, S.; Ali, M.T.; Halim, M.H.A. Gain enhancement of rectangular microstrip patch antenna using air gap at 2.4 GHz. *Int. J. Nanoelectron. Mater.* **2020**, *13*, 211–224.
11. Mekki, A.S.; Hamidon, M.N.; Ismail, A.; Alhawari, A.R.H. Gain Enhancement of a Microstrip Patch Antenna Using a Reflecting Layer. *Int. J. Antennas Propag.* **2015**, *2015*, 975263. [CrossRef]
12. Kapoor, A.; Mishra, R.; Kapoor, A.; Kumar, P. Compact wideband-printed antenna for sub-6 GHz fifth-generation applications. *Int. J. Smart Sens. Intell. Syst.* **2020**, *13*, 1–10. [CrossRef]
13. Kulkarni, J.; Sim, C. Wideband CPW-Fed Oval-Shaped Monopole Antenna for Wi-Fi 5 and Wi-Fi 6 Applications. *Prog. Electromagn. Res. C* **2021**, *107*, 173–182. [CrossRef]
14. Liu, X.; Wang, H.; Yang, X.; Wang, J. Quad-Band Circular Polarized Antenna for GNSS, 5G and WIFI-6E Applications. *Electronics* **2022**, *11*, 1133. [CrossRef]
15. Hu, Z.; Xiao, Z.; Jiang, S.; Song, R.; He, D. A Dual-Band Conformal Antenna Based on Highly Conductive. *Materials* **2021**, *14*, 5087. [CrossRef] [PubMed]
16. Analysis, C.M. Design of a 5G Sub-6 GHz Vehicular Cellular Antenna Characteristic Mode Analysis. *Sensors* **2022**, *22*, 8862.
17. Lawoye, T.O.; Kumar, P. A High Gain Antenna with DGS for Sub-6 GHz 5G Communications. *Adv. Electromagn.* **2022**, *11*, 41–50.
18. Patch, A.S.D.; Array, A.; Application, W. A Simple Dual-Polarized Patch Antenna Array for Wi-Fi 6/6E Application. *IEEE Trans. Antennas Propag.* **2022**, *70*, 11143–11148.
19. Kumar, A.; Althuwayb, A.A.; Al-hasan, M.J. Wideband Triple Resonance Patch Antenna for 5G Wi-Fi Spectrum. *Prog. Electromagn. Res. Lett.* **2020**, *93*, 89–97. [CrossRef]
20. Sim, C.; Chen, C.-C.; Zhang, X.Y.; Lee, Y.-L.; Chiang, C.-Y. Very Small-Size Uniplanar Printed Monopole Antenna for Dual-Band WLAN Laptop Computer Applications. *IEEE Trans. Antennas Propag.* **2017**, *65*, 2916–2922. [CrossRef]
21. Parizi, S.A.R. Bandwidth Enhancement Techniques. *Trends Res. Microstrip Antennas* **2017**, 9–11. [CrossRef]
22. Balanis, C.A. Antenna theory: A review. *Proc. IEEE* **1992**, *80*, 7–23. [CrossRef]

sensors

MDPI

Article

Flexible Antenna with Circular/Linear Polarization for Wideband Biomedical Wireless Communication

Mohammed E. Yassin [1], Khaled F. A. Hussein [2], Qammer H. Abbasi [3,*], Muhammad A. Imran [3] and Shaimaa A. Mohassieb [1,3]

[1] Electronics and Communications Engineering Department, Akhbar Elyom Academy, 6th of October City 12573, Egypt; m.ezzat@akhbaracademy.edu.eg (M.E.Y.); s.mohassieb@akhbaracademy.edu.eg (S.A.M.)
[2] Microwave Engineering Department, Electronics Research Institute (ERI), Cairo 11843, Egypt; fkhalid@eri.sci.eg
[3] James Watt School of Engineering, University of Glasgow, Glasgow G12 8QQ, UK; muhammad.imran@glasgow.ac.uk
* Correspondence: qammer.abbasi@glasgow.ac.uk

Abstract: A wideband low-profile radiating G-shaped strip on a flexible substrate is proposed to operate as biomedical antenna for off-body communication. The antenna is designed to produce circular polarization over the frequency range 5–6 GHz to communicate with WiMAX/WLAN antennas. Furthermore, it is designed to produce linear polarization over the frequency range 6–19 GHz for communication with the on-body biosensor antennas. It is shown that an inverted G-shaped strip produces circular polarization (CP) of the opposite sense to that produced by G-shaped strip over the frequency range 5–6 GHz. The antenna design is explained and its performance is investigated through simulation, as well as experimental measurements. This antenna can be viewed as composed of a semicircular strip terminated with a horizontal extension at its lower end and terminated with a small circular patch through a corner-shaped strip extension at its upper end to form the shape of "G" or inverted "G". The purpose of the corner-shaped extension and the circular patch termination is to match the antenna impedance to 50 Ω over the entire frequency band (5–19 GHz) and to improve the circular polarization over the frequency band (5–6 GHz). To be fabricated on only one face of the flexible dielectric substrate, the antenna is fed through a co-planar waveguide (CPW). The antenna and the CPW dimensions are optimized to obtain the most optimal performance regarding the impedance matching bandwidth, 3dB axial ratio (AR) bandwidth, radiation efficiency, and maximum gain. The results show that the achieved 3dB-AR bandwidth is 18% (5–6 GHz). Thus, the proposed antenna covers the 5GHz frequency band of the WiMAX/WLAN applications within its 3dB-AR frequency band. Furthermore, the impedance matching bandwidth is 117% (5–19 GHz) which enables low-power communication with the on-body sensors over this wide range of the frequency. The maximum gain and radiation efficiency are 5.37 dBi and 98%, respectively. The overall antenna dimensions are 25 × 27 × 0.13 mm^3 and the bandwidth-dimension ratio (BDR) is 1733.

Keywords: circular polarized antenna; compact antenna; flexible antenna; wideband antenna

Citation: Yassin, M.E.; Hussein, K.F.A.; Abbasi, Q.H. ; Imran, M.A.; Mohassieb, S.A. Flexible Antenna with Circular/Linear Polarization for Wideband Biomedical Wireless Communication. *Sensors* **2023**, *23*, 5608. https://doi.org/10.3390/s23125608

Academic Editor: Domenico Ciuonzo

Received: 12 May 2023
Revised: 2 June 2023
Accepted: 13 June 2023
Published: 15 June 2023

1. Introduction

Recently, a lot of research work has been focused on the wideband flexible antennas for biomedical applications [1–7]. For example, in the application of wireless body area network (WBAN), on-body or implantable biosensors are placed at different positions of the patient body to provide biotelemetry data continuously, at regular time intervals [5,6], or whenever it is required. The biotelemetry data transmitted by the biosensor antennas are collected by a nearby (central) antenna of the WBAN (may be either on-body or off-body antenna). To be available for a wider area network, the central antenna retransmits the collected data to a nearby base station, which is often a WiMAX or Wi-Fi (WLAN) antenna. Figure 1 shows an example of WBAN for a patient in the Intensive Care Unit (ICU) where the G-shaped antenna proposed in the present work is employed as a central off-body antenna to perform the dual function of collecting the biotelemetry data from the biosensors distributed on (or implantable inside) the patient body and then retransmitting them to a WiMAX/WLAN antenna. The minimization of the power consumption is a vital requirement in the biosensor antennas to produce low-power transceivers in biomedical applications, such as the WBAN. To minimize the power required for communication within the nodes of the WBAN, the ultra-wideband (UWB) antennas are preferred to narrowband antennas for their low-power spectral density that is restricted to 41.3 dBm/MHz [8,9]. For this reason, the proposed G-shaped antenna is designed to operate in the "on-body" mode with linear polarization over the wide frequency band (6–19 GHz). In this mode, the central antenna operates with linear polarization to communicate with the biosensor antennas for collecting biotelemetry data. On the other hand, the G-shaped antenna is designed to operate in "off-body" mode for re-transmitting the collected data to the WiMAX/WLAN antenna in the frequency band 5–6 GHz. As shown in Figure 2, the WiMAX/WLAN antenna that may have arbitrary polarization (vertical or horizontal). To reduce the power loss caused by misalignment with the WiMAX/WLAN antenna, the G-shaped antenna is designed to produce circular polarization while operating in the "off-body" mode (5–6 GHz) and linear polarization while operating in the "on-body" mode (6–19 GHz).

The proposed antenna is designed to communicate not only with the biosensors on the body but also with the surrounding medical apparatus attached to the body in a health care system and, also, to communicate with the WLAN access points. To allow better mobility of the patient and the attached medical equipment, this antenna is designed to operate near body and not stuck to the body, as shown in Figure 1. Moreover, this antenna operates in the off-body mode (while communicating with the WLAN) emitting relatively higher level of the power than that emitted in the on-body mode operation. To reduce the specific absorption rate (SAR) in the off-body mode it is preferred not to be placed directly on the patient body. Because of these reasons, the proposed antenna is designed to be placed in a region central to the body sensors, the surrounding medical equipment, and the wireless LAN access point rather than to be integrated into the patient body.

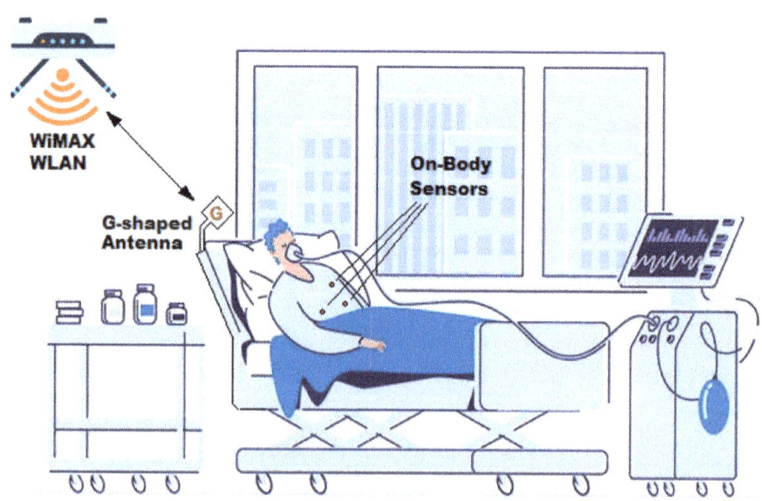

Figure 1. Wireless body area network in intensive care unit where the proposed G-shaped antenna is employed as central off-body antenna for transmitting the biotelemetry data to a nearby WiMAX/WLAN base station antenna.

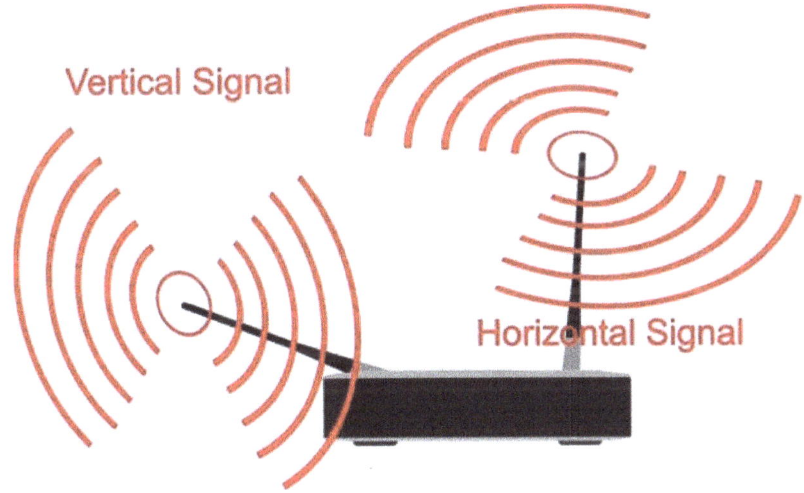

Figure 2. WiMAX/WLAN access point with antennas of arbitrary polarization.

Polarization is a critical and significant feature of antennas in modern wireless communication technology [10]. The orientation of the electric field vector determines the polarization of the antenna, which can be either linear, circular, or elliptical [11]. Antennas with circular polarization (CP) radiate electromagnetic energy in a circular spiral pattern, where the two perpendicular field components are nearly equal in magnitude and in phase quadrature [12]. The enhancement of CP bandwidth is a significant challenge for antenna designers looking to create compact antennas without compromising their performance in wireless communication technologies, such as 5G, Wi-Fi, and WiMax [13].

Lately, different antenna types have been employed to attain circular polarization, including patch antennas, slot antennas, and spiral antennas [14–19]. These antennas offer several advantages over traditional linearly polarized antennas, including improved signal quality, lower interference, and increased capacity in wireless communication systems [20–22]. In addition to the different antenna types, CP antennas can also be classified based on their feeding networks. Single-feed CP antennas, which utilize a single feed point, are simpler to design and fabricate but have limited bandwidth [23–25]. On the other hand, multiple-feed CP antennas, which use multiple feed points, can achieve wider bandwidths but are more complex to design and manufacture [26,27]. Recently, the single-feed co-planar waveguide (CPW) has become the preferred feeding method for wideband CP antennas [28–30]. Various types of CPW antennas have been documented to enhance CP performance, including monopole, patch, and slot antennas [31–34]. These antennas offer wideband circular polarization with high gain and radiation efficiency, making them ideal for use in modern wireless communication technologies.

The flexibility of a circularly polarized (CP) antenna's substrate can significantly impact its performance [35]. When the antenna is bent or deformed, the polarization purity of the transmitted and received signals can be affected, resulting in a decrease in antenna efficiency and signal quality. However, using a flexible substrate in the design of CP antennas can help mitigate these issues by allowing the antenna to conform to curved or irregular surfaces, reducing the impact of substrate deformation on antenna performance [36,37]. In recent years, there has been a growing interest in designing circularly polarized antennas using flexible substrates to improve their performance under bending conditions. Flexible substrates can conform to curved or irregular surfaces and provide better integration with conformal devices, making them ideal for use in wearable or flexible communication systems [38]. The use of flexible substrates in circularly polarized antennas has been shown to improve radiation performance under bending conditions and reduce the impact of substrate deformation on antenna performance [39,40]. This makes them particularly useful for applications such as smart textiles and Internet of Things (IoT) devices, where antennas need to be compact, low-profile, and able to withstand deformation. As such, research into the design of circularly polarized antennas using flexible substrates has become an active area of investigation in the field of wireless communication technology [41–43].

The proposed G-shaped and inverted G-shaped printed strip CP antennas are designed on a low-loss flexible Roger RO3003^{TM} that exhibits good performance even under bending. The effects of bending the substrate on the antenna parameters were fully investigated by numerical simulations. For enhancing the antenna performance, the antenna is constricted as semicircular wide strip terminated by a small circular patch and fed through a CPW. The geometric dimensions have been optimized for enhanced performance regarding the impedance matching bandwidth, 3dB axial ratio (AR) bandwidth, radiation efficiency, and gain.

2. Antenna Design

The WiMAX/WLAN antennas may have arbitrary polarization depending on their orientation as shown in Figure 2. To reduce the losses resulting from misalignment of the WBAN central antenna with the WiMAX/WLAN antenna during the off-body mode of operation (5 GHz band), it is proposed that the G-shaped antenna produces circular polarization over the frequency range 5–6 GHz. During the on-body mode of operation (6–19 GHz), it is preferred to operate with linear polarization as it is easier to align the central antenna with the on-body biosensor antennas. The functions assigned to the proposed antenna and the required polarization during the two modes of operation are listed in Table 1.

Table 1. Operations assigned to the G-shaped strip antenna proposed for WBAN in ICU.

Mode of Operation	Function of the Antenna	Frequency Range	Polarization
On-body	Communicate with the biosensor antennas for biotelemetry	6–19 GHz	Linear
Off-body	Retransmit biotelemetry to WiMAX/WLAN antennas	5–6 GHz	Circular

The G-shaped and inverted G-shaped strip antennas are designed to operate over the frequency band (5–19 GHz). The main issue of the proposed antenna design is to fulfill the requirements of the dual function to produce dual polarization (circular/linear) over a wide frequency band with good impedance matching and high radiation efficiency. The G-shaped radiating strip is a turn-like antenna, i.e., it is similar to a helix or spiral of one turn. Therefore, it can produce both circular and linear polarization by adjusting its dimensions relative to the wavelength. The frequency band 5–6 GHz is actually the unique frequency band that is commonly dedicated for WiMAX and WLAN applications together. Therefore, the frequency band 5–6 GHz is selected for circular polarization. The CST® Studio Suite 3D EM simulator (CST-MWS) is used to design the antenna and evaluate its performance. The geometry of the inverted G-shaped antenna is presented in Figure 3. The inverted-G shaped antenna produces right-hand (CP) in the (+ve z-direction) and left-hand (CP) in the (-ve z-direction) whereas the G-shaped antenna produces left-hand (CP) in the (+ve z-direction) and right-hand (CP) in the (-ve z-direction).

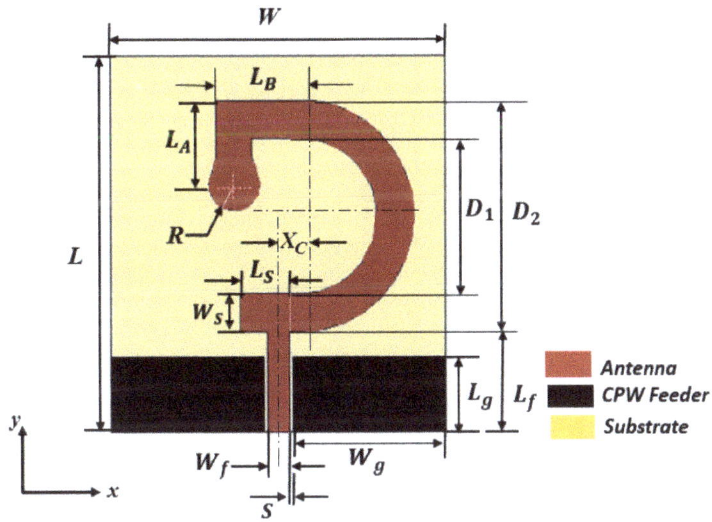

Figure 3. Geometry of the inverted G-shaped strip antenna showing the dimensional parameters.

Both the G-shaped and inverted G-shaped antennas can be viewed as composed of semicircular strip connected, at its upper end, to horizontal strip extension of length L_B followed by vertical strip extension of length L_A to form a right-angle corner-shaped extension that is terminated by a small circular patch of radius R. At its lower end, the semicircular strip is terminated with a horizontal strip extension of length L_S and width W_S, as shown in Figure 3. The semicircular strip can be described by the inner and outer diameters D_1 and D_2, respectively. The antenna is excited through a CPW feeding line. The CPW has a length of L_g, the width of its central strip is W_f, and the width of each side slot is S. The characteristic impedance of the CPW is obtained based on W_f and S that are calculated using the CPW design equations [44,45]. The length of the CPW central strip extension to connect the antenna to the feed line is L_f. The antenna and the CPW feeder are printed on a flexible Rogers RO3003^{TM} substrate which has a dielectric constant $\epsilon_r = 3, loss\ tangent\ tan\delta = 0.001$, and thickness $h = 0.13$ mm. The total size of the printed antenna is $L \times W \times h$. The best dimensions of the proposed antenna are shown in Table 2.

Table 2. Optimum dimensions of the proposed CP antenna.

Parameter	W	L	D_1	D_2	R	L_S	h	W_S
Value (mm)	25	27	11.2	16.6	1.9	4.6	0.13	2.7
Parameter	L_A	L_B	W_g	L_g	L_f	W_f	S	X_C
Value (mm)	6	6.4	11.4	5.4	7.2	1.6	0.3	1.8

2.1. Evolution of the Antenna Design

The proposed antenna design has evolved in three main steps as presented in Figure 4a. Other details of the antenna design have been achieved through other minor steps of the design process. However, the major three steps are explained in the present section.

The first step is a trial to produce circular polarization by using a semicircular strip radiator with small horizontal extensions at the end points of the circular arc, as shown in the geometry of Antenna#1. From the curves of $\mid S_{11} \mid$ and AR against the frequency Figure 4b,c, it seems that the antenna impedance matching is realized over some frequency bands within the desired wideband and, on the other hand, the AR seems to be improved showing a minimum near 5 GHz. However, neither the desired impedance matching nor the 3dB-AR is achieved.

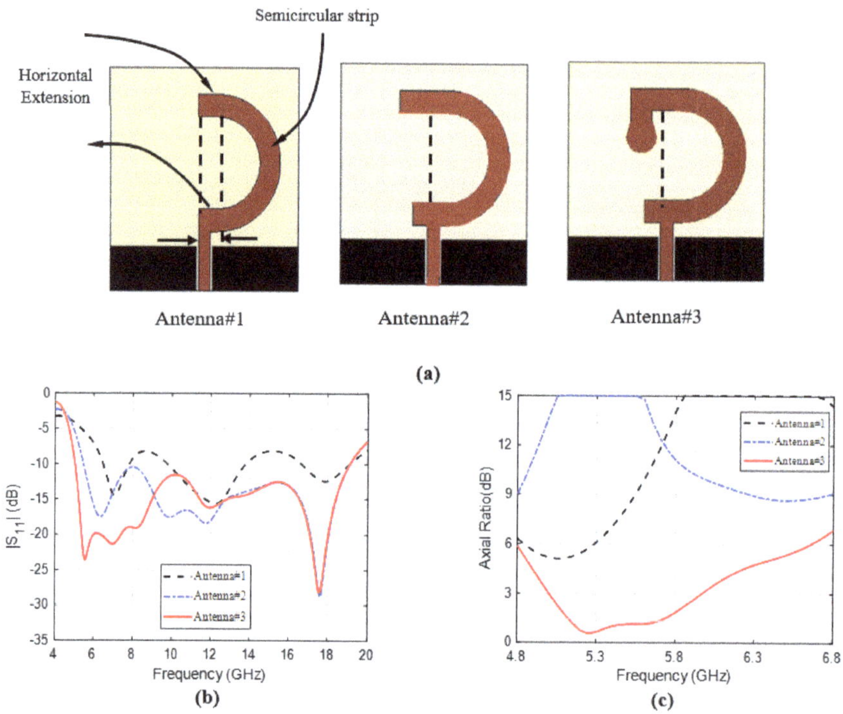

Figure 4. Evolution of the Inverted G-shape CP Antenna Design: (**a**) Evolution steps. (**b**) The reflection coefficient $| S_{11} |$ graph of each design step. (**c**) The AR graph of each design step.

In the next step of the antenna design, the end points of the radiating strip are further extended to obtain the geometry of Antenna#2. This step of the design results in a great improvement of the impedance matching bandwidth as shown in Figure 4b but, however, the AR is badly affected as shown in Figure 4c. In the third and final major steps of the design process, the strip radiator is extended at its upper end by a vertical extension terminated with a small circular strip as shown in the geometry of Antenna#3. The final design achieves the designed bandwidth of impedance matching, and 3dB-AR as shown in Figure 4b,c.

2.2. *Investigation of Optimal Dimensions*

The G-shaped antenna being proposed consists of five parts, (i) the semicircular strip, (ii) the vertical extension of the central strip of the CPW region, (iii) the horizontal extension at the lower end of the semicircular strip, (iv) the corner extension at the upper end of the semicircular strip, and (v) the small circular patch that terminates the corner extension of the semicircular patch. It is required to obtain the widest impedance matching and 3dB-AR bandwidth and to enhance the radiation efficiency. For this purpose, A thorough investigation of various parameters has been conducted to determine the optimal dimensions of the antenna. The effects of various dimensional parameters on $| S_{11} |$ and AR are numerically investigated in the following subsections.

2.2.1. Effect of the Diameters of the Semicircular Strip

The effects of changing the outer and inner diameters, D_2 and D_1, respectively, are depicted in the present section. The effects of changing the outer diameters D_2 on $| S_{11} |$ and the AR is shown in Figure 5. It is evident in Figure 5a that increasing D_2 leads to a decrease in the lower and higher frequency bands and the entire impedance matching band is shifted towards the left. However, Figure 5b shows that the best performance regarding the 3dB-AR bandwidth is achieved when D_2 = 16.6 mm.

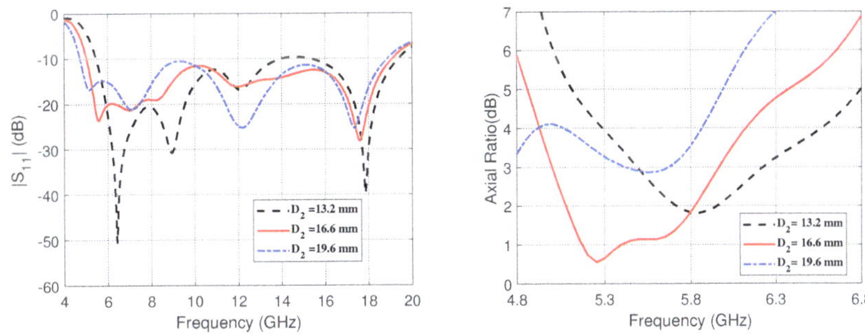

(**a**) (**b**)

Figure 5. Effect of changing the outer diameter, D_2, of the semicircular strip on (**a**) The magnitude of the reflection coefficient , $|S_{11}|$, as a function of frequency (4–20 GHz). (**b**) The AR over the frequency range (4.8–6.8 GHz).

On the other hand, the effects of changing the inner diameter, D_1, of the semicircular strip on the impedance matching and 3dB-AR bandwidths are presented in Figure 6. Figure 6a shows that increasing D_1 decreases the lower frequency at which the impedance bandwidth matches whereas the higher frequency seems to insensitive to such changes of D_1. The proposed antenna performs optimally in terms of the 3dB-AR bandwidth when D_1 = 11.2 mm, as illustrated in Figure 6b.

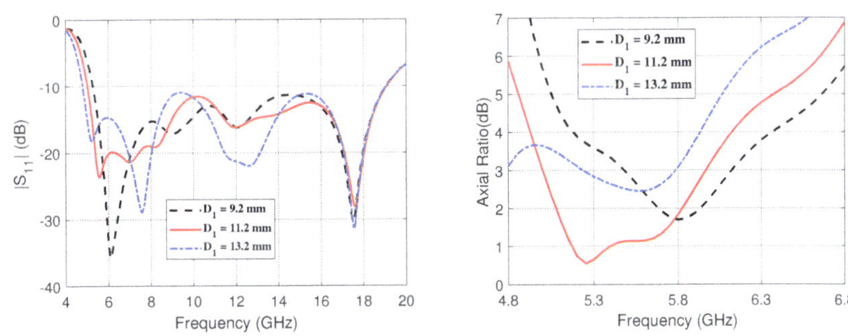

(**a**) (**b**)

Figure 6. Effect of changing the outer diameter, D_1, of the semicircular strip on (**a**) the magnitude of the reflection coefficient , $|S_{11}|$, as a function of frequency (4–20 GHz), and (**b**) the AR over the frequency range (4.8–6.8 GHz).

2.2.2. Effects of the Dimensions of the Corner-Shaped Extension of the Curved Strip

The effects of changing the vertical and horizontal dimensions, L_A and L_B, respectively, of the corner-shaped extension of the curved strip on the reflection coefficient magnitude and the AR are investigated in the present section. The effects of changing vertical length, L_A, on the frequency response of the reflection coefficient magnitude, $|S_{11}|$, and the AR are presented in Figure 7, respectively. It is shown in Figure 7a that the impedance matching frequency band seems to be slightly affected at its lower frequency by changing L_A. On the other hand, the AR seems to be strongly dependent on L_A, as shown in Figure 7b. It is clear that setting L_A = 6 mm gives the best 3dB-AR bandwidth.

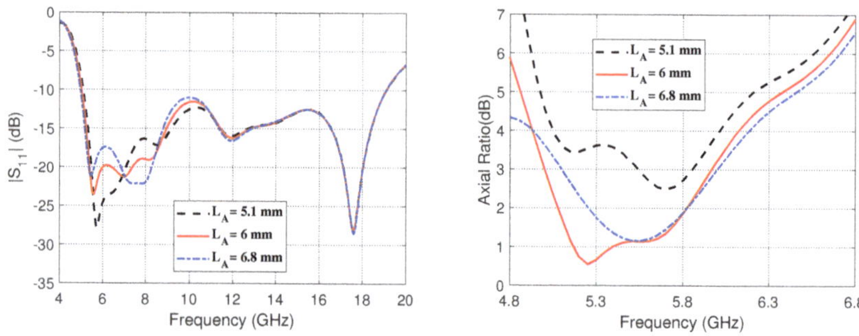

(**a**) (**b**)

Figure 7. Effect of changing the vertical length, L_A, of the corner-shaped extension of the curved strip on (**a**) The magnitude of the reflection coefficient , | S_{11} |, as a function of frequency (4–20 GHz). (**b**) The AR over the frequency range (4.8–6.8 GHz).

The effects of changing horizontal length, L_B, on the frequency response of the reflection coefficient magnitude, | S_{11} |, and the AR are presented in Figure 8, respectively. It is shown in Figure 8a that the impedance matching frequency band seems to be slightly affected at its lower frequency by changing L_B. On the other hand, the AR seems to be strongly dependent on L_B as shown in Figure 8b. It is clear that setting L_B = 6.4 mm gives the best 3dB-AR bandwidth.

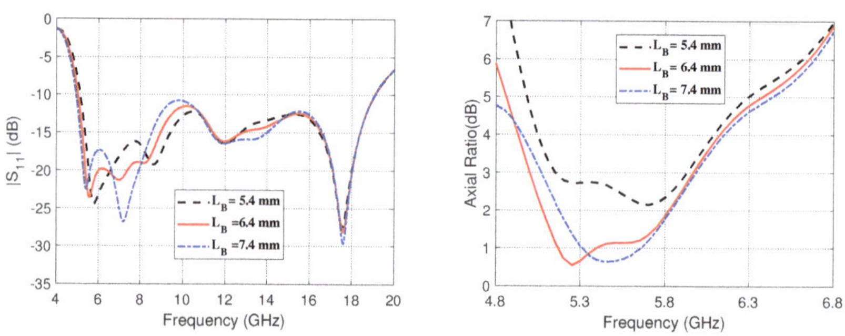

(**a**) (**b**)

Figure 8. Effect of changing the horizontal length, L_B, of the corner-shaped extension of the curved strip on (**a**) The magnitude of the reflection coefficient , | S_{11} |, as a function of frequency (4–20 GHz). (**b**) The AR over the frequency range (4.8–6.8 GHz).

2.2.3. Influence of the Radius of the Circular Patch Termination

The impact of altering the radius, *R*, of the small circular patch termination of the G-shaped strip antenna on the frequency dependence of the magnitude of the reflection coefficient, | S_{11} |, and the AR are presented in Figure 9. It is shown in Figure 9a that the impedance matching frequency band seems to be slightly affected at its lower frequency by changing *R*. On the other hand, the AR seems to be strongly dependent on *R* as shown in Figure 9b. It is clear that setting *R* = 1.9 mm gives the best 3dB-AR bandwidth.

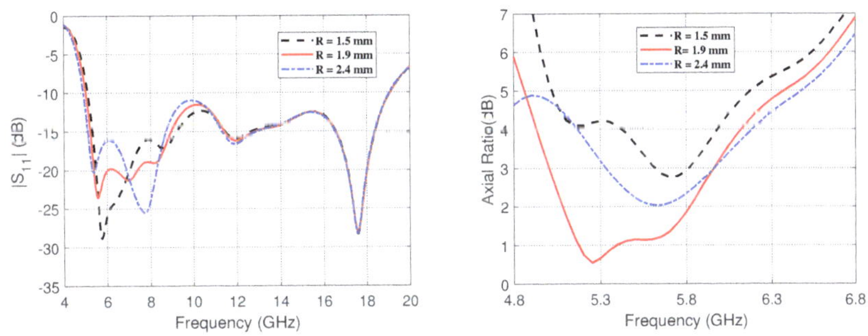

(a) (b)

Figure 9. The impact of altering the radius, R, of the circular patch termination on (**a**) The reflection coefficient magnitude, $|S_{11}|$, over the frequency range (4–20 GHz). (**b**) The AR over the frequency range (4.8–6.8 GHz).

2.2.4. The Impact of Altering the Length of the CPW Central Strip Extension

The impact of altering the length L_f of the CPW central strip extension on the magnitude of the reflection coefficient, $|S_{11}|$, and the AR over the applicable frequency ranges are shown in Figure 10. Figure 10a shows that the lower frequency of the impedance matching bandwidth slightly decreases with increasing L_f. However, the widest bandwidth is obtained by setting L_f = 7.2 mm. This value of L_f results in a 3dB-AR frequency range covering 5 GHz to 6 GHz as shown in Figure 10b.

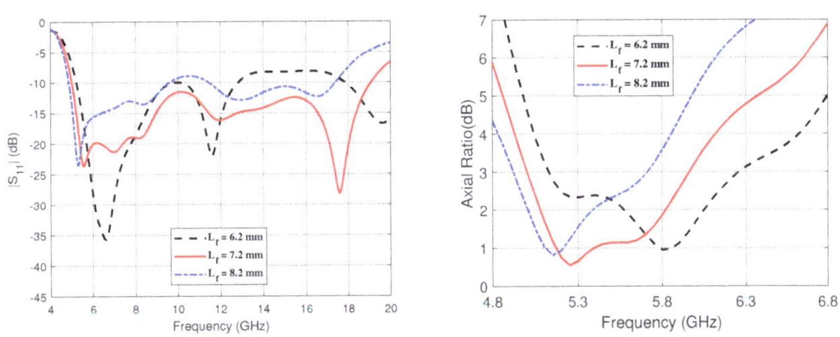

(a) (b)

Figure 10. The impact of altering the length, L_f, of the central strip extension on (**a**) The reflection coefficient magnitude, $|S_{11}|$, over the frequency range (4–20 GHz). (**b**) The AR as a function of frequency (4.8–6.8 GHz).

2.2.5. Optimum Dimensional Parameters

The parametric sweeps presented in Sections 2.2.1–2.2.4 are examples for the complete investigation of various parameters that has been conducted to determine the optimal values for the geometrical design parameters of the inverted G-shaped antenna. The results of this investigation are presented in Table 2. Furthermore, it should be noted that the dimensions of the CPW region are adjusted to improve the impedance matching where the analytic rules in [44,45] have been used as initial values and then optimized by EM simulation.

3. Effect of Bend Stresses on the Inverted G-Shape CP Antenna Characteristics

Owing to its flexible structure, the proposed antenna shall preserve its high performance even while being subjected to bend strains (to some extents) in the different directions.

For studying the effects of the bend stresses on the characteristics of the proposed antenna, it is subjected to different bend angles (β) and the corresponding frequency dependencies of the reflection coefficient, $| S_{11} |$ and AR are investigated. The bending is applied in one of two perpendicular directions, the bend angle β_L that is applied in the longitudinal plane parallel to the feeding line as shown in Figure 11a or the bend angle β_T that is applied in the transverse plane perpendicular to the feed line, as shown in Figure 11b.

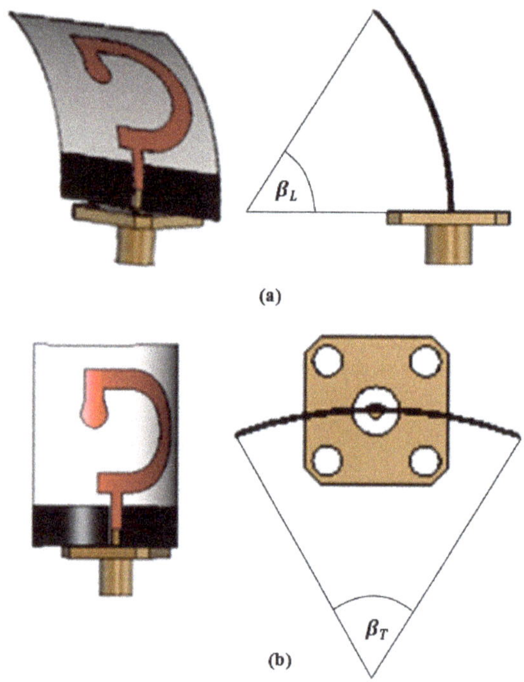

Figure 11. The proposed antenna subjected to bend stresses in (**a**) The longitudinal plane, (**b**) The transverse plane.

3.1. Effects of Bending the Antenna in the Longitudinal Plane on the Impedance Matching and Axial Ratio

The frequency dependence of the reflection coefficient, $| S_{11} |$, for different values of the longitudinal bend angle, β_L is presented in Figure 12. It is shown that for longitudinal bend angles less than 55°, the dependence of $| S_{11} |$ on the frequency is almost unchanged keeping the impedance matching bandwidth almost the same for $\beta_L < 55°$. However, for $\beta_L \geq 55°$, the value of $| S_{11} |$ starts to increase above -10 dB leading to decrease the impedance matching bandwidth.

The frequency dependence of the axial ratio for different values of the longitudinal bend angle, β_L is presented in Figure 13. It is shown that, with increasing β_L the frequency band for 3dB-AR is slightly shifted towards the left keeping the 3dB-AR bandwidth almost unchanged keeping the impedance matching bandwidth almost the same for $\beta_L < 45°$. However, for $\beta_L \geq 45°$, the value of the axial ratio increases above 3 dB leading to lose the circular polarization over a significant part of the frequency band.

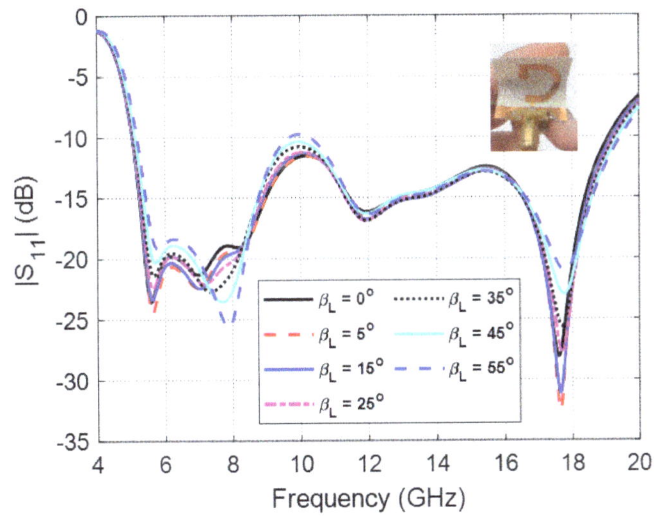

Figure 12. Frequency dependence of the reflection coefficient, $| S_{11} |$, for different values of the longitudinal bend angle, β_L.

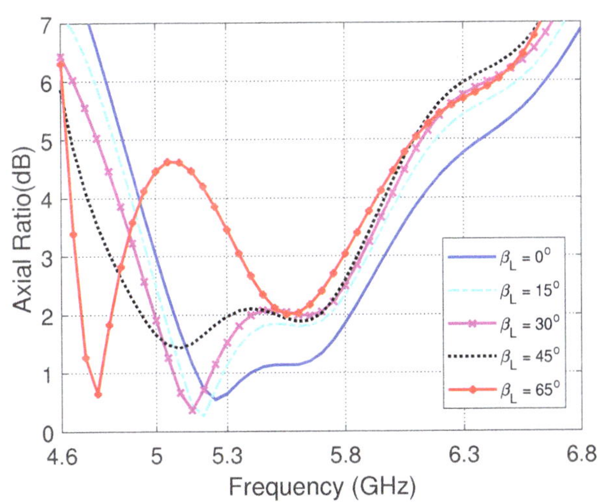

Figure 13. Frequency dependence of the AR for different values of the longitudinal bend angle, β_L.

3.2. Effects of Bending the Antenna in the Transverse Plane on the Impedance Matching and Axial Ratio

The frequency dependence of the reflection coefficient, $| S_{11} |$, for different values of the transverse bend angle, β_T is presented in Figure 14. It is shown that for $\beta_T < 25°$, the dependence of $| S_{11} |$ on the frequency is slightly affected but, however, the impedance matching bandwidth is almost the same. For $\beta_T \geq 25°$, the value of $| S_{11} |$ starts to increase above -10 dB leading to decrease the impedance matching bandwidth.

The frequency dependence of the axial ratio for different values of the transverse bend angle, β_T is presented in Figure 15. It is shown that, with increasing β_T the 3dB-AR bandwidth is continuously decreased as the lower frequency is increased and the higher frequency is decreased. For $\beta_T \geq 25°$, the minimum value of the axial ratio increases above 3 dB leading to lose the circular polarization over the entire frequency band.

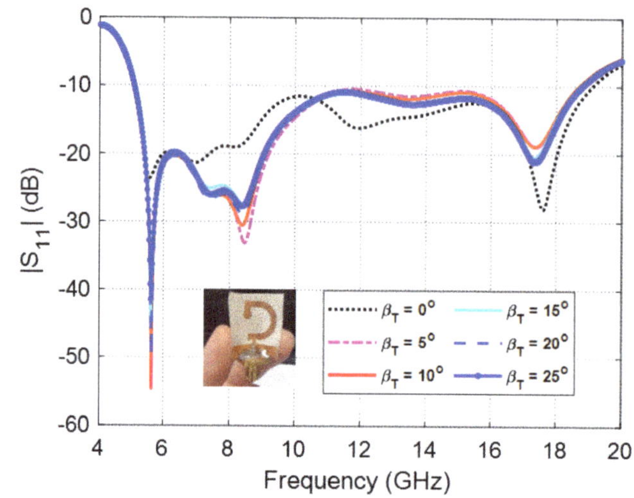

Figure 14. Frequency dependence of the reflection coefficient, $| S_{11} |$, for different values of the transverse bend bend angle, β_T.

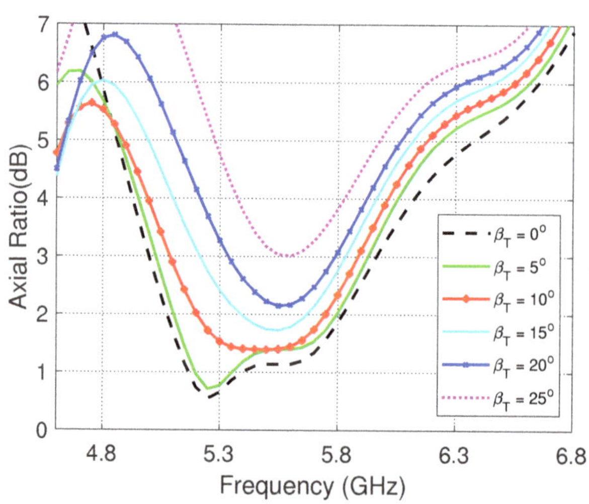

Figure 15. Frequency dependence of the AR for different values of the transverse bend angle, β_T.

3.3. *Effect of Bending the Antenna on the Impedance Matching Bandwidth*

The dependence of the percent impedance matching bandwidth on both the longitudinal and transverse bend angles, β_L and β_T, respectively, is presented in Figure 16a. It is shown that the impedance matching bandwidth is almost independent of the bend angles as long as both β_L and β_T are less than 55°. Increasing any of the two bend angles above 55° causes a dramatic drop of the percent bandwidth of impedance matching.

3.4. Effect of Bending the Antenna on the 3dB Axial Ratio Bandwidth

The dependence of the percent bandwidth of 3dB-AR on both the longitudinal and transverse bend angles, β_L and β_T, respectively, is presented in Figure 16b. It is shown that the impedance matching bandwidth is almost independent of the longitudinal bend angle as long as $\beta_L \leq 45°$. Increasing β_L above 45° causes a dramatic drop of the percent bandwidth 3dB-AR. On the other hand, increasing the transverse bend angle β_T beyond 5° leads to a fast decay of the 3dB-AR bandwidth until the circular polarization is completely lost for $\beta_T \geq 25°$.

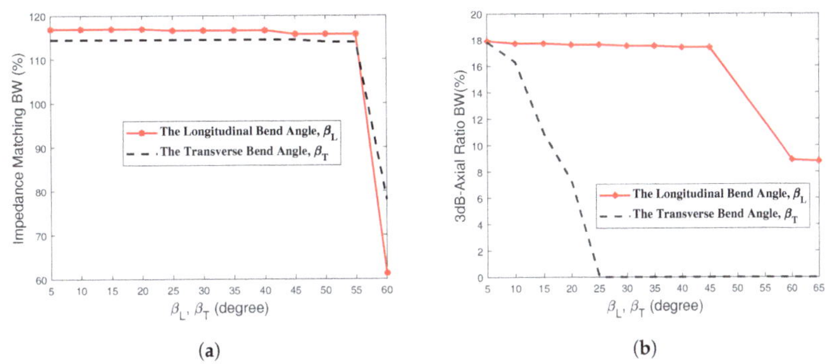

Figure 16. Dependence of the percentage (**a**) impedance matching bandwidth and (**b**) 3dB-AR bandwidth on the bend angles, β_L and β_T.

Thus, it can be concluded that the proposed antenna preserves the impedance matching bandwidth and the 3dB-AR bandwidth for longitudinal and transverse bend angles less than 55° (i.e., for $\beta_L < 55°$ and $\beta_T < 55°$). However, to preserve the achieved antenna performance regarding the 3dB-AR bandwidth, it is recommended to keep the longitudinal bend angle less than 45° and to keep the transverse bend angle less than 5° (i.e., $\beta_L < 45°$ and $\beta_T < 5°$) otherwise the 3dB-AR bandwidth will be badly affected.

4. Mechanism of Circular Polarization

The way that the proposed inverted G-shaped strip antenna produces circular polarization can be demonstrated by showing the surface current distribution on the semi-circle strip radiator, as depicted in Figure 17. The surface current distributions are presented at sequential orthogonal phases 0°, 90°, 180°, and 270° which correspond to time delays 0, $1/4T$, $1/2T$, and $3/4T$, respectively, where T is the periodic time at different frequencies, such as 5, 5.5, and 6 GHz.

It is shown that the surface current on the inverted-G-shaped antenna is circulating in the counter clockwise direction, thereby producing RHCP in the +ve z-direction. As the dielectric substrate is very thin relative to the wavelength (thickness = 0.002λ) it can be considered transparent to the wave at this frequency. Hence, the amount of power radiated towards the lower half of the space (-ve z-direction) is almost equal to the amount of power radiated towards the upper half of the space (+ve z-direction) but with LHCP.

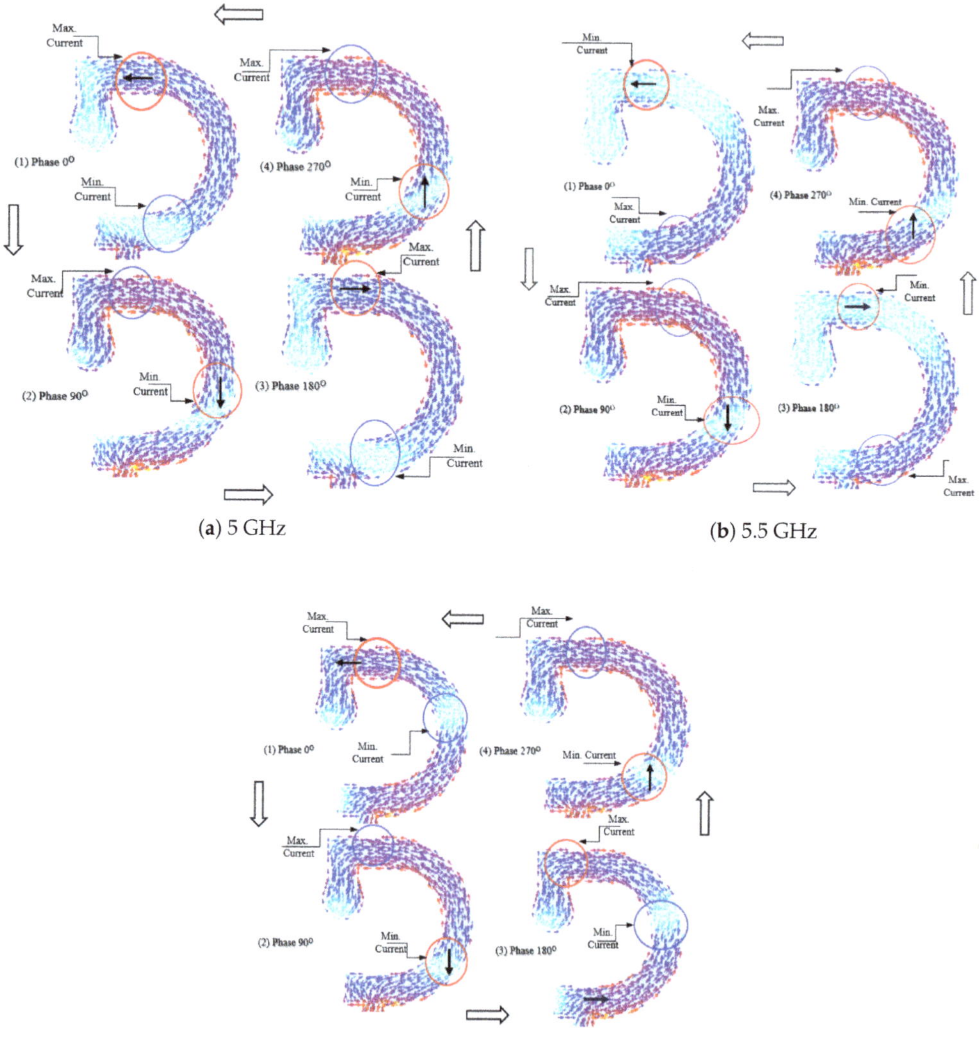

(**a**) 5 GHz

(**b**) 5.5 GHz

(**c**) 6 GHz

Figure 17. Surface current distributions on the surface of the inverted G-shaped strip antenna at different frequencies range at sequential orthogonal phases (sequential time delays). (**a**) 5 GHz. (**b**) 5.5 GHz, (**c**) 6 GHz.

5. Procedure for Measurement of the Radiation Patterns, Gain, and Efficiency

This section is concerned with explaining the procedure used to measure the radiation pattern, gain, and efficiency of the proposed antenna. The radiation pattern measurement setup presented in Figure 18, where the antenna under-test is zoomed-in while being mounted on the fixture tool during measurement is established for this purpose. A reference-gain cross-polarized horn antenna is used for gain measurement over the entire frequency band (4–20 GHz). This antenna can be used to separately measure the horizontally and the vertically polarized fields. During measurement, the reference antenna is maintained oriented to the antenna under test (AUT) which is placed on the rotator for complete rotation in the azimuth and elevation planes.

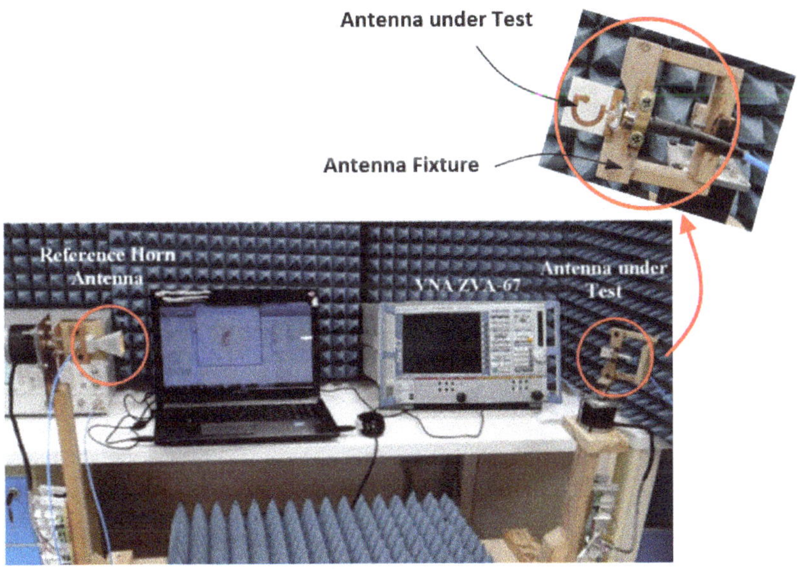

Figure 18. Experimental setup for measurement of the gain, radiation patterns, and antenna efficiency.

The reference antenna is connected to port 2 of the VNA whereas the AUT is connected to port 1. While rotating the AUT in the desired plane, the readings of the transmission scattering parameter S_{21} are uniformly acquired over the entire frequency band (4–20 GHz) with the preset resolution. A Matlab® program on the laptop plays the role of a central controller and data processor. It controls the rotation of the AUT and the data acquisition of the VNA and stores the S_{21} data during measurement. When the antenna rotation is completed to cover the entire space $(0 \leq \theta \leq \pi$ and $0 \leq \phi \leq 2\pi)$ with appropriate angular resolution, where θ and ϕ are spherical angular coordinates related to the coordinate system presented in Figure 3. The stored S_{21} data are processed, the gain is calculated and the radiation patterns at all the frequencies can be drawn. The radiation efficiency over the entire frequency band is also measured. A novel method followed to measure, the radiated field, gain, and radiation efficiency is described as follows.

The effective aperture area of the reference-gain horn antenna (employed as a receiver during measurement) can be expressed as follows

$$A_{REF} = \frac{\lambda^2}{4\pi} G_{REF} \tag{1}$$

where λ is the operating wavelength and G_{REF} is the gain of the reference-gain horn antenna.

Let P_0 be the power output of the wave generator used for transmission during measurement, and P_{RL} be the power returned to the wave generator (the transmitter) due to the impedance mismatch between the AUT and the wave generator. The following expression can be used to evaluate P_{RL}.

$$\frac{P_{RL}}{P_0} = |S_{11}|^2 \tag{2}$$

The power accepted by the AUT excitation port can be expressed as follows:

$$P_A = P_0 - P_{RL} = P_0\left(1 - |S_{11}|^2\right) \tag{3}$$

5.1. Measurement of the Radiation Pattern

Let $P_R(\theta,\phi)$ be the power received at the reference-gain horn antenna when the rotator directs the AUT at the direction(θ,ϕ) . The following expression can be used to evaluate $P_R(\theta,\phi)$.

$$\frac{P_R(\theta,\phi)}{P_0} = \frac{|S_{21}(\theta,\phi)|^2}{1-|S_{22}|^2} \tag{4}$$

where $S_{21}(\theta,\phi)$ is the mutual S-parameter between ports 1 and 2 of the VNA when the rotator directs the AUT towards the direction(θ,ϕ) during measurement and S_{22} is the reflection coefficient measured at port 2 to which the reference-gain antenna is connected during measurement. Thus, the normalized power patterns can be expressed as follows:

$$\hat{P}_R(\theta,\phi) = \frac{|S_{21}(\theta,\phi)|^2}{\max\left(|S_{21}(\theta,\phi)|^2\right)} \tag{5}$$

The far field radiation pattern can be expressed as follows

$$|E(\theta,\phi)| = \sqrt{2\zeta P_R(\theta,\phi)} \tag{6}$$

where ζ is the intrinsic wave impedance of free space.

Employing (3), the far field radiation pattern:

$$|E(\theta,\phi)| = \sqrt{2\zeta}\frac{|S_{21}(\theta,\phi)|}{\sqrt{1-|S_{22}|^2}} \tag{7}$$

The normalized radiation pattern can be evaluated as follows:

$$\hat{E}(\theta,\phi) = \frac{|S_{21}(\theta,\phi)|}{\max\left(|S_{21}(\theta,\phi)|\right)} \tag{8}$$

5.2. Measurement of the Antenna Gain

The power density at the reference-gain antenna when the AUT is directed at (θ,ϕ) can be expressed as follows:

$$\rho(\theta,\phi) = \frac{P_A G_A(\theta,\phi)}{4\pi D^2} \tag{9}$$

where $G_A(\theta,\phi)$ is the gain of the AUT in the direction (θ,ϕ), D is the distance between the transmitting and receiving antennas during measurement. The power received at the reference-gain antenna when the AUT is directed at (θ,ϕ) can be expressed as follows.

$$P_R(\theta,\phi) = \rho(\theta,\phi)\frac{A_{REF}}{\eta_{REF}} \tag{10}$$

where η_{REF} is the radiation efficiency of the reference-gain antenna. It is used in the denominator of the right-hand side of (10) to compensate for the losses of the reference-gain antenna other than the return loss.

Substituting from (9) into (10), the power at the receiving antenna can be expressed as follows:

$$P_R(\theta,\phi) = \frac{P_A G_A(\theta,\phi)}{4\pi D^2}\frac{A_{REF}}{\eta_{REF}} \tag{11}$$

Making use of (1), (3), and (11), the power received at the reference-gain antenna can be expressed as follows:

$$P_R(\theta,\phi) = P_0\left(1-|S_{11}|^2\right)G_A(\theta,\phi)\frac{\lambda^2}{(4\pi D)^2}\frac{G_{REF}}{\eta_{REF}} \tag{12}$$

Thus, the AUT antenna gain, $G_A(\theta,\phi)$, can be expressed as follows:

$$G_A(\theta,\phi) = \frac{P_R(\theta,\phi)}{P_0\left(1-|S_{11}|^2\right)}\frac{(4\pi D)^2}{\lambda^2}\frac{\eta_{REF}}{G_{REF}} \tag{13}$$

Substituting from (4) into (13), the following expression is obtained.

$$G_A(\theta,\phi) = \frac{|S_{21}(\theta,\phi)|^2}{\left(1-|S_{11}|^2\right)\left(1-|S_{22}|^2\right)}\frac{(4\pi D)^2}{\lambda^2}\frac{\eta_{REF}}{G_{REF}} \tag{14}$$

Note that S_{11}, S_{22}, and $S_{21}(\theta,\phi)$ are measured by the VNA and all the other parameters on the right-hand side of (14) are known before carrying out the measurement procedure. Hence, the expression (14) can be used for measuring the gain of the AUT.

The realized antenna gain of the AUT can be expressed as follows:

$$G_R(\theta,\phi) = \left(1-|S_{11}|^2\right)G_A(\theta,\phi) = \frac{|S_{21}(\theta,\phi)|^2}{\left(1-|S_{22}|^2\right)}\frac{(4\pi D)^2}{\lambda^2}\frac{\eta_{REF}}{G_{REF}} \tag{15}$$

5.3. *Measurement of the Antenna Efficiency*

The power received by the reference-gain horn antenna can be obtained by calculating the following double integral

$$P_{Rad} = \int_0^{2\pi}\int_0^{\pi}\rho(\theta,\phi)D^2\sin\theta d\theta d\phi \tag{16}$$

where $\rho(\theta,\phi)$ is the power density at the location of the reference-horn antenna (the receiver) when the AUT is directed at (θ,ϕ) by the rotator.

The total power radiated by the AUT can be calculated using the following double integral.

$$P_{Rad} = \frac{4\pi D^2}{\lambda^2}\frac{\eta_{REF}}{G_{REF}}\int_0^{2\pi}\int_0^{\pi}P_R(\theta,\phi)\sin\theta d\theta d\phi \tag{17}$$

5.3.1. Total Antenna Efficiency

Dividing both sides of (17) by P_0, the following expression is obtained for the total efficiency of the AUT.

$$\eta_{\text{Total}} = \frac{P_{Rad}}{P_0} = \frac{4\pi D^2}{\lambda^2}\frac{\eta_{REF}}{G_{REF}}\int_0^{2\pi}\int_0^{\pi}\frac{P_R(\theta,\phi)}{P_0}\sin\theta d\theta d\phi \tag{18}$$

Substituting from (4) into (18), the following expression is obtained.

$$\eta_{\text{Total}} = \frac{P_{Rad}}{P_0} = \frac{1}{1-|S_{22}|^2}\frac{4\pi D^2}{\lambda^2}\frac{\eta_{REF}}{G_{REF}}\int_0^{2\pi}\int_0^{\pi}|S_{21}(\theta,\phi)|^2\sin\theta d\theta d\phi \tag{19}$$

5.3.2. Antenna Radiation Efficiency

The radiation efficiency of the AUT can be expressed as follows.

$$\eta_{Rad} = \frac{P_{Rad}}{P_0 - P_{RL}} \tag{20}$$

The expression (20) can be reformulated as follows.

$$\eta_{\text{Rad}} = \frac{P_{Rad}}{P_0}\left(1 - \frac{P_{RL}}{P_0}\right)^{-1} \quad (21)$$

Making use of (2) and (19), the expression (21) can be reformulated so that radiation efficiency can be calculated as follows.

$$\eta_{Rad} = \frac{1}{\left(1 - |S_{22}|^2\right)\left(1 - |S_{11}|^2\right)} \frac{4\pi D^2}{\lambda^2} \frac{\eta_{REF}}{G_{REF}} \int_0^{2\pi} \int_0^{\pi} |S_{21}(\theta, \phi)|^2 \sin\theta d\theta d\phi \quad (22)$$

6. Experimental Results and Discussions

In this section, the antenna fabrication is described, and the experimental assessment of the antenna performance are presented and compared to the simulation results.

6.1. Fabrication and Measurements of Reflection Coefficient

For experimental investigations, both the G-shaped and inverted G-shaped antennas are fabricated. The antenna prototypes are presented in Figure 19. Each of them is subjected to experimental measurement of the input impedance and radiation characteristics.

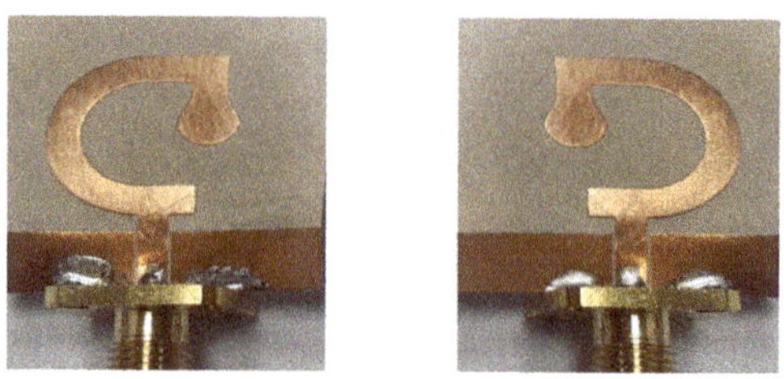

(a) (b)

Figure 19. A photograph of the fabricated antennas. (**a**) G-shaped strip antenna. (**b**) Inverted G-shaped strip antenna.

To measure the reflection coefficient magnitude, $|S_{11}|$, the antenna is connected to Rohde and Schwarz vector network analyzer (VNA) model ZVA 67, as shown in Figure 20. The frequency dependence of $|S_{11}|$ is presented in Figure 21a where the experimental and simulation results come in good agreement with each other, indicating that the impedance bandwidth is about 14 GHz (5–19 GHz). The change in the AR over the frequency range of (4.8–6.8 GHz) is shown in Figure 21b. The measurement results are consistent with the simulation results, indicating that the 3dB-AR bandwidth is around 1 GHz (5–6 GHz), i.e., 18% percentage fractional bandwidth.

The antenna is designed to produce circular polarization over the frequency range 5–6 GHz (for the 5GHz band of the WLAN/WiMAX applications) with left-hand sense in the directions of one half-space and right-hand polarization in the opposite directions of the other half-space. On the other hand, it is designed to produce linear polarization over the frequency range 6–19 GHz. This makes the proposed antenna more able to operate in the applications requiring wideband with varying polarization characteristics.

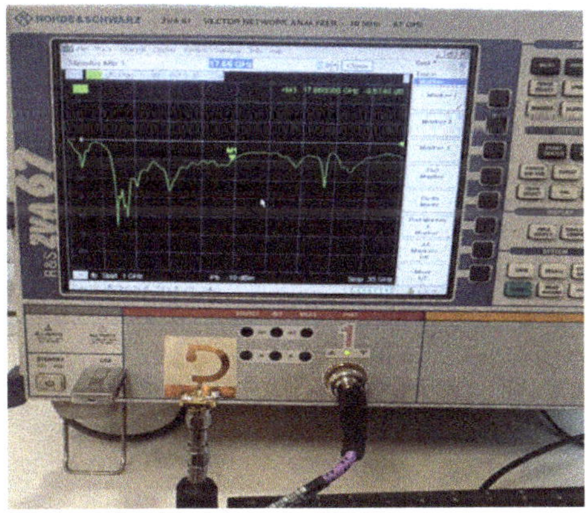

Figure 20. The fabricated prototype of the inverted G-strip antenna is connected to the VNA to measure the reflection coefficient magnitude, $| S_{11} |$, at the antenna feeding port.

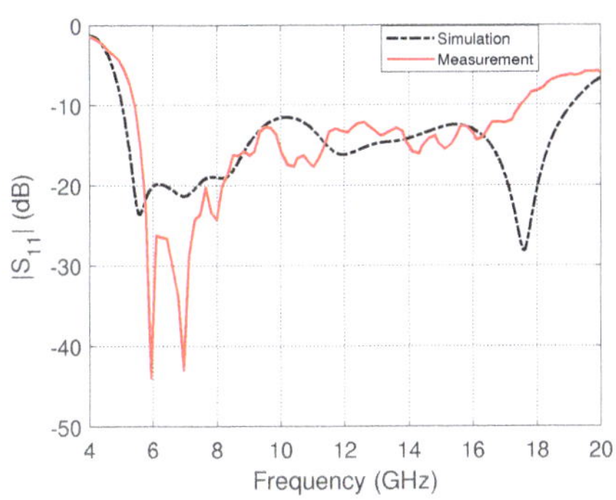

(a)

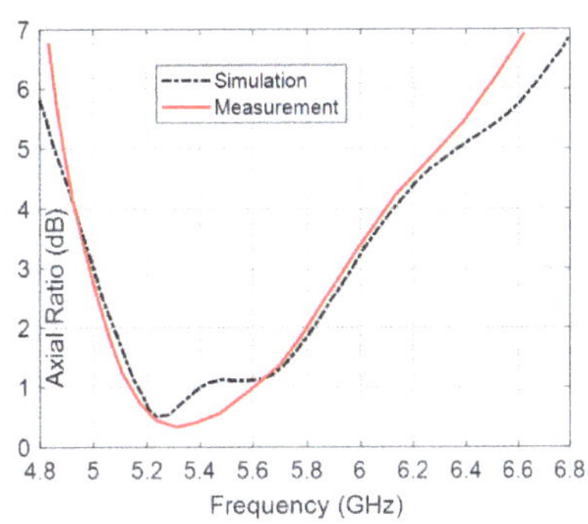

(b)

Figure 21. Frequency response as obtained by simulation compared to that obtained by measurement for the inverted G-shaped strip antenna (**a**) The $| S_{11} |$. (**b**) The AR.

The frequency bands allocated for the WiMAX and WLAN applications are listed in the in Table 3. The proposed antenna is well-suited to operate in the 5 GHz band allocated for both applications [46].

Table 3. The WiMAX and WLAN frequency bands application.

Name of the Allocated Band	WiMAX	WLAN
700 MHz band	470–862 MHz	–
1.4 GHz band	1.390–1.435 GHz	–
2.3 GHz band	2.300–2.400 GHz	–
2.4 GHz band	2.498–2.800 GHz	–
2.5 GHz band	–	2.400–2.484
3.5 GHz band	3.300–3.800 GHz	–
3.6 GHz band	–	3.675–3.890
5 GHz band	4.900–5.980 GHz	4.915–5.925

6.2. Gain and Radiation Efficiency

The dependence of the maximum gain generated by the inverted G-shaped antenna on the frequency is shown in Figure 22a. Both simulation and experimental results indicate that the maximum gain varies between 3.5 dBi and 5.37 dBi over the frequency band of impedance matching (5–19 GHz) and between 3.5 dBi and 3.8 dBi over the 3dB-AR frequency band (5–6 GHz). This means that the maximum gain produced by the proposed G-shaped and inverted G-shaped antennas is relatively stable over the frequency band of concern.

The results of the simulation and experimental testing of the radiation efficiency of the inverted G-shaped antenna, which were conducted over the frequency range of (4–20 GHz), are presented in Figure 22b and demonstrate good agreement. Owing to the coplanar structure and the high quality and small thickness of the substrate material, the radiation efficiency is maintained above 98% over almost the entire frequency band of operation (5–19 GHz).

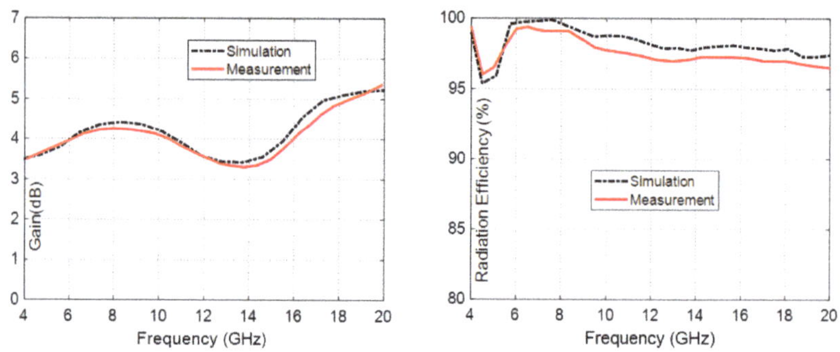

(**a**) (**b**)

Figure 22. The inverted G-shaped antenna exhibits a dependency on frequency, as obtained by simulation and measurement over the frequency range (4–20 GHz) for (**a**) The Gain. (**b**) The radiation efficiency.

6.3. Radiation Patterns

6.3.1. Circularly Polarized Radiated Fields

The measured far field radiation patterns of the circularly polarized components E-left and E-right generated by the inverted G-shaped antenna have been added and compared to those obtained by simulation in the plane $\phi = 0°$ at 5, 5.5, and 6 GHz, as shown in Figures 23–25, respectively.

It is shown that the radiated field of the inverted G-shaped antenna is dominated by RHCP component in upper half-space (+ve z-direction) and LHCP component in the lower half of the space (-ve z-direction).

6.3.2. Linearly Polarized Radiated Fields

The measured far field radiation patterns of the total electric field (Linearly Polarized) produced by the inverted G-shaped in the plane $\phi = 0°$ at 9, 12, 15, and 18 GHz, respectively, as shown in Figure 26. The measured and the simulated radiation patterns of the total radiated field at multiple frequencies over the frequency range 6–19 GHz show good agreement.

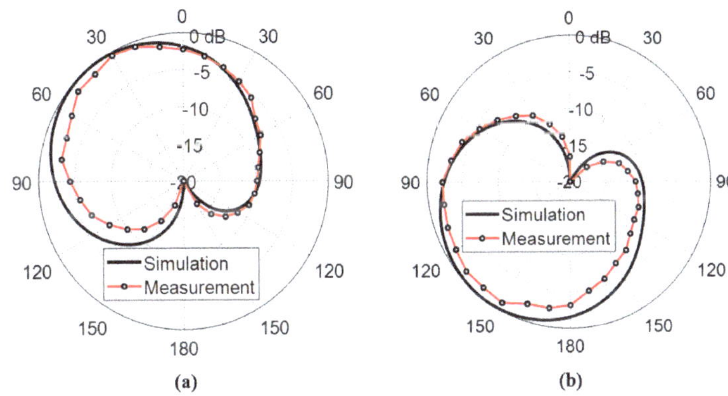

Figure 23. Radiation patterns of (**a**) RHCP and (**b**) LHCP fields produced by the inverted G-shaped antenna in the plane $\phi = 0°$ at the 5 GHz.

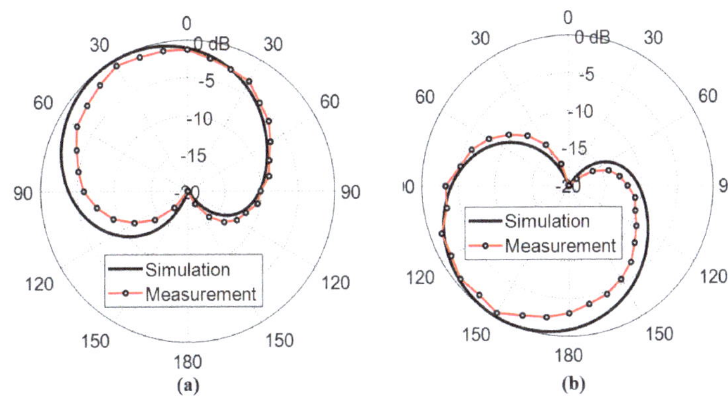

Figure 24. Radiation patterns of (**a**) RHCP and (**b**) LHCP fields produced by the inverted G-shaped antenna in the plane $\phi = 0°$ at the 5.5 GHz.

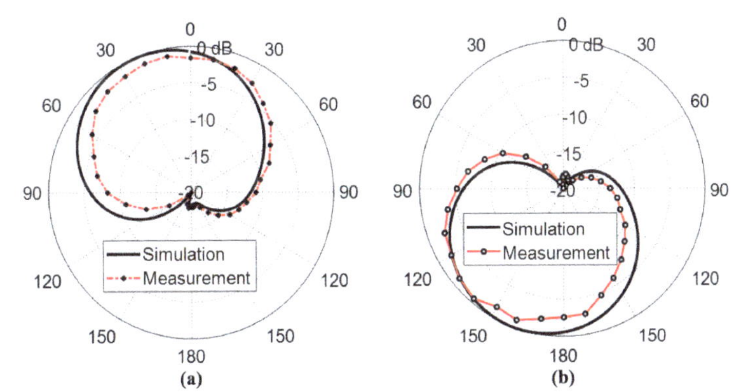

Figure 25. Radiation patterns of (**a**) RHCP and (**b**) LHCP fields produced by the inverted G-shaped antenna in the plane $\phi = 0°$ at the 6 GHz.

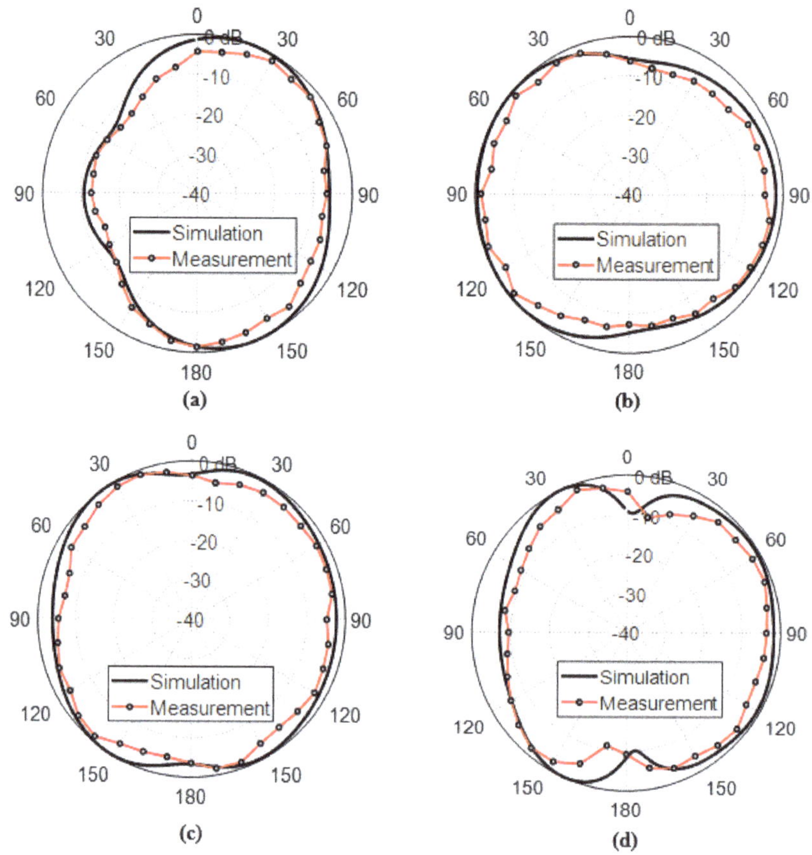

Figure 26. Radiation patterns of the total electric field produced by the inverted G-shaped antenna (Linearly Polarized) in the plane $\phi = 0°$ (**a**) 9 GHz, (**b**) 12 GHz, (**c**) 15 GHz, and (**d**) 18 GHz.

6.4. Summary of Comparative Performance

To demonstrate the contribution of the present work in the context of the published work with similar interest, some performance measures of the proposed G-shaped antenna are provided in Table 4 in comparison to those of other CP antenna designs proposed in some recent publications.

Table 4 includes designs for rigid and flexible antennas. The flexibility of the antenna is required especially for the medical applications where a central Wireless Body Area Network (WBAN) antenna should be conformal to the surface on which it is mounted. Such a type of antenna has dual vital functions of communication with the on-body biosensor antennas and the off-body LAN antennas. In the off-body mode this antenna communicates with the WiMAX/WLAN antenna to send the data collected form the body sensors to the LAN so that the doctors can take care of their patients remotely.

By comparison with the other antennas listed in Table 4, it becomes clear that the antenna proposed in the present work is the unique antenna that is flexible, operating with circular polarization over one of the frequency bands of the WiMAX/WLAN applications (4.9–6.0 GHz; see Table 3), and operating with linear polarization over the very wide band (6–19 GHz). The circular polarization is necessary while communicating with the LAN antenna as the patient movements result in misalignment between the attached central antenna and the WiMAX/WLAN antenna, which would degrade the signal level if the antenna were linearly polarized.

Table 4. A comparison of the performance measures of the G-shaped antenna with those of other antenna designs that have been recently published.

Work	Dimensions (mm × mm)	Frequency Band (GHz)	% BW for Impedance Matching	Radiation Efficiency	Gain (dBi)	% BW for 3dB-AR	Substrate
[1]	35 × 35	5.6–6.1	6.6%	NA	7.2	3.8%	Flexible
[4]	65 × 65	1.4–1.8	25.4%	NA	−1.25	30.1%	Flexible
[21]	20 × 20	5.2–6.8	13.7%	NA	2.94	6.8%	Flexible
[30]	13 × 13.4	4.8–16	118%	NA	3.33	55.8%	Rigid FR-4
[47]	29 × 46.5	2.1–3.6	60%	97%	4.5	13%	Flexible
[48]	32 × 38	5.5–7	24%	NA	NA	16.6%	Flexible
[49]	3 × 5	3.4–7	69%	85%	6.5	21.8%	Rigid FR-4
[50]	50 × 50	1.8–2.4	37.4%	NA	3.95	15.5%	Rigid FR-4
[51]	31 × 38	3.8–4.5	17.53%	91%	3.1	10.47%	Flexible
[52]	39 × 39	2.22–9.92	126.5%	80%	3.98	73.3%	Rigid FR-4
Present	25 × 27	4.9–19	117%	98%	5.37	18%	Flexible

On the other hand, the proposed antenna can use the remaining part of the operational band (6–19 GHz) for high data rate and wideband communication with the on-body (and in-body) biosensor antennas attached to (or implantable in) the patient body through linear polarization. During this mode of operation, the alignment between the proposed antenna and the biosensor antennas can be easily maintained. Regarding the above requirements (flexibility, circular polarization in the band 4.9–6.0 GHz, and linear polarization over the wideband 6–19 GHz) can be uniquely satisfied by the antenna proposed in the present work when compared to the other antennas listed in Table 4.

7. Conclusions

A wideband low-profile G-shaped strip antenna on a flexible substrate has been proposed for biomedical for the application of WBAN to work as off-body communication. The antenna produces circular polarization over the frequency band 5–6 GHz to communicate with WiMAX/WLAN antennas and to produce linear polarization over the frequency band 6–19 GHz for communication with the biosensor antennas. It has been shown that an inverted G-shaped strip produces CP of the opposite sense to that produced by G-shaped strip over the frequency range 5–6 GHz. The antenna structure is composed of a semicircular strip terminated with a horizontal extension at its lower end and terminated with a small circular patch through a corner-shaped strip extension at its upper end to form the shape of "G" or inverted "G". The corner-shaped extension and the circular patch termination has been added to the antenna geometry for impedance matching over the entire frequency band (5–19 GHz) and to improve the circular polarization over the frequency band (5–6 GHz). The antenna has been fed through a CPW to allow the fabrication on only one face of the flexible dielectric substrate. The dimensions of the antenna as well as the CPW have been optimized to obtain the best performance regarding the impedance matching bandwidth, 3dB AR bandwidth, radiation efficiency, and maximum gain. The results have shown that the achieved 3dB-AR bandwidth is 18% (5–6 GHz) to cover the 5 GHz frequency band of the WiMAX/WLAN applications within its 3dB-AR frequency band. Furthermore, the impedance matching bandwidth has been shown to reach 117% (5–19 GHz) which enables low-power communication with the biosensors over this wide range of the frequency. It has been shown that the maximum gain and radiation efficiency are 5.37 dBi and 98%, respectively. The antenna dimensions are 25 × 27 × 0.13 mm^3 and the BDR is 1733.

Author Contributions: Conceptualization, M.E.Y. and S.A.M.; methodology, M.E.Y.; software, M.E.Y.; validation, S.A.M. and K.F.A.H.; formal analysis, K.F.A.H.; investigation, S.A.M.; resources, M.A.I.; data curation, Q.H.A.; writing—original draft preparation, S.A.M.; writing—review and editing, K.F.A.H. and S.A.M.; visualization, Q.H.A.; supervision, Q.H.A.; project administration, M.A.I.; funding acquisition, M.A.I. All authors have read and agreed to the published version of the manuscript.

Funding: This work was supported in parts by The Royal Academy of Engineering Industrial Fellowship: IF2223-201 and The EPSRC Research Council [grant no EP/R511705/1].

Data Availability Statement: Not applicable.

Acknowledgments: The authors are grateful to the support in parts by The Royal Academy of Engineering Industrial Fellowship: IF2223-201 and The EPSRC Research Council [grant no EP/R511705/1].

Conflicts of Interest: The authors declare no conflict of interest.

Abbreviations

The following abbreviations are used in this manuscript:

CP	Circularly Polarized
CPW	Co-planar Waveguide
WBAN	Wireless Body Area Network
UWB	Ultra-wideband
RHCP	Right-hand CP
LHCP	Left-hand CP
AR	Axial Ratio
BDR	Bandwidth-Dimension Ratio
VNA	Vector Network Analyzer

References

1. Yang, H.C.; Liu, X.Y.; Fan, Y.; Tentzeris, M.M. Flexible circularly polarized antenna with axial ratio bandwidth enhancement for off-body communications. *IET Microwaves Antennas Propag.* **2021**, *15*, 754–767. [CrossRef]
2. Ullah, U.; Mabrouk, I.B.; Koziel, S. A compact circularly polarized antenna with directional pattern for wearable off-body communications. *IEEE Antennas Wirel. Propag. Lett.* **2019**, *18*, 2523–2527. [CrossRef]
3. Anderson, R.M.; Turner, H.C.; Farrell, S.H.; Truscott, J.E. Flexible antennas: A review. *Micromachines* **2020**, *11*, 847.
4. Ahmad, S.; Khabba, A.; Ghaffar, A.; Li, X.J. A Compact Wideband Flexible Circularly Polarized Implantable Antenna for Biotelemetry Applications. In Proceedings of IEEE International Symposium on Antennas and Propagation and USNC-URSI Radio Science Meeting (APS/URSI), Marina Bay Sands, Singapore, 4–10 December 2021; pp. 695–696.
5. Särestöniemi, M.; Sonkki, M.; Myllymäki, S.; Pomalaza-Raez, C. Wearable flexible antenna for UWB on-body and implant communications. *Telecom* **2021**, *2*, 19. [CrossRef]
6. Farahat, A.E.; Hussein, K.F.A. Wearable Button-Like Dual-Band Central Antenna for Wireless Bodyarea Networks. *Prog. Electromagn. Res. B* **2021**, *90*, 21–41. [CrossRef]
7. Farahat, A.E.; Hussein, K.F.A. Dual-band dual-mode monopole/spiral central antenna for WBAN applications. *Microw. Opt. Technol. Lett.* **2022**, *64*, 1401–1406. [CrossRef]
8. Poffelie, L.A.Y.; Soh, P.J.; Yan, S.; Vandenbosch, G.A.E. A high-fidelity all-textile UWB antenna with low back radiation for off-body WBAN applications. *IEEE Trans. Antennas Propag.* **2015**, *64*, 757–760. [CrossRef]
9. Federal Communications Commission. *FCC Report and Order for Part 15 Acceptance of Ultra Wideband (UWB) Systems from 3.1 to 10.6 GHz*; FCC: Washington, DC, USA, 2002; pp. 1–10.
10. Balanis, C.A. *Antenna Theory and Design*, 3rd ed.; Wiley: New York, NY, USA, 2005.
11. Shichang, S.S.; Luo, Q.; Zhu, F. *Circularly Polarized Antennas*, Wiley-IEEE Press: Hoboken, NJ, USA, 2014.
12. Nadeem, I.; Alibakhshikenari, M.; Babaeian. F.; Althuwayb, A.; Virdee, B.; Azpilicueta, L.; Khan, S.; Huynen, I.; Falcone, F.; Denidni, T.; et al. A comprehensive survey on 'circular polarized antennas' for existing and emerging wireless communication technologies. *J. Phys. Appl. Phys.* **2022**, *55*, 033002. [CrossRef]
13. Trinh, L.H.; Truong, N.V.; Ferrero, F. Low cost circularly polarized antenna for IoT space applications. *Electronics* **2020**, *9*, 1564. [CrossRef]
14. Mondal, T.; Maity, S.; Ghatak, R.; Bhadra, C.; Sekhar, R. Design and analysis of a wideband circularly polarised perturbed psi-shaped antenna. *IET Microwaves Antennas Propag.* **2018**, *12*, 1582–1586. [CrossRef]
15. Yassin, M.E.; Mohamed, H.A.; Abdallah, E.A.F.; El-Hennawy, H.S. Circularly polarized wideband-to-narrowband switchable antenna. *IEEE Access* **2019**, *7*, 36010–36018. [CrossRef]

16. Deegwal, J.K; Sharma, V. Dual Band Circular Polarized Printed Dipole Antenna for S and C Band Wireless Applications. *Prog. Electromagn. Res. C* **2020**, *105*, 129–146.
17. Zhai, H.; Yang, D.; Xi, L.; Feng, D. A new CPW-fed broadband circularly polarized printed monopole antenna for UWB application. *Microw. Opt. Technol. Lett.* **2018**, *60*, 364–369. [CrossRef]
18. Saraswat, K.; Harish, A.R. Dual-band CP coplanar waveguide-fed split-ring resonator-loaded G-shaped slot antenna with wide-frequency ratio. *IET Microwaves Antennas Propag.* **2018**, *12*, 1920–1925. [CrossRef]
19. Li, G.; Zhai, H.; Ma, X.; Li, T.; Liang, C. Design of a CPW-fed slot antenna with small size and ultra-broadband circularly polarized radiation. *J. Electromagn. Waves Appl.* **2014**, *28*, 1212–1220. [CrossRef]
20. Liu, X.; Wang, H.; Yang, X.; Wang, J. Quad-band circular polarized antenna for GNSS, 5G and WIFI-6E applications. *Electronics* **2022**, *11*, 1133. [CrossRef]
21. Babu, B.A.; Madhav, B.T.P.; Vineel, B.; Chandini, G.; Amrutha, C.; Rao, M.C. Design and Analysis of a Circularly polarized flexible, compact and transparent antenna for Vehicular Communication Applications. *J. Physics Conf. Ser.* **2021**, *1804*, 012192. [CrossRef]
22. Al-Gburi, A.J.A.; Zakaria, Z.; Alsariera, H.; Akbar, M.F.; Ibrahim, I.M.; Ahmad, K.S.; Ahmad, S.; Al-Bawri, S.S. Broadband Circular Polarised Printed Antennas for Indoor Wireless Communication Systems: A Comprehensive Review. *Micromachines* **2022**, *13*, 1048. [CrossRef]
23. Banerjee, U.; Karmakar, A..; Saha, A. A review on circularly polarized antennas, trends and advances. *Int. J. Microw. Wirel. Technol.* **2020**, *12*, 922–943. [CrossRef]
24. Reddy, V.; Sarma, N. Single feed circularly polarized poly fractal antenna for wireless applications. *Int. J. Comput. Inf. Eng.* **2014**, *11*, 1710–1713.
25. Sharma, V.; Jhajharia, T. Square slot antenna for wide circularly polarized bandwidth and axial ratio beamwidth. *Electr. Control. Commun. Eng.* **2021**, *17*, 1–11. [CrossRef]
26. Kumar, M.; Nath, V. A circularly polarized printed elliptical wide-slot antenna with high bandwidth-dimension-ratio for wireless applications. *Wirel. Netw.* **2020**, *17*, 5485–5499. [CrossRef]
27. AboEl-Hassan, M.; Hussein, K.F.; Awadalla, K.H. A novel microstrip antenna with L-shaped slots for circularly polarized satellite applications. *Microw. Opt. Technol. Lett.* **2020**, *62*, 839–844. [CrossRef]
28. Chen, Z.N.; Qing, X. Slotted microstrip antennas for circular polarization with compact size. *IEEE Antennas Propag. Mag.* **2013**, *55*, 124–137.
29. Ahmad, S.; Manzoor, B.; Naseer, S.; Ghaffar, A.; Hussein, M. A Flexible Broadband CPW-Fed Circularly Polarized Biomedical Implantable Antenna with Enhanced Axial Ratio Bandwidth. *Res. Sq.* **2021**, *preprint*. [CrossRef]
30. Chaudhary, P.; Kumar, A.; Mittra, R. Quadrilateral-shaped wideband circularly polarized CPW-fed monopole antenna. In Proceedings of the URSI Asia-Pacific Radio Science Conference (AP-RASC), New Delhi, India, 9–15 March 2019; pp. 1–4.
31. Ullah, U.; Koziel, S. A novel coplanar-strip-based excitation technique for design of broadband circularly polarization antennas with wide 3 dB axial ratio beamwidth. *IEEE Trans. Antennas Propag.* **2019**, *67*, 4224–4229. [CrossRef]
32. Kumar, M.; Nath, V. Circularly polarized microstrip-line-fed antenna with rotated elliptical slot serving satellite communications. *IEEE Trans. Antennas Propag.* **2020**, *110*, 1443–1458. [CrossRef]
33. Liang, C.-F.; Lyu, Y.-P.; Chen, D.; Cheng, C.-H. Wideband Circularly Polarized Stacked Patch Antenna Based on TM_{11} and TM_{10}. *IEEE Trans. Antennas Propag.* **2021**, *70*, 2459–2467. [CrossRef]
34. Kumar, R.; Chaudhary, R.K. Investigation of higher order modes excitation through F-shaped slot in rectangular dielectric resonator antenna for wideband circular polarization with broadside radiation characteristics. *Int. J. Microw.-Comput.-Aided Eng.* **2018**, *28*, e21281. [CrossRef]
35. Sayem, A.S.M.; Simorangkir, R.B.V.B.; Esselle, K.P.; Lalbakhsh, A.; Gawade, D.R.; O'Flynn, B.; Buckley, J.L. Flexible and transparent circularly polarized patch antenna for reliable unobtrusive wearable wireless communications. *Sensors* **2022**, *22*, 1276. [CrossRef] [PubMed]
36. Xia, Y.; Yuan, M.; Dobrea, A.; Li, C.; Heidari, H.; Mottram, N.; Ghannam, R. Reconfigurable Wearable Antenna for 5G Applications using Nematic Liquid Crystals. *arXiv* **2022**, arXiv:2212.08622.
37. Mostafa, A.; Ibrahim, F.; Ahmed, H.; Yassin, M. Wideband Energy Harvesting System Based on Circularly Polarized Monopole Antenna. *Int. Undergrad. Res. Conf.* **2021**, *5*, 403–408.
38. Li, J.; Jiang, Y.; Zhao, X. Circularly polarized wearable antenna based on NinjaFlex-embedded conductive fabric. *Int. J. Antennas Propag.* **2019**, *2019*, 3059480. [CrossRef]
39. Fernández-Prades, C.; Rogier, H.; Collado, A.; Tentzeris, M.M. Flexible substrate antennas. *Int. J. Antennas Propag.* **2012**, *2012* , 1–2. [CrossRef]
40. Virothu, S.; Anuradha, M.S. Flexible CP diversity antenna for 5G cellular Vehicle-to-Everything applications. *AEU-Int. J. Electron. Commun.* **2022**, *152*, 154248. [CrossRef]
41. Birwal, A.; Singh, S.; Kanaujia, B.K.; Kumar, S. Broadband CPW-fed circularly polarized antenna for IoT-based navigation system. *Int. J. Microw. Wirel. Technol.* **2019**, *11*, 835–843. [CrossRef]
42. Kirtania, S.G.; Younes, B.A.; Hossain, A.R.; Karacolak, T.; Sekhar, P.K. CPW-fed flexible ultra-wideband antenna for IoT applications. *Micromachines* **2021**, *12*, 453. [CrossRef]
43. Zhang, R.; Liu, J.; Wang, Y.; Luo, Z.; Zhang, B.; Duan, J. Flexible wearable composite antennas for global wireless communication systems. *Sensors* **2021**, *21*, 6083. [CrossRef]

44. Fahmy, W.M.; Farahat, A.E.; Hussein, K.F.A.; Ammar, A.A. High Q-factor bandstop filter based on CPW resonator broadside-coupled to CPW through-line. *Prog. Electromagn. Res. B* **2020**, *86*, 121–138. [CrossRef]
45. Fahmy, W.M.; Farahat, A.E.; Hussein, K.F.A.; Ammar, A.A. Dual-Band Bandpass Filter Optimized for High Q-Factor. *Appl. Comput. Electromagn. Soc. J. (ACES)* **2021**, *36*, 398–410. [CrossRef]
46. Available online: https://wireless-instruments.com/lte-wimax-bands/ (accessed on 11 May 2023).
47. Ahmad, S.; Cherif, N.; Naseer, S.; Ijaz, U.; Faouri, Y.S.; Ghaffar, A.; Hussein, M. A wideband circularly polarized CPW-fed substrate integrated waveguide based antenna array for ISM band applications. *Heliyon* **2022**, *8*, e10058. [CrossRef]
48. Venkateswara Rao, M.; Madhav, B.T.P.; Anilkumar, T.; Prudhvinadh, B. Circularly polarized flexible antenna on liquid crystal polymer substrate material with metamaterial loading. *Microw. Opt. Technol. Lett.* **2020**, *62*, 866–874. [CrossRef]
49. Pakkathillam, J.K.; Kanagasabai, M. Circularly polarized broadband antenna deploying fractal slot geometry. *IEEE Antennas Wirel. Propag. Lett.* **2015**, *14*, 1286–1289. [CrossRef]
50. Kumar, Y.; Gangwar, R.K.; Kanaujia, B.K. Characterization of CP radiations in a planar monopole antenna using tuning fork fractal slot for LTE band13/Wi-Max and Wi-Fi applications. *IEEE Access* **2020**, *8*, 127123–127133. [CrossRef]
51. Hossain, A.R.; Karacolak, T. CPW-Fed Compact Circularly Polarized Flexible Antenna for C Band Applications. In Proceedings of the 2023 United States National Committee of URSI National Radio Science Meeting (USNC-URSI NRSM), Online, 10–14 January 2023; pp. 246–247.
52. Samsuzzaman, M.; Islam, M.T. Circularly polarized broadband printed antenna for wireless applications. *Sensors* **2018**, *18*, 4261. [CrossRef] [PubMed]

sensors

Article

Linear Antenna Array Sectorized Beam Scanning Approaches Using Element Position Perturbation in the Azimuth Plane

Safaa I. Abd Elrahman [1], Ahmed M. Elkhawaga [2,*], Amr H. Hussein [2,3,*] and Abd Elhameed A. Shaalan [1,*]

1 Electronics and Communication Engineering Department, Faculty of Engineering, Zagazig University, Zagazig 44519, Egypt; safaa9ibrahim@gmail.com
2 Electronics and Electrical Communications Engineering Department, Faculty of Engineering, Tanta University, Tanta 31527, Egypt
3 Electronics and Electrical Communications Engineering Department, Faculty of Engineering, Horus University Egypt, New Damietta 34518, Egypt
* Correspondence: ahmed.elkhawaga@f-eng.tanta.edu.eg (A.M.E.); amrvips@yahoo.com (A.H.H.); shaalan_zag2010@yahoo.com (A.E.A.S.)

Abstract: In this paper, two sector beam scanning approaches (BSAs) based on element position perturbations (EPPs) in the azimuth plane are introduced. In EPP-BSA, the elements' excitations are kept constant and the elements' positions in the direction normal to the array line are changed according to a predetermined EPP pattern. The magnitude and repetition rate of the selected EPP pattern determines the steering angle of the main beam. However, EPP-BSA results in a wide scanning range with a significant increase in the side lobe level (SLL). To mitigate this drawback, a reduction in the SLL of the array pattern is firstly performed using the single convolution/genetic algorithm (SC/GA) technique and then perturbing the elements' positions in the azimuth plane. This combination between SLL reduction and EPP-BSA (SLL/EPP-BSA) results in a smaller scanning range with a relatively constant half power beamwidth (HPBW) and a much lower SLL. In addition, keeping the synthesized excitation coefficients constant without adding progressive phase shifters facilitates the manufacturing process and reduces the cost of the feeding network. Furthermore, a planar antenna array thinning approach is proposed to realize the EPP-BSA. The results are realized using the computer simulation technology (CST) microwave studio software package, which provides users with an optimized modeling environment and results in realizable and realistic designs.

Keywords: array thinning; beam scanning approach (BSA); element position perturbations (EPP); linear antenna array (LAA); side lobe level (SLL); half power beamwidth (HPBW); genetic algorithm (GA); planar antenna array (PAA)

Citation: Abd Elrahman, S.I.; Elkhawaga, A.M.; Hussein, A.H.; Shaalan, A.E.A. Linear Antenna Array Sectorized Beam Scanning Approaches Using Element Position Perturbation in the Azimuth Plane. *Sensors* **2023**, *23*, 6557. https://doi.org/10.3390/s23146557

Academic Editor: Manuel Sierra Castañer

Received: 1 June 2023
Revised: 8 July 2023
Accepted: 19 July 2023
Published: 20 July 2023

1. Introduction

In both military and civilian applications, such as wireless communication systems, wireless power transmission systems, and radar systems, phased antenna arrays have various benefits including flexible beam scanning, high tracking accuracy, and widespread use [1]. Flexible beam-scanning antenna design is currently a hot topic of research because of the emergence of contemporary applications that need antenna arrays with variable scanning capabilities [2]. Automotive radars, weather observations, aviation surveillance [3], and satellite communication [4] are just a few of these numerous applications.

In [5], a (2×3) phased antenna array is developed using a rectangular microstrip patch antenna on a FR-4 dielectric substrate. The elements are excited with constant amplitude and different phases to steer the beam within a scanning range of 49° (from −25° to +24°). In [6], four leaf-shaped bow-tie slot antennas in a linear array are used for beam scanning within a scanning range from 9° to 13°. In [7], the two-element antenna array is fed through two transmission lines that are inclined at an angle from a common base line. The main beam can be directed from one antenna's normal to the other antenna's normal by adjusting the excitations of the two antennas. The scan angle depended on changing the excitation coefficient in every scan angle or changing the tilt angle. In [8], a broadband reconfigurable antenna that can guide the main beam in three distinct directions has been introduced. Besides the main radiating patch, three parasitic tiny patches are included to control the main beam steering. Three beam-steering orientations of the radiation pattern are offered by PIN diode switching. In [9], the same concept of switched parasitic patches at 60 GHz for WLAN applications has been introduced for main beam steering. In [10], a sparse substrate integrated waveguide (SIW) slot antenna array has been introduced. The genetic algorithm was utilized to optimize the position of each element over the aperture of 4.5 λ_o, where λ_o is the free space wavelength. The sparse array consists of array elements, phase shifters, and power dividers on a substrate integrated with an SIW. This sparse antenna can scan the beam in the range of $\pm 15°$. In [11], a millimeter-wave (MMW) multi-folded frequency-dependent progressive phase shifter based on a substrate integrated waveguide (SIW) has been introduced. By combining the antenna element and the phase shifter together, the array can achieve a wide scanning range from $-18°$ to $32°$.

On the other hand, array thinning has been utilized for large array synthesis to achieve desired radiation patterns with a minimum array size and cost [12–16]. In [12], a numerical stochastic optimization technique that is known as multiagent genetic algorithm (MAGA) has been used for planar array thinning and synthesis for both low SLL and high directivity. In [13], discrete particle swarm optimization (DPSO) has been used for large array thinning. The peak side lobe level (PSLL) performance of the algorithm is verified through several representative examples, demonstrating its effectiveness and robustness.

In [14], the binary genetic algorithm (BGA) has been used for thinned planar array design, where only the outer sub-planar array elements with small amplitude excitations are optimized for thinning. The BGA has been used to remove a number of useless antenna elements while keeping the inner sub-planar elements with high amplitude excitations unchanged. In [15], a GA in combination with array thinning and tapering techniques has been introduced to minimize the SLL. Thinning is applied to idle some elements of the array, and tapering is used to achieve a superior SLL. In [16], a hybrid approach combining difference sets (DS) and discrete Fourier transform (DFT) techniques has been introduced for the synthesis of thinned planar antenna arrays. DS is used to establish pattern synthesis rules, while DFT is employed to find the nonuniform excitations of a planar antenna array. The proposed method extracts excitations using cyclic shifts from a suitable DS sequence to achieve antenna array thinning with a reduced PSLL.

In this paper, two beamforming approaches for sectorized beam scanning of linear antenna arrays based on the element position perturbation (EPP) technique denoted as EPP-BSA and SLL/EPP-BSA are introduced. In both techniques, the elements' excitations are kept constant while the positions of the elements are perturbed in the azimuth plane according to a predefined perturbation pattern. In this work, the cosine wave pattern is utilized for this purpose. The proposed EPP-BSA provides wider scanning ranges than the proposed SLL/EPP-BSA but with much higher SLLs at the endpoints of the scanning range. In this regard, the proposed SLL/EPP-BSA may be useful in interference-sensitive applications that require small scanning sectors. The proposed techniques are based on modifying the EPP pattern rather than using phase shifters while maintaining the elements excitations. This simplifies the production process and lowers the cost of the feeding network. Furthermore, the proposed technique eliminates interference and coupling between the main beam and the back lobe at any scanning angle by maintaining a 180 spacing angle between them. Furthermore, a proposed PAA thinning approach is proposed to realize the proposed EPP-BSA approach. It allows for precise array thinning to select the desired active "ON" antenna elements and turn off the other elements to configure the designed EPP pattern to achieve a specific scan angle with the desired radiation properties. The results are realized using the computer simulation technology (CST) microwave studio software package, which is highly matched to the MATLAB simulations.

The main contributions of this work can be summarized as follows:

I. The proposed techniques provide continuous beam steering with high accuracy rather than switched beam steering or discrete beam steering techniques.
II. The required beam steering is simply achieved by only adjusting the amplitude and repetition rate of the utilized perturbation pattern that is fully controlled.
III. The elements' excitations (magnitude and phase) are kept constant throughout the entire steering range. That completely avoids the design of complex and expensive feeding networks. On the other hand, the existing beam steering techniques are mainly based on utilizing n-bit phase shifters that are complex and expensive. Also, the accuracy of main beam steering depends on the phase quantization errors of the utilized phase shifters arising from the limited number of bits per phase shifter.
IV. Although the EPP has been employed for pattern nulling or side lobe cancellation, it has been utilized in a novel manner for achieving continuous beam steering techniques that can be applicable for beam steering of millimeter wave patch antenna arrays as described in the paper.
V. With an HPBW decrease of around 3° at the extremities of the steering range, the first technique, EPP-BSA, offers a wide steering range with a width of 102.72°. While at the extremes of the steering range, it experiences a high SLL of −2.668 dB.
VI. The second technique, SLL/EPP-BSA, provides a steering range of width 61.46° with fixed HPBW and $SLL < -10$ dB over the entire steering range.
VII. A proposed PAA thinning approach is introduced to realize the proposed EPP-BSA technique. It allows for precise array thinning to select the desired active "ON" antenna elements and turn off "OFF" the other elements to configure the designed EPP pattern to steer the main beam in a specific direction with the desired radiation properties.

2. Proposed EPP-Based Beam Scanning Approaches

This section introduces two array beam scanning approaches that employ element position perturbation (EPP) in the azimuth plane. In these techniques, the array elements are distributed using well-known waveform patterns such as sine, cosine, and triangle waves. In the first approach, the original excitations of the elements are kept constant. While the amplitude and repetition rate of the chosen waveform pattern along the array line define the element position and the main beam's steering angle. So, it can be denoted as the EPP-based scanning approach (EPP-BSA). However, the EPP-BSA results in a wide scanning range with a significant increase in the side lobe level (SLL).

To mitigate the SLL problem of the EPP-BSA, a second array beam scanning approach is introduced. In this approach, SLL reduction of the array pattern is firstly performed using the single convolution/genetic algorithm (SC/GA) technique introduced in [17] and then performance-optimized EPP in the azimuth plane. This combination between SLL reduction and EPP (SLL/EPP) results in a limited scanning range but with a relatively constant HPBW and minimal variations in the synthesized SLL. The proposed SLL/EPP-based beam scanning approach can be denoted as the SLL/EPP-BSA.

2.1. Proposed EPP-BSA

To derive a closed form expression of the synthesized array factor applying EPP, we considered a linear antenna array (LAA) whose elements distributions are shown in Figure 1 before and after EPP.

Consider a uniform LAA consisting of N antenna elements with uniform element spacing d as shown in Figure 2, its array factor $AF(\theta)$ is given by [18]:

$$AF(\theta) = \sum_{n=1}^{N} a_n \, e^{jk(n-1)d\cos\theta} \tag{1}$$

where a_n is the excitation coefficient of the nth antenna elements, $k = 2\pi/\lambda$ is the wave number, and λ is the wavelength.

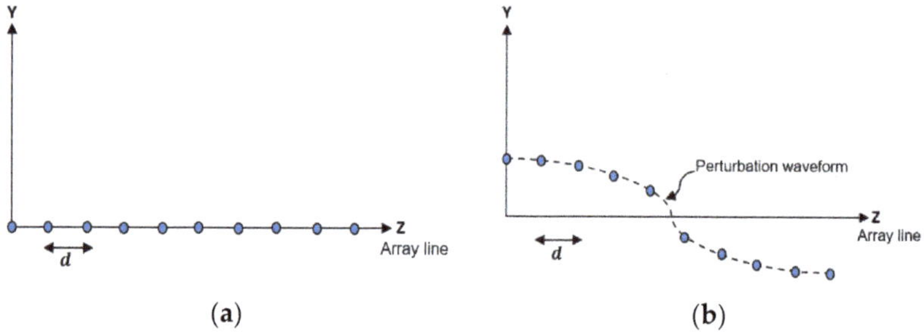

Figure 1. Linear antenna array (**a**) without EPP and (**b**) with EPP.

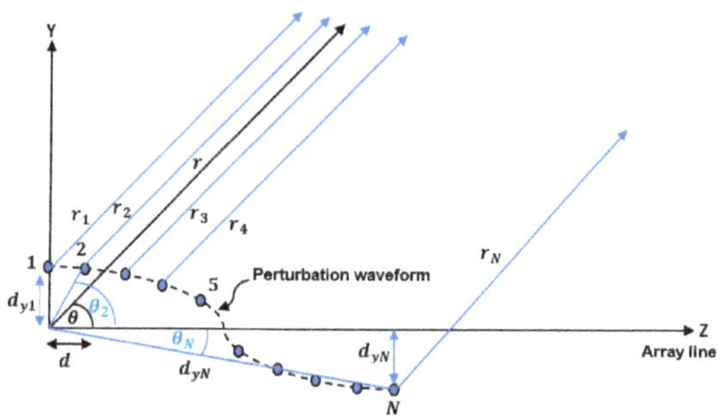

Figure 2. Geometry of a linear array of elements aligned along the Z-axis and has position perturbation along the Y-axis.

If the positions of the N antenna elements of the linear antenna array are perturbed along the Y-axis as shown in Figure 2, the array factor of Equation (1) should be modified. The total electric field E_t at the far field point is the superposition of the individual electric fields arising from each antenna element such that:

$$E_t = E_{1+}E_2 + \cdots\cdots\cdots + E_N \tag{2}$$

where

$$E_1 = a_1 \frac{E_0}{r_1} e^{-jkr_1} \tag{3}$$

$$E_2 = a_2 \frac{E_0}{r_2} e^{-jkr_2} \tag{4}$$

$$E_N = a_N \frac{E_0}{r_{N-1}} e^{-jkr_{N-1}} \tag{5}$$

Then, the total field can be expressed as:

$$E_t = \left(\frac{E_0}{r_1} a_1 e^{-jkr_1} + \frac{E_0}{r_2} a_2 e^{-jkr_2} + \cdots + a_N \frac{E_0}{r_{N-1}} e^{-jkr_{N-1}} \right) \tag{6}$$

where r_n, $n = 1, 2, 3, \ldots, N$ is the distances from the nth antenna element to the far field point. For far field approximation we use the following assumptions:

$$\frac{1}{r_1} = \frac{1}{r_2} = \frac{1}{r_3} = \cdots = \frac{1}{r_N} = \frac{1}{r} \tag{7}$$

$$r_1 \parallel r_2 \parallel r_3 \parallel \cdots \parallel r_N \tag{8}$$

where r is the distance from the origin or reference point to the far field point and $\|$ is the parallel operator. And, θ is the angle between the r vector and the Z-axis. d is the distance between successive elements on the Z-axis, d_{yn} is the distance of the nth element in the Y-direction, d_n is distance between the nth element and origin, and θ_n is the angle between d_n and the Z axis as shown in Figure 3.

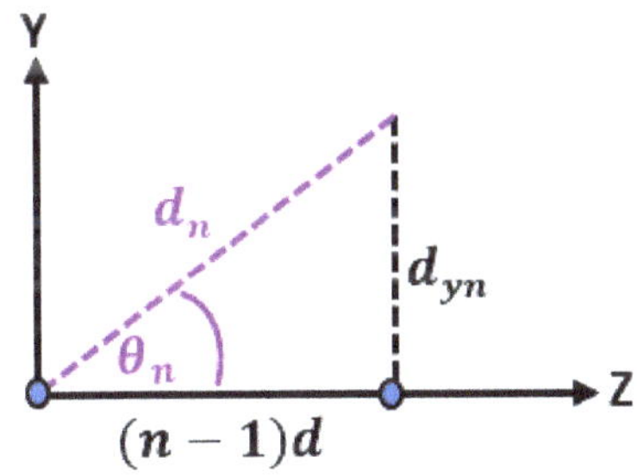

Figure 3. The triangular relation between the element position in the Y-direction and the distance from the reference antenna element.

Based on Figure 3, the angle θ_n can be calculated as follows:

$$sin\theta_n = \frac{d_{yn}}{d_n} \tag{9}$$

$$cos\theta_n = \frac{(n-1)\,d}{d_n} \tag{10}$$

To determine r_1 for the first antenna element consider the array geometry shown in Figure 4. In this case, r_1 can be expressed as:

$$r_1 = r - d_{y1}\sin\theta \tag{11}$$

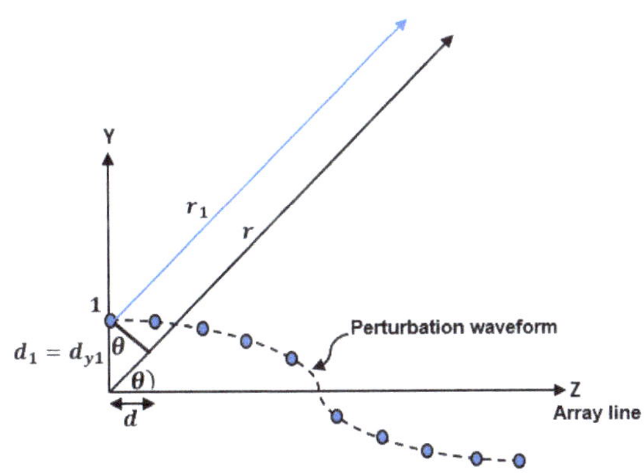

Figure 4. Geometry of a linear array of elements aligned along the Z-axis and has position perturbation along the Y-axis indicating the far field radiation from the first antenna element.

On the other hand, r_2 of the second antenna element can be determined as shown in Figure 5 such that: $r_2 = r - d_2\cos(\theta_2 - \theta)$

$$r_2 = r - d_2\cos\theta\cos\theta_2 - d_2\sin\theta\sin\theta_2 \tag{12}$$

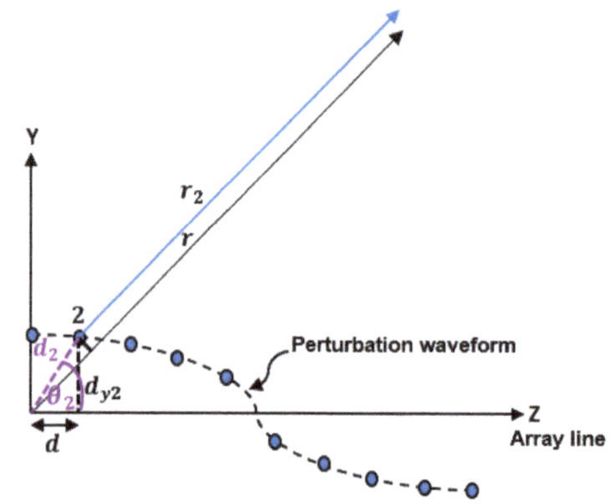

Figure 5. Geometry of a linear array of elements aligned along the *Z*-axis and has position perturbation along the *Y*-axis indicating the far field radiation from the second antenna element.

Substituting $cos\theta_2 = \frac{d}{d_2}$ and $sin\theta_2 = \frac{d_{y2}}{d_2}$ into Equation (12) then

$$r_2 = r - d_2 \cos\theta \times \frac{d}{d_2} - d_2 \sin\theta \times \frac{d_{y2}}{d_2} r_2 = r - d\cos\theta - d_{y_2} \sin\theta \tag{13}$$

By the same way, r_3 of the third antenna element can be determined as:

$$r_3 = r - 2d\cos\theta - d_{y_3} \sin\theta \tag{14}$$

In general, the r_n of the nth antenna element can be expressed as:

$$r_n = r - (n-1)d\cos\theta - d_{y_n} \sin\theta \tag{15}$$

Substituting the far field approximations in Equation (6), the total electric field E_t can be rewritten as:

$$E_t = \frac{E_0}{r} e^{-jkr} \left(a_1 e^{jkd_{y_1} \sin\theta} + a_2 e^{jkd\cos\theta + jkd_{y_2} \sin\theta} + \cdots + a_N e^{jk(N-1)d\cos\theta + jkd_{y_N} \sin\theta} \right) \tag{16}$$

Then,

$$E_t = \frac{E_0}{r} e^{-jkr} \times AF(\theta) \tag{17}$$

Accordingly, the modified array factor can be expressed as:

$$AF(\theta) = a_1 e^{jkd_{y_1} \sin\theta} + a_2 e^{jkd\cos\theta + jkd_{y_2} \sin\theta} + \cdots + a_N e^{jk(N-1)d\cos\theta + jkd_{y_N} \sin\theta} \tag{18}$$

In other words, it can be written as:

$$AF(\theta) = \sum_{n=1}^{N} a_n e^{(jk(n-1)d\cos\theta + jkd_{y_n} \sin\theta)} \tag{19}$$

It is clear that $AF(\theta)$ is a function of the element position perturbation, $d^{y_n} = \left[d_{y1} \; d_{y2} \ldots \; d_{y_N} \right]$ in the *Y*-direction.

2.1.1. Simulation Results of the Proposed EPP-BSA

In this section, several simulations are carried out to verify the proposed approach and give a recommendation for the best case. Consider a broadside uniform linear antenna array (ULAA) consisting of $N = 8$ elements with uniform element spacing $d = \lambda/2$ whose elements are arranged along the *Z*-axis. It is required to scan the array main beam to the right and left of the broadside direction. This is achieved by performing EPP in the Y-direction following a known waveform pattern. In this case, the cosine waveform is utilized for the EPP where the cosine wave amplitude and period are expressed in terms of the operating wavelength λ. The half period H of the cosine wave along the Z-axis is chosen related to the array length $L_h = (N-1) \times d = 3.5\ \lambda$.

In this case, the element position in the Y-direction along the array line (Z-direction) is given by:

$$d^{y_n} = \pm A_c \times cos\left(\frac{z \times \pi}{H}\right) \quad (20)$$

where A_c is the amplitude of the cosine wave, z is the element position on the Z-axis, and H is the half period of cosine that is chosen to be $H \geq L_h$. Both the amplitude of the cosine wave A_c and the half period H control the scanning angle of the array main beam.

The beam scanning is performed at different compression ratios R that are defined as:

$$R = \frac{L_h}{H}, \ H \geq L_h \quad (21)$$

It is noticed that as H increases more than L_h, the array size in the Y-direction is reduced and becomes compact, so R is denoted as the compression ratio. Consequently, there are different test cases for beam scanning for different values of compression ratio.

Test case (1): $R = 1$

In this case, the cosine wave half period is chosen to be equal to the array size, $H = L_h = 3.5\ \lambda$, hence $R = 1$. Accordingly, the EPPs of the array elements for A_c changing from positive to negative and A_c changing from negative to positive are shown in Figures 6a and 6b, respectively.

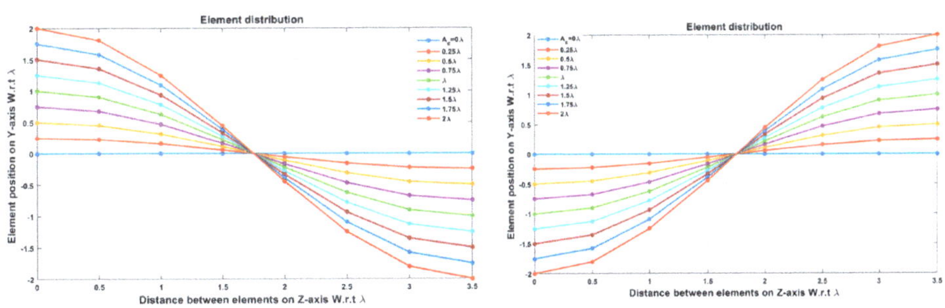

(a) (b)

Figure 6. The EPPs of the array elements at compression ratio $R = 1$ for (**a**) A_c is changing from positive to negative and (**b**) A_c is changing from negative to positive.

When the cosine wave amplitude is changing from positive to negative and its magnitude is changing from $A_c = 0$ to $A_c = 0.19\ \lambda$, the array main beam is scanned from 0° to 7.01° with HPBW and the SLL is less than −10 dB. However, as the cosine wave amplitude increases from than 0.19 λ to 1.5 λ, the scan angle is increased from 7.01° to 44.6°, the SLL is increased more than −10 dB, and the HPBW is decreased by 3°. However, as the cosine wave amplitude increases from than 1.5 λ to 2 λ, the scan angle is increased from 44.6° to 52.9°. But, the grating lobe appears, and the SLL is increased by more than −10 dB as listed in Table 1. The synthesized radiation patterns using the EPP technique are shown in Figure 7.

Table 1. The relation between the amplitude changes of the cosine waveform and the scanning angle, HPBW, SLL, and maximum scanning range θ_r for 8-element uniform LAA at compression ratio $R = 1$.

Figure	Cosine Wave Amplitude A_c	Main Beam Direction θ_o, in °	SLL (dB)	HPBW in °	Maximum Scanning Range
Figure 7a	0	90°	−12.8 dB	12.8°	
Figure 7b	0.19 λ	82.99°	−10.09 dB	12.67°	
Figure 7c	0.25 λ	80.92°	−9.346 dB	12.65°	
Figure 7d	0.5 λ	72.16°	−6.82 dB	12.33°	
Figure 7e	0.75 λ	63.91°	−4.963 dB	11.85°	
Figure 7f	λ	56.69°	−3.618 dB	11.1°	$-52.91° \leq \theta_r \leq 52.91°$
Figure 7g	1.25 λ	50.5°	−2.671 dB	10.45°	
Figure 7h	1.5 λ	45.34°	−2.002 dB	9.8°	
Figure 7i	1.75 λ	40.7°	−1.454 dB	9.3°	
Figure 7j	2 λ	37.09°	−1.233 dB	8.75°	

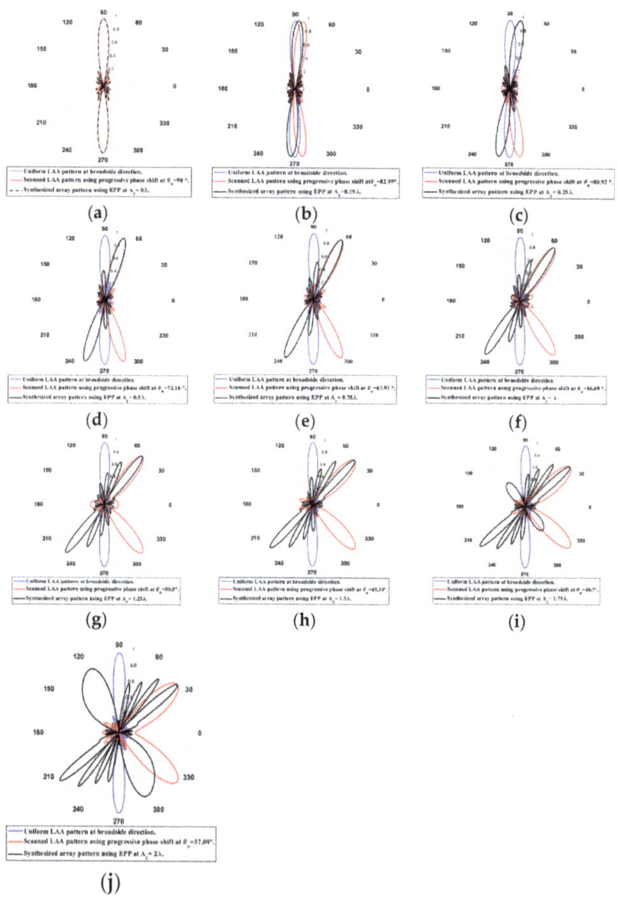

Figure 7. Polar plots of the synthesized radiation patterns using the EPP-BSA approach for 8-element LAA at compression ratio $R = 1$ and (**a**) $A_c = 0$, (**b**) $A_c = 0.19\ \lambda$, (**c**) $A_c = 0.25\ \lambda$, (**d**) $A_c = 0.5\ \lambda$, (**e**) $A_c = 0.75\ \lambda$, (**f**) $A_c = \lambda$, (**g**) $A_c = 1.25\ \lambda$, (**h**) $A_c = 1.5\ \lambda$, (**i**) $A_c = 1.75\ \lambda$, and (**j**) $A_c = 2\ \lambda$ and changing from positive to negative.

On the other hand, when the cosine wave amplitude is changing from negative to positive and its magnitude is changing from $A_c = 0$ to $A_c = 0.19\lambda$, the array main beam is scanned from 0° to −7.01° with a constant HPBW and the SLL is less than −10 dB. However, as the cosine wave amplitude increases from 0.19 λ to 1.5 λ, the scan angle is increased from −7.01° to −44.6°, the SLL is increased by more than −10 dB, and the HPBW is decreased by 3°. But, as the cosine wave amplitude increases from 1.5 λ to 2 λ, the scan angle is increased from −44.6° to −52.9° and the grating lobe appears with increased SLL as shown in Figure 8. The achieved maximum scanning range is $-52.91° \leq \theta_r \leq 52.91°$ around the broadside direction.

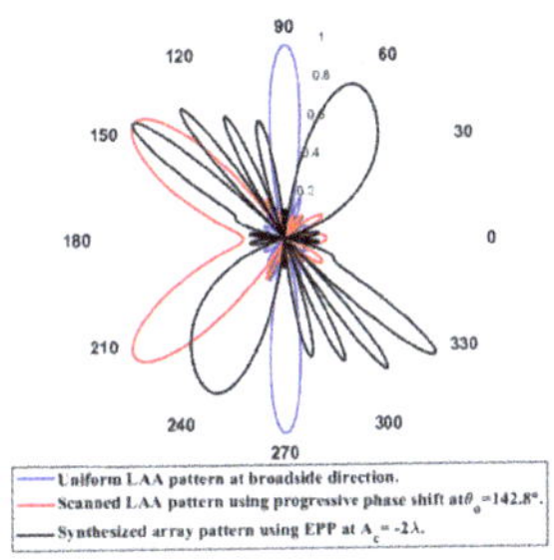

Figure 8. Polar plot of the synthesized radiation pattern using the EPP-BSA approach for 8-element ULAA at compression ratio $R = 1$ for $A_c = 2\ \lambda$ and changing from negative to positive.

Unlike traditional progressive phase shift scanning, as the scanning angle increases away from the broadside direction, the SLL increases, and the HPBW is significantly increased many folds of its value in the broadside case as shown in Figures 7 and 8. However, in the proposed EPP-BSA approach, as the amplitude of the cosine wave A_c increases, the HPBW of the main beam decreases as the length or size of the array projection on the plane normal to the main beam direction θ_o is increased.

Test case (2): $R = 0.875$

In this case, the cosine wave half period is chosen to be greater than the array size, $H > L_h$ and equals $H = 4\ \lambda$, hence $R = 3.5\ \lambda/4\ \lambda = 0.875$. Accordingly, the EPPs of the array elements for A_c changing from positive to negative and A_c changing from negative to positive are shown in Figures 9a and 9b, respectively.

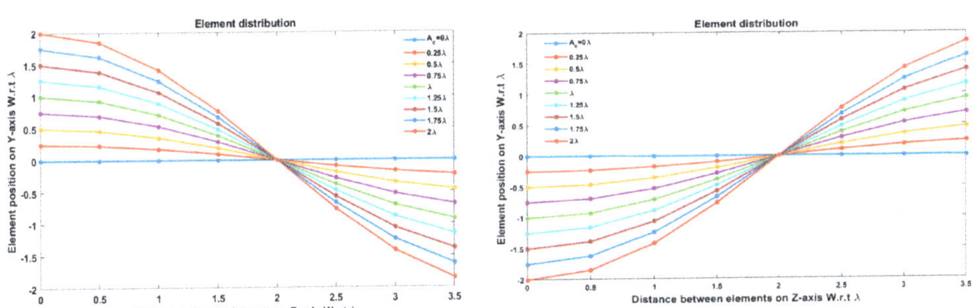

(**a**) (**b**)

Figure 9. The EPPs of the array elements at compression ratio $R = 0.875$ for (**a**) A_c is changing from positive to negative and (**b**) A_c is changing from negative to positive.

When the cosine wave amplitude is changing from positive to negative and its magnitude is changing from $A_c = 0$ to $A_c = 0.25\ \lambda$, the array main beam is scanned from 0° to 8.56° with a constant HPBW and the SLL is less than −10 dB. However, as the cosine wave amplitude increases from 0.25 λ to 2 λ, the scan angle is increased from 8.56° to 51.36°, the SLL is increased by more than −10 dB, and the HPBW is decreased by 3.6° as listed in Table 2. The synthesized radiation patterns using the EPP technique are shown in Figure 10. It is clear that as the amplitude of the cosine waveform increases, the scanning angle increases away from the broadside direction, the HPBW decreases, and the SLL increases as shown in Figure 11.

Table 2. The relation between the amplitude changes of the cosine waveform and the scanning angle, HPBW, SLL, and maximum scanning range θ_r for 8-element ULAA at compression ratio $R = 0.875$.

Figure	Cosine Wave Amplitude A_c	Main Beam Direction θ_o, in °	SLL (dB)	HPBW in °	Maximum Scanning Range
Figure 10a	0	90°	−12.8 dB	12.8°	
Figure 10b	0.25 λ	81.44°	−10.09 dB	12.6°	
Figure 10c	0.5 λ	73.19°	−7.88 dB	12.4°	
Figure 10d	0.75 λ	65.46°	−6.181 dB	12°	
Figure 10e	λ	58.24°	−4.917 dB	11.2°	$-51.36° \leq \theta_r \leq 51.36°$
Figure 10f	1.25 λ	52.05°	−4.009 dB	10.8°	
Figure 10g	1.5 λ	46.86°	−3.386 dB	10.5°	
Figure 10h	1.75 λ	42.25°	−2.949 dB	9.7°	
Figure 10i	2 λ	38.64°	−2.668 dB	9.2°	

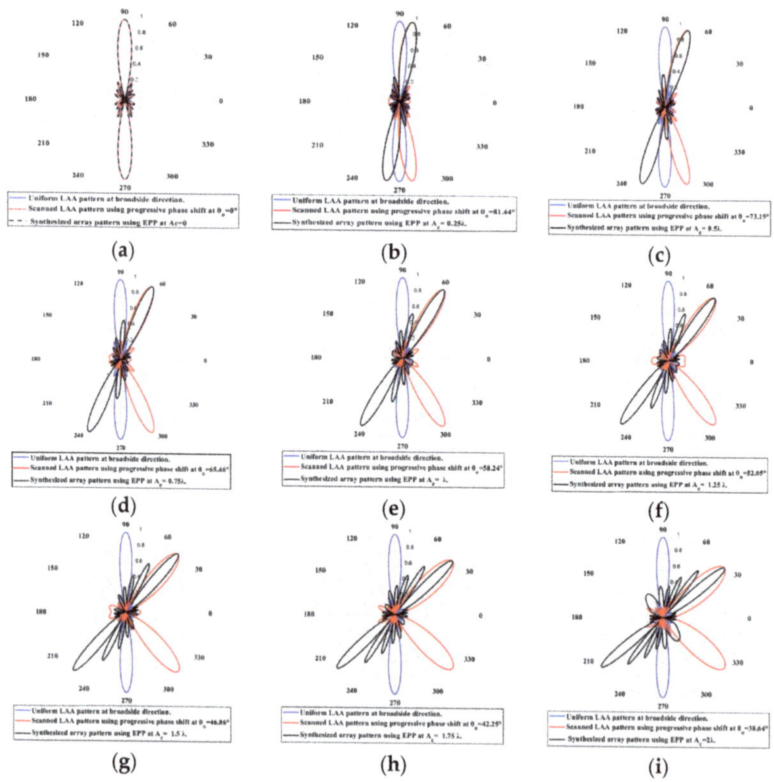

Figure 10. Polar plots of the synthesized radiation patterns using the EPP-BSA approach for an 8-element ULAA at compression ratio $R = 0.875$ and (**a**) $A_c = 0$, (**b**) $A_c = 0.25\ \lambda$, (**c**) $A_c = 0.5\ \lambda$, (**d**) $A_c = 0.75\ \lambda$, (**e**) $A_c = \lambda$, (**f**) $A_c = 1.25\ \lambda$, (**g**) $A_c = 1.5\ \lambda$, (**h**) $A_c = 1.75\ \lambda$, and (**i**) $A_c = 2\ \lambda$ and changing from positive to negative.

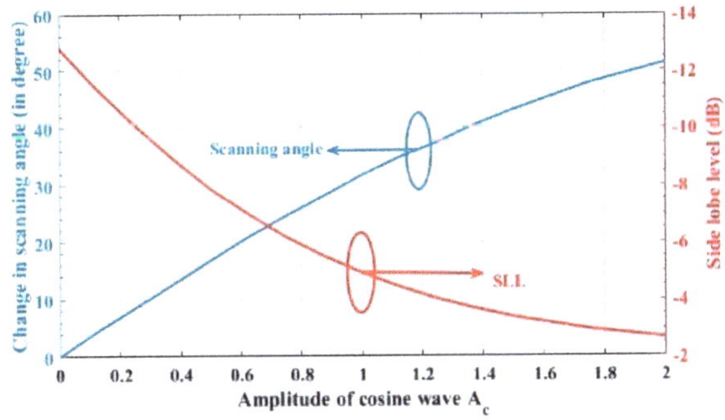

Figure 11. Relation between changes in the main beam scanning angle around the broadside direction and side lobe level changes.

On the other hand, when the cosine wave amplitude is changing from negative to positive and its magnitude is changing from $A_c = 0$ to $A_c = 0.25\ \lambda$, the array main beam is scanned from $0°\ to - 8.56°$ with a constant HPBW and the SLL is less than -10 dB. However, as the cosine wave amplitude increases from $0.25\ \lambda$ to $2\ \lambda$, the scan angle is increased from $-8.56°$ to $-51.36°$, the SLL is increased by more than -10 dB, and the HPBW is decreased by 3.6° as shown in Figure 12. The achieved maximum scanning range is $-51.36° \leq \theta_r \leq 51.36°$ around the broadside direction.

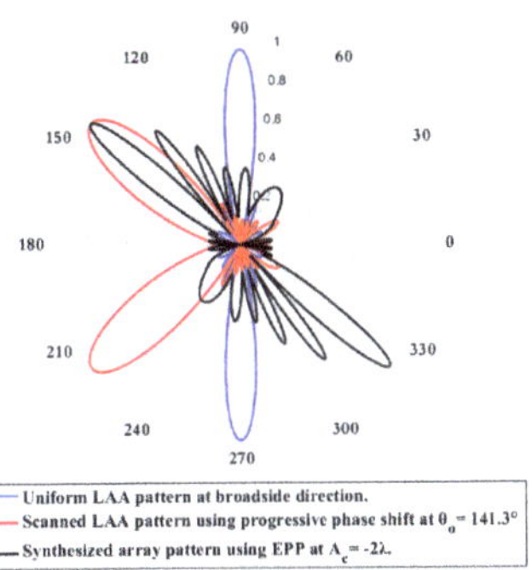

Figure 12. Polar plot of the synthesized radiation pattern using the EPP-BSA approach for 8-element ULAA at compression ratio $R = 0.875$ for $A_c = 2\ \lambda$ and changing from negative to positive.

Test case (3): $R = 0.7778$

In this case, the cosine wave half period is chosen to be greater than the array size, $H > L_h$ and equals $H = 4\ \lambda$, hence $R = 4.5\ \lambda / 4\ \lambda = 0.7778$. Accordingly, the EPPs of the array elements for A_c changing from positive to negative and A_c changing from negative to positive are shown in Figures 13a and 13b, respectively.

When the cosine wave amplitude is changing from positive to negative and its magnitude is changing from $A_c = 0$ to $A_c = 0.3\ \lambda$, the array main beam is scanned from 0° to 9.08° with a constant HPBW and the SLL is less than -10 dB. However, as the cosine wave amplitude increases from $0.3\ \lambda$ to $2\ \lambda$, the scan angle is increased from 9.08° to 48.78°, the SLL is increased by more than -10 dB, and the HPBW is decreased by 1.38° as listed in Table 3. The synthesized radiation patterns using the EPP technique are shown in Figure 14. On the other hand, when the cosine wave amplitude is changing from negative to positive and its magnitude is changing from $A_c = 0$ to $A_c = 0.3\ \lambda$, the array main beam is scanned from 0° to $-9.08°$ with a constant HPBW and the SLL is less than -10 dB. However, as the cosine wave amplitude increases from than $0.3\ \lambda$ to $2\ \lambda$, the scan angle is increased from $-9.08°$ to $-48.78°$, the SLL is increased by more than -10 dB, and the HPBW is decreased by 1.38° as shown in Figure 15. The achieved maximum scanning range is $-48.78° \leq \theta_r \leq 48.78°$ around the broadside direction.

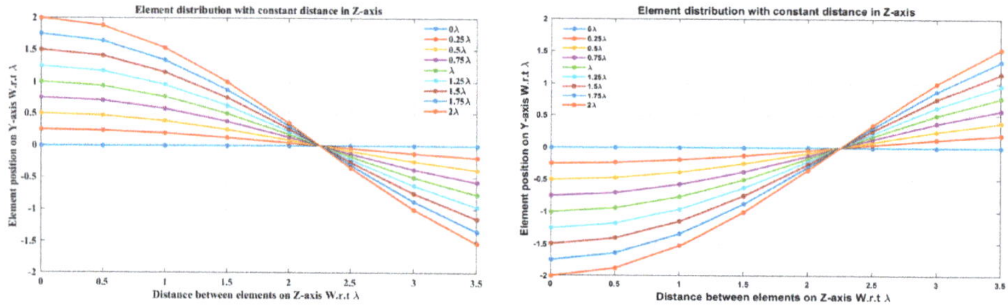

(a) (b)

Figure 13. The EPPs of the array elements at compression ratio $R = 0.7778$ for (**a**) A_c is changing from positive to negative and (**b**) A_c is changing from negative to positive.

Table 3. The relation between the amplitude changes of the cosine waveform and the scanning angle, HPBW, SLL, and maximum scanning range θ_r for 8-element ULAA at compression ratio $R = 0.7778$.

Figure	Cosine Wave Amplitude A_c	Main Beam Direction θ_o, in °	SLL (dB)	HPBW in °	Maximum Scanning Range
Figure 14a	0	90°	−12.8 dB	12.8°	
Figure 14b	0.25 λ	82.47°	−10.75 dB	12.7°	
Figure 14c	0.27 λ	81.44°	−10.55 dB	12.71°	
Figure 14d	0.3 λ	80.92°	−10.03 dB	12.68°	
Figure 14e	0.5 λ	74.74°	−8.679 dB	12.52°	
Figure 14f	0.75 λ	68.03°	−6.93 dB	12.21°	$-48.78° \leq \theta_r \leq 48.78°$
Figure 14g	λ	61.33°	−5.567 dB	11.96°	
Figure 14h	1.25 λ	55.14°	−4.547 dB	11.77°	
Figure 14i	1.5 λ	49.99°	−3.842 dB	11.61°	
Figure 14j	1.75 λ	45.34°	−3.366 dB	11.5°	
Figure 14k	2 λ	41.22°	−3.097 dB	11.42°	

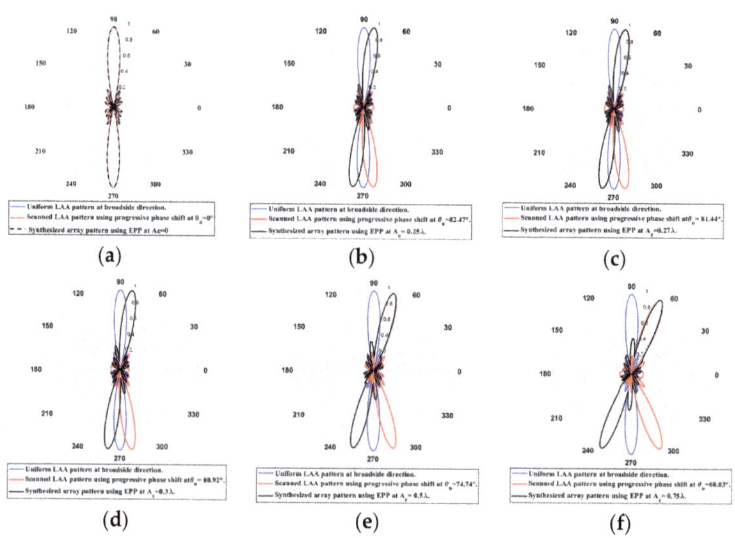

Figure 14. *Cont.*

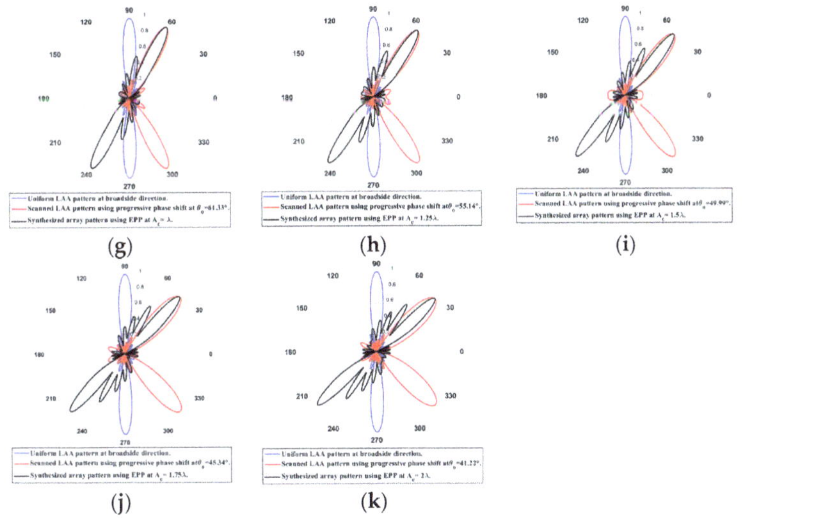

Figure 14. Polar plots of the synthesized radiation patterns using the EPP-BSA approach for an 8-element ULAA at compression ratio $R = 0.7778$ at (a) $A_c = 0$, (b) $A_c = 0.25\ \lambda$, (c) $A_c = 0.27\ \lambda$, (d) $A_c = 0.3\ \lambda$ (e) $A_c = 0.5\ \lambda$, (f) $A_c = 0.75\ \lambda$, (g) $A_c = \lambda$, (h) $A_c = 1.25\ \lambda$, (i) $A_c = 1.5\ \lambda$, (j) $A_c = 1.75\ \lambda$, and (k) $A_c = 2\ \lambda$ and changing from positive to negative.

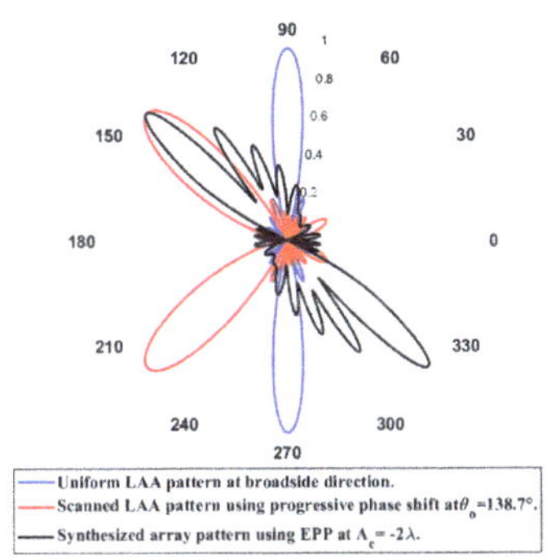

Figure 15. Polar plot of the synthesized radiation pattern using the EPP-BSA approach for 8-element ULAA at compression ratio $R = 0.7778$ for $A_c = 2\ \lambda$ and changing from negative to positive.

Test case (4): $R = 0.7$

In this case, the cosine wave half period is chosen to be equal to $H = 5\ \lambda$, and the array length equals $L_h = 3.5\ \lambda$, hence $R = 0.7$. Accordingly, the EPPs of the array elements for A_c changing from positive to negative and A_c changing from negative to positive are shown in Figures 16a and 16b, respectively.

When the cosine wave amplitude is changing from positive to negative and its magnitude is changing from $A_c = 0$ to $A_c = 0.39\ \lambda$, the array main beam is scanned from 0° to 10.68° with a constant HPBW and the SLL is less than −10 dB. However, as the cosine wave amplitude increases from 0.39 λ to 2 λ, the scan angle is increased from 10.68° to 45.69°, the SLL is increased by more than −10 dB, and the HPBW is decreased by 1.1° as listed in Table 4. The synthesized radiation patterns using the EPP technique are shown in Figure 17.

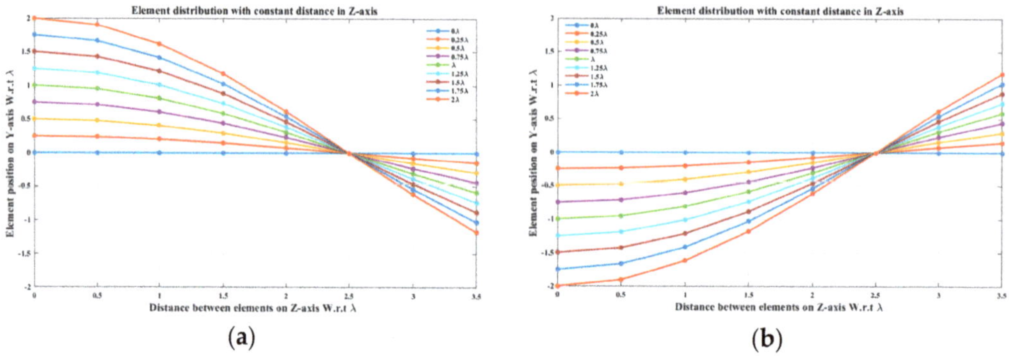

Figure 16. The EPPs of the array elements at compression ratio $R = 0.7$ for (**a**) A_c is changing from positive to negative and (**b**) A_c is changing from negative to positive.

Table 4. The relation between the amplitude changes of the cosine waveform and the scanning angle, HPBW, SLL, and maximum scanning range θ_r for 8-element ULAA at compression ratio $R = 0.7$.

Figure	Cosine Wave Amplitude A_c	Main Beam Direction θ_o, in $^\circ$	SLL (dB)	HPBW in $^\circ$	Maximum Scanning Range
Figure 17a	0	90°	−12.8 dB	12.8°	
Figure 17b	0.25 λ	82.99°	−11.2 dB	12.74°	
Figure 17c	0.35 λ	80.41°	−10.36 dB	12.66°	
Figure 17d	0.39 λ	79.38°	−10.05 dB	12.66°	
Figure 17e	0.5 λ	76.28°	−9.198 dB	12.65°	
Figure 17f	0.75 λ	70.1°	−7.39 dB	12.5°	$-45.69^\circ \leq \theta_r \leq 45.69^\circ$
Figure 17g	λ	63.91°	−5.869 dB	12.42°	
Figure 17h	1.25 λ	58.24°	−4.68 dB	12.4°	
Figure 17i	1.5 λ	53.08°	−3.791 dB	12.7°	
Figure 17j	1.75 λ	48.44°	−3.185 dB	13.4°	
Figure 17k	2 λ	44.31°	−2.819 dB	13.9°	

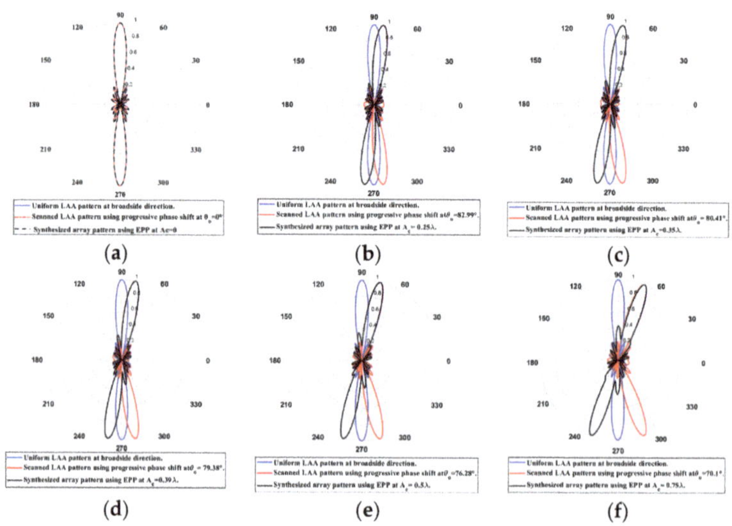

Figure 17. *Cont.*

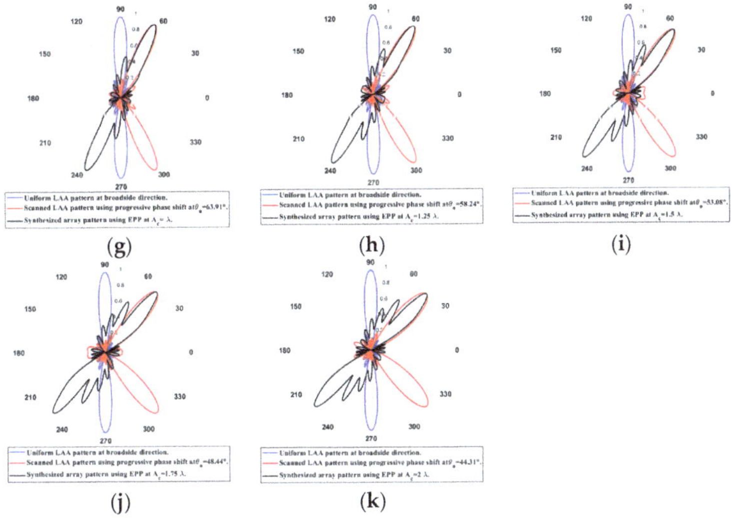

Figure 17. Polar plots of the synthesized radiation patterns using the EPP techniques for 8-element ULAA at compression ratio $R = 0.7$ at: (**a**) $A_c = 0$, (**b**) $A_c = 0.25\ \lambda$, (**c**) $A_c = 0.35\ \lambda$, (**d**) $A_c = 0.39\ \lambda$, (**e**) $A_c = 0.5\ \lambda$, (**f**) $A_c = 0.75\ \lambda$, (**g**) $A_c = \lambda$, (**h**) $A_c = 1.25\ \lambda$, (**i**) $A_c = 1.5\ \lambda$, (**j**) $A_c = 1.75\ \lambda$, and (**k**) $A_c = 2\ \lambda$ and changing from positive to negative.

On the other hand, when the cosine wave amplitude is changing from negative to positive and its magnitude is changing from $A_c = 0$ to $A_c = 0.39\ \lambda$, the array main beam is scanned from $0^\circ\ to - 10.68^\circ$ with a constant HPBW and the SLL is less than -10 dB. However, as the cosine wave amplitude increases from $0.39\ \lambda$ to $2\ \lambda$, the scan angle is increased from -10.68° to -45.69°, the SLL is increased by more than -10 dB, and the HPBW is decreased by 1.1° as shown in Figure 18. The achieved maximum scanning range is $-45.69^\circ \leq \theta_r \leq 45.69^\circ$ around the broadside direction.

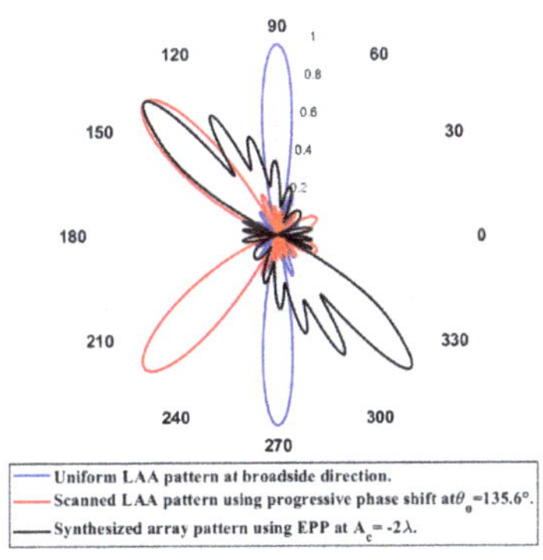

Figure 18. Polar plot of the synthesized radiation pattern using the EPP-BSA approach for 8-element ULAA at compression ratio $R = 0.7$ for $A_c = 2\ \lambda$ and changing from negative to positive.

Test case (5): $R = 0.5$

In this case, the cosine wave half period is chosen to be equal to $H = 7\ \lambda$, and the array length equals $L_h = 3.5\ \lambda$, hence $R = 0.5$. Accordingly, the EPPs of the array elements for A_c changing from positive to negative and A_c changing from negative to positive are shown in Figures 19a and 19b, respectively.

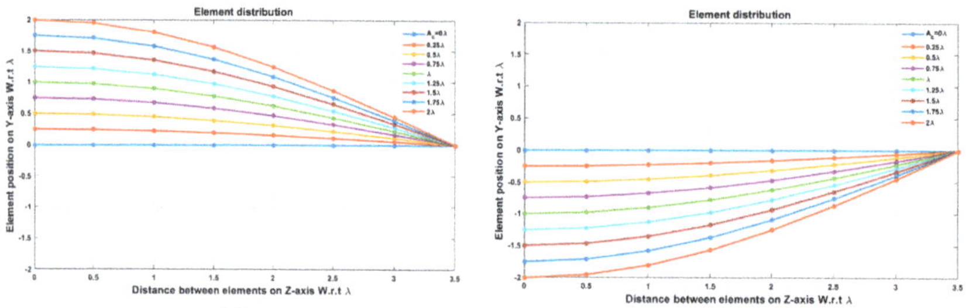

(**a**) (**b**)

Figure 19. The EPPs of the array elements at compression ratio $R = 0.5$ for (**a**) A_c is changing from positive to negative and (**b**) A_c is changing from negative to positive.

When the cosine wave amplitude is changing from positive to negative and its magnitude is changing from $A_c = 0$ to $A_c = 0.65\ \lambda$, the array main beam is scanned from 0°to 10.83° with a constant HPBW and the SLL is less than −10 dB. However, as the cosine wave amplitude increases from 0.65 λ to 2 λ, the scan angle is increased from 10.83° to 31.25 °, the SLL is increased by more than −10 dB, and the HPBW is decreased by 2.83° as listed in Table 5. The synthesized radiation patterns using the EPP technique are shown in Figure 20.

Table 5. The relation between the amplitude changes of the cosine waveform and the scanning angle, HPBW, SLL, and maximum scanning range θ_r for 8-element ULAA at compression ratio $R = 0.5$.

Figure	Cosine Wave Amplitude A_c	Main Beam Direction θ_o, in °	SLL (dB)	HPBW in °	Maximum Scanning Range
Figure 19a	0	90°	−12.8 dB	12.8°	
Figure 19b	0.25 λ	85.57°	−12.08 dB	12.79°	
Figure 19c	0.5 λ	81.44°	−10.89 dB	12.79°	
Figure 19d	0.6 λ	79.89°	−10.32 dB	12.77°	
Figure 19e	0.65 λ	79.17°	−10.02 dB	12.76°	
Figure 19f	0.75 λ	77.32°	−9.416 dB	12.8°	$-31.25° \leq \theta_r \leq 31.25°$
Figure 19g	λ	73.71°	−7.874 dB	12.89°	
Figure 19h	1.25 λ	69.58°	−6.407 dB	13°	
Figure 19i	1.5 λ	65.97°	−5.087 dB	13.328°	
Figure 19j	1.75 λ	62.36°	−3.915 dB	13.94°	
Figure 19k	2 λ	58.75°	−2.929 dB	15.63°	

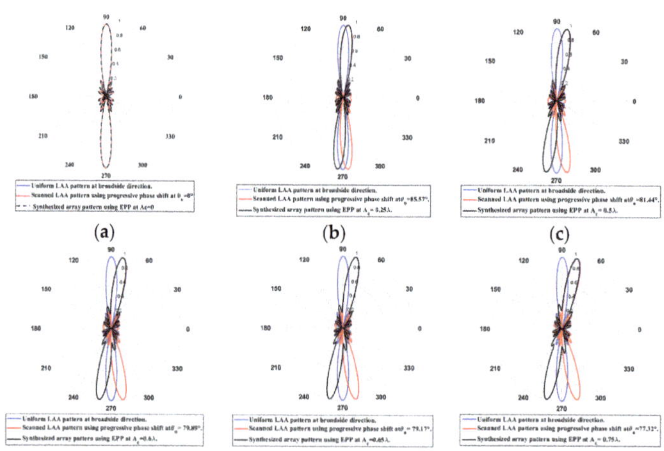

(**a**) (**b**) (**c**)

(**d**) (**e**) (**f**)

Figure 20. *Cont.*

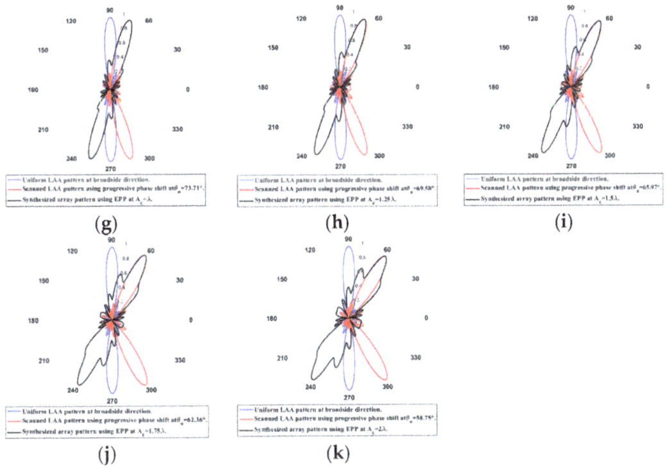

Figure 20. Polar plots of the synthesized radiation patterns using the EPP-BSA approach for 8-element uniform LAA at compression ratio equal to 0.5 at: (**a**) $A_c = 0$, (**b**) $A_c = 0.25\ \lambda$, (**c**) $A_c = 0.5\ \lambda$, (**d**) $A_c = 0.6\ \lambda$, (**e**) $A_c = 0.65\ \lambda$, (**f**) $A_c = 0.75\ \lambda$, (**g**) $A_c = \lambda$, (**h**) $A_c = 1.25\ \lambda$, (**i**) $A_c = 1.5\ \lambda$, (**j**) $A_c = 1.75\ \lambda$, and (**k**) $A_c = 2\ \lambda$ and changing from positive to negative.

On the other hand, when the cosine wave amplitude is changing from negative to positive and its magnitude is changing from $A_c = 0$ to $A_c = 0.65\ \lambda$, the array main beam is scanned from 0° to − 10.83° with a constant HPBW and the SLL is less than −10 dB. However, as the cosine wave amplitude increases from 0.65 λ to 2 λ , the scan angle is increased from −10.83° to − 31.25°, the SLL is increased by more than −10 dB, and the HPBW is decreased by 2.83° as shown in Figure 21. The achieved maximum scanning range is $-31.25° \leq \theta_r \leq 31.25°$ around the broadside direction.

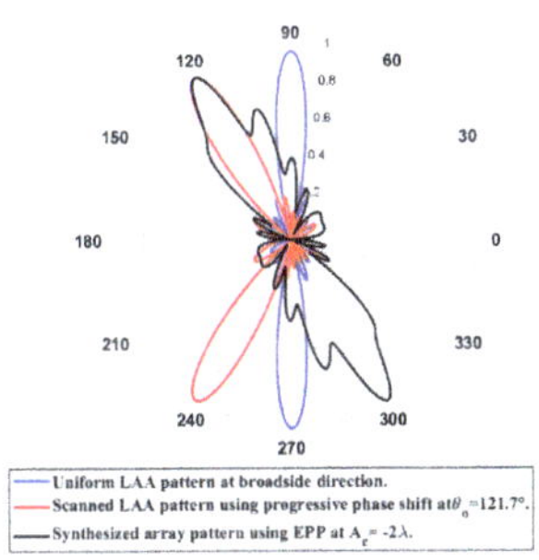

Figure 21. Polar plot of the synthesized radiation pattern using the EPP-BSA approach for 8-element uniform LAA due to compression ratio equal to 0.5 for $A_c = 2\ \lambda$ and changing from negative to positive.

2.1.2. Comparison between the Five Test Cases of Compression Ratio

The results of the aforementioned five test cases of the compression ratio R are summarized in Table 6. It is clear that for test cases 1, 2, and 3, as the cosine wave amplitude increases, the scanning angle increases, and the HPBW decreases. However, test case 2 at $R = 0.875$, provides the largest decrease in the HPBW. On the other hand, for test cases 4 and 5, as the cosine wave amplitude increases, the scanning angle increases giving rise to a larger scanning range than in the previous test cases, and the HPBW increases. But, the null between the first side lobe and main beam is relatively high, resulting in stronger interference. As a consequence, we can conclude that test case 2 with $R = 0.875$ yields the best results in terms of HPBW and scanning range.

Table 6. Comparison between the five test cases of compression ratio. The optimal value is in bold.

No.	Compression Ratio	Cosine Wave Amplitude A_c	Scan Angle Range with Respect to Broadside Direction	SLL	HPBW
1	$R = 1$	0 to 0.19 λ	From 0° to 7.01°	< -10 dB	Constant
		0.19λ to 1.5 λ	From 7.01° to 52.91°	> -10 dB	Decreased by 3°
2	**$R = 0.875$**	**0 to 0.25 λ**	From **0°** to **8.56°**	$<$ **−10 dB**	Constant
		0.25 λ to 2 λ	From **8.5°** to **51.36°**	$>$ **−10 dB**	Decreased by **3.6°**
3	$R = 0.7778$	0 to 0.3 λ	From 0° to 9.08°	< -10 dB	Constant
		0.3 λ to 2 λ	From 9.08° to 48.78°	> -10 dB	Decreased by 1.38°
4	$R = 0.7$	0 to 0.39 λ	From 0° to 10.62°	< -10 dB	Constant
		0.39 λ to 2 λ	From 10.62° to 45.69°	> -10 dB	Increase by 1.1°
5	$R = 0.5$	0 to 0.65 λ	From 0° to 10.83°	< -10 dB	Constant
		0.65 λ to 2 λ	From 10.62° to 31.25°	> -10 dB	Increased by 2.83°

2.1.3. CST Realization of the Proposed EPP-BSA

In this section, the proposed EPP-BSA is realistically validated for actual antenna elements rather than isotropic antennas. The antenna array is created using the CST microwave studio software package utilizing a dipole element, the dimensions of which, as well as the H-plane and E-plane patterns, are shown in Figure 22. The simulated scattering parameter (reflection coefficient) $|S_{11}|$ of the dipole antenna is illustrated in Figure 23, showing a resonance frequency $f_o = 1$ GHz.

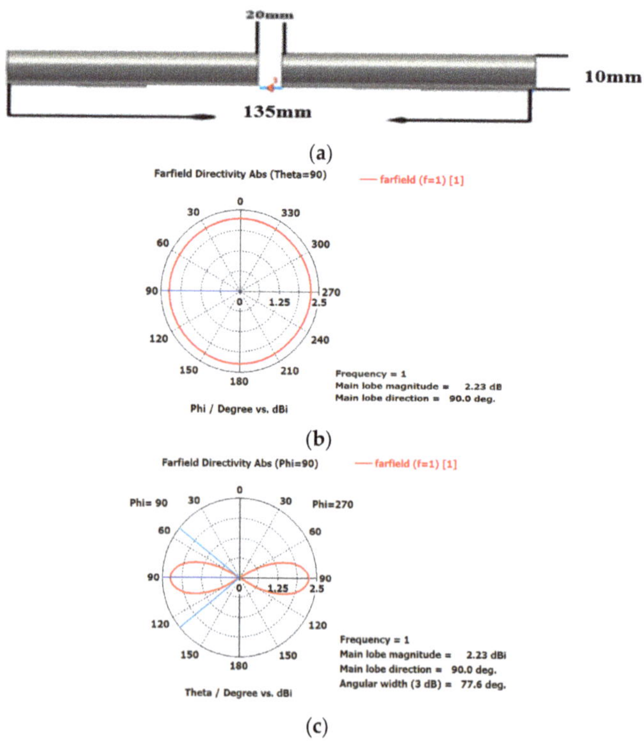

Figure 22. (**a**) Dimensions of dipole antenna, (**b**) H-plane pattern, and (**c**) E-plane pattern.

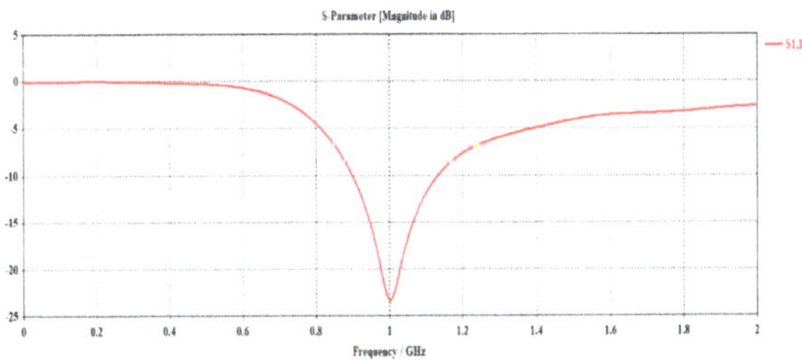

Figure 23. The scattering parameter (reflection coefficient) $|S_{11}|$ against the frequency for the designed dipole antenna.

Considering the recommended test case 2 with a compression ratio $R = 0.875$, which yields the best results in terms of HPBW and scanning range, we created the synthesized eight-element antenna arrays applying the proposed EPP-BSA approach. The synthesized antenna arrays and the associated 3D radiation patterns are shown in Figure 24, while the polar plots of the synthesized radiation patterns using the CST software package are shown in Figure 25. The CST simulation results for A_c changing from 0 to 2 λ are recorded in Table 7 compared to the MATLAB simulation results indicating the resultant scan angle, SLL, and HPBW. The comparison shows a high matching between the CST and MATLAB simulation results with minimal fractal changes because the CST acts as a real environment and considers the mutual coupling between antenna elements in the synthesized array structures. This demonstrates the ability of practical validation of the proposed EPP-BSA approach.

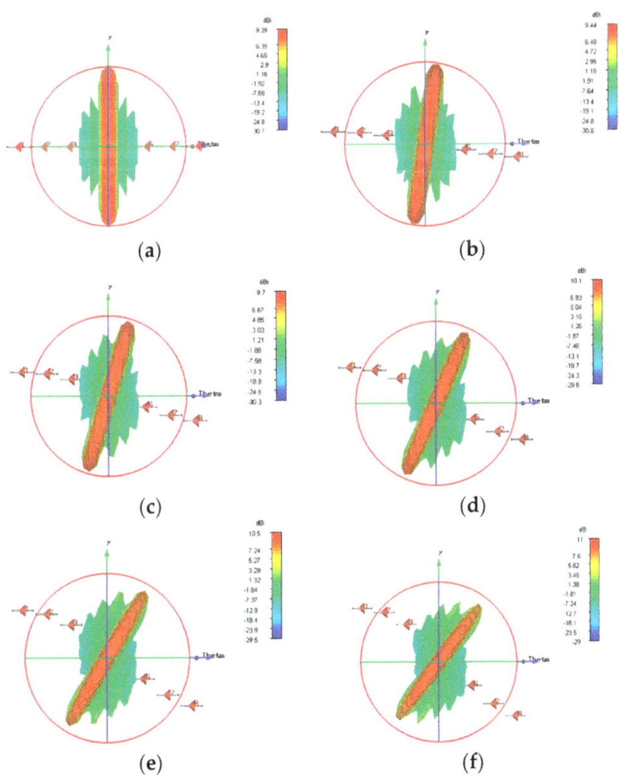

Figure 24. *Cont.*

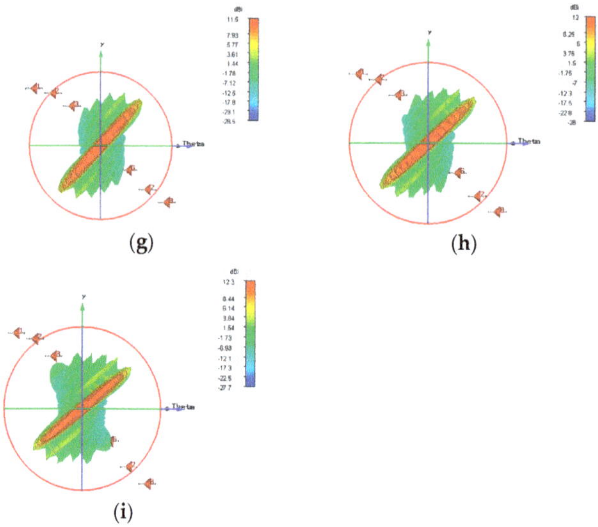

Figure 24. The 3D radiation pattern plots of the synthesized antenna arrays using CST software package applying the proposed EPP-BSA approach for an 8-element ULAA for $R = 0.875$ at: (**a**) $A_c = 0$, (**b**) $A_c = 0.25\ \lambda$, (**c**) $A_c = 0.5\ \lambda$, (**d**) $A_c = 0.75\ \lambda$, (**e**) $A_c = 1\ \lambda$, (**f**) $A_c = 1.251\ \lambda$, (**g**) $A_c = 1.5\ \lambda$, (**h**) $A_c = 1.75\ \lambda$, (**i**) $A_c = 2\ \lambda$, and changing from positive to negative.

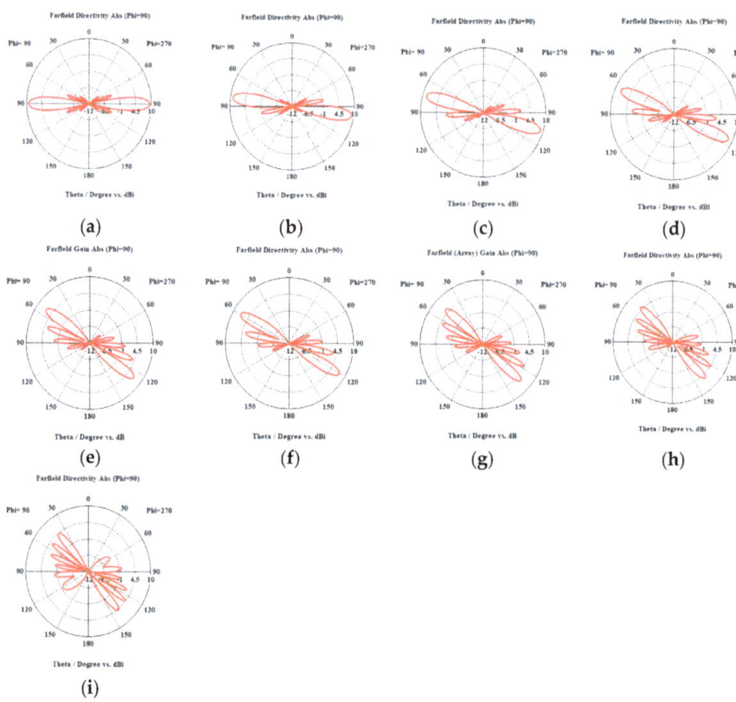

Figure 25. Polar plots of the synthesized antenna arrays radiation patterns using CST software package applying the proposed EPP-BSA approach for an 8-element ULAA for $R = 0.875$ at: (**a**) $A_c = 0$, (**b**) $A_c = 0.25\lambda$, (**c**) $A_c = 0.5\ \lambda$, (**d**) $A_c = 0.75\ \lambda$, (**e**) $A_c = 1\ \lambda$, (**f**) $A_c = 1.251\ \lambda$, (**g**) $A_c = 1.5\ \lambda$, (**h**) $A_c = 1.75\ \lambda$, (**i**) $A_c = 2\ \lambda$, and changing from positive to negative.

Table 7. The CST simulation results for A_c changing from 0 to 2 λ compared to the MATLAB simulation results indicating the resultant scan angle, SLL, and HPBW.

Cosine Wave Amplitude A_c	CST Results			MATLAB Results		
	Scan Angle	HPBW	SLL	Scan Angle	HPBW	SLL
$A_c = 0$	0°	12.3°	−13 dB	0°	12.8°	−12.8 dB
0.25 λ	8°	12.2°	−12.2 dB	8.56°	12.6°	−10.09 dB
0.5 λ	17°	12°	−7.5 dB	16.81°	12.4°	−7.88 dB
0.75 λ	24°	11.6°	−5.6 dB	24.54°	12°	−6.181 dB
λ	31°	11.2°	−4.1 dB	31.76°	11.2°	−4.917 dB
1.25 λ	37°	10.7°	−3.2 dB	37.95°	10.8°	−4.009 dB
1.5 λ	43°	10.2°	−2.1 dB	43.14°	10.5°	−3.386 dB
1.75 λ	47°	9.7°	−1.3 dB	47.75°	9.7°	−2.949 dB
2 λ	51°	9.1°	−2.1 dB	51.36°	9.2°	−2.668 dB

2.2. Proposed SLL/EPP-BSA

To mitigate the SLL problem of the EPP-BSA, the SLL/EPP-based beam scanning approach, which is denoted as SLL/EPP-BSA, is introduced. In this approach, SLL reduction of the array pattern is firstly performed using our single convolution/genetic algorithm (SC/GA) technique introduced in [13] and then performance-optimized EPP in the azimuth plane. This combination between the SLL reduction and the EPP (SLL/EPP) results in a wide scanning range from the broadside to the end-fire direction with a relatively constant HPBW and minimal variations in the synthesized SLL. In addition, keeping the synthesized excitation coefficients constant without adding progressive phase shifters facilitates the manufacturing process and reduces the cost of the feeding network. The SC/GA SLL reduction technique is utilized as it provides a twofold decrease in the SLL. Consider the ULAA configuration shown in Figure 1 whose array factor $AF(\theta)$ is defined by Equation (1), where a_n is the excitation coefficient of the nth antenna element. The SC is used to determine the synthesized excitation coefficients, while the GA is utilized to determine the element spacing between the antenna array elements. The SC/GA technique can be summarized as follows:

For N-element ULAA, the $N \times 1$ excitation vector $\mathrm{A}_{N\times 1} = [a_1\, a_2\, a_3\, a_4 \ldots a_N]$ is convolved by itself such that the resultant 1D convolution vector $\mathbb{C}_{(2N-1)\times 1}$ can be expressed as:

$$\mathbb{C}_{(2N-1)\times 1} = \mathrm{A}_{N\times 1} * \mathrm{A}_{N\times 1} \tag{22}$$

However, the size of the resultant excitation vector $\mathbb{C}_{I\times 1} = \mathbb{C}_{(2N-1)\times 1}$ from the single convolution process is much larger than the size of the original excitation vector $\mathrm{A}_{N\times 1}$. In order to synthesis the array factor $AF_S(\theta)$ using a reduced number of antenna elements, we divided the vector $\mathbb{C}_{I\times 1} = \mathbb{C}_{(2N-1)\times 1}$ into two vectors $\mathbb{C}_{(N\times 1)_O}$ and $\mathbb{C}_{(N-1)\times 1_E}$ that contain the odd and even excitations to implement the synthesized array factors $AF_{SO}(\theta)$ and $AF_{SE}(\theta)$, respectively.

The odd excitation vector can be determined as follows:

$$\mathbb{C}_{(N\times 1)_O} = \mathbb{C}_{I\times 1}|_{odd} \tag{23}$$

where the elements $\mathbb{C}_O(n,1)$ of the vector $\mathbb{C}_{(N\times 1)_O}$ can be obtained from the elements $\mathbb{C}(i,1)$ of the vector $\mathbb{C}_{I\times 1}$ such that:

$$\mathbb{C}_O(n,1) = \mathbb{C}(2i-1,1) \tag{24}$$

where $1 \leq (i = n) \leq \frac{I+1}{2}$

The even excitation vector can be determined as follows:

$$\mathbb{C}_{(N-1)\times 1_E} = \mathbb{C}_{I\times 1}|_{even} \quad (25)$$

where the elements $\mathbb{C}_E(n,1)$ of the vector $\mathbb{C}_{(N-1)\times 1_E}$ can be obtained from the elements $\mathbb{C}(i,1)$ of the vector $\mathbb{C}_{I\times 1}$ such that:

$$\mathbb{C}_E(n,1) = \mathbb{C}(2i,1) \quad (26)$$

where $1 \leq (i=n) \leq \left(\frac{I+1}{2}\right) - 1$

Accordingly, the synthesized array factor using odd excitation coefficients $AF_{SO}(\theta)$ is given by:

$$AF_{SO}(\theta) = \sum_{n=1}^{\frac{I+1}{2}} \mathbb{C}_O(n,1)e^{(jk(n-1)d_s\cos\theta)} \quad (27)$$

While the synthesized array factor using even excitation coefficients $AF_{SE}(\theta)$ is given by:

$$AF_{SE}(\theta) = \sum_{n=1}^{(\frac{I+1}{2})-1} \mathbb{C}_E(n,1)e^{(jk(n-1)d_s\cos\theta)} \quad (28)$$

where d_s is the synthesized element spacing determined by the GA [13]. The GA optimizes the element spacing such that the synthesized array factors $AF_{SO}(\theta)$ and $AF_{SE}(\theta)$ provide the same HPBW as the original array factor $AF(\theta)$. It is worth mentioning that both synthesized $AF_{SO}(\theta)$ and $AF_{SE}(\theta)$ provide a twofold decrease in the SLL compared to the original array factor $AF(\theta)$.

After SLL reduction, the synthesized excitation coefficients $\mathbb{C}_{(N\times 1)_O}$ and $\mathbb{C}_{(N-1)\times 1_E}$ are kept constant and the EPP is performed for perturbing the elements' positions in the azimuth plane to steer the main beam to the desired direction.

Accordingly, the synthesized array factor using odd excitation coefficients $AF_{SO}(\theta)$ and EPP can be expressed as:

$$AF_{SO-EPP}(\theta) = \sum_{n=1}^{\frac{I+1}{2}} \mathbb{C}_O(n,1)e^{(jk(n-1)d_s\cos\theta + jkd_{y_n}\sin\theta)} \quad (29)$$

While the synthesized array factor using even excitation coefficients $AF_{SE}(\theta)$ and EPP can be expressed as:

$$AF_{SE-EPP}(\theta) = \sum_{n=1}^{(\frac{I+1}{2})-1} \mathbb{C}_E(n,1)e^{(jk(n-1)d_s\cos\theta + jkd_{y_n}\sin\theta)} \quad (30)$$

where d_{y_n} is related to the synthesized element spacing d_s according to Figure 26 shown below.

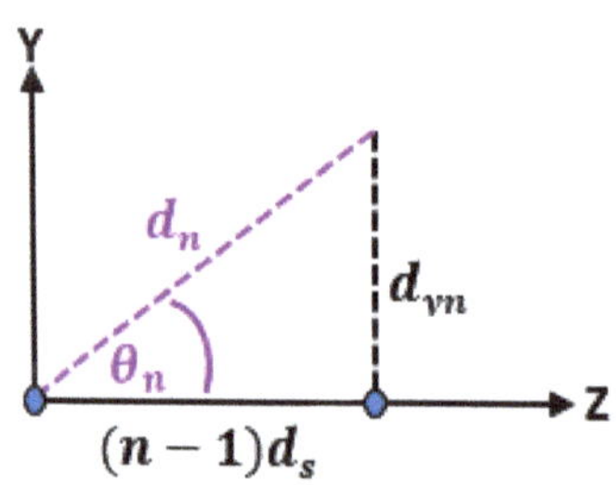

Figure 26. The triangular relation between the element position d_{y_n} in the Y-direction and the distance from the reference antenna element.

2.2.1. Simulation Results of the Proposed SLL/EPP-BSA

Consider a broadside ULAA consisting of $N = 8$ elements with uniform element spacing $d = \lambda/2$ whose elements are arranged along the Z-axis. Firstly, the ULAA is synthesized using the SC/GA for SLL reduction. The synthesized array factors $AF_{SO}(\theta)$

and $AF_{SE}(\theta)$ using odd and even excitations compared to the original array factor $AF(\theta)$ are shown in Figures 27 and 28, respectively. The synthesized excitation coefficients $\mathbb{C}_{(N\times1)_O}$ and $\mathbb{C}_{(N-1)\times1_E}$, element spacing d_s, SLL, and HPBW are listed in Table 8.

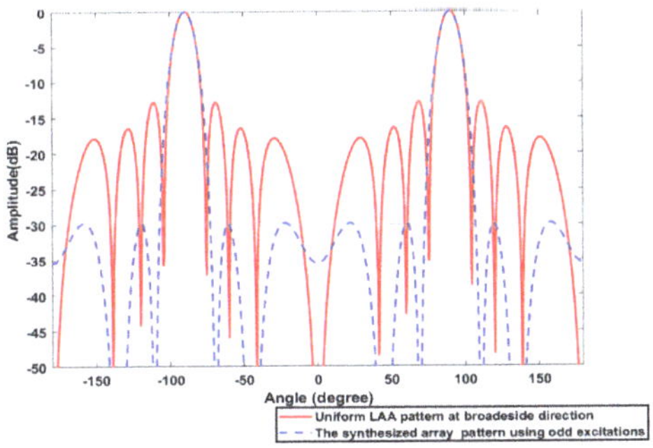

Figure 27. The synthesized array factor $AF_{SO}(\theta)$ using odd excitations compared to the original array factor $AF(\theta)$ of the 8-element ULAA.

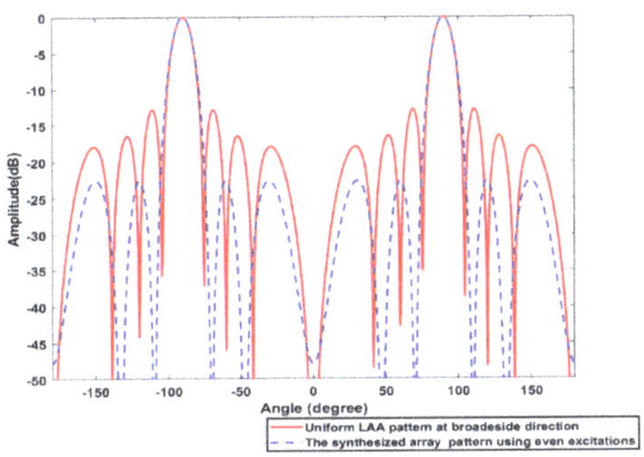

Figure 28. The synthesized array factor $AF_{SE}(\theta)$ using even excitations compared to the original array factor $AF(\theta)$ of the 8-element ULAA.

Table 8. The synthesized excitation coefficients $\mathbb{C}_{(N\times1)_O}$ and $\mathbb{C}_{(N-1)\times1_E}$, element spacing d_s, SLL, and HPBW.

Original ULAA		The Synthesized Array Using Odd Excitations		The Synthesized Array Using Even Excitations	
N = 8 elements		N = 8 elements		N = 7 elements	
$d_s \backslash \lambda = 0.5$		$d_s \backslash \lambda = 0.705$		$d_s \backslash \lambda = 0.735$	
a_1	1	a_1	1	a_1	2
a_2	1	a_2	3	a_2	4
a_3	1	a_3	5	a_3	6
a_4	1	a_4	7	a_4	8
a_5	1	a_5	7	a_5	6
a_6	1	a_6	5	a_6	4
a_7	1	a_7	3	a_7	2
a_8	1	a_8	1		
$HPBW = 12.8°$		$HPBW = 12.8°$		$HPBW = 12.8°$	
$SLL = -12.8$ dB		$SLL = -29.79$ dB		$SLL = -22.61$ dB	

2.2.2. Simulation Results of the Proposed SLL/EPP-BSA for Five Test Case

Test case (1): $R = 1$

In this case, the cosine wave half period is chosen to be equal to the array length, $H = L_h = 4.935\ \lambda$, hence $R = 1$. Accordingly, the EPPs of the synthesized array elements for A_c changing from positive to negative are shown Figure 29. The relation between the amplitude changes of the cosine waveform and the scanning angle, HPBW, SLL, and maximum scanning range for the eight-element ULAA at a compression ratio $R = 1$ are listed in Table 9. When the cosine wave amplitude is changing from positive to negative and its magnitude is changing from $A_c = 0$ to $A_c = \lambda$, the array main beam tilted from a broadside direction of $\theta^\circ = 90^\circ$ to $\theta^\circ = 62.36^\circ$ with minimal changes in the HPBW that is increased by 0.12° at $\theta^\circ = 62.36^\circ$ compared to the broadside direction, while the SLL increased from -29.79 dB to -10.53 dB as the main beam is titled away from the broadside direction to $\theta^\circ = 62.36^\circ$. The achieved maximum scanning range is $-27.64^\circ \leq \theta_r \leq 27.64^\circ$ around the broadside direction. The polar plot of the synthesized radiation pattern using the SLL/EPP-BSA approach at compression ratio $R = 1$ and $A_c = \lambda$ is shown in Figure 30. This technique might be beneficial in mobile network applications when coverage is limited, and the coverage area is divided into sectors.

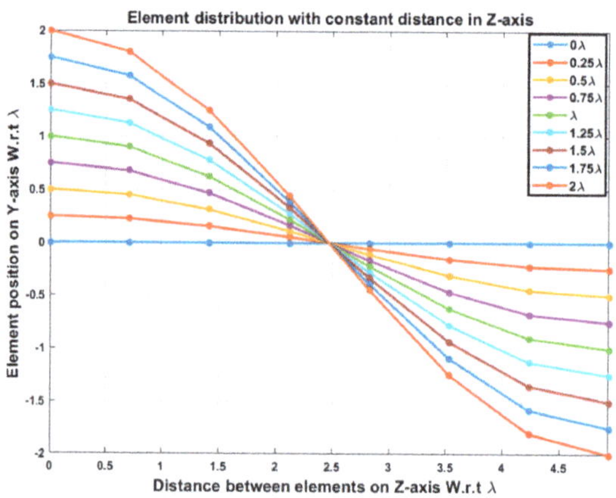

Figure 29. The EPPs of the synthesized array elements at compression ratio $\boldsymbol{R} = 1$ for $\boldsymbol{A_c}$ is changing from positive to negative.

Table 9. The relation between the amplitude changes of the cosine waveform and the scanning angle, HPBW, SLL, and maximum scanning range θ_r for 8-element ULAA at compression ratio $R = 1$.

Cosine Wave Amplitude A_c	Main Beam Direction θ_o, in °	SLL (dB)	HPBW in °	Maximum Scanning Range
0	90°	−29.79 dB	12.8°	$-27.64^\circ \leq \theta_r \leq 27.64^\circ$
0.25 λ	82.99°	−22.94 dB	12.78°	
0.5 λ	75.77°	−16.46 dB	12.83°	
0.75 λ	69.06°	−12.84 dB	12.89°	
λ	62.36°	−10.53 dB	12.92°	
1.25 λ	Grating lobe appearance			
1.5 λ				
1.75 λ				
2 λ				

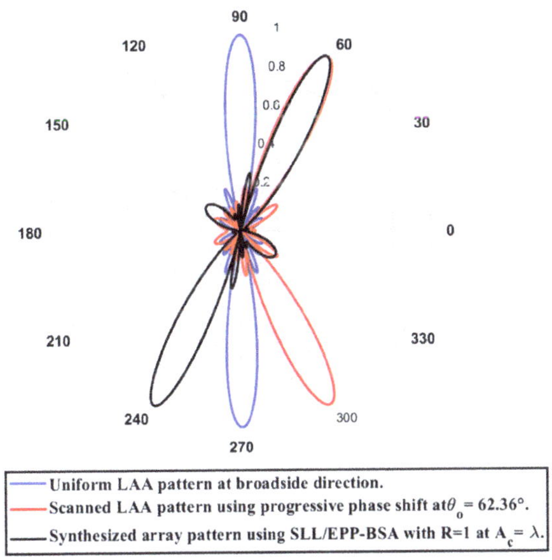

Figure 30. Polar plot of the synthesized radiation pattern using the SLL/EPP-BSA approach at compression ratio $R = 1$ and $A_c = \lambda$.

Test case (2): $R = 0.875$

In this case, the cosine wave half period is chosen to be greater than the array length, $H > L_h$ and equals $H = 5.64\ \lambda$, hence $R = 4.935\ \lambda/5.64\ \lambda = 0.875$. Accordingly, after applying the EPPs of the synthesized array elements for A_c changing from positive and the cosine wave amplitude is changing from positive to negative and its magnitude is changing from $A_c = 0$ to $A_c = 1.25\ \lambda$, the array main beam tilted from a broadside direction of $\theta^\circ = 90^\circ$ to $\theta^\circ = 59.27^\circ$ with minimal changes in the HPBW that decreased by 0.34° at $\theta^\circ = 59.27^\circ$ compared to the broadside direction, while the SLL increased from −29.79 dB to −10.83 dB as the main beam titled away from the broadside direction to $\theta^\circ = 59.27^\circ$. The achieved maximum scanning range is $-30.73^\circ \leq \theta_r \leq 30.73^\circ$ around the broadside direction. The relation between the amplitude changes of the cosine waveform and the scanning angle, HPBW, SLL, and maximum scanning range for the eight-element ULAA at a compression ratio $R = 0.875$ are listed in Table 10. The polar plot of the synthesized radiation pattern using the SLL/EPP-BSA approach at compression ratio $R = 0.875$ and $A_c = 1.25\ \lambda$ is shown in Figure 31.

Table 10. The relation between the amplitude changes of the cosine waveform and the scanning angle, HPBW, SLL, and maximum scanning range θ_r for 8-element ULAA at compression ratio $R = 0.875$.

Cosine Wave Amplitude A_c	Main Beam Direction θ_o, in °	SLL (dB)	HPBW in °	Maximum Scanning Range
0	90°	−29.79 dB	12.8°	
0.25 λ	83.5°	−25.55 dB	12.76°	
0.5 λ	76.8°	−18.78 dB	12.77°	$-30.73^\circ \leq \theta_r \leq 30.73^\circ$
0.75 λ	70.61°	−14.96 dB	12.75°	
λ	64.94°	−12.5 dB	12.67°	
1.25 λ	59.27°	−10.83 dB	12.46°	
1.5 λ				
1.75 λ	Grating lobe appearance			
2 λ				

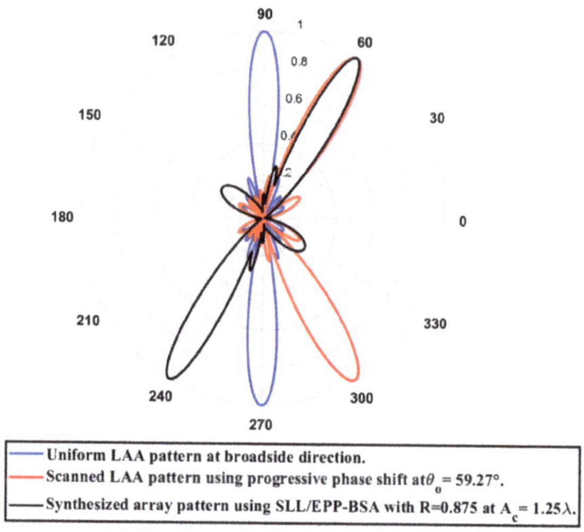

Figure 31. Polar plot of the synthesized radiation pattern using the SLL/EPP-BSA approach at compression ratio $R = 0.875$ and $A_c = 1.25\ \lambda$.

Test case (3): $R = 0.7778$

In this case, the cosine wave half period is chosen to be greater than the array length, $H > L_h$ and equals $H = 6.345\ \lambda$, hence $R = 4.935\lambda/6.345\lambda = 0.7778$. Accordingly, after applying the EPPs of the synthesized array elements for A_c changing from positive and the cosine wave amplitude is changing from positive to negative and its magnitude is changing from $A_c = 0$ to $A_c = 1.25\lambda$, the array main beam tilted from the broadside direction of $\theta^\circ = 90^\circ$ to $\theta^\circ = 62.36^\circ$ with minimal changes in the HPBW that is increased by 0.23° at $\theta^\circ = 62.36^\circ$ compared to the broadside direction, while the SLL is increased from −29.79 dB to −19.1 dB as the main beam is titled away from the broadside direction to $\theta^\circ = 62.36^\circ$. The achieved maximum scanning range is $-27.64^\circ \leq \theta_r \leq 27.64^\circ$ around the broadside direction. The relation between the amplitude changes of the cosine waveform and the scanning angle, HPBW, SLL, and maximum scanning range for the eight-element ULAA at a compression ratio $R = 0.7778$ are listed in Table 11. The polar plot of the synthesized radiation pattern using the SLL/EPP-BSA approach at compression ratio $R = 0.7778$ and $A_c = 1.25\ \lambda$ is shown in Figure 32.

Table 11. The relation between the amplitude changes of the cosine waveform and the scanning angle, HPBW, SLL, and maximum scanning range θ_r for 8-element ULAA at compression ratio $R = 0.7778$.

Cosine Wave Amplitude A_c	Main Beam Direction θ_o, in °	SLL (dB)	HPBW in °	Maximum Scanning Range
0	90°	−29.79 dB	12.8°	$-27.64^\circ \leq \theta_r \leq 27.64^\circ$
0.25 λ	84.02°	−28.9 dB	12.78°	
0.5 λ	78.35°	−27.14 dB	12.88°	
0.75 λ	72.67°	−24.59 dB	12.98°	
λ	67.52°	−21.74 dB	13.06°	
1.25 λ	62.36°	−19.1 dB	13.03°	
1.5 λ	Grating lobe appearance			
1.75 λ				
2 λ				

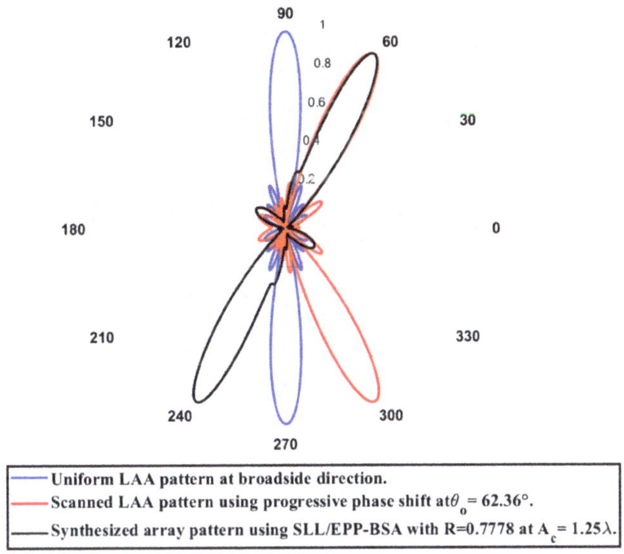

Figure 32. Polar plot of the synthesized radiation pattern using the SLL/EPP-BSA approach at compression ratio $R = 0.7778$ and $A_c = 1.25\ \lambda$.

Test case (4): $R = 0.7$

In this case, the cosine wave half period is chosen to be greater than the array length, $H > L_h$ and equals $H = 7.05\ \lambda$, and the array length equal $L_h = 4.935\ \lambda$, hence $R = 0.7$. Accordingly, after applying the EPPs of the synthesized array elements for A_c changing from positive and the cosine wave amplitude is changing from positive to negative and its magnitude is changing from $A_c = 0$ to $A_c = 1.5\lambda$, the array main beam tilted from the broadside direction of $\theta^\circ = 90^\circ$ to $\theta^\circ = 60.81^\circ$ with minimal changes in the HPBW that is increased by 0.89° at $\theta^\circ = 60.81$ compared to the broadside direction, while SLL is increased from −29.79 dB to −12.68 dB as the main beam is titled away from the broadside direction to $\theta^\circ = 60.81^\circ$. The achieved maximum scanning range is $-29.19^\circ \leq \theta_r \leq 29.19^\circ$ around the broadside direction. The relation between the amplitude changes of the cosine waveform and the scanning angle, HPBW, SLL, and maximum scanning range for the eight-element ULAA at a compression ratio $R = 0.7$ are listed in Table 12. The polar plot of the synthesized radiation pattern using the SLL/EPP-BSA approach at compression ratio $R = 0.7$ and $A_c = 1.5\ \lambda$ is shown in Figure 33.

Table 12. The relation between the amplitude changes of the cosine waveform and the scanning angle, HPBW, SLL, and maximum scanning range θ_r for 8-element ULAA at compression ratio $R = 0.7$.

Cosine Wave Amplitude A_c	Main Beam Direction θ_o, in °	SLL (dB)	HPBW in °	Maximum Scanning Range
0	90°	−29.79 dB	12.8°	$-29.19^\circ \leq \theta_r \leq 29.19^\circ$
0.25 λ	85.05°	−29.28 dB	12.82°	
0.5 λ	79.89°	−28.21 dB	12.95°	
0.75 λ	74.74°	−26.2 dB	13.17°	
λ	70.1°	−23.44 dB	13.42°	
1.25 λ	65.46°	−18.94 dB	13.64°	
1.5 λ	60.81°	−12.68 dB	13.69°	
1.75 λ	Grating lobe appearance			
2 λ				

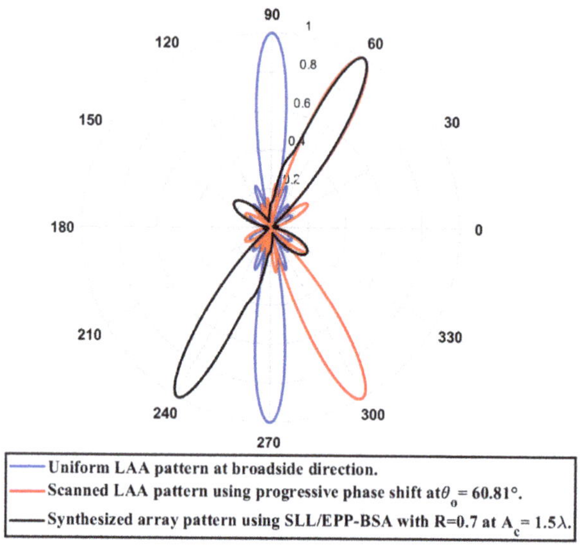

Figure 33. Polar plot of the synthesized radiation pattern using the SLL/EPP-BSA approach at compression ratio $R = 0.7$ and $A_c = 1.5\,\lambda$.

Test case (5): $R = 0.5$

In this case, the cosine wave half period is chosen to be greater than the array length, $H = 9.87\,\lambda$, and the array length equal $L_h = 4.935\,\lambda$, hence $R = 0.5$. Accordingly, after applying the EPPs of the synthesized array elements for A_c changing from positive and the cosine wave amplitude is changing from positive to negative and its magnitude is changing from $A_c = 0$ to $A_c = 2\,\lambda$, the array main beam tilted from the broadside direction of $\theta^\circ = 90^\circ$ to $\theta^\circ = 65.97^\circ$ with minimal changes in the HPBW that is increased by 3.02° at $\theta^\circ = 65.97^\circ$ compared to the broadside direction, while the SLL is increased from −29.79 dB to −27.35 dB as the main beam is titled away from the broadside direction to $\theta^\circ = 65.97^\circ$. The achieved maximum scanning range is $-24.03^\circ \leq \theta_r \leq 24.03^\circ$ around the broadside direction. The relation between the amplitude changes of the cosine waveform and the scanning angle, HPBW, SLL, and maximum scanning range for the eight-element ULAA at a compression ratio $R = 0.5$ are listed in Table 13. The polar plot of the synthesized radiation pattern using the SLL/EPP-BSA approach at compression ratio $R = 0.5$ and $A_c = 2\,\lambda$ is shown in Figure 34.

Table 13. The relation between the amplitude changes of the cosine waveform and the scanning angle, HPBW, SLL, and maximum scanning range θ_r for 8-element ULAA at compression ratio $R = 0.5$.

Cosine Wave Amplitude A_c	Main Beam Direction θ_o, in °	SLL (dB)	HPBW in °	Maximum Scanning Range
0	90°	−29.79 dB	12.8°	
0.25 λ	87.11°	−29.77 dB	12.81°	
0.5 λ	84.02°	−29.82 dB	12.94°	
0.75 λ	80.92°	−29.67 dB	13.16°	
λ	77.83°	−29.05 dB	13.48°	$-24.03^\circ \leq \theta_r \leq 24.03^\circ$
1.25 λ	74.74°	−28.06 dB	13.9°	
1.5 λ	72.16°	−30.35 dB	14.46°	
1.75 λ	69.06°	−28.86 dB	15.18°	
2 λ	65.97°	−27.35 dB	15.82°	

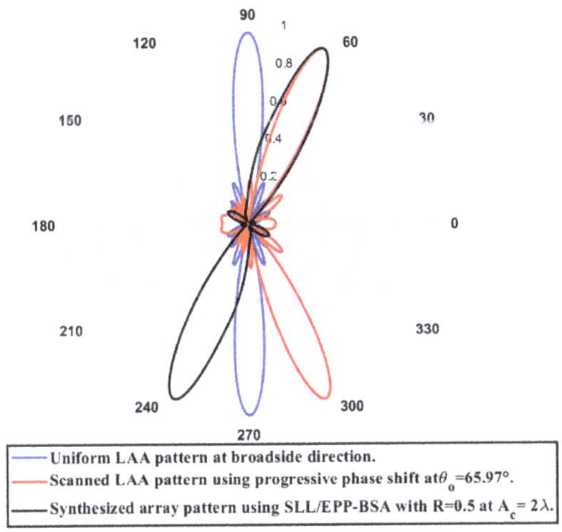

Figure 34. Polar plot of the synthesized radiation pattern using the SLL/EPP-BSA approach at compression ratio $R = 0.5$ and $A_c = 2\ \lambda$.

2.2.3. Comparison between EPP/BSA and SLL/EPP-BSA

Based on prior simulations, we can conclude that the proposed EPP-BSA provides wider scanning ranges than the proposed SLL/EPP-BSA but with much higher SLLs at the endpoints of the scanning ranges. In this regard, the proposed SLL/EPP-BSA may be useful in interference-sensitive applications that require small scanning sectors.

3. Realization of EPP-BSA Using Planar Antenna Array Thinning

It is possible to design a PAA with a large number of antennas packed into a tiny space on a chip to create a system for millimeter-wave applications when the size of the antenna element is very small. Consider the test case of the $M =$ eight-element LAA with uniform element spacing $d_z = 0.5\ \lambda$ and size $L_h = 3.5\ \lambda$, where the cosine wave half period is chosen to be greater than the array size, $H > L_h$ and equals $H = 4\lambda$, and $R = 3.5\ \lambda / 4\ \lambda = 0.875$. For cosine wave amplitude A_c changing from $2\ \lambda$ to $-2\ \lambda$ in the Y-direction, its span will be $2A_c = 4\ \lambda$. The realization of the eight-element LAA sector beam scanning using the proposed EPP-BSA approach can be implemented by constructing a PAA of dimensions $(4\ \lambda \times 3.5\ \lambda)$ in the Y-direction and the Z-direction, respectively. Taking a uniform element spacing in the Y-direction to be $d_y = 0.1\ \lambda$ and a uniform element spacing in the Z-direction to be $d_z = 0.5\ \lambda$, a $\left(M_y \times M_z\right) = (41 \times 8)$ planar antenna array (PAA) can be constructed as shown in Figure 35, where M_y is the number of elements in the Y-direction that can be calculated as follows:

$$M_y = (2A_c \,/\, 0.1\ \lambda) + 1 = (4\lambda \,/\, 0.1\lambda) + 1 = 41\ elements \tag{31}$$

M_z is the number of elements in the Z-direction that can be calculated as follows:

$$M_z = M = 8\ elements \tag{32}$$

Rather than using a rotating motor or progressive phase shift techniques for beam scanning, the proposed PAA configuration allows for precise array thinning to select the desired active "ON" eight elements and turn off the other elements to configure the designed EPP pattern to achieve a specific scan angle with the desired radiation properties. However, there may be some errors due to the mismatch between the positions of the elements of the actual antenna array and the positions of the selected active eight elements on the PAA for the desired EPP pattern. So that a position quantization is performed to select the nearest element on the PAA that matches the corresponding element on the actual antenna array in order to eliminate these inaccuracies as much as feasible. Table 14 shows a comparison between the actual element positions in the Y-direction compared to the corresponding quantized element positions in the Y-direction at fixed element spacing in the Z-direction.

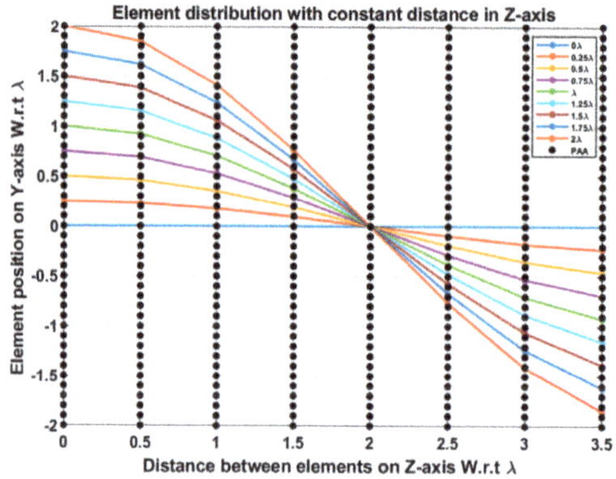

Figure 35. Element distribution of the proposed (41 × 8) PAA configuration and the element distributions of the $M = 8$ elements of the actual array using the proposed EPP-BSA approach at uniform element spacing $d_z = 0.5\ \lambda$ on the Z-axis and A_c changing from positive to negative.

Table 14. A comparison between the actual elements' positions in Y-direction compared to the corresponding quantized elements' positions in Y-direction at fixed element spacing in Z-direction for the realization of EPP-BSA approach using the proposed PAA thinning.

Cosine Wave Amplitude A_c	Element Position on Y-Direction/λ	Number of Antenna Element on Z-Direction							
		1	2	3	4	5	6	7	8
$0\ \lambda$	Exact position	0	0	0	0	0	0	0	0
	Quantized position	0	0	0	0	0	0	0	0
$0.5\ \lambda$	Exact position	0.5	0.46	0.35	0.19	0	−0.19	−0.35	−0.46
	Quantized position	0.5	0.5	0.4	0.2	0	−0.2	−0.4	−0.5
λ	Exact position	1	0.9	0.7	0.4	0	−0.4	−0.7	−0.9
	Quantized position	1	0.9	0.7	0.4	0	−0.4	−0.7	−0.9
$2\ \lambda$	Exact position	2	1.848	1.414	0.7654	0	−0.7654	−1.414	−1.848
	Quantized position	2	1.8	1.4	0.7	0	−0.7	−1.4	−1.8

The proposed PAA thinning technique is superior to utilizing progressive phase shifters to scan the main beam from the broadside to the desired direction since the proposed technique is based on modifying the EPP pattern rather than adjusting the phase shifter. This simplifies the production process while also lowering the cost of the feeding network. Furthermore, the proposed technique eliminates interference and coupling between the main beam and the back lobe at any scanning angle by maintaining a 180 spacing angle between them. However, when utilizing the progressive phase shift technique, the spacing angle between the main beam and the back lobe decreases as the scan angle increases, resulting in increased interference and coupling between them, which indeed increases the HPBW as the main beam approaches the end-fire direction.

The implementation of the proposed (41 × 8) PAA configuration using the CST microwave studio is shown in Figure 36. The array is realized using the dipole antenna illustrated in Figure 22 with resonance frequency $f_o = 1$ GHz. Table 15 shows a comparison between the CST simulation results and MATLAB simulation results of the proposed EPP/BSA approach using the proposed PAA thinning and actual EPP/BSA indicating the resultant scan angle, SLL, and HPBW at $R = 0.875$.

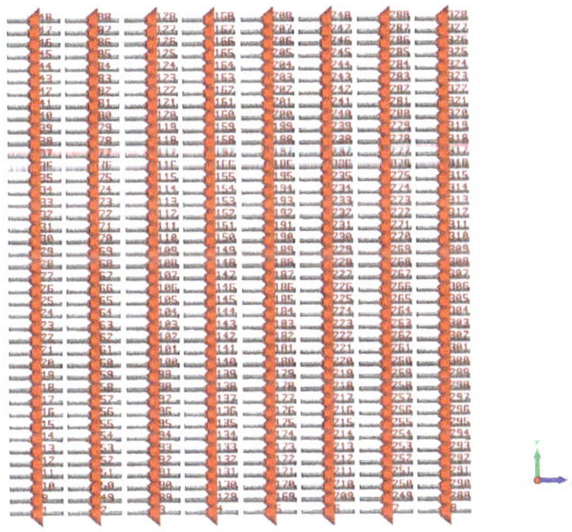

Figure 36. The implementation of the proposed (41 × 8) PAA configuration using CST microwave studio utilizing a dipole antenna with resonance frequency $f_o = 1$ GHz.

Table 15. A comparison between the CST simulation results and MATLAB simulation results of the proposed EPP/BSA approach using the proposed PAA thinning and actual EPP/BSA indicating the resultant scan angle, SLL, and HPBW at $R = 0.875$.

Cosine Wave Amplitude A_c	CST Simulation Results Using PAA Thinning			CST Simulation Results Using Actual EPP/BSA			MATLAB Simulation Results Using Actual EPP/BSA		
	Scan Angle	HPBW	SLL	Scan Angle	HPBW	SLL	Scan Angle	HPBW	SLL
0 λ	0°	12.6°	−13.6 dB	0°	12.3°	−13 dB	0°	12.8°	−12.8 dB
0.5 λ	17°	12.2°	−6.5 dB	17°	12°	−7.5 dB	16.81°	12.4°	−7.88 dB
λ	30°	10.9°	−2.8 dB	31°	11.2°	−4.1 dB	31.76°	11.2°	−4.917 dB
2 λ	50°	8.7°	−1.1 dB	51°	9.1°	−2.1 dB	51.36°	9.2°	−2.668 dB

Figure 37 shows the CST simulated 3D radiation patterns using the PAA thinning for implementation of the proposed EPP-BSA approach for the eight-element ULAA at $R = 0.875$, $A_c = 0$, $A_c = 0.5\ \lambda$, $A_c = \lambda$, $A_c = 2\ \lambda$, and changing from positive to negative. While the corresponding polar plots are shown in Figure 38.

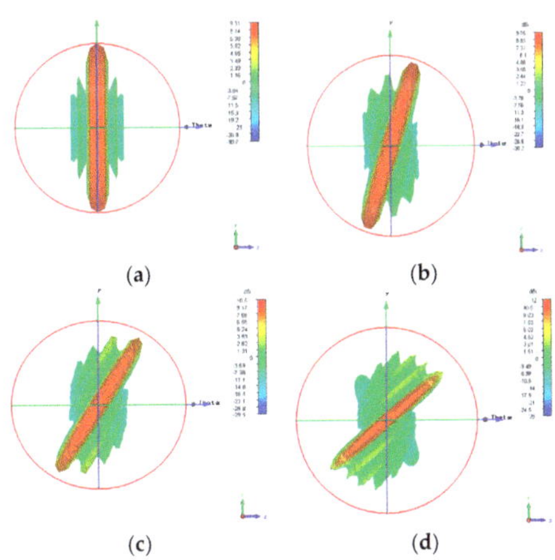

Figure 37. CST-simulated 3D radiation patterns using the PAA thinning for implementation of the proposed EPP-BSA approach for 8-element ULAA at $R = 0.875$: (**a**) $A_c = 0$, (**b**) $A_c = 0.5\ \lambda$, (**c**) $A_c = \lambda$, (**d**) $A_c = 2\ \lambda$, and changing from positive to negative.

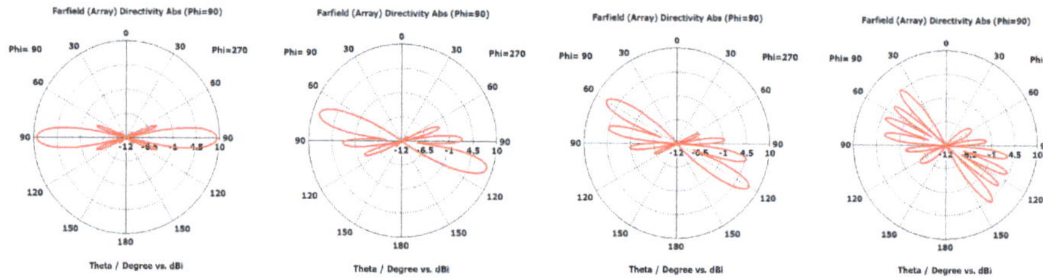

(a) (b) (c) (d)

Figure 38. CST simulated polar plots of the radiation patterns using the PAA thinning for implementation of the proposed EPP-BSA approach for 8-element ULAA at $R = 0.875$: (**a**) $A_c = 0$, (**b**) $A_c = 0.5\,\lambda$, (**c**) $A_c = \lambda$, (**d**) $A_c = 2\,\lambda$, and changing from positive to negative.

4. Comparison with State-of-the-Art Works

In this section, the proposed beam steering techniques are compared with the state-of-the-art works. The comparison is performed in terms of the main concept of the utilized technique, beam steering range, maximum SLL, HPBW, and the significant remarks (drawbacks and advantages) as mentioned in Table 16.

Table 16. Comparison with state-of-the-art works.

Ref.	Beam Steering Range (BSR)	Maximum SLL	HPBW
[19]	Relatively continuous beam steering from $-50°$ to $50°$ by switching eight different sampling states of the switching mechanism.	Poor SLL performance as the realized SLL of the array is higher than the ideal case due to the phase quantization errors that reaches 22.5° in this 3-bit phase shifter.	Whenever the main beam's direction changes, the HPBW correspondingly changes.
[20]	Switched beam steering at four distinct angles : $(-41°, -12°, +15°, \text{and} + 48°)$.	High SLL within the range from -9 dB to -4 dB.	Whenever the main beam's direction changes, the HPBW correspondingly changes.
[21]	Switched beam steering at four distinct angles : $(+10°, -38°, +38°, \text{and} - 10°)$.	Good SLL performance that equal $(-13\text{ dB}, -8\text{ dB}, -8\text{ dB}, -13\text{ dB}, \text{and} - 6\text{ dB})$ at the directions $(+10°, -38°, +38°, \text{and} - 10°)$, respectively.	The HPBW is changed by 4.1° from 23.62° to 27.51° when the main beam is steered from $+10°$ to $+38°$ or from $-10°$ to $-38°$, respectively.
[22]	Continuous beam steering from $-30°$ to $30°$.	Maximum SLL equals -9.2 dB	The HPBW is changed by less than 1°.
[23]	Continuous beam steering from $-20°$ to $20°$.	Not available	Not available
[24]	Switched beam steering from $-19°$ to $19°$.	Not available	Not available
This work	Continuous beam steering from $-51.36°$to$51.36°$.	High SLL $= -2.668$ dB at the edges of the steering range.	The HPBW is decreased by 3° at the edges of the steering range.
	Continuous beam steering from $-30.73°$ to $30.73°$.	Good SLL < -10 dB over the entire steering range.	Constant HPBW over the entire steering range.

In [19], the main beam steering is performed using a three-bit phase shifter. The steering network was formed from a uniform eight-branch power divider with each terminal feeding a microstrip line equipped with a switching mechanism that enabled three-bit periodic phase shifting for beam steering. The antenna array elements are fed by the network via coupled feeding. While the main drawbacks of the work are: (i) A complicated feeding network where the progressive phase shift of each antenna element is needed to be quantized and mapped from 0° to $-360°$ in the eight different states of the individual 3-bit phase shifter. (ii) Each three-bit phase shifter requires eight PIN diodes, DC coupling capacitors, and a DC biasing network that complicates the design and increases the feeding network cost. (iii) The realized SLL of the array is higher than the ideal case due to the phase quantization errors that reaches 22.5° in this quantized three-bit phase shifter. (v) In contrast to the ideal situation, when phase quantization was not used, this phase inaccuracy is the reason for the main beam direction mismatch. (iv) The array gain changes with the change in the main beam direction.

In [20], a switched beam antenna array has been introduced. In order to produce four distinct main beam directions, the antenna's phase shifter made use of a 4×4 planar Butler matrix with phase variations of ±135° and ±45° between its outputs. While the main drawbacks of the work are: (i) It did not offer continuous beam scanning; it just offered switched beams in limited and distinct directions. (ii) The beamwidth of the four main beams ranges from 38.5° to 55.4° in the H-plane and 26° to 31.8° in the E-plane, respectively. (iii) The proposed method suffers from significant HPBW broadening. (v) The feeding network design is complicated. In [21], the authors presented a switched beam antenna array. The main beam steering is performed using a 4 × 4 Butler matrix phase-shifter. While the main drawbacks of the work are: (i) It did not offer continuous beam scanning; it just offered switched beams in limited and distinct directions. (ii) The feeding network design is simple.

In [22], beam steering has been performed by utilizing a metasurface lens array that is fed by a phased array with fewer phase shifters. While the main drawbacks of the work are: (i) The design of the antenna array is complicated. (ii) It requires phase shifters. (iii) The integration of the lens array and the phased array is a cumbersome process and greatly affects the impedance matching of the array system. In [23], analog beam steering has been performed based on Huygens' metasurface pixel technique without using external phase shifters to the array elements. The required phase shifts between antenna elements were generated by integrating conventional probe-fed microstrip patch antenna arrays with anisotropic Huygens metasurface. In [24], the grid array antenna structure has been used to introduce switched beam steering. By switching the excitation of the elements at various locations, the array main beam can be steered, while it suffers from limited accuracy.

5. Conclusions

In this paper, two sector beam scanning approaches based on the EPP technique, denoted as the EPP-BSA and the SLL/EPP-BSA, are introduced. Based on prior simulations, we can conclude that the proposed EPP-BSA provides wider scanning ranges than the proposed SLL/EPP-BSA but with much higher SLLs at the endpoints of the scanning range. In this regard, the proposed SLL/EPP-BSA may be useful in interference-sensitive applications that require small scanning sectors. The proposed techniques are based on modifying the EPP pattern rather than using phase shifters while maintaining the elements excitations. This simplifies the production process and lowers the cost of the feeding network. Furthermore, the proposed technique eliminates interference and coupling between the main beam and the back lobe at any scanning angle by maintaining a 180 spacing angle between them. Furthermore, a proposed PAA thinning approach is proposed to realize the proposed EPP-BSA approach. It allows for precise array thinning to select the desired active "ON" antenna elements and turn off the other elements to configure the designed EPP pattern to achieve a specific scan angle with the desired radiation properties. The results are realized using the computer simulation technology (CST) microwave studio software package, which is highly matched to the MATLAB simulations.

Author Contributions: Methodology, A.M.E.; Software, S.I.A.E.; Writing—review & editing, A.H.H.; Supervision, A.E.A.S. All authors have read and agreed to the published version of the manuscript.

Funding: This research received no external funding.

Informed Consent Statement: Not applicable.

Conflicts of Interest: The authors declare no conflict of interest.

References

1. Mailloux, R.J. *Phased Array Antenna Handbook*, 2nd ed.; Artech House: Norwood, MA, USA, 2008.
2. Shaker, J.; Chaharmir, M.R.; Cuhaci, M.; Ittipiboon, A. Reflectarray Research at The Communications Research Centre Canada. *IEEE Antennas Propag. Mag.* **2008**, *50*, 31–52. [CrossRef]
3. Mirmozafari, M.; Saeedi, S.; Zhang, G.; Rahmat-Samii, Y. A Crossed Dipole Phased Array Antenna Architecture with Enhanced Polarization and Isolation Characteristics. *IEEE Trans. Antennas Propag.* **2020**, *68*, 4469–4478. [CrossRef]

4. Chahat, N.; Decrossas, E.; Gonzalez-Ovejero, D.; Yurduseven, O.; Radway, M.J.; Hodges, R.E.; Estabrook, P.; Baker, J.D.; Bell, D.J.; Cwik, T.A.; et al. Advanced Cubesat Antennas for Deep Space and Earth Science Missions: A Review. *IEEE Antennas Propag. Mag.* **2019**, *61*, 37–46. [CrossRef]
5. Sadman, A.M.; Hossam-E-Haider, M. Design of a 2 × 3 Microstrip Patch Phased Array Antenna for GNSS Augmentation. In Proceedings of the 2020 23rd International Conference on Computer and Information Technology (ICCIT), Dhaka, Bangladesh, 19 December 2020; pp. 1–6.
6. Hor, M.; Hikage, T.; Yamamoto, M. Leaf-Shaped Bowtie Slot Antenna Array for Application in Beam Scanning. In Proceedings of the 2020 International Symposium on Antennas and Propagation (ISAP), Osaka, Japan, 25–28 January 2021; pp. 777–778.
7. Sayidmarie, K.H.; Shakeeb, A.R. Array Beam Scanning and Switching by Variation of Amplitude-Only Excitations. *Int. J. Electromagn. Appl.* **2017**, *7*, 25–30.
8. Li, Z.; Bahceci, I.; Cetiner, B.A. Broadband beam-steering reconfigurable antenna. *Microwave Opt. Technol. Lett.* **2017**, *59*, 63–65. [CrossRef]
9. Khalat, A.; Towfiq, M.A.; Cetiner, B.A.; Ceylan, O.; Biyikli, N. A 60 GHz beam-steering reconfigurable antenna. In Proceedings of the 2016 IEEE International Symposium on Antennas and Propagation (APSURSI), Fajardo, PR, USA, 28 June–1 July 2016; pp. 1279–1280.
10. Li, Y.; Li, Y. Investigation on SIW Slot Antenna Array with Beam Scanning Ability. *Int. J. Antennas Propag.* **2019**, *2019*, 8293624. [CrossRef]
11. Cheng, X.; Yao, Y.; Tomura, T.; Hirokawa, J.; Yu, T.; Yu, J.; Chen, X. Millimeter-Wave Frequency Beam Scanning Array with A Phase Shifter Based on Substrate-Integrated-Waveguide. *IEEE Access* **2018**, *6*, 47408–47414. [CrossRef]
12. Hong, Y.; Yang, S.; Ma, Y.; Chen, Y.; Qu, S.W. Thinned Planar Array Synthesis Based on Multiagent Genetic Algorithm. In Proceedings of the IEEE MTT-S International Wireless Symposium (IWS), Guangzhou, China, 19–22 May 2019; pp. 1–3.
13. Cai, W.; Ji, L.; Guo, C.; Mei, K.; Zeng, H. Constrained Planar Array Thinning Based on Discrete Particle Swarm Optimization with Hybrid Search Strategies. *Sensors* **2022**, *22*, 7656. [CrossRef] [PubMed]
14. Mohammed, J.R. A Method for Thinning Useless Elements in The Planar Antenna Arrays. *Progress Electromag. Res. Lett.* **2021**, *97*, 105–113. [CrossRef]
15. Gupta, P.; Gupta, V. Thinned 8 × 8 Planar Antenna Array with Reduced Side Lobe Levels For 5g Applications. *Wirel. Personal Commun.* **2021**, *119*, 639–655. [CrossRef]
16. Nath, G.V.; Subhashini, K.R. A Hybrid Discrete Fourier Transform Based Difference Set Approach for Reduction in Peak Sidelobe Level of Planar Antenna Array. *Int. J. RF Microwave Comput.-Aided Eng.* **2022**, *32*, e23381. [CrossRef]
17. Elkhawaga, A.M.; Mahmoud, N.M.; Abd Elnaby, M.M.; Hussein, A.H. New Hybrid Approaches for SLL Reduction and Size Miniaturization of Planar Antenna Arrays Based on 2D Convolution and Genetic Algorithm. *Alex. Eng. J.* **2022**, *61*, 7233–7264. [CrossRef]
18. Balanis, C.A. *Antenna Theory: Analysis and Design*, 3rd ed.; Wiley: New York, NY, USA, 2005.
19. Uddin, M.N.; Tarek, M.N.; Islam, M.K.; Alwan, E.A. A Reconfigurable Beamsteering Antenna Array at 28 GHz Using a Corporate-fed 3-Bit Phase Shifter. *IEEE Open J. Antennas Propag.* **2023**, *4*, 126–140. [CrossRef]
20. Nguyen, T.; Duyen, B.T.; Thao, H.T. A Compact Switched Beam Antenna Array for ISM Band. *J. Sci. Technol. Issue Inform. Commun. Technol.* **2021**, *19*, 1–5. [CrossRef]
21. Reddy, M.H.; Siddle, D.; Sheela, D. Design and Implementation Of A Beam-Steering Antenna Array Using Butler Matrix Feed Network For X-Band Applications. *AEU-Int. J. Electron. Commun.* **2022**, *147*, 154147–154160. [CrossRef]
22. Xu, R.; Chen, Z.N. A Compact Beamsteering Metasurface Lens Array Antenna with Low-Cost Phased Array. *IEEE Trans. Antennas Propag.* **2020**, *69*, 1992–2002. [CrossRef]
23. Nguyen, Q.; Zaghloul, A.I. Design of Beam Steering Patch Arrays Using Self-Phased Metasurface Pixels. In Proceedings of the 2020 IEEE International Symposium on Antennas and Propagation and North American Radio Science Meeting, Montreal, QC, Canada, 5–10 July 2020; pp. 909–910.
24. Qing, X. Switchable Beam Steering Antenna for Ka-Band Airborne Applications. In Proceedings of the 2020 IEEE International Conference on Computational Electromagnetics (ICCEM), Singapore, 24–26 August 2020; pp. 201–202.

electronics

MDPI

Communication

Design of a Crossed Dielectric Resonator-Loaded, Dual-Band Dual-Polarized Differential Patch Antenna with Improved Port Isolation and Gain

Dongdong Wang [1,*], Yudong Liu [2] and Jia Liang [3,4]

1 The 15th Research Institute of China Electronics Technology Group Corporation, Beijing 100076, China
2 China Academy of Launch Vehicle Technology, Beijing 100076, China; heaton02@163.com
3 The Institute of Information and Navigation, Air Force Engineering University, Xi'an 710077, China; sai_liang@126.com
4 The Collaborative Innovation Center of Information Sensing and Understanding, Xi'an 710077, China
* Correspondence: darrenwang1983@163.com

Abstract: To meet the urgent requirement for more channel capacity in modern wireless communication systems, antennas with more operation bands are demanded. However, large amounts of antennas suffer from low radiation gains and low port isolation levels. In view of this, a differentially fed, dual-wideband, dual-polarized patch antenna is proposed in this paper. Compared with conventional crossed-feeding structures, the proposed crossed dielectric resonator (CDR) can provide extra resonances with improved isolation levels and radiation gain. Further, four shorting pins are introduced to the radiating patch to help improve the impedance-matching performance. In addition, the proposed antenna also has a very compact size of $0.46\lambda \times 0.46\lambda \times 0.12\lambda$. Finally, a prototype of the proposed antenna is fabricated to validate the design concept. The measured results show that the proposed antenna generates dual wide bands of 1.86–2.52 GHz and 3.26–3.72 GHz for $|S_{11}| < -10$ dB. High radiation gains of 8.9 ± 0.9 dBi and 10.8 ± 1.2 dBi are also obtained, as well as high port isolation levels of better than 38.4 dB and 36.2 dB at the two bands. The excellent performance of the proposed antenna makes it a promising candidate for 4G/5G wireless communication systems.

Keywords: crossed dielectric resonator; differential antenna; dual-band antenna; dual-polarized antenna; port isolation; radiation gain

Citation: Wang, D.; Liu, Y.; Liang, J. Design of a Crossed Dielectric Resonator-Loaded, Dual-Band Dual-Polarized Differential Patch Antenna with Improved Port Isolation and Gain. *Electronics* **2023**, *12*, 3570. https://doi.org/10.3390/electronics12173570

Academic Editors: Raed A. Abd-Alhameed, Chan Hwang See and Naser Ojaroudi Parchin

Received: 11 July 2023
Revised: 12 August 2023
Accepted: 15 August 2023
Published: 24 August 2023

1. Introduction

As modern wireless communication devices such as base stations get concentrated in crowded spaces, electromagnetic environments get worse, which leads to unexpected interference between different wireless communication systems. One of the solutions is integrating different functions into one wireless communication device. As a result, antennas are demanded to have wider operation bands or multiple operation bands [1–3].

In multi-band antenna designs, parasitic structures [4–7] and slots [8–11] are commonly used to generate multiple operation bands. For example, fan-shaped parasitic patches are introduced to circularly polarized patches [4], and additional resonant mode can be excited at a lower band with linearly polarized characteristics. In [5–7], stacked patches are utilized to generate extra resonant modes for dual-band operation. In [8] and [9], symmetric slots etching at the edges of radiating patches contribute to a higher resonant mode at the upper band. Except for the above designs, multi-band characteristics can also be obtained by loading shorting pins or stubs [11–13]. For example, a dual-band microstrip patch antenna is designed in [12] using four shorting pins and a slot, and the radiation gain reaches 10.2 and 10.0 dBi at the two bands. In recent years, dielectric resonators have been investigated in antenna designs because of their stable, resonant characteristics [14–18].

For example, the higher-order resonance in [14] is obtained by loading copper tape on the surface of a defected DRA. In [15], a coupling slot and a metal strip are used to excite HEM_{111} and HEM_{113} modes with orthogonal polarizations. In [16], higher-order resonant modes are excited using crossed DRA differentially fed by probes, and two operation bands are obtained. In [17], dual-operation bands can also be excited by etching a cylindrical hole in the middle of a cylindrical DRA.

However, very few studies concentrate simultaneously on gain and isolation improvement for dual-band antennas. In view of this, a differentially fed, dual-band, dual-polarized differential patch antenna is developed in this paper. By introducing a crossed dielectric resonator as the feeding structure, extra resonant modes are excited with better impedance matching performance, as well as improved port isolation levels. In addition, the shorted patch antenna can not only excite extra resonances but also work as a 1×2 array at the two bands. Thus, the realized gain in the main radiating direction is relatively higher than in other designs. Measured results show that the proposed antenna has two wide operation bands of 1.86–2.52 GHz and 3.26–3.72 GHz for $|S_{11}| < -10$ dB with high port isolation levels of higher than 38.4 dB and 36.2 dB at the two operation bands. The antenna configuration and detailed design process and analysis are presented in Section 2. The experiment results are listed in Section 3.

2. Antenna Design Process and Analysis

Figure 1 shows the configuration of the dual-band dual-polarized differential antenna. The proposed antenna consists of a CDR with relative permittivity of 9.2, an F4B substrate (the relative permittivity is 2.65, the loss tangent is 0.002), four printed patches, four shorting pins, as well as a metal ground. The shorted patches are printed on the top of the F4B substrate. The CDR is differentially fed by two pairs of coaxial cables with inner conductors of the coaxial cables connected to each edge of the CDR. When port1+ and port1− are differentially fed, vertical polarization is obtained. On the other hand, horizontal polarization can be obtained when port2+ and port2− are differentially excited. Due to the symmetry of the antenna structure, the two differential ports have similar frequency responses. Thus, only the simulated and measured results when port1+ and port1− are differentially fed are displayed in the following analysis.

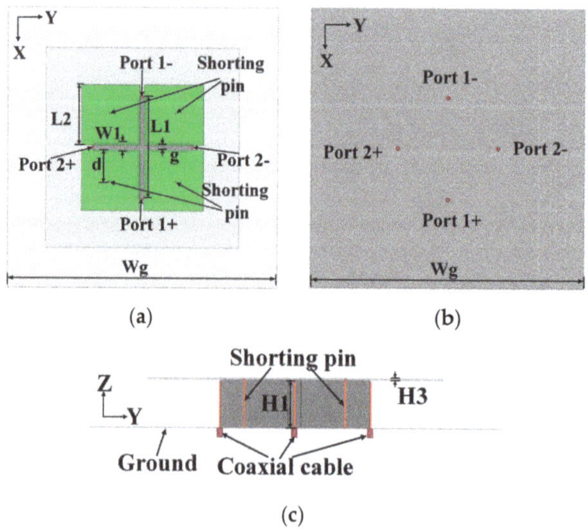

Figure 1. Configuration of the proposed antenna: (**a**) top view, (**b**) back view, and (**c**) front view. The optimized parameters are as follows: L1 = 50.5, L2 = 30.2, W1 = 4.3, g = 2.1, Wg = 140.0, H1 = 15.0, H3 = 0.5 (Unit: mm).

The evolution process of the proposed antenna is given in Figure 2 to show the detailed design concept. Ant.1 is a conventional patch antenna differentially excited by a bent metal strip. In Ant.2, the metal strip is replaced by a CDR. The proposed antenna has four shorting pins loading on the radiating patches of Ant.2.

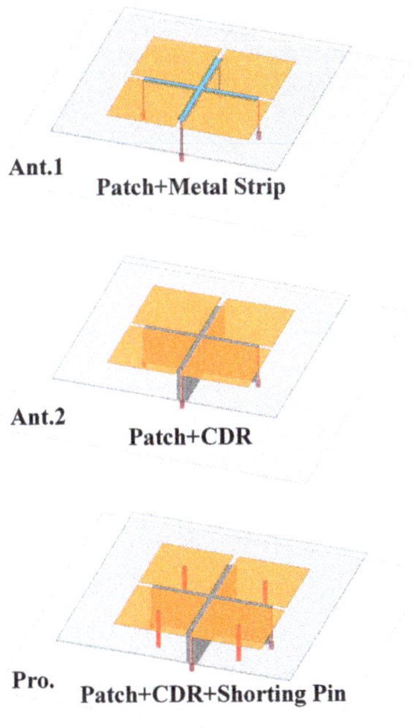

Figure 2. Evolution process of the proposed antenna.

The simulated S-parameters and radiation gains for Ant.1 and Ant.2 are illustrated in Figure 3. When Ant.1 is differentially fed, two resonances are obtained at about 1.92 GHz and 3.96 GHz. The isolation level is better than 37.5 dB and 25.2 dB at the two bands, while the realized gains at the two bands are 8.8 and 9.0 dBi, respectively. When the crossed strip is replaced by the CDR in Ant.2, the original two resonant modes move to lower bands, and another resonant mode is generated at about 2.48 GHz. In addition, the isolation level of Ant.2 is 17.5 dB and 16.9 dB higher than that of Ant.1 at the two bands.

Current distributions on Ant.1 and Ant.2 at the two frequencies are shown in Figure 4 for better demonstration. The dominant current of Ant.1 at 1.92 GHz flows along the edges of radiation patches that are close to the feeding structure. The currents flow along the whole feeding structure at 3.96 GHz, which indicates a poor isolation level between the two differential ports. Additionally, the dominant current path of Ant.1 at 1.92 GHz is about twice the current path at the upper band, which corresponds to two different resonant modes. As the dominant currents flow along X-axis, unidirectional radiation patterns can be obtained for Ant.1 at the upper band. When the crossed metal strip is replaced by the CDR in Ant.2, another resonant mode occurs at 2.48 GHz. The current distributions are like Ant.1 at the lower band, as Figure 4 shows. When differentially fed, the resonating mode for the CDR at 3.64 GHz is fixed along the two ports, and the orthogonal mode can hardly be excited, which leads to isolation improvement between the two pairs of differential ports. Additionally, the effective currents flow along the outer side of the radiating patches, which means that the antenna can work as a 1×2 array, and radiation gains are relatively higher than that of other patch antennas. Meanwhile, as the impedance matching gets better, the peak realized gains at the two bands increase by 1 dB and 1.2 dB, respectively.

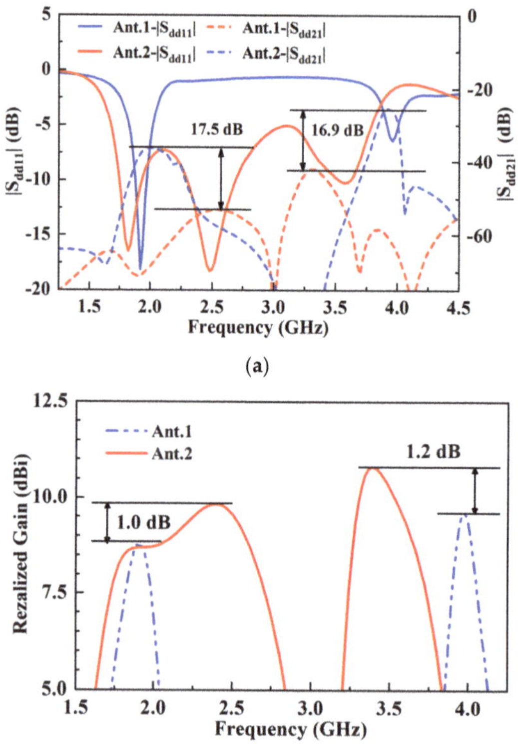

(**b**)

Figure 3. Simulated (**a**) S-parameters and (**b**) realized gains of Ant.1 and Ant.2.

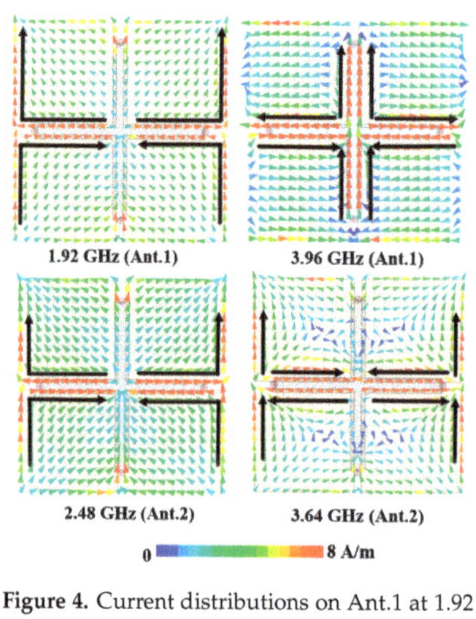

Figure 4. Current distributions on Ant.1 at 1.92 GHz and 3.96 GHz and on Ant.2 at 2.48 GHz and 3.64 GHz.

Figure 5 illustrates the input impedance for Ant.1 and Ant.2. It can be clearly seen from the real part that Ant.1 has two resonances. The four resonances of Ant.2 indicate that the CDR offers one more resonance at the upper band. In view of this, a parameter study is developed to validate the influence of the CDR on the impedance-matching performance of Ant.2, as shown in Figures 6 and 7. It can be inferred that both the radiating patch and the CDR provide one resonance at the each operation band. As the additional resonance at the upper band is introduced by the CDR, the resonant frequency is more sensitive to the length and width of the CDR. When L1 or W1 increases, the resonant mode of the CDR at the upper band moves to low-frequency band and fuses with the patch mode. Thus, the impedance matching performance also gets better as L1 or W1 increase.

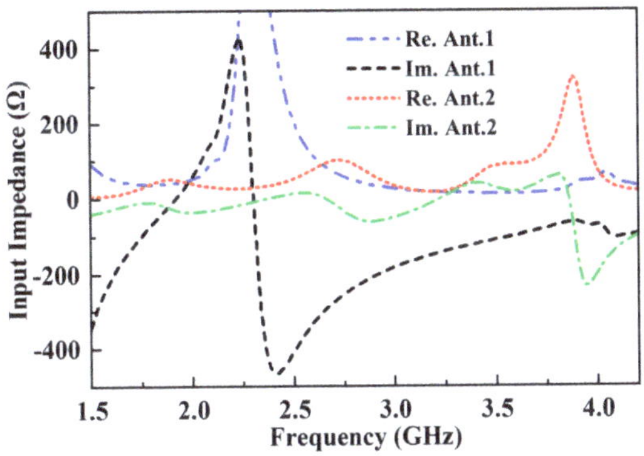

Figure 5. Input impedance for Ant.1 and Ant.2.

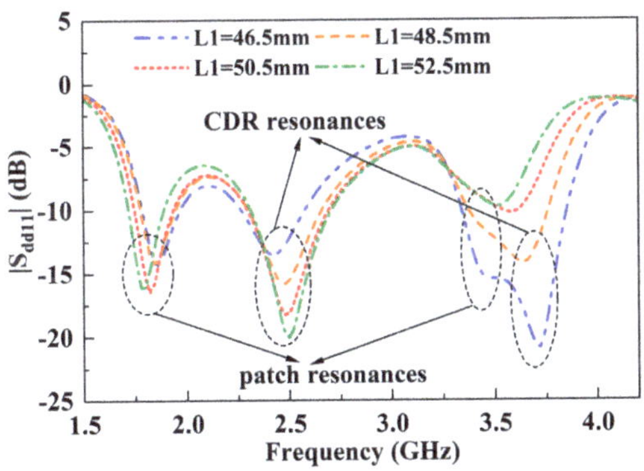

Figure 6. S-parameter for Ant.2 varied with length of the CDR.

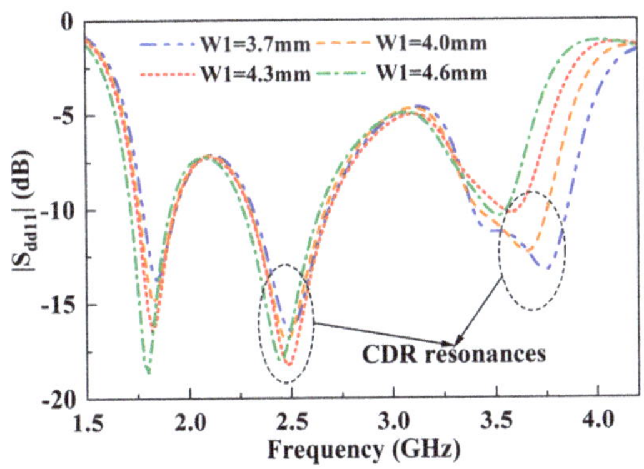

Figure 7. S-parameter for Ant.2 varied with width of the CDR.

However, the impedance-matching performance of Ant.2 is still not satisfied. In view of this, four shorting pins are loaded on the radiating patches, which can effectively tune the impedance to a matching status, as shown in Figure 8. As a result, wideband performance with dual resonances is obtained at the lower and upper bands. Additionally, radiation gains of the proposed antenna increase by 0.3 dB and 1.3 dB at the two bands owing to the improvement of impedance matching performance.

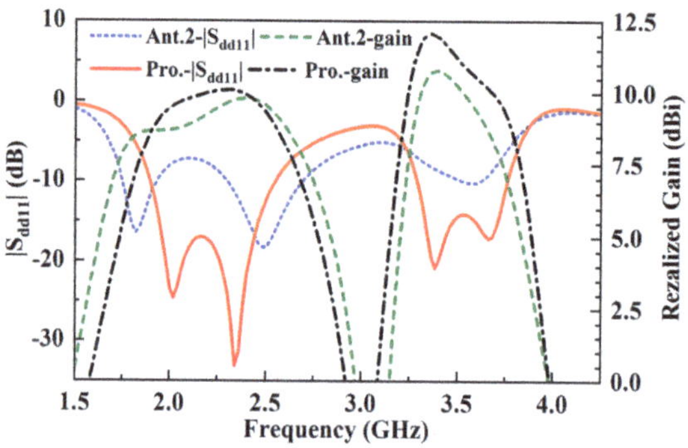

Figure 8. S-parameter and realized gains for Ant.2 and the proposed antenna.

To further clarify the influence of the shorting pins, the input impedance of Ant.2 and the final design are investigated, as shown in Figure 9. The four shorting pins have the function of inductors at the lower band, mainly because the length of the shorting pin is 0.09 wavelength. Thus, the capacitive character of Ant.2 can be neutralized by the shorting pins at the lower band. As a result, the impedance matching performance is improved at the lower band. Additionally, as the dominant current path is increased by the shorting pin, resonance for the patch at the lower band moves to a higher frequency band and fuses with the CDR mode. On the other hand, the shorting pin at the upper band is about 0.2 wavelength, which means that it can be used as an open-ended structure approximately. Thus, the shoring pin presents a capacitive character with a low real part. As Ant.2 shows the inductive imaginary part at the upper band with high impedance, the introduction of four shorting pins can help to tune the impedance to a matching status. At last, the proposed antenna exhibits dual operation bands of 1.89 GHz–2.55 GHz and 3.30 GHz–3.76 GHz for $|Sdd_{11}| < -10$ dB. The realized gains at the two bands are 9.38 ± 0.74 dBi and 10.63 ± 1.44 dBi, respectively. To better understand the influence of shorting pins in the proposed antenna, a parameter study on the position of the shorting pins is made, as Figure 10 shows. As the shorting pin can be regarded as an open-ended structure at the upper band, the dominant current path gets longer at the upper band when d increases, which can be inferred from Figure 4. At the same time, the resonant frequency at the upper band moves towards the CDR mode at about 3.3 GHz. Moreover, as the shorting pin can improve the resonant frequency of patch mode at the lower band, the two resonances get close when d increases. At last, d is chosen to be 15.0 mm as a compromise.

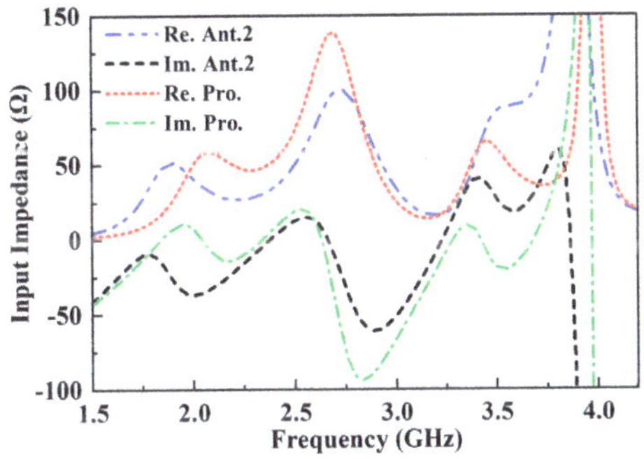

Figure 9. Input impedance for Ant.2 and the proposed antenna.

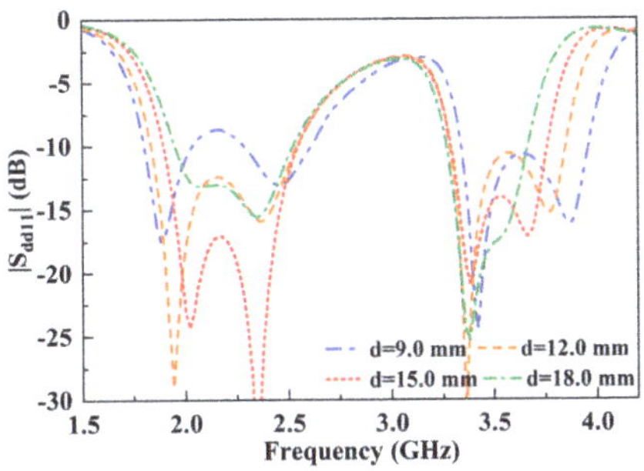

Figure 10. S-parameter for the proposed antenna varied with distance from the shorting pin to patch edge.

3. Experimental Results

A prototype of the proposed dual-band dual-polarized antenna is manufactured and measured, as shown in Figure 11. Two pairs of 3 dB couplers are designed to test the S-parameters and realized gains of the two ports at both lower and upper bands. Thus, the difference between simulated and measured results mainly comes from manufacturing and testing errors, as well as the introduction of 3 dB couplers.

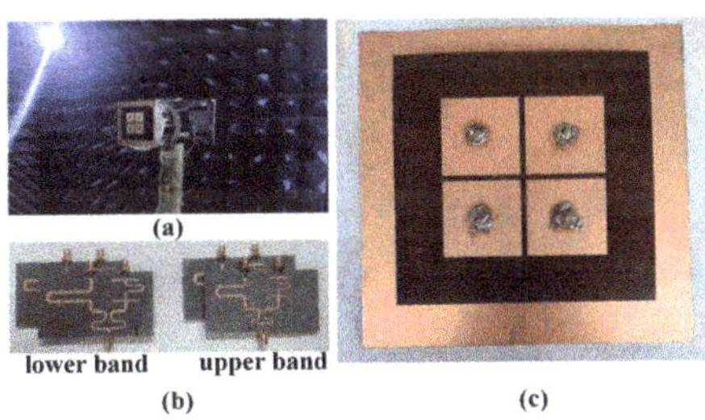

Figure 11. (a) testing environment, (b) 3 dB couplers, and (c) prototype of the proposed antenna.

Figure 12 shows the measured S-parameters of the proposed antenna. The proposed antenna operates at 1.86 GHz to 2.52 GHz and 3.26 GHz to 3.72 GHz for $|S_{11}| < -10$ dB. Besides, the measured port isolation levels at the two operation bands are higher than 38.4 dB and 36.2 dB, respectively. As Figure 13 shows, the measured radiation gains are 8.9 ± 0.9 dBi and 10.8 ± 1.2 dBi at the two bands, which are close to the simulated results. In addition, insertion loss of the antenna is added due to the introduction of 3 dB couplers. The measured gains are a bit lower than the simulated ones, as well as the radiation efficiency. The measured radiation patterns of the proposed antenna at 2.2 GHz and 3.5 GHz are plotted in Figure 14. Stable and unidirectional radiation patterns can also be observed with low cross-polarization levels at the two working bands.

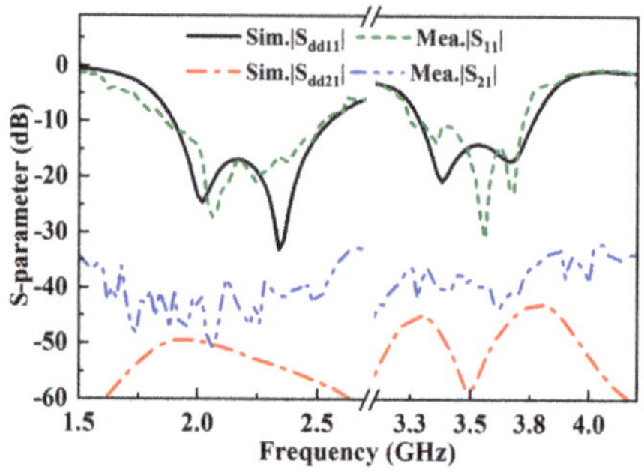

Figure 12. Simulated and measured S-parameters of the proposed antenna.

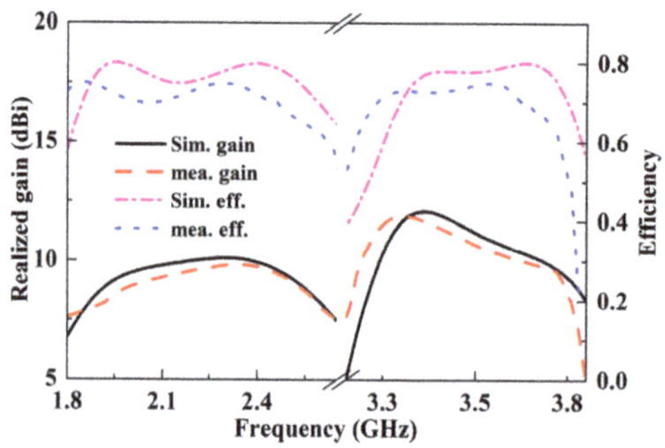

Figure 13. Simulated and measured realized gains and radiation efficiencies of the proposed antenna.

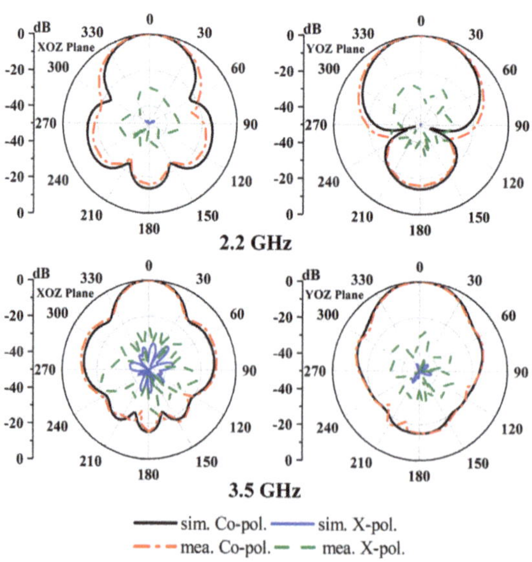

Figure 14. Simulated and measured radiation patterns of the antenna at different frequencies.

4. Discussion

A comparison between the proposed antenna and reference antennas is made, as Table 1 shows (λc is the wavelength at the center frequency of the lower band). The proposed antenna shows a wider operation band than the other references except for [6] at the upper band. Although the proposed antenna has a higher profile than that of [6,7], the realized gains at the two bands are the highest, especially in designs using dielectric resonators. In general, the proposed dual-band dual-polarized antenna reveals a low profile, wide bandwidth, and high radiation gains.

Table 1. Comparison of dual-band antennas.

Ref.	Radiator Size (λ_c)	Bandwidth	Polar	Peak Gain (dBi)	Iso. (dB)
[6]	0.29 × 0.29 × 0.097	22%/19.3%	RHCP/LP	8.5/10.8	>40.2/>32.8
[7]	0.48 × 0.48 × 0.085	11.4%/5.1%	±45°/±45°	8.3/10.5	>40/-
[8]	0.49 × 0.49 × 0.13	12.3%/7.6%	±45°/±45°	8.7/9.8	>37/>38
[12]	0.138^2 × 0.025	1.6%/2.2%	LP/LP	10.2/10.0	-/-
[15]	0.136^2 × 0.31	18.38%/7.01%	LP/LP	6.44/8.98	>36/-
[17]	0.21^2 × 0.17	10.1%/3.7%	LP/LP	1.18/0	-/-
Pro.	0.46 × 0.46 × 0.12	30.1% 13.2%	VP/HP VP/HP	9.8 12.0	>38.4 >36.2

5. Conclusions

In this paper, a differentially fed, dual-band, dual-polarized patch antenna is proposed. By introducing CDR as the feeding structure of the patch antenna, two additional resonances are obtained, as well as high port isolation levels and gains. Further, the impedance matching performance is greatly enhanced by loading four shorting pins. Measured results show that the proposed antenna has dual wideband from 1.86 GHz to 2.52 GHz and 3.26 GHz to 3.72 GHz for $|S_{11}| < -10$ dB, which can cover B1/B34/B39/B40 band for 4G wireless communication and N1/N7/N25/N34/N38/N39/N40/N41/N78/N84 band for 5G wireless communication. In addition, the measured gains at the two operation bands are 8.9 ± 0.9 dBi and 10.8 ± 1.2 dBi, respectively, which is 1–2 dB higher than that of conventional dual-band dual-polarized antenna designs. Moreover, the radiator has a very compact size of 0.46λ × 0.46λ × 0.12λ, which indicates a promising candidate for 4G/5G wireless communication systems, such as base stations.

Author Contributions: Methodology, D.W.; software simulation and validation, D.W. and Y.L.; fabrication and testing, J.L.; writing—original draft preparation, D.W. All authors have read and agreed to the published version of the manuscript.

Funding: This research received no external funding.

Data Availability Statement: All data have been included in the study.

Conflicts of Interest: The authors declare no conflict of interest.

References

1. Li, Y.; Zhao, Z.; Tang, Z.; Yin, Y. Differentially-Fed, Wideband Dual-Polarized Filtering Antenna With Novel Feeding Structure for 5G Sub-6 GHz Base Station Applications. *IEEE Access* **2019**, *7*, 184718–184725. [CrossRef]
2. Shi, C.; Zou, J.; Gao, J.; Liu, C. Gain Enhancement of a Dual-Band Antenna with the FSS. *Electronics* **2022**, *11*, 2882. [CrossRef]
3. Wen, L.H.; Gao, S.; Luo, Q.; Yang, Q.; Hu, W.; Yin, Y.; Ren, X.; Wu, J. A Wideband Differentially Fed Dual-Polarized Antenna with Wideband Harmonic Sup-pression. *IEEE Trans. Antennas Propag.* **2019**, *67*, 6176–6181. [CrossRef]
4. Lin, J.; Qian, Z.; Cao, W.; Shi, S.; Wang, Q.; Zhong, W. A Low-Profile Dual-Band Dual-Mode and Dual-Polarized Antenna Based on AMC. *IEEE Antennas Wirel. Propag. Lett.* **2017**, *16*, 2473–2476. [CrossRef]

5. Zhu, X.-Q.; Guo, Y.-X.; Wu, W. Miniaturized Dual-Band and Dual-Polarized Antenna for MBAN Applications. *IEEE Trans. Antennas Propag.* **2016**, *64*, 2805–2814. [CrossRef]
6. Wang, K.; Liang, X.; Zhu, W.; Geng, J.; Li, J.; Ding, Z.; Jin, R. A Dual-Wideband Dual-Polarized Aperture-Shared Patch Antenna with High Isolation. *IEEE Antennas Wirel. Propag. Lett.* **2018**, *17*, 735–738. [CrossRef]
7. Feng, B.; Li, L.; Cheng, J.-C.; Sim, C.-Y. A Dual-Band Dual-Polarized Stacked Microstrip Antenna with High-Isolation and Band-Notch Characteristics for 5G Microcell Communications. *IEEE Trans. Antennas Propag.* **2019**, *67*, 4506–4516. [CrossRef]
8. Li, Y.; Zhao, Z.; Tang, Z.; Yin, Y. Differentially Fed, Dual-Band Dual-Polarized Filtering Antenna with High Selectivity for 5G Sub-6 GHz Base Station Applications. *IEEE Trans. Antennas Propag.* **2019**, *68*, 3231–3236. [CrossRef]
9. Cui, Y.; Wang, X.; Shen, G.; Li, R. A Triband SIW Cavity-Backed Differentially Fed Dual-Polarized Slot Antenna for WiFi/5G Applications. *IEEE Trans. Antennas Propag.* **2020**, *68*, 8209–8214. [CrossRef]
10. Liu, N.-W.; Zhu, L.; Liu, Z.-X.; Liu, Y. Dual-Band Single-Layer Microstrip Patch Antenna with Enhanced Bandwidth and Beamwidth Based on Reshaped Multiresonant Modes. *IEEE Trans. Antennas Propag.* **2019**, *67*, 7127–7132. [CrossRef]
11. Olan-Nuñez, K.N.; Murphy-Arteaga, R.S. Dual-Band Antenna on 3D-Printed Substrate for 2.4/5.8 GHz ISM-Band Applications. *Electronics* **2023**, *12*, 2368. [CrossRef]
12. Zhang, X.; Zhu, L. Dual-Band High-Gain Differentially Fed Circular Patch Antenna Working in TM11 and TM12 Modes. *IEEE Trans. Antennas Propag.* **2018**, *66*, 3160–3165. [CrossRef]
13. Liu, Z.-X.; Zhu, L.; Liu, N.-W. Dual-Band Dual-Mode Patch Antenna with High-Gain and Wide-Beam Radiations in Two Respective Bands. *IEEE Trans. Antennas Propag.* **2021**, *69*, 8058–8068. [CrossRef]
14. Feng, L.Y.; Leung, K.W. Wideband Dual-Frequency Antenna with Large Frequency Ratio. *IEEE Trans. Antennas Propag.* **2019**, *67*, 1981–1986. [CrossRef]
15. Sun, Y.X.; Leung, K.W. Dual-Band and Wideband Dual-Polarized Cylindrical Dielectric Resonator Antennas. *IEEE Antennas Wirel. Propag. Lett.* **2013**, *12*, 384–387. [CrossRef]
16. Tang, H.; Chen, J.-X.; Yang, W.-W.; Zhou, L.-H.; Li, W. Differential Dual-Band Dual-Polarized Dielectric Resonator Antenna. *IEEE Trans. Antennas Propag.* **2017**, *65*, 855–860. [CrossRef]
17. Hu, P.F.; Pan, Y.M.; Leung, K.W.; Zhang, X.Y. Wide-/Dual-Band Omnidirectional Filtering Dielectric Resonator Antennas. *IEEE Trans. Antennas Propag.* **2018**, *66*, 2622–2627. [CrossRef]
18. Khan, S.; Khan, M.A.; Anab, M.; Marwat, S.N.K.; Jan, N.; Ghoniem, R.M. Wideband Singly Fed Compact Circularly Polarized Rectangular Dielectric Resonator Antenna for X-Band Wireless Applications. *Electronics* **2022**, *11*, 3281. [CrossRef]

electronics

MDPI

Article

W-Band Broadband Circularly Polarized Reflectarray Antenna

Zhicheng Wang [1,2], Rui Zhang [1,*], Wenke Song [1,2], Bingchuan Xie [1,2], Xiaobo Lin [1,2], Haixuan Li [1,2] and Lu Tian [1,2]

1 Aerospace Information Research Institute, Chinese Academy of Sciences, Beijing 100190, China; wangzhicheng21@mails.ucas.ac.cn (Z.W.); songwenke19@mails.ucas.ac.cn (W.S.); xiebingchuan@163.com (B.X.); linxiaobo20@mails.ucas.ac.cn (X.L.); lihaixuan22@mails.ucas.ac.cn (H.L.); tianlu22@mails.ucas.ac.cn (L.T.)

2 The School of Electronic, Electrical and Communication Engineering, University of Chinese Academy of Sciences, Beijing 100408, China

* Correspondence: ruizhang@mail.ie.ac.cn

Abstract: We propose a W-band circularly polarized reflectarray antenna in this article, which contains a single-layer reflectarray and a linearly polarized horn feed. To realize the proposed antenna, we designed a novel W-band multi-resonant element containing a Rogers RT5880 substrate with copper patches printed on its both surfaces. Then, a circular reflectarray with 12.25 λ_0 (39.1 mm) aperture diameter was designed based on the proposed multi-resonant element, whose center frequency is 94 GHz. We further fabricated the proposed circular reflectarray and tested its performance. In the measured results, we can see that the obtained 1 dB gain bandwidth is 19.1% (91~109 GHz) and the obtained 2 dB gain bandwidth can reach as wide as 27.6% (89~115 GHz). Moreover, the 3 dB axial ratio bandwidth is 13.8% (89~102 GHz). The measured gain of our proposed reflectarray antenna at 94 GHz is 29.1 dBi and the corresponding aperture efficiency can reach as high as 52.0%. Those results show that our proposed antenna may be prospective in wireless communication applications due to its strengths in broadband and high aperture efficiency.

Keywords: W-band; reflectarray; broadband; circularly polarized

Citation: Wang, Z.; Zhang, R.; Song, W.; Xie, B.; Lin, X.; Li, H.; Tian, L. W-Band Broadband Circularly Polarized Reflectarray Antenna. *Electronics* **2023**, *12*, 3849. https://doi.org/10.3390/electronics12183849

Academic Editor: Dimitra I. Kaklamani

Received: 12 July 2023
Revised: 18 August 2023
Accepted: 7 September 2023
Published: 12 September 2023

1. Introduction

In the 1960s, Berry et al. firstly proposed the concept of the reflectarray antenna [1]. However, due to the large volume and heavy weight, the reflectarray antenna at that time was not highly valued. This was until the 1970s, with the development of printed circuit board (PCB) technology, when scholars began to use printed microstrip elements produced by PCB technology to design reflectarray antennas. In 1978, C.S. Malagisi firstly proposed the use of microstrip patch elements as phase modulation units for reflectarray antennas, which have advantages such as light weight, low profile, and easy conformability [2]. In 1987, R.E. Munson applied for a US patent for a microstrip reflectarray antenna, which gave rise to the concept of the microstrip reflectarray antenna (RA) [3].

RAs, as substitutes for traditional reflectors, have received widespread attention in recent years. Many studies have investigated the design of circularly polarized reflectarrays [4–19]. The advantages of circularly polarized electromagnetic waves include the ability to overcome multipath effects and the interference of rain and fog and to solve polarization matching problems. Therefore, circularly polarized antennas are widely used in deep space exploration, remote sensing and telemetry, satellite communication, and other fields.

According to the type of feed, circularly polarized reflectarray antennas (CPRAs) can be divided into two categories: CPRAs with circularly polarized (CP) feeds or with linearly polarized (LP) feeds. CPRAs with CP feeds are mostly based on rotating elements [4–8], and the required phase compensation is obtained by changing the rotation angle of the rotating elements. However, as the frequency of electromagnetic waves increases, especially when working in the millimeter wave, terahertz, or higher bands, the difficulty in designing and processing CP feeds increases sharply. Therefore, scholars have begun to study CPRAs with LP feeds. Multiple types of elements have been reported for CPRAs with LP feeds, including square patch and cross slot dual-layer elements [9], dual-layer rectangular patch elements [10], dual-layer T-shaped elements [11], split square ring elements [12], dual-layer Jerusalem-shaped elements [13], and double-layered elliptical microstrip patch elements [14].

The state of the art on CPRAs has made significant steps in recent years, such as dual-band CPRAs [15–17], circularly polarized reflect-transmit-array antennas [18], and beam-scanning CPRAs [19]. However, due to the high machining accuracy required for W-band RAs, there are only a few studies on W-band CPRAs, and most of them remain in the simulation stage without fabrication and measurement. Therefore, the study of W-band CPRAs is of great significance.

In our study, a W-band single-layer broadband circularly polarized reflectarray antenna was proposed, which contains a low-profile reflectarray and a linearly polarized horn antenna. Firstly, we designed and analyzed a single-layer multi-resonant element for our proposed CPRA, including a Rogers RT5880 substrate with copper patches printed on both its surfaces. The simulation results showed that elements can independently control the reflected wave phase in two orthogonal directions by changing the corresponding parameters and can achieve a smooth phase shift range of over 360° in both orthogonal directions. Then, we designed and manufactured a planar circular reflectarray consisting of 437 elements and further verified of our proposed element.

2. The Design of the Reflectarray Element

The reflectarray of CPRAs with an LP feed needs the ability to convert LP incident waves into CP reflected waves. Based on electromagnetic theory, CP waves are composed of two spatially orthogonal LP waves with equal amplitudes and phase differences of $\pi/2$. Therefore, the designed reflectarray should be able to generate different reflective phases for LP incident waves in two orthogonal directions to obtain a 90° phase difference between the two orthogonal parts of the reflected wave to obtain CP waves. Meanwhile, each reflectarray element is necessary to bring the appropriate reflective phase to compensate the spatial phase delay, so as to imitate the ability of the traditional reflector.

Among the basic structures of reflectarray elements, the dipole has the ability to change the reflective phase of the LP incident wave in the corresponding direction by adjusting the length, while almost having no effect on the reflective phase of the LP incident wave in its orthogonal direction. Meanwhile, the structure of the dipole is simple, small, and easy to integrate into multi-resonant structures. Thus, the dipole has advantages in the high-precision processing of W-band antennas. The combination of dipoles can be used to design W-band circularly polarized reflectarray elements.

As shown in Figure 1, a reflectarray element is formed by four dipoles. The l_x and l_y can be independently changed, so the element is not a completely symmetrical structure. The substrate material is Rogers RT5880, with a thickness of 0.508 mm.

The simulation results of this element, where the incident wave is an LP wave in the X direction, are shown in Figure 2. It can be seen that for an LP incident wave paralleled to the X axis, the phase shift of the reflected wave is more sensitive to the changes in the parameter l_x, while the change in the parameter l_y has little effect on its reflective phase. We can assume that the reflective phase of the LP incident wave paralleled to the X axis is only determined by l_x. Similarly, the reflective phase of the LP incident wave paralleled to the Y axis is only determined by l_y. Therefore, the reflective phases of the LP incident waves paralleled to the X axis or Y axis can be independently obtained by changing the values of l_x or l_y.

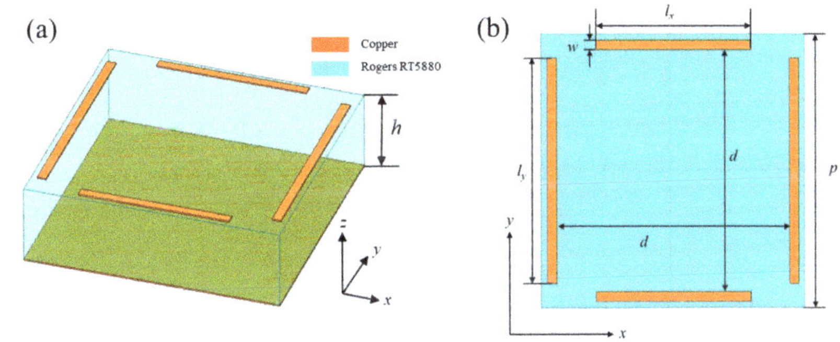

Figure 1. The structure of the element formed by four dipoles. (**a**) Perspective view, (**b**) top view.

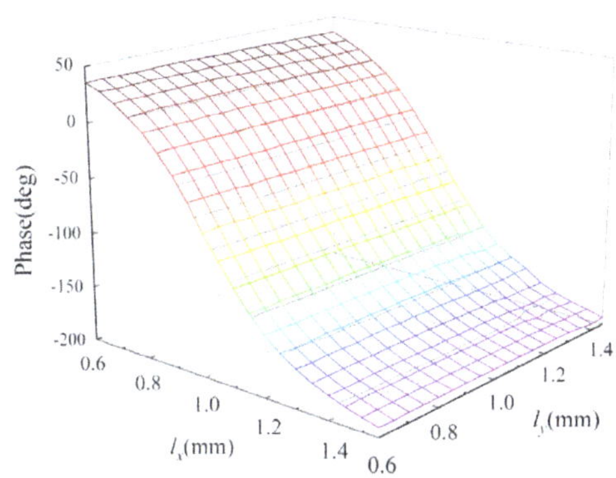

Figure 2. Simulation results of the mentioned element with E_i^x at 94 GHz.

However, this single-resonant element cannot meet the demands of designing RAs because its reflective phase shift range is only about 250°. To change the resonance characteristic of the element, four additional dipoles were added into our proposed element to obtain a single-layer multi-resonant element.

Figure 3 shows the structure of our proposed single-layer multi-resonant element. It contains a dielectric substrate (Rogers RT5880, blue part) with metal patches (copper, orange parts) printed on its both surfaces. To show three typical resonant forms of the element more intuitively, the simulated surface current distributions of the multi-resonant element with E_i^x at three different frequency points are shown in Figure 4. When the frequency is 84 GHz, the surface current of the outer dipole is strong, while the inner dipole is weak. When the frequency is 94 GHz, the surface current distribution intensity of the inner and outer dipoles is equivalent. When the frequency is 104 GHz, the surface current of the outer dipole is weak, while the inner dipole is strong. We can see the variation in surface current distributions with different frequencies. As the frequency increases, the surface current of the outer dipole gradually decreases, while the surface current of the inner dipole gradually increases.

In the realization of reflectarrays, element spacing is an important factor. Too large element spacing will lead to the generation of grid lobes in the antenna pattern. We can use Equation (1) to calculate the proper element spacing p so as to avoid this problem [20]:

$$p \leq \frac{\lambda_0}{1+\sin\theta} \tag{1}$$

where λ_0 is the electromagnetic wave length at the center frequency and θ is the maximum angle of incidence from the feed.

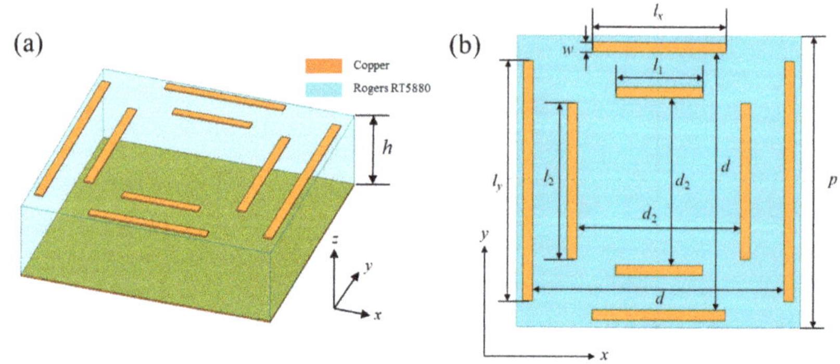

Figure 3. (**a**) Perspective view and (**b**) top view of the single-layer multi-resonant element.

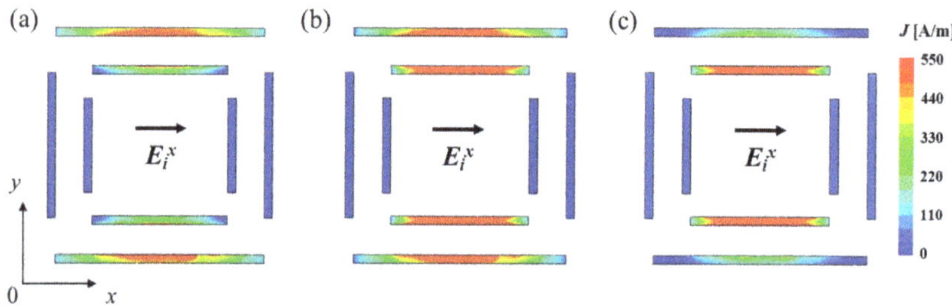

Figure 4. Simulated surface current distributions of the multi-resonant element with E_i^x at (**a**) 84 GHz, (**b**) 94 GHz, and (**c**) 104 GHz.

With the center frequency of 94 GHz, considering the limitation in the range of phase shift and practical manufacture, we chose an element spacing of 0.53 λ_0. Therefore, the parameter p is 1.7 mm.

Microstrip reflectarrays are fabricated by PCB technology. The minimum line spacing is an important limiting factor in the process of PCB technology. In addition, the thicknesses of the boards used in the fabrication process have some fixed standards. We chose a substrate with a thickness of 0.508 mm. Then, we proposed a coefficient k, where d_2 is equal to $k \times d$, l_1 is equal to $k \times l_x$, and l_2 is equal to $k \times l_y$, and further optimized the structure parameters by using simulation software. To design the reflectarray elements, two requirements should be met. Firstly, the phase shift range of the element needs to exceed 360°. Moreover, the phase shift curve needs to be flat and have good linearity, which can effectively expand the bandwidth of the antenna. In our proposed element, the parameter k is a significant structure parameter because it can affect the coupling between the outer and inner dipoles. The parameter k could affect the distance and strength of the resonance points of the proposed element. With a decrease in parameter k, the distance between the inner and outer dipoles increases, so that the coupling effect weakens. It also leads to the increase in the spacing between resonance points and the decrease in intensity of the resonance points. The parameter w is the width of each dipole and it also has a significant impact on the reflection performance of the element. With a decrease in parameter w, the dipoles become thinner, so that the position of each resonance point moves towards the right in the coordinate axis. Meanwhile, the strength of resonance points increases, but the distance between the resonance points does not change much. The parameter d has little effect on the position of the resonance point, it only slightly affects the strength of the resonance point. However, if the parameter d is too small, it will affect the allowable size range of l_x or l_y. If the parameter d is too large, it will exacerbate the coupling between the adjacent elements, resulting in a change in the actual reflection phase shift of the element and further causing the decreased gain of the antenna. Therefore, in the design of multi-resonant broadband reflectarray elements, the selection of the parameter d should be moderate. We can see the optimized parameters of the single-layer multi-resonant element in Table 1.

Table 1. The dimensions of the single-layer multi-resonant element.

Parameter	Value	Parameter	Value
p	1.7 mm	w	0.06 mm
d	1.5 mm	d_2	0.65 d
l_x	0.6~1.532 mm	l_1	0.65 l_x
l_y	0.6~1.532 mm	l_2	0.65 l_y

The master–slave boundary method with Floquet port excitations is a commonly used method to build and simulate the reflectarray element. Figure 5 shows the reflective phase shift performance of the proposed reflectarray element with E_i^X. It can be seen that for incident waves paralleled to the X axis, the phase shift of the reflected wave is more sensitive to the changes in the parameter l_x, while the change in the parameter l_y has little effect on its reflective phase. As shown in Figure 6a, when l_y is 1 mm and l_x varies from 0.600 to 1.532 mm, at the center frequency of 94 GHz, the phase shift range of the reflected wave is larger than 360° and the amplitude is larger than −0.34 dB. The frequency sweep results ranging from 84 to 110 GHz are shown in Figure 6b. We can see that when the frequency changes, the reflection characteristics of the element also change. The reflection phase shift range of the element is positively correlated with the frequency. As the frequency decreases, the reflection phase shift curve tends to be flat and the reflection phase shift range decreases. Conversely, the reflection phase shift range increases. However, the whole change is not significant and the reflection phase shift curves are approximately parallel. It means that, within a relatively wide frequency band, the element can provide a reflection phase shift that is not significantly different from that required. Our proposed element has good broadband performance due to its approximately identical phase compensation characteristic at different frequencies.

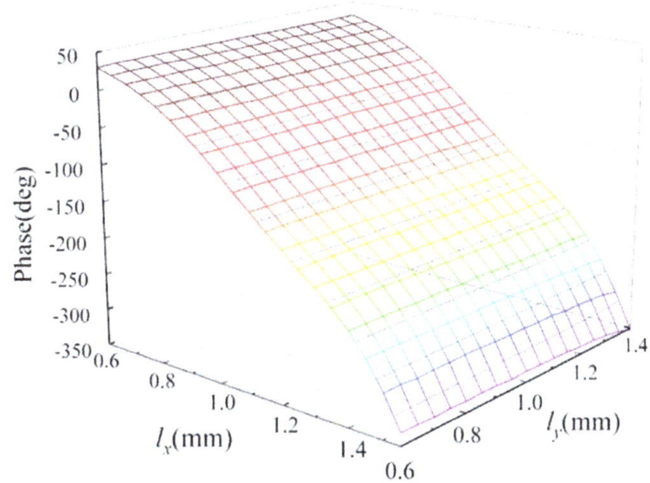

Figure 5. The reflective phase shift characteristics of our proposed element with E_i^X at 94 GHz.

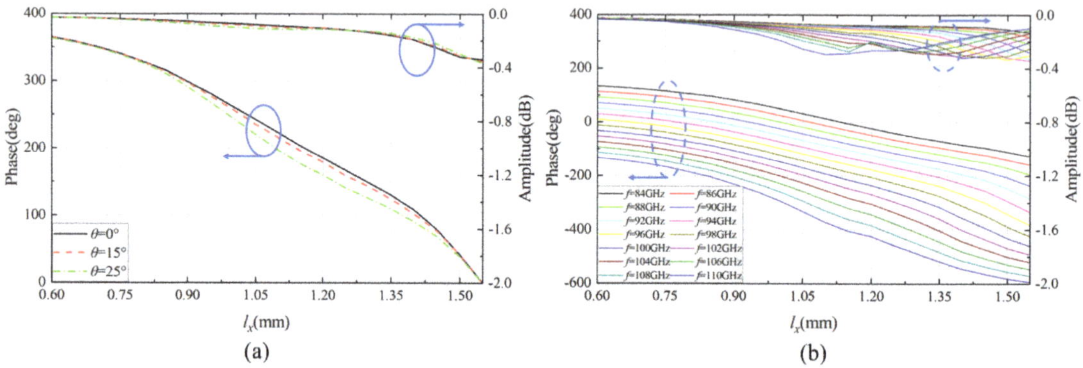

Figure 6. The characteristics of our proposed element when l_y is 1 mm. (**a**) Different incidence angles. (**b**) Different frequencies. (The blue arrow shows the coordinate axis which is applicable to the curve in the blue circle).

3. The Design and Realization of RA

Choosing optimum phase distribution in the reflectarray aperture is the key to obtaining an RA with high gain and aperture efficiency. We can use Equation (2) to obtain the optimum value of required phase ϕ_i of each element in the coordinate plane across the reflectarray:

$$\phi_i = k(R_i - \vec{r_i} \times \hat{r_0}) + \phi_0 \tag{2}$$

where k is the wavenumber, R_i is the distance between the phase center and the ith element, $\vec{r}_i$ is the position vector of the ith element, $\hat{r}_0$ is the unit vector in the main beam direction, and ϕ_0 is the relative phase [21].

To design the aperture phase distribution, firstly, we should choose the proper focal distance F and aperture size D. The focus-to-diameter ratio (F/D) is a key factor to obtain a higher aperture efficiency of the RA. In our research, the feed we chose is a horn antenna with a 17 dBi gain at the center frequency of 94 GHz. Based on reflectarray theory, spillover efficiency η_s and illumination efficiency η_i are the most important factors affecting the value of aperture efficiency η_a. And the increased F/D value leads to lower spillover efficiency η_s and higher illumination efficiency η_i [22]. Therefore, the proper F/D value can be chosen as 1.116, which means that the maximum angle of incidence from the feed is about 24.13°, as shown in Figure 7. To ensure a sufficiently high gain of the antenna, we chose the planar circular reflectarray containing 437 elements; thus, the aperture diameter D is 12.25 λ_0 (39.1 mm) and the focal distance F is 47.43 mm.

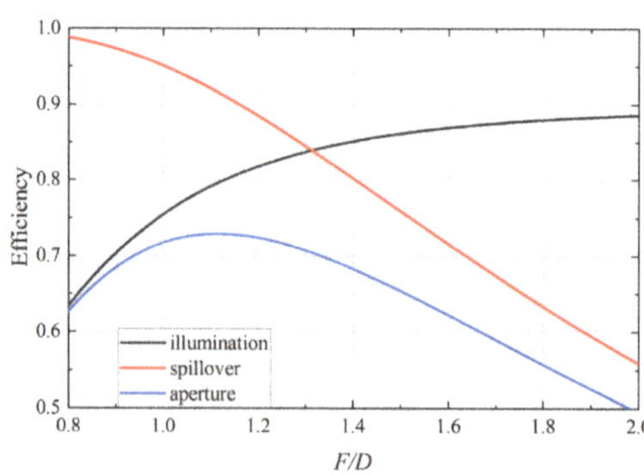

Figure 7. The efficiencies versus the value of F/D.

After determining the F/D value, the optimum required phases of each element in the coordinate plane across the reflectarray can be calculated by Equation (2). The required compensation phases of the X-direction incident wave differ by 90° from the Y-direction incident wave, as shown in Figure 8, because the reflectarray needs the ability to convert LP incident waves into CP reflected waves.

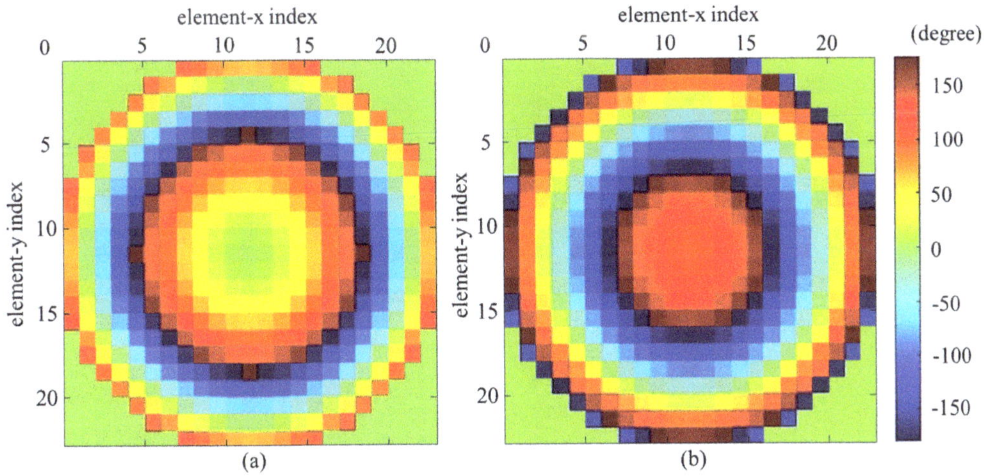

Figure 8. The required compensation phase of (**a**) the X-direction incident wave and (**b**) the Y-direction incident wave.

Then, we simulated, fabricated, and tested our proposed RA. To test the gain performance of RAs, a far-field measurement system with a 1 GHz frequency resolution was used. As shown in Figure 9a, the reflectarray and the feed horn are fixed by the plastic bracket and then connected to the frequency multiplier and the vector network analyzer.

Figure 10 shows a good consistency between the simulated and measured results at the three different frequency points. We can see that, at the center frequency of 94 GHz, the value of the measured side-lobe in the E plane is as low as −16.2 dB and in the H plane it is −15.7 dB, and the value of measured cross-polarization is as low as −17.4 dB. Moreover, the maximum measured side-lobe is −14.5 dB in the H plane at 89 GHz. The larger connecting flange and the occlusion by metal waveguides are significant reasons leading to the differences between the radiation patterns of simulated and fabricated RAs.

Moreover, Figure 11 shows the result of the axial ratio and antenna gain of our proposed RA. At 94 GHz, the measured gain reached 29.1 dBi and the corresponding aperture efficiency reached 52.0%. The 1 dB gain bandwidths and 2 dB gain bandwidths of the designed RA are 19.1% and 27.6%, respectively. The 3 dB axial ratio bandwidth is 13.8%, ranging from 89 GHz to 102 GHz. We found that the small differences between the antenna gain and axial ratio of our proposed RA may be caused by the feed blocking and the discrepancies between the fabricated and simulated reflectarrays. The PCB process used in our proposed reflectarray has a fabrication tolerance of −0.006~0 mm. Under the condition of a manufacturing error of −0.006 mm, the average phase compensation error of the element is 2.30° and the max error is about 5.04°. Due to the phase compensation errors, the gain of the RA will decrease and the axial ratio will also change.

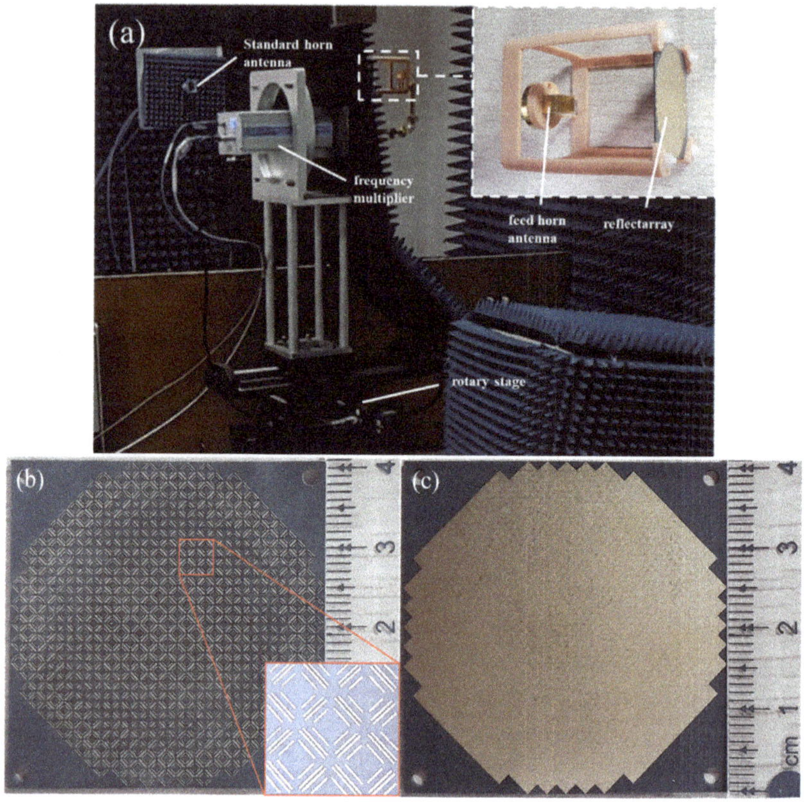

Figure 9. (**a**) Measured setup of the proposed reflectarray antenna. (**b**) Top and (**c**) bottom layer of the fabricated antenna sample.

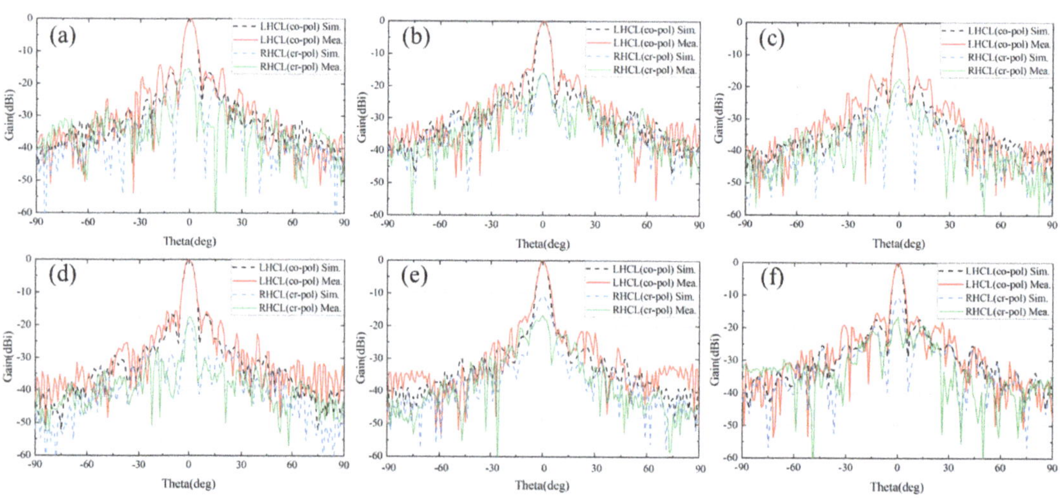

Figure 10. Simulated and measured radiation patterns of the proposed RA. (**a**) E plane and (**b**) H plane at 89 GHz; (**c**) E plane and (**d**) H plane at 94 GHz; (**e**) E plane and (**f**) H plane at 102 GHz.

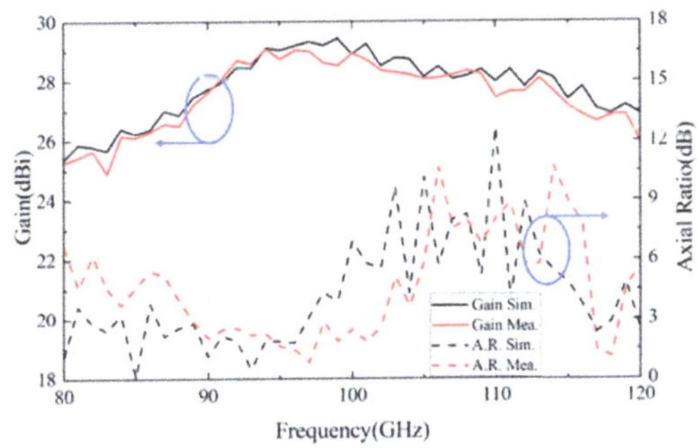

Figure 11. Measured and simulated antenna gain or axial ratio within frequency band of 80–120 GHz. (The blue arrow shows the coordinate axis which is applicable to the curve in the blue circle).

A comparison between our work and related RAs in the W-band or higher bands is shown in Table 2. Our proposed CPRA converts the LP wave emitted by the feed into the CP wave. The incident of our feed is at a 45° angle to the coordinate axis. The incident wave is divided into two orthogonal components and then converted into a CP wave by changing the phase difference between the two components. Therefore, this type of CPRA has a gain 1.5 dBi lower than the LPRA with same aperture and feed, and the aperture efficiency is correspondingly much lower. Moreover, due to the high machining accuracy required for W-band RAs, there are a few studies about W-band CPRAs, and most of them remain in the simulation stage without fabrication and measurement. Considering this characteristic, our proposed CPRA has undergone processing and measuring. It has advantages in the comprehensive performance of the polarization mode, aperture efficiency, and bandwidth.

Table 2. Comparison of related RAs in W-band or higher bands.

	This Work	[23]	[24]	[22]	[25]	[26]
f_0 (GHz)	94	88	94	94	90	165
Polarization	CP	CP	CP	LP	LP	LP
Feed polarization	LP	LP	CP	LP	LP	LP
Gain (dBi)	29.1	28.4	30.6	31.1	28.5	33.2
Aperture Size (λ_0)	12.25 (circular)	17.6 (circular)	16.14 (rectangular)	13.78 (rectangular)	12 (circular)	23.76 (circular)
AE (%)	52	22.6	45.56	53.4	55.3	32.5
1 dB gain BW (%)	19.1	6.81	--	25.5	28.8	24.2
2 dB gain BW (%)	27.6	--	22.22	--	--	--
3 dB AR BW (%)	13.8	6.81	22.22	--	--	--
Measured (Y/N)	Y	N	N	Y	Y	Y

4. Conclusions

In this article, a W-band circularly polarized reflectarray antenna is proposed. This antenna is composed of a single-layer planar reflectarray fed with a linearly polarized horn antenna. The single-layer planar reflectarray has a 39.1 mm diameter circular aperture containing 437 elements. And we further fabricated and tested it to verify its performance. Our measurement results indicate that the measured peak gain of our proposed RA can reach 29.1 dBi and the corresponding aperture efficiency is 52.0%. The 1 dB gain bandwidth and the 2 dB gain bandwidth are as wide as 19.3% (91–109 GHz) and 27.6% (89–115 GHz), respectively, and the 3 dB axial ratio bandwidth is 13.8% (89~102 GHz). These results show that our proposed antenna may be prospective in wireless communication applications due to its strengths in broadband and high aperture efficiency.

Author Contributions: Conceptualization, Z.W.; methodology, Z.W. and W.S.; validation, Z.W.; writing—original draft preparation, Z.W.; writing—review, R.Z, W.S., B.X., X.L., H.L. and L.T.; supervision, R.Z.; funding acquisition, R.Z. All authors have read and agreed to the published version of the manuscript.

Funding: This research was funded by the National Magnetic Confinement Fusion Energy Research and Development Program of China, grant number 2018YFE0305100.

Data Availability Statement: All data have been included in the paper.

Conflicts of Interest: The authors declare no conflict of interest.

References

1. Berry, D.; Malech, R.; Kennedy, W. The reflectarray antenna. *IEEE Trans. Antennas Propag.* **1963**, *AP-11*, 645–651. [CrossRef]
2. Malagisi, C.S. Microstrip disc element reflectarray. *Electron. Aerosp. Syst. Conv.* **1978**, *12*, 186–192.
3. Munson, R.E.; Haddad, H.; Hanlen, J. Microstrip Reflectarray Antenna for Satellite Communication and RCS Enhancement or Reduction. U.S. Patent 4684952, 4 August 1987.
4. Strassner, B.; Han, C.; Chang, K. Circularly polarized reflectarray with microstrip ring elements having variable rotation angles. *IEEE Trans. Antennas Propag.* **2004**, *52*, 1122–1125. [CrossRef]
5. Han, C.; Chang, K. Ka-band reflectarray using ring elements. *Electron. Lett.* **2003**, *39*, 491–493. [CrossRef]
6. Yu, A.; Yang, F.; Elsherbeni, A.Z.; Huang, J. An X-band circularly polarized reflectarray using split square ring elements and the modified element rotation technique. In Proceedings of the 2008 IEEE Antennas and Propagation Society International Symposium, San Diego, CA, USA, 5–11 July 2008; pp. 1–4.
7. Yu, A.; Yang, F.; Elsherbeni, A.Z.; Huang, J. Design and measure-ment of a circularly polarized Ka-band reflectarray antenna. In Proceedings of the 2009 3rd European Conference on Antennas and Propagation, Berlin, Germany, 23–27 March 2009; pp. 2769–2773.
8. Zhao, M.-Y.; Zhang, G.-Q.; Lei, X.; Wu, J.-M.; Shang, J.-Y. Design of new single-layer multiple-resonance broadband circularly polarized reflectarrays. *IEEE Antennas Wirel. Propag. Lett.* **2013**, *12*, 356–359. [CrossRef]
9. Chaharmir, M.R.; Shaker, J.; Cuhaci, M.; Sebak, A. Circularly polarised reflectarray with cross-slot of varying arms on ground plane. *Electron. Lett.* **2002**, *38*, 1492–1493. [CrossRef]
10. Wu, Z.H.; Zhang, W.X.; Liu, Z.G.; Shen, W. Circularly polarized reflectarray with linearly polarised feed. *Electron. Lett.* **2005**, *41*, 387–388. [CrossRef]
11. Ren, L.-S.; Jiao, Y.-C.; Li, F.; Zhao, J.-J.; Zhao, G. A dual-layer T-shaped element for broadband circularly polarized reflectarray with linearly polarized feed. *IEEE Antennas Wirel. Propag. Lett.* **2011**, *10*, 407–410.
12. Yu, A.; Yang, F.; Elsherbeni, A.Z.; Huang, J.; Kim, Y. Anoffset-fed X-band reflectarray antenna using a modified element rotation technique. *IEEE Trans. Antennas Propag.* **2012**, *60*, 1619–1624. [CrossRef]
13. Zheng, Q.; Guo, C.; Qu, Y.; Ding, J. Design of broadband circularly polarized reflectarray with linearly polarized feed. In Proceedings of the 2016 Asia-Pacific Microwave Conference (APMC), New Delhi, India, 5–9 December 2016; pp. 1–3.
14. Lee, S.R.; Lim, E.H.; Lo, F.L.; Ng, W.H. Circularly Polarized Elliptical Microstrip Patch Reflectarray. *IEEE Antennas Wirel. Propag. Lett.* **2017**, *65*, 4322–4327. [CrossRef]
15. Yang, S.; Yan, Z.; Cai, M.; Li, X. Low-Profile Dual-Band Circularly Polarized Antenna Combining Transmitarray and Reflectarray for Satellite Communications. *IEEE Trans. Antennas Propag.* **2022**, *70*, 5983–5988. [CrossRef]
16. Jiang, P.; Jiang, W.; Hu, W.; Gong, S. A Mesh-Type Shared-Aperture Dual-Band Circularly Polarized Transmit-Reflect-Array Antenna. *IEEE Trans. Antennas Propag.* **2023**, *71*, 1590–1601. [CrossRef]
17. Mei, X.; Gu, P.; Tao, X.; Ding, D. Single-Layer Dual-Band Circularly Polarized Reflectarray Antenna. In Proceedings of the 2022 International Applied Computational Electromagnetics Society Symposium (ACES-China), Xuzhou, China, 9–12 December 2022; pp. 1–3. [CrossRef]
18. Feng, J.; Yan, Z.; Yang, S.; Fan, F.; Zhang, T.; Liu, X.; Zhao, X.; Chen, Q. Reflect–Transmit-Array Antenna with Independent Dual Circularly Polarized Beam Control. *IEEE Antennas Wirel. Propag. Lett.* **2023**, *22*, 89–93. [CrossRef]
19. Al-Nuaimi, M.K.T.; Mahmoud, A.; Hong, W.; He, Y. Design of Single-Layer Circularly Polarized Reflectarray with Efficient Beam Scanning. *IEEE Antennas Wirel. Propag. Lett.* **2020**, *19*, 1002–1006. [CrossRef]
20. Josefsson, L.; Persson, P. *Conformal Array Antenna Theory and Design*; John Wiley & Sons: Hoboken, NJ, USA, 2006.
21. Song, W.; Xue, Q.; Cai, Y.; Guo, N.; Liu, K.; Li, S.; Ding, H. A Single-Layer Reflect-Transmit-Array Antenna with Polarization-Dependent Operation. *IEEE Access* **2021**, *9*, 167928–167935. [CrossRef]
22. Wang, Z.; Zhang, R.; Song, W.; Lin, X.; Xie, B.; Wang, J.; Zhao, R. W-Band Single-Layer Broadband Reflectarray Antenna. *IEEE Access* **2023**, *11*, 66309–66317. [CrossRef]
23. Sun, Y.; Qi, S.-S.; Huang, H.; Zhou, T.; Wu, W. Broadband Single-Layer Circularly Polarized Reflectarray Antenna at W Band. In Proceedings of the 2019 International Conference on Microwave and Millimeter Wave Technology (ICMMT), Guangzhou, China, 19–22 May 2019; pp. 1–3.

24. Al-Nuaimi, M.K.T.; Huang, G.-L. Design of CP Reflectarray for 80–100 GHz Band Based on Pancharatnam-Berry Phase Theory. In Proceedings of the 2021 IEEE MTT-S International Microwave Workshop Series on Advanced Materials and Processes for RF and THz Applications (IMWS-AMP), Chongqing, China, 15–17 November 2021; pp. 185–186. [CrossRef]
25. Choi, E.-C.; Nam, S. W-Band Low Phase Sensitivity Reflectarray Antennas with Wideband Characteristics Considering the Effect of Angle of Incidence. *IEEE Access* **2020**, *8*, 111064–111073. [CrossRef]
26. Miao, Z.-W.; Hao, Z.-C.; Yuan, Q. Design and Implementation of a G-Band Silicon Based Single-Layer Reflectarray Antenna. *IEEE Antennas Wirel. Propag. Lett.* **2017**, *16*, 2191–2194. [CrossRef]

electronics MDPI

Article

Optimal Pattern Synthesis of Linear Array Antennas Using the Nonlinear Chaotic Grey Wolf Algorithm

Kunxia Zhao, Yan Liu * and Kui Hu

School of Physics and Electronic Information, Yunnan Normal University, Kunming 650000, China; kunxiazhao@163.com (K.Z.); 18468288015@163.com (K.H.)
* Correspondence: liuyan1@ynnu.edu.cn

Abstract: The grey wolf optimization (GWO) algorithm is a new nature-inspired meta-heuristic algorithm inspired by the social hierarchy and hunting behavior of grey wolves. In this paper, the GWO algorithm is improved to overcome previous shortcomings of being easily trapped in local optima and having a low convergence rate. The proposed enhancement of the GWO algorithm utilizes logistic-tent double mapping to generate initialized populations, which enhances its global search capability and convergence rate. This improvement is called the nonlinear chaotic grey wolf optimization (NCGWO) algorithm. The performance of the NCGWO algorithm was evaluated with four representative benchmark functions. Then, the NCGWO algorithm was applied to perform an optimal pattern synthesis of linear array antennas (LAAs) using two distinct approaches: optimizing the amplitudes of the antenna currents while preserving uniform spacing and optimizing the positions of the antennas while assuming uniform excitation. To validate the effectiveness of the proposed approach, the results obtained by the NCGWO algorithm were compared with those obtained by other intelligent algorithms. Additionally, the NCGWO algorithm was applied to a more complex planar antenna array to further validate its performance. Our results demonstrate that the NCGWO algorithm exhibits superior performance regarding electromagnetic optimization problems compared to widely recognized algorithms.

Keywords: nonlinear chaotic grey wolf optimization (NCGWO) algorithm; logistic-tent double mapping; array antennas; pattern synthesis

Citation: Zhao, K.; Liu, Y.; Hu, K. Optimal Pattern Synthesis of Linear Array Antennas Using the Nonlinear Chaotic Grey Wolf Algorithm. *Electronics* **2023**, *12*, 4087. https://doi.org/10.3390/electronics12194087

Academic Editors: Raed A. Abd-Alhameed, Athanasios D. Panagopoulos, Chan Hwang See and Naser Ojaroudi Parchin

Received: 27 July 2023
Revised: 12 September 2023
Accepted: 25 September 2023
Published: 29 September 2023

1. Introduction

Array antennas are indispensable in present-day Wi-Fi verbal exchange systems. Compared to a single antenna, array antennas have greater directivities, greater gains, and higher competencies in beam control and null placement [1], which make them a crucial component in implementing fifth-generation (5G) and satellite communication systems [2]. Since environmental friendliness has become a significant theme of the current era, technological innovation is the ultimate solution to a green and sustainable system. A new track is opened by array antennas. For example, compared with the U.S. Arecibo array, the 500 m Aperture Spherical Radio Telescope of China has a 64% larger aperture, 2.5 times higher sensitivity, and 10 times higher comprehensive performance, and it can detect up to 13.7 billion light years away. Consequently, the efficiency enhancements of array antennas are apparent.

In a broader context, array antenna elements can be structured in various configurations: linear, planar, and three-dimensional. When focusing on the pattern synthesis for linear array antennas (LAAs), there are two different types of LAAs: a uniform array with equal spacing and a nonuniform array. Furthermore, in the realm of planar arrays, the rectangular configuration stands out as a prevalent choice due to its geometric attributes and inherent traits [3]. Placing null placements in particular directions and reducing the sidelobe level (SLL) are essential methods to prevent the interference of array antennas [4].

However, in the pattern synthesis of array antennas, attaining a combination of low SLL, null control, and narrow main beam is challenging due to their inherent mutual constraints. The mutual coupling of array elements can affect the efficiency of array antennas, so the coupling effect between the elements should also be considered in the design to ensure better performance of the array antennas [2,5,6].

Previously, the conventional Taylor's method was employed for pattern synthesis with low SLL [7]. With the development of mathematics, several evolutionary algorithms have proven effective in the electromagnetic field, many of which are applied in the pattern synthesis of array antennas, such as the grey wolf optimization (GWO) algorithm [8], particle swarm optimization (PSO) algorithm [9], biogeography-based optimization (BBO) algorithm [10], firefly algorithm (FA) [11], genetic algorithm (GA) [12], difference evolution (DE) algorithm [13], exponential chaotic differential evolution (ECDE) [14], grasshopper optimization algorithm (GOA) [15], quantum particle swarm optimization (QPSO) algorithm [16], spider monkey optimization (SMO) algorithm [17], mayfly algorithm (MA) [18], wind-driven optimization (WDO) algorithm [19], invasive weed (IWO) algorithm [20], and atom search optimization (ASO) algorithm [21]. These algorithms have demonstrated commendable performance within their capabilities; however, the growing demand for anti-interference in information transmission has underscored the necessity for continuous research in the realm of intelligent antenna optimization. Hence, research on the enhancement of the current and new optimization methods has become active recently.

The GWO algorithm was employed to optimize LAAs to synthesize optimal radiation patterns. The primary objective was to achieve an array pattern with a minimal SLL and place nulls in the specified directions. The GWO algorithm was successfully utilized to accomplish this goal [22]. The Improved Grey Wolf (IGWO) algorithm was previously utilized to optimize a uniform LAA for optimal pattern synthesis [23]. In this study, the original GWO algorithm has been reworked to an NCGWO algorithm, logistic-tent double mapping is adopted to generate initialized populations, and a convergence factor formula of the Gaussian function is used to enhance the overall performance of the GWO algorithm. The work proposes using the NCGWO algorithm for the pattern synthesis of LAAs and planar arrays.

The remainder of this study is structured as follows: Section 2 provides a comprehensive description of the GWO algorithm. Section 3 outlines the enhancements and specific formulations of the NCGWO algorithm. Section 4 compares various algorithms based on test functions. Then, Sections 5 and 6 describe the array factor expressions of array antennas and data results, respectively. Finally, Section 7 draws the conclusion.

2. GWO Algorithm

The GWO algorithm imitates the observed natural hierarchy of leadership and hunting mechanism in grey wolves. Figure 1 shows the hierarchy of grey wolves. There are three steps in hunting: searching for prey, encircling the prey, and attacking the prey.

Figure 1. Hierarchy of grey wolves.

The α wolf is the leader, and its decision making encompasses the hunting behavior, habitat selection, and food distribution. Other wolves obey the orders of the α wolf. The second layer of the pyramid is called the β wolf, which is the successor of the α wolf. The third level is δ, which follows the commands of the α and β wolves. Older α and β wolves also descend to the level of δ wolves. The bottom layer, i.e., the ω wolf, is responsible for balancing the relationships within the population. This hierarchical structure ensures efficient and effective decision making within the pack. Grey wolves are known for their cooperative hunting behavior, where the β wolf and δ wolf engage in pursuing, encircling, and attacking prey under the guidance of the α wolf [23]. Figure 2 illustrates how an ω wolf adjusts its location within a 2D space by following the guidance of the α, β, and δ wolves.

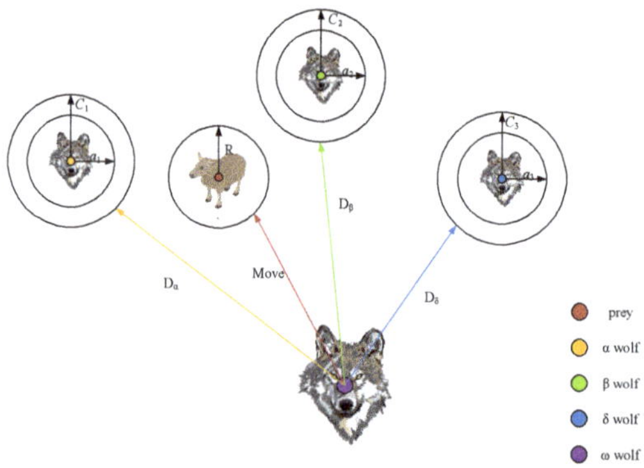

Figure 2. Wolves in 2D space.

2.1. Social Hierarchy

To optimize the search using the GWO algorithm, the α solution is designated as the most appropriate solution, which is followed by the β and δ solutions. Any other potential solution is categorized as an ω solution to align with the progression of the three prior solutions since the search process of the algorithm is guided by them [23].

2.2. Encircling Prey

Upon the detection of a prey ($t = 1$), the first iteration commences. Subsequently, the ω wolves, which are guided by the α, β, and δ wolves, pursue and ultimately surround the prey. The following equations describe the encircling behavior:

$$\vec{D}_\alpha = \left|\vec{C}_1 \bullet \vec{X}_\alpha - \vec{X}(t)\right| \tag{1}$$

$$\vec{D}_\beta = \left|\vec{C}_2 \bullet \vec{X}_\beta - \vec{X}(t)\right| \tag{2}$$

$$\vec{D}_\delta = \left|\vec{C}_3 \bullet \vec{X}_\delta - \vec{X}(t)\right| \tag{3}$$

where $\vec{D}_i(i = \alpha, \beta, \delta)$ is the wolf's distance from the prey; the location vector of the prey is defined by $\vec{X}_i(i = \alpha, \beta, \delta)$, the location vector of a grey wolf at iteration t is $\vec{X}(t)$, and the coefficient vector $\vec{C}_i(i = 1, 2, 3)$ represents random disturbances in the positions of the wolves. These coefficients are calculated as follows:

$$\vec{C}_i = 2\vec{r}_i \tag{4}$$

where $\vec{r}_i$ is a random vector in [0, 1].

2.3. Hunting and Attacking Prey

Grey wolves have a remarkable ability to discern the location of their prey and orchestrating its encirclement. In the pursuit and assault of a prey, the role of the leader is typically assumed by the α, which is occasionally accompanied by the participation of the β and δ. Conventionally, it is widely acknowledged that the α, β, and δ individuals are the most experienced hunters and have superior knowledge regarding potential prey positions. Other members of the wolf pack follow the directives given by the α, β, and δ. As a result, the foremost three optimal solutions are conserved, which compels the remaining search agents to adjust their positions by the coordinates of the leading search agent. To elucidate this concept, the following formula is proposed [8].

The positions of the α, β, and δ wolves are individually computed as follows:

$$\vec{X}_1 = \vec{X}_\alpha - \vec{A}_1 \bullet \vec{D}_\alpha \tag{5}$$

$$\vec{X}_2 = \vec{X}_\beta - \vec{A}_2 \bullet \vec{D}_\beta \tag{6}$$

$$\vec{X}_3 = \vec{X}_\delta - \vec{A}_3 \bullet \vec{D}_\delta \tag{7}$$

where $\vec{X}_i(i = 1,2,3)$ are the positions of the α, β, and δ wolves; $\vec{X}_i(i = \alpha,\beta,\delta)$ is the position vector of the prey; $\vec{A}_i(i = 1,2,3)$ is the vector coefficient; and $\vec{D}_i(i = \alpha,\beta,\delta)$ is the wolf's distance from the prey.

$\vec{A}$ is the vector coefficient; when $|\vec{A}| > 1$, the grey wolf expands its search to find food, which corresponds to a global search; when $|\vec{A}| < 1$, the grey wolf narrows the envelope, which corresponds to a local search. $\vec{A}$. is calculated as follows:

$$\vec{A} = 2a\vec{r} - a \tag{8}$$

where $\vec{r}$ is a vector of random values of 0–1, a is the convergence factor, and a is a linearly decreasing variable that starts at 2 and approaches 0 with each iteration. This is represented in the GWO algorithm as follows:

$$a = 2 - 2\left(\frac{t}{T_{\max}}\right) \tag{9}$$

where t is the number of current iterations, and $T_{\max}$ is the maximum number of iterations. In the original GWO, the positions of the grey population in the next generation can be obtained by the average positions of the α, β, and δ wolves. These are calculated as follows:

$$\vec{X}(t+1) = \frac{\vec{X}_1 + \vec{X}_2 + \vec{X}_3}{3} \tag{10}$$

3. Nonlinear Chaotic Grey Wolf Optimization (NCGWO) Algorithm

The NCGWO algorithm is based on the GWO algorithm with two improvements: the initial population is generated by introducing logistic-tent double mapping and the convergence factor based on the Gaussian function is enhanced. This section describes the improvements and specific expressions of the NCGWO algorithm.

3.1. Logistic-Tent Double Mapping

Logistic mapping is a single-peaked mapping, which perfectly inherits the advantages of randomness and the initial value of sensitivity inherent in chaos [24]. The logistic mapping is expressed as follows:

$$X_{n+1} = UX_n(1 - X_n) \tag{11}$$

where X_n is the iteration result $X_n \in (0,1)$, and U is a parameter $U \in (0,4)$.

The tent chaotic system is a segmented linear one-dimensional mapping with chaotic properties, uniform probability density and power spectral density, and a relatively good autocorrelation [25]. It is mathematically derived as follows:

$$X_{n+1} = \begin{cases} ax_n, 0 \leq x_n < 0.5 \\ a(1-x_n), 0.5 \leq x_n \leq 1 \end{cases} \tag{12}$$

where x is the system variable and $x \in [0,1]$, and a control parameter is between (0 and 1) and it is $a \neq 0.5$ when the system is chaotic.

Compared to logistic mapping, tent mapping has a uniform probability density, a uniform power spectral density, ideal correlation properties, and a faster iteration speed [26].

This study uses the logistic-tent double mapping [27], which combines the complex chaotic dynamics properties of the logistic mapping with a faster iteration speed, greater autocorrelation, and the applicability to many sequences of the tent chaotic system. Its mathematical formulation is as follows:

$$X_{n+1} = \begin{cases} \left[rx_n(1-x_n) + \frac{(4-r)}{2}x_n\right]\mathrm{mod}1, \mathrm{if}x_n < 0.5 \\ \left[rx_n(1-x_n) + \frac{(4-r)(1-x_n)}{2}\right]\mathrm{mod}1, \mathrm{if}x_n \geq 0.5 \end{cases} \tag{13}$$

where X is the system variable, r is the control parameter $x \in [0,1]$, and $r \in [0,4]$. In this study, the logistic-tent double mapping is used for initialization, so that the initial solutions are distributed as uniformly as possible in the solution space, as shown in Figure 3.

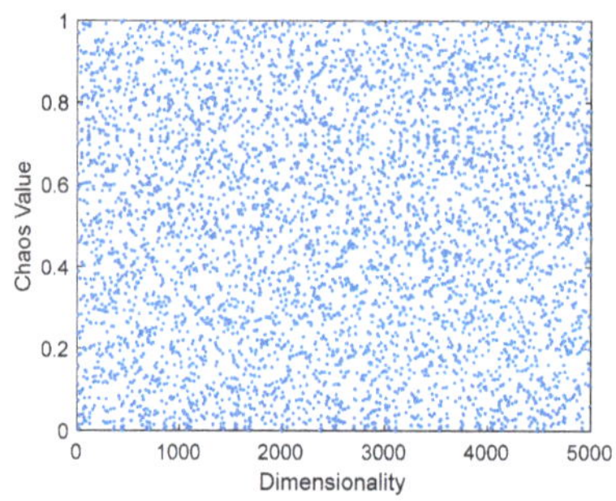

Figure 3. Population distribution of chaotic mapping.

3.2. Nonlinear Variable Speed Convergence Factor

All population intelligence optimization algorithms have the problem of balancing the global search ability and local search ability. A strong global search ability can ensure the diversity of the population, and a strong local search ability can ensure a locally accurate search and accelerate the convergence of the algorithm. Therefore, the importance of balancing the global search ability and local search ability in the GWO algorithm is self-evident.

The convergence factor of the GWO algorithm linearly decreases from 2 to 0 over successive iterations. This decrease during continuous convergence renders a strategy of linearly decreasing the convergence factor that is inadequate for holistically capturing the intricate optimization search dynamics, as noted by [28]. Accordingly, this study introduces a convergence factor by exponential attenuation:

$$a = He^{-(t^2/2M^2)} - (H-2) \tag{14}$$

where e is the natural exponent; t is the current iteration number; M is the maximum number of iterations; H is the coefficient vector, which is used to adjust the convergence factor to make it nonlinearly decrease from 2 to 0. Figure 4 shows the nonlinear relationship between the improved convergence coefficient and the number of iterations.

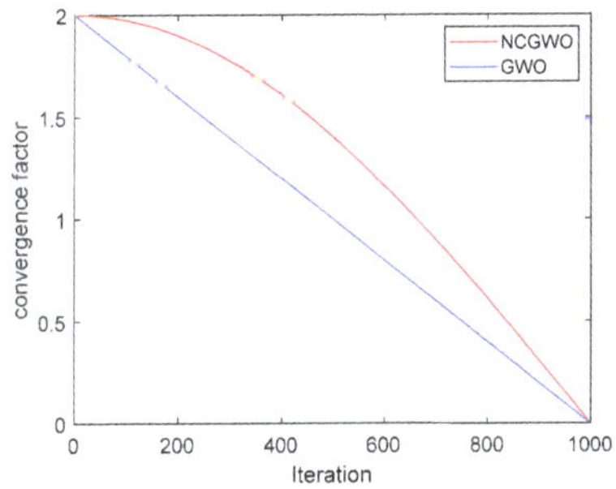

Figure 4. Convergence coefficients.

Based on the data plotted in Figure 4, when the number of iterations increases, the improved convergence coefficient nonlinearly decreases. The decay rate is lower during the initial stages of the algorithm to enhance the search for the global optimal solution, while it accelerates in the later stages to achieve a more precise localization of the local optimal solution. This approach effectively balances the search capabilities for both global and local solutions [29].

Based on this description, Table 1 succinctly outlines the disparities between NCGWO and GWO algorithms.

Table 1. Comparison of the GWO and NCGWO improvement points.

	NCGWO	GWO
Number 1	Logistic-tent double mapping generating initial populations	Randomly generated initial populations
Formula	$X_{n+1} = \begin{cases} \left[r x_n(1-x_n) + \frac{(4-r)}{2} x_n \right] \text{mod} 1, \text{if} x_n < 0.5 \\ \left[r x_n(1-x_n) + \frac{(4-r)(1-x_n)}{2} \right] \text{mod} 1, \text{if} x_n \geq 0.5 \end{cases}$	$X_{n+1} = rand(X_n)$
Reason 1	The GWO algorithm suffers from a lack of diversity in the initial population due to the random generation of individuals. This limitation hampers the algorithm's search flexibility. To address this, logistic-tent dual chaotic mapping methods are employed to enhance the initial population's diversity, optimizing the global search process and increasing the algorithm's search flexibility.	
Number 2	Nonlinear variable speed convergence factor	Linear convergence factor
Formula	$a = He^{-(t^2/2M^2)} - (H-2)$	$a = 2 - 2\left(\frac{t}{T_{\max}}\right)$
Reason 2	The a plays an important role in balancing the exploration and exploitation of the candidate's individual search. The search process of the GWO algorithm is nonlinear and complicated; the linear convergence factor cannot balance the global and local search ability. The better performance would be obtained if there was a nonlinear variable speed convergence factor.	

3.3. NCGWO Algorithm Steps

Figure 5 shows the NCGWO implementation steps. Initially, the initial population is generated, and parameters are defined for the fitness value calculation, where the best values are stored as α, β, and δ. Subsequently, the positions are updated, which ultimately leads to the output of the optimal solution X_α.

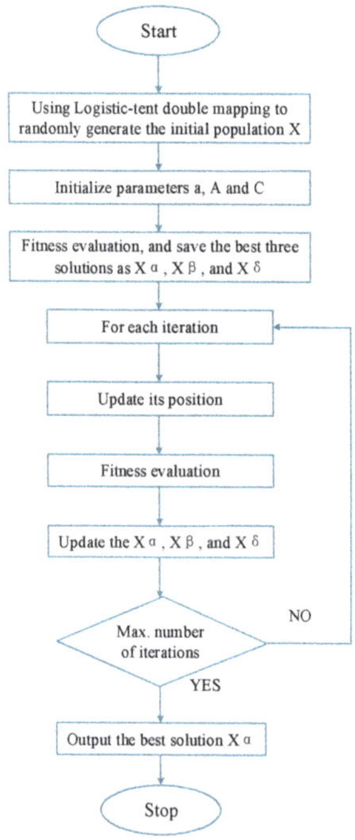

Figure 5. Flow chart of the nonlinear chaotic grey wolf optimization (NCGWO) algorithm.

4. NCGWO Performance Analysis

The overall performance of NCGWO algorithm used to be first examined on four popular benchmark functions, which are used as metrics to consider the performances of optimization algorithms in many instances. The results were compared to those of well-known metaheuristic optimization methods such as GWO, PSO, and GA [30] (Table 2). Table 3 compares the other algorithms on the benchmark functions. Figure 6 shows the 2D graphics.

Figure 6a,b are unimodal functions, each of which contains only one optimum. Figure 6a is smooth, unimodal, strongly convex, and symmetric. Figure 6b has a very narrow ridge, the tip of the ridge is very sharp, and it runs around a parabola. Figure 6c,d are multimodal functions with many local optima but only one global optimum. Figure 6c has a large search space and many local minima, and there are millions of locally optimal solutions in this interval alone. Figure 6d is the term of the summation that produces a parabola, while the local optima are above the parabola level [31].

Table 2. Benchmark function and their properties.

F	Expression	Boundaries	Optimal Solution i.e., $f(x^*)$
F1	$f(x) = \sum_{i=1}^{D} x_i^2$	$-100 \leq x_i \leq 100$	$x^* = [0,0,\ldots,0] f(x^*) = 0$
F2	$f(x) = \sum_{i=1}^{D-1} \left[100(x_{i+1} - x_i^2) + (1 - x_i)^2\right]$	$-30 \leq x_i \leq 30$	$x^* = [1,1,\ldots,1] f(x^*) = 0$
F3	$f(x) = \sum_{i=1}^{D} [x_i^2 - 10\cos(2\pi x_i) + 10]$	$-5.12 \leq x_i \leq 5.12$	$x^* = [0,0,\ldots,0] f(x^*) = 0$
F4	$f(x) = \frac{1}{4000}\sum_{i=1}^{D} x_i^2 - \prod_{i=1}^{D} \cos\left(\frac{x_i}{\sqrt{i}}\right) + 1$	$-100 \leq x_i \leq 100$	$x^* = [0,0,\ldots,0] f(x^*) = 0$

Table 3. Result comparison with 50 independent runs.

Objective Function	Algorithm	Min	Mean	SD
$f_1(x)$	NCGWO	7.6893×10^{-99}	2.6133×10^{-100}	1.4032×10^{-99}
	GWO	2.60×10^{-73}	3.55×10^{-71}	1.18×10^{-70}
	PSO	5.01×10^{-17}	5.96×10^{-15}	2.54×10^{-14}
	GA	2.23×10^{-31}	1.3×10^{-3}	0.105
$f_2(x)$	NCGWO	27.10	26.01	0.59
	GWO	34.45	36.40	0.75
	PSO	10.574	54.55	56.00
	GA	23.89	284.04	255.57
$f_3(x)$	NCGWO	0	0	0
	GWO	0	2.27×10^{-15}	1.11×10^{-14}
	PSO	21.89	77.89	24.37
	GA	11.94	35.97	12.26
$f_4(x)$	NCGWO	0	0	0
	GWO	0	6.24×10^{-4}	2.52×10^{-3}
	PSO	0	1.48×10^{-2}	2.38×10^{-2}
	GA	0	1.7×10^{-3}	1.46×10^{-2}

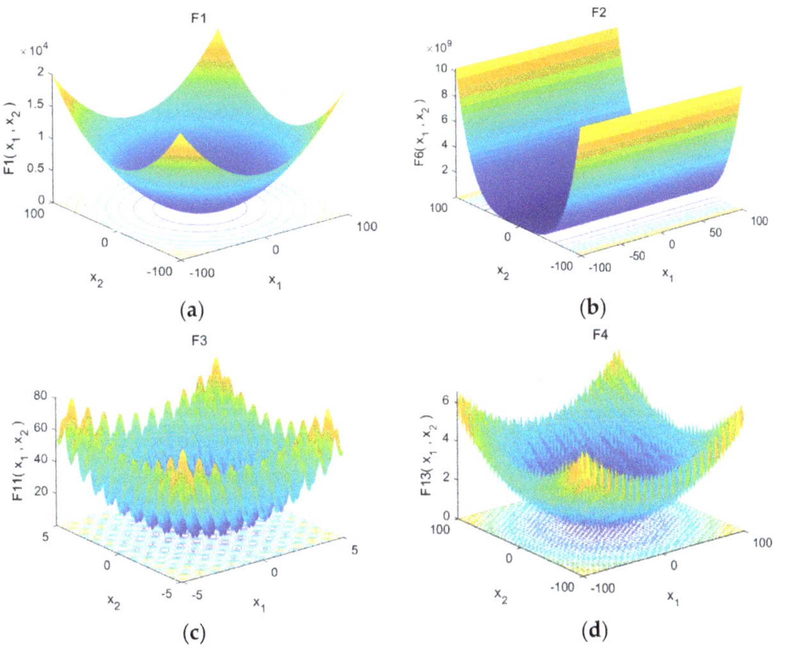

Figure 6. Illustration of the two-dimensional versions of the test functions. (**a**,**b**) are unimodal functions; (**c**,**d**) are multimodal functions.

In this test case, each algorithm was independently executed 50 times, and the population size and maximum number of iterations were set to 100 and 1000, respectively. Table 3 shows that the NCGWO algorithm outperformed the other algorithms on test functions F1, F3, and F4, while the PSO algorithm showed a better performance on F2. However, the NCGWO algorithm consistently demonstrated superior performance across all test functions (Table 3). Therefore, we conclude that the NCGWO algorithm outperforms the GWO algorithm.

5. Array Antenna

This section presents the model and array factor of LAAs and planar arrays. The problem of optimizing the PSLL and objective function is detailed in [32].

5.1. Linear Array Antennas

In this study, a uniform LAA was employed, where an equi-spaced array of $2N$ electric dipoles was symmetrically positioned along the x-axis (Figure 7a). The electric field in the far-field region for an electric dipole is given in Ref. [32]:

$$E(\theta) = j\frac{Idl}{4\pi}\left(\frac{e^{-jkR}}{R}\right)\eta_0 k \sin\theta \tag{15}$$

where Idl is the excitation amplitude, η_0 is the impedance of free space, k is the free space wave number $k = 2\pi/\lambda$, and λ is the wavelength. Here, we can ignore $\left(e^{-jkR}/R\right)$, the far-field expression.

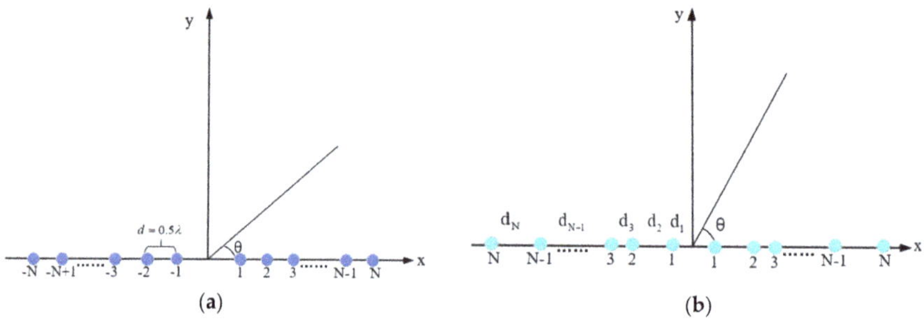

Figure 7. (**a**) Uniform LAA geometry; (**b**) sparse LAA geometry.

For array antennas, the field pattern $f(\theta)$ is expressed as a product of $E(\theta)$ and the array factor $F(\theta)$:

$$f(\theta) = E(\theta) \times F(\theta) \tag{16}$$

where the array factor of the far-field uniform linear array antenna is expressed in decibels using an equation in Ref. [33], as follows:

$$F(\theta) = 2\sum_{n=1}^{N} I_n \cos(kd\cos\theta + \delta_n) \tag{17}$$

The excitation amplitude of the Nth element is represented by I, k is the wave number $k = 2\pi/\lambda$,λ is the wavelength in free space, and distance d between arrays was set to 0.5λ. The variable θ is the angle between the ray direction and the array axis, and δ_n is the phase of the Nth element, which was set to $\delta_n = 0$.

For the sparse linear array antenna in Figure 7b, the inter-element spacing was optimized, while the amplitude and phase were assumed to be 1 and 0, respectively. The equation for the sparse LAA array factor in Ref. [33] is

$$F(\theta) = 2\sum_{n=1}^{N} \cos(kd\cos\theta) \tag{18}$$

For an even-numbered array, Equation (16) is applicable. If it is an odd-numbered array, the array factor can be obtained by adding 1 to Equation (18).

5.2. Planar Array Antenna

Planar array antennas enhance the antenna directionality compared to LAAs. In Figure 8, a rectangular planar array consists of $n = n_x \bullet n_y$ identical antenna units, where dx and dy are the intervals between antenna units along the x and y axes, respectively. The equation for the planar array antenna pattern is expressed in Ref. [34] as follows:

$$F(\theta, \varphi) = F_x(\theta, \varphi) \bullet F_y(\theta, \varphi) \tag{19}$$

$$F(\theta,\varphi)=\sum_{n=1}^{N} I_n e^{j2\pi d_x \sin\theta\cos\varphi} e^{j2\pi d_y \sin\theta\sin\varphi} \quad (20)$$

where N is the total number of antenna elements in the array; θ is the elevation angle with respect to the z-axis; φ is the azimuth angle with respect to the x-axis; I is the excitation amplitude of the Nth element; d_x and d_y are the intervals between antenna units along the x- and y-axes, respectively; $d_x = d_y = 0.5\lambda$; λ is the wavelength.

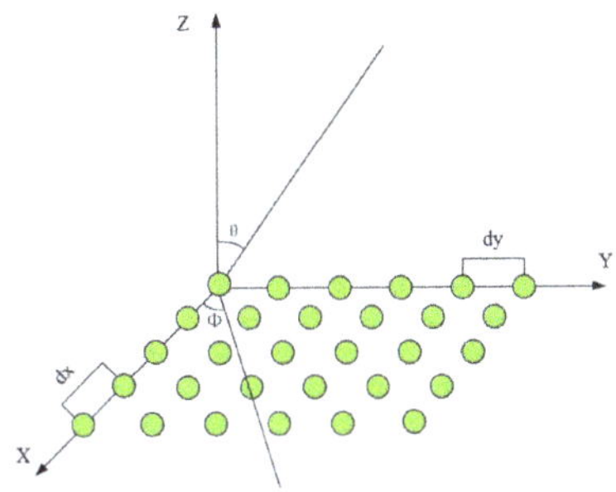

Figure 8. Planar array antennas geometry.

5.3. Pattern Synthesis of Array Antennas

The objective is to minimize the decrease in SLL below the predicted value while maintaining a constant main lobe width of the array and achieving deep nulls in specific directions. To achieve this objective, the following fitness function was used [23]:

$$Fit = \eta_1|PSLL - TPSL| + \eta_2\left|\max\{F(\theta_K)\}_{k=1,\ldots N_0} - DDNL\right| \quad (21)$$

where $PSLL$ is the peak sidelobe level; $TPSL$ is the target peak sidelobe level; $F(\theta_k)$ is the direction of the kth deep null; $DDNL$ is the desired deep null level; η_1 and η_2 are the weighting factors and subject to the condition $\eta_1 + \eta_2 = 1$.

6. Data Analysis

6.1. Reducing the First Sidelobe Level (SLL) Nearest to the Main Lobe of a Uniform LAA

In the first example, the NCGWO algorithm was utilized to synthesize the pattern of a uniform LAA. The performance of the NCGWO algorithm was compared with that of the IGWO [23], GWO, PSO, and MA [35] algorithms. Assuming zero-phase excitation, the amplitudes of the excitation current were optimized for LAA with $2N = 10$ to achieve the desired peak sidelobe level (PSLL). The main lobe width was set to 28°, while the sidelobe regions were $\theta = [0^\circ, 76^\circ] \cup [104^\circ, 180^\circ]$. Meanwhile, the first sidelobe, which is closest to the main lobe, was minimized within the region defined by $\theta = [69^\circ, 76^\circ] \cup [104^\circ, 111^\circ]$. The NCGWO algorithm was executed with a population size of 20 for 1000 iterations, which is consistent with previous studies [34].

Figure 9a depicts the array pattern obtained using the NCGWO algorithm, which was compared with the results of the IGWO, GWO, PSO, and MA algorithms. Figure 9b illustrates the iteration curves that correspond to the GWO, IGWO, and NCGWO algorithms. Evidently, the NCGWO algorithm converges to the optimal value within approximately 100 generations, whereas the GWO and IGWO algorithms require approximately 400 generations to achieve the same outcome. This outcome underscores the superior performance of the NCGWO algorithm compared to the GWO and IGWO algorithms. Figure 10 shows the 3D radiation patterns of reducing the 10-element LAA's first SLL before and after the NCGWO optimization, which clearly demonstrates the effectiveness of the NCGWO algorithm in significantly reducing the first SLL.

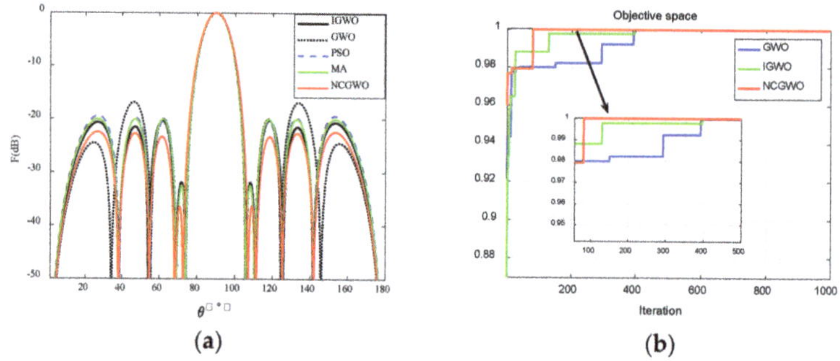

Figure 9. (a) Radiation patterns of a 10-element linear array antenna SLL; (b) iteration curve.

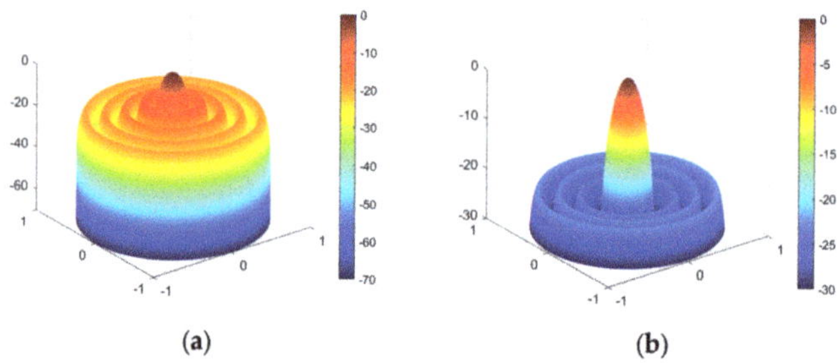

Figure 10. 3D radiation patterns of a 10-element linear array antenna SLL near the main lobe minimization: (a) before and (b) after.

Table 4 shows the optimization results in the example. The NCGWO algorithm achieved a minimized nearest SLL of −36.4027 dB and a running time of 0.8290 s. This minimized nearest SLL was lower than those of GWO, IGWO, PSO, and MA by 4.5627, 4.4898, 3.9827, and 3.9527 dB, respectively. The PSLL achieved by the NCGWO algorithm was −22.4420 dB, which was 5.585, 2.1659, 3.0379, and 2.4053 dB lower than those of GWO, IGWO, PSO, and MA, respectively.

Table 4. Results for a 10-element linear array antenna.

Algorithm	Optimized Amplitude Current	Peak SLL (dB)	Near SLL (dB)	Execution Time (s)
GWO	0.9531, 0.9452, 0.5549, 0.4634, 0.5083	−16.8570	−31.8400	3.1800
IGWO	1.0000, 0.8376, 0.6409, 0.4349, 0.4768	−20.2761	−31.9129	2.4520
PSO	0.9242, 0.7997, 0.6030, 0.3814, 0.4766	−19.4041	−32.4200	2.9300
MA	1.0000, 0.8585, 0.6383, 0.4253, 0.5070	−20.0367	−32.4500	9.3700
NCGWO	1.0000, 0.8645, 0.6455, 0.4105, 0.3999	−22.4420	−36.4027	0.8290

6.2. Array Factor of a 20-Element Uniform Linear Array Antenna with a Lower Notch

In the second example, an LAA with 20 elements was considered to decrease the SLL in the sidelobe regions of $\theta = [0^\circ, 82^\circ] \cup [98^\circ, 180^\circ]$ with lower notches in the specific directions of $\theta = [50^\circ, 60^\circ] \cup [120^\circ, 130^\circ]$. The NCGWO algorithm was executed with a population size of 30 and at most 500 iterations, consistent with previous studies [36].

Figure 11a depicts the LAA pattern synthesized using the NCGWO algorithm and compares it with the results of the GWO, IGWO, GOA, and SMO algorithms. Figure 11b illustrates the iteration curves that correspond to the GWO, IGWO, and NCGWO algorithms. Although all three algorithms obtain the optimal value within 500 generations, a discernible disparity is evident in the figure. Specifically, the NCGWO algorithm demonstrates a more favorable optimal value than GWO and IGWO. Figure 12 illustrates the

3D radiation patterns of a 20-element LAA before and after the NCGWO optimization. NCGWO significantly reduces the PSLL and the notch.

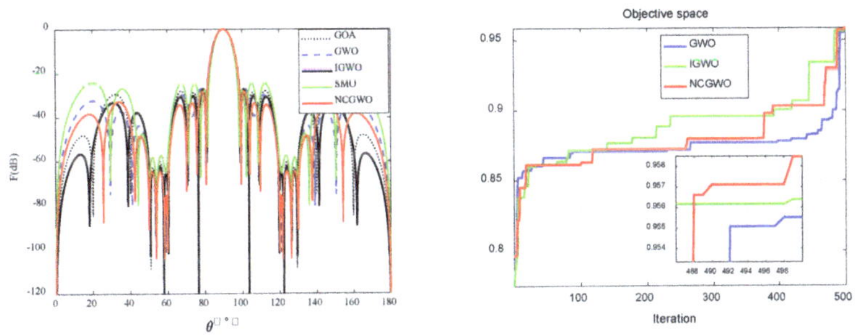

(a) (b)

Figure 11. (**a**) Radiation patterns of the SLL of a 20-element linear array antenna with a lower notch; (**b**) iteration curve.

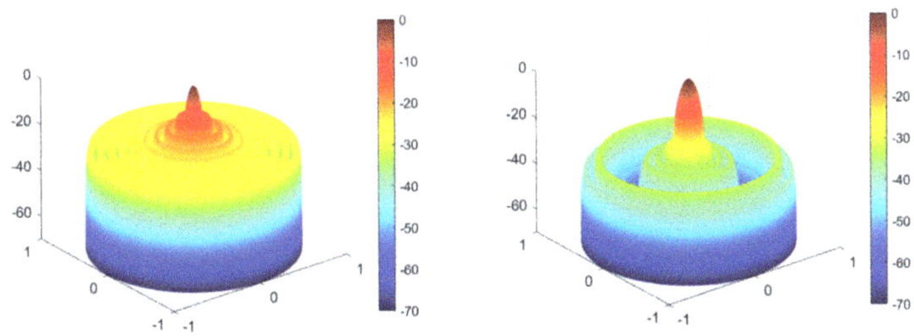

(a) (b)

Figure 12. 3D radiation patterns of the SLL of a 20-element linear array antenna with a lower notch: (**a**) before and (**b**) after.

Table 5 shows the optimization results. The results demonstrate that the NCGWO algorithm achieved a PSLL of −28.1781 dB, and the PSLL of the notch was −64.3883 dB. Compared with the reported results of the GWO, IGWO, GOA, and SMO algorithms [36], the PSLL of the NCGWO algorithm was 1.2140, 1.1637, 4.0781, and 0.4781 dB lower, respectively, and the PSLL of the notch was 1.7452, 2.2447, 7.6883, and 3.1883 dB lower, respectively. Previous studies [35] did not provide information on the running time, and the present study independently executed the algorithm 50 times and calculated the average. Table 5 shows that the NCGWO algorithm requires less time than other algorithms.

Table 5. Results for a 20-element linear array antenna.

Algorithm	Optimized Amplitude Current	Peak SLL (dB)	Notch Depth (dB)	Execution Time (s)
GWO	0.8785, 0.9057, 0.8858, 0.7978, 0.6418 0.5180, 0.3663, 0.3101, 0.1407, 0.0939	−26.9677	62.6431	3.075
IGWO	0.9795, 0.9716, 0.9386, 0.7924, 0.7202 0.5613, 0.5115, 0.3950, 0.2175, 0.1190	−27.0144	−62.1436	2.5920
SMO	1.0000, 0.9990, 1.0000, 0.8360, 0.6430 0.6540, 0.4770, 0.5970, 0.2580, 0.2150	−24.1000	−56.7000	9.612
GOA	1.0000, 0.9860, 0.9900, 0.7960, 0.7360 0.5630, 0.5270, 0.4470, 0.2430, 0.1510	−27.7000	−61.2000	7.714
NCGWO	1.0000, 0.9636, 0.9417, 0.7958, 0.7000 0.5752, 0.4474, 0.3626, 0.1699, 0.1008	−28.1781	−64.3883	1.998

In this design example, the NCGWO algorithm exhibited significant advantages in suppressing the PSLL and achieving a greater depth in the notch. The NCGWO algorithm also demonstrated better optimization capabilities than other algorithms.

6.3. Ten-Element Sparse LAA with Constraint on the Minimum Element Spacing

The fourth example focuses on a 10-element sparse LAA with the objective of minimizing the PSLL. The main lobe width was configured to be 22°, while the sidelobe regions were designated as $\theta = [0^{\circ}, 79^{\circ}] \cup [101^{\circ}, 180^{\circ}]$. The minimum element spacing was 0.5λ. The NCGWO algorithm was executed with a population size of 40 and at most 1000 iterations, which is consistent with previous studies [37,38].

Figure 13a depicts the sparse LAA pattern using the NCGWO algorithm and a comparison with the results of the GWO, IGWO, WDO, and ACO algorithms. Figure 13b presents the iteration curves of the GWO, IGWO, and NCGWO algorithms. The NCGWO algorithm outperforms both IGWO and GWO, as it can more quickly converge towards optimal values. Figure 14 shows the 3D radiation patterns of the 10-element sparse LAA before and after the NCGWO algorithm, where the NCGWO algorithm effectively reduces the PSLL in an intuitive manner.

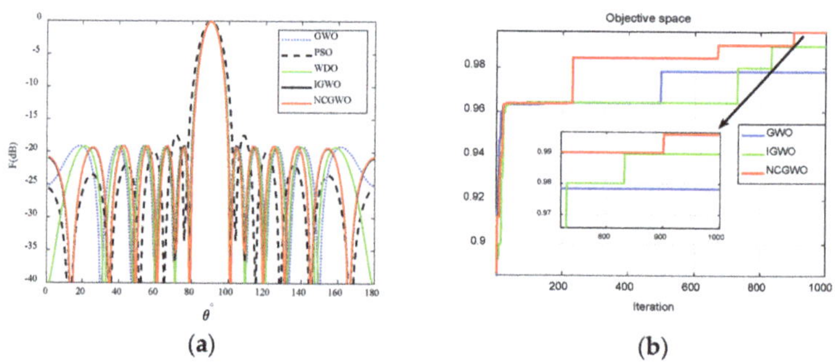

Figure 13. (**a**) Radiation patterns of the SLL of a 10-element sparse linear array antenna; (**b**) iteration curve.

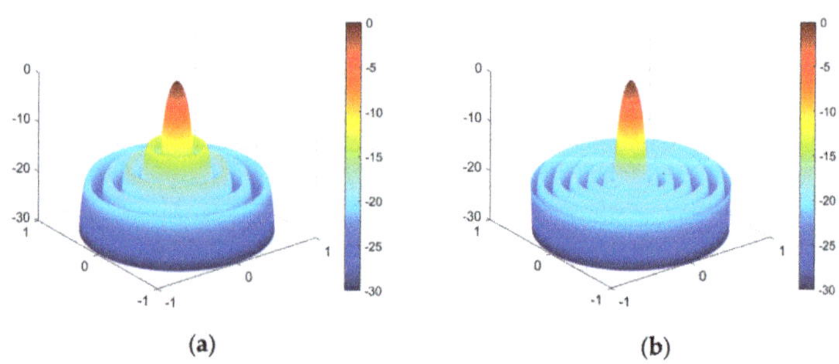

Figure 14. 3D radiation patterns of the SLL of a 10-element sparse linear array antenna: (**a**) before and (**b**) after.

Table 6 shows the optimization results. The PSLL of the NCGWO algorithm was −19.0848 dB, which was 0.0148, 0.0236, 1.6748, and 0.0348 dB lower than those of the GWO, IGWO, PSO, and WDO algorithms, respectively. Previous studies [35,36] did not provide information on the running time; therefore, the present study independently executed the algorithm 50 times and calculated the average. As shown in Table 6, the NCGWO algorithm requires less time than other algorithms.

Table 6. Comparison of results of algorithms for 10-element sparse linear array antennas.

Algorithm	Optimized Element Position	Peak SLL (dB)	Execution Time (s)
GWO	0.2204, 0.7078, 1.2023, 1.8266, 2.5468	−19.0700	4.9820
IGWO	0.2234, 0.7196, 1.2224, 1.8604, 2.5966	−19.0712	4.3400
PSO	0.2515, 0.5550, 1.0650, 1.5000, 2.1100	−17.4100	9.5150
WDO	0.2233, 0.7179, 1.2221, 1.8591, 2.5936	−19.0500	6.2090
NCGWO	0.2298, 0.7401, 1.2613, 1.9205, 2.6785	−19.0848	3.3540

6.4. Sparse LAA with Constraints on the PSLL Minimization and Null

The fifth example focuses on a 32-element LAA to minimize the PSLL while aiming for nulls at $\theta = 81^{\circ}$ and $\theta = 99^{\circ}$. The main lobe width is configured as 10°, and the sidelobe regions are designated as $\theta = \left[0^{\circ}, 84^{\circ}\right] \cup \left[96^{\circ}, 180^{\circ}\right]$. To attain the lowest null and enable a fair comparison with other algorithms, the population size of the NCGWO algorithm was set to 30, and the maximum number of iterations was set to 1000.

Figure 15a illustrates the sparse LAA pattern using the NCGWO algorithm and compares it with the results of the GWO, IGWO, WDO, and ACO algorithms [39,40]. Figure 15b displays the iterative curves of GWO, IGWO, and NCGWO. The NCGWO algorithm performs better than its IGWO and GWO counterparts due to its rapid convergence and attainment of optimal values. Figure 16 shows the 3D radiation patterns of the 32-element sparse LAA before and after the NCGWO algorithm optimization. NCGWO was successful in greatly reducing the PSLL and lowering the null.

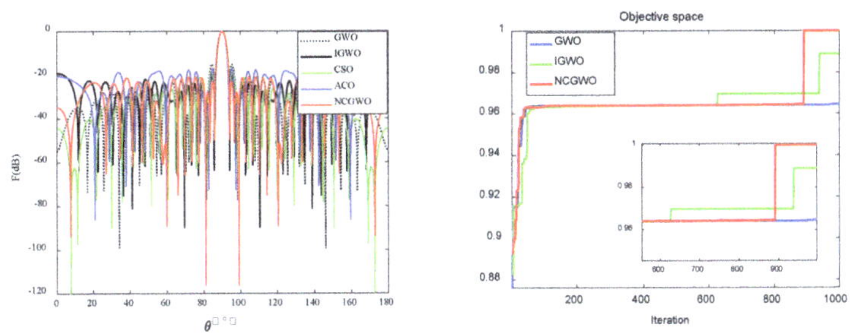

(a) (b)

Figure 15. (**a**) Radiation patterns of the SLL minimization and null of 32-element sparse linear array antennas; (**b**) iteration curve.

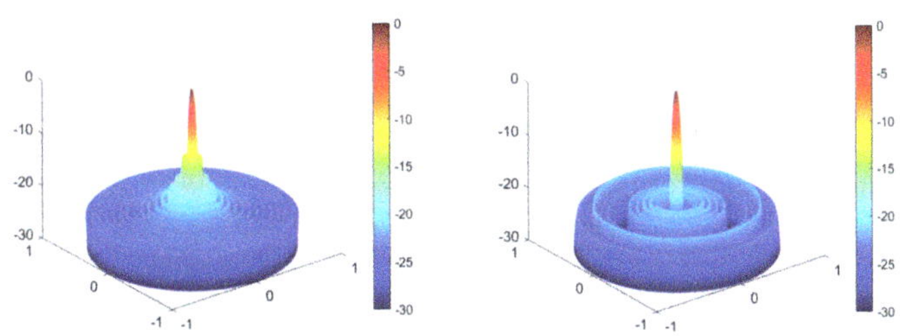

(a) (b)

Figure 16. 3D radiation patterns of the SLL minimization and null of 32-element sparse linear array antennas: (**a**) before and (**b**) after.

Table 7 presents the optimization results. In this example, the ACO, CSO, IGWO and GWO algorithm achieved nulls of −50 dB, −80 dB, −108.2477 dB, and −106 dB, respectively. The proposed NCGWO algorithm achieved the deepest null of −116 dB,

8. Mirjalili, S.; Mirjalili, S.M.; Lewis, A. Grey wolf optimizer. *Expert Syst. Appl.* **2016**, *47*, 106–119. [CrossRef]
9. Güneş, F.; Tokan, F. Pattern search optimization with applications on synthesis of linear antenna arrays. *Expert Syst. Appl.* **2010**, *37*, 4698–4705. [CrossRef]
10. Singh, U.; Kamal, T. Design of non-uniform circular antenna arrays using biogeography-based optimisation. *IET Microw. Antennas Propag.* **2011**, *5*, 1365–1370. [CrossRef]
11. Kaur, K.; Banga, V.K. 'Optimization of linear antenna array using firefly algorithm. *Int. J. Sci. Eng. Res.* **2013**, *4*, 601–606.
12. Zhang, Z.; Li, T.; Yuan, F.; Yin, L. Synthesis of Linear Antenna Array Using Genetic Algorithm to Control Side Lobe Level. In *Computer Engineering and Networking*; Springer: Cham, Switzerland, 2013; pp. 39–46.
13. Yang, S.; Gan, Y.B.; Qing, A. Antenna-array pattern nulling using a differential evolution algorithm. *Int. J. RF Microw. Comput. Eng.* **2003**, *14*, 57–63. [CrossRef]
14. Abdelkader, E.M.; Moselhi, O.; Marzouk, M.; Zayed, T. An exponential chaotic differential evolution algorithm for optimizing bridge maintenance plans. *Autom. Constr.* **2022**, *134*, 104107. [CrossRef]
15. Saremi, S.; Mirjalili, S.; Lewis, A. Grasshopper optimisation algorithm: Theory and application. *Adv. Eng. Softw.* **2017**, *105*, 30–47. [CrossRef]
16. Xia, G.; Han, Z.; Zhao, B.; Wang, X. Local path planning for unmanned surface vehicle collision avoidance based on modified quantum particle swarm optimization. *Complexity* **2020**, *2020*, 3095426. [CrossRef]
17. Al-Azza, A.A.; Al-Jodah, A.A.; Harackiewicz, F.J. Spider monkey optimization: A novel technique for antenna optimization. *IEEE Antennas Wirel. Propag. Lett.* **2015**, *15*, 1016–1019. [CrossRef]
18. Zervoudakis, K.; Tsafarakis, S. A mayfly optimization algorithm. *Comput. Ind. Eng.* **2020**, *145*, 106559. [CrossRef]
19. Mahto, S.K.; Choubey, A.; Suman, S. Linear array synthesis with minimum side lobe level and null control using wind driven optimization. In Proceedings of the 2015 International Conference on Signal Processing and Communication Engineering Systems, Guntur, India, 2–3 January 2015; pp. 191–195.
20. Guo, H.; Li, J.; Hao, H.; Song, P.; Zhang, L.; Zhang, X. Synthesis of Linear Antenna Array for Wireless Power Transmission. *Synthesis* **2023**, *2023*, 6–27. [CrossRef]
21. Pavani, T.; Padmavathi, K.; Kumari, C.U.; Ushasree, A. Design of array antennas via atom search optimization. *Mater. Today Proc.* **2023**, *80*, 2051–2054. [CrossRef]
22. Lakhlef, N.; Oudira, H.; Dumond, C. Failure Correction of Linear Antenna Array using Grey Wolf Optimization. In Proceedings of the 2020 6th IEEE Congress on Information Science and Technology (CiSt), Agadir-Essaouira, Morocco, 5–12 June 2021; pp. 384–388.
23. Zhao, K.X.; Liu, Y.; Hu, K. Optimal Pattern Synthesis of Array Antennas Using Improved Grey Wolf Algorithm. In Proceedings of the 2022 IEEE 12th International Conference on Electronics Information and Emergency Communication (ICEIEC), Beijing, China, 15–17 July 2022; pp. 172–175.
24. Persohn, K.J.; Povinelli, R.J. Analyzing logistic map pseudorandom number generators for periodicity induced by finite precision floating-point representation. *Chaos Solitons Fractals* **2012**, *45*, 238–245. [CrossRef]
25. Nezhad, S.Y.D.; Safdarian, N.; Zadeh, S.A.H. New method for fingerprint images encryption using DNA sequence and chaotic tent map. *Optik* **2020**, *224*, 165661. [CrossRef]
26. Hua, Z.; Zhu, Z.; Yi, S.; Zhang, Z.; Huang, H. Cross-plane colour image encryption using a two-dimensional logistic tent modular map. *Inf. Sci.* **2021**, *546*, 1063–1083. [CrossRef]
27. Lawnik, M. Combined logistic and tent map. *J. Phys. Conf. Ser.* **2018**, *1141*, 012132. [CrossRef]
28. Panda, G.; Pradhan, P.M.; Majhi, B. IIR system identification using cat swarm optimization. *Expert Syst. Appl.* **2011**, *38*, 12671–12683. [CrossRef]
29. Tan, F.; Zhao, J.; Wang, Q. A grey wolf optimization algorithm with improved nonlinear convergence. *Microelectron. Comput.* **2019**, *36*, 89–95.
30. Li, X.; Luk, K.M. The grey wolf optimizer and its applications in electromagnetics. *IEEE Trans. Antennas Propag.* **2019**, *68*, 2186–2197. [CrossRef]
31. El-Kenawy, E.-S.M.; Mirjalili, S.; Alassery, F.; Zhang, Y.-D.; Eid, M.M.; El-Mashad, S.Y.; Aloyaydi, B.A.; Ibrahim, A.; Abdelhamid, A.A. Novel meta-heuristic algorithm for feature selection, unconstrained functions and engineering problems. *IEEE Access* **2022**, *10*, 40536–40555. [CrossRef]
32. Raji, M.F.; Zhao, H.; Monday, H.N. Fast optimization of sparse antenna array using numerical Green's function and genetic algorithm. *Int. J. Numer. Model. Electron. Netw. Devices Fields* **2020**, *33*, e2544. [CrossRef]
33. Guo, Q.; Wang, Y.; Li, Y.; Qi, L.; Chernogor, L.F. Pattern synthesis of nonuniform linear antenna array based on FFDM. *Digit. Signal Process.* **2023**, *137*, 104022. [CrossRef]
34. Boopalan, N.; Ramasamy, A.K.; Nagi, F. A Comparison of Faulty Antenna Detection Methodologies in Planar Array. *Appl. Sci.* **2023**, *13*, 3695. [CrossRef]
35. Owoola, E.O.; Xia, K.; Wang, T.; Umar, A.; Akindele, R.G. Pattern synthesis of uniform and sparse linear antenna array using mayfly algorithm. *IEEE Access* **2021**, *9*, 77954–77975. [CrossRef]
36. Wang, H.; Liu, C.; Wu, H.; Li, B.; Xie, X. Optimal pattern synthesis of linear antenna array and broadband design of whip antenna using grasshopper optimization algorithm. *Int. J. Antennas Propag.* **2020**, *2020*, 5904018. [CrossRef]

37. Khodier, M.M.; Christodoulou, C.G. Linear antenna array geometry synthesis with minimum sidelobe level and null control using particle swarm optimization. *IEEE Trans. Antennas Propag.* **2005**, *53*, 2674–2679. [CrossRef]
38. Bayraktar, Z.; Komurcu, M.; Bossard, J.A.; Werner, D.H. The wind driven optimization technique and its application in electromagnetics. *IEEE Trans. Antennas Propag.* **2013**, *61*, 2745–2757. [CrossRef]
39. Pappula, L.; Ghosh, D. Linear antenna array synthesis using cat swarm optimization. *AEU Int. J. Electron. Commun.* **2014**, *68*, 540–549. [CrossRef]
40. Aboul-Seoud, A.K.; Mahmoud, A.K.; Hafez, A.E.D.S. A sidelobe level reduction (SLL) for planar array antennas with −30 dB attenuators weight precision. *Aerosp. Sci. Technol.* **2010**, *14*, 316–320. [CrossRef]

MDPI
St. Alban-Anlage 66
4052 Basel
Switzerland
www.mdpi.com

MDPI Books Editorial Office
E-mail: books@mdpi.com
www.mdpi.com/books

www.ingramcontent.com/pod-product-compliance
Lightning Source LLC
Chambersburg PA
CBHW061446220926
51850CB00004B/23
9783725806713